Law of Sines

$$\frac{\sin A}{a} = \frac{\sin B}{b} = \frac{\sin C}{c}$$

Area of a Triangle

$$\text{Area} = \frac{1}{2}bc \sin A$$

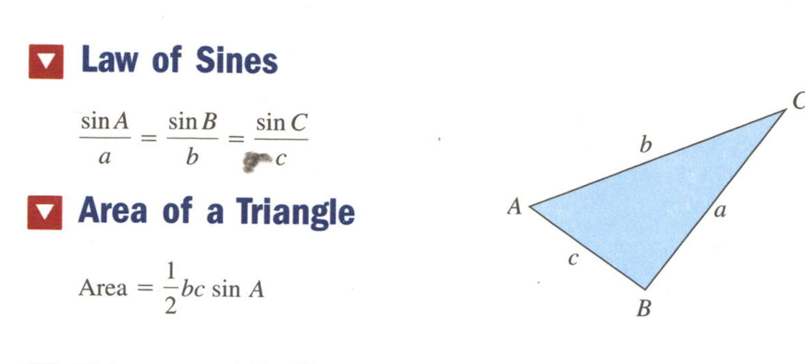

Law of Cos

$a^2 = b^2 + c^2 -$

$b^2 = a^2 + c^2 - 2ac \cos B$

$c^2 = a^2 + b^2 - 2ab \cos C$

Trigonometric Functions of a Real Number

For any real number t and point $P(x, y)$ on the unit circle associated with t:

$\cos t = x$ $\qquad \sin t = y$ $\qquad \tan t = \frac{y}{x};\ x \neq 0$

$\sec t = \frac{1}{x};\ x \neq 0$ $\qquad \csc t = \frac{1}{y};\ y \neq 0$ $\qquad \cot t = \frac{x}{y};\ y \neq 0$

Trigonometry and the Coordinate Plane

For $P(x, y)$ a point on the terminal side of an angle θ in standard position:

$\cos\theta = \frac{x}{r}$ $\qquad \sin\theta = \frac{y}{r}$ $\qquad \tan\theta = \frac{y}{x},\ x \neq 0$

$\sec\theta = \frac{r}{x},\ x \neq 0$ $\qquad \csc\theta = \frac{r}{y},\ y \neq 0$ $\qquad \cot\theta = \frac{x}{y},\ y \neq 0$

Right Triangle Trigonometry

For right $\triangle ABC$ with indicated sides **adj**acent and **opp**osite to acute angle θ:

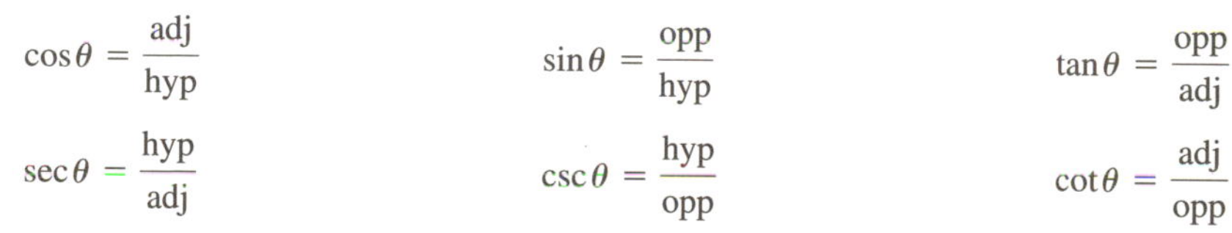

$\cos\theta = \frac{\text{adj}}{\text{hyp}}$ $\qquad \sin\theta = \frac{\text{opp}}{\text{hyp}}$ $\qquad \tan\theta = \frac{\text{opp}}{\text{adj}}$

$\sec\theta = \frac{\text{hyp}}{\text{adj}}$ $\qquad \csc\theta = \frac{\text{hyp}}{\text{opp}}$ $\qquad \cot\theta = \frac{\text{adj}}{\text{opp}}$

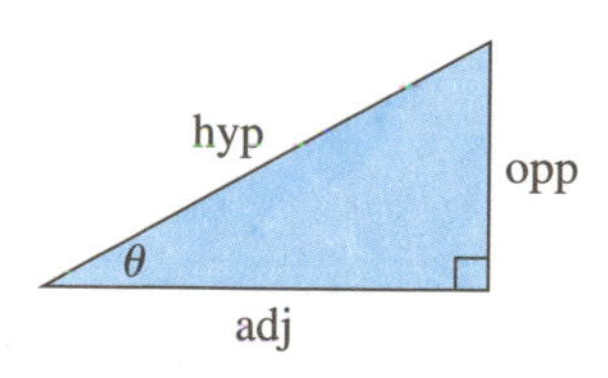

Special Triangles and Special Angles

θ	$\sin\theta$	$\cos\theta$	$\tan\theta$	$\csc\theta$	$\sec\theta$	$\cot\theta$
$0° = 0$	0	1	0	—	1	—
$30° = \frac{\pi}{6}$	$\frac{1}{2}$	$\frac{\sqrt{3}}{2}$	$\frac{1}{\sqrt{3}}$	2	$\frac{2}{\sqrt{3}}$	$\sqrt{3}$
$45° = \frac{\pi}{4}$	$\frac{\sqrt{2}}{2}$	$\frac{\sqrt{2}}{2}$	1	$\sqrt{2}$	$\sqrt{2}$	1
$60° = \frac{\pi}{3}$	$\frac{\sqrt{3}}{2}$	$\frac{1}{2}$	$\sqrt{3}$	$\frac{2}{\sqrt{3}}$	2	$\frac{1}{\sqrt{3}}$
$90° = \frac{\pi}{2}$	1	0	—	1	—	1

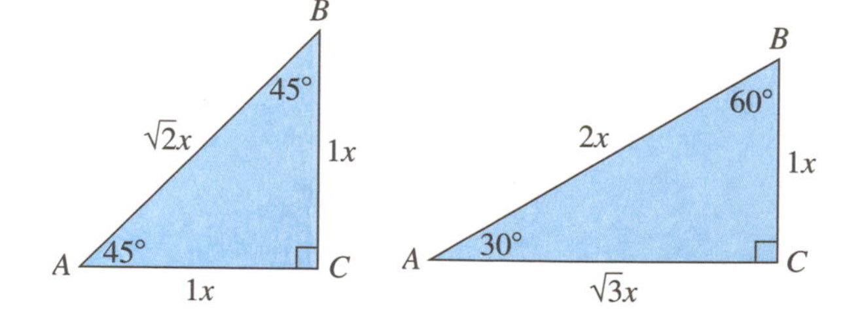

Degree and Radian Conversions

degrees to radians: multiply by $\frac{\pi}{180°}$ (degrees cancel) $\qquad$ radians to degrees: multiply by $\frac{180°}{\pi}$ (radians cancel)

Arcs and Sectors

For a circle of radius r and angle θ in radians:

arc length: $s = r\theta$

area of sector: $A = \frac{1}{2}r^2\theta$

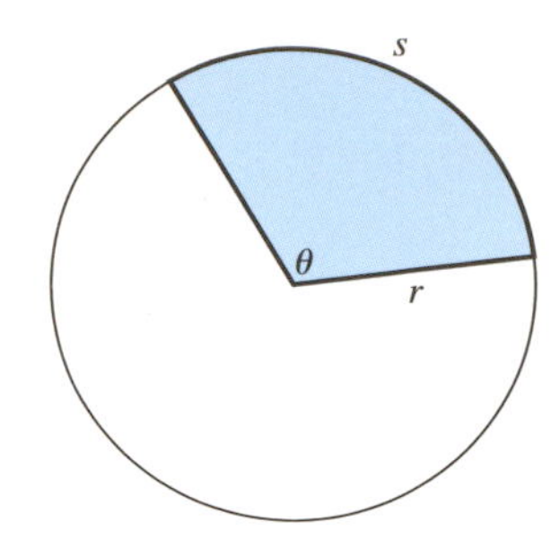

Graphs of the Trigonometric Functions

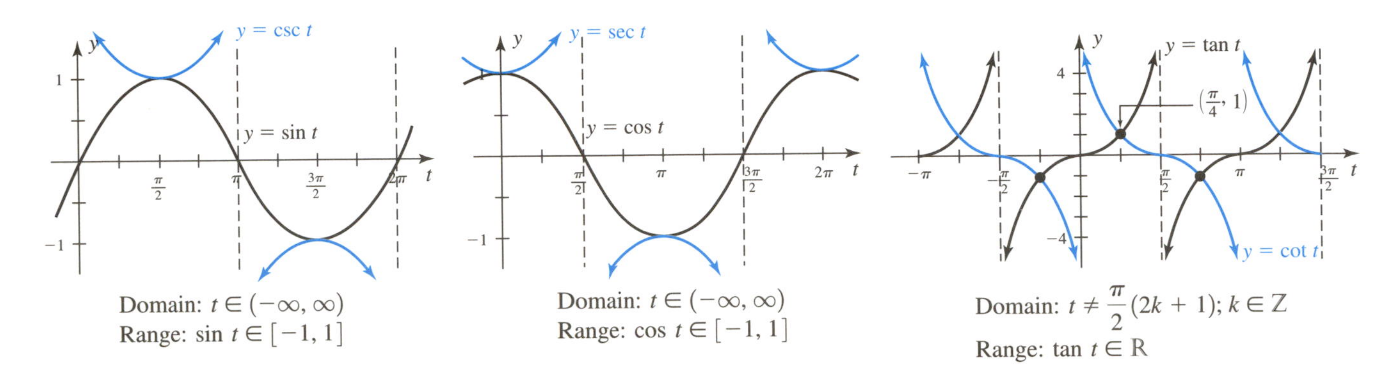

Domain: $t \in (-\infty, \infty)$
Range: $\sin t \in [-1, 1]$

Domain: $t \in (-\infty, \infty)$
Range: $\cos t \in [-1, 1]$

Domain: $t \neq \frac{\pi}{2}(2k + 1); k \in \mathbb{Z}$
Range: $\tan t \in \mathbb{R}$

Transformations of Basic Trig Graphs

Given Function

$y = f(x)$

Transformation of $y = f(x)$

$$y = Af\left[B\left(x \pm \frac{C}{B}\right)\right] \pm D$$

- A: north/south reflections; vertical stretches and compressions
- $\pm \frac{C}{B}$: horizontal shift, opposite direction of sign
- $\pm D$: vertical shift, same direction as sign

For $y = A\sin\left[B\left(x \pm \frac{C}{B}\right)\right] \pm D$ we have: amplitude: $|A|$, period: $\frac{2\pi}{B}$, horizontal shift: $\frac{C}{B}$, vertical shift: D

The Inverse Trigonometric Functions

For $y = \sin t$ with $t \in \left[-\frac{\pi}{2}, \frac{\pi}{2}\right]$ and $y \in [1,\ 1]$, the inverse function is $y = \sin^{-1}t$, where $t \in [1, 1]$ and $y \in \left[-\frac{\pi}{2}, \frac{\pi}{2}\right]$.

For $y = \cos t$ with $t \in [0,\ \pi]$ and $y \in [1,\ 1]$, the inverse function is $y = \cos^{-1}t$, where $t \in [1, 1]$ and $y \in [0,\ \pi]$.

For $y = \tan t$ with $t \in \left(-\frac{\pi}{2}, \frac{\pi}{2}\right)$ and $y \in \mathbb{R}$, the inverse function is $y = \tan^{-1}t$, where $t \in \mathbb{R}$ and $y \in \left(-\frac{\pi}{2}, \frac{\pi}{2}\right)$.

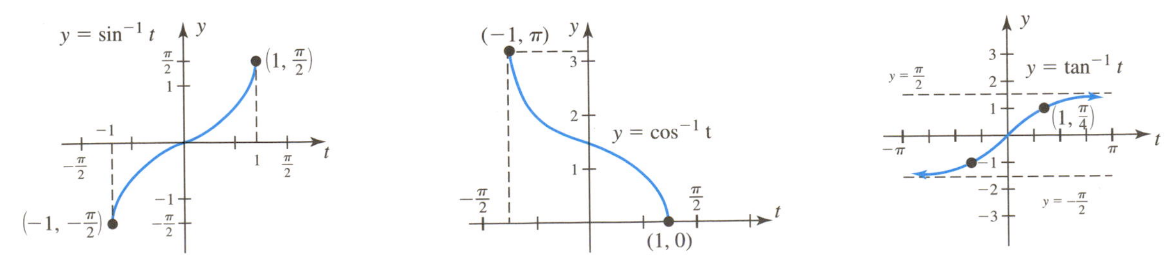

Trigonometry

COBURN'S PRECALCULUS SERIES

COLLEGE ALGEBRA

Coburn's *College Algebra* offers an energizing and engaging new look at conventional topics, while introducing new topics involving the use of data, technology, and alternative approaches to traditional methods. All are supported by carefully crafted exercise sets, designed to develop long-term retention, cement connections, and foster an appreciation of mathematics and its power.
(ISBN–13: 978-0-07-290119-1; ISBN–10: 0-07-290119-5)

TRIGONOMETRY

New in 2007, Coburn's *Trigonometry* offers an exciting and innovative look at a study of trigonometry with a special emphasis on applications using a wealth of modern, up-to-date topics. Students and instructors will enjoy the careful development of each concept, and the support each idea receives in the exercise sets.
(ISBN–13: 978-0-07-291005-6; ISBN–10: 0-07-291005-4)

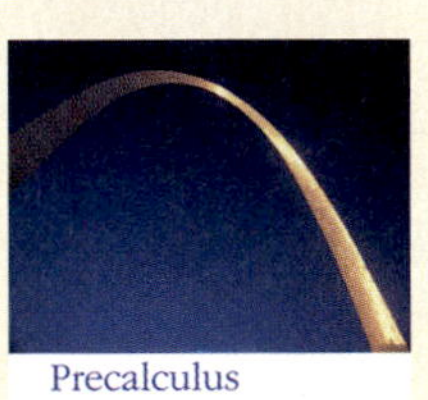

PRECALCULUS

Coburn's *Precalculus* combines all of the essential elements of his *College Algebra* and *Trigonometry* texts, while maintaining the easy-reading style and the quality, quantity, and variety of exercises that his books are becoming known for. The author's ability to maintain a conversational approach, with no compromise in mathematical integrity, shines through offering the ideal preparation for a study of calculus.
(ISBN–13: 978-0-07-290469-7; ISBN–10: 0-07-290469-0)

Trigonometry

John W. Coburn

St. Louis Community College at Florissant Valley

Boston Burr Ridge, IL Dubuque, IA New York San Francisco St. Louis
Bangkok Bogotá Caracas Kuala Lumpur Lisbon London Madrid Mexico City
Milan Montreal New Delhi Santiago Seoul Singapore Sydney Taipei Toronto

The McGraw·Hill Companies

TRIGONOMETRY

This book is printed on acid-free paper.

1 2 3 4 5 6 7 8 9 0 DOW/DOW 0 9 8 7 6

ISBN 978–0–07–291005–6
MHID 0–07–291005–4

ISBN 978–0–07–331266–8 (Instructor's Edition)
MHID 0–07–331266–5

Publisher: *Elizabeth J. Haefele*
Sponsoring Editor: *Dawn R. Bercier*
Director of Development: *David Dietz*
Developmental Editor: *Lindsay Roth*
Marketing Director: *Ryan Blankenship*
Senior Project Manager: *Vicki Krug*
Senior Production Supervisor: *Sherry L. Kane*
Lead Media Project Manager: *Stacy A. Patch*
Media Producer: *Amber M. Huebner*
Senior Designer: *David W. Hash*
Cover/Interior Designer: *Maureen McCutcheon*
(USE) Cover Image: *The Gateway Arch, Saint Louis, Missouri, ©Gary Cralle/Getty Images*
Senior Photo Research Coordinator: *John C. Leland*
Photo Research: *Emily Tietz*
Supplement Producer: *Melissa M. Leick*
Compositor: *Techbooks*
Typeface: 10.5/12 *Times Roman*
Printer: *R. R. Donnelley Willard, OH*

Chapter 1 Opener: © Brand X Pictures/PunchStock; p. 32: © Royalty-Free/CORBIS

Chapter 2 Opener: © Andrew Michael/Stone/Getty Images; p. 93: © Royalty-Free/CORBIS; p. 147: © Royalty-Free/CORBIS

Chapter 3 Opener: © Bettmann/CORBIS; p. 204: © Tony Freeman/Photo Edit

Chapter 4 Opener: © Adam Jones / Visuals Unlimited; p. 228: © Keystone, Denis Balibouse, Pool/AP Photo; p. 275: © John Wang/Getty Images; p. 287: © Dennis MacDonald/Photo Edit

Chapter 5 Opener: © AGE fotostock / SuperStock; p. 350: © Steve Raymer/CORBIS; p. 367: © Maanwar Mirza/Reuters/CORBIS

Chapter 6 Opener: © RubberBall / SuperStock; p. 439: © H. Wiesenhofer/PhotoLink/Getty Images; p. 461: © Cindy Charles/Photo Edit

Library of Congress Cataloging-in-Publication Data

Coburn, John W.
Trigonometry / John W. Coburn. – 1st ed.
p. cm. – (Coburn's precalculus series)
Includes index.
ISBN 978–0–07–291005–6 --- ISBN 0–07–291005–4 (acid-free paper)
1. Trigonometry–Textbooks. I. Title.

QA531.C63 2008
516.24–dc22 2006047031

www.mhhe.com

ABOUT THE AUTHOR

John Coburn grew up in the Hawaiian Islands, the seventh of sixteen children. In 1979 he received a bachelor's degree in education from the University of Hawaii. After being lured into the business world for a number of years, he returned to his first love, accepting a teaching position in high school mathematics where he was recognized as Teacher of the Year in 1987. Soon afterward, John decided to seek a master's degree, which he received two years later from the University of Oklahoma. For the last seventeen years, he has been teaching mathematics at the Florissant Valley campus of St. Louis Community College, where he is now a full professor. During his tenure there he has received numerous nominations as an outstanding teacher by the local chapter of Phi Theta Kappa, and was recognized as Post–Secondary Teacher of the Year in 2004 by the Mathematics Educators of Greater St. Louis (MEGSL). He has made numerous presentations at local, state, and national conferences on a wide variety of topics. His other loves include his family, music, athletics, games, and all things beautiful. We hope this love of life comes through in his writing, and serves to make the learning experience an interesting and engaging one for all students.

DEDICATION

To my wife and best friend Helen, whose love, support, and willingness to sacrifice never faltered.

Contents

Preface

FROM THE AUTHOR

I was raised on the island of Oahu, and was a boy of four when Hawaii celebrated its statehood. From Laie Elementary to my graduation from the University of Hawaii, my educational experience was hugely cosmopolitan. Every day was filled with teachers and fellow students from every race, language, culture, and country imaginable, and this experience made an indelible impression on my view of the world. I can only hope that this exposure to different ideas and new perspectives contributed to an ability to connect with a diverse audience. It has certainly instilled the desire to communicate effectively with students from all walks of life—students like yours. Even my home experience helped to mold my thinking in this direction, because my education at home was closely connected to my public education. You see, Mom and Dad were both teachers. Mom taught English and Dad, as fate would have it, held advanced degrees in physics, chemistry, and . . . mathematics. But where my father was well known, well respected, and a talented mathematician, I was no prodigy and had to work very hard to see the connections so necessary for success in mathematics. In many ways, my writing is born of this experience, as it seemed to me that many texts offered too scant a framework to build concepts, too terse a development to make connections, and insufficient support in their exercise sets to develop long-term retention or foster a love of mathematics. To this end I've adopted a mantra of sorts, that being, "If you want more students to reach the top, you gotta put a few more rungs on the ladder." These are some of the things that have contributed to the text's unique and engaging style, and I hope in the end, to its widespread appeal.

Chapter Overview

The organization and pedagogy of each chapter support an approach sustained throughout the text, that of laying a firm foundation, building a solid framework, and providing strong connections. In the end, you'll have a beautiful, strong, and lasting structure, designed to support further learning opportunities. Each chapter also offers *Mid-Chapter Checks,* and contains the features *Reinforcing Basic Concepts* and *Strengthening Core Skills,* all designed to support student efforts and build long-term retention. The *Summary and Concept Reviews* offer on-the-spot, structured review exercises, while the *Mixed Review* gives students the opportunity to decide among available solution strategies. All *Practice Tests* have been carefully crafted to match the tone, type, and variety of exercises introduced in the chapter, with the *Cumulative Reviews* closely linked to the *Maintaining Your Skills* feature found in every section. Finally, the *Calculator Exploration and Discovery* feature, well . . . it does just that, offering students the opportunity to go beyond what is possible with paper and pencil alone.

Section Overview

Every section begins by putting some perspective on upcoming material while placing it in the context of the "larger picture." Objectives for the section are clearly laid out. The *Point of Interest* features were carefully researched and help to color the mathematical landscape, or make it more closely connected. The exposition has a smooth and conversational style, and includes helpful hints, mathematical connections, cautions, and opportunities for further exploration. Examples were carefully chosen to weave a tight-knit fabric, and everywhere possible, to link concepts and topics under discussion to real-world experience. A wealth of exercises support the section's main ideas, and due to their range of difficulty, there is very strong support for weaker students, while advanced students are challenged to reach even further. Each exercise set includes the following categories: *Concepts and Vocabulary; Developing Your Skills; Working with Formulas; Applications; Writing, Research, and Decision Making; Extending the Concept;* and *Maintaining Your Skills;* all carefully planned, sequenced, and thought out. The majority of reviewers seem to think that the applications were first-rate, a staple of this text, and one of its strongest, most appealing features.

Technology Overview

Writing a text that recognizes the diversity that exists among teaching methods and philosophies was a very difficult task. While the majority of the text can in fact be taught with minimal calculator use, there is an abundance of resources for teachers that advocate its total integration into the curriculum. Almost every section contains a detailed *Technology Highlight,* every chapter a *Calculator Exploration and Discovery* feature, and calculator use is demonstrated at appropriate times and in appropriate ways throughout. For the far greater part, instructors can use graphing and calculating technology where and how they see fit and feel supported by the text. Additionally, there are a number of on-line features and supplements that encourage further mathematical exploration, additional support for the use of graphing and programming technology, with substantive and meaningful student collaborations using the *Mathematics in Action* features available at www.mhhe.com/coburn.

Summary and Conclusion

You have in your hands a powerful tool with numerous features. All of your favorite and familiar components are there, to be used in support of your own unique style, background, and goals. The additional features are closely linked and easily accessible, enabling you to try new ideas and extend others. It is our hope that this textbook and its optional supplements provide all the tools you need to teach the course you've always wanted to teach. Writing these texts was one of the most daunting and challenging experiences of my life, particularly with an 8-year-old daughter often sitting in my lap as I typed, and the twins making off with my calculators so they could draw pretty graphs. But as you might imagine, in undertaking an endeavor of this scope and magnitude, I was blessed to experience the thrill of discovery and rediscovery a thousand times. I'd like to conclude by soliciting your help. As hard as we've worked on this project, and as proud as our McGraw-Hill team is of the result, we know there is room for improvement. Our reviewers have proven many times over there is a wealth of untapped ideas, new perspectives, and alternative approaches that can help bring a new and higher level of clarity to the teaching and learning of mathematics. Please let us know how we can make a good thing better.

ACKNOWLEDGMENTS

I first want to express a deep appreciation for the guidance, comments, and suggestions offered by those who reviewed various portions of the manuscript. I found their collegial exchange of ideas and experience very refreshing, instructive, and sometimes chastening, but always helping to create a better learning tool for our students.

Rosalie Abraham
Florida Community College at Jacksonville

Jay Abramson
Arizona State University

Omar Adawi
Parkland College

Carolyn Autrey
University of West Georgia

Jannette Avery
Monroe Community College

Adele Berger
Miami Dade College

Jean Bevis
Georgia State University

Patricia Bezona
Valdosta State University

Patrick Bibby
Miami Dade College

Elaine Bouldin Tenpenny
Middle Tennessee State University

Anna Butler
East Carolina University

Cecil Coone
Southwest Tennessee Community College

Charles Cooper
University of Central Oklahoma

Sally Copeland
Johnson County Community College

Nancy Covey Jenkins
Strayer University

Julane Crabtree
Johnson County Community College

Steve Cunningham
San Antonio College

Tina Deemer
University of Arizona

Jennifer Dollar
Grand Rapids Community College

Patricia Ellington
University of Texas at Arlington

Angela Everett
Chattanooga State Technical Community College

Gerry Fitch
Louisiana State University

James Gilbert
Mississippi Gulf Coast Community College

Ilene Grant
Georgia Perimeter College

Jim Hardman
Sinclair Community College

Brenda Helms
Mississippi Gulf Coast Community College

Laura Hillerbrand
Broward Community College

Linda Hurst
Central Texas College

John Kalliongis
Saint Louis University

Fritz Keinert
Iowa State University

Thomas Keller
Southwest Texas State University

Marlene Kovaly
Florida Community College at Jacksonville

Betty Larson
South Dakota State University

Denise LeGrand
University of Arkansas at Little Rock

Lisa Mantini
Oklahoma State University

Nancy Matthews
University of Oklahoma

Thomas McMillan
University of Arkansas at Little Rock

Owen Mertens
Southwest Missouri State University

James Miller
West Virginia University

Christina Morian
Lincoln University

Jeffrey O'Connell
Ohlone College

Debra Otto
University of Toledo

Luke Papademas
DeVry University–Chicago

Frank Pecchioni
Jefferson Community College

Greg Perkins
Hartnell College

Shahla Peterman
University of Missouri

Jeanne Pirie
Erie Community College

David Platt
Front Range Community College

Evelyn Pupplo-Cody
Marshall University

Lori Pyle
University of Central Florida

Linda Reist
Macomb Community College

Ira Lee Riddle
Pennsylvania State University–Abington

Kathy Rodgers
University of Southern Indiana

Behnaz Rouhani
Georgia Perimeter College

David Schultz
Mesa Community College

John Seims
Mesa Community College–Red Mountain Campus

Delphy Shaulis
University of Colorado

Jean Shutters
Harrisburg Area Community College

Albert Simmons
Ozarks Technical Community College

Mohan Tikoo
Southeast Missouri State University

Diane Trimble
Tulsa Community College–West Campus

Anthony Vance
Austin Community College

Arun Verma
Hampton University

Erin Wall
College of the Redwoods

Anna Wlodarczyk
Florida International University

Kevin Yokoyama
College of the Redwoods

I would also like to thank those who participated in various precalculus symposia and offered valuable advice.

Robert Anderson
University of Wisconsin–Eau Claire

George Avirappattu
Kean University

Rajilakshmi Baradwaj
University of Maryland–Baltimore County

Judy Barclay
Cuesta College

Donna Beatty
Ventura College

Kim Bennekin
Georgia Perimeter College

Branson Brade
Houston Community College

Beverly Broomell
Suffolk County Community College

Mary Cottier
St. Phillips College

Donna Densmore
Bossier Parish Community College

Hamidulla Farhat
Hampton University

Patricia Foard
South Plains College

Bill Forrest
Baton Rouge Community College

Nancy Forrester
Northeast State Community College

Marc Grether
University of North Texas

Steve Grosteffon
Santa Fe Community College

Ali Hajjafar
University of Akron

Sharon Hamsa
Longview Community College

Janice Hector
De Anza Community College

Ellen Hill
Minnesota State University–Moorhead

Teresa Houston
Eastern Mississippi Community College

Tim Howard
Columbus State University

Miles Hubbard
St. Cloud State University

Fred Keene
Pasadena City College

Fritz Keinert
Iowa State University

Paul Kimble
Southwest Texas Junior College

Tor Kwembe
Jackson State University

Marie Larsen
Cuesta College

Danny Lau
Gainesville State College

Kathryn Lavelle
Westchester Community College

Mitch Levy
Broward Community College

Manoug Manougian
University of South Florida, Tampa

Nancy Matthews
University of Oklahoma

Steve Matusow
Evergreen Community College

Ram Mohapatra
University of Central Florida

Scott Mortensen
Dixie State College

James Newsom
Tidewater Community College

Curtis Paul
Moorpark Community College

Michael Rosenthal
Florida International University

Beverly Reed
Kent State University

Geoffrey Schulz
Community College of Philadelphia

Rebecca Sellers
Jefferson State Community College

John Smith
Hawaii Pacific University

Dave Sobecki
Miami University

Eleanor Storey
Front Range Community College

Scott Sykes
University of West Georgia

Linda Tucker
Rose State College

Anthony Vance
Austin Community College

Michele Wallace
Washington State University

Larissa Williamson
University of Florida

Randall Wills
Southeastern Louisiana University

Nate Wilson
St. Louis Community College

Additional gratitude goes to Carrie Green for her careful accuracy checking and helpful suggestions. Thank you to Rosemary Karr and Lesley Seale for authoring the solutions manuals. Rosemary is owed a special debt of gratitude for her tireless attention to detail and her willingness to go above and beyond the call of duty. I would especially like to thank John Leland and Emily Tietz for their efforts in securing just the right photos; Vicki Krug (whose motto is undoubtedly *From Panta Rhei to Fait Accompli*) for her uncanny ability to bring innumerable parts from all directions into a unified whole; Patricia Steele, a copy editor *par excellance* who can tell an en dash from a minus sign at 50 paces; Dawn Bercier for her enthusiasm in marketing the Coburn series; Suzanne Alley for her helpful suggestions, infinite patience, and steady hand in bringing the manuscript to completion; and Steve Stembridge, whose personal warmth, unflappable manner, and down-to-earth approach to problem solving kept us all on time and on target. In truth, my hat is off to all the fine people at McGraw-Hill for their continuing support and belief in this series. A final word of thanks must go to Rick Armstrong, whose depth of knowledge, experience, and mathematical connections seems endless; Anne Marie Mosher for her contributions to various features of the text and to J. D. Herdlick, Richard Pescarino, and the rest of my colleagues at St. Louis Community College whose friendship, encouragement, and love of mathematics makes going to work each day a joy.

A COMMITMENT TO ACCURACY

You have a right to expect an accurate textbook, and McGraw-Hill invests considerable time and effort to make sure that we deliver one. Listed below are the many steps we take to make sure this happens.

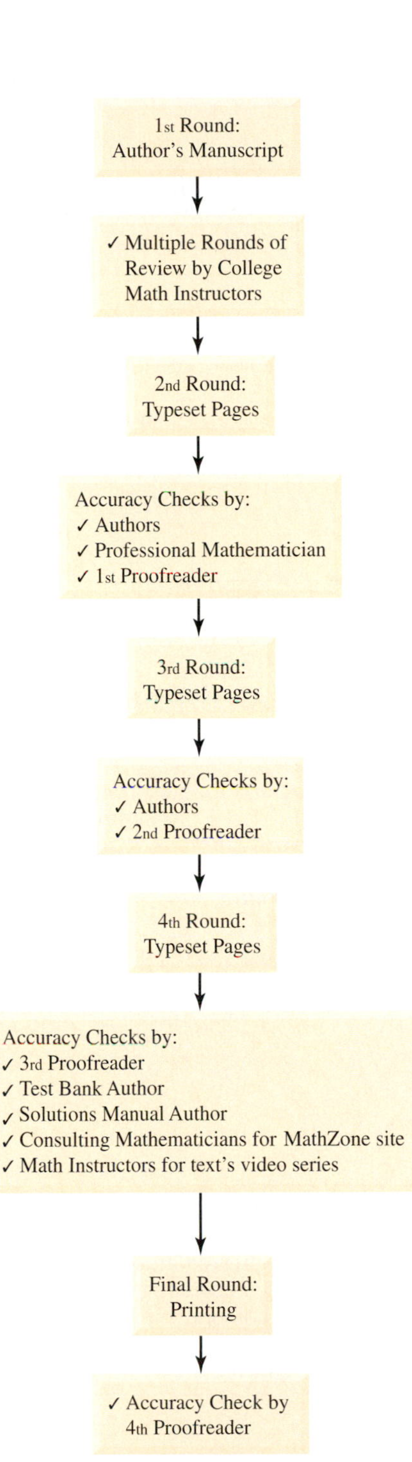

OUR ACCURACY VERIFICATION PROCESS

First Round

Step 1: Numerous **college math instructors** review the manuscript and report on any errors that they may find, and the authors make these corrections in their final manuscript.

Second Round

Step 2: Once the manuscript has been typeset, the **authors** check their manuscript against the first page proofs to ensure that all illustrations, graphs, examples, exercises, solutions, and answers have been correctly laid out on the pages, and that all notation is correctly used.

Step 3: An outside, **professional mathematician** works through every example and exercise in the page proofs to verify the accuracy of the answers.

Step 4: A **proofreader** adds a triple layer of accuracy assurance in the first pages by hunting for errors, then a second, corrected round of page proofs is produced.

Third Round

Step 5: The **author team** reviews the second round of page proofs for two reasons: 1) to make certain that any previous corrections were properly made, and 2) to look for any errors they might have missed on the first round.

Step 6: A **second proofreader** is added to the project to examine the new round of page proofs to double check the author team's work and to lend a fresh, critical eye to the book before the third round of paging.

Fourth Round

Step 7: A **third proofreader** inspects the third round of page proofs to verify that all previous corrections have been properly made and that there are no new or remaining errors.

Step 8: Meanwhile, in partnership with **independent mathematicians,** the text accuracy is verified from a variety of fresh perspectives:

- The **test bank author** checks for consistency and accuracy as they prepare the computerized test item file.
- The **solutions manual author** works every single exercise and verifies their answers, reporting any errors to the publisher.
- A **consulting group of mathematicians,** who write material for the text's MathZone site, notifies the publisher of any errors they encounter in the page proofs.
- A video production company employing **expert math instructors** for the text's videos will alert the publisher of any errors they might find in the page proofs.

Final Round

Step 9: The **project manager,** who has overseen the book from the beginning, performs a **fourth proofread** of the textbook during the printing process, providing a final accuracy review.

⇒ What results is a mathematics textbook that is as accurate and error-free as is humanly possible, and our authors and publishing staff are confident that our many layers of quality assurance have produced textbooks that are the leaders of the industry for their integrity and correctness.

Guided Tour

Laying a Firm Foundation . . .

OUTSTANDING EXAMPLES

Abundant examples carefully prepare the students for homework and exams. Easily located on the page, Coburn's numerous worked examples expose the learner to more exercise types than most other texts.

Now Try boxes immediately follow most examples to guide the students to specific matched and structured exercises they can try for practice and further understanding.

EXAMPLE 1 Use the symmetry of the unit circle and reference arcs of standard values to complete Table 2.3.

Table 2.3

t	π	$\frac{7\pi}{6}$	$\frac{5\pi}{4}$	$\frac{4\pi}{3}$	$\frac{3\pi}{2}$	$\frac{5\pi}{3}$	$\frac{7\pi}{4}$	$\frac{11\pi}{6}$	2π
$\sin t$									

Solution: Symmetry shows that for any odd multiple of $t = \frac{\pi}{4}$, function values will be $\pm\frac{\sqrt{2}}{2}$ depending on the quadrant of the terminal side. Similarly, for any reference arc of $\frac{\pi}{6}$, $\sin t = \pm\frac{1}{2}$, while for a reference arc of $\frac{\pi}{3}$, $\sin t = \pm\frac{\sqrt{3}}{2}$. With these, we complete the table as shown in Table 2.4.

Table 2.4

t	π	$\frac{7\pi}{6}$	$\frac{5\pi}{4}$	$\frac{4\pi}{3}$	$\frac{3\pi}{2}$	$\frac{5\pi}{3}$	$\frac{7\pi}{4}$	$\frac{11\pi}{6}$	2π
$\sin t$	0	$-\frac{1}{2}$	$-\frac{\sqrt{2}}{2}$	$-\frac{\sqrt{3}}{2}$	-1	$-\frac{\sqrt{3}}{2}$	$-\frac{\sqrt{2}}{2}$	$-\frac{1}{2}$	0

NOW TRY EXERCISES 7 AND 8

EXAMPLE 1 Use algebra to write four additional identities that belong to the Pythagorean family.

Solution: Starting with $\sin^2\theta + \cos^2\theta = 1$,

$\sin^2\theta + \cos^2\theta = 1$ — original identity
- $\sin^2\theta = 1 - \cos^2\theta$ — subtract $\cos^2\theta$
- $\sin\theta = \pm\sqrt{1 - \cos^2\theta}$ — take square root

$\sin^2\theta + \cos^2\theta = 1$ — original identity
- $\cos^2\theta = 1 - \sin^2\theta$ — subtract $\sin^2\theta$
- $\cos\theta = \pm\sqrt{1 - \sin^2\theta}$ — take square root

For the identities involving a radical, the choice of sign will depend on the quadrant of the terminal side.

NOW TRY EXERCISES 9 AND 10

Annotations located to the right of the solution sequence help the students recognize which property or procedure is being applied.

GRAPHICAL SUPPORT

The analysis of $y = 2.5\sin\left[\frac{\pi}{4}(t + 3)\right] + 6$ from Example 5(b) can be verified on a graphing calculator. Enter the function as Y_1 on the Y= screen and set an appropriate window size using the information gathered. Press the TRACE key and −3 ENTER and the calculator gives the average value $y = 6$ as output. Repeating this for $x = 5$ shows one complete cycle has been completed.

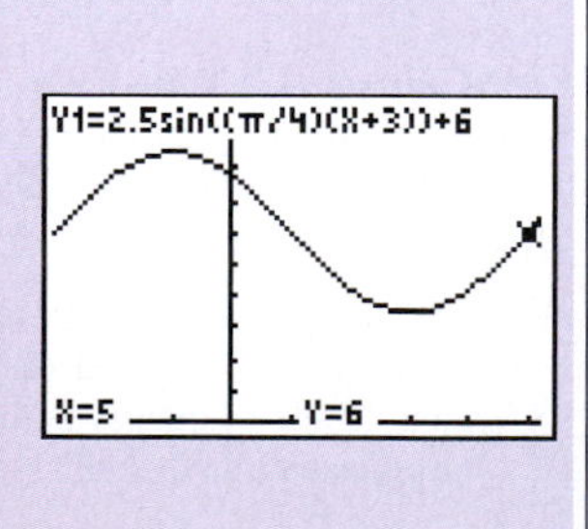

Graphical Support Boxes, located after selected examples, visually reinforce algebraic concepts with a corresponding graphing calculator example.

Building a Solid Framework . . .

SECTION EXERCISES

Concepts and Vocabulary exercises help students recall and retain important mathematical terms, building the solid vocabulary they need to verbalize and understand algebraic concepts.

CONCEPTS AND VOCABULARY

Fill in each blank with the appropriate word or phrase. Carefully reread the section if needed.

1. The double-angle identities can be derived using the ______ identities with $\alpha = \beta$. For $\cos(2\theta)$ we expand $\cos(\alpha + \beta)$ using ______.
2. If θ is in QI, $0 < \theta < 90°$. This means 2θ must be in ______ or ______, since ______ $< 2\theta <$ ______. If θ is in QIII then $180° < \theta < 270°$ and $\frac{\theta}{2}$ must be in ______ since ______ $< \frac{\theta}{2} <$ ______.
3. Multiple-angle identities can also be derived using the sum and difference identities. For $\sin(3x)$ use $\sin($____ + ____$)$.
4. The half-angle identities are closely related to the ______ ______ identities and can be developed using algebra and a change of ______.

DEVELOPING YOUR SKILLS

Find exact values for $\sin(2\theta)$, $\cos(2\theta)$, and $\tan(2\theta)$ using the information given.

7. $\sin\theta = \frac{5}{13}$; θ in QII

8. $\cos\theta = -\frac{21}{29}$; θ in QII

9. $\cos\theta = -\frac{9}{41}$; θ in QII

10. $\sin\theta = -\frac{63}{65}$; θ in QIII

11. $\tan\theta = \frac{13}{84}$; θ in QIII

12. $\sec\theta = \frac{53}{28}$; θ in QI

Developing Your Skills exercises help students reinforce what they have learned by offering plenty of practice with increasing levels of difficulty.

Working with Formulas exercises demonstrate how equations and functions model the real world by providing contextual applications of well-known formulas.

Graphing Calculator icons appear next to examples and exercises where important concepts can be supported by use of graphing technology.

WORKING WITH FORMULAS

73. Force required to maintain equilibrium using a screw jack: $F = \frac{Wk}{c}\tan(p - \theta)$

The force required to maintain equilibrium when a screw jack is used can be modeled by the formula shown, where p is the pitch angle of the screw, W is the weight of the load, θ is the angle of friction, with k and c being constants related to a particular jack. Simplify the formula using the difference formula for tangent given $p = \frac{\pi}{6}$ and $\theta = \frac{\pi}{4}$.

WRITING, RESEARCH, AND DECISION MAKING

109. As mentioned in the *Point of Interest,* many methods have been used for angle measure over the centuries, some more logical or meaningful than what is popular today. Do some research on the evolution of angle measure, and compare/contrast the benefits and limitations of each method. In particular, try to locate information on the history of degrees, radians, mils, and gradients, and identify those still in use.

110. Use the diagram given to develop the fixed ratios for the sides of a 30-60-90 triangle. Ancient geometers knew that a hexagon (six sides) could be inscribed in a circle by laying out six consecutive chords equal in length to the radius ($r = 10$ cm for illustration). After connecting the diagonals of the hexagon, six equilateral triangles are formed with sides of 10 cm.

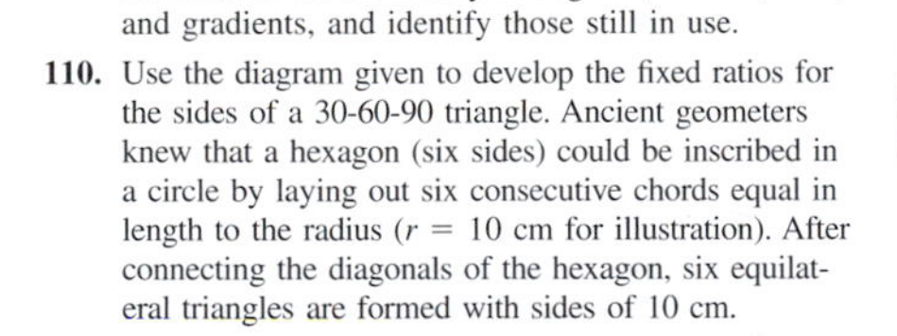

Exercise 110

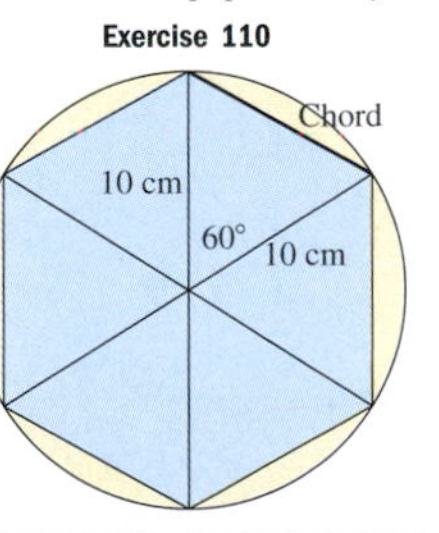

Writing, Research, and Decision Making exercises encourage students to communicate their understanding of the topics at hand or explore topics of interest in greater depth.

Wait, There's More!

- **Technology Highlights,** located before most section exercise sets, assist those interested in exploring a section topic with a graphing calculator.
- **Extending the Concept** exercises are designed to be more challenging, requiring synthesis of related concepts or the use of higher-order thinking skills.
- **Maintaining Your Skills** exercises review topics from previous chapters, helping students to retain concepts and keep skills sharp.

Mid-Chapter Checks assess student progress before they continue to the second half of the chapter.

Reinforcing Basic Concepts immediately follow the Mid-Chapter Check. This feature extends and explores a chapter topic in greater detail.

END-OF-CHAPTER MATERIAL

The **Summary and Concept Review,** located at the end of Chapters 1–6, lists key concepts and is organized by section. This format provides additional practice exercises and makes it easy for students to review the terms and concepts they will need prior to a quiz or exam.

Mixed Review exercises offer more practice on topics from the entire chapter, are arranged in random order, and require students to identify problem types and solution strategies on their own.

The **Practice Test** gives students the opportunity to check their mastery and prepare for classroom quizzes, tests, and other assessments.

Cumulative Reviews help students retain previously learned skills and concepts by revisiting important ideas from earlier chapters.

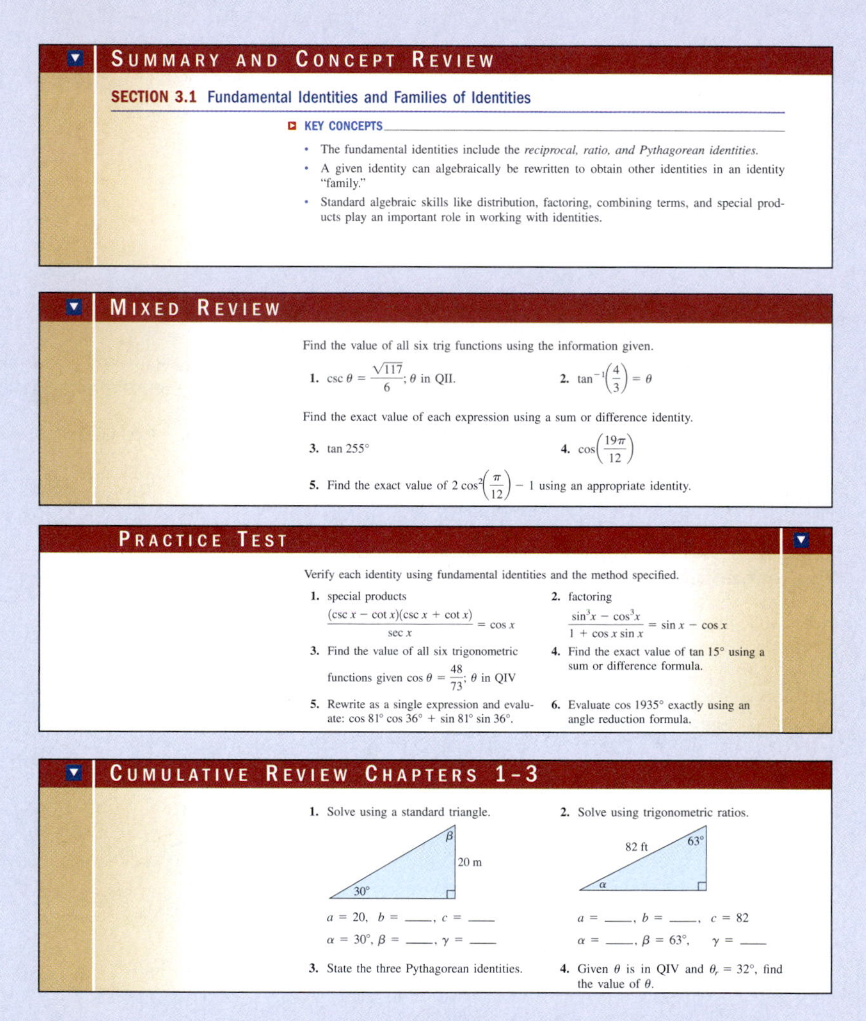

MID-CHAPTER CHECK

1. Which function, $y = \tan t$ or $y = \cot t$, is decreasing on its domain? Which function, $y = \cos t$ or $y = \sin t$, begins at (0, 1) in the interval $t \in [0, 2\pi)$?
2. State the period of $y = \sin\left(\frac{\pi}{2}t\right)$. Where will the max/min values occur in the primary interval?
3. Evaluate without using a calculator: (a) $\cot 60°$ and (b) $\sin\left(\frac{7\pi}{4}\right)$.
4. Evaluate using a calculator: (a) $\sec\left(\frac{\pi}{12}\right)$ and (b) $\tan 4.3$.
5. Which of the six trig functions are even functions?
6. State the domain of $y = \sin t$ and $y = \tan t$.

SUMMARY AND CONCEPT REVIEW

SECTION 3.1 Fundamental Identities and Families of Identities

KEY CONCEPTS

- The fundamental identities include the *reciprocal, ratio, and Pythagorean identities.*
- A given identity can algebraically be rewritten to obtain other identities in an identity "family."
- Standard algebraic skills like distribution, factoring, combining terms, and special products play an important role in working with identities.

MIXED REVIEW

Find the value of all six trig functions using the information given.

1. $\csc\theta = \frac{\sqrt{117}}{6}$; θ in QII.
2. $\tan^{-1}\left(\frac{4}{3}\right) = \theta$

Find the exact value of each expression using a sum or difference identity.

3. $\tan 255°$
4. $\cos\left(\frac{19\pi}{12}\right)$
5. Find the exact value of $2\cos^2\left(\frac{\pi}{12}\right) - 1$ using an appropriate identity.

PRACTICE TEST

Verify each identity using fundamental identities and the method specified.

1. special products
$\frac{(\csc x - \cot x)(\csc x + \cot x)}{\sec x} = \cos x$
2. factoring
$\frac{\sin^3 x - \cos^3 x}{1 + \cos x \sin x} = \sin x - \cos x$
3. Find the value of all six trigonometric functions given $\cos\theta = \frac{48}{73}$; θ in QIV
4. Find the exact value of tan 15° using a sum or difference formula.
5. Rewrite as a single expression and evaluate: $\cos 81° \cos 36° + \sin 81° \sin 36°$.
6. Evaluate cos 1935° exactly using an angle reduction formula.

CUMULATIVE REVIEW CHAPTERS 1–3

1. Solve using a standard triangle.

$a = 20$, $b =$ ___, $c =$ ___
$\alpha = 30°$, $\beta =$ ___, $\gamma =$ ___

2. Solve using trigonometric ratios.

$a =$ ___, $b =$ ___, $c = 82$
$\alpha =$ ___, $\beta = 63°$, $\gamma =$ ___

3. State the three Pythagorean identities.
4. Given θ is in QIV and $\theta_r = 32°$, find the value of θ.

Wait, There's *Still* More!

- The **Calculator Exploration and Discovery** feature is designed to extend the borders of a student's mathematical understanding using the power of graphing and calculating technology.
- **Strengthening Core Skills** exercises help students strengthen skills that form the bedrock of mathematics and lead to continued success.

Providing Strong Connections . . .

THROUGH APPLICATIONS!

The **Index of Applications** is located immediately after the Guided Tour and is organized by discipline to help identify applications relevant to a particular field.

Meaningful Applications—over 500 carefully chosen applications explore a wide variety of interests and illustrate how mathematics is connected to other disciplines and the world around us.

82. Fluid mechanics: In studies of fluid mechanics, the equation $\gamma_1 V_1 \sin\alpha = \gamma_2 V_2 \sin(\alpha - \beta)$ sometimes arises. Use a difference identity to show that if $\gamma_1 V_1 = \gamma_2 V_2$, the equation is equivalent to $\cos\beta - \cot\alpha \sin\beta = 1$.

83. Art and mathematics: When working in two-point geometric perspective, artists must scale their work to fit on the paper or canvas they are using. In doing so, the equation $\frac{A}{B} = \frac{\tan\theta}{\tan(90° - \theta)}$ arises. Rewrite the expression on the right in terms of sine and cosine, then use the difference identities to show the equation can be rewritten as $\frac{A}{B} = \tan^2\theta$.

84. Traveling waves: If two waves of the same frequency, velocity, and amplitude are traveling along a string in opposite directions, they can be represented by the equations ...d $Y_2 = A\sin(kx + \omega t)$. Use the sum and difference formulas ...ult $Y_R = Y_1 + Y_2$ of these waves can be expressed as

91. Temperature and altitude: The temperature (in degrees Fahrenheit) at a given altitude can be approximated by the function $f(x) = -\frac{7}{2}x + 59$, where $f(x)$ represents the temperature and x represents the altitude in thousands of feet. (a) What is the approximate temperature at an altitude of 35,000 ft (normal cruising altitude for commercial airliners)? (b) Find $f^{-1}(x)$, then input your answer from part (a) and comment on the result. (c) If the temperature outside a weather balloon is $-18°$F, what is the approximate altitude of the balloon?

92. Fines for speeding: In some localities, there is a set formula to determine the amount of a fine for exceeding posted speed limits. Suppose the amount of the fine for exceeding a 50 mph speed limit was given by the function $f(x) = 12x - 560$ where $f(x)$ represents the fine in dollars for a speed of x mph. (a) What is the fine for traveling 65 mph through this speed zone? (b) Find $f^{-1}(x)$, then input your answer from

Looking for Interactive Applications? Look Online!

The **Mathematics in Action** activities, located at www.mhhe.com/coburn, enable students to work collaboratively as they manipulate applets that apply mathematical concepts in real-world contexts.

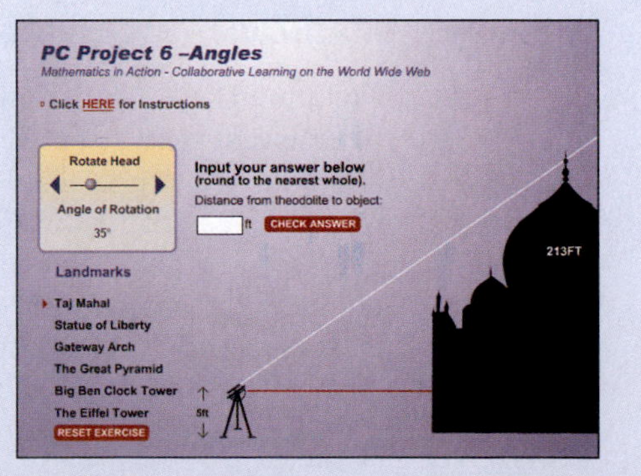

Concepts for Calculus icons identify concepts or skills that a student will likely see in a first semester calculus course.

SUPPLEMENTS FOR INSTRUCTORS

Instructor's Edition
ISBN–13: 978-0-07-313704-9 (ISBN–10: 0-07-313704-9)

In the Instructor's Edition (IE), **exercise answers appear in the back of the IE as an appendix.**

Instructor's Solutions Manual
ISBN–13: 978-0-07-330791-6 (ISBN–10: 0-07-330791-2)

Authored by Rosemary Karr and Lesley Seale, the *Instructor's Solutions Manual* contains detailed, **worked-out solutions** to all exercises in the text.

Instructor's Testing and Resource CD-ROM
ISBN–13: 978-0-07-330792-3 (ISBN–10: 0-07-330792-0)

This cross-platform CD-ROM provides a wealth of resources for the instructor. Among the supplements featured on the CD-ROM is a **computerized test bank** utilizing Brownstone Diploma® algorithm-based testing software to quickly create customized exams. This user-friendly program enables instructors to search for questions by topic, format, or difficulty level; to edit existing questions or to add new ones; and to scramble questions and answer keys for multiple versions of a single test. Hundreds of text-specific, open-ended and multiple-choice questions are included in the question bank. Sample chapter tests, midterms, and final exams in Microsoft Word® and PDF formats are also provided.

Video Lectures on DVD
ISBN–13: 978-0-07-330794-7 (ISBN–10: 0-07-330794-7)

In the videos, qualified teachers work through selected problems from the textbook, following the solution methodology employed in the text. The video series is available on DVD or VHS videocassette, or online as an assignable element of MathZone (see section on MathZone). The DVDs are closed-captioned for the hearing impaired, subtitled in Spanish, and meet the Americans with Disabilities Act Standards for Accessible Design. Instructors can use them as resources in a learning center, for online courses, and/or to provide extra help to students who require extra practice.

MathZone—www.mathzone.com

McGraw-Hill's **MathZone 3.0** is a complete **Web-based tutorial and course management system** for mathematics and statistics, designed for greater ease of use than any other system available. Available with selected McGraw-Hill textbooks, the system enables instructors to **create and share courses and assignments** with colleagues, adjunct faculty members, and teaching assistants with only a few mouse clicks. All **assignments, exercises, e-Professor multimedia tutorials, video lectures, and NetTutor® live tutors** follow the textbook's learning objectives and problem-solving style and notation. Using MathZone's **assignment builder,** instructors can **edit questions and algorithms, import their own content,** and **create**

announcements and due dates for homework and quizzes. MathZone's **automated grading function** reports the results of easy-to-assign algorithmically generated homework, quizzes, and tests. All student activity within MathZone is recorded and available through a **fully integrated gradebook** that can be downloaded to Microsoft Excel®. (See "Supplements for the Student" for descriptions of the elements of MathZone.)

ALEKS

ALEKS (**A**ssessment and **LE**arning in **K**nowledge **S**paces) is an artificial-intelligence-based system for mathematics learning, available over the Web 24/7. Using unique adaptive questioning, ALEKS accurately assesses what topics each student knows and then determines exactly what each student is ready to learn next. ALEKS interacts with the students much as a skilled human tutor would, moving between explanation and practice as needed, correcting and analyzing errors, defining terms and changing topics on request, and helping them master the course content more quickly and easily. Moreover, the new ALEKS 3.0 now links to text-specific videos, multimedia tutorials, and textbook pages in PDF format. ALEKS also offers a robust classroom management system that enables instructors to monitor and direct student progress toward mastery of curricular goals. See www.highed.aleks.com for more information.

SUPPLEMENTS FOR STUDENTS

Student's Solutions Manual
ISBN-13: 978-0-07-291766-6 (ISBN-10: 0-07-291766-0)

Authored by Rosemary Karr and Lesley Seale, the *Student's Solutions Manual* contains detailed, **worked-out solutions** to all the problems in the Mid-Chapter Checks, Reinforcing Basic Concepts, Summary and Concept Review Exercises, Practice Tests, Cumulative Reviews, and Strengthening Core Skills. Also included are **worked-out solutions** for odd-numbered exercises of the section exercises and the mixed reviews. The steps shown in solutions are carefully matched to the style of solved examples in the textbook.

MathZone—www.mathzone.com

McGraw-Hill's MathZone is a powerful Web-based tutorial for homework, quizzing, testing, and multimedia instruction. Also available in CD-ROM format, MathZone offers:

- **Practice exercises** based on the text and generated in an unlimited quantity for as much practice as needed to master an objective.
- **Video clips** of classroom instructors showing how to solve exercises from the text; **e-Professor** animations that take a student through step-by-step instructions (delivered on-screen and narrated by a teacher on audio) for solving exercises from the textbook; the user controls the pace of the explanations and can review as needed.
- **NetTutor,** which offers personalized instruction by live tutors familiar with the textbook's objectives and problem-solving methods.

Every assignment, exercise, video lecture, and e-Professor is derived from the textbook.

Video Lectures on DVD

ISBN-13: 978-0-07-330794-7 (ISBN-10: 0-07-330794-7)

The video series is based on exercises from the textbook. Each presenter works through selected problems, following the solution methodology employed in the text. The video series is available on DVD or online as part of MathZone. The DVDs are closed-captioned for the hearing impaired, subtitled in Spanish, and meet the Americans with Disabilities Act Standards for Accessible Design.

NetTutor

Available through MathZone, NetTutor is a revolutionary system that enables students to interact with a live tutor over the Web. NetTutor's Web-based, graphical chat capabilities enable students and tutors to use mathematical notation and even to draw graphs as they work through a problem together. Students can also submit questions and receive answers, browse previously answered questions, and view previous sessions. Tutors are familiar with the textbook's objectives and problem-solving styles.

ALEKS

(**A**ssessment and **LE**arning in **K**nowledge **S**paces) is an artificial intelligence-based system for mathematics learning, available online 24/7. ALEKS interacts with the student much as a skilled human tutor would, moving between explanation and practice as needed, helping you master the course content more quickly and easily. NEW! ALEKS 3.0 now links to text-specific videos, multimedia tutorials, and textbook pages in PDF format. See www.highed.aleks.com for more information.

Index of Applications

MEDICINE/NURSING/NUTRITION/DIETETICS/HEALTH

METEOROLOGY

PHYSICS/ASTRONOMY/PLANETARY STUDIES

SPORTS/LEISURE

TRANSPORTATION

Chapter 1

An Introduction to Trigonometry

Chapter Outline

Preview

In this chapter we'll draw on numerous concepts from previous courses as we begin a study of trigonometry. The classical approach involves **static trigonometry** and a study of right triangles, the word *static* meaning, "standing or fixed in one place" (Merriam Webster). We'll also look at **dynamic trigonometry**—the trigonometry of objects in motion or in the process of change. Sometimes you'll still be able to see a "right triangle connection" in the applications we consider, but more often the applications will be free from any concept of angle measure as we focus on the trigonometry of real numbers. All in all, we hope to create a framework that unites algebra, geometry, and trigonometry, as it was indeed their union that enabled huge advances in knowledge, technology, and an understanding of things that are.

1.1 Angle Measure, Special Triangles, and Special Angles

LEARNING OBJECTIVES

In Section 1.1 you will learn how to:

A. **Use the vocabulary associated with a study of angles and triangles**

B. **Find fixed ratios of the sides of special triangles using a polygon inscribed in a circle**

C. **Use radians for angle measure and compute circular arc length and area using radians**

D. **Convert between degrees and radians for nonstandard angles**

E. **Solve applications involving angular velocity and linear velocity using radians**

INTRODUCTION

Trigonometry, like her sister science geometry, has its origins deeply rooted in the practical and scientific use of measurement and proportion. In this section we'll look at the fundamental concepts on which trigonometry is based, which we hope will lead to a better understanding and a greater appreciation of the wonderful science that trigonometry has become.

POINT OF INTEREST

While angle measure based on a 360° circle has been almost universally accepted for centuries, its basic construct is contrived, artificial, and no better or worse than other measures proposed and used over time (stadia, gons, cirs, points, mils, gradients, and so on). In the 1870s, mathematician Thomas Muir and James Thomson (the brother of Lord Kelvin) advocated the need for a new unit, stating the measure of an angle in terms of a circle's inherent characteristics, rather than arbitrarily declared numbers like 360 (degrees), 400 (gradients), or 1000 (mils). The new measure, which after some deliberation they called *radians,* conveniently expressed the standard angles used since antiquity $\left(30° = \frac{\pi}{6}, 45° = \frac{\pi}{4}, 60° = \frac{\pi}{3}, \text{etc.}\right)$, while contributing to the simplification of many mathematical formulas and procedures. Using radians also allows a clearer view of the trigonometric functions as *functions of a real number,* rather than merely functions of an angle, thus extending their influence on both pure and applied mathematics.

A. Angle Measure in Degrees

Beginning with the common notion of a straight line, a **ray** is a half line, or all points extending from a single point, in a single direction. An **angle** is the joining of two rays at a common endpoint called the **vertex.** Arrowheads are used to indicate the half lines continue forever and can be extended if necessary. Angles can be named using a single letter at the vertex, the letters from the rays forming the sides, or by a single Greek letter, with the favorites being **alpha** α, **beta** β, **gamma** γ, and **theta** θ. The symbol $\angle$ is often used to designate an angle (see Figure 1.1).

Figure 1.1

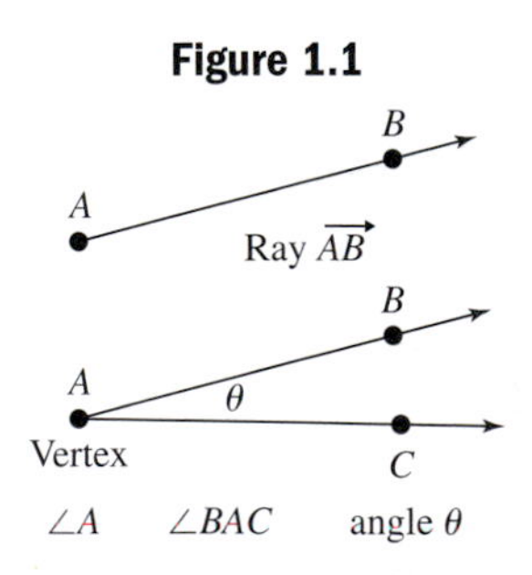

Euclid (325–265 B.C.), often thought of as the *father of geometry,* described an angle as "the inclination of one to another of two lines which meet in a plane." This *amount of inclination* gives rise to the common notion of angle measure in degrees, often measured with a semicircular **protractor** like the one shown in Figure 1.2. The notation for degrees is

Figure 1.2

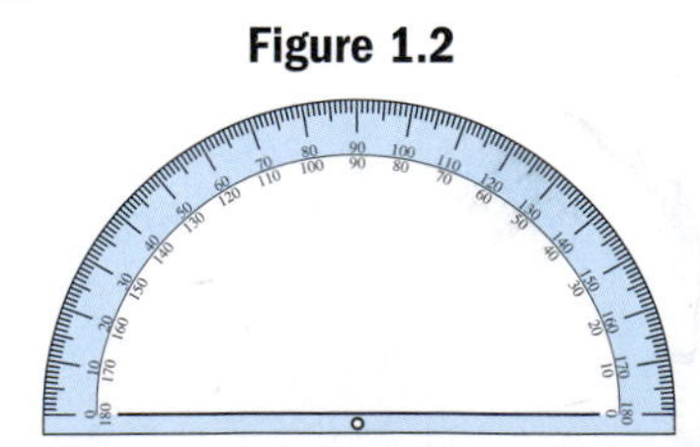

the ° symbol. By definition $1°$ is $\frac{1}{360}$ of a full rotation, so this protractor can be used to measure any angle from $0°$ (where the two rays are on top of each other) to $180°$ (where they form a straight line). An angle measuring $180°$ is called a **straight angle,** while an angle that measures $90°$ is called a **right angle.** Two angles that sum to $90°$ are said to be **complementary,** while two that sum to $180°$ are called **supplementary** angles.

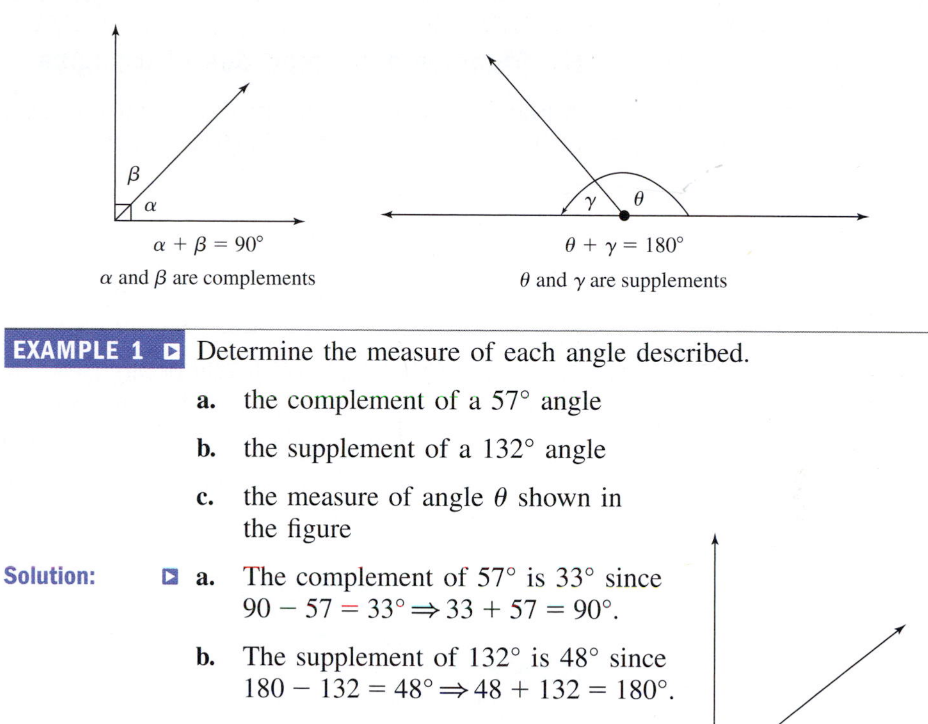

$\alpha + \beta = 90°$

α and β are complements

$\theta + \gamma = 180°$

θ and γ are supplements

EXAMPLE 1 Determine the measure of each angle described.

a. the complement of a $57°$ angle

b. the supplement of a $132°$ angle

c. the measure of angle θ shown in the figure

Solution:

a. The complement of $57°$ is $33°$ since $90 - 57 = 33° \Rightarrow 33 + 57 = 90°$.

b. The supplement of $132°$ is $48°$ since $180 - 132 = 48° \Rightarrow 48 + 132 = 180°$.

c. Since θ and $39°$ are complements, $\theta = 90 - 39 = 51°$.

NOW TRY EXERCISES 7 THROUGH 10

In a study of trigonometry, it helps to further classify the various angles we encounter. An angle greater than $0°$ but less than $90°$ is called an **acute** angle. An angle greater than $90°$ but less than $180°$ is called an **obtuse** angle. For very fine measurements, each degree is divided into 60 smaller parts called **minutes,** and each minute into 60 smaller parts called **seconds.** This means that a minute is $\frac{1}{60}$ of a degree, while a second is $\frac{1}{3600}$ of a degree. The angle whose measure is "sixty-one degrees, eighteen minutes, and forty-five seconds" is written as $61° \, 18' \, 45''$. The degrees-minutes-seconds (DMS) method of measuring angles is commonly used in aviation and navigation, while in other areas **decimal degrees** such as $61.3125°$ are preferred. You will sometimes be asked to convert between the two.

EXAMPLE 2 Convert as indicated.

a. $61° \, 18' \, 45''$ to decimal degrees

b. $142.215°$ to DMS

Solution:

a. Since $1' = \frac{1}{60}$ of a degree and $1'' = \frac{1}{3600}$ of a degree, we have $\left(61 + \frac{18}{60} + \frac{45}{3600}\right)° = 61.3125°$.

b. For the conversion to DMS we write the fractional part separate from the whole number part to compute the number of minutes it represents: $142° + 0.215° = 142° + 0.215(60)'$ or $142°\ 12.9'$. We then extend the process to find the number of seconds represented by 0.9 minutes. $142°\ 12.9' = 142°\ 12' + 0.9(60)'' = 142°\ 12'\ 54''$.

NOW TRY EXERCISES 11 THROUGH 26

B. Triangles and Properties of Triangles

A triangle is a closed plane figure with three straight sides and three angles. Regardless of its size or orientation, triangles have the following properties.

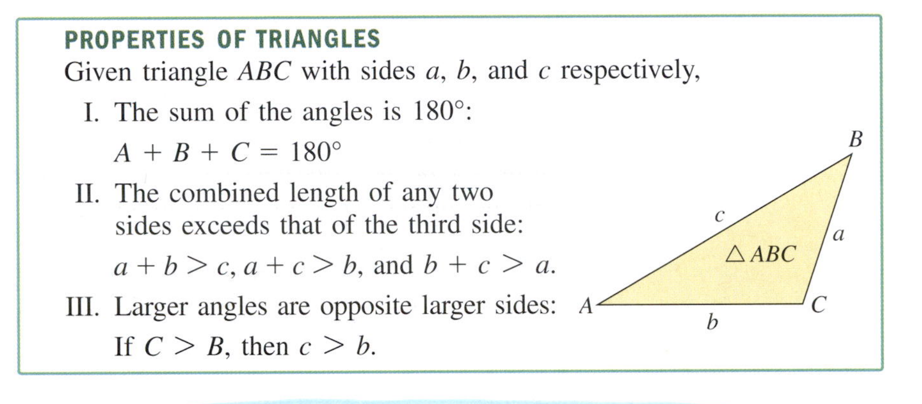

PROPERTIES OF TRIANGLES

Given triangle ABC with sides a, b, and c respectively,

I. The sum of the angles is 180°:
$A + B + C = 180°$

II. The combined length of any two sides exceeds that of the third side:
$a + b > c$, $a + c > b$, and $b + c > a$.

III. Larger angles are opposite larger sides:
If $C > B$, then $c > b$.

Two triangles are **similar triangles** if corresponding angles are equal. Since antiquity it's been known that *if two triangles are similar, corresponding sides are proportional.* This relationship, used extensively by the engineers of virtually all ancient civilizations, undergirds our study of trigonometry. Example 3 illustrates how proportions and similar triangles are often used. Note that *corresponding sides* are those opposite the equal angles from each triangle.

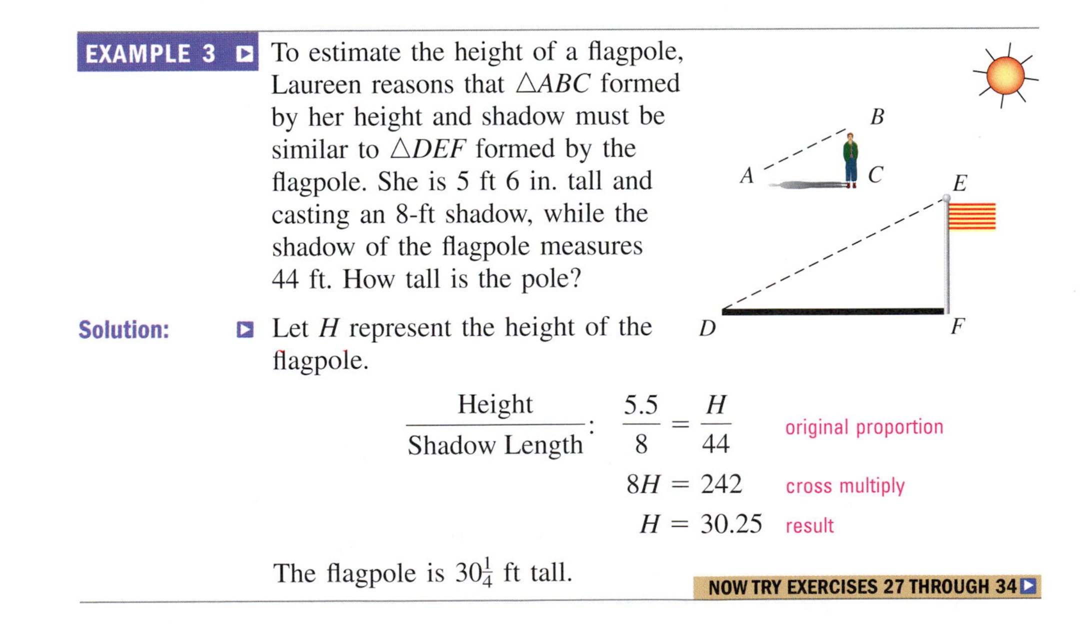

EXAMPLE 3 To estimate the height of a flagpole, Laureen reasons that $\triangle ABC$ formed by her height and shadow must be similar to $\triangle DEF$ formed by the flagpole. She is 5 ft 6 in. tall and casting an 8-ft shadow, while the shadow of the flagpole measures 44 ft. How tall is the pole?

Solution: Let H represent the height of the flagpole.

$$\frac{\text{Height}}{\text{Shadow Length}}: \quad \frac{5.5}{8} = \frac{H}{44} \quad \text{original proportion}$$

$$8H = 242 \quad \text{cross multiply}$$

$$H = 30.25 \quad \text{result}$$

The flagpole is $30\frac{1}{4}$ ft tall.

NOW TRY EXERCISES 27 THROUGH 34

Figure 1.3

Figure 1.3 shows Laureen standing along the shadow of the flagpole, again illustrating the proportional relationships that exist. Early mathematicians quickly recognized the power of these relationships, realizing if the angle of inclination and the related fixed proportions were known, they had the ability to find mountain heights, the widths of lakes, and even the ability to estimate distances to the Sun and Moon. What was needed was an accurate and systematic method of finding these "fixed proportions" for various angles, so they could be applied more widely. In support of this search, two special triangles were used. These triangles, commonly called **45-45-90** and **30-60-90** triangles, are *special* because no estimation or interpolation is needed to find the relationships between their sides. Consider a circle with $r = 10$ cm circumscribed about a square. By drawing the diagonals of the square, four right triangles are formed, each with a base and height of 10 cm (see Figure 1.4). The hypotenuse of each is a *chord of the circle*, or a line segment whose endpoints lie on the circle. Due to the symmetry, the two nonright angles are equal and measure 45°, and the Pythagorean theorem shows the length of the chord (hypotenuse) is $\sqrt{200} = 10\sqrt{2}$. This 45-45-90 triangle has sides in the proportion $10{:}10{:}10\sqrt{2}$. Actually, the relationship easily generalizes for circles of any radius and the result is stated here.

Figure 1.4

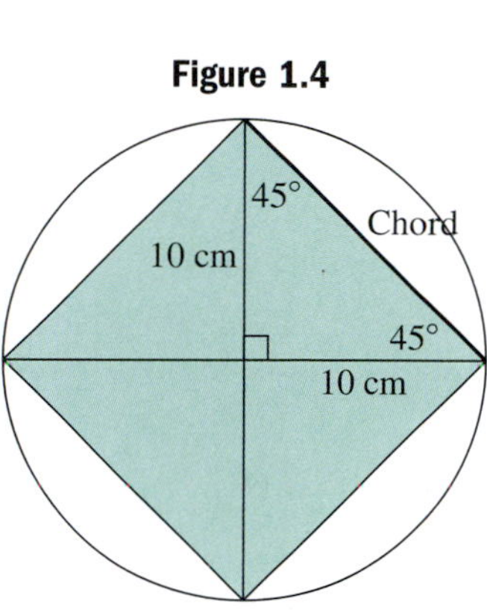

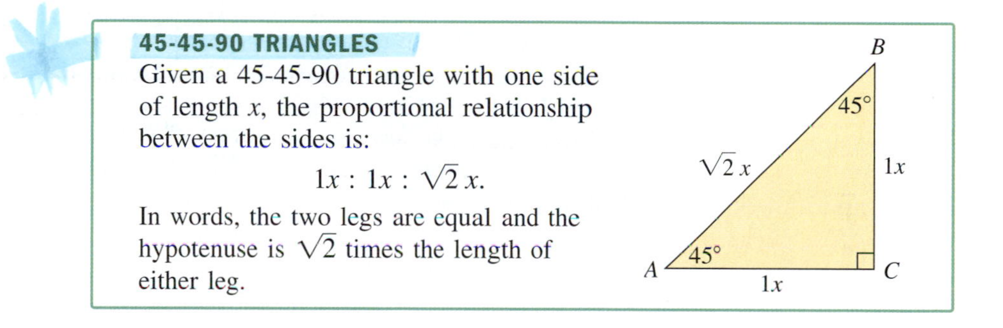

45-45-90 TRIANGLES

Given a 45-45-90 triangle with one side of length x, the proportional relationship between the sides is:

$$1x : 1x : \sqrt{2}\,x.$$

In words, the two legs are equal and the hypotenuse is $\sqrt{2}$ times the length of either leg.

The proportional relationship for a 30-60-90 triangle is developed in Exercise 110, and the result is stated here.

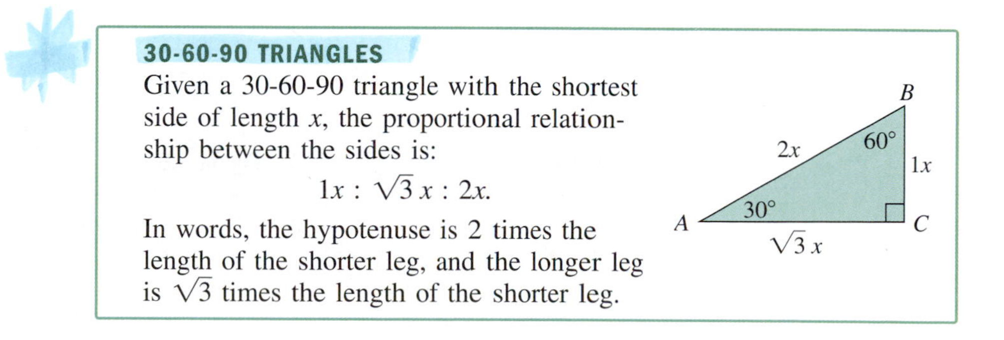

30-60-90 TRIANGLES

Given a 30-60-90 triangle with the shortest side of length x, the proportional relationship between the sides is:

$$1x : \sqrt{3}\,x : 2x.$$

In words, the hypotenuse is 2 times the length of the shorter leg, and the longer leg is $\sqrt{3}$ times the length of the shorter leg.

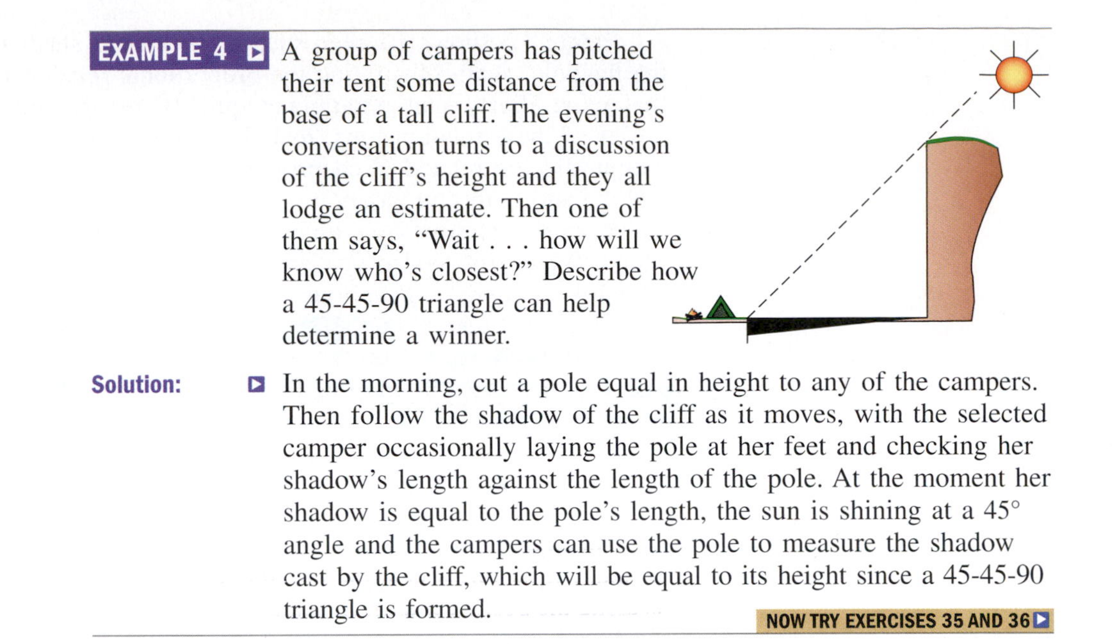

EXAMPLE 4 A group of campers has pitched their tent some distance from the base of a tall cliff. The evening's conversation turns to a discussion of the cliff's height and they all lodge an estimate. Then one of them says, "Wait . . . how will we know who's closest?" Describe how a 45-45-90 triangle can help determine a winner.

Solution: In the morning, cut a pole equal in height to any of the campers. Then follow the shadow of the cliff as it moves, with the selected camper occasionally laying the pole at her feet and checking her shadow's length against the length of the pole. At the moment her shadow is equal to the pole's length, the sun is shining at a 45° angle and the campers can use the pole to measure the shadow cast by the cliff, which will be equal to its height since a 45-45-90 triangle is formed.

NOW TRY EXERCISES 35 AND 36

C. Angle Measure in Radians; Arc Length and Area

As stated in the *Point of Interest,* there are numerous ways to measure angles and these can actually be used interchangeably. However, where degree measure has its roots in measuring "the amount of inclination" between two rays, angle measure in radians uses a slightly different perspective. For two rays joined at a vertex, angle measure can also be considered as the *amount of rotation* from a fixed ray called the **initial side,** to the rotated ray called the **terminal side.** This allows for the possibility of angles greater than 180°, and for positive or negative angles, depending on the direction of rotation. Angles formed by a counterclockwise rotation are considered **positive angles,** and angles formed by a clockwise rotation are **negative angles.** We can then name an angle of any size, including those greater than 360° where the amount of rotation exceeds one revolution. See Figures 1.5 through 1.10.

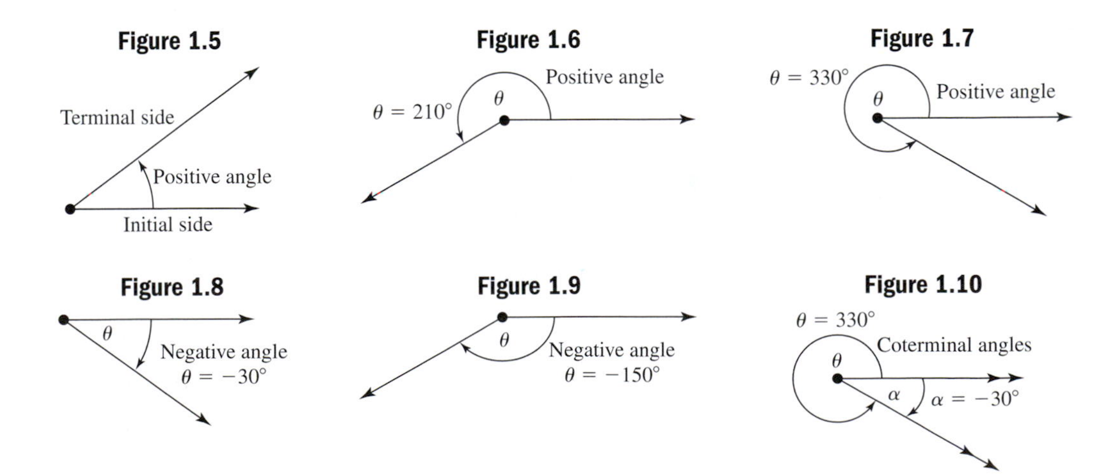

Note in Figure 1.10 that angle $\theta = 330°$ and angle $\alpha = -30°$ share the same initial and terminal sides and are called **coterminal angles.** Coterminal angles will always differ by 360°, meaning for any integer k, angles θ and $\theta + 360k$ will be coterminal.

EXAMPLE 5 Find two positive angles and two negative angles that are coterminal with 60°.

Solution: For $k = -2$, $60° + 360(-2) = -660°$.
For $k = -1$, $60° + 360(-1) = -300°$.
For $k = 1$, $60° + 360(1) = 420°$.
For $k = 2$, $60° + 360(2) = 780°$.

NOW TRY EXERCISES 37 THROUGH 40

An angle is said to be in **standard position** in the xy-plane if its vertex is at the origin and the initial side is along the positive x-axis. In standard position, the terminal sides of 90°, 180°, 270°, and 360° angles coincide with one of the axes and are called **quadrantal angles.** To help develop these ideas further, we use a **central circle,** or a circle in the xy-plane with its center at the origin. A **central angle** is an angle whose vertex is at the center of the circle. For central angle θ intersecting the circle at points B and C, we say circular arc BC, denoted $\overset{\frown}{BC}$, **subtends** $\angle BAC$, as shown in Figure 1.11. The letter s is commonly used to represent arc length, and if we define **1 radian** (abbreviated rad) to be the measure of an angle subtended by an arc equal in length to the radius, then $\theta = 1$ rad when $s = r$ (see Figure 1.12). We can then find the radian measure of any central angle by dividing the length of the subtended arc by r: $\frac{s}{r} = \theta$ radians. Multiplying both sides by r gives a formula for the length of any arc subtended on a circle of radius r: $s = r\theta$ if θ is in radians.

Figure 1.11

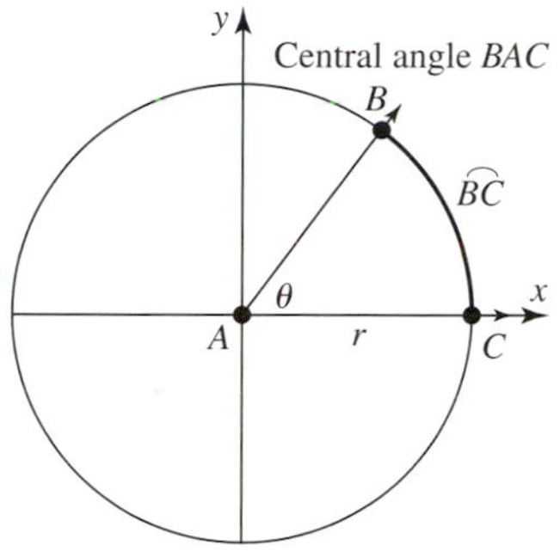

Figure 1.12

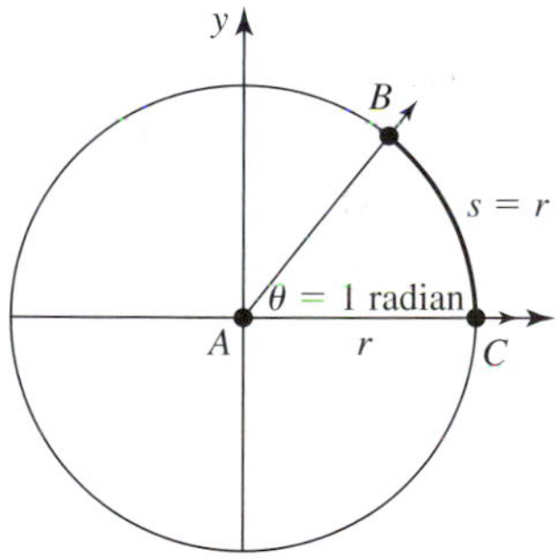

RADIANS
1 radian is the measure of a central angle subtended by an arc that is equal in length to the radius: $\theta = 1$ when $s = r$.

ARC LENGTH
If θ is a central angle on a circle of radius r, then the length of the subtended arc s is given by $s = r\theta$, *provided θ is in radians.*

EXAMPLE 6 If the circle in Figure 1.12 has radius $r = 10$ cm, what is the length of the arc subtended by an angle of 3.5 rad?

Solution: Using the formula $s = r\theta$ with $r = 10$ and $\theta = 3.5$ gives

$$s = 10(3.5) \quad \text{substitute 10 for } r \text{ and 3.5 for } \theta$$
$$s = 35 \quad \text{result}$$

The subtended arc has a length of 35 cm.

NOW TRY EXERCISES 41 THROUGH 52

Using a central angle θ measured in radians, we can also develop a formula for the **area of a circular sector** (a pie slice) using a proportion. The arc encompassing the 360° of a whole circle is identical to its circumference, and has length $C = 2\pi r$. While you've likely not considered this before, note the formula can be written as $C = 2\pi \cdot r$, which implies that the radius, or an arc of length r, can be wrapped around the circumference of the circle $2\pi \approx 6.28$ times, as illustrated in Figure 1.13. This shows the radian measure of a full 360° rotation is 2π: 2π rad $= 360°$. This can be verified as before, using the relation $\theta \text{ radians} = \frac{s}{r} = \frac{2\pi r}{r} = 2\pi$. The ratio of the area of a sector to the total area will be identical to the ratio of the subtended angle to one full rotation. Using $\mathcal{A}$ to represent the area of the sector, we have $\frac{\mathcal{A}}{\pi r^2} = \frac{\theta}{2\pi}$ and solving for $\mathcal{A}$ gives $\mathcal{A} = \frac{1}{2}r^2\theta$.

Figure 1.13

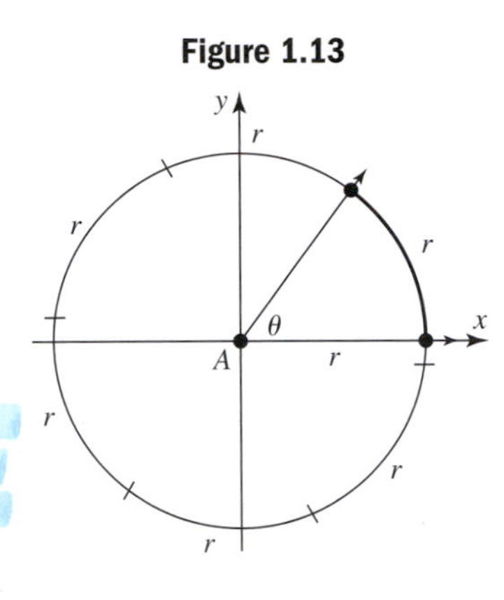

AREA OF A SECTOR

If θ is a central angle on a circle of radius r, then the area of the sector formed by angle θ is

$\mathcal{A} = \frac{1}{2}r^2\theta$, *provided θ is in radians.*

EXAMPLE 7 What is the area of the circular sector formed by a central angle of $\frac{3\pi}{4}$, if the radius of the circle is 72 ft? Round to tenths.

Solution: Using the formula $\mathcal{A} = \frac{1}{2}r^2\theta$ we have

$$\mathcal{A} = \left(\frac{1}{2}\right)(72)^2\left(\frac{3\pi}{4}\right) \quad \text{substitute 72 for } r, \frac{3\pi}{4} \text{ for } \theta$$
$$= 1944\pi \text{ ft}^2 \quad \text{result}$$

The area of this sector is approximately 6107.3 ft^2.

NOW TRY EXERCISES 53 THROUGH 64

D. Converting Between Degrees and Radians

In addition to its use in developing formulas for arc length and the area of a sector, the relation 2π rad $= 360°$ enables us to state the radian measures of the standard angles using a simple division. For π rad $= 180°$ we have division by 2: $\frac{\pi}{2} = 90°$ division by 3: $\frac{\pi}{3} = 60°$

division by 4: $\frac{\pi}{4} = 45°$ division by 6: $\frac{\pi}{6} = 30°$.

See Figure 1.14. The radian measures of these standard angles play a major role in this chapter, and you are encouraged to become very familiar with them. Additional conversions can quickly be found using multiples of these four. For example, multiplying both sides of $\frac{\pi}{3} = 60°$ by two gives $\frac{2\pi}{3} = 120°$. The relationship π radians $= 180°$ also gives the factors needed for converting from degrees to radians or from radians to degrees, even if θ is a nonstandard angle. Dividing by π we have $1 \text{ rad} = \frac{180°}{\pi}$, while division by $180°$ shows $1° = \frac{\pi}{180}$ rad. Multiplying a given angle by the appropriate conversion factor gives an equivalent measure.

Figure 1.14

90° or $\frac{\pi}{2}$; 60° or $\frac{\pi}{3}$; 45° or $\frac{\pi}{4}$; 30° or $\frac{\pi}{6}$; 0°; 180° π; C; 270° $\frac{3\pi}{2}$

WORTHY OF NOTE

We will often use the convention that unless degree measure is explicitly implied or noted with the ° symbol, radian measure is being used. In other words, $\theta = \frac{\pi}{2}$, $\theta = 2$, and $\theta = 32.76$ all indicate angles measured in radians.

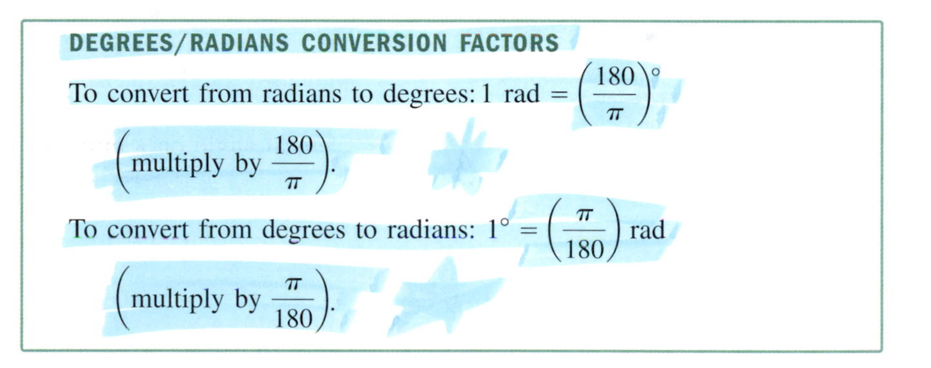

DEGREES/RADIANS CONVERSION FACTORS

To convert from radians to degrees: $1 \text{ rad} = \left(\frac{180}{\pi}\right)°$ $\left(\text{multiply by } \frac{180}{\pi}\right)$.

To convert from degrees to radians: $1° = \left(\frac{\pi}{180}\right)$ rad $\left(\text{multiply by } \frac{\pi}{180}\right)$.

EXAMPLE 8 Convert each angle as indicated: (a) 75° to radians and (b) $\frac{\pi}{24}$ rad to degrees.

Solution: **a.** For degrees to radians, use the conversion factor $\frac{\pi}{180}$:

$$75° = 75 \cdot 1° = 75 \cdot \frac{\pi}{180} = \frac{5\pi}{12} \text{ rad}$$

b. For radians to degrees, use the conversion factor $\frac{180}{\pi}$:

$$\frac{\pi}{24} \text{ rad} = \frac{\pi}{24} \cdot 1 \text{ rad} = \frac{\pi}{24} \cdot \left(\frac{180}{\pi}\right)° = 7.5°$$

NOW TRY EXERCISES 65 THROUGH 92

One example where these conversions are useful is in applications involving longitude and latitude (see Figure 1.15). The **latitude** of a fixed point on the Earth's surface tells how many degrees north or south of the equator the point is, as measured from the center of the Earth. The **longitude** of a fixed point on the Earth's surface tells how many degrees east or west of the Greenwich Meridian (in England) the point is, as measured along the equator to the north/south line going through the point.

Figure 1.15

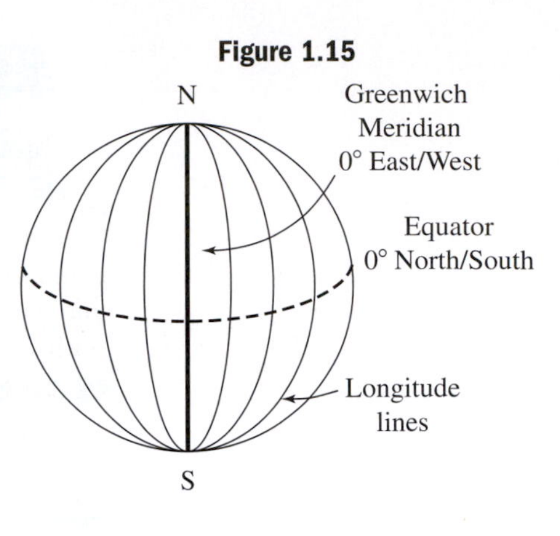

EXAMPLE 9 The cities of Quito, Ecuador, and Macapá, Brazil, both lie very near the equator, at a latitude of 0°. However, Quito is at approximately 78° west longitude, while Macapá is at 51° west longitude. Assuming the Earth has a radius of 3960 mi, how far apart are these cities?

WORTHY OF NOTE

Note that $r = 3960$ mi was used because Quito and Macapá *are both on the equator.* For other cities sharing the same longitude but not on the equator, the radius of the Earth *at that longitude* must be used. See Exercise 112.

Solution: First we note that $(78 - 51)° = 27°$ of longitude separate the two cities. Using the conversion factor $1° = \left(\frac{\pi}{180}\right)$ rad, we find the equivalent radian measure is $27\left(\frac{\pi}{180}\right) = \frac{3\pi}{20}$. The arc length formula gives

$$s = r\theta \quad \text{arc length formula; } \theta \text{ in radians}$$

$$= 3960\left(\frac{3\pi}{20}\right) \quad \text{substitute 3960 for } r \text{ and } \frac{3\pi}{20} \text{ for } \theta$$

$$= 594\pi \quad \text{result}$$

Quito and Macapá are approximately 1866 mi apart (see *Worthy of Note* in the margin).

NOW TRY EXERCISES 93 THROUGH 96

E. Angular and Linear Velocity

When the formula for uniform motion (Distance = Rate × Time: $D = RT$) is expressed in terms of R, we have $R = \frac{D}{T}$ where the rate is measured as a distance per unit time. We can also apply this concept to a *rate of rotation* per unit time, called **angular velocity.** In this context, we use the symbol ω to represent the rate of rotation and θ to represent the angle through which the terminal side has moved, measured in radians. The result is a formula for angular velocity: $\omega = \frac{\theta}{t}$. For instance, a bicycle wheel turning at 150 revolutions per minute (rpm) or $\frac{150 \text{ revolutions}}{1 \text{ min}}$ has an angular velocity of $\omega = \frac{150(2\pi)}{1\text{ min}} = \frac{300\pi}{1\text{ min}}$, since 1 revolution $= 2\pi$. We can also use $\omega = \frac{\theta}{t}$ to find the **linear velocity** of this bicycle (the speed of the bicycle). In the context of angular motion, the formula $R = \frac{D}{T}$ becomes $V = \frac{s}{t}$, and since $s = r\theta$, $V = \frac{r\theta}{t}$. The formula can also be written directly in terms of ω, since $\omega = \frac{\theta}{t}$: $V = r\omega$. In summary, $V = r\omega$ gives the linear velocity of a point on the circumference of a rotating circular object, as long as ω is in radians/unit time.

ANGULAR AND LINEAR VELOCITY

Given a central circle of radius r with point P on the circumference, and central angle θ in radians with P on the terminal side. If P moves along the circumference at a uniform rate:

(1) The rate at which θ changes is called the *angular velocity* ω,

$$\omega = \frac{\theta}{t}.$$

(2) The rate at which the *position* of P changes is called the *linear velocity* V,

$$V = \frac{r\theta}{t} \quad \Rightarrow \quad V = r\omega.$$

EXAMPLE 10 The wheels on a racing bicycle have a radius of 13 in. How fast is the cyclist traveling in miles per hour, if the wheels are turning at 150 rpm?

Solution: Earlier we noted 150 rpm $= \frac{300\pi}{1\text{ min}}$. Using the formula $V = r\omega$ gives a linear velocity of $V = (13\text{ in.})\frac{300\pi}{1\text{ min}} \approx \frac{12{,}252.2\text{ in.}}{1\text{ min}}$.

To convert this to miles per hour we convert minutes to hours (1 hr = 60 min) and inches to miles (1 mi = 5280 × 12 in.):

$\left(\frac{12{,}252.2\text{ in.}}{1\text{ min}}\right)\left(\frac{60\text{ min}}{1\text{ hr}}\right)\left(\frac{1\text{ mi}}{63{,}360\text{ in.}}\right) \approx 11.6$ mph. The bicycle is traveling about 11.6 mph.

NOW TRY EXERCISES 99 THROUGH 102

TECHNOLOGY HIGHLIGHT

Decimal Degree and Radian Conversions

The keystrokes shown apply to a TI-84 Plus model. Please consult your manual or our Internet site for other models.

Most graphing calculators are programmed to compute conversions involving angle measure. On the TI-84 Plus, this is accomplished using the "°" feature for degree to radian conversions *while in radian* MODE, and the "r" feature for radian to degree conversions *while in degree* MODE. Both are found on the **ANGLE** submenu located at 2nd APPS, as is the **4:▶DMS** feature used for conversion to the degrees, minutes, seconds format (see Figure 1.16). We'll illustrate by converting both standard and nonstandard angles. To convert 180°, 72°, and −45° to radians, be sure you are in radian MODE then enter 180 2nd APPS ENTER (the **1:**° feature is the default), then ENTER once again to execute the operation. The screen shows a value of 3.141592654, which we expected since 180° = π. For 72° and −45°, we simply recall 180° (2nd ENTER) and overwrite the desired value (see Figure 1.17). For radian-to-degree conversions, be sure you are in degree MODE and complete the conversions in a similar manner, using 2nd APPS **3:**ʳ instead of **1:**°.

Figure 1.16

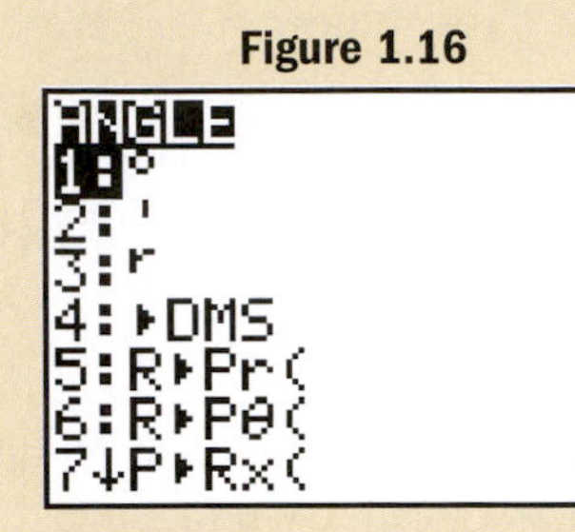

Figure 1.17

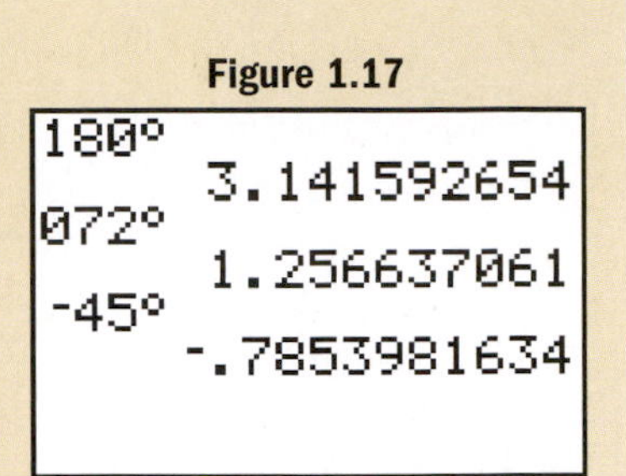

Exercise 1: Use your graphing calculator to convert the radian measures $\frac{\pi}{2}$, π, $\frac{25\pi}{12}$, and 2.37 to degrees, then verify each using standard angles or a conversion factor.

Exercise 2: Experiment with the 2nd APPS **3:▶DMS** feature, and use it to convert 108.716° to the DMS format. Verify the result manually.

1.1 EXERCISES

CONCEPTS AND VOCABULARY

Fill in each blank with the appropriate word or phrase. Carefully reread the section if needed.

1. ______________ angles sum to 90°. Supplementary angles sum to ____°. Acute angles are ___ than 90°. Obtuse angles are _______ than 90°.

2. The expression "theta equals two degrees" is written ______ using the "°" notation. The expression, "theta equals two radians" is simply written _____.

3. The formula for arc length is s = __. The area of a sector is A = _____. For both formulas, θ must be in _______.

4. If θ is not a standard angle, multiply by _____ to convert radians to degrees. To convert degrees to radians, multiply by _____.

5. Discuss/explain the difference between angular velocity and linear velocity. In particular, why does one depend on the radius while the other does not?

6. Discuss/explain the difference between 1° and 1 radian. Exactly what is a radian? Without any conversions, explain why an angle of 4 rad terminates in QIII.

DEVELOPING YOUR SKILLS

Determine the measure of the angle described or indicated.

7. a. The complement of a 12.5° angle **b.** The supplement of a 149.2° angle

8. a. The complement of a 62.4° angle **b.** The supplement of a 74.7° angle

9. The measure of angle α

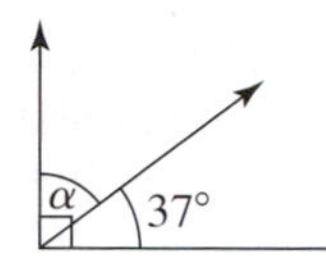

10. The measure of angle β

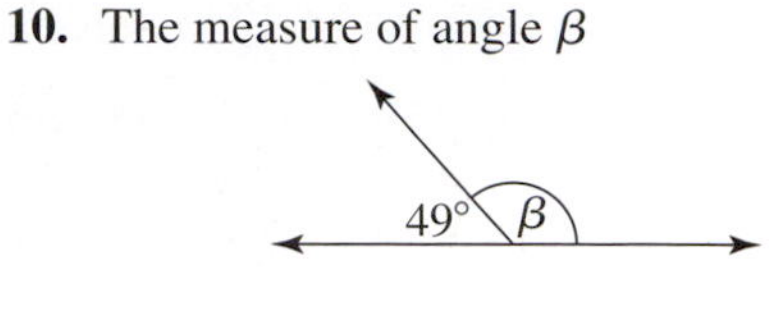

Convert from DMS (degree/minute/sec) notation to decimal degrees. Round to hundredths of a degree.

11. $42°30'$ **12.** $125°45'$ **13.** $67°22'42''$ **14.** $9°08'58''$
15. $285°00'48''$ **16.** $312°00'24''$ **17.** $45°45'45''$ **18.** $30°30'30''$

Convert the angles from decimal degrees to DMS (degree/minute/sec) notation.

19. 20.25° **20.** 40.75° **21.** 67.30° **22.** 83.5°
23. 275.33° **24.** 330.45° **25.** 5.4525° **26.** 12.3275°

27. Is the triangle shown possible? Why/why not?

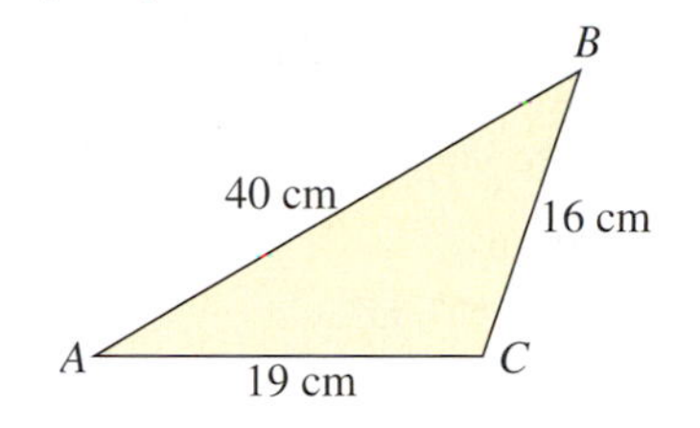

28. Is the triangle below possible? Why/why not?

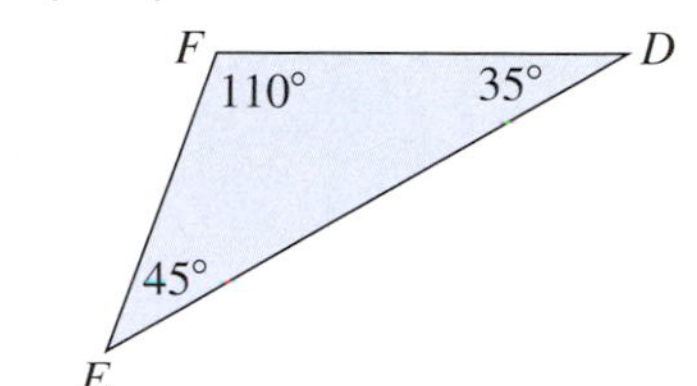

Determine the measure of the angle indicated.

29. angle α

30. angle β

31. $\angle A$

32. $\angle B$

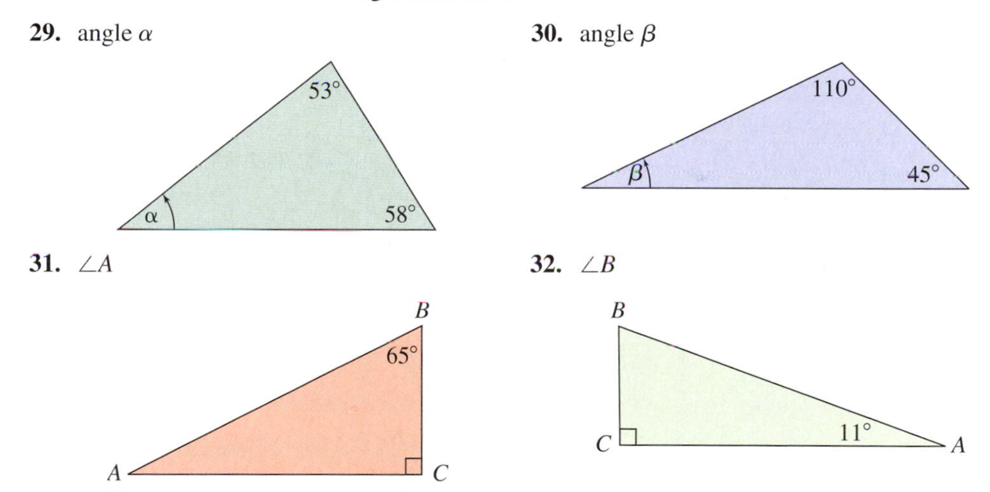

33. Similar triangles: A helicopter is hovering over a crowd of people watching a police standoff in a parking garage across the street. Stewart notices the shadow of the helicopter is lagging approximately 50 m behind a point directly below the helicopter. If he is 2 m tall and is casting a shadow of 1.6 m at this time, what is the altitude of the helicopter?

34. Similar triangles: Near Fort Macloud, Alberta (Canada), there is a famous cliff known as *Head Smashed in Buffalo Jump.* The area is now a Canadian National Park, but at one time the Native Americans hunted buffalo by steering a part of the herd over the cliff. While visiting the park late one afternoon, Denise notices that its shadow reaches 201 ft from the foot of the cliff, at the same time she is casting a shadow of $12'1''$. If Denise is $5'4''$ tall, what is the height of the cliff?

Solve using special triangles. Answer in both exact and approximate form.

35. Special triangles: A ladder-truck arrives at a high-rise apartment complex where a fire has broken out. If the maximum length the ladder extends is 82 ft and the angle of inclination is 45°, how high up the side of the building does the ladder reach? Assume the ladder is mounted atop a 10 ft high truck.

36. Special triangles: A heavy-duty ramp is used to winch heavy appliances from street level up to a warehouse loading dock. If the ramp is 7.5 ft high and the incline is 15 ft long, (a) what angle α does the dock make with the street? (b) How long is the base of the ramp?

Find two positive angles and two negative angles that are coterminal with the angle given. Answers may vary.

37. $\theta = 75°$ **38.** $\theta = 225°$ **39.** $\theta = -45°$ **40.** $\theta = -60°$

Use the formula for arc length to find the value of the remaining unknown: $s = r\theta$.

41. $\theta = 3.5$; $r = 280$ m
42. $\theta = 2.3$; $r = 129$ cm
43. $s = 2007$ mi; $r = 2676$ mi
44. $s = 4435.2$ km; $r = 12{,}320$ km
45. $\theta = \frac{3\pi}{4}$; $s = 4146.9$ yd
46. $\theta = \frac{11\pi}{6}$; $s = 28.8$ nautical miles
47. $\theta = \frac{4\pi}{3}$; $r = 2$ mi
48. $\theta = \frac{3\pi}{2}$; $r = 424$ in.
49. $s = 252.35$ ft; $r = 980$ ft
50. $s = 942.3$ mm; $r = 1800$ mm
51. $\theta = 320°$; $s = 52.5$ km
52. $\theta = 202.5°$; $s = 7627$ m

Use the formula for area of a circular sector to find the value of the remaining unknown: $A = \frac{1}{2}r^2\theta$.

53. $\theta = 5$; $r = 6.8$ km
54. $\theta = 3$; $r = 45$ mi
55. $A = 1080\ \text{mi}^2$; $r = 60$ mi
56. $A = 437.5\ \text{cm}^2$; $r = 12.5$ cm
57. $\theta = \frac{7\pi}{6}$; $A = 16.5\ \text{m}^2$
58. $\theta = \frac{19\pi}{12}$; $A = 753\ \text{cm}^2$

Find the angle, radius, arc length, and/or area as needed, until all values are known.

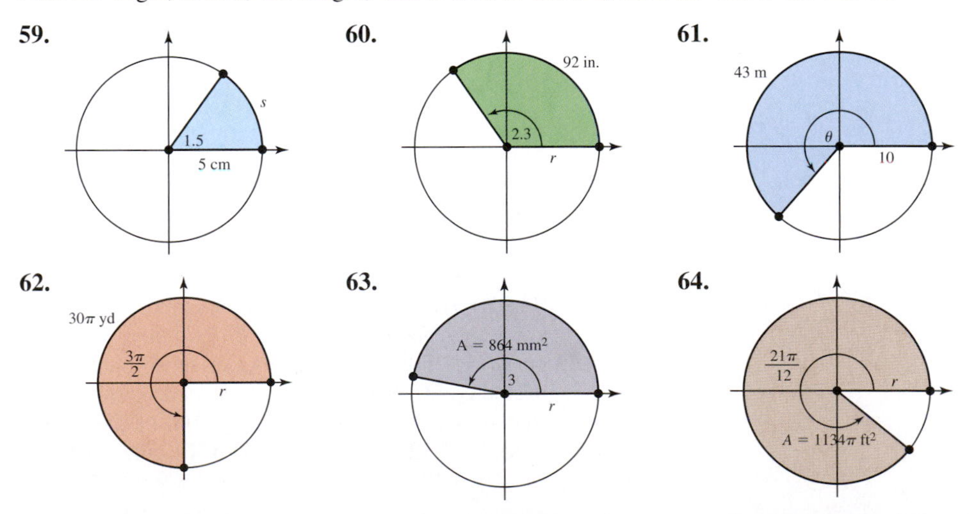

Convert the following degree measures to radians in exact form, without the use of a calculator.

65. $\theta = 360°$ **66.** $\theta = 180°$ **67.** $\theta = 45°$ **68.** $\theta = 30°$

69. $\theta = 210°$ **70.** $\theta = 330°$ **71.** $\theta = -120°$ **72.** $\theta = -225°$

Convert each degree measure to radians. Round to the nearest ten-thousandth.

73. $\theta = 27°$ **74.** $\theta = 52°$ **75.** $\theta = 227.9°$ **76.** $\theta = 154.4°$

Convert each radian measure to degrees, without the use of a calculator.

77. $\theta = \frac{\pi}{3}$ **78.** $\theta = \frac{\pi}{4}$ **79.** $\theta = \frac{\pi}{6}$ **80.** $\theta = \frac{\pi}{2}$

81. $\theta = \frac{2\pi}{3}$ **82.** $\theta = \frac{5\pi}{6}$ **83.** $\theta = 4\pi$ **84.** $\theta = 6\pi$

Convert each radian measure to degrees. Round to the nearest tenth.

85. $\theta = \frac{11\pi}{12}$ **86.** $\theta = \frac{17\pi}{36}$ **87.** $\theta = 3.2541$ **88.** $\theta = 1.0257$

89. $\theta = 3$ **90.** $\theta = 5$ **91.** $\theta = -2.5$ **92.** $\theta = -3.7$

93. Arc length: The city of Pittsburgh, Pennsylvania, is directly north of West Palm Beach, Florida. Pittsburg is at 40.3° north latitude, while West Palm Beach is at 26.4° north latitude. Assuming the Earth has a radius of 3960 mi, how far apart are these cities?

94. Arc length: Both Libreville, Gabon, and Jamame, Somalia, lie near the equator, but on opposite ends of the African continent. If Libreville is at 9.3° east longitude and Jamame is 42.5° east longitude, how wide is the continent of Africa at the equator?

95. Area of a sector: A water sprinkler is set to shoot a stream of water 12 m long and rotate through an angle of 40°. (a) What is the area of the lawn it waters? (b) For $r = 12$ m, what angle is required to water twice as much area? (c) For $\theta = 40°$, what range for the water stream is required to water twice as much area?

96. Area of a sector: A motion detector can detect movement up to 25 m away through an angle of 75°. (a) What area can the motion detector monitor? (b) For $r = 25$ m, what angle is required to monitor 50% more area? (c) For $\theta = 75°$, what range is required for the detector to monitor 50% more area?

WORKING WITH FORMULAS

97. Relationships in a right triangle: $h = \frac{ab}{c}$, $m = \frac{b^2}{c}$, **and** $n = \frac{a^2}{c}$

Given $\angle C$ is a right angle, and h is the altitude of $\triangle ABC$, then h, m, and n can all be expressed directly in terms of a, b, and c by the relationships shown here. Compute the value of h, m, and n for a right triangle with sides of 8, 15, and 17 cm.

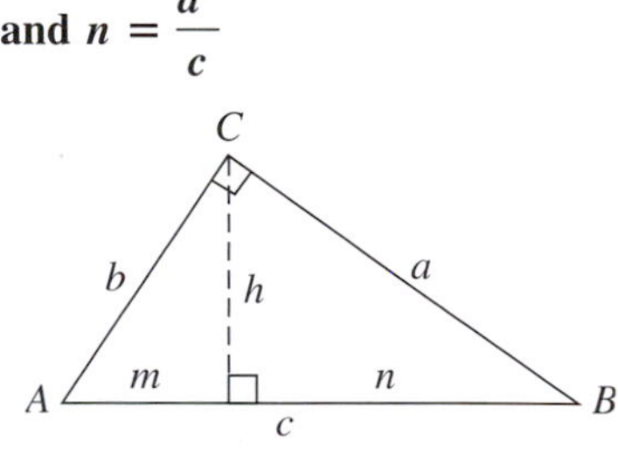

98. The height of an equilateral triangle: $H = \frac{\sqrt{3}}{2}S$

Given an equilateral triangle with sides of length S, the height of the triangle is given by the formula shown. Once the height is known the area of the triangle can easily be found (also see Exercise 97). The Gateway Arch in St. Louis, Missouri, is actually composed of stainless steel sections that are equilateral triangles. At the base of the arch the length of the sides is 54 ft. The smallest cross section at the top of the arch has sides of 17 ft. Find the area of these cross sections.

APPLICATIONS

99. Riding a round-a-bout: At the park two blocks from our home, the kid's round-a-bout has a radius of 56 in. About the time the kids stop screaming, "Faster, Daddy, faster!" I estimate the round-a-bout is turning at $\frac{3}{4}$ revolutions per second. (a) What is the related angular velocity? (b) What is the linear velocity (in miles per hour) of Eli and Reno, who are "hanging on for dear life" at the rim of the round-a-bout?

100. Carnival rides: At carnivals and fairs, the *Gravity Drum* is a popular ride. People stand along the wall of a circular drum with radius 12 ft, which begins spinning very fast, pinning them against the wall. The drum is then turned on its side by an armature, with the riders screaming and squealing with delight. As the drum is raised to a near-vertical position, it is spinning at a rate of 35 rpm. (a) What is the related angular velocity? (b) What is the linear velocity (in miles per hour) of a person on this ride?

101. Speed of a winch: A winch is being used to lift a turbine off the ground so that a tractor-trailer can back under it and load it up for transport. The winch drum has a radius of 3 in. and is turning at 20 rpm. Find (a) the angular velocity of the drum, (b) the linear velocity of the turbine in feet per second as it is being raised, and (c) how long it will take to get the load to the desired height of 6 ft (ignore the fact that the cable may wind over itself on the drum).

102. Speed of a current: An instrument called a *fluviometer* is used to measure the speed of flowing water, like that in a river or stream. A cruder method involves placing a paddle wheel in the current, and using the wheel's radius and angular velocity to calculate the speed of water flow. If the paddle wheel has a radius of 5.6 ft and is turning at 30 rpm, find (a) the angular velocity of the wheel and (b) the linear velocity of the water current in miles per hour.

On topographical maps, each concentric figure (a figure within a figure) represents a fixed change in elevation (the vertical change) according to a given *contour interval.* The *measured distance* on the map from point A to point B indicates the horizontal distance or the horizontal change between point A and a location directly beneath point B, according to a given *scale of distances.*

Exercises 103 and 104

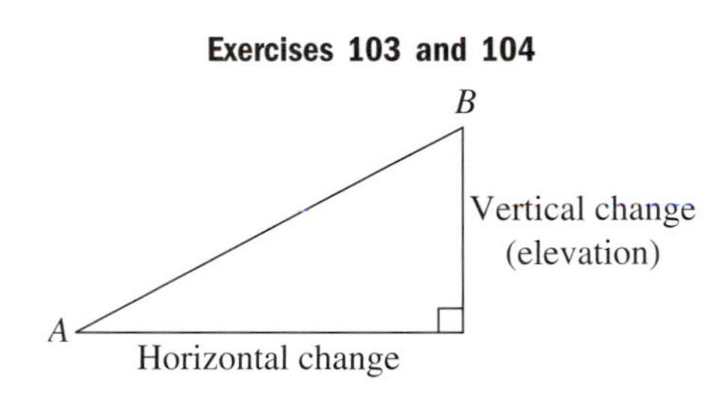

Exercise 103

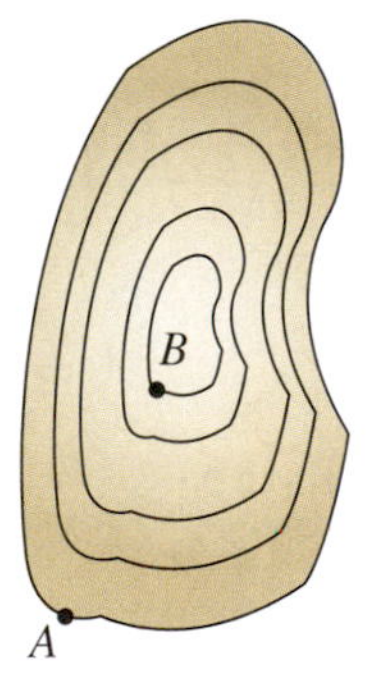

103. Special triangles: In the figure shown, the *contour interval* is 1:250 (each concentric line indicates an increase of 250 m in elevation) and the scale of distances is 1 cm = 625 m. (a) Find the change of elevation from A to B; (b) use a proportion to find the horizontal change between points A and B if the measured distance between them is 1.6 cm; and (c) Draw the corresponding right triangle and use a special triangle relationship to find the length of the trail up the mountain side that connects A and B.

104. Special triangles: As part of park maintenance, the 2 by 4 handrail alongside a mountain trail leading to the summit of Mount Marilyn must be replaced. In the figure, the *contour interval* is 1:200 (each concentric line indicates an increase of 200 m in elevation) and the scale of distances is 1 cm = 400 m. (a) Find the change of elevation from A to B; (b) use a proportion to find the horizontal change between A and B if the measured distance between them is 4.33 cm; and (c) draw the corresponding right triangle and use a special triangle relationship to find the length needed to replace the handrail (recall that $\sqrt{3} \approx 1.732$).

Exercise 104

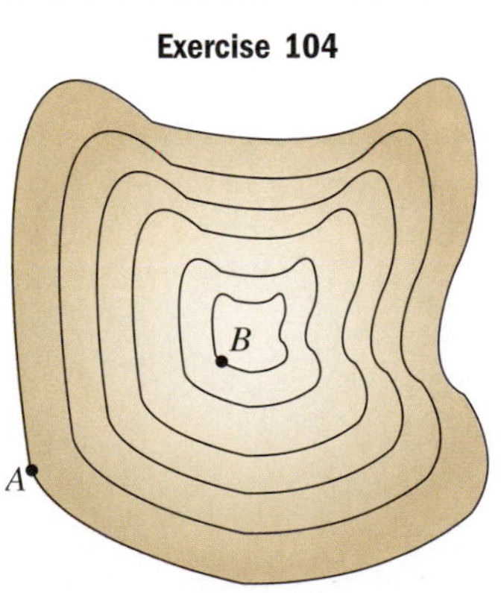

105. Special triangles: Two light planes are flying in formation at 100 mph, doing some reconnaissance work. As they approach a mountain, one pilot breaks to the left at an angle of 45° to the other plane and increases speed to 141.4 mph. Assuming they keep the same altitude and the first plane continues to fly at 100 mph, use a special triangle to find the distance between them after 0.5 hr. Note $\sqrt{2} \approx 1.414$.

106. Special triangles: Two ships are cruising together on the open ocean at 10 nautical miles per hour. One of them turns to make a 90° angle with the first and increases speed to 17.32 nautical miles per hour, heading for port. Assuming the first ship continues traveling at 10 knots, use a special triangle to find the distance between the ships in 1 hr, just as the second ship reaches port. Note $\sqrt{3} \approx 1.732$.

107. Angular and linear velocity: The planet Jupiter's largest moon, Ganymede, rotates around the planet at a distance of about 656,000 miles, in an orbit that is perfectly circular. If the moon completes one rotation about Jupiter in 7.15 days, (a) find the angle θ that the moon moves through in 1 day, in both degrees and radians, (b) find the angular velocity of the moon in radians per hour, and (c) find the moon's linear velocity in miles per second as it orbits Jupiter.

108. Angular and linear velocity: The planet Neptune has a very low eccentricity and an orbit that is nearly circular. It orbits the Sun at a distance of 4497 million kilometers and completes one revolution every 165 yr. (a) Find the angle θ that the planet moves through in one year in both degrees and radians and (b) find the linear velocity (km/hr) as it orbits the Sun.

WRITING, RESEARCH, AND DECISION MAKING

109. As mentioned in the *Point of Interest,* many methods have been used for angle measure over the centuries, some more logical or meaningful than what is popular today. Do some research on the evolution of angle measure, and compare/contrast the benefits and limitations of each method. In particular, try to locate information on the history of degrees, radians, mils, and gradients, and identify those still in use.

Exercise 110

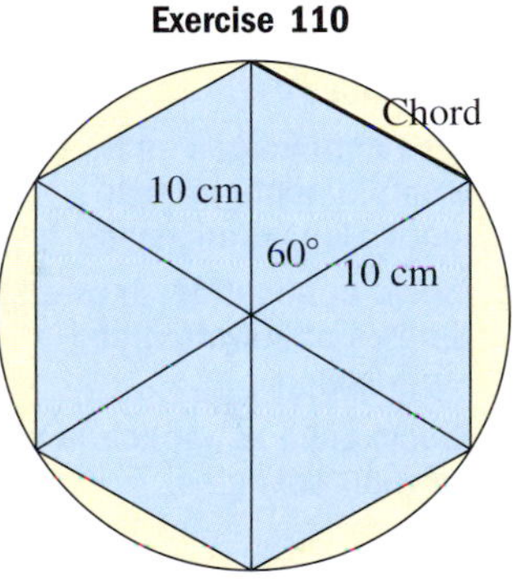

110. Use the diagram given to develop the fixed ratios for the sides of a 30-60-90 triangle. Ancient geometers knew that a hexagon (six sides) could be inscribed in a circle by laying out six consecutive chords equal in length to the radius ($r = 10$ cm for illustration). After connecting the diagonals of the hexagon, six equilateral triangles are formed with sides of 10 cm.

EXTENDING THE CONCEPT

111. The Duvall family is out on a family bicycle ride around Creve Couer Lake. The adult bikes have a pedal sprocket with a 4-in. radius, wheel sprocket with 2-in. radius, and tires with a 13-in. radius. The kids' bikes have pedal sprockets with a 2.5-in. radius, wheel sprockets with 1.5-in. radius, and tires with a 9-in. radius. (a) If adults and kids both pedal at 50 rpm, how far ahead (in yards) are the adults after 2 min? (b) If adults pedal at 50 rpm, how fast do the kids have to pedal to keep up?

112. Suppose two cities *not on the equator* shared the same latitude but different longitude. To find the distance between them we can no longer use a circumference of $C \approx 24{,}881$ mi or a radius of $R = 3960$ mi, since this only applies for east/west measurements along the equator. For the circumference c and the related "radius" r at other latitudes, consider the diagram shown. (a) Use the diagram to find a formula for r in terms of R and h; (b) given $h \approx 1054$ mi at $47.2°$ north latitude, use the formula to help find the distance between Zurich, Switzerland, and Budapest, Hungary if both are at $47.2°$ north latitude, with Zurich at $8.3°$ east longitude, and Budapest at $19°$ east longitude; (c) given $h \approx 1446$ mi at $39.4°$ north latitude, use the formula to find the distance between Denver, Colorado, and Indianapolis, Indiana, if both are at $39.4°$ N latitude, with Denver at $105°$ west longitude and Indianapolis at $86°$ west longitude.

Note: Later in this chapter we'll see how basic trigonometry can be used to find h.

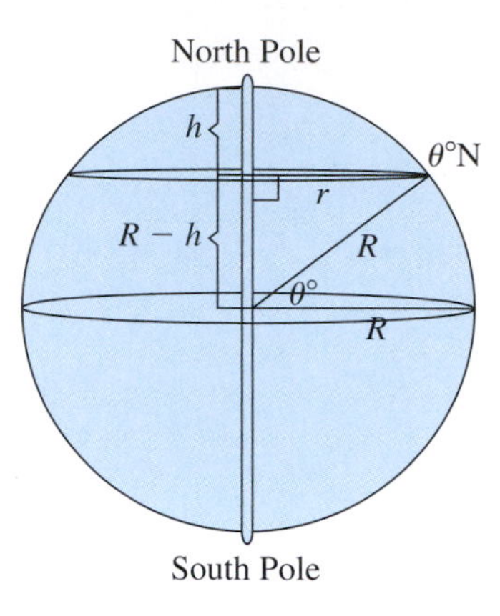

1.2 The Trigonometry of Right Triangles

LEARNING OBJECTIVES

In Section 1.2 you will learn how to:

A. Find values of the six trigonometric functions from their ratio definition

B. Solve right triangles given one angle and one side

C. Solve a right triangle given two sides

D. Use cofunctions and complements to write equivalent expressions

E. Solve applications involving angles of elevation and depression

F. Solve general applications of right triangles

INTRODUCTION

Over a long period of time, what began as a study of chord lengths by Hipparchus, Ptolemy, Aryabhata, and others, became a systematic application of the ratios of the sides of a right triangle (see *Point of Interest* following). In this section we develop the sine, cosine, and tangent functions from a right triangle perspective, and explore certain relationships that exist between them. Surprisingly, this view of the trig functions also leads to a number of significant applications.

POINT OF INTEREST

The study of chords and half-chords is closely linked to the modern concept of the sine ratio: $\frac{\text{opp}}{\text{hyp}}$. In the centuries following Ptolemy (~85–165 A.D.), numerous tables of these chord and half-chord lengths were published, and with greater and greater accuracy. Through a mistranslation, the Hindu word for half-chord became the Latin word *sinus,* from which we have the *sine* function. Around 1620 Edmund Gunter suggested *cosine*, short for "*complement of sine a*" because $\text{sine}\,\theta = \text{cosine}(90° - \theta)$.

A. Trigonometric Ratios and Their Values

In Section 1.1, we looked at applications involving 45-45-90 and 30-60-90 triangles, using the fixed ratios that exist between their sides. To apply this concept more generally using other right triangles, each side is given a specific name using its location relative to a specified angle. For the 30-60-90 triangle in Figure 1.18(a), the side **opposite (opp)** and the side **adjacent (adj)** are named with respect to the 30° angle, with the **hypotenuse (hyp)** always across from the right angle. Likewise for the 45-45-90 triangle in Figure 1.18(b). Using these designations to name the various ratios, we can develop a systematic method for applying the concept. Note that the x's "cancel" in each ratio, reminding us the ratios are independent of the triangle's size (if two triangles are similar, the ratio of corresponding sides is constant).

Figure 1.18

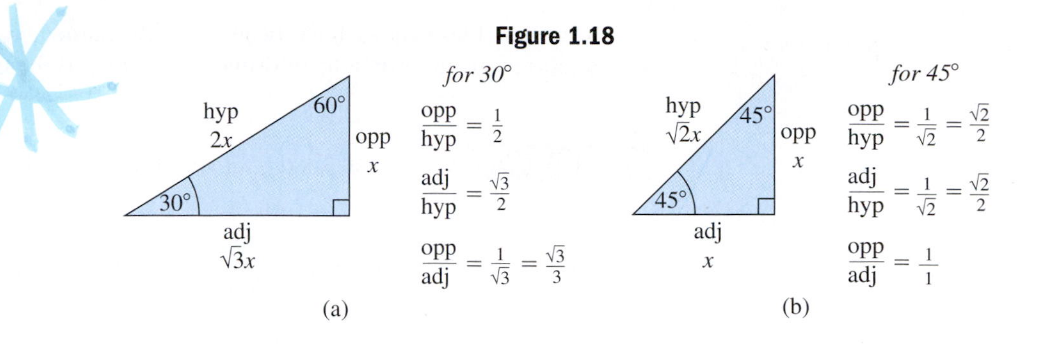

As hinted at in the *Point of Interest,* ancient mathematicians were able to find values for the ratios corresponding to *any acute angle* in a right triangle, and it soon became apparent that *naming* each ratio would be helpful. These ratios are $\frac{\text{opp}}{\text{hyp}} \rightarrow$ **sine,** $\frac{\text{adj}}{\text{hyp}} \rightarrow$ **cosine,** and $\frac{\text{opp}}{\text{adj}} \rightarrow$ **tangent.** Since each ratio depends on the measure of an acute angle θ, they are often referred to as **functions of an acute angle** and written in function form.

$$\text{sine}\,\theta = \frac{\text{opp}}{\text{hyp}} \qquad \text{cosine}\,\theta = \frac{\text{adj}}{\text{hyp}} \qquad \text{tangent}\,\theta = \frac{\text{opp}}{\text{adj}}$$

The reciprocal ratios, for example, $\frac{\text{hyp}}{\text{opp}}$ instead of $\frac{\text{opp}}{\text{hyp}}$, also play a significant role in this view of trigonometry, and are likewise given a name:

$$\text{cosecant}\,\theta = \frac{\text{hyp}}{\text{opp}} \qquad \text{secant}\,\theta = \frac{\text{hyp}}{\text{adj}} \qquad \text{cotangent}\,\theta = \frac{\text{adj}}{\text{opp}}$$

The definitions hold regardless of the triangle's orientation (how it is drawn) or which of the acute angles is used.

In actual use, each function name is written in the abbreviated form as $\sin\theta$, $\cos\theta$, $\tan\theta$, $\csc\theta$, $\sec\theta$, and $\cot\theta$, respectively. Over the course of this study, you will see many connections between this view of trigonometry and the unit circle approach (studied in Section 1.4). In particular, note that based on these designations, we have

$$\sin\theta = \frac{1}{\csc\theta} \qquad \cos\theta = \frac{1}{\sec\theta} \qquad \tan\theta = \frac{1}{\cot\theta}$$

$$\csc\theta = \frac{1}{\sin\theta} \qquad \sec\theta = \frac{1}{\cos\theta} \qquad \cot\theta = \frac{1}{\tan\theta}$$

WORTHY OF NOTE

Over the years, a number of memory tools have been invented to help students recall these ratios correctly. One such tool is the acronym SOH CAH TOA, from the first letter of the function and the corresponding ratio. It is often recited as, "Sit On a Horse, Canter Away Hurriedly, To Other Adventures." Try making up a memory tool of your own.

In general:

TRIGONOMETRIC FUNCTIONS OF AN ACUTE ANGLE

$$\sin\alpha = \frac{b}{c} \qquad \sin\beta = \frac{a}{c}$$

$$\cos\alpha = \frac{a}{c} \qquad \cos\beta = \frac{b}{c}$$

$$\tan\alpha = \frac{b}{a} \qquad \tan\beta = \frac{a}{b}$$

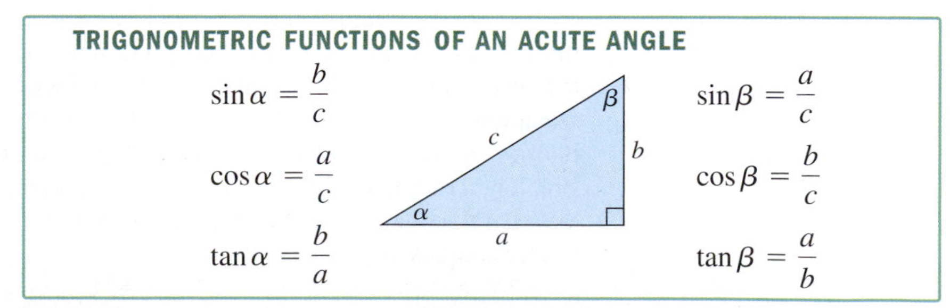

Now that these ratios have been formally named, we can state the value of all six functions given sufficient information about a right triangle.

EXAMPLE 1 Given that $\sin\theta = \frac{4}{7}$, find the value of the remaining trig functions of θ.

Solution: For $\sin\theta = \dfrac{4}{7} = \dfrac{\text{opp}}{\text{hyp}}$, we draw a triangle with a side of 4 units opposite a designated angle θ, and label a hypotenuse of 7 (see the figure). Using the Pythagorean theorem we find the length of the adjacent side: adj $= \sqrt{7^2 - 4^2} = \sqrt{33}$. The ratios are

$$\sin\theta = \frac{4}{7} \qquad \cos\theta = \frac{\sqrt{33}}{7} \qquad \tan\theta = \frac{4}{\sqrt{33}}$$

$$\csc\theta = \frac{7}{4} \qquad \sec\theta = \frac{7}{\sqrt{33}} \qquad \cot\theta = \frac{\sqrt{33}}{4}$$

NOW TRY EXERCISES 7 THROUGH 12

Note that due to similar triangles, identical results would be obtained using any ratio of sides that is equivalent to $\frac{4}{7}$. In other words, $\frac{2}{3.5} = \frac{4}{7} = \frac{8}{14} = \frac{16}{28}$ and so on, will all give the same value for $\sin\theta$.

B. Solving Right Triangles Given One Angle and One Side

Example 1 gave values of the trig functions for an *unknown angle* θ. Using standard triangles, we can state the value of each trig function for 30°, 45°, and 60° based on the related ratio (see Table 1.1).

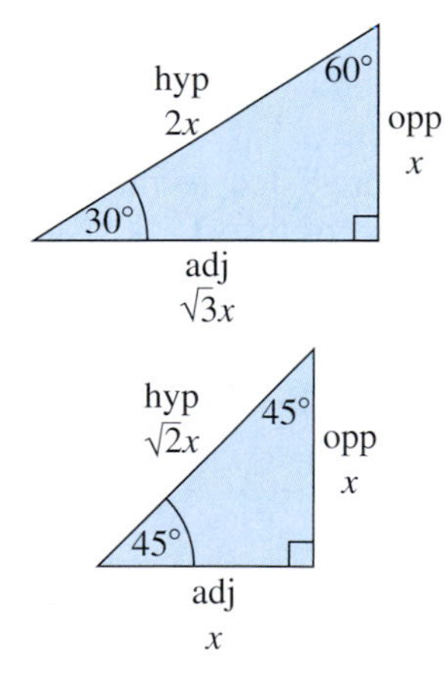

Table 1.1

θ	$\sin\theta$	$\cos\theta$	$\tan\theta$	$\csc\theta$	$\sec\theta$	$\cot\theta$
30°	$\frac{1}{2}$	$\frac{\sqrt{3}}{2}$	$\frac{1}{\sqrt{3}} = \frac{\sqrt{3}}{3}$	2	$\frac{2}{\sqrt{3}} = \frac{2\sqrt{3}}{3}$	$\sqrt{3}$
45°	$\frac{\sqrt{2}}{2}$	$\frac{\sqrt{2}}{2}$	1	$\sqrt{2}$	$\sqrt{2}$	1
60°	$\frac{\sqrt{3}}{2}$	$\frac{1}{2}$	$\sqrt{3}$	$\frac{2}{\sqrt{3}} = \frac{2\sqrt{3}}{3}$	2	$\frac{1}{\sqrt{3}} = \frac{\sqrt{3}}{3}$
90°	1	0	undefined	1	undefined	0

To **solve a right triangle** means to find the measure of all three angles and all three sides. This is accomplished using any combination of the Pythagorean theorem, the properties of triangles, and the trigonometric ratios. We will adopt the convention of naming the angles with a capital letter at the vertex or using a Greek letter on the interior. Each side is labeled using the related small case letter from the angle opposite. The complete solution should be organized in table form as in Example 2. Note the quantities shown in **bold** were given, and the remaining values were found using the techniques mentioned.

EXAMPLE 2 ▶ Solve the triangle given.

Solution: ▶ Applying the sine ratio (since the side opposite 30° is given), we have: $\sin 30° = \frac{\text{opp}}{\text{hyp}}$.

For side c:

$$\sin 30° = \frac{17.9}{c} \qquad \sin 30° = \frac{\text{opposite}}{\text{hypotenuse}}$$

$$c \sin 30° = 17.9 \qquad \text{multiply by } c$$

$$c = \frac{17.9}{\sin 30°} \qquad \text{divide by } \sin 30° = \frac{1}{2}$$

$$= 35.8 \qquad \text{result}$$

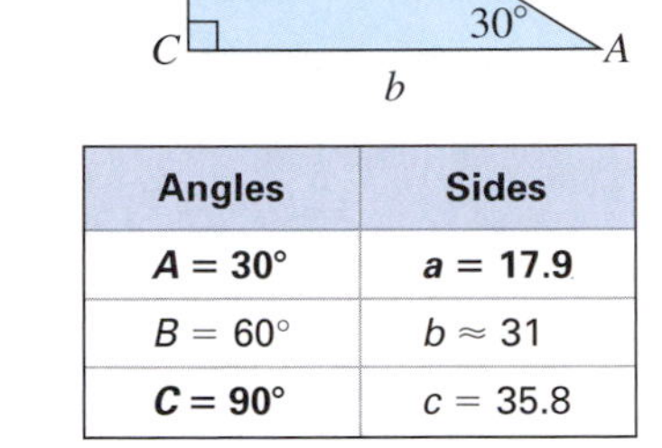

Using the Pythagorean theorem we find side $b \approx 31$, and since $\angle A$ and $\angle B$ are complements, $B = 60°$. Note the results would have been identical if the standard ratios from the 30-60-90 triangle were applied. The hypotenuse is twice the shorter side: $c = 2(17.9) = 35.8$, and the longer side is $\sqrt{3}$ times the shorter: $b = 17.9(\sqrt{3}) \approx 31$.

Angles	Sides
$A = 30°$	**$a = 17.9$**
$B = 60°$	$b \approx 31$
$C = 90°$	$c = 35.8$

NOW TRY EXERCISES 13 THROUGH 16 ▶

As mentioned in the *Point of Interest,* what was begun by Hipparchus and Ptolemy in their study of chords, gradually became a complete table of values for the sine, cosine, and tangent of acute angles (and half-angles). Prior to the widespread availability of handheld calculators, these tables were used to find $\sin\theta$, $\cos\theta$, and $\tan\theta$ for nonstandard angles. Table 1.2 shows the sine of 49°30′ is approximately 0.7604.

Table 1.2
$\sin\theta$

θ	0′	10′	20′	30′
45°	0.7071	0.7092	0.7112	0.7133
46	0.7193	0.7214	0.7234	0.7254
47	0.7314	0.7333	0.7353	0.7373
48	0.7431	0.7451	0.7470	0.7490
49	0.7547	0.7566	0.7585	0.7604

Today these trig values are programmed into your calculator and we can retrieve them with the push of a button (or two). To find the sine of a nonstandard angle like 48°, make sure your calculator is in degree **MODE**, then press the **SIN** key, 48, and **ENTER**.

EXAMPLE 3 ▶ Solve the triangle shown in the figure.

Solution: ▶ We know $\angle B = 58°$ since $A + B = 90°$. We can find length b using the tangent function:

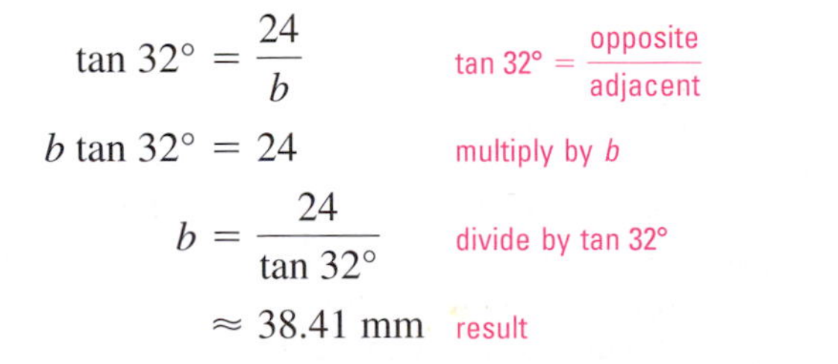

$$\tan 32° = \frac{24}{b} \qquad \tan 32° = \frac{\text{opposite}}{\text{adjacent}}$$

$$b \tan 32° = 24 \qquad \text{multiply by } b$$

$$b = \frac{24}{\tan 32°} \qquad \text{divide by } \tan 32°$$

$$\approx 38.41 \text{ mm} \qquad \text{result}$$

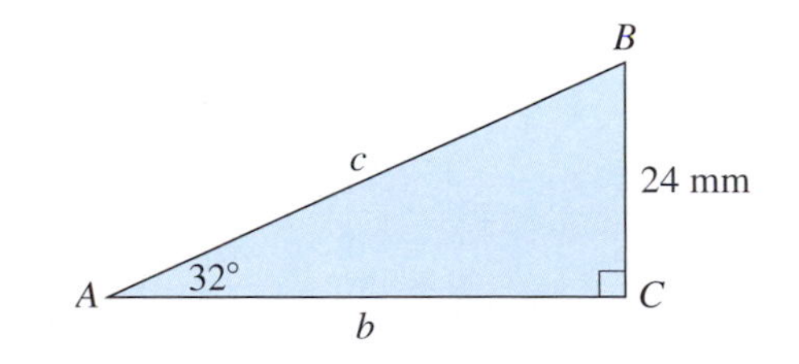

We can find the length c by simply applying the Pythagorean theorem, or by using another trig ratio and a known angle.

For side c:

$$\sin 32° = \frac{24}{c} \qquad \sin 32° = \frac{\text{opposite}}{\text{hypotenuse}}$$

$$c \sin 32° = 24 \qquad \text{multiply by } c$$

$$c = \frac{24}{\sin 32°} \qquad \text{divide by } \sin 32°$$

$$\approx 45.29 \text{ mm} \qquad \text{result}$$

Angles	Sides
$A = 32°$	$a = 24$
$B = 58°$	$b \approx 38.41$
$C = 90°$	$c \approx 45.29$

The complete solution is shown in the table.

NOW TRY EXERCISES 17 THROUGH 22

When the trig functions are applied to solve a triangle, any of the relationships available can be employed: (1) angles must sum to 180°, (2) Pythagorean theorem, (3) standard triangles, and (4) the trigonometric functions of an acute angle. However, the resulting equation must have only one unknown or it cannot be used. For the triangle shown in Figure 1.19, we cannot begin with the Pythagorean theorem since sides a and b are unknown, and tan 51° is unusable for the same reason. Since the hypotenuse is given, we could begin with $\cos 51° = \frac{b}{152}$ and first solve for b, or with $\sin 51° = \frac{a}{152}$ and solve for a, then work out a complete solution. Verify that $a \approx 118.13$ ft and $b \approx 95.66$ ft.

Figure 1.19

C. Solving Right Triangles Given Two Sides

The partial table for $\sin\theta$ given earlier was also used in times past to find an angle whose sine was known, meaning if $\sin\theta \approx 0.7604$, then θ must be 49.5°. This is shown on the last line of Table 1.2. The modern notation for "an angle whose sine is known" is $\theta = \mathbf{sin^{-1}}\boldsymbol{x}$ or $\theta = \mathbf{arcsin}\ \boldsymbol{x}$, where x is the known value for $\sin\theta$. The values for the acute angles $\theta = \sin^{-1}x$, $\theta = \cos^{-1}x$, and $\theta = \tan^{-1}x$ are also programmed into your calculator and are generally accessed using the INV or 2nd keys with the related SIN, COS, or TAN key. With these we are completely equipped to find all six measures of a right triangle, given any three. As an alternative to naming the angles with a capital letter, we sometimes use a Greek letter at the interior of the vertex.

EXAMPLE 4 Solve the triangle given in the figure.

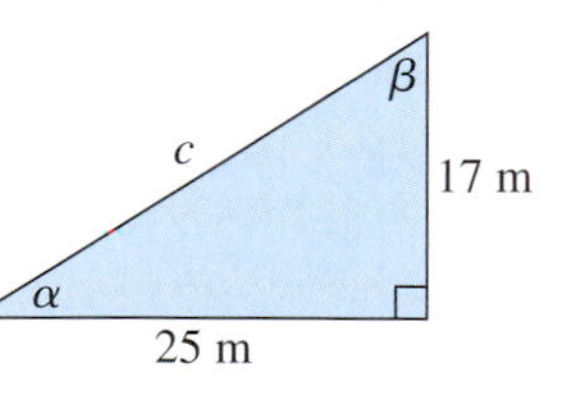

Solution: Since the hypotenuse is unknown, we cannot begin with the sine or cosine ratios. The opposite and adjacent sides *are* known, so we use $\tan\alpha$. For $\tan\alpha = \frac{17}{25}$ we find $\alpha = \tan^{-1}\left(\frac{17}{25}\right) \approx 34.2°$. Since α and β are complements, $\beta \approx 90 - 34.2 = 55.8°$. The Pythagorean theorem shows the hypotenuse is about 30.23 m.

Angles	Sides
$\alpha \approx 34.2°$	$a = 17$
$\beta \approx 55.8°$	$b = 25$
$\gamma = 90°$	$c \approx 30.23$

NOW TRY EXERCISES 23 THROUGH 54

We could have started Example 4 by using the Pythagorean theorem to find the hypotenuse, then $\sin^{-1}\left(\frac{17}{30.23}\right)$ or $\cos^{-1}\left(\frac{25}{30.23}\right)$ to find α. Either expression gives a value very near 34.2°.

D. Using Cofunctions and Complements to Write Equivalent Expressions

In Figure 1.20, $\angle\alpha$ and $\angle\beta$ must be complements since we have a right triangle, and the sum of the three angles must be 180°. The complementary angles in a right triangle have a unique relationship that is often used. Specifically $\alpha + \beta = 90°$ means $\beta = 90° - \alpha$. Note that $\sin\alpha = \frac{a}{c}$ and $\cos\beta = \frac{a}{c}$. This means $\sin\alpha = \cos\beta$ or $\sin\alpha = \cos(90° - \alpha)$ by substitution. In words, "The sine of an angle is equal to the cosine of its complement." For this reason sine and cosine are called **cofunctions** (hence the name **co**sine), as are secant/cosecant, along with tangent/cotangent. As a test, we use a calculator to check the statement $\sin 52.3° = \cos(90 - 52.3)°$

Figure 1.20

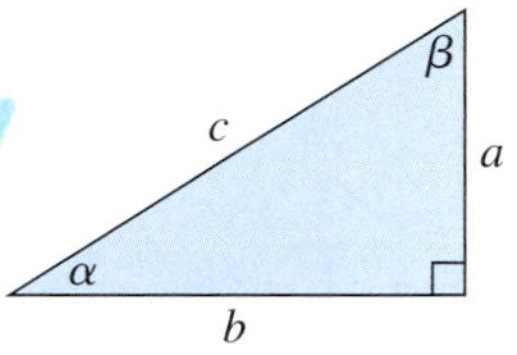

$$\sin 52.3° = \cos 37.7°$$
$$0.791223533 = 0.791223533 \checkmark$$

SUMMARY OF COFUNCTIONS

sine and cosine	tangent and cotangent	secant and cosecant
$\sin\theta = \cos(90 - \theta)$	$\tan\theta = \cot(90 - \theta)$	$\sec\theta = \csc(90 - \theta)$
$\cos\theta = \sin(90 - \theta)$	$\cot\theta = \tan(90 - \theta)$	$\csc\theta = \sec(90 - \theta)$

For use in Example 5 and elsewhere in the text, note the expression $\cos^2 75°$ is simply a more convenient way of writing $(\cos 75°)^2$.

EXAMPLE 5 Given $\sin 15° = \frac{1}{2}\sqrt{2 - \sqrt{3}}$ in exact form, find the exact value of $4\cos^2 75°$ using a cofunction. Check the result using a calculator.

Solution: Using $\cos 75° = \sin(90° - 75°) = \sin 15°$ gives

$$4\cos^2 75° = 4\sin^2 15° \quad \text{cofunctions}$$
$$= 4\left(\frac{1}{2}\sqrt{2 - \sqrt{3}}\right)^2 \quad \text{substitute known value}$$
$$= 4\left[\frac{1}{4}(2 - \sqrt{3})\right] \quad \text{square as indicated}$$
$$= 2 - \sqrt{3} \quad \text{result}$$

Using a calculator, we verify $4\cos^2 75° \approx 0.2679491924 \approx 2 - \sqrt{3}$.

NOW TRY EXERCISES 55 THROUGH 68

E. Applications Using Angles of Elevation/Depression

While the name seems self-descriptive, in more formal terms an **angle of elevation** is defined to be the acute angle formed by a **horizontal line of orientation** (parallel to level ground) and the line of sight (see Figure 1.21). An **angle of depression** is likewise defined but involves a line of sight that is below the horizontal line of orientation (Figure 1.22).

Figure 1.21 **Figure 1.22**

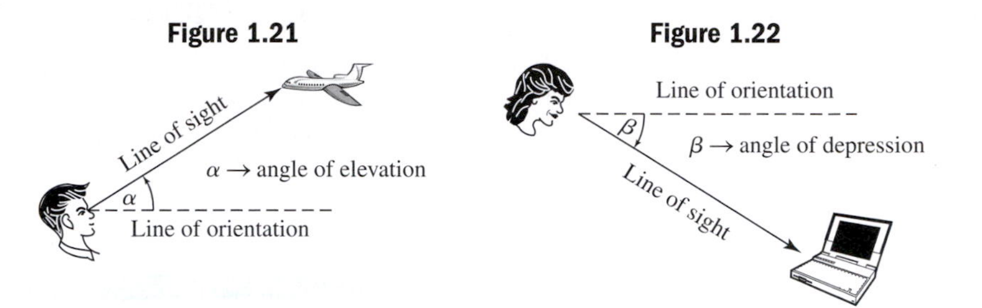

Angles of elevation/depression make distance and length computations of all magnitudes a relatively easy matter and are extensively used by surveyors, engineers, astronomers, and even the casual observer who is familiar with the basics of trigonometry.

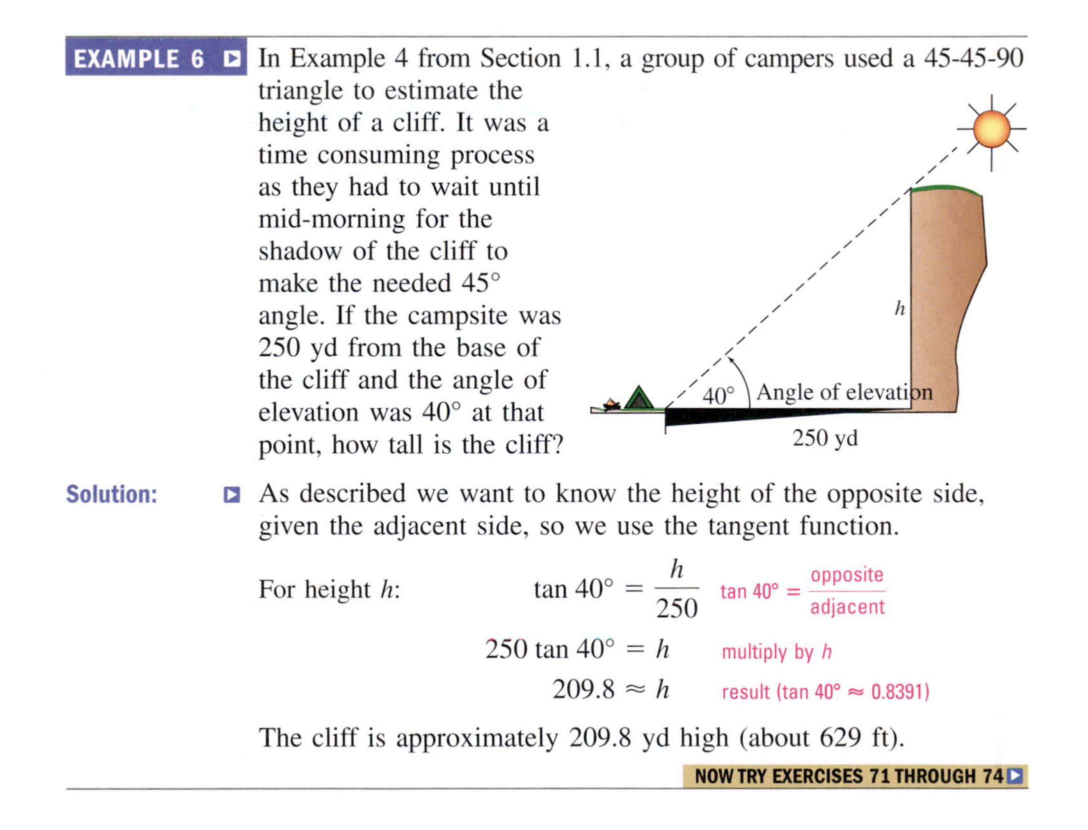

EXAMPLE 6 In Example 4 from Section 1.1, a group of campers used a 45-45-90 triangle to estimate the height of a cliff. It was a time consuming process as they had to wait until mid-morning for the shadow of the cliff to make the needed 45° angle. If the campsite was 250 yd from the base of the cliff and the angle of elevation was 40° at that point, how tall is the cliff?

Solution: As described we want to know the height of the opposite side, given the adjacent side, so we use the tangent function.

For height h:

$$\tan 40^\circ = \frac{h}{250} \qquad \tan 40^\circ = \frac{\text{opposite}}{\text{adjacent}}$$

$$250 \tan 40^\circ = h \qquad \text{multiply by } h$$

$$209.8 \approx h \qquad \text{result } (\tan 40^\circ \approx 0.8391)$$

The cliff is approximately 209.8 yd high (about 629 ft).

NOW TRY EXERCISES 71 THROUGH 74

Closely related to angles of depression/elevation are acute angles of rotation from a fixed orientation to a fixed line of sight. In this case, the movement is simply horizontal rather than vertical.

EXAMPLE 7 To thwart drivers who have radar detection equipment, a state trooper takes up a hidden position 50 ft from the roadway. Using a sighting device she finds the angle between her position and a road sign in the distance is 79°. She then uses a stop watch to determine how long it takes a vehicle to pass her location and reach the road sign. In quick succession—an 18-wheeler, a truck, and a car pass her position, with the time each takes to travel this distance noted. Find the speed of each vehicle in miles per hour if (a) the 18-wheeler takes 2.7 sec, (b) the truck takes 2.3 sec, and (c) the car takes 1.9 sec.

50 ft
79°

Solution: We begin by finding the distance traveled by each vehicle. Using $\tan 79° = \frac{d}{50}$ gives $d = 50 \tan 79° \approx 257$ ft. To convert a rate given in feet per second to miles per hour, recall there are 5280 feet in 1 mi and 3600 sec in 1 hr.

a. 18-wheeler: $\left(\frac{257\text{ ft}}{2.7\text{ sec}}\right)\left(\frac{1\text{ mi}}{5280\text{ ft}}\right)\left(\frac{3600\text{ sec}}{1\text{ hr}}\right) \approx \left(\frac{65\text{ mi}}{1\text{ hr}}\right)$
The 18-wheeler is traveling approximately 65 mph.

b. Using the same calculation with 2.3 sec shows the truck was going about 76 mph.

c. At 1.9 sec, the car was traveling about 92 mph.

NOW TRY EXERCISES 75 AND 76

F. General Applications of Right Triangles

In their widest and most beneficial use, the trig functions of acute angles are used in the context of other problem-solving skills, such as drawing a diagram, labeling unknowns, working the solution out in stages, and so on. Example 8 serves to illustrate some of these combinations.

EXAMPLE 8 From his hotel room window on the sixth floor, Singh notices some window washers high above him on the hotel across the street. Curious as to their height above ground, he quickly estimates the buildings are 50 ft apart, the angle of elevation to the workers is about 80°, and the angle of depression to the base of the hotel is about 50°. (a) How high above ground is the window of Singh's hotel room? (b) How high above ground are the workers?

Solution: **a.** To find the height of the window we'll use the tangent ratio, since the adjacent side of the angle is known, and the opposite side is the height we desire.

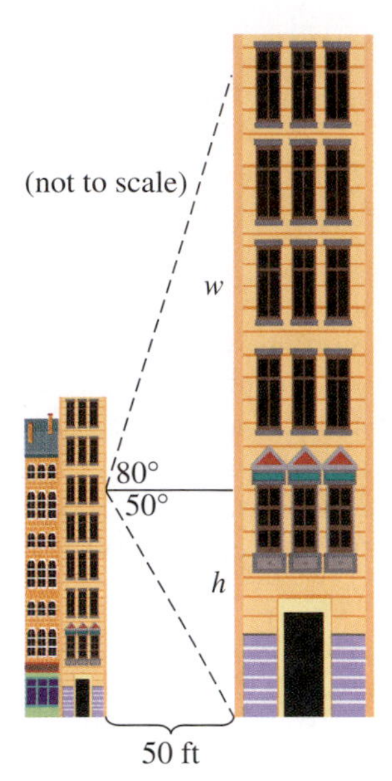

For the height h: $\tan 50° = \frac{h}{50}$ $\tan 50° = \frac{\text{opposite}}{\text{adjacent}}$

$50 \tan 50° = h$ solve for h

$59.6 \approx h$ result ($\tan 50° \approx 1.1918$)

The window is approximately 59.6 ft above ground.

b. For the workers w: $\tan 80° = \frac{w}{50}$ $\tan 80° = \frac{\text{opposite}}{\text{adjacent}}$

$50 \tan 80° = h$ solve for h

$283.6 \approx h$ result ($\tan 80° \approx 5.6713$)

The workers are approximately $283.6 + 59.6 = 343.2$ ft above ground.

NOW TRY EXERCISES 77 THROUGH 80

There are a number of additional, interesting applications in the exercise set.

TECHNOLOGY HIGHLIGHT

Using the Storage and Recall Features of a Graphing Calculator

The keystrokes shown apply to a TI-84 Plus model. Please consult your manual or our Internet site for other models.

Computations involving the trig ratios often produce irrational numbers. Sometimes the number is used numerous times in an application, and it helps to store the value in a memory location so it can instantly be recalled without having to look it up, recompute its value, or enter it digit by digit. Storage locations are also used when writing programs for your graphing calculator. Suppose the value $\frac{1 + \sqrt{5}}{2} \approx 1.6180339887$ were to be used repeatedly. You could store this value in the *temporary memory location* X,T,θ,n using the keystrokes $\frac{1 + \sqrt{5}}{2}$ STO▸ X,T,θ,n , or in a permanent memory location using the ALPHA locations A through Z. In temporary storage, the value could potentially be overwritten once you leave the home screen. To use a stored value, we simply bring up the location name to the home screen.

EXAMPLE 1 Save the value of $\frac{1 + \sqrt{5}}{2}$ in location X,T,θ,n , then investigate the relationship between (a) this number and its reciprocal and (b) this number and its square. What do you notice? Why does the value of x^{-1} seem to be off by one decimal place?

Figure 1.23

```
(1/2)(1+√(5))→X
      1.618033989
X-1
      .6180339887
X²
      2.618033989
```

Solution: After storing the number, as shown in Figure 1.23, we find its reciprocal is equal to the original number minus 1, while its square is equal to the original number plus 1. The value of x^{-1} appears off due to rounding.

Now suppose we wanted to investigate the trigonometric formula for a triangle's area: $\mathcal{A} = \frac{1}{2}ab\sin\theta$, where b is the base, a is the length of one side, and θ is the angle between them. The formula has four unknowns, $\mathcal{A}$, a, b, and θ. The idea here is to let θ represent the independent variable x, and $\mathcal{A}$ the dependent variable y, then evaluate the function for different values of a and b. On the Y= screen, enter the formula as $Y_1 = 0.5\text{ AB}\sin \text{X}$. On the home screen, store a value of 2 in location A and 12 in location B using 2 STO▸ ALPHA MATH and 12 STO▸ ALPHA APPS, respectively, as shown in Figure 1.24. The final result can be computed on the home screen (or using the TABLE feature) by supplying a value to memory location X,T,θ,n, then calling up Y_1

Figure 1.24

```
2→A
                2
12→B
               12
30→X
               30
Y1(X)
```

(VARS ▸ (Y-VARS) 1:Function ENTER) and evaluating $Y_1(X)$. As Figure 1.24 shows, we used 30 STO▸ X,T,θ,n, and the area of a triangle with $a = 2$, $b = 12$, and $\theta = 30°$ is 6 units^2 (although the display couldn't hold all of the information without scrolling).

Exercise 1: Evaluate the area formula once again, using $a = 5$ and $b = 5\sqrt{3}$. What values of X will give an area greater than 10 units^2?

1.2 EXERCISES

CONCEPTS AND VOCABULARY

Fill in each blank with the appropriate word or phrase. Carefully reread the section if needed.

1. The phrase, "an angle whose tangent is known," is written notationally as __________.
2. Given $\sin\theta = \frac{7}{24}$, $\csc\theta =$ __________ because they are __________.
3. The sine of an angle involves the ratio of the __________ side to the __________.
4. The cosine of an angle involves the ratio of the __________ side to the __________.
5. Discuss/explain exactly what is meant when you are asked to "solve a triangle." Include an illustrative example.
6. Given an acute angle and the length of the adjacent leg, which four (of the six) trig functions could be used to begin solving the triangle?

DEVELOPING YOUR SKILLS

Use the function value given to determine the value of the other five trig functions of the acute angle θ. Answer in exact form (a diagram will help).

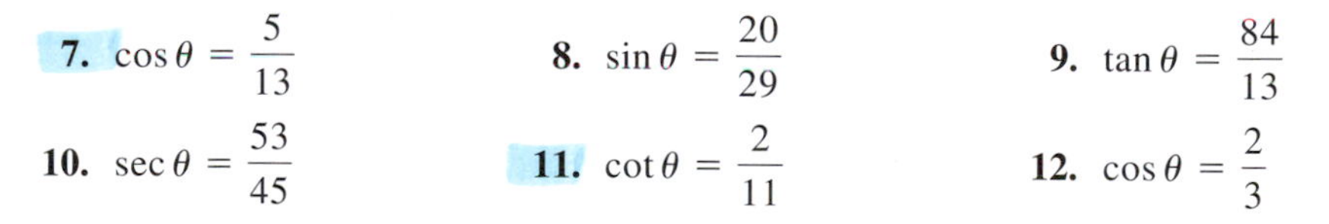

7. $\cos\theta = \dfrac{5}{13}$ **8.** $\sin\theta = \dfrac{20}{29}$ **9.** $\tan\theta = \dfrac{84}{13}$

10. $\sec\theta = \dfrac{53}{45}$ **11.** $\cot\theta = \dfrac{2}{11}$ **12.** $\cos\theta = \dfrac{2}{3}$

Solve each triangle using trig functions of an acute angle θ. Give a complete answer (in table form) using exact values.

13. **14.**

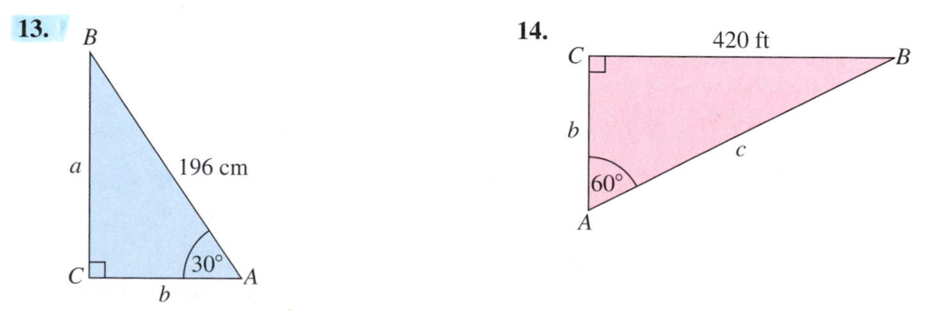

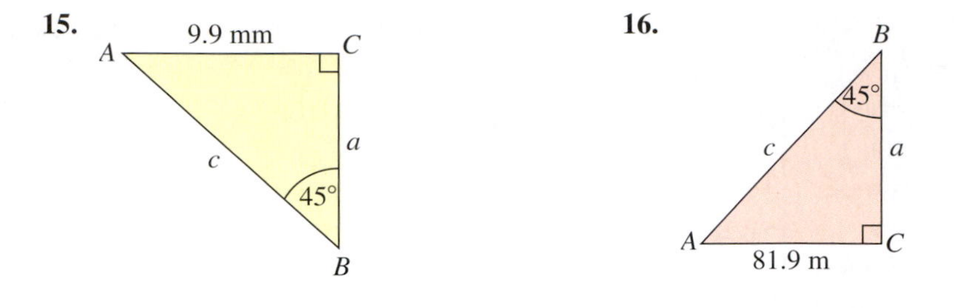

Solve the triangles shown and write answers in table form. Round sides to the nearest 100th of a unit. Double check that angles sum to 180° and that the three sides satisfy (approximately) the Pythagorean theorem.

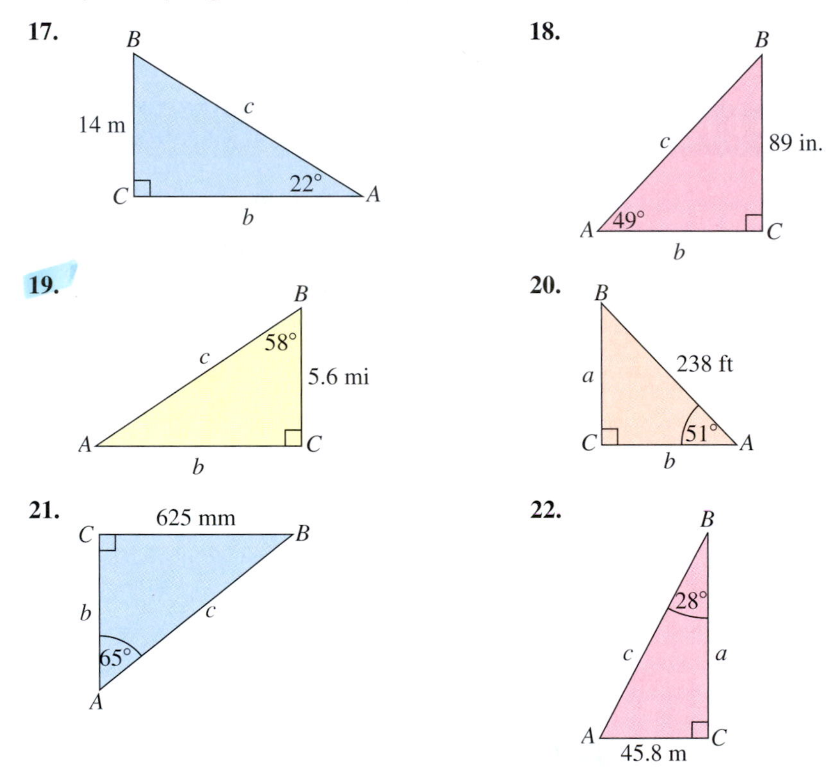

Use a calculator to find the indicated ratios for each angle, rounded to four decimal places.

23. sin 27° **24.** cos 72° **25.** tan 40° **26.** cot 57.3°

27. sec 40.9° **28.** csc 39° **29.** sin 65° **30.** tan 84.1°

Use a calculator to find the angle whose corresponding ratio is given. Round to the nearest 10th of a degree. For Exercises 31 through 38, note the relationship to Exercises 23 through 30.

31. $\sin A = 0.4540$ **32.** $\cos B = 0.3090$ **33.** $\tan \theta = 0.8391$

34. $\cot A = 0.6420$ **35.** $\sec B = 1.3230$ **36.** $\csc \beta = 1.5890$

37. $\sin A = 0.9063$ **38.** $\tan B = 9.6768$ **39.** $\tan \alpha = 0.9903$

40. $\cot \alpha = 0.9903$ **41.** $\sin \alpha = 0.9903$ **42.** $\tan \alpha = 3.1245$

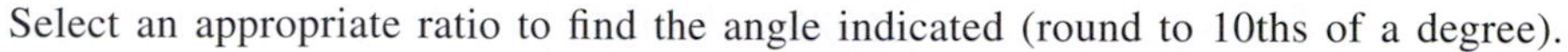

Select an appropriate ratio to find the angle indicated (round to 10ths of a degree).

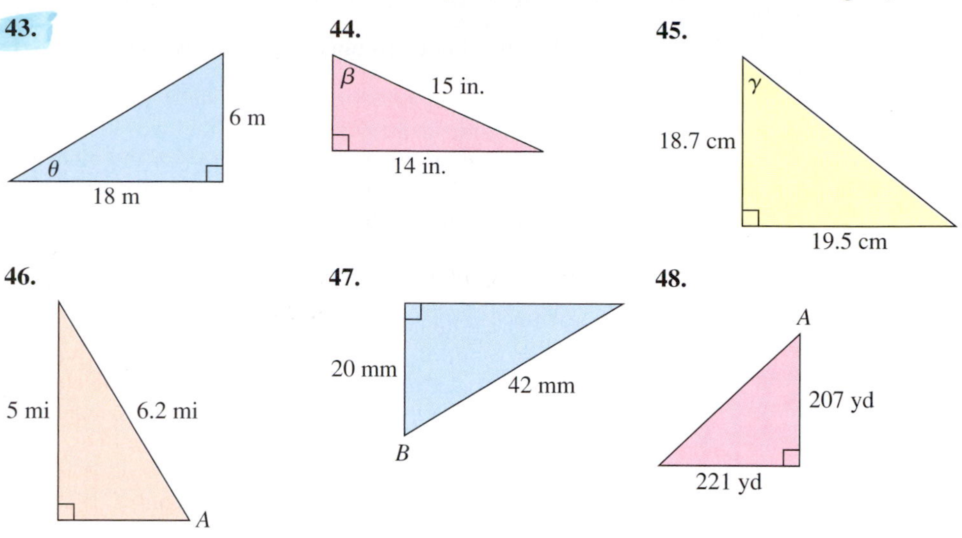

Exercises 49 to 54

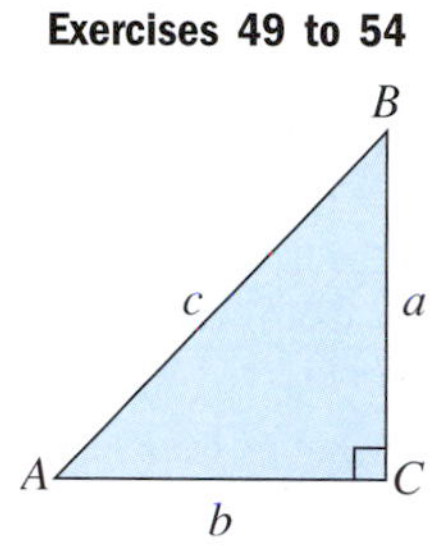

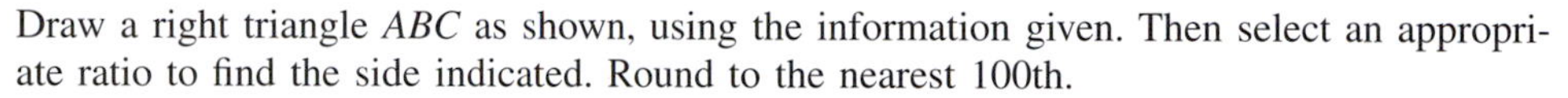

Draw a right triangle *ABC* as shown, using the information given. Then select an appropriate ratio to find the side indicated. Round to the nearest 100th.

49. $\angle A = 25°$
$c = 52$ mm
find side a

50. $\angle B = 55°$
$b = 31$ ft
find side c

51. $\angle A = 32°$
$a = 1.9$ mi
find side b

52. $\angle B = 29.6°$
$c = 9.5$ yd
find side a

53. $\angle A = 62.3°$
$b = 82.5$ furlongs
find side c

54. $\angle B = 12.5°$
$a = 32.8$ km
find side b

Use a calculator to evaluate each pair of functions and comment on what you notice.

55. sin 25°, cos 65° **56.** sin 57°, cos 33° **57.** tan 5°, cot 85° **58.** sec 40°, csc 50°

Based on your observations in Exercises 55 to 58, fill in the blank so that the ratios given are equal.

59. sin 47°, cos ____ **60.** cos ____, sin 12° **61.** cot 69°, tan ____ **62.** csc 17°, sec ____

Complete the following tables without referring to the text or using a calculator.

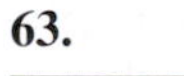

63.

θ	sin θ	cos θ	tan θ	sin(90 − θ)	cos(90 − θ)	tan(90 − θ)	csc θ	sec θ	cot θ
30°									

64.

θ	sin θ	cos θ	tan θ	sin(90 − θ)	cos(90 − θ)	tan(90 − θ)	csc θ	sec θ	cot θ
45°									

Evaluate the expressions below without a calculator, using the cofunction relationship and the following exact forms: $\sin 18° = \frac{1}{4}(\sqrt{5} - 1)$; $\tan 75° = 2 + \sqrt{3}$; and $\sin 22.5 = \frac{1}{2}\sqrt{2 - \sqrt{2}}$

65. $4 \cos 72°$ **66.** $\cot^2 15°$ **67.** $\cos 67.5°$ **68.** $4 \cos^2 67.5°$

WORKING WITH FORMULAS

69. The sine of an angle between two sides of a triangle: $\sin\theta = \frac{2A}{ab}$

If the area A and two sides a and b of a triangle are known, the sine of the angle between the two sides is given by the formula shown. Find the angle θ for the triangle to the right given $A \approx 38.9$, and use it to solve the triangle. (*Hint:* Apply the same concept to angle γ or β.)

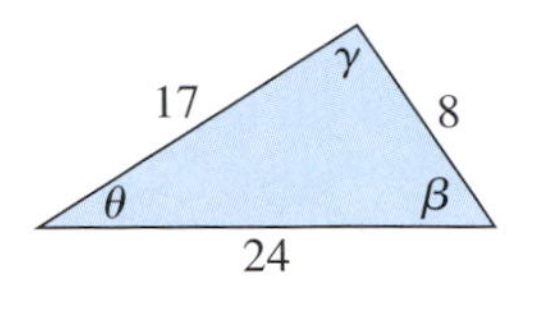

70. Illumination of a surface: $E = \frac{I\cos\theta}{d^2}$

The illumination E of a surface by a light source is a measure of the luminous flux per unit area that reaches the surface. The value of E [in lumens (lm) per square foot] is given by the formula shown, where d is the distance from the light source (in feet), I is the intensity of the light [in candelas (cd)], and θ is the angle the light flux makes with the vertical. For reading a book, an illumination E of at least 18 lm/ft^2 is recommended. Assuming the open book is lying on a horizontal surface, how far away should a light source be placed if it has an intensity of 90 cd (about 75 W) and the light flux makes an angle of 65° with the book's surface (i.e., $\theta = 25°$)?

APPLICATIONS

71. Angle of elevation: For a person standing 100 m from the center of the base of the Eiffel Tower, the angle of elevation to the top of the tower is 71.6°. How tall is the Eiffel Tower?

72. Angle of depression: A person standing near the top of the Eiffel Tower notices a car wreck some distance from the tower. If the angle of depression from the person's eyes to the wreck is 32°, how far away is the accident? See Exercise 71.

73. Angle of elevation: In 2001, the tallest building in the world was the Petronas Tower I in Kuala Lumpur, Malaysia. For a person standing 25.9 ft from the base of the tower, the angle of elevation to the top of the tower is 89°. How tall is the Petronas tower?

74. Angle of depression: A person standing on the top of the Petronas Tower I looks out across the city and pinpoints her residence. If the angle of depression from the person's eyes to her home is 5°, how far away (in feet and in miles) is the residence? See Exercise 73.

75. Acute angle of rotation: While standing near the edge of a farmer's field, Johnny watches a crop-duster dust the farmer's field for insect control. Curious as to the plane's speed during each drop, Johnny attempts an estimate using the angle of rotation from one end of the field to the other, while standing 50 ft from one corner. Using a stopwatch he finds the plane makes each pass in 2.35 sec. If the angle of rotation was 83°, how fast (in miles per hour) is the plane flying as it applies the insecticide?

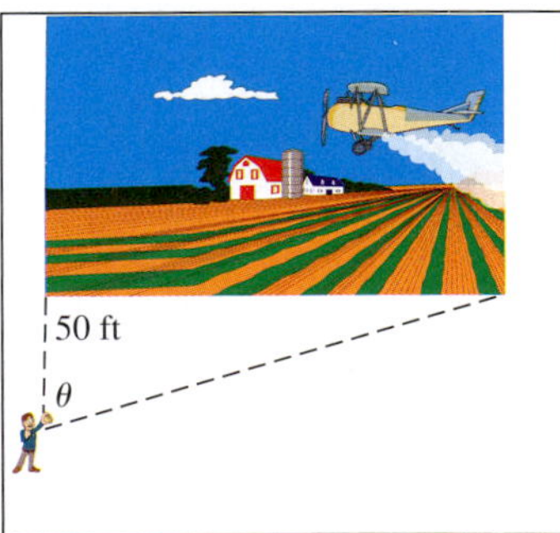

76. Acute angle of rotation: While driving to their next gig, Josh and the boys get stuck in a line of cars at a railroad crossing as the gates go down. As the sleek, speedy express train approaches, Josh decides to pass the time estimating its speed. He spots a large oak tree beside the track some distance away, and figures the angle of rotation from the crossing to the tree is about 80°. If their car is 60 ft from the crossing and it takes the train 3 sec to reach the tree, how fast is the train moving in miles per hour?

77. A local Outdoors Club has just hiked to the south rim of a large canyon, when they spot some climbers attempting to scale the northern face. Knowing the distance between the northern and southern faces of the canyon is approximately 175 yd, they attempt to compute the distance remaining for the climbers to reach the top of the northern rim. Using a homemade transit, they sight an angle of depression of 55° to the bottom of the north face, and angles of elevation of 24° and 30° to the climbers and top of the northern rim respectively. (a) How high is the southern rim of the canyon? (b) How high is the northern rim? (c) How much further until the climbers reach the top?

78. From her elevated observation post 300 ft away, a naturalist spots a troop of baboons high up in a tree. Using the small transit attached to her telescope, she finds the angle of depression to the bottom of this tree is 14°, while the angle of elevation to the top of the tree is 25°. The angle of elevation to the troop of baboons is 21°. Use this information to find (a) the height of the observation post, (b) the height of the baboon's tree, and (c) the height of the baboons above ground.

79. Angle of elevation: The tallest free-standing tower in the world is the CNN Tower in Toronto, Canada. The tower includes a rotating restaurant high above the ground. From a distance of 500 ft the angle of elevation to the pinnacle of the tower is 74.6°. The angle of elevation to the restaurant from the same vantage point is 66.5°. How tall is the CNN Tower? How far below the pinnacle of the tower is the restaurant located?

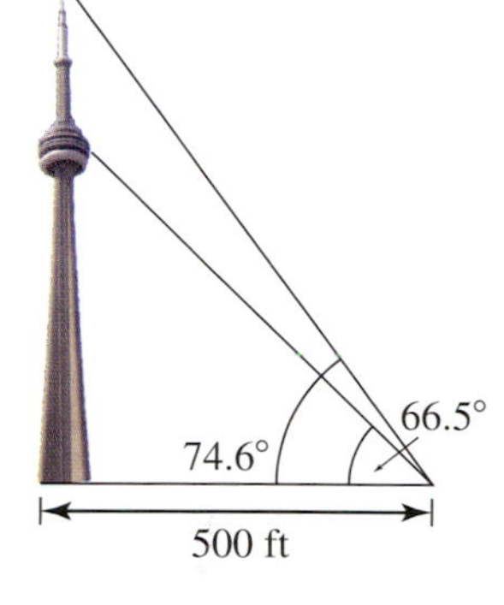

80. Angle of elevation: In August 2004, Taipei 101 captured the record as the world's tallest building, according to the Council on Tall Buildings and Urban Habitat [Source: www.ctbuh.org]. Measured at a point 108 m from its base, the angle of elevation to the top of the spire is 78°. From a distance of about 95 m, the angle of elevation to the top of the roof is also 78°. How tall is Taipei 101 from street level to the top of the spire? How tall is the spire itself?

Alternating current: In AC (alternating current) applications, the relationship between measures known as the impedance (Z), resistance (R), and the phase angle (θ) can be demonstrated using a right triangle. Both the resistance and the impedance are measured in ohms (Ω).

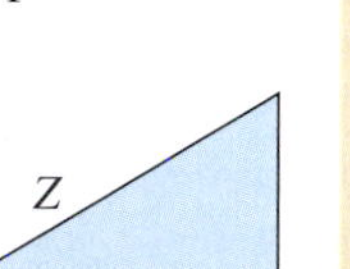

81. Find the impedance Z if the phase angle θ is 34°, and the resistance R is 320 Ω.

82. Find the phase angle θ if the impedance Z is 420 Ω, and the resistance R is 290 Ω.

83. Contour maps: In the figure shown, the *contour interval* is 175 m (each concentric line represents an increase of 175 m in elevation), and the scale of horizontal distances is 1 cm = 500 m. (a) Find the vertical change from A to B (the increase in elevation); (b) use a proportion to find the horizontal change between points A and B if the measured distance on the map is 2.4 cm; and (c) draw the corresponding right triangle and use it to estimate the length of the trail up the mountain side that connects A and B, then use trig to compute the approximate angle of incline as the hiker climbs from point A to point B.

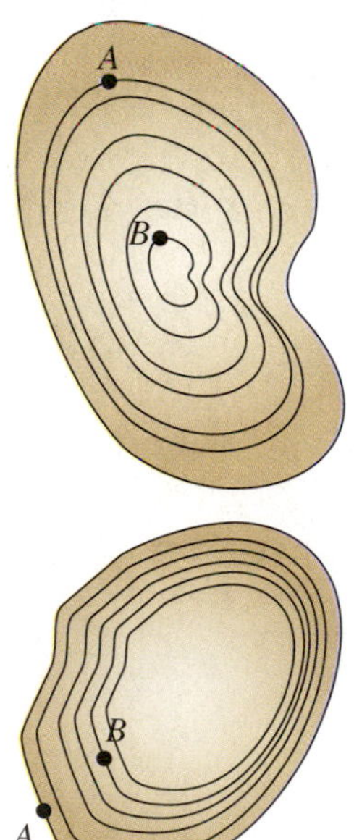

84. Contour maps: In the figure shown, the *contour interval* is 150 m (each concentric line represents an increase of 150 m in elevation), and the scale of horizontal distances is 1 cm = 250 m. (a) Find the vertical change from A to B (the increase in elevation); (b) use a proportion to find the horizontal change between points A and B if the measured distance on the map is 4.5 cm; and (c) draw the corresponding right triangle and use it to estimate the length of the trail up the mountain side that connects A and B, then use trig to compute the approximate angle of incline as the hiker climbs from point A to point B.

85. Height of a rainbow: While visiting the Lapahoehoe Memorial on the island of Hawaii, Bruce and Carma see a spectacularly vivid rainbow arching over the bay. Bruce speculates the rainbow is 500 ft away, while Carma estimates the angle of elevation to the highest point of the rainbow is about 27°. What was the approximate height of the rainbow?

86. High-wire walking: As part of a circus act, a high-wire walker not only "walks the wire," she walks a wire that is *set at an incline of* 10° to the horizontal! If the length of the (inclined) wire is 25.39 m, (a) how much higher is the wire set at the destination pole than at the departure pole? (b) How far apart are the poles?

87. Diagonal of a cube: A cubical box has a diagonal measure of 35 cm. (a) Find the dimensions of the box and (b) the angle θ that the diagonal makes at the lower corner of the box.

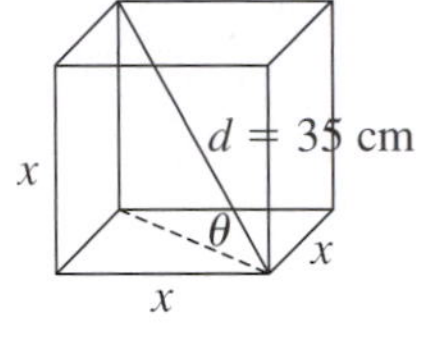

Exercise 88

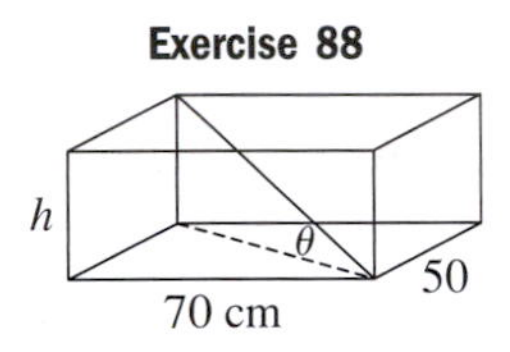

88. Diagonal of a rectangular parallelepiped: A rectangular box has a width of 50 cm and a length of 70 cm. (a) Find the height h that ensures the diagonal across the middle of the box will be 90 cm and (b) the angle θ that the diagonal makes at the lower corner of the box.

WRITING, RESEARCH, AND DECISION MAKING

89. As you can see from the preceding collection of exercises, trigonometry has many intriguing, useful, and sometimes fun applications. Create two application exercises of your own that are modeled on those here, or better yet come up with an original! Ask a fellow student to solve them and if you both agree they are "*good*" exercises, see if your instructor will consider using one of them on the next quiz or test (you never know).

90. An *angle of repose* is the angle at which the very top layer of elements begins sliding down the sides of a conical pile as more material is added to the top. In other words, the height and width of the conical pile may change, but the angle of repose will remain fairly constant and depends on the type of material being dumped. Angles of repose are important to the study and prevention of avalanches. Using the library, Internet, or other available resources, do some further research on angles of repose and their applications. Write up a short summary of what you find and include a few illustrative examples.

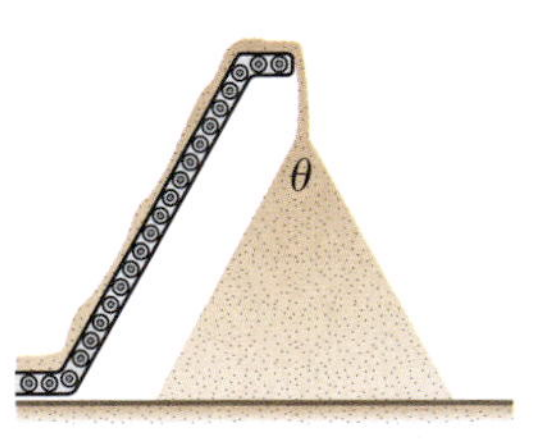

EXTENDING THE CONCEPT

91. The formula $h = \dfrac{d}{\cot u - \cot v}$ can be used to calculate the height h of a building when distance x is unknown but distance d is known (see the diagram). Use the ratios for $\cot u$ and $\cot v$ to derive the formula (note x is "absent" from the formula).

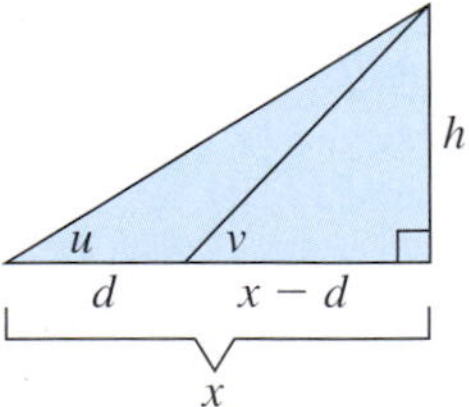

92. Use the diagram given to derive a formula for the height h of the taller building in terms of the height x of the shorter building and the ratios for tan u and tan v. Then use the formula to find h given the shorter building is 75 m tall with $u = 40°$ and $v = 50°$.

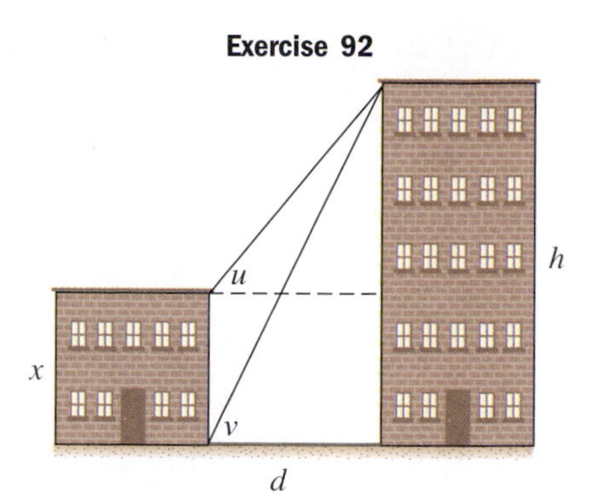

93. Aristarchus of Samos (~310–230 B.C.) was a Greek astronomer/mathematician. He appears to be among the first to realize that when the moon is in its first quarter, the triangle formed by the Sun, the Earth, and the Moon (ΔEMS in the figure) must be a right triangle. Although he did not have trigonometry or even degrees at his disposal, in effect he estimated $\angle MES$ to be 87° and used this right triangle to reckon how many times further the Sun was from the Earth, than the Moon was from the Earth (the true angle is much closer to 89.85°). Using 240,000 mi as the distance $\overline{EM}$ from the Earth to the Moon, (a) find Aristarchus' original estimate of the Sun's distance from the Earth, (b) compute the difference between Aristarchus' estimate and the improved estimate using 89.85°; and (c) determine why the error is so large for a mere 2.85° difference.

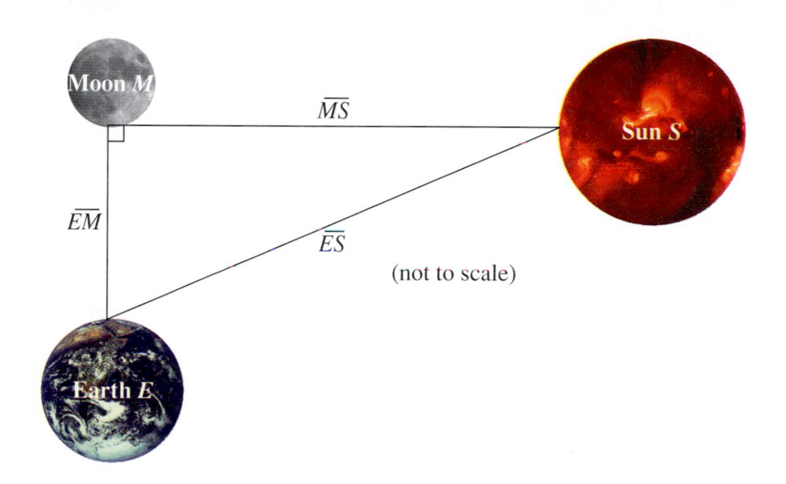

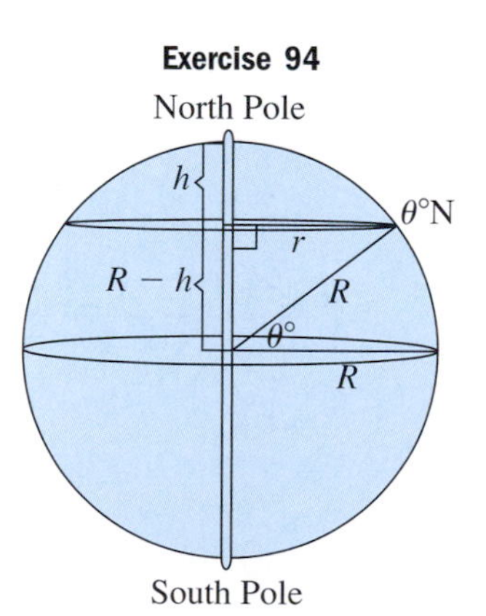

94. The radius of the Earth at the equator (0° N latitude) is approximately 3960 mi. Beijing, China, is located at 39.5° N latitude, 116° E longitude. Philadelphia, Pennsylvania, is located at the same latitude, but at 75° W longitude. (a) Use the diagram given and a cofunction relationship to find the radius r of the Earth (parallel to the equator) at this latitude; (b) use the arc length formula to compute the *shortest distance* between these two cities along this latitude; and (c) if the supersonic Concorde flew a direct flight between Beijing and Philadelphia along this latitude, approximate the flight time assuming a cruising speed of 1250 mph. (Note: The shortest distance is actually found by going over the North Pole.)

MAINTAINING YOUR SKILLS

95. (1.1) Convert from DMS to decimal degrees.

$132°42'54''$

96. (1.1) Convert from decimal degrees to DMS

36.4525°

97. (1.1) Convert θ to degrees.

$$\theta = \frac{4\pi}{3}$$

98. (1.1) Convert to radians. Leave the result in terms of π.

$\theta = 150°$

99. (1.1) Find (a) the radian measure of θ, and (b) the area of the circular sector.

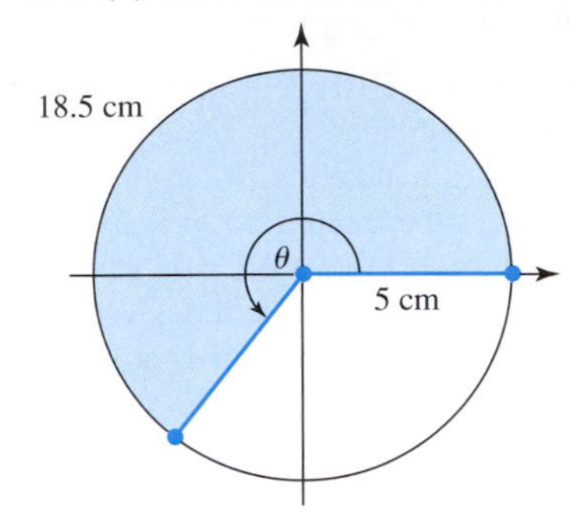

100. (1.1) A windshield wiper has a radius of 15 in. and sweeps through an angle of 105° as it wipes the windshield. (a) To the nearest whole, what is the area of the windshield it wipes? (b) To the nearest whole, what angle (in degrees) would be needed for the wiper to wipe an area of $\mathcal{A} = 226\text{ in}^2$?

Mid-Chapter Check

1. The city of Las Vegas, Nevada, is located at $36°06'36''$ north latitude, $115°04'48''$ west longitude. Convert both measures to (a) decimal degrees and (b) radians (round to four decimal places).
2. Referring to Exercise 1, how far north of the equator is Las Vegas given that the radius of the Earth is 3,960 mi?
3. Find the angle in radians subtended by the arc shown in the figure, then determine the area of the sector.

Exercise 3

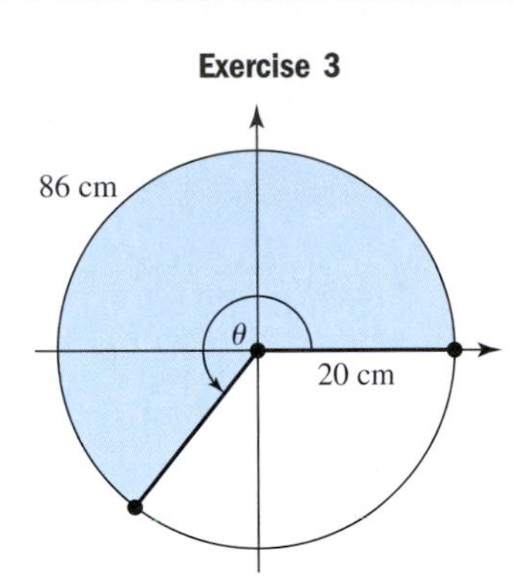

4. Evaluate using a calculator. Be sure your calculator is in degree **MODE** or radian **MODE** as needed. Round to four decimal places.

   a. $\tan 53°$ **b.** $\sin 128.4°$

   c. $\cos\left(\frac{7\pi}{4}\right)$ **d.** $\tan 4.4$

5. Solve the triangle shown in the diagram.
6. Given $\tan 75° = 2 + \sqrt{3}$, what is the value of $\cot 15°$? Why?
7. Name one positive angle and one negative angle that are coterminal with the following angles. Answers may vary.

   a. $18°$ **b.** $\frac{5\pi}{6}$

Exercise 5

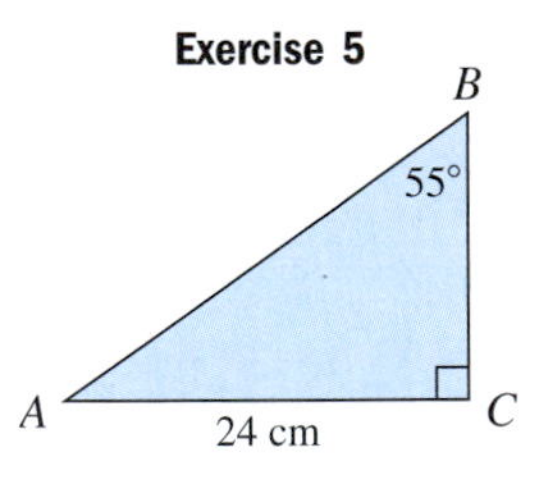

8. Complete the table from memory.

θ	sin θ	cos θ	tan θ	csc θ	sec θ	cot θ
30°						
45°						
60°						

9. At a kid's carnival, the merry-go-round makes 10 revolutions every minute and the outer-most riders are a distance of 15 ft from the center. (a) What is the angular velocity of these outer-most riders? (b) What is the linear velocity of these riders in miles per hour?

10. At a high school gym, sightings are taken from the basketball half-court line to help determine the height of the backboard. The angle of elevation to the top of the backboard is 18°, while the angle of elevation to the bottom of the backboard is 13.4°. If the half-court line is 40 ft away, how tall is the backboard? Answer in feet and inches to the nearest inch.

REINFORCING BASIC CONCEPTS

The Area of a Triangle

While you're likely familiar with the most common formula for the area of a triangle, $\mathcal{A} = \frac{1}{2}bh$, you might be surprised to learn that over 20 different formulas exist for computing this area. Many of these involve very basic trigonometric ideas, and we'll use some of these ideas to develop two additional formula types here. For $\mathcal{A} = \frac{1}{2}bh$, recall that b represents the length of a designated base of the triangle, and h represents the length of the altitude drawn to that base (Figure 1.25). If the height h of the triangle isn't given, but two sides a and b and the angle C between them are known (Figure 1.26), h can be found using a sine ratio: $\sin C = \frac{h}{a}$, giving $a \sin C = h$.

Figure 1.25

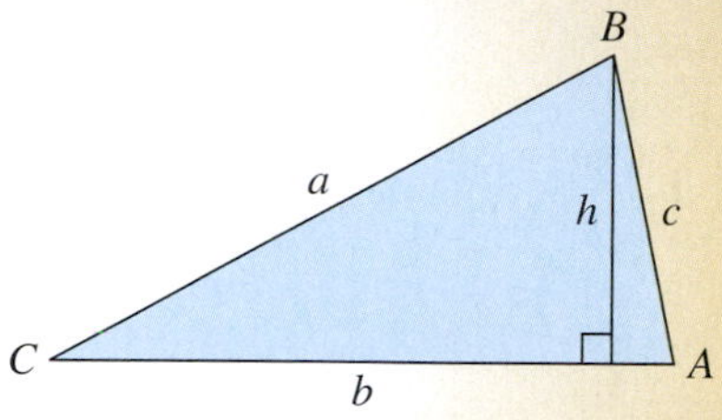

Figure 1.26

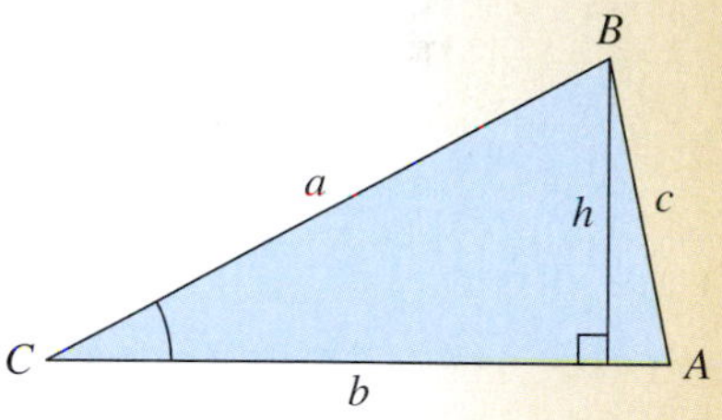

Substituting $a \sin C$ for h in the formula $\mathcal{A} = \frac{1}{2}bh$ gives $\mathcal{A} = \frac{1}{2}ba \sin C = \frac{1}{2}ab \sin C$. By designating the other sides as "the base," and drawing altitudes from the opposite vertex (Figures 1.27 and 1.28), the formulas $\mathcal{A} = \frac{1}{2}bc \sin A$ and $\mathcal{A} = \frac{1}{2}ac \sin B$ are likewise obtained.

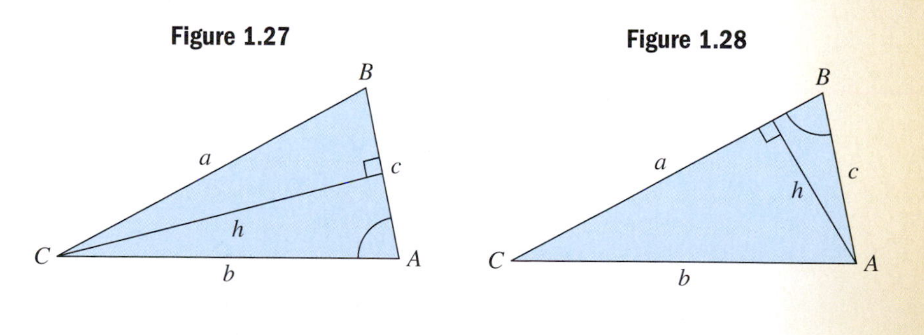

Figure 1.27 **Figure 1.28**

AREA GIVEN TWO SIDES AND AN INCLUDED ANGLE (SAS)

1. $\mathcal{A} = \frac{1}{2}ab \sin C$ 2. $\mathcal{A} = \frac{1}{2}bc \sin A$ 3. $\mathcal{A} = \frac{1}{2}ac \sin B$

In words, the formulas say the area of a triangle is equal to one-half the product of two sides times the sine of the angle between them.

ILLUSTRATION 1 Find the area of $\triangle ABC$ given $a = 16.2$ cm, $b = 25.6$ cm, and $C = 28.3°$.

Solution: Since sides a, b, and angle C are given, we apply the first formula.

$$\mathcal{A} = \frac{1}{2}ab\sin C \quad \text{area formula}$$

$$= \frac{1}{2}(16.2)(25.6)\sin 28.3° \quad \text{substitute 16.2 for } a\text{, 25.6 for } b\text{, and } 28.3° \text{ for } C$$

$$\approx 98.3 \text{ cm}^2 \quad \text{result}$$

The area of this triangle is approximately 98.3 cm^2.

Using the SAS formulas, a formula requiring two angles and one side can be developed. Observe that formula 1 can be written in terms of b: $b = \frac{2\mathcal{A}}{c \sin A}$; and formula 3 can be written in terms of a: $a = \frac{2\mathcal{A}}{c \sin B}$. Substituting $\frac{2\mathcal{A}}{c \sin A}$ for b and $\frac{2\mathcal{A}}{c \sin B}$ for a in formula 2 gives the following:

$$\mathcal{A} = \frac{1}{2}ab\sin C \quad \text{given formula}$$

$$2\mathcal{A} = \frac{2\mathcal{A}}{c\sin B}\cdot\frac{2\mathcal{A}}{c\sin A}\sin C \quad \text{substitute } \frac{2\mathcal{A}}{c\sin B} \text{ for } a\text{, } \frac{2\mathcal{A}}{c\sin A} \text{ for } b$$

$$c^2 \sin A \sin B = 2\mathcal{A} \sin C \quad \text{multiply by } c\sin A \text{ and } c\sin B\text{, divide by } 2\sin C$$

$$\frac{c^2 \sin A \sin B}{2\sin C} = \mathcal{A} \quad \text{solve for } \mathcal{A}$$

Note that if any two angles A and B are given, the third can easily be found by subtracting the sum of these two from 180°: $C = 180° - (A + B)$. As with the previous formula, versions relying on side b or side a can also be found.

AREA GIVEN TWO ANGLES AND ANY SIDE (AAS/ASA)

1. $\mathcal{A} = \dfrac{c^2 \sin A \sin B}{2\sin C}$ 2. $\mathcal{A} = \dfrac{a^2 \sin B \sin C}{2\sin A}$ 3. $\mathcal{A} = \dfrac{b^2 \sin A \sin C}{2\sin B}$

In Chapter 3, yet another useful relationship for the area of a triangle will be developed, called Heron's formula. The formula requires only the length of the three sides, and can be developed both algebraically and using trigonometry. Use the appropriate formula to find the area of the following triangles. Round to the nearest tenth.

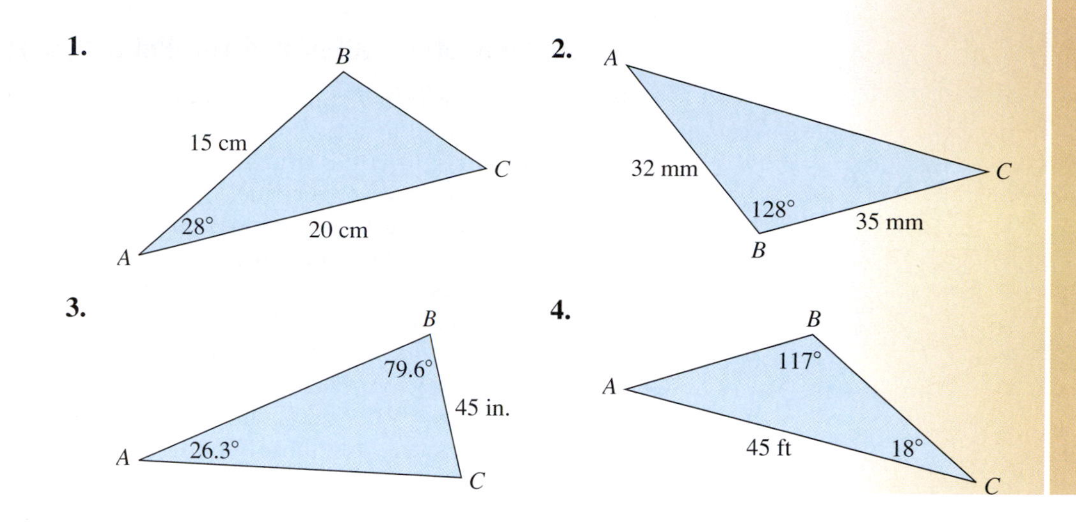

1.3 Trigonometry and the Coordinate Plane

LEARNING OBJECTIVES

In Section 1.3 you will learn how to:

A. Define the trigonometric functions using the coordinates of a point in QI

B. Use reference angles to evaluate the trig functions for any angle

C. Solve applications using the trig functions of any angle

INTRODUCTION

This section tends to bridge the study of *static trigonometry* and the angles of a right triangle, with the study of *dynamic trigonometry* and the unit circle. This is accomplished by noting that the domain of the trig functions from a triangle point of view *need not be restricted to acute angles.* We'll soon see that the domain can be extended to include trig functions of any angle, a view that greatly facilitates our work in Chapter 5, where many applications involve angles greater than 90°.

POINT OF INTEREST

Science and astronomy flowered brilliantly in India from 500 to 1200 A.D. This period saw publication of *Surya Siddhanta* (Knowledge from the Sun), a remarkable treatise on astronomy. Significant advances were also made in mathematics, including the work of Bhaskara (1114–1185 A.D.). Although he did not state it expressly, Bhaskara recognized that the sides of all right triangles are in proportion $2x:x^2 - 1:x^2 + 1$ for some real number x. For the Pythagorean triple (3, 4, 5), $x = 2$, while for (28, 45, 53), $x = \frac{7}{2}$. For more on Pythagorean triples, see the *Technology Extension: Generating Pythagorean Triples* at www.mhhe.com/coburn.

A. Trigonometric Ratios and the Point $P(x, y)$

Regardless of where a right triangle is situated or how it is oriented, each trig function is defined as a given ratio of sides with respect to a given angle. In this light, consider a 30-60-90 triangle placed in the first quadrant with the 30° angle at the origin and the longer side along the x-axis. From our previous review of similar triangles, the trig ratios will have the same value regardless of the triangle's size so for convenience, we'll use a hypotenuse of 10. This gives sides of 5, $5\sqrt{3}$, and 10, and from the diagram in Figure 1.29 we note the point (x, y) marking the vertex of the 60° angle has coordinates $(5\sqrt{3}, 5)$. Further, the diagram shows that sin 30°, cos 30°, and tan 30° can all be expressed in terms of these coordinates since $\frac{\text{opposite}}{\text{hypotenuse}} = \frac{5}{10} = \frac{y}{r}$ (sine), $\frac{\text{adjacent}}{\text{hypotenuse}} = \frac{5\sqrt{3}}{10} = \frac{x}{r}$ (cosine), and $\frac{\text{opposite}}{\text{adjacent}} = \frac{5}{5\sqrt{3}} = \frac{y}{x}$ (tangent), where r is the length of the hypotenuse. Each result reduces to the more familiar values seen earlier in our study of right triangle trigonometry: $\sin 30° = \frac{1}{2}$, $\cos 30° = \frac{\sqrt{3}}{2}$, and $\tan 30° = \frac{1}{\sqrt{3}} = \frac{\sqrt{3}}{3}$. This suggests we might be able to define the six trig functions in terms of x, y, and r, where $r = \sqrt{x^2 + y^2}$ (the distance from the point to the origin). The slope of the line coincident with the hypotenuse has $\frac{\text{rise}}{\text{run}} = \frac{1}{\sqrt{3}} = \frac{\sqrt{3}}{3}$, and since the line goes through the origin its equation must be $y = \frac{\sqrt{3}}{3}x$. Any point (x, y) on this line will be at the 60° vertex of a right triangle formed by drawing a perpendicular line from the point (x, y) to the x-axis. As Example 1 shows, we obtain the standard values for sin 30°, cos 30°, and tan 30° regardless of the point chosen.

Figure 1.29

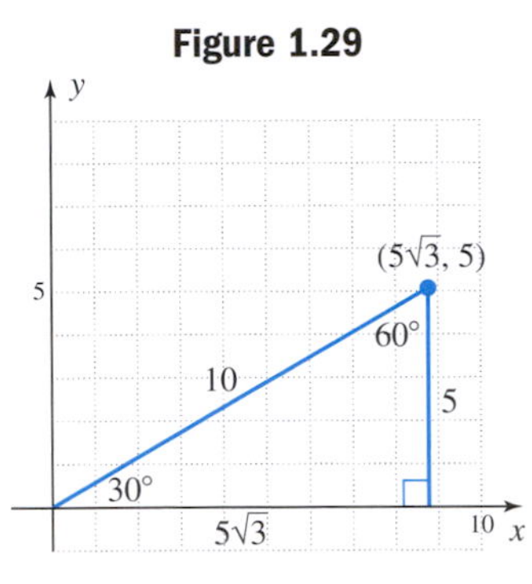

EXAMPLE 1 Pick an arbitrary point in QI that satisfies $y = \frac{\sqrt{3}}{3}x$, then draw the corresponding right triangle and evaluate sin 30°, cos 30°, and tan 30°.

Solution: The coefficient of x has a denominator of 3, so we choose a multiple of 3 for convenience. For $x = 6$ we have $y = \frac{\sqrt{3}}{3}(6) = 2\sqrt{3}$. As seen in the figure, the point $(6, 2\sqrt{3})$ is on the line and at the vertex of the 60° angle. Evaluating the trig functions at 30°, we obtain:

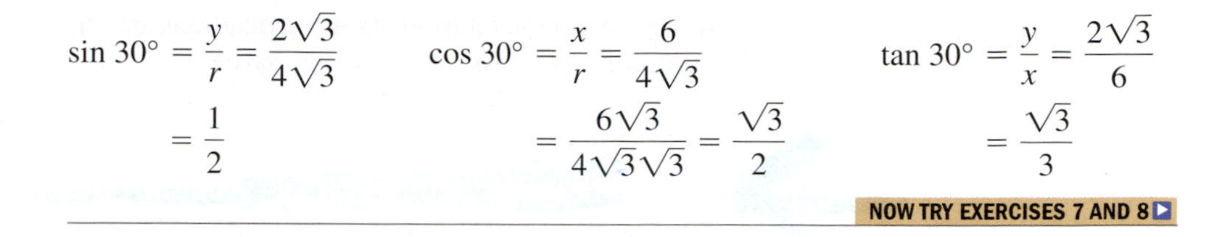

$$\sin 30° = \frac{y}{r} = \frac{2\sqrt{3}}{4\sqrt{3}} = \frac{1}{2} \qquad \cos 30° = \frac{x}{r} = \frac{6}{4\sqrt{3}} = \frac{6\sqrt{3}}{4\sqrt{3}\sqrt{3}} = \frac{\sqrt{3}}{2} \qquad \tan 30° = \frac{y}{x} = \frac{2\sqrt{3}}{6} = \frac{\sqrt{3}}{3}$$

NOW TRY EXERCISES 7 AND 8

In general, consider *any* two points (x, y) and (X, Y) on an arbitrary line $y = kx$, at corresponding distances r and R from the origin (Figure 1.30). Because the triangles formed are similar, we have $\frac{y}{x} = \frac{Y}{X}$, $\frac{x}{r} = \frac{X}{R}$, and so on, and we conclude that the value of the trig functions are indeed independent of the point chosen.

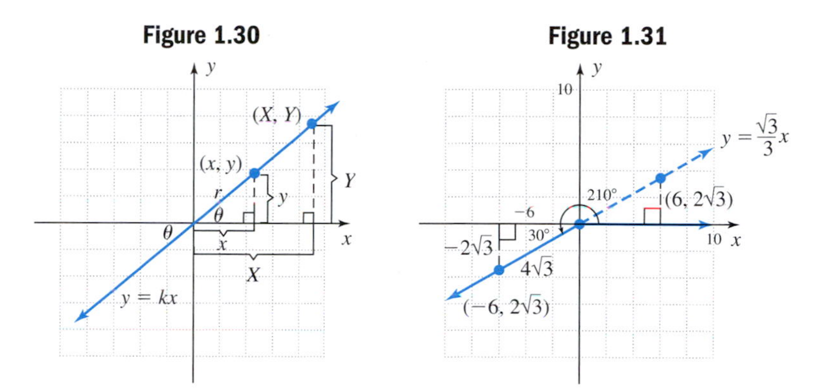

Figure 1.30 **Figure 1.31**

Viewing the trig functions in terms of x, y, and r produces significant and powerful results. In Figure 1.31, we note the line $y = \frac{\sqrt{3}}{3}x$ from Example 1 also extends into QIII, and *creates another* 30° *angle whose vertex is at the origin* (since vertical angles are equal). The sine, cosine, and tangent functions can still be evaluated for this angle, but in QIII both x and y are negative. It's here that our view of angles as a rotation bridges the right triangle view of trigonometry with the unit circle view in Section 1.4. If we consider the angle in QIII to be a positive rotation of 210° (180° + 30°), we can evaluate the trig functions using the values of x, y, and r from any point on the terminal side, since these are fixed by the 30° angle created, and they're *equivalent to those in QI except for their sign:*

Figure 1.32

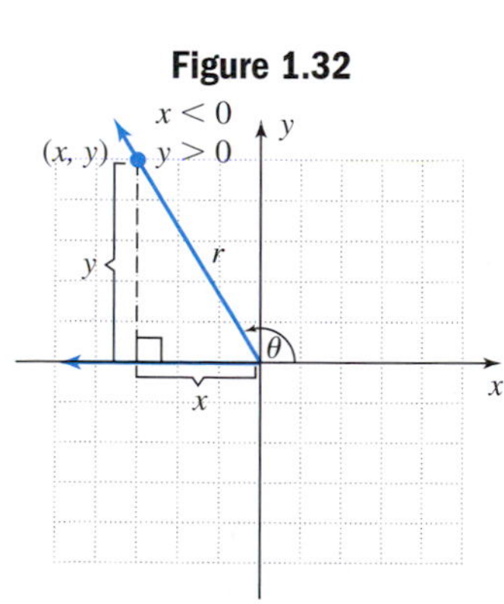

$$\sin 210° = \frac{y}{r} = \frac{-2\sqrt{3}}{4\sqrt{3}} = -\frac{1}{2} \qquad \cos 210° = \frac{x}{r} = \frac{-6}{4\sqrt{3}} = -\frac{\sqrt{3}}{2} \qquad \tan 210° = \frac{y}{x} = \frac{-2\sqrt{3}}{-6} = \frac{\sqrt{3}}{3}$$

For *any* rotation θ and a point (x, y) on the terminal side, the distance r can be found using $r = \sqrt{x^2 + y^2}$ and the six trig functions defined in terms of x, y, and r. Figure 1.32 shows a rotation that terminates in QII, where x is negative and y is positive. Correctly evaluating the trig functions of any angle depends heavily on the quadrant of the terminal side, since this will dictate the signs for x and y. Students are strongly encouraged to make these quadrant and sign observations the *first step* in any solution process. In summary, we have

TRIGONOMETRIC FUNCTIONS OF ANY ANGLE

Given $P(x, y)$ is any point on the terminal side of angle θ in standard position, with $r = \sqrt{x^2 + y^2}$ the distance from the origin to (x, y). The six trigonometric functions of θ are

$$\sin\theta = \frac{y}{r} \qquad \cos\theta = \frac{x}{r} \qquad \tan\theta = \frac{y}{x},\ x \neq 0$$

$$\csc\theta = \frac{r}{y},\ y \neq 0 \qquad \sec\theta = \frac{r}{x},\ x \neq 0 \qquad \cot\theta = \frac{x}{y},\ y \neq 0$$

EXAMPLE 2 Find the value of the six trigonometric functions given $P(-5, 5)$ is on the terminal side of angle θ in standard position.

Solution: For $P(-5, 5)$ we have $x < 0$ and $y > 0$ so the terminal side is in QII. Solving for r yields $r = \sqrt{(-5)^2 + (5)^2} = \sqrt{50} = 5\sqrt{2}$. For $x = -5$, $y = 5$, and $r = 5\sqrt{2}$, we obtain

$$\sin\theta = \frac{y}{r} = \frac{5}{5\sqrt{2}} = \frac{\sqrt{2}}{2} \qquad \cos\theta = \frac{x}{r} = \frac{-5}{5\sqrt{2}} = -\frac{\sqrt{2}}{2} \qquad \tan\theta = \frac{y}{x} = \frac{5}{-5} = -1$$

The remaining functions can be evaluated using the reciprocals.

$$\csc\theta = \frac{2}{\sqrt{2}} = \sqrt{2} \qquad \sec\theta = -\frac{2}{\sqrt{2}} = -\sqrt{2} \qquad \cot\theta = -1$$

Note the connection between these results and the standard values for $\theta = 45°$.

NOW TRY EXERCISES 9 THROUGH 24

EXAMPLE 3 Given that $P(x, y)$ is a point on the terminal side of angle θ in standard position, find the value of $\sin\theta$, $\cos\theta$, and $\tan\theta$ if (a) the terminal side is in QII and coincident with the line $y = -\frac{12}{5}x$ and (b) the terminal side is in QIV and coincident with the line $y = -\frac{12}{5}x$.

Solution: **a.** Select any convenient point in QII that satisfies this equation. We select $x = -5$ since x is negative in QII, which gives $y = 12$ and the point $(-5, 12)$. Solving for r gives $r = \sqrt{(-5)^2 + (12)^2} = 13$. The ratios are

$$\sin\theta = \frac{y}{r} = \frac{12}{13} \qquad \cos\theta = \frac{x}{r} = \frac{-5}{13} \qquad \tan\theta = \frac{y}{x} = \frac{12}{-5}$$

b. In QIV we select $x = 10$ since x is positive in QIV, giving $y = -24$ and the point $(10, -24)$. Next we find r: $r = \sqrt{(10)^2 + (-24)^2} = 26$. The ratios are

$$\sin\theta = \frac{y}{r} = \frac{-24}{26} = -\frac{12}{13} \qquad \cos\theta = \frac{x}{r} = \frac{10}{26} = \frac{5}{13} \qquad \tan\theta = \frac{y}{x} = \frac{-24}{10} = -\frac{12}{5}$$

NOW TRY EXERCISES 25 THROUGH 28

Figure 1.33

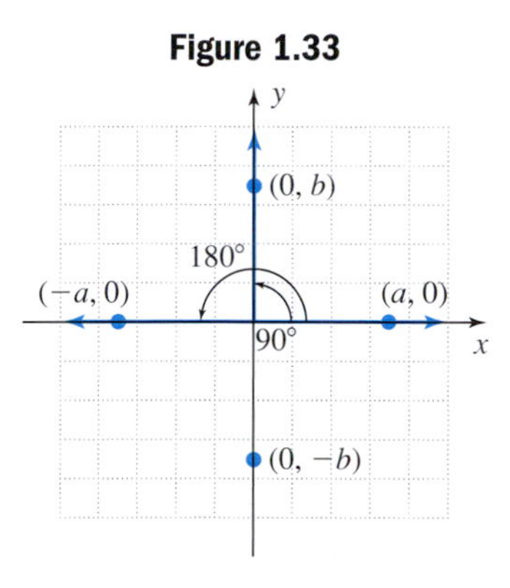

In Example 3, note the ratios are the same in QII and QIV *except for their sign.* We will soon use this observation to great advantage.

Now that we've defined the trig functions in terms of ratios involving x, y, and r, the question arises as to their value at the quadrantal angles. For 90° and 270°, any point on the terminal side of the angle has an x-*value* of zero, meaning tan 90°, sec 90°, tan 270°, and sec 270° are all undefined since $x = 0$ is in the denominator. Similarly, at 180° and 360°, the y-value of any point on the terminal side is zero, so cot 180°, csc 180°, cot 360°, and csc 360° are also undefined (see Figure 1.33).

EXAMPLE 4 Evaluate the six trig functions for $\theta = 270°$.

Solution: Here, θ is the quadrantal angle whose terminal side separates QIII and QIV. Since the evaluation is independent of the point chosen on this side, we choose $(0, -1)$ for convenience, giving $r = 1$. For $x = 0$, $y = -1$, and $r = 1$ we obtain

$$\sin\theta = \frac{-1}{1} = -1 \qquad \cos\theta = \frac{0}{-1} = 0 \qquad \tan\theta = \frac{-1}{0}\ (\textit{undefined})$$

The remaining ratios can be evaluated using the reciprocal relationships.

$$\csc\theta = -1 \qquad \sec\theta = \frac{-1}{0}\ (\textit{undefined}) \qquad \cot\theta = \frac{0}{-1} = 0$$

NOW TRY EXERCISES 29 AND 30

Results for the quadrantal angles are summarized in Table 1.3.

Table 1.3

θ	$\sin\theta = \frac{y}{r}$	$\cos\theta = \frac{x}{r}$	$\tan\theta = \frac{y}{x}$	$\csc\theta = \frac{r}{y}$	$\sec\theta = \frac{r}{x}$	$\cot\theta = \frac{x}{y}$
$0° \to (1, 0)$	0	1	0	undefined	1	undefined
$90° \to (0, 1)$	1	0	undefined	1	undefined	0
$180° \to (-1, 0)$	0	−1	0	undefined	−1	undefined
$270° \to (0, -1)$	−1	0	undefined	−1	undefined	0

B. Reference Angles and the Trig Functions of Any Angle

For any angle θ in standard position, the acute angle θ_r formed by the terminal side and the nearest x-axis is called the **reference angle.** Several examples of this concept are illustrated in Figures 1.34 through 1.37 for $\theta > 0$ and a point (x, y) on the terminal side.

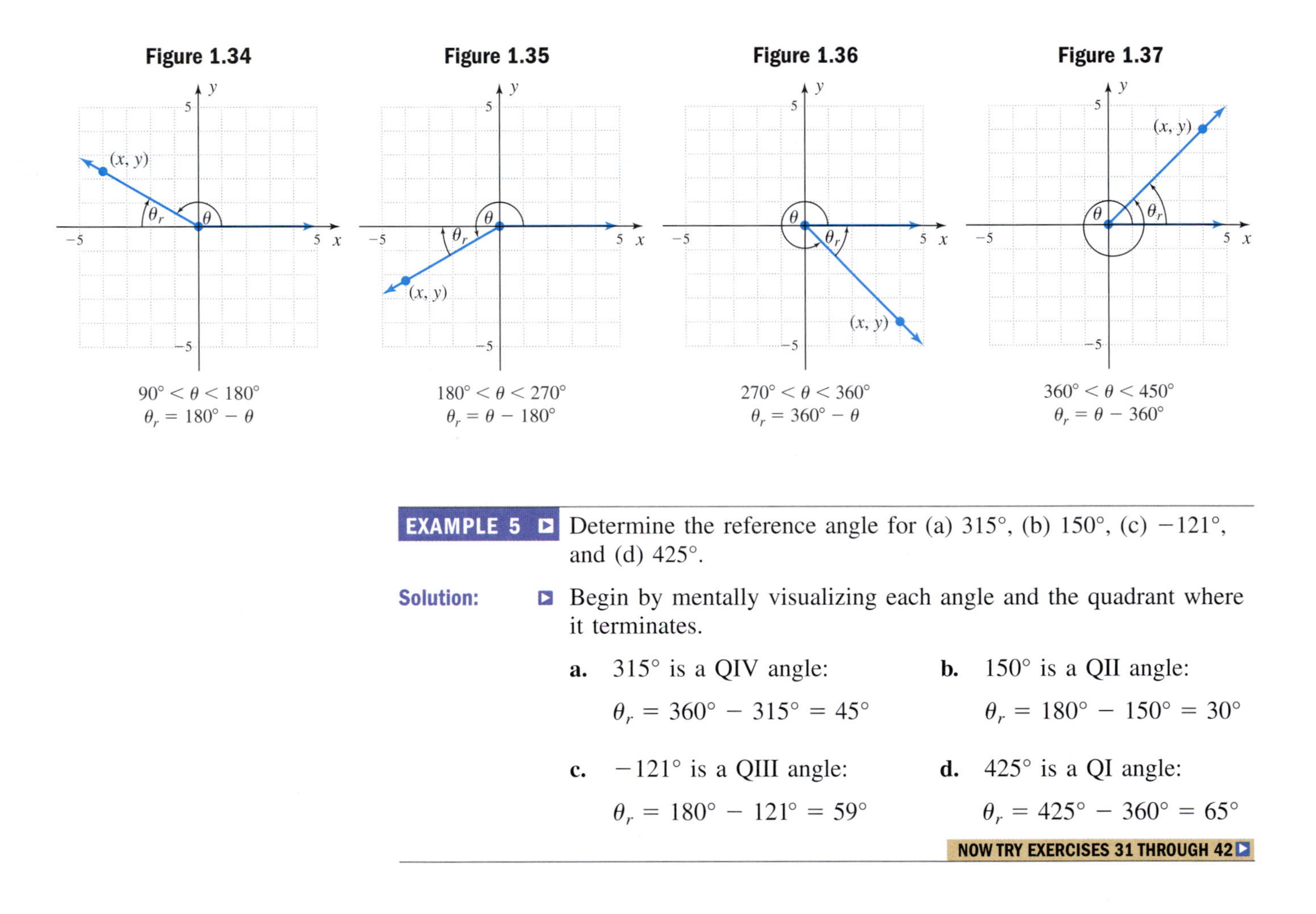

Figure 1.34 $90° < \theta < 180°$, $\theta_r = 180° - \theta$

Figure 1.35 $180° < \theta < 270°$, $\theta_r = \theta - 180°$

Figure 1.36 $270° < \theta < 360°$, $\theta_r = 360° - \theta$

Figure 1.37 $360° < \theta < 450°$, $\theta_r = \theta - 360°$

EXAMPLE 5 Determine the reference angle for (a) 315°, (b) 150°, (c) −121°, and (d) 425°.

Solution: Begin by mentally visualizing each angle and the quadrant where it terminates.

a. 315° is a QIV angle:

$\theta_r = 360° - 315° = 45°$

b. 150° is a QII angle:

$\theta_r = 180° - 150° = 30°$

c. −121° is a QIII angle:

$\theta_r = 180° - 121° = 59°$

d. 425° is a QI angle:

$\theta_r = 425° - 360° = 65°$

NOW TRY EXERCISES 31 THROUGH 42

The reference angles from Examples 5(a) and 5(b) were standard angles, which means we automatically know the absolute value of the trig ratios using θ_r. The best way to remember the signs of the trig functions is to keep in mind that sine is associated with y, cosine with x, and tangent with both x and y. In addition, there are several mnemonic devices (memory tools) to assist you. One is to use the first letter of the function that is positive in each quadrant and create a catchy acronym. For instance **ASTC → All Students Take Classes** (see Figure 1.38). Note that a trig function and its reciprocal function will always have the same sign.

Figure 1.38

Quadrant II **S**ine is positive	Quadrant I **A**ll are positive
Tangent is positive Quadrant III	**C**osine is positive Quadrant IV

EXAMPLE 6 ▶ Use a reference angle to evaluate $\sin\theta$, $\cos\theta$, and $\tan\theta$ for $\theta = 315°$.

Solution: ▶ The terminal side is in QIV where x is positive and y is negative. With $\theta_r = 45°$, we have:

$$\sin 315° = -\frac{\sqrt{2}}{2} \qquad \cos 315° = \frac{\sqrt{2}}{2}$$

$$\tan 315° = -1$$

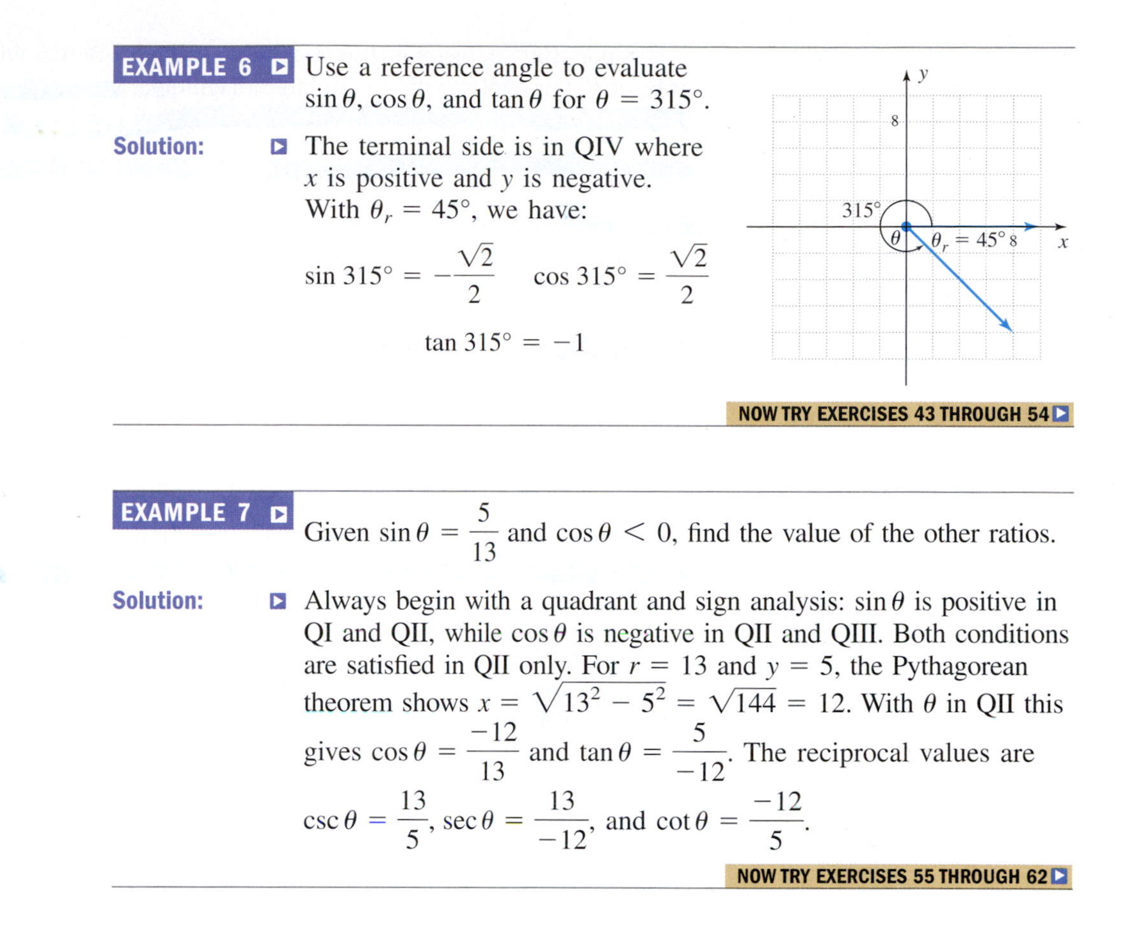

NOW TRY EXERCISES 43 THROUGH 54 ▶

EXAMPLE 7 ▶ Given $\sin\theta = \frac{5}{13}$ and $\cos\theta < 0$, find the value of the other ratios.

Solution: ▶ Always begin with a quadrant and sign analysis: $\sin\theta$ is positive in QI and QII, while $\cos\theta$ is negative in QII and QIII. Both conditions are satisfied in QII only. For $r = 13$ and $y = 5$, the Pythagorean theorem shows $x = \sqrt{13^2 - 5^2} = \sqrt{144} = 12$. With θ in QII this gives $\cos\theta = \frac{-12}{13}$ and $\tan\theta = \frac{5}{-12}$. The reciprocal values are $\csc\theta = \frac{13}{5}$, $\sec\theta = \frac{13}{-12}$, and $\cot\theta = \frac{-12}{5}$.

NOW TRY EXERCISES 55 THROUGH 62 ▶

In our everyday experience, there are many actions and activities where angles greater than or equal to 360° are applied. Some common instances are a professional basketball player that "does a three-sixty" (360°) while going to the hoop, a diver that completes a "two-and-a-half" (900°) off the high board, and a skater that executes a perfect triple axel ($3\frac{1}{2}$ turns or 1260°). As these examples suggest, angles greater than 360° must still terminate in one of the four quadrants, allowing a reference angle to be found and the functions to be evaluated for any angle *regardless of size.* Recall that two angles in standard position that share the same terminal side are called coterminal angles. Figure 1.39 illustrates that $\alpha = 135°$, $\beta = -225°$, and $\theta = 495°$ are all coterminal, with *all three having a reference angle of* 45°.

Figure 1.39

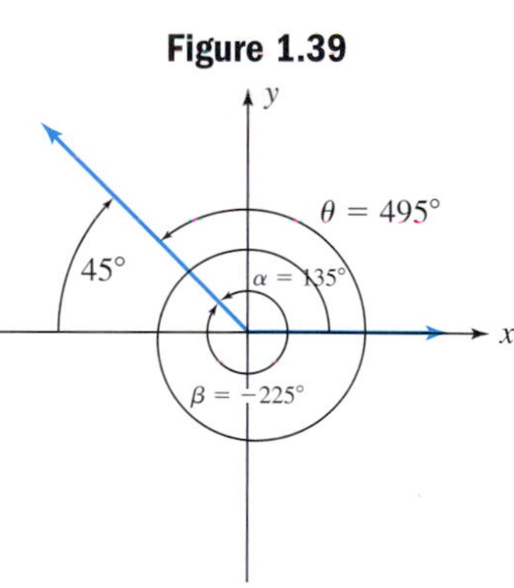

EXAMPLE 8 ▶ Evaluate sin 135°, cos −225°, and tan 495°.

Solution: ▶ The angles are coterminal and terminate in QII, where $x < 0$ and $y > 0$. With $\theta_r = 45°$ we have $\sin 135° = \frac{\sqrt{2}}{2}$, $\cos -225° = -\frac{\sqrt{2}}{2}$, and $\tan 495° = -1$.

NOW TRY EXERCISES 63 THROUGH 74 ▶

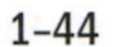

Since 360° is one full rotation, all angles $\theta + 360°k$ will be coterminal for any integer k. For angles with a very large magnitude, we can find the quadrant of the terminal side by subtracting as many integer multiples of 360° as needed from the angle. For $\alpha = 1908°$, $\frac{1908}{360} = 5.3$ and $1908 - 360(5) = 108°$. This angle is in QII with $\theta_r = 72°$. See Exercises 75 through 90.

Figure 1.40

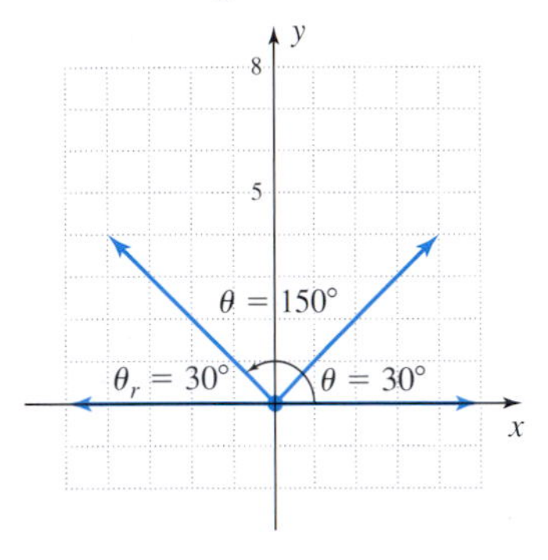

C. Applications of the Trig Functions of Any Angle

One of the most basic uses of coterminal angles is determining all values of θ that satisfy a stated relationship. For example, by now you are aware that if $\sin\theta = \frac{1}{2}$ (positive one-half), then $\theta = 30°$ or $\theta = 150°$ (see Figure 1.40). But this is also true for all angles coterminal with these two, and we would write the solutions as $\theta = 30° + 360°k$ and $\theta = 150° + 360°k$ for all integers k.

EXAMPLE 9 Find all angles satisfying the relationship. Answer in degrees.

a. $\cos\theta = -\frac{\sqrt{2}}{2}$ **b.** $\tan\theta = -1.4654$

Solution: **a.** Cosine is negative in QII and QIII. Recognizing $\cos 45° = \frac{\sqrt{2}}{2}$, we reason $\theta_r = 45°$ and two solutions are $\theta = 135°$ from QII and $\theta = 225°$ from QIII. For all values of θ satisfying the relationship, we have $\theta = 135° + 360°k$ and $\theta = 225° + 360°k$. See Figure 1.41.

Figure 1.41

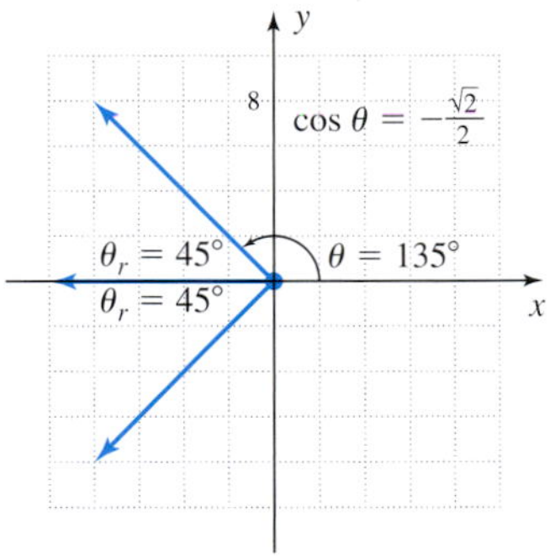

b. Tangent is negative in QII and QIV. For -1.4654 we again find θ_r using a calculator: 2nd TAN -1.4654 ENTER shows $\tan^{-1}(-1.4654) \approx -55.7$, so $\theta_r = 55.7°$. Two solutions are $\theta = 180° - 55.7° = 124.3°$ from QII, and in QIV $\theta = 360° - 55.7° = 304.3°$. The result is $\theta = 124.3° + 360°k$ and $\theta = 304.3° + 360°k$. Note that these can be combined into the single statement $\theta = 124.3° + 180°k$. See Figure 1.42.

Figure 1.42

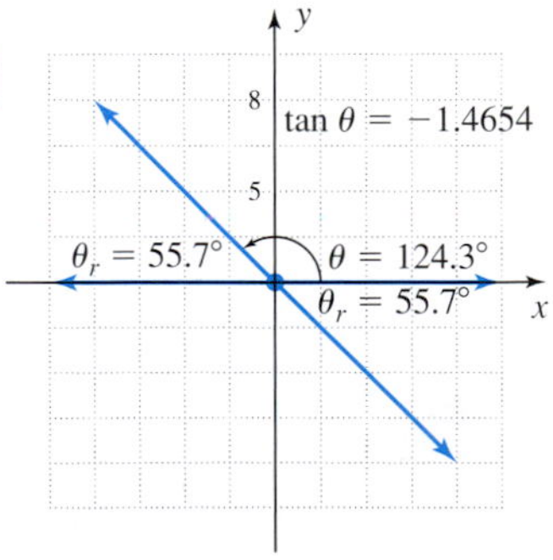

NOW TRY EXERCISES 93 THROUGH 100

We close this section with an additional application of the concepts related to trigonometric functions of any angle.

EXAMPLE 10 A radar operator calls the captain over to her screen saying, "Sir, we have an unidentified bogey at heading 20° (20° east of due north or a standard 70° rotation). I think it's a UFO." The captain asks, "What makes you think so?" To which the sailor replies, "Because it's at 25,000 ft and not moving!" Name all angles for which the UFO causes a "blip" to occur on the radar screen.

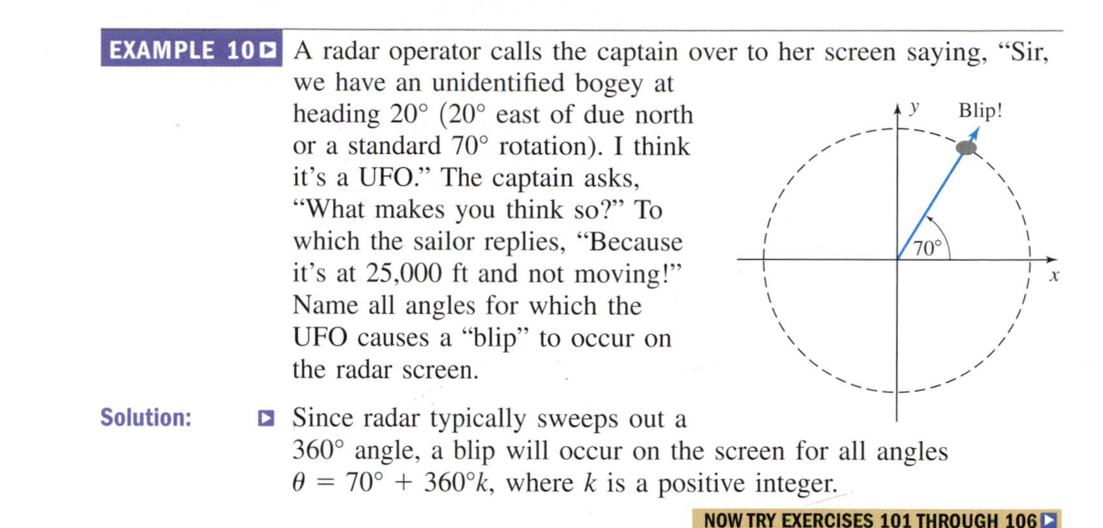

Solution: Since radar typically sweeps out a 360° angle, a blip will occur on the screen for all angles $\theta = 70° + 360°k$, where k is a positive integer.

NOW TRY EXERCISES 101 THROUGH 106

TECHNOLOGY HIGHLIGHT

x, y, r, and functions of any angle

The keystrokes shown apply to a TI-84 Plus model. Please consult your manual or our Internet site for other models.

Graphing calculators offer a number of features that can assist a study of the trig functions of any angle. On the TI-84 Plus, the keystrokes 2nd APPS **(ANGLE)** will bring up the menu shown in Figure 1.43. Options 1 through 4 are basically used for angle conversions (DMS degrees to decimal degrees, degrees to radians, and so on). Of interest to us here are options 5 and 6, which can be used to determine the radius r (option 5) or the angle θ (option 6) related to a given point (x, y). For $(-12, 35)$, CLEAR the home screen and press 2nd APPS **(ANGLE) 5:R▶Pr(**, which will place the option on the home screen. This feature supplies the left parenthesis of the ordered pair, and you simply complete it: **5:R▶Pr(−12, 35)**. As shown in Figure 1.44, the calculator returns 37, and $(-12)^2 + (35)^2 = (37)^2$✓. To find the related angle, it is assumed that θ is in standard position and (x, y) is on the terminal side. Pressing 2nd APPS **(ANGLE) 6:R▶Pθ(** and completing the ordered pair as before, shows the corresponding angle is approximately 108.9° (Figure 1.44). Note this is a QII angle as expected, since $x < 0$ and $y > 0$. Use these features to complete the following exercises.

Figure 1.43

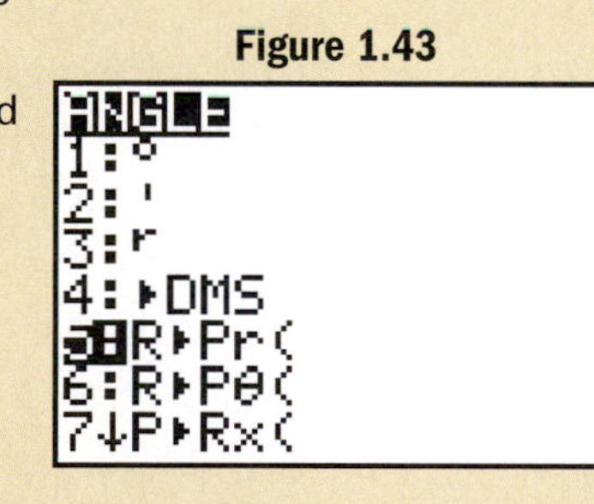

Figure 1.44

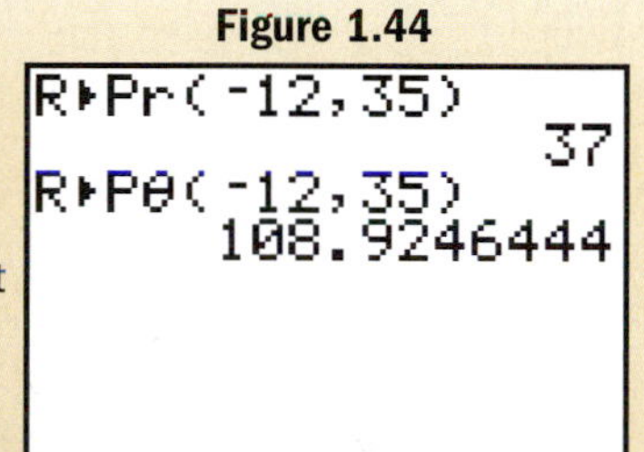

Exercise 1: Find the radius corresponding to the point $(-5, 5\sqrt{3})$.

Exercise 2: Find the angle corresponding to $(-28, -45)$, then use a calculator to evaluate $\sin\theta$, $\cos\theta$, and $\tan\theta$ for this angle. Compare each result to the values given by $\sin\theta = \frac{y}{r}$, $\cos\theta = \frac{x}{r}$, and $\tan\theta = \frac{y}{x}$.

1.3 EXERCISES

CONCEPTS AND VOCABULARY

Fill in each blank with the appropriate word or phrase. Carefully reread the section if needed.

1. An angle is in standard position if its vertex is at the ________ and the initial side is along the ________.
2. A(n) ________ angle is one where the ________ side is coincident with one of the coordinate axes.
3. Angles formed by a counterclockwise rotation are ________ angles. Angles formed by a ________ rotation are negative angles.
4. For any angle θ, its reference angle θ_r is the positive ________ angle formed by the ________ side and the nearest x-axis.
5. Discuss the similarities and differences between the trigonometry of right triangles and the trigonometry of *any* angle.
6. Let $T(x)$ represent any one of the six basic trig functions. Explain why the equation $T(x) = k$ will always have exactly two solutions in $[0, 2\pi)$ if x is not a quadrantal angle.

DEVELOPING YOUR SKILLS

7. Draw a 30-60-90 triangle with the 60° angle at the origin and the short side along the x-axis. Determine the slope and equation of the line coincident with the hypotenuse, then pick any point on this line and evaluate $\sin 60°$, $\cos 60°$, and $\tan 60°$. Comment on what you notice.
8. Draw a 45-45-90 triangle with a 45° angle at the origin and one side along the x-axis. Determine the slope and equation of the line coincident with the hypotenuse, then pick any point on this line and evaluate $\sin 45°$, $\cos 45°$, and $\tan 45°$. Comment on what you notice.

Find the value of the six trigonometric functions given $P(x, y)$ is on the terminal side of angle θ, with θ in standard position.

9. $(8, 15)$
10. $(7, 24)$
11. $(-20, 21)$
12. $(-3, -1)$
13. $(7.5, -7.5)$
14. $(9, -9)$
15. $(4\sqrt{3}, 4)$
16. $(-6, 6\sqrt{3})$
17. $(2, 8)$
18. $(6, -15)$
19. $(-3.75, -2.5)$
20. $(6.75, 9)$
21. $\left(-\frac{5}{9}, \frac{2}{3}\right)$
22. $\left(\frac{3}{4}, -\frac{7}{16}\right)$
23. $\left(\frac{1}{4}, -\frac{\sqrt{5}}{2}\right)$
24. $\left(-\frac{\sqrt{3}}{5}, \frac{22}{25}\right)$

Graph each linear equation. Then state the quadrants it traverses and pick one point on the line from each quadrant and evaluate the functions $\sin\theta$, $\cos\theta$ and $\tan\theta$ using these points.

25. $y = \frac{3}{4}x$
26. $y = \frac{5}{12}x$
27. $y = -\frac{\sqrt{3}}{3}x$
28. $y = -\frac{\sqrt{3}}{2}x$
29. Evaluate the six trig functions in terms of x, y, and r for $\theta = 90°$.
30. Evaluate the six trig functions in terms of x, y, and r for $\theta = 180°$.

Name the reference angle θ_r for the angle θ given.

31. $\theta = 120°$
32. $\theta = 210°$
33. $\theta = 135°$
34. $\theta = 315°$
35. $\theta = -45°$
36. $\theta = -240°$
37. $\theta = 112°$
38. $\theta = 179°$
39. $\theta = 500°$
40. $\theta = 750°$
41. $\theta = -168.4°$
42. $\theta = -328.2°$

State the quadrant of the terminal side of θ, using the information given.

43. $\sin\theta > 0$, $\cos\theta < 0$
44. $\cos\theta < 0$, $\tan\theta < 0$
45. $\tan\theta < 0$, $\sin\theta > 0$
46. $\sec\theta > 0$, $\tan\theta > 0$

Find the exact value of $\sin\theta$, $\cos\theta$, and $\tan\theta$ using reference angles.

47. $\theta = 330°$ **48.** $\theta = 390°$ **49.** $\theta = -45°$ **50.** $\theta = -120°$

51. $\theta = 240°$ **52.** $\theta = 315°$ **53.** $\theta = -150°$ **54.** $\theta = -210°$

For the information given, find the related values of x, y, and r. Clearly indicate the quadrant of the terminal side of θ, then state the values of the six trig functions of θ.

55. $\cos\theta = \frac{4}{5}$ and $\sin\theta < 0$

56. $\tan\theta = -\frac{12}{5}$ and $\cos\theta > 0$

57. $\csc\theta = -\frac{37}{35}$ and $\tan\theta > 0$

58. $\sin\theta = -\frac{20}{29}$ and $\cot\theta < 0$

59. $\csc\theta = 3$ and $\cos\theta > 0$

60. $\csc\theta = -2$ and $\cos\theta > 0$

61. $\sin\theta = -\frac{7}{8}$ and $\sec\theta < 0$

62. $\cos\theta = \frac{5}{12}$ and $\sin\theta < 0$

Find two positive and two negative angles that are coterminal with the angle given. Answers will vary.

63. 52° **64.** 12° **65.** 87.5° **66.** 22.8°

67. 225° **68.** 175° **69.** −107° **70.** −215°

Evaluate in exact form as indicated.

71. $\sin 120°$, $\cos -240°$, $\tan 480°$

72. $\sin 225°$, $\cos 585°$, $\tan -495°$

73. $\sin -30°$, $\cos -390°$, $\tan 690°$

74. $\sin 210°$, $\cos 570°$, $\tan -150°$

Find the exact value of $\sin\theta$, $\cos\theta$, and $\tan\theta$ using reference angles.

75. $\theta = 600°$ **76.** $\theta = 480°$ **77.** $\theta = -840°$

78. $\theta = -930°$ **79.** $\theta = 570°$ **80.** $\theta = 495°$

81. $\theta = -1230°$ **82.** $\theta = 3270°$

For each exercise, state the quadrant of the terminal side and the sign of the function in that quadrant. Then evaluate the expression using a calculator. Round to four decimal places.

83. $\sin 719°$ **84.** $\cos 528°$ **85.** $\tan -419°$

86. $\sec -621°$ **87.** $\csc 681°$ **88.** $\tan 995°$

89. $\cos 805°$ **90.** $\sin 772°$

WORKING WITH FORMULAS

91. The area of a parallelogram: $A = ab\sin\theta$

The area of a parallelogram is given by the formula shown, where a and b are the lengths of the sides and θ is the angle between them. Use the formula to complete the following: (a) find the area of a parallelogram with sides $a = 9$ and $b = 21$, given $\theta = 50°$. (b) What is the smallest integer value of θ where the area is greater than 150 units^2? (c) State what happens when $\theta = 90°$. (d) How can you find the area of a triangle using this formula?

92. The angle between two lines in the plane: $\tan\theta = \frac{m_2 - m_1}{1 + m_2 m_1}$

Given line 1 and line 2 with slopes m_1 and m_2, respectively, the angle between the two lines is given by the formula shown. Find the angle θ if the equation of line 1 is $y_1 = \frac{3}{4}x + 2$ and line 2 has equation $y = -\frac{2}{3}x + 5$.

APPLICATIONS

Find all angles satisfying the stated relationship. For standard angles, express your answer in exact form. For nonstandard values, use a calculator and round function values to tenths.

93. $\cos\theta = \frac{1}{2}$ **94.** $\sin\theta = \frac{\sqrt{2}}{2}$ **95.** $\sin\theta = -\frac{\sqrt{3}}{2}$ **96.** $\tan\theta = -\frac{\sqrt{3}}{1}$

97. $\sin\theta = 0.8754$ **98.** $\cos\theta = 0.2378$ **99.** $\tan\theta = -2.3512$ **100.** $\cos\theta = -0.0562$

101. Nonacute angles: At a recent carnival, one of the games on the midway was played using a large spinner that turns clockwise. On Jorge's spin the number 25 began at the 12 o'clock (top/center) position, returned to this position five times during the spin and stopped at the 3 o'clock position. What angle θ did the spinner spin through? Name all angles that are coterminal with θ.

Exercise 101

102. Nonacute angles: One of the four blades on a ceiling fan has a decal on it and begins at a designated "12 o'clock" position. Turning the switch on and then immediately off, causes the blade to make over three complete, counterclockwise rotations, with the blade stopping at the 8 o'clock position. What angle θ did the blade turn through? Name all angles that are coterminal with θ.

Exercise 103

103. High dives: As part of a diving competition, David executes a perfect reverse two-and-a-half flip. Does he enter the water feet first or head first? Through what angle did he turn from takeoff until the moment he entered the water?

104. Gymnastics: While working out on a trampoline, Charlene does three complete, forward flips and then belly-flops on the trampoline before returning to the upright position. What angle did she turn through from the flip to the belly-flop?

Exercise 105

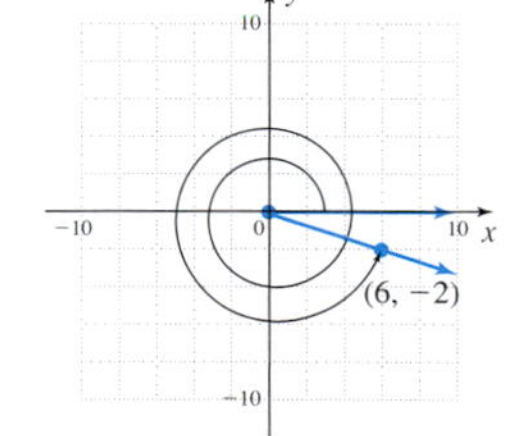

105. Spiral of Archimedes: The graph shown is called the spiral of Archimedes. Through what angle θ has the spiral turned, given the spiral terminates at $(6, -2)$ as indicated?

106. Involute of a circle: The graph shown is called the involute of a circle. Through what angle θ has the involute turned, given the graph terminates at $(-4, -3.5)$ as indicated?

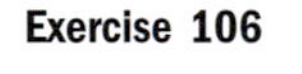
Exercise 106

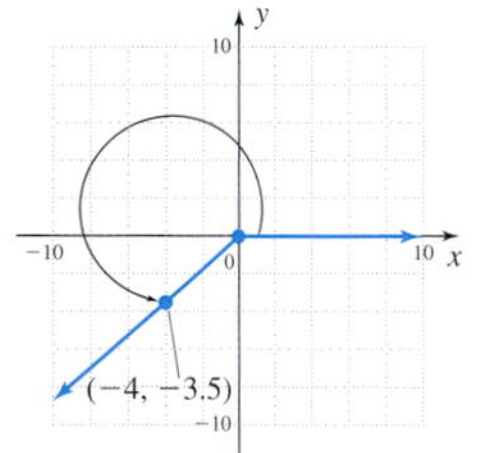

Area bounded by chord and circumference: Find the area of the shaded region, rounded to the nearest 100th. Note the area of a triangle is one-half the area of a parallelogram (see Exercise 91).

107. **108.**

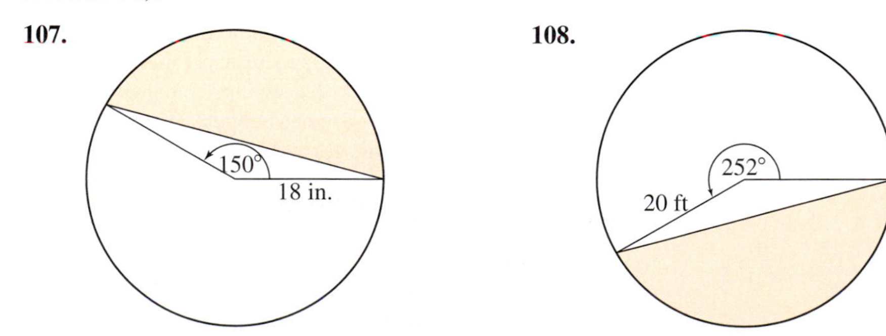

WRITING, RESEARCH, AND DECISION MAKING

109. Each of our eyes can individually see through a cone approximately 150° wide, called the **cone of vision.** When the cone from the right and left eyes overlap, the eyes can gather light (*see*) from almost 180° (including the peripheral vision), although the clearest focus is within a central 60° cone. Use the Internet or the resources of a local library to investigate and research *cones of vision.* Try to determine what trigonometric principles are involved in studies of the eye, and its ability to see and interpret light. Prepare a short summary on what you find.

110. In an elementary study of trigonometry, the hands of a clock are often studied because of the angle relationship that exists between the hands. For example, at 3 o'clock, the angle between the two hands is a right angle and measures 90°. (a) What is the angle between the two hands at 1 o'clock? 2 o'clock? Explain why. (b) What is the angle between the two hands at 6:30? 7:00? 7:30? Explain why. (c) Name four times at which the hands will form a 45° angle. (d) What other questions of interest can you think of regarding the angle between the hands of a clock?

EXTENDING THE CONCEPT

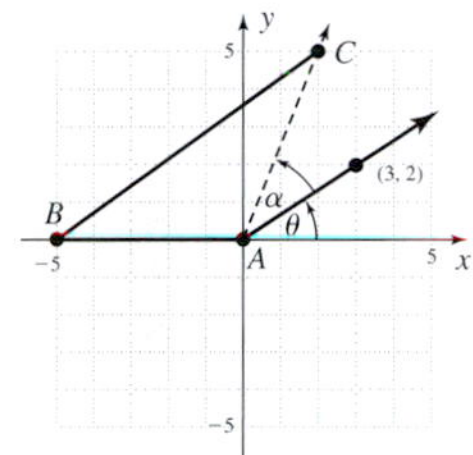

111. In the diagram shown, the indicated ray is of arbitrary length. (a) Through what additional angle α would the ray have to be rotated to create triangle ABC? (b) What will be the length of side AC once the triangle is complete?

112. Referring to Exercise 102, suppose the fan blade had a radius of 20 in. and is turning at a rate of 12 revolutions per second. (a) Find the angle the blade turns through in 3 sec. (b) Find the circumference of the circle traced out by the tip of the blade. (c) Find the total distance traveled by the blade tip in 10 sec. (d) Find the speed, in miles per hour, that the tip of the blade is traveling.

MAINTAINING YOUR SKILLS

113. (1.1) Complete the table from memory or mental calculation only.

degrees	30°	45°	60°	90°	180°	270°	360°
radians							

114. (1.2) Given $\tan 40° \approx 0.8391$, what is the value of $\cot 50°$? Why?

115. (1.2) Solve the given triangle using a calculator. Express all angles and sides to the nearest tenth of a unit. Answer in table form.

116. (1.2) Solve the given triangle using special angles. Express all angles and sides in exact form. Answer in table form.

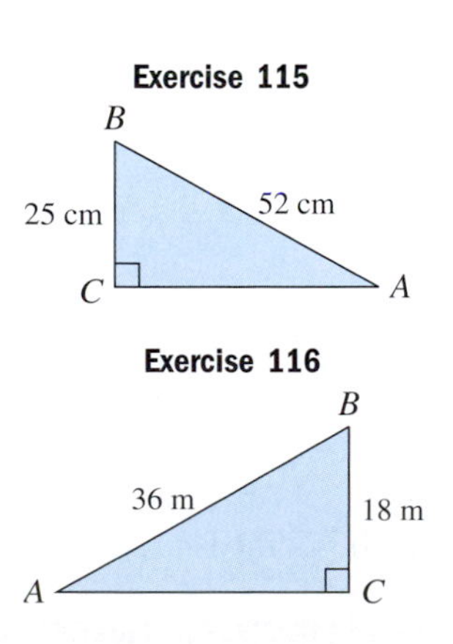

117. (1.1) For emissions testing, automobiles are held stationary while a heavy roller installed in the floor allows the wheels to turn freely. If the large wheels of a customized pickup have a radius of 18 in. and are turning at 300 revolutions per minute, what speed is the odometer of the truck reading in miles per hour?

118. (1.2) Jazon is standing 117 ft from the base of the Washington Monument in Washington, D.C. If his eyes are 5 ft above level ground and he must hold his head at a 78° angle from horizontal to see the top of the monument (the angle of elevation of 78°), estimate the height of the monument. Answer to the nearest tenth of a foot.

1.4 Unit Circles and the Trigonometry of Real Numbers

LEARNING OBJECTIVES

In Section 1.4 you will learn how to:

- **A.** Locate points on a unit circle and use symmetry to locate other points
- **B.** Use standard triangles to find points on a unit circle and locate other points using symmetry
- **C.** Define the six trig functions in terms of a point on the unit circle
- **D.** Define the six trig functions in terms of a real number *t*
- **E.** Find the real number *t* corresponding to given values of sin *t*, cos *t*, and tan *t*

INTRODUCTION

In this section we introduce the **trigonometry of real numbers,** a view of trig that exists free of the traditional right triangle view. In fact, the ultimate value of the trig functions lies not in its classical study, but in the input/output nature of the trig functions and the cyclic values they generate. These functions have powerful implications in our continuing study, and important applications in this course and those that follow.

POINT OF INTEREST

As functions of a real number, trigonometry has applications in some surprising areas: (a) blood pressure, (b) predator/prey models, (c) electric generators, (d) tidal motion, (e) meteorology, (f) planetary studies, (g) engine design, and (h) intensity of light/sound. Actually the list is endless, but perhaps these are sufficient to engage and intrigue, drawing us into our current study.

A. The Unit Circle

Figure 1.45

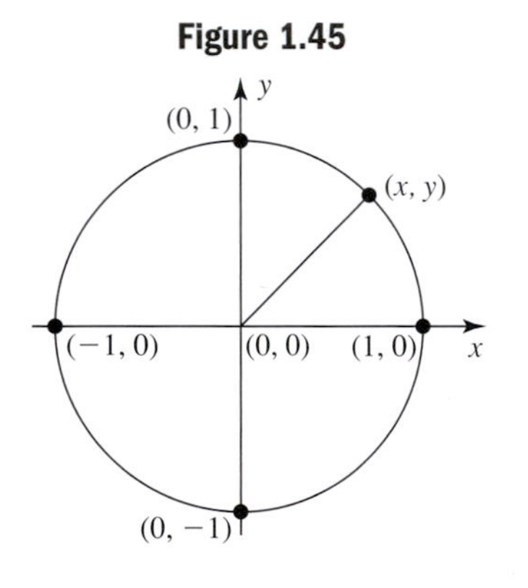

A circle is defined as the set of all points in a plane that are a *fixed distance* called the **radius** from a *fixed point* called the **center.** Since the definition involves distance, we can construct the general equation of a circle using the distance formula. Assume the center has coordinates (h, k) and let (x, y) represent any point on the graph. Since the distance between these points is the radius r, the distance formula yields $\sqrt{(x-h)^2 + (y-k)^2} = r$. Squaring both sides gives $(x-h)^2 + (y-k)^2 = r^2$. For central circles both h and k are zero, and the result is the equation for a **central circle** of radius r: $x^2 + y^2 = r^2$. The **unit circle** is defined as a central circle with radius 1 unit: $x^2 + y^2 = 1$. As such, the figure can easily be graphed by drawing a circle through the four **quadrantal points** $(1, 0)$, $(-1, 0)$, $(0, 1)$, and $(0, -1)$ as in Figure 1.45. To find other points on the circle, we simply select any value of x, where $|x| < 1$, then substitute and solve for y; or any value of y, where $|y| < 1$, then solve for x.

EXAMPLE 1 Find a point on the unit circle given $y = \frac{1}{2}$ with (x, y) in QII.

Solution: Using the equation of a unit circle, we have

$$x^2 + y^2 = 1 \quad \text{unit circle equation}$$

$$x^2 + \left(\frac{1}{2}\right)^2 = 1 \quad \text{substitute } \frac{1}{2} \text{ for } y$$

$$x^2 + \frac{1}{4} = 1 \quad \left(\frac{1}{2}\right)^2 = \frac{1}{4}$$

$$x^2 = \frac{3}{4} \quad \text{subtract } \frac{1}{4}$$

$$x = \pm\frac{\sqrt{3}}{2} \quad \text{result}$$

With (x, y) in QII, we choose $x = -\frac{\sqrt{3}}{2}$. The point is $\left(-\frac{\sqrt{3}}{2}, \frac{1}{2}\right)$.

NOW TRY EXERCISES 7 THROUGH 18

Additional points on the unit circle can be found using symmetry. The simplest examples come from the quadrantal points, where $(1, 0)$ and $(-1, 0)$ are on opposite sides of the y-axis, and $(0, 1)$ and $(0, -1)$ are on opposite sides of the x-axis. In general, if a and b are positive real numbers and (a, b) is on the unit circle, then $(-a, b)$, $(a, -b)$, and $(-a, -b)$ are also on the circle *because a circle is symmetric to both axes and the origin*! For the point $\left(-\frac{\sqrt{3}}{2}, \frac{1}{2}\right)$ from Example 1, three other points can quickly be located, since the coordinates will differ only in sign. They are $\left(-\frac{\sqrt{3}}{2}, -\frac{1}{2}\right)$ in QIII, $\left(\frac{\sqrt{3}}{2}, -\frac{1}{2}\right)$ in QIV, and $\left(\frac{\sqrt{3}}{2}, \frac{1}{2}\right)$ in QI. See Figure 1.46.

Figure 1.46

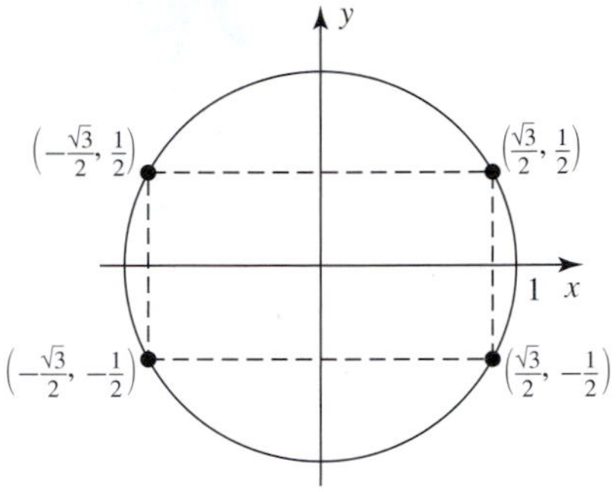

EXAMPLE 2 Name the quadrant containing $\left(-\frac{3}{5}, -\frac{4}{5}\right)$ and verify it's on a unit circle. Then use symmetry to find three other points on the circle.

Solution: Since both coordinates are negative, $\left(-\frac{3}{5}, -\frac{4}{5}\right)$ is in QIII. Substituting into the equation for a unit circle yields

$$x^2 + y^2 = 1 \quad \text{unit circle equation}$$

$$\left(\frac{-3}{5}\right)^2 + \left(\frac{-4}{5}\right)^2 = 1 \quad \text{substitute } \tfrac{-3}{5} \text{ for } x \text{ and } \tfrac{-4}{5} \text{ for } y$$

$$\frac{9}{25} + \frac{16}{25} = 1 \quad \text{simplify}$$

$$\frac{25}{25} = 1 \quad \text{result checks}$$

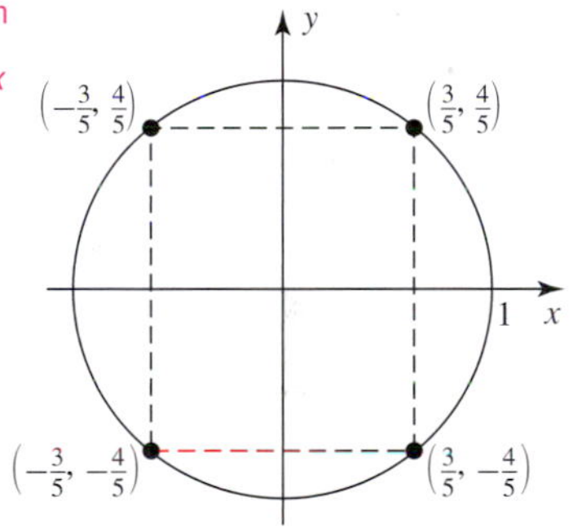

Since $\left(\frac{-3}{5}, \frac{-4}{5}\right)$ is on the unit circle, $\left(\frac{3}{5}, \frac{-4}{5}\right)$ $\left(\frac{-3}{5}, \frac{4}{5}\right)$, and $\left(\frac{3}{5}, \frac{4}{5}\right)$ are also on the circle due to symmetry (see figure).

NOW TRY EXERCISES 19 THROUGH 26

B. Standard Triangles and the Unit Circle

The standard triangles can also be used to find points on a unit circle. As usually written, the triangles state a proportional relationship between their sides after assigning a value of 1 to the shortest sides. However, precisely due to this proportional relationship, *we can divide all sides by the length of the hypotenuse,* giving *it* a length of 1 unit (see Figures 1.47 and 1.48). We then place the triangle within the unit circle, and reflect it from quadrant to quadrant to find additional points. We use *the sides* of the triangle to determine the absolute value of each coordinate, and *the quadrant* to give each coordinate the appropriate sign. Note the standard triangles are now expressed in radians.

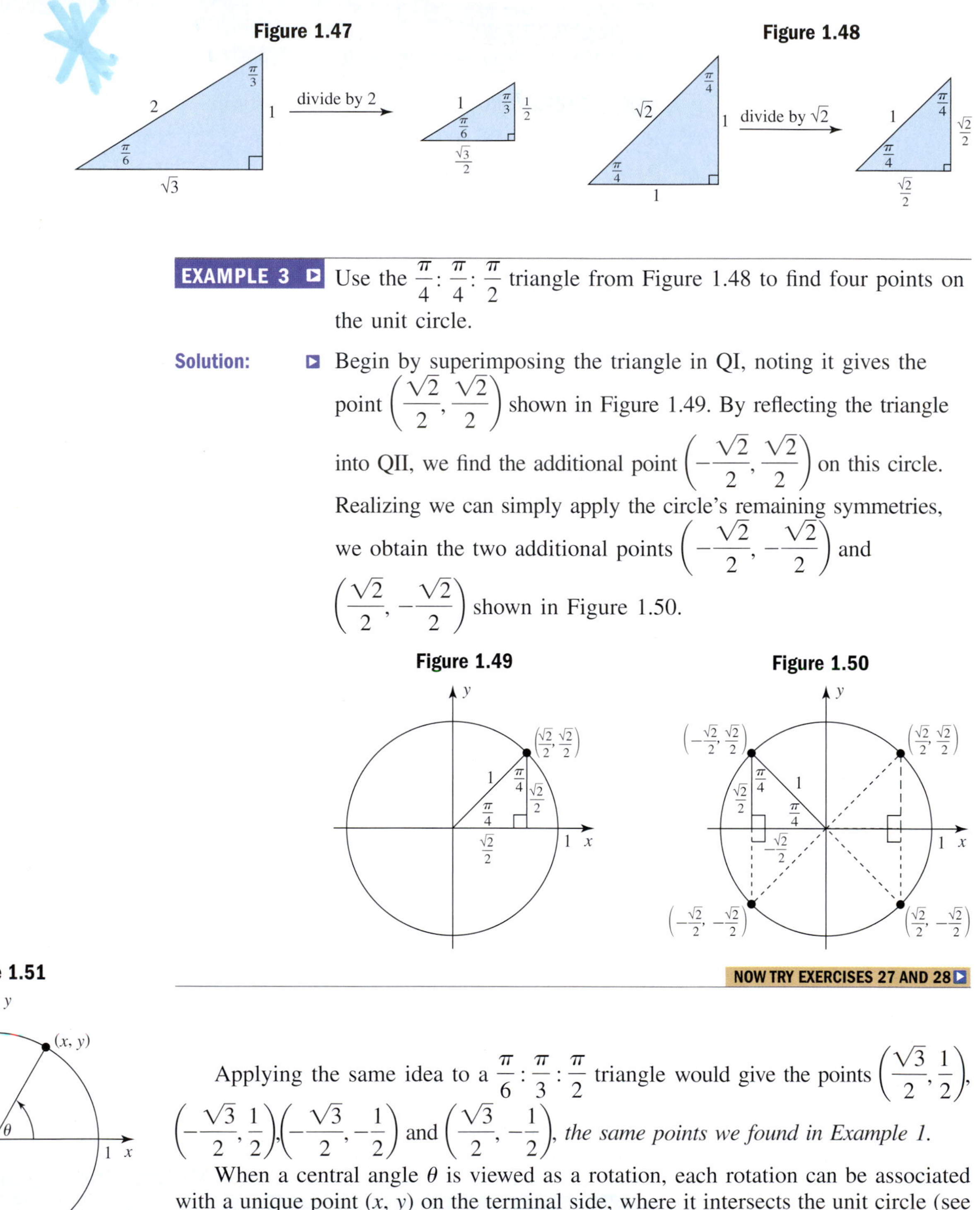

EXAMPLE 3 Use the $\frac{\pi}{4}:\frac{\pi}{4}:\frac{\pi}{2}$ triangle from Figure 1.48 to find four points on the unit circle.

Solution: Begin by superimposing the triangle in QI, noting it gives the point $\left(\frac{\sqrt{2}}{2}, \frac{\sqrt{2}}{2}\right)$ shown in Figure 1.49. By reflecting the triangle into QII, we find the additional point $\left(-\frac{\sqrt{2}}{2}, \frac{\sqrt{2}}{2}\right)$ on this circle. Realizing we can simply apply the circle's remaining symmetries, we obtain the two additional points $\left(-\frac{\sqrt{2}}{2}, -\frac{\sqrt{2}}{2}\right)$ and $\left(\frac{\sqrt{2}}{2}, -\frac{\sqrt{2}}{2}\right)$ shown in Figure 1.50.

NOW TRY EXERCISES 27 AND 28

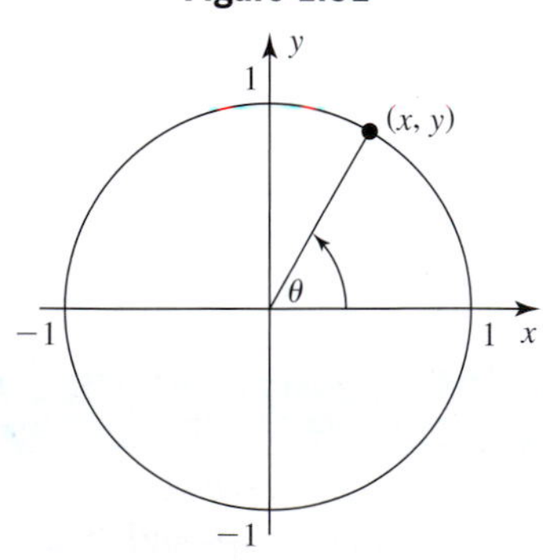

Figure 1.51

Applying the same idea to a $\frac{\pi}{6}:\frac{\pi}{3}:\frac{\pi}{2}$ triangle would give the points $\left(\frac{\sqrt{3}}{2}, \frac{1}{2}\right)$, $\left(-\frac{\sqrt{3}}{2}, \frac{1}{2}\right)$, $\left(-\frac{\sqrt{3}}{2}, -\frac{1}{2}\right)$ and $\left(\frac{\sqrt{3}}{2}, -\frac{1}{2}\right)$, *the same points we found in Example 1.*

When a central angle θ is viewed as a rotation, each rotation can be associated with a unique point (x, y) on the terminal side, where it intersects the unit circle (see

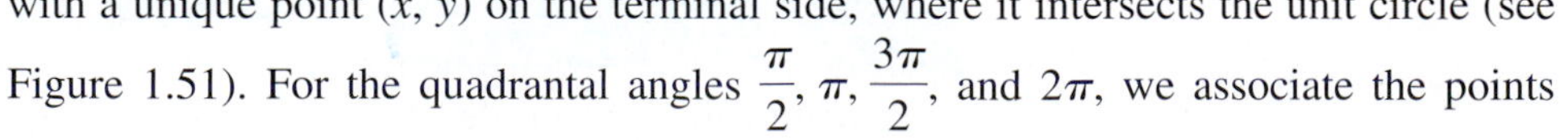

Figure 1.51). For the quadrantal angles $\frac{\pi}{2}$, π, $\frac{3\pi}{2}$, and 2π, we associate the points

(0, 1), $(-1, 0)$, $(0, -1)$, and (0, 0), respectively. When this rotation results in a standard angle θ, the association can be found using a standard triangle in a manner similar to Example 3. Figure 1.52 shows we associate the point $\left(\frac{\sqrt{3}}{2}, \frac{1}{2}\right)$ with $\theta = \frac{\pi}{6}$, $\left(\frac{\sqrt{2}}{2}, \frac{\sqrt{2}}{2}\right)$ with $\theta = \frac{\pi}{4}$, and by reorienting the $\frac{\pi}{6} : \frac{\pi}{3} : \frac{\pi}{2}$ triangle, $\left(\frac{1}{2}, \frac{\sqrt{3}}{2}\right)$ is associated with a rotation of $\theta = \frac{\pi}{3}$.

Figure 1.52

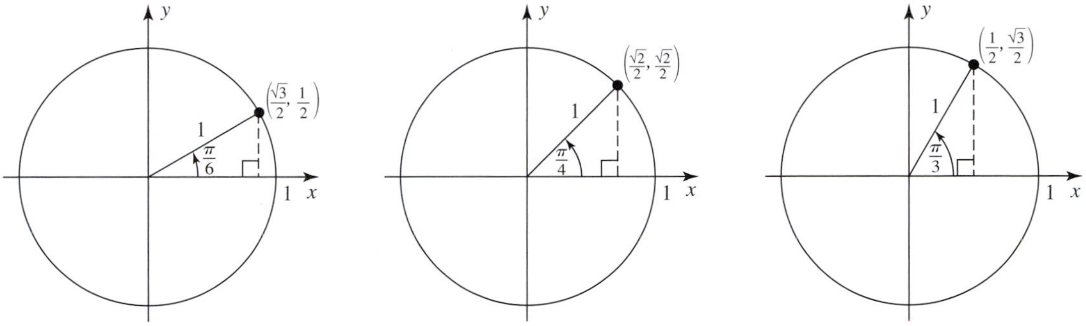

For standard rotations from $\theta = 0$ to $\theta = \frac{\pi}{2}$ we have the following:

Rotation θ	0	$\frac{\pi}{6}$	$\frac{\pi}{4}$	$\frac{\pi}{3}$	$\frac{\pi}{2}$
Associated point (x, y)	(0, 0)	$\left(\frac{\sqrt{3}}{2}, \frac{1}{2}\right)$	$\left(\frac{\sqrt{2}}{2}, \frac{\sqrt{2}}{2}\right)$	$\left(\frac{1}{2}, \frac{\sqrt{3}}{2}\right)$	(0, 1)

Each of these points gives rise to three others using the symmetry of the circle. With this symmetry and a reference angle θ_r, we can associate additional points on a unit circle for $\theta > \frac{\pi}{2}$. Several examples of the reference angle concept are shown in Figure 1.53 for θ in radians.

Figure 1.53

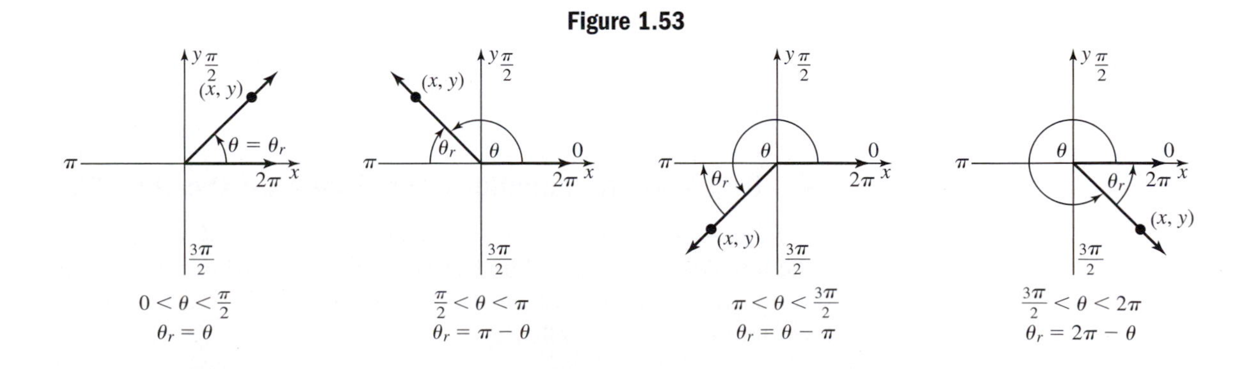

Due to the symmetries of the circle, reference angles of $\frac{\pi}{6}$, $\frac{\pi}{4}$, and $\frac{\pi}{3}$ serve to fix the absolute value of the coordinates for x and y, so all that remains is to *use the appropriate sign for each coordinate* (r is always positive). This depends solely on the quadrant of the terminal side.

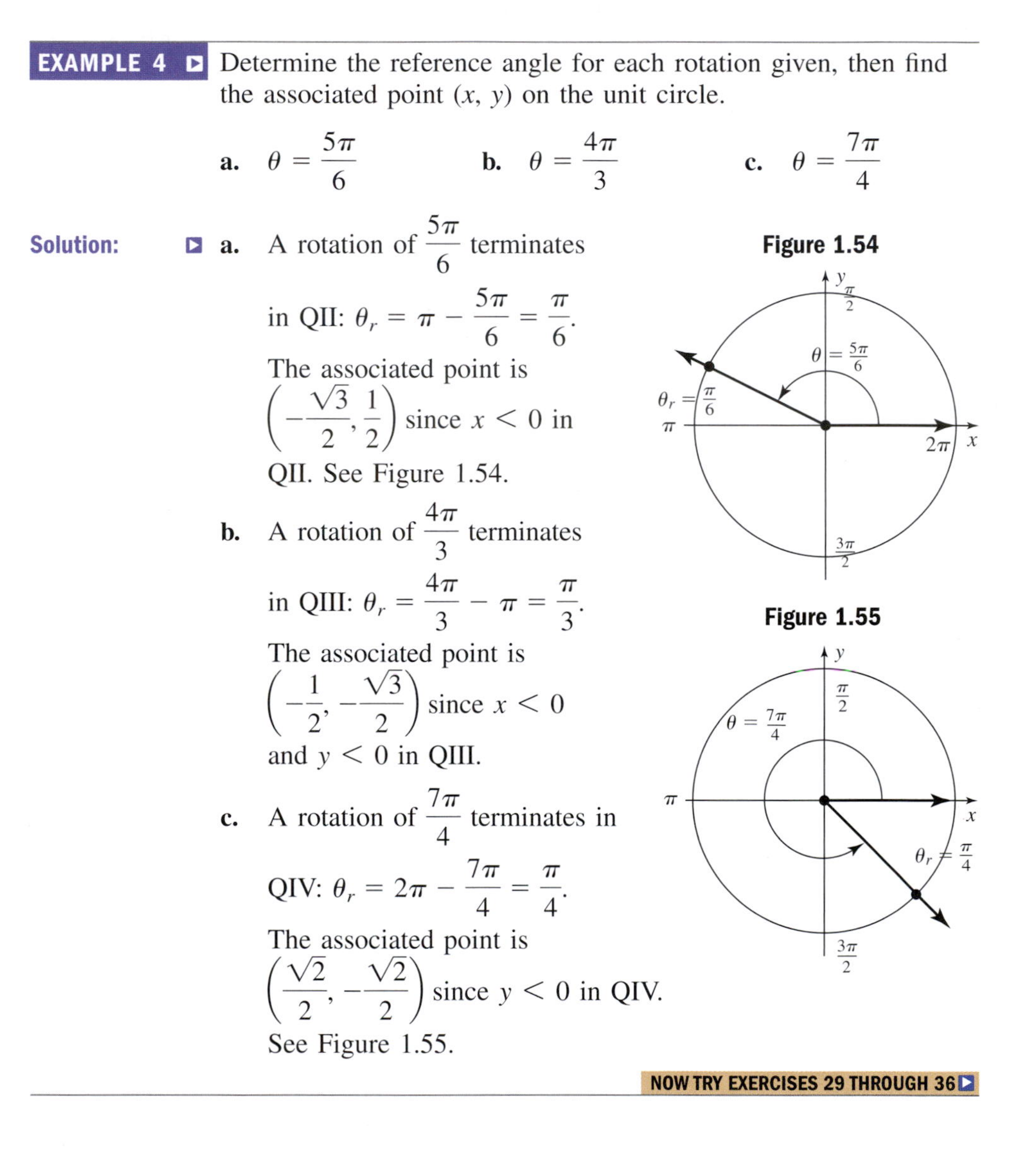

EXAMPLE 4 Determine the reference angle for each rotation given, then find the associated point (x, y) on the unit circle.

a. $\theta = \frac{5\pi}{6}$ **b.** $\theta = \frac{4\pi}{3}$ **c.** $\theta = \frac{7\pi}{4}$

Solution: **a.** A rotation of $\frac{5\pi}{6}$ terminates in QII: $\theta_r = \pi - \frac{5\pi}{6} = \frac{\pi}{6}$. The associated point is $\left(-\frac{\sqrt{3}}{2}, \frac{1}{2}\right)$ since $x < 0$ in QII. See Figure 1.54.

b. A rotation of $\frac{4\pi}{3}$ terminates in QIII: $\theta_r = \frac{4\pi}{3} - \pi = \frac{\pi}{3}$. The associated point is $\left(-\frac{1}{2}, -\frac{\sqrt{3}}{2}\right)$ since $x < 0$ and $y < 0$ in QIII.

c. A rotation of $\frac{7\pi}{4}$ terminates in QIV: $\theta_r = 2\pi - \frac{7\pi}{4} = \frac{\pi}{4}$. The associated point is $\left(\frac{\sqrt{2}}{2}, -\frac{\sqrt{2}}{2}\right)$ since $y < 0$ in QIV. See Figure 1.55.

Figure 1.54

Figure 1.55

NOW TRY EXERCISES 29 THROUGH 36

C. Trigonometric Functions and Points on the Unit Circle

We can now define the six trig functions in terms of a point (x, y) on the unit circle associated with a given rotation θ, with the use of right triangles quickly fading from view. For this reason they are sometimes called the **circular functions.** We define them as follows:

THE CIRCULAR FUNCTIONS

For any rotation θ and point $P(x, y)$ on the unit circle associated with θ,

$$\cos\theta = x \qquad \sin\theta = y \qquad \tan\theta = \frac{y}{x}; x \neq 0$$

$$\sec\theta = \frac{1}{x}; x \neq 0 \qquad \csc\theta = \frac{1}{y}; y \neq 0 \qquad \cot\theta = \frac{x}{y}; y \neq 0$$

Recall that once $\sin\theta$, $\cos\theta$, and $\tan\theta$ are known, the values of $\csc\theta$, $\sec\theta$, and $\cot\theta$ follow automatically since a number and its reciprocal always have the same sign. See Figure 1.56.

Figure 1.56

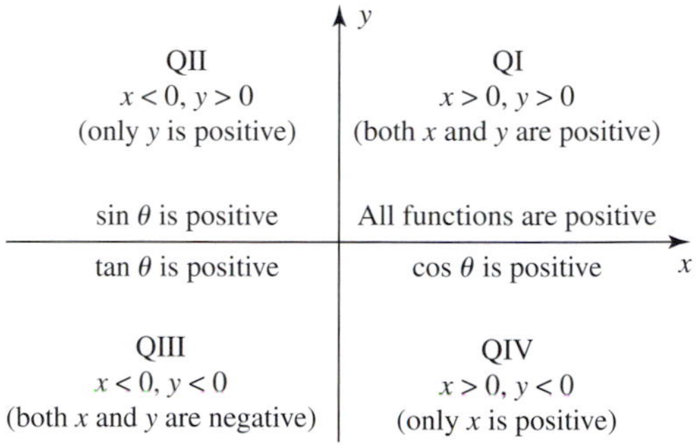

QII $x < 0, y > 0$ (only y is positive) $\sin\theta$ is positive	QI $x > 0, y > 0$ (both x and y are positive) All functions are positive
$\tan\theta$ is positive QIII $x < 0, y < 0$ (both x and y are negative)	$\cos\theta$ is positive QIV $x > 0, y < 0$ (only x is positive)

EXAMPLE 5 Evaluate the six trig functions for $\theta = \frac{5\pi}{4}$.

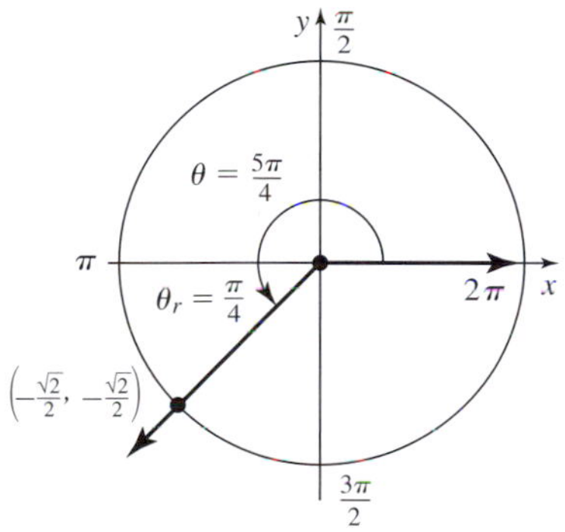

Solution: A rotation of $\frac{5\pi}{4}$ terminates in QIII, so $\theta_r = \frac{5\pi}{4} - \pi = \frac{\pi}{4}$. The associated point is $\left(-\frac{\sqrt{2}}{2}, -\frac{\sqrt{2}}{2}\right)$ since $x < 0$ and $y < 0$ in QIII. This yields

$$\cos\left(\frac{5\pi}{4}\right) = -\frac{\sqrt{2}}{2} \qquad \sin\left(\frac{5\pi}{4}\right) = -\frac{\sqrt{2}}{2} \qquad \tan\left(\frac{5\pi}{4}\right) = 1$$

Noting the reciprocal of $-\frac{\sqrt{2}}{2}$ is $-\sqrt{2}$ after rationalizing, we have

$$\sec\left(\frac{5\pi}{4}\right) = -\sqrt{2} \qquad \csc\left(\frac{5\pi}{4}\right) = -\sqrt{2} \qquad \cot\left(\frac{5\pi}{4}\right) = 1$$

NOW TRY EXERCISES 37 THROUGH 40

Figure 1.57

D. The Trigonometry of Real Numbers

Defining the trig functions in terms of a point on the unit circle is precisely what we needed to work with them as functions of a real number. This is because when $r = 1$ and θ is in radians, *the length of the subtended arc is numerically the same as the measure of the angle:* $s = (1)\theta \rightarrow s = \theta$! This means we can view any function of θ as a like function of arc length s, where $s \in \mathbb{R}$. As a compromise the variable t is commonly used, with t representing *either* the amount of rotation *or* the length of the arc. As such we will assume t is a dimensionless quantity, although there are other reasons for this assumption. In Figure 1.57, a rotation of $\theta = \frac{3\pi}{4}$ is subtended by an arc length of $s = \frac{3\pi}{4}$

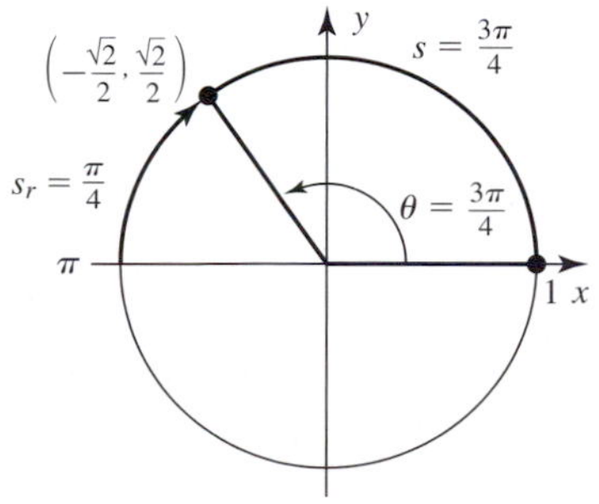

WORTHY OF NOTE

Once again, to evaluate sec *t*, csc *t*, and cot *t*, we simply reciprocate the values found for cos *t*, sin *t*, and tan *t*, respectively. For Example 6(a), these reciprocals resulted in a radical denominator, and the expression was rationalized to give the values shown.

(about 2.356 units). The reference angle for θ is $\frac{\pi}{4}$, which we will now refer to as a **reference arc.** As you work through the remaining examples and the exercises that follow, it will often help to draw a quick sketch similar to that in Figure 1.57 to determine the quadrant of the terminal side, the reference arc, and the signs of each function.

EXAMPLE 6 Evaluate the six trig functions for the given value of *t*.

a. $t = \frac{11\pi}{6}$ **b.** $t = \frac{3\pi}{2}$

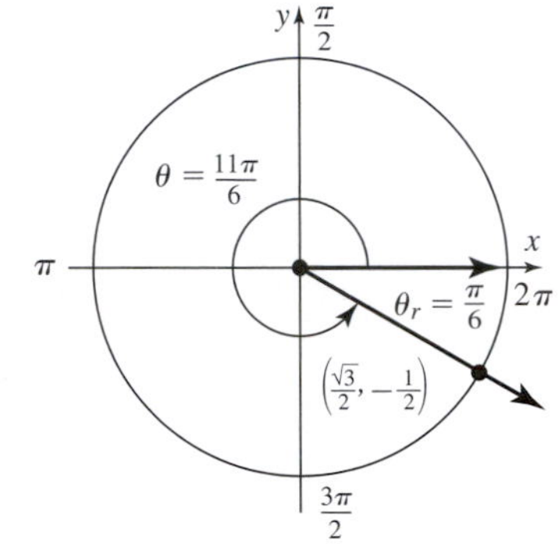

Solution: **a.** For $t = \frac{11\pi}{6}$, the arc terminates in QIV, where $x > 0$ and $y < 0$. The reference arc is $\frac{\pi}{6}$ and from our previous work we know the corresponding point (x, y) is $\left(\frac{\sqrt{3}}{2}, -\frac{1}{2}\right)$. This gives:

$$\cos\left(\frac{11\pi}{6}\right) = \frac{\sqrt{3}}{2} \quad \sin\left(\frac{11\pi}{6}\right) = -\frac{1}{2} \quad \tan\left(\frac{11\pi}{6}\right) = -\frac{\sqrt{3}}{3}$$

$$\sec\left(\frac{11\pi}{6}\right) = \frac{2\sqrt{3}}{3} \quad \csc\left(\frac{11\pi}{6}\right) = -2 \quad \cot\left(\frac{11\pi}{6}\right) = -\sqrt{3}$$

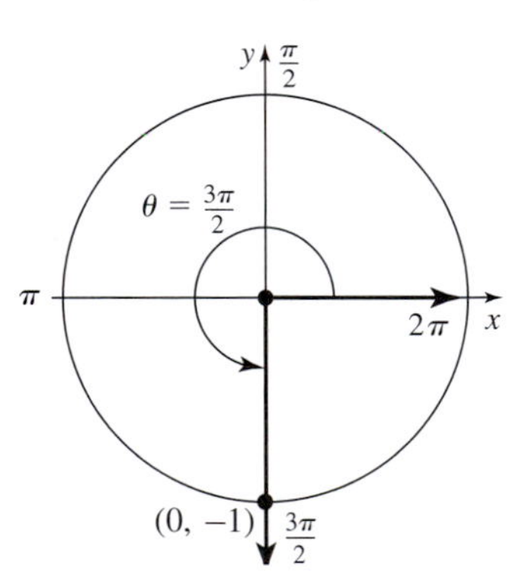

b. $t = \frac{3\pi}{2}$ is a quadrantal angle and the associated point is $(0, -1)$. This yields:

$$\cos\left(\frac{3\pi}{2}\right) = 0 \quad \sin\left(\frac{3\pi}{2}\right) = -1 \quad \tan\left(\frac{3\pi}{2}\right) = \text{undefined}$$

$$\sec\left(\frac{3\pi}{2}\right) = \text{undefined} \quad \csc\left(\frac{3\pi}{2}\right) = -1 \quad \cot\left(\frac{3\pi}{2}\right) = 0$$

NOW TRY EXERCISES 41 THROUGH 44

As Example 6(b) indicates, as functions of a real number the concept of domain comes into play. From their definition it is apparent there are no restrictions on the domain of cosine and sine, but the domains of the other functions must be restricted to exclude division by zero. For functions with *x* in the denominator, we cast out the odd multiples of $\frac{\pi}{2}$, since the *x*-coordinate of the related quadrantal points is zero: $\frac{\pi}{2} \rightarrow (0, 1), \frac{3\pi}{2} \rightarrow (0, -1)$, and so on. The excluded values can be stated as $t \neq \frac{\pi}{2} + \pi k$ for all integers *k*. For functions with *y* in the denominator, we cast out all multiples of π ($t \neq \pi k$ for all integers *k*) since the *y*-coordinate of these points is zero: $0 \rightarrow (1, 0)$, $\pi \rightarrow (-1, 0)$, $2\pi \rightarrow (1, 0)$, and so on.

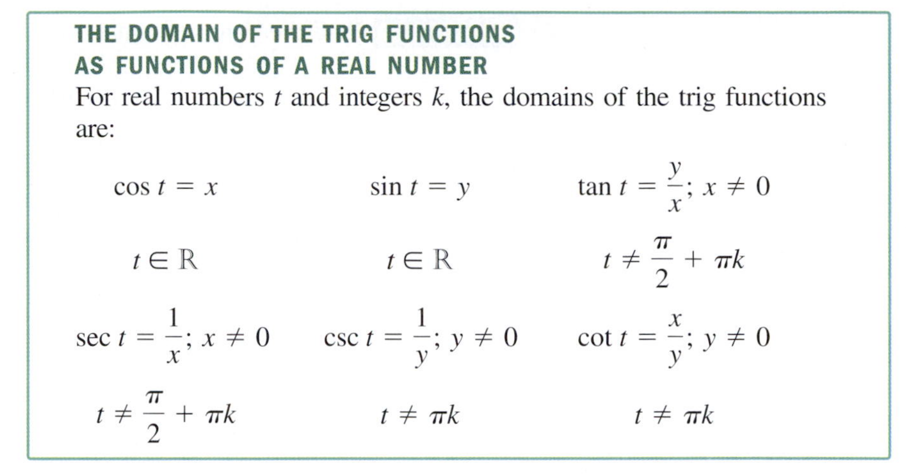

THE DOMAIN OF THE TRIG FUNCTIONS AS FUNCTIONS OF A REAL NUMBER

For real numbers t and integers k, the domains of the trig functions are:

$\cos t = x$	$\sin t = y$	$\tan t = \frac{y}{x};\ x \neq 0$
$t \in \mathbb{R}$	$t \in \mathbb{R}$	$t \neq \frac{\pi}{2} + \pi k$
$\sec t = \frac{1}{x};\ x \neq 0$	$\csc t = \frac{1}{y};\ y \neq 0$	$\cot t = \frac{x}{y};\ y \neq 0$
$t \neq \frac{\pi}{2} + \pi k$	$t \neq \pi k$	$t \neq \pi k$

For any point (x, y) on the unit circle, the definition of the trig functions can still be applied even if t is unknown.

EXAMPLE 7 Given $\left(\frac{-7}{25}, \frac{24}{25}\right)$ is a point on the unit circle corresponding to a real number t, find the value of all six trig functions of t.

Solution: Using the definitions from the previous box we have $\cos t = \frac{-7}{25}$, $\sin t = \frac{24}{25}$, and $\tan t = \frac{\sin t}{\cos t} = \frac{24}{-7}$. The values of the reciprocal functions are then $\sec t = \frac{25}{-7}$, $\csc t = \frac{25}{24}$, and $\cot t = \frac{-7}{24}$.

NOW TRY EXERCISES 45 THROUGH 70

E. Finding a Real Number *t* Whose Function Value Is Known

In Example 7, we were able to apply the definition of the trig functions even though t was unknown. In many cases, however, we need to *find* the related value of t. For instance, what is the value of t given $\cos t = -\frac{\sqrt{3}}{2}$ with t in QII?

Exercises of this type fall into two broad categories: (1) you recognize the given number as one of the standard values: $\pm\left\{0, \frac{1}{2}, \frac{\sqrt{2}}{2}, \frac{\sqrt{3}}{2}, \frac{\sqrt{3}}{3}, \sqrt{3}, 1\right\}$; or (2) you don't. If you recognize a standard value, you can often name the real number t after a careful consideration of the related quadrant and required sign. The diagram in Figure 1.58 reviews these standard values for $0 \leq t \leq \frac{\pi}{2}$ but remember—all other standard values can be found using reference arcs and the symmetry of the circle.

Figure 1.58

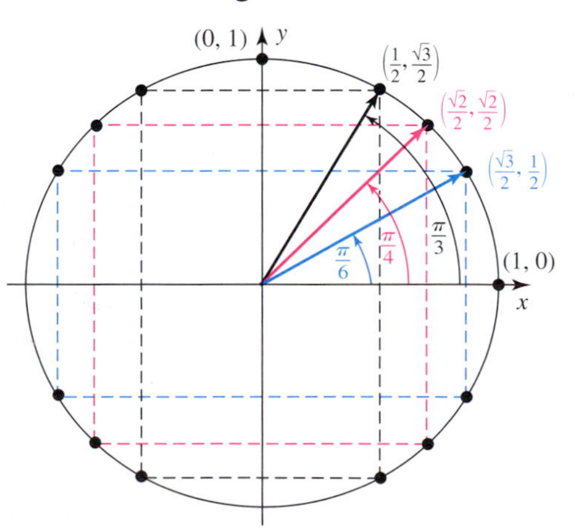

EXAMPLE 8 Find the value of t that corresponds to the following functions:

a. $\cos t = -\frac{\sqrt{2}}{2}$; t in QII **b.** $\tan t = \sqrt{3}$; t in QIII

Solution:

a. The cosine function is negative in QII and QIII where $x < 0$. We recognize $-\frac{\sqrt{2}}{2}$ as a standard value for sine and cosine, related to certain multiples of $t = \frac{\pi}{4}$. In QII, we have $t = \frac{3\pi}{4}$.

b. The tangent function is positive in QI and QIII where x and y have like signs. We recognize $\sqrt{3}$ as a standard value for tangent and cotangent, related to certain multiples of $t = \frac{\pi}{3}$. For tangent in QIII, we have $t = \frac{8\pi}{6} = \frac{4\pi}{3}$.

NOW TRY EXERCISES 71 THROUGH 86

If you don't recognize the given value, most calculators are programmed to give a reference value for t that corresponds to the given function value, using the $\sin^{-1}$, $\cos^{-1}$, and $\tan^{-1}$ keys. However, the quadrant, signs, and reference arcs must still be considered to correctly name t.

EXAMPLE 9 Find the value of t that corresponds to the following functions:

a. $\sin t = \frac{\sqrt{13}}{7}$; t in QI **b.** $\tan t = \frac{4}{3}$; t in QIII

Solution:

a. The sine function is positive in QI and QII where $y > 0$. With your calculator in radian mode, the expression $\sin^{-1}\left(\frac{\sqrt{13}}{7}\right)$ [2nd SIN $(\sqrt{13})/7$ ENTER] gives $t \approx 0.5411$ to four decimal places, which is the desired QI value since $0 < 0.5411 < \frac{\pi}{2} \approx 1.57$.

b. The tangent function is positive in QI and QIII. Since $\tan t = \frac{4}{3}$ is likewise not a standard value, we use 2nd TAN 4/3 ENTER which gives $t \approx 0.9273$. We note this is the value from QI, since $0 < 0.9273 < \frac{\pi}{2} \approx 1.57$. The QIII value is then $t = \pi + 0.9273 \approx 4.0689$.

NOW TRY EXERCISES 87 THROUGH 102

Using radian measure and the unit circle is much more than a simple convenience to trigonometry. Whether the unit is 1 cm, 1 m, 1 km, or even 1 light-year, using 1 unit designations serves to simplify a great many practical applications, including those involving the arc length formula, $s = r\theta$. See Exercises 105 through 112.

TECHNOLOGY HIGHLIGHT

Graphing the Unit Circle

The keystrokes shown apply to a TI-84 Plus model. Please consult your manual or our Internet site for other models.

When using a graphing calculator to study the unit circle, it's important to keep two things in mind. First, most graphing calculators are only capable of graphing *functions*, which means we must modify the equation of the circle (and relations like ellipses, hyperbolas, horizontal parabolas, and so on) before it can be graphed. Second, most standard viewing windows have the x- and y-values preset at $[-10, 10]$ even though the calculator screen is 94 pixels wide and 64 pixels high. This tends to compress the y-values and give a skewed image of the graph. Consider the equation $x^2 + y^2 = 1$. From our work in this section, we know this is the equation of a circle centered at (0, 0) with radius $r = 1$. For the calculator to graph this relation, we must define it in two pieces, each of which is a function, by solving for y:

$$x^2 + y^2 = 1 \quad \text{original equation}$$
$$y^2 = 1 - x^2 \quad \text{isolate } y^2$$
$$y = \pm\sqrt{1 - x^2} \quad \text{solve for } y$$

Note that we can separate this result into two parts, each of which is a function. The graph of $Y_1 = \sqrt{1 - x^2}$ gives the "upper half" of the circle, while $Y_2 = -\sqrt{1 - x^2}$ gives the "lower half." We can enter these on the Y= screen as shown, using the expression $-Y_1$ instead of reentering the entire expression. The function variables Y_1, Y_2, Y_3, and so on, can be accessed using VARS ► **(Y-VARS)** ENTER **(1:Function)**. Graphing Y_1 and Y_2 on the standard screen, the result appears very small and more elliptical than circular (Figure 1.59). One way to fix this (there are many others), is to use the ZOOM **4:ZDecimal** option, which places the tic marks equally spaced on both axes, instead of trying to force both to display points from −10 to 10. Using this option gives the screen shown in Figure 1.60. An even better graph can be obtained using the ZOOM **2:Zoom In** option (or by manually resetting the window size). Using the TRACE feature enables us to view points on the unit circle, but recall that this image is actually the union of two graphs and we may need to jump between the upper and lower halves using the up ▲ or down ▼ arrows.

Figure 1.59

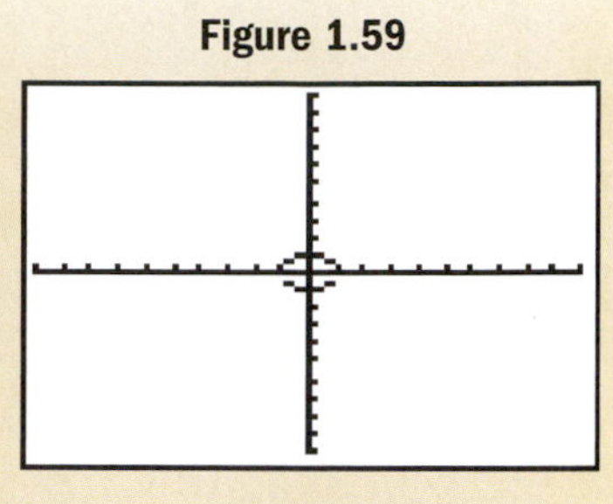

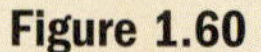

Figure 1.60

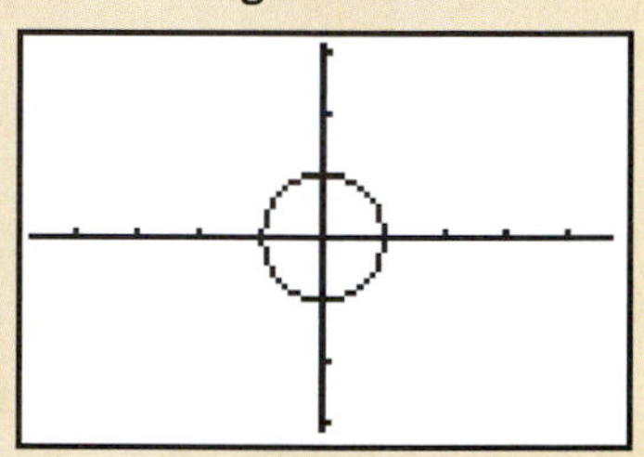

Exercise 1: Use the TRACE feature to verify the point (0.6, 0.8) is on the unit circle, as well as the other three related points given by symmetry (as shown in Example 2).

Exercise 2: Use the 2nd TRACE **(CALC)** feature to evaluate the function at $\frac{\sqrt{2}}{2}$. What do you notice about the output? For cos t or sin t, what value of t can we associate with this point?

Exercise 3: What other standard values can you identify as you TRACE around the circle?

1.4 EXERCISES

CONCEPTS AND VOCABULARY

Fill in each blank with the appropriate word or phrase. Carefully reread the section if needed.

1. A central circle is symmetric to the ______ axis, the ______ axis and to the ______.

2. Since $\left(\frac{5}{13}, -\frac{12}{13}\right)$ is on the unit circle, the point ________ in QII is also on the circle.

3. On a unit circle, $\cos t =$ ______, $\sin t =$ ______, and $\tan t =$ ______; while $\frac{1}{x} =$ ______, $\frac{1}{y} =$ ______, and $\frac{x}{y} =$ ______.

4. On a unit circle with θ in radians, the length of a(n) ________ ______ is numerically the same as the measure of the ______, since for $s = r\theta$, $s = \theta$ when $r = 1$.

5. Discuss/explain how knowing only one point on the unit circle, actually gives the location of four points. Why is this helpful to a study of the circular functions?

6. A student is asked to find t using a calculator, given $\sin t \approx 0.5592$ with t in QII. The answer submitted is $t = \sin^{-1} 0.5592 \approx 34°$. Discuss/explain why this answer is not correct. What is the correct response?

DEVELOPING YOUR SKILLS

Given the point is on a unit circle, complete the ordered pair (x, y) for the quadrant indicated. For Exercises 7 to 14, answer in radical form as needed. For Exercises 15 to 18, round results to four decimal places.

7. $(x, -0.8)$; QIII **8.** $(-0.6, y)$; QII **9.** $\left(\frac{5}{13}, y\right)$; QIV **10.** $\left(x, -\frac{8}{17}\right)$; QIV

11. $\left(\frac{\sqrt{11}}{6}, y\right)$; QI **12.** $\left(x, -\frac{\sqrt{13}}{7}\right)$; QIII **13.** $\left(-\frac{\sqrt{11}}{4}, y\right)$; QII **14.** $\left(x, \frac{\sqrt{6}}{5}\right)$; QI

15. $(x, -0.2137)$; QIII **16.** $(0.9909, y)$; QIV **17.** $(x, 0.1198)$; QII **18.** $(0.5449, y)$; QI

Verify the point given is on a unit circle, then use symmetry to find three more points on the circle. Results for Exercises 19 to 22 are exact, results for Exercises 23 to 26 are approximate.

19. $\left(-\frac{\sqrt{3}}{2}, \frac{1}{2}\right)$ **20.** $\left(\frac{\sqrt{7}}{4}, -\frac{3}{4}\right)$ **21.** $\left(\frac{\sqrt{11}}{6}, -\frac{5}{6}\right)$

22. $\left(-\frac{\sqrt{6}}{3}, -\frac{\sqrt{3}}{3}\right)$ **23.** $(0.3325, 0.9431)$ **24.** $(0.7707, -0.6372)$

25. $(0.9937, -0.1121)$ **26.** $(-0.2029, 0.9792)$

27. Use a $\frac{\pi}{6} : \frac{\pi}{3} : \frac{\pi}{2}$ triangle with a hypotenuse of length 1 to verify that $\left(\frac{1}{2}, \frac{\sqrt{3}}{2}\right)$ is a point on the unit circle.

28. Use the results from Exercise 27 to find three additional points on the circle and name the quadrant of each point.

Find the reference angle associated with each rotation, then find the associated point (x, y).

29. $\theta = \frac{5\pi}{4}$ **30.** $\theta = \frac{5\pi}{3}$ **31.** $\theta = -\frac{5\pi}{6}$ **32.** $\theta = -\frac{7\pi}{4}$

33. $\theta = \frac{11\pi}{4}$ **34.** $\theta = \frac{11\pi}{3}$ **35.** $\theta = \frac{25\pi}{6}$ **36.** $\theta = \frac{39\pi}{4}$

Use the symmetry of the circle and reference angles as needed to state the exact value of the trig functions for the given angle, without the use of a calculator. A diagram may help.

37. a. $\sin\left(\frac{\pi}{4}\right)$ **b.** $\sin\left(\frac{3\pi}{4}\right)$ **c.** $\sin\left(\frac{5\pi}{4}\right)$ **d.** $\sin\left(\frac{7\pi}{4}\right)$

e. $\sin\left(\frac{9\pi}{4}\right)$ **f.** $\sin\left(-\frac{\pi}{4}\right)$ **g.** $\sin\left(-\frac{5\pi}{4}\right)$ **h.** $\sin\left(-\frac{11\pi}{4}\right)$

38. a. $\tan\left(\frac{\pi}{3}\right)$ **b.** $\tan\left(\frac{2\pi}{3}\right)$ **c.** $\tan\left(\frac{4\pi}{3}\right)$ **d.** $\tan\left(\frac{5\pi}{3}\right)$

e. $\tan\left(\frac{7\pi}{3}\right)$ **f.** $\tan\left(-\frac{\pi}{3}\right)$ **g.** $\tan\left(-\frac{4\pi}{3}\right)$ **h.** $\tan\left(-\frac{10\pi}{3}\right)$

39. **a.** $\cos \pi$ **b.** $\cos 0$ **c.** $\cos\left(\frac{\pi}{2}\right)$ **d.** $\cos\left(\frac{3\pi}{2}\right)$

40. **a.** $\sin \pi$ **b.** $\sin 0$ **c.** $\sin\left(\frac{\pi}{2}\right)$ **d.** $\sin\left(\frac{3\pi}{2}\right)$

Use the symmetry of the circle and reference arcs as needed to state the exact value of the trig functions for the given real number, without the use of a calculator. A diagram may help.

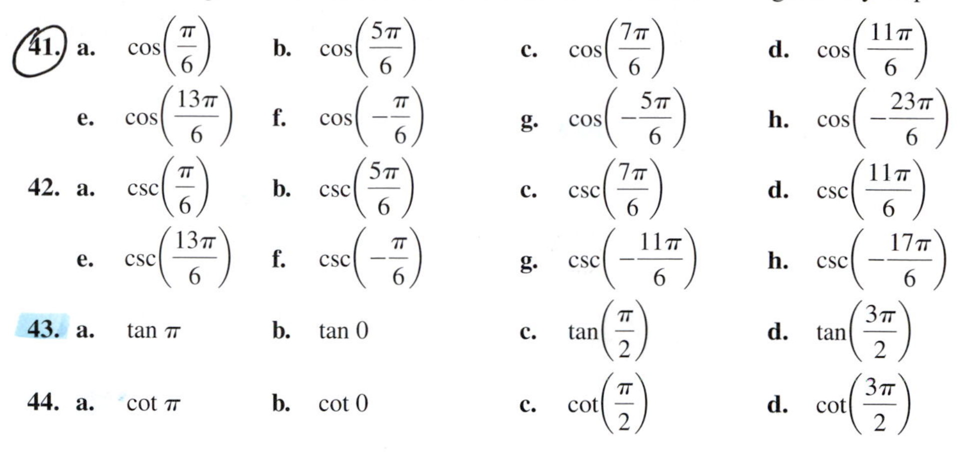

41. **a.** $\cos\left(\frac{\pi}{6}\right)$ **b.** $\cos\left(\frac{5\pi}{6}\right)$ **c.** $\cos\left(\frac{7\pi}{6}\right)$ **d.** $\cos\left(\frac{11\pi}{6}\right)$

e. $\cos\left(\frac{13\pi}{6}\right)$ **f.** $\cos\left(-\frac{\pi}{6}\right)$ **g.** $\cos\left(-\frac{5\pi}{6}\right)$ **h.** $\cos\left(-\frac{23\pi}{6}\right)$

42. **a.** $\csc\left(\frac{\pi}{6}\right)$ **b.** $\csc\left(\frac{5\pi}{6}\right)$ **c.** $\csc\left(\frac{7\pi}{6}\right)$ **d.** $\csc\left(\frac{11\pi}{6}\right)$

e. $\csc\left(\frac{13\pi}{6}\right)$ **f.** $\csc\left(-\frac{\pi}{6}\right)$ **g.** $\csc\left(-\frac{11\pi}{6}\right)$ **h.** $\csc\left(-\frac{17\pi}{6}\right)$

43. **a.** $\tan \pi$ **b.** $\tan 0$ **c.** $\tan\left(\frac{\pi}{2}\right)$ **d.** $\tan\left(\frac{3\pi}{2}\right)$

44. **a.** $\cot \pi$ **b.** $\cot 0$ **c.** $\cot\left(\frac{\pi}{2}\right)$ **d.** $\cot\left(\frac{3\pi}{2}\right)$

Given (x, y) is a point on a unit circle corresponding to t, find the value of all six circular functions of t.

45.

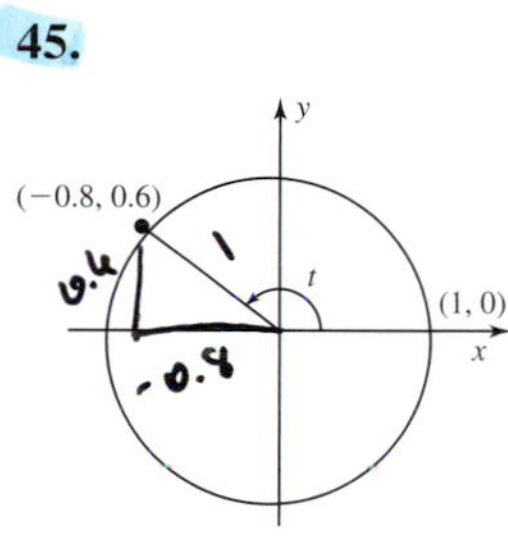

46.

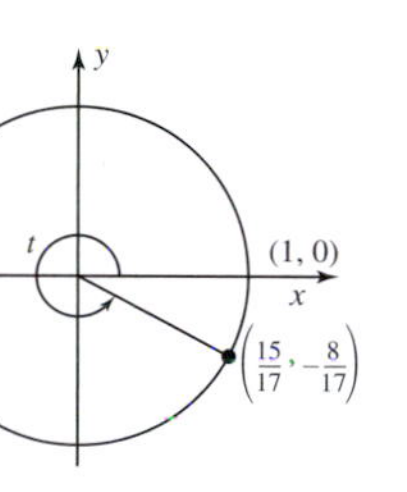

47.

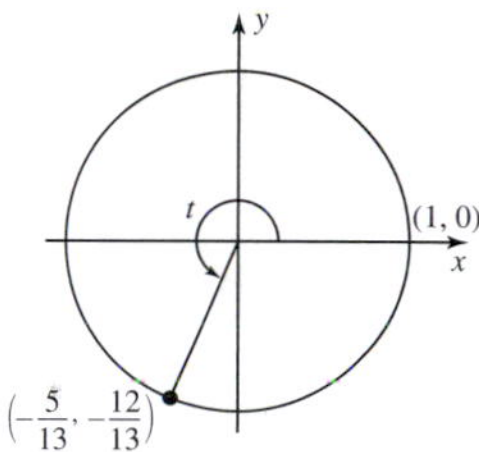

48.

y, t, (1, 0), x, $\left(-\frac{24}{25}, -\frac{7}{25}\right)$

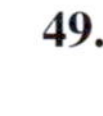

49.

y, $\left(\frac{5}{6}, \frac{\sqrt{11}}{6}\right)$, t, (1, 0), x

50.

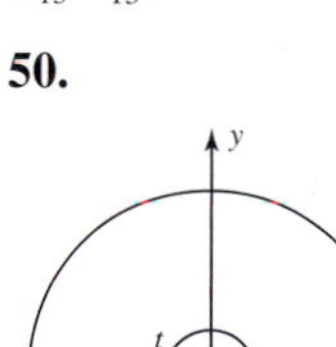

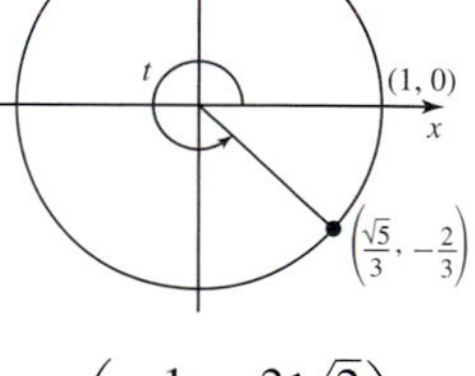

51. $\left(-\frac{2}{5}, \frac{\sqrt{21}}{5}\right)$ **52.** $\left(\frac{\sqrt{7}}{4}, -\frac{3}{4}\right)$ **53.** $\left(-\frac{1}{3}, -\frac{2\sqrt{2}}{3}\right)$

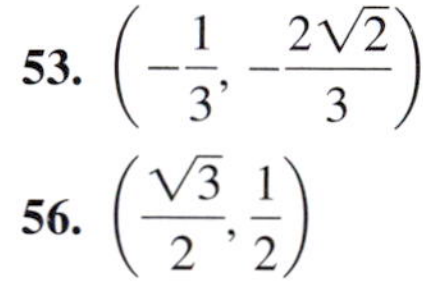

54. $\left(-\frac{2\sqrt{6}}{5}, -\frac{1}{5}\right)$ **55.** $\left(\frac{1}{2}, \frac{\sqrt{3}}{2}\right)$ **56.** $\left(\frac{\sqrt{3}}{2}, \frac{1}{2}\right)$

57. $\left(-\frac{\sqrt{2}}{2}, \frac{\sqrt{2}}{2}\right)$ **58.** $\left(\frac{\sqrt{2}}{3}, -\frac{\sqrt{7}}{3}\right)$

On a unit circle, the real number t can represent either the amount of rotation or the *length of the arc* when associating t with a point (x, y) on the circle. In the circle diagram on p. 62, the real number t in radians is marked off along the circumference. For Exercises 59

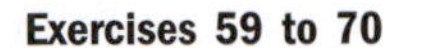
Exercises 59 to 70

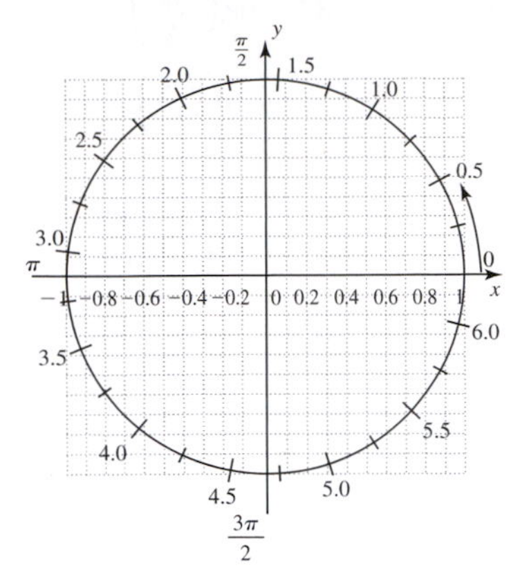

through 70, name the quadrant in which t terminates and use the figure to estimate function values to one decimal place (use a straightedge). Check results using a calculator.

59. sin 0.75 **60.** cos 2.75 **61.** cos 5.5 **62.** sin 4.0

63. tan 0.8 **64.** sec 3.75 **65.** csc 2.0 **66.** cot 0.5

67. $\cos\left(\frac{5\pi}{8}\right)$ **68.** $\sin\left(\frac{5\pi}{8}\right)$ **69.** $\tan\left(\frac{8\pi}{5}\right)$ **70.** $\sec\left(\frac{8\pi}{5}\right)$

Without using a calculator, find the value of t in $[0, 2\pi)$ that corresponds to the following functions.

71. $\sin t = \frac{\sqrt{3}}{2}$; t in QII **72.** $\cos t = \frac{1}{2}$; t in QIV **73.** $\cos t = -\frac{\sqrt{3}}{2}$; t in QIII

74. $\sin t = -\frac{1}{2}$; t in QIV **75.** $\tan t = -\sqrt{3}$; t in QII **76.** $\sec t = -2$; t in QIII

77. $\sin t = 1$; t is quadrantal **78.** $\cos t = -1$; t is quadrantal

Without using a calculator, find the two values of t (where possible) in $[0, 2\pi)$ that make each equation true.

79. $\sec t = -\sqrt{2}$ **80.** $\csc t = -\frac{2}{\sqrt{3}}$ **81.** $\tan t$ undefined **82.** $\csc t$ undefined

83. $\cos t = -\frac{\sqrt{2}}{2}$ **84.** $\sin t = \frac{\sqrt{2}}{2}$ **85.** $\sin t = 0$ **86.** $\cos t = -1$

Use a calculator to find the value of t in $[0, 2\pi)$ that satisfies the equation and condition given. Round to four decimal places.

87. $\cos t = \frac{\sqrt{5}}{4}$; t in QIV **88.** $\sin t = \frac{\sqrt{6}}{5}$; t in QII **89.** $\tan t = -\frac{7}{8}$; t in QIV

90. $\sec t = -\frac{7}{4}$; t in QIII **91.** $\sin t = -0.9872$; t in QIII **92.** $\cos t = -0.5467$; t in QII

93. $\cot t = 6.4521$; t in QI **94.** $\csc t = 2.2551$; t in QII

Find an additional value of t (to four decimal places) in $[0, 2\pi)$ that makes the equation true.

95. $\sin 0.8 \approx 0.7174$ **96.** $\cos 2.12 \approx -0.5220$ **97.** $\cos 4.5 \approx -0.2108$

98. $\sin 5.23 \approx -0.8690$ **99.** $\tan 0.4 \approx 0.4228$ **100.** $\sec 5.7 \approx 1.1980$

101. Given $\left(\frac{3}{4}, -\frac{4}{5}\right)$ is a point on the unit circle that corresponds to t. Find the coordinates of the point corresponding to (a) $-t$ and (b) $t + \pi$.

102. Given $\left(-\frac{7}{25}, \frac{24}{25}\right)$ is a point on the unit circle that corresponds to t. Find the coordinates of the point corresponding to (a) $-t + \pi$ and (b) $t - \pi$.

WORKING WITH FORMULAS

103. From Pythagorean triples to points on the unit circle: $(x, y, r) \rightarrow \left(\frac{x}{r}, \frac{y}{r}, 1\right)$

While not strictly a "formula," there is a simple algorithm for rewriting any Pythagorean triple as a triple with hypotenuse 1. This enables us to identify certain points on a unit circle, and to evaluate the six trig functions of the acute angles. Here are two common and two not-so-common Pythagorean triples. Rewrite each as a triple with hypotenuse 1, verify $\left(\frac{x}{r}, \frac{y}{r}\right)$ is a point on the unit circle, and evaluate the six trig functions using this point.

a. (5, 12, 13) **b.** (7, 24, 25) **c.** (12, 35, 37) **d.** (9, 40, 41)

104. The sine and cosine of $(2k + 1)\frac{\pi}{4}$

In the solution to Example 8(a), we mentioned $\pm\frac{\sqrt{2}}{2}$ were standard values for sine and cosine, "related to certain multiples of $\frac{\pi}{4}$." Actually, we meant "odd multiples of $\frac{\pi}{4}$." The odd multiples of $\frac{\pi}{4}$ are given by the "formula" shown, where k is any integer. (a) What multiples of $\frac{\pi}{4}$ are generated by $k = -3, -2, -1, 0, 1, 2, 3$? (b) Find similar formulas for Example 8(b), where $\sqrt{3}$ is a standard value for tangent and cotangent, "related to certain multiples of $\frac{\pi}{6}$."

APPLICATIONS

105. Laying new sod: When new sod is laid, a heavy roller is used to press the sod down to ensure good contact with the ground beneath. The radius of the roller is 1 ft. (a) Through what angle (in radians) has the roller turned after being pulled across 5 ft of yard? (b) What angle must the roller turn through to press a length of 30 ft?

Exercise 106

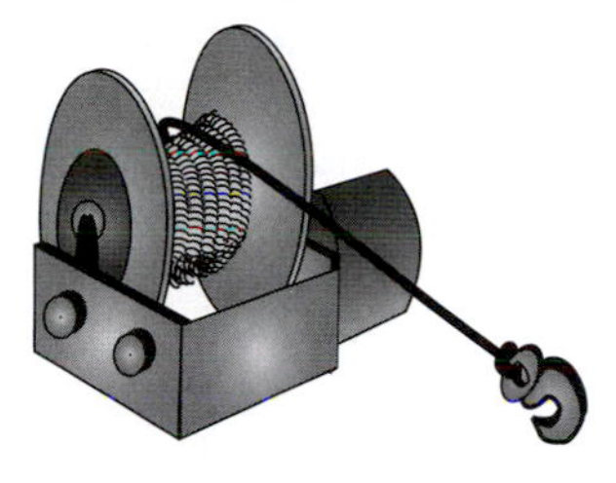

106. Cable winch: A large winch with a radius of 1 ft winds in 3 ft of cable. (a) Through what angle (in radians) has it turned? (b) What angle must it turn through in order to winch in 12.5 ft of cable?

107. Wiring an apartment: In the wiring of an apartment complex, electrical wire is being pulled from a spool with radius 1 decimeter (1 dm = 10 cm). (a) What length (in decimeters) is removed as the spool turns through 5 rad? (b) How many decimeters are removed in one complete turn ($t = 2\pi$) of the spool?

108. Barrel races: In the barrel races popular at some family reunions, contestants stand on a hard rubber barrel with a radius of 1 cubit (1 cubit = 18 in.), and try to "walk the barrel" from the start line to the finish line without falling. (a) What distance (in cubits) is traveled as the barrel is walked through an angle of 4.5 rad? (b) If the race is 25 cubits long, through what angle will the winning barrel walker walk the barrel?

Interplanetary measurement: In around 1905, astronomers decided to begin using what are called astronomical units or AU to study the distances between the celestial bodies of our solar system. It represents the average distance between the Earth and the Sun, which is about 93 million miles. Pluto is roughly 39.24 AU from the Sun.

109. If the Earth travels through an angle of 2.5 rad about the Sun, (a) what distance in astronomical units (AU) has it traveled? (b) How many AU does it take for one complete orbit around the Sun?

110. Since Jupiter is actually the middle (fifth of nine) planet from the Sun, suppose astronomers had decided to use *its* average distance from the Sun as 1 AU. In this case 1 AU would be 480 million miles. If Jupiter travels through an angle of 4 rad about the Sun, (a) what distance in the "new" astronomical units (AU) has it traveled? (b) How many of the new AU does it take to complete one-half an orbit about the Sun? (c) What distance in the new AU is the dwarf planet Pluto from the Sun?

111. Compact disk circumference: A standard compact disk has a radius of 6 cm. Call this length "1 unit." Draw a long, straight line on a blank sheet of paper, then carefully roll the compact disk along this line without slippage, through one full revolution (2π rad) and mark the spot. Take an accurate measurement of the resulting line segment. Is the result equal to 2π "units" ($2\pi \times 6$ cm)?

Exercise 112

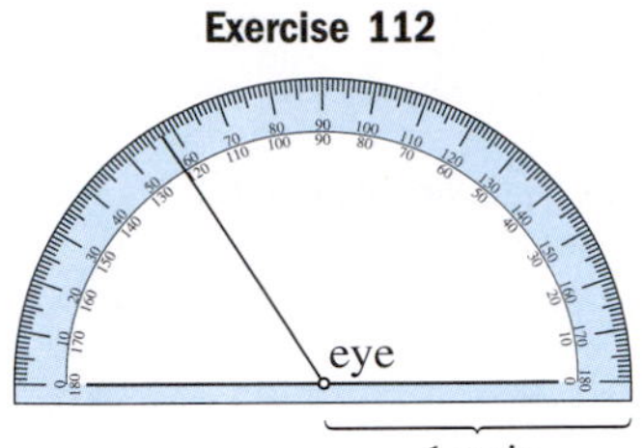

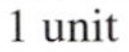

112. Verifying $s = r\theta$: On a protractor, carefully measure the distance from the middle of the protractor's eye to the edge of the protractor along the 0° mark, to the nearest half-millimeter. Call this length "1 unit." Then use a ruler to draw a straight line on a blank sheet of paper, and with the protractor on edge, start the zero degree mark at one end of the line, carefully roll the protractor until it reaches 1 radian (57.3°), and mark this spot. Now measure the length of the line segment created. Is it very close to 1 "unit" long?

WRITING, RESEARCH, AND DECISION MAKING

113. In this section, we discussed the *domain* of the circular functions, but said very little about their *range*. Review the concepts presented here and determine the range of $y = \cos t$ and $y = \sin t$. In other words, what are the largest and smallest output values we can expect?

114. Since $\tan t = \dfrac{\sin t}{\cos t}$, what can you say about the range of the tangent function?

EXTENDING THE CONCEPT

Use the radian grid with Exercises 59–70 to answer Exercises 115 and 116.

115. Given $\cos(2t) = -0.6$ with the terminal side of the arc in QII, (a) what is the value of $2t$? (b) What quadrant is t in? (c) What is the value of $\cos t$? (d) Does $\cos(2t) = 2\cos t$?

116. Given $\sin(2t) = -0.8$ with the terminal side of the arc in QIII, (a) what is the value of $2t$? (b) What quadrant is t in? (c) What is the value of $\sin t$? (d) Does $\sin(2t) = 2\sin t$?

117. Find the measure of all sides and all angles (in radians) for the triangles given. Round the angles to four decimal places and sides to the nearest tenth. (a) How are the two triangles related? (b) Which can be used to identify points on the unit circle? (c) What is the value of t in $\sin t = \frac{28}{53}$?

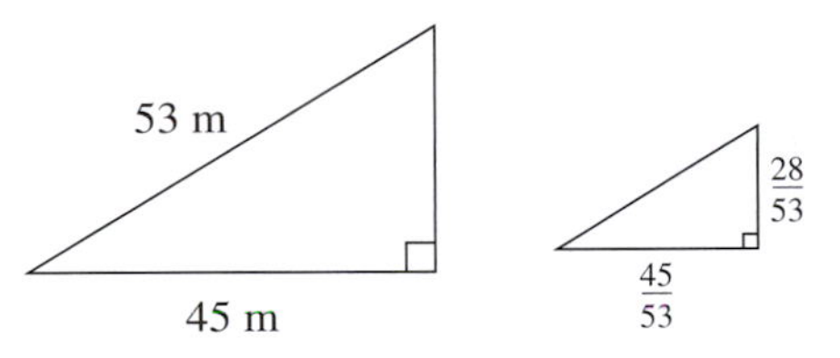

MAINTAINING YOUR SKILLS

118. (1.1) The armature for the rear windshield wiper has a length of 24 in., with a rubber wiper blade that is 20 in. long. What area of my rear windshield is cleaned as the armature swings back-and-forth through an angle of 110°?

119. (1.1) The boxes used to ship some washing machines are perfect cubes with edges measuring 38 in. Use a special triangle to find the length of the diagonal d of one side, and the length of the interior diagonal D (through the middle of the box).

120. (1.3) Given $\sin\theta = \frac{21}{29}$ and $\cos\theta < 0$, find the value of the other five trig functions of θ.

121. (1.2) From the far end of a 50 ft pool, the angle of elevation to the highest diving board is 19.8°. To the nearest foot, how high is the board?

122. (1.3) For $\theta = \dfrac{11\pi}{6}$, use a reference angle θ_r to evaluate the six trig functions of θ.

123. (1.2) Triangle ABC has sides of $a = 12$ m, $b = 35$ m, and $c = 37$ m. Verify that this is a right triangle and find the measure of angles A, B, and C, rounded to the nearest tenth.

SUMMARY AND CONCEPT REVIEW

SECTION 1.1 Angle Measure, Special Triangles, and Special Angles

KEY CONCEPTS

- An angle is defined as the joining of two rays at a common endpoint called the vertex.
- Angle measure can be viewed as the amount of inclination between two intersecting lines, or the amount of rotation from a fixed (initial) side to a terminal side.
- Angles are often named using a capital letter at the vertex or using Greek letters in the interior. The most common are α (alpha), β (beta), γ (gamma), and θ (theta).
- An angle in standard position has its vertex at the origin and its initial side coterminal with the positive x-axis.
- Two angles in standard position are coterminal if they have the same terminal side.
- A counterclockwise rotation gives a positive angle, a clockwise rotation gives a negative angle.
- One (1°) degree is defined to be $\frac{1}{360}$ of a full revolution. One (1) radian is the measure of a central angle subtended by an arc equal in length to the radius.
- Degrees can be divided into a smaller unit called minutes: $1^\circ = 60'$; minutes can be divided into a smaller unit called seconds: $1' = 60''$. This implies $1^\circ = 3600''$.
- Straight angles measure 180°; right angles measure 90°.
- Two angles are complementary if they sum to 90° and supplementary if they sum to 180°.
- Properties of triangles: (I) the sum of the angles is 180°; (II) the combined length of any two sides must exceed that of the third side and; (III) larger angles are opposite larger sides.
- Given two triangles, if all three corresponding angles are equal, the triangles are said to be similar. If two triangles are similar, then corresponding sides are in proportion.
- If $\triangle ABC$ is a right triangle with hypotenuse c, then $a^2 + b^2 = c^2$ (Pythagorean theorem). For $\triangle ABC$ with longest side c, if $a^2 + b^2 = c^2$, then $\triangle ABC$ is a right triangle (converse of Pythagorean theorem).
- In a 45-45-90 triangle, the sides are in the proportion $1x : 1x : \sqrt{2}x$.
- In a 30-60-90 triangle, the sides are in the proportion $1x : \sqrt{3}x : 2x$.
- Since $C = 2\pi r$, there are 2π rad in a complete revolution (the radius can be laid out along the circumference 2π times).
- The formula for arc length in degrees: $s = \left(\frac{\theta}{360^\circ}\right)2\pi r$; in radians: $s = r\theta$.
- The formula for the area of a circular sector in degrees: $\mathcal{A} = \left(\frac{\theta}{360^\circ}\right)\pi r^2$; in radians: $\mathcal{A} = \frac{1}{2}r^2\theta$.
- To convert degree measure to radians, multiply by $\frac{\pi}{180}$; for radians to degrees, multiply by $\frac{180}{\pi}$.
- Standard conversions should be committed to memory: $30^\circ = \frac{\pi}{6}$, $45^\circ = \frac{\pi}{4}$, $60^\circ = \frac{\pi}{3}$, $90^\circ = \frac{\pi}{2}$.
- A location north or south of the equator is given in degrees latitude; a location east or west of the Greenwich Meridian is given in degrees longitude.
- Angular velocity is a rate of rotation per unit time: $\omega = \frac{\theta}{t}$. Linear velocity is a change in position per unit time: $V = \frac{\theta r}{t}$ or $V = r\omega$.

EXERCISES

1. Convert $147°36'48''$ to decimal degrees.
2. Convert $32.87°$ to degrees, minutes, and seconds.
3. All of the right triangles given are similar. Find the dimensions of the largest triangle.

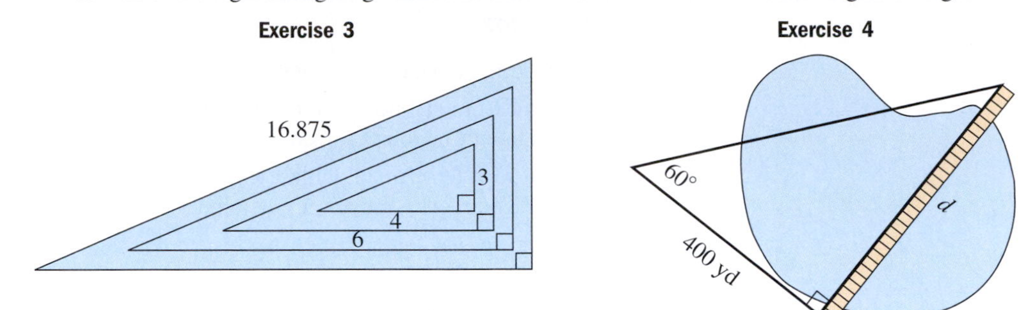

4. Use special angles/special triangles to find the length of the bridge needed to cross the lake shown in the figure.
5. Convert to degrees: $\frac{2\pi}{3}$.
6. Convert to radians: $210°$.
7. Find the arc length if $r = 5$ and $\theta = 57°$.
8. Evaluate without using a calculator: $\sin\left(\frac{7\pi}{6}\right)$.

Find the angle, radius, arc length, and/or area as needed, until all values are known.

9. **10.** **11.**

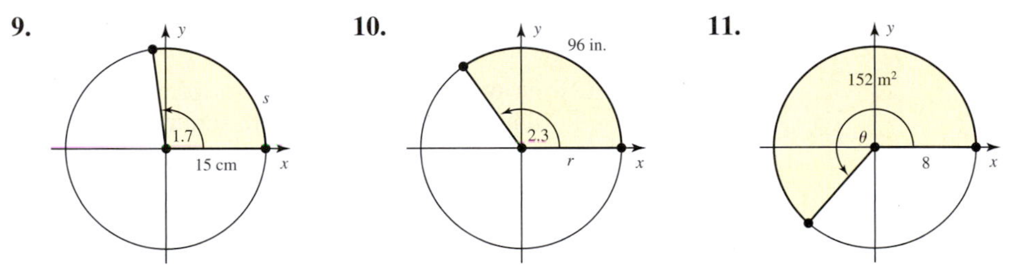

12. With great effort, 5-year-old Mackenzie has just rolled her bowling ball down the lane, and it is traveling painfully slow. So slow, in fact, that you can count the number of revolutions the ball makes using the finger holes as a reference. (a) If the ball is rolling at 1.5 revolutions per second, what is the angular velocity? (b) If the ball's radius is 5 in., what is its linear velocity in feet per second? (c) If the distance to the first pin is 60 feet and the ball is true, how long until it hits?

SECTION 1.2 The Trigonometry of Right Triangles

KEY CONCEPTS

- The sides of a triangle can be named according to their location with respect to a given angle.

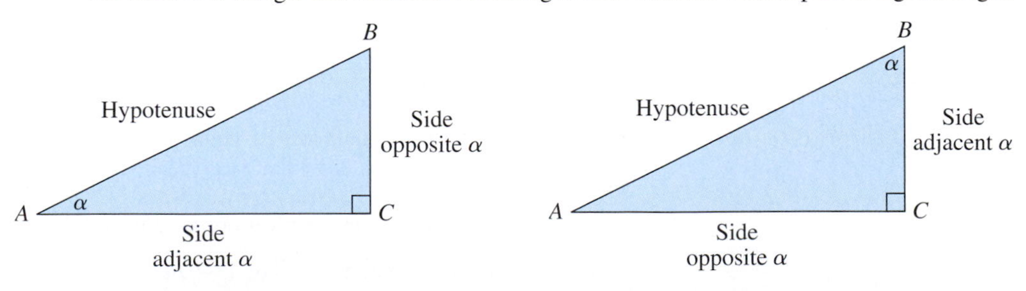

- The ratios of two sides with respect to a given angle are named as follows:

$$\sin\alpha = \frac{\text{opposite}}{\text{hypotenuse}} \qquad \cos\alpha = \frac{\text{adjacent}}{\text{hypotenuse}} \qquad \tan\alpha = \frac{\text{opposite}}{\text{adjacent}}$$

- The reciprocal of the ratios above play a vital role and are likewise given special names:

$$\csc\alpha = \frac{\text{hypotenuse}}{\text{opposite}} \qquad \sec\alpha = \frac{\text{hypotenuse}}{\text{adjacent}} \qquad \cot\alpha = \frac{\text{adjacent}}{\text{opposite}}$$

$$\csc\alpha = \frac{1}{\sin(\alpha)} \qquad \sec\alpha = \frac{1}{\cos(\alpha)} \qquad \cot\alpha = \frac{1}{\tan(\alpha)}$$

- The cofunctions of these ratios are so named because each function of α is equal to the cofunction of its complement. For instance, the complement of sine is *co*sine and $\sin\alpha = \cos(90° - \alpha)$.
- To solve a right triangle means to apply any combination of the sine, cosine, and tangent ratios, along with the Pythagorean theorem, until all three sides and all three angles are known.
- An angle of elevation is the angle formed by a horizontal line of orientation (parallel to level ground) and the line of sight. An angle of depression is likewise formed, but with the line of sight below the line of orientation.

EXERCISES

13. Use a calculator to solve for A:

a. $\cos 37° = A$ **b.** $\cos A = 0.4340$

14. Rewrite each expression in terms of a cofunction.

a. $\tan 57.4°$ **b.** $\sin(19°30'15'')$

Solve each triangle. Round angles to the nearest tenth and sides to the nearest hundredth.

15. **16.**

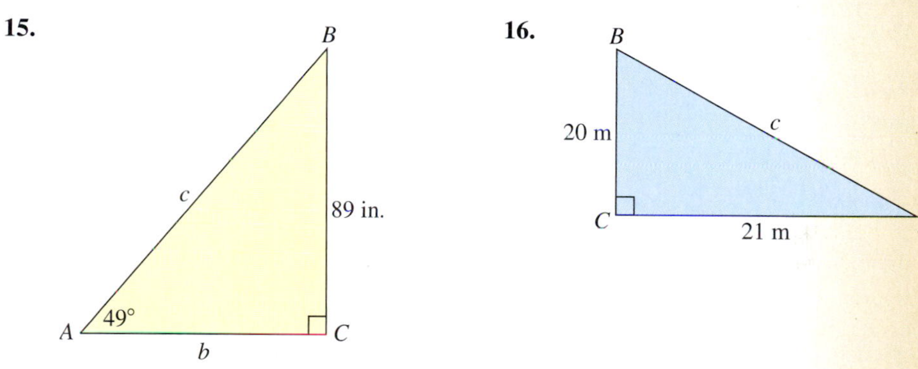

17. Josephine is to weld a support to a 20-m ramp so that the incline is exactly 15°. What is the height h of the support that must be used?

18. From the observation deck of a seaside building 480 m high, Armando sees two fishing boats in the distance, along a single line of sight. The angle of depression to the nearer boat is 63.5°, while for the boat farther away the angle is 45°. (a) How far out to sea is the nearer boat? (b) How far apart are the two boats?

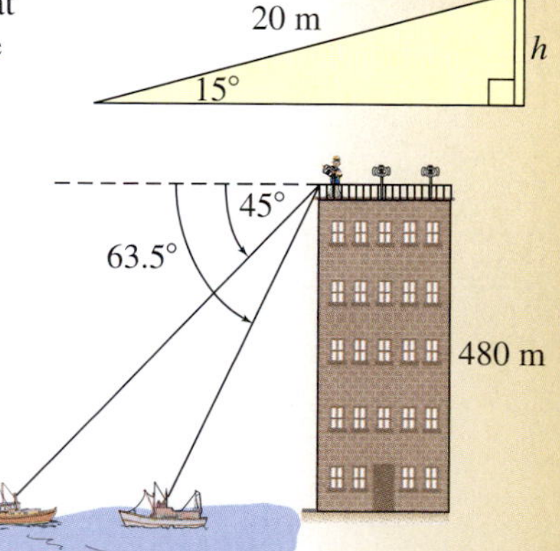

19. A slice of bread is roughly 14 cm by 10 cm. A sandwich is made and then cut diagonally. What two acute angles are formed?

SECTION 1.3 Trigonometry and the Coordinate Plane

KEY CONCEPTS

- In standard position, the terminal sides of 90°, 180°, 270°, and 360° angles coincide with one of the axes and are called quadrantal angles.
- By placing a right triangle in the coordinate plane with one acute angle at the origin and one side along the x-axis, we note the trig functions can be defined in terms of a point $P(x, y)$.
- Given $P(x, y)$ is any point on the terminal side of an angle θ in standard position, with $r = \sqrt{x^2 + y^2}$ the distance from the origin to this point. The six trigonometric functions of θ are

$$\sin\theta = \frac{y}{r} \qquad \cos\theta = \frac{x}{r} \qquad \tan\theta = \frac{y}{x} \qquad \csc\theta = \frac{r}{y} \qquad \sec\theta = \frac{r}{x} \qquad \cot\theta = \frac{x}{y}$$
$$\qquad\qquad\qquad\qquad x \neq 0 \qquad\quad y \neq 0 \qquad\quad x \neq 0 \qquad\quad y \neq 0$$

- A reference angle θ_r is defined to be the acute angle formed by the terminal side of a given angle θ and the nearest x-axis.
- Reference angles can be used to evaluate the trig functions of any angle, since the values are fixed by the ratio of sides and the signs are dictated by the quadrant of the terminal side.
- If a specific value of a trig function is known, the related angle θ can be found using a reference arc/angle, or the $\sin^{-1}$, $\cos^{-1}$, or $\tan^{-1}$ features of a calculator for nonstandard values.
- If θ is a solution to $\sin\theta = k$, then $\theta + 360k$ is also a solution for any integer k.

EXERCISES

20. Find two positive angles and two negative angles that are coterminal with $\theta = 207°$.

21. Name the reference angle for the angles given:
$\theta = -152°$ $\quad\theta = 521°$ $\quad\theta = 210°$

22. Find the value of the six trigonometric functions, given $P(x, y)$ is on the terminal side of angle θ in standard position.
a. $P(-12, 35)$ **b.** $(12, -18)$

23. Find the value of x, y, and r using the information given, and state the quadrant of the terminal side of θ. Then state the values of the six trig functions of θ.
a. $\cos\theta = \frac{4}{5}$; $\sin\theta < 0$
b. $\tan\theta = -\frac{12}{5}$; $\cos\theta > 0$

24. Find all angles satisfying the stated relationship. For standard angles, express your answer in exact form. For nonstandard angles, use a calculator and round to the nearest tenth.
a. $\tan\theta = -1$ **b.** $\cos\theta = \frac{\sqrt{3}}{2}$ **c.** $\tan\theta = 4.0108$ **d.** $\sin\theta = -0.4540$

SECTION 1.4 Unit Circles and the Trigonometry of Real Numbers

KEY CONCEPTS

- A central unit circle refers to a circle of radius 1 with its center at the origin, and can be graphed by drawing a circle through its quadrantal points: (1, 0), (0, 1), (−1, 0), and (0, −1).
- A central circle is symmetric to both axes and the origin. This means that if (a, b) is a point on the circle, then $(-a, b)$, $(-a, -b)$, and $(a, -b)$ are also on the circle and satisfy the equation of the circle.

- Points on a unit circle can be located using a right triangle of radius 1, with the vertex of an acute angle at the center. The coordinates of the point are (x, y), where x and y are the lengths of the legs.
- On a unit circle with θ in radians, the length of a subtended arc is numerically the same as the subtended angle, making the arc a "circular number line" and allowing us to treat the trig functions as functions of a real number.
- As functions of a real number, we refer to a reference arc s_r rather than a reference angle θ_r.
- For any real number t and a point on the unit circle associated with t, we have:

$$\cos t = x \qquad \sin t = y \qquad \tan t = \frac{y}{x} \qquad \sec t = \frac{1}{x} \qquad \csc t = \frac{1}{y} \qquad \cot t = \frac{x}{y}$$
$$\qquad\qquad\qquad\qquad x \neq 0 \qquad\quad x \neq 0 \qquad\quad y \neq 0 \qquad\quad y \neq 0$$

- The domain of each trig function must exclude division by zero (see box on page 57).
- Given the specific value of any function, the related real number t or angle θ can be found using a reference arc/angle, or the $\sin^{-1}$, $\cos^{-1}$, or $\tan^{-1}$ features of a calculator for nonstandard values.

EXERCISES

25. Given $\left(\frac{\sqrt{13}}{7}, y\right)$ is on a unit circle, find y if the point is in QIV, then use the symmetry of the circle to locate three other points.

26. Given $\left(\frac{3}{4}, -\frac{\sqrt{7}}{4}\right)$ is on the unit circle, find the value of all six trig functions of t without the use of a calculator.

27. Without using a calculator, find two values in $[0, 2\pi)$ that make the equation true: $\csc t = \frac{2}{\sqrt{3}}$.

28. Use a calculator to find the value of t that corresponds to the situation described: $\cos t = -0.7641$ with t in QII.

29. A crane used for lifting heavy equipment has a winch-drum with a 1-yd radius. (a) If 59 ft of cable has been wound in while lifting some equipment to the roof-top of a building, what angle has the drum turned through? (b) What angle must the drum turn through to wind in 75 ft of cable?

MIXED REVIEW

1. The shortest side of a 30-60-90 triangle has a length of $5\sqrt{3}$ cm. (a) What is the length of the longer side? (b) What is the length of the hypotenuse?

2. Name two values in $[0, 2\pi]$ where $\tan t = 1$.

3. Name two values in $[0, 2\pi]$ where $\cos t = -\frac{1}{2}$.

Exercise 7

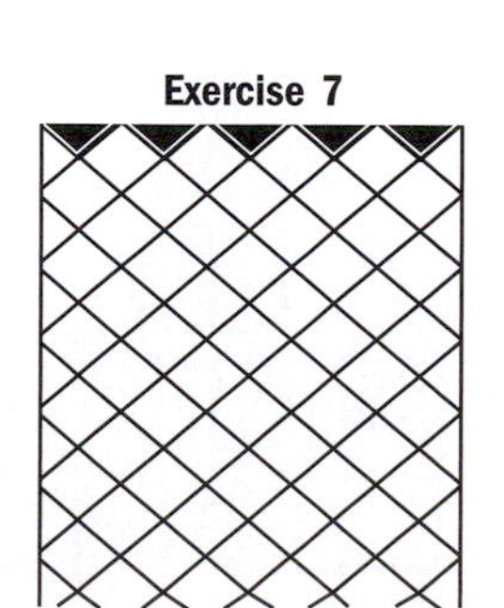

4. Given $\sin \theta = \frac{8}{\sqrt{185}}$ with θ in QII, state the value of the other five trig functions.

5. Convert to DMS form: $220.813\overline{8}°$.

6. Find two negative angles and two positive angles that are coterminal with (a) 57° and (b) 135°.

7. To finish the top row of the tile pattern on our bathroom wall, 12″ × 12″ tiles must be cut diagonally. Use a standard triangle to find the length of each cut and the width of the wall covered by tiles.

8. The service door into the foyer of a large office building is 36″ wide by 78″ tall. The building manager has ordered a large wall painting 85″ × 85″ to add some atmosphere to the foyer area. (a) Can the painting be brought in the service door? (b) If so, at what two integer-valued angles (with respect to level ground) could the painting be tilted?

9. Find the arc length and area of the shaded sector.

Exercise 9

$(-4\sqrt{3}, -4)$

10. Given $\sin\theta = \frac{16}{63}$, what is the value of $\cos(90 - \theta)$? Why?

11. Convert from DMS to decimal degrees: 86° 54′ 54″.

12. Name the reference angle θ_r for the angle θ given.

a. 735° **b.** −135° **c.** $\frac{5\pi}{6}$ **d.** $-\frac{5\pi}{3}$

13. Find the value of all six trig functions of θ, given the point $(15, -8)$ is on the terminal side.

14. Verify that $\left(-\frac{\sqrt{2}}{2}, \frac{\sqrt{2}}{2}\right)$ is a point on the unit circle and find the value of all six trig functions at this point.

Exercise 15

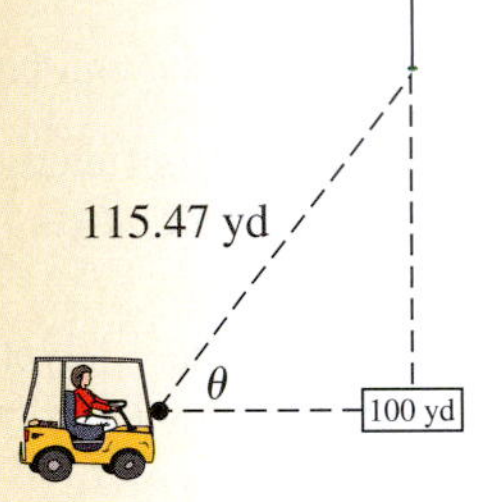

15. On your approach shot to the ninth green, the Global Positioning System (GPS) your cart is equipped with tells you the pin is 115.47 yd away. The distance plate states the straight line distance to the hole is 100 yd (see the diagram). Relative to a parallel line between the plate and the hole, at what complementary angle θ should you hit the shot?

16. The electricity supply lines to the top of Lone Eagle Plateau must be replaced, and the new lines will be run in conduit buried slightly beneath the surface. The scale of elevation is 1:150 (each concentric line indicates an increase in 150 ft of elevation), and the scale of horizontal distance is 1 in. = 200 ft. (a) Find the increase in elevation from point A to point B, (b) use a proportion to find the horizontal distance from A to B if the measured distance on the map is $2\frac{1}{4}$ inches, (c) draw the corresponding right triangle and use it to estimate the length of conduit needed from A to B and the angle of incline the installers will experience while installing the conduit.

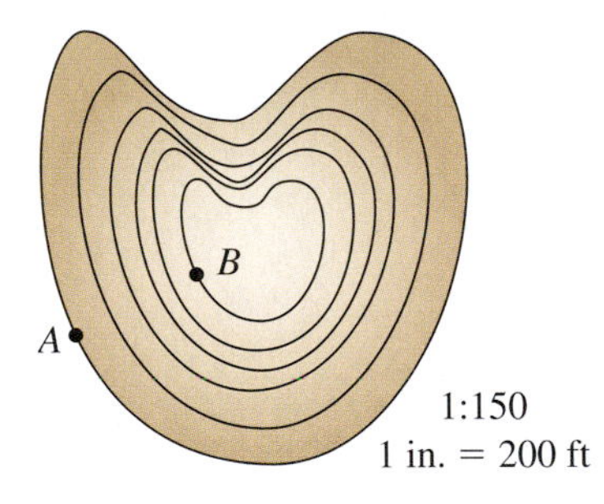

17. A salad spinner consists of a colander basket inside a large bowl, and is used to wash and dry lettuce and other salad ingredients. In vigorous use, the spinner is turned at about 3 revolutions per second. (a) Find the angular velocity and (b) find the linear velocity of a point of the circumference if the basket has a 20 cm radius.

18. Use a reference angle to find the value of (a) tan 780° and (b) $\cos\left(\frac{37\pi}{4}\right)$.

19. Virtually everyone is familiar with the Statue of Liberty in New York Bay, but fewer know that America is home to a second "Statue of Liberty" standing proudly atop the iron dome of the Capitol Building. From a distance of 600 ft, the angle of elevation from ground level to the top of the statue (from the east side) is 25.60°. The angle of elevation to the base of the statue is 24.07°. How tall is *Freedom,* the name sculptor Thomas Crawford gave this statue?

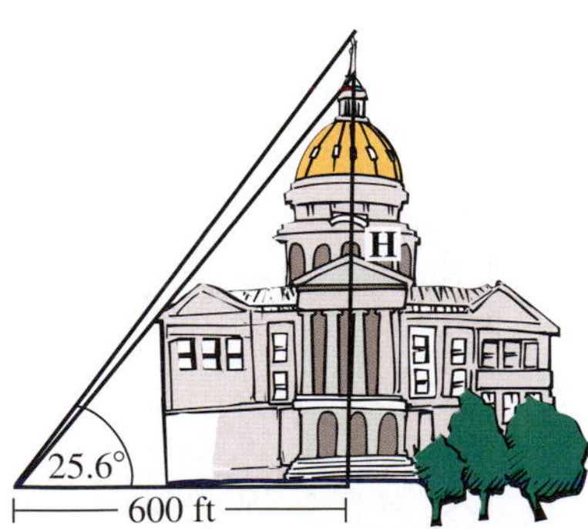

20. Given $\cos\theta = \frac{15}{17}$ and $\sin\theta = -\frac{8}{17}$, state the value of $\tan\theta$, $\cot\theta$, $\csc\theta$, and $\sec\theta$.

PRACTICE TEST

1. State the complement and supplement of a 35° angle.
2. Find two negative angles and two positive angles that are coterminal with $\theta = 30°$. Many solutions are possible.
3. Name the reference angle of each angle θ given.

   a. 225° **b.** −510° **c.** $\frac{7\pi}{6}$ **d.** $\frac{25\pi}{3}$

4. Convert from DMS to decimal degrees or decimal degrees to DMS as indicated.

   a. 100°45′18″ to decimal degrees

   b. 48.2125° to DMS

5. Four Corners USA is the point at which Utah, Colorado, Arizona, and New Mexico meet. Using the southern border of Colorado, the western border of Kansas, and the point P where Colorado, Nebraska, and Kansas meet, very nearly approximates a 30-60-90 triangle. If the western border of Kansas is 215 mi long, (a) what is the distance from Four Corners USA to point P? (b) How long is Colorado's southern border?

Exercise 5

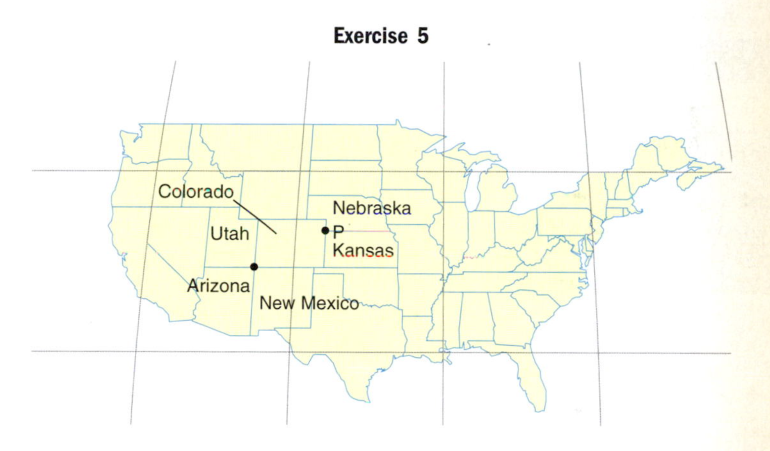

6. Complete the table using reference angles and exact values. If a function is undefined, so state.

t	sin t	cos t	tan t	csc t	sec t	cot t
0						
$\frac{2\pi}{3}$						
$\frac{7\pi}{6}$						
$\frac{5\pi}{4}$						
$\frac{5\pi}{3}$						
$\frac{7\pi}{4}$						
$\frac{13\pi}{6}$						

7. Given $\cos\theta = \frac{2}{5}$ and $\tan\theta < 0$, find the value of the other five trig functions of θ.

8. Verify that $\left(\frac{1}{3}, -\frac{2\sqrt{2}}{3}\right)$ is a point on the unit circle, then find the value of all six trig functions associated with this point.

9. In order to take pictures of a dance troupe as it performs, a camera crew rides in a cart on tracks that trace a circular arc. The radius of the arc is 75 ft, and from end to end the cart sweeps out an angle of 172.5° in 20 seconds. Use this information to find (a) the length of the track in feet and inches, (b) the angular velocity of the cart, and (c) the linear velocity of the cart in both ft/sec and mph.

Exercise 10

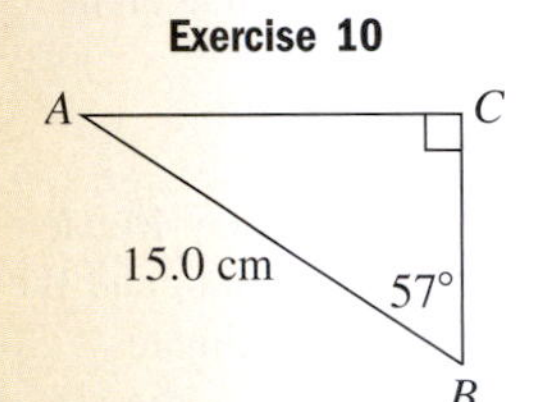

10. Solve the triangle shown. Answer in table form.

11. The "plow" is a yoga position in which a person lying on their back brings their feet up, over, and behind their head and touches them to the floor. If distance from hip to shoulder (at the right angle) is 57 cm and from hip to toes is 88 cm, find the distance from shoulders to toes and the angle formed at the hips.

Exercise 11

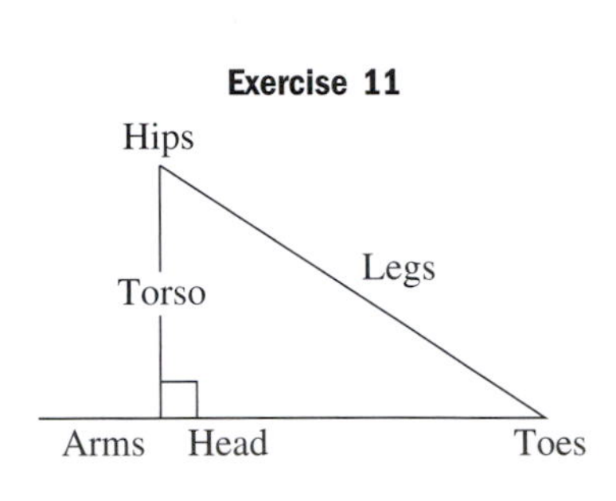

Exercise 12

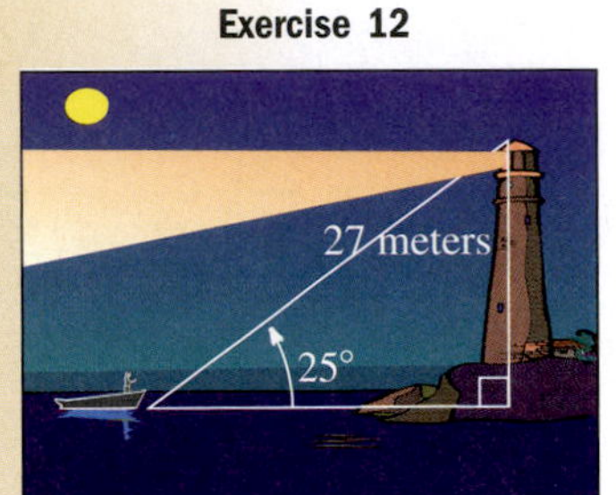

12. While doing some night fishing, you round a peninsula and a tall light house comes into view. Taking a sighting, you find the angle of elevation to the top of the lighthouse is 25°. If the lighthouse is known to be 27 m tall, how far away from shore are you?

13. Find the value of $t \in [0, 2\pi)$ satisfying the conditions given.

a. $\sin t = -\frac{1}{2}$, t in QIII **b.** $\sec t = \frac{2\sqrt{3}}{3}$, t in QIV **c.** $\tan t = -1$, t in QII

14. Memphis, Tennessee, is directly north of New Orleans, Louisiana, at 90° W longitude. Find the approximate distance between the cities in kilometers, given the Earth has a radius of 6373 km and Memphis is at 35° N latitude, while New Orleans is at 29.6° N latitude.

15. Given that $\left(\frac{20}{29}, \frac{21}{29}\right)$ is a point on the central unit circle, use the symmetry of the circle to name three other points on the circle.

16. Show that the length of chord $\mathcal{L}$ in the diagram is given by $\mathcal{L} = 2r$, where R is the radius of the circle and $r = R\cos\theta$. If the radius R is 15.7 cm and $\theta = 33°$, how long is the chord?

Exercise 16

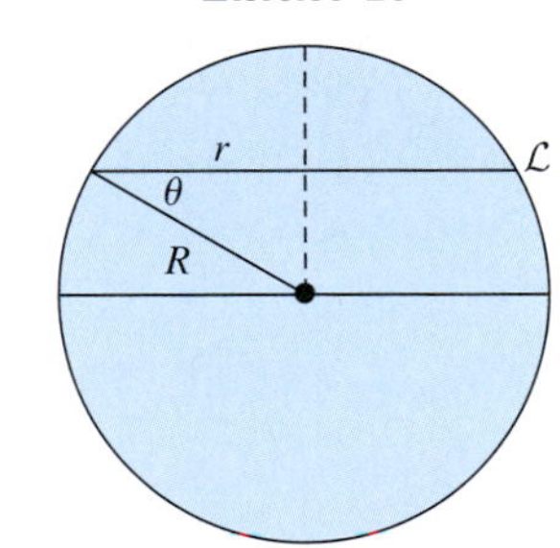

17. An athlete throwing the shot-put begins his first attempt facing due east, completes three and one-half turns and launches the shot facing due west. What angle did her body turn through?

18. On a calm, clear day, a helium balloon is released and begins rising. (a) How high is the balloon if the angle of elevation is 38° from a distance of 50 ft? (b) If a second reading of 56° is taken five seconds later, how high is the balloon? (c) How fast is the balloon rising in miles per hour (round to tenths)?

19. Given $\sin 32° \approx 0.53$, find the value of $\sin 148°$, $\sin 212°$, and $\sin 328°$ without using a calculator.

20. Find the value of t satisfying the given conditions.

a. $\sin t = -0.7568$; t in QIII

b. $\sec t = -1.5$; t in QII

Calculator Exploration and Discovery

Signs, Quadrants, and Reference Arcs

The keystrokes shown apply to a TI-84 Plus model. Please consult your manual or our Internet site for other models.

Graphing calculators can help us visualize the concept of reference arcs and angles, and better understand their connection to the circular functions. Just prior to Example 1 from Section 1.3, we placed a 30-60-90 triangle in QI, with the 30° angle at the origin and the longer side along the x-axis. We later noted that the slope of the line coincident with the hypotenuse must be $m = \frac{1}{\sqrt{3}} = \frac{\sqrt{3}}{3}$, since the $\frac{\text{rise}}{\text{run}}$ ratio was identical to the ratio of the shorter side to the longer side of the triangle. The equation of this line is then $y = \frac{\sqrt{3}}{3}x$. From elementary geometry, we know that vertical angles are equal, and so the acute angle in QIII formed by this line and the negative x-axis must also measure 30°, and is the reference angle for 210° (see Figure 1.61). By reorienting the same triangle in QII, two additional 30° angles can be formed, where the line coincident with this hypotenuse has an equation of $y = -\frac{\sqrt{3}}{3}x$. These are the 30° reference angles for 150° and 330°. From our work in Section 1.4, the equation of a unit circle is given by $x^2 + y^2 = 1$, which is $y = \pm\sqrt{1 - x^2}$ in function form. By graphing these functions on a graphing calculator, we can verify two important concepts: (1) that the point on a unit circle associated with $30° = \frac{\pi}{6}$ is $\left(\frac{\sqrt{3}}{2}, \frac{1}{2}\right)$ and (2) that the coordinates of the points associated with reference angles/arcs of similar measure are identical except for their sign. Using a graphing calculator, enter $Y_1 = \sqrt{1 - x^2}$ (the upper half of a unit circle), $Y_2 = \frac{\sqrt{3}}{3}x$, $Y_3 = -\sqrt{1 - x^2}$ (the lower half of a unit circle), and $Y_4 = -\frac{\sqrt{3}}{3}x$ (see Figure 1.62). Graphing the functions using the ZOOM **4:ZDecimal** feature will produce a "friendly" and square viewing window, but with a relatively small unit circle. Using the ZOOM **2:Zoom In** feature will give a better view, with the final window size depending on the settings of your zoom factors. Our settings were at XFactor = 2 and YFactor = 2, which produced the screen shown in Figure 1.63 (to access the zoom factors, press ZOOM ► **4:SetFactors**). Now comes the fun part. To find the point on the unit circle corresponding to $30° = \frac{\pi}{6}$, we need only find where the line intersects the unit circle in QI. This is accomplished using the keystrokes 2nd TRACE **(CALC) 5:Intersect,** and identifying the graphs we're interested in. After doing so, the calculator returns the values shown in Figure 1.64, which are indeed the equivalent of $\left(\frac{\sqrt{3}}{2}, \frac{1}{2}\right)$. Using the down arrow ▼ at this point will "jump the cursor" to the point where $Y_3 = -\sqrt{1 - x^2}$ and $Y_4 = -\frac{\sqrt{3}}{3}x$ intersect, where we note *the output values remain the same* except that

Figure 1.61

Figure 1.62

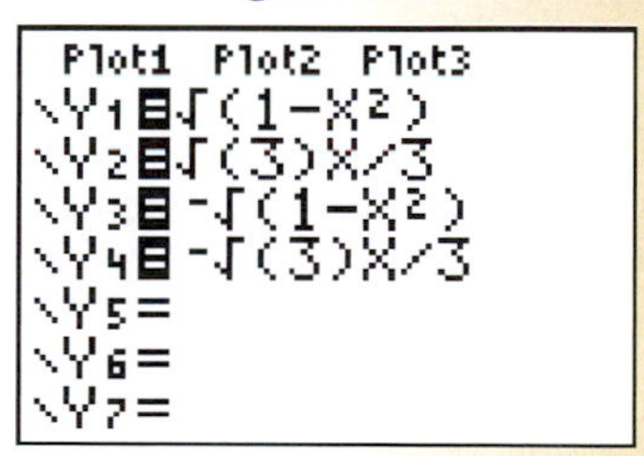

Figure 1.63

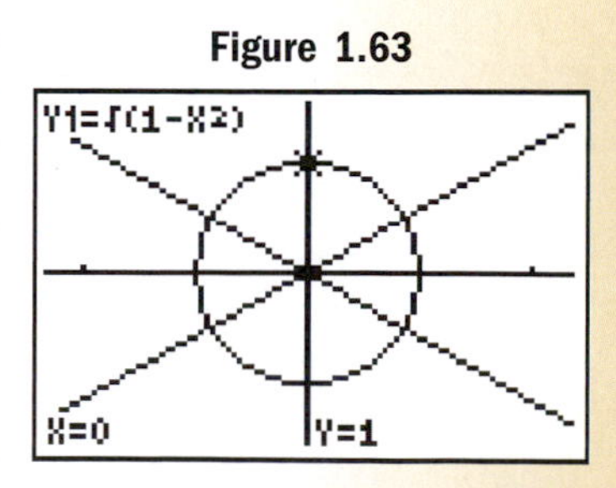

in QIV, the y-coordinate is negative (see Figure 1.65). In addition, since the calculator stores the last used x-value in the temporary location X, T, θ, n we can find the point of intersection in QIII $\left(Y_3 = -\sqrt{1-x^2} \text{ with } Y_2 = \frac{\sqrt{3}}{3}x\right)$ by simply using the keystrokes (–) X, T, θ, n ENTER, and the point of intersection in QII $\left(Y_1 = \sqrt{1-x^2} \text{ with } Y_4 = -\frac{\sqrt{3}}{3}x\right)$ using the up arrow ▲ (press the up and down arrows repeatedly for effect). Exercises of this type help to reinforce the value of reference angles and arcs, and give visual support for the unit circle definition of the trig functions, as we can more clearly see, for example, that for $\cos t = x$: cos 30° in QI and cos 330° in QIV are both equal to $\frac{\sqrt{3}}{2}$, while cos 150° in QII and cos 210° in QIII are both equal to $-\frac{\sqrt{3}}{2}$.

Figure 1.64

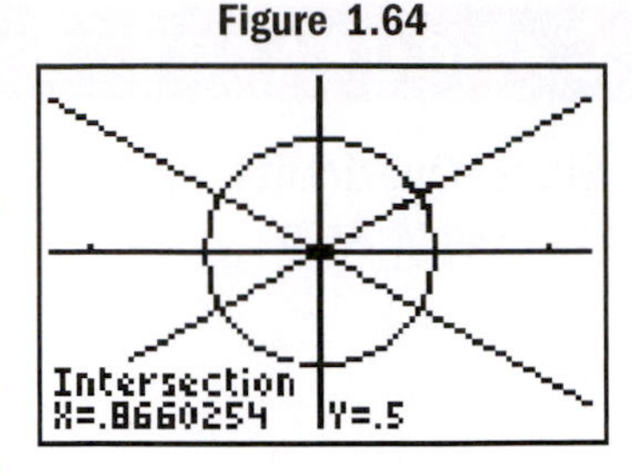

Figure 1.65

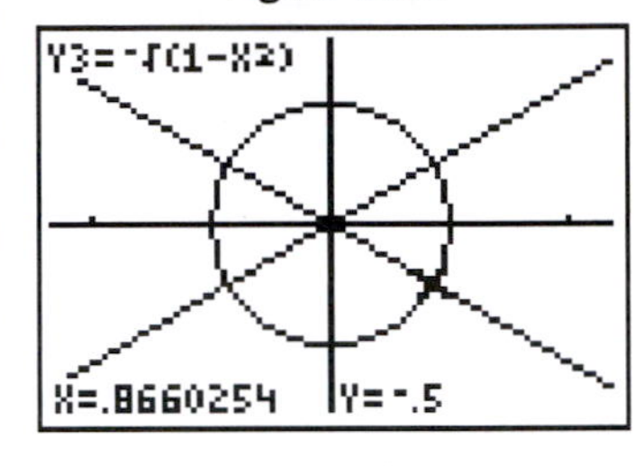

Exercise 1: Place a 45-45-90 triangle in QI, with a 45° angle at the origin and one side along the positive x-axis, then answer the following questions.

- **a.** What is the equation of the line coincident with the hypotenuse?
- **b.** What acute angle is formed in QIII by the line from (a) and the negative x-axis? Why?
- **c.** The acute angle from (b) is the reference angle for what positive angle θ $(0 < \theta < 360°)$?
- **d.** If the 45-45-90 triangle were reoriented in QII, what would be the equation of the line coincident with the hypotenuse?
- **e.** Using the triangle and line from (d), reference angles of $\theta_r = 45°$ in QII and QIV are created. Find θ for each of these angles $(0 < \theta < 360°)$.
- **f.** Use a graphing calculator to determine where the lines from (a) and (d) intersect with the unit circle, and use the results to find/verify the value of sin 45°, sin 135°, sin 225° and sin 315°.

Exercise 2: Repeat Exercise 1 using a 30-60-90 triangle with the *shorter side* along the x-axis. Use the results to find the value of tan 60°, tan 120°, tan 240° and tan 300°.

Strengthening Core Skills

More on Radians

To increase your understanding and appreciation of radian measure, consider the protractor shown in Figure 1.66, which is marked in both degrees and radians. Besides the obvious conversions it illustrates, $15° = \frac{\pi}{12}$, $30° = \frac{\pi}{6}$, and so on, we note that either system is adequate for measuring the amount of rotation for an angle in standard position. As drawn, this would be particularly true for multiples of $15° = \frac{\pi}{12}$. However, angles measured in radians have a twofold advantage over those measured in degrees.

First, the formula $s = r\theta$ (θ in radians) enables us to find the length of the related circular arc directly, without having to determine a proportional part of the 360° in a full

Figure 1.66

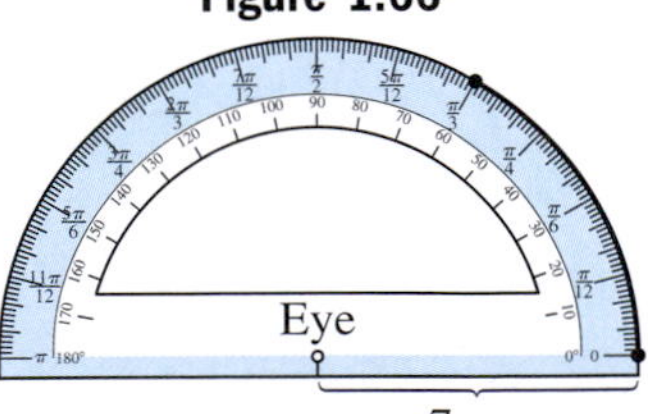

circle. For the protractor shown, $r = 7$ cm and using an angle of $60° = \frac{\pi}{3}$, we have $s = 7\left(\frac{\pi}{3}\right) \approx 7.33$ cm. To verify calculations of this kind, take any semicircular protractor you have available and determine its radius r. On a sheet of paper, stand the protractor on the 0° end, mark where the 0° meets the paper and roll the protractor along a straight line to $60° = \frac{\pi}{3}$ and make another mark. Then draw a line segment between these two marks and measure its length—the result will be very close to $s = r\left(\frac{\pi}{3}\right)$. Repeat this process for other angles measured in radians.

Figure 1.67

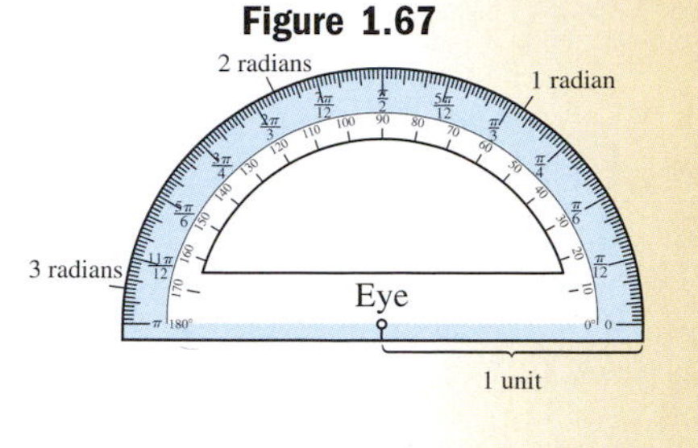

Second, if we also mark the protractor in *unit radians*, as shown in Figure 1.67 $\left(\text{note } \frac{\pi}{3} = 1.047 \text{ or just more than 1 rad}\right)$, the measure of the arc will be numerically equal to the measure of the angle. If you repeat the previous experiment and "roll the protractor" to $t = 2$ rad, the resulting line segment will be exactly twice as long as the radius. Try it! This again shows that we can view t as either an angle in radians *or* as the length of the related arc. More importantly, when this "unit protractor" is seen as the upper half of a unit circle (see Figure 1.68), we are also reminded of why we can view the trig functions as functions of a real number. Specifically, this is because the arc length t, $t \in \mathbb{R}$, acts as a "circular number line" that associates any real number t with a unique point (x, y) on the unit circle. For $\cos t = x$ and $\sin t = y$, the trig functions are now indeed functions of any real number, since the real number line, circular or otherwise, is infinite in length and extends in both a positive and negative direction (note that this view of trigonometry *is independent of the right triangle view*). Using the grid provided in Figure 1.68, we estimate that an arc length of $t = 1.25$ units corresponds to the point (0.32, 0.95) on the unit circle. After verifying $(0.32)^2 + (0.95)^2 \approx 1$, we use a calculator to support our findings and sure enough, $\cos 1.25 = 0.3153223624$ and $\sin 1.25 = 0.9489846194$. Use this information to complete the following exercises.

Figure 1.68

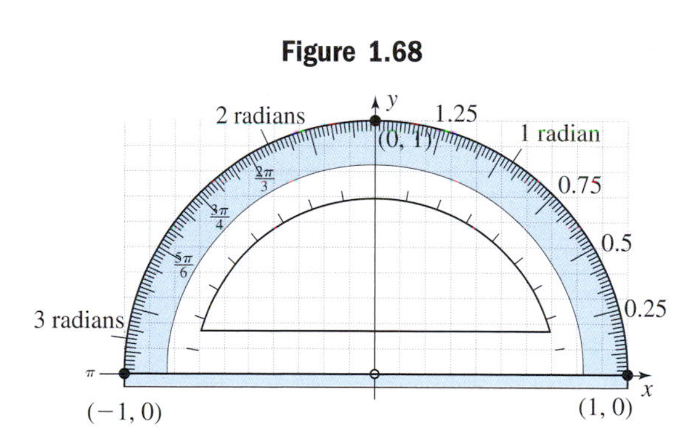

1. Estimate the point (x, y) on the unit circle associated with the values of t indicated, then verify that $x^2 + y^2 \approx 1$. Finally, use a calculator to show $x \approx \cos t$ and $y \approx \sin t$.

a. $t = 0.25$ **b.** $t = 0.5$ **c.** $t = 0.75$ **d.** $t = 1$

2. Estimate the point (x, y) on the unit circle associated with the values of t indicated, then verify that $x^2 + y^2 \approx 1$. Finally, use standard values to show $x \approx \cos t$ and $y \approx \sin t$.

a. $t = \frac{2\pi}{3}$ **b.** $t = \frac{3\pi}{4}$ **c.** $t = \frac{5\pi}{6}$ **d.** $t = \pi$

Chapter 2

Trigonometric Graphs and Models

Chapter Outline

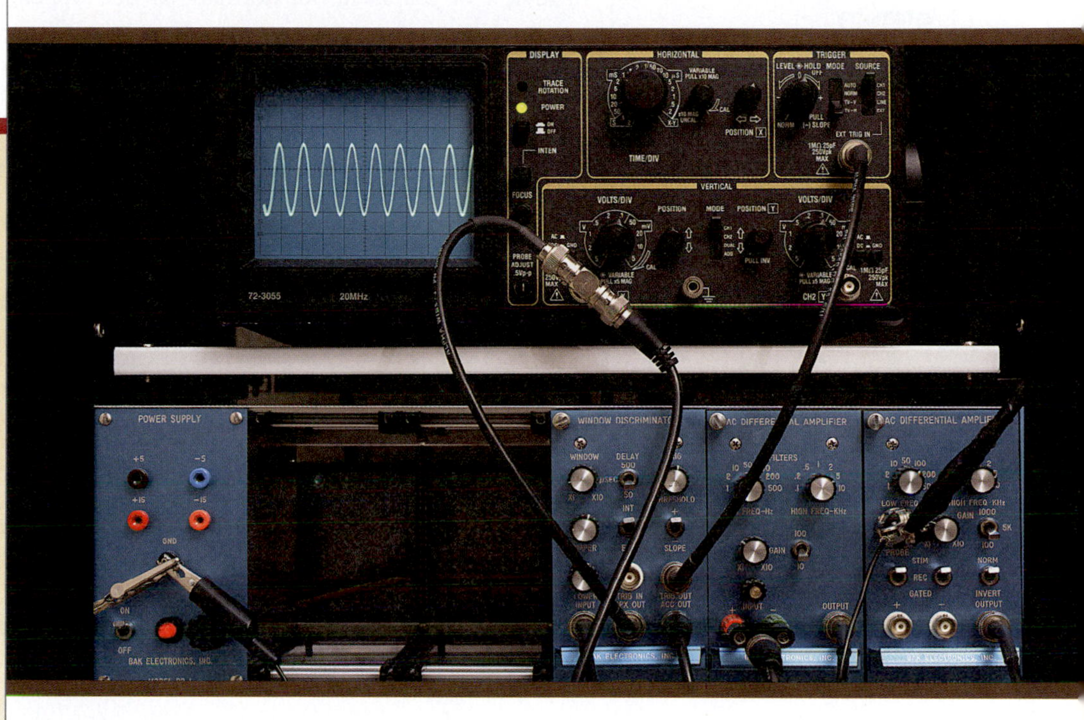

Preview

While written records of trigonometry are much more recent that those of geometry and algebra, the *roots* of trigonometry are likely just as ancient. Peering into the night-time heavens, ancient scholars noticed patterns among the celestial bodies, giving rise to the desire to model their regular reoccurrence. But even at the beginning of the modern age, with the tools of geometry at hand and the study of algebra maturing, astronomers were unable to come up with accurate models. This had to wait for the study of periodic functions to mature. Via the popular media, many people have an awareness of **sinusoidal graphs** like the image shown on the oscilloscope in the figure, and a basic understanding of cycles, periods, and wave phenomenon.

2.1 Graphs of the Sine and Cosine Functions

LEARNING OBJECTIVES

In Section 2.1 you will learn how to:

A. **Graph $f(t) = \sin t$ using standard values and symmetry**

B. **Graph $f(t) = \cos t$ using standard values and symmetry**

C. **Graph sine and cosine functions with various amplitudes and periods**

D. **Investigate graphs of the reciprocal functions $f(t) = \csc(Bt)$ and $f(t) = \sec(Bt)$**

E. **Write the equation for a given graph**

INTRODUCTION

As with the graphs of other functions, trigonometric graphs contribute a great deal toward the understanding of each trig function and its applications. For now, our primary interest is the general shape of each basic graph and some of the transformations that can be applied. We will also learn to analyze each graph, and to capitalize on the features that enable us to apply the functions as real-world models. For a review of graphical transformations, see Appendix II.

POINT OF INTEREST

The close of the third century B.C. marked the end of the glory years of Grecian mathematics. Lacking the respect his predecessors had for math, science, and art, Ptolemy VII exiled all scholars who would not swear loyalty to him. As things turned out, Alexandria's loss was the rest of Asia Minor's gain, and mathematical and scientific knowledge spread. According to one Athenaeus of Naucratis, "The king sent many Alexandrians into exile, filling the islands and towns with . . . philologists, philosophers, mathematicians, musicians, painters, physicians and other professional men. The refugees, reduced by poverty to teaching what they knew, instructed many other men."

A. Graphing $f(t) = \sin t$

From our work in previous sections, we have the values for $y = \sin t$ in the interval $\left[0, \frac{\pi}{2}\right]$ shown in Table 2.1.

Figure 2.1

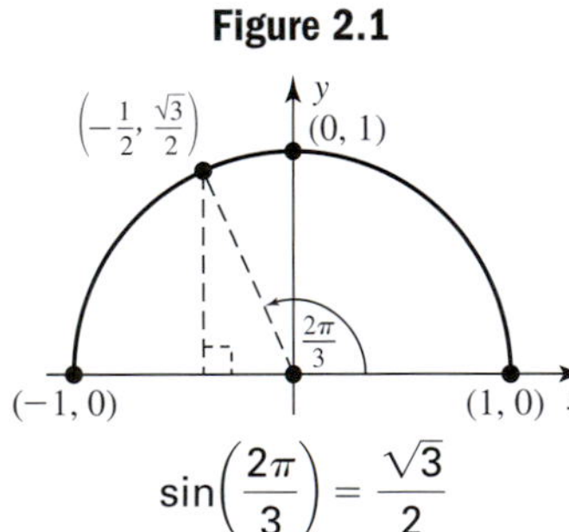

$$\sin\left(\frac{2\pi}{3}\right) = \frac{\sqrt{3}}{2}$$

Table 2.1

t	0	$\frac{\pi}{6}$	$\frac{\pi}{4}$	$\frac{\pi}{3}$	$\frac{\pi}{2}$
$\sin t$	0	$\frac{1}{2}$	$\frac{\sqrt{2}}{2}$	$\frac{\sqrt{3}}{2}$	1

Figure 2.2

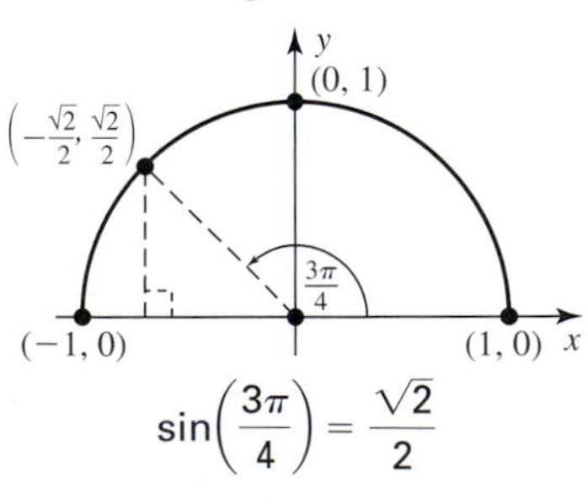

$$\sin\left(\frac{3\pi}{4}\right) = \frac{\sqrt{2}}{2}$$

Observe that in this interval (representing Quadrant I), sine values are increasing from 0 to 1. From $\frac{\pi}{2}$ to π (Quadrant II), standard values taken from the unit circle show sine values are decreasing from 1 to 0, *but through the same output values as in QI.* See Figures 2.1 through 2.3.

With this information we can extend our table of values through π, noting that $\sin \pi = 0$

Figure 2.3

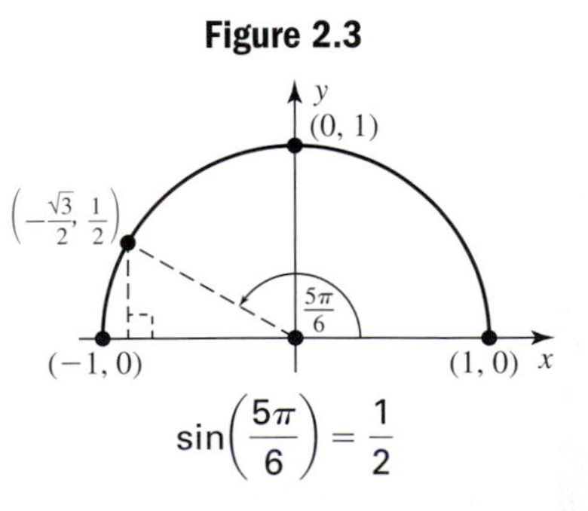

$$\sin\left(\frac{5\pi}{6}\right) = \frac{1}{2}$$

(see Table 2.2). Note that both the table and unit circle show the range of the sine function is $y \in [-1, 1]$.

Table 2.2

t	0	$\frac{\pi}{6}$	$\frac{\pi}{4}$	$\frac{\pi}{3}$	$\frac{\pi}{2}$	$\frac{2\pi}{3}$	$\frac{3\pi}{4}$	$\frac{5\pi}{6}$	π
$\sin t$	0	$\frac{1}{2}$	$\frac{\sqrt{2}}{2}$	$\frac{\sqrt{3}}{2}$	1	$\frac{\sqrt{3}}{2}$	$\frac{\sqrt{2}}{2}$	$\frac{1}{2}$	0

Using the symmetry of the circle and the fact that y is negative in QIII and QIV, we can complete the table for values between π and 2π.

EXAMPLE 1 Use the symmetry of the unit circle and reference arcs of standard values to complete Table 2.3.

Table 2.3

t	π	$\frac{7\pi}{6}$	$\frac{5\pi}{4}$	$\frac{4\pi}{3}$	$\frac{3\pi}{2}$	$\frac{5\pi}{3}$	$\frac{7\pi}{4}$	$\frac{11\pi}{6}$	2π
$\sin t$									

Solution: Symmetry shows that for any odd multiple of $t = \frac{\pi}{4}$, function values will be $\pm\frac{\sqrt{2}}{2}$ depending on the quadrant of the terminal side. Similarly, for any reference arc of $\frac{\pi}{6}$, $\sin t = \pm\frac{1}{2}$, while for a reference arc of $\frac{\pi}{3}$, $\sin t = \pm\frac{\sqrt{3}}{2}$. With these, we complete the table as shown in Table 2.4.

Table 2.4

t	π	$\frac{7\pi}{6}$	$\frac{5\pi}{4}$	$\frac{4\pi}{3}$	$\frac{3\pi}{2}$	$\frac{5\pi}{3}$	$\frac{7\pi}{4}$	$\frac{11\pi}{6}$	2π
$\sin t$	0	$-\frac{1}{2}$	$-\frac{\sqrt{2}}{2}$	$-\frac{\sqrt{3}}{2}$	-1	$-\frac{\sqrt{3}}{2}$	$-\frac{\sqrt{2}}{2}$	$-\frac{1}{2}$	0

NOW TRY EXERCISES 7 AND 8

Noting that $\frac{1}{2} = 0.5$, $\frac{\sqrt{2}}{2} \approx 0.71$, and $\frac{\sqrt{3}}{2} \approx 0.87$, we plot these points and connect them with a smooth curve to graph $y = \sin t$ for $t \in [0, 2\pi]$. The first five plotted points are labeled in Figure 2.4.

Figure 2.4

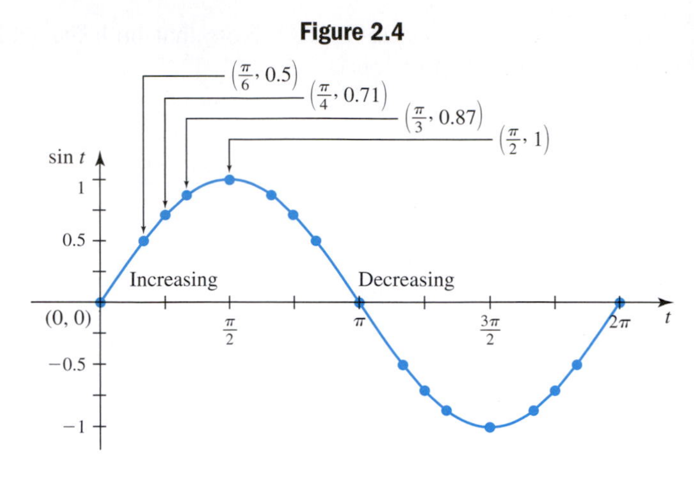

Expanding the table from 2π to 4π using reference arcs and the unit circle shows that function values begin to repeat. For example, $\sin\left(\frac{13\pi}{6}\right) = \sin\left(\frac{\pi}{6}\right)$ since $\theta_r = \frac{\pi}{6}$; $\sin\left(\frac{9\pi}{4}\right) = \sin\left(\frac{\pi}{4}\right)$ since $\theta_r = \frac{\pi}{4}$, and so on. Functions that cycle through a set pattern of values are said to be **periodic functions.**

> **PERIODIC FUNCTIONS**
> A function f is said to be periodic if there is a positive number P such that $f(t + P) = f(t)$ for all t in the domain. The smallest number P for which this occurs is called the **period** of f.

For the sine function we have $\sin t = \sin(t + 2\pi)$, as in $\sin\left(\frac{13\pi}{6}\right) = \sin\left(\frac{\pi}{6} + 2\pi\right)$ and $\sin\left(\frac{9\pi}{4}\right) = \sin\left(\frac{\pi}{4} + 2\pi\right)$, with the idea extending to all other real numbers t: $\sin t = \sin(t + 2\pi k)$ for all integers k. The sine function is periodic with period $P = 2\pi$.

Although we initially focused on positive values of t in $[0, 2\pi]$, $t < 0$ and $k < 0$ are certainly possibilities and we note the graph of $y = \sin t$ extends infinitely in both directions (see Figure 2.5).

Figure 2.5

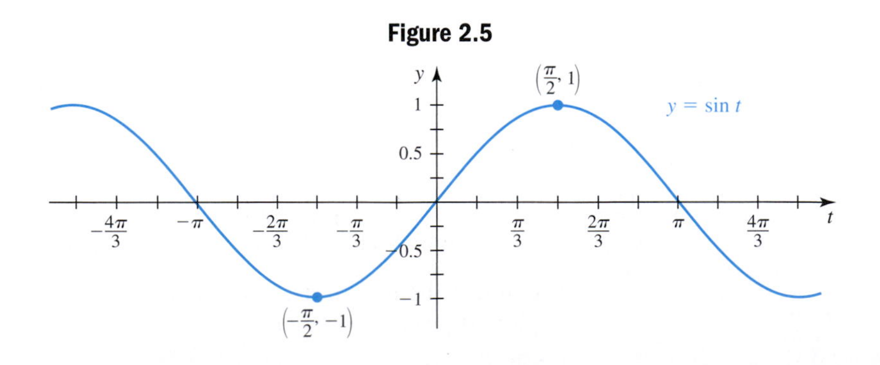

From Figure 2.5 we also note that $y = \sin t$ is an odd function. For instance, we have $\sin\left(-\frac{\pi}{2}\right) = -\sin\left(\frac{\pi}{2}\right) = -1$ and in general $\sin(-t) = -\sin t$ for all $t \in \mathbb{R}$. As a handy reference, the following box summarizes the main characteristics of $y = \sin t$ using features discussed here and ideas from Section 1.4.

CHARACTERISTICS OF $f(t) = \sin t$

Unit Circle Definition	Domain	Symmetry	Maximum values	Increasing: $t \in (0, 2\pi)$
$\sin t = y$	$t \in \mathbb{R}$	Odd: $\sin(-t) = -\sin t$	$\sin t = 1$ at $t = \frac{\pi}{2} + 2\pi k$	$t \in \left(0, \frac{\pi}{2}\right) \cup \left(\frac{3\pi}{2}, 2\pi\right)$
Period	**Range**	**Zeroes**	**Minimum values**	**Decreasing: $t \in (0, 2\pi)$**
2π	$y \in [-1, 1]$	$t = k\pi, k \in \mathbb{Z}$	$\sin t = -1$ at $t = \frac{3\pi}{2} + 2\pi k$	$t \in \left(\frac{\pi}{2}, \frac{3\pi}{2}\right)$

Many of the transformations applied to algebraic graphs can also be applied to trigonometric graphs (see Appendix II). These transformations may stretch, reflect, or translate the graph, but it will still retain its basic shape. In numerous applications it will help if you're able to draw a quick, accurate sketch of the transformations involving $f(t) = \sin t$. To assist this effort, we'll begin with the standard interval $t \in [0, 2\pi]$, combine the characteristics just listed with some simple geometry, and offer the following four-step process. Steps I through IV are illustrated in Figures 2.6 through 2.9.

Step I: Draw the y-axis, mark zero halfway up, with -1 and 1 an equal distance from this zero. Then draw an extended t-axis and tick mark 2π to the extreme right.

Step II: Mark halfway between the y-axis and 2π and label it "π," mark halfway between π on either side and label the marks $\frac{\pi}{2}$ and $\frac{3\pi}{2}$. Halfway between these you can draw additional tick marks to represent the remaining multiples of $\frac{\pi}{4}$.

Step III: Next, lightly draw a rectangular frame, which we'll call the **reference rectangle,** $P = 2\pi$ units wide and 2 units tall, centered on the t-axis and with the y-axis along one side.

Step IV: Knowing $y = \sin t$ is positive and increasing in QI; that the range is $[-1, 1]$; that the zeroes are 0, π, and 2π; and that maximum and minimum values *occur halfway between the zeroes* (since there is no horizontal shift), we can draw a reliable graph of $y = \sin t$ by partitioning the reference rectangle into four equal parts to locate the zeroes and max/min values. We will call this partitioning of the reference rectangle the **rule of fourths,** since we are then scaling the t-axis in increments of $\frac{P}{4}$.

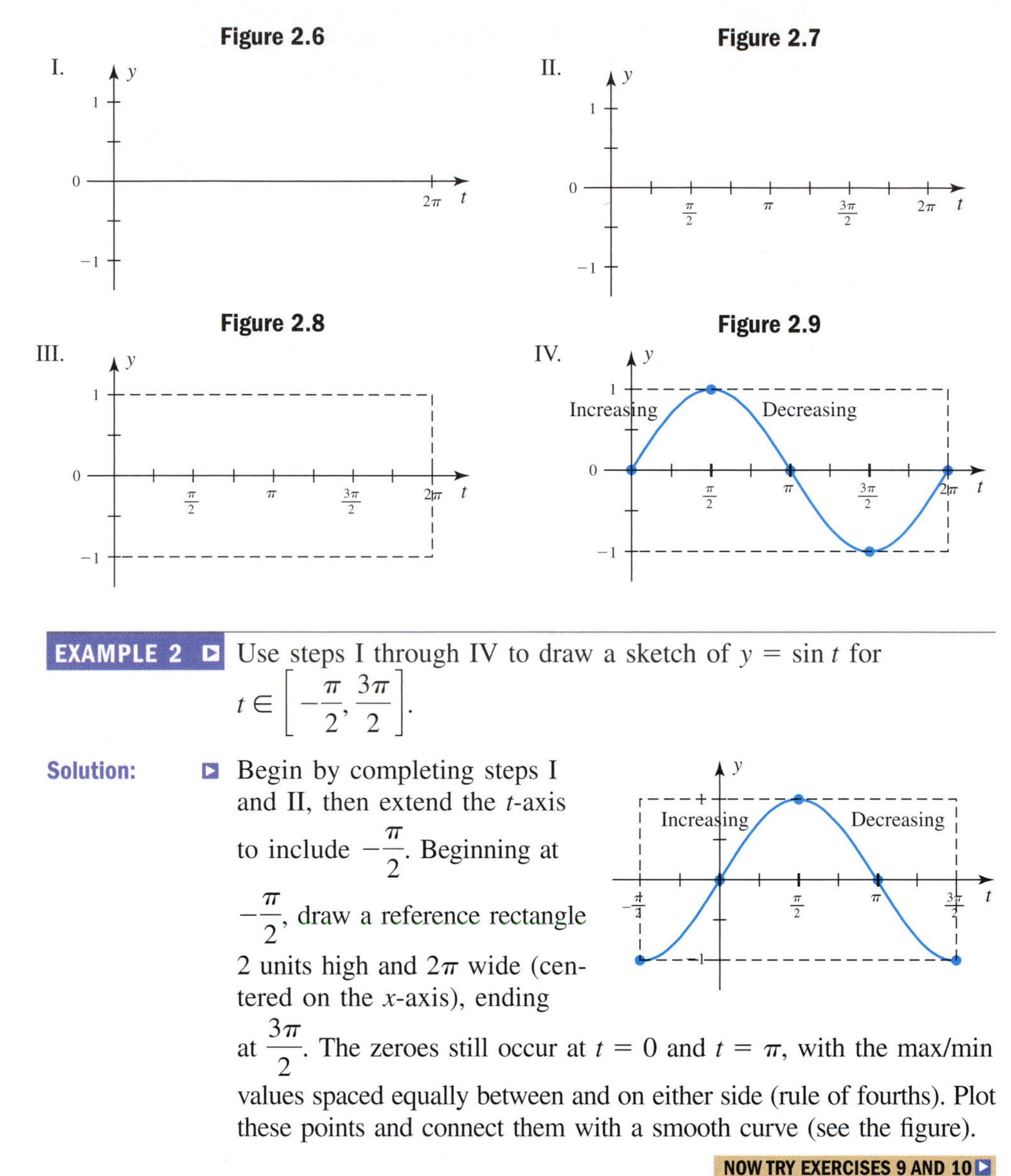

EXAMPLE 2 Use steps I through IV to draw a sketch of $y = \sin t$ for $t \in \left[-\frac{\pi}{2}, \frac{3\pi}{2}\right]$.

Solution: Begin by completing steps I and II, then extend the t-axis to include $-\frac{\pi}{2}$. Beginning at $-\frac{\pi}{2}$, draw a reference rectangle 2 units high and 2π wide (centered on the x-axis), ending at $\frac{3\pi}{2}$. The zeroes still occur at $t = 0$ and $t = \pi$, with the max/min values spaced equally between and on either side (rule of fourths). Plot these points and connect them with a smooth curve (see the figure).

NOW TRY EXERCISES 9 AND 10

B. Graphing $f(t) = \cos t$

With the graph of $f(t) = \sin t$ established, sketching the graph of $f(t) = \cos t$ is a very natural next step. First, note that when $t = 0$, $\cos t = 1$ so the graph of $y = \cos t$ will begin at (0, 1) in the interval $[0, 2\pi]$. Second, we've seen $\left(\pm\frac{1}{2}, \pm\frac{\sqrt{3}}{2}\right)$, $\left(\pm\frac{\sqrt{3}}{2}, \pm\frac{1}{2}\right)$ and $\left(\pm\frac{\sqrt{2}}{2}, \pm\frac{\sqrt{2}}{2}\right)$ are all points on the unit circle since they satisfy $x^2 + y^2 = 1$. Since $\cos t = x$ and $\sin t = y$, the equation $\cos^2 t + \sin^2 t = 1$ can be obtained by direct substitution. This means if $\sin t = \pm\frac{1}{2}$, then $\cos t = \pm\frac{\sqrt{3}}{2}$ and vice versa, with the signs taken from the appropriate quadrant. The table of values for cosine then becomes a simple extension of the table for sine, as shown in Table 2.5 for $t \in [0, \pi]$.

Table 2.5

t	0	$\frac{\pi}{6}$	$\frac{\pi}{4}$	$\frac{\pi}{3}$	$\frac{\pi}{2}$	$\frac{2\pi}{3}$	$\frac{3\pi}{4}$	$\frac{5\pi}{6}$	π
sin t	0	$\frac{1}{2} = 0.5$	$\frac{\sqrt{2}}{2} \approx 0.71$	$\frac{\sqrt{3}}{2} \approx 0.87$	1	$\frac{\sqrt{3}}{2} \approx 0.87$	$\frac{\sqrt{2}}{2} \approx 0.71$	$\frac{1}{2} = 0.5$	0
cos t	**1**	$\frac{\sqrt{3}}{2} \approx 0.87$	$\frac{\sqrt{2}}{2} \approx 0.71$	$\frac{1}{2} = 0.5$	**0**	$-\frac{1}{2} = -0.5$	$-\frac{\sqrt{2}}{2} \approx -0.71$	$-\frac{\sqrt{3}}{2} \approx -0.87$	**−1**

The same values can be taken from the unit circle, but this view requires much less effort and easily extends to values of t in $[\pi, 2\pi]$. Using the points from Table 2.5 and its extension through $[\pi, 2\pi]$, we can draw the graph of $y = \cos t$ for $t \in [0, 2\pi]$ and identify where the function is increasing and decreasing in this interval. See Figure 2.10.

Figure 2.10

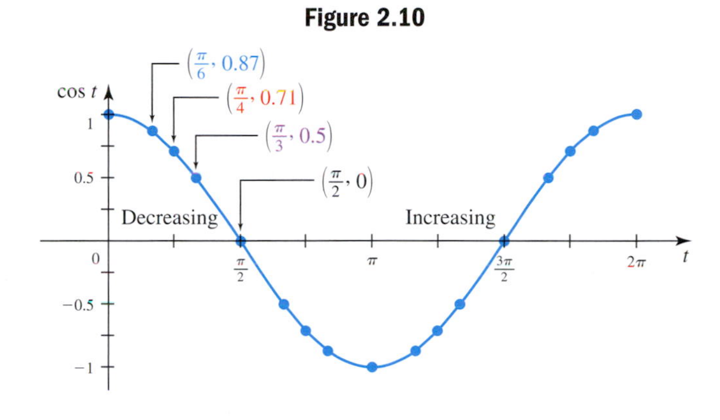

The function is decreasing for $t \in (0, \pi)$, and increasing for $t \in (\pi, 2\pi)$.

The end result appears to be the graph of $y = \sin t$, shifted to the left $\frac{\pi}{2}$ units, meaning $\left(\frac{-\pi}{2}, 0\right)$ is on the graph. This is in fact the case, and is a relationship we will prove in Chapter 3. Like $y = \sin t$, the function $y = \cos t$ is periodic with period $P = 2\pi$, and extends infinitely in both directions.

Finally, we note that cosine is an **even function,** meaning $\cos(-t) = \cos t$ for all t in the domain. For instance, $\cos\left(-\frac{\pi}{2}\right) = \cos\left(\frac{\pi}{2}\right) = 0$. Here is a summary of important characteristics of the cosine function.

CHARACTERISTICS OF $f(t) = \cos t$

Unit Circle Definition	Domain	Symmetry	Maximum values	Increasing: $t \in (0, 2\pi)$
$\cos t = x$	$t \in \mathbb{R}$	Even: $\cos(-t) = \cos t$	$\cos t = 1$ at $t = 2\pi k$	$t \in (\pi, 2\pi)$
Period	**Range**	**Zeroes**	**Minimum values**	**Decreasing: $t \in (0, 2\pi)$**
2π	$y \in [-1, 1]$	$t = \frac{\pi}{2} + \pi k;\ k \in \mathbb{Z}$	$\cos t = -1$ at $t = \pi + 2\pi k$	$t \in (0, \pi)$

EXAMPLE 3 Draw a sketch of $y = \cos t$ for $t \in \left[-\pi, \frac{3\pi}{2}\right]$.

Solution: Using steps I through IV (frame, scaling, zeroes, and max/min values) produces this graph for $y = \cos t$ in $\left[-\pi, \frac{3\pi}{2}\right]$. Note the reference rectangle goes from $-\pi$ to π (2π units in length), and the graph was then extended to the interval $\left[-\pi, \frac{3\pi}{2}\right]$.

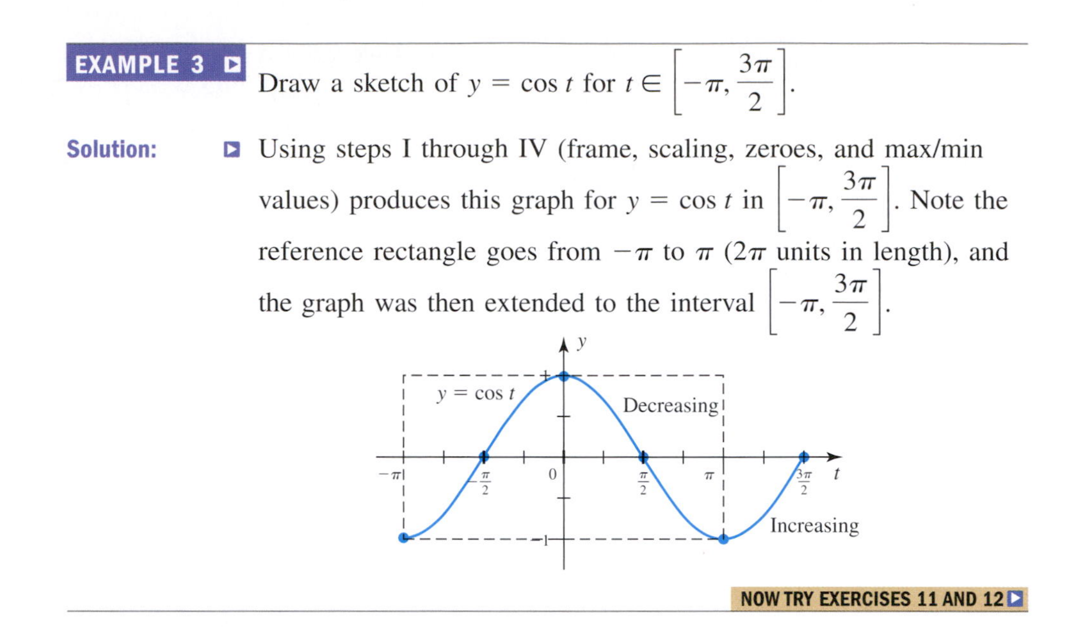

NOW TRY EXERCISES 11 AND 12

WORTHY OF NOTE

Note that the equations $y = A \sin t$ and $y = A \cos t$ both indicate y is a function of t, with no reference to the unit circle definitions $\cos t = x$ and $\sin t = y$.

C. Graphing $y = A \sin(Bt)$ and $y = A \cos(Bt)$

In many applications, trig functions have maximum and minimum values other than 1 and -1, and periods other than 2π. For instance, in tropical regions the maximum and minimum temperatures may vary by no more than 20°, while for desert regions this difference may be 40° or more. This variation is modeled by the *amplitude* of sine and cosine functions.

Amplitude and the Coefficient A ($B = 1$)

For functions of the form $y = A \sin t$ and $y = A \cos t$, let M represent the *M*aximum value and m the *m*inimum value of the functions. Then the quantity $\frac{M + m}{2}$ gives the **average value** of the function, while $\frac{M - m}{2}$ gives the **amplitude** of the function. Amplitude is the maximum displacement from the average value in the positive or negative direction. It is represented by $|A|$, with A playing a role similar to that seen for algebraic graphs ($Af(x)$ vertically stretches or compresses the graph of f, and reflects it across the t-axis if $A < 0$). Graphs of the form $y = \sin t$ (and $y = \cos t$) can quickly be sketched with any amplitude by noting that the *zeroes of the function remain fixed* (since $\sin t = 0$ implies $A \sin t = 0$), and that the *maximum and minimum values are A and $-A$* respectively (since $\sin t = 1$ or -1 implies $A \sin t = A$ or $-A$). Connecting the points that result with a smooth curve will complete the graph.

EXAMPLE 4 Draw a sketch of $y = 4 \sin t$ for $t \in [0, 2\pi]$.

Solution: This graph has the same zeroes as $y = \sin t$, but the maximum value is $4 \sin\left(\frac{\pi}{2}\right) = 4(1) = 4$, with a minimum value of

$4\sin\left(\frac{3\pi}{2}\right) = 4(-1) = -4$ (the amplitude is $|A| = 4$). Connecting these points with a "sine curve" gives the graph shown ($y = \sin t$ is also shown for comparison).

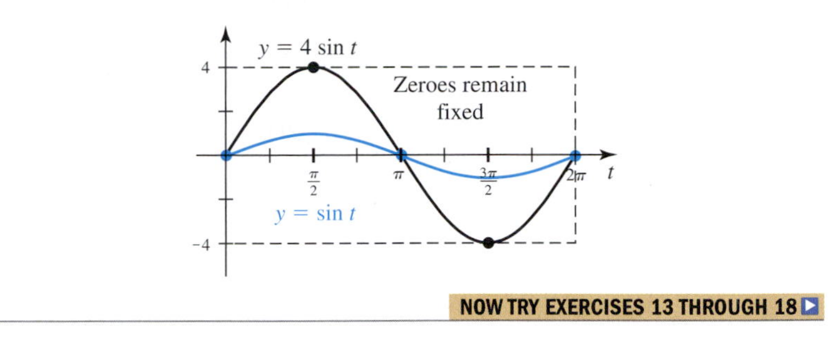

NOW TRY EXERCISES 13 THROUGH 18

Period and the Coefficient B

While basic sine and cosine functions have a period of 2π, in many applications the period may be very long (tsunami's) or very short (electromagnetic waves). For the equations $y = A\sin(Bt)$ and $y = A\cos(Bt)$, the period depends on the value of B. To see why, consider the function $y = \cos(2t)$ and Table 2.6. Multiplying input values by 2 means each cycle will be completed twice as fast. The table shows that $y = \cos(2t)$ completes a full cycle in $[0, \pi]$, giving a period of $P = \pi$ (Figure 2.11, red graph).

Table 2.6

t	0	$\frac{\pi}{4}$	$\frac{\pi}{2}$	$\frac{3\pi}{4}$	π
$2t$	0	$\frac{\pi}{2}$	π	$\frac{3\pi}{2}$	2π
$\cos(2t)$	1	0	-1	0	1

Dividing input values by 2 (or multiplying by $\frac{1}{2}$) will cause the function to complete a cycle only half as fast, doubling the time required to complete a full cycle. Table 2.7 shows $y = \cos\left(\frac{1}{2}t\right)$ completes only one-half cycle in 2π (Figure 2.11, blue graph).

Table 2.7

t	0	$\frac{\pi}{4}$	$\frac{\pi}{2}$	$\frac{3\pi}{4}$	π	$\frac{5\pi}{4}$	$\frac{3\pi}{2}$	$\frac{7\pi}{4}$	2π
$\frac{1}{2}t$	0	$\frac{\pi}{8}$	$\frac{\pi}{4}$	$\frac{3\pi}{8}$	$\frac{\pi}{2}$	$\frac{5\pi}{8}$	$\frac{3\pi}{4}$	$\frac{7\pi}{8}$	π
$\cos\left(\frac{1}{2}t\right)$	1	0.92	$\frac{\sqrt{2}}{2}$	0.38	0	-0.38	$-\frac{\sqrt{2}}{2}$	-0.92	-1

The graphs of $y = \cos t$, $y = \cos(2t)$, and $y = \cos\left(\frac{1}{2}t\right)$ shown in Figure 2.11 clearly illustrate this relationship and how the value of B affects the period of a graph.

To find the period for arbitrary values of B, the formula $P = \dfrac{2\pi}{B}$ is used. Note for $y = \cos(2t)$, $B = 2$, and $P = \dfrac{2\pi}{2} = \pi$, as shown. For $y = \cos\left(\dfrac{1}{2}t\right)$, $B = \dfrac{1}{2}$, and $P = \dfrac{2\pi}{1/2} = 4\pi$.

Figure 2.11

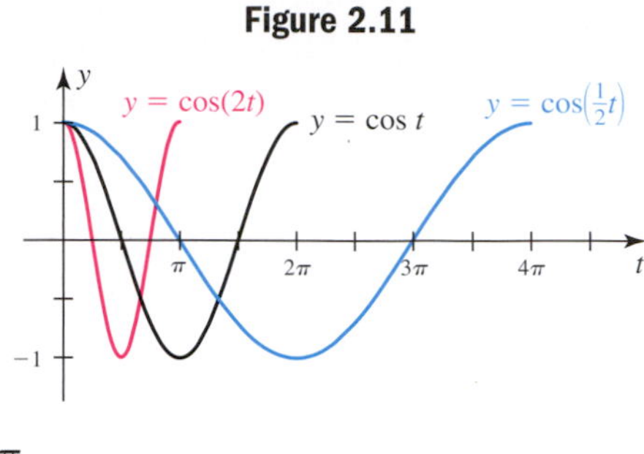

PERIOD FORMULA FOR SINE AND COSINE

For B a real number and functions $y = A\sin(Bt)$ and $y = A\cos(Bt)$, the period is given by $P = \dfrac{2\pi}{B}$.

To sketch these functions for periods other than 2π, we still use a reference rectangle of height $2A$ and length P, then break the enclosed t-axis in four equal parts to help find and graph the zeroes and max/min values. In general, if the period is "very large" one full cycle is appropriate for the graph. If the period is very small, graph at least two cycles.

Note the value of B in Example 5 includes a factor of π. This actually happens quite frequently in applications of the trig functions.

EXAMPLE 5 Draw a sketch of $y = -2\cos(0.4\pi t)$ for $t \in [-\pi, 2\pi]$.

Solution: The amplitude is $|A| = 2$, so the reference rectangle will be $2(2) = 4$ units high. Since $A < 0$ the *graph will be vertically reflected across the t-axis*. The period is $P = \dfrac{2\pi}{0.4\pi} = 5$ (note the factors of π reduce to 1), so the frame will be 5 units in length. Breaking the t-axis into four parts within the frame gives $\left(\frac{1}{4}\right)5 = \frac{5}{4}$ units, indicating that we should scale the t-axis in multiples of $\frac{1}{4}$. In cases where the π factor reduces, we scale the t-axis as a "standard" number line, and *estimate the location of multiples of* π. For practical reasons, we first draw the unreflected graph (shown in blue) for guidance in drawing the reflected graph.

NOW TRY EXERCISES 19 THROUGH 30

D. Graphs of $y = \csc(Bt)$ and $y = \sec(Bt)$

The graphs of these reciprocal functions follow quite naturally from the graphs of $y = A\sin(Bt)$ and $y = A\cos(Bt)$, by using these observations: (1) you cannot divide by zero, (2) the reciprocal of a very small number is a very large number (and vice versa), and (3) the reciprocal of 1 is 1. Just as with rational functions, division by zero creates a vertical asymptote, so the graph of $y = \csc t = \frac{1}{\sin t}$ will have a vertical asymptote at every point where $\sin t = 0$. This occurs at $t = \pi k$, where k is an integer $(\ldots -2\pi, -\pi, 0, \pi, 2\pi, \ldots)$. Further, the graph of $y = \csc(Bt)$ will share the maximums and minimums of $y = \sin(Bt)$, since the reciprocal of 1 and -1 are still 1 and -1. Finally, due to observation 2, the graph of the cosecant function will be increasing when the sine function is decreasing, and decreasing when the sine function is increasing. In most cases, we graph $y = \csc(Bt)$ by drawing a sketch of $y = \sin(Bt)$, then using the preceding observations as demonstrated in Example 6. In doing so, we discover that the period of the cosecant function is also 2π.

EXAMPLE 6 Graph the function $y = \csc t$ for $t \in [0, 4\pi]$.

Solution: The related sine function is $y = \sin t$, which means we'll draw a rectangular frame $2A = 2$ units high. The period is $P = \frac{2\pi}{1} = 2\pi$, so the reference frame will be 2π units in length. Breaking the t-axis into four parts within the frame means each tick mark will be $\left(\frac{1}{4}\right)\left(\frac{2\pi}{1}\right) = \frac{\pi}{2}$ units apart, with the asymptotes occurring at 0, π, and 2π. A partial table and the resulting graph are shown.

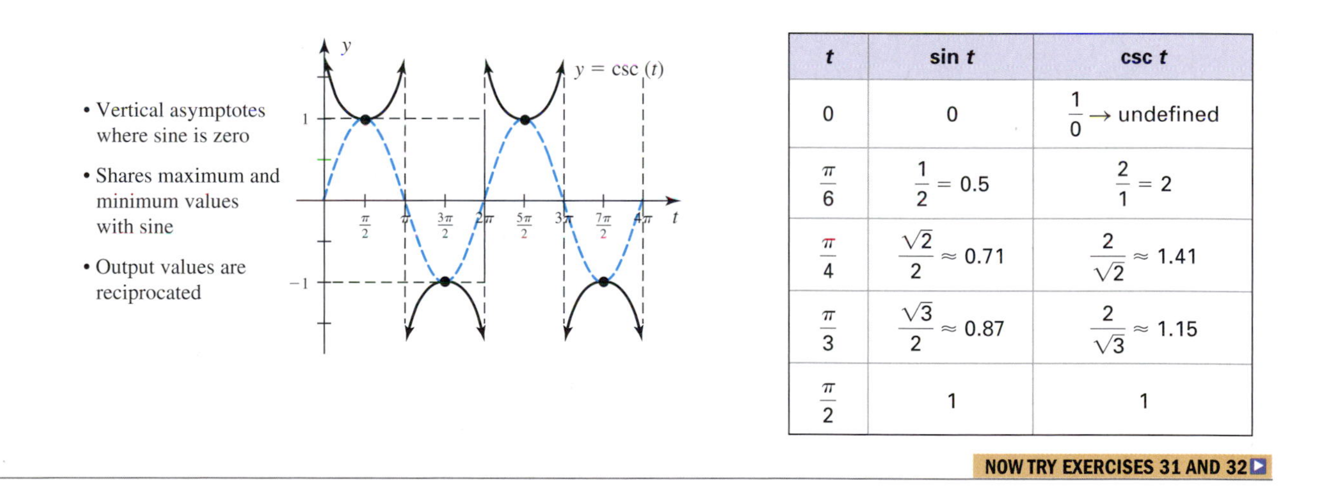

t	$\sin t$	$\csc t$
0	0	$\frac{1}{0} \rightarrow$ undefined
$\frac{\pi}{6}$	$\frac{1}{2} = 0.5$	$\frac{2}{1} = 2$
$\frac{\pi}{4}$	$\frac{\sqrt{2}}{2} \approx 0.71$	$\frac{2}{\sqrt{2}} \approx 1.41$
$\frac{\pi}{3}$	$\frac{\sqrt{3}}{2} \approx 0.87$	$\frac{2}{\sqrt{3}} \approx 1.15$
$\frac{\pi}{2}$	1	1

NOW TRY EXERCISES 31 AND 32

Similar observations can be made regarding $y = \sec(Bt)$ and its relationship to $y = \cos(Bt)$ (see Exercises 8, 33, and 34). The most important characteristics of the cosecant and secant functions are summarized in the following box. For these functions, there is no discussion of amplitude, and no mention is made of their zeroes since neither graph intersects the t-axis.

CHARACTERISTICS OF $y = \csc t$ and $y = \sec t$

$y = \csc t$		$y = \sec t$	
Unit Circle Definition	Period	Unit Circle Definition	Period
$\csc t = \frac{1}{y}$	2π	$\sec t = \frac{1}{x}$	2π
Domain	Range	Domain	Range
$t \neq k\pi$	$y \in (-\infty, -1] \cup [1, \infty)$	$t \neq \frac{\pi}{2} + \pi k$	$y \in (-\infty, -1] \cup [1, \infty)$

E. Writing Equations from Graphs

Mathematical concepts are best reinforced by working with them in both "forward and reverse." Where graphs are concerned, this means we should attempt to find the equation of a given graph, rather than only using an equation to sketch the graph. Exercises of this type require that you become very familiar with the graph's basic characteristics and how each is expressed as part of the equation.

EXAMPLE 7 The graph shown here is of the form $y = A\sin(Bt)$. Find the value of A and B.

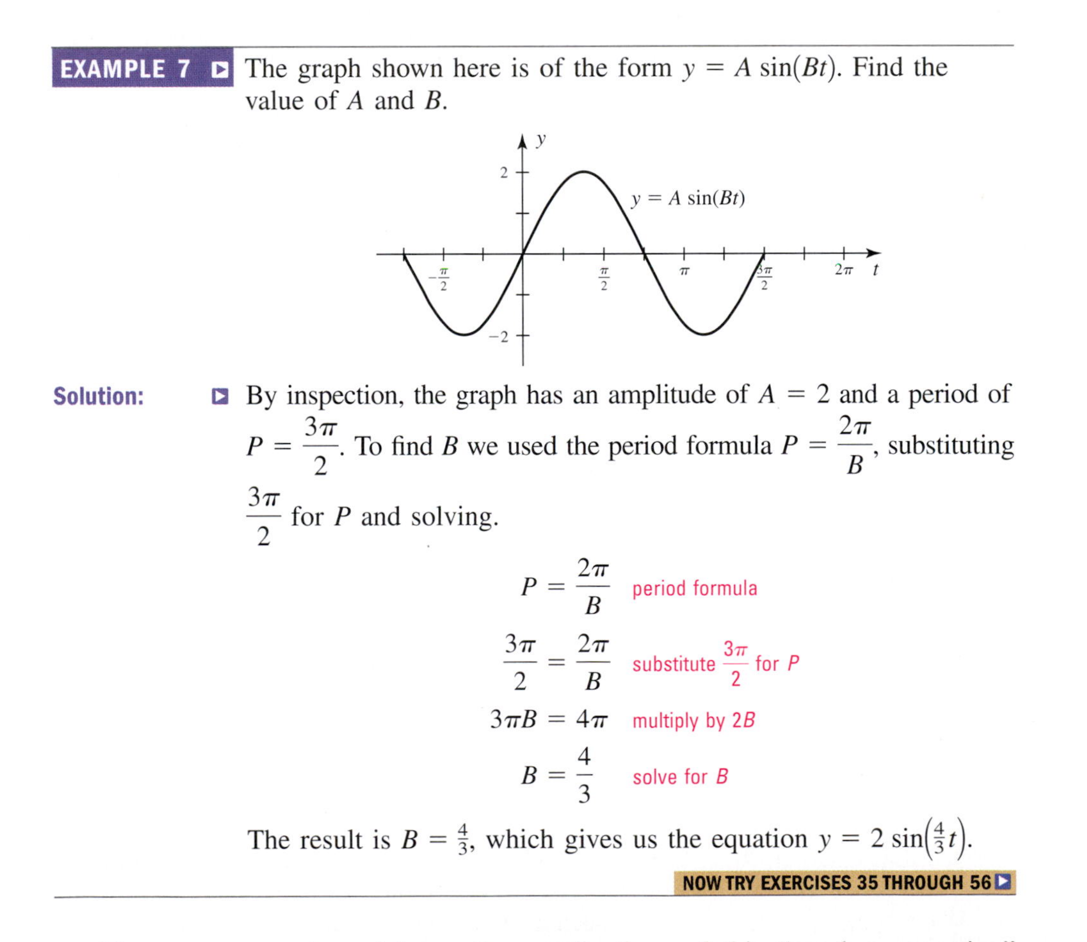

Solution: By inspection, the graph has an amplitude of $A = 2$ and a period of $P = \frac{3\pi}{2}$. To find B we used the period formula $P = \frac{2\pi}{B}$, substituting $\frac{3\pi}{2}$ for P and solving.

$$P = \frac{2\pi}{B} \quad \text{period formula}$$

$$\frac{3\pi}{2} = \frac{2\pi}{B} \quad \text{substitute } \frac{3\pi}{2} \text{ for } P$$

$$3\pi B = 4\pi \quad \text{multiply by } 2B$$

$$B = \frac{4}{3} \quad \text{solve for } B$$

The result is $B = \frac{4}{3}$, which gives us the equation $y = 2\sin\left(\frac{4}{3}t\right)$.

NOW TRY EXERCISES 35 THROUGH 56

There are a number of interesting applications of this "graph to equation" process in the exercise set. See Exercises 59 to 70.

TECHNOLOGY HIGHLIGHT

Exploring Amplitudes and Periods

The keystrokes shown apply to a TI-84 Plus model. Please consult your manual or our Internet site for other models.

In practice, trig applications offer an immense range of coefficients, creating amplitudes that are sometimes very large and sometimes extremely small, as well as periods ranging from nanoseconds, to many years. This *Technology Highlight* is designed to help you use the calculator more effectively in the study of these functions. To begin, we note the TI-84 Plus offers a preset ZOOM option that automatically sets a window size convenient to many trig graphs. The resulting WINDOW after pressing ZOOM **7:ZTrig** is shown in Figure 2.12 for a calculator set in **Radian** MODE.

Figure 2.12

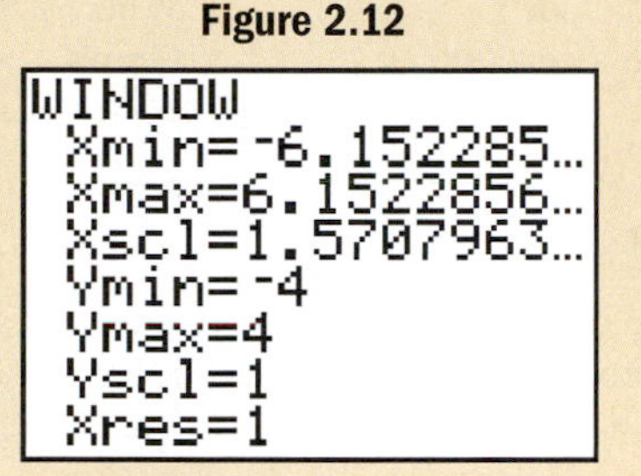

One important concept of Section 2.1 is that a change in amplitude will not change the location of the zeroes or max/min values. On the Y= screen, enter $Y_1 = \frac{1}{2}\sin x$, $Y_2 = \sin x$, $Y_3 = 2\sin x$, and $Y_4 = 4\sin x$, then use ZOOM **7:ZTrig** to graph the functions. As you see in Figure 2.13, each graph rises to the expected amplitude, while "holding on" to the zeroes (graph the functions in **Simul**taneous MODE).

Figure 2.13

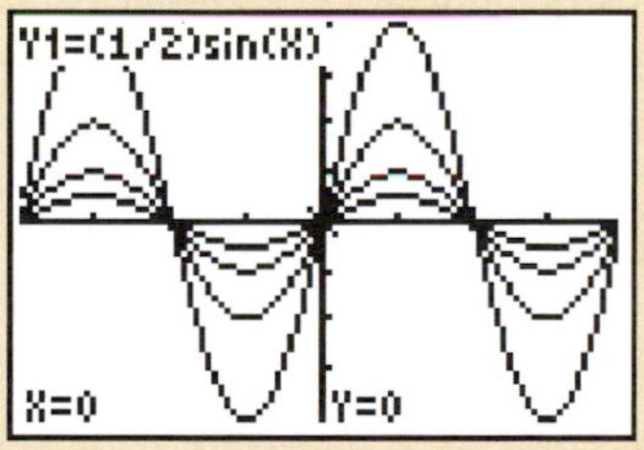

To explore concepts related to the coefficient B and its effect on the period of a trig function, enter $Y_1 = \sin\left(\frac{1}{2}x\right)$ and $Y_2 = \sin(2x)$ on the Y= screen and graph using ZOOM **7:ZTrig**. While the result is "acceptable," the graphs are difficult to read and compare, so we manually change the window size to obtain a better view (Figure 2.14).

Figure 2.14

After pressing GRAPH we can use TRACE and the up or down arrows to help identify each function.

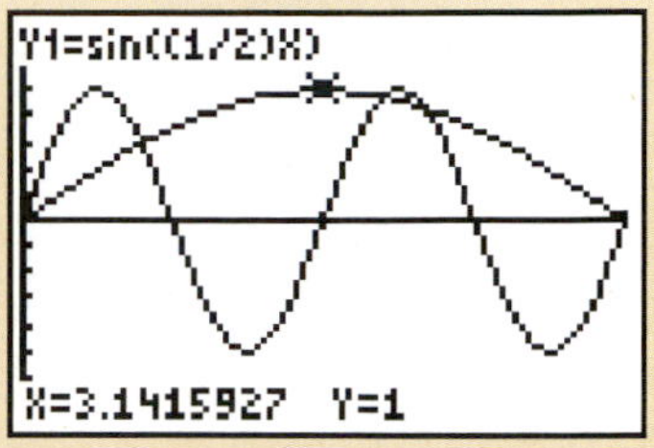

A true test of effective calculator use comes when the amplitude or period is a very large or very small number. For instance, the tone you hear while pressing "5" on your telephone is actually a combination of the tones modeled by $Y_1 = \sin[2\pi(770)t]$ and $Y_2 = \sin[2\pi(1336)t]$. Graphing these functions requires a careful analysis of the period, otherwise the graph can appear either garbled, misleading, or difficult to read—try graphing Y_1 on the ZOOM **7:ZTrig** or ZOOM **6:ZStandard** screens (see Figure 2.15). First note the amplitude is $A = 1$, and the period is $P = \frac{2\pi}{2\pi 770}$ or $\frac{1}{770}$. With a period this short, even graphing the function from Xmin $= -1$ to Xmax $= 1$ gives a distorted graph.

Figure 2.15

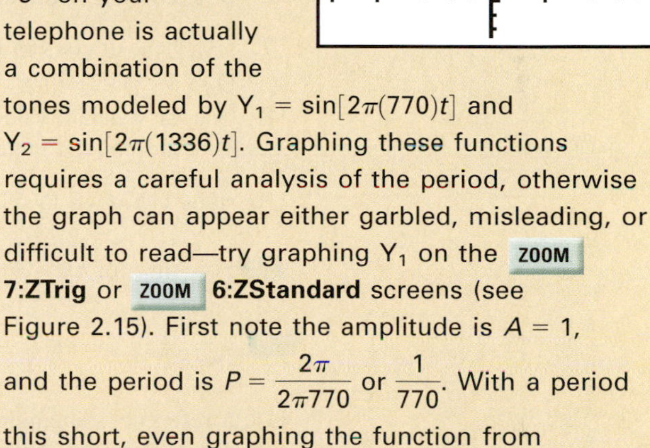

Setting Xmin to $-1/770$, Xmax to $1/770$, and Xscl to $(1/770)/10$ gives the graph in Figure 2.16, which can be used to investigate characteristics of the function.

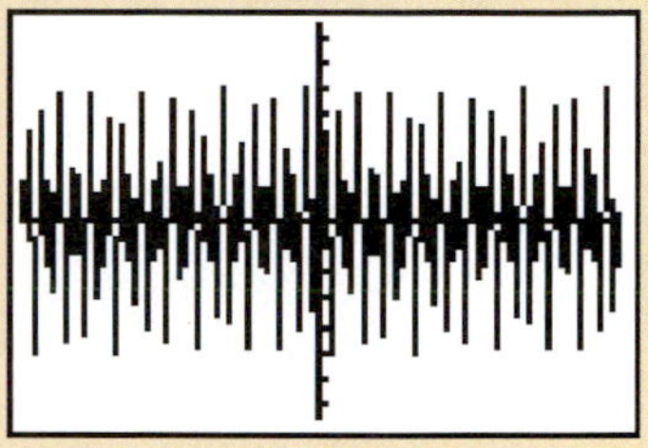

Figure 2.16

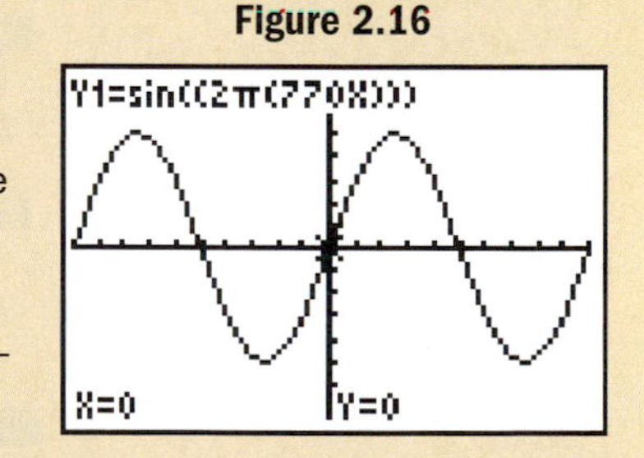

Exercise 1: Graph the second tone $Y_2 = \sin[2\pi(1336)t]$ mentioned here and find its value at $t = 0.00025$ sec.

Exercise 2: Graph the function $Y_1 = 950\sin(0.005t)$ on a "friendly" window and find the value at $x = 550$.

2.1 EXERCISES

CONCEPTS AND VOCABULARY

Fill in each blank with the appropriate word or phrase. Carefully reread the section if needed.

1. For the sine function, output values are __________ in the interval $\left[0, \frac{\pi}{2}\right]$.

2. For the cosine function, output values are __________ in the interval $\left[0, \frac{\pi}{2}\right]$.

3. For the sine and cosine functions, the domain is __________ and the range is __________.

4. The amplitude of sine and cosine is defined to be the maximum __________ from the __________ value in the positive and negative directions.

5. Discuss/describe the four-step process outlined in this section for the graphing of basic trig functions. Include a worked-out example and a detailed explanation.

6. Discuss/explain how you would determine the domain and range of $y = \sec x$. Where is this function undefined? Why? Graph $y = 2\sec(2t)$ using $y = 2\cos(2t)$. What do you notice?

DEVELOPING YOUR SKILLS

7. Use the symmetry of the unit circle and reference arcs of standard values to complete a table of values for $y = \cos t$ in the interval $t \in [\pi, 2\pi]$.

8. Use the standard values for $y = \cos t$ for $t \in [\pi, 2\pi]$ to create a table of values for $y = \sec t$ on the same interval.

Use steps I through IV given in this section to draw a sketch of

9. $y = \sin t$ for $t \in \left[-\frac{3\pi}{2}, \frac{\pi}{2}\right]$

10. $y = \sin t$ for $t \in [-\pi, \pi]$

11. $y = \cos t$ for $t \in \left[-\frac{\pi}{2}, 2\pi\right]$

12. $y = \cos t$ for $t \in \left[-\frac{\pi}{2}, \frac{5\pi}{2}\right]$

Use a reference rectangle and the *rule of fourths* to draw an accurate sketch of the following functions through two complete cycles—one where $t > 0$, and one where $t < 0$. Clearly state the amplitude and period as you begin.

13. $y = 3\sin t$

14. $y = 4\sin t$

15. $y = -2\cos t$

16. $y = -3\cos t$

17. $y = \frac{1}{2}\sin t$

18. $y = \frac{3}{4}\sin t$

19. $y = -\sin(2t)$

20. $y = -\cos(2t)$

21. $y = 0.8\cos(2t)$

22. $y = 1.7\sin(4t)$

23. $f(t) = 4\cos\left(\frac{1}{2}t\right)$

24. $y = -3\cos\left(\frac{3}{4}t\right)$

25. $f(t) = 3\sin(4\pi t)$

26. $g(t) = 5\cos(8\pi t)$

27. $y = 4\sin\left(\frac{5\pi}{3}t\right)$

28. $y = 2.5\cos\left(\frac{2\pi}{5}t\right)$

29. $f(t) = 2\sin(256\pi t)$

30. $g(t) = 3\cos(184\pi t)$

Draw the graph of each function by first sketching the related sine and cosine graphs, and applying the observations made in this section.

31. $y = 3\csc t$

32. $g(t) = 2\csc(4t)$

33. $y = 2\sec t$

34. $f(t) = 3\sec(2t)$

Clearly state the amplitude and period of each function, then match it with the corresponding graph.

35. $y = -2\cos(4t)$

36. $y = 2\sin(4t)$

37. $y = 3\sin(2t)$

38. $y = -3\cos(2t)$

39. $y = 2\csc\left(\frac{1}{2}t\right)$

40. $y = 2\sec\left(\frac{1}{4}t\right)$

41. $f(t) = \frac{3}{4}\cos(0.4t)$

42. $g(t) = \frac{7}{4}\cos(0.8t)$

43. $y = \sec(8\pi t)$

44. $y = \csc(12\pi t)$

45. $y = 4\sin(144\pi t)$

46. $y = 4\cos(72\pi t)$

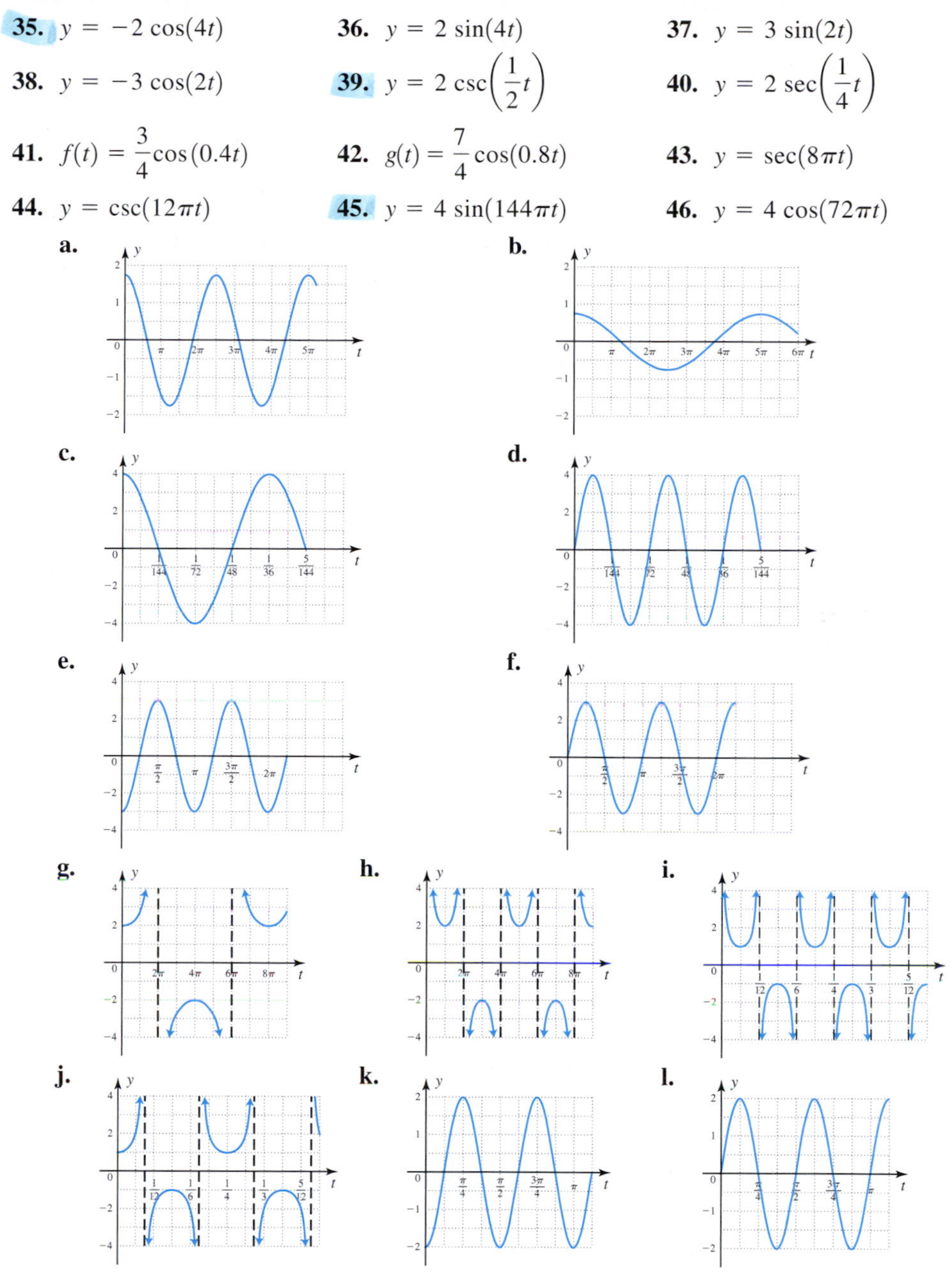

The graphs shown are of the form $y = A\cos(Bt)$ or $y = A\csc(Bt)$. Use the characteristics illustrated for each graph to determine its equation.

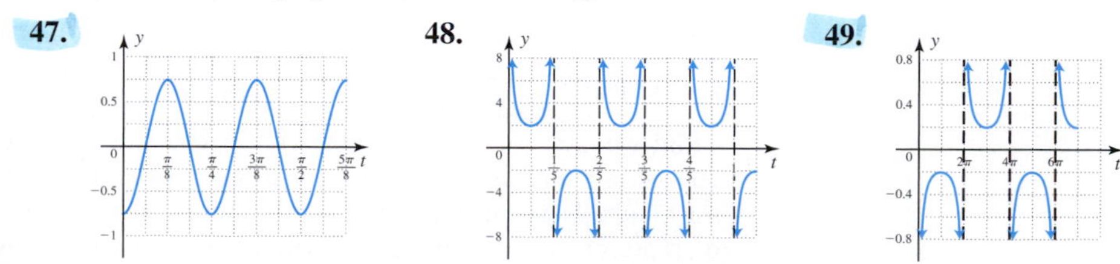

50. **51.** **52.**

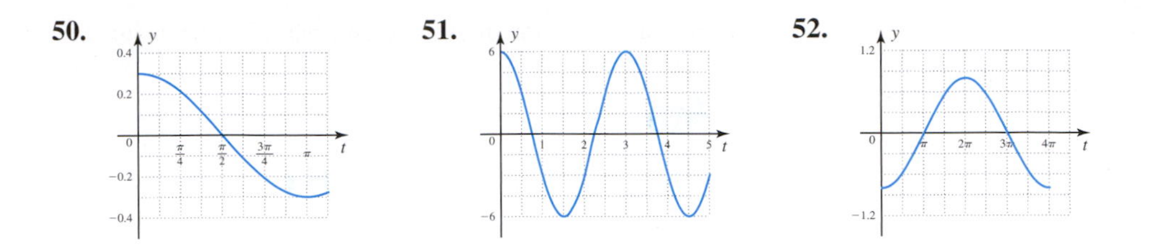

Match each graph to its equation, then graphically estimate the points of intersection. Confirm or contradict your estimate(s) by substituting the values into the given equations using a calculator.

53. $y = -\cos x;\ y = \sin x$

54. $y = -\cos x;\ y = \sin(2x)$

55. $y = -2\cos x;\ y = 2\sin(3x)$

56. $y = 2\cos(2\pi x);\ y = -2\sin(\pi x)$

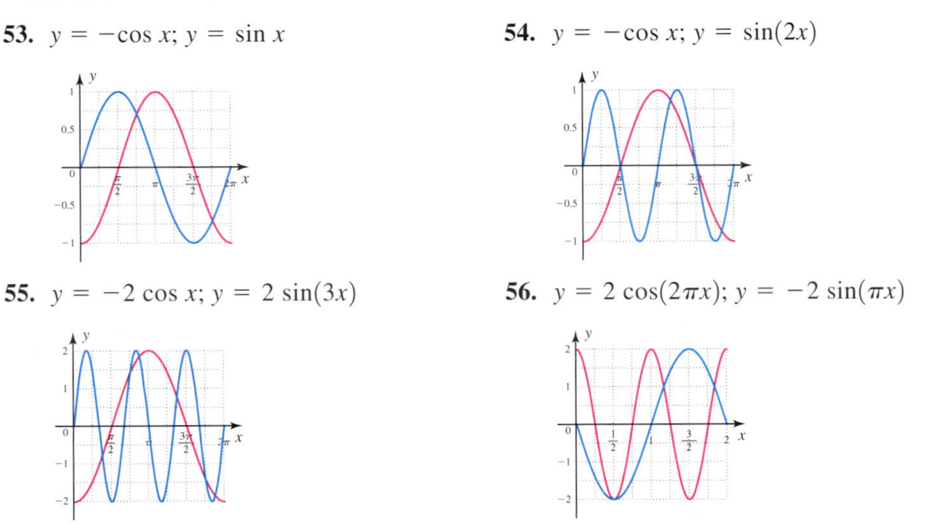

WORKING WITH FORMULAS

57. The Pythagorean theorem in trigonometric form: $\sin^2\theta + \cos^2\theta = 1$

The formula shown is commonly known as a Pythagorean identity and is introduced more formally in Chapter 3. It is derived by noting that on a unit circle, $\cos t = x$ and $\sin t = y$, while $x^2 + y^2 = 1$. Given that $\sin t = \frac{15}{113}$, use the formula to find the value of $\cos t$ in Quadrant I. What is the Pythagorean triple associated with these values of x and y?

Exercise 58

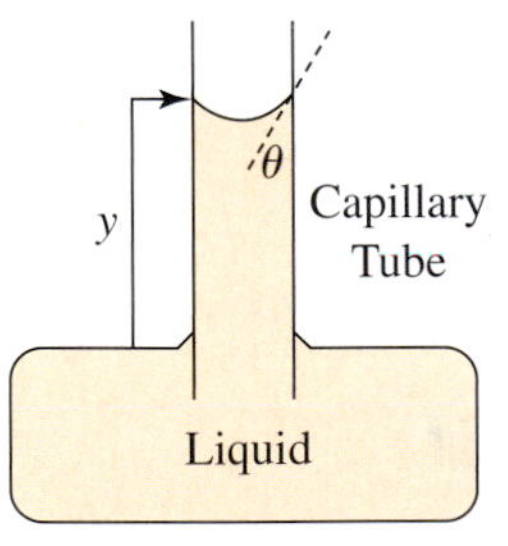

58. Hydrostatics, surface tension, and contact angles: $y = \dfrac{2\gamma\cos\theta}{kr}$

The height that a liquid will rise in a capillary tube is given by the formula shown, where r is the radius of the tube, θ is the contact angle of the liquid with the side of the tube (the meniscus), γ is the surface tension of the liquid-vapor film, and k is a constant that depends on the weight-density of the liquid. How high will the liquid rise given that the surface tension γ has a value of 0.2706, the tube has radius $r = 0.2$ cm, the contact angle θ is 22.5°, and the constant $k = 1.25$?

APPLICATIONS

Tidal waves: Tsunamis, also known as tidal waves, are ocean waves produced by earthquakes or other upheavals in the Earth's crust and can move through the water undetected for hundreds of miles at great speed. While traveling in the open ocean, these waves can be represented by a sine graph with a very long wavelength (period) and a very small amplitude. Tsunami waves only attain a monstrous size as they approach the shore, and represent a very different phenomenon than the ocean swells created by heavy winds over an extended period of time.

59. A graph modeling a tsunami wave is given in the figure. (a) What is the height of the tsunami wave (from crest to trough)? Note that $h = 0$ is considered the level of a calm ocean. (b) What is the tsunami's wavelength? (c) Find the equation for this wave.

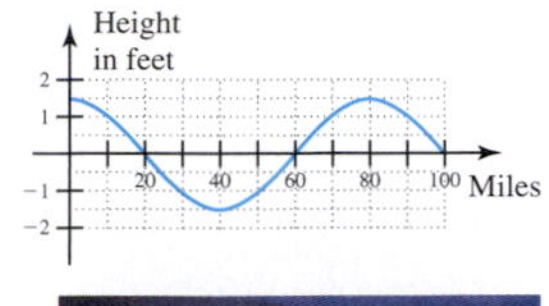

60. A heavy wind is kicking up ocean swells approximately 10 ft high (from crest to trough), with wavelengths of 250 ft. (a) Find the equation that models these swells. (b) Graph the equation. (c) Determine the height of a wave measured 200 ft from the trough of the previous wave.

Sinusoidal models: The sine and cosine functions are of great importance to meteorological studies, as when modeling the temperature based on the time of day, the illumination of the Moon as it goes through its phases, or even the prediction of tidal motion.

61. The graph given shows the deviation from the average daily temperature for the hours of a given day, with $t = 0$ corresponding to 6 A.M. (a) Use the graph to determine the related equation. (b) Use the equation to find the deviation at $t = 11$ (5 P.M.) and confirm that this point is on the graph. (c) If the average temperature for this day was 72°, what was the temperature at midnight?

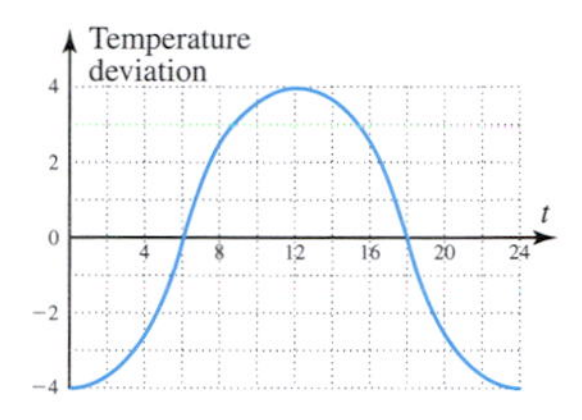

62. The equation $y = 7\sin\left(\frac{\pi}{6}t\right)$ models the height of the tide along a certain coastal area, as compared to average sea level. Assuming $t = 0$ is midnight, (a) graph this function over a 12-hr period. (b) What will the height of the tide be at 5 A.M.? (c) Is the tide rising or falling at this time?

Sinusoidal movements: Many animals exhibit a wavelike motion in their movements, as in the tail of a shark as it swims in a straight line or the wingtips of a large bird in flight. Such movements can be modeled by a sine or cosine function and will vary depending on the animal's size, speed, and other factors.

63. The graph shown models the position of a shark's tail at time t, as measured to the left (negative) and right (positive) of a straight line along its length. (a) Use the graph to determine the related equation. (b) Is the tail to the right, left, or at center when $t = 6.5$ sec? How far? (c) Would you say the shark is "swimming leisurely," or "chasing its prey"? Justify your answer.

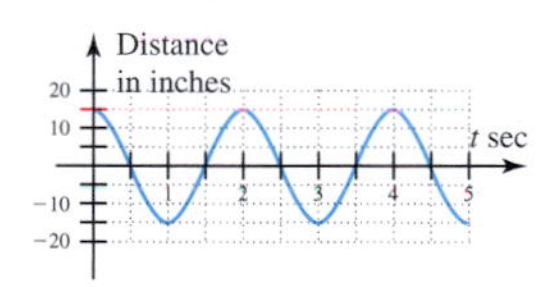

64. The State Fish of Hawaii is the *humuhumunukunukuapua'a*, a small colorful fish found abundantly in coastal waters. Suppose the tail motion of an adult fish is modeled by the equation $d(t) = \sin(15\pi t)$ with $d(t)$ representing the position of the fish's tail at time t, as measured in inches to the left (negative) or right (positive) of a straight line along its length. (a) Graph the equation over two periods. (b) Is the tail to the left or right of center at $t = 2.7$ sec? How far? (c) Would you say this fish is "swimming leisurely," or "running for cover"? Justify your answer.

Kinetic energy: The kinetic energy a planet possesses as it orbits the Sun can be modeled by a cosine function. When the planet is at its apogee (greatest distance from the Sun), its kinetic energy is at its lowest point as it slows down and "turns around" to head back toward the Sun. The kinetic energy is at its highest when the planet "whips around the Sun" to begin a new orbit.

65. Two graphs are given here. (a) Which of the graphs could represent the kinetic energy of a planet orbiting the Sun if the planet is at its perigee (closest distance to the Sun) when $t = 0$? (b) For what value(s) of t does this planet possess 62.5% of its maximum kinetic energy with the kinetic energy increasing? (c) What is the orbital period of this planet?

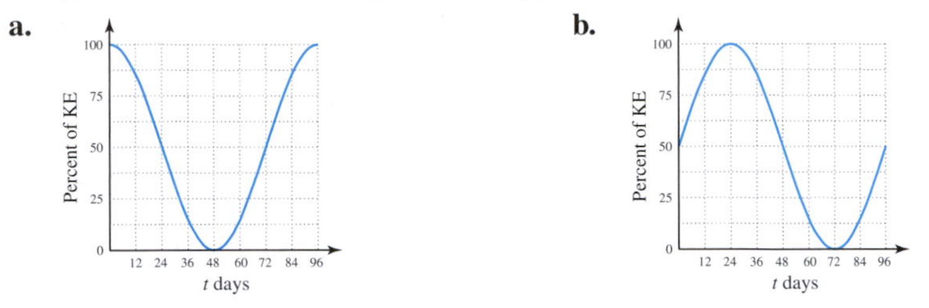

66. The *potential energy* of the planet is the antipode of its kinetic energy, meaning when kinetic energy is at 100%, the potential energy is 0%, and when kinetic energy is at 0% the potential energy is at 100%. (a) How is the graph of the kinetic energy related to the graph of the potential energy? In other words, what transformation could be applied to the kinetic energy graph to obtain the potential energy graph? (b) If the kinetic energy is at 62.5% and increasing [as in Graph 65(b)], what can be said about the potential energy in the planet's orbit at this time?

Visible light: One of the narrowest bands in the electromagnetic spectrum is the region involving visible light. The wavelengths (periods) of visible light vary from 400 nanometers (purple/violet colors) to 700 nanometers (bright red). The approximate wavelengths of the other colors are shown in the diagram.

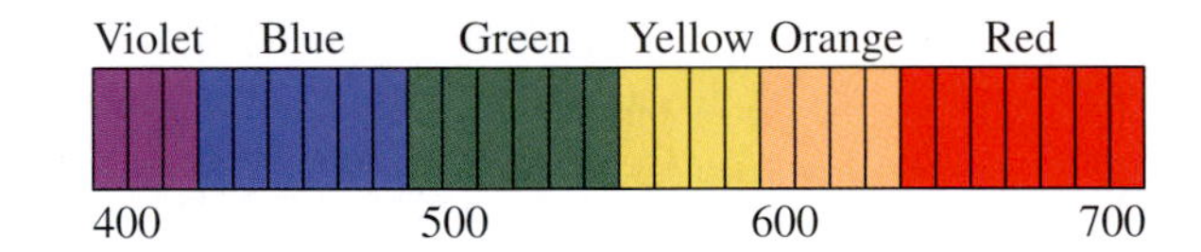

67. The equations for the colors in this spectrum have the form $y = \sin(\gamma t)$, where $\frac{2\pi}{\gamma}$ gives the length of the sine wave. (a) What color is represented by the equation $y = \sin\left(\frac{\pi}{240}t\right)$? (b) What color is represented by the equation $y = \sin\left(\frac{\pi}{310}t\right)$?

68. Name the color represented by each of the graphs (a) and (b) here and write the related equation.

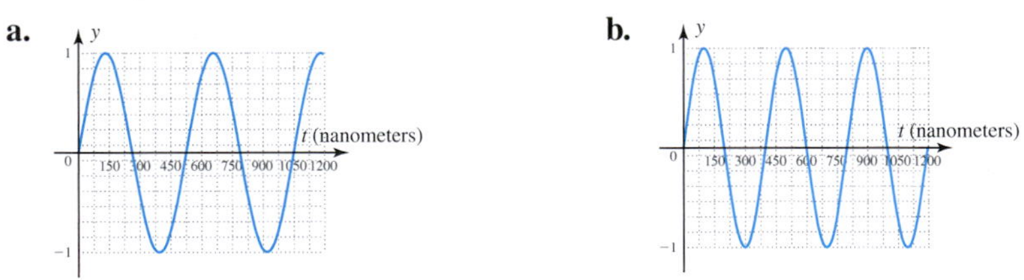

Alternating current: Surprisingly, even characteristics of the electric current supplied to your home can be modeled by sine or cosine functions. For alternating current (AC), the amount of current I (in amps) at time t can be modeled by $I = A\sin(\omega t)$, where A represents the maximum current that is produced, and ω is related to the frequency at which the generators turn to produce the current.

Exercise 69

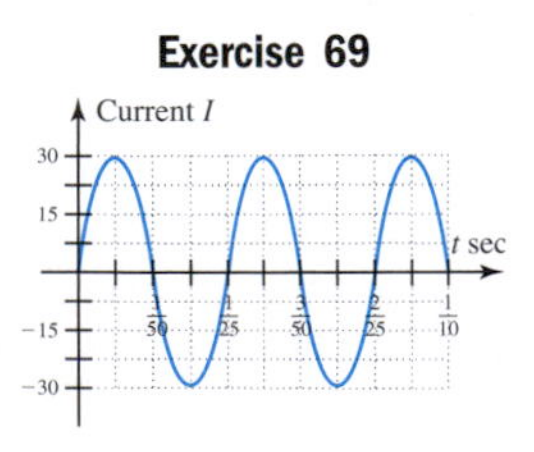

69. Find the equation of the household current modeled by the graph, then use the equation to determine I when $t = 0.045$ sec. Verify that the resulting ordered pair is on the graph.

70. If the *voltage* produced by an AC circuit is modeled by the equation $E = 155\sin(120\pi t)$, (a) what is the period and amplitude of the related graph? (b) What voltage is produced when $t = 0.2$?

WRITING, RESEARCH, AND DECISION MAKING

71. For $y = A\sin(Bx)$ and $y = A\cos(Bx)$, the expression $\frac{M + m}{2}$ gives the average value of the function, where M and m represent the maximum and minimum values respectively. What was the average value of every function graphed in this section? Compute a table of values for the function $y = 2\sin t + 3$, and note its maximum and minimum values. What is the average value of this function? What transformation has been applied to change the average value of the function? Can you name the average value of $y = -2\cos t + 1$ by inspection? How is the amplitude related to this average value? (*Hint:* Graph the horizontal line $y = \frac{M + m}{2}$ on the same grid.)

72. Use the Internet or the resources of a local library to do some research on tsunamis (see Exercises 59 and 60). Attempt to find information on (a) exactly how they are generated; (b) some of the more notable tsunamis in history; (c) the average amplitude of a tsunami; (d) their average wavelength or period; and (e) the speeds they travel in the open ocean. Prepare a short summary of what you find.

73. To understand where the period formula $P = \frac{2\pi}{B}$ came from, consider that if $B = 1$, the graph of $y = \sin(Bt) = \sin(1t)$ completes one cycle from $1t = 0$ to $1t = 2\pi$. If $B \neq 1$, $y = \sin(Bt)$ completes one cycle from $Bt = 0$ to $Bt = 2\pi$. Discuss how this observation validates the period formula.

74. Horizontal stretches and compressions are remarkably similar for all functions. Use a graphing calculator to graph the functions $Y_1 = \sin x$ and $Y_2 = \sin(2x)$ on a ZOOM **7:ZTrig** screen (graph Y_1 in **bold**). Note that Y_2 completes two periods in the time it takes Y_1 to complete 1 period (the graph of Y_2 is horizontally compressed). Now enter $Y_1 = (x^2 - 1)(x^2 - 4)$ and $Y_2 = ([2x]^2 - 1)([2x]^2 - 4)$, substituting $2x$ for x as before. After graphing these on a ZOOM **4:ZDecimal** screen, what do you notice? How do these algebraic graphs compare to the trigonometric graphs?

EXTENDING THE CONCEPT

75. The tone you hear when pressing the digit "9" on your telephone is actually a combination of two separate tones, which can be modeled by the functions $f(t) = \sin[2\pi(852)t]$ and $g(t) = \sin[2\pi(1477)t]$. Which of the two functions has the shortest period? By carefully scaling the axes, graph the function having the shorter period using the steps I through IV discussed in this section.

76. Consider the functions $f(x) = \sin(|t|)$ and $g(x) = |\sin t|$. For one of these functions, the portion of the graph below the x-axis is reflected above the x-axis, creating a series of "humps." For the other function, the graph obtained for $t > 0$ is reflected across the y-axis to form the graph for $t < 0$. After a thoughtful contemplation (without actually graphing), try to decide which description fits $f(x)$ and which fits $g(x)$, justifying your thinking. Then confirm or contradict your guess using a graphing calculator.

MAINTAINING YOUR SKILLS

77. (1.3) Given $\sin 1.12 \approx 0.9$, find an additional value of t in $[0, 2\pi)$ that makes the equation $\sin t \approx 0.9$ true.

78. (1.1) Invercargill, New Zealand, is at $46°14'24''$ south latitude. If the Earth has a radius of 3960 mi, how far is Invercargill from the equator?

79. (1.3) Given $\cos t = \frac{28}{53}$ with $\tan t < 0$: (a) find the related values of x, y, and r; (b) state the quadrant of the terminal side; and (c) give the value of the other five trig functions of t.

80. (1.1) Use a standard triangle to calculate the distance from the ball to the pin on the seventh hole, given the ball is in a straight line with the 100-yd plate, as shown in the figure.

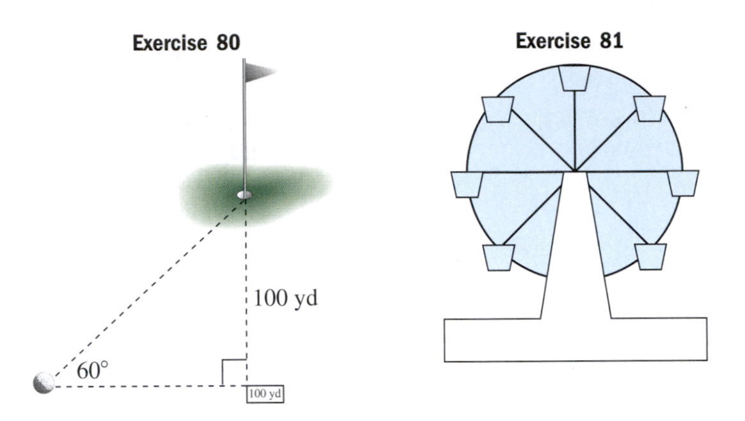

81. (1.1) The Ferris wheel shown has a radius of 25 ft and is turning at a rate of 14 rpm. (a) What is the angular velocity in radians? (b) What distance does a seat on the rim travel as the Ferris wheel turns through an angle of 225°? (c) What is the linear velocity (in miles per hour) of a person sitting in a seat at the rim of the Ferris wheel?

82. (1.2) The world's tallest unsupported flagpole was erected in 1985 in Vancouver, British Columbia. Standing 60 ft from the base of the pole, the angle of elevation is 78°. How tall is the pole?

2.2 Graphs of the Tangent and Cotangent Functions

LEARNING OBJECTIVES

In Section 2.2 you will learn how to:

A. **Graph $y = \tan t$ using asymptotes, zeroes, and the ratio $\frac{\sin t}{\cos t}$**

B. **Graph $y = \cot t$ using asymptotes, zeroes, and the ratio $\frac{\cos t}{\sin t}$**

C. **Identify and discuss important characteristics of $y = \tan t$ and $y = \cot t$**

D. **Graph $y = A\tan(Bt)$ and $y = A\cot(Bt)$ with various values of A and B**

E. **Solve applications of $y = \tan t$ and $y = \cot t$**

INTRODUCTION

Unlike the other four trig functions, tangent and cotangent have no maximum or minimum value on any open interval of their domain. However, it is precisely this unique feature that adds to their value as mathematical models. Collectively, the six functions give scientists the tools they need to study, explore, and investigate a wide range of phenomena, extending our understanding of the world around us.

POINT OF INTEREST

The sine and cosine functions evolved from studies of the chord lengths within a circle and their applications to astronomy. Although we know today they are related to the tangent and cotangent functions, it appears the latter two developed quite independently of this context. When time was measured by the length of the shadow cast by a vertical stick, observers noticed the shadow was extremely long at sunrise, cast no shadow at noon, and returned to extreme length at sunset. Records tracking shadow length in this way date back as far as 1500 B.C. in ancient Egypt, and are the precursor of our modern tangent and cotangent functions.

A. The Graph of $y = \tan t$

Like the secant and cosecant functions, tangent is defined in terms of a ratio that creates asymptotic behavior at the zeroes of the denominator. In terms of the unit circle, $\tan t = \frac{y}{x}$, which means in $[-\pi, 2\pi]$, vertical asymptotes occur at $t = -\frac{\pi}{2}$, $t = \frac{\pi}{2}$, and $\frac{3\pi}{2}$, since the x-coordinate on the unit circle is zero. We further note $\tan t = 0$ when the y-coordinate is zero, so the function will have t-intercepts at $t = -\pi, 0, \pi$, and 2π in the same interval. This produces the framework for graphing the tangent function shown in Figure 2.17.

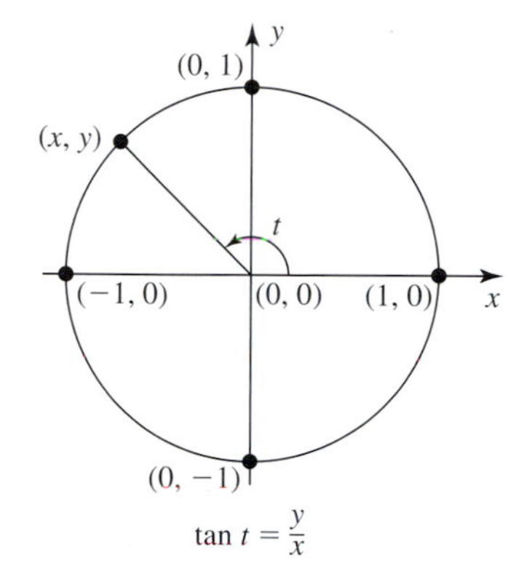

Figure 2.17

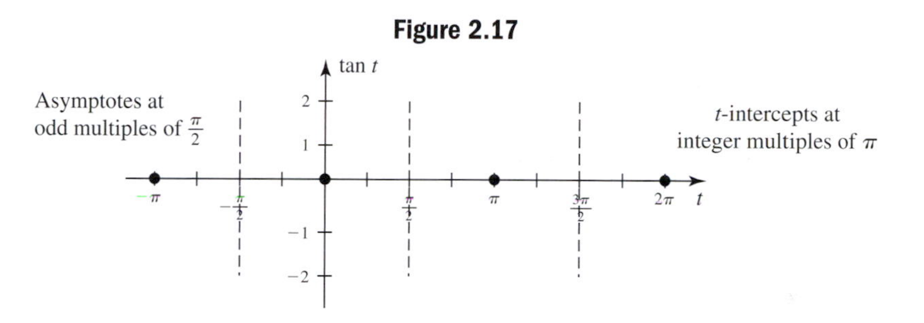

Knowing the graph must go through these zeroes and approach the asymptotes, we are left with determining the *direction of the approach.* This can be discovered by noting that in QI, the y-coordinates of points on the unit circle start at 0 and increase, while the x-values start at 1 and decrease. This means the ratio $\frac{y}{x}$ defining $\tan t$ is increasing, and in fact becomes infinitely large as t gets very close to $\frac{\pi}{2}$. A similar observation can be made for a negative rotation of t in QIV. Using the additional points provided by $\tan\left(-\frac{\pi}{4}\right) = -1$ and $\tan\left(\frac{\pi}{4}\right) = 1$, we find the graph of $\tan t$ is increasing throughout the interval $\left(-\frac{\pi}{2}, \frac{\pi}{2}\right)$ and that the function has a period of π. We also note $y = \tan t$ is an odd function (symmetric about the origin), since $\tan(-t) = -\tan t$ as evidenced by the two points just computed. The completed graph is shown in Figure 2.18 with the primary interval in red.

Figure 2.18

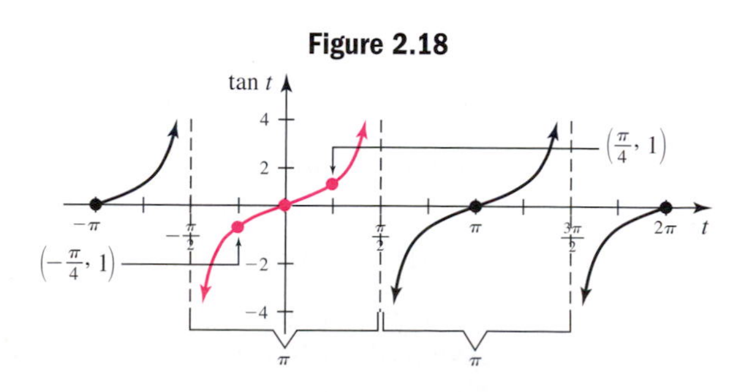

The graph can also be developed by noting the ratio relationship that exists between $\sin t$, $\cos t$, and $\tan t$. In particular, since $\sin t = y$, $\cos t = x$, and $\tan t = \frac{y}{x}$, we have $\tan t = \frac{\sin t}{\cos t}$ by direct substitution. These and other relationships between the trig functions will be fully explored in Chapter 3.

EXAMPLE 1 Complete Table 2.8 shown for $\tan t = \frac{y}{x}$ using the values given for $\sin t$ and $\cos t$.

Table 2.8

t	0	$\frac{\pi}{6}$	$\frac{\pi}{4}$	$\frac{\pi}{3}$	$\frac{\pi}{2}$	$\frac{2\pi}{3}$	$\frac{3\pi}{4}$	$\frac{5\pi}{6}$	π
$\sin t = y$	0	$\frac{1}{2}$	$\frac{\sqrt{2}}{2}$	$\frac{\sqrt{3}}{2}$	1	$\frac{\sqrt{3}}{2}$	$\frac{\sqrt{2}}{2}$	$\frac{1}{2}$	0
$\cos t = x$	1	$\frac{\sqrt{3}}{2}$	$\frac{\sqrt{2}}{2}$	$\frac{1}{2}$	0	$-\frac{1}{2}$	$-\frac{\sqrt{2}}{2}$	$-\frac{\sqrt{3}}{2}$	-1
$\tan t = \frac{y}{x}$									

Solution: For the noninteger values of x and y, the "twos will cancel" each time we compute $\frac{y}{x}$. This means we can simply list the ratio of numerators. The results are shown in Table 2.9.

Table 2.9

t	0	$\frac{\pi}{6}$	$\frac{\pi}{4}$	$\frac{\pi}{3}$	$\frac{\pi}{2}$	$\frac{2\pi}{3}$	$\frac{3\pi}{4}$	$\frac{5\pi}{6}$	π
$\sin t = y$	0	$\frac{1}{2}$	$\frac{\sqrt{2}}{2}$	$\frac{\sqrt{3}}{2}$	1	$\frac{\sqrt{3}}{2}$	$\frac{\sqrt{2}}{2}$	$\frac{1}{2}$	0
$\cos t = x$	1	$\frac{\sqrt{3}}{2}$	$\frac{\sqrt{2}}{2}$	$\frac{1}{2}$	0	$-\frac{1}{2}$	$-\frac{\sqrt{2}}{2}$	$-\frac{\sqrt{3}}{2}$	-1
$\tan t = \frac{y}{x}$	0	$\frac{1}{\sqrt{3}} \approx 0.58$	1	$\sqrt{3} \approx 1.7$	undefined	$-\sqrt{3}$	-1	$-\frac{1}{\sqrt{3}}$	0

NOW TRY EXERCISES 7 AND 8

Additional values can be found using a calculator as needed. For future use and reference, it will help to recognize the approximate decimal equivalent of all standard values and radian angles. In particular, note that $\sqrt{3} \approx 1.73$ and $\frac{1}{\sqrt{3}} \approx 0.58$. See Exercises 9 through 14.

B. The Graph of $y = \cot t$

Since the cotangent function is also defined in terms of a ratio, it too displays asymptotic behavior at the zeroes of the denominator, with t-intercepts at the zeroes of the numerator. Like the tangent function, $\cot t = \frac{x}{y}$ can be written in terms of $\cos t = x$ and $\sin t = y$: $\cot t = \frac{\cos t}{\sin t}$, and the graph obtained by plotting points as in Example 2.

EXAMPLE 2 Complete a table of values for $\cot t = \frac{x}{y}$ for $t \in [0, \pi]$ using its ratio relationship with cos t and sin t. Use the results to verify the location of the asymptotes and plotted points, then graph the function for $t \in (-\pi, 2\pi)$.

Solution: The completed table is shown here. In this interval, the cotangent function has asymptotes at 0 and π since $y = 0$ at these points, and has a t-intercept at $\frac{\pi}{2}$ since $x = 0$. The graph shown in Figure 2.19 was completed using the period $P = \pi$.

t	0	$\frac{\pi}{6}$	$\frac{\pi}{4}$	$\frac{\pi}{3}$	$\frac{\pi}{2}$	$\frac{2\pi}{3}$	$\frac{3\pi}{4}$	$\frac{5\pi}{6}$	π
$\sin t = y$	0	$\frac{1}{2}$	$\frac{\sqrt{2}}{2}$	$\frac{\sqrt{3}}{2}$	1	$\frac{\sqrt{3}}{2}$	$\frac{\sqrt{2}}{2}$	$\frac{1}{2}$	0
$\cos t = x$	1	$\frac{\sqrt{3}}{2}$	$\frac{\sqrt{2}}{2}$	$\frac{1}{2}$	0	$-\frac{1}{2}$	$-\frac{\sqrt{2}}{2}$	$-\frac{\sqrt{3}}{2}$	-1
$\cot t = \frac{x}{y}$	undefined	$\sqrt{3}$	1	$\frac{1}{\sqrt{3}}$	0	$-\frac{1}{\sqrt{3}}$	-1	$-\sqrt{3}$	undefined

Figure 2.19

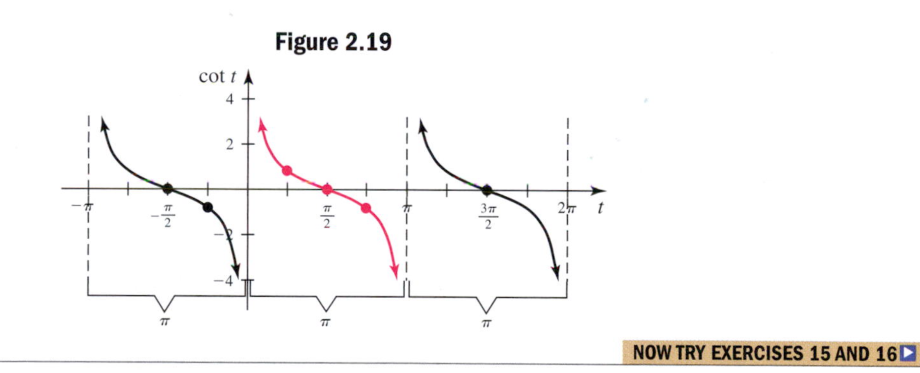

NOW TRY EXERCISES 15 AND 16

C. Characteristics of $y = \tan t$ and $y = \cot t$

The most important characteristics of the tangent and cotangent functions are summarized in the following box. There is no discussion of amplitude, maximum, or minimum values, since maximum or minimum values do not exist. For future use and reference, perhaps the most significant characteristic distinguishing tan t from cot t is that *tan t increases,* while *cot t decreases* over their respective domains. Also note that due to symmetry, the zeroes of each function are always located halfway between the asymptotes.

CHARACTERISTICS OF $y = \tan t$ and $y = \cot t$

$y = \tan t$

Unit Circle Definition	Domain	Range
$\tan t = \frac{y}{x}$	$t \neq \frac{(2k+1)\pi}{2}$ $k \in \mathbb{Z}$	$y \in (-\infty, \infty)$
Period	**Behavior**	**Symmetry**
π	increasing	Odd $\tan(-t) = -\tan t$

$y = \cot t$

Unit Circle Definition	Domain	Range
$\cot t = \frac{x}{y}$	$t \neq k\pi$ $k \in \mathbb{Z}$	$y \in (-\infty, \infty)$
Period	**Behavior**	**Symmetry**
π	decreasing	Odd $\cot(-t) = -\cot t$

EXAMPLE 3 Given $\tan\left(\frac{\pi}{6}\right) = \frac{1}{\sqrt{3}}$, what can you say about $\tan\left(\frac{7\pi}{6}\right)$, $\tan\left(\frac{13\pi}{6}\right)$, and $\tan\left(-\frac{5\pi}{6}\right)$?

Solution: Each of the arguments differs by a multiple of π: $\tan\left(\frac{7\pi}{6}\right) = \tan\left(\frac{\pi}{6} + \pi\right)$, $\tan\left(\frac{13\pi}{6}\right) = \tan\left(\frac{\pi}{6} + 2\pi\right)$ and $\tan\left(-\frac{5\pi}{6}\right) = \tan\left(\frac{\pi}{6} - \pi\right)$. Since the period of the tangent function is $P = \pi$, all of these expressions have a value of $\frac{1}{\sqrt{3}}$.

NOW TRY EXERCISES 17 THROUGH 22

Since the tangent function is more common than the cotangent, many needed calculations will first be done using the tangent function and its properties, then reciprocated. For instance, to evaluate $\cot\left(-\frac{\pi}{6}\right)$ we reason that $\cot t$ is an odd function, so $\cot\left(-\frac{\pi}{6}\right) = -\cot\left(\frac{\pi}{6}\right)$. Since cotangent is the reciprocal of tangent and $\tan\left(\frac{\pi}{6}\right) = \frac{1}{\sqrt{3}}$, $-\cot\left(\frac{\pi}{6}\right) = -\sqrt{3}$. See Exercises 23 and 24.

D. Graphing $y = A\tan(Bt)$ and $y = A\cot(Bt)$

The Coefficient A: Vertical Stretches and Compressions

For the tangent and cotangent functions, the role of coefficient A is best seen through an analogy from basic algebra (the concept of amplitude is foreign to these functions). Consider the graph of $y = x^3$ (Figure 2.20), which you may recall has the appearance of a vertical propeller. Comparing the parent function $y = x^3$ with functions $y = Ax^3$, the graph is stretched vertically if $|A| > 1$ (see Figure 2.21) and compressed if $0 < |A| < 1$. In the latter case the graph becomes very "flat" near the zeroes, as shown in Figure 2.22.

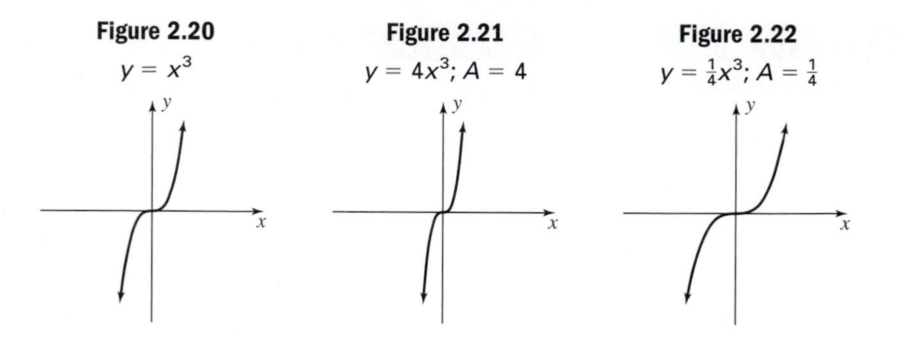

Figure 2.20 $y = x^3$

Figure 2.21 $y = 4x^3;\ A = 4$

Figure 2.22 $y = \frac{1}{4}x^3;\ A = \frac{1}{4}$

While ***cubic functions are not asymptotic,*** they are a good illustration of A's effect on the tangent and cotangent functions. Fractional values of A ($|A| < 1$) compress the graph, flattening it out near its zeroes. Numerically, this is because a fractional part of a small quantity is an even smaller quantity. For instance, compare $\tan\left(\frac{\pi}{6}\right)$ with $\frac{1}{4}\tan\left(\frac{\pi}{6}\right)$. To two decimal places, $\tan\left(\frac{\pi}{6}\right) = 0.57$, while $\frac{1}{4}\tan\left(\frac{\pi}{6}\right) = 0.14$, so the graph must be "nearer the t-axis" at this value.

EXAMPLE 4 Draw a "comparative sketch" of $y = \tan t$ and $y = \frac{1}{4}\tan t$ on the same axis and discuss similarities and differences. Use the interval $t \in [-\pi, 2\pi]$.

Solution: Both graphs will maintain their essential features (zeroes, asymptotes, period increasing, and so on). However, the graph of $y = \frac{1}{4}\tan t$ is vertically compressed, causing it to flatten out near its zeroes and changing how the graph approaches its asymptotes in each interval.

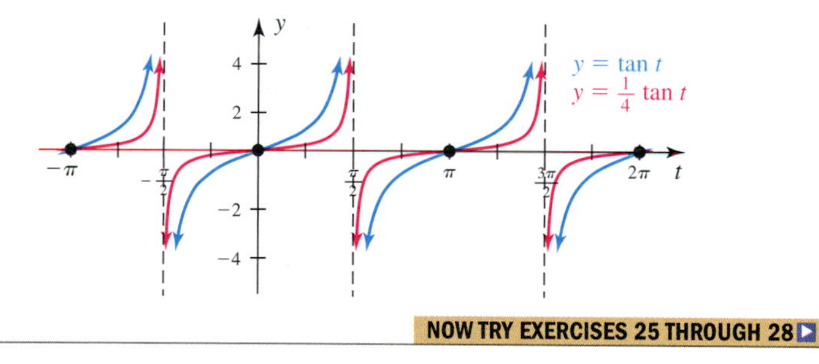

NOW TRY EXERCISES 25 THROUGH 28

WORTHY OF NOTE

It may be easier to interpret the phrase "twice as fast" as $2P = \pi$ and "one-half as fast" as $\frac{1}{2}P = \pi$. In each case, solving for P gives the correct interval for the period of the new function.

The Coefficient B: The Period of Tangent and Cotangent

Like the other trig functions, the value of B has a material impact on the period of the function, and with the same effect. The graph of $y = \cot(2t)$ completes a cycle twice as fast as $y = \cot t\left(P = \frac{\pi}{2} \text{ versus } P = \pi\right)$, while $y = \cot\left(\frac{1}{2}t\right)$ completes a cycle one-half as fast ($P = 2\pi$ versus $P = \pi$).

This type of reasoning leads us to a **period formula** for tangent and cotangent, namely, $P = \frac{\pi}{B}$, where B is the coefficient of the input variable.

WORTHY OF NOTE

Similar to the four-step process used to graph sine and cosine functions, we can graph tangent and cotangent functions using a rectangle $P = \frac{\pi}{B}$ units in length and $2A$ units high, centered on the primary interval. After dividing the length of the rectangle into fourths, the t-intercept will always be the halfway point, with y-values of $|A|$ occuring at the $\frac{1}{4}$ and $\frac{3}{4}$ marks. See Example 5.

EXAMPLE 5 ▶ Sketch the graph of $y = 3\cot(2t)$ over the interval $[-\pi, \pi]$.

Solution: ▶ For $y = 3\cot(2t)$, $A = 3$ which results in a vertical stretch, and $B = 2$ which gives a period of $\frac{\pi}{2}$. The function is still undefined at $t = 0$ and is asymptotic there, then at all integer multiples of $P = \frac{\pi}{2}$. Selecting the inputs $t = \frac{\pi}{8}$ and $t = \frac{3\pi}{8}$ yields the points $\left(\frac{\pi}{8}, 3\right)$ and $\left(\frac{3\pi}{8}, -3\right)$, which we'll use along with the period and symmetry of the function to complete the graph:

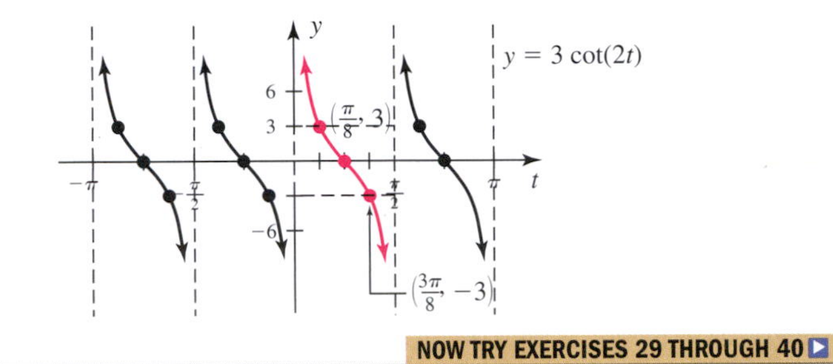

NOW TRY EXERCISES 29 THROUGH 40 ▶

As with the trig functions from Section 2.1, it is possible to determine the equation of a tangent or cotangent function from a given graph. Where previously we noted the amplitude, period, and max/min values to obtain our equation, here we first determine the period of the function by calculating the "distance" between asymptotes, then choose any convenient point on the graph (other than an x-intercept) and substitute in the equation to solve for A.

EXAMPLE 6 ▶ Find the equation of the graph, given it's of the form $y = A\tan(Bt)$.

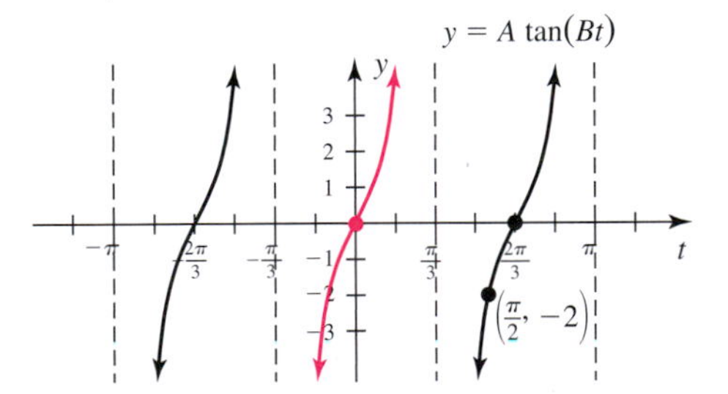

Solution: ▶ Using the interval centered at the origin and the asymptotes at $t = -\frac{\pi}{3}$ and $t = \frac{\pi}{3}$, we find the period of the function is $P = \frac{\pi}{3} - \left(-\frac{\pi}{3}\right) = \frac{2\pi}{3}$. To find the value of B we substitute in $P = \frac{\pi}{B}$ and find $B = \frac{3}{2}$. This gives the equation $y = A\tan\left(\frac{3}{2}t\right)$.

To find A, we take the point $\left(\frac{\pi}{2}, -2\right)$ given, and use $t = \frac{\pi}{2}$ with $y = -2$ to solve for A:

$$y = A\tan\left(\frac{3}{2}t\right) \qquad \text{substitute } \frac{3}{2} \text{ for } B$$

$$-2 = A\tan\left[\left(\frac{3}{2}\right)\left(\frac{\pi}{2}\right)\right] \qquad \text{substitute } -2 \text{ for } y \text{ and } \frac{\pi}{2} \text{ for } t$$

$$-2 = A\tan\left(\frac{3\pi}{4}\right) \qquad \text{multiply}$$

$$A = \frac{-2}{\tan\left(\frac{3\pi}{4}\right)} \qquad \text{solve for } A$$

$$= 2 \qquad \text{result}$$

The equation of the graph is $y = 2\tan\left(\frac{3}{2}t\right)$.

NOW TRY EXERCISES 41 THROUGH 46

E. Applications of Tangent and Cotangent Functions

We end this section with one example of how tangent and cotangent functions can be applied. Numerous others can be found in the exercise set.

EXAMPLE 7 One evening, in port during a *Semester at Sea,* Marlon is debating a project choice for his Precalculus class. Looking out his porthole, he notices a revolving light turning at a constant speed near the corner of a long warehouse. The light throws its beam along the length of the warehouse, then disappears into the air, and then returns time and time again. Suddenly—Marlon has his project. He notes the time it takes the beam to traverse the warehouse wall is very close to 4 sec, and in the morning he measures the wall's length at 127.26 m. His project? Modeling the distance of the beam from the corner of the warehouse with a tangent function. Can you help?

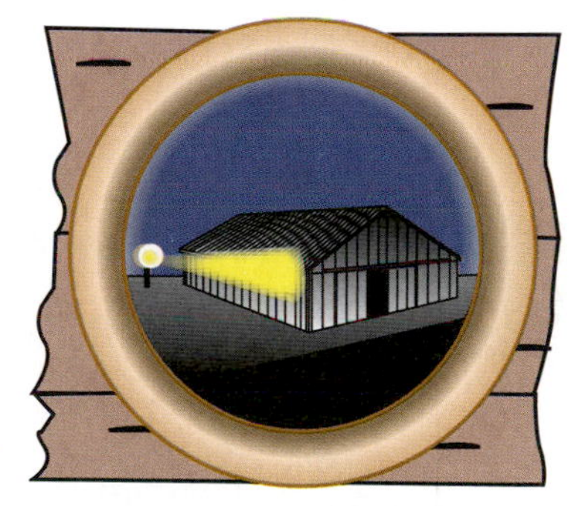

Solution: The equation model will have the form $D(t) = A\tan(Bt)$, where $D(t)$ is the distance (in meters) of the beam from the corner after t sec. The distance along the wall is measured in positive values so we're using only $\frac{1}{2}$ the period of the function, giving $\frac{1}{2}P = 4$ (the beam "disappears" at $t = 4$) so $P = 8$. Substitution in the period formula gives $B = \frac{\pi}{8}$ and the equation $D = A\tan\left(\frac{\pi}{8}t\right)$.

Knowing the beam travels 127.26 m in about 4 sec (when it disappears into infinity), we'll use $t = 3.9$ and 127.26 for D in order to solve for A and complete our equation model (see note following this example).

$$A \tan\left(\frac{\pi}{8}t\right) = D \quad \text{equation model}$$

$$A \tan\left[\frac{\pi}{8}(3.9)\right] = 127.26 \quad \text{substitute 127.26 for } D \text{ and 3.9 for } t$$

$$A = \frac{127.26}{\tan\left[\frac{\pi}{8}(3.9)\right]} \quad \text{solve for } A$$

$$\approx 5 \quad \text{result}$$

One equation modeling the distance of the beam from the corner of the warehouse is $D(t) = 5\tan\left(\frac{\pi}{8}t\right)$.

NOW TRY EXERCISES 49 THROUGH 52

For Example 7, we should note the choice of 3.9 for t was very arbitrary, and while we obtained an "acceptable" model, different values of A would be generated for other choices. For instance, $t = 3.95$ gives $A \approx 2.5$, while $t = 3.99$ gives $A \approx 0.5$. The true value of A depends on the distance of the light from the corner of the warehouse wall. In any case, it's interesting to note that at $t = 2$ sec (one-half the time it takes the beam to disappear), the beam has traveled only 5 m from the corner of the building: $D(2) = 5\tan\left(\frac{\pi}{4}\right) = 5$ m. Although the light is rotating at a constant angular speed, the speed of the beam along the wall increases *dramatically* as t gets close to 4 sec.

TECHNOLOGY HIGHLIGHT

Zeroes, Asymptotes, and the Tangent/Cotangent Functions

The keystrokes shown apply to a TI-84 Plus model. Please consult your manual or our Internet site for other models.

In this *Technology Highlight* we'll explore the tangent and cotangent functions from the perspective of their ratio definition. While we could easily use $Y_1 = \tan x$ to generate and explore the graph, we would miss an opportunity to note the many important connections that emerge from a ratio definition perspective. To begin, enter $Y_1 = \sin x$, $Y_2 = \cos x$, and $Y_3 = \frac{Y_1}{Y_2}$, as shown in Figure 2.23 [recall that function variables are accessed using VARS ▶ **(Y-VARS)** ENTER **(1:Function)]**. Note that Y_2 has been disabled by overlaying the cursor on the equal sign and pressing ENTER. In addition, note the slash next to Y_1 is more **bold** than the other slashes. The TI-84 Plus offers options that help distinguish between graphs when more than one is being displayed on the GRAPH

Figure 2.23

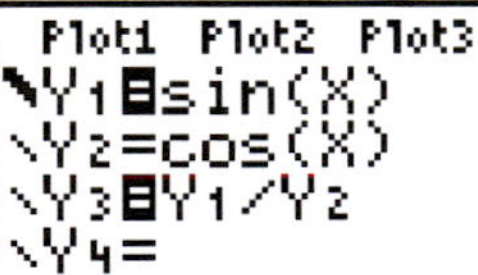

screen, and we selected a **bold** line for Y_1 by moving the cursor to the far left position and repeatedly pressing ENTER until the desired option appeared. Pressing ZOOM **7:ZTrig** at this point produces the screen shown in Figure 2.24, where we immediately note that tan x is zero everywhere that sin x is zero. This is hardly surprising since $\tan x = \frac{\sin x}{\cos x}$, but is a point that is often overlooked. Going back to the Y= and disabling Y_1 while enabling Y_2 will produce the graph shown in Figure 2.25 where similar observations can be made.

Figure 2.24

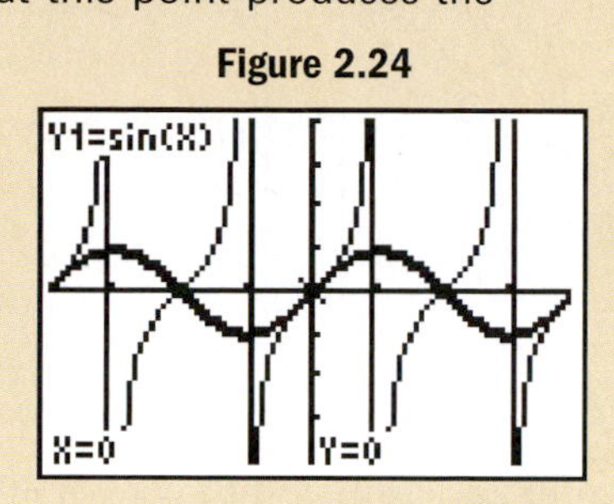

Figure 2.25

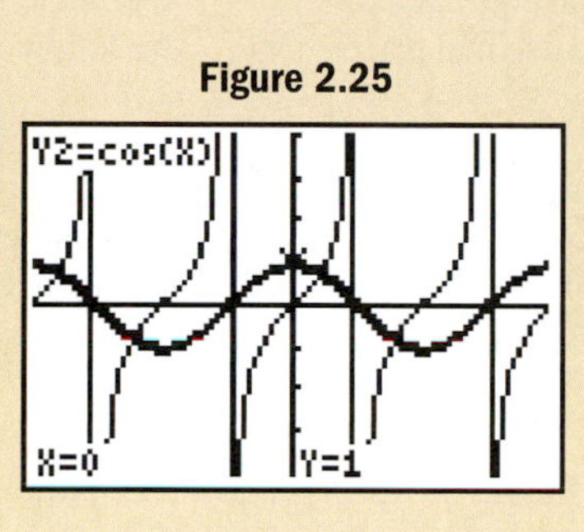

Exercise 1: What do you notice about the zeroes of cos x as they relate to the graph of $Y_3 = \tan x$?

Exercise 2: Going back to the graph of Y_1 and Y_3, from Y_3 we note the tangent function is increasing everywhere it is defined. What do you notice about the increasing/decreasing intervals for sin x as they relate to tan x? What do you notice about the intervals where each function is positive or negative?

Exercise 3: Go to the Y= screen and change Y_3 from $\frac{Y_1}{Y_2}$ (tangent) to $\frac{Y_2}{Y_1}$ (cotangent), then graph Y_2 and Y_3 on the same screen. From the graph of Y_3 we note the cotangent function is decreasing everywhere it is defined. What do you notice about the increasing/decreasing intervals for cos x as they relate to cot x? What do you notice about the intervals where each function is positive or negative?

2.2 EXERCISES

CONCEPTS AND VOCABULARY

Fill in each blank with the appropriate word or phrase. Carefully reread the section if needed.

1. The period of $y = \tan t$ and $y = \cot t$ is ________. To find the period of $y = \tan(Bt)$ and $y = \cot(Bt)$, the formula ________ is used.

2. The function $y = \tan t$ is ________ everywhere it is defined. The function $y = \cot t$ is ________ everywhere it is defined.

3. Tan t and cot t are ______ functions, so $f(-t) =$ ________. If $\tan\left(-\frac{11\pi}{12}\right)$, ≈ 0.268, then $\tan\left(\frac{11\pi}{12}\right) \approx$ ______.

4. The asymptotes of $y =$ ______ are located at odd multiples of $\frac{\pi}{2}$. The asymptotes of $y =$ ______ are located at integer multiples of π.

5. Discuss/explain how you can obtain a table of values for $y = \cot t$ (a) given the values for $y = \sin t$ and $y = \cos t$, and (b) given the values for $y = \tan t$.

6. Explain/discuss how the zeroes of $y = \sin t$ and $y = \cos t$ are related to the graphs of $y = \tan t$ and $y = \cot t$. How can these relationships help graph functions of the form $y = A\tan(Bt)$ and $y = A\cot(Bt)$?

DEVELOPING YOUR SKILLS

Use the values given for sin t and cos t to complete the tables.

7.

t	π	$\frac{7\pi}{6}$	$\frac{5\pi}{4}$	$\frac{4\pi}{3}$	$\frac{3\pi}{2}$
$\sin t = y$	0	$-\frac{1}{2}$	$-\frac{\sqrt{2}}{2}$	$-\frac{\sqrt{3}}{2}$	-1
$\cos t = x$	-1	$-\frac{\sqrt{3}}{2}$	$-\frac{\sqrt{2}}{2}$	$-\frac{1}{2}$	0
$\tan t = \frac{y}{x}$					

8.

	$\frac{3\pi}{2}$	$\frac{5\pi}{3}$	$\frac{7\pi}{4}$	$\frac{11\pi}{6}$	2π
$\sin t = y$	-1	$-\frac{\sqrt{3}}{2}$	$-\frac{\sqrt{2}}{2}$	$-\frac{1}{2}$	0
$\cos t = x$	0	$\frac{1}{2}$	$\frac{\sqrt{2}}{2}$	$\frac{\sqrt{3}}{2}$	1
$\tan t = \frac{y}{x}$					

9. Without reference to a text or calculator, attempt to name the decimal equivalent of the following values to one decimal place.

$$\frac{\pi}{2} \quad \frac{\pi}{4} \quad \frac{\pi}{6} \quad \sqrt{2} \quad \frac{\sqrt{2}}{2} \quad \frac{2}{\sqrt{3}}$$

10. Without reference to a text or calculator, attempt to name the decimal equivalent of the following values to one decimal place.

$$\frac{\pi}{3} \quad \pi \quad \frac{3\pi}{2} \quad \sqrt{3} \quad \frac{\sqrt{3}}{2} \quad \frac{1}{\sqrt{3}}$$

11. State the value of each expression without the use of a calculator.

a. $\tan\left(-\frac{\pi}{4}\right)$ **b.** $\cot\left(\frac{\pi}{6}\right)$

c. $\cot\left(\frac{3\pi}{4}\right)$ **d.** $\tan\left(\frac{\pi}{3}\right)$

12. State the value of each expression without the use of a calculator.

a. $\cot\left(\frac{\pi}{2}\right)$ **b.** $\tan \pi$

c. $\tan\left(-\frac{5\pi}{4}\right)$ **d.** $\cot\left(-\frac{5\pi}{6}\right)$

13. State the value of each expression without the use of a calculator, given $t \in [0, 2\pi)$ terminates in the quadrant indicated.

a. $\tan t = -1$, t in QIV

b. $\cot t = \sqrt{3}$, t in QIII

c. $\cot t = -\frac{1}{\sqrt{3}}$, t in QIV

d. $\tan t = -1$, t in QII

14. State the value of each expression without the use of a calculator, given $t \in [0, 2\pi)$ terminates in the quadrant indicated.

a. $\cot t = 1$, t in QI

b. $\tan t = -\sqrt{3}$, t in QII

c. $\tan t = \frac{1}{\sqrt{3}}$, t in QI

d. $\cot t = 1$, t in QIII

Use the values given for sin t and cos t to complete the tables.

15.

t	π	$\frac{7\pi}{6}$	$\frac{5\pi}{4}$	$\frac{4\pi}{3}$	$\frac{3\pi}{2}$
$\sin t = y$	0	$-\frac{1}{2}$	$-\frac{\sqrt{2}}{2}$	$-\frac{\sqrt{3}}{2}$	-1
$\cos t = x$	-1	$-\frac{\sqrt{3}}{2}$	$-\frac{\sqrt{2}}{2}$	$-\frac{1}{2}$	0
$\cot t = \frac{x}{y}$					

16.

	$\frac{3\pi}{2}$	$\frac{5\pi}{3}$	$\frac{7\pi}{4}$	$\frac{11\pi}{6}$	2π
$\sin t = y$	-1	$-\frac{\sqrt{3}}{2}$	$-\frac{\sqrt{2}}{2}$	$-\frac{1}{2}$	0
$\cos t = x$	0	$\frac{1}{2}$	$\frac{\sqrt{2}}{2}$	$\frac{\sqrt{3}}{2}$	1
$\cot t = \frac{x}{y}$					

17. Given $t = \frac{11\pi}{24}$ is a solution to $\tan t \approx 7.6$, use the period of the function to name three additional solutions. Check your answer using a calculator.

18. Given $t = \frac{7\pi}{24}$ is a solution to $\cot t \approx 0.77$, use the period of the function to name three additional solutions. Check your answer using a calculator.

19. Given $t \approx 1.5$ is a solution to $\cot t = 0.07$, use the period of the function to name three additional solutions. Check your answers using a calculator.

20. Given $t \approx 1.25$ is a solution to $\tan t = 3$, use the period of the function to name three additional solutions. Check your answers using a calculator.

Verify the value shown for t is a solution to the equation given, then use the period of the function to name all real roots. Check two of these roots on a calculator.

21. $t = \frac{\pi}{10}$; $\tan t \approx 0.3249$

22. $t = -\frac{\pi}{16}$; $\tan t \approx -0.1989$

23. $t = \frac{\pi}{12}$; $\cot t = 2 + \sqrt{3}$

24. $t = \frac{5\pi}{12}$; $\cot t = 2 - \sqrt{3}$.

Graph each function over the interval indicated, noting the period, asymptotes, zeroes, and value of A. Include a comparative sketch of $y = \tan t$ or $y = \cot t$ as indicated.

25. $f(t) = 2 \tan t$; $[-2\pi, 2\pi]$

26. $g(t) = \frac{1}{2} \tan t$; $[-2\pi, 2\pi]$

27. $h(t) = 3 \cot t$; $[-2\pi, 2\pi]$

28. $r(t) = \frac{1}{4} \cot t$; $[-2\pi, 2\pi]$

Graph each function over the interval indicated, noting the period, asymptotes, zeroes, and value of A.

29. $y = \tan(2t)$; $\left[-\frac{\pi}{2}, \frac{\pi}{2}\right]$

30. $y = \tan\left(\frac{1}{4}t\right)$; $[-4\pi, 4\pi]$

31. $y = \cot(4t)$; $\left[-\frac{\pi}{4}, \frac{\pi}{4}\right]$

32. $y = \cot\left(\frac{1}{2}t\right)$; $[-2\pi, 2\pi]$

33. $y = 2 \tan(4t)$; $\left[-\frac{\pi}{4}, \frac{\pi}{4}\right]$

34. $y = 4 \tan\left(\frac{1}{2}t\right)$; $[-2\pi, 2\pi]$

35. $y = 5 \cot\left(\frac{1}{3}t\right)$; $[-3\pi, 3\pi]$

36. $y = \frac{1}{2} \cot(2t)$; $\left[-\frac{\pi}{2}, \frac{\pi}{2}\right]$

37. $y = 3 \tan(2\pi t)$; $\left[-\frac{1}{2}, \frac{1}{2}\right]$

38. $y = 4 \tan\left(\frac{\pi}{2}t\right)$; $[-2, 2]$

39. $f(t) = 2 \cot(\pi t)$; $[-1, 1]$

40. $p(t) = \frac{1}{2} \cot\left(\frac{\pi}{4}t\right)$; $[-4, 4]$

Find the equation of each graph, given it is of the form $y = A \tan(Bt)$.

41.

42.

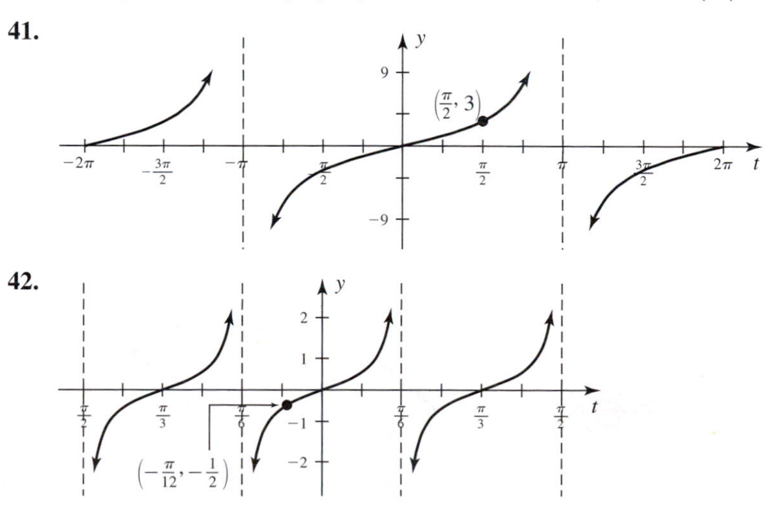

Find the equation of each graph, given it is of the form $y = A\cot(Bt)$.

43.

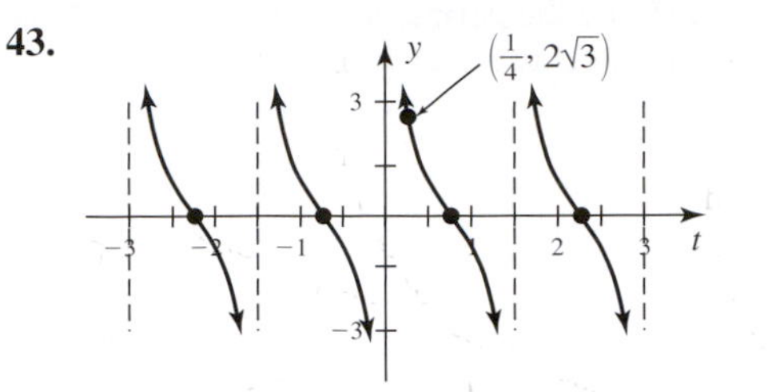

44.

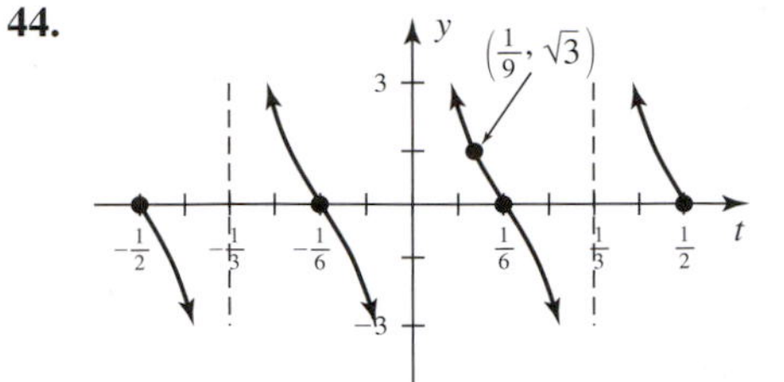

45. Given that $t = -\frac{\pi}{8}$ and $t = -\frac{3\pi}{8}$ are solutions to $\cot(3t) = \tan t$, use a graphing calculator to find two additional solutions in $[0, 2\pi]$.

46. Given $t = \frac{1}{6}$ is a solution to $\tan(2\pi t) = \cot(\pi t)$, use a graphing calculator to find two additional solutions in $[-1, 1]$.

WORKING WITH FORMULAS

47. Position of an image reflected from a spherical lens: $\tan\theta = \frac{h}{s - k}$

The equation shown is used to help locate the position of an image reflected by a spherical mirror, where s is the distance of the object from the lens along a horizontal axis, θ is the angle of elevation from this axis, h is the altitude of the right triangle indicated, and k is distance from the lens to the foot of altitude h. Find the distance k given $h = 3$ mm, $\theta = \frac{\pi}{24}$, and that the object is 24 mm from the lens.

48. The height of an object calculated from a distance: $h = \frac{d}{\cot u - \cot v}$

The height h of a tall structure can be computed using two angles of elevation measured some distance apart along a straight line with the object. This height is given by the formula shown, where d is the distance between the two points from which angles u and v were measured. Find the height h of a building if $u = 40°$, $v = 65°$, and $d = 100$ ft.

APPLICATIONS

Tangent function data models: Model the data in Exercises 49 and 50 using the function $y = A\tan(Bx)$. State the period of the function, the location of the asymptotes, the value of A, and name the point (x, y) used to calculate A (answers may vary). Use your equation model to evaluate the function at $x = -2$ and $x = 2$. What observations can you make? Also see Exercise 58.

49.

Input	Output	Input	Output
−6	−∞	1	1.4
−5	−20	2	3
−4	−9.7	3	5.2
−3	−5.2	4	9.7
−2	−3	5	20
1	−1.4	6	∞
0	0		

50.

Input	Output	Input	Output
−3	−∞	0.5	6.4
−2.5	−91.3	1	13.7
−2	−44.3	1.5	23.7
−1.5	−23.7	2	44.3
−1	−13.7	2.5	91.3
−0.5	−6.4	3	∞
0	0		

Exercise 51

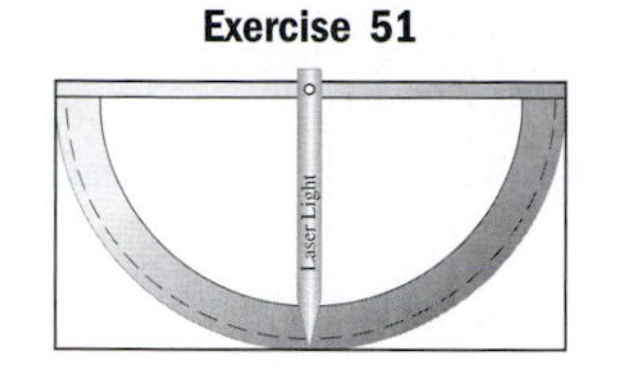

51. As part of a lab setup, a laser pen is made to swivel on a large protractor as illustrated in the figure. For their lab project, students are asked to take the instrument to one end of a long hallway and measure the distance of the projected beam relative to the angle the pen is being held, and collect the data in a table. Use the data to find a function of the form $y = A\tan(B\theta)$. State the period of the function, the location of the asymptotes, the value of A, and name the point (θ, y) you used to calculate A (answers may vary). Based on the result, can you approximate the length of the laser pen? Note that in degrees, the period formula for tangent is $P = \dfrac{180°}{B}$.

θ (degrees)	Distance (cm)
0	0
10	2.1
20	4.4
30	6.9
40	10.1
50	14.3
60	20.8
70	33.0
80	68.1
89	687.5

52. Use the equation model obtained in Exercise 51 to compare the values given by the equation with the actual data. As a percentage, what was the largest deviation between the two?

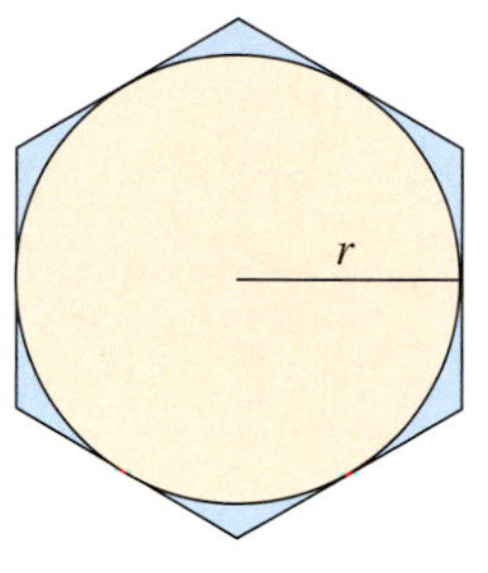

53. Circumscribed polygons: The *perimeter* of a regular polygon circumscribed about a circle of radius r is given by $P = 2nr\tan\left(\dfrac{\pi}{n}\right)$, where n is the number of sides $(n \geq 3)$ and r is the radius of the circle. Given $r = 10$ cm, (a) What is the perimeter of the circle? (b) What is the perimeter of the polygon when $n = 4$? Why? (c) Calculate the perimeter of the polygon for $n = 10, 20, 30$, and 100. What do you notice?

54. Circumscribed polygons: The area of a regular polygon circumscribed about a circle of radius r is given by $A = nr^2\tan\left(\dfrac{\pi}{n}\right)$, where n is the number of sides $(n \geq 3)$ and r is the radius of the circle. Given $r = 10$ cm,

a. What is the area of the circle?

b. What is the area of the polygon when $n = 4$? Why?

c. Calculate the area of the polygon for $n = 10, 20, 30$, and 100. What do you notice?

Coefficients of friction: Pulling someone on a sled is much easier during the winter than in the summer, due to a phenomenon known as the *coefficient of friction*. The friction between the sled's skids and the snow is much lower than the friction between the skids and the dry ground or pavement. Basically, the coefficient of friction is defined by the relationship $\mu = \tan\theta$, where θ is the angle at which a block composed of one material will slide down an inclined plane made of another material, with a constant velocity. Coefficients of friction have been established experimentally for many materials and a short list is shown here.

Material	Coefficient
steel on steel	0.74
copper on glass	0.53
glass on glass	0.94
copper on steel	0.68
wood on wood	0.5

55. Graph the function $\mu = \tan\theta$ with θ in degrees over the interval $[0°, 60°]$ and use the graph to estimate solutions to the following. Confirm or contradict your estimates using a calculator.

a. A block of copper is placed on a sheet of steel, which is slowly inclined. Is the block of copper moving when the angle of inclination is 30°? At what angle of inclination will the copper block be moving with a constant velocity down the incline?

b. A block of copper is placed on a sheet of cast-iron. As the cast-iron sheet is slowly inclined, the copper block begins sliding at a constant velocity when the angle of inclination is approximately 46.5°. What is the coefficient of friction for copper on cast-iron?

c. Why do you suppose coefficients of friction greater than $\mu = 2.5$ are extremely rare? Give an example of two materials that likely have a high μ-value.

56. Graph the function $\mu = \tan\theta$ with θ in radians over the interval $\left[0, \frac{5\pi}{12}\right]$ and use the graph to estimate solutions to the following. Confirm or contradict your estimates using a calculator.

a. A block of glass is placed on a sheet of glass, which is slowly inclined. Is the block of glass moving when the angle of inclination is $\frac{\pi}{4}$? What is the smallest angle of inclination for which the glass block will be moving with a constant velocity down the incline (rounded to four decimal places)?

b. A block of Teflon is placed on a sheet of steel. As the steel sheet is slowly inclined, the Teflon block begins sliding at a constant velocity when the angle of inclination is approximately 0.04. What is the coefficient of friction for Teflon on steel?

c. Why do you suppose coefficients of friction less than $\mu = 0.04$ are extremely rare for two solid materials? Give an example of two materials that likely have a very low μ value.

57. Tangent lines: The actual definition of the word *tangent* comes from the Latin *tangere*, meaning "to touch." In mathematics, a tangent line touches the graph of a circle at only one point and function values for $\tan\theta$ are obtained from the length of the line segment tangent to a unit circle.

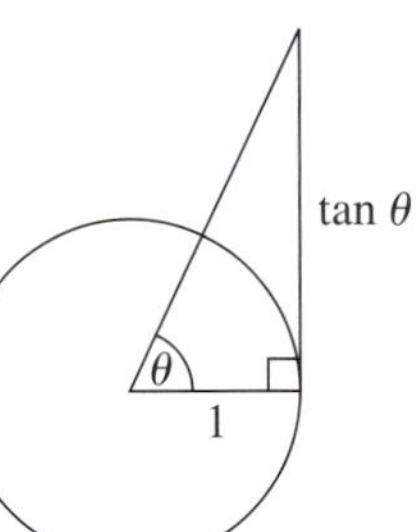

a. What is the length of the line segment when $\theta = 80°$?

b. If the line segment is 16.35 units long, what is the value of θ?

c. Can the line segment ever be greater than 100 units long? Why or why not?

d. How does your answer to (c) relate to the asymptotic behavior of the graph?

WRITING, RESEARCH, AND DECISION MAKING

58. Rework Exercises 49 and 50, obtaining a new equation for the data using a different ordered pair to compute the value of A. What do you notice? Try yet another ordered pair and calculate A once again for another equation Y_2. Complete a table of values using the given inputs, with the outputs of the three equations generated (original, Y_1, and Y_2). Does any one equation seem to model the data better than the others? Are all of the equation models "acceptable"? Please comment.

59. The golden ratio $\frac{-1+\sqrt{5}}{2}$ has long been thought to be the most pleasing ratio in art and architecture. It is commonly believed that many forms of ancient architecture were constructed using this ratio as a guide. The ratio actually turns up in some surprising places, far removed from its original inception as a line segment cut in "mean and extreme" ratio. Do some research on the golden ratio and some of the equations that have been used to produce it. Given $x = 0.6662394325$, try to find a connection between $y = \cos x$, $y = \tan x$, and $y = \sin x$.

EXTENDING THE CONCEPT

60. Regarding Example 7, we can use the standard distance/rate/time formula $D = RT$ to compute the average velocity of the beam of light along the wall in any interval of time: $R = \frac{D}{T}$. For example, using $D(t) = 5\tan\left(\frac{\pi}{8}t\right)$, the average velocity in the interval $[0, 2]$

is $\frac{D(2) - D(0)}{2 - 0} = 2.5$ m/sec. Calculate the average velocity of the beam in the time intervals [2, 3], [3, 3.5], and [3.5, 3.8] sec. What do you notice? How would the average velocity of the beam in the interval [3.9, 3.99] sec compare?

61. Determine the slope of the line drawn *through* the parabola (called a **secant** line) in Figure I. Use the same method (any two points on the line) to calculate the slope of the line drawn **tangent** to the parabola in Figure II. Compare your calculations to the tangent of the angles α and β that each line makes with the x-axis. What can you conclude? Write a formula for the point/slope equation of a line using $\tan\theta$ instead of m.

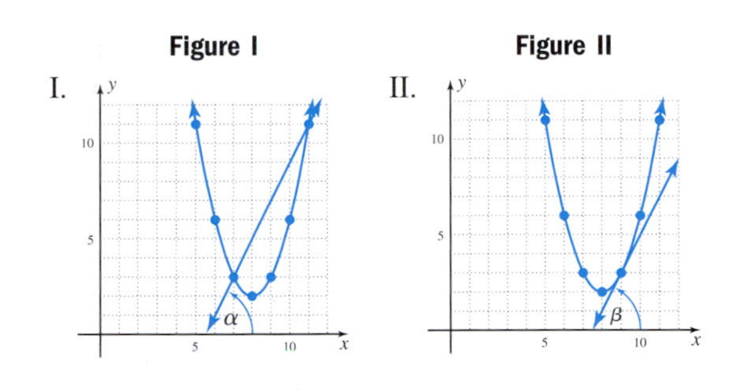

MAINTAINING YOUR SKILLS

62. (1.2) A tent rope is 4 ft long and attached to the tent wall 2 ft above the ground. How far from the tent is the stake holding the rope?

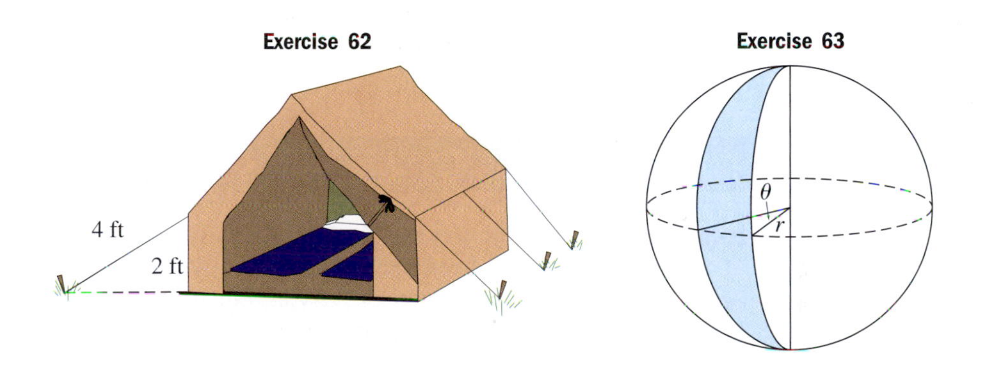

63. (1.1) A lune is a section of surface area on a sphere, which is subtended by an angle θ at the circumference. For θ *in radians,* the surface area of a lune is $A = 2r^2\theta$, where r is the radius of the sphere. Find the area of a lune on the surface of the Earth which is subtended by an angle of 15°. Assume the radius of the Earth is 6373 km.

64. (1.4) Use a reference arc to determine the value of $\cos\left(\frac{29\pi}{6}\right)$.

65. (1.4) State the points on the unit circle that correspond to $t = 0, \frac{\pi}{4}, \frac{\pi}{2}, \pi, \frac{3\pi}{4}, \frac{3\pi}{2}$, and 2π. What is the value of $\tan\left(\frac{\pi}{2}\right)$? Why?

66. (1.3) Given $\sin 212° \approx -0.53$, find another angle θ in [0°, 360°) that satisfies $\sin\theta \approx -0.53$ without using a calculator.

67. (2.1) At what value(s) of t does the horizontal line $y = 3$ intersect the graph of $y = -3\sin t$?

MID-CHAPTER CHECK

1. Which function, $y = \tan t$ or $y = \cot t$, is decreasing on its domain? Which function, $y = \cos t$ or $y = \sin t$, begins at (0, 1) in the interval $t \in [0, 2\pi)$?
2. State the period of $y = \sin\left(\frac{\pi}{2}t\right)$. Where will the max/min values occur in the primary interval?
3. Evaluate without using a calculator: (a) cot 60° and (b) $\sin\left(\frac{7\pi}{4}\right)$.
4. Evaluate using a calculator: (a) $\sec\left(\frac{\pi}{12}\right)$ and (b) tan 4.3.
5. Which of the six trig functions are even functions?
6. State the domain of $y = \sin t$ and $y = \tan t$.
7. Name the location of the asymptotes and graph $y = 3\tan\left(\frac{\pi}{2}t\right)$ for $t \in [-2\pi, 2\pi]$.
8. Clearly state the amplitude and period, then sketch the graph: $y = -3\cos\left(\frac{\pi}{2}t\right)$.
9. On a unit circle, if arc t has length 5.94, (a) in what quadrant does it terminate? (b) What is its reference arc? (c) Of sin t, cos t, and tan t, which are negative for this value of t?
10. For the graph given here, (a) clearly state the amplitude and period; (b) find the equation of the graph; (c) graphically find $f(\pi)$ and then confirm/contradict your estimation using a calculator.

Exercise 10

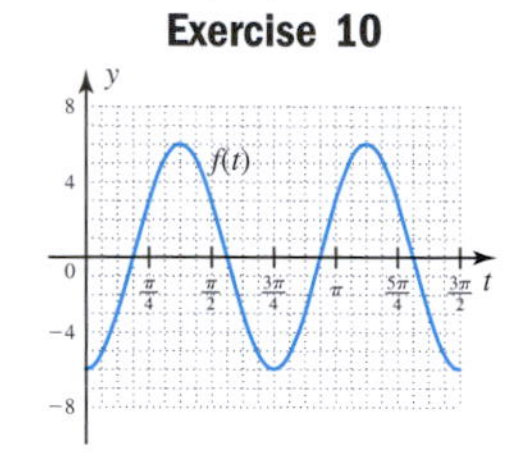

REINFORCING BASIC CONCEPTS

Trigonometric Potpourri

This *Reinforcing Basic Concepts* is simply a collection of patterns, observations, hints, and reminders connected with an introduction to trigonometry. Individually the points may seem trivial, but taken together they tend to reinforce the core fundamentals of trig, enabling you to sequence and store the ideas in your own way. Having these basic elements available for instant retrieval builds a stronger bridge to future concepts, assists in the discovery of additional connections, and enables a closer tie between these concepts and the real-world situations to which they will be applied. Just a little work now pays big dividends later. As Louis Pasteur once said, "Fortune favors a prepared mind."

Exercise 1

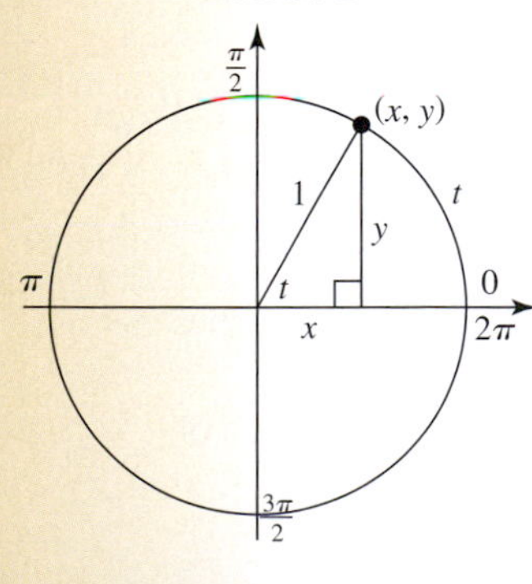

1. The collection begins with an all-encompassing view of the trig functions, as seen by the imposition of a right triangle in a unit circle, on the coordinate grid. This allows all three approaches to trigonometry to be seen at one time:

$$\text{right triangle: } \cos t = \frac{\text{adj}}{\text{hyp}} \qquad \text{any angle: } \cos t = \frac{x}{r}$$

$$\text{real number: } \cos t = x$$

Should you ever forget the association of x *with cosine and* y *with sine,* just remember that c comes before s in the alphabet in the same way the x comes before y: $\cos t \rightarrow x$ and $\sin t \rightarrow y$.

2. *Know the standard angles and the standard values.* As mentioned earlier, they are used repeatedly throughout higher mathematics to introduce new concepts and skills without the clutter and distraction of large decimal values. It is interesting to note the standard values (from QI) can always be recreated using the *pattern of fourths*. Simply write the fractions with integer numerators from $\frac{0}{4}$ through $\frac{4}{4}$, and take their square root:

$$\sqrt{\frac{0}{4}} = 0 \qquad \sqrt{\frac{1}{4}} = \frac{1}{2} \qquad \sqrt{\frac{2}{4}} = \frac{\sqrt{2}}{2} \qquad \sqrt{\frac{3}{4}} = \frac{\sqrt{3}}{2} \qquad \sqrt{\frac{4}{4}} = 1$$

3. There are many *decimal equivalents* in elementary mathematics that contribute to concept building. In the same way you recognize the decimal values of $\frac{1}{4}$, $\frac{1}{2}$, $\frac{3}{4}$, and others, it helps *tremendously* to recognize or "know" decimal equivalents for values that are commonly used in a study of trig. They help to identify the quadrant of an arc/angle's terminal side when t is expressed in radians, they assist in graphing and estimation skills, and are used extensively throughout this text and in other areas of mathematics. In terms of π these are $\frac{\pi}{2}$, π, $\frac{3\pi}{2}$, and 2π. In terms of radicals they are $\sqrt{2}$, $\frac{\sqrt{2}}{2}$, $\sqrt{3}$, and $\frac{\sqrt{3}}{2}$.

Knowing $\pi \approx 3.14$ leads directly to $\frac{\pi}{2} \approx 1.57$ and $2\pi \approx 6.28$, while adding the decimal values for π and $\frac{\pi}{2}$ gives $\frac{3\pi}{2} \approx 4.71$. Further, since $\sqrt{2} \approx 1.41$, $\frac{\sqrt{2}}{2} \approx 0.7$, and since $\sqrt{3} \approx 1.73$, $\frac{\sqrt{3}}{2} \approx 0.87$.

4. To specifically remember what standard value is associated with a standard angle or arc, recall that

a. If t is a quadrantal arc/angle, it's easiest to use the coordinates (x, y) of a point on a unit circle.

b. If t is any odd multiple of $\frac{\pi}{4}$, $\sin t$ and $\cos t$ must be $-\frac{\sqrt{2}}{2}$ or $\frac{\sqrt{2}}{2}$, with the choice depending on the quadrant of the terminal side. The values of the other functions can be found using these.

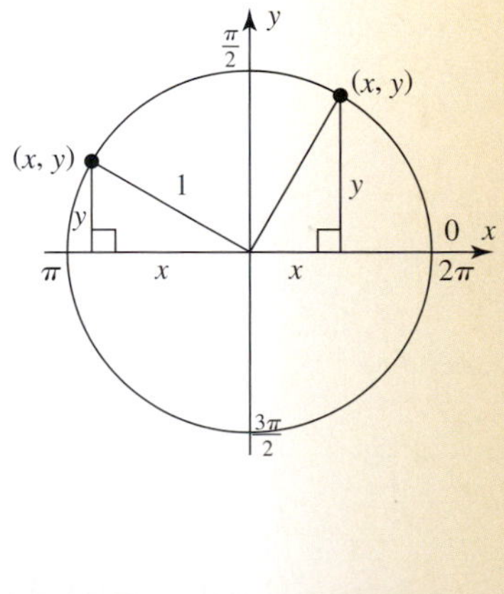

c. If t is a multiple of $\frac{\pi}{6}$ (excluding the quadrantal angles), the value for sine and cosine must be either $\pm\frac{1}{2}$ or $\pm\frac{\sqrt{3}}{2}$, depending on the quadrant of the terminal side. If there's any hesitation about which value applies to sine and which to cosine, mental imagery can once again help. Since $\frac{\sqrt{3}}{2} \approx 0.87 > \frac{1}{2} = 0.5$, we simply apply the larger value to the larger arc/angle. Note that for the triangle drawn in QI $\left(\frac{\pi}{3} = 60° \text{ angle at vertex}\right)$, y is obviously longer than x, meaning the association must be $\sin t = \frac{\sqrt{3}}{2}$ and $\cos t = \frac{1}{2}$. For the triangle in QII $\left(\frac{5\pi}{6} = 150° \text{ whose reference angle is } 30°\right)$, x is longer than y, meaning the association must be $\cos t = -\frac{\sqrt{3}}{2}$ and $\sin t = \frac{1}{2}$.

5. Although they are often neglected or treated lightly in a study of trig, the secant, cosecant, and cotangent functions play an integral role in more advanced mathematics classes. Be sure you're familiar with their reciprocal relationship to the more common cosine, sine, and tangent functions:

$$\sec t = \frac{1}{\cos t},\quad \cos t = \frac{1}{\sec t},\quad \sec t \cos t = 1;\quad \cos\left(\frac{2\pi}{3}\right) = -\frac{1}{2} \rightarrow \sec\left(\frac{2\pi}{3}\right) = -2$$

$$\csc t = \frac{1}{\sin t},\quad \sin t = \frac{1}{\csc t},\quad \csc t \sin t = 1;\quad \sin\left(\frac{2\pi}{3}\right) = \frac{\sqrt{3}}{2} \rightarrow \csc\left(\frac{2\pi}{3}\right) = \frac{2}{\sqrt{3}}$$

$$\cot t = \frac{1}{\tan t},\quad \tan t = \frac{1}{\cot t},\quad \cot t \tan t = 1;\quad \tan 60° = \sqrt{3} \rightarrow \cot 60° = \frac{1}{\sqrt{3}}$$

6. Finally, the need to be very familiar with the basic graphs of the trig functions would be hard to overstate. As with transformations of the toolbox functions (from algebra), transformations of the basic trig graphs are a huge help to the understanding and solution of trig equations and inequalities, as well as to their application in the context of real-world phenomena.

2.3 Transformations and Applications of Trigonometric Graphs

LEARNING OBJECTIVES

In Section 2.3 you will learn how to:

A. Apply vertical translations in context

B. Apply horizontal translations in context

C. Solve applications involving harmonic motion

INTRODUCTION

From your algebra experience, you may remember beginning with a study of linear graphs, then moving on to quadratic graphs and their characteristics. By combining and extending the knowledge you gained, you were able to investigate and understand a variety of polynomial graphs—along with some powerful applications. A study of trigonometry follows a similar pattern, and by "combining and extending" our understanding of the basic trig graphs, we'll look at some powerful applications in *this* section.

POINT OF INTEREST

In some coastal areas, tidal motion is simple, predictable, and under perfect conditions can be approximated using a single sine function. In other areas, however, a combination of factors make the motion much more complex, although still predictable. The mathematical model for these tides requires more than a single trig function, a concept known as the *addition of ordinates.* Similar concepts also apply in areas such as sound waves and musical tones.

Figure 2.26

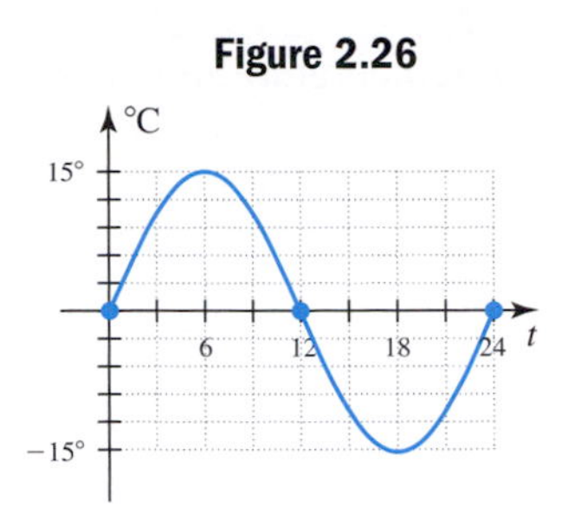

A. Vertical Translations: $y = A \sin(Bt) + D$

On any given day, outdoor temperatures tend to follow a **sinusoidal pattern,** or a pattern that can be modeled by a sine function. As the sun rises, the morning temperature begins to warm and rise until reaching its high in the late afternoon, then begins to cool during the early evening and nighttime hours until falling to its nighttime low just prior to sunrise. Next morning, the cycle begins again. In the northern latitudes where the winters are very cold, it's not unreasonable to assume an average daily temperature of 0°C (32°F), and a temperature graph in degrees Celsius that looks like the one in Figure 2.26. For the moment, we'll assume that $t = 0$ corresponds to 12:00 noon.

EXAMPLE 1 Use the graph in Figure 2.26 to: (a) state the amplitude and period of the function, (b) estimate the temperature at 9:00 P.M., and (c) estimate the number of hours the temperature is below -12°C.

Solution:

a. By inspection we see the amplitude of the graph is $|A| = 15$, since this is the maximum displacement from the average value. As the context and graph indicate, the period is 24 hr.

b. Since 9:00 P.M. corresponds to $t = 9$, we read along the t-axis to 9, then move up to the graph to approximate the related temperature value on the °C axis (see figure). It appears the temperature will be close to 11°C.

c. In part (b) we're given the time and asked for a temperature. Here we're given a temperature and asked for a time. Begin on the °C axis at -12°, move horizontally until you intersect the graph, then move upward to the t-axis. The temperature is below 12°C for about 5 hr (from $t \approx 15.5$ to $t \approx 20.5$.)

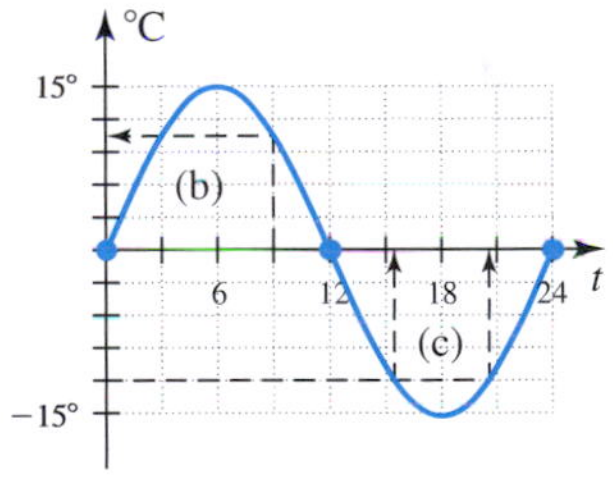

NOW TRY EXERCISES 7 THROUGH 10

If you live in a more temperate area, the daily temperatures still follow a sinusoidal pattern, but the average temperature could be much higher. This is an example of a **vertical shift,** and is the role D plays in the equation $y = A \sin(Bt) + D$. All other aspects of a graph remain the same; it is simply shifted D units up if $D > 0$ and D units down if $D < 0$. As in Section 2.1, for maximum value M and minimum value m, $\frac{M - m}{2}$ gives the amplitude A of a sine curve, while $\frac{M + m}{2}$ gives the **average value** D. From Example 1 we have $\frac{15 + (-15)}{2} = 0$ and $\frac{15 - (-15)}{2} = 15$✓.

EXAMPLE 2 On a fine day in Galveston, Texas, the high temperature might be about 85°F with an overnight low of 61°F. (a) Find a sinusoidal equation model for the daily temperature; (b) sketch the graph; and (c) approximate what time(s) of day the temperature is 65°F. Assume $t = 0$ corresponds to 12:00 noon.

Solution:

a. We first note the period is still $P = 24$, giving $B = \frac{\pi}{12}$, and the equation model will have the form $y = A\sin\left(\frac{\pi}{12}t\right) + D$. Using $\frac{M + m}{2} = \frac{85 + 61}{2}$, we find the *average value* $D = 73$, with amplitude $A = \frac{85 - 61}{2} = 12$. The resulting equation is $y = 12\sin\left(\frac{\pi}{12}t\right) + 73$.

b. To sketch the graph, use a reference rectangle $2A = 24$ units tall and $P = 24$ units wide, along with the *rule of fourths* to locate zeroes and max/min values (see Figure 2.27). Then lightly sketch a sine curve through these points and within the rectangle as in Figure 2.28. This is the graph of $y = 12\sin\left(\frac{\pi}{12}t\right) + 0$.

Figure 2.27 **Figure 2.28**

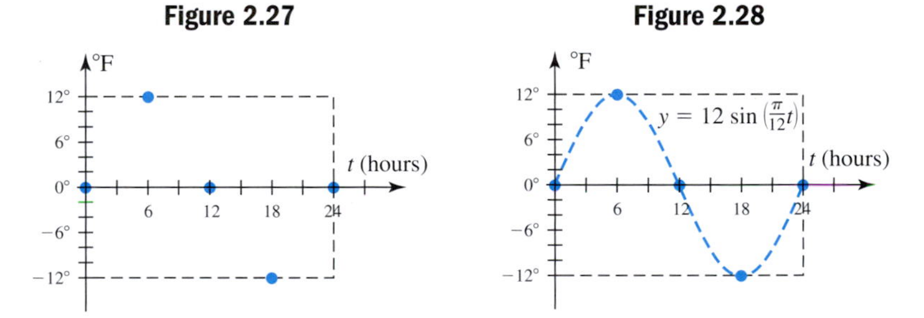

Using an appropriate scale, shift the rectangle and plotted points vertically upward 73 units and carefully draw the finished graph through the points and within the rectangle (see Figure 2.29).

Figure 2.29

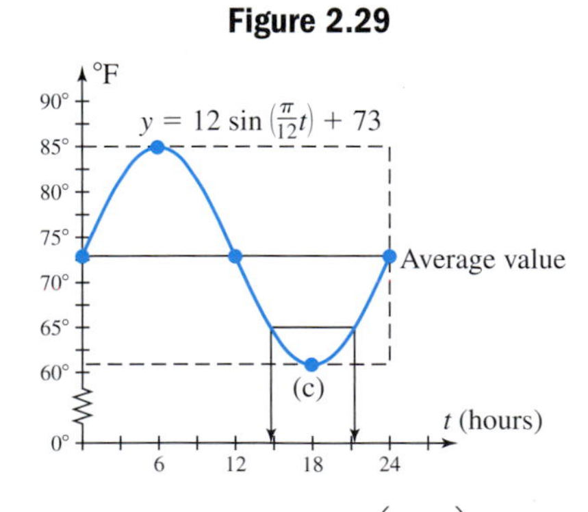

This gives the graph of $y = 12\sin\left(\frac{\pi}{12}t\right) + 73$. Note the brokenline notation "≷" in Figure 2.29 indicates that certain values along an axis are unused (in this case, we skipped 0° to 60°), and we began scaling the axis with the values needed.

c. As indicated in Figure 2.29, the temperature hits 65° twice, at about 15 and 21 hr after 12:00 noon, or at 3:00 A.M. and 9:00 A.M. Verify by computing $f(15)$ and $f(21)$.

NOW TRY EXERCISES 11 THROUGH 18

Sinusoidal graphs actually include both sine and cosine graphs, the difference being that sine graphs begin at the average value, while cosine graphs begin at the maximum value. Sometimes it's more advantageous to use one over the other, but equivalent forms can easily be found. In Example 3, a cosine function is used to model an animal population that fluctuates sinusoidally due to changes in food supplies.

EXAMPLE 3 The population of a certain animal species can be modeled by the function $P(t) = 1200\cos\left(\frac{\pi}{5}t\right) + 9000$, where $P(t)$ represents the population in year t. Use the model to: (a) find the period of the function; (b) graph the function over one period; (c) find the maximum and minimum values; and (d) estimate the number of years the population is less than 8000.

Solution:

a. Since $B = \frac{\pi}{5}$, the period is $P = \frac{2\pi}{\pi/5} = 10$, meaning the population of this animal species fluctuates over a 10-yr cycle.

b. Use a reference rectangle ($2A = 2400$ by $P = 10$ units) and the *rule of fourths* to locate zeroes and max/min values, then sketch the unshifted graph $y = 1200\cos\left(\frac{\pi}{5}t\right)$. With $P = 10$, these occur at $t = 0, 2.5, 5, 7.5$, and 10 (see Figure 2.30). Shift this graph upward 9000 units (using an appropriate scale) to obtain the graph of $P(t)$ shown in Figure 2.31.

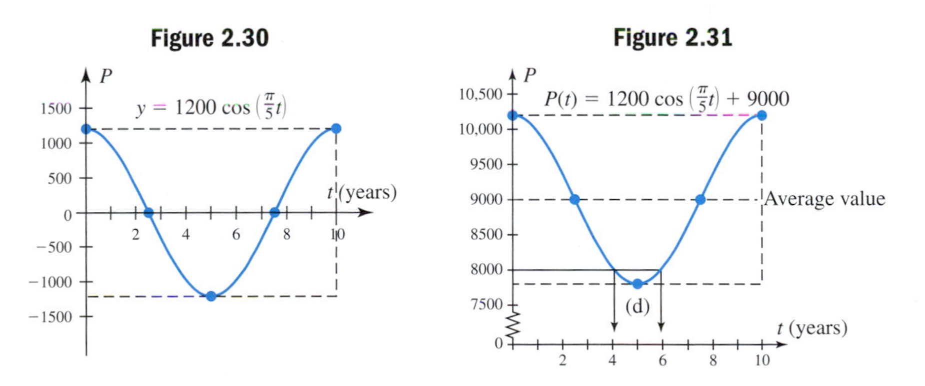

c. The maximum value is $9000 + 1200 = 10{,}200$ and the minimum value is $9000 - 1200 = 7800$.

d. As determined from the graph, the population drops below 8000 animals for approximately 2 yr. Verify by computing $P(4)$ and $P(6)$.

NOW TRY EXERCISES 19 AND 20

B. Horizontal Translations: $y = A\sin(Bt + C) + D$

In some cases, scientists would rather "benchmark" their study of sinusoidal phenomenon by placing the average value at $t = 0$ instead of a maximum value (as in Example 3), or by placing the maximum or minimum value at $t = 0$ instead of the average value (as in Example 1). Rather than make additional studies or recompute using available data, *we can simply shift these graphs using a horizontal translation*. To help understand how, consider the graph of $y = x^2$. The graph is a parabola, concave up, with a vertex at the origin. Comparing this function with $y_1 = (x - 3)^2$ and $y_2 = (x + 3)^2$, we note y_1 is simply the parent graph shifted 3 units right, and y_2 is the parent graph shifted 3 units left ("opposite the sign"). See Figures 2.32 through 2.34.

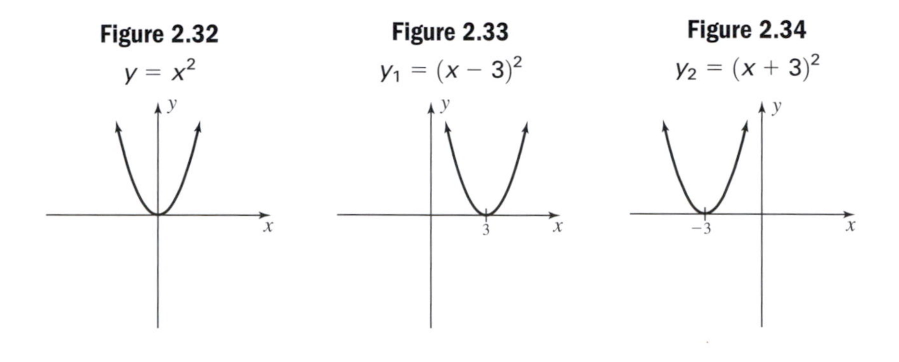

Figure 2.32 $y = x^2$

Figure 2.33 $y_1 = (x - 3)^2$

Figure 2.34 $y_2 = (x + 3)^2$

While ***quadratic functions have no maximum value if A > 0,*** these graphs are a good reminder of how a basic graph can be horizontally shifted. We simply *replace the independent variable x with $(x \pm h)$ or t with $(t \pm h)$*, where h is the desired shift and the sign is chosen depending on the direction of the shift.

EXAMPLE 4 Use a horizontal translation to shift the graph from Example 3 so that the average population value begins at $t = 0$. Verify the result on a graphing calculator, then find a sine function that gives the same graph as the shifted cosine function.

Solution: For $P(t) = 1200\cos\left(\frac{\pi}{5}t\right) + 9000$ from Example 3, the average value first occurs at $t = 2.5$. For the average value to occur at $t = 0$, we can shift the graph to the right 2.5 units. Replacing t with $(t - 2.5)$ gives

$$P(t) = 1200\cos\left[\frac{\pi}{5}(t - 2.5)\right] + 9000.$$

Y2=1200cos((π/5)(X-2.5))
X=.05319149 Y=9040.098

A graphing calculator shows the desired result is obtained (see figure), and it appears to be a sine function with the same amplitude and period. The equation is $y = 1200\sin\left(\frac{\pi}{5}t\right) + 9000$.

NOW TRY EXERCISES 21 AND 22

Equations like $P(t) = 1200\cos\left[\frac{\pi}{5}(t - 2.5)\right] + 9000$ from Example 4 are said to be written in **shifted form,** since we can easily tell the magnitude and direction of the shift in this form. To obtain the **standard form** we *distribute the value of B*: $P(t) = 1200\cos\left(\frac{\pi}{5}t - \frac{\pi}{2}\right) + 9000$. In general, the *standard form* of a sinusoidal equation (using *either* a cosine or sine function) is written $y = A\sin(Bt \pm C) + D$, with the *shifted form* found by factoring out B from the argument:

$$y = A\sin(Bt \pm C) + D \rightarrow y = A\sin\left[B\left(t \pm \frac{C}{B}\right)\right] + D$$

In either case, C gives what is known as the **phase shift** of the function, and is used to discuss how far (as an angle measure) a given function is "out of phase" with a reference function. In the latter case, $\frac{C}{B}$ is simply the horizontal shift of the function and gives the magnitude and direction of the shift (opposite the sign).

CHARACTERISTICS OF SINUSOIDAL MODELS

Given the basic graph of $y = \sin t$, the graph can be transformed and written as $y = A\sin(Bt)$, where

1. $|A|$ gives the *amplitude* of the graph, or the maximum displacement from the average value.
2. B is related to the *period P* of the graph according to the ratio $P = \frac{2\pi}{B}$, giving the interval required for one complete cycle.

The graph of $y = A\sin(Bt)$ can be translated and written in the following forms:

Standard form	Shifted form
$y = A\sin(Bt \pm C) + D$	$y = A\sin\left[B\left(t \pm \frac{C}{B}\right)\right] + D$

3. In either case, C is called the *phase shift* of the graph, while the ratio $\pm\frac{C}{B}$ gives the magnitude and direction of the *horizontal shift* (opposite the given sign).
4. D gives the *vertical shift* of the graph, and the location of the average value. The shift will be in the same direction as the given sign.

It's important that you don't confuse the standard form with the shifted form. Each has a place and purpose, but the horizontal shift can be identified only by focusing on the change in an independent variable. Even though the equations $y = 4(x + 3)^2$ and $y = (2x + 6)^2$ are equivalent, only the first explicitly shows that $y = 2x^2$ has been shifted three units left. Likewise $y = \sin[2(t + 3)]$ and $y = \sin(2t + 6)$ are equivalent, but only the first explicitly gives the horizontal shift (three units left). Applications involving a horizontal shift come in an infinite variety, and the shifts are generally not uniform or standard. Knowing where each cycle begins and ends is a helpful part of sketching a graph of the equation model. The **primary interval** for a sinusoidal graph can be found by solving the inequality $0 \le Bt \pm C < 2\pi$, with the reference rectangle and *rule of fourths* giving the zeroes, max/min values, and a sketch of the graph in this interval. The graph can then be extended as needed in either direction, then shifted vertically D units.

EXAMPLE 5 For each function, identify the amplitude, period, horizontal shift, vertical shift (average value), and endpoints of the primary interval.

a. $y = 120\cos\left[\frac{\pi}{8}(t - 6)\right] + 350$

b. $y = 2.5\sin\left(\frac{\pi}{4}t + \frac{3\pi}{4}\right) + 6$

Solution:

a. The equation shows amplitude $|A| = 120$, with an average value of $D = 350$. This indicates the maximum value will be $y = 120(1) + 350 = 470$, with a minimum of $y = 120(-1) + 350 = 230$. Since $B = \frac{\pi}{8}$, the period is $P = \frac{2\pi}{\pi/8} = 16$. With the argument in factored form, we note the horizontal shift is 6 units to the right (6 units opposite the sign). For the endpoints of the primary interval we solve $0 \le \frac{\pi}{8}(t - 6) < 2\pi$, giving $6 \le t < 22$.

b. The equation shows amplitude $|A| = 2.5$, with an average value of $D = 6$. The maximum value will be $y = 2.5(1) + 6 = 8.5$, with a minimum of $y = 2.5(-1) + 6 = 3.5$. With $B = \frac{\pi}{4}$, the period is $P = \frac{2\pi}{\pi/4} = 8$. To find the horizontal shift, we factor out $\frac{\pi}{4}$ from the argument to place the equation in shifted form: $\left(\frac{\pi}{4}t + \frac{3\pi}{4}\right) = \frac{\pi}{4}(t + 3)$. The horizontal shift is 3 units left (3 units opposite the sign). For the endpoints of the primary interval we solve $0 \le \frac{\pi}{4}(t + 3) < 2\pi$, which gives $-3 \le t < 5$.

NOW TRY EXERCISES 23 THROUGH 34

GRAPHICAL SUPPORT

The analysis of $y = 2.5\sin\left[\frac{\pi}{4}(t + 3)\right] + 6$ from Example 5(b) can be verified on a graphing calculator. Enter the function as Y_1 on the Y= screen and set an appropriate window size using the information gathered. Press the TRACE key and -3 ENTER and the calculator gives the average value $y = 6$ as output. Repeating this for $x = 5$ shows one complete cycle has been completed.

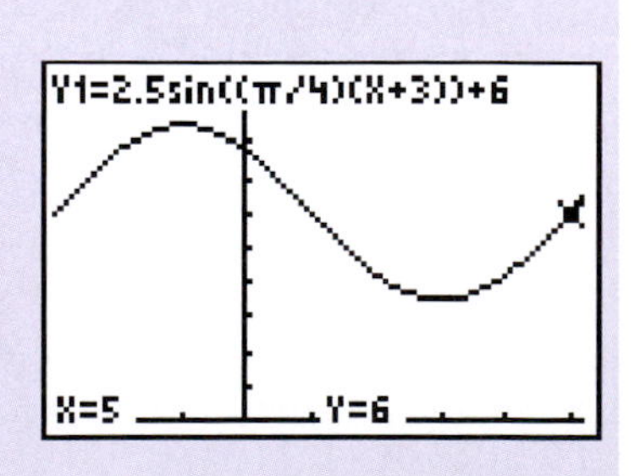

To help gain a better understanding of sinusoidal functions, their graphs, and the role the coefficients A, B, C, and D play, it's often helpful to reconstruct the equation of a given graph.

EXAMPLE 6 Determine the equation of the given graph using a sine function.

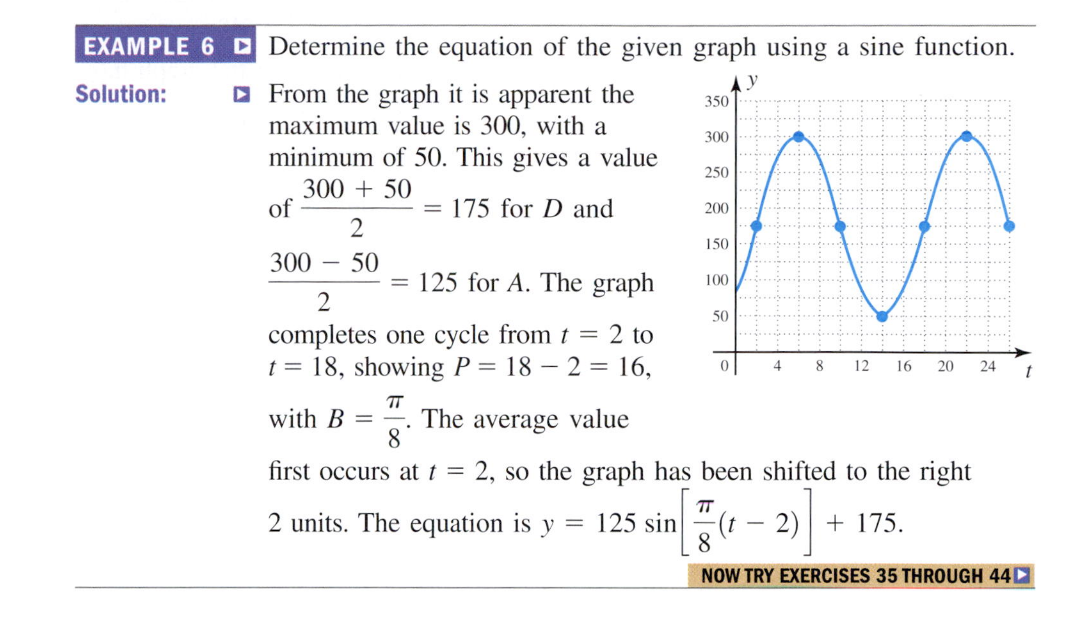

Solution: From the graph it is apparent the maximum value is 300, with a minimum of 50. This gives a value of $\frac{300 + 50}{2} = 175$ for D and $\frac{300 - 50}{2} = 125$ for A. The graph completes one cycle from $t = 2$ to $t = 18$, showing $P = 18 - 2 = 16$, with $B = \frac{\pi}{8}$. The average value first occurs at $t = 2$, so the graph has been shifted to the right 2 units. The equation is $y = 125 \sin\left[\frac{\pi}{8}(t - 2)\right] + 175$.

NOW TRY EXERCISES 35 THROUGH 44

C. Simple Harmonic Motion: $y = A\sin(Bt)$ or $y = A\cos(Bt)$

The periodic motion of springs, tides, sound, and other phenomena all exhibit what is known as **harmonic motion,** which can be modeled using sinusoidal functions.

Harmonic Models—Springs

Consider a spring hanging from a beam with a weight attached to one end. When the weight is at rest, we say it is in **equilibrium,** or has zero displacement from center. Stretching the spring and then releasing it causes the weight to "bounce up and down," with its displacement from center neatly modeled over time by a sine wave (see Figure 2.35).

Figure 2.35 **Figure 2.36**

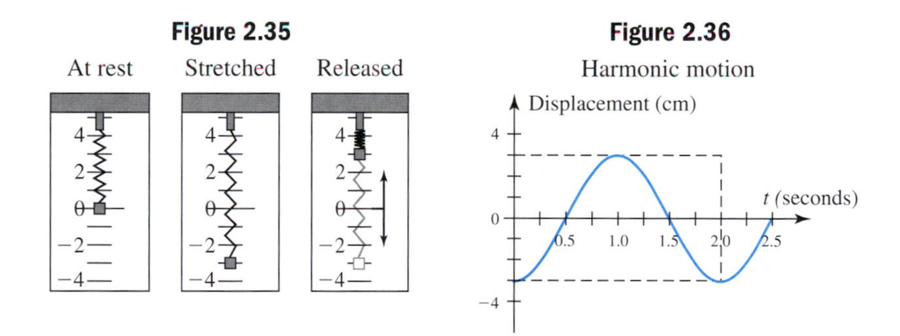

For objects in harmonic *motion* (there are other harmonic models), the input variable t is always a time unit (seconds, minutes, days, etc.), so in addition to the period of the sinusoid, we are very interested in its **frequency—**the number of cycles it completes per unit time (see Figure 2.36). Since the period gives the time required to complete one cycle, the frequency f is given by $f = \frac{1}{P} = \frac{B}{2\pi}$.

EXAMPLE 7 For the harmonic motion modeled by the sinusoid in Figure 2.36, (a) find an equation of the form $y = A\cos(Bt)$; (b) determine the frequency; and (c) use the equation to find the position of the weight at $t = 1.8$ sec.

Solution:

a. By inspection the graph has an amplitude $|A| = 3$ and a period $P = 2$. After substitution into $P = \frac{2\pi}{B}$, we obtain $B = \pi$ and the equation $y = -3\cos(\pi t)$.

b. Frequency is the reciprocal of the period so $f = \frac{1}{2}$, showing one-half a cycle is completed each second (as the graph indicates).

c. Evaluating the model at $t = 1.8$ gives $y = -3\cos[\pi(1.8)] \approx -2.43$, meaning the weight is 2.43 cm below the equilibrium point at this time.

NOW TRY EXERCISES 47 THROUGH 50

Harmonic Models—Sound Waves

A second example of harmonic motion is the production of sound. For the purposes of this study, we'll look at musical notes. The vibration of matter produces a **pressure wave** or **sound energy,** which in turn vibrates the eardrum. Through the intricate structure of the middle ear, this sound energy is converted into mechanical energy and sent to the inner ear where it is converted to nerve impulses and transmitted to the brain. If the sound wave has a high frequency, the eardrum vibrates with greater frequency, which the brain interprets as a "high-pitched" sound. The *intensity* of the sound wave can also be transmitted to the brain via these mechanisms, and if the arriving sound wave has a high amplitude, the eardrum vibrates more forcefully and the sound is interpreted as "loud" by the brain. These characteristics are neatly modeled using $y = A\sin(Bt)$. For the moment we will focus on the frequency, keeping the amplitude constant at $A = 1$.

The musical note known as A_4 or "the A above middle C" is produced with a frequency of 440 vibrations per second, or 440 hertz (Hz) (this is the note most often used in the tuning of pianos and other musical instruments). For any given note, the same note one octave higher will have double the frequency, and the same note one octave lower will have one-half the frequency. In addition, with $f = \frac{1}{P}$ the value of $B = 2\pi\left(\frac{1}{P}\right)$ can always be expressed as $B = 2\pi f$, so A_4 has the equation $y = \sin[440(2\pi t)]$ (after rearranging the factors). The same note one octave lower is A_3 and has the equation $y = \sin[220(2\pi t)]$, with one-half the frequency. To draw the representative graphs, we must scale the t-axis in very small increments (seconds $\times$ 10^{-3}) since $P = \frac{1}{440} \approx 0.0023$ for A_4, and $P = \frac{1}{220} \approx 0.0045$ for A_3. Both are graphed in

Figure 2.37, where we see that the higher note completes two cycles in the same interval that the lower note completes one.

Figure 2.37

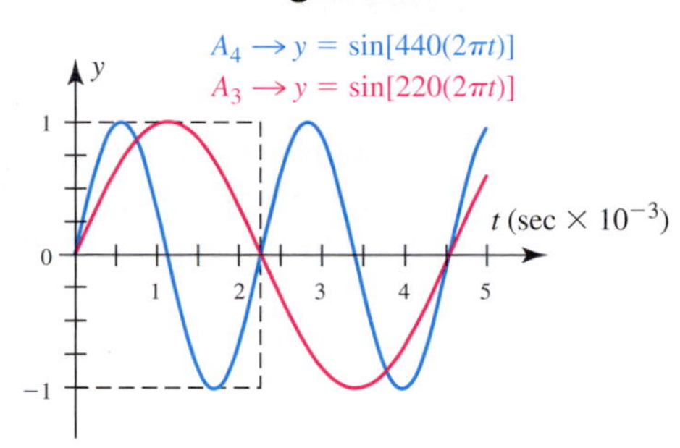

EXAMPLE 8 The table here gives the frequencies for three octaves of the 12 "chromatic" notes with frequencies between 110 Hz and 840 Hz. Two of the 36 notes are graphed in the figure. Which two?

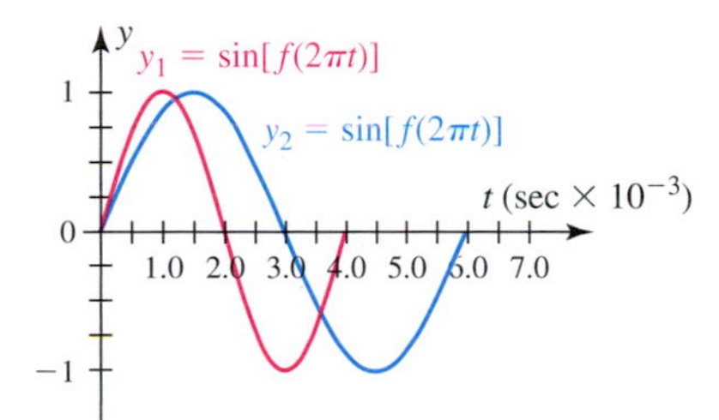

	Frequency by Octave		
Note	**Octave 3**	**Octave 4**	**Octave 5**
A	110.00	220.00	440.00
A#	116.54	233.08	466.16
B	123.48	**246.96**	493.92
C	130.82	261.64	523.28
C#	138.60	277.20	554.40
D	146.84	293.68	587.36
D#	155.56	311.12	622.24
E	**164.82**	329.24	659.28
F	174.62	349.24	698.48
F#	185.00	370.00	740.00
G	196.00	392.00	784.00
G#	207.66	415.32	830.64

Solution: Since amplitudes are equal, the only difference is the frequency and period of the notes. It appears that y_1 has a period of about 0.004 sec, giving a frequency of $\frac{1}{0.004} = 250$ Hz—very likely a B_4 (in bold). The graph of y_2 has a period of about 0.006, for a frequency of $\frac{1}{0.006} \approx 167$ Hz—probably an E_3 (in bold).

NOW TRY EXERCISES 51 THROUGH 54

TECHNOLOGY HIGHLIGHT

Locating Zeroes/Roots/x-intercepts on a Graphing Calculator

The keystrokes shown apply to a TI-84 Plus model. Please consult your manual or our Internet site for other models.

As you know, the zeroes of a function are *input* values that cause an *output* of zero, and are analogous to the roots of an equation. Graphically both show up as *x*-intercepts and once a function is graphed on the *TI-84 Plus,* they can be located (if they exist) using the 2nd CALC **2:zero** feature. This feature is similar to the **3:minimum** and **4:maximum** features, in that we have the calculator search a specified interval by giving a **left bound** and a **right bound**. To illustrate, enter $Y_1 = 3\sin\left(\frac{\pi}{2}x\right) - 1$ on the Y= screen and graph it using the ZOOM **7:ZTrig** option. The resulting graph shows there are six zeroes in this interval and we'll locate the first root to the left of zero. Knowing the ZOOM **7:Trig** option uses tick marks that are spaced every $\frac{\pi}{2}$ units, this root is in the interval $\left(-\pi, -\frac{\pi}{2}\right)$. After pressing 2nd CALC **2:zero** the calculator returns you to the graph, and requests a "Left Bound," asking you to narrow down the interval it has to search (see Figure 2.38). We enter $-\pi$ (press ENTER) as already discussed, and the calculator marks this choice at the top of the screen with a "▶" marker (pointing to the right), then asks you to enter a "Right Bound." After entering $-\frac{\pi}{2}$, the calculator marks this with a "◀" marker and asks for a "Guess." This option is primarily used when the interval selected has more than one zero and you want to guide the calculator toward a particular zero. Often we will simply bypass it by pressing ENTER once again (see Figure 2.39). The calculator searches the specified interval until it locates a zero (Figure 2.40) or displays an error message indicating it was unable to comply (no zeroes in the interval). It's important to note that the "solution" displayed is actually just a very accurate estimate. Use these ideas to locate the zeroes of the following functions in $[0, \pi]$.

Figure 2.38

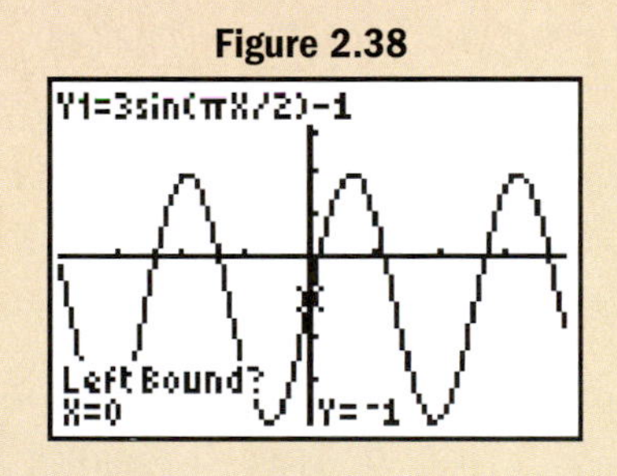

Figure 2.39

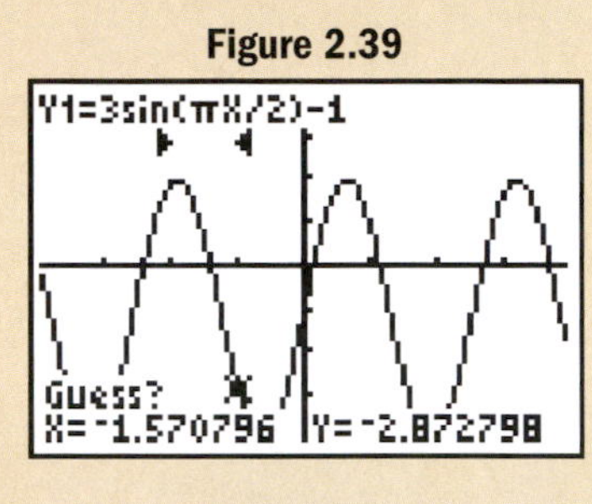

Figure 2.40

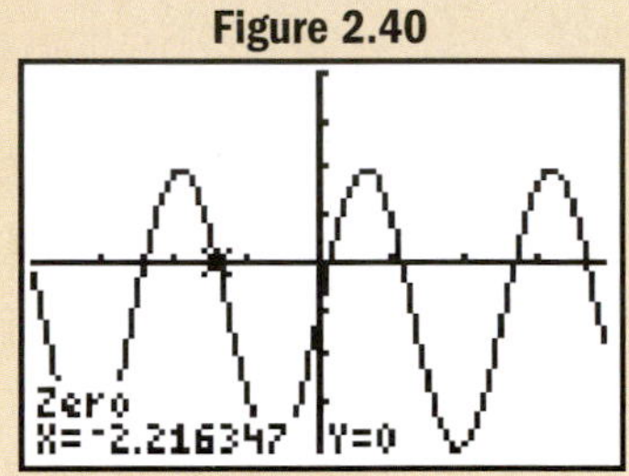

Exercise 1: $y = -2\cos(\pi t) + 1$

Exercise 2: $y = 0.5\sin[\pi(t - 2)]$

Exercise 3: $y = \frac{3}{2}\tan(2x) - 1$

Exercise 4: $y = x^3 - \cos x$

2.3 EXERCISES

CONCEPTS AND VOCABULARY

Fill in each blank with the appropriate word or phrase. Carefully reread the section if needed.

1. A sinusoidal wave or pattern is one that can be modeled by functions of the form ________ or ________.

2. The graph of $y = \sin x + k$ is the graph of $y = \sin x$ shifted ________ k units. The graph of $y = \sin(x - h)$ is the graph of $y = \sin x$ shifted ________ h units.

3. To find the primary interval of a sinusoidal graph, solve the inequality ________.

4. Given the period P, the frequency is ________, and given the frequency f, the value of B is ________.

5. Explain/discuss the difference between the *standard form* of a sinusoidal equation, and the *shifted form*. How do you obtain one from the other? For what benefit?

6. Write out a step-by-step procedure for sketching the graph of $y = 30\sin\left(\frac{\pi}{2}t - \frac{1}{2}\right) + 10$. Include use of the reference rectangle, primary interval, zeroes, max/mins, and so on. Be complete and thorough.

DEVELOPING YOUR SKILLS

Use the graphs given to (a) state the amplitude A and period P of the function; (b) estimate the value at $x = 14$; and (c) estimate the interval in $[0, P]$, where $f(x) \geq 20$.

7. 8.

Use the graphs given to (a) state the amplitude A and period P of the function; (b) estimate the value at $x = 2$; and (c) estimate the interval in $[0, P]$, where $f(x) \leq -100$.

9. 10.

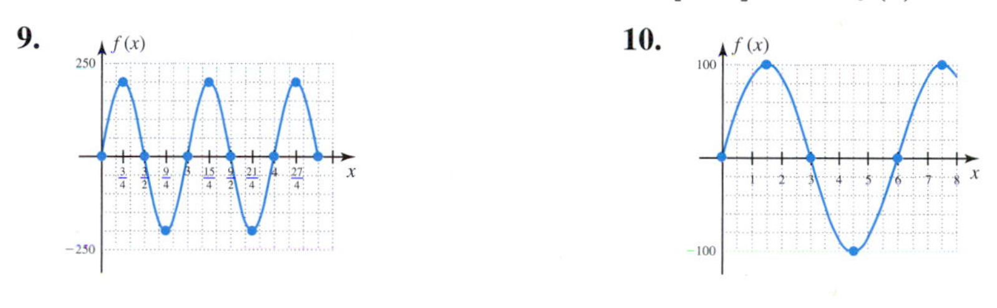

Use the information given to write a sinusoidal equation and sketch its graph. Recall $B = \frac{2\pi}{P}$.

11. Max: 100, min: 20, $P = 30$

12. Max: 95, min: 40, $P = 24$

13. Max: 20, min: 4, $P = 360$

14. Max: 12,000, min: 6500, $P = 10$

Use the information given to write a sinusoidal equation, sketch its graph, and answer the question posed.

15. In Geneva, Switzerland, the daily temperature in January ranges from an average high of 39°F to an average low of 29°F. (a) Find a sinusoidal equation model for the daily temperature; (b) sketch the graph; and (c) approximate the time(s) each January day the temperature reaches the freezing point (32°F). Assume $t = 0$ corresponds to noon.

   Source: *2004 Statistical Abstract of the United States,* Table 1331.

16. In Nairobi, Kenya, the daily temperature in January ranges from an average high of 77°F to an average low of 58°F. (a) Find a sinusoidal equation model for the daily temperature; (b) sketch the graph; and (c) approximate the time(s) each January day the temperature reaches a comfortable 72°F. Assume $t = 0$ corresponds to noon.

   Source: *2004 Statistical Abstract of the United States,* Table 1331.

17. In Oslo, Norway, the number of hours of daylight reaches a low of 6 hr in January, and a high of nearly 18.8 hr in July. (a) Find a sinusoidal equation model for the

number of daylight hours each month; (b) sketch the graph; and (c) approximate the number of *days* each year there are more than 15 hr of daylight. Use 1 month $\approx$ 30.5 days. Assume $t = 0$ corresponds to January 1.

Source: www.visitnorway.com/templates.

18. In Vancouver, British Columbia, the number of hours of daylight reaches a low of 8.3 hr in January, and a high of nearly 16.2 hr in July. (a) Find a sinusoidal equation model for the number of daylight hours each month; (b) sketch the graph; and (c) approximate the number of *days* each year there are more than 15 hr of daylight. Use 1 month $\approx$ 30.5 days. Assume $t = 0$ corresponds to January 1.

Source: www.bcpassport.com/vital/temp.

19. Recent studies seem to indicate the population of North American porcupine (*Erethizon dorsatum*) varies sinusoidally with the solar (sunspot) cycle due to its effects on Earth's ecosystems. Suppose the population of this species in a certain locality is modeled by the function $P(t) = 250\cos\left(\frac{2\pi}{11}t\right) + 950$, where $P(t)$ represents the population of porcupines in year t. Use the model to (a) find the period of the function; (b) graph the function over one period; (c) find the maximum and minimum values; and (d) estimate the number of years the population is less than 740 animals.

Source: Ilya Klvana, McGill University (Montreal), Master of Science thesis paper, November 2002.

20. The population of mosquitoes in a given area is primarily influenced by precipitation, humidity, and temperature. In tropical regions, these tend to fluctuate sinusoidally in the course of a year. Using trap counts and statistical projections, fairly accurate estimates of a mosquito population can be obtained. Suppose the population in a certain region was modeled by the function $P(t) = 50\cos\left(\frac{\pi}{26}t\right) + 950$, where $P(t)$ was the mosquito population (in thousands) in week t of the year. Use the model to (a) find the period of the function; (b) graph the function over one period; (c) find the maximum and minimum population values; and (d) estimate the number of weeks the population is less than 915,000.

21. Use a horizontal translation to shift the graph from Exercise 19 so that the average population of the North American porcupine begins at $t = 0$. Verify results on a graphing calculator, then find a sine function that gives the same graph as the shifted cosine function.

22. Use a horizontal translation to shift the graph from Exercise 20 so that the average population of mosquitoes begins at $t = 0$. Verify results on a graphing calculator, then find a sine function that gives the same graph as the shifted cosine function.

Identify the amplitude (A), period (P), horizontal shift (HS), vertical shift (VS), and endpoints of the primary interval (PI) for each function given.

23. $y = 120\sin\left[\frac{\pi}{12}(t - 6)\right]$ **24.** $y = 560\sin\left[\frac{\pi}{4}(t + 4)\right]$ **25.** $h(t) = \sin\left(\frac{\pi}{6}t - \frac{\pi}{3}\right)$

26. $r(t) = \sin\left(\frac{\pi}{10}t - \frac{2\pi}{5}\right)$ **27.** $y = \sin\left(\frac{\pi}{4}t - \frac{\pi}{6}\right)$ **28.** $y = \sin\left(\frac{\pi}{3}t + \frac{5\pi}{12}\right)$

29. $f(t) = 24.5\sin\left[\frac{\pi}{10}(t - 2.5)\right] + 15.5$ **30.** $g(t) = 40.6\sin\left[\frac{\pi}{6}(t - 4)\right] + 13.4$

31. $g(t) = 28\sin\left(\frac{\pi}{6}t - \frac{5\pi}{12}\right) + 92$ **32.** $f(t) = 90\sin\left(\frac{\pi}{10}t - \frac{\pi}{5}\right) + 120$

33. $y = 2500\sin\left(\frac{\pi}{4}t + \frac{\pi}{12}\right) + 3150$ **34.** $y = 1450\sin\left(\frac{3\pi}{4}t + \frac{\pi}{8}\right) + 2050$

Find the equation of the graph given. Write answers in the form $y = A\sin(Bt + C) + D$.

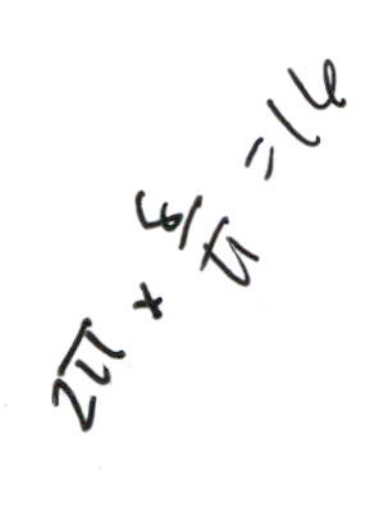

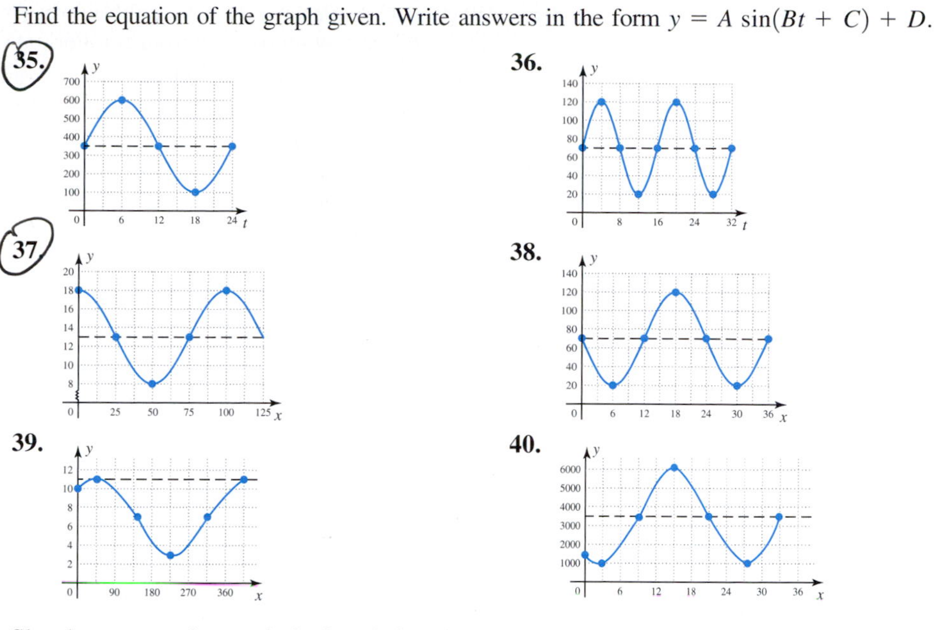

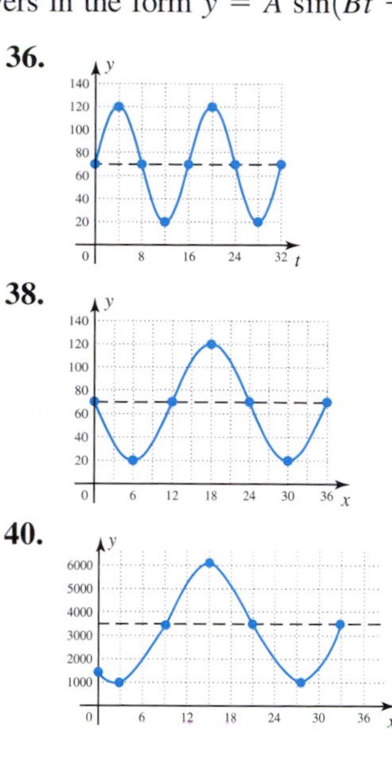

Sketch one complete period of each function.

41. $f(t) = 25\sin\left[\frac{\pi}{4}(t - 2)\right] + 55$

42. $g(t) = 24.5\sin\left[\frac{\pi}{10}(t - 2.5)\right] + 15.5$

43. $h(t) = 1500\sin\left(\frac{\pi}{8}t + \frac{\pi}{4}\right) + 7000$

44. $p(t) = 350\sin\left(\frac{\pi}{6}t + \frac{\pi}{3}\right) + 420$

WORKING WITH FORMULAS

45. The relationship between the coefficient B, the frequency f, and the period P

In many applications of trigonometric functions, the equation $y = A\sin(Bt)$ is written as $y = A\sin[(2\pi f)t]$, where $B = 2\pi f$. Justify the new equation using $f = \frac{1}{P}$ and $P = \frac{2\pi}{B}$. In other words, explain how $A\sin(Bt)$ becomes $A\sin[(2\pi f)t]$, as though you were trying to help another student with the ideas involved.

46. Number of daylight hours: $D(t) = \frac{K}{2}\sin\left[\frac{2\pi}{365}(t - 79)\right] + 12$

The number of daylight hours for a particular day of the year is modeled by the formula given, where $D(t)$ is the number of daylight hours on day t of the year and K is a constant related to the total variation of daylight hours, latitude of the location, and other factors. For the city of Reykjavik, Iceland, $K \approx 17$, while for Detroit, Michigan, $K \approx 6$. How many hours of daylight will each city receive on June 30 (the 182nd day of the year)?

APPLICATIONS

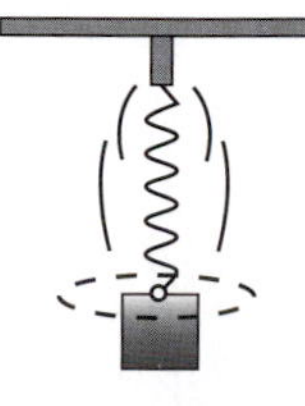

47. Harmonic motion: A weight on the end of a spring is oscillating in harmonic motion. The equation model for the oscillations is $d(t) = 6\sin\left(\frac{\pi}{2}t\right)$, where d is the distance (in centimeters) from the equilibrium point in t sec.

a. What is the period of the motion? What is the frequency of the motion?

b. What is the displacement from equilibrium at $t = 2.5$? Is the weight moving toward the equilibrium point or away from equilibrium at this time?

c. What is the displacement from equilibrium at $t = 3.5$? Is the weight moving toward the equilibrium point or away from equilibrium at this time?

d. How far does the weight move between $t = 1$ and $t = 1.5$ sec? What is the average velocity for this interval? Do you expect a greater or lesser velocity for $t = 1.75$ to $t = 2$? Explain why.

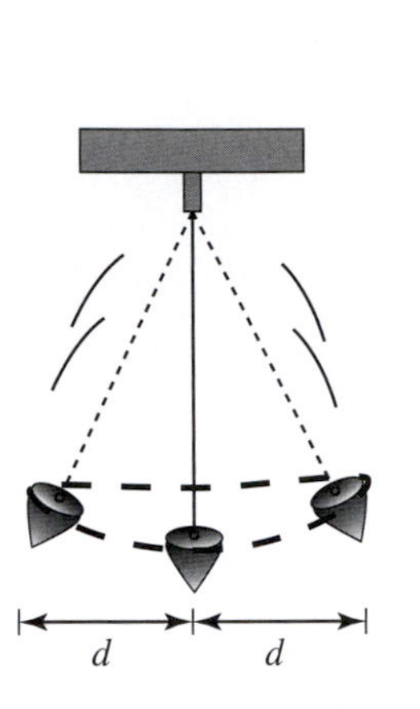

48. Harmonic motion: The bob on the end of a 24-in. pendulum is oscillating in harmonic motion. The equation model for the oscillations is $d(t) = 20\cos(4t)$, where d is the distance (in inches) from the equilibrium point, t sec after being released from one side.

a. What is the period of the motion? What is the frequency of the motion?

b. What is the displacement from equilibrium at $t = 0.25$ sec? Is the weight moving toward the equilibrium point or away from equilibrium at this time?

c. What is the displacement from equilibrium at $t = 1.3$ sec? Is the weight moving toward the equilibrium point or away from equilibrium at this time?

d. How far does the bob move between $t = 0.25$ and $t = 0.35$ sec? What is its average velocity for this interval? Do you expect a greater velocity for the interval $t = 0.55$ to $t = 0.6$? Explain why.

49. Harmonic motion: A simple pendulum 36 in. in length is oscillating in harmonic motion. The bob at the end of the pendulum swings through an arc of 30 in. (from the far left to the far right, or one-half cycle) in about 0.8 sec. What is the equation model for this harmonic motion?

50. Harmonic motion: As part of a study of wave motion, the motion of a floater is observed as a series of uniform ripples of water move beneath it. By careful observation, it is noted that the floater bobs up and down through a distance of 2.5 cm every $\frac{1}{3}$ sec. What is the equation model for this harmonic motion?

51. Sound waves: Two of the musical notes from the chart on page 123 are graphed in the figure. Use the graphs given to determine which two.

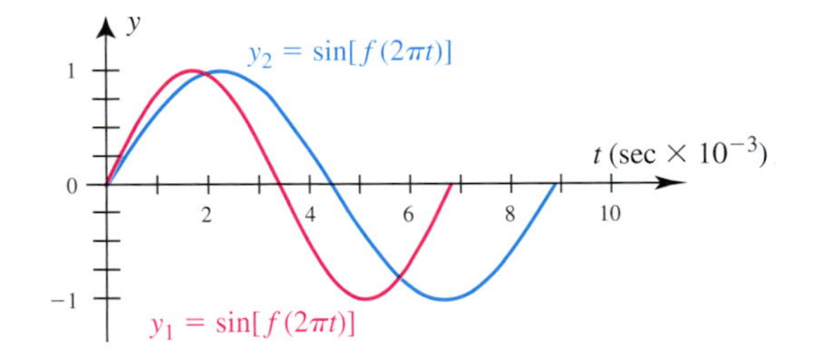

52. Sound waves: Two chromatic notes *not on the chart from page 123* are graphed in the figure. Use the graphs and the discussion regarding octaves to determine which two. Note the scale of the t-axis *has been changed* to hundredths of a second.

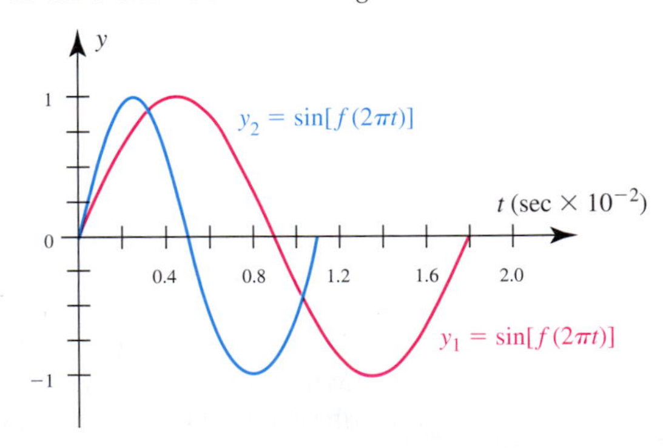

Sound waves: Use the chart on page 123 to draw graphs representing each pair of these notes. Write the equation for each note in the form $y = \sin[f(2\pi t)]$ and clearly state the period of each note.

53. notes D_3 and G_4

54. the notes A_5 and $C\#_3$

Temperature models: Use the information given to determine the amplitude, period, average value (AV), and horizontal shift. Then write the equation and sketch the graph. Assume each context is sinusoidal.

55. During a typical January (summer) day in Buenos Aires, Argentina, the daily high temperature is 84°F and the daily low is 64°F. Assume the low temperature occurs around 6:00 A.M. Assume $t = 0$ corresponds to midnight.

56. In Moscow, Russia, a typical January day brings a high temperature of 21°F and a low of 11°F. Assume the high temperature occurs around 4:00 P.M. Assume $t = 0$ corresponds to midnight.

Daylight hours model: Solve using a graphing calculator and the formula given in Exercise 46.

57. For the city of Caracas, Venezuela, $K \approx 1.3$, while for Tokyo, Japan, $K \approx 4.8$.

a. How many hours of daylight will each city receive on January 15th (the 15th day of the year)?

b. Graph the equations modeling the hours of daylight on the same screen. Then determine (i) what days of the year these two cities will have the same number of hours of daylight, and (ii) the number of days each year that each city receives 11.5 hr or less of daylight.

58. For the city of Houston, Texas, $K \approx 3.8$, while for Pocatello, Idaho, $K \approx 6.2$.

a. How many hours of daylight will each city receive on December 15 (the 349th day of the year)?

b. Graph the equations modeling the hours of daylight on the same screen. Then determine (i) how many days each year Pocatello receives more daylight than Houston, and (ii) the number of days each year that each city receives 13.5 hr or more of daylight.

WRITING, RESEARCH, AND DECISION MAKING

59. Some applications of sinusoidal graphs use variable amplitudes and variable frequencies, rather than constant values of f and P as illustrated in this section. The classic example is AM (amplitude modulated) and FM (frequency modulated) radio waves. Use the Internet, an encyclopedia, or the resources of a local library to research the similarities and differences between AM and FM radio waves, and include a sketch of each type of wave. What are the advantages and disadvantages of each? Prepare a short summary on what you find. For other applications involving variable amplitudes, see the *Calculator Exploration and Discovery* feature at the end of this chapter.

60. A laser beam is actually a very narrow beam of light with a constant frequency and very high intensity (LASER is an acronym for *light amplified* by *stimulated emission* of *radiation*). Lasers have a large and growing number of practical applications, including repair of the eye's retina, sealing of blood vessels during surgery, treatment of skin cancer, drilling small holes, and making precise measurements. In the study of various kinds of laser light, sinusoidal equations are used to model certain characteristics of the light emitted. Do some further reading and research on lasers, and see if you can locate and discuss some of these sinusoidal models.

EXTENDING THE CONCEPT

61. The formulas we use in mathematics can sometimes seem very mysterious. We know they "work," and we can graph and evaluate them—but where did they come from? Consider the formula for the number of daylight hours from Exercise 46: $D(t) = \frac{K}{2}\sin\left[\frac{2\pi}{365}(t - 79)\right] + 12.$

a. We know that the addition of 12 represents a vertical shift, but what does a vertical shift of 12 mean *in this context*?

b. We also know the factor $(t - 79)$ represents a phase shift of 79 to the right. But what does a horizontal (phase) shift of 79 mean *in this context*?

c. Finally, the coefficient $\frac{K}{2}$ represents a change in amplitude, but what does a change of amplitude mean *in this context*? Why is the coefficient bigger for the northern latitudes?

62. Use a graphing calculator to graph the equation $f(x) = \frac{3x}{2} - 2\sin(2x) - 1.5$.

a. Determine the interval between each peak of the graph. What do you notice?

b. Graph $g(x) = \frac{3x}{2} - 1.5$ on the same screen and comment on what you observe.

c. What would the graph of $f(x) = -\frac{3x}{2} + 2\sin(2x) + 1.5$ look like? What is the x-intercept?

MAINTAINING YOUR SKILLS

63. (1.1) In what quadrant does the angle $t = 3.7$ terminate? What is the reference arc?

64. (1.3) The planet Venus orbits the Sun in a path that is nearly circular, with a radius of 67.2 million miles. Through what angle of its orbit has the planet moved after traveling 168 million miles?

65. (1.3) Given $\sin\theta = -\frac{5}{12}$ with $\tan\theta < 0$, find the value of all six trig functions of θ.

66. (1.4) While waiting for John in the parking lot of the hotel, Rick passes the time by estimating the angle of elevation of the hotel's exterior elevator. By careful observation, he notes that from an angle of elevation of 20°, it takes the elevator 5 sec until the angle grows to 52° and reaches the penthouse floor. If his car is parked 150 ft from the hotel, (a) how far does the elevator move during this time? (b) What is the average speed (in miles per hour) of the elevator?

67. (1.1) The vertices of a triangle have coordinates $(-2, 6)$, $(1, 10)$, and $(2, 3)$. Verify that a right triangle is formed and find the measures of the two acute angles.

68. (2.1/2.2) Draw a quick sketch of $y = 2\sin x$, $y = 2\cos x$, and $y = 2\tan x$ for $x \in \left[-\frac{\pi}{2}, 2\pi\right]$.

2.4 Trigonometric Models

LEARNING OBJECTIVES

In Section 2.4 you will learn how to:

A. Create a trigonometric model from critical points or data

B. Create a sinusoidal model from data using regression

INTRODUCTION

In the most common use of the word, a cycle is any series of events or operations that occur in a predictable pattern and return to a starting point. This includes things as diverse as the wash cycle on a washing machine and the powers of i. There are a number of common events that occur in *sinusoidal* cycles, or events that can be modeled by a sine wave. As in Section 2.3, these include monthly average temperatures, monthly average daylight hours, and harmonic motion, among many others. Less well-known applications include alternating current, biorhythm theory, and animal populations that fluctuate over a known period of years. In this section, we develop two methods for creating a sine model. The first uses information about the critical points, where the cycle reaches its maximum or minimum values. The second uses a set of data and the ability of a calculator to run a sinusoidal regression.

POINT OF INTEREST

The *concept* of modeling data with a function is actually very ancient. Early function models consisted of a table of values, as in tracking the length of a gnomon's shadow (the protruding piece on a sundial) each hour of the day—the precursor of the tangent function. Other tables kept track of the number of daylight hours in each month, a phenomenon neatly modeled by a sine function.

A. Critical Points and Sinusoidal Models

Although future courses will define them more precisely, we will consider **critical points** to be *inputs* where a function attains a minimum or maximum value. If an event or phenomenon is known to behave sinusoidally (regularly fluctuating between a maximum and minimum), we can create an acceptable model of the form $y = A\sin(Bx + C) + D$ given these **critical points** (the critical value and the resulting max/min) and the period. For instance, many weather patterns have a period of 12 months. Using the formula $P = \frac{2\pi}{B}$, we find $B = \frac{2\pi}{P}$ and substituting 12 for P gives $B = \frac{\pi}{6}$ (always the case for phenomena with a 12-month cycle). The maximum value of $A\sin(Bx + C) + D$ will always occur when $\sin(Bx + C) = 1$, and the minimum at $\sin(Bx + C) = -1$, giving this system of equations: max value $M = A(1) + D$ and min value $m = A(-1) + D$. Solving the system for A and D gives $A = \frac{M - m}{2}$ and $D = \frac{M + m}{2}$ as before. To find C, assume the maximum and minimum values occur at (x_2, M) and (x_1, m), respectively. We can substitute the values computed for A, B, and D in $y = A\sin(Bx + C) + D$, along with either

(x_2, M) or (x_1, m), and solve for C. Using the minimum value (x_1, m), where $x = x_1$ and $y = m$, we have:

$$y = A\sin(Bx + C) + D \quad \text{sinusoidal equation model}$$
$$m = A\sin(Bx_1 + C) + D \quad \text{substitute } m \text{ for } y \text{ and } x_1 \text{ for } x$$
$$\frac{m - D}{A} = \sin(Bx_1 + C) \quad \text{isolate sine function}$$

Fortunately, for sine models constructed from critical points we have $\frac{y - D}{A} \rightarrow \frac{m - D}{A}$, which is always equal to -1 (see Exercise 33). This gives a simple result for C, since $-1 = \sin(Bx_1 + C)$ leads to $\frac{3\pi}{2} = Bx_1 + C$ or $C = \frac{3\pi}{2} - Bx_1$. Exercises 7 through 12 offer abundant practice with these ideas. Note how they're used in the context of an application.

EXAMPLE 1 When the Spirit and Odyssey Rovers landed on Mars (January 2004), there was a renewed public interest in studying the planet. Of particular interest were the polar ice caps, which are now thought to hold frozen water, especially the northern cap. The Martian ice caps expand and contract with the seasons, just as they do here on Earth but there are about 687 days in a Martian year, making each Martian "month" just over 57 days long (1 Martian day $\approx$ 1 Earth day). At its smallest size, the northern ice cap covers an area of roughly 0.17 million square miles. At the height of winter, the cap covers about 3.7 million square miles (an area about the size of the 50 United States). Suppose these occur at the beginning of month 4 ($x = 4$) and month 10 ($x = 10$) respectively. (a) Use this information to create a sinusoidal model of the form $f(x) = A\sin(Bx + C) + D$, (b) use the model to predict the area of the ice cap in the eighth Martian month, and (c) use a graphing calculator to determine the number of months the cap covers less than 1 million mi^2.

WORTHY OF NOTE

For a complete review of solving equations graphically using a graphing calculator, see Appendix II.

Solution: **a.** Assuming a "12-month" weather pattern, $P = 12$ and $B = \frac{\pi}{6}$. The maximum and minimum points are (10, 3.7) and (4, 0.17). Using this information, $D = \frac{3.7 + 0.17}{2} = 1.935$ and $A = \frac{3.7 - 0.17}{2} = 1.765$. Using $C = \frac{3\pi}{2} - Bx_1$, gives $C = \frac{3\pi}{2} - \frac{\pi}{6}(4) = \frac{5\pi}{6}$. The equation model is $f(x) = 1.765\sin\left(\frac{\pi}{6}x + \frac{5\pi}{6}\right) + 1.935$, where $f(x)$ represents millions of square miles in month x.

b. For the size of the cap in month 8 we evaluate the function at $x = 8$.

$$f(x) = 1.765 \sin\left(\frac{\pi}{6}x + \frac{5\pi}{6}\right) + 1.935 \quad \text{sinusoidal equation model}$$

$$f(8) = 1.765 \sin\left[\frac{\pi}{6}(8) + \frac{5\pi}{6}\right] + 1.935 \quad \text{substitute } x = 8$$

$$= 2.8175 \quad \text{evaluate using a calculator}$$

In month 8, the polar ice cap will cover about 2,817,500 mi^2.

c. Of the many options available, we opt to solve by locating the points where $Y_1 = 1.765 \sin\left(\frac{\pi}{6}x + \frac{5\pi}{6}\right) + 1.935$ and $Y_2 = 1$ intersect. After entering the functions on the Y= screen, we set an appropriate window. Here we set $x \in [0, 12]$ and $y \in [-1, 5]$ for a window with a frame around the output values. On the TI-84 Plus, press 2nd TRACE **(CALC) 5:intersect** to find the intersection points. To four decimal places they occur at $x = 2.0663$ and $x = 5.9337$. The ice cap at the northern pole of Mars has an area of less than 1 million mi^2 from early in the second month to late in the fifth month. The second intersection is shown in the figure.

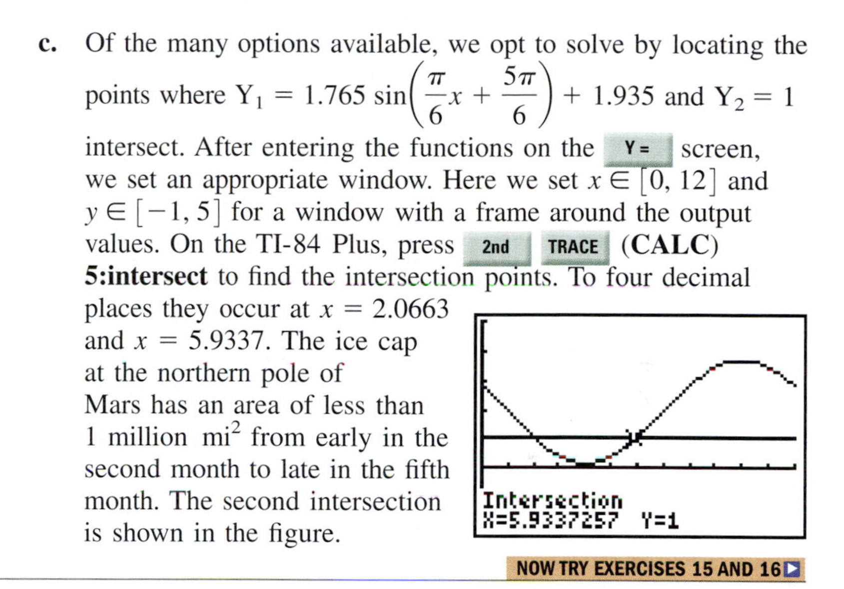

NOW TRY EXERCISES 15 AND 16

While this form of "equation building" can't match the accuracy of a regression model (computed from a larger set of data), it does lend insight as to how sinusoidal functions work. The equation will always contain the maximum and minimum values, and using the period of the phenomena, we can create a smooth sine wave that "fills in the blanks" between these critical points.

EXAMPLE 2 Naturalists have found that many animal populations, such as the arctic lynx, some species of fox, and certain rabbit breeds, tend to fluctuate sinusoidally over 10-year periods. Suppose that an extended study of a lynx population began in 1990, and in the third year of the study, the population had fallen to a minimum of 2500. In the eighth year the population hit a maximum of 9500. (a) Use this information to create a sinusoidal model of the form $P(x) = A \sin(Bx + C) + D$, (b) use the model to predict the lynx population in the year 1996, and (c) use a graphing calculator to determine the number of years the lynx population is above 8000 in a 10-year period.

Solution:

a. Since $P = 10$, we have $B = \frac{2\pi}{10} = \frac{\pi}{5}$. Using 1990 as year zero, the minimum and maximum populations occur at (3, 2500) and (8, 9500). From the information given, $D = \frac{9500 + 2500}{2} = 6000$, and $A = \frac{9500 - 2500}{2} = 3500$. Using the minimum value we have $C = \frac{3\pi}{2} - \frac{\pi}{5}(3) = \frac{9\pi}{10}$, giving an equation model of $P(x) = 3500 \sin\left(\frac{\pi}{5}x + \frac{9\pi}{10}\right) + 6000$, where $P(x)$ represents the lynx population in year x.

b. For the population in 1996 we evaluate the function at $x = 6$.

$$P(x) = 3500 \sin\left(\frac{\pi}{5}x + \frac{9\pi}{10}\right) + 6000 \quad \text{sinusoidal function model}$$

$$P(6) = 3500 \sin\left[\frac{\pi}{5}(6) + \frac{9\pi}{10}\right] + 6000 \quad \text{substitute } x = 6$$

$$\approx 7082 \quad \text{evaluate using a calculator}$$

In 1996, the lynx population was about 7082.

c. Using a graphing calculator and the functions $Y_1 = 3500 \sin\left(\frac{\pi}{5}x + \frac{9\pi}{10}\right) + 6000$ and $Y_2 = 8000$, we attempt to find points of intersection. Enter the functions (press Y=) and set the viewing window (we used $x \in [0, 12]$ and $y \in [0, 10{,}000]$). On the TI-84 Plus, press 2nd TRACE (CALC) **5:intersect** to find where Y_1 and Y_2 intersect. To four decimal places this occurs at $x = 6.4681$ and $x = 9.5319$. The lynx population exceeded 8000 for roughly 3 yr. The first intersection is shown.

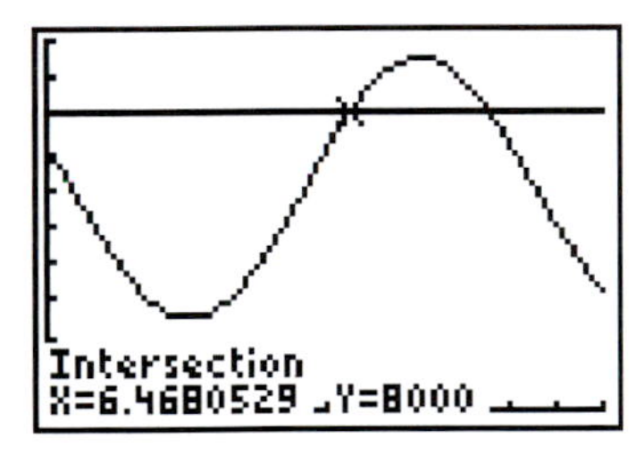

NOW TRY EXERCISES 17 AND 18

This type of equation building isn't limited to the sine function, in fact there are many situations where a sine model cannot be applied. Consider the length of the shadows cast by a flagpole or radio tower as the Sun makes its way across the sky. The shadow's length follows a regular pattern (shortening then lengthening) and "always returns to a starting point," yet when the Sun is low in the sky the shadow becomes (theoretically) infinitely long, unlike the output values from a sine function. In this case, an equation involving $\tan x$ might provide a good model, although the data will vary greatly depending on latitude. We'll attempt to model the data using $y = A\tan(Bx \pm C)$, with the D-term absent since a vertical shift in this context has no meaning. Recall that the period of the tangent function is $P = \frac{\pi}{|B|}$ and that $\pm\frac{C}{B}$ gives the magnitude and direction of the horizontal shift, in a direction opposite the sign.

EXAMPLE 3 The data given tracks the length of a gnomon's shadow for the 12 daylight hours at a certain location near the equator (positive and negative values indicate lengths before noon and after noon). Assume $t = 0$ represents 6:00 A.M. (a) Use the data to find an equation model of the form $L(t) = A\tan(Bt \pm C)$. (b) Graph the function and scatter-plot. (c) Find the shadow's length at 4:30 P.M. (d) If the shadow is 6.1 cm long, what time in the morning is it?

Hour of the Day	Length (cm)	Hour of the Day	Length (cm)
0	∞	7	−2.1
1	29.9	8	−4.6
2	13.9	9	−8.0
3	8.0	10	−13.9
4	4.6	11	−29.9
5	2.1	12	$-\infty$
6	0		

Solution:

a. We begin by noting this phenomenon has a period of $P = 12$. Using the period formula for tangent we solve for B: $P = \frac{\pi}{B}$ gives $12 = \frac{\pi}{B}$, so $B = \frac{\pi}{12}$. Since we want (6, 0) to be the "center" of the function [instead of (0, 0)], we desire a horizontal shift 6 units to the right. Using the ratio $\frac{C}{B}$ $\left(\text{with } B = \frac{\pi}{12}\right)$ gives $-6 = \frac{12C}{\pi}$ so $C = -\frac{\pi}{2}$. To find A we use the equation built so far: $L(t) = A\tan\left(\frac{\pi}{12}t - \frac{\pi}{2}\right)$, and *any data point* to solve for A.

Using (3, 8) we obtain $8 = A\tan\left[\frac{\pi}{12}(3) - \frac{\pi}{2}\right]$:

$$8 = A\tan\left(-\frac{\pi}{4}\right) \quad \text{simplify argument}$$

$$-8 = A \quad \text{solve for } A\text{: } \tan\left(-\frac{\pi}{4}\right) = -1$$

The equation model is $L(t) = -8\tan\left(\frac{\pi}{12}t - \frac{\pi}{2}\right)$.

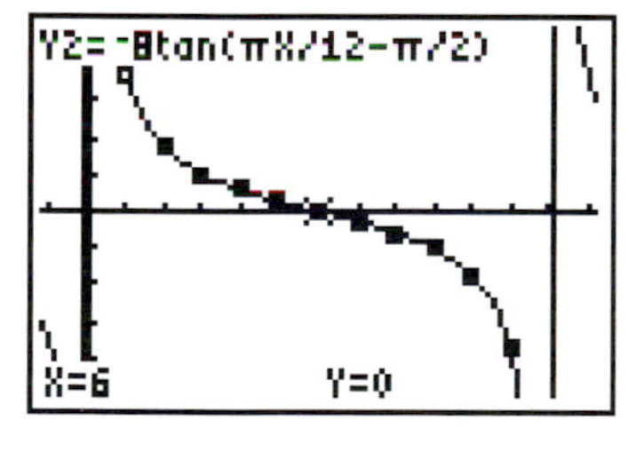

b. The scatter-plot and graph are shown in the figure to the left.

c. 4:30 P.M. indicates $t = 10.5$. Evaluating $L(10.5)$ gives

$$L(t) = -8\tan\left(\frac{\pi}{12}t - \frac{\pi}{2}\right) \quad \text{function model}$$

$$L(10.5) = -8\tan\left[\frac{\pi}{12}(10.5) - \frac{\pi}{2}\right] \quad \text{substitute 10.5 for } t$$

$$= -8\tan\left(\frac{3\pi}{8}\right) \quad \text{simplify argument}$$

$$\approx -19.31 \quad \text{result}$$

At 4:30 P.M., the shadow has a length of $|-19.31| = 19.31$ cm.

d. Substituting 6.1 for $L(t)$ and solving for t graphically gives the graph shown, where we note the day is about 3.5 hr old—it is about 9:30 A.M.

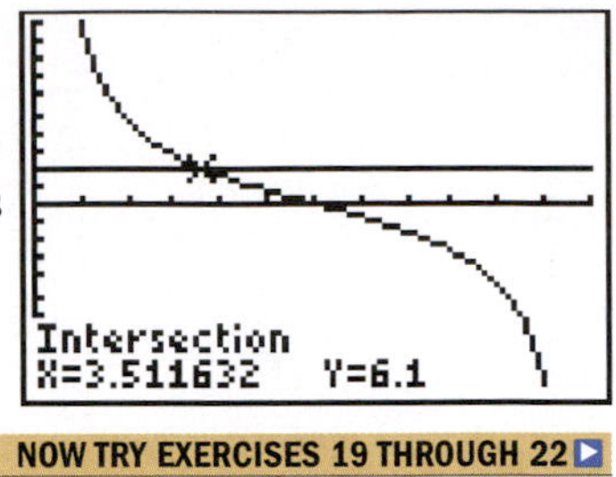

NOW TRY EXERCISES 19 THROUGH 22

B. Data and Sinusoidal Regression

Most graphing calculators are programmed to handle numerous forms of polynomial and nonpolynomial regression, including **sinusoidal regression.** The sequence of steps used is the same regardless of the form chosen, and the fundamentals are reviewed in Appendix III. Exercises 23 through 26 offer further practice with regression fundamentals. Example 4 illustrates their use in context.

EXAMPLE 4 The data shown give the record high temperature for selected months for Bismarck, North Dakota. (a) Use the data to draw a scatter-plot, then find a sinusoidal regression model and graph both on the same screen. (b) Use the equation model to estimate the record high temperatures for months 2, 6, and 8. (c) Determine what month gives the largest difference between the actual data and the computed results.

Month (Jan → 1)	Temp. (°F)	Month (Jan → 1)	Temp. (°F)
1	63	9	105
3	81	11	79
5	98	12	65
7	109		

Source: NOAA Comparative Climate Data 2004.

Solution: **a.** Entering the data and running the regression results in the coefficients shown in Figure 2.41. After entering the equation in Y_1 and pressing ZOOM **9:Zoom Stat** we obtain the graph shown in Figure 2.42.

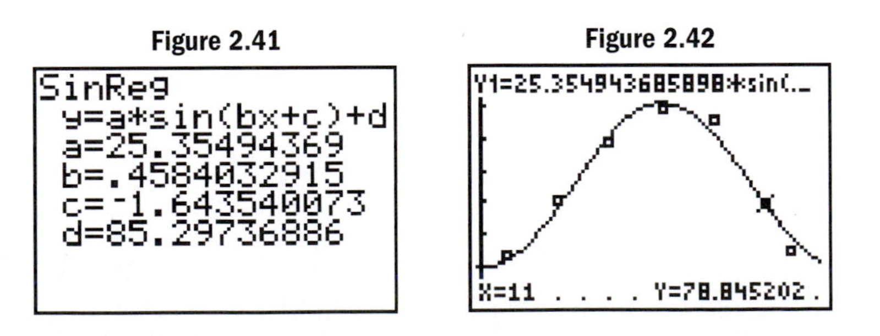

Figure 2.43

L2	L3	L4
63	61.805	1.1953
81	78.575	2.4248
98	100.61	-2.611
109	110.65	-1.652
105	100.83	4.167
79	78.845	.1548
65	68.661	-3.661

L4(1)=1.195267717...

b. Using $x = 2$, $x = 6$, and $x = 8$ as inputs, projects record high temperatures of 68.5°, 108.0°, and 108.1°, respectively.

c. In the header of L3, use Y_1(L1) ENTER to evaluate the regression model using the inputs from L1, and place the results in L3. Entering L2 − L3 in the header of L4 gives the results shown in Figure 2.43 and we note the largest difference occurs in September—about 4°.

NOW TRY EXERCISES 27 AND 28

Weather patterns differ a great deal depending on the locality. For example, the annual rainfall received by Seattle, Washington, far exceeds that received by Cheyenne, Wyoming. Our final example compares the two amounts and notes an interesting fact about the relationship.

EXAMPLE 5 The average monthly rainfall (in inches) for Cheyenne, Wyoming, and Seattle, Washington, is shown in the table. (a) Use the data to find a sinusoidal regression model for the average monthly rainfall in each city. Enter or paste the equation for Cheyenne in Y_1 and the equation for Seattle in Y_2. (b) Graph both equations on the same screen (without the scatter-plots) and use TRACE or 2nd TRACE **(CALC) 5:intersect** to help estimate the number of months Cheyenne receives more rainfall than Seattle.

Source: NOAA Comparative Climate Data 2004.

Month (Jan. → 1)	WY Rain	WA Rain
1	0.45	5.13
2	0.44	4.18
3	1.05	3.75
4	1.55	2.59
5	2.48	1.77
6	2.12	1.49
7	2.26	0.79
8	1.82	1.02
9	1.43	1.63
10	0.75	3.19
11	0.64	5.90
12	0.46	5.62

Solution:

a. Setting the calculator in **Float 0 1 2 3 4 5 6 7 8 9** MODE and running sinusoidal regressions gives the equations shown in Figure 2.44.

b. Both graphs are shown in Figure 2.45. Using the TRACE feature, we find the graphs intersect at approximately (4.7, 2.0) and (8.4, 1.7). Cheyenne receives more rain than Seattle for about $8.4 - 4.7 = 3.7$ months of the year.

Figure 2.44 **Figure 2.45**

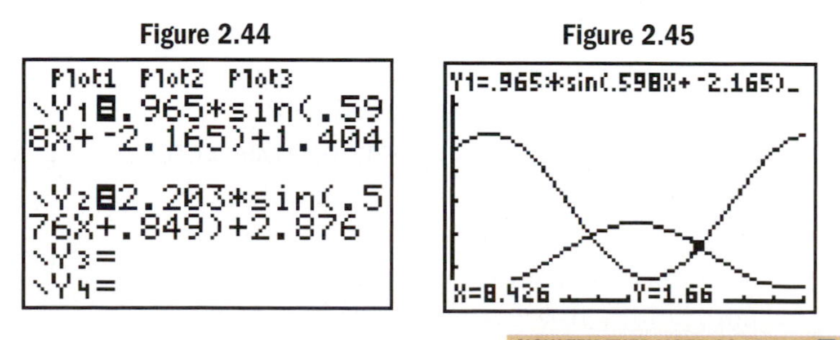

NOW TRY EXERCISES 29 AND 30

The exercise set offers additional significant and interesting applications, and many more can be found by searching the Internet for data generated by sinusoidal phenomena.

TECHNOLOGY HIGHLIGHT

Working with Data—Common "Breaks and Fixes"

The keystrokes illustrated refer to a TI-84 Plus model. For other models please consult your manual or visit our Internet site.

In the process of using a calculator to explore data sets, extensive work is done with lists, regressions, equations, tables, and graphs, with these and other features combined. With so many ideas working together, it helps to be aware of the most common mistakes and how to correct them. The following list is not exhaustive, but could prove helpful when using graphing and calculating technology in support of your studies.

1. **Break:** You accidently skip a data entry; **Fix:** 2nd DEL (**INS**).
   The calculator will insert a new position between the cursor location and the previous datum.
2. **Break:** You enter data in the wrong list, say L2 instead of L1; **Fix:** Use L1 = L2.
   Place the cursor in the heading of List1 and press 2nd 2 (L2) ENTER. The calculator will automatically copy the contents of L2 into L1. You can then overwrite or DEL the entries in L2.
3. **Break:** You accidently delete a List name; **Fix:** Reset List names using **5:SetUpEditor.**
   Using STAT **5:SetUpEditor** will reset list names to the default L1 through L6, without affecting the data in any list. Use with caution in case you have Lists stored under custom names.
4. **Break:** You get the message **ERR:DIM MISMATCH; Fix:** several.
   This usually occurs when the lists used to run a regression have an unequal number of entries (one was overlooked or accidently deleted). The primary fix is to double-check the Lists and find the missing entry. If you are investigating the regression *equation* and no longer need a scatter-plot, try turning the plots off (2nd Y= (**STATPLOT**)) so the calculator won't recognize the lists.
5. **Break:** The axes do not show up on the GRAPH screen; **Fix:** 2nd zoom **AXESON**.
   Some programs run on the TI will "turn-off" the axes so they don't interfere with how the program is displayed. Just turn them back on.
6. **Break:** You get the message **ERR:WINDOW RANGE; Fix:** reset window size.
   This occurs if you accidently enter Ymax < Ymin or Xmax < Xmin (other situations as well). Simply double-check and correct the entries for window size.
7. **Break:** You get the message **ERR:domain; Fix:** check context/adjust data.
   Most of the time this happens when the form of regression you've chosen can't use zero entries. Rethink the context of the data and try another form of regression, or replace the zero with 0.001. (Use with caution! $0 \neq 0.001$ and sometimes results can be significantly affected.)

Exercise 1: In Example 4, suppose the entry 98° was accidently skipped when inputting the data. Do you place the cursor on the 81 or the 109 before pressing 2nd DEL (INS)?

Exercise 2: While working through Example 5, the temperatures for Seattle were accidently listed in L1 instead of L3 where you wanted them. How do you move the entries to L3?

2.4 EXERCISES

CONCEPTS AND VOCABULARY

Fill in each blank with the appropriate word or phrase. Carefully reread the section if needed.

1. For $y = A\sin(Bx + C) + D$, the maximum value occurs when ________ $= 1$, leaving $y =$ ________.

2. For $y = A\sin(Bx + C) + D$, the minimum value occurs when ________ $= -1$, leaving $y =$ ________.

3. For $y = A\sin(Bx + C) + D$, the value of C is found using $C = \frac{3\pi}{2} - Bx$, where x is a ________ point.

4. Any phenomenon with sinusoidal behavior regularly fluctuates between a ________ and ________ value.

5. Rework Example 1, but this time use the maximum value in the calculation for C. Do the equations differ? Discuss/explain why either critical point can be used to build the equation model.

6. Discuss/explain: (a) How is data concerning average rainfall related to the average discharge rate of rivers? (b) How is data concerning average daily temperature related to water demand? Are both sinusoidal with the same period? Are there other associations you can think of?

DEVELOPING YOUR SKILLS

Find the sinusoidal equation for the information as given.

7. minimum value at (9, 25); maximum value at (3, 75); period: 12 min

8. minimum value at (4.5, 35); maximum value at (1.5, 121); period: 6 yr

9. minimum value at (15, 3); maximum value at (3, 7.5); period: 24 hr

10. minimum value at (3, 3.6); maximum value at (7, 12); period: 8 hr

11. minimum value at (5, 279); maximum value at (11, 1285); period: 12 yr

12. minimum value at (6, 8280); maximum value at (22, 23,126); period: 32 yr

WORKING WITH FORMULAS

13. **Orbiting distance north or south of the equator: $D(t) = A\cos(Bt)$**

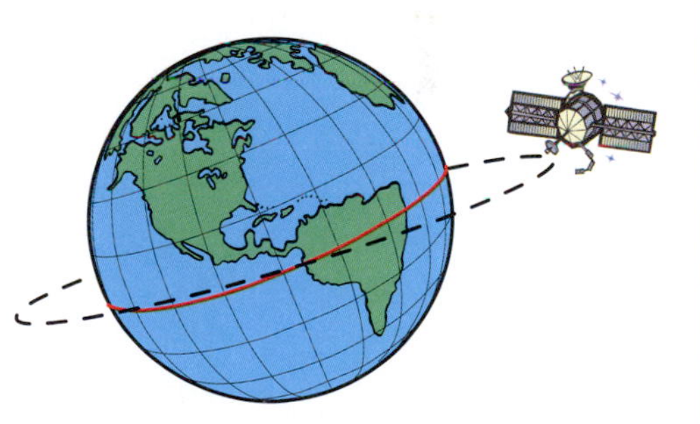

Unless a satellite is placed in a strict equatorial orbit, its distance north or south of the equator will vary according to the sinusoidal model shown, where $D(t)$ is the distance t min after entering orbit. Negative values indicate it is south of the equator, and the distance D is actually a two-dimensional distance, as seen from a vantage point in outer space. The value of B depends on the speed of the satellite and the time it takes to complete one orbit, while $|A|$ represents the maximum distance from the equator. (a) Find the equation model for a satellite whose maximum distance north of the equator is 2000 miles and that completes one orbit every 2 hr ($P = 120$). (b) How many minutes after entering orbit is the satellite directly above the equator $[D(t) = 0]$? (c) Is the satellite north or south of the equator 257 min after entering orbit? How far north or south?

14. **Biorhythm theory: $P(d) = 50\sin(Bd) + 50$**

Advocates of biorhythm theory believe that human beings are influenced by certain biological cycles that begin at birth, have different periods, and continue throughout life. The classical cycles and their periods are physical potential (23 days), emotional potential (28 days), and intellectual potential (33 days). On any given day of life, the percent of potential in these three areas is purported to be modeled by the function shown, where $P(d)$ is the percent of available potential on day d of life. Find the value of B for each of the physical, emotional, and intellectual potentials and use it to see what the theory has to say about your potential today. Use day $d = 365.25(\text{age}) +$ days since last birthday.

APPLICATIONS

15. The U.S. National Oceanic and Atmospheric Administration (USNOAA) keeps temperature records for most major U.S. cities. For Phoenix, Arizona, they list an average high temperature of 65.0°F for the month of January (month 1) and an average high temperature of 104.2°F for July (month 7). Assuming January and July are the coolest and warmest months of the year, (a) build a sinusoidal function model for temperatures in Phoenix, and (b) use the model to find the average high temperature in September. (c) If a person has a tremendous aversion to temperatures over 95°, during what months should they plan to vacation elsewhere?

16. Much like the polar ice cap on Mars, the sea ice that surrounds the continent of Antarctica (the Earth's southern polar cap) varies seasonally, from about 8 million mi^2 in September to about 1 million mi^2 in March. Use this information to (a) build a sinusoidal equation that models the advance and retreat of the sea ice, and (b) determine the size of the ice sheet in May. (c) Find the months of the year that the sea ice covers more than 6.75 million mi^2.

Exercise 16

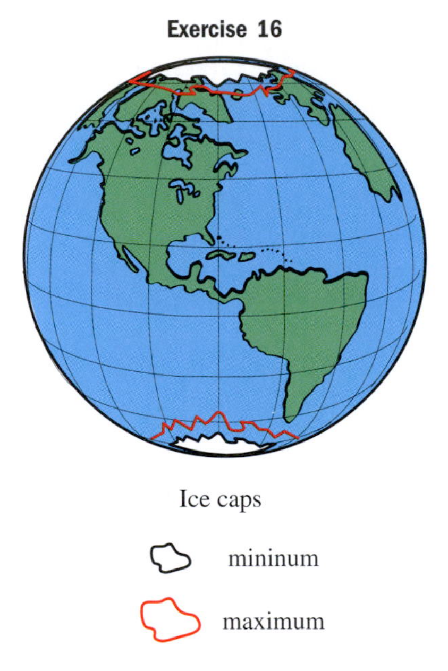

17. A phenomenon is said to be *circadian* if it occurs in 24-hr cycles. A person's body temperature is circadian, since there are normally small, sinusoidal variations in body temperature from a low of 98.2°F to a high of 99°F throughout a 24-hr day. Use this information to (a) build the circadian equation for a person's body temperature, given $t = 0$ corresponds to midnight and that a person usually reaches their minimum temperature at 5 A.M.; (b) find the time(s) during a day when a person reaches "normal" body temperature (98.6°); and (c) find the number of hours each day that body temperature is 98.4°F or less.

18. For an internal combustion engine, the position of a piston in the cylinder can be modeled by a sinusoidal function. For a particular engine size and idle speed, the piston head is 0 in. from the top of the cylinder (the minimum value) when $t = 0$ at the beginning of the intake stroke, and reaches a maximum distance of 4 in. from the top of the cylinder (the maximum value) when $t = \frac{1}{48}$ sec at the beginning of the compression stroke. Following the compression stroke is the power stroke ($t = \frac{2}{48}$), the exhaust stroke ($t = \frac{3}{48}$), and the intake stroke ($t = \frac{4}{48}$), after which it all begins again. Given the period of a four-stroke engine under these conditions is $P = \frac{1}{24}$ second, (a) find the sinusoidal equation modeling the position of the piston, and (b) find the distance of the piston from the top of the cylinder at $t = \frac{1}{9}$ sec. Which stroke is the engine in at this moment?

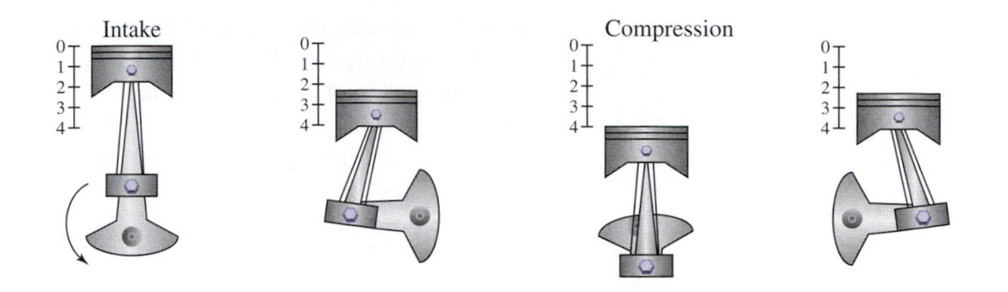

Use the data given to find an equation model of the form $f(x) = A\tan(Bx + C)$. Then graph the function and scatter plot to help find (a) the output for $x = 2.5$ and (b) the value of x where $f(x) = 16$.

19.

x	y	x	y
0	∞	7	−1.4
1	20	8	−3
2	9.7	9	−5.2
3	5.2	10	−9.7
4	3	11	−20
5	1.4	12	$-\infty$
6	0		

20.

x	y	x	y
0	∞	7	6.4
1	−91.3	8	13.7
2	−44.3	9	23.7
3	−23.7	10	44.3
4	−13.7	11	91.3
5	−6.4	12	∞
6	0		

Exercise 21

Distance Traveled (mi)	Height (cm)
0	0
3	1
6	1.8
9	2.8
12	4.2
15	6.3
18	10
21	21
24	∞

21. While driving toward a Midwestern town on a long, flat stretch of highway, I decide to pass the time by measuring the apparent height of the tallest building in the downtown area as I approach. At the time the idea occurred to me, the buildings were barely visible. Three miles later I hold a 30 cm ruler up to my eyes at arm's length, and the apparent height of the tallest building is 1 cm. After three more miles the apparent height is 1.8 cm. Measurements are taken every 3 mi until I reach town and are shown in the table (assume I was 24 mi from the parking garage when I began this activity). (a) Use the data to come up with a tangent function model of the building's apparent height after traveling a distance of x mi closer. (b) What was the apparent height of the building at after I had driven 19 mi? (c) How many miles had I driven when the apparent height of the building took up all 30 cm of my ruler?

22. The **theory of elastic rebound** has been used by seismologists to study the cause of earthquakes. As seen in the figure, the Earth's crust is stretched to a breaking point by the slow movement of one tectonic plate in a direction opposite the other along a fault line, and when the rock snaps—each half violently rebounds to its original alignment causing the Earth to quake.

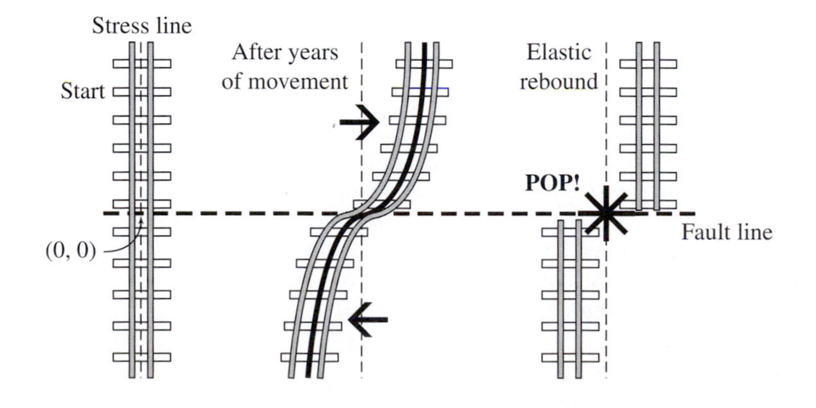

x	y	x	y
−4.5	−61	1	2.1
−4	−26	2	6.8
−3	−14.8	3	15.3
−2	−7.2	4	25.4
−1	−1.9	4.5	59
0	0		

Suppose the *misalignment* of these plates through the stress and twist of crustal movement can be modeled by a tangent graph, where x represents the horizontal distance from the original stress line, and y represents the vertical distance from the fault line. Assume a "period" of 10.2 m. (a) Use the data from the table on page 141 to come up with a trigonometric model of the deformed stress line. (b) At a point 4.8 m along the fault line, what is the distance to the deformed stress line (moving parallel to the original stress line)? (c) At what point along the fault line is the vertical distance to the deformed stress line 50 m?

For the following sets of data (a) find a sinusoidal regression equation using your calculator; (b) construct an equation manually using the period and maximum/minimum values; and (c) graph both on the same screen, then use a TABLE to find the largest difference between output values.

23.

Day of Month	Output
1	15
4	41
7	69
10	91
13	100
16	90
19	63
22	29
25	5
28	2
31	18

24.

Day of Month	Output
1	179
4	201
7	195
10	172
13	145
16	120
19	100
22	103
25	124
28	160
31	188

25.

Month (Jan. = 1)	Output
1	16
2	19
3	21
4	22
5	21
6	19
7	16
8	13
9	11
10	10
11	11
12	13

26.

Month (Jan. = 1)	Output
1	86
2	96
3	99
4	95
5	83
6	72
7	56
8	48
9	43
10	49
11	58
12	73

27. The highest temperature of record for the even months of the year are given in the table for the city of Pittsburgh, Pennsylvania. (a) Use the data to draw a scatter-plot, then find a sinusoidal regression model and graph both on the same screen. (b) Use the equation to estimate the record high temperature for the odd-numbered months. (c) What month shows the largest difference between the actual data and the computed results?

Source: 2004 Statistical Abstract of the United States, Table 378.

Month (Jan → 1)	High Temp. (°F)
2	76
4	89
6	98
8	100
10	87
12	74

28. The average discharge rate of the Alabama River is given in the table for the odd-numbered months of the year. (a) Use the data to draw a scatter-plot, then find a sinusoidal regression model and graph both on the same screen. (b) Use the equation to estimate the flow rate for the even-numbered months. (c) Use the graph and equation to estimate the number of days per year the flow rate is below 500 m^3/sec.

Source: Global River Discharge Database Project; www.rivdis.sr.unh.edu.

Month (Jan → 1)	Rate (m^3/sec)
1	1569
3	1781
5	1333
7	401
9	261
11	678

29. The average monthly rainfall (in inches) for Reno, Nevada, is shown in the table. (a) Use the data to find a sinusoidal regression model for the monthly rainfall. (b) Graph this equation model and the rainfall equation model for Cheyenne, Wyoming (from Example 5), on the same screen, and estimate the number of months that Reno gets more rainfall than Cheyenne.

Month (Jan → 1)	Reno Rainfall	Month (Jan → 1)	Reno Rainfall
1	1.06	7	0.24
2	1.06	8	0.27
3	0.86	9	0.45
4	0.35	10	0.42
5	0.62	11	0.80
6	0.47	12	0.88

Source: NOAA Comparative Climate Data 2004.

30. The number of daylight hours per month (as measured on the 15th of each month) is shown in the table for the cities of Beaumont, Texas, and Minneapolis, Minnesota. (a) Use the data to find a sinusoidal regression model of the daylight hours for each city. (b) Graph both equations on the same screen and use the graphs to estimate the *number of days* each year that Beaumont receives more daylight than Minneapolis (use 1 month = 30.5 days).

Source: www.encarta.msn.com/media_701500905/Hours_of_Daylight_by_Latitude.html.

Exercise 30

Month (Jan → 1)	TX Sunlight	MN Sunlight
1	10.4	9.1
2	11.2	10.4
3	12.0	11.8
4	12.9	13.5
5	14.4	16.2
6	14.1	15.7
7	13.9	15.2
8	13.3	14.2
9	12.4	12.6
10	11.5	11.0
11	10.7	9.6
12	10.2	8.7

31. The table given indicates the percent of the Moon that is illuminated for the days of a particular month, at a given latitude. (a) Use a graphing calculator to find a sinusoidal regression model. (b) Use the model to determine what percent of the Moon is illuminated on day 20. (c) Use the maximum and minimum values with the period and an appropriate horizontal shift to create your own model of the data. How do the values for A, B, C, and D compare?

32. The mood of persons with SAD syndrome (seasonal affective disorder) often depends on the weather. Victims of SAD are typically more despondent in rainy weather than when the Sun is out, and more comfortable in the daylight hours than at night. The table shows the average number of daylight hours for Vancouver, British Columbia, for 12 months of a year. (a) Use a calculator to find a sinusoidal regression model. (b) Use the model to estimate the *number of days* per year (use 1 month $\approx$ 30.5 days) with more than 14 hr of daylight. (c) Use the maximum and minimum values with the period and an appropriate horizontal shift to *create* a model of the data. How do the values for A, B, C, and D compare?

Source: Vancouver Climate at www.bcpassport.com/vital.

Exercise 31

Day	% Illum.	Day	% Illum.
1	28	19	34
4	55	22	9
7	82	25	0
10	99	28	9
13	94	31	30
16	68		

Exercise 32

Month	Hours	Month	Hours
1	8.3	7	16.2
2	9.4	8	15.1
3	11.0	9	13.5
4	12.9	10	11.7
5	14.6	11	9.9
6	15.9	12	8.5

WRITING, RESEARCH, AND DECISION MAKING

33. For the equations from Examples 1 and 2, use the minimum value (x, m) to show that $\frac{y - D}{A} \rightarrow \frac{m - D}{A}$ is equal to -1. Then verify this relationship in general by substituting $\frac{M - m}{2}$ for A, $\frac{M + m}{2}$ for D.

34. Unlike the rainfall data from Example 5, precipitation models are often valid for only part of the year, usually during an annual rainy season. Using the Internet or the resources of a local library, do some research on rainfall patterns and construct a sinusoidal exercise of your own.

EXTENDING THE CONCEPT

35. A *dampening factor* is any function whose product with a sinusoidal function causes a systematic change in amplitude. In the graph shown, $y = \sin(3x)$ is dampened by the function $y = -\frac{1}{4}x + 2$. Notice the peaks and valleys of the sine graph are points on the graph of this line. The table given shows points of intersection for $y = \sin(3x)$ and another dampening factor. Use the regression capabilities of a graphing calculator to find its equation and graph both functions on the same screen.

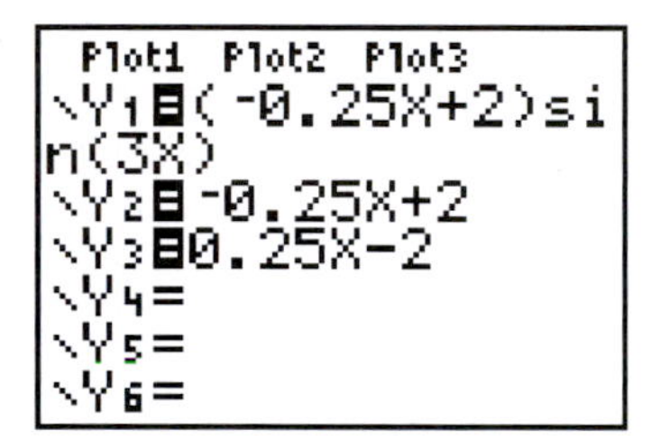

x	y	x	y
$-\frac{7\pi}{4}$	4.64	$\frac{5\pi}{4}$	1.33
$-\frac{3\pi}{4}$	3.14	$\frac{9\pi}{4}$	1.02
$\frac{\pi}{4}$	2.04	$\frac{13\pi}{4}$	1.09

MAINTAINING YOUR SKILLS

36. (2.2) Draw a quick sketch of $y = -\tan x$ for $x \in \left[-\frac{3\pi}{2}, 2\pi\right]$.

37. (1.3) What four values of t satisfy the equation $|\cos t| = \frac{1}{2}$?

38. (2.3) Determine the equation of the graph shown, given it's of the form $y = A\cos(Bt \pm C) + D$.

Exercise 38

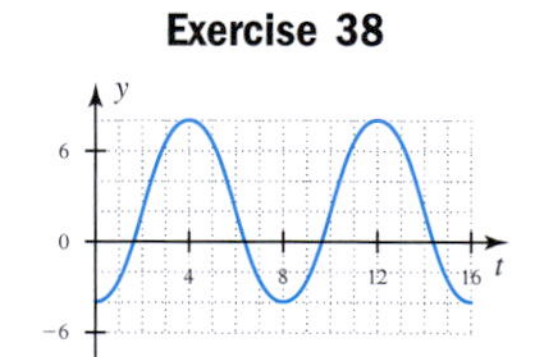

39. (2.2) The graph of $y = \sec x$ is shifted to the right $\frac{\pi}{3}$ units. What is the equation of the shifted graph?

40. (1.2) Clarke is standing between two tall buildings. The angle of elevation to the top of the building to her north is 60°, while the angle of elevation to the top of the building to her south is 70°. If she is 400 m from the northern building and 200 m from the southern building, which one is taller?

41. (1.1) Lying on his back in three inches of newly fallen snow, Mitchell stretches out his arms and moves them back and forth (like a car's wiper blades) to form a snow angel. If his arms are 24 in. long and move through an angle of 35° at the shoulders, what is the area of the angel's wings that are formed?

SUMMARY AND CONCEPT REVIEW

SECTION 2.1 Graphs of the Sine and Cosine Functions

KEY CONCEPTS

- On a unit circle, $\cos t = x$ and $\sin t = y$. Graphing these functions using the x- and y-coordinates of points on the unit circle, results in a periodic, wavelike graph with domain $(-\infty, \infty)$.

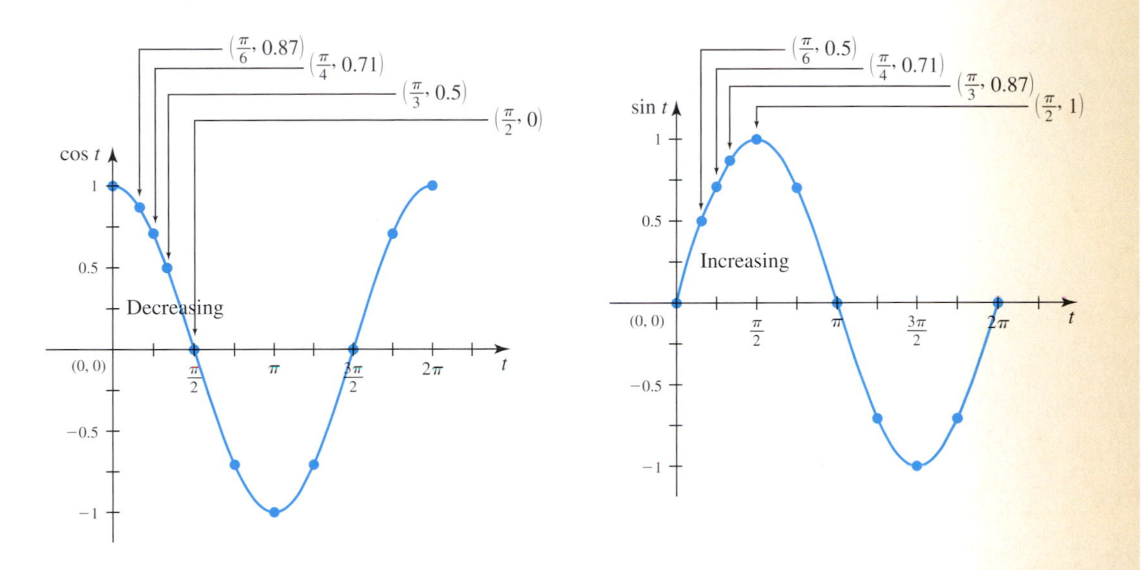

- The characteristics of each graph play a vital role in their contextual application, and these are summarized on pages 81 and 83.
- The amplitude of a sine or cosine graph is the maximum displacement from the average value. For $y = A\sin(Bt)$ and $y = A\cos(Bt)$, where no vertical shift has been applied, the average value is $y = 0$ (x-axis).
- When sine and cosine functions are applied in context, they often have amplitudes other than 1 and periods other than 2π. For $y = A\sin(Bt)$ and $y = A\cos(Bt)$, $|A|$ gives the amplitude; $P = \frac{2\pi}{B}$ gives the period.
- If $|A| > 1$, the graph is vertically stretched, if $0 < |A| < 1$ the graph is vertically compressed, and if $A < 0$ the graph is reflected across the x-axis, just as with algebraic functions.
- If $B > 1$, the graph is horizontally compressed (the period is smaller/shorter); if $B < 1$ the graph is horizontally stretched (the period is larger/longer).
- To graph $y = A\sin(Bt)$ or $A\cos(Bt)$, draw a reference rectangle $2A$ units high and $P = \frac{2\pi}{B}$ units wide, then use the *rule of fourths* to locate zeroes and max/min values. Connect these points with a smooth curve.
- The graph of $y = \sec t = \frac{1}{\cos t}$ will be asymptotic everywhere $\cos t = 0$, increasing where $\cos t$ is decreasing, and decreasing where $\cos t$ is increasing. It will also "share" the max/min values of $\cos t$.

- The graph of $y = \csc t = \dfrac{1}{\sin t}$ will be asymptotic everywhere $\sin t = 0$, increasing where $\sin t$ is decreasing, and decreasing where $\sin t$ is increasing. It will also "share" the max/min values of $\sin t$.

EXERCISES

Use a reference rectangle and the *rule of fourths* to draw an accurate sketch of the following functions through at least one full period. Clearly state the amplitude (as applicable) and period as you begin.

1. $y = 3 \sin t$

2. $y = 3 \sec t$

3. $y = -\cos(2t)$

4. $y = 1.7 \sin(4t)$

5. $f(t) = 2 \cos(4\pi t)$

6. $g(t) = 3 \sin(398\pi t)$

The given graphs are of the form $y = A \sin(Bt)$ and $y = A \csc(Bt)$. Determine the equation of each graph.

7. **8.**

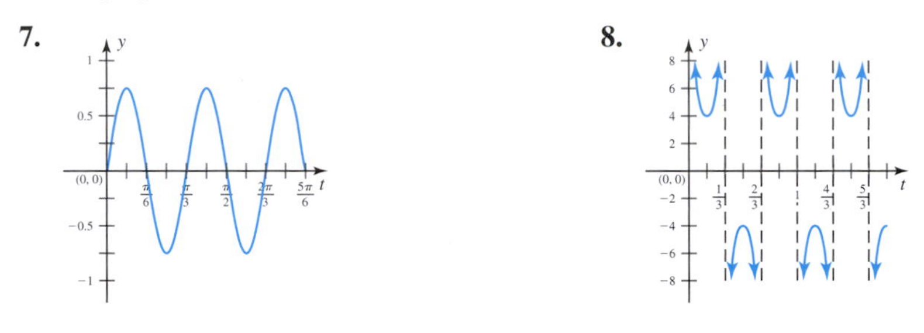

9. Referring to the chart of colors visible in the electromagnetic spectrum (page 94), what color is represented by the equation $y = \sin\left(\dfrac{\pi}{270}t\right)$? By $y = \sin\left(\dfrac{\pi}{320}t\right)$?

SECTION 2.2 Graphs of the Tangent and Cotangent Functions

KEY CONCEPTS

- Since $\tan t$ is defined in terms of the ratio $\dfrac{y}{x}$, the graph will be asymptotic everywhere $x = 0$ on the unit circle, meaning all odd multiples of $\dfrac{\pi}{2}$.
- Since $\cot t$ is defined in terms of the ratio $\dfrac{x}{y}$, the graph will be asymptotic everywhere $y = 0$ on the unit circle, meaning all integer multiples of π.
- The graph of $y = \tan t$ is increasing everywhere it is defined; the graph of $y = \cot t$ is decreasing everywhere it is defined.
- The characteristics of each graph play a vital role in their contextual application, and these are summarized on page 100.

$\frac{\pi}{2}$ (0, 1)
(x, y)
π (−1, 0)
(1, 0) 0 x
$\frac{3\pi}{2}$ (0, −1)

- For the more general tangent and cotangent graphs $y = A\tan(Bt)$ and $y = A\cot(Bt)$, if $|A| > 1$, the graph is vertically stretched, if $0 < |A| < 1$ the graph is vertically compressed, and if $A < 0$ the graph is reflected across the x-axis, just as with algebraic functions.
- If $B > 1$, the graph is horizontally compressed (the period is smaller/shorter); if $B < 1$ the graph is horizontally stretched (the period is larger/longer).
- To graph $y = A\tan(Bt)$, note $A\tan(Bt)$ is still zero at $t = 0$. Compute the period $P = \frac{\pi}{B}$ and draw asymptotes a distance of $\frac{P}{2}$ on either side of the y-axis (zeroes occur halfway between asymptotes). Plot other convenient points and use the symmetry and period of $\tan t$ to complete the graph.
- To graph $y = A\cot(Bt)$, note it is asymptotic at $t = 0$. Compute the period $P = \frac{\pi}{B}$ and draw asymptotes a distance P on either side of the y-axis (zeroes occur halfway between asymptotes). Other convenient points along with the symmetry and period can be used to complete the graph.

EXERCISES

10. State the value of each expression without the aid of a calculator:
$\tan\left(\frac{7\pi}{4}\right)$ $\qquad$ $\cot\left(\frac{\pi}{3}\right)$

11. State the value of each expression without the aid of a calculator, given that t terminates in QII.
$\tan^{-1}(-\sqrt{3})$ $\qquad$ $\cot^{-1}\left(-\frac{1}{\sqrt{3}}\right)$

12. Graph $y = 6\tan\left(\frac{1}{2}t\right)$ in the interval $[-2\pi, 2\pi]$.

13. Graph $y = \frac{1}{2}\cot(2\pi t)$ in the interval $[-1, 1]$.

14. Use the period of $y = \cot t$ to name three additional solutions to $\cot t = 0.0208$, given $t = 1.55$ is a solution. Many solutions are possible.

15. Given $t = 0.4444$ is a solution to $\cot^{-1}(t) = 2.1$, use an analysis of signs and quadrants to name an additional solution in $[0, 2\pi)$.

16. Find the height of Mount Rushmore, using the formula $h = \frac{d}{\cot u - \cot v}$ and the values shown.

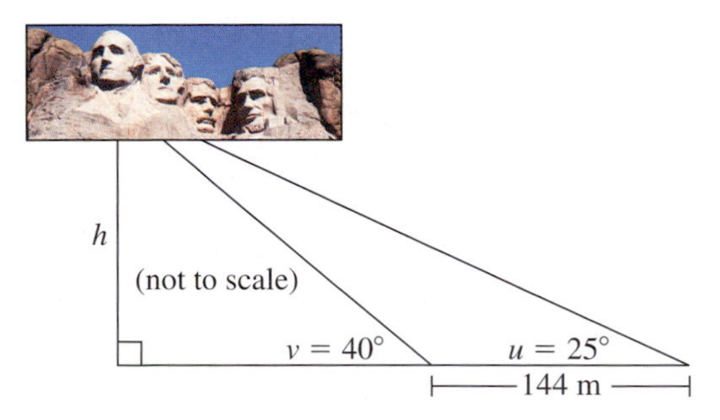

17. Model the data in the table using a tangent function. Clearly state the period, the value of A, and the location of the asymptotes.

Input	Output	Input	Output
−6	$-\infty$	1	1.4
−5	−19.4	2	3
−4	−9	3	5.2
−3	−5.2	4	9
−2	−3	5	19.4
1	−1.4	6	∞
0	0		

SECTION 2.3 Transformations and Applications of Trigonometric Graphs

KEY CONCEPTS

- Many everyday phenomena follow a sinusoidal pattern, or a pattern that can be modeled by a sine or cosine function (e.g., daily temperatures, hours of daylight, and some animal populations).
- To obtain accurate equation models of sinusoidal phenomena, we often must use vertical and horizontal shifts of a basic function.
- The equation $y = A\sin(Bt \pm C) + D$ is called the *standard form* of a general sinusoid. The equation $y = A\sin\left[B\left(t \pm \frac{C}{B}\right)\right] + D$ is called the *shifted form* of a general sinusoid.
- In either form, D represents the average value of the function and a vertical shift D units upward if $D > 0$, D units downward if $D < 0$. For a maximum value M and minimum value m, $\frac{M + m}{2} = D$, $\frac{M - m}{2} = A$.
- The shifted form $y = A\sin\left[B\left(t \pm \frac{C}{B}\right)\right] + D$ enables us to quickly identify the horizontal shift of the function: $\frac{C}{B}$ units in a direction opposite the given sign.
- To graph a shifted sinusoid, locate the primary interval by solving $0 \le Bt + C < 2\pi$, then use a reference rectangle along with the *rule of fourths* to sketch the graph in this interval. The graph can then be extended as needed in either direction, then shifted vertically D units.
- One basic application of sinusoidal graphs involves phenomena in harmonic motion, or motion that can be modeled by functions of the form $y = A\sin(Bt)$ or $y = A\cos(Bt)$ (with no horizontal or vertical shift).

EXERCISES

For each equation given, (a) identify/clearly state the amplitude, period, horizontal shift, and vertical shift; then (b) graph the equation using the primary interval, a reference rectangle, and *rule of fourths*.

18. $y = 240\sin\left[\frac{\pi}{6}(t - 3)\right] + 520$

19. $y = 3.2\cos\left(\frac{\pi}{4}t + \frac{3\pi}{2}\right) + 6.4$

For each graph given, identify the amplitude, period, horizontal shift, and vertical shift, and give the equation of the graph.

20. **21.**

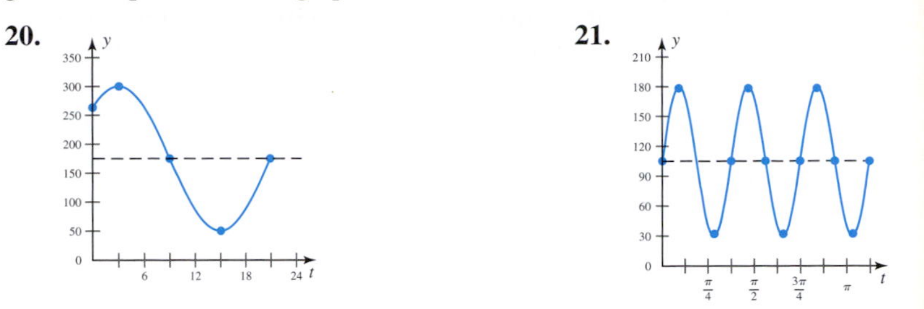

22. Monthly precipitation in Cheyenne, Wyoming, can be modeled by a sine function, by using the average precipitation for July (2.26 in.) as a maximum (actually slightly higher in May), and the average precipitation for February (0.44 in.) as a minimum. Assume $t = 0$ corresponds to March. (a) Use the information to construct a sinusoidal model, and (b) use the model to estimate the inches of precipitation Cheyenne receives in August ($t = 5$) and December ($t = 9$).

Source: 2004 Statistical Abstract of the United States, Table 380.

SECTION 2.4 Trigonometric Models

KEY CONCEPTS

- If the period P and critical points (X, M) and (x, m) of a sinusoidal function are known, a model of the form $y = A\sin(Bx + C) + D$ can be obtained:

$$B = \frac{2\pi}{P} \qquad A = \frac{M - m}{2} \qquad D = \frac{M + m}{2} \qquad C = \frac{3\pi}{2} - Bx$$

- If an event or phenomenon is known to be sinusoidal, a graphing calculator can be used to find an equation model if at least four data points are given.

EXERCISES

For the following sets of data, (a) find a sinusoidal regression equation using your calculator; (b) construct an equation manually using the period and maximum/minimum values; and (c) graph both on the same screen, then use a TABLE to find the largest difference between output values.

23.

Day of Year	Output	Day of Year	Output
1	4430	184	90
31	3480	214	320
62	3050	245	930
92	1890	275	2490
123	1070	306	4200
153	790	336	4450

24.

Day of Month	Output	Day of Month	Output
1	69	19	98
4	78	22	92
7	84	25	85
10	91	28	76
13	96	31	67
16	100		

25. The record high temperature for the even months of the year are given in the table for the city of Juneau, Alaska. (a) Use a graphing calculator to find a sinusoidal regression model. (b) Use the equation to estimate the record high temperature for the month of July. (c) Compare the actual data to the results produced by the regression model. What comments can you make about the accuracy of the model?

Source: 2004 Statistical Abstract of the United States, Table 378.

Month (Jan. → 1)	High Temp. (°F)
2	57
4	72
6	86
8	83
10	61
12	54

MIXED REVIEW

Exercises 1 and 2

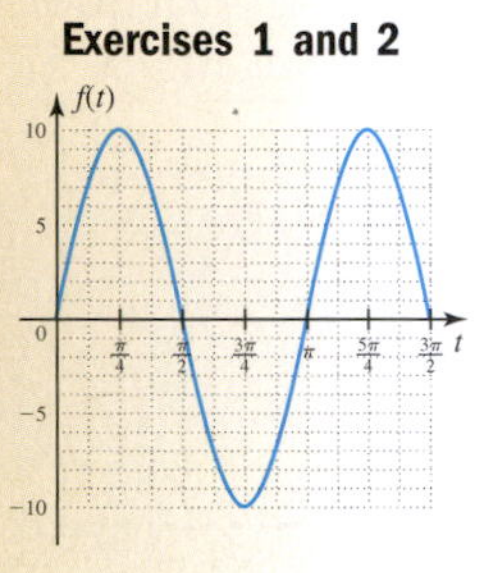

Exercises 4 and 5

State the equation of the function $f(t)$ shown using a

1. sine function

2. cosine function

3. The average high temperature for a Midwestern town is given in the table for selected days of the year. (a) Find an appropriate regression model. (b) Use the model to predict the temperature on the 100th day of the year. (c) Calculate the average value of these average temperatures. How does the result compare to your graph and equation?

Exercise 3

Day of Year	Average High Temp. (°F)
0	25
30	26
60	35
90	49
120	65
150	79
180	85
210	83
240	72
270	57
300	41
330	30
360	25

State the equation of the function $g(t)$ shown using a

4. tangent function

5. cotangent function

6. The data given in the table tracks the length of the shadow cast by an obelisk for the 12 daylight hours at a certain location (positive and negative values indicate lengths before noon and afternoon, respectively). Assume $t = 0$ corresponds to 6:00 A.M. (a) Use the data to construct an equation modeling the length L of the shadow and graph the function. (b) Use the function to find the shadow's length at 1:30 P.M. (d) If the shadow is "-52 m" long, what time is it?

Hour of Day	Length (m)	Hour of Day	Length (m)
0	∞	7	-10.7
1	149.3	8	-23.1
2	69.3	9	-40
3	40.0	10	-69.3
4	23.1	11	-149.3
5	10.7	12	$-\infty$
6	0		

Use a reference rectangle and the *rule of fourths* to graph the following functions in $[0, 2\pi)$.

7. $y = -2 \csc t$

8. $y = 3 \sec t$

9. $f(t) = \cos(2t) - 1$

10. $g(t) = -\sin(3t) + 1$

11. $h(t) = \dfrac{3}{2}\sin\left(\dfrac{\pi}{2}t\right)$

12. $p(t) = 6\cos(\pi t)$

13. Monthly precipitation in Minneapolis, Minnesota, can be modeled by a sine function, by using the average precipitation for August (4.05 in.) as a maximum (actually slightly higher in June), and the average precipitation for February (0.79 in.) as a minimum. Assume $t = 0$ corresponds to April. (a) Use the information to construct a sinusoidal model, and (b) use the model to approximate the inches of precipitation Minneapolis receives in July ($t = 3$) and December ($t = 8$).

Source: 2004 Statistical Abstract of the United States, Table 380.

14. Which of the following functions indicates that the graph of $y = \sin(2t)$ has been shifted horizontally $\frac{\pi}{4}$ units left?

a. $y = \sin\left[2\left(t + \frac{\pi}{4}\right)\right]$ **b.** $y = \sin\left(2t + \frac{\pi}{4}\right)$

c. $y = \sin\left[2\left(t - \frac{\pi}{4}\right)\right]$ **d.** $y = \sin(2t) - \frac{\pi}{4}$

15. Given $t = 2.85$ is an approximate solution to $\tan t = -0.3$, use an analysis of signs and quadrants to name another solution in $[0, 2\pi)$.

16. Solve each equation in $[0, 2\pi)$ without the use of a calculator. If the expression is undefined, so state.

a. $x = \sin\left(-\frac{\pi}{4}\right)$ **b.** $\sec x = \sqrt{2}$ **c.** $\cot\left(\frac{\pi}{2}\right) = x$

d. $\cos \pi = x$ **e.** $\csc x = \frac{2\sqrt{3}}{3}$ **f.** $\tan\left(\frac{\pi}{2}\right) = x$

State the amplitude, period, horizontal shift, vertical shift, and endpoints of the primary interval, then sketch the graph using a reference rectangle and the *rule of fourths.*

17. $y = 5\cos(2t) - 8$ **18.** $y = \frac{7}{2}\sin\left[\frac{\pi}{2}(x - 1)\right]$

State the period and phase shift, then sketch the graph of each function.

19. $y = 2\tan\left(\frac{1}{4}t\right)$ **20.** $y = 3\sec\left(x - \frac{\pi}{2}\right)$

PRACTICE TEST

1. Complete the table using exact values, including the point on the unit circle associated with t.

t	0	$\frac{\pi}{6}$	$\frac{\pi}{4}$	$\frac{\pi}{2}$	$\frac{2\pi}{3}$	$\frac{5\pi}{6}$	$\frac{5\pi}{4}$	$\frac{4\pi}{3}$	$\frac{3\pi}{2}$
sin t									
cos t									
tan t									
csc t									
sec t									
cot t									
$P(x, y)$									

2. State the value of each expression without the use of a calculator:

a. $\cos\left(\frac{3\pi}{2}\right)$ **b.** $\sin \pi$ **c.** $\tan\left(\frac{5\pi}{4}\right)$ **d.** $\sec\left(\frac{\pi}{6}\right)$

3. State the value of each expression without the use of a calculator:

a. $\tan^{-1}(\sqrt{3})$; QI **b.** $\cos^{-1}\left(-\frac{1}{2}\right)$; QII

c. $\sin^{-1}\left(-\frac{\sqrt{3}}{2}\right)$; QIII **d.** $\sec^{-1}(\sqrt{2})$; QIV

State the amplitude (if it exists) and period of each function, then sketch the graph in the interval indicated using a reference rectangle and the *rule of fourths.*

4. $y = 3\sin\left(\frac{\pi}{5}t\right),\ t \in [0, 10)$ **5.** $y = 2\sec(2t),\ t \in [0, 2\pi)$

6. $y = 4\tan(3t),\ t \in [0, \pi)$

State the amplitude (if it exists), period, horizontal shift, and vertical shift of each function, then sketch the graph in the interval indicated using a reference rectangle and the *rule of fourths.*

7. $y = 12\sin\left[3\left(t - \frac{\pi}{6}\right)\right] + 19,\ t \in [0, 2\pi)$ **8.** $y = \frac{3}{4}\cos\left(\frac{1}{2}\pi t\right) - \frac{1}{2},\ t \in [-2, 6)$

9. $y = 2\cot\left[\frac{1}{3}\pi\left(t - \frac{3}{2}\right)\right],\ t \in [0, 6)$

10. The revenue for Otake's Mower Repair is very seasonal, with business in the summer months far exceeding business in the winter months. Monthly revenue for the company can be modeled by the function $R(x) = 7.5\cos\left(\frac{\pi}{6}x + \frac{4\pi}{3}\right) + 12.5$, where $R(x)$ is the average revenue (in thousands of dollars) for month x ($x = 1 \rightarrow$ Jan). (a) What is the average revenue for September? (b) For what months of the year is revenue at least \$12,500?

11. Given $t = 4.25$ is a solution to $\tan t \approx 2$, use the period $P = \pi$ to find two other solutions.

12. Given $t \approx 1.4602$ is a solution to $\tan^{-1}(t) = 9$, find another solution using a reference angle/arc.

13. Although Reno, Nevada, is a very arid city, the amount of monthly precipitation tends to follow a sinusoidal pattern. (a) Find an equation model for the precipitation in Reno, given the annual high is 1.06 in., while annual low is 0.24 in. and occurs in the month of July. (b) In what month does the annual high occur?

14. The lowest record temperature for the even months of the year are given in the table for the city of Denver, Colorado. (a) Use a graphing calculator to find a sinusoidal regression model, and (b) use the equation to estimate the record low temperature for the odd numbered months.

Source: 2004 Statistical Abstract of the United States, Table 379.

Exercise 14

Month (Jan. → 1)	High Temp. (°F)
2	−30
4	−2
6	30
8	41
10	3
12	−25

Exercise 15

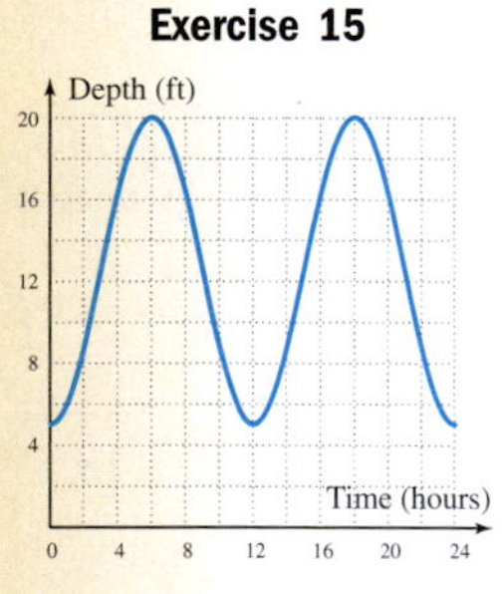

Exercise 16

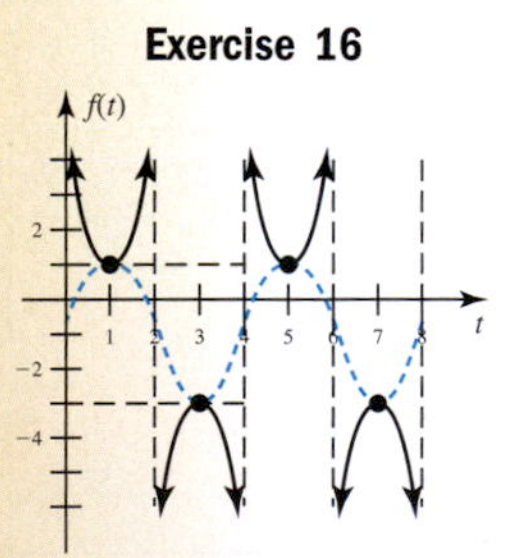

15. Due to tidal motions, the depth of water in Brentwood Bay varies sinusoidally as shown in the diagram, where time is in hours and depth is in feet. Find an equation that models the depth of water at time t.

16. Determine the equation of the graph shown, given it is of the form $y = A\csc(Bt) + D$.

Match each equation with its corresponding graph.

17. $y = 3\sin\left[2\left(t - \frac{\pi}{6}\right)\right]$

18. $y = 3\sin(\pi t) - 1$

19. $y = -3\sin(\pi t) + 1$

20. $y = 3\sin\left(2t - \frac{\pi}{6}\right)$

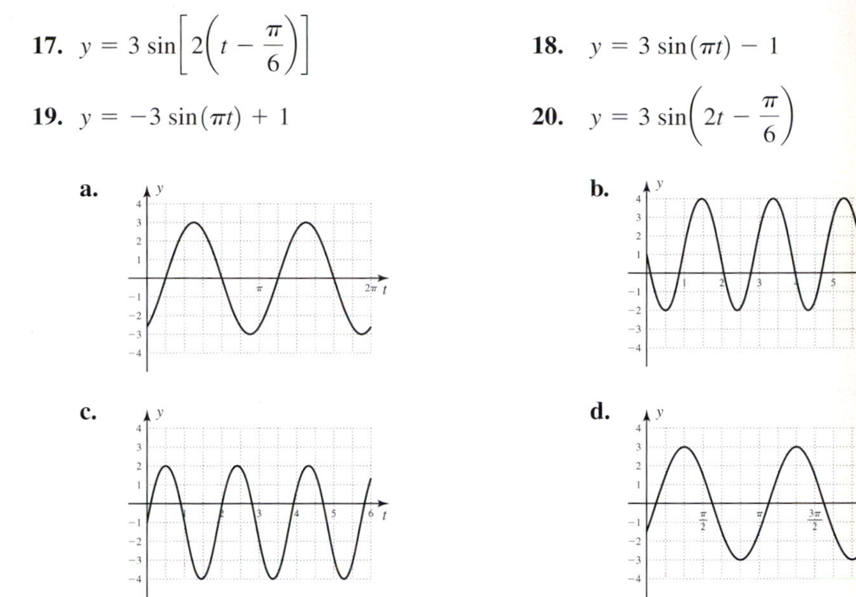

Calculator Exploration and Discovery

Variable Amplitudes and Modeling the Tides

As mentioned in the *Point of Interest* from Section 2.3, tidal motion is often too complex to be modeled by a single sine function. In this *Exploration and Discovery,* we'll look at a method that combines two sine functions to help model a tidal motion with variable amplitude. In the process, we'll use much of what we know about the amplitude, horizontal shifts, and vertical shifts of a sine function. The graph in Figure 2.46 shows three days of tidal motion for Davis Inlet, Canada.

Figure 2.46

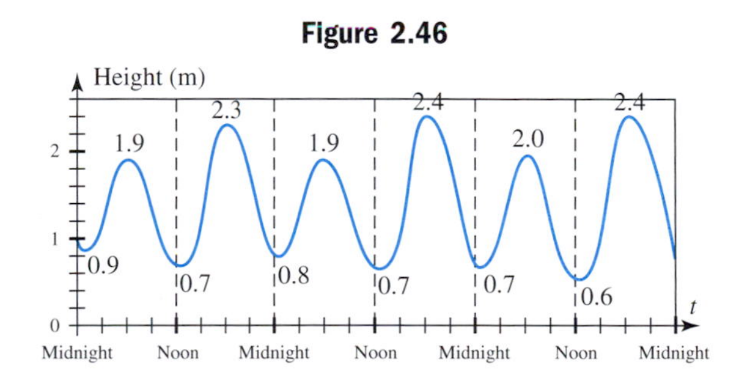

As you can see, the amplitude of the graph varies, and there is no *single* sine function that can serve as a model. However, notice that the amplitude *varies predictably,* and that the high tides and low tides can independently be modeled by a sine function. To simplify our exploration, we will use the assumption that tides have an exact 24-hr period (close, but no), that variations between high and low tides takes place every 12 hr (again close but not exactly true), and the variation between the "low-high" (1.9 m) and the "high-high" (2.4 m) is uniform. A similar assumption is made for the low tides. The result is the graph in Figure 2.47.

Figure 2.47

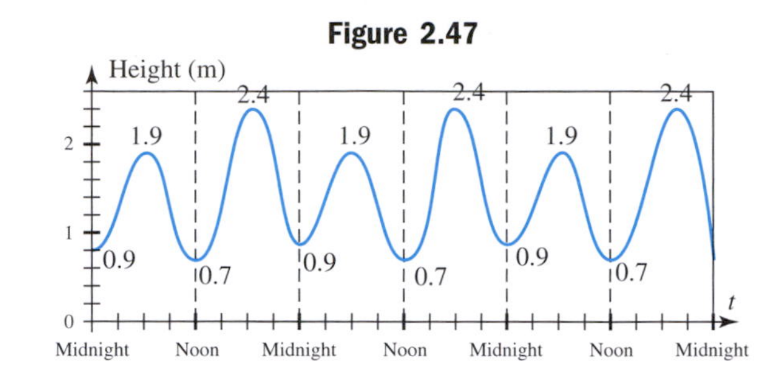

Figure 2.48

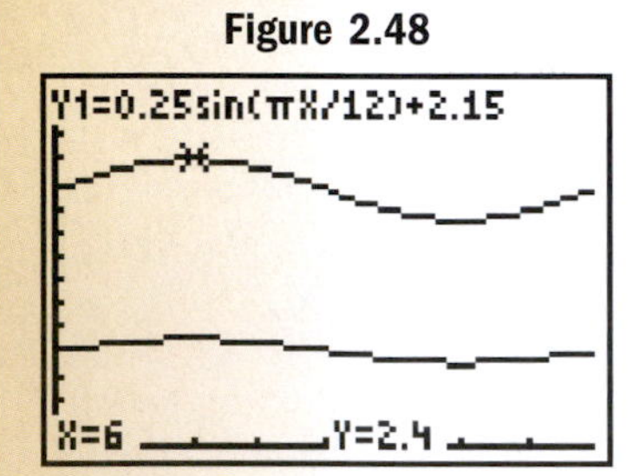

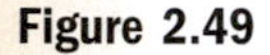

Figure 2.49

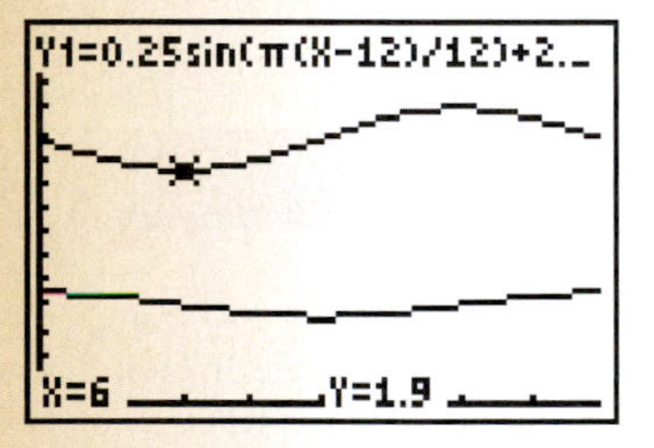

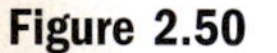

Figure 2.50

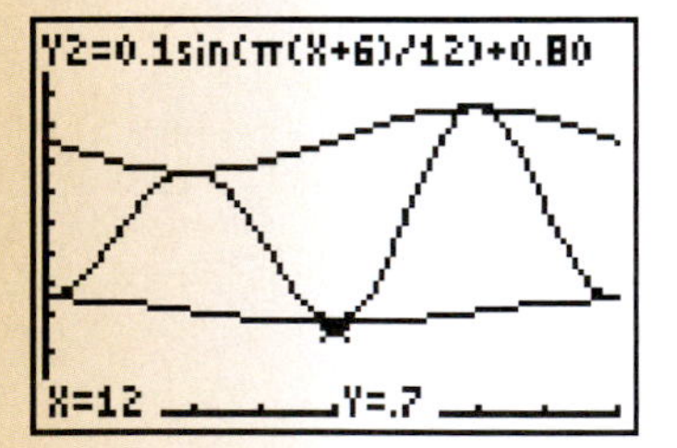

First consider the high tides, which vary from a maximum of 2.4 to a minimum of 1.9. Using the ideas from Section 2.3 to construct an equation model gives $A = \frac{2.4 - 1.9}{2} = 0.25$ and $D = \frac{2.4 + 1.9}{2} = 2.15$. With a period of $P = 24$ hr we obtain the equation $Y_1 = 0.25\sin\left(\frac{\pi}{12}x\right) + 2.15$. Using 0.9 and 0.7 as the maximum and minimum low tides, similar calculations yield the equation $Y_2 = 0.1\sin\left(\frac{\pi}{12}x\right) + 0.8$ (verify this). Graphing these two functions over a 24-hr period yields the graph in Figure 2.48, where we note the high and low values are correct, but the two functions are in phase with each other. As can be determined from Figure 2.47, we want the high tide model to start at the average value and decrease, and the low tide equation model to start at high-low and decrease. Replacing x with $x - 12$ in Y_1 and x with $x + 6$ in Y_2 accomplishes this result (see Figure 2.49). Now comes the fun part! Since Y_1 represents the low/high maximum values for high tide, and Y_2 represents the low/high minimum values for low tide, *the amplitude and average value for the tidal motion at Davis Inlet are* $A = \frac{Y_1 - Y_2}{2}$ *and* $D = \frac{Y_1 + Y_2}{2}$! By entering $Y_3 = \frac{Y_1 - Y_2}{2}$ and $Y_4 = \frac{Y_1 + Y_2}{2}$, the equation for the tidal motion (with its variable amplitude) will have the form $Y_5 = Y_3\sin(Bx \pm C) + Y_4$, where the values of B and C must be determined. The key here is to note there is only a 12-hr difference between the changes in amplitude, so $P = 12$ (instead of 24) and $B = \frac{\pi}{6}$ for this function. Also, from the graph (Figure 2.47) we note the tidal motion begins at a minimum and increases, indicating a shift of 3 units to the right is required. Replacing x with $x - 3$ gives the equation modeling these tides, and the final equation is $Y_5 = Y_3\sin\left[\frac{\pi}{6}(x - 3)\right] + Y_4$. Figure 2.50 gives a screen shot of Y_1, Y_2, and Y_5 in the interval [0, 24]. The tidal graph from Figure 2.47 is shown in Figure 2.51 with Y_3 and Y_4 superimposed on it.

Figure 2.51

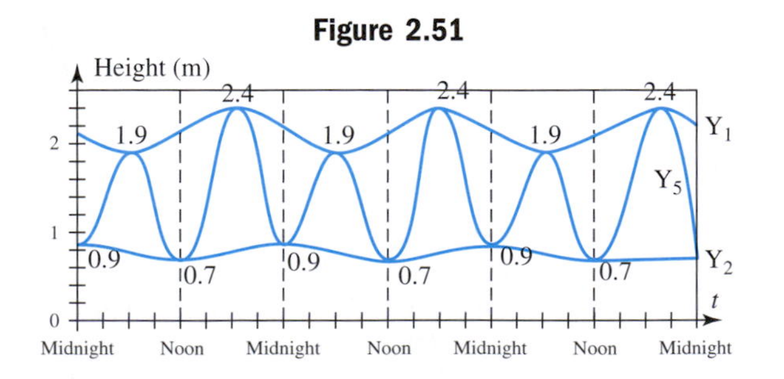

Exercise 1: The website www.tides.com/tcpred.htm offers both *t*ide and *c*urrent *pred*ictions for various locations around the world, in both numeric and graphical form. In addition, data for the "two" high tides and "two" low tides are clearly highlighted. Select a coastal area where tidal motion is similar to that of Davis Inlet, and repeat this exercise. Compare your model to the actual data given on the website. How good was the fit?

STRENGTHENING CORE SKILLS

Standard Angles, Reference Angles, and the Trig Functions

A review of the main ideas discussed in this chapter indicates there are four of what might be called "core skills." These are skills that (a) play a fundamental part in the acquisition of concepts, (b) hold the overall structure together as we move from concept to concept, and (c) are ones we return to again and again throughout our study. The first of these is *(1) knowing the standard angles and standard values.*

The standard angles/values brought us to the trigonometry of any angle, forming a strong bridge to the second core skill: *(2) using reference angles to determine the value of the trig functions in each quadrant.* For any angle θ where $\theta_r = 30°$, $\sin\theta = \frac{1}{2}$ or $\sin\theta = -\frac{1}{2}$ since the *ratio is fixed* but the sign *depends on the quadrant of* θ : $\sin 30° = \frac{1}{2}$ [QI], $\sin 150° = \frac{1}{2}$ [QII], $\sin 210° = -\frac{1}{2}$ [QIII], $\sin 330° = -\frac{1}{2}$ [QIV], and so on (see Figure 2.52).

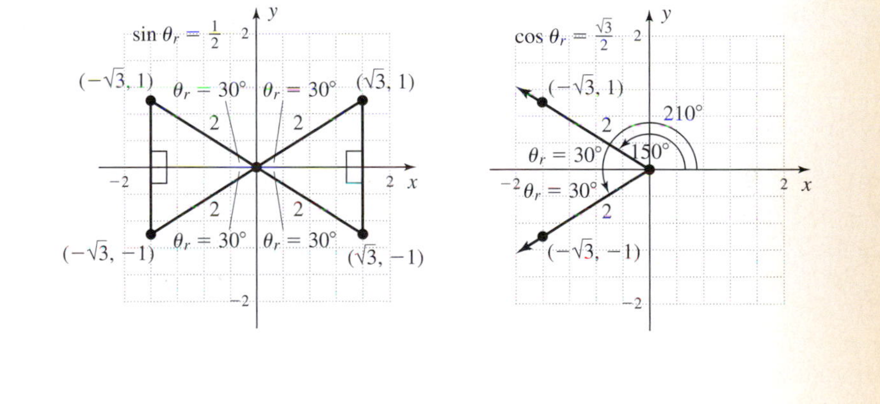

In turn, the reference angles led us to a third core skill, helping us realize that if θ was not a quadrantal angle, *(3) equations like* $\cos(\theta) = -\dfrac{\sqrt{3}}{2}$ *must have two solutions in* $[0, 360°)$. The solutions in $[0, 360°)$ are $\theta = 150°$ and $\theta = 210°$ (see Figure 2.53).

Of necessity, this brings us to the fourth core skill, *(4) effective use of a calculator.* The standard angles are a wonderful vehicle for introducing the basic ideas of trigonometry, and actually occur quite frequently in real-world applications. But by far, most of the values we encounter will be nonstandard values where θ_r must be found using a calculator.

The *Summary and Concept Review Exercises,* as well as the *Practice Test* offer ample opportunities to refine these skills.

Exercise 1: Fill in the table from memory.

t	0	$\frac{\pi}{6}$	$\frac{\pi}{4}$	$\frac{\pi}{3}$	$\frac{\pi}{2}$	$\frac{2\pi}{3}$	$\frac{3\pi}{4}$	$\frac{5\pi}{6}$	π	$\frac{7\pi}{6}$	$\frac{5\pi}{4}$
$\sin t = y$											
$\cos t = x$											
$\tan t = \frac{y}{x}$											

Exercise 2: Solve each equation in $[0, 2\pi)$ without the use of a calculator.

a. $2 \sin t + \sqrt{3} = 0$

b. $-3\sqrt{2} \cos t + 4 = 1$

c. $-\sqrt{3} \tan t + 2 = 1$

d. $2 \sin t = 1$

Exercise 3: Solve each equation in $[0, 2\pi)$ using a calculator and rounding answers to four decimal places.

a. $\sqrt{6} \sin t - 2 = 1$

b. $-3\sqrt{2} \cos t + \sqrt{2} = 0$

c. $3 \tan t + \frac{1}{2} = -\frac{1}{4}$

d. $2 \sec t = -5$

CUMULATIVE REVIEW CHAPTERS 1–2

1. Use a reference rectangle and the *rule of fourths* to graph $y = 2.5 \sin\left(\frac{\pi}{2}t\right)$ in $[-2, 6)$.

2. State the period, horizontal shift, and x-intercepts of the function $y = -\tan\left(2x + \frac{\pi}{2}\right)$ in $\left[-\frac{\pi}{2}, \pi\right)$. Then state whether the function is increasing or decreasing and sketch its graph.

3. Given that $\tan \theta = \frac{80}{39}$, draw a right triangle that corresponds to this ratio, then use the Pythagorean theorem to find the length of the missing side. Finally, find the two acute angles.

4. Without a calculator, what values in $[0, 2\pi)$ make the equation true: $\sin t = -\frac{\sqrt{3}}{2}$?

5. Given $\left(\frac{3}{4}, -\frac{\sqrt{7}}{4}\right)$ is a point on the unit circle corresponding to t, find all six trig functions of t.

Exercise 6

12 cm

56°

6. During a storm, a tall tree snapped, as shown in the diagram. Find the tree's original height. Round to tenths of a meter.

7. Find the measure of angle α for the triangle shown and convert it to

a. decimal degrees **b.** radians

Exercise 7

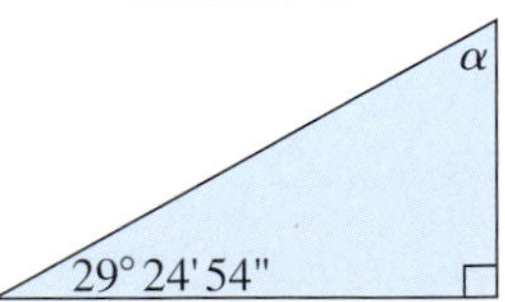

8. Verify the point $\left(\frac{8}{17}, \frac{15}{17}\right)$ is on a unit circle, and use the circle's symmetry to find three other points on the circle.

9. For $\theta = 729.5°$, find the value of $\sin\theta$, $\cos\theta$, and $\tan\theta$ using reference angles and a calculator. Round to four decimal places.

10. Given $t = \frac{11\pi}{6}$, use reference arcs to state the value of all six trig functions of t without using a calculator.

11. The world's tallest indoor waterfall is in Detroit, Michigan, in the lobby of the International Center Building. Standing 66 ft from the base of the falls, the angle of elevation is 60°. How tall is the waterfall?

12. It's a warm, lazy Saturday and Hank is watching a county maintenance crew mow the park across the street. He notices the mower takes 29 sec to pass through 77° of rotation from one end of the park to the other. If the corner of the park is 60 ft directly across the street from his house, (a) how wide is the park? (b) How fast (in mph) does the mower travel as it cuts the grass?

13. The conveyor belt used in many grocery check-out lines is on rollers that have a 2-in. radius and turn at 252 rpm. (a) What is the angular velocity of the rollers? (b) How fast are your groceries moving (in mph) when the belt is moving?

14. The planet Mars has a near-circular orbit (its eccentricity is 0.093—the closer to zero, the more circular the orbit). It orbits the Sun at an average distance of 142,000,000 mi and takes 687 days to complete one circuit. (a) Find the angular velocity of its orbit in radians per year, and (b) find the planet's linear velocity as it orbits the Sun, in miles per second.

15. Find $f(\theta)$ for all six trig functions, given the point $P(-9, 40)$ is a point on the terminal side of the angle. Then find the angle θ in degrees, rounded to tenths.

16. Given $t = 5.37$, (a) in what quadrant does the arc terminate? (b) What is the reference arc? (c) Find the value of $\sin t$ rounded to four decimal places.

17. A jet-stream water sprinkler shoots water a distance of 15 m and turns back and forth through an angle of $t = 1.2$ rad. (a) What is the length of the arc that the sprinkler reaches? (b) What is the area in m^2 of the yard that is watered?

18. Determine the equation of the graph shown given it is of the form $y = A\tan(Bt)$.

19. Determine the equation of the graph shown given it is of the form $y = A\sin(Bt \pm C) + D$.

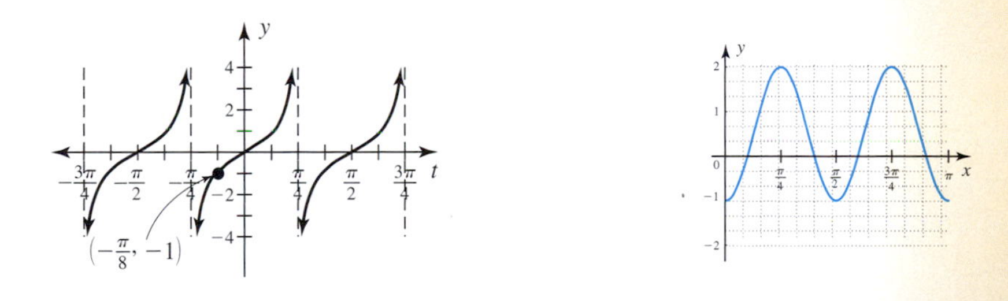

20. In London, the average temperature on a summer day ranges from a high of 72°F to a low of 56°F. Use this information to write a sinusoidal equation model, assuming the low temperature occurs at 6:00 A.M. Clearly state the amplitude, average value, period, and horizontal shift.

Source: 2004 Statistical Abstract of the United States, Table 1331.

21. State true or false and explain your response: $(\sin t)(\sec t) = 1$.

22. Complete the following statement: The cofunction of sine is cosine because. . . .

23. Find the value of all six trig functions given $\tan t = -\frac{68}{51}$ and $\sin t > 0$.

24. The profits of Red-Bud Nursery can be modeled by a sinusoid, with profit peaking twice each year. Given profits reach a yearly low of $4000 in mid-January (month 1.5), and a yearly high of $14,000 in mid-April (month 4.5), (a) construct an equation for their yearly profits. (b) Use the model to find their profits for August. (c) Name the other month at which profit peaks.

25. The Earth has a radius of 3960 mi. Mexico City, Mexico, is located at approximately 19° N latitude, and is almost exactly due south of Hutchinson, Kansas, located at about 38° N latitude. How many miles separate the two cities?

Chapter 3

Identities: Their Place, Purpose, and Application

Chapter Outline

Preview

In this chapter we focus on the threefold purpose of studying identities. The first involves their puzzle-like nature and the innate curiosity that drives many people to *take things apart and put things back together.* The second involves the maturing and refinement of manipulative skills that, even with the ready availability of calculators, continue to play a vital role in the study of mathematics. The third is a practical application of these identities to rewrite, simplify, and solve equations that would otherwise remain unsolvable. It is important to realize that identities were not created to challenge the ancient scholars, these scholars were challenged to *create the identities* as they sought to study and understand the solar system, and to apply what they learned in very practical ways.

3.1 Fundamental Identities and Families of Identities

LEARNING OBJECTIVES

In Section 3.1 you will learn how to:

A. Use fundamental identities to help understand and recognize identity "families"

B. Verify other identities using the fundamental identities and basic algebra skills

C. Use fundamental identities to express a given trig function in terms of the other five

D. Use counterexamples and contradictions to show an equation is not an identity

INTRODUCTION

In this section we begin laying the foundation necessary to work with identities successfully. The cornerstone of this effort is a healthy respect for the fundamental identities and vital role they play. Students are strongly encouraged to do more than memorize them—they should be *internalized,* meaning they must become a natural and instinctive part of your core mathematical knowledge.

POINT OF INTEREST

The word *identity* is likely from the Latin root *identidem,* which means "many times," or "over and over (repeatedly)." The connotation is that no matter what values you substitute, you will get equal results time and time again. Some identities are trivial, like $x = x$, while others are based on algebraic properties: $2(x + 3) = 2x + 6$. Some algebraic identities rely on recognition or frequency of use to "see" their validity: $x^3 + 8 = (x + 2)(x^2 - 2x + 4)$, while others must be rewritten or simplified before the equality can clearly be seen: $\frac{x^3 + 8}{x^2 - 2x + 4} = x + 2$. The same is true for trigonometric identities.

A. Fundamental Identities and Identity Families

Successfully working with all other forms of identities will depend a great deal on your familiarity with the *fundamental identities*. In fact, the word *fundamental* itself means, "a basis or foundation supporting an essential structure or function" (*Merriam-Webster's Collegiate Dictionary,* 11th ed.). For convenience and review, the definition of the trig functions *and the identities that result* from them are listed here.

TRIG FUNCTIONS AND FUNDAMENTAL IDENTITIES

For any rotation θ of a central angle, and the point $P(x, y)$ on the unit circle associated with θ, we have $x^2 + y^2 = 1$ with

$$\cos\theta = x \qquad \sin\theta = y \qquad \tan\theta = \frac{y}{x};\ x \neq 0$$

$$\sec\theta = \frac{1}{x};\ x \neq 0 \qquad \csc\theta = \frac{1}{y};\ y \neq 0 \qquad \cot\theta = \frac{x}{y};\ y \neq 0$$

The Fundamental Identities That Result

Reciprocal Identities: $\sin\theta = \frac{1}{\csc\theta}$ $\quad\cos\theta = \frac{1}{\sec\theta}$ $\quad\tan\theta = \frac{1}{\cot\theta}$

Ratio Identities: $\tan\theta = \frac{\sin\theta}{\cos\theta}$ $\quad\tan\theta = \frac{\sec\theta}{\csc\theta}$ $\quad\cot\theta = \frac{\cos\theta}{\sin\theta}$

Pythagorean Identities: $\sin^2\theta + \cos^2\theta = 1$ $\quad\tan^2\theta + 1 = \sec^2\theta$ $\quad 1 + \cot^2\theta = \csc^2\theta$

Identities due to Symmetry: $\sin(-\theta) = -\sin\theta$ $\quad\cos(-\theta) = \cos\theta$ $\quad\tan(-\theta) = -\tan\theta$

These identities seem to naturally separate themselves into the four groups or families listed, with each group having additional relationships that can be inferred from the definitions. For instance, since $\sin\theta$ is the reciprocal of $\csc\theta$, $\csc\theta$ must be the reciprocal of $\sin\theta$. Similar statements can be made regarding $\cos\theta$ and $\sec\theta$ as well as $\tan\theta$ and $\cot\theta$. Recognizing these additional "family members" enlarges the number of identities you can work with, and will help you use them more efficiently. In particular, since they *are* reciprocals: $\sin\theta\csc\theta = 1$, $\cos\theta\sec\theta = 1$, $\tan\theta\cot\theta = 1$. See Exercises 7 and 8.

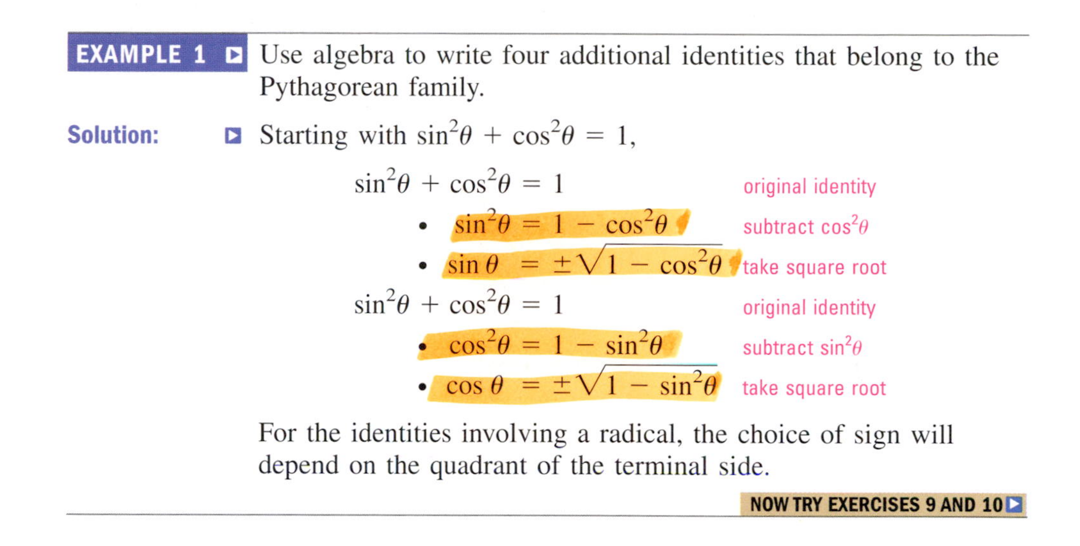

EXAMPLE 1 Use algebra to write four additional identities that belong to the Pythagorean family.

Solution: Starting with $\sin^2\theta + \cos^2\theta = 1$,

$$\sin^2\theta + \cos^2\theta = 1 \qquad \text{original identity}$$
$$\bullet \ \sin^2\theta = 1 - \cos^2\theta \qquad \text{subtract } \cos^2\theta$$
$$\bullet \ \sin\theta = \pm\sqrt{1 - \cos^2\theta} \qquad \text{take square root}$$
$$\sin^2\theta + \cos^2\theta = 1 \qquad \text{original identity}$$
$$\bullet \ \cos^2\theta = 1 - \sin^2\theta \qquad \text{subtract } \sin^2\theta$$
$$\bullet \ \cos\theta = \pm\sqrt{1 - \sin^2\theta} \qquad \text{take square root}$$

For the identities involving a radical, the choice of sign will depend on the quadrant of the terminal side.

NOW TRY EXERCISES 9 AND 10

The four additional Pythagorean identities are marked with a "•" in Example 1. The fact that each of them represents an equality gives us more options when attempting to verify or prove more complex identities. For instance, since $\cos^2\theta = 1 - \sin^2\theta$, we can replace $\cos^2\theta$ with $1 - \sin^2\theta$, or replace $1 - \sin^2\theta$ with $\cos^2\theta$, *any time they occur in an expression.* Note there are many other members of this family, since similar steps can be performed on the other Pythagorean identities. In fact, each of the fundamental identities can be similarly rewritten and there are a variety of exercises at the end of this section for practice.

B. Verifying an Identity Using Algebra

An identity is an equation that is true for all values of the input variable where the functions involved are defined. This means we cannot *prove* an equation is an identity by repeatedly substituting input values and obtaining a true equation. This would be a finite exercise and we might easily miss a value or even a range of values for which the equation is false. Instead we attempt to rewrite one side of the equation until we obtain a perfect match with the other side, so that there can be no doubt. As hinted at earlier, this is done using basic algebra skills combined with the fundamental identities and the substitution principle. For now we'll focus on verifying identities by simplifying expressions using algebra. In Section 3.2 we'll introduce some guidelines and ideas that will help you verify a wider range of identities.

EXAMPLE 2 Use algebra to verify the identity: $\sin\theta(\csc\theta - \sin\theta) = \cos^2\theta$.

Solution: To begin, use the distributive property to simplify the left-hand side.

$$\sin\theta(\csc\theta - \sin\theta) = \cos^2\theta \quad \text{original equation}$$
$$\sin\theta\csc\theta - \sin^2\theta = \quad \text{distribute}$$
$$1 - \sin^2\theta = \quad \text{substitute 1 for } \sin\theta\csc\theta$$
$$\cos^2\theta = \quad 1 - \sin^2\theta = \cos^2\theta$$

Since we were able to transform the left-hand side into a duplicate of the right, there can be no doubt the original equation is an identity.

NOW TRY EXERCISES 11 THROUGH 20

The verification of identities is truly a union of algebraic skills with the fundamental identities. Often we must *factor* an expression, rather than multiply, to begin the verification process.

EXAMPLE 3 Verify the identity: $1 = \cot^2x\sec^2x - \cot^2x$.

Solution: The left side is as simple as it gets. The terms on the right side have a common factor and we begin there.

$$1 = \cot^2x\sec^2x - \cot^2x \quad \text{original equation}$$
$$= \cot^2x(\sec^2x - 1) \quad \text{factor out } \cot^2x$$
$$= \cot^2x\tan^2x \quad \text{substitute } \tan^2x \text{ for } \sec^2x - 1$$
$$= (\cot x\tan x)^2 \quad \text{power property of exponents}$$
$$= 1^2 \quad \cot x\tan x = 1$$

NOW TRY EXERCISES 21 THROUGH 28

Examples 2 and 3 show you can begin the verification process on either the left or right side of the equation, whichever seems more convenient. Example 4 shows how the special products $(A + B)(A - B) = A^2 - B^2$ and/or $(A + B)^2 = A^2 + 2AB + B^2$ can be used in the verification process.

EXAMPLE 4 Use a special product to verify the identity $(\sin x - \cos x)^2 = 1 + 2\sin(-x)\cos x$.

Solution: By squaring the left-hand side, we might be able to use a Pythagorean identity and we begin there.

$$(\sin x - \cos x)^2 = 1 + 2\sin(-x)\cos x \quad \text{original equation}$$
$$\sin^2x - 2\sin x\cos x + \cos^2x = \quad \text{expand binomial}$$
$$1 - 2\sin x\cos x = \quad \text{substitute 1 for } \sin^2(x) + \cos^2(x)$$

At this point we appear to be off by a sign, but quickly recall that sine is an odd function and $-\sin(x) = \sin(-x)$. After rewriting the expression on the left, a direct substitution shows the equation is indeed an identity.

$$1 + 2(-\sin x)\cos x = \quad \text{rewrite expression}$$

$$1 + 2\sin(-x)\cos x = \quad \text{substitute } \sin(-x) \text{ for } -\sin(x)$$

NOW TRY EXERCISES 29 THROUGH 34

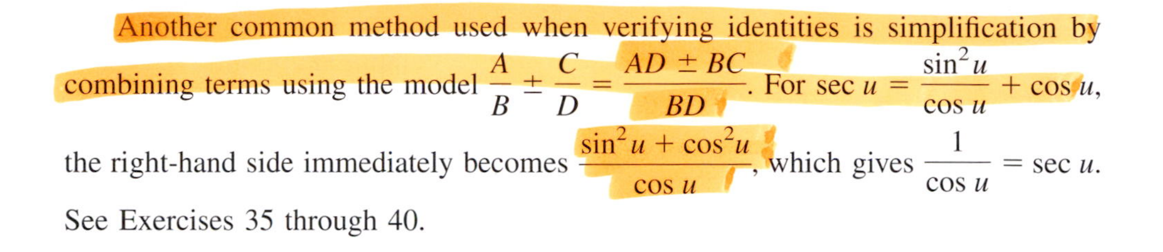

Another common method used when verifying identities is simplification by combining terms using the model $\frac{A}{B} \pm \frac{C}{D} = \frac{AD \pm BC}{BD}$. For $\sec u = \frac{\sin^2 u}{\cos u} + \cos u$, the right-hand side immediately becomes $\frac{\sin^2 u + \cos^2 u}{\cos u}$, which gives $\frac{1}{\cos u} = \sec u$. See Exercises 35 through 40.

C. Writing One Function in Terms of Another

Any one of the six trigonometric functions can be written in terms of any of the other functions using fundamental identities. The process involved offers practice in working with identities, highlights how each function is related to the other, and has practical applications in verifying more complex identities.

EXAMPLE 5 Write the function $\cos x$ in terms of the tangent function.

Solution: We begin by noting the two functions share common ground via $\sec x$, since $1 + \tan^2 x = \sec^2 x$ and $\cos x = \frac{1}{\sec x}$. Starting with $\sec^2 x$ we have

$$\sec^2 x = 1 + \tan^2 x \quad \text{Pythagorean identity}$$

$$\sec x = \pm\sqrt{1 + \tan^2 x} \quad \text{square roots}$$

We can now substitute $\pm\sqrt{1 + \tan^2 x}$ for $\sec x$ in $\cos x = \frac{1}{\sec x}$.

$$\cos x = \frac{1}{\pm\sqrt{1 + \tan^2 x}} \quad \text{substitute } \pm\sqrt{1 + \tan^2 x} \text{ for } \sec x$$

NOW TRY EXERCISES 41 THROUGH 46

WORTHY OF NOTE

It is important to note the stipulation "valid where both are defined" does not preclude a difference in the domains of each function. The result of Example 5 is indeed an identity, even though the expressions have unequal domains.

Example 5 also highlights two very important points. First, the sign of the radical is dependent on the terminal side of the angle. If the terminal side is in QII or QIII, we choose the negative sign since $\cos x < 0$ in those quadrants. Second, recall that identities are valid *only where both expressions are defined.* The expression on the right is undefined for all odd multiples of $\frac{\pi}{2}$.

D. Showing an Equation Is Not an Identity

To show an equation is *not* an identity, we need only find a single value for which the functions involved are defined but the equation is *false*. This can often be done by trial and error, or even by inspection. To illustrate the process, we'll use two common misconceptions that arise in working with identities.

EXAMPLE 6 Show the equations given are *not* identities.

a. $\sin(2x) = 2\sin x$ **b.** $\cos(\alpha + \beta) = \cos\alpha + \cos\beta$

Solution: **a.** The assumption here seems to be that we can factor out the coefficient from the argument. By inspection we note the amplitude of $\sin(2x)$ is $A = 1$, while the amplitude of $2\sin x$ is $A = 2$. This means they cannot possibly be equal for all values of x, although they are equal for integer multiples of π. Verify using $x = \frac{\pi}{6}$ or other standard values.

GRAPHICAL SUPPORT

While not a definitive method of proof, a graphing calculator can be used to investigate whether an equation is an identity. Since the left and right members of the equation must be equal for all values (where they are defined), their graphs must be identical. Graphing the functions from Example 6(a) as Y_1 and Y_2 shows the equation $\sin(2x) = 2\sin x$ is definitely *not* an identity.

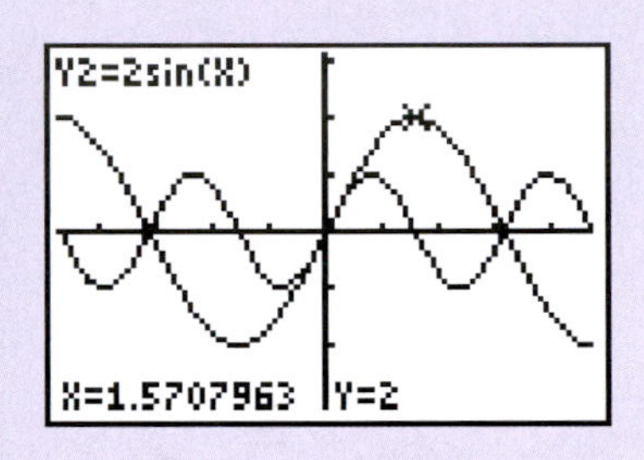

b. The assumption here is that we can distribute function values. This is similar to saying $\sqrt{x + 4} = \sqrt{x} + 2$, a statement obviously false for all values except $x = 0$. Here we'll substitute convenient values to prove the equation false, namely, $\alpha = \frac{3\pi}{4}$ and $\beta = \frac{\pi}{4}$.

$$\cos\left(\frac{3\pi}{4} + \frac{\pi}{4}\right) = \cos\left(\frac{3\pi}{4}\right) + \cos\left(\frac{\pi}{4}\right)$$ substitute $\frac{\pi}{3}$ for α and $\frac{\pi}{4}$ for β

$$\cos\pi = -\frac{\sqrt{2}}{2} + \frac{\sqrt{2}}{2}$$ simplify

$$-1 \neq 0$$ result is false

NOW TRY EXERCISES 47 THROUGH 53

Similar to our work in Chapter 2, given the value of $\cot t$ and the quadrant of t, the fundamental identities enable us to find the value of the other five functions of t. In fact, this is generally true for any trig function and real number or angle t. For instance, given $\cot t = \frac{-9}{40}$ and t in QIV, we know $\tan t = -\frac{40}{9}$, since they are reciprocals. Since $\sec^2 t = 1 + \tan^2 t$, the value of $\sec t$ can be found by substitution, yielding $\pm\frac{41}{9}$ after doing the algebra. With secant positive in QIV, we have $\sec t = \frac{41}{9}$, which automatically gives $\cos t = \frac{9}{41}$. The value of $\sin t$ and $\csc t$ can now be found using the Pythagorean and reciprocal identities. See Exercises 54 through 62.

TECHNOLOGY HIGHLIGHT

Equations and Identities on a Graphing Calculator

The keystrokes shown apply to a TI-84 Plus model. Please consult your manual or our Internet site for other models.

Earlier we stated, "An identity is an equation that is true for all values of the input variable where the functions are defined." To better understand exactly what this means, consider the equation $\frac{\cos x}{1 + \sin x} = \sec x - \tan x$. Even though the right-hand side of this equation is undefined at $x = \frac{\pi}{2}$, and both sides are undefined at $\frac{3\pi}{2}$, the equation is still an identity since it is a true equation for *all other values of x* (see Figure 3.1). On the other hand, the equation $\sin\left(\frac{\theta}{2}\right) = \frac{\sin \theta}{\sin 2}$ is *not* an identity (thougth it is sometimes mistaken for one), even though it is a true equation for $\theta = 2\pi k$ and $\theta \approx \pm 2.1976 + 2\pi k$, where k is an integer. These solutions appear graphically as points of intersection (see Figure 3.2). Finally, we consider the equation $2x^2 - 1 = \cos(2\cos^{-1}x)$. Judging from the graph shown in Figure 3.3, it appears the equation is true for all real values of x, since we presumably see a single graph. Actually the equation is only true when $x \in [-1, 1]$, for reasons we'll discover later in this chapter. While it *is an identity,* it is only valid where both functions are defined.

Figure 3.1

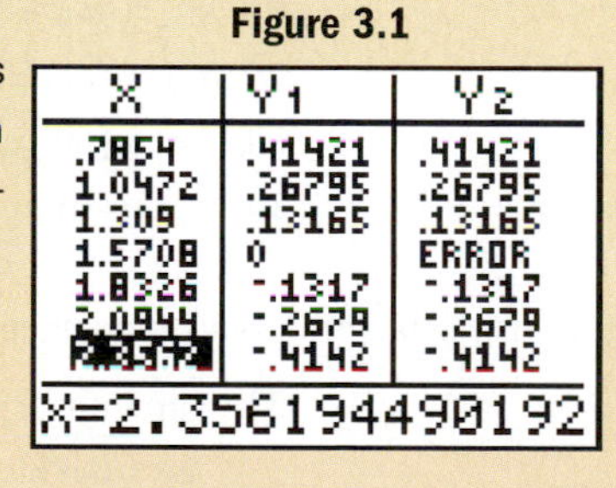

X	Y1	Y2
.7854	.41421	.41421
1.0472	.26795	.26795
1.309	.13165	.13165
1.5708	0	ERROR
1.8326	-.1317	-.1317
2.0944	-.2679	-.2679
2.3562	-.4142	-.4142

X=2.356194490192

Figure 3.2

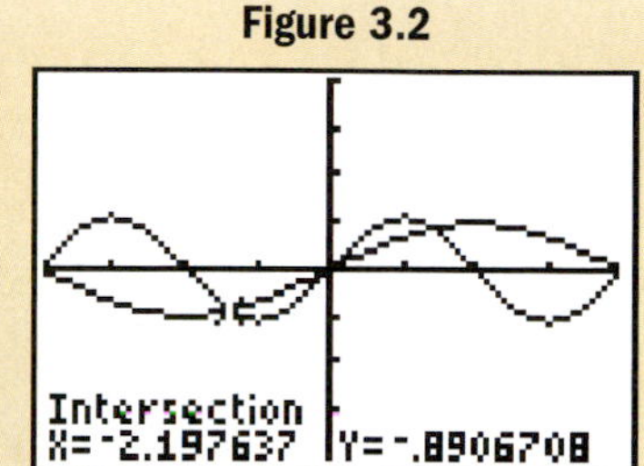

Figure 3.3

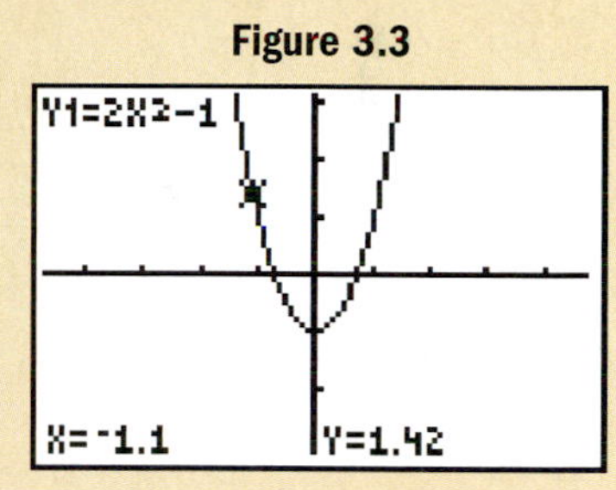

Use a graphing calculator and graphs or tables to determine which of the following equations are not identities.

Exercise 1: $\cos x \cot x = \csc x - \sin x$

Exercise 2: $\tan^2 x + \cot^2 x = 1$

Exercise 3: $\frac{1}{\csc x + \sec x} = \sin x + \cos x$

Exercise 4: $2x\sqrt{1 - x^2} = \sin(2\cos^{-1}x)$

3.1 EXERCISES

CONCEPTS AND VOCABULARY

Fill in each blank with the appropriate word or phrase. Carefully reread the section if needed.

1. Three fundamental ratio identities are $\tan \theta = \frac{?}{\cos \theta}$, $\tan \theta = \frac{?}{\csc \theta}$, and $\cot \theta = \frac{?}{\sin \theta}$.

2. When verifying an identity, we can work on either the _____ or _____ side of an equation, or on both sides __________ of each other.

3. To show an equation is *not an identity,* we must find at least ____ value(s) where both sides of the equation are defined, but which makes the equation ____.

4. Using a calculator we find $\sec^2 45° =$ ____ and $3\tan 45° - 1 =$ ____. We also find $\sec^2 225° =$ ____ and $3\tan 225° - 1 =$ ____. Is the equation $\sec^2\theta = 3\tan\theta - 1$ an identity?

5. Use the pattern $\frac{A}{B} \pm \frac{C}{D} = \frac{AD \pm BC}{BD}$ to add the following terms, and comment on this process versus "finding a common denominator:" $\frac{\cos x}{\sin x} - \frac{\sin x}{\sec x}$.

6. Name at least four algebraic skills that are used with the fundamental identities in order to rewrite a trigonometric expression. Use algebra to quickly rewrite $(\sin x + \cos x)^2$.

DEVELOPING YOUR SKILLS

Starting with the ratio identity given, use substitution and fundamental identities to write four new identities belonging to the ratio family.

7. $\tan x = \frac{\sin x}{\cos x}$

8. $\cot x = \frac{\cos x}{\sin x}$

Starting with the Pythagorean identity given, use algebra to write four additional identities belonging to the Pythagorean family.

9. $1 + \tan^2 x = \sec^2 x$

10. $1 + \cot^2 x = \csc^2 x$

Verify the equation is an identity using multiplication and fundamental identities.

11. $\sin x \cot x = \cos x$

12. $\cos x \tan x = \sin x$

13. $\sec^2 x \cot^2 x = \csc^2 x$

14. $\csc^2 x \tan^2 x = \sec^2 x$

15. $\cos x(\sec x - \cos x) = \sin^2 x$

16. $\tan x(\cot x + \tan x) = \sec^2 x$

17. $\sin x(\csc x - \sin x) = \cos^2 x$

18. $\cot x(\tan x + \cot x) = \csc^2 x$

19. $\tan x(\csc x + \cot x) = \sec x + 1$

20. $\cot x(\sec x + \tan x) = \csc x + 1$

Verify the equation is an identity using factoring and fundamental identities.

21. $\tan^2 x \csc^2 x - \tan^2 x = 1$

22. $\sin^2 x \cot^2 x + \sin^2 x = 1$

23. $\frac{\sin x \cos x + \sin x}{\cos x + \cos^2 x} = \tan x$

24. $\frac{\sin x \cos x + \cos x}{\sin x + \sin^2 x} = \cot x$

25. $\frac{1 + \sin x}{\cos x + \cos x \sin x} = \sec x$

26. $\frac{1 + \cos x}{\sin x + \cos x \sin x} = \csc x$

27. $\frac{\sin x \tan x + \sin x}{\tan x + \tan^2 x} = \cos x$

28. $\frac{\cos x \cot x + \cos x}{\cot x + \cot^2 x} = \sin x$

Verify the equation is an identity using special products and fundamental identities.

29. $\frac{(\sin x + \cos x)^2}{\cos x} = \sec x + 2\sin x$

30. $\frac{(1 + \tan x)^2}{\sec x} = \sec x + 2\sin x$

31. $(1 + \sin x)[1 + \sin(-x)] = \cos^2 x$

32. $(\sec x + 1)[\sec(-x) - 1] = \tan^2 x$

33. $\frac{(\csc x - \cot x)(\csc x + \cot x)}{\tan x} = \cot x$

34. $\frac{(\sec x + \tan x)(\sec x - \tan x)}{\csc x} = \sin x$

Verify the equation is an identity using fundamental identities and $\frac{A}{B} \pm \frac{C}{D} = \frac{AD \pm BC}{BD}$ to combine terms.

35. $\frac{\cos^2 x}{\sin x} + \frac{\sin x}{1} = \csc x$

36. $\frac{\sec \alpha}{1} - \frac{\tan^2 \alpha}{\sec \alpha} = \cos \alpha$

37. $\frac{\tan x}{\csc x} - \frac{\sin x}{\cos x} = \frac{\sin x - 1}{\cot x}$

38. $\frac{\cot x}{\sec x} - \frac{\cos x}{\sin x} = \frac{\cos x - 1}{\tan x}$

39. $\frac{\sec x}{\sin x} - \frac{\csc x}{\sec x} = \tan x$

40. $\frac{\csc x}{\cos x} - \frac{\sec x}{\csc x} = \cot x$

Write the given function entirely in terms of the second function indicated.

41. $\tan x$ in terms of $\sin x$

42. $\tan x$ in terms of $\sec x$

43. $\sec x$ in terms of $\cot x$

44. $\sec x$ in terms of $\sin x$

45. $\cot x$ in terms of $\sin x$

46. $\cot x$ in terms of $\csc x$

Show that the following equations *are not identities.*

47. $\sin\left(\theta + \frac{\pi}{3}\right) = \sin \theta + \sin\left(\frac{\pi}{3}\right)$

48. $\cos\left(\frac{\pi}{4}\right) + \cos \theta = \cos\left(\frac{\pi}{4} + \theta\right)$

49. $\cos(2\theta) = 2 \cos \theta$

50. $\tan(2\theta) = 2 \tan \theta$

51. $\tan\left(\frac{\theta}{4}\right) = \frac{\tan \theta}{\tan 4}$

52. $\cos^2\theta - \sin^2\theta = -1$

53. $\sqrt{\sin^2 x - 9} = \sin x - 3$

For the function $f(\theta)$ and the quadrant in which θ terminates, state the value of the other five trig functions.

54. $\cos \theta = -\frac{20}{29}$ with θ in QII

55. $\sin \theta = \frac{12}{37}$ with θ in QII

56. $\tan \theta = \frac{15}{8}$ with θ in QIII

57. $\sec \theta = \frac{45}{27}$ with θ in QIV

58. $\cot \theta = \frac{x}{5}$ with θ in QI

59. $\csc \theta = \frac{7}{x}$ with θ in QII

60. $\sin \theta = -\frac{7}{13}$ with θ in QIII

61. $\cos \theta = \frac{23}{25}$ with θ in QIV

62. $\sec \theta = -\frac{9}{7}$ with θ in QII

WORKING WITH FORMULAS

63. The area of regular polygon: $A = \left(\frac{nx^2}{4}\right)\frac{\cos(\frac{\pi}{n})}{\sin(\frac{\pi}{n})}$

The area of a regular polygon is given by the formula shown, where n represents the number of sides and x is the length of each side.

a. Rewrite the formula in terms of a single trig function.

b. Verify the formula for a square with sides of 8 m.

c. Find the area of a dodecagon (12 sides) with 10-in. sides.

64. The illuminance of a point on a surface by a source of light: $E = \dfrac{I\cos\theta}{r^2}$

The illuminance E (in lumens/m^2) of a point on a horizontal surface is given by the formula shown, where I is the intensity of the light source in lumens, r is the distance in meters from the light source to the point, and θ is the complement of the angle α (in degrees) made by the light source and the horizontal surface. Calculate the illuminance if $I = 800$ lumens, and the flashlight is held so that the distance r is 2 m while the angle α is 40°.

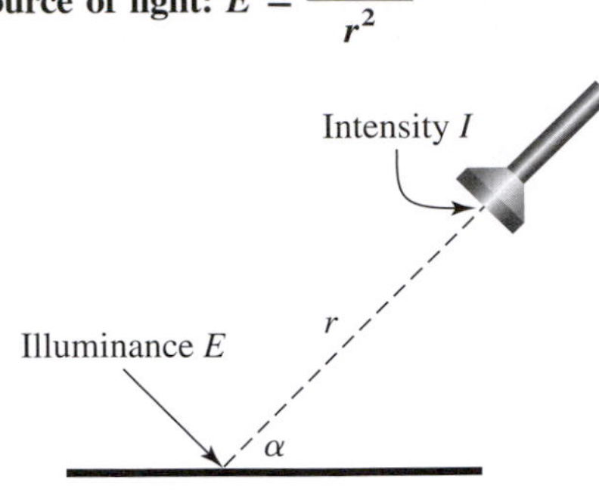

APPLICATIONS

Writing a given expression in an alternative form is an idea used at all levels of mathematics. In future classes, it is often helpful to decompose a power into smaller powers (as in writing A^3 as $A \cdot A^2$) or to rewrite an expression using known identities so that it can be factored.

65. Show that $\cos^3 x$ can be written as $\cos x(1 - \sin^2 x)$.

66. Show that $\tan^3 x$ can be written as $\tan x(\sec^2 x - 1)$.

67. Show that $\tan x + \tan^3 x$ can be written as $\tan x(\sec^2 x)$.

68. Show that $\cot^3 x$ can be written as $\cot x(\csc^2 x - 1)$.

69. Show $\tan^2 x \sec x - 4\tan^2 x$ can be factored into $(\sec x - 4)(\sec x - 1)(\sec x + 1)$.

70. Show $2\sin^2 x\cos x - \sqrt{3}\sin^2 x$ can be factored into $(1 - \cos x)(1 + \cos x)(2\cos x - \sqrt{3})$.

71. Show $\cos^2 x \sin x - \cos^2 x$ can be factored into $-1(1 + \sin x)(1 - \sin x)^2$.

72. Show $2\cot^2 x\csc x + 2\sqrt{2}\cot^2 x$ can be factored into $2(\csc x + \sqrt{2})(\csc x - 1)(\csc x + 1)$.

Many applications of fundamental identities involve geometric figures, as in Exercises 73 and 74.

Exercise 73

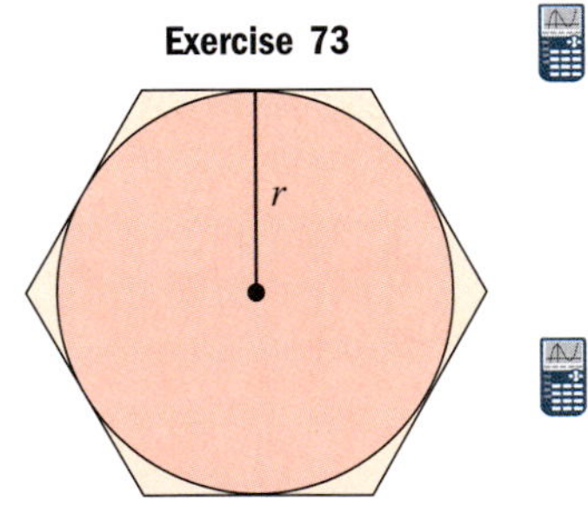

73. Area of a polygon: The area of a regular polygon that has been circumscribed about a circle of radius r (see figure) is given by the formula $A = nr^2\dfrac{\sin\left(\frac{\pi}{n}\right)}{\cos\left(\frac{\pi}{n}\right)}$, where n represents the number of sides. (a) Rewrite the formula in terms of a single trig function; (b) verify the formula for a square circumscribed about a circle with radius 4 m; and (c) find the area of a dodecagon (12 sides) circumscribed about the same circle.

74. Perimeter of a polygon: The perimeter of a regular polygon circumscribed about a circle of radius r is given by the formula $P = 2nr\dfrac{\sin\left(\frac{\pi}{n}\right)}{\cos\left(\frac{\pi}{n}\right)}$, where n represents the number of sides. (a) Rewrite the formula in terms of a single trig function; (b) verify the formula for a square circumscribed about a circle with radius 4 m; and (c) Find the perimeter of a dodecagon (12 sides) circumscribed about the same circle.

75. Angle of intersection: At their point of intersection, the angle θ between any two nonparallel lines satisfies the relationship $(m_2 - m_1)\sin\theta = \cos\theta + m_1 m_2\cos\theta$, where m_1 and m_2 represent the slopes of the two lines. Rewrite the equation in terms of a single trig function.

76. Angle of intersection: Use the result of Exercise 75 to find the angle between the lines $Y_1 = \frac{2}{5}x - 3$ and $Y_2 = \frac{7}{3}x + 1$.

77. Angle of intersection: Use the result of Exercise 75 to find the angle between the lines $Y_1 = 3x - 1$ and $Y_2 = -2x + 7$.

WRITING, RESEARCH, AND DECISION MAKING

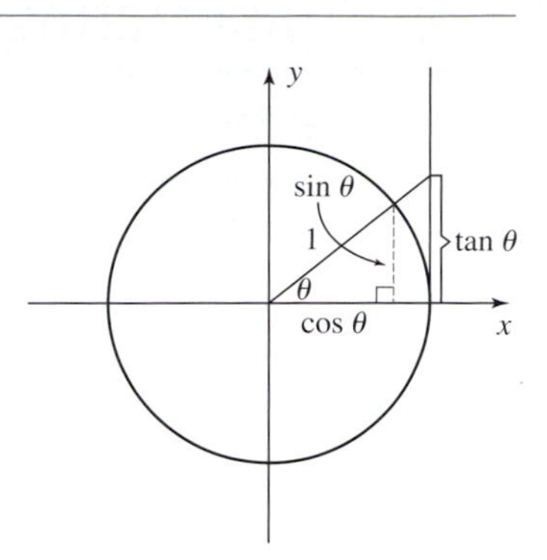

78. The word *tangent* literally means "to touch," which in mathematics we take to mean *touches in only and exactly one point.* In the figure, the circle has a radius of 1 and the vertical line is "tangent" to the circle at the x-axis. The figure can be used to verify the Pythagorean identity for sine and cosine, as well as the ratio identity for tangent. Discuss/demonstrate how.

79. During the development of the extensive chord tables mentioned in Chapter 1, a function called the ***versed sine*** was heavily used and, in fact, is still used by some surveyors and, engineers today. Use the Internet or the resources of a local library to do some research on the versed sine and its relationship to chords, half-chords, and the sine and cosine functions. Include several examples of its use, including how it was used to find exact forms for 15° and 7.5°.

EXTENDING THE CONCEPT

80. Use factoring and fundamental identities to help find the x-intercepts of f in $[0, 2\pi)$. $f(\theta) = -2\sin^4\theta + \sqrt{3}\sin^3\theta + 2\sin^2\theta - \sqrt{3}\sin\theta$.

81. Regarding Exercises 73 and 74, note the area of the polygon is greater than the area of the circle, and the perimeter of the polygon is greater than the perimeter of the circle. When comparing the area and perimeter of each: (a) how many sides must the polygon have for their areas to differ by less than 5%? (b) How many sides must the polygon have for their perimeters to differ by less than 2%?

MAINTAINING YOUR SKILLS

82. (1.4) Verify the point $\left(-\frac{3}{4}, \frac{\sqrt{7}}{4}\right)$ is on the unit circle and use it to give the value of the six trig functions of θ.

83. (1.2) Standing 265 ft from the base of the Strastosphere Tower in Las Vegas, Nevada, the angle of elevation to the top of the tower is about 77°. Approximate the height of the tower to the nearest foot.

84. (2.3) The equation of the function graphed in the figure is of the form $y = \cot(Bx)$. What is the value of B?

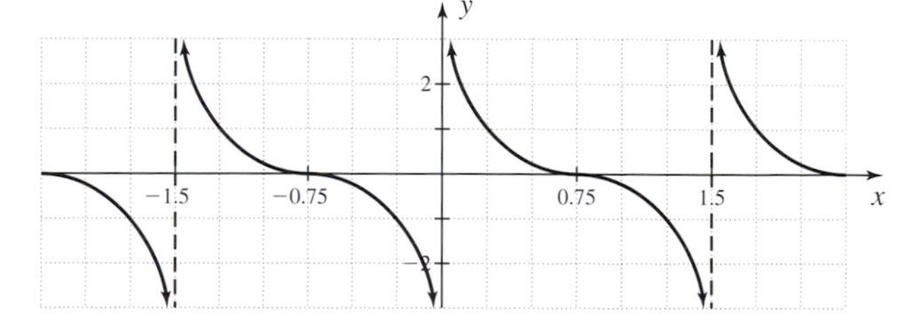

85. (1.3) Find $f(\theta)$ for all six trig functions, given $P(-16, -63)$ is on the terminal side of θ.

86. (1.1) The Earth has a radius of 3960 mi. Charlotte, North Carolina, is located at 35° N latitude, 80.5° W longitude. Amarillo, Texas, is located at 35° N latitude, 101.5° W longitude. How many miles separate the two cities?

87. (2.1) Use a reference rectangle and the *rule of fourths* to sketch the graph of $y = 2\sin(2t)$ for $t \in [0, 2\pi)$.

3.2 Constructing and Verifying Identities

LEARNING OBJECTIVES

In Section 3.2 you will learn how to:

A. Create and verify a new identity

B. Verify general identities

INTRODUCTION

In Section 3.1, our primary goal was to illustrate how basic algebra skills are used to help rewrite trigonometric expressions. In Section 3.2 we'll sharpen and refine these skills so they can be applied more generally, and develop the ability to verify a much wider range of identities.

POINT OF INTEREST

Around A.D. 505 the Indian mathematician Varahamihira worked in the city of Ujjain, one of the leading centers of mathematical ideas in India. Most of Varahamihira's work centered on astronomy and the computation/compilation of tables related to movements of heavenly bodies. He made several important mathematical discoveries, including the identities (written in modern notation) $\sin^2 x + \cos^2 x = 1$ (Pythagorean identity), $\sin x = \cos\left(\frac{\pi}{2} - x\right)$ (cofunction identity), and $\sin^2 x = \frac{1 - \cos(2x)}{2}$ (half-angle identity—introduced in Section 3.4).

A. Creating and Verifying New Identities

In Example 2 of Section 3.1, we showed that $\sin\theta(\csc\theta - \sin\theta) = \cos^2\theta$ was an identity by transforming the left side into $\cos^2\theta$. One of the subtleties involved in verifying an identity is that *the steps must be reversible*. In other words, we should be able to rewrite $\cos^2\theta$ using algebra and the fundamental identities until it becomes $\sin\theta(\csc\theta - \sin\theta)$. To gain a better insight into identities and how they can be verified, we will use the idea of reversibility to create and then verify random identities. This may seem unusual and very arbitrary to you (actually, it *is*), ***and results could vary from student to student.*** But you are asked to keep the underlying message in mind, rather than any specific steps. When working with identities that message is—if two things are equal, one can be substituted for the other at any time and the result will be equivalent. If there is ever any doubt, graphical support for the intermediate stages and the final result can be employed, as noted in Section 3.1.

EXAMPLE 1 Starting with the expression $\csc x + \cot x$, use fundamental identities to rewrite the expression and create a new identity. Then verify the identity by reversing the steps.

Solution:

$\csc x + \cot x$ — original expression

$= \dfrac{1}{\sin x} + \dfrac{\cos x}{\sin x}$ — substitute reciprocal and ratio identities

$= \dfrac{1 + \cos x}{\sin x}$ — write as a single term

Resulting identity: $\csc x + \cot x = \frac{1 + \cos x}{\sin x}$

Verify identity: Working with the right-hand side, we reverse each step with a view toward the original expression.

$$\csc x + \cot x = \frac{1 + \cos x}{\sin x} \qquad \text{identity}$$

$$= \frac{1}{\sin x} + \frac{\cos x}{\sin x} \qquad \text{rewrite left side as individual terms}$$

$$= \csc x + \cot x \qquad \text{substitute reciprocal and ratio identities}$$

NOW TRY EXERCISES 7 THROUGH 9

For Example 1, we stopped "creating" identities after a two-step process. In reality, the substitutions and rewriting could be sustained for three or more steps using algebra combined with the fundamental identities.

EXAMPLE 2 Starting with the expression $2 \tan x \sec x$, use fundamental identities to rewrite the expression and create a new identity. Then verify the identity by reversing the steps.

Solution:

$$2 \tan x \sec x \qquad \text{original expression}$$

$$= 2 \cdot \frac{\sin x}{\cos x} \cdot \frac{1}{\cos x} \qquad \text{substitute ratio and reciprocal identities}$$

$$= \frac{2 \sin x}{\cos^2 x} \qquad \text{multiply}$$

$$= \frac{2 \sin x}{1 - \sin^2 x} \qquad \text{substitute } 1 - \sin^2 x \text{ for } \cos^2 x$$

Resulting identity: $2 \tan x \sec x = \frac{2 \sin x}{1 - \sin^2 x}$ identity

Verify identity: Working with the right-hand side, we reverse each step with a view toward the original expression.

$$2 \tan x \sec x = \frac{2 \sin x}{1 - \sin^2 x} \qquad \text{identity}$$

$$= \frac{2 \sin x}{\cos^2 x} \qquad \text{substitute } \cos^2 x \text{ for } 1 - \sin^2 x$$

$$= 2 \cdot \frac{\sin x}{\cos x} \cdot \frac{1}{\cos x} \qquad \text{rewrite using } \cos^2 x = \cos x \cdot \cos x$$

$$= 2 \tan x \sec x \qquad \text{substitute ratio and reciprocal identities}$$

NOW TRY EXERCISES 10 THROUGH 12

As a reminder, support for the "resulting identity" in Examples 1 and 2 can be found using the GRAPH and/or TABLE features of a graphing calculator. See the Technology Highlight just prior to the Exercise Set.

B. Verifying Identities

We're now ready to put these ideas, and the ideas from Section 3.1, to work for us. When verifying identities we attempt to mold, change, or rewrite one side of the equality until we obtain a match with the other side. What follows is a collection of the ideas and methods we've observed so far, which we'll call the *Guidelines for Verifying Identities*. One small piece of advice before we begin—there really is no *right* place to start. Think things over for a moment, then attempt a substitution, simplification, or operation and just see where it leads. If you hit a dead end, that's okay! Just back up and try something else.

GUIDELINES FOR VERIFYING IDENTITIES

1. Work on only one side of the identity.
   - We cannot assume the equation is true, so properties of equality cannot be applied.
   - We verify the identity by changing the form of one side until we get a match with the other.
2. Work with the more complex side, as it is easier to reduce/simplify than to "build."
3. If an expression contains more than one term, it is often helpful to combine terms using $\frac{A}{B} \pm \frac{C}{D} = \frac{AD \pm BC}{BD}$.
4. Converting all functions to sines and cosines or to x's and y's (from the unit circle definition) can be helpful.
5. Apply basic algebra skills as appropriate: distribute, factor, multiply by a conjugate, and so on.
6. *Know the fundamental identities inside out, upside down, and backward—they are the key!*

WORTHY OF NOTE

When verifying identities, it is actually permissible to work on each side of the equality *independently*, in the effort to create a "match." But properties of equality can never be used, since we cannot assume an equality exists.

Note how these ideas are employed in Examples 3 through 5, particularly the frequent use of fundamental identities.

EXAMPLE 3 Verify the identity: $\sin^2 x \tan^2 x = \tan^2 x - \sin^2 x$.

Solution: As a general rule, the side with the greater number of terms or the side with rational terms is considered more complex, so we begin with the right-hand side.

$$\begin{aligned}
\sin^2 x \tan^2 x &= \tan^2 x - \sin^2 x && \text{original identity} \\
&= \frac{\sin^2 x}{\cos^2 x} - \frac{\sin^2 x}{1} && \text{ratio identity (write in terms of sines and cosines)} \\
&= \frac{\sin^2 x}{1}\left(\frac{1}{\cos^2 x} - 1\right) && \text{factor out } \sin^2 x \\
&= \sin^2 x(\sec^2 x - 1) && \text{reciprocal identity} \\
&= \sin^2 x \tan^2 x && \text{substitute } \tan^2 x \text{ for } \sec^2 x - 1
\end{aligned}$$

NOW TRY EXERCISES 13 THROUGH 18

Example 3 involved *factoring* out a common expression. Just as often, we'll need to *multiply* numerators and denominators by a common expression, as in Example 4.

EXAMPLE 4 Verify the identity: $\frac{\cos x}{1 + \sec x} = \frac{1 - \cos x}{\tan^2 x}$.

Solution: Both sides of the identity have a single term and one is really no more complex than the other. As a matter of choice, we begin on the left side.

$$\frac{\cos x}{1 + \sec x} = \frac{1 - \cos x}{\tan^2 x}$$ original identity

$$\left(\frac{\cos x}{1 + \sec x}\right)\left(\frac{1 - \sec x}{1 - \sec x}\right) =$$ multiply above and below by the conjugate

$$\frac{\cos x - 1}{1 - \sec^2 x} =$$ $\cos x \sec x = 1$; $(A + B)(A - B) = A^2 - B^2$

$$\frac{\cos x - 1}{-\tan^2 x} =$$ Pythagorean identity (watch signs)

$$\frac{1 - \cos x}{\tan^2 x} =$$ multiply above and below by -1

NOW TRY EXERCISES 19 THROUGH 22

Example 4 highlights the need to be very familiar with the fundamental identities. The related Pythagorean identity is $1 + \tan^2 x = \sec^2 x$, so to replace $1 - \sec^2 x$, we had to use $-\tan^2 x$, not simply $\tan^2 x$.

As noted in the guidelines, combining rational terms is often productive. Once again students are encouraged to work with the pattern $\frac{A}{B} \pm \frac{C}{D} = \frac{AD \pm BC}{BD}$ until the method comes quickly and naturally.

EXAMPLE 5 Verify the identity: $\frac{\sec x}{\sin x} - \frac{\sin x}{\sec x} = \frac{\tan^2 x + \cos^2 x}{\tan x}$.

Solution: We begin with the left-hand side.

$$\frac{\sec x}{\sin x} - \frac{\sin x}{\sec x} = \frac{\tan^2 x + \cos^2 x}{\tan x}$$ original identity

$$\frac{\sec^2 x - \sin^2 x}{\sin x \sec x} =$$ combine terms: $\frac{A}{B} - \frac{C}{D} = \frac{AD - BC}{BD}$

$$\frac{(1 + \tan^2 x) - (1 - \cos^2 x)}{\frac{\sin x}{1}\left(\frac{1}{\cos x}\right)} =$$ Pythagorean identities; ratio identity

$$\frac{\tan^2 x + \cos^2 x}{\tan x} =$$ result

NOW TRY EXERCISES 23 THROUGH 28

Identities come in an infinite variety and it would be impossible to illustrate all variations. Using the general ideas and skills presented should prepare you to verify any of those given in the exercise set, as well as those you encounter in your future studies. See Exercises 29 through 58.

TECHNOLOGY HIGHLIGHT

Identities and Graphical Tests

The keystrokes shown apply to a TI-84 Plus model. Please consult your manual or our Internet site for other models.

Unless you are told to verify a known identity, how would you know ahead of time that it was possible? What if there was a misprint on a test, quiz, or homework assignment and what you were trying to prove wasn't possible? In cases like these, a *graphical test* can be used. While not fool-proof, seeing if the graphs appear identical can either *suggest* the identity is true, or definitely show it is not. When testing identities, it helps to ENTER the left-hand side of the equation as Y_1 on the Y= screen, and the right-hand side as Y_2. We can then test whether an identity relationship might exist by activating the two relations in question, graphing both relations, and noting whether two graphs or a single graph appears on the GRAPH screen. After entering the six functions shown in Figure 3.4, leave Y_1 and Y_2 active, then deactivate Y_3 through Y_6 by overlaying the equal sign with the cursor and pressing ENTER. In this way, we'll test each pair to see if their graphs appear identical. Of course, the calculator's TABLE feature could also be used. Using the calculator's ZOOM **7:ZTrig** feature, we obtain the graph shown in Figure 3.5, which indicates that an identity relationship likely exists between Y_1 and Y_2. After deactivating Y_1 and Y_2, we activate Y_3 and Y_4 and again note the likelihood of an identity relationship, since only one graph can be seen. However, when testing Y_5 and Y_6, we note the existence of two distinct graphs (see Figure 3.6), and conclude that $\sin(2x) \neq 2\sin x$—even though $\sin(2x) = 2\sin x$ for integer multiples of π (the points of intersection). Use this technique to help determine which of the following equations are not identities.

Figure 3.4

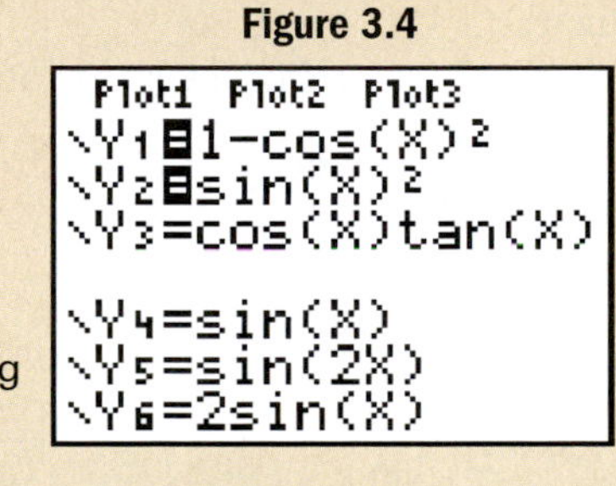

Figure 3.5

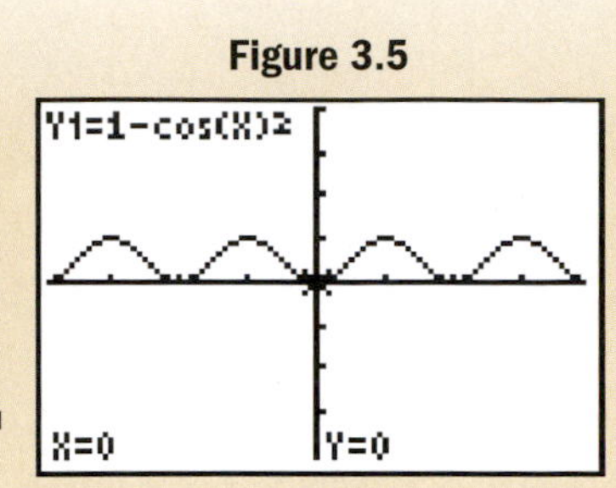

Figure 3.6

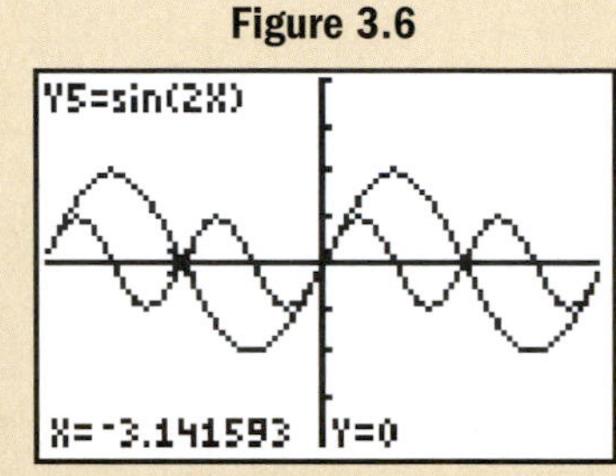

Exercise 1: $\dfrac{1 - \sin^2\theta}{\cos\theta} = \cos\theta$

Exercise 2: $\cos(2x) = 1 - 2\sin^2 x$

Exercise 3: $\dfrac{\cos x}{1 + \sin x} = \dfrac{1 - \sin x}{\cos x}$

Exercise 4: $\dfrac{\cos x}{1 - \sin x} = \sec x - \tan x$

3.2 EXERCISES

CONCEPTS AND VOCABULARY

Fill in each blank with the appropriate word or phrase. Carefully reread the section if needed.

1. The process used to verify an identity must be __________, meaning you can retrace your steps back to the original expression.

2. When attempting to verify an identity, if an expression contains more than one rational term it is often helpful to ________ them using the least common denominator.

3. To verify an identity, always begin with the more ________ expression, since it is easier to ________ than to ________.

4. Converting all terms to functions of ______ and ______, or writing each function in terms of its ______ ______ definition may help verify an identity.

5. Discuss/explain why you must not add, subtract, multiply, or divide both sides of the equation when verifying identities.

6. Discuss/explain the difference between operating on both sides of an equation (see Exercise 5) and working on each side independently.

DEVELOPING YOUR SKILLS

Use algebra and the fundamental identities to rewrite the given expression and create a new identity relationship. Then verify your identity by reversing the steps. Answers will vary.

7. $\sec x + \tan x$

8. $(\cos x + \sin x)^2$

9. $(1 - \sin^2 x)\sec x$

10. $2 \cot x \csc x$

11. $\dfrac{\sin x - \sin x \cos x}{\sin^2 x}$

12. $(\cos x + \sin x)(\cos x - \sin x)$

Verify that the following equations are identities.

13. $\cos^2 x \tan^2 x = 1 - \cos^2 x$

14. $\sin^2 x \cot^2 x = 1 - \sin^2 x$

15. $\tan x + \cot x = \sec x \csc x$

16. $\cot x \cos x = \csc x - \sin x$

17. $\dfrac{\cos x}{\tan x} = \csc x - \sin x$

18. $\dfrac{\sin x}{\cot x} = \sec x - \cos x$

19. $\dfrac{\cos \theta}{1 - \sin \theta} = \sec \theta + \tan \theta$

20. $\dfrac{\sin \theta}{1 - \cos \theta} = \csc \theta + \cot \theta$

21. $\dfrac{1 - \sin x}{\cos x} = \dfrac{\cos x}{1 + \sin x}$

22. $\dfrac{1 - \cos x}{\sin x} = \dfrac{\sin x}{1 + \cos x}$

23. $\dfrac{\csc x}{\cos x} - \dfrac{\cos x}{\csc x} = \dfrac{\cot^2 x + \sin^2 x}{\cot x}$

24. $\dfrac{1}{\cos^2 x} + \dfrac{1}{\sin^2 x} = \csc^2 x \sec^2 x$

25. $\dfrac{\sin x}{1 + \sin x} - \dfrac{\sin x}{1 - \sin x} = -2 \tan^2 x$

26. $\dfrac{\cos x}{1 + \cos x} - \dfrac{\cos x}{1 - \cos x} = -2 \cot^2 x$

27. $\dfrac{\cot x}{1 + \csc x} - \dfrac{\cot x}{1 - \csc x} = 2 \sec x$

28. $\dfrac{\tan x}{1 + \sec x} - \dfrac{\tan x}{1 - \sec x} = 2 \csc x$

29. $\dfrac{\sec^2 x}{1 + \cot^2 x} = \tan^2 x$

30. $\dfrac{\csc^2 x}{1 + \tan^2 x} = \cot^2 x$

31. $\sin^2 x(\cot^2 x - \csc^2 x) = -\sin^2 x$

32. $\cos^2 x(\tan^2 x - \sec^2 x) = -\cos^2 x$

33. $\cos x \cot x + \sin x = \csc x$

34. $\sin x \tan x + \cos x = \sec x$

35. $\dfrac{\sec x}{\cot x + \tan x} = \sin x$

36. $\dfrac{\csc x}{\cot x + \tan x} = \cos x$

37. $\dfrac{\sin x - \csc x}{\csc x} = -\cos^2 x$

38. $\dfrac{\cos x - \sec x}{\sec x} = -\sin^2 x$

39. $\dfrac{1}{\csc x - \sin x} = \tan x \sec x$

40. $\dfrac{1}{\sec x - \cos x} = \cot x \csc x$

41. $\dfrac{1 + \sin x}{1 - \sin x} = (\tan x + \sec x)^2$

42. $\dfrac{1 - \cos x}{1 + \cos x} = (\csc x - \cot x)^2$

43. $\dfrac{\cos x - \sin x}{1 - \tan x} = \dfrac{\cos x + \sin x}{1 + \tan x}$

44. $\dfrac{1 - \cot x}{1 + \cot x} = \dfrac{\sin x - \cos x}{\sin x + \cos x}$

45. $\dfrac{\tan^2 x - \cot^2 x}{\tan x - \cot x} = \csc x \sec x$

46. $\dfrac{\cot x - \tan x}{\cot^2 x - \tan^2 x} = \sin x \cos x$

47. $\dfrac{\cot x}{\cot x + \tan x} = 1 - \sin^2 x$

48. $\dfrac{\tan x}{\cot x + \tan x} = 1 - \cos^2 x$

49. $\dfrac{\sec^4 x - \tan^4 x}{\sec^2 x + \tan^2 x} = 1$

50. $\dfrac{\csc^4 x - \cot^4 x}{\csc^2 x + \cot^2 x} = 1$

51. $\dfrac{\cos^4 x - \sin^4 x}{\cos^2 x} = 2 - \sec^2 x$

52. $\dfrac{\sin^4 x - \cos^4 x}{\sin^2 x} = 2 - \csc^2 x$

53. $(\sec x + \tan x)^2 = \dfrac{(\sin x + 1)^2}{\cos^2 x}$

54. $(\csc x + \cot x)^2 = \dfrac{(\cos x + 1)^2}{\sin^2 x}$

55. $\dfrac{\cos x}{\sin x} + \dfrac{\sin x}{\cos x} + \dfrac{\csc x}{\sec x} = \dfrac{\sec x + \cos x}{\sin x}$

56. $\dfrac{\cos x}{\sin x} + \dfrac{\sin x}{\cos x} + \dfrac{\sec x}{\csc x} = \dfrac{\csc x + \sin x}{\cos x}$

57. $\dfrac{\sin^4 x - \cos^4 x}{\sin^3 x + \cos^3 x} = \dfrac{\sin x - \cos x}{1 - \sin x \cos x}$

58. $\dfrac{\sin^4 x - \cos^4 x}{\sin^3 x - \cos^3 x} = \dfrac{\sin x + \cos x}{1 + \sin x \cos x}$

WORKING WITH FORMULAS

59. Distance to top of movie screen: $d^2 = (20 + x\cos\theta)^2 + (20 - x\sin\theta)^2$

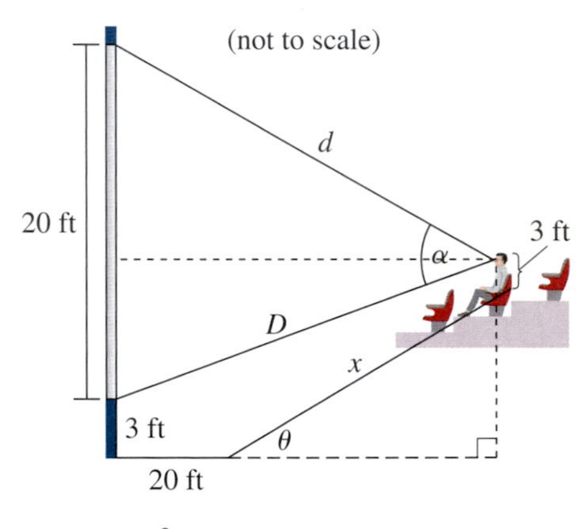

At a theater, the optimum viewing angle depends on a number of factors, like the height of the screen, the incline of the auditorium, the location of a seat, the height of your eyes while seated, and so on. One of the measures needed to find the "best" seat is the distance from your eyes to the top of the screen. For a theater with the dimensions shown, this distance is given by the formula here (x is the diagonal distance from the horizontal floor to your seat). (a) Show the formula is equivalent to $800 + 40x(\cos\theta - \sin\theta) + x^2$. (b) Find the distance d if $\theta = 18°$ and you are sitting in the eighth row with the rows spaced 3 ft apart.

60. The area of triangle ABC: $A = \dfrac{c^2 \sin A \sin B}{2 \sin C}$

If one side and three angles of a triangle are known, its area can be computed using this formula, where side c is opposite angle C. Find the area of the triangle shown in the diagram.

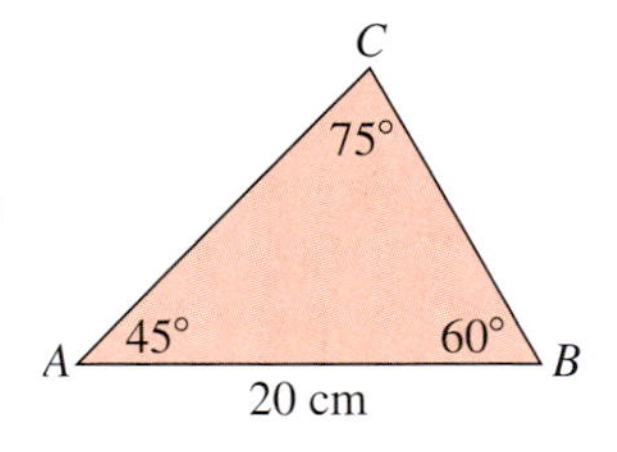

APPLICATIONS

61. Pythagorean theorem: For the triangle shown, (a) find an expression for the length of the hypotenuse in terms of tan x and cot x, then determine the length of the hypotenuse when $x = 1.5$ rad; (b) show the expression you found in part (a) is equivalent to $h = \sqrt{\csc x \sec x}$ and recompute the length of the hypotenuse using this expression. Did the answers match?

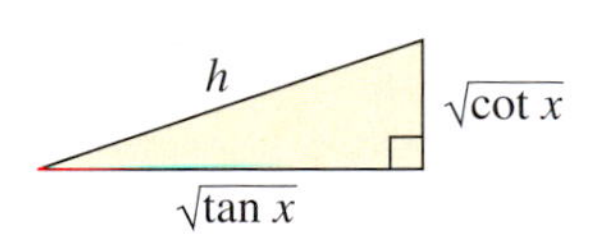

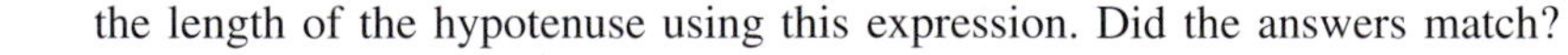

62. Pythagorean theorem: For the triangle shown, (a) find an expression for the area of the triangle in terms of cot x and cos x, then determine its area given $x = \frac{\pi}{6}$; (b) show the expression you found in part (a) is equivalent to $A = \frac{1}{2}(\csc x - \sin x)$ and recompute the area using this expression. Did the answers match?

Exercise 62

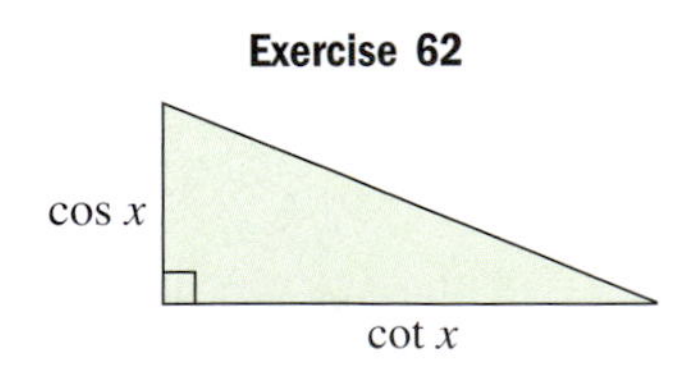

63. Viewing distance: Referring to Exercise 59, find a formula for D—the distance from this patron's eyes to the *bottom* of the movie screen. Simplify the result using a Pythagorean identity, then find the value of D.

64. Viewing distance: Referring to Exercises 59 and 63, once d and D are known, the viewing angle α (the angle subtended by the movie screen and the viewer's eyes) can be found using the formula $\cos\alpha = \dfrac{d^2 + D^2 - 20^2}{2dD}$. Find the value of $\cos\alpha$ for this particular theater, person, and seat.

65. Intensity of light: In a study of the luminous intensity of light, the expression $\sin\alpha = \dfrac{I_1\cos\theta}{\sqrt{(I_1\cos\theta)^2 + (I_2\sin\theta)^2}}$ can occur. Simplify the equation for the moment $I_1 = I_2$.

66. Intensity of light: Referring to Exercise 65, find the angle θ given $I_1 = I_2$ and $\alpha = 60°$.

WRITING, RESEARCH, AND DECISION MAKING

67. Just as the points $P(x, y)$ on the unit circle $x^2 + y^2 = 1$ are used to name the circular trigonometric functions, the points $P(x, y)$ on the unit hyperbola $x^2 - y^2 = 1$ are used to name what are called the **hyperbolic trigonometric functions.** The hyperbolic functions are used extensively in many of the applied sciences. The identities for these functions have many similarities to those for the circular functions, but also have some significant differences. Using the Internet or the resources of a library, do some research on the functions $\sinh t$, $\cosh t$, and $\tanh t$, where t is any real number. In particular, see how the Pythagorean identities compare/contrast between the two forms of trigonometry.

68. Hipparchus and Ptolemy, the forefathers of modern trigonometry, appear to have made use of the Pythagorean identities, though not in the form we know them today. In the writings of Varahamihira (in about A.D. 505) we find the identity $\sin^2\theta + \cos^2\theta = 1$ given in prose form. The Arab mathematician Abu'l-Wefa (~A.D. 940–998) gives a discussion of the Pythagorean identity involving secant and tangent, with references to other identities as well. Do some research on Abu'l-Wefa and report on the scope of his contributions to trigonometry. In particular, what other identities does he mention?

EXTENDING THE CONCEPT

69. Verify the identity $\dfrac{\sin^6 x - \cos^6 x}{\sin^4 x - \cos^4 x} = 1 - \sin^2 x\cos^2 x$.

70. Use factoring to show the equation is an identity: $\sin^4 x + 2\sin^2 x\cos^2 x + \cos^4 x = 1$.

MAINTAINING YOUR SKILLS

71. (1.4) Verify that $\left(\dfrac{\sqrt{7}}{4}, \dfrac{3}{4}\right)$ is a point on the unit circle, then state the values of $\sin t$, $\cos t$, and $\tan t$ associated with this point.

72. (1.1) Find the area of the circular segment and the length of the arc subtended by $\theta = 57°$ for $r = 2$ ft.

Exercise 74

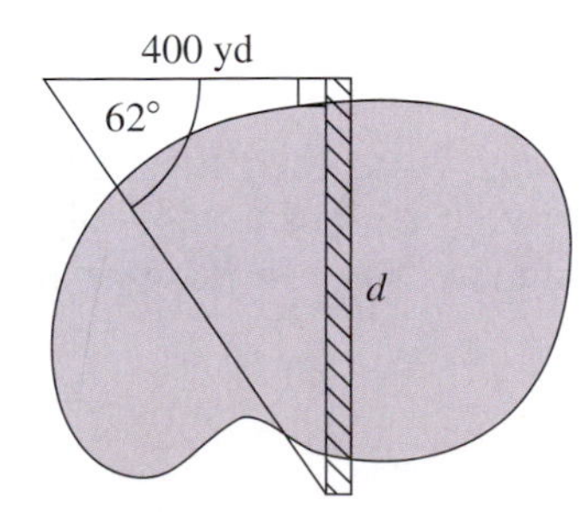

73. (1.1) Name the standard ratio of the sides for a 45-45-90 and 30-60-90 triangle. Include a diagram.

74. (1.3) Use an appropriate trig ratio to find the length of the bridge needed to cross the lake shown in the figure.

75. (2.3) Graph using transformations of a basic function: $y = -2\cos\left(t + \dfrac{\pi}{4}\right)$; $t \in [0, 2\pi)$.

76. (2.2) State the domain of the function: $y = \tan t$.

Mid-Chapter Check

Verify the identities by computing the product.

1. $\sin x(\csc x - \sin x) = \cos^2 x$

2. $(1 + \sec t)(1 - \sec t) = \tan^2 t$

Verify the identities by factoring the left-hand side.

3. $\cos^2 x - \cot^2 x = -\cos^2 x \cot^2 x$

4. $1 - \sin^4 t = (1 + \sin^2 t)\cos^2 t$

Verify the identities by combining the terms on the left.

5. $\dfrac{2\sin x}{\sec x} - \dfrac{\cos x}{\csc x} = \cos x \sin x$

6. $\dfrac{1 - \cos t}{\cos t} + \dfrac{\sec t - 1}{\sec t} = \sec t - \cos t$

Verify the identities by writing all expressions in terms of sine and cosine.

7. $\csc^2 x - \cot^2 x = 1$

8. $\dfrac{\tan x}{\sec x} = \sin x$

Show the equations given are not identities.

9. $1 + \sec^2 x = \tan^2 x$

10. $\cos^2 t = \sin^2 t - 1$

Reinforcing Basic Concepts

Understanding Identities

Sometimes a tactile verification of identities can help us understand and appreciate them. For the following exercises you will need a cm/mm ruler and a protractor.

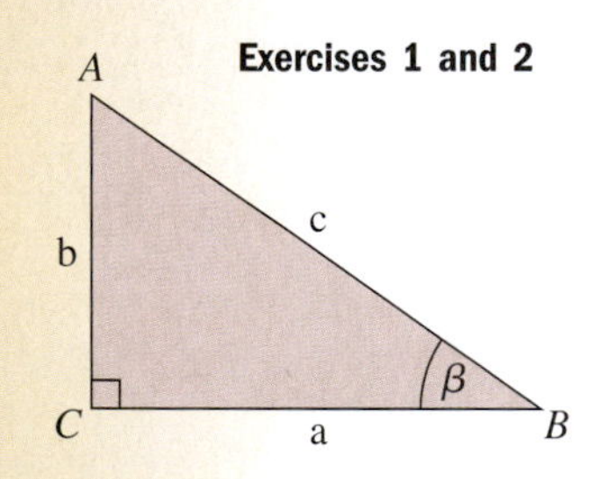

Exercise 1: As carefully as you can (very carefully), measure the opposite and adjacent sides of each triangle to the nearest half-millimeter (mm) with respect to the angle α or β given. Then use the Pythagorean theorem to *compute* the length of the hypotenuse. Then *measure* the hypotenuse and compare your measured result with your computed result. What was the percent difference between the two measurements?

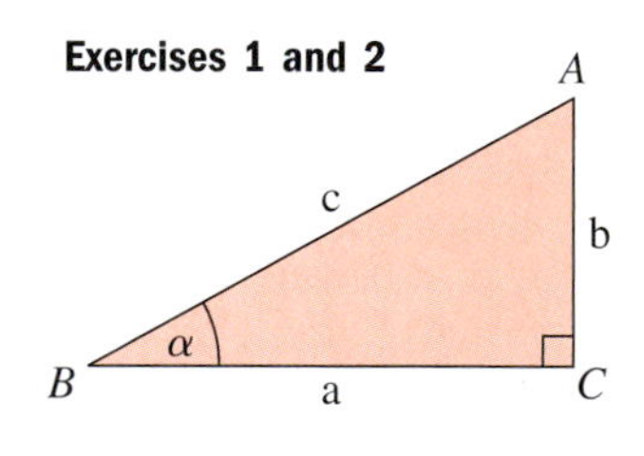

Exercise 2: Using the measured lengths, write the values for $\sin\alpha$, $\cos\alpha$, $\tan\alpha$, $\sin\beta$, $\cos\beta$, and $\tan\beta$ for each triangle. Compute $\sin^2\alpha + \cos^2\alpha$ and $\sin^2\beta + \cos^2\beta$. What do you notice?

Exercise 3: Is it true that $\tan\alpha = \dfrac{\sin\alpha}{\cos\alpha}$? $\tan\beta = \dfrac{\sin\beta}{\cos\beta}$?

As an additional tool to help *verify* identities, we can write each function in terms of its unit circle definition, enabling us to work through a proof in more algebraic terms. This sometimes makes the process less cumbersome. Recall that for $P(x, y)$ on the unit circle, $x^2 + y^2 = 1$.

Verify the identity: $\cos^2\theta - \sin^2\theta = \dfrac{1 - \tan^2\theta}{1 + \tan^2\theta}$.

We begin with the rational term on the right.

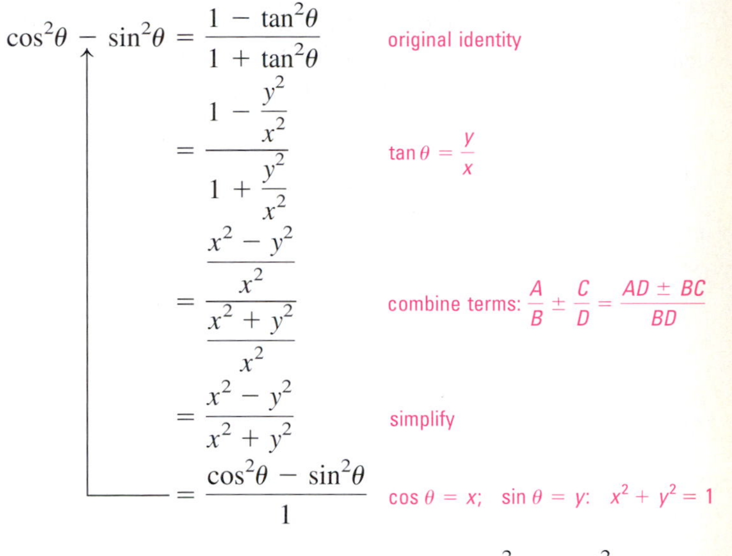

$$\cos^2\theta - \sin^2\theta = \frac{1 - \tan^2\theta}{1 + \tan^2\theta} \qquad \text{original identity}$$

$$= \frac{1 - \frac{y^2}{x^2}}{1 + \frac{y^2}{x^2}} \qquad \tan\theta = \frac{y}{x}$$

$$= \frac{\frac{x^2 - y^2}{x^2}}{\frac{x^2 + y^2}{x^2}} \qquad \text{combine terms: } \frac{A}{B} \pm \frac{C}{D} = \frac{AD \pm BC}{BD}$$

$$= \frac{x^2 - y^2}{x^2 + y^2} \qquad \text{simplify}$$

$$= \frac{\cos^2\theta - \sin^2\theta}{1} \qquad \cos\theta = x;\ \sin\theta = y;\ x^2 + y^2 = 1$$

Exercise 4: Use this method to verify the identity $\dfrac{\csc^2 x - \cot^2 x}{\cot^2 x} = \sec^2 x - 1$.

3.3 The Sum and Difference Identities

LEARNING OBJECTIVES

In Section 3.3 you will learn how to:

A. **Develop and use sum and difference identities for cosine**

B. **Use the sum and difference identities to verify the cofunction identities and angle reduction formulas**

C. **Develop and use sum and difference identities for sine and tangent**

D. **Use the sum and difference identities to write the difference quotient for trig functions**

INTRODUCTION

The sum and difference formulas for sine and cosine have a long and ancient history. Originally developed to help map the motion of celestial bodies, they were used centuries later to develop the derivatives of the trig functions, advance complex number theory, and study wave motion in different mediums. These identities are also used to find exact results (in radical form) for many nonstandard angles, a result of great importance to the ancient astronomers and still of notable mathematical significance today.

POINT OF INTEREST

Early in the second century, Claudius Ptolemy [tō-la-mē] wrote the *Almagest,* a work considering motions of the five known planets. To accurately map and predict these motions, Ptolemy computed and compiled a table of chords (the ancient precursor of our modern trig tables). Of interest here is the trigonometric identities he developed to help compile the table.

It had long been known that for a quadrilateral inscribed in a semicircle, $\overline{AB}\cdot\overline{CD} + \overline{BC}\cdot\overline{AD} = \overline{AC}\cdot\overline{BD}$. Using a diameter of 120 units (as Ptolemy did) and writing the relationship in terms of the *chord function* he used, we have chord(β)chord($180° - \alpha$) + 120chord($\alpha - \beta$) = chord(α)chord($180° - \beta$). Solving for 120chord($\alpha - \beta$) gives 120chord($\alpha - \beta$) = chord(α)chord($180° - \beta$) − chord(β) chord($180° - \alpha$), which closely resembles the difference formula for sine in use today: $\sin(u - v) = \sin u \cos v - \cos u \sin v$.

A. The Sum and Difference Identities for Cosine

On a unit circle with center C, consider the point A on the terminal side of angle α, and point B on the terminal side of angle β, as shown in Figure 3.7. Since $r = 1$, the coordinates of A and B are $(\cos\alpha, \sin\alpha)$ and $(\cos\beta, \sin\beta)$, respectively. Using the distance formula, we find that $\overline{AB}$ is equal to

Figure 3.7

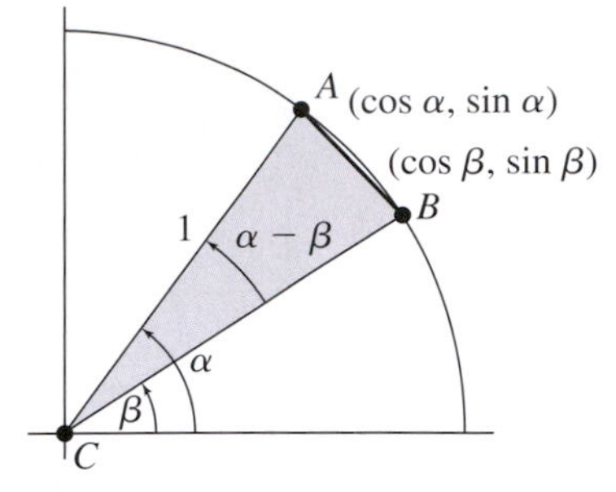

$$
\begin{aligned}
\overline{AB} &= \sqrt{(\cos\alpha - \cos\beta)^2 + (\sin\alpha - \sin\beta)^2} \\
&= \sqrt{\cos^2\alpha - 2\cos\alpha\cos\beta + \cos^2\beta + \sin^2\alpha - 2\sin\alpha\sin\beta + \sin^2\beta} && \text{binomial squares} \\
&= \sqrt{(\cos^2\alpha + \sin^2\alpha) + (\cos^2\beta + \sin^2\beta) - 2\cos\alpha\cos\beta - 2\sin\alpha\sin\beta} && \text{regroup} \\
&= \sqrt{2 - 2\cos\alpha\cos\beta - 2\sin\alpha\sin\beta} && \cos^2 u + \sin^2 u = 1
\end{aligned}
$$

With no loss of generality, we can rotate sector CAB clockwise, until side $\overline{CB}$ coincides with the x-axis. This creates new coordinates of $(1, 0)$ for B, and new coordinates of $(\cos(\alpha - \beta), \sin(\alpha - \beta))$ for A, *but the distance* $\overline{AB}$ *remains unchanged!* (see Figure 3.8). Recomputing the distance gives

Figure 3.8

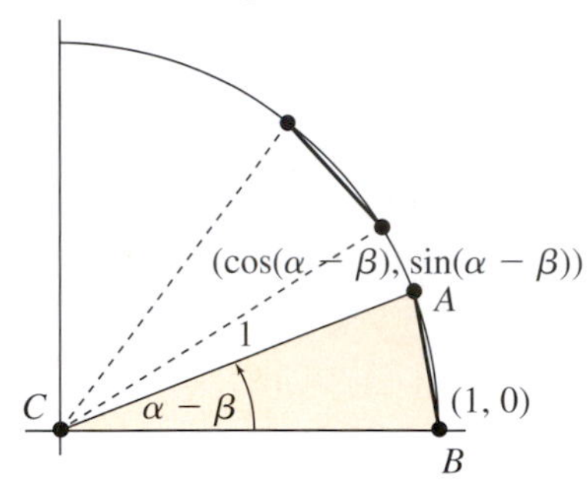

$$
\begin{aligned}
\overline{AB} &= \sqrt{[\cos(\alpha - \beta) - 1]^2 + [\sin(\alpha - \beta) - 0]^2} \\
&= \sqrt{\cos^2(\alpha - \beta) - 2\cos(\alpha - \beta) + 1 + \sin^2(\alpha - \beta)} \\
&= \sqrt{[\cos^2(\alpha - \beta) + \sin^2(\alpha - \beta)] - 2\cos(\alpha - \beta) + 1} \\
&= \sqrt{2 - 2\cos(\alpha - \beta)}
\end{aligned}
$$

Since both expressions represent the same distance, we can set them equal to each other and solve for $\cos(\alpha - \beta)$.

$$
\begin{aligned}
\sqrt{2 - 2\cos(\alpha - \beta)} &= \sqrt{2 - 2\cos\alpha\cos\beta - 2\sin\alpha\sin\beta} && \overline{AB} = \overline{AB} \\
2 - 2\cos(\alpha - \beta) &= 2 - 2\cos\alpha\cos\beta - 2\sin\alpha\sin\beta && \text{square both sides} \\
-2\cos(\alpha - \beta) &= -2\cos\alpha\cos\beta - 2\sin\alpha\sin\beta && \text{subtract 2} \\
\cos(\alpha - \beta) &= \cos\alpha\cos\beta + \sin\alpha\sin\beta && \text{divide both sides by } -2
\end{aligned}
$$

The result is called the **difference identity for cosine.** The **sum identity for cosine** follows immediately, by substituting $-\beta$ for β.

$$
\begin{aligned}
\cos(\alpha - \beta) &= \cos\alpha\cos\beta + \sin\alpha\sin\beta && \text{difference identity} \\
\cos(\alpha - [-\beta]) &= \cos\alpha\cos(-\beta) + \sin\alpha\sin(-\beta) && \text{substitute } -\beta \text{ for } \beta \\
\cos(\alpha + \beta) &= \cos\alpha\cos\beta - \sin\alpha\sin\beta && \cos(-\beta) = \cos\beta;\ \sin(-\beta) = -\sin\beta
\end{aligned}
$$

EXAMPLE 1 Use the sum and difference identities for cosine to find exact values for (a) $\cos 15° = \cos(45° - 30°)$, and (b) $\cos 75° = \cos(45° + 30°)$. Check results on a calculator.

Solution: Each involves a direct application of the related identity, using standard values.

WORTHY OF NOTE

Note that $\cos 45° - \cos 30° = \frac{\sqrt{2}}{2} - \frac{\sqrt{3}}{2}$ is a negative number very near zero, while *cos 15° is a positive number and very near 1*. This definitely shows $\cos(45° - 30°) \neq \cos(45°) - (30°)$. Likewise, since $\frac{\cos 30°}{2} < \cos 30°$, the idea that $\frac{\cos 30°}{2} = \cos 15°$ is also false.

a.

$$\cos(\alpha - \beta) = \cos\alpha\cos\beta + \sin\alpha\sin\beta \quad \text{difference identity}$$
$$\cos(45° - 30°) = \cos 45°\cos 30° + \sin 45°\sin 30° \quad \alpha = 45°, \beta = 30°$$
$$= \left(\frac{\sqrt{2}}{2}\right)\left(\frac{\sqrt{3}}{2}\right) + \left(\frac{\sqrt{2}}{2}\right)\left(\frac{1}{2}\right) \quad \text{standard values}$$
$$\cos 15° = \frac{\sqrt{6} + \sqrt{2}}{4} \quad \text{combine terms}$$

To 10 decimal places, $\cos 15° = 0.9659258263$.

b.

$$\cos(\alpha + \beta) = \cos\alpha\cos\beta - \sin\alpha\sin\beta \quad \text{sum identity}$$
$$\cos(45° + 30°) = \cos 45°\cos 30° - \sin 45°\sin 30° \quad \alpha = 45°, \beta = 30°$$
$$= \left(\frac{\sqrt{2}}{2}\right)\left(\frac{\sqrt{3}}{2}\right) - \left(\frac{\sqrt{2}}{2}\right)\left(\frac{1}{2}\right) \quad \text{standard values}$$
$$\cos 75° = \frac{\sqrt{6} - \sqrt{2}}{4} \quad \text{combine terms}$$

To 10 decimal places, $\cos 75° = 0.2588190451$.

NOW TRY EXERCISES 7 THROUGH 12

These identities are listed here using the "$\pm$" and "$\mp$" notation to avoid needless repetition. In their application, use both upper symbols or both lower symbols depending on whether you're evaluating the sum or difference of two angles. As with the other identities, these can be rewritten to form other members of the identity family, as when they are used to consolidate a larger expression. This is shown in Example 2.

THE SUM AND DIFFERENCE IDENTITIES FOR COSINE

cosine family: $\cos(\alpha \pm \beta) = \cos\alpha\cos\beta \mp \sin\alpha\sin\beta$ — functions repeat, signs alternate

$\cos\alpha\cos\beta \mp \sin\alpha\sin\beta = \cos(\alpha \pm \beta)$ — can be used to *expand* or *contract*

EXAMPLE 2 Write as a single expression in cosine: $\cos 12°\cos 3° - \sin 12°\sin 3°$

Solution: Since the functions repeat and are expressed as a difference, we use the sum identity for cosine to rewrite the difference as a single expression.

$$\cos\alpha\cos\beta - \sin\alpha\sin\beta = \cos(\alpha + \beta) \quad \text{sum identity for cosine}$$
$$\cos 12°\cos 3° - \sin 12°\sin 3° = \cos(12° + 3°) \quad \alpha = 12°, \beta = 3°$$

The expression is equal to $\cos 15°$, which we know has a value of $\frac{\sqrt{6} + \sqrt{2}}{4}$.

NOW TRY EXERCISES 13 THROUGH 16

The sum and difference identities can be used to evaluate the cosine of the sum of two angles, even when they are not adjacent, or even expressed in terms of cosine.

Figure 3.9

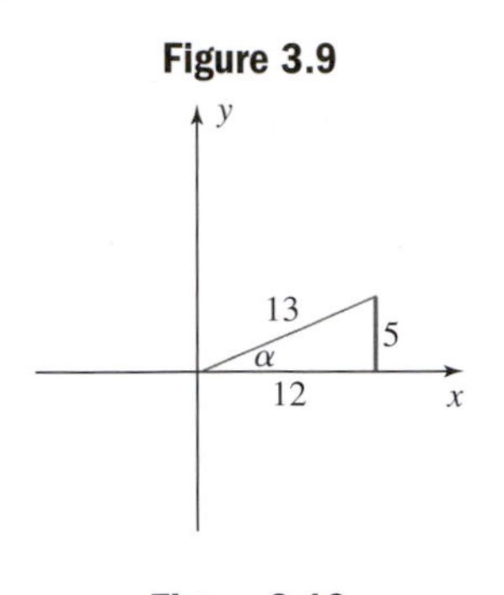

Figure 3.10

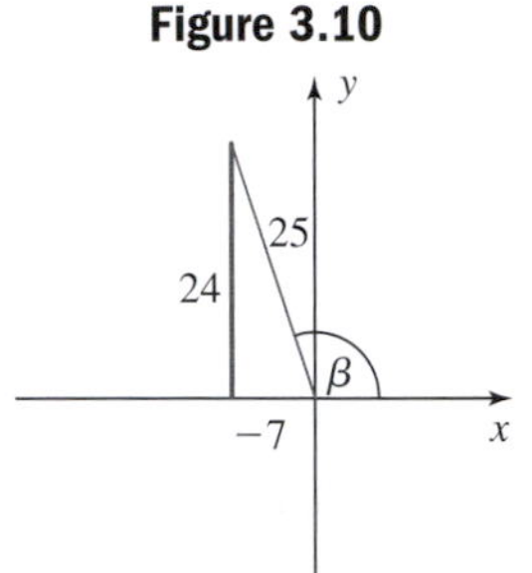

EXAMPLE 3 Given $\sin\alpha = \frac{5}{13}$ with the terminal side in QI, and $\tan\beta = -\frac{24}{7}$ with the terminal side in QII. Compute the value of $\cos(\alpha + \beta)$.

Solution: To use the sum formula we need the value of $\cos\alpha$, $\sin\alpha$, $\cos\beta$, and $\sin\beta$. Using the given information about the quadrants along with the Pythagorean theorem, we draw the triangles shown in Figures 3.9 and 3.10, yielding the values that follow.

$$\cos\alpha = \frac{12}{13}\text{ (QI)},\ \sin\alpha = \frac{5}{13}\text{ (QI)},\ \cos\beta = -\frac{7}{25}\text{ (QII)},\text{ and }\sin\beta = \frac{24}{25}\text{ (QII)}$$

Using $\cos(\alpha + \beta) = \cos\alpha\cos\beta - \sin\alpha\sin\beta$ gives this result:

$$\begin{aligned}\cos(\alpha + \beta) &= \left(\frac{12}{13}\right)\left(-\frac{7}{25}\right) - \left(\frac{5}{13}\right)\left(\frac{24}{25}\right)\\ &= -\frac{84}{325} - \frac{120}{325}\\ &= -\frac{204}{325}\end{aligned}$$

NOW TRY EXERCISES 17 THROUGH 18

WORTHY OF NOTE

It is worth pointing out that if we approximate the values of α and β using tables or a calculator, we find $\alpha \approx 22.62°$ and $\beta \approx 106.26°$. Sure enough, $\cos(22.62° + 106.26°) \approx -\frac{204}{325}$!

B. The Cofunction Identities and Angle Reduction Formulas

The cofunction identities were actually introduced in Section 1.2 by noting relationships between functions using a right triangle. In this section we demonstrate that $\cos\left(\frac{\pi}{2} - t\right) = \sin t$ and $\sin\left(\frac{\pi}{2} - t\right) = \cos t$. The first result follows directly. For the second we use $\cos\left(\frac{\pi}{2} - t\right) = \sin t$ and replace t with the real number $\frac{\pi}{2} - s$.

$$\cos\left(\frac{\pi}{2} - t\right) = \sin t \qquad \text{cofunction identity for cosine}$$

$$\cos\left(\frac{\pi}{2} - \left[\frac{\pi}{2} - s\right]\right) = \sin\left(\frac{\pi}{2} - s\right) \qquad \text{substitute } \frac{\pi}{2} - s \text{ for } t$$

$$\cos s = \sin\left(\frac{\pi}{2} - s\right) \qquad \text{result}$$

This establishes the cofunction relationship for sine: $\sin\left(\frac{\pi}{2} - s\right) = \cos s$ for any real number s. See Exercises 19 through 24.

The sum and difference identities can also be used to establish the **angle reduction formulas.** These will be of great use in our study of complex numbers in trigonometric form (Chapter 5), and other places. The reduction formulas use the period of a function to reduce large angles to an angle in $[0, 360°)$ or $[0, 2\pi)$ having an equivalent function value. Using a sum identity, we can show $\cos(t + 2\pi k) = \cos t$ and $\cos(\theta + 360°k) = \cos\theta$ for k an integer. Here we'll verify that $\cos(t + 2\pi k) = \cos t$.

$$\begin{aligned}\cos(t + 2\pi k) &= \cos t\cos(2\pi k) - \sin t\sin(2\pi k) && \text{sum identity for cosine}\\ &= \cos t(1) - \sin t(0) && \cos(2\pi k) = 1;\ \sin(2\pi k) = 0\\ &= \cos t\end{aligned}$$

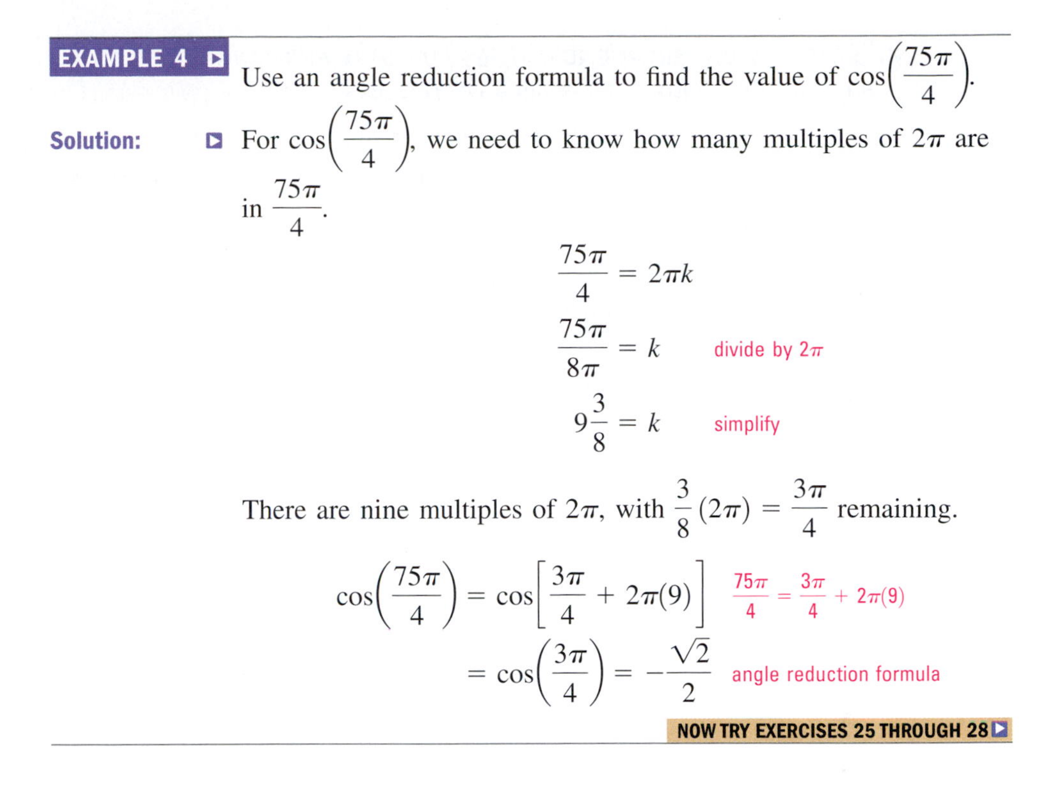

EXAMPLE 4 ▷ Use an angle reduction formula to find the value of $\cos\left(\frac{75\pi}{4}\right)$.

Solution: ▷ For $\cos\left(\frac{75\pi}{4}\right)$, we need to know how many multiples of 2π are in $\frac{75\pi}{4}$.

$$\frac{75\pi}{4} = 2\pi k$$

$$\frac{75\pi}{8\pi} = k \qquad \text{divide by } 2\pi$$

$$9\frac{3}{8} = k \qquad \text{simplify}$$

There are nine multiples of 2π, with $\frac{3}{8}(2\pi) = \frac{3\pi}{4}$ remaining.

$$\cos\left(\frac{75\pi}{4}\right) = \cos\left[\frac{3\pi}{4} + 2\pi(9)\right] \qquad \frac{75\pi}{4} = \frac{3\pi}{4} + 2\pi(9)$$

$$= \cos\left(\frac{3\pi}{4}\right) = -\frac{\sqrt{2}}{2} \qquad \text{angle reduction formula}$$

NOW TRY EXERCISES 25 THROUGH 28 ▷

Similar relationships exist for $\sin(\theta + 360°k)$ and $\sin(t + 2\pi k)$ since cosine and sine have the same period: $\sin(\theta + 360°k) = \sin\theta$ and $\sin(t + 2\pi k) = \sin t$.

C. The Sum and Difference Identities for Sine and Tangent

The sum and difference identities for sine can easily be developed using cofunction identities. Since $\sin t = \cos\left(\frac{\pi}{2} - t\right)$, we need only rename t as the sum $(\alpha + \beta)$ or the difference $(\alpha - \beta)$ and work from there.

$$\sin t = \cos\left(\frac{\pi}{2} - t\right) \qquad \text{cofunction identity}$$

$$\sin(\alpha + \beta) = \cos\left[\frac{\pi}{2} - (\alpha + \beta)\right] \qquad \text{substitute } (\alpha + \beta) \text{ for } t$$

$$= \cos\left[\left(\frac{\pi}{2} - \alpha\right) - \beta\right] \qquad \text{regroup argument}$$

$$= \cos\left(\frac{\pi}{2} - \alpha\right)\cos\beta + \sin\left(\frac{\pi}{2} - \alpha\right)\sin\beta \qquad \text{apply difference identity for cosine}$$

$$\sin(\alpha + \beta) = \sin\alpha\cos\beta + \cos\alpha\sin\beta \qquad \text{result}$$

The difference identity for sine is likewise developed. The sum and difference identities for tangent can be derived using ratio identities and their derivation is left as an exercise (see Exercise 80).

THE SUM AND DIFFERENCE IDENTITIES FOR SINE AND TANGENT

sine family: $\sin(\alpha \pm \beta) = \sin\alpha\cos\beta \pm \cos\alpha\sin\beta$ — functions alternate, signs repeat

$\sin\alpha\cos\beta \pm \cos\alpha\sin\beta = \sin(\alpha \pm \beta)$ — can be used to *expand* or *contract*

tangent family: $\tan(\alpha \pm \beta) = \dfrac{\tan\alpha \pm \tan\beta}{1 \mp \tan\alpha\tan\beta}$ — sum of tangents in numerator; product of tangents in denominator

$\dfrac{\tan\alpha \pm \tan\beta}{1 \mp \tan\alpha\tan\beta} = \tan(\alpha \pm \beta)$ — can be used to *expand* or *contract*

EXAMPLE 5A Write as a single expression in sine: $\sin(2t)\cos t + \cos(2t)\sin t$.

Solution: Since the functions in the expression alternate and are expressed as a sum, we use the sum identity for sine:

$$\sin\alpha\cos\beta + \cos\alpha\sin\beta = \sin(\alpha + \beta) \quad \text{sum identity for sine}$$

$$\sin(2t)\cos t + \cos(2t)\sin t = \sin(2t + t) \quad \text{substitute } 2t \text{ for } \alpha \text{ and } t \text{ for } \beta$$

The expression is equal to $\sin(3t)$.

EXAMPLE 5B Use the sum or difference identity for tangent to find the exact value of $\tan\frac{11\pi}{12}$.

Solution: Since an exact value is requested, $\frac{11\pi}{12}$ must be the sum or difference of two standard angles. A casual inspection reveals $\frac{11\pi}{12} = \frac{2\pi}{3} + \frac{\pi}{4}$. This gives

$$\tan(\alpha + \beta) = \frac{\tan\alpha + \tan\beta}{1 - \tan\alpha\tan\beta} \quad \text{sum identity for tangent}$$

$$\tan\left(\frac{2\pi}{3} + \frac{\pi}{4}\right) = \frac{\tan\left(\frac{2\pi}{3}\right) + \tan\left(\frac{\pi}{4}\right)}{1 - \tan\left(\frac{2\pi}{3}\right)\tan\left(\frac{\pi}{4}\right)} \quad \alpha = \frac{2\pi}{3}, \beta = \frac{\pi}{4}$$

$$= \frac{-\sqrt{3} + 1}{1 - (-\sqrt{3})(1)} \quad \tan\left(\frac{2\pi}{3}\right) = -\sqrt{3}, \tan\left(\frac{\pi}{4}\right) = 1$$

$$= \frac{1 - \sqrt{3}}{1 + \sqrt{3}} \quad \text{simplify expression}$$

NOW TRY EXERCISES 29 THROUGH 50

D. Sum/Difference Identities and the Difference Quotient

The difference quotient for any function $f(x)$ is defined to be $\dfrac{f(x + h) - f(x)}{h}$. As innocent as it looks, this expression lays the foundation for the study of differential calculus. In Example 6, we explore one of the more notable applications of the sum/difference identities, the simplification of the difference quotient as it applies to trig functions. While we lack the tools needed to simplify the result completely, using the difference identity to obtain the form shown will help create a stronger tie to future course work.

EXAMPLE 6 ▷ Given $f(x) = \cos x$, show that the difference quotient results in the expression $\cos x \dfrac{\cos h - 1}{h} - \sin x \dfrac{\sin h}{h}$.

Solution: ▷ For $f(x) = \cos x$, $f(x + h) = \cos(x + h) \rightarrow \cos x \cos h - \sin x \sin h$

$$\frac{f(x+h) - f(x)}{h} = \frac{\cos(x+h) - \cos x}{h} \quad \text{difference quotient}$$

$$= \frac{\cos x \cos h - \sin x \sin h - \cos x}{h} \quad \text{sum identity}$$

$$= \frac{\cos x \cos h - \cos x - \sin x \sin h}{h} \quad \text{rewrite terms}$$

$$= \frac{\cos x(\cos h - 1) - \sin x \sin h}{h} \quad \text{factor}$$

$$= \frac{\cos x(\cos h - 1)}{h} - \frac{\sin x \sin h}{h} \quad \text{write as separate expressions}$$

$$= \cos x \frac{\cos h - 1}{h} - \sin x \frac{\sin h}{h} \quad \text{rewrite terms}$$

NOW TRY EXERCISES 75 THROUGH 79 ▷

TECHNOLOGY HIGHLIGHT

Relationships Between the Sum/Difference Formulas

The keystrokes shown apply to a TI-84 Plus model. Please consult your manual or our Internet site for other models.

In Example 3, we found that for $\sin\alpha = \frac{5}{13}$ in QI and $\tan\beta = -\frac{24}{7}$ in QII, $\cos(\alpha + \beta) = -\frac{204}{325}$. Once this value has been found, the values of $\sin(\alpha + \beta)$ and $\tan(\alpha + \beta)$ can easily be found without having to apply the respective sum identity. Using the Pythagorean identity $1 - \cos^2\theta = \sin^2\theta$ (where $\alpha + \beta = \theta$) we have $\sin^2\theta = 1 - (-\frac{204}{325})^2$ yielding $\sin^2\theta = 0.6060023669$ (see Figure 3.11). To compute $\sqrt{0.6060023669}$, we use $\sqrt{\text{Ans}}$ where "Ans" is the result of the most recent calculation and can be accessed using 2nd (−). Knowing the answer must be rational (discuss why), we convert this value to fraction form using MATH ENTER (**1:▶Frac**) ENTER giving $\sin\theta = \sin(\alpha + \beta) = \frac{253}{325}$ ($\alpha + \beta$ is in QII where $\sin\theta > 0$). At this point the value of $\tan(\alpha + \beta)$ is even easier to calculate, since $\dfrac{\sin\theta}{\cos\theta} = \tan\theta$ for all θ (including $\theta = \alpha + \beta$). Since the denominators from $\sin(\alpha + \beta)$ and $\cos(\alpha + \beta)$ are identical, they reduce to 1 leaving the ratio of numerators, giving $\tan\theta = \tan(\alpha + \beta) = -\frac{253}{204}$. Use these ideas to complete the following exercises.

Figure 3.11

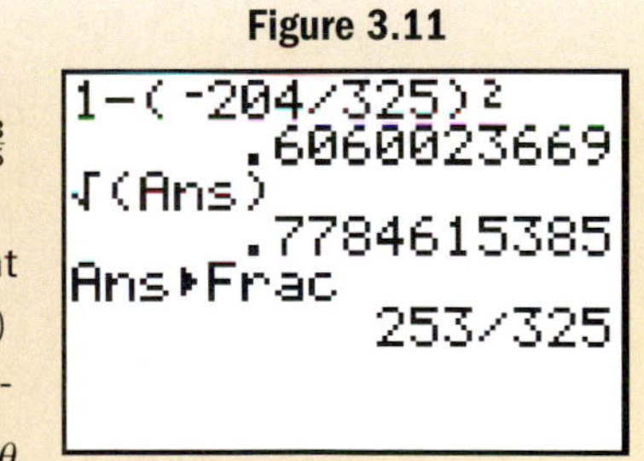

Exercise 1: For $\tan\alpha = \frac{40}{9}$ in QIII and $\cos\beta = \frac{12}{37}$ in QIV, find $\cos(\alpha - \beta)$, $\sin(\alpha - \beta)$, and $\tan(\alpha - \beta)$.

Exercise 2: For $\cos\alpha = -\frac{8}{17}$ in QIII and $\sin\beta = \frac{20}{29}$ in QII, find $\sin(\alpha + \beta)$, $\cos(\alpha + \beta)$, and $\tan(\alpha + \beta)$.

3.3 EXERCISES

CONCEPTS AND VOCABULARY

Fill in each blank with the appropriate word or phrase. Carefully reread the section if needed.

1. Since $\tan 45° + \tan 60° > 1$, we know $\tan 45° + \tan 60° = \tan 105°$ is _____ since $\tan\theta < 0$ in QII.

2. The sum/difference identities can be used to rewrite a sum or difference as a _____ expression.

3. For the cosine sum/difference identities, the functions _____ and the signs _____.

4. For the sine sum/difference identities, the functions _____ and the signs _____.

5. Discuss/explain how the angle reduction formula can be applied to $\cos\theta$, where $\theta = \frac{21\pi}{2}$.

6. Compare/contrast the cofunction identities as they relate to right triangle trig and the trigonometry of any angle.

DEVELOPING YOUR SKILLS

Find the exact value of the expression given using a sum or difference identity. Some simplifications may involve using symmetry and the formulas for negatives.

7. $\cos 105°$ **8.** $\cos 135°$ **9.** $\cos\left(\frac{7\pi}{12}\right)$ **10.** $\cos\left(-\frac{5\pi}{12}\right)$

Use sum/difference identities to verify that both expressions give the same result.

11. a. $\cos(45° + 30°)$ **b.** $\cos(120° - 45°)$

12. a. $\cos\left(\frac{\pi}{6} - \frac{\pi}{4}\right)$ **b.** $\cos\left(\frac{\pi}{4} - \frac{\pi}{3}\right)$

Rewrite as a single expression in cosine.

13. $\cos(7\theta)\cos(2\theta) + \sin(7\theta)\sin(2\theta)$

14. $\cos\left(\frac{\theta}{3}\right)\cos\left(\frac{\theta}{6}\right) - \sin\left(\frac{\theta}{3}\right)\sin\left(\frac{\theta}{6}\right)$

Find the exact value of the given expressions.

15. $\cos 183° \cos 153° + \sin 183° \sin 153°$

16. $\cos\left(\frac{7\pi}{36}\right)\cos\left(\frac{5\pi}{36}\right) - \sin\left(\frac{7\pi}{36}\right)\sin\left(\frac{5\pi}{36}\right)$

17. For $\sin\alpha = -\frac{4}{5}$ with terminal side in QIV and $\tan\beta = -\frac{5}{12}$ with terminal side in QII, find $\cos(\alpha + \beta)$.

18. For $\sin\alpha = \frac{112}{113}$ with terminal side in QII and $\sec\beta = -\frac{89}{39}$ with terminal side in QII, find $\cos(\alpha - \beta)$.

Use a cofunction identity to write an equivalent expression.

19. $\cos 57°$ **20.** $\sin 18°$ **21.** $\tan\left(\frac{5\pi}{12}\right)$

22. $\sec\left(\frac{\pi}{10}\right)$ **23.** $\sin\left(\frac{\pi}{6} - \theta\right)$ **24.** $\cos\left(\frac{\pi}{3} + \theta\right)$

Use an angle reduction formula to find the exact value of each expression.

25. $\cos 1665°$

26. $\cos\left(\frac{91\pi}{6}\right)$

27. $\sin\left(\frac{41\pi}{6}\right)$

28. $\sin 2385°$

Rewrite as a single expression.

29. $\sin(3x)\cos(5x) + \cos(3x)\sin(5x)$

30. $\sin\left(\frac{x}{2}\right)\cos\left(\frac{x}{3}\right) - \cos\left(\frac{x}{2}\right)\sin\left(\frac{x}{3}\right)$

31. $\dfrac{\tan(5\theta) - \tan(2\theta)}{1 + \tan(5\theta)\tan(2\theta)}$

32. $\dfrac{\tan\left(\frac{x}{2}\right) + \tan\left(\frac{x}{8}\right)}{1 - \tan\left(\frac{x}{2}\right)\tan\left(\frac{x}{8}\right)}$

Find the exact value of the given expressions.

33. $\sin 137° \cos 47° - \cos 137° \sin 47°$

34. $\sin\left(\frac{11\pi}{24}\right)\cos\left(\frac{5\pi}{24}\right) + \cos\left(\frac{11\pi}{24}\right)\sin\left(\frac{5\pi}{24}\right)$

35. $\dfrac{\tan\left(\frac{11\pi}{21}\right) - \tan\left(\frac{4\pi}{21}\right)}{1 + \tan\left(\frac{11\pi}{21}\right)\tan\left(\frac{4\pi}{21}\right)}$

36. $\dfrac{\tan\left(\frac{3\pi}{20}\right) + \tan\left(\frac{\pi}{10}\right)}{1 - \tan\left(\frac{3\pi}{20}\right)\tan\left(\frac{\pi}{10}\right)}$

37. For $\cos\alpha = -\frac{7}{25}$ with terminal side in QII and $\cot\beta = \frac{15}{8}$ with terminal side in QIII, find

a. $\sin(\alpha + \beta)$ **b.** $\tan(\alpha + \beta)$

38. For $\csc\alpha = \frac{29}{20}$ with terminal side in QI and $\cos\beta = -\frac{12}{37}$ with terminal side in QII, find

a. $\sin(\alpha - \beta)$ **b.** $\tan(\alpha - \beta)$

Find the exact value of the expression given using a sum or difference identity. Some simplifications may involve using symmetry and the formulas for negatives.

39. $\sin 105°$

40. $\sin(-75°)$

41. $\sin\left(\frac{5\pi}{12}\right)$

42. $\sin\left(\frac{11\pi}{12}\right)$

43. $\tan 150°$

44. $\tan 75°$

45. $\tan\left(\frac{2\pi}{3}\right)$

46. $\tan\left(-\frac{\pi}{12}\right)$

Use sum/difference identities to verify that both expressions give the same result.

47. **a.** $\sin(45° - 30°)$ **b.** $\sin(135° - 120°)$

48. **a.** $\sin\left(\frac{\pi}{3} - \frac{\pi}{4}\right)$ **b.** $\sin\left(\frac{\pi}{4} - \frac{\pi}{6}\right)$

49. Find $\sin 255°$ given $150° + 105° = 255°$. See Exercises 7 and 39.

50. Find $\cos\left(\frac{19\pi}{12}\right)$ given $2\pi - \frac{5\pi}{12} = \frac{19\pi}{12}$. See Exercises 10 and 41.

51. Given α and β are acute angles with $\sin\alpha = \frac{12}{13}$ and $\tan\beta = \frac{35}{12}$, find

a. $\sin(\alpha + \beta)$ **b.** $\cos(\alpha - \beta)$ **c.** $\tan(\alpha + \beta)$

52. Given α and β are acute angles with $\cos\alpha = \frac{8}{17}$ and $\sec\beta = \frac{25}{7}$, find

a. $\sin(\alpha + \beta)$ **b.** $\cos(\alpha - \beta)$ **c.** $\tan(\alpha + \beta)$

53. Given α and β are obtuse angles with $\sin\alpha = \frac{28}{53}$ and $\cos\beta = -\frac{13}{85}$, find

a. $\sin(\alpha - \beta)$ **b.** $\cos(\alpha + \beta)$ **c.** $\tan(\alpha - \beta)$

54. Given α and β are obtuse angles with $\tan\alpha = -\frac{60}{11}$ and $\sin\beta = \frac{35}{37}$, find

a. $\sin(\alpha - \beta)$ **b.** $\cos(\alpha + \beta)$ **c.** $\tan(\alpha - \beta)$

55. Use the diagram indicated to compute the following:

a. $\sin A$ **b.** $\cos A$ **c.** $\tan A$

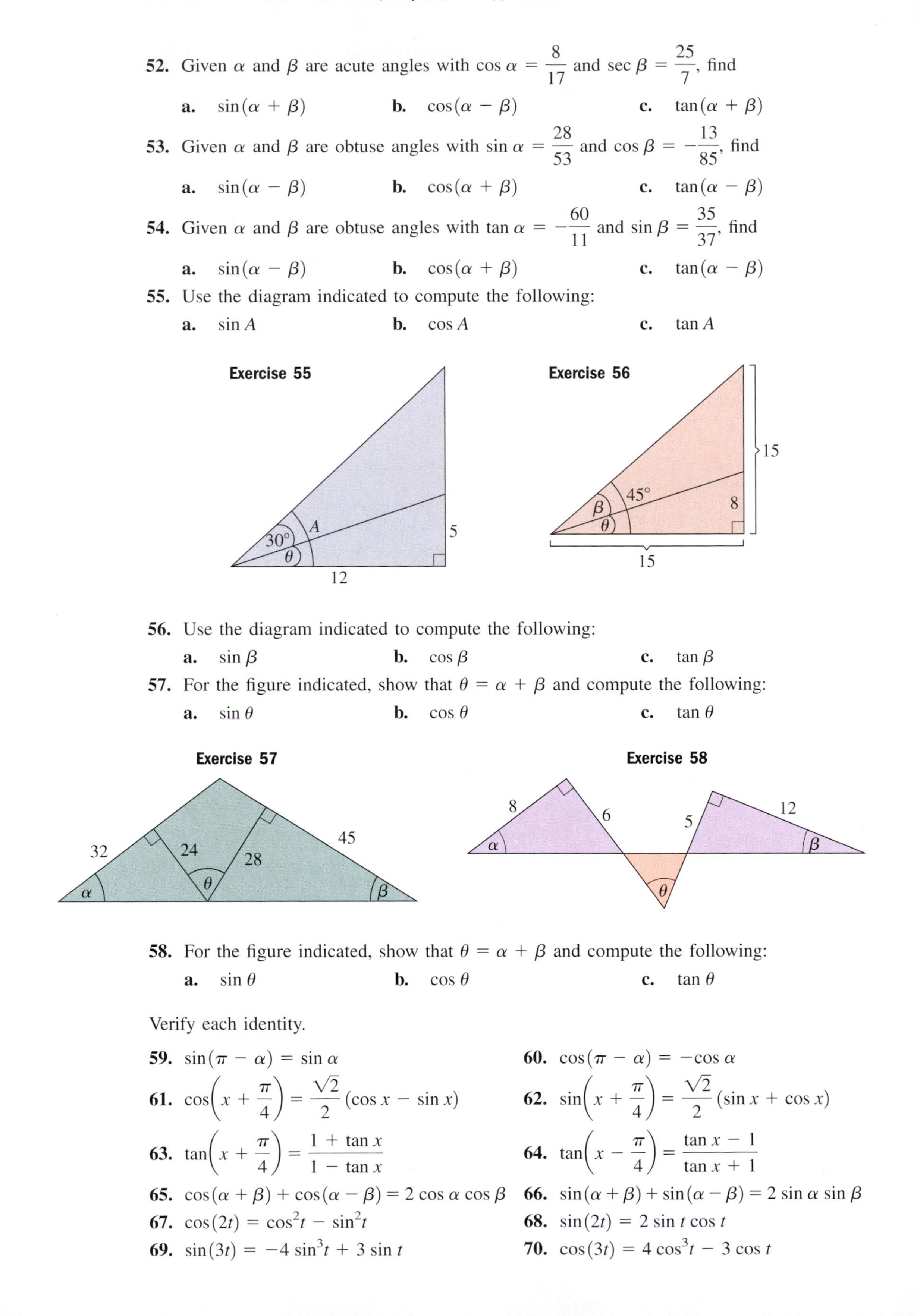

56. Use the diagram indicated to compute the following:

a. $\sin\beta$ **b.** $\cos\beta$ **c.** $\tan\beta$

57. For the figure indicated, show that $\theta = \alpha + \beta$ and compute the following:

a. $\sin\theta$ **b.** $\cos\theta$ **c.** $\tan\theta$

58. For the figure indicated, show that $\theta = \alpha + \beta$ and compute the following:

a. $\sin\theta$ **b.** $\cos\theta$ **c.** $\tan\theta$

Verify each identity.

59. $\sin(\pi - \alpha) = \sin\alpha$

60. $\cos(\pi - \alpha) = -\cos\alpha$

61. $\cos\left(x + \frac{\pi}{4}\right) = \frac{\sqrt{2}}{2}(\cos x - \sin x)$

62. $\sin\left(x + \frac{\pi}{4}\right) = \frac{\sqrt{2}}{2}(\sin x + \cos x)$

63. $\tan\left(x + \frac{\pi}{4}\right) = \frac{1 + \tan x}{1 - \tan x}$

64. $\tan\left(x - \frac{\pi}{4}\right) = \frac{\tan x - 1}{\tan x + 1}$

65. $\cos(\alpha + \beta) + \cos(\alpha - \beta) = 2\cos\alpha\cos\beta$

66. $\sin(\alpha + \beta) + \sin(\alpha - \beta) = 2\sin\alpha\sin\beta$

67. $\cos(2t) = \cos^2 t - \sin^2 t$

68. $\sin(2t) = 2\sin t\cos t$

69. $\sin(3t) = -4\sin^3 t + 3\sin t$

70. $\cos(3t) = 4\cos^3 t - 3\cos t$

71. Use a difference identity to show that $\cos\left(x - \frac{\pi}{4}\right)$ is equivalent to $\frac{\sqrt{2}}{2}(\cos x + \sin x)$.

72. Use sum/difference identities to show that $\sin\left(x + \frac{\pi}{4}\right) + \sin\left(x - \frac{\pi}{4}\right)$ is equivalent to $\sqrt{2}\sin x$.

WORKING WITH FORMULAS

73. Force required to maintain equilibrium using a screw jack: $F = \frac{Wk}{c}\tan(p - \theta)$

The force required to maintain equilibrium when a screw jack is used can be modeled by the formula shown, where p is the pitch angle of the screw, W is the weight of the load, θ is the angle of friction, with k and c being constants related to a particular jack. Simplify the formula using the difference formula for tangent given $p = \frac{\pi}{6}$ and $\theta = \frac{\pi}{4}$.

74. Brewster's law: reflection and refraction of unpolarized light: $\tan\theta_p = \frac{n_2}{n_1}$

Brewster's law of optics states that when unpolarized light strikes a dielectric surface, the transmitted light rays and the reflected light rays are perpendicular to each other. The proof of Brewster's law involves the expression $n_1 \sin\theta_p = n_2 \sin\left(\frac{\pi}{2} - \theta_p\right)$. Use the difference identity for sine to verify that this expression leads to Brewster's law.

APPLICATIONS

In Exercises 75 to 78, the difference quotient $\frac{f(x+h) - f(x)}{h}$ has been applied to the expression given. Simplify the resulting expression as much as possible using a sum identity.

75. $\sin\left(\frac{\pi}{2}\right)$; $\frac{\sin\left(\frac{\pi}{2} + h\right) - \sin\left(\frac{\pi}{2}\right)}{h}$

76. $\cos\left(\frac{\pi}{2}\right)$; $\frac{\cos\left(\frac{\pi}{2} + h\right) - \cos\left(\frac{\pi}{2}\right)}{h}$

77. $\tan\left(\frac{\pi}{4}\right)$; $\frac{\tan\left(\frac{\pi}{4} + h\right) - \tan\left(\frac{\pi}{4}\right)}{h}$

78. $\tan\pi$; $\frac{\tan(\pi + h) - \tan\pi}{h}$

79. Difference quotient: Given $f(x) = \sin x$, show that the difference quotient results in the expression $\sin x \frac{\cos h - 1}{h} + \cos x\left(\frac{\sin h}{h}\right)$.

80. Difference identity: Derive the difference identity for tangent using $\tan(\alpha - \beta) = \frac{\sin(\alpha - \beta)}{\cos(\alpha - \beta)}$. (*Hint:* After applying the difference identities, divide the numerator and denominator by $\cos\alpha\cos\beta$.)

81. AC circuits: In a study of AC circuits, the equation $R = \frac{\cos s \cos t}{\overline{\omega}C\sin(s + t)}$ sometimes arises. Use a sum identity and algebra to show this equation is equivalent to $R = \frac{1}{\overline{\omega}C(\tan s + \tan t)}$.

82. **Fluid mechanics:** In studies of fluid mechanics, the equation $\gamma_1 V_1 \sin\alpha = \gamma_2 V_2 \sin(\alpha - \beta)$ sometimes arises. Use a difference identity to show that if $\gamma_1 V_1 = \gamma_2 V_2$, the equation is equivalent to $\cos\beta - \cot\alpha\sin\beta = 1$.

83. **Art and mathematics:** When working in two-point geometric perspective, artists must scale their work to fit on the paper or canvas they are using. In doing so, the equation $\frac{A}{B} = \frac{\tan\theta}{\tan(90° - \theta)}$ arises. Rewrite the expression on the right in terms of sine and cosine, then use the difference identities to show the equation can be rewritten as $\frac{A}{B} = \tan^2\theta$.

84. **Traveling waves:** If two waves of the same frequency, velocity, and amplitude are traveling along a string in opposite directions, they can be represented by the equations $Y_1 = A\sin(kx - \omega t)$ and $Y_2 = A\sin(kx + \omega t)$. Use the sum and difference formulas for sine to show the result $Y_R = Y_1 + Y_2$ of these waves can be expressed as $Y_R = 2A\sin(kx)\cos(\omega t)$.

WRITING, RESEARCH, AND DECISION MAKING

Exercise 86

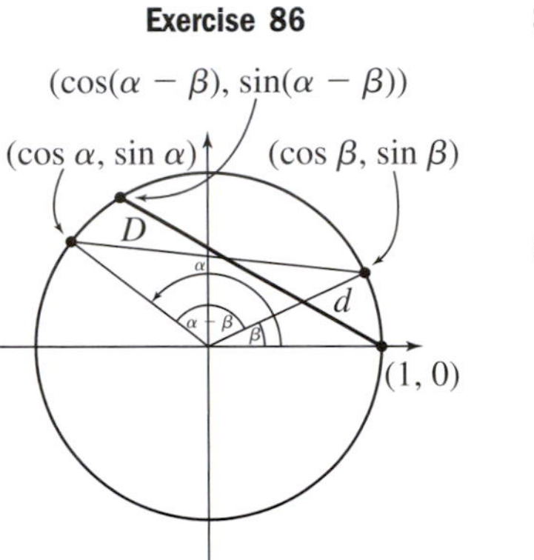

85. The identity $\sin^2\alpha + \cos^2\alpha = 1$ is the standard Pythagorean identity for sine and cosine. Knowing that cosine is the cofunction of sine, show that if α and β are complementary, $\sin^2\alpha + \sin^2\beta = 1$. Explain/discuss how complementary angles, cofunctions, and sum/difference identities can be used to verify a similar relationship for the other Pythagorean identities.

86. An alternative method of proving the difference formula for cosine uses a unit circle and the fact that equal arcs are subtended by equal chords ($D = d$ in the diagram). Using a combination of algebra, the distance formula, and a Pythagorean identity, show that $\cos(\alpha - \beta) = \cos\alpha\cos\beta + \sin\alpha\sin\beta$ (start by computing D^2 and d^2). Then discuss/explain how the sum identity can be found using the fact that $\beta = -(-\beta)$.

EXTENDING THE CONCEPT

87. **A proof without words:** Verify the Pythagorean theorem for each right triangle in the diagram, then discuss/explain how the diagram offers a proof of the sum identities for sine and cosine. Be detailed and thorough.

88. Show that $A_1\cos(Bx) + A_2\sin(Bx)$ is equal to $A_1\sec C\cos(Bx - C)$ given that $\tan C = \frac{A_2}{A_1}$. (*Hint:* Solve for A_2 and substitute.)

Exercise 87

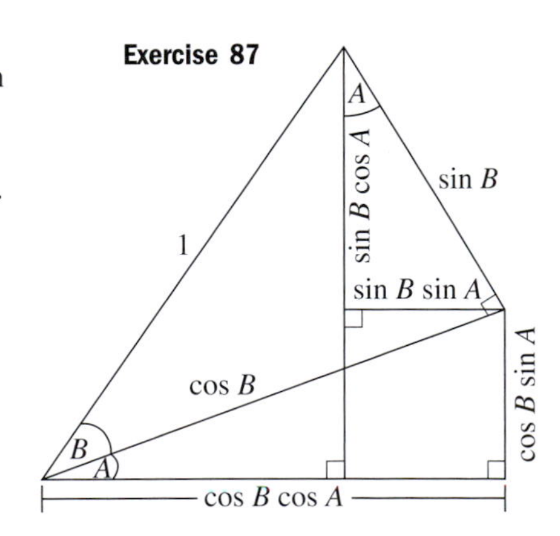

MAINTAINING YOUR SKILLS

89. (2.1/2.2) State the period of the functions given:

a. $y = 3\sin\left(\frac{\pi}{8}x - \frac{\pi}{3}\right)$

b. $y = 4\tan\left(2x + \frac{\pi}{4}\right)$

90. (3.1) State the three Pythagorean identities.

91. (1.2) Solve the triangles given using only special ratio/special triangle values.

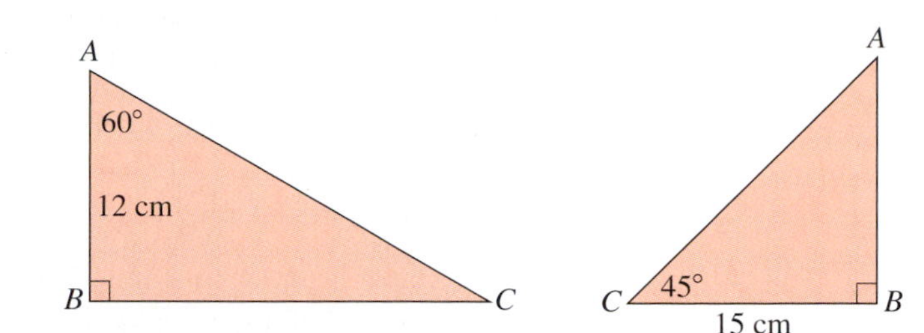

92. (1.2) Clarence the Clown is about to be shot from a circus cannon to a safety net on the other side of the main tent. If the cannon is 30 ft long and must be aimed at 40° for Clarence to hit the net, the end of the cannon must be how high from ground level?

93. (1.1) The area of a circular sector is $A = \frac{125\pi}{3}r^2$. If $r = 25$ m, what angle is subtended?

94. (2.4) Temperatures in Baghdad, Iraq, range from an average low of 58°F in January (month 1) to an average high of 110°F in July (month 7). Use this information to construct a sinusoidal model that approximates the average temperatures in Baghdad for the other 10 months.

3.4 The Double-Angle, Half-Angle, and Product-to-Sum Identities

LEARNING OBJECTIVES

In Section 3.4 you will learn how to:

A. Derive and use the double-angle identities for cosine, tangent, and sine

B. Develop and use the power reduction and half-angle identities

C. Derive and use the product-to-sum and sum-to-product identities

D. Solve applications using these identities

INTRODUCTION

The derivation of the sum and difference identities in Section 3.3 was a "watershed event" in the study of identities. By making various substitutions, they lead us very naturally to many new identity families, giving us a heightened ability to simplify expressions, solve equations, find exact values, and model real-world phenomena. In fact, many of the identities are applied in very practical ways, as in a study of projectile motion and the conic sections (Chapter 6). In addition, one of the most profound principles discovered in the eighteenth and nineteenth centuries was that electricity, light, and sound could all be studied using sinusoidal waves. These waves often interact with each other, creating the phenomena known as reflection, diffraction, superposition, interference, standing waves, and others. The product-to-sum and sum-to-product identities play a fundamental role in the investigation and study of these phenomena.

POINT OF INTEREST

Like many elements of trigonometry, the half-angle identities reach far back into antiquity. Ptolemy used them implicitly in compiling his table of chords, and there is evidence that Hipparchus of Nicea (~180–125 B.C.) was aware of them 200 years earlier. We say *implicitly* because chords are related to the sine

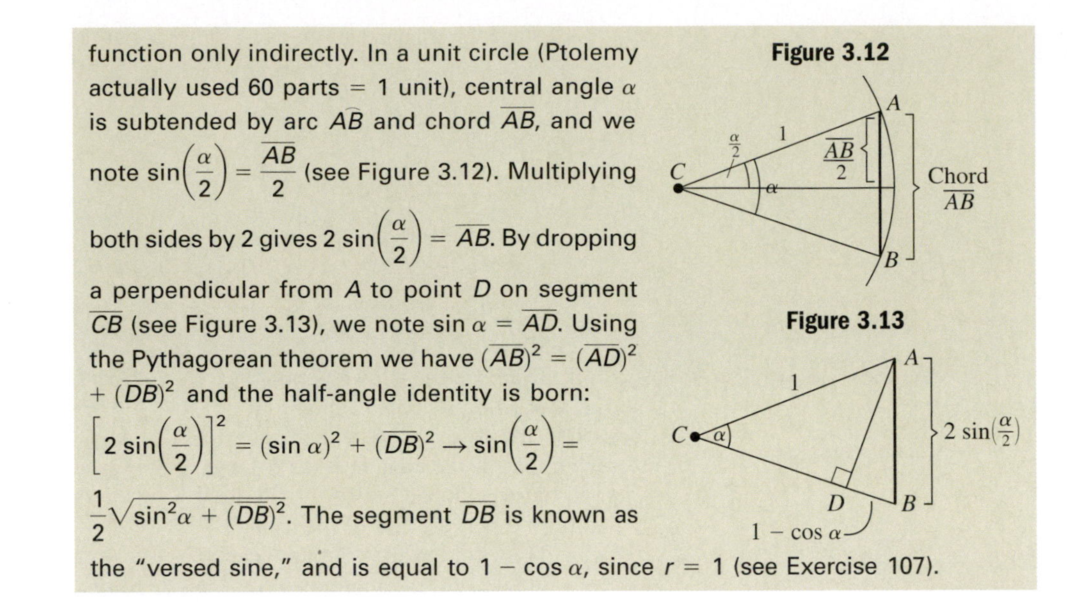

function only indirectly. In a unit circle (Ptolemy actually used 60 parts = 1 unit), central angle α is subtended by arc $\overset{\frown}{AB}$ and chord $\overline{AB}$, and we note $\sin\left(\frac{\alpha}{2}\right) = \frac{\overline{AB}}{2}$ (see Figure 3.12). Multiplying both sides by 2 gives $2 \sin\left(\frac{\alpha}{2}\right) = \overline{AB}$. By dropping a perpendicular from A to point D on segment $\overline{CB}$ (see Figure 3.13), we note $\sin \alpha = \overline{AD}$. Using the Pythagorean theorem we have $(\overline{AB})^2 = (\overline{AD})^2 + (\overline{DB})^2$ and the half-angle identity is born: $\left[2 \sin\left(\frac{\alpha}{2}\right)\right]^2 = (\sin \alpha)^2 + (\overline{DB})^2 \rightarrow \sin\left(\frac{\alpha}{2}\right) = \frac{1}{2}\sqrt{\sin^2\alpha + (\overline{DB})^2}$. The segment $\overline{DB}$ is known as the "versed sine," and is equal to $1 - \cos \alpha$, since $r = 1$ (see Exercise 107).

A. The Double-Angle Identities

The double-angle identities for sine, cosine, and tangent can all be derived using the related sum identities with two equal angles ($\alpha = \beta$). We'll illustrate the process here for the cosine of twice an angle.

$$\cos(\alpha + \beta) = \cos \alpha \cos \beta - \sin \alpha \sin \beta \qquad \text{sum identity for cosine}$$
$$\cos(\alpha + \alpha) = \cos \alpha \cos \alpha - \sin \alpha \sin \alpha \qquad \text{assume } \alpha = \beta \text{ and substitute } \alpha \text{ for } \beta$$
$$\cos(2\alpha) = \cos^2\alpha - \sin^2\alpha \qquad \text{simplify—double-angle identity for cosine}$$

Using the Pythagorean identity $\cos^2\alpha + \sin^2\alpha = 1$, we can easily find two additional members of this family, which are often quite useful. For $\cos^2\alpha = 1 - \sin^2\alpha$ we have

$$\cos(2\alpha) = \cos^2\alpha - \sin^2\alpha \qquad \text{double-angle identity for cosine}$$
$$= (1 - \sin^2\alpha) - \sin^2\alpha \qquad \text{substitute } 1 - \sin^2\alpha \text{ for } \cos^2\alpha$$
$$\cos(2\alpha) = 1 - 2\sin^2\alpha \qquad \text{double-angle in terms of sine}$$

Using $\sin^2\alpha = 1 - \cos^2\alpha$ we obtain an additional form:

$$\cos(2\alpha) = \cos^2\alpha - \sin^2\alpha \qquad \text{double-angle identity for cosine}$$
$$= \cos^2\alpha - (1 - \cos^2\alpha) \qquad \text{substitute } 1 - \sin^2\alpha \text{ for } \cos^2\alpha$$
$$\cos(2\alpha) = 2\cos^2\alpha - 1 \qquad \text{double-angle in terms of cosine}$$

The derivations of $\sin(2\alpha)$ and $\tan(2\alpha)$ are likewise developed and are asked for in Exercise 109. The double-angle identities are collected here for your convenience.

THE DOUBLE-ANGLE IDENTITIES

cosine family: $\cos(2\alpha) = \cos^2\alpha - \sin^2\alpha$
$= 1 - 2\sin^2\alpha$
$= 2\cos^2\alpha - 1$

sine family: $\sin(2\alpha) = 2\sin\alpha\cos\alpha$

tangent family: $\tan(2\alpha) = \dfrac{2\tan\alpha}{1 - \tan^2\alpha}$

EXAMPLE 1 Given $\sin\alpha = \frac{5}{8}$, find the value of $\cos(2\alpha)$.

Solution: Using the double-angle identity involving sine, we find

$$\begin{aligned}\cos(2\alpha) &= 1 - 2\sin^2\alpha && \text{double-angle in terms of sine}\\ &= 1 - 2\left(\frac{5}{8}\right)^2 && \text{substitute } \tfrac{5}{8} \text{ for } \sin\alpha\\ &= 1 - \frac{25}{32} && 2\left(\tfrac{5}{8}\right)^2 = \tfrac{25}{32}\\ &= \frac{7}{32} && \text{result}\end{aligned}$$

If $\sin\alpha = \frac{5}{8}$, then $\cos(2\alpha) = \frac{7}{32}$.

NOW TRY EXERCISES 7 THROUGH 20

Note that since $\sin\alpha > 0$, α must be in QI or QII. For α in QI, $0 < \alpha < \frac{\pi}{2}$ implies that 2α is in QII: $0 < 2\alpha < \pi$ (by multiplication). For α in QII, $\frac{\pi}{2} < \alpha < \pi$ implies that 2α is in QIII or QIV: $\pi < 2\alpha < 2\pi$. Since $\cos(2\alpha)$ turned out to be positive, we know $\sin\alpha = \frac{5}{8}$ is large enough that its double angle 2α must in fact be in QIV.

Like the fundamental identities, the double-angle identities can be used to verify or develop others. In Example 2, we explore one of many **multiple-angle identities,** verifying that $\cos(3\theta)$ can be rewritten as $4\cos^3\theta - 3\cos\theta$.

EXAMPLE 2 Verify that $\cos(3\theta) = 4\cos^3\theta - 3\cos\theta$ is an identity.

Solution: Use the sum identity for cosine, with $\alpha = 2\theta$ and $\beta = \theta$. Note that our goal is an expression using cosines only, with no multiple angles.

$$\begin{aligned}\cos(\alpha + \beta) &= \cos\alpha\cos\beta - \sin\alpha\sin\beta && \text{sum identity for cosine}\\ \cos(2\theta + \theta) &= \cos(2\theta)\cos\theta - \sin(2\theta)\sin\theta && \text{substitute } 2\theta \text{ for } \alpha \text{ and } \theta \text{ for } \beta\end{aligned}$$

$$\begin{aligned}
\cos(3\theta) &= (2\cos^2\theta - 1)\cos\theta - (2\sin\theta\cos\theta)\sin\theta && \text{substitute for } \cos(2\theta) \text{ and } \sin(2\theta)\\
&= 2\cos^3\theta - \cos\theta - 2\cos\theta\sin^2\theta && \text{multiply}\\
&= 2\cos^3\theta - \cos\theta - 2\cos\theta(1 - \cos^2\theta) && \text{substitute } 1 - \cos^2\theta \text{ for } \sin^2\theta\\
&= 2\cos^3\theta - \cos\theta - 2\cos\theta + 2\cos^3\theta && \text{multiply}\\
&= 4\cos^3\theta - 3\cos\theta && \text{combine terms}
\end{aligned}$$

NOW TRY EXERCISES 21 AND 22

EXAMPLE 3 Find the exact value of $\sin 22.5° \cos 22.5°$.

Solution: A product of sines and cosines having the same argument hints at the double-angle identity for sine. Using $\sin(2\alpha) = 2\sin\alpha\cos\alpha$ and dividing by 2 gives

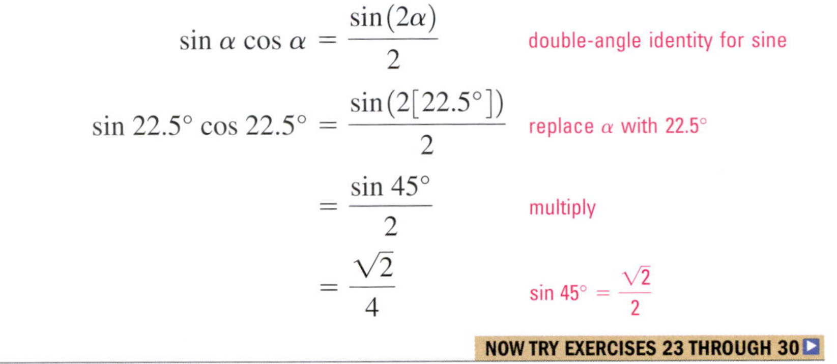

$$\begin{aligned}
\sin\alpha\cos\alpha &= \frac{\sin(2\alpha)}{2} && \text{double-angle identity for sine}\\
\sin 22.5°\cos 22.5° &= \frac{\sin(2[22.5°])}{2} && \text{replace } \alpha \text{ with } 22.5°\\
&= \frac{\sin 45°}{2} && \text{multiply}\\
&= \frac{\sqrt{2}}{4} && \sin 45° = \frac{\sqrt{2}}{2}
\end{aligned}$$

NOW TRY EXERCISES 23 THROUGH 30

B. The Power Reduction and Half-Angle Identities

Expressions having a trigonometric function raised to a power actually occur quite frequently in various applications. We can rewrite even powers of these trig functions in terms of an expression containing only cosine to the power 1, using what are called the **power reduction identities.** This makes the expression easier to use and evaluate. It can legitimately be argued that the power reduction identities are actually members of the double-angle family, as all three are a direct consequence. To find identities for $\cos^2 x$ and $\sin^2 x$, we solve the related double-angle identity involving $\cos(2x)$.

$$\begin{aligned}
1 - 2\sin^2\alpha &= \cos(2\alpha) && \cos(2\alpha) \text{ in terms of sine}\\
-2\sin^2\alpha &= \cos(2\alpha) - 1 && \text{subtract 1}\\
\sin^2\alpha &= \frac{1 - \cos(2\alpha)}{2} && \text{power reduction identity for sine}
\end{aligned}$$

Using the same approach for $\cos^2\alpha$ gives $\cos^2\alpha = \frac{1 + \cos(2\alpha)}{2}$. The identity for $\tan^2\alpha$ can be derived from $\tan(2\alpha) = \frac{2\tan\alpha}{1 - \tan^2\alpha}$ (see Exercise 110), but in this case it's easier to use the identity $\tan^2 u = \frac{\sin^2 u}{\cos^2 u}$. The result is $\frac{1 - \cos(2\alpha)}{1 + \cos(2\alpha)}$.

THE POWER REDUCTION IDENTITIES

$$\cos^2\alpha = \frac{1+\cos(2\alpha)}{2} \qquad \sin^2\alpha = \frac{1-\cos(2\alpha)}{2} \qquad \tan^2\alpha = \frac{1-\cos(2\alpha)}{1+\cos(2\alpha)}$$

EXAMPLE 4 Write $8\sin^4 x$ in terms of an expression containing only cosines to the power 1.

Solution:

$$8\sin^4 x = 8(\sin^2 x)^2 \qquad \text{original expression}$$

$$= 8\left[\frac{1-\cos(2x)}{2}\right]^2 \qquad \text{substitute } \frac{1-\cos(2x)}{2} \text{ for } \sin^2 x$$

$$= 2[1 - 2\cos(2x) + \cos^2(2x)] \qquad \text{multiply}$$

$$= 2\left[1 - 2\cos(2x) + \frac{1-\cos(4x)}{2}\right] \qquad \text{substitute } \frac{1-\cos(4x)}{2} \text{ for } \cos^2(2x)$$

$$= 2 - 4\cos(2x) + 1 - \cos(4x) \qquad \text{multiply}$$

$$= 3 - 4\cos(2x) - \cos(4x) \qquad \text{result}$$

NOW TRY EXERCISES 31 THROUGH 36

The half-angle identities follow directly from those above, using algebra and a simple change of variable. For $\cos^2\alpha = \frac{1+\cos(2\alpha)}{2}$, we first take square roots and obtain $\cos\alpha = \pm\sqrt{\frac{1+\cos(2\alpha)}{2}}$. Using the substitution $u = 2\alpha$ gives $\alpha = \frac{u}{2}$, and making these substitutions results in the half-angle identity for cosine: $\cos\left(\frac{u}{2}\right) = \pm\sqrt{\frac{1+\cos u}{2}}$, where the radical's sign depends on the quadrant in which $\frac{u}{2}$ terminates. Using the same substitution for sine gives $\sin\left(\frac{u}{2}\right) = \pm\sqrt{\frac{1-\cos u}{2}}$, and for the tangent identity, $\tan\left(\frac{u}{2}\right) = \pm\sqrt{\frac{1-\cos u}{1+\cos u}}$. In the case of $\tan\left(\frac{u}{2}\right)$, we can actually develop identities that are free of radicals by rationalizing the denominator or numerator. We'll illustrate the former, leaving the latter as an exercise (see Exercise 108).

$$\tan\left(\frac{u}{2}\right) = \pm\sqrt{\frac{(1-\cos u)(1-\cos u)}{(1+\cos u)(1-\cos u)}} \qquad \text{multiply by conjugate}$$

$$= \pm\sqrt{\frac{(1-\cos u)^2}{1-\cos^2 u}} \qquad \text{rewrite}$$

$$= \pm\sqrt{\frac{(1-\cos u)^2}{\sin^2 u}} \qquad \text{Pythagorean identity}$$

$$= \pm\left|\frac{1-\cos u}{\sin u}\right| \qquad \sqrt{x^2} = |x|$$

Since $1 - \cos u > 0$ and $\sin u$ has the same sign as $\tan\left(\frac{u}{2}\right)$ for all u, the relationship can simply be written $\tan\left(\frac{u}{2}\right) = \frac{1-\cos u}{\sin u}$.

THE HALF-ANGLE IDENTITIES

$$\cos\left(\frac{u}{2}\right) = \pm\sqrt{\frac{1 + \cos u}{2}} \qquad \sin\left(\frac{u}{2}\right) = \pm\sqrt{\frac{1 - \cos u}{2}} \qquad \tan\left(\frac{u}{2}\right) = \pm\sqrt{\frac{1 - \cos u}{1 + \cos u}}$$

$$\tan\left(\frac{u}{2}\right) = \frac{1 - \cos u}{\sin u} \qquad \tan\left(\frac{u}{2}\right) = \frac{\sin u}{1 + \cos u}$$

EXAMPLE 5 Use the half-angle identities to find exact values for (a) sin 15° and (b) tan 15°.

Solution: Noting that 15° is one-half the standard angle 30°, we can find each value by applying the respective half-angle identity with $u = 30°$ in Quadrant I.

a.
$$\sin\left(\frac{30}{2}\right) = \sqrt{\frac{1 - \cos 30}{2}} = \sqrt{\frac{1 - \frac{\sqrt{3}}{2}}{2}}$$
$$\sin 15° = \frac{\sqrt{2 - \sqrt{3}}}{2}$$

b.
$$\tan\left(\frac{30}{2}\right) = \frac{1 - \cos 30}{\sin 30}$$
$$\tan 15° = \frac{1 - \frac{\sqrt{3}}{2}}{\frac{1}{2}} = 2 - \sqrt{3}$$

NOW TRY EXERCISES 37 THROUGH 48

EXAMPLE 6 Given $\cos\theta = -\frac{7}{25}$ and θ in QIII, find exact values of $\sin\left(\frac{\theta}{2}\right)$ and $\cos\left(\frac{\theta}{2}\right)$.

Solution: With θ in QIII $\rightarrow \pi < \theta < \frac{3\pi}{2}$, we know $\frac{\theta}{2}$ must be in QII $\rightarrow \frac{\pi}{2} < \frac{\theta}{2} < \frac{3\pi}{4}$ and we choose our signs accordingly: $\sin\left(\frac{\theta}{2}\right) > 0$ and $\cos\left(\frac{\theta}{2}\right) < 0$.

$$\sin\left(\frac{\theta}{2}\right) = \sqrt{\frac{1 - \cos\theta}{2}} = \sqrt{\frac{1 - \left(-\frac{7}{25}\right)}{2}} = \sqrt{\frac{16}{25}} = \frac{4}{5}$$

$$\cos\left(\frac{\theta}{2}\right) = -\sqrt{\frac{1 + \cos\theta}{2}} = -\sqrt{\frac{1 + \left(-\frac{7}{25}\right)}{2}} = -\sqrt{\frac{9}{25}} = -\frac{3}{5}$$

NOW TRY EXERCISES 49 THROUGH 64

C. The Product-to-Sum Identities

As mentioned in the introduction, the product-to-sum and sum-to-product identities are of immense importance to the study of any phenomenon that travels in waves, like light and sound. In fact, the tones you hear as you dial a telephone are actually the sum of two sound waves interacting with each other. Each derivation of a product-to-sum identity is very similar (see Exercise 111), and we illustrate by deriving the identity for $\cos\alpha\cos\beta$. Beginning with the sum and difference identities for cosine, we have

$$\begin{aligned} \cos\alpha\cos\beta + \sin\alpha\sin\beta &= \cos(\alpha-\beta) && \text{cosine of a difference} \\ +\ \cos\alpha\cos\beta - \sin\alpha\sin\beta &= \cos(\alpha+\beta) && \text{cosine of a sum} \\ 2\cos\alpha\cos\beta &= \cos(\alpha-\beta)+\cos(\alpha+\beta) && \text{combine equations} \\ \cos\alpha\cos\beta &= \frac{1}{2}[\cos(\alpha-\beta)+\cos(\alpha+\beta)] && \text{divide by 2} \end{aligned}$$

The identities from this family are listed here.

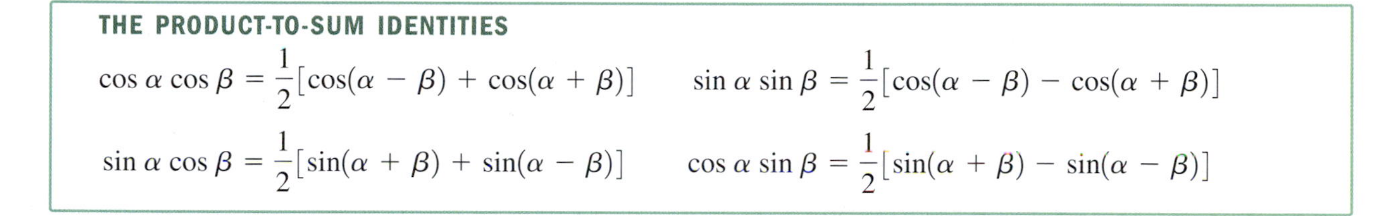

THE PRODUCT-TO-SUM IDENTITIES

$$\cos\alpha\cos\beta = \frac{1}{2}[\cos(\alpha-\beta)+\cos(\alpha+\beta)] \qquad \sin\alpha\sin\beta = \frac{1}{2}[\cos(\alpha-\beta)-\cos(\alpha+\beta)]$$

$$\sin\alpha\cos\beta = \frac{1}{2}[\sin(\alpha+\beta)+\sin(\alpha-\beta)] \qquad \cos\alpha\sin\beta = \frac{1}{2}[\sin(\alpha+\beta)-\sin(\alpha-\beta)]$$

EXAMPLE 7 Write the product $2\cos(27t)\cos(15t)$ as the sum of two cosine functions.

Solution: This is a direct application of the product-to-sum identity, with $\alpha = 27t$ and $\beta = 15t$.

$$\begin{aligned} \cos\alpha\cos\beta &= \frac{1}{2}[\cos(\alpha-\beta)+\cos(\alpha+\beta)] && \text{product-to-sum identity} \\ 2\cos(27t)\cos(15t) &= 2\left(\frac{1}{2}\right)[\cos(27t-15t)+\cos(27t+15t)] && \text{substitute} \\ &= \cos(12t)+\cos(42t) && \text{result} \end{aligned}$$

NOW TRY EXERCISES 65 THROUGH 73

There are times we find it necessary to "work in the other direction," writing a sum of two trig functions as a product. This family of identities can be derived from the product-to-sum identities using a change of variable. We'll illustrate the process for $\sin u + \sin v$. You are asked for the derivation of $\cos u + \cos v$ in Exercise 112. To begin, we use $2\alpha = u + v$ and $2\beta = u - v$. This creates the sum $2\alpha + 2\beta = 2u$ and the difference $2\alpha - 2\beta = 2v$, yielding $\alpha + \beta = u$ and $\alpha - \beta = v$, respectively.

Dividing the original expressions by 2 gives $\alpha = \frac{u + v}{2}$ and $\beta = \frac{u - v}{2}$, which all together make the derivation a matter of direct substitution. Using these values in any product-to-sum identity gives the related sum-to-product, as shown here.

$$\sin\alpha\cos\beta = \frac{1}{2}[\sin(\alpha + \beta) + \sin(\alpha - \beta)]$$ product-to-sum identity (sum of sines)

$$\sin\left(\frac{u + v}{2}\right)\cos\left(\frac{u - v}{2}\right) = \frac{1}{2}(\sin u + \sin v)$$ substitute $\frac{u+v}{2}$ for α, $\frac{u-v}{2}$ for β, substitute u for $\alpha + \beta$ and v for $\alpha - \beta$

$$2\sin\left(\frac{u + v}{2}\right)\cos\left(\frac{u - v}{2}\right) = \sin u + \sin v$$ multiply by 2

The sum-to-product identities are listed next.

THE SUM-TO-PRODUCT IDENTITIES

$$\cos u + \cos v = 2\cos\left(\frac{u + v}{2}\right)\cos\left(\frac{u - v}{2}\right) \qquad \sin u + \sin v = 2\sin\left(\frac{u + v}{2}\right)\cos\left(\frac{u - v}{2}\right)$$

$$\sin u - \sin v = 2\cos\left(\frac{u + v}{2}\right)\sin\left(\frac{u - v}{2}\right) \qquad \cos u - \cos v = -2\sin\left(\frac{u + v}{2}\right)\sin\left(\frac{u - v}{2}\right)$$

EXAMPLE 8 Given $y_1 = \sin(12\pi t)$ and $y_2 = \sin(10\pi t)$, express the sum $y_1 + y_2$ as a product of trigonometric functions.

Solution: This is a direct application of the sum-to-product identity $\sin u + \sin v$, with $u = 12\pi t$ and $v = 10\pi t$.

$$\sin u + \sin v = 2\sin\left(\frac{u + v}{2}\right)\cos\left(\frac{u - v}{2}\right)$$ sum-to-product identity

$$\sin(12\pi t) + \sin(10\pi t) = 2\sin\left(\frac{12\pi t + 10\pi t}{2}\right)\cos\left(\frac{12\pi t - 10\pi t}{2}\right)$$ substitute $12\pi t$ for u and $10\pi t$ for v

$$= 2\sin(11\pi t)\cos(\pi t)$$ substitute

NOW TRY EXERCISES 74 THROUGH 82

D. Applications of Identities

In more advanced mathematics courses, rewriting an expression using identities enables the extension or completion of a task that would otherwise be very difficult (or even impossible). In addition, there are a number of practical applications in the physical sciences.

EXAMPLE 9 A projectile is any object that is thrown, shot, kicked, dropped, or otherwise given an initial velocity, but lacking a continuing source of propulsion. If air resistance is ignored, the range of projectile depends only on its initial velocity v and the angle θ at which it is propelled. This phenomenon is modeled by the function $r(\theta) = \frac{1}{16}v^2\sin\theta\cos\theta$.

a. Use an identity to show this function is equivalent to $r(\theta) = \frac{1}{32}v^2\sin(2\theta)$.

b. If the projectile is thrown with an initial velocity of $v = 96$ ft/sec, how far will it travel if $\theta = 15°$?

c. From the result of part (a), determine what angle θ will give the maximum range for the projectile.

Solution:

a. Note that we can use a double-angle identity if we rewrite the coefficient. Writing $\frac{1}{16}$ as $2\left(\frac{1}{32}\right)$ and commuting the factors gives $r(\theta) = \left(\frac{1}{32}\right)v^2(2\sin\theta\cos\theta) = \left(\frac{1}{32}\right)v^2\sin(2\theta)$.

b. With $v = 96$ ft/sec and $\theta = 15°$, the formula gives $r(15°) = \left(\frac{1}{32}\right)(96)^2\sin 30°$. The projectile travels a horizontal distance of 144 ft.

c. For any initial velocity v, $r(\theta)$ will be maximized when $\sin(2\theta)$ is a maximum. This occurs when $\sin(2\theta) = 1$, meaning $2\theta = 90°$ and $\theta = 45°$. The maximum range is achieved when the projectile is released at an angle of 45°.

NOW TRY EXERCISES 103 AND 104

GRAPHICAL SUPPORT

The result in Example 9(c) can be verified graphically by assuming an initial velocity of 96 ft/sec and entering the function $r(\theta) = \frac{1}{32}(96)^2\sin(2\theta) = 288\sin(2\theta)$ as Y_1 on a graphing calculator. With an amplitude of 288 and results confined to the first quadrant, we set an appropriate window, graph the function, and use the 2nd TRACE (**CALC 4:maximum**) feature. As shown in the figure, the max occurs at $\theta = 45°$.

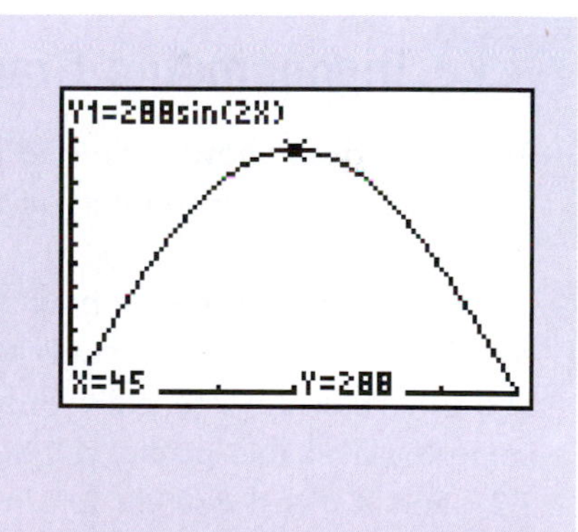

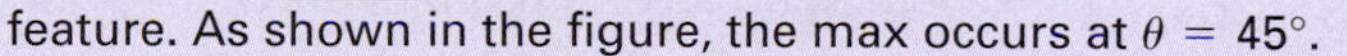

Each tone you hear on a touch-tone phone is actually the combination of precisely two sound waves with different frequencies (frequency f is defined as $f = \frac{B}{2\pi}$). This is why the tones you hear sound identical, regardless of what phone you use. The sum-to-product and product-to-sum formulas help us to understand, study, and use sound in very powerful and practical ways, like sending faxes and using other electronic media.

EXAMPLE 10 On a touch-tone phone, the sound created by pressing 5 is produced by combining a sound wave with frequency 1336 cycles/sec, with another wave having frequency 770 cycles/sec. Their respective equations are $y_1 = \cos(2\pi\, 1336t)$ and $y_2 = \cos(2\pi\, 770t)$, with the resultant wave being $y = y_1 + y_2$ or $y = \cos(2672\pi t) + \cos(1540\pi t)$. Rewrite this sum as a product.

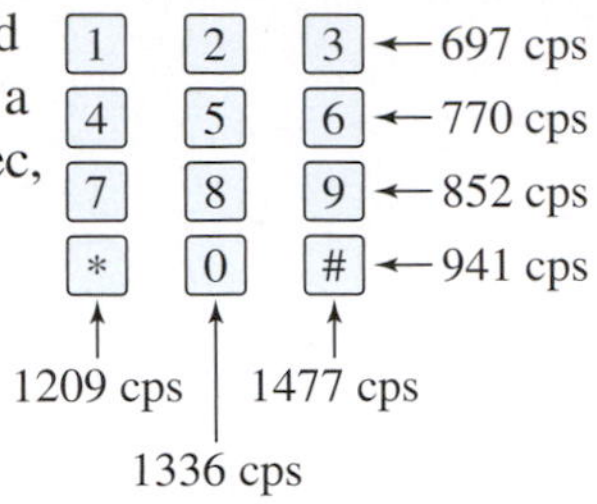

1	2	3	← 697 cps
4	5	6	← 770 cps
7	8	9	← 852 cps
*	0	#	← 941 cps
1209 cps	1336 cps	1477 cps	

Solution: This is a direct application of the sum-to-product identity, with $u = 2672\pi t$ and $v = 1540\pi t$. Computing one-half the sum/difference of u and v gives $\frac{2672\pi t + 1540\pi t}{2} = 2106\pi t$ and $\frac{2672\pi t - 1540\pi t}{2} = 566\pi t$.

$$\cos u + \cos v = 2\cos\left(\frac{u+v}{2}\right)\cos\left(\frac{u-v}{2}\right)$$ sum-to-product identity

$$\cos(2672\pi t) + \cos(1540\pi t) = 2\cos(2106\pi t)\cos(566\pi t)$$ substitute $2672\pi t$ for u and $1540\pi t$ for v

NOW TRY EXERCISES 105 AND 106

Note we can identify the button pressed when the wave is written as a sum. If we have only the resulting wave (written as a product), the product-to-sum formula must be used to identify which button was pressed.

TECHNOLOGY HIGHLIGHT

Trigonometric Graphs and ZoomSto/ZoomRcl

The keystrokes shown apply to a TI-84 Plus model. Please consult your manual or our Internet site for other models.

In Chapter 2, we saw how the ZOOM **7:ZTrig** screen could be used to study trigonometric graphs. The calculator automatically creates a friendly window using the available pixels, setting Xmin ≈ −360° or -2π and Xmax ≈ 360° or 2π. In addition, it sets Ymin = −4 and Ymax = 4 since a large number of trig graphs fall within this amplitude range. However, a great deal of study, investigation, and comparison is done in the interval $[0, 2\pi]$ and with sine and cosine graphs having a much smaller amplitude. In these cases, it helps to use the ZOOM ► **(MEMORY)** feature of the TI-84 Plus. To use this feature, you set the window to the desired settings and press ZOOM ► **(MEMORY) 2:ZoomSto,** which will store these settings for future use. The window size will be changed each time ZOOM **7:ZTrig,** ZOOM **4:ZDecimal,** or ZOOM **7:ZStandard** is used, but you can always return to the stored settings using ZOOM ► **(MEMORY) 3:ZoomRcl.** Suppose we wanted to investigate the graph of $Y_1 = \cos(2x)$, along with the three related identities developed in this section. Using ZOOM **7:ZTrig** in degree mode gives the window shown in Figure 3.14. If our particular interest was [0, 360] we could reset Xmin = 0, Xmax = 360, and Xscl = 45. Using the preset Ymin and Ymax values leaves too much "wasted space," so we reset these to Ymin = −2 and

Figure 3.14

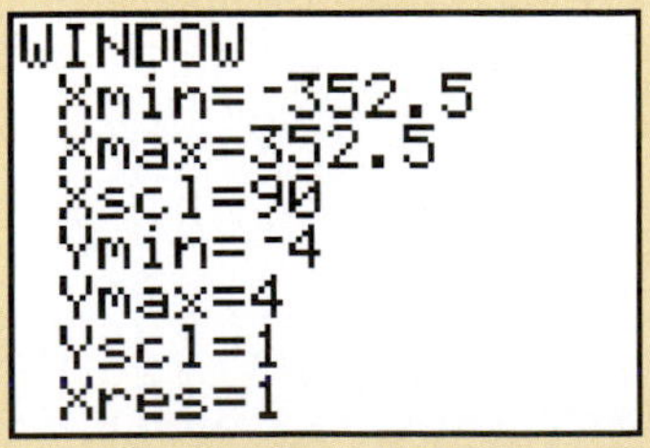

```
WINDOW
 Xmin=-352.5
 Xmax=352.5
 Xscl=90
 Ymin=-4
 Ymax=4
 Yscl=1
 Xres=1
```

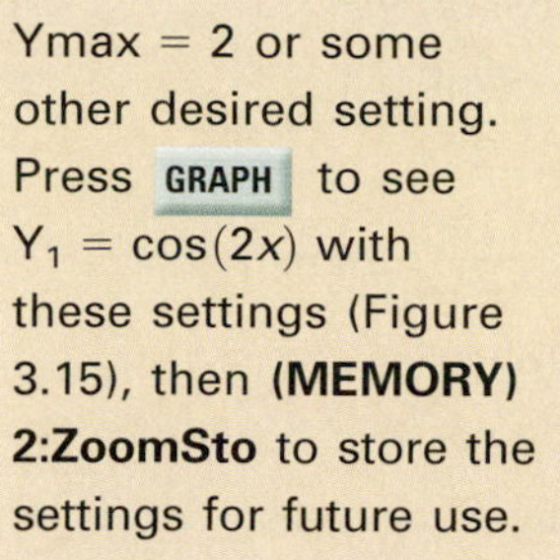

Ymax = 2 or some other desired setting. Press GRAPH to see $Y_1 = \cos(2x)$ with these settings (Figure 3.15), then **(MEMORY) 2:ZoomSto** to store the settings for future use.

Figure 3.15

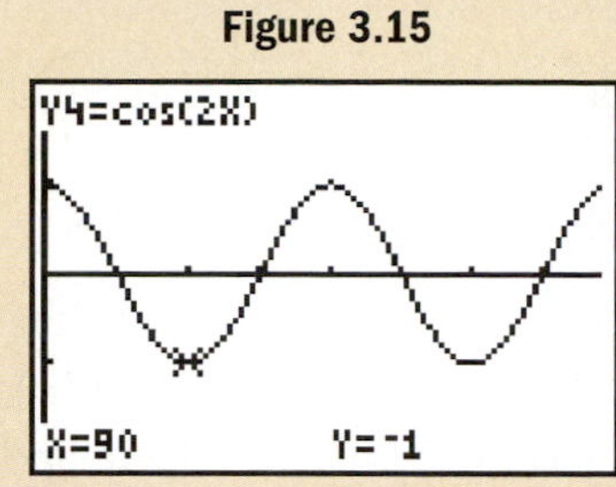

Exercise 1: Enter $Y_2 = \cos^2x - \sin^2x$ and $Y_3 = 1 - 2\sin^2x$, then graph Y_1, Y_2, and Y_3 on the ZOOM **7:ZTrig** screen. What do you notice? Now use the keystrokes ZOOM ▶ **(MEMORY) 3:ZoomRcl** to graph these functions using the stored settings. Can you distinguish between the graphs?

Exercise 2: Enter $Y_1 = \tan^2x$ and $Y_2 = \dfrac{1 - \cos(2x)}{1 + \cos(2x)}$.

Use the ZOOM **7:ZTrig** screen to graphically verify that $Y_1 = Y_2$ is an identity, then take a closer look at the graphs using ZOOM ▶ **(MEMORY) 3:ZoomRcl.** Use this screen to approximate the values of x for which $\tan^2x = 1$.

3.4 EXERCISES

CONCEPTS AND VOCABULARY

Fill in each blank with the appropriate word or phrase. Carefully reread the section if needed.

1. The double-angle identities can be derived using the ________ identities with $\alpha = \beta$. For $\cos(2\theta)$ we expand $\cos(\alpha + \beta)$ using ________.

2. If θ is in QI, $0 < \theta < 90°$. This means 2θ must be in ______ or ______, since ______ $< 2\theta <$ ______. If θ is in QIII then $180° < \theta < 270°$ and $\frac{\theta}{2}$ must be in ______ since ______ $< \frac{\theta}{2} <$ ______.

3. Multiple-angle identities can also be derived using the sum and difference identities. For $\sin(3x)$ use sin(____ + ____).

4. The half-angle identities are closely related to the ________ ________ identities and can be developed using algebra and a change of ________.

5. Explain/discuss how the three different identities for $\tan\left(\frac{u}{2}\right)$ are related. Verify that $\dfrac{1 - \cos x}{\sin x} = \dfrac{\sin x}{1 + \cos x}$.

6. In Example 8, we were given $\cos\theta = -\frac{7}{25}$ and θ in QIII. Discuss how the result would differ if we stipulate that θ is in QII instead.

DEVELOPING YOUR SKILLS

Find exact values for $\sin(2\theta)$, $\cos(2\theta)$, and $\tan(2\theta)$ using the information given.

7. $\sin\theta = \frac{5}{13}$; θ in QII

8. $\cos\theta = -\frac{21}{29}$; θ in QII

9. $\cos\theta = -\frac{9}{41}$; θ in QII

10. $\sin\theta = -\frac{63}{65}$; θ in QIII

11. $\tan\theta = \frac{13}{84}$; θ in QIII

12. $\sec\theta = \frac{53}{28}$; θ in QI

13. $\sin\theta = \frac{48}{73}$; $\cos\theta < 0$

14. $\cos\theta = -\frac{8}{17}$; $\tan\theta > 0$

15. $\csc\theta = \frac{5}{3}$; $\sec\theta < 0$

16. $\cot\theta = -\frac{80}{39}$; $\cos\theta > 0$

Find exact values for $\sin\theta$, $\cos\theta$, and $\tan\theta$ using the information given.

17. $\sin(2\theta) = \frac{24}{25}$; 2θ in QII

18. $\sin(2\theta) = -\frac{240}{289}$; 2θ in QIII

19. $\cos(2\theta) = -\frac{41}{841}$; 2θ in QII

20. $\cos(2\theta) = \frac{120}{169}$; 2θ in QIV

21. Verify the following identity: $\sin(3\theta) = 3\sin\theta - 4\sin^3\theta$

22. Verify the following identity: $\cos(4\theta) = 8\cos^4\theta - 8\cos^2\theta + 1$

Use a double-angle identity to find exact values for the following expressions.

23. $\cos 75° \sin 75°$

24. $\cos^2 15° - \sin^2 15°$

25. $1 - 2\sin^2\left(\frac{\pi}{8}\right)$

26. $2\cos^2\left(\frac{\pi}{12}\right) - 1$

27. $\frac{2\tan 22.5°}{1 - \tan^2 22.5°}$

28. $\frac{2\tan(\frac{\pi}{12})}{1 - \tan^2(\frac{\pi}{12})}$

29. Use a double-angle identity to rewrite $9\sin(3x)\cos(3x)$ as a single function. [*Hint:* $9 = \frac{9}{2}(2)$.]

30. Use a double-angle identity to rewrite $2.5 - 5\sin^2 x$ as a single term. [*Hint:* Factor out a constant.]

Rewrite in terms of an expression containing only cosines to the power 1.

31. $\sin^2 x \cos^2 x$

32. $\sin^4 x \cos^2 x$

33. $3\cos^4 x$

34. $\cos^4 x \sin^4 x$

35. $2\sin^6 x$

36. $4\cos^6 x$

Use a half-angle identity to find exact values for $\sin\theta$, $\cos\theta$, and $\tan\theta$ for the given value of θ.

37. $\theta = 22.5°$

38. $\theta = 75°$

39. $\theta = \frac{\pi}{12}$

40. $\theta = \frac{5\pi}{12}$

41. $\theta = 67.5°$

42. $\theta = 112.5°$

43. $\theta = \frac{3\pi}{8}$

44. $\theta = \frac{11\pi}{12}$

Use the results of Exercises 37–40 and a half-angle identity to find the exact value.

45. $\sin 11.25°$

46. $\tan 37.5°$

47. $\sin\left(\frac{\pi}{24}\right)$

48. $\cos\left(\frac{5\pi}{24}\right)$

Use a half-angle identity to rewrite each expression as a single, nonradical function.

49. $\sqrt{\frac{1 + \cos 30°}{2}}$

50. $\sqrt{\frac{1 - \cos 45°}{2}}$

51. $\sqrt{\frac{1 - \cos(4\theta)}{1 + \cos(4\theta)}}$

52. $\frac{1 - \cos(6x)}{\sin(6x)}$

53. $\frac{\sin(2x)}{1 + \cos(2x)}$

54. $\frac{\sqrt{2}(1 + \cos x)}{1 + \cos x}$

Find exact values for $\sin\left(\frac{\theta}{2}\right)$, $\cos\left(\frac{\theta}{2}\right)$, and $\tan\left(\frac{\theta}{2}\right)$ using the information given.

55. $\sin\theta = \frac{12}{13}$; θ is obtuse

56. $\cos\theta = -\frac{8}{17}$; θ is obtuse

57. $\cos\theta = -\frac{4}{5}$; θ in QII

58. $\sin\theta = -\frac{7}{25}$; θ in QIII

59. $\tan\theta = -\frac{35}{12}$; θ in QII

60. $\sec\theta = -\frac{65}{33}$; θ in QIII

61. $\sin\theta = \frac{15}{113}$; θ is acute

62. $\cos\theta = \frac{48}{73}$; θ is acute

63. $\cot\theta = \frac{21}{20}; \pi < \theta < \frac{3\pi}{2}$

64. $\csc\theta = \frac{41}{9}; \frac{\pi}{2} < \theta < \pi$

Write each product as a sum using the product-to-sum identities.

65. $\sin(-4\theta)\sin(8\theta)$

66. $\cos(15\alpha)\sin(-3\alpha)$

67. $2\cos\left(\frac{7t}{2}\right)\cos\left(\frac{3t}{2}\right)$

68. $2\sin\left(\frac{5t}{2}\right)\sin\left(\frac{9t}{2}\right)$

69. $2\cos(1979\pi t)\cos(439\pi t)$

70. $2\cos(2150\pi t)\cos(268\pi t)$

Find the exact value using product-to-sum identities.

71. $2\cos 15°\sin 135°$

72. $\sin\left(\frac{7\pi}{8}\right)\cos\left(\frac{\pi}{8}\right)$

73. $\sin\left(\frac{7\pi}{12}\right)\sin\left(-\frac{\pi}{12}\right)$

Write each sum as a product using the sum-to-product identities.

74. $\cos(9h) + \cos(4h)$

75. $\sin(14k) + \sin(41k)$

76. $\sin\left(\frac{11x}{8}\right) - \sin\left(\frac{5x}{8}\right)$

77. $\cos\left(\frac{7x}{6}\right) - \cos\left(\frac{5x}{6}\right)$

78. $\cos(697\pi t) + \cos(1447\pi t)$

79. $\cos(852\pi t) + \cos(1209\pi t)$

Find the exact value using sum-to-product identities.

80. $\cos 75° + \cos 15°$

81. $\sin\left(\frac{17\pi}{12}\right) - \sin\left(\frac{13\pi}{12}\right)$

82. $\sin\left(\frac{11\pi}{12}\right) + \sin\left(\frac{7\pi}{12}\right)$

Verify the following identities.

83. $\frac{2\sin x\cos x}{\cos^2 x - \sin^2 x} = \tan(2x)$

84. $\frac{1 - \sin^2 x}{2\sin x\cos x} = \cot(2x)$

85. $(\sin x + \cos x)^2 = 1 + \sin(2x)$

86. $(\sin^2 x - 1)^2 = \sin^4 x + \cos(2x)$

87. $\cos(8\theta) = \cos^2(4\theta) - \sin^2(4\theta)$

88. $\sin(4x) = 4\sin x\cos x(1 - 2\sin^2 x)$

89. $\frac{\cos(2\theta)}{\sin^2\theta} = \cot^2\theta - 1$

90. $\csc^2\theta - 2 = \frac{\cos(2\theta)}{\sin^2\theta}$

91. $\tan(2\theta) = \frac{2}{\cot\theta - \tan\theta}$

92. $\cot\theta - \tan\theta = \frac{2\cos(2\theta)}{\sin(2\theta)}$

93. $\tan x + \cot x = 2\csc(2x)$

94. $\csc(2x) = \frac{1}{2}\csc x\sec x$

95. $\cos^2\left(\frac{x}{2}\right) - \sin^2\left(\frac{x}{2}\right) = \cos x$

96. $1 - 2\sin^2\left(\frac{x}{4}\right) = \cos\left(\frac{x}{2}\right)$

97. $1 - \sin^2(2\theta) = 1 - 4\sin^2\theta + 4\sin^4\theta$

98. $2\cos^2\left(\frac{x}{2}\right) - 1 = \cos x$

99. $\frac{\sin(120\pi t) + \sin(80\pi t)}{\cos(120\pi t) - \cos(80\pi t)} = -\cot(20\pi t)$

100. $\frac{\sin m + \sin n}{\cos m + \cos n} = \tan\left(\frac{m + n}{2}\right)$

WORKING WITH FORMULAS

101. Supersonic speeds, the sound barrier, and Mach numbers: $\mathcal{M} = \csc\left(\frac{\theta}{2}\right)$

The speed of sound varies with temperature and altitude. At 32°F, sound travels about 742 mi/hr at sea level. A jet-plane flying faster than the speed of sound (called supersonic speed) has "broken the sound barrier." The plane projects three-dimensional sound waves about the nose of the craft that form the shape of a cone. The cone intersects the Earth along a hyperbolic path, with a sonic boom being heard by anyone along this path. The ratio of the plane's speed to the speed of sound is called its Mach number $\mathcal{M}$, meaning a plane flying at $\mathcal{M} = 3.2$ is traveling 3.2 times the speed of sound. This Mach number can be determined using the formula given here, where θ is the vertex angle of the cone described. For the following exercises, use the formula to find $\mathcal{M}$ or θ as required. For parts (a) and (b), answer in exact form (using a half-angle identity) and approximate form.

a. $\theta = 30°$ **b.** $\theta = 45°$ **c.** $\mathcal{M} = 2$

102. Malus's law: $I = I_0 \cos^2\theta$
When a beam of plane-polarized light with intensity I_0 hits an analyzer, the intensity I of the transmitted beam of light can be found using the formula shown, where θ is the angle formed between the transmission axes of the polarizer and the analyzer. Find the intensity of the beam when $\theta = 15°$ and $I_0 = 300$ candelas (cd). Answer in exact form (using a power reduction identity) and approximate form.

APPLICATIONS

Range of a projectile: Exercises 103 and 104 refer to Example 9. In Example 9, we noted that the range of a projectile was maximized at $\theta = 45°$. If $\theta > 45°$ or $\theta < 45°$, the projectile falls short of its maximum potential distance. In Exercises 103 and 104 assume that the projectile has an initial velocity of 96 ft/sec.

103. Compute how many feet short of maximum the projectile falls if (a) $\theta = 22.5°$ and (b) $\theta = 67.5°$. Answer in both exact and approximate form.

104. Use a calculator to compute how many feet short of maximum the projectile falls if (a) $\theta = 40°$ and $\theta = 50°$ and (b) $\theta = 37.5°$ and $\theta = 52.5°$. Do you see a pattern? Discuss/explain what you notice and experiment with other values to confirm your observations.

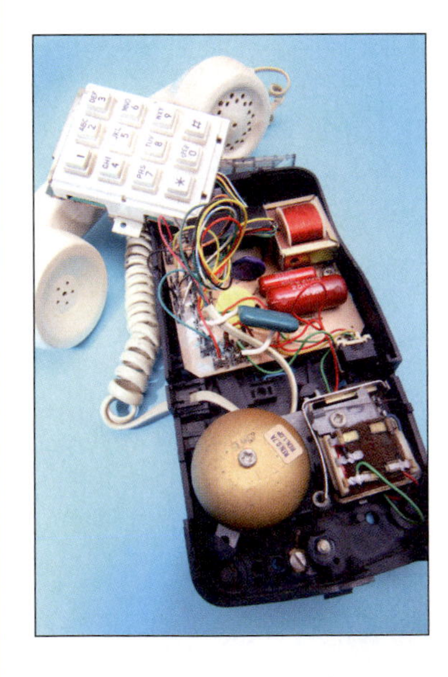

Touch-tone phones: The diagram given in Example 10 shows the various frequencies used to create the tones for a touch-tone phone. One button is randomly pressed and the resultant wave is modeled by $y(t)$ shown. Use a product-to-sum identity to write the expression as a sum and determine the button pressed.

105. $y(t) = 2\cos(2150\pi t)\cos(268\pi t)$

106. $y(t) = 2\cos(1906\pi t)\cos(512\pi t)$

Working with identities

107. From the diagram in the *Point of Interest,* show the Pythagorean theorem holds for the smaller triangle. In other words, show $\sin^2\alpha + (1 - \cos\alpha)^2 = \left[2\sin\left(\frac{\alpha}{2}\right)\right]^2$.

108. Show that $\tan\left(\frac{u}{2}\right) = \pm\sqrt{\frac{1 - \cos u}{1 + \cos u}}$ is equivalent to $\frac{\sin u}{1 + \cos u}$ by rationalizing the numerator.

109. Derive the identity for $\sin(2\alpha)$ and $\tan(2\alpha)$ using $\sin(\alpha + \beta)$ and $\tan(\alpha + \beta)$, where $\alpha = \beta$.

110. Derive the identity for $\tan^2(\theta)$ using $\tan(2\alpha) = \frac{2\tan(\alpha)}{1 - \tan^2(\alpha)}$.

111. Derive the product-to-sum identity for $\sin\alpha \sin\beta$.

112. Derive the sum-to-product identity for $\cos u + \cos v$.

113. Clock angles: Kirkland City has a large clock atop city hall, with a minute hand that is 3 ft long. Claire and Monica independently attempt to devise a function that will track the distance between the tip of the minute hand at t minutes between the hours, and the tip of the minute hand when it is in the vertical position as shown. Claire finds the function $d(t) = \left|6\sin\left(\frac{\pi t}{60}\right)\right|$, while Monica devises $d(t) = \sqrt{18\left[1 - \cos\left(\frac{\pi t}{30}\right)\right]}$. Use the identities from this section to show the functions are equivalent.

114. Origami: The Japanese art of origami involves the repeated folding of a single piece of paper to create various art forms. When the upper right corner of a rectangular 21.6-cm by 28-cm piece of paper is folded down until the corner is flush with the other side, the length L of the fold is related to the angle θ by $L = \frac{10.8}{\sin\theta\cos^2\theta}$. (a) Show this is equivalent to $L = \frac{21.6\sec\theta}{\sin(2\theta)}$, (b) find the length of the fold if $\theta = 30°$, and (c) find the angle θ if $L = 28.8$ cm.

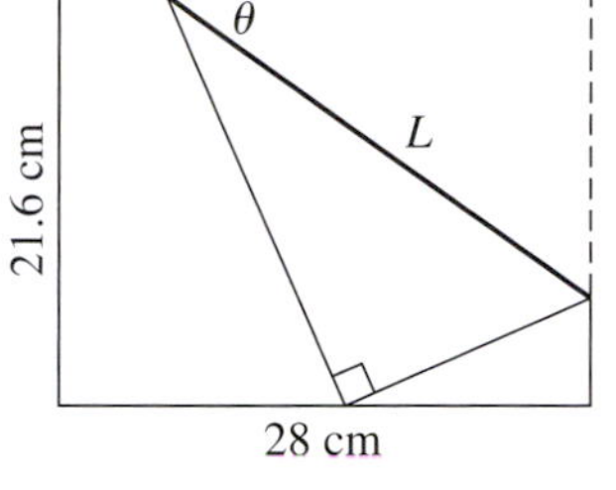

115. Machine gears: A machine part involves two gears. The first has a radius of 2 cm and the second a radius of 1 cm, so the smaller gear turns twice as fast as the larger gear. Let θ represent the angle of rotation in the larger gear, measured from a vertical and downward starting position. Let P be a point on the circumference of the smaller gear, starting at the vertical and downward position. Four engineers working on an improved design for this component devise functions that track the height of point P above the horizontal plane shown, for a rotation of $\theta°$ by the larger gear. The functions they develop are: Engineer A: $f(\theta) = \sin(2\theta - 90°) + 1$; Engineer B: $g(\theta) = 2\sin^2\theta$; Engineer C: $k(\theta) = 1 + \sin^2\theta - \cos^2\theta$; and Engineer D: $h(\theta) = 1 - \cos(2\theta)$. Use any of the identities you've learned so far to show these four functions are equivalent.

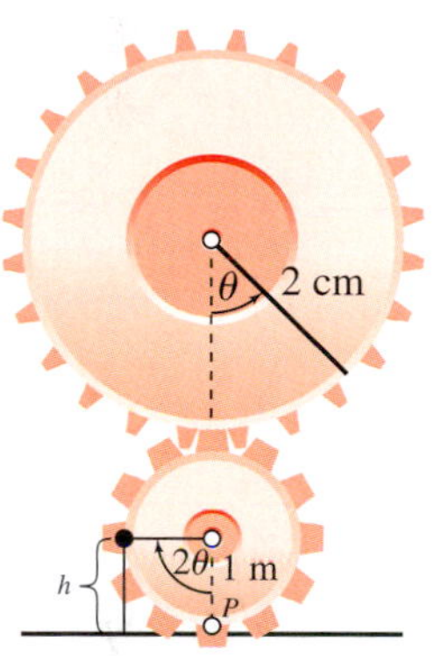

116. Working with identities: Compute the value of sin 15° two ways, first using the half-angle identity for sine, and second using the difference identity for sine. (a) Find a decimal approximation for each to show the results are equivalent and (b) verify algebraically that they are equivalent. (*Hint:* Square both sides.)

117. Working with identities: Compute the value of cos 15° two ways, first using the half-angle identity for cosine, and second using the difference identity for cosine. (a) Find a decimal approximation for each to show the results are equivalent and (b) verify algebraically that they are equivalent. (*Hint:* Square both sides.)

WRITING, RESEARCH, AND DECISION MAKING

118. As it turns out, the sine of any angle that is a multiple of 3° can be written in exact form using only rational numbers and square roots. Some are very simple, like the standard angles, while others involve numerous terms and radicals. Only the multiples of 15° can be expressed without using nested radicals (one radical inside another), with the exception of sin 18° and sin 54°. Do some research and study with regard to the sine of these angles. In particular, find a closed form expression for sin 15° and sin 75°—which can be done using the ideas of this section, as well as closed form expressions for sin 18° and sin 54°, which use a combination of geometry and algebra.

119. Can you find three distinct, real numbers whose sum is equal to their product? A little known fact from trigonometry stipulates that for any triangle, the sum of the tangents of the angles is equal to the products of their tangents. Use a calculator to test this statement, recalling the three angles must sum to 180°. Our website at www.mhhe.com/coburn shows a method that enables you to verify the statement using tangents that are all rational values.

EXTENDING THE CONCEPT

Exercise 120

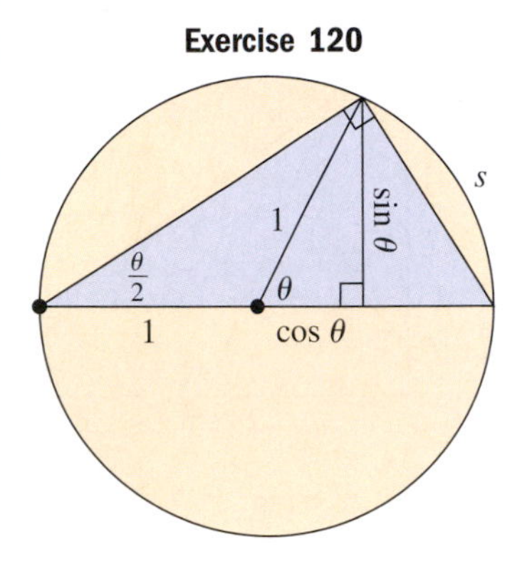

120. A proof without words: From elementary geometry we have the following: (a) an angle inscribed in a semicircle is a right angle, and (b) the measure of an inscribed angle (vertex on the circumference) is one-half the measure of its intercepted arc. Discuss/explain how the unit-circle diagram offers a proof that $\tan\left(\frac{x}{2}\right) = \frac{\sin x}{1 + \cos x}$. Be detailed and thorough.

121. Using $\theta = 30°$ and repeatedly applying the half-angle identity for cosine, show that $\cos 3.75°$ is equal to $\frac{\sqrt{2 + \sqrt{2 + \sqrt{2 + \sqrt{3}}}}}{2}$. Verify the result using a calculator, then use the patterns noted to write the value of cos 1.875° in closed form (also verify this result). As θ becomes very small, what appears to be happening to the value of $\cos \theta$?

MAINTAINING YOUR SKILLS

122. (1.3) Use a calculator to find two values of t that satisfy $\sin t \approx 0.9889$. Round to four decimal places.

123. (2.2) Given $Y_1 = \tan \alpha$ and $Y_2 = \cot \alpha$, which function is increasing? Which is defined for $\alpha \in (0, \pi)$?

124. (1.1) The hypotenuse of a certain right triangle is twice the shortest side. Solve the triangle.

125. (1.4) Verify that $\left(\frac{16}{65}, \frac{63}{65}\right)$ is on the unit circle, then find $\tan \theta$ and $\sec \theta$ to verify $1 + \tan^2\theta = \sec^2\theta$.

126. (2.3) Write the equation of the function graphed in terms of a sine function of the form $y = A \sin(Bx + C) + D$.

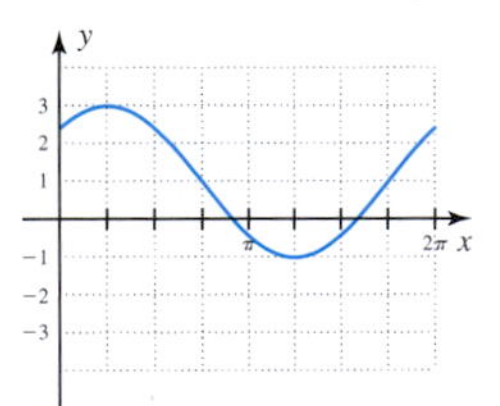

127. (3.3) Use the sum identity for cosine to find the exact value of cos(105°).

SUMMARY AND CONCEPT REVIEW

SECTION 3.1 Fundamental Identities and Families of Identities

KEY CONCEPTS

- The fundamental identities include the *reciprocal, ratio, and Pythagorean identities.*
- A given identity can algebraically be rewritten to obtain other identities in an identity "family."
- Standard algebraic skills like distribution, factoring, combining terms, and special products play an important role in working with identities.

- The pattern $\frac{A}{B} \pm \frac{C}{D} = \frac{AD \pm BC}{BD}$ gives an efficient method for combining rational terms.
- Using fundamental identities, a given trig function can be expressed in terms of any other trig function.
- Once the value of a given trig function is known, the value of the other five can be uniquely determined using fundamental identities, *if the quadrant of the terminal side is known.*
- To show an equation is not an identity, find any one value where the expressions are defined but the equation is false, or graph both functions on a calculator to see if the graphs are identical.

EXERCISES

Verify using the method specified and fundamental identities.

1. multiplication
$\sin x(\csc x - \sin x) = \cos^2 x$

2. factoring
$\frac{\tan^2 x \csc x + \csc x}{\sec^2 x} = \csc x$

3. special products
$\frac{(\sec x - \tan x)(\sec x + \tan x)}{\csc x} = \sin x$

4. combine terms using $\frac{A}{B} \pm \frac{C}{D} = \frac{AD \pm BC}{BD}$
$\frac{\sec^2 x}{\csc x} - \sin x = \frac{\tan^2 x}{\csc x}$

Find the value of all six trigonometric functions using the information given.

5. $\cos\theta = -\frac{12}{37}$; θ in QIII

6. $\sec\theta = \frac{25}{23}$; θ in QIV

SECTION 3.2 Constructing and Verifying Identities

KEY CONCEPTS

- The steps used to verify an identity must be reversible.
- If two expressions are equal, one may be substituted for the other and the result will be equivalent.
- To verify an identity we mold, change, substitute, and rewrite one side until we "match" the other side.
- Verifying identities often involves a combination of algebraic skills with the fundamental trig identities.

A collection and summary of the *Guidelines for Verifying Identities* can be found on page 172.

EXERCISES

Rewrite each expression to create a new identity, then verify the identity by reversing the steps.

7. $\csc x + \cot x$

8. $\frac{\cos x - \sin x \cos x}{\cos^2 x}$

Verify that each equation is an identity.

9. $\frac{\csc^2 x\,(1 - \cos^2 x)}{\tan^2 x} = \cot^2 x$

10. $\frac{\cot x}{\sec x} - \frac{\csc x}{\tan x} = \cot x(\cos x - \csc x)$

11. $\dfrac{\sin^4 x - \cos^4 x}{\sin x \cos x} = \tan x - \cot x$

12. $\dfrac{(\sin x + \cos x)^2}{\sin x \cos x} = \csc x \sec x + 2$

SECTION 3.3 The Sum and Difference Identities

KEY CONCEPTS

The sum and difference identities can be used to

- Find exact values for nonstandard angles that are a sum or difference of two standard angles.
- Verify the cofunction identities and to rewrite a given function in terms of its cofunction.
- Find coterminal angles in $[0, 360°)$ for very large angles (the angle reduction formulas).
- Evaluate the difference quotient for $\sin x$, $\cos x$, and $\tan x$.
- Rewrite a sum as a single expression: $\cos\alpha\cos\beta + \sin\alpha\sin\beta = \cos(\alpha - \beta)$.

The sum and difference identities for sine and cosine can be remembered by noting

- For $\cos(\alpha \pm \beta)$, the function repeats and the signs alternate:
  $\cos(\alpha \pm \beta) = \cos\alpha\cos\beta \mp \sin\alpha\sin\beta$
- For $\sin(\alpha \pm \beta)$ the signs repeat and the functions alternate:
  $\sin(\alpha \pm \beta) = \sin\alpha\cos\beta \pm \cos\alpha\sin\beta$

EXERCISES

Find exact values for the following expressions using sum and difference formulas.

13. a. $\cos 75°$ **b.** $\tan\left(\dfrac{\pi}{12}\right)$

14. a. $\tan 15°$ **b.** $\sin\left(-\dfrac{\pi}{12}\right)$

Evaluate exactly using sum and difference formulas.

15. a. $\cos 109° \cos 71° - \sin 109° \sin 71°$ **b.** $\sin 139° \cos 19° - \cos 139° \sin 19°$

Rewrite as a single expression using sum and difference formulas.

16. a. $\cos(3x)\cos(-2x) - \sin(3x)\sin(-2x)$ **b.** $\sin\left(\dfrac{x}{4}\right)\cos\left(\dfrac{3x}{8}\right) + \cos\left(\dfrac{x}{4}\right)\sin\left(\dfrac{3x}{8}\right)$

Evaluate exactly using sum and difference formulas, by reducing the angle to an angle in $[0, 360°)$ or $[0, 2\pi)$.

17. a. $\cos 1170°$ **b.** $\sin\left(\dfrac{57\pi}{4}\right)$

Use a cofunction identity to write an equivalent expression for the one given.

18. a. $\cos\left(\dfrac{x}{8}\right)$ **b.** $\sin\left(x - \dfrac{\pi}{12}\right)$

19. Verify that both expressions yield the same result using sum and difference formulas.
$\tan 15° = \tan(45° - 30°)$ and $\tan 15° = \tan(135° - 120°)$.

20. Use sum and difference formulas to verify the following identity.

$$\cos\left(x + \frac{\pi}{6}\right) + \cos\left(x - \frac{\pi}{6}\right) = \sqrt{3}\cos x$$

SECTION 3.4 The Double-Angle, Half-Angle, and Product-to-Sum Identities

KEY CONCEPTS

- When multiple angle identities (identities involving $n\theta$) are used to find exact values, the terminal side of θ must be determined so the appropriate sign can be used.
- The power reduction identities for $\cos^2x$ and $\sin^2x$ are closely related to the double-angle identities, and can be derived directly from $\cos(2x) = 2\cos^2x - 1$ and $\cos(2x) = 1 - 2\sin^2x$.
- The half-angle identities can be developed from the power reduction identities by using a change of variable and taking square roots. The sign is then chosen based on the quadrant of the half angle.
- The product-to-sum and sum-to-product identities can be derived using the sum and difference formulas, and have important applications in many areas of science.

EXERCISES

Find exact values for $\sin(2\theta)$, $\cos(2\theta)$, and $\tan(2\theta)$ using the information given.

21. a. $\cos\theta = \frac{13}{85}$; θ in QIV **b.** $\csc\theta = -\frac{29}{20}$; θ in QIII

Find exact values for $\sin\theta$, $\cos\theta$, and $\tan\theta$ using the information given.

22. a. $\cos(2\theta) = -\frac{41}{841}$; θ in QII **b.** $\sin(2\theta) = -\frac{336}{625}$; θ in QII

Find exact values using the appropriate double-angle identity.

23. a. $\cos^2 22.5° - \sin^2 22.5°$ **b.** $1 - 2\sin^2\left(\frac{\pi}{12}\right)$

Find exact values for $\sin\theta$ and $\cos\theta$ using the appropriate half-angle identity.

24. a. $\theta = 67.5°$ **b.** $\theta = \frac{5\pi}{8}$

Find exact values for $\sin\left(\frac{\theta}{2}\right)$ and $\cos\left(\frac{\theta}{2}\right)$ using the given information.

25. a. $\cos\theta = \frac{24}{25}$; $0° < \theta < 360°$; θ in QIV **b.** $\csc\theta = -\frac{65}{33}$; $-90° < \theta < 0$; θ in QIV

26. Verify the equation is an identity.

$$\frac{\cos(3\alpha) - \cos\alpha}{\cos(3\alpha) + \cos\alpha} = \frac{2\tan^2\alpha}{\sec^2\alpha - 2}$$

27. Solve using a sum-to-product formula.

$$\cos(3x) + \cos x = 0$$

28. The area of an isosceles triangle (two equal sides) is given by the formula

$A = x^2\sin\left(\frac{\theta}{2}\right)\cos\left(\frac{\theta}{2}\right)$, where the equal sides have length x and the vertex angle measures $\theta°$. (a) Use this formula and the half-angle identities to find the area of an isosceles triangle with vertex angle $\theta = 30°$ and equal sides of 12 cm. (b) Use substitution and a double-angle identity to verify that $x^2\sin\left(\frac{\theta}{2}\right)\cos\left(\frac{\theta}{2}\right) = \frac{1}{2}x^2\sin\theta$, then recompute the triangle's area. Do the results match?

MIXED REVIEW

Find the value of all six trig functions using the information given.

1. $\csc\theta = \frac{\sqrt{117}}{6}$; θ in QII
2. $\tan^{-1}\left(\frac{4}{3}\right) = \theta$

Find the exact value of each expression using a sum or difference identity.

3. $\tan 255°$
4. $\cos\left(\frac{19\pi}{12}\right)$
5. Find the exact value of $2\cos^2\left(\frac{\pi}{12}\right) - 1$ using an appropriate identity.
6. Find the exact value of $2\sin\left(\frac{\pi}{8}\right)\cos\left(\frac{\pi}{8}\right)$ using an appropriate identity.

Verify that each equation is an identity.

7. $\dfrac{1 - \cos^2\theta + \sin^2\theta}{\tan^2\theta} = 1 + \cos(2\theta)$
8. $\dfrac{(\cos t + \sin t)^2}{\tan t} = \cot t + 2\cos^2 t$

Find exact values for $\sin\left(\frac{x}{2}\right)$ and $\cos\left(\frac{x}{2}\right)$ using the information given.

9. $\sin x = \frac{-6}{7.5}$; $540° < x < 630°$
10. $\sec x = \frac{11.7}{4.5}$; $0 < x < \frac{\pi}{2}$

Verify the following identities *using a sum formula.*

11. $\sin(2\alpha) = 2\sin\alpha\cos\alpha$
12. $\cos(2\alpha) = \cos^2\alpha - \sin^2\alpha$

Find the value of each expression using sum-to-product and half-angle identities (without using a calculator).

13. $\sin 172.5° - \sin 52.5°$
14. $\cos 172.5° + \cos 52.5°$

Use the product-to-sum formulas to find the exact value of

15. $\sin\left(\frac{13\pi}{24}\right)\cos\left(\frac{7\pi}{24}\right)$
16. $\sin\left(\frac{13\pi}{24}\right)\sin\left(\frac{7\pi}{24}\right)$
17. Verify the identity: $\sin(\alpha + \beta)\sin(\alpha - \beta) = \sin^2\alpha - \sin^2\beta$
18. Verify the identity: $\left[\cos\left(\frac{\theta}{2}\right) - \sin\left(\frac{\theta}{2}\right)\right]^2 = 1 - \sin\theta$

The horizontal distance an object will travel when it is projected at angle θ with initial velocity v is given by the equation $R = \frac{1}{16}v^2\sin\theta\cos\theta$.

19. Use an identity to show this equation can be written as $R = \frac{1}{32}v^2 \sin(2\theta)$.

20. Use this equation to show why the horizontal distance traveled by the object is the same for any two complementary angles.

θ

R

PRACTICE TEST

Verify each identity using fundamental identities and the method specified.

1. special products
$$\frac{(\csc x - \cot x)(\csc x + \cot x)}{\sec x} = \cos x$$

2. factoring
$$\frac{\sin^3 x - \cos^3 x}{1 + \cos x \sin x} = \sin x - \cos x$$

3. Find the value of all six trigonometric functions given $\cos\theta = \frac{48}{73}$; θ in QIV

4. Find the exact value of tan 15° using a sum or difference formula.

5. Rewrite as a single expression and evaluate: $\cos 81° \cos 36° + \sin 81° \sin 36°$.

6. Evaluate cos 1935° exactly using an angle reduction formula.

7. Use sum and difference formulas to verify
$$\sin\left(x + \frac{\pi}{4}\right) - \sin\left(x - \frac{\pi}{4}\right) = \sqrt{2}\cos x.$$

8. Find exact values for $\sin\theta$, $\cos\theta$, and $\tan\theta$ given $\cos(2\theta) = -\frac{161}{289}$; θ in QI

9. Use a double-angle identity to evaluate $2\cos^2 75° - 1$.

10. Find exact values for $\sin\left(\frac{\theta}{2}\right)$ and $\cos\left(\frac{\theta}{2}\right)$ given $\tan\theta = \frac{12}{35}$; θ in QI

11. The area of a triangle is given geometrically as $A = \frac{1}{2}$ base · height. The trigonometric formula for the triangle's area is $\boldsymbol{A = \frac{1}{2}bc\sin\alpha}$, where α is the angle formed by the sides b and c. In a certain triangle, $b = 8$, $c = 10$, and $\alpha = 22.5°$. Use the formula for A given here and a half-angle identity to find the area of the triangle in exact form.

b a α c

12. The equation $Ax^2 + Bxy + Cy^2 = 0$ can be written in an alternative form that makes it easier to graph. This is done by eliminating the mixed xy-term using the relation $\tan(2\theta) = \frac{B}{A - C}$ to find θ. We can then find values for $\sin\theta$ and $\cos\theta$, which are used in a conversion formula. Find $\sin\theta$ and $\cos\theta$ for $17x^2 + 5\sqrt{3}xy + 2y^2 = 0$, assuming 2θ in QI.

Verify that each equation is an identity.

13. $\frac{\tan\theta + \cot\theta}{\sin\theta\cos\theta} = \csc^2\theta\sec^2\theta$

14. $-2\cos^4\theta + 3\cos^2\theta - 1 = \sin^2\theta\cos(2\theta)$

Find exact values for the following expressions using sum and difference formulas.

15. $\tan(15°)$

16. $\cos(15°) - \sin(15°)$

17. Find exact values for $\sin(2\theta)$, $\cos(2\theta)$, and $\tan(2\theta)$ using the information given.

$\sin\theta = -\frac{12}{37}$; θ in QIII

18. Verify that the following equation is an identity:

$$\frac{\csc^2 x - 2}{2\cot^2 x - \csc^2 x} = 1$$

19. One of the buttons on a telephone is pressed, and the equation of the resultant wave is modeled by $y(t) = \cos(2\pi 1336t) + \cos(2\pi 941t)$. (a) Use the diagram on page 200 to determine which button on the touch-tone phone was pressed. (b) Use a sum-to-product identity to write this sum as a product.

20. Write the product as a sum using a product-to-sum identity: $2\cos(1979\pi t)\cos(439\pi t)$.

CALCULATOR EXPLORATION AND DISCOVERY

Seeing the Beats as the Beats Go On

The keystrokes shown apply to a TI-84Plus model. Please consult your manual or our Internet site for other models.

When two sound waves of slightly different frequencies are combined, the resultant wave varies periodically in amplitude over time. These amplitude pulsations are called **beats.** In this *Exploration and Discovery,* we'll look at ways to "see" the beats more clearly on a graphing calculator, by representing sound waves very simplistically as $Y_1 = \cos(mt)$ and $Y_2 = \cos(nt)$ and noting a relationship between m, n, and the number of beats in $[0, 2\pi]$. Using a sum-to-product formula, we can represent the resultant wave as a single term. For $Y_1 = \cos(12t)$ and $Y_2 = \cos(8t)$ the result is

$$\cos(12t) + \cos(8t) = 2\cos\left(\frac{12t + 8t}{2}\right)\cos\left(\frac{12t - 8t}{2}\right)$$
$$= 2\cos(10t)\cos(2t)$$

The window used and resulting graph are shown in Figures 3.16 and 3.17, and it appears that "silence" occurs four times in this interval—where the graph of the combined waves is tangent to (bounces off of) the x-axis. This indicates a total of four beats. Note the number of beats is equal to the difference $m - n$: $12 - 8 = 4$. Further experimentation will show this is not a coincidence, and this enables us to construct two additional functions that will *frame these pulsations* and make them easier to see. Since the maximum amplitude of the resulting wave is 2, we use functions of the form $\pm 2\cos\left(\frac{k}{2}x\right)$ to construct the frame, where k is the number of beats in the interval $(m - n = k)$. For $Y_1 = \cos(12t)$ and $Y_2 = \cos(8t)$, we have $k = \frac{12 - 8}{2} = 2$ and the functions we use will be $Y_2 = 2\cos(2x)$ and $Y_3 = -2\cos(2x)$ as shown in Figure 3.18. The result is shown in Figure 3.19, where the frame clearly shows the four beats or more precisely, the four moments of silence. As an alternative to this analysis, we could rewrite the expression as $2\cos(2t)\cos(10t)$ and view $A = 2\cos(2t)$ (the slower frequency) as the varying or "modulating" amplitude for $\cos(10t)$ to produce this frame.

Figure 3.16

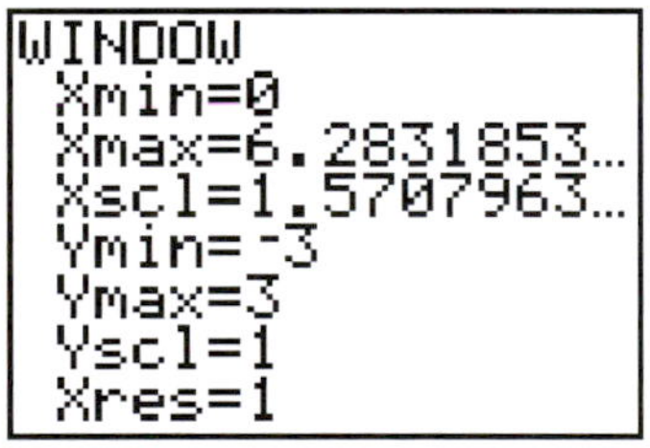

```
WINDOW
 Xmin=0
 Xmax=6.2831853…
 Xscl=1.5707963…
 Ymin=-3
 Ymax=3
 Yscl=1
 Xres=1
```

Figure 3.17

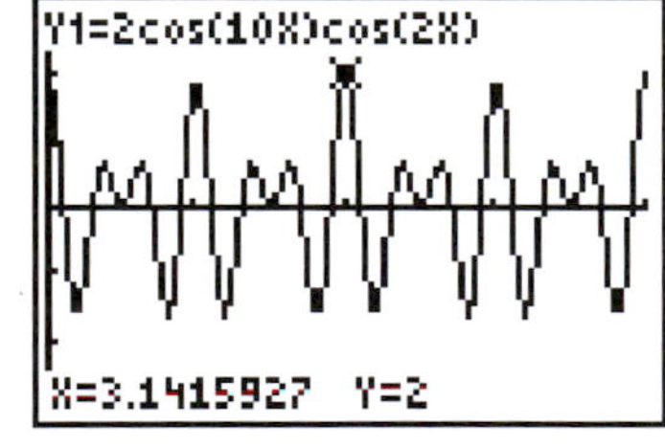

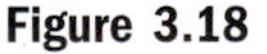

Figure 3.18

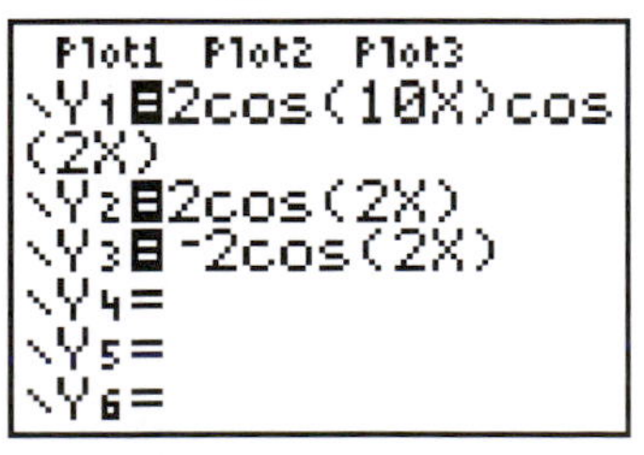

For each exercise, (a) express the sum $Y_1 + Y_2$ as a product, (b) graph Y_R on a graphing calculator for $x \in [0, 2\pi]$ and identify the number of beats in this interval, and (c) determine what value of k in $\pm 2\cos\left(\frac{k}{2}x\right)$ would be used to frame the resultant Y_R, then enter these as Y_2 and Y_3 to check the result.

Figure 3.19

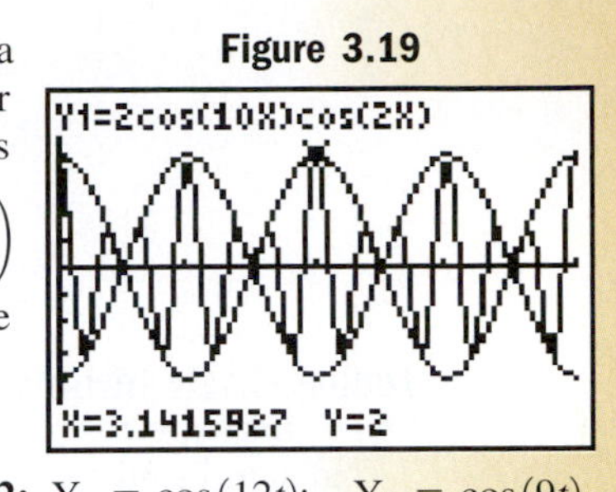

Exercise 1: $Y_1 = \cos(14t)$; $Y_2 = \cos(8t)$ **Exercise 2:** $Y_1 = \cos(12t)$; $Y_2 = \cos(9t)$

STRENGTHENING CORE SKILLS

Identities-Connections and Relationships

It is a well-known fact that information is retained longer and used more effectively when it is organized, sequential, and connected. In this *Strengthening Core Skills (SCS),* we attempt to do just that with our study of identities. In flowchart form we'll show that the entire range of identities has only two tiers, and that the fundamental identities and the sum and difference identities are really the keys to the entire range of identities. Beginning with the right triangle definition of sine, cosine, and tangent, the **reciprocal identities** and **ratio identities** are more semantic (word related) than mathematical, and the **Pythagorean identities** follow naturally from the properties of right triangles. These form the first tier.

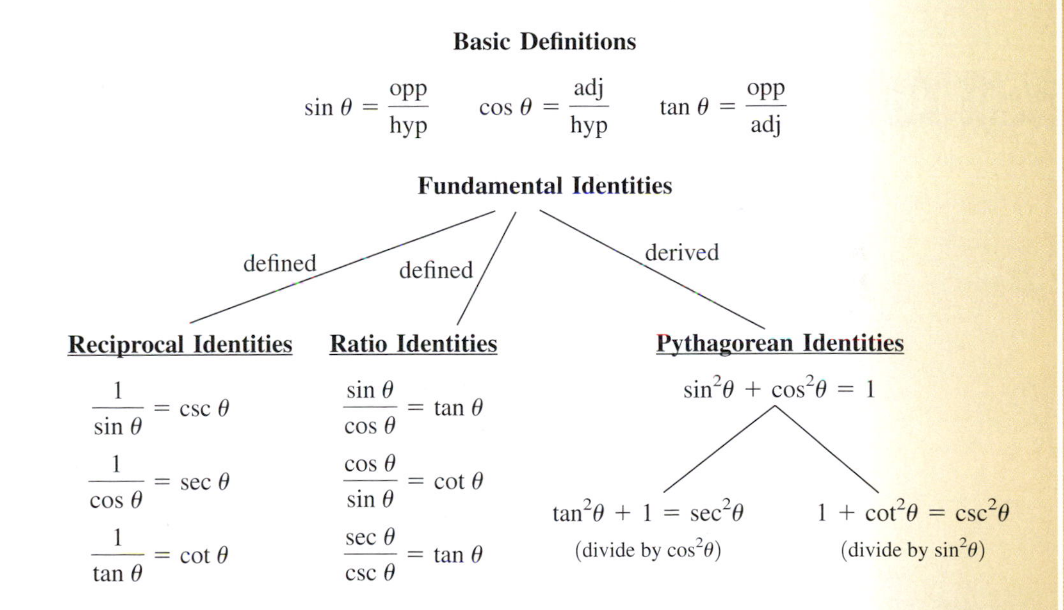

The reciprocal and ratio identities are actually *defined*, while the Pythagorean identities are *derived* from these two families. In addition, the identity $\sin^2\theta + \cos^2\theta = 1$ is the only Pythagorean identity we actually need to memorize; the other two follow by division of $\cos^2\theta$ and $\sin^2\theta$ as indicated.

In virtually the same way, the sum and difference identities for sine and cosine are the only identities that need to be memorized, as all other identities in the second tier flow from these.

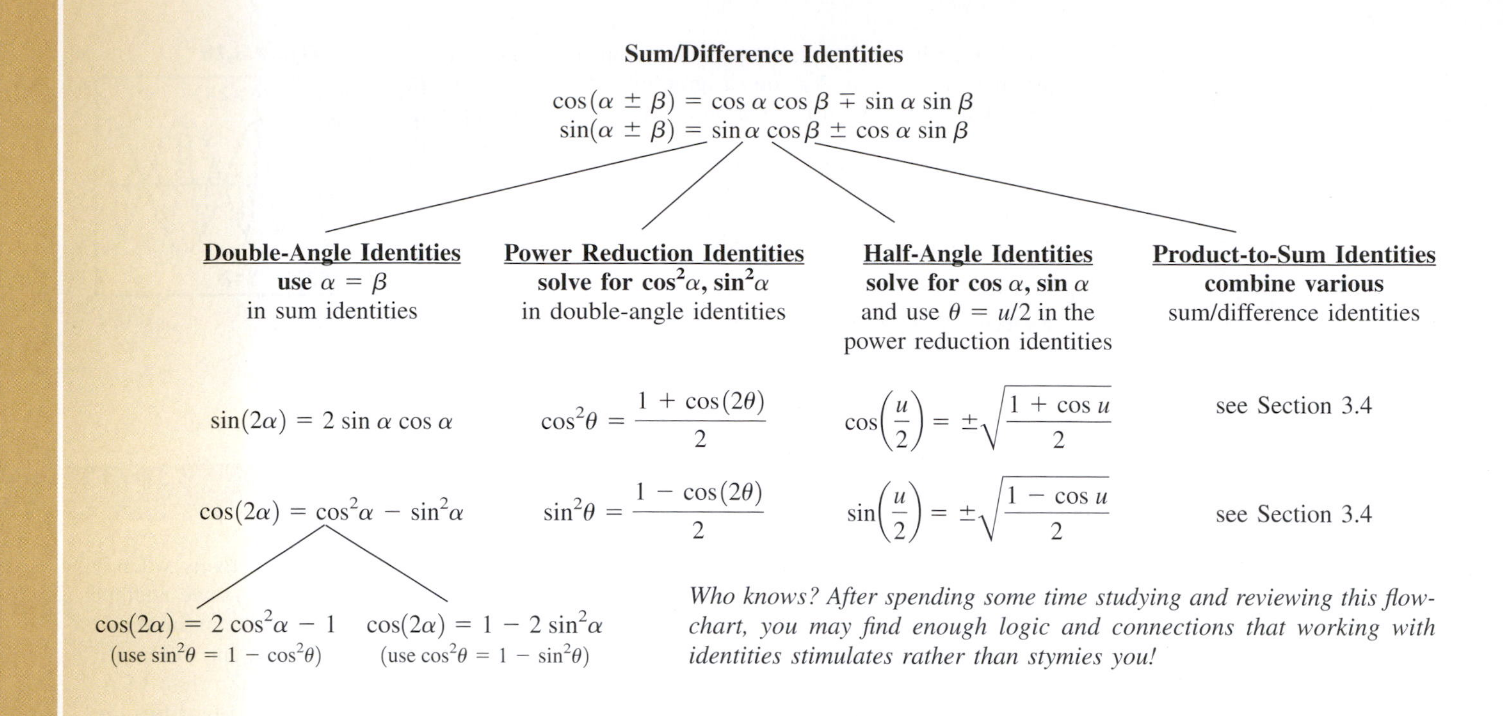

Who knows? After spending some time studying and reviewing this flowchart, you may find enough logic and connections that working with identities stimulates rather than stymies you!

Cumulative Review Chapters 1–3

1. Solve using a standard triangle.

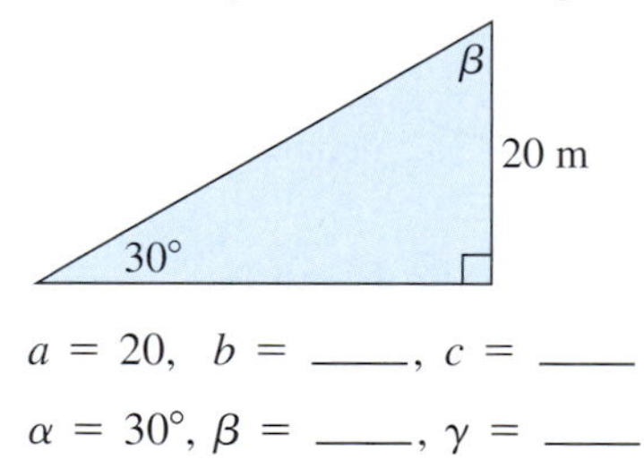

2. Solve using trigonometric ratios.

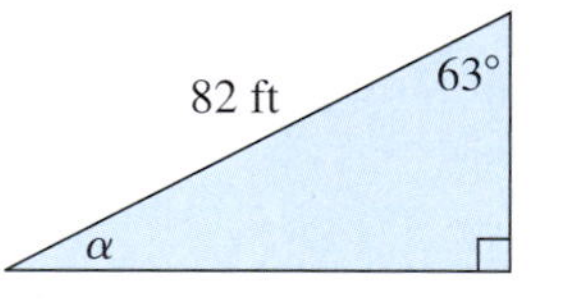

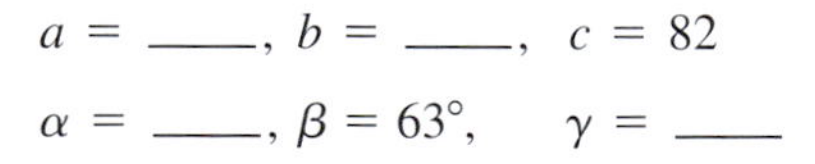

3. State the three Pythagorean identities.

4. Given θ is in QIV and $\theta_r = 32°$, find the value of θ.

5. The rollers on a conveyor belt have a radius of 6 in. and are turning at 300 rpm. How fast (in feet per second) is an object on the conveyor belt moving?

6. State the value of all six trig functions given $\tan\alpha = -\frac{3}{4}$ with $\cos\alpha > 0$.

7. Sketch the graph of $y = 3\cos\left(2x - \frac{\pi}{4}\right)$ using a reference rectangle and the *rule of fourths.*

8. Given $\cos 53° \approx 0.6$ and $\cos 37° \approx 0.8$ approximate the value of $\cos 16°$ without using a calculator.

9. Verify that the following is an identity:
$$\cos^2\left(\frac{\alpha}{2}\right) = \frac{\sec\alpha + 2 + \cos\alpha}{2\sec\alpha + 2}$$

10. Determine the equation of the function shown, given it is of the form $y = A\cos[B(t + C)] + D$.

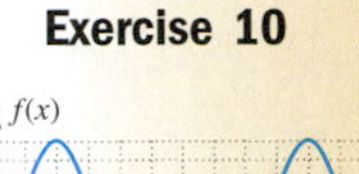

Exercise 10

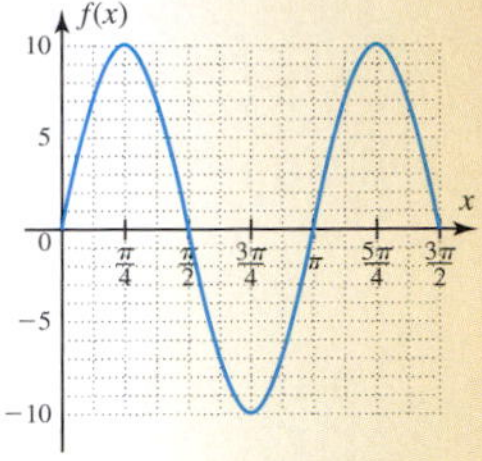

11. State the double-angle identities for sine and cosine.

12. Find $\sin(\alpha + \beta)$ and $\cos(\alpha + \beta)$ given $\cos\alpha = \frac{7}{25}$ in QIV and $\csc\beta = -\frac{13}{5}$ in QIII.

13. Use the sum identities to find exact values for $\sin 195°$ and $\cos 195°$.

14. Use a product-to-sum formula to find an exact value for $\cos\left(\frac{\pi}{12}\right)\sin\left(\frac{5\pi}{12}\right)$.

15. The approximate number of hours of daylight for Juneau, Alaska, is given month by month in the table to the right. Use the data to find an appropriate regression equation, then answer the following questions.

Month (Jan. → 1)	Daylight (hr)	Month (Jan. → 1)	Daylight (hr)
1	5.5	7	18.1
2	8	8	17
3	11	9	14.5
4	13.5	10	12.5
5	16.5	11	9.5
6	18	12	7.5

   a. Approximately how many daylight hours were there on April 15 ($t = 4.5$)?

   b. Approximate the dates between which there are over 15 hr of daylight.

16. Road Gang 52 out of Oklahoma City, Oklahoma, is responsible for rebuilding dangerous roadbeds for the rail line between Denton, Texas, and Arkansas City, Kansas. Due to favorable weather conditions, the gang can refurbish a high of 0.8 mi of track per day in June, but only 0.62 mi of track in the cold of December. (a) Use this information to build a sinusoidal equation that models the amount of track Gang 52 can rebuild each month. (b) How many months of the year can Gang 52 rebuild more than 0.75 mi of track per day?

Verify the identities.

17. $\cot x\left(\tan x - \frac{\sin x}{\cos^3 x}\right) = \tan^2 x$

18. $\cos^2 x - \csc^2 x\cos^2 x = -\frac{\cos^4 x}{\sin^2 x}$

19. The cities of Kiev and Leningrad have played significant roles in the history of Russia. Kiev is located at 50.3° N latitude and 30.3° E longitude. Leningrad is located at 59.6° N latitude and at roughly the same longitude. If the radius of the Earth is 6372 km, how many kilometers separate the two cities?

20. The planet Venus completes one orbit of the Sun in roughly 225 days. Find its linear velocity in miles per second if the radius of its orbit is approximately 67 million miles.

Chapter 4

Inverse Functions and Trigonometric Equations

Chapter Outline

Preview

This chapter will unify much of what we've learned so far, and lead us to some intriguing, sophisticated, and surprising applications of trigonometry. The definition of the trig functions (Chapter 1) helped us see and study a number of new relationships. Their graphs (Chapter 2) gave us insights into how the functions are related to each other, and enabled us to study periodic phenomena. The identities (Chapter 3) were used to simplify complex expressions and to show how trig functions often work together to model natural events. In this chapter we'll see how each of these are combined to solve trigonometric equations and aid in a study of additional trigonometric models.

4.1 One-to-One and Inverse Functions

LEARNING OBJECTIVES

In Section 4.1 you will learn how to:

A. Identify one-to-one functions

B. Investigate inverse functions using ordered pairs

C. Find inverse functions using an algebraic method

D. Graph a function and its inverse on the same grid

INTRODUCTION

Throughout the algebra sequence, inverse operations are used to solve basic equations. To solve the equation $2x - 3 = -8$ we add 3 to both sides, then divide by 2 since subtraction and addition are inverse operations, as are division and multiplication. In this section, we introduce the idea of an *inverse function*, or one function that "undoes" the operations of another.

POINT OF INTEREST

In the old children's bedtime story, Hansel and Gretel lay out a trail of bread crumbs as they walk into the forest, hoping to eventually follow the trail back home. Although they were foiled in this attempt (birds ate the crumbs), the idea of retracing your steps to get home is a familiar theme. In a related way, an inverse function helps us to "find our way back" to the variable, and thereby solve equations.

A. Identifying One-to-One Functions

From your earlier work you may recall that if every vertical line crosses the graph of a relation in at most one point, the relation is a function. In other words, each first coordinate x must correspond to only one second coordinate y. Consider the graphs of $y = 2x + 3$ and $y = x^2$ given in Figures 4.1 and 4.2, respectively.

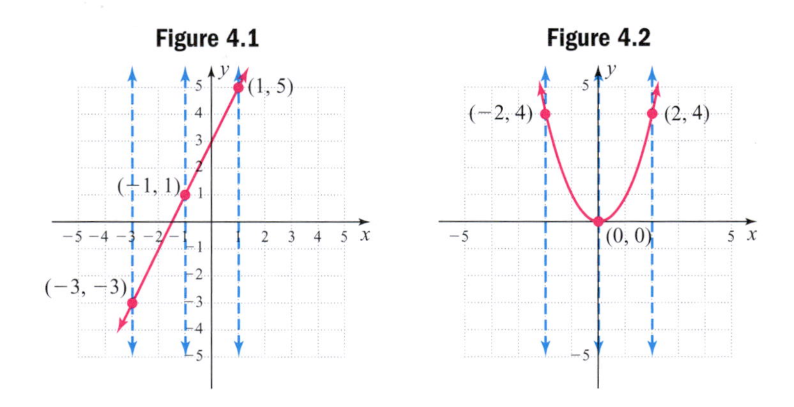

The dashed, vertical lines indicated on each graph clearly show that each x corresponds to only one y. For $y = 2x + 3$, the points $(-3, -3)$, $(-1, 1)$, and $(1, 5)$ are indicated, while for $y = x^2$ we have $(-2, 4)$, $(0, 0)$, and $(2, 4)$. Both are functions, but the points from $y = 2x + 3$ have one characteristic those from $y = x^2$ do not—*each second coordinate y corresponds to a unique first coordinate x*. Note the output "4" from the range of $y = x^2$ corresponds to both -2 and 2 from the domain. If each element from the range of a function corresponds to a unique element of the domain, the function is said to be **one-to-one.** Identifying one-to-one functions is an important part of finding inverse functions.

Figure 4.3

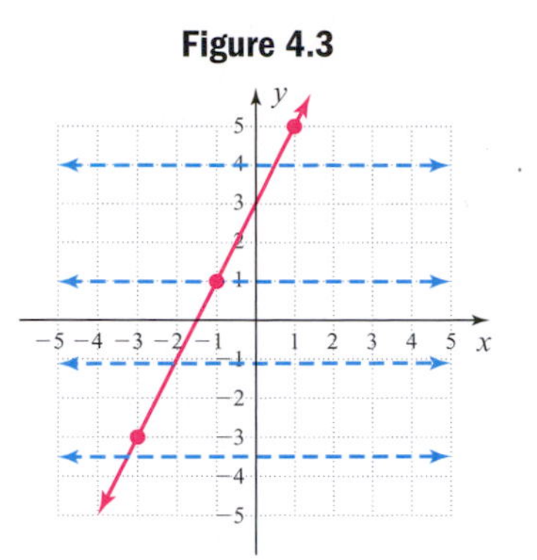

Figure 4.4

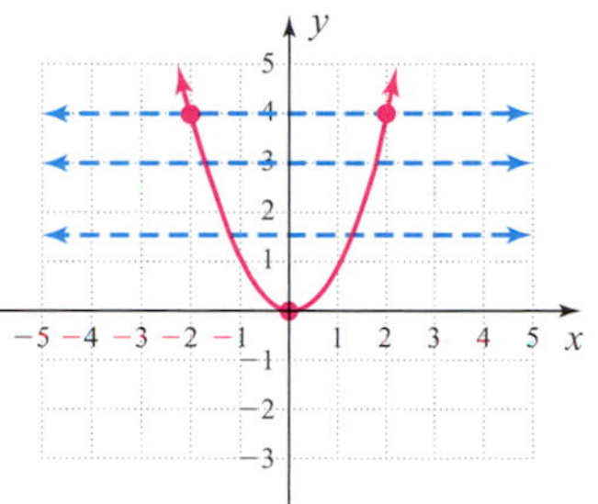

ONE-TO-ONE FUNCTIONS

A function f with domain D and range R is said to be one-to-one if no two elements in D correspond to the same element in R:
If $f(x_1) = f(x_2)$, then $x_1 = x_2$. If $f(x_1) \neq f(x_2)$, then $x_1 \neq x_2$.

From this definition we conclude the graph of a one-to-one function must not only pass the vertical line test to show that each x corresponds to only one y, it must also pass a **horizontal line test** to show that each y also corresponds to only one x.

HORIZONTAL LINE TEST

If every horizontal line intersects the graph of a function in at most one point, the function is one-to-one.

Notice the graph of $y = 2x + 3$ (Figure 4.3) passes the horizontal line test, while the graph of $y = x^2$ (Figure 4.4) does not.

EXAMPLE 1 Use the horizontal line test to determine whether each graph given here is the graph of a one-to-one function.

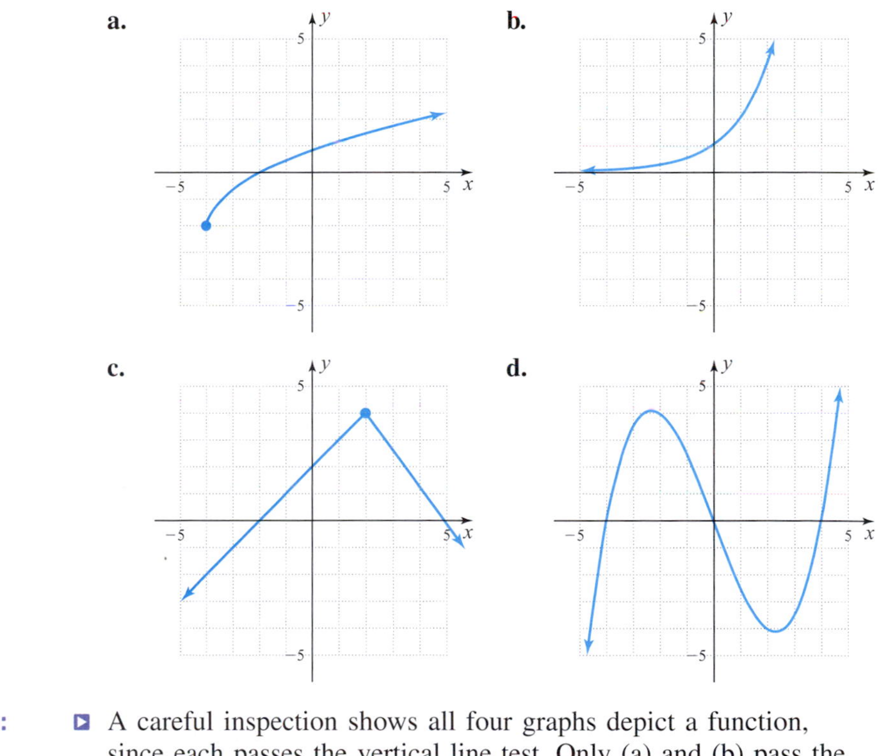

Solution: A careful inspection shows all four graphs depict a function, since each passes the vertical line test. Only (a) and (b) pass the horizontal line test and are *one-to-one* functions.

NOW TRY EXERCISES 7 THROUGH 28

If the function is given in ordered pair form, we simply check to see that no given second coordinate is paired with more than one first coordinate (see exercises 16 through 19).

B. Inverse Functions and Ordered Pairs

Consider the linear function $f(x) = x + 3$. This function simply "adds 3 to the input value," and some of the ordered pairs generated are $(-7, -4)$, $(-5, -2)$, $(0, 3)$, and $(2, 5)$. On an intuitive level, we might say the inverse of this function would have to "undo" the addition of 3, and $g(x) = x - 3$ is a likely candidate. Some ordered pairs for g are $(-4, -7)$, $(-2, -5)$, $(3, 0)$, and $(5, 2)$. Note that if you interchange the x- and y-coordinates of f, you get exactly the coordinates of the points from g! This shows how the second function "undoes" the operations of the first and vice versa, an observation that will help lead us to the general definition of an **inverse function.** For now, if f is a one-to-one function with ordered pairs (a, b), then the inverse of f is the one-to-one function with ordered pairs of the form (b, a). The inverse function is denoted f^{-1} and is read "f inverse," or "the inverse of f." For the function $f(x) = x + 3$ as given, we have $f^{-1}(x) = x - 3$.

CAUTION

The notation $f^{-1}(x)$ is simply a way of denoting an inverse function and has nothing to do with exponential properties. In particular, $f^{-1}(x)$ does *not* mean $\frac{1}{f(x)}$.

It's important to note that if a function is not one-to-one, no inverse function exists since the interchange of x- and y-coordinates will result in a nonfunction. For example, interchanging the coordinates of $(-2, 4)$, $(0, 0)$, and $(2, 4)$ from $y = x^2$ results in $(4, -2)$, $(0, 0)$, and $(4, 2)$, and we have one x-value being mapped to two y-values, in violation of the function definition.

EXAMPLE 2 Find the inverse of each one-to-one function given:

a. $f(x) = \{(-4, 13), (-1, 7), (0, 5), (2, 1), (5, -5), (8, -11)\}$

b. $p(x) = \frac{2}{3}x$

Solution:

a. The inverse function for part (a) can be found by simply interchanging the x- and y-coordinates: $f^{-1}(x) = \{(13, -4), (7, -1), (5, 0), (1, 2), (-5, 5), (-11, 8)\}$.

b. For part (b), we reason the inverse function for p must undo the multiplication of $\frac{2}{3}$ and $q(x) = \frac{3}{2}x$ is a good possibility. Creating a few ordered pairs for p yields $(-3, -2)$, $(-1, -\frac{2}{3})$, $(0, 0)$, $(2, \frac{4}{3})$, and $(6, 4)$. After interchanging the x- and y-coordinates, we check to see if $(-2, -3)$, $(-\frac{2}{3}, -1)$, $(0, 0)$, $(\frac{4}{3}, 2)$, and $(4, 6)$ satisfy q. Since this is the case, we assume $q(x) = \frac{3}{2}x$ is a likely candidate for $p^{-1}(x)$, deferring a formal proof until later in the section.

NOW TRY EXERCISES 29 THROUGH 40

One important consequence of the relationship between a function and its inverse is that the *domain* of the function becomes the range of the inverse, and the *range* of the function becomes the domain of the inverse (since all input and output values are interchanged).

INVERSE FUNCTIONS

Given f is a one-to-one function with domain D and range R, the inverse function is denoted f^{-1} and has domain R and range D, where $f^{-1}(y) = x$ implies $f(x) = y$, and $f(x) = y$ implies $f^{-1}(y) = x$, for all y in R.

Using this definition, we more clearly see that if $f(-4) = 13$, as in Example 2(a), $f^{-1}(13) = -4$.

C. Finding Inverse Functions Using an Algebraic Method

The fact that interchanging x- and y-values helps determine an inverse function can be generalized to develop an **algebraic method** for finding inverses. Instead of interchanging specific x- and y-values, we actually *interchange the x and y variables* themselves, then solve the equation for y. The process is summarized here.

ALGEBRAIC METHOD FOR FINDING THE INVERSE OF A ONE-TO-ONE FUNCTION

1. Use y instead of $f(x)$.
2. Interchange x and y.
3. Solve the equation for y.
4. The result gives the inverse function: substitute $f^{-1}(x)$ for y.

EXAMPLE 3 Use the algebraic method to find the inverse function for $f(x) = \sqrt[3]{x+5}$.

Solution:

$$f(x) = \sqrt[3]{x+5} \quad \text{given function}$$
$$y = \sqrt[3]{x+5} \quad \text{use } y \text{ instead of } f(x)$$
$$x = \sqrt[3]{y+5} \quad \text{interchange } x \text{ and } y$$
$$x^3 = y + 5 \quad \text{cube both sides}$$
$$x^3 - 5 = y \quad \text{solve for } y$$
$$x^3 - 5 = f^{-1}(x) \quad \text{the result is } f^{-1}(x)$$

For $f(x) = \sqrt[3]{x+5}$, $f^{-1}(x) = x^3 - 5$.

NOW TRY EXERCISES 41 THROUGH 48

Actually, there is a conclusive way to *prove* that one function is the inverse of another. Consider the result of Example 3, where we saw that for $f(x) = \sqrt[3]{x+5}$, the inverse function was $f^{-1}(x) = x^3 - 5$. Substituting 3 into f gives 2, and substituting 2 into f^{-1} returns us to 3. Along these same lines, substituting an arbitrary value v in f will yield $\sqrt[3]{v+5}$, and substituting $\sqrt[3]{v+5}$ into f^{-1} should return us to v. This indicates that by **composing** f and f^{-1}, we can conclusively verify whether or not one function is the inverse of another (see Appendix IV for a review of composition, if needed).

INVERSE FUNCTIONS

If f is a one-to-one function, then the inverse of f is the function f^{-1} such that $(f \circ f^{-1})(x) = x$ and $(f^{-1} \circ f)(x) = x$.
Note the composition must be verified both ways.

EXAMPLE 4 Use the algebraic method to find the inverse function for $f(x) = \sqrt{x + 2}$. Then verify that you've found the correct inverse.

Solution: We know f is a square root function, with domain $x \in [-2, \infty)$ and range $y \in [0, \infty)$. This is important since the *domain and range values will be interchanged for the inverse function*. The domain of f^{-1} will be $x \in [0, \infty)$ and its range $y \in [-2, \infty)$.

$f(x) = \sqrt{x + 2}$	given function; $x \geq -2$
$y = \sqrt{x + 2}$	use y instead of $f(x)$
$x = \sqrt{y + 2}$	interchange x and y
$x^2 = y + 2$	solve for y (square both sides)
$x^2 - 2 = y$	subtract 2
$f^{-1}(x) = x^2 - 2$	the result is $f^{-1}(x)$; D: $x \in [0, \infty)$, R: $y \in [-2, \infty)$

Verify:

$(f \circ f^{-1})(x) = f[f^{-1}(x)]$	$f^{-1}(x)$ is an input for f
$= \sqrt{f^{-1}(x) + 2}$	f adds 2 to inputs, then takes square root
$= \sqrt{(x^2 - 2) + 2}$	substitute $x^2 - 2$ for $f^{-1}(x)$
$= \sqrt{x^2}$	simplify
$= x$ ✓	since the domain of $f^{-1}(x)$ is $x \in [0, \infty)$

Verify:

$(f^{-1} \circ f)(x) = f^{-1}[f(x)]$	$f(x)$ is an input for f^{-1}
$= [f(x)]^2 - 2$	f^{-1} squares inputs, then subtracts 2
$= [\sqrt{x + 2}]^2 - 2$	$f(x) = \sqrt{x + 2}$
$= x + 2 - 2$	simplify
$= x$ ✓	result

NOW TRY EXERCISES 49 THROUGH 74

Figure 4.5

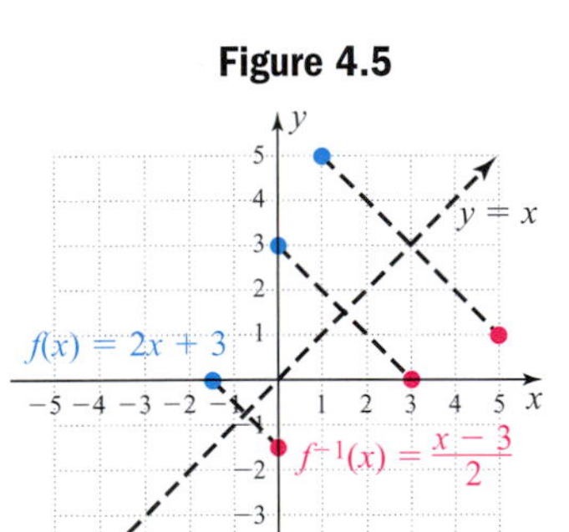

Figure 4.6

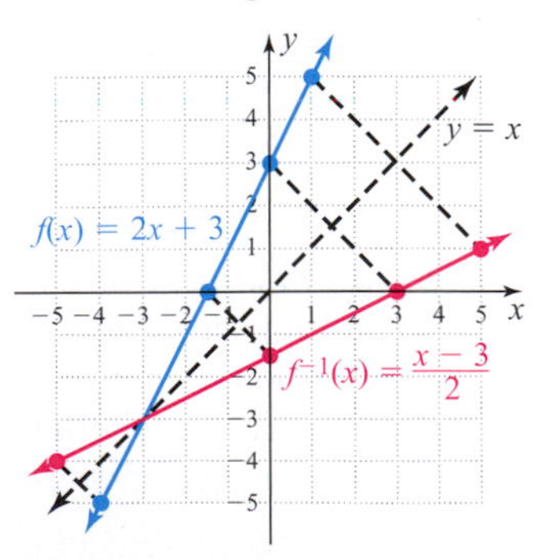

D. The Graph of a Function and Its Inverse

When a function and its inverse are graphed on the same grid, an interesting and useful relationship is noted—they are reflections across the line $y = x$ (the identity function). Consider the function $f(x) = 2x + 3$, and its inverse function $f^{-1}(x) = \frac{x - 3}{2} = \frac{1}{2}x - \frac{3}{2}$. The intercepts of f are $(0, 3)$ and $(-\frac{3}{2}, 0)$ and the points $(-4, -5)$ and $(1, 5)$ are on the graph. The intercepts for f^{-1} are $(0, -\frac{3}{2})$ and $(3, 0)$ with both $(-5, -4)$ and $(5, 1)$ on its graph (note the interchange of coordinates once again). When these points are plotted on a coordinate grid (Figure 4.5), we see they are symmetric to the line $y = x$. When both graphs are drawn (Figure 4.6), this relationship is seen even more clearly, with the graphs intersecting on the line of symmetry at $(-3, -3)$.

EXAMPLE 5 In Example 4, we found the inverse function for $f(x) = \sqrt{x+2}$ was $f^{-1}(x) = x^2 - 2, x \geq 0$. Plot these functions on the same grid and comment on how the graphs are related. Also state where they intersect.

Solution: The graph of f is a square root function with the node at $(-2, 0)$, a y-intercept of $(0, \sqrt{2})$, and an x-intercept of $(-2, 0)$ (Figures 4.7 and 4.8 in blue). The graph of $x^2 - 2, x \geq 0$ is the right-hand branch of a parabola, with y-intercept at $(0, -2)$ and an x-intercept at $(\sqrt{2}, 0)$ (Figures 4.7 and 4.8 in red).

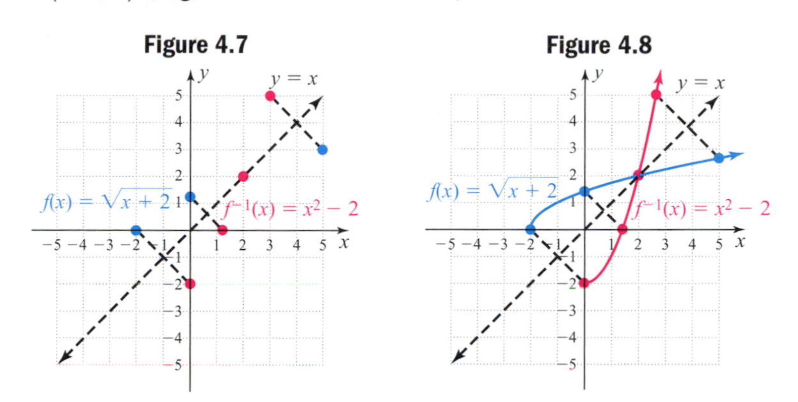

Their graphs are symmetric to the line $y = x$ and intersect on the line of symmetry at $(2, 2)$.

NOW TRY EXERCISES 75 THROUGH 82

EXAMPLE 6 Given the graph shown in Figure 4.9, use the grid in Figure 4.10 to draw a graph of the inverse function.

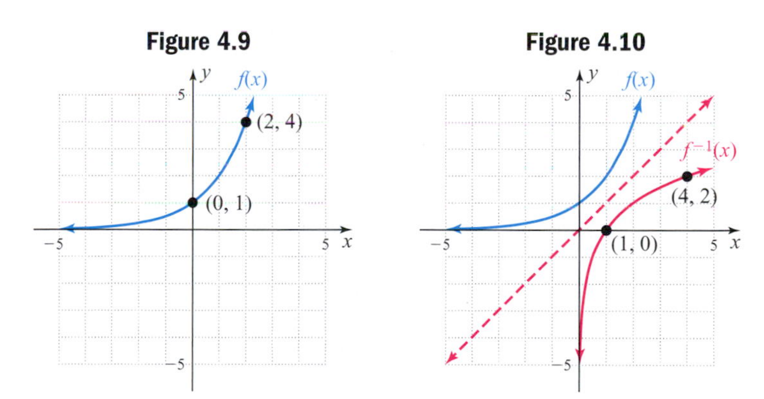

Solution: From the graph, the domain of f appears to be $x \in \mathbb{R}$ and the range is $y \in (0, \infty)$. This means the domain of f^{-1} will be $x \in (0, \infty)$ and the range will be $y \in \mathbb{R}$. The points $(0, 1)$, $(1, 2)$, and $(2, 4)$ seem to be on the graph of f. To sketch f^{-1}, draw the line $y = x$, interchange the x- and y-coordinates of the selected points, and use the domain and range boundaries as a guide. The resulting graph is that of f^{-1} (shown in red).

NOW TRY EXERCISES 83 THROUGH 88

A summary of important points is given here.

FUNCTIONS AND INVERSE FUNCTIONS

1. If a function passes the horizontal line test, it is a one-to-one function.
2. If a function f is one-to-one, the inverse function f^{-1} exists.
3. The domain of f^{-1} is the range of f, and the range of f^{-1} is the domain of f.
4. For a one-to-one function f and its inverse function f^{-1}, $(f \circ f^{-1})(x) = x$ and $(f^{-1} \circ f)(x) = x$.
5. The graphs of f and f^{-1} are symmetric with respect to the line $y = x$.

TECHNOLOGY HIGHLIGHT

Using a Graphing Calculator to Investigate Inverse Functions

The keystrokes shown apply to a TI-84 Plus model. Please consult your manual or our Internet site for other models.

Many of the important points from this section can be illustrated and verified using a graphing calculator. To begin, enter $Y_1 = x^3$ and $Y_2 = \sqrt[3]{x}$ (which are clearly inverse functions) on the Y= screen, then press ZOOM **4:ZDecimal** to graph these equations on a friendly window. The vertical and horizontal propeller functions appear on the screen and seem to be reflections across the line $y = x$ as expected. To verify, use the TABLE feature with the inputs $x = 2$ and $x = 8$. As illustrated in Figure 4.11, the calculator shows the point (2, 8) is on the graph of Y_1, and the point (8, 2) is on the graph of Y_2. As another check, we can have the calculator locate points of intersection. Recall this is done using the keystrokes 2nd TRACE (CALC) **5**, moving the cursor to a location near the desired point of intersection, then pressing ENTER ENTER ENTER. As shown in Figure 4.12, the graphs intersect at (1, 1), which is clearly on the line $y = x$. Finally, we could just return to the Y= screen, enter $Y_3 = x$ and GRAPH all three functions. The line $y = x$ shows a beautiful symmetry between the graphs. Work through the following exercises, then use a graphing calculator to check your results as illustrated in this *Technology Highlight.*

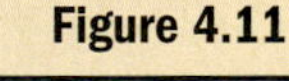
Figure 4.11

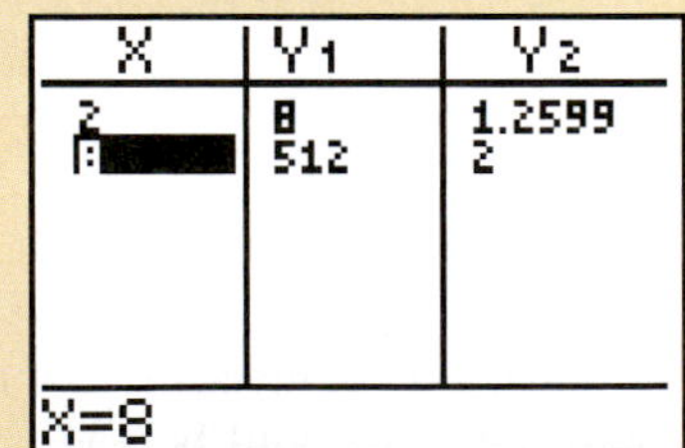

Figure 4.12

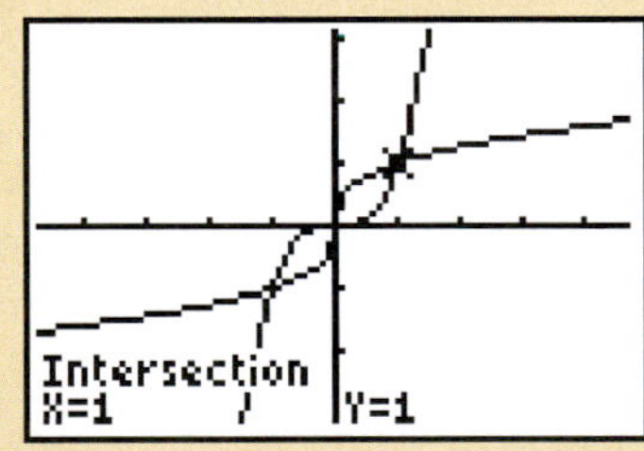

Exercise 1: Given $f(x) = 2x + 1$, find the inverse function $f^{-1}(x)$, then verify they are inverses by (a) using ordered pairs and (b) showing the point of intersection is on the line $y = x$.

Exercise 2: Given $f(x) = x^2 + 1$; $x \geq 0$, find the inverse function $f^{-1}(x)$, then verify they are inverses by (a) using ordered pairs and (b) showing each is a reflection of the other across the line $y = x$.

4.1 EXERCISES

CONCEPTS AND VOCABULARY

Fill in each blank with the appropriate word or phrase. Carefully reread the section if needed.

1. A function is one-to-one if each ______ coordinate corresponds to exactly ______ first coordinate.

2. If every ______ line intersects the graph of a function in at most ______ point, the function is one-to-one.

3. A certain function is defined by the ordered pairs $(-2, -11)$, $(0, -5)$, $(2, 1)$, and $(4, 19)$. The inverse function is ______________.

4. To find f^{-1} using the algebraic method, we (1) use ______ instead of $f(x)$, (2) ______ x and y, (3) ______ for y and replace y with $f^{-1}(x)$.

5. State true or false and explain why: *To show that g is the inverse function for f, simply show that* $(f \circ g)(x) = x$. Include an example in your response.

6. Discuss/explain why no inverse function exists for $f(x) = (x + 3)^2$ and $g(x) = \sqrt{4 - x^2}$. How would the domain of each function have to be restricted to allow for an inverse function?

DEVELOPING YOUR SKILLS

Determine whether each graph given is the graph of a one-to-one function. If not, give examples of how the definition of one-to-oneness is violated.

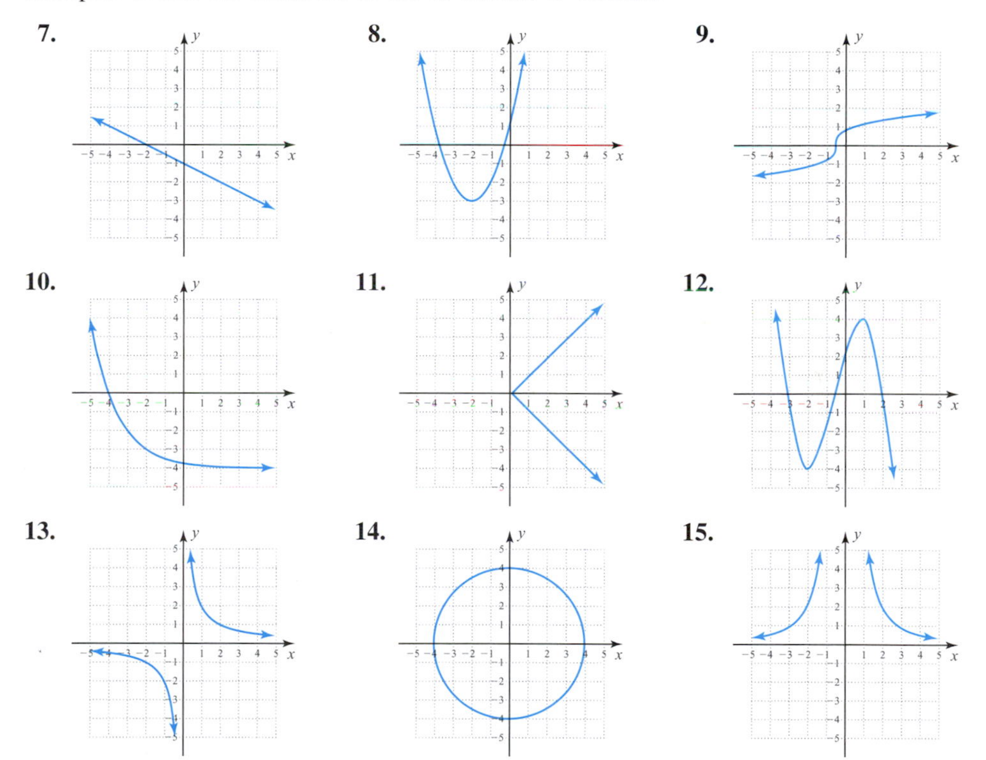

Determine whether the functions given are one-to-one. If not, state why.

16. $\{(-7, 4), (-1, 9), (0, 5), (-2, 1), (5, -5)\}$

17. $\{(9, 1), (-2, 7), (7, 4), (3, 9), (2, 7)\}$

18. $\{(-6, 1), (4, -9), (0, 11), (-2, 7), (-4, 5), (8, 1)\}$

19. $\{(-6, 2), (-3, 7), (8, 0), (12, -1), (2, -3), (1, 3)\}$

Determine if the functions given are one-to-one by noting the function family to which each belongs and mentally picturing the shape of the graph. If a function is not one-to-one, discuss how the definition of one-to-oneness is violated.

20. $f(x) = 3x - 5$ **21.** $g(x) = (x + 2)^3 - 1$ **22.** $h(x) = -|x - 4| + 3$

23. $p(t) = 3t^2 + 5$ **24.** $s(t) = \sqrt{2t - 1} + 5$ **25.** $r(t) = \sqrt[3]{t + 1} - 2$

26. $y = 3$ **27.** $y = -2x$ **28.** $y = x$

For Exercises 29 to 32, find the inverse function of the one-to-one functions given.

29. $f(x) = \{(-2, 1), (-1, 4), (0, 5), (2, 9), (5, 15)\}$

30. $g(x) = \{(-2, 30), (-1, 11), (0, 4), (1, 3), (2, 2)\}$

31. $v(x)$ is defined by the ordered pairs shown.

32. $w(x)$ is defined by the ordered pairs shown.

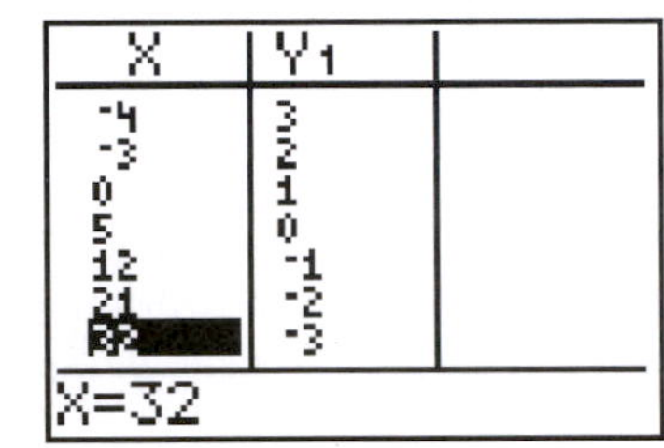

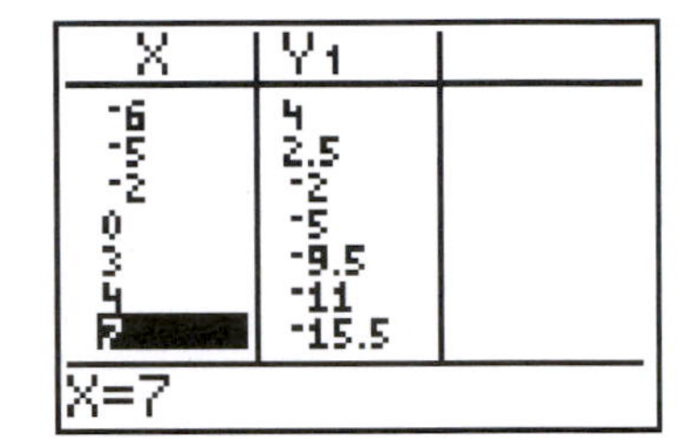

Determine a likely candidate for the inverse function by reasoning and test points.

33. $f(x) = x + 5$ **34.** $g(x) = x - 4$ **35.** $p(x) = -\frac{4}{5}x$ **36.** $r(x) = \frac{3}{4}x$

37. $f(x) = 4x + 3$ **38.** $g(x) = 5x - 2$ **39.** $Y_1 = \sqrt[3]{x - 4}$ **40.** $Y_2 = \sqrt[3]{x + 2}$

Find the inverse of each function given, then compute at least five ordered pairs and check the result. Note that choices of ordered pairs will vary.

41. $f(x) = 2x + 7$ **42.** $f(x) = 5x - 4$ **43.** $f(x) = \sqrt{x - 2}$ **44.** $f(x) = \sqrt{x + 3}$

45. $f(x) = x^2 + 3;\ x \geq 0$ **46.** $f(x) = x^2 - 4;\ x \geq 0$ **47.** $f(x) = x^3 + 1$ **48.** $f(x) = x^3 - 2$

For each function $f(x)$ given, prove (using a composition) that $g(x) = f^{-1}(x)$.

49. $f(x) = -2x + 5,\ g(x) = \dfrac{x - 5}{-2}$

50. $f(x) = 3x - 4,\ g(x) = \dfrac{x + 4}{3}$

51. $f(x) = \sqrt[3]{x + 5},\ g(x) = x^3 - 5$

52. $f(x) = \sqrt{<x - 4},\ g(x) = x^3 + 4$

53. $f(x) = \frac{2}{3}x - 6,\ g(x) = \frac{3}{2}x + 9$

54. $f(x) = \frac{4}{5}x + 6,\ g(x) = \frac{5}{4}x - \frac{15}{2}$

55. $f(x) = x^2 - 3;\ x \geq 0,\ g(x) = \sqrt{x + 3}$

56. $f(x) = x^2 + 8;\ x \geq 0,\ g(x) = \sqrt{x - 8}$

Find the inverse of each function $f(x)$ given, then prove (by composition) your inverse function is correct. Note the domain of f is all real numbers.

57. $f(x) = 3x - 5$ **58.** $f(x) = 5x + 4$ **59.** $f(x) = \dfrac{x - 5}{2}$ **60.** $f(x) = \dfrac{x + 4}{3}$

61. $f(x) = \frac{1}{2}x - 3$ **62.** $f(x) = \frac{2}{3}x + 1$ **63.** $f(x) = x^3 + 3$ **64.** $f(x) = x^3 - 4$

65. $f(x) = \sqrt[3]{2x + 1}$ **66.** $f(x) = \sqrt[3]{3x - 2}$ **67.** $f(x) = \dfrac{(x - 1)^3}{8}$ **68.** $f(x) = \dfrac{(x + 3)^3}{-27}$

Find the inverse of each function given, then prove (by composition) your inverse function is correct. Note the implied domain of each function and use it to state any necessary restrictions on the inverse.

69. $f(x) = \sqrt{3x + 2}$ **70.** $g(x) = \sqrt{2x - 5}$ **71.** $p(x) = 2\sqrt{x - 3}$

72. $q(x) = 4\sqrt{x + 1}$ **73.** $v(x) = x^2 + 3; x \geq 0$ **74.** $w(x) = x^2 - 1; x \geq 0$

Plot each function $f(x)$ and its inverse $f^{-1}(x)$ on the same grid and "dash-in" the line $y = x$. Note how the graphs are related. Then verify the "inverse function" relationship using a composition.

75. $f(x) = 4x + 1; f^{-1}(x) = \frac{x - 1}{4}$

76. $f(x) = 2x - 7; f^{-1}(x) = \frac{x + 7}{2}$

77. $f(x) = \sqrt[3]{x + 2}; f^{-1}(x) = x^3 - 2$

78. $f(x) = \sqrt[3]{x - 7}; f^{-1}(x) = x^3 + 7$

79. $f(x) = 0.2x + 1; f^{-1}(x) = 5x - 5$

80. $f(x) = \frac{2}{9}x + 4; f^{-1}(x) = \frac{9}{2}x - 18$

81. $f(x) = (x + 2)^2; x \geq -2;$
$f^{-1}(x) = \sqrt{x} - 2$

82. $f(x) = (x - 3)^2; x \geq 3;$
$f^{-1}(x) = \sqrt{x} + 3$

Determine the domain and range for each function whose graph is given, and use this information to state the domain and range of the inverse function. Then sketch in the line $y = x$, estimate the location of two or more points on the graph, and use these to graph $f^{-1}(x)$ on the same grid.

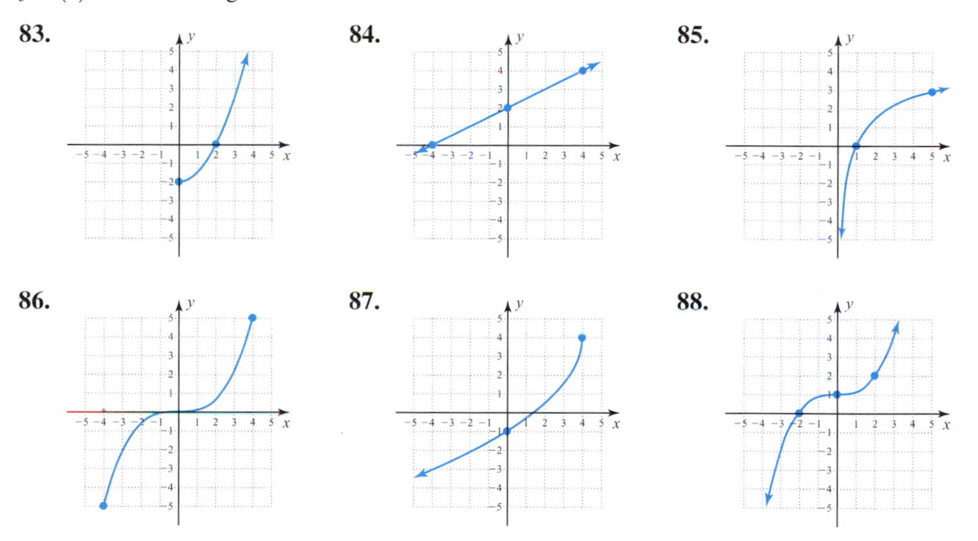

WORKING WITH FORMULAS

89. The height of a projected image: $f(x) = \frac{1}{2}x - 8.5$

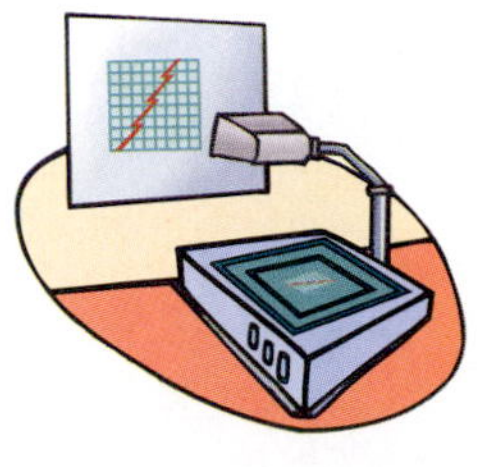

The height of an image projected on a screen by an overhead projector is given by the formula shown, where $f(x)$ represents the actual height of the image on the projector (in centimeters) and x is the distance of the projector from the screen (in centimeters). (a) When the projector is 80 cm from the screen, how large is the image? (b) Show that the inverse function is $f^{-1}(x) = 2x + 17$, then input your answer from part (a) and comment on the result. What information does the inverse function give?

90. The radius of a sphere: $r(V) = \sqrt[3]{\frac{3V}{4\pi}}$

In generic form, the radius of a sphere is given by the formula shown, where $r(V)$ represents the radius and V represents the volume of the sphere in cubic units. (a) If a weather balloon that is roughly spherical holds 14,130 in^3 of air, what is the radius of the balloon (use $\pi \approx 3.14$)? (b) Show that the inverse function is $f^{-1}(r) = \frac{4}{3}\pi r^3$, then input your answer from part (a) and comment on the result. What information does the inverse function give?

APPLICATIONS

91. Temperature and altitude: The temperature (in degrees Fahrenheit) at a given altitude can be approximated by the function $f(x) = -\frac{7}{2}x + 59$, where $f(x)$ represents the temperature and x represents the altitude in thousands of feet. (a) What is the approximate temperature at an altitude of 35,000 ft (normal cruising altitude for commercial airliners)? (b) Find $f^{-1}(x)$, then input your answer from part (a) and comment on the result. (c) If the temperature outside a weather balloon is $-18°$F, what is the approximate altitude of the balloon?

92. Fines for speeding: In some localities, there is a set formula to determine the amount of a fine for exceeding posted speed limits. Suppose the amount of the fine for exceeding a 50 mph speed limit was given by the function $f(x) = 12x - 560$ where $f(x)$ represents the fine in dollars for a speed of x mph. (a) What is the fine for traveling 65 mph through this speed zone? (b) Find $f^{-1}(x)$, then input your answer from part (a) and comment on the result. (c) If a fine of \$172 were assessed, how fast was the driver going through this speed zone?

93. Effect of gravity: Due to the effect of gravity, the distance an object has fallen after being dropped is given by the function $f(x) = 16x^2; x \geq 0$, where $f(x)$ represents the distance in feet after x sec. (a) How far has the object fallen 3 sec after it has been dropped? (b) Find $f^{-1}(x)$, then input your answer from part (a) and comment on the result. (c) If the object is dropped from a height of 784 ft, how many seconds until it hits the ground (stops falling)?

94. Area and radius: In generic form, the area of a circle is given by $f(r) = \pi r^2$, where $f(r)$ represents the area in square units for a circle with radius r. (a) A pet dog is tethered to a stake in the backyard. If the tether is 10 ft long, how much area does the dog have to roam (use $\pi \approx 3.14$)? (b) Find $f^{-1}(r)$, then input your answer from part (a) and comment on the result. (c) If the owners want to allow the dog 1256 ft^2 of area to live and roam, how long a tether should be used?

95. Volume of a cone: In generic form, the volume of an equipoise cone (height equal to radius) is given by $f(h) = \frac{1}{3}\pi h^3$, where $f(h)$ represents the volume in units^3 and h represents the height of the cone. (a) Find the volume of such a cone if $r = 30$ ft (use $\pi \approx 3.14$). (b) Find $f^{-1}(h)$, then input your answer from part (a) and comment on the result. (c) If the volume of water in the cone is 763.02 ft^3, how deep is the water at its deepest point?

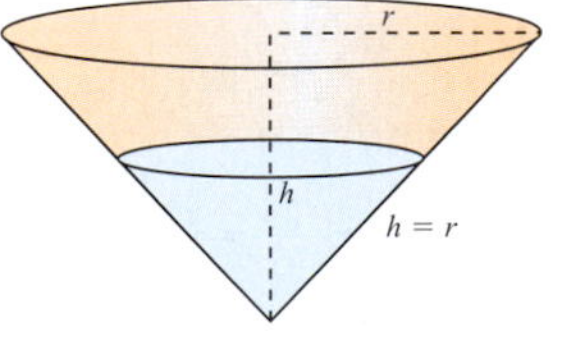

96. **Wind power:** The power delivered by a certain wind-powered generator can be modeled by the function $f(x) = \frac{x^3}{2500}$, where $f(x)$ is the horsepower (hp) delivered by the generator and x represents the speed of the wind in miles per hour. (a) Use the model to determine how much horsepower is generated by a 30 mph wind. (b) The person monitoring the output of the generators (wind generators are usually erected in large numbers) would like a function that gives the wind speed based on the horsepower readings on the gauges in the monitoring station. For this purpose, find $f^{-1}(x)$. Check your work by using your answer from part (a) as an input in $f^{-1}(x)$ and comment on the result. (c) If gauges show 25.6 hp is being generated, how fast is the wind blowing?

WRITING, RESEARCH, AND DECISION MAKING

97. The volume of an equipoise cylinder (height equal to radius) is given by $f(x) = \pi x^3$, where $f(x)$ represents the volume and x represents the height (or radius) of the cylinder. (a) What radius is needed to produce cans with volume $V = 392.5\text{ cm}^3$ (use $\pi \approx 3.14$)? (b) If a can manufacturer will be producing many different sizes based on customer need, would it make more sense to give the metalworkers a formula for the radius required based on required volume? Find $f^{-1}(x)$ to produce this formula, then input $V = 392.5\text{ cm}^3$ and comment on the result. (c) Which formula found the radius more efficiently?

98. Inverse functions can be illustrated in very practical ways by retracing sequences we use everyday. Consider this sequence: (a) leave house, (b) take keys from pocket, (c) open car door, (d) get into driver's seat, (e) insert keys in ignition, (f) turn car on, and (g) drive to work. To get back into your home after work would require that you "undo" each step in this sequence and in reverse order. Think of another everyday sequence that has at least three steps, and list what must be done to undo each step. Comment on how this might relate to finding the inverse function for $f(x) = 2x^2 - 3$.

99. The function $f(x) = \frac{1}{x}$ is one of the few functions that is its own inverse. This means the ordered pairs (a, b) and (b, a) must satisfy both f and f^{-1}. (a) Find f^{-1} using the algebraic method to verify that $f(x) = f^{-1}(x) = \frac{1}{x}$. (b) Graph the function $f(x) = \frac{1}{x}$ using a table of integers from -4 to 4. Note that for any ordered pair (a, b) on f, the ordered pair (b, a) is also on f. (c) State where the graph of $y = x$ will intersect the graph of this function and discuss why.

EXTENDING THE CONCEPT

100. Which of the following is the inverse function for $f(x) = \frac{2}{3}\left(x - \frac{1}{2}\right)^5 + \frac{4}{5}$?

a. $\sqrt[5]{\frac{1}{2}\left(x - \frac{2}{3}\right)} - \frac{4}{5}$ **b.** $\frac{3}{2}\sqrt[5]{(x - 2)} - \frac{5}{4}$

c. $\frac{3}{2}\sqrt[5]{\left(x + \frac{1}{2}\right)} - \frac{5}{4}$ **d.** $\sqrt[5]{\frac{3}{2}\left(x - \frac{4}{5}\right)} + \frac{1}{2}$

101. Suppose a function is defined as $f(x) =$ *the exponent that goes on 9 to obtain x*. For example, $f(81) = 2$ since 2 is the exponent that goes on 9 to obtain 81, and $f(3) = \frac{1}{2}$ since $\frac{1}{2}$ is the exponent that goes on 9 to obtain 3. Determine the value of each of the following:

a. $f(1)$ **b.** $f(729)$ **c.** $f^{-1}(2)$ **d.** $f^{-1}\left(\frac{1}{2}\right)$

MAINTAINING YOUR SKILLS

102. (3.3) Verify the following is an identity: $\tan^2\theta \sin^2\theta = \tan^2\theta - \sin^2\theta$.

103. (3.4) Given $\tan(2\beta) = \frac{7}{24}$, with 2β in QI, use double-angle formulas to find exact values for $\cos \beta$ and $\sin \beta$.

104. (2.1) Write the equation of the function graphed in two ways—in terms of a sine function then in terms of a cosine function.

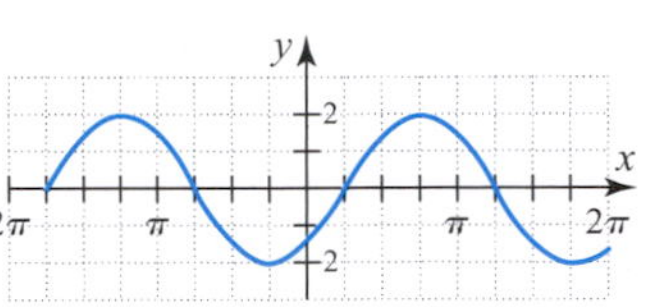

105. (1.3) Find $f(\theta)$ for all six trig functions, given $P(-8, 15)$ is on the terminal side of the angle.

106. (1.2) Standing 9 km (9000 meters) from the base of Mount Fuji (Japan), the angle of elevation to the summit is 22° 46′. Is Mount Fuji taller than Mount Hood (Oregon), which stands 3428 m high?

107. (1.1) The area of a circular sector is $A = \frac{125\pi}{3}$ m^2. If $r = 10$ m, what angle is subtended?

4.2 The Inverse Trig Functions and Their Applications

LEARNING OBJECTIVES

In Section 4.2 you will learn how to:

A. **Find and graph the inverse sine function and evaluate related expressions**

B. **Find and graph the inverse cosine and tangent functions and evaluate related expressions**

C. **Apply the definition and notation of inverse trig functions to simplify compositions**

D. **Write a given expression in trigonometric form using a substitution**

E. **Find and graph inverse functions for sec x, csc x, and cot x**

INTRODUCTION

While we usually associate the number π with the features of a circle, it also occurs in some "odd" places, such as the study of normal (bell) curves, Bessel functions, Fourier series, Laplace transforms, and infinite series. In much the same way, the trigonometric functions are surprisingly versatile, finding their way into a study of complex numbers and vectors, the simplification of algebraic expressions, and finding the area under certain curves—applications that are hugely important in a continuing study of mathematics. As you'll see, a study of the inverse trig functions helps support these fascinating applications.

POINT OF INTEREST

Prior to the widespread availability of handheld calculators, students of mathematics had to use published tables for $\sin\theta$, $\cos\theta$, and $\tan\theta$ to find their function values. A small portion of the table for $\sin\theta$ is shown here, where we note that sin 46° is approximately 0.7193 (shaded). The tables were also used to find an angle whose sine was known, meaning if $\sin\theta \approx 0.7604$, then θ must be $49.5° = 49°30'$ (also shaded). The modern notation for "an angle whose sine is known" is $\theta = \sin^{-1}x$ or $\theta = \arcsin x$, where x is the known value for $\sin\theta$.

sin θ

deg \ min	0′	10′	20′	30′	40′	50′
45°	0.7071	0.7092	0.7112	0.7133	0.7153	0.7173
46	0.7193	0.7214	0.7234	0.7254	0.7274	0.7294
47	0.7314	0.7333	0.7353	0.7373	0.7392	0.7412
48	0.7431	0.7451	0.7470	0.7490	0.7509	0.7528
49	0.7547	0.7566	0.7585	0.7604	0.7623	0.7642

A. The Inverse Sine Function

In Section 4.1 we established that only one-to-one functions have an inverse. All six trig functions fail the horizontal line test and are not one-to-one as given. However, by suitably restricting the domain, a one-to-one function can be defined that makes finding an inverse possible. For the sine function, it seems natural to choose the interval $\left[-\frac{\pi}{2}, \frac{\pi}{2}\right]$ since it is centrally located and the sine function attains all possible output values in this interval. A graph of $y = \sin x$ is shown in Figure 4.13, with the portion corresponding to $x \in \left[-\frac{\pi}{2}, \frac{\pi}{2}\right]$ colored in red. Note the range is still $y \in [-1, 1]$ (Figure 4.14).

Figure 4.13 Figure 4.14

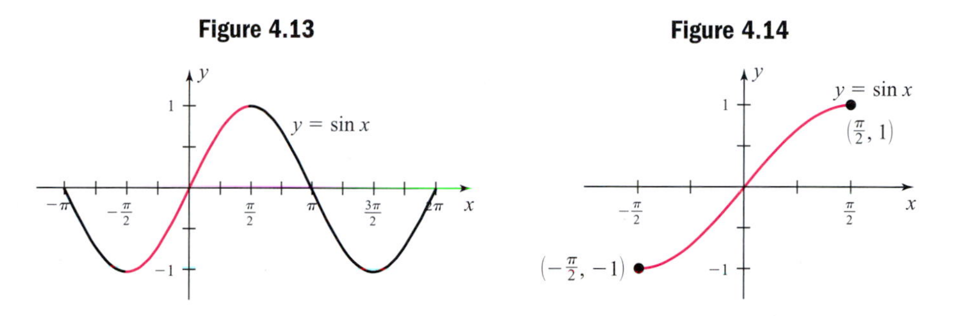

We can obtain an implicit equation for the inverse of $y = \sin x$ by interchanging x- and y-values, obtaining $x = \sin y$. By accepted convention, the *explicit* form of the inverse sine function is written $y = \sin^{-1}x$ or $y = \arcsin x$. Since domain and range values have been interchanged, the domain of $y = \sin^{-1}x$ is $x \in [-1, 1]$ and the range is $y \in \left[-\frac{\pi}{2}, \frac{\pi}{2}\right]$. The graph of $y = \sin^{-1}x$ can be found by reflecting the portion in red across the line $y = x$ and using the endpoints of the domain and range (see Figure 4.15).

Figure 4.15

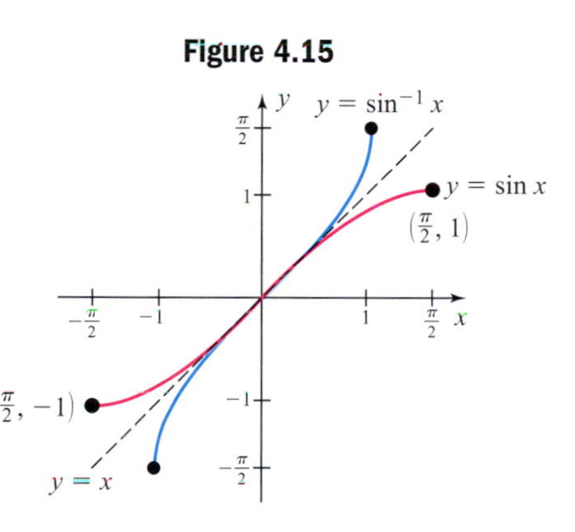

THE INVERSE SINE FUNCTION

For $y = \sin x$ with domain $x \in \left[-\frac{\pi}{2}, \frac{\pi}{2}\right]$ and range $y \in [-1, 1]$, the inverse sine function is

$$\mathbf{y = \sin^{-1}x \text{ or } y = \arcsin x,}$$

with domain $x \in [-1, 1]$ and range $y \in \left[-\frac{\pi}{2}, \frac{\pi}{2}\right]$.

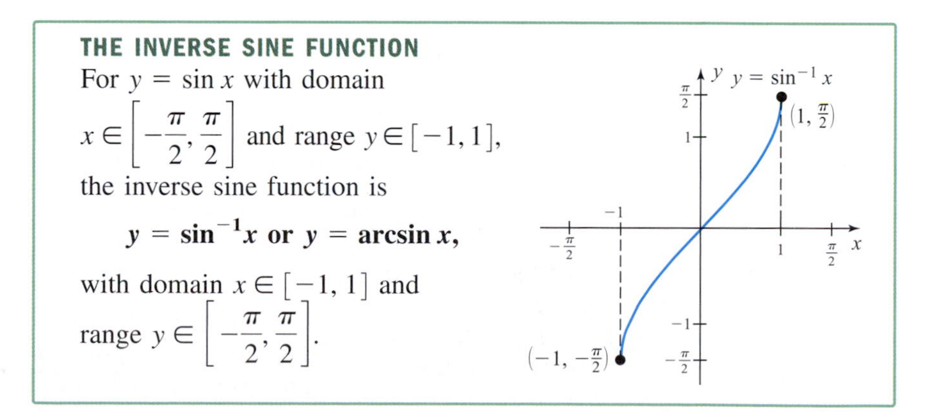

From the implicit form $x = \sin y$, we learn to interpret the inverse function as, "y is the number or angle whose sine is x." Learning to read and interpret the explicit form in this way will be helpful. That is, $y = \sin^{-1}x$ means "y is the number or angle whose sine is x."

$$\mathbf{y = \sin^{-1}x \leftrightarrow x = \sin y \qquad x = \sin y \leftrightarrow y = \sin^{-1}x}$$

EXAMPLE 1 ▶ Evaluate the inverse sine function for the values given:

a. $y = \sin^{-1}\left(\frac{\sqrt{3}}{2}\right)$ **b.** $y = \arcsin\left(-\frac{1}{2}\right)$ **c.** $y = \sin^{-1}2$

Solution: ▶ For $x \in [-1, 1]$ and $y \in \left[-\frac{\pi}{2}, \frac{\pi}{2}\right]$,

a. $y = \sin^{-1}\left(\frac{\sqrt{3}}{2}\right)$: y is the number or angle whose sine is $\frac{\sqrt{3}}{2} \rightarrow \sin y = \frac{\sqrt{3}}{2}$, so $\sin^{-1}\left(\frac{\sqrt{3}}{2}\right) = \frac{\pi}{3}$.

b. $y = \arcsin\left(-\frac{1}{2}\right)$: y is the number or angle whose sine is $-\frac{1}{2} \rightarrow \sin y = -\frac{1}{2}$, so $\arcsin\left(-\frac{1}{2}\right) = -\frac{\pi}{6}$.

c. $y = \sin^{-1}(2)$: y is the number or angle whose sine is $2 \rightarrow \sin y = 2$. Since $2 \notin [-1, 1]$, $\sin^{-1}(2)$ is undefined.

NOW TRY EXERCISES 7 THROUGH 12 ▶

Table 4.1

x	$\sin x$
$-\frac{\pi}{2}$	-1
$-\frac{\pi}{3}$	$-\frac{\sqrt{3}}{2}$
$-\frac{\pi}{4}$	$-\frac{\sqrt{2}}{2}$
$-\frac{\pi}{6}$	$-\frac{1}{2}$
0	0
$\frac{\pi}{6}$	$\frac{1}{2}$
$\frac{\pi}{4}$	$\frac{\sqrt{2}}{2}$
$\frac{\pi}{3}$	$\frac{\sqrt{3}}{2}$
$\frac{\pi}{2}$	1

In Examples 1a and 1b, note that the equations $\sin y = \frac{\sqrt{3}}{2}$ and $\sin y = -\frac{1}{2}$ each have an infinite number of solutions, but only one solution in $\left[-\frac{\pi}{2}, \frac{\pi}{2}\right]$.

When x is one of the standard values $\left(0, \frac{1}{2}, \frac{\sqrt{3}}{2}, 1\text{, and so on}\right)$, $y = \sin^{-1}x$ can be evaluated by reading a standard table "in reverse." For $y = \arcsin(-1)$, we locate the number -1 in the right-hand column of Table 4.1, and note the "number or angle whose sine is -1," is $-\frac{\pi}{2}$. If x is between -1 and 1 but is not a standard value, we can use the $\sin^{-1}$ function on a calculator, which is most often the 2nd or INV function for SIN.

EXAMPLE 2 ▶ Evaluate each inverse sine function twice. First in radians rounded to four decimal places, then in degrees to the nearest tenth.

a. $y = \sin^{-1}0.8492$ **b.** $y = \arcsin(-0.2317)$

Solution: ▶ For $x \in [-1, 1]$, we evaluate $y = \sin^{-1}x$.

a. $y = \sin^{-1}0.8492$: With the calculator in radian MODE, use the keystrokes 2nd SIN 0.8492) ENTER. We find $\sin^{-1}(0.8492) \approx 1.0145$ radians. In degree MODE, the same sequence of keystrokes gives $\sin^{-1}(0.8492) \approx 58.1°$ (note that 1.0145 rad $\approx 58.1°$).

b. $y = \arcsin(-0.2317)$: In radian MODE, we find $\sin^{-1}(-0.2317) \approx -0.2338$ rad. In degree MODE, $\sin^{-1}(-0.2317) \approx -13.4°$.

NOW TRY EXERCISES 13 THROUGH 16

From our work in Section 4.1, we know that if f and g are inverses, $(f \circ g)(x) = x$ and $(g \circ f)(x) = x$. For $f(x) = \sin x$ and $g(x) = \sin^{-1}x$ this means $\sin(\sin^{-1}x) = x$ for $x \in [-1, 1]$ and $\sin^{-1}(\sin x) = x$ for $x \in \left[-\frac{\pi}{2}, \frac{\pi}{2}\right]$.

INVERSE FUNCTION PROPERTIES FOR SINE

For $f(x) = \sin x$ and $g(x) = \sin^{-1}x$:

I. $(f \circ g)(x) = \sin(\sin^{-1}x) = x$ for $x \in [-1, 1]$

and

II. $(g \circ f)(x) = \sin^{-1}(\sin x) = x$ for $x \in \left[-\frac{\pi}{2}, \frac{\pi}{2}\right]$

EXAMPLE 3 Evaluate each expression and verify the result on a calculator.

a. $\sin\left[\sin^{-1}\left(\frac{1}{2}\right)\right]$ **b.** $\arcsin\left[\sin\left(\frac{\pi}{4}\right)\right]$ **c.** $\sin^{-1}\left[\sin\left(\frac{5\pi}{6}\right)\right]$

Solution:

a. $\sin\left[\sin^{-1}\left(\frac{1}{2}\right)\right] = \frac{1}{2}$, since $\frac{1}{2} \in [-1, 1]$ — Property I

b. $\arcsin\left[\sin\left(\frac{\pi}{4}\right)\right] = \frac{\pi}{4}$, since $\frac{\pi}{4} \in \left[-\frac{\pi}{2}, \frac{\pi}{2}\right]$ — Property II

c. $\sin^{-1}\left[\sin\left(\frac{5\pi}{6}\right)\right] \neq \frac{5\pi}{6}$, since $\frac{5\pi}{6} \notin \left[-\frac{\pi}{2}, \frac{\pi}{2}\right]$.

This doesn't mean the expression cannot be evaluated, only that we cannot use Property II. Since $\sin\left(\frac{5\pi}{6}\right) = \sin\left(\frac{\pi}{6}\right)$ and $\frac{\pi}{6} \in \left[-\frac{\pi}{2}, \frac{\pi}{2}\right]$, $\sin^{-1}\left[\left(\sin\frac{5\pi}{6}\right)\right] = \sin^{-1}\left[\sin\left(\frac{\pi}{6}\right)\right] = \frac{\pi}{6}$.

The calculator verification for each is shown in Figures 4.16 and 4.17. Note $\frac{\pi}{6} \approx 0.5236$ and $\frac{\pi}{4} \approx 0.7854$.

Figure 4.16

Parts (a) and (b)

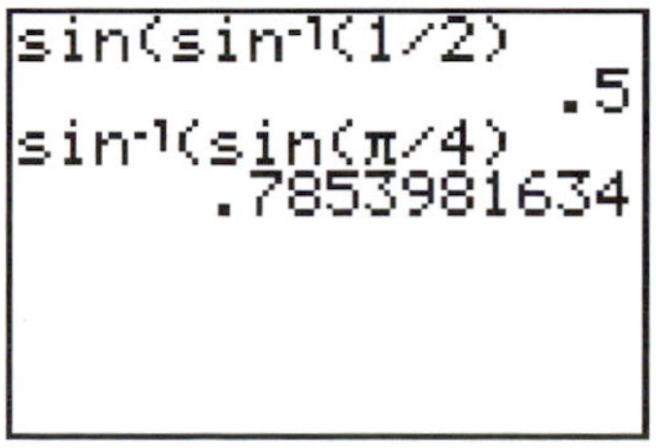

```
sin(sin⁻¹(1/2)
                .5
sin⁻¹(sin(π/4)
      .7853981634
```

Figure 4.17

Part (c)

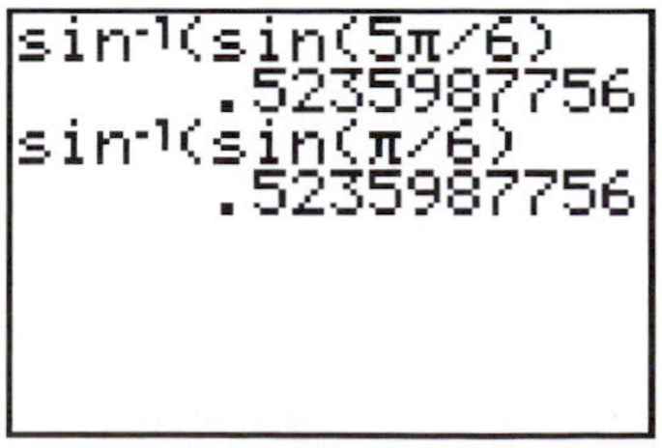

```
sin⁻¹(sin(5π/6)
      .5235987756
sin⁻¹(sin(π/6)
      .5235987756
```

NOW TRY EXERCISES 17 THROUGH 24

B. The Inverse Cosine and Inverse Tangent Functions

Like the sine function, the cosine function is not one-to-one and its domain must also be restricted to develop an inverse function. For convenience we choose the interval $x \in [0, \pi]$ since it is again somewhat central and takes on all of its values in this interval. A graph of the cosine function, with the interval corresponding to $x \in [0, \pi]$ shown in red, is given in Figure 4.18. Note the range is still $y \in [-1, 1]$ (Figure 4.19).

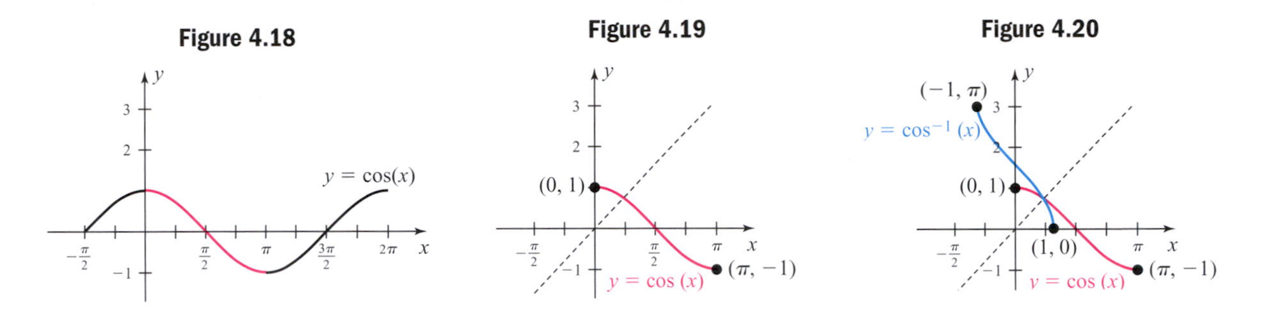

For the implicit equation of inverse cosine, $y = \cos x$ becomes $x = \cos y$, with the corresponding explicit forms being $y = \cos^{-1}x$ and $y = \arccos x$. By reflecting the graph of $y = \cos x$ across the line $y = x$, we obtain the graph of $y = \cos^{-1}x$ shown in Figure 4.20.

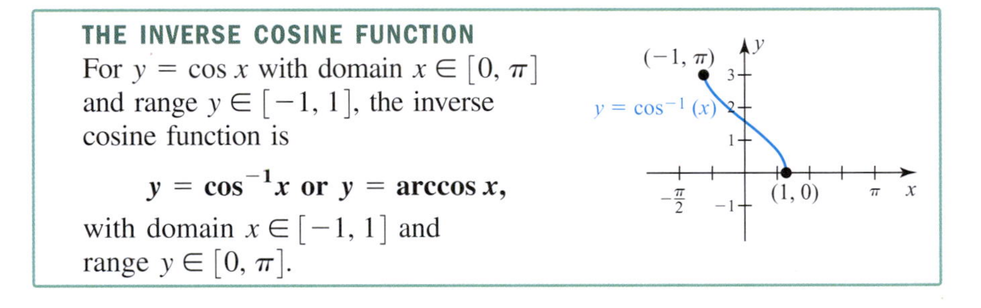

THE INVERSE COSINE FUNCTION

For $y = \cos x$ with domain $x \in [0, \pi]$ and range $y \in [-1, 1]$, the inverse cosine function is

$$\mathbf{y = \cos^{-1}x \text{ or } y = \arccos x,}$$

with domain $x \in [-1, 1]$ and range $y \in [0, \pi]$.

EXAMPLE 4 Evaluate the inverse cosine function for the values given:

a. $y = \cos^{-1}0$ **b.** $y = \arccos\left(-\frac{\sqrt{3}}{2}\right)$ **c.** $y = \cos^{-1}\pi$

Solution: For $x \in [-1, 1]$, we evaluate $y = \cos^{-1}x$.

a. $y = \cos^{-1}0$: y is the number or angle whose cosine is $0 \rightarrow \cos y = 0$. This shows $\cos^{-1}0 = \frac{\pi}{2}$.

b. $y = \arccos\left(-\frac{\sqrt{3}}{2}\right)$: y is the number or angle whose cosine is $-\frac{\sqrt{3}}{2} \rightarrow \cos y = -\frac{\sqrt{3}}{2}$. This shows $\arccos\left(-\frac{\sqrt{3}}{2}\right) = \frac{5\pi}{6}$.

c. $y = \cos^{-1}\pi$: y is the number or angle whose cosine is $\pi \rightarrow \cos y = \pi$. Since $\pi \notin [-1, 1]$, $\cos^{-1}\pi$ is undefined.

NOW TRY EXERCISES 25 THROUGH 34

Knowing that $y = \cos x$ and $y = \cos^{-1}x$ are inverse functions enables us to state inverse function properties similar to those for sine.

INVERSE FUNCTION PROPERTIES FOR COSINE

For $f(x) = \cos x$ and $g(x) = \cos^{-1}x$:

I. $(f \circ g)(x) = \cos(\cos^{-1}x) = x$ for $x \in [-1, 1]$

and

II. $(g \circ f)(x) = \cos^{-1}(\cos x) = x$ for $x \in [0, \pi]$

EXAMPLE 5 Evaluate each expression.

a. $\cos[\cos^{-1}(0.73)]$ **b.** $\arccos\left[\cos\left(\frac{\pi}{12}\right)\right]$ **c.** $\cos^{-1}\left[\cos\left(\frac{4\pi}{3}\right)\right]$

Solution:

a. $\cos[\cos^{-1}(0.73)] = 0.73$, since $0.73 \in [-1, 1]$ Property I

b. $\arccos\left[\cos\left(\frac{\pi}{12}\right)\right] = \frac{\pi}{12}$, since $\frac{\pi}{12} \in [0, \pi]$ Property II

c. $\cos^{-1}\left[\cos\left(\frac{4\pi}{3}\right)\right] \neq \frac{4\pi}{3}$, since $\frac{4\pi}{3} \notin [0, \pi]$.

This expression cannot be evaluated using Property II. Since $\cos\left(\frac{4\pi}{3}\right) = \cos\left(\frac{2\pi}{3}\right)$ and $\frac{2\pi}{3} \in [0, \pi]$, $\cos^{-1}\left[\cos\left(\frac{4\pi}{3}\right)\right] = \cos^{-1}\left[\cos\left(\frac{2\pi}{3}\right)\right] = \frac{2\pi}{3}$.

The results can also be verified using a calculator.

NOW TRY EXERCISES 35 THROUGH 42

For the tangent function, we likewise restrict the domain to obtain a one-to-one function, with the most common choice being $x \in \left(-\frac{\pi}{2}, \frac{\pi}{2}\right)$. The corresponding range is $y \in \mathbb{R}$. The *implicit* equation for the inverse tangent function is $x = \tan y$ with the explicit forms $y = \tan^{-1}x$ or $y = \arctan x$. With the domain and range interchanged, the domain of $y = \tan^{-1}x$ is $x \in \mathbb{R}$, and the range is $y \in \left(-\frac{\pi}{2}, \frac{\pi}{2}\right)$. The graph of $y = \tan x$ for $x \in \left(-\frac{\pi}{2}, \frac{\pi}{2}\right)$ is shown in red in Figure 4.21, with the inverse function $y = \tan^{-1}x$ shown in blue in Figure 4.22.

Figure 4.21

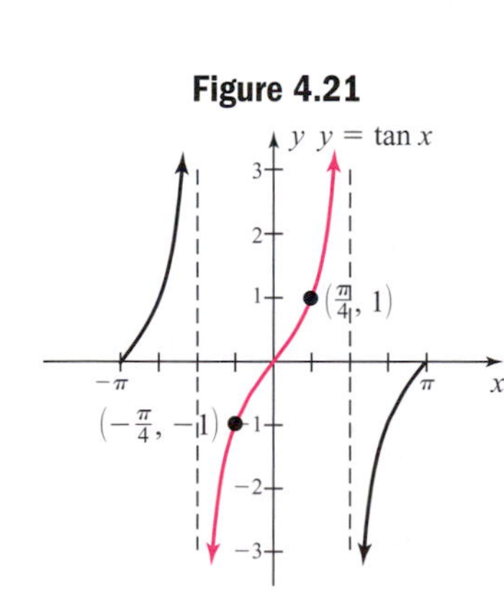

Figure 4.22

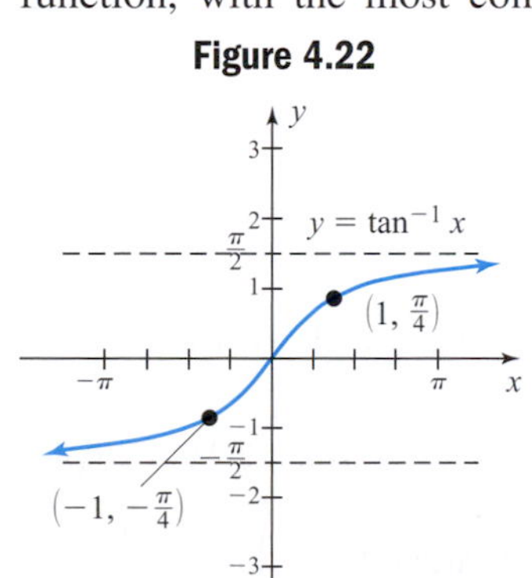

THE INVERSE TANGENT FUNCTION

For $y = \tan x$ with domain $x \in \left(-\frac{\pi}{2}, \frac{\pi}{2}\right)$ and range $y \in \mathbb{R}$, the inverse tangent function is

$$\mathbf{y = \tan^{-1} x} \text{ or } \mathbf{y = \arctan x},$$

with domain $x \in \mathbb{R}$ and range $y \in \left(-\frac{\pi}{2}, \frac{\pi}{2}\right)$.

INVERSE FUNCTION PROPERTIES FOR TANGENT

For $f(x) = \tan x$ and $g(x) = \tan^{-1} x$:

I. $(f \circ g)(x) = \tan(\tan^{-1}x) = x$ for $x \in \mathbb{R}$

and

II. $(g \circ f)(x) = \tan^{-1}(\tan x) = x$ for $x \in \left(-\frac{\pi}{2}, \frac{\pi}{2}\right)$.

EXAMPLE 6 ▶ Evaluate each expression.

a. $\tan^{-1}(-\sqrt{3})$ **b.** $\arctan[\tan(-0.89)]$

Solution: ▶ For $x \in \mathbb{R}$ and $y \in \left(-\frac{\pi}{2}, \frac{\pi}{2}\right)$,

a. $\tan^{-1}(-\sqrt{3}) = -\frac{\pi}{3}$

b. $\arctan[\tan(-0.89)] = -0.89$, since $-0.89 \in \left(-\frac{\pi}{2}, \frac{\pi}{2}\right)$ Property II

NOW TRY EXERCISES 43 THROUGH 52 ▶

C. Using the Inverse Trig Functions to Evaluate Compositions

In the context of angle measure, the expression $y = \sin^{-1}\left(-\frac{1}{2}\right)$ represents the *angle y* whose sine is $-\frac{1}{2}$. It seems natural to ask, "What happens if we take the tangent of this angle?" In other words, what does the expression $\tan\left[\sin^{-1}\left(-\frac{1}{2}\right)\right]$ mean? Similarly, if $y = \cos\left(\frac{\pi}{3}\right)$ represents a real number between -1 and 1, how do we compute $\sin^{-1}\left[\cos\left(\frac{\pi}{3}\right)\right]$? Expressions like these occur in many fields of study.

EXAMPLE 7 ▶ Simplify each expression: (a) $\tan\left[\arcsin\left(-\frac{1}{2}\right)\right]$ (b) $\sin^{-1}\left[\cos\left(\frac{\pi}{3}\right)\right]$

Solution: ▶ **a.** In Example 1 we found $\arcsin\left(-\frac{1}{2}\right) = -\frac{\pi}{6}$. Substituting $-\frac{\pi}{6}$ for $\arcsin\left(-\frac{1}{2}\right)$ gives $\tan\left(-\frac{\pi}{6}\right) = -\frac{\sqrt{3}}{3}$, showing $\tan\left[\arcsin\left(-\frac{1}{2}\right)\right] = -\frac{\sqrt{3}}{3}$.

b. For $\sin^{-1}\left[\cos\left(\frac{\pi}{3}\right)\right]$, we begin with the inner function $\cos\left(\frac{\pi}{3}\right) = \frac{1}{2}$. Substituting $\frac{1}{2}$ for $\cos\left(\frac{\pi}{3}\right)$ gives $\sin^{-1}\left(\frac{1}{2}\right)$. With the appropriate checks satisfied we have $\sin^{-1}\left(\frac{1}{2}\right) = \frac{\pi}{6}$, showing $\sin^{-1}\left[\cos\left(\frac{\pi}{3}\right)\right] = \frac{\pi}{6}$.

NOW TRY EXERCISES 53 THROUGH 64

If the argument is not a standard value and we need the answer in exact form, we can draw the triangle described by the inner expression using the definition of the trigonometric functions as ratios. In other words, for either y or $\theta = \sin^{-1}\left(\frac{8}{17}\right)$, we draw a triangle with hypotenuse 17 and side 8 opposite θ to model the statement, "an angle whose sine is $\frac{8}{17} = \frac{\text{opp}}{\text{hyp}}$" (see Figure 4.23). Using the Pythagorean theorem, we find the adjacent side is 15 and can now name any of the other trig functions.

Figure 4.23

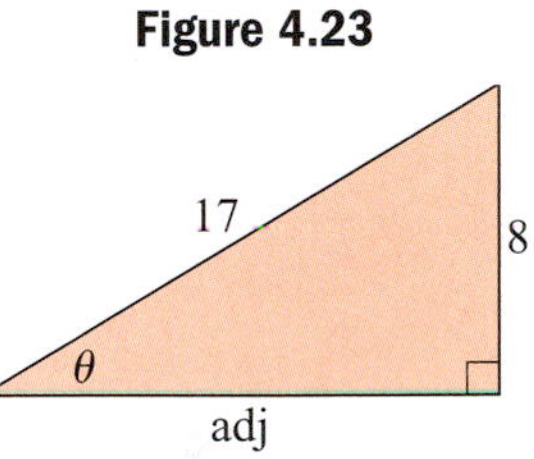

WORTHY OF NOTE

To verify the result of Example 8, we can actually find the value of $\sin^{-1}\left(-\frac{8}{17}\right)$ on a calculator, then take the tangent of the result. See Figure 4.25.

Figure 4.25

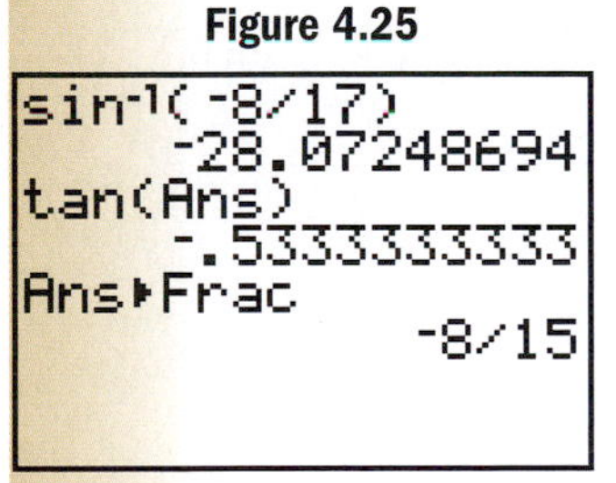

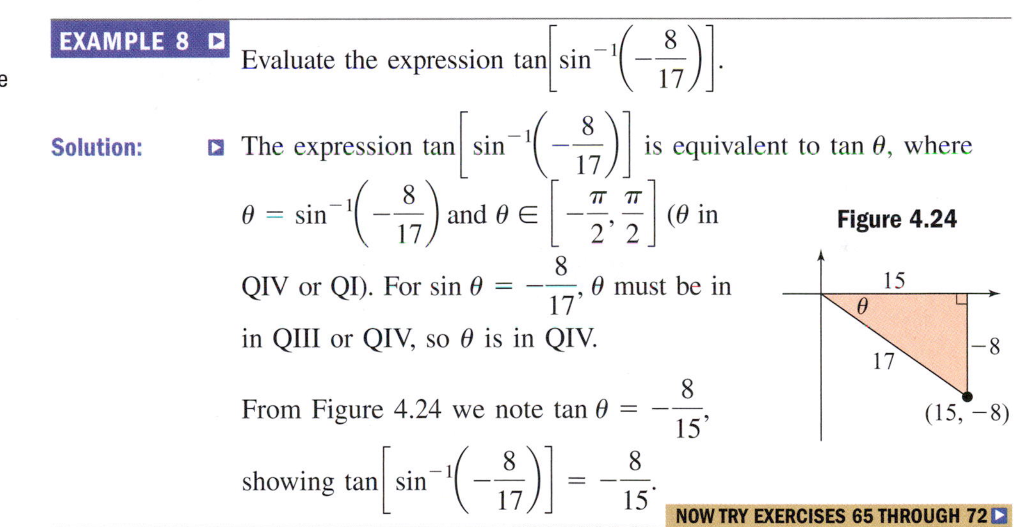

EXAMPLE 8 Evaluate the expression $\tan\left[\sin^{-1}\left(-\frac{8}{17}\right)\right]$.

Solution: The expression $\tan\left[\sin^{-1}\left(-\frac{8}{17}\right)\right]$ is equivalent to $\tan\theta$, where $\theta = \sin^{-1}\left(-\frac{8}{17}\right)$ and $\theta \in \left[-\frac{\pi}{2}, \frac{\pi}{2}\right]$ (θ in QIV or QI). For $\sin\theta = -\frac{8}{17}$, θ must be in in QIII or QIV, so θ is in QIV.

From Figure 4.24 we note $\tan\theta = -\frac{8}{15}$, showing $\tan\left[\sin^{-1}\left(-\frac{8}{17}\right)\right] = -\frac{8}{15}$.

Figure 4.24

NOW TRY EXERCISES 65 THROUGH 72

These ideas apply even when one side of the triangle is unknown. In other words, we can still draw a triangle for $\theta = \cos^{-1}\left(\frac{x}{\sqrt{x^2 + 16}}\right)$, since "$\theta$ is an angle whose cosine is $\frac{x}{\sqrt{x^2 + 16}} = \frac{\text{adj}}{\text{hyp}}$."

EXAMPLE 9 Evaluate the expression $\tan\left[\cos^{-1}\left(\frac{x}{\sqrt{x^2+16}}\right)\right](x > 0)$.

Solution: Rewrite $\tan\left[\cos^{-1}\left(\frac{x}{\sqrt{x^2+16}}\right)\right]$ as $\tan\theta$, where $\theta = \cos^{-1}\left(\frac{x}{\sqrt{x^2+16}}\right)$.

Draw a triangle with side x adjacent to θ and a hypotenuse of $\sqrt{x^2+16}$. The Pythagorean theorem gives $x^2 + \text{opp}^2 = (\sqrt{x^2+16})^2$, which leads to $\text{opp}^2 = (x^2+16) - x^2$ giving $\text{opp} = \sqrt{16} = 4$. This shows $\tan\theta = \tan\left[\cos^{-1}\left(\frac{x}{\sqrt{x^2+16}}\right)\right] = \frac{4}{x}$ (see Figure 4.26).

Figure 4.26

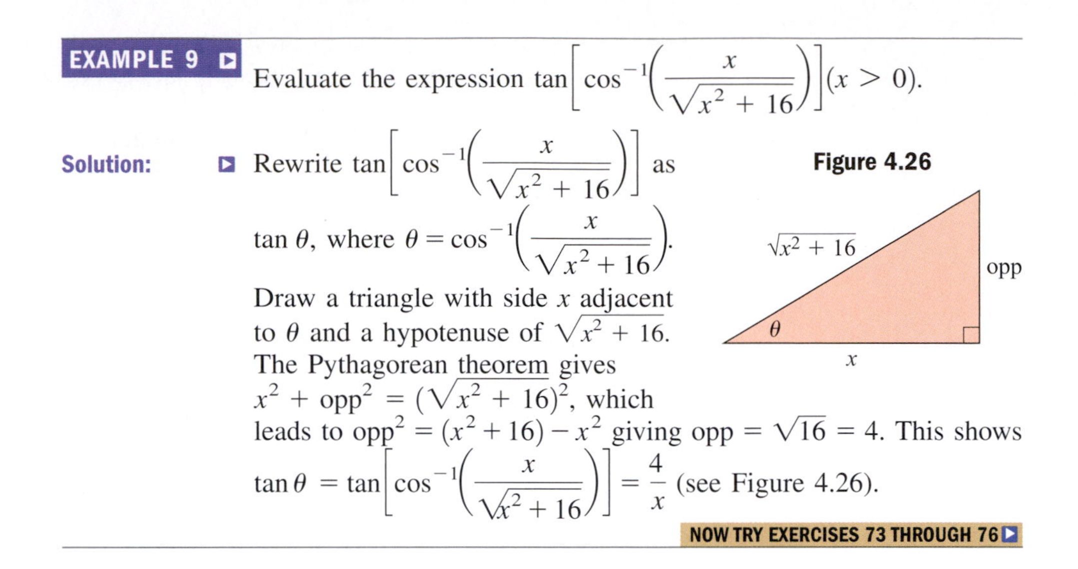

NOW TRY EXERCISES 73 THROUGH 76

D. Trig Substitutions and Trigonometric Forms

In mathematics, we often encounter expressions that are difficult to use or understand in their given form, and attempt to write the expression in a simpler but equivalent form. For instance, solving $x^{\frac{2}{3}} + 3x^{\frac{1}{3}} - 10 = 0$ may seem challenging at first, but after substituting $u = x^{\frac{1}{3}}(u^2 = x^{\frac{2}{3}})$ the equation is easily factorable: $u^2 + 3u - 10 = 0$. We are then able to find solutions in u and "unsubstitute" to write the final answer in terms of the original variable x (the solutions are $x = -125$ and $x = 8$). In much the same way, the equation $y = \frac{x}{\sqrt{9-x^2}}$ appears somewhat daunting at first, but is easily simplified using the substitution $x = 3\sin\theta$. When doing so, we restrict the domain so that $x = 3\sin\theta$ is one-to-one: $\theta \in \left[-\frac{\pi}{2}, \frac{\pi}{2}\right]$.

EXAMPLE 10 Simplify $y = \frac{x}{\sqrt{9-x^2}}$ using the substitution $x = 3\sin\theta$.

Solution:

$$y = \frac{3\sin\theta}{\sqrt{9-(3\sin\theta)^2}} \quad \text{substitute } 3\sin\theta \text{ for } x$$

$$= \frac{3\sin\theta}{\sqrt{9-9\sin^2\theta}} \quad (3\sin\theta)^2 = 9\sin^2\theta$$

$$= \frac{3\sin\theta}{\sqrt{9(1-\sin^2\theta)}} \quad \text{factor}$$

$$= \frac{3\sin\theta}{\sqrt{9\cos^2\theta}} \quad 1-\sin^2\theta = \cos^2\theta$$

$$= \frac{3\sin\theta}{3\cos\theta} \quad \sqrt{9\cos^2\theta} = 3\cos\theta \text{ since } -\frac{\pi}{2} < \theta < \frac{\pi}{2}$$

$$= \tan\theta; \text{ where } x = 3\sin\theta$$

NOW TRY EXERCISES 77 THROUGH 80

Using the notation for inverse functions, we can write the result from Example 10 in terms of x and use a calculator to compare it with the original function. For $x = 3\sin\theta$ we have $\frac{x}{3} = \sin\theta$ or $\theta = \sin^{-1}\left(\frac{x}{3}\right)$. Substituting into $y = \tan\theta$, we obtain $y = \tan\left[\sin^{-1}\left(\frac{x}{3}\right)\right]$. Place the calculator in radian MODE, then enter $Y_1 = \frac{x}{\sqrt{9 - x^2}}$ and $Y_2 = \tan\left[\sin^{-1}\left(\frac{x}{3}\right)\right]$ on the Y= screen. Use TblStart $= -3$, since $-3 \leq 3\sin\theta \leq 3$. The resulting table is shown in Figure 4.27, which seems to indicate the functions are indeed equivalent.

Figure 4.27

X	Y1	Y2
-3	ERROR	ERROR
-2.9	-3.775	-3.775
-2.8	-2.6	-2.6
-2.7	-2.065	-2.065
-2.6	-1.737	-1.737
-2.5	-1.508	-1.508
-2.4	-1.333	-1.333

X=-2.4

E. The Inverse Functions for Secant, Cosecant, and Cotangent

As we did for the other functions, we restrict the domain of the secant, cosecant, and cotanget functions to obtain a one-to-one function that is invertible (an inverse can be found). Once again the choice is arbitrary, and some restricted domains are easier to work with than others in more advanced mathematics. For $y = \sec x$ we've chosen the "most intuitive" restriction, one that seems more centrally located (nearer the origin). The graph of $y = \sec x$ is reproduced here, along with its inverse function (see Figures 4.28 and 4.29). The domain, range, and graphs of the functions $y = \csc^{-1}x$ and $y = \cot^{-1}x$ are asked for in the Exercises (see Exercise 104).

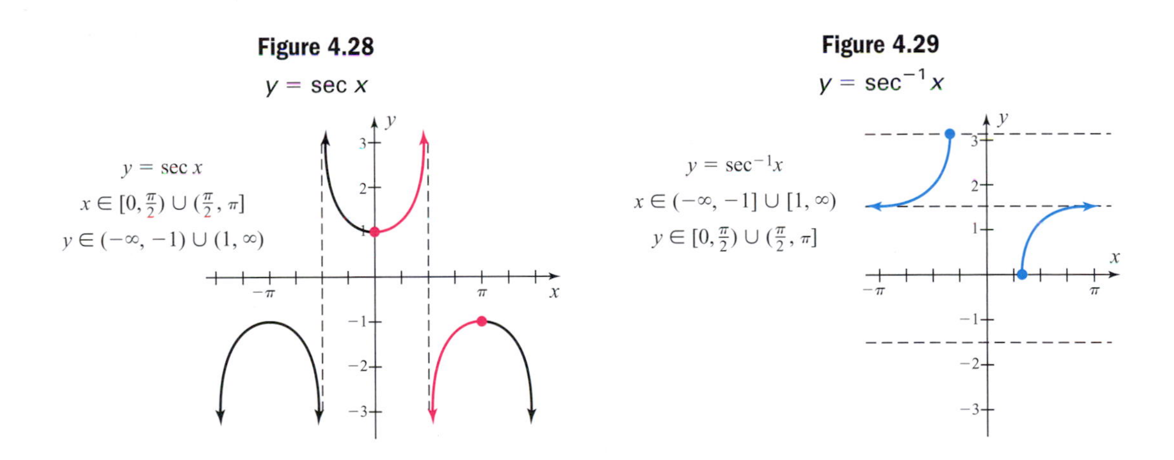

Figure 4.28 $y = \sec x$

Figure 4.29 $y = \sec^{-1}x$

The functions $y = \sec^{-1}x$, $y = \csc^{-1}x$, and $y = \cot^{-1}x$ can be evaluated by noting their relationship to $y = \cos^{-1}x$, $y = \sin^{-1}x$, and $y = \tan^{-1}x$, respectively. For $y = \sec^{-1}x$, we have $\sec y = x \rightarrow \frac{1}{\cos y} = x$, which leads to $\cos y = \frac{1}{x} \rightarrow y = \cos^{-1}\left(\frac{1}{x}\right)$. That is, to find the value of $y = \sec^{-1}x$, evaluate $y = \cos^{-1}\left(\frac{1}{x}\right)$. Values for $\csc^{-1}x$ and $\cot^{-1}x$ can be similarly found.

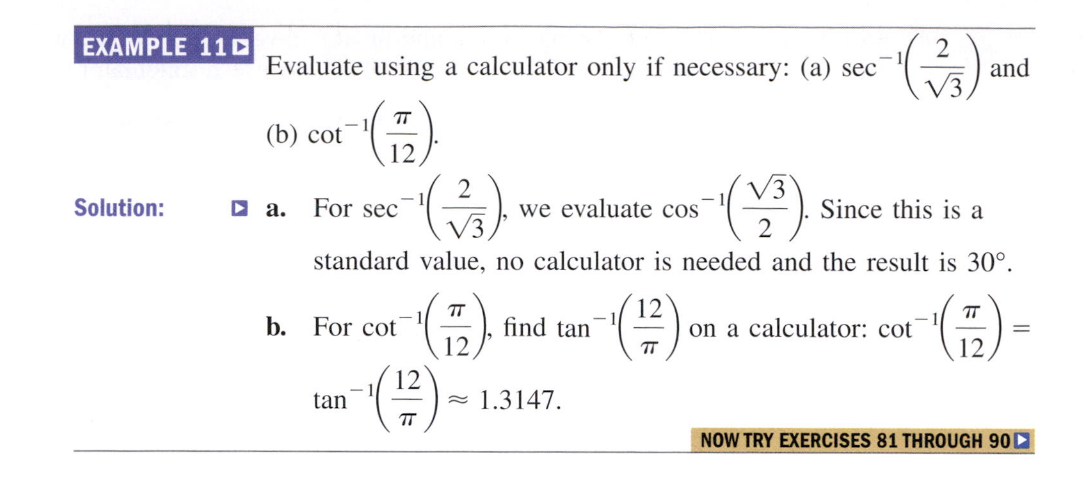

EXAMPLE 11 Evaluate using a calculator only if necessary: (a) $\sec^{-1}\left(\frac{2}{\sqrt{3}}\right)$ and (b) $\cot^{-1}\left(\frac{\pi}{12}\right)$.

Solution:

a. For $\sec^{-1}\left(\frac{2}{\sqrt{3}}\right)$, we evaluate $\cos^{-1}\left(\frac{\sqrt{3}}{2}\right)$. Since this is a standard value, no calculator is needed and the result is 30°.

b. For $\cot^{-1}\left(\frac{\pi}{12}\right)$, find $\tan^{-1}\left(\frac{12}{\pi}\right)$ on a calculator: $\cot^{-1}\left(\frac{\pi}{12}\right) = \tan^{-1}\left(\frac{12}{\pi}\right) \approx 1.3147$.

NOW TRY EXERCISES 81 THROUGH 90

TECHNOLOGY HIGHLIGHT

More on Inverse Functions

The keystrokes illustrated refer to a TI-84 Plus model. For other models please consult your manual or visit our Internet site.

The domain and range of the inverse functions for sine, cosine, and tangent are preprogrammed into most graphing calculators, making them an ideal tool for reinforcing the concepts involved, and *reminding us* of the concepts involved. In particular, $\sin x = y$ implies that $\sin^{-1} y = x$ only if $-90° \leq y \leq 90°$ and $-1 \leq x \leq 1$. To get a stark reminder of this fact we'll use the TABLE feature of the grapher. Begin by going to the TBLSET screen (2nd WINDOW) and set TblStart = 90 with ΔTbl = −30. After placing the calculator in degree MODE , go to the Y= screen and input $Y_1 = \sin x$, $Y_2 = \sin^{-1} x$, and $Y_3 = Y_2(Y_1)$ (the composition $Y_2 \circ Y_1$). Then disable Y_2 [turn it off—Y_3 will read it anyway) so that both Y_1 and Y_3 will be displayed simultaneously on the TABLE screen. Now let's see what the TABLE has to say to us. Pressing 2nd GRAPH brings up the TABLE shown in Figure 4.30, where we note the inputs are standard angles, the outputs in Y_1 are the (expected) standard values, and the outputs in Y_3 return the original standard values. In other words, $Y_1 = \sin x$ is giving the standard values and the inverse function $Y_2(Y_1)$ is doing its job, returning us to the standard angles. Now scroll upward until 180° is at the top of the X column (Figure 4.31), and note that $Y_3 = Y_2(Y_1)$ continues to return standard angles in the interval $[-90°, 90°]$—a stark reminder that while the expression $\sin 150° = 0.5$, $\sin^{-1}(\sin 150°) \neq 150°$. Further, we are reminded that while $\sin^{-1}(\sin 150°)$ can still be evaluated, it cannot be evaluated directly using the inverse function properties. Use these ideas to complete the following exercises.

Figure 4.30

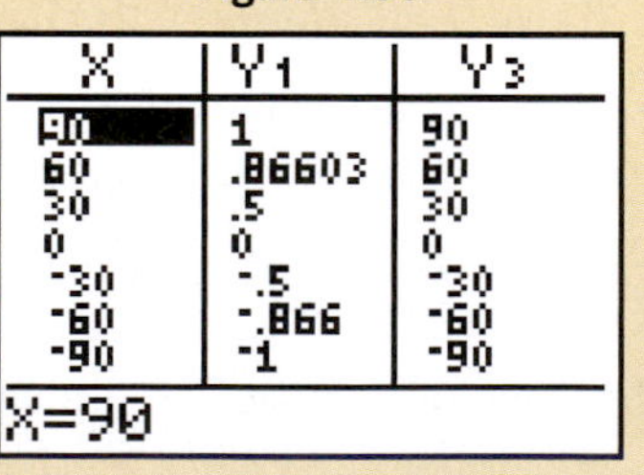

X	Y1	Y3
90	1	90
60	.86603	60
30	.5	30
0	0	0
-30	-.5	-30
-60	-.866	-60
-90	-1	-90

X=90

Figure 4.31

X	Y1	Y3
180	0	0
150	.5	30
120	.86603	60
90	1	90
60	.86603	60
30	.5	30
0	0	0

X=180

Exercise 1: Go through an exercise similar to the one here using $Y_1 = \cos x$ and $Y_2 = \cos^{-1} x$. Remember to modify the TBLSET to accommodate the restricted domain for cosine.

Exercise 2: Complete parts (a) and (b) using the TABLE from Exercise 1. Complete parts (c) and (d) without a calculator.

a. $\cos^{-1}(\cos 150)$ **b.** $\cos^{-1}(\cos 210)$

c. $\cos^{-1}(\cos 120)$ **d.** $\cos^{-1}(\cos 240)$

4.2 EXERCISES

CONCEPTS AND VOCABULARY

Fill in each blank with the appropriate word or phrase. Carefully reread the section if needed.

1. All six trigonometric functions fail the __________ ____ test and therefore are not ____ to ____.

2. The two most common ways of writing the inverse function for $y = \sin x$ are __________ and __________.

3. The domain for the inverse sine function is ______ and the range is ______.

4. The domain for the inverse cosine function is ______ and the range is ______.

5. Most calculators do not have a key for evaluating an expression like $\sec^{-1}5$. Explain how it is done using the COS key.

6. Discuss/explain what is meant by the *implicit form* of an inverse function and the *explicit form*. Give algebraic and trigonometric examples.

DEVELOPING YOUR SKILLS

The tables here show values of $\sin\theta$, $\cos\theta$, and $\tan\theta$ for $\theta \in [-180° \text{ to } 210°]$. The restricted domain used to develop the inverse functions is shaded. Use the information from these tables to complete the exercises that follow.

$y = \sin\theta$

θ	$\sin\theta$	θ	$\sin\theta$
−180°	0	30°	$\frac{1}{2}$
−150°	$-\frac{1}{2}$	60°	$\frac{\sqrt{3}}{2}$
−120°	$-\frac{\sqrt{3}}{2}$	90°	1
−90°	−1	120°	$\frac{\sqrt{3}}{2}$
−60°	$-\frac{\sqrt{3}}{2}$	150°	$\frac{1}{2}$
−30°	$-\frac{1}{2}$	180°	0
0	0	210°	$-\frac{1}{2}$

$y = \cos\theta$

θ	$\cos\theta$	θ	$\cos\theta$
−180°	−1	30°	$\frac{\sqrt{3}}{2}$
−150°	$-\frac{\sqrt{3}}{2}$	60°	$\frac{1}{2}$
−120°	$-\frac{1}{2}$	90°	0
−90°	0	120°	$-\frac{1}{2}$
−60°	$\frac{1}{2}$	150°	$-\frac{\sqrt{3}}{2}$
−30°	$\frac{\sqrt{3}}{2}$	180°	−1
0	1	210°	$-\frac{\sqrt{3}}{2}$

$y = \tan\theta$

θ	$\tan\theta$	θ	$\tan\theta$
−180°	0	30°	$\frac{\sqrt{3}}{3}$
−150°	$\frac{\sqrt{3}}{3}$	60°	$\sqrt{3}$
−120°	$\sqrt{3}$	90°	—
−90°	—	120°	$-\sqrt{3}$
−60°	$-\sqrt{3}$	150°	$-\frac{\sqrt{3}}{3}$
−30°	$-\frac{\sqrt{3}}{3}$	180°	0
0	0	210°	$\sqrt{3}$

Use the preceding tables to fill in each blank (principal values only).

7.

$\sin 0 = 0$	$\sin^{-1}0 =$ ____
$\sin\left(\frac{\pi}{6}\right) =$ ____	$\arcsin\left(\frac{1}{2}\right) = \frac{\pi}{6}$
$\sin\left(-\frac{5\pi}{6}\right) = -\frac{1}{2}$	$\sin^{-1}\left(-\frac{1}{2}\right) =$ ____
$\sin\left(-\frac{\pi}{2}\right) = -1$	$\sin^{-1}(-1) =$ ____

8.

$\sin \pi = 0$	$\sin^{-1}0 =$ ____
$\sin 120° = \frac{\sqrt{3}}{2}$	$\sin^{-1}\left(\frac{\sqrt{3}}{2}\right) =$ ____
$\sin(-60°) = -\frac{\sqrt{3}}{2}$	$\arcsin\left(-\frac{\sqrt{3}}{2}\right) =$ ____
$\sin 180° =$ ____	$\arcsin 0 = 0$

Evaluate without the aid of calculators or tables, *keeping the domain and range of each function in mind.* Answer in radians.

9. $\sin^{-1}\left(\frac{\sqrt{2}}{2}\right)$ **10.** $\arcsin\left(\frac{\sqrt{3}}{2}\right)$ **11.** $\sin^{-1}1$ **12.** $\arcsin\left(-\frac{1}{2}\right)$

Evaluate using a calculator, *keeping the domain and range of each function in mind.* Answer in radians to the nearest ten-thousandth *and* in degrees to the nearest tenth.

13. $\arcsin 0.8892$ **14.** $\arcsin\left(\frac{7}{8}\right)$ **15.** $\sin^{-1}\left(\frac{1}{\sqrt{7}}\right)$ **16.** $\sin^{-1}\left(\frac{1-\sqrt{5}}{2}\right)$

Evaluate each expression.

17. $\sin\left[\sin^{-1}\left(\frac{\sqrt{2}}{2}\right)\right]$ **18.** $\sin\left[\arcsin\left(\frac{\sqrt{3}}{2}\right)\right]$ **19.** $\arcsin\left[\sin\left(\frac{\pi}{3}\right)\right]$

20. $\sin^{-1}(\sin 30°)$ **21.** $\sin^{-1}(\sin 135°)$ **22.** $\arcsin\left[\sin\left(\frac{-2\pi}{3}\right)\right]$

23. $\sin(\sin^{-1} 0.8205)$ **24.** $\sin\left[\arcsin\left(\frac{3}{5}\right)\right]$

Use the tables given prior to Exercise 7 to fill in each blank (principal values only).

25.

$\cos 0 = 1$	$\cos^{-1}1 =$ ____
$\cos\left(\frac{\pi}{6}\right) =$ ____	$\arccos\left(\frac{\sqrt{3}}{2}\right) = \frac{\pi}{6}$
$\cos 120° = -\frac{1}{2}$	$\arccos\left(-\frac{1}{2}\right) =$ ____
$\cos \pi = -1$	$\cos^{-1}(-1) =$ ____

26.

$\cos(-60°) = \frac{1}{2}$	$\cos^{-1}\left(\frac{1}{2}\right) =$ ____
$\cos\left(-\frac{\pi}{6}\right) = \frac{\sqrt{3}}{2}$	$\cos^{-1}\left(\frac{\sqrt{3}}{2}\right) =$ ____
$\cos(-120°) =$ ____	$\arccos\left(-\frac{1}{2}\right) = 120°$
$\cos(2\pi) = 1$	$\cos^{-1}1 =$ ____

Evaluate without the aid of calculators or tables. Answer in radians.

27. $\cos^{-1}\left(\frac{1}{2}\right)$ **28.** $\arccos\left(-\frac{\sqrt{3}}{2}\right)$ **29.** $\cos^{-1}(-1)$ **30.** $\arccos(0)$

Evaluate using a calculator. Answer in radians to the nearest ten-thousandth, degrees to the nearest tenth.

31. $\arccos 0.1352$ **32.** $\arccos\left(\frac{4}{7}\right)$ **33.** $\cos^{-1}\left(\frac{\sqrt{5}}{3}\right)$ **34.** $\cos^{-1}\left(\frac{\sqrt{6}-1}{5}\right)$

Evaluate each expression.

35. $\arccos\left[\cos\left(\frac{\pi}{4}\right)\right]$ **36.** $\cos^{-1}(\cos 60°)$ **37.** $\cos(\cos^{-1}0.5560)$

38. $\cos\left[\arccos\left(-\frac{8}{17}\right)\right]$ **39.** $\cos\left[\cos^{-1}\left(-\frac{\sqrt{2}}{2}\right)\right]$ **40.** $\cos\left[\arccos\left(\frac{\sqrt{3}}{2}\right)\right]$

41. $\cos^{-1}\left[\cos\left(\frac{5\pi}{4}\right)\right]$ **42.** $\arccos(\cos 44.2°)$

Use the tables presented before Exercise 7 to fill in each blank. Convert from radians to degrees as needed.

43.

$\tan 0 = 0$	$\tan^{-1} 0 = $ ____
$\tan\left(-\frac{\pi}{3}\right) = $ ____	$\arctan(-\sqrt{3}) = -\frac{\pi}{3}$
$\tan 30° = \frac{\sqrt{3}}{3}$	$\arctan\left(\frac{\sqrt{3}}{3}\right) = $ ____
$\tan\left(\frac{\pi}{3}\right) = $ ____	$\tan^{-1}(\sqrt{3}) = $ ____

44.

$\tan(-150°) = \frac{\sqrt{3}}{3}$	$\tan^{-1}\left(\frac{\sqrt{3}}{3}\right) = $ ____
$\tan \pi = 0$	$\tan^{-1} 0 = $ ____
$\tan 120° = -\sqrt{3}$	$\arctan(-\sqrt{3}) = $ ____
$\tan\left(\frac{\pi}{4}\right) = $ ____	$\arctan 1 = \frac{\pi}{4}$

Evaluate without the aid of calculators or tables.

45. $\tan^{-1}\left(-\frac{\sqrt{3}}{3}\right)$ **46.** $\arctan(-1)$ **47.** $\arctan(\sqrt{3})$ **48.** $\tan^{-1} 0$

Evaluate using a calculator, *keeping the domain and range of each function in mind.* Answer in radians to the nearest ten-thousandth *and* in degrees to the nearest tenth.

49. $\tan^{-1}(-2.05)$ **50.** $\tan^{-1}(0.3267)$ **51.** $\arctan\left(\frac{29}{21}\right)$ **52.** $\arctan(-\sqrt{6})$

Simplify each expression without using a calculator.

53. $\sin^{-1}\left[\cos\left(\frac{2\pi}{3}\right)\right]$ **54.** $\cos^{-1}\left[\sin\left(-\frac{\pi}{3}\right)\right]$ **55.** $\tan\left[\arccos\left(\frac{\sqrt{3}}{2}\right)\right]$

56. $\sec\left[\arcsin\left(\frac{1}{2}\right)\right]$ **57.** $\csc\left[\sin^{-1}\left(\frac{\sqrt{2}}{2}\right)\right]$ **58.** $\cot\left[\cos^{-1}\left(-\frac{1}{2}\right)\right]$

59. $\arccos[\sin(-30°)]$ **60.** $\arcsin(\cos 135°)$

Explain why the following expressions are not defined.

61. $\tan(\sin^{-1} 1)$ **62.** $\cot(\arccos 1)$ **63.** $\sin^{-1}\left[\csc\left(\frac{\pi}{4}\right)\right]$ **64.** $\cos^{-1}\left[\sec\left(\frac{2\pi}{3}\right)\right]$

Use the diagrams below to write the value of: (a) $\sin\theta$, (b) $\cos\theta$, and (c) $\tan\theta$.

65. **66.** **67.** **68.**

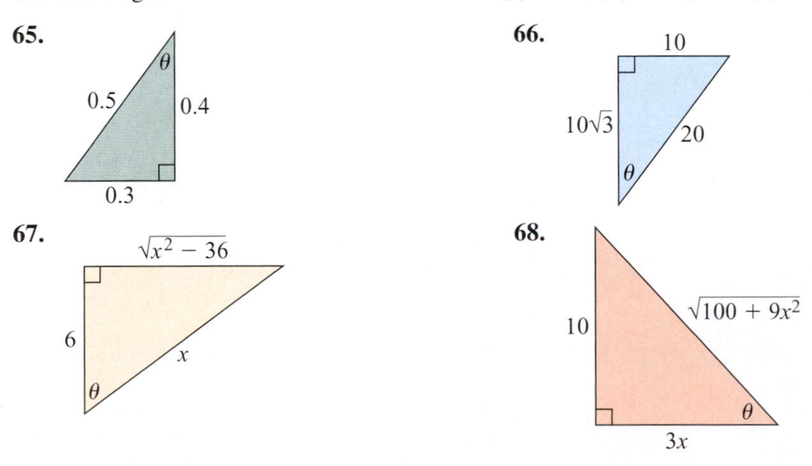

Evaluate each expression by drawing a right triangle and labeling the sides.

69. $\sin\left[\cos^{-1}\left(-\frac{7}{25}\right)\right]$ **70.** $\cos\left[\sin^{-1}\left(-\frac{11}{61}\right)\right]$ **71.** $\sin\left[\tan^{-1}\left(\frac{\sqrt{5}}{2}\right)\right]$

72. $\tan\left[\cos^{-1}\left(\frac{\sqrt{23}}{12}\right)\right]$ **73.** $\cot\left[\arcsin\left(\frac{3x}{5}\right)\right]$ **74.** $\tan\left[\operatorname{arcsec}\left(\frac{5}{2x}\right)\right]$

75. $\cos\left[\sin^{-1}\left(\frac{x}{\sqrt{12+x^2}}\right)\right]$ **76.** $\tan\left[\sec^{-1}\left(\frac{\sqrt{9+x^2}}{x}\right)\right]$

Simplify the expression using the substitution given and state the domain for which the new expression is valid. Then use the TABLE feature of a graphing calculator to show the expressions are equivalent.

77. $\frac{\sqrt{25+x^2}}{x}$; $x = 5\tan\theta$ **78.** $\frac{\sqrt{36-x^2}}{x}$; $x = 6\sin\theta$ **79.** $\frac{\sqrt{16-x^2}}{x}$; $x = 4\cos\theta$ **80.** $\frac{\sqrt{49+x^2}}{x}$; $x = 7\tan\theta$

Use the tables given prior to Exercise 7 to help fill in each blank.

81.

$\sec 0 = 1$	$\sec^{-1}1 =$ ____
$\sec\left(\frac{\pi}{3}\right) =$ ____	$\operatorname{arcsec} 2 = \frac{\pi}{3}$
$\sec(-30°) = \frac{2}{\sqrt{3}}$	$\operatorname{arcsec}\left(\frac{2}{\sqrt{3}}\right) =$ ____
$\sec(\pi) =$ ____	$\sec^{-1}(-1) =$ ____

82.

$\sec(-60°) = 2$	$\operatorname{arcsec} 2 =$ ____
$\sec\left(\frac{7\pi}{6}\right) = -\frac{2}{\sqrt{3}}$	$\operatorname{arcsec}\left(-\frac{2}{\sqrt{3}}\right) =$ ____
$\sec(-360°) = 1$	$\operatorname{arcsec} 1 =$ ____
$\sec(60°) =$ ____	$\sec^{-1}2 = 60°$

Evaluate using a calculator only as necessary.

83. $\operatorname{arccsc} 2$ **84.** $\csc^{-1}\left(-\frac{2}{\sqrt{3}}\right)$ **85.** $\cot^{-1}\sqrt{3}$ **86.** $\operatorname{arccot}(-1)$

87. $\operatorname{arcsec} 5.789$ **88.** $\cot^{-1}\left(-\frac{\sqrt{7}}{2}\right)$ **89.** $\sec^{-1}\sqrt{7}$ **90.** $\operatorname{arccsc} 2.9875$

WORKING WITH FORMULAS

91. The force normal to an object on an inclined plane: $F_N = mg\cos\theta$

When an object is on an inclined plane, the **normal force** is the force acting perpendicular to the plane and away from the force of gravity, and is measured in a unit called **newtons (N).** The magnitude of this force depends on the angle of incline of the plane according to the formula above, where m is the mass of the object in kilograms and g is the force of gravity (9.8 m/sec^2). Given $m = 225$ g, find (a) F_N for $\theta = 15°$ and $\theta = 45°$ and (b) θ for $F_N = 1$ N and $F_N = 2$ N.

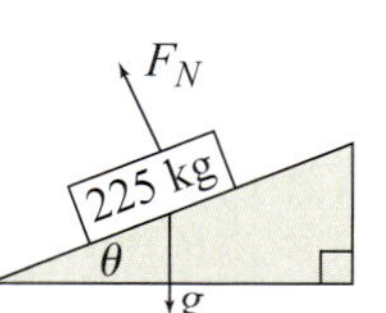

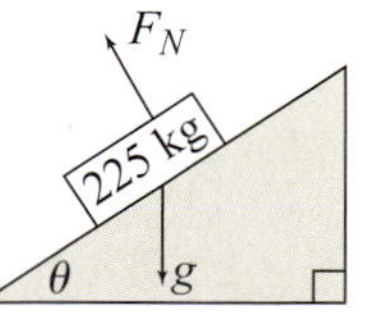

92. Heat flow on a cylindrical pipe:

$$T = (T_0 - T_R)\sin\left(\frac{y}{\sqrt{x^2+y^2}}\right) + T_R;\ y \geq 0$$

When a circular pipe is exposed to a fan-driven source of heat, the temperature of the air reaching the pipe is greatest at the point nearest to the source (see diagram). As you move around the circumference of the pipe away from the source, the temperature of the air reaching the pipe gradually decreases. One possible model of this phenomenon is given by the formula shown, where T is the temperature of the air at a point (x, y)

Exercise 92

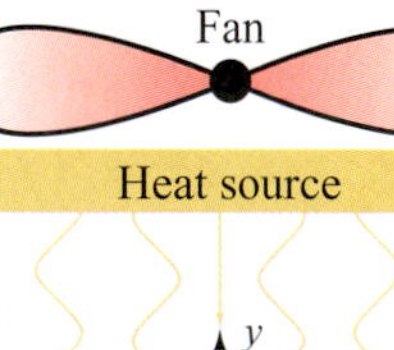

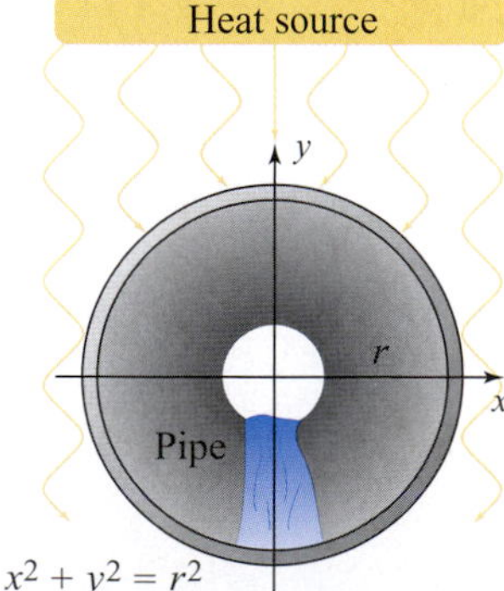

on the circumference of a pipe with outer radius $r = \sqrt{x^2 + y^2}$, T_0 is the temperature of the air at the source, and T_R is the surrounding room temperature. Assuming $T_0 = 220°\text{F}$, $T_R = 72°$ and $r = 5$ cm: (a) Find the temperature of the air at the points (0, 5), (3, 4), (4, 3), (4.58, 2), and (4.9, 1). (b) Why is the temperature decreasing for this sequence of points? (c) Simplify the formula using $r = 5$ and use it to find two points on the pipe's circumference where the temperature of the air is 113°.

APPLICATIONS

Exercise 93

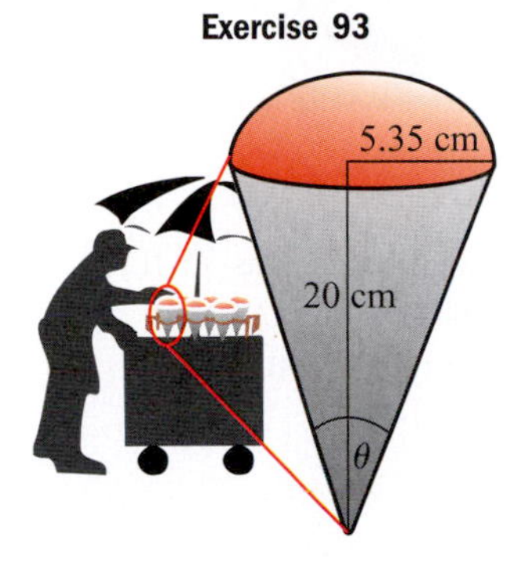

93. Snowcone dimensions: *Made in the Shade Snowcones* sells a colossal size cone that uses a conical cup holding 20 oz of ice and liquid. The cup is 20 cm tall and has a radius of 5.35 cm. Find the angle θ formed by a cross-section of the cup.

94. Avalanche conditions: Winter avalanches occur for many reasons, one being the slope of the mountain. Avalanches seem to occur most often for slopes between 35° and 60° (snow gradually slides off steeper slopes). The slopes at a local ski resort have an average rise of 2000 ft for each horizontal run of 2559 ft. Is this resort prone to avalanches? Find the angle θ and respond.

Exercise 94

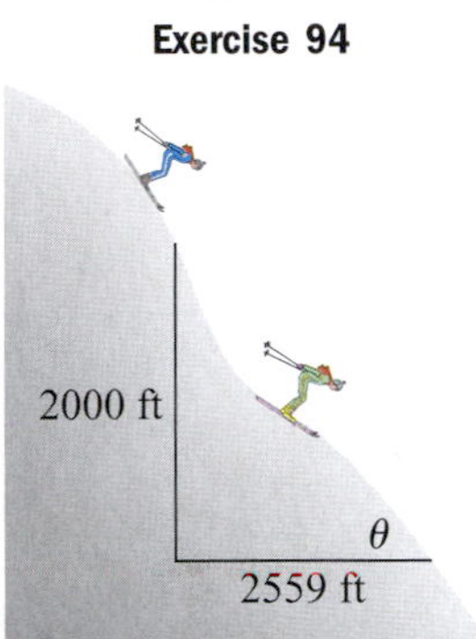

Exercise 95

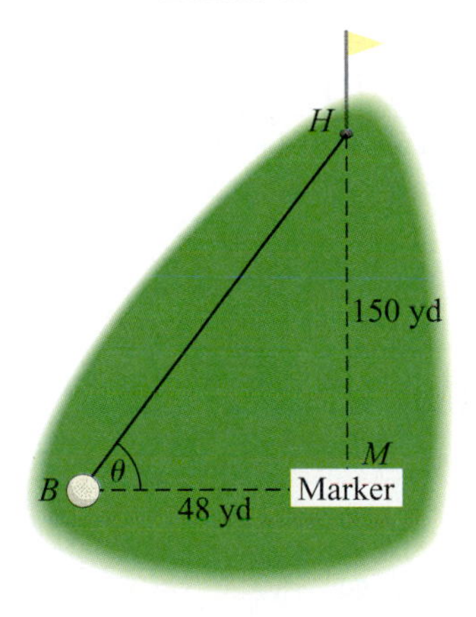

95. Distance to hole: A popular story on the PGA Tour has Gerry Yang, Tiger Woods' teammate at Stanford and occasional caddie, using the Pythagorean theorem to find the distance Tiger needed to reach a particular hole. Suppose you notice a marker in the ground stating that the straight line distance from the marker to the hole (H) is 150 yd. If your ball B is 48 yd from the marker (M) and angle BMH is a right angle, determine the angle θ and *your* straight line distance from the hole.

96. Ski jumps: At a waterskiing contest on a large lake, skiers use a ramp rising out of the water that is 30 ft long and 10 ft high at the high end. What angle θ does the ramp make with the lake?

Exercise 96

θ 10 ft 30 ft

97. For $Y_1 = \dfrac{x}{\sqrt{100 - x^2}}$: (a) find an equivalent expression Y_2 using the substitution $x = 10 \sin \theta$; (b) express Y_2 in terms of x using $x = 10 \sin \theta$ (solve for θ) and substitution; and (c) use the TABLE feature of a graphing calculator to show that $Y_1 = Y_2$ for $x \in (-10, 10)$.

98. For $Y_1 = \dfrac{\sqrt{x^2 - 1}}{|x|}$: (a) find an equivalent expression Y_2 using the substitution $x = \sec \theta$; (b) express Y_2 in terms of x using $x = \sec \theta$ (solve for θ) and substitution; and (c) use the TABLE feature of a graphing calculator to show that $Y_1 = Y_2$ for $x \neq 0$.

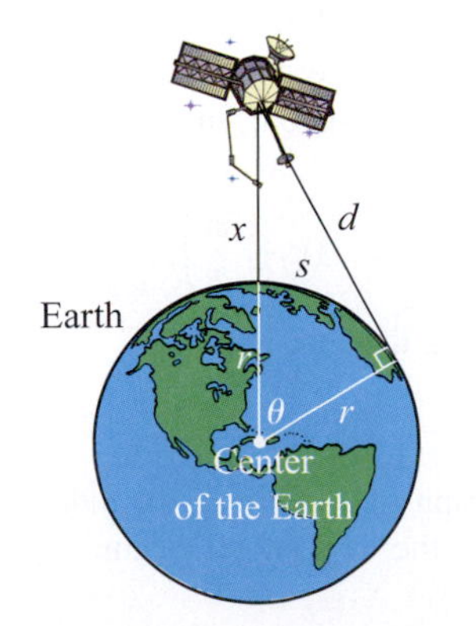

WRITING, RESEARCH, AND DECISION MAKING

Consider a satellite orbiting at an altitude of x mi above the Earth. The distance d from the satellite to the horizon and the length s of the corresponding arc of the Earth are shown in the diagram.

99. To find the distance d we use the formula $d = \sqrt{2rx + x^2}$. Show how this formula was developed using the Pythagorean theorem.

100. Find a formula for the angle θ in terms of r and x, then a formula for the arc length s.

101. If the Earth has a radius of 3960 mi and the satellite is orbiting at an altitude of 150 mi, (a) what is the measure of angle θ? (b) How much longer is d than s?

EXTENDING THE CONCEPT

A projectile is any object that is shot, thrown, slung, or otherwise projected and has no continuing source of propulsion. The horizontal and vertical position of the projectile depends on its initial velocity, angle of projection, and height of release (air resistance is neglected). The horizontal position of the projectile is given by $x = v_0\cos\theta\, t$, while its vertical position is modeled by $y = y_0 + v_0\sin\theta\, t - 16t^2$, where y_0 is the height it is projected from, θ is the projection angle, and t is the elapsed time in seconds.

102. A circus clown is shot out of a specially made cannon at an angle of 55°, with an initial velocity of 85 ft/sec, and the end of the cannon 10 ft high.

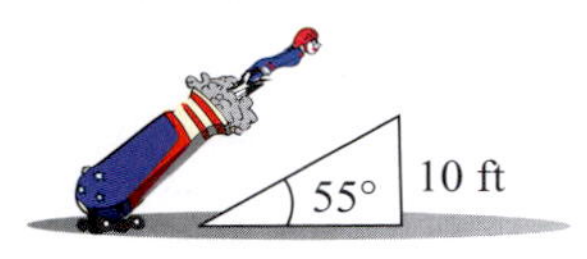

a. Find the position of the safety net (distance from the cannon and height from the ground) if the clown hits the net after 4.3 sec.

b. Find the angle at which the clown was shot if the initial velocity was 75 ft/sec and hits a net which is placed 175.5 ft away after 3.5 sec.

Exercise 103

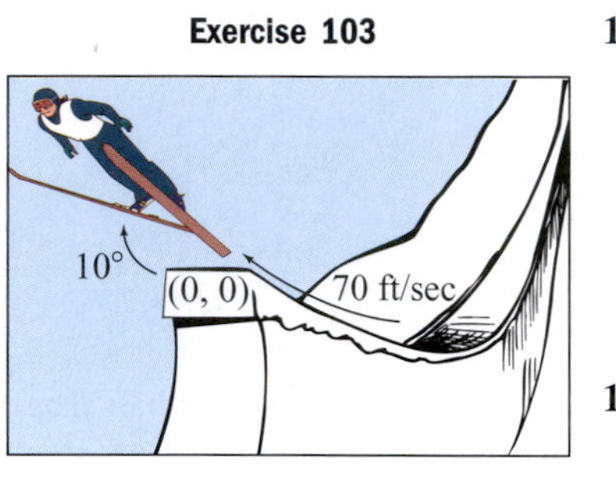

103. A winter ski jumper leaves the ski-jump with an initial velocity of 70 ft/sec at an angle of 10°. Assume the jump-off point has coordinates (0, 0).

a. What is the horizontal position of the skier after 6 sec?

b. What is the vertical position of the skier after 6 sec?

c. What diagonal distance (down the mountain side) was traveled if the skier touched down after being airborne for 6 sec?

104. Suppose the domain of $y = \csc x$ was restricted to $x \in \left[-\frac{\pi}{2}, 0\right) \cup \left(0, \frac{\pi}{2}\right]$, and the domain of $y = \cot x$ to $x \in (0, \pi)$. **(a)** Would these functions then be one-to-one? **(b)** What are the corresponding ranges? **(c)** State the domain and range of $y = \csc^{-1}x$ and $y = \cot^{-1}x$. **(d)** Graph each function.

MAINTAINING YOUR SKILLS

105. (3.4) Use the triangle given with a double-angle identity to find the exact value of $\sin(2\theta)$.

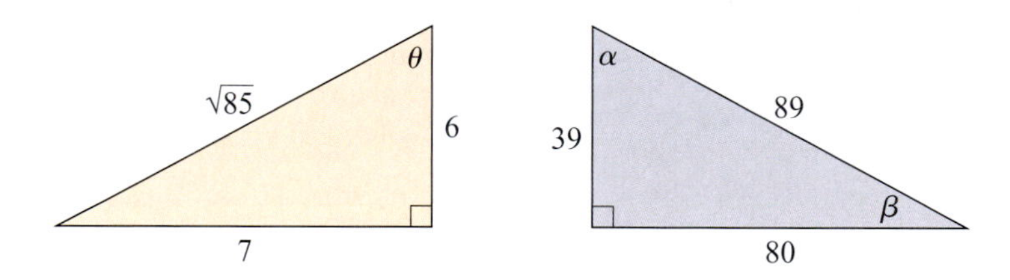

106. (3.3) Use the triangle given with a sum identity to find the exact value of $\sin(\alpha + \beta)$.

107. (1.1) Charlene just bought her daughter a battery operated jeep. If the wheels have a radius of 4 in. and are turning at 3.5 revolutions per second at top speed, find the top speed of the vehicle in miles per hour, rounded to the nearest tenth.

108. (2.1) State the amplitude, period, and horizontal shift for $y = 5\sin\left(3x - \frac{\pi}{2}\right)$.

109. (1.3) Evaluate $\sin\theta$, $\cos\theta$, and $\tan\theta$ if the terminal side is along the line $y = \frac{12}{5}x$ with θ in QI.

110. (2.1) Explain how the graph of $y = \sec x$ is related to the graph of $y = \cos x$. Include a discussion of the domain and range of $y = \sec x$ and where the asymptotes occur.

4.3 Solving Basic Trig Equations

LEARNING OBJECTIVES

In Section 4.3 you will learn how to:

A. Use a graph to gain information about principal roots, roots in $[0, 2\pi)$, and roots in $\mathbb{R}$

B. Use inverse functions to solve trig equations for the principal root

C. Solve trig equations for roots in $[0, 2\pi)$ or $[0, 360°)$

D. Solve trig equations for roots in $\mathbb{R}$

E. Solve trig equations using fundamental identities

F. Solve trig equations using graphing technology

INTRODUCTION

In this section we'll take the elements of basic equation solving and use them to help solve **trig equations,** or equations containing trigonometric functions. All of the algebraic techniques previously used can be applied to these equations, including the properties of equality and all forms of factoring (common terms, difference of squares, etc.). As with polynomial equations, we continue to be concerned with the *number of solutions* as well as with the *solutions themselves*, but there is one major difference. There is no "algebra" that can transform a function like $\sin x = \frac{1}{2}$ into $x = \textit{solution}$. For that we rely on the inverse trig functions from Section 4.2.

POINT OF INTEREST

The algebraic equation $\frac{3x}{5} = 1$ is of a very simple form, and is solved quite easily with solution $x = \frac{5}{3}$. Likewise the trig equation $2\sin x = 2$ can be solved with little difficulty, as we recognize that for $\sin x = 1$, $x = \frac{\pi}{2}$. But the equation $2\sin x + \frac{3x}{5} = 2$, containing both of these expressions, poses an immense challenge to would-be solvers—it is not possible to "get x alone on one side" using elementary means. In ancient times, equations of this type were "solved" by constructing a table of values, in an exhaustive search for all numbers x that made the equation true. Today we can quickly find approximate solutions on a graphing calculator (see Example 9).

A. The Principal Root, Roots in $[0, 2\pi)$, and Real Roots

One of the greatest advantages a student can bring to a study of trig equations is the ability to think in pictures. In a study of polynomial equations, making a connection between the degree of an equation, its graph, and its possible roots, helped give insights as to the number, location, and nature of the roots. Similarly, keeping graphs of basic trig functions *constantly* in mind helps you gain information regarding the solutions to trig equations.

This is one reason we carefully developed the transformations of these graphs in Section 2.3. When solving trig equations, we refer to the solution found using $\sin^{-1}$, $\cos^{-1}$, and $\tan^{-1}$ as the **principal root.** You will alternatively be asked to find (1) the principal root, (2) solutions in $[0, 2\pi)$ or $[0°, 360°)$, or (3) solutions from the set of real numbers $\mathbb{R}$. For convenience, graphs of the basic sine, cosine, and tangent functions are repeated in Figures 4.32 through 4.34. Take a mental snapshot of them and keep them close at hand.

WORTHY OF NOTE

For the polynomial equation $y = -2(x - 5)^2 + 1$, knowing it has degree two (a parabola), is stretched vertically ($|a| = 2$) and is concave down ($a < 0$) with vertex at (5, 1), enables us to conclude that there are two positive real roots.

Figure 4.33

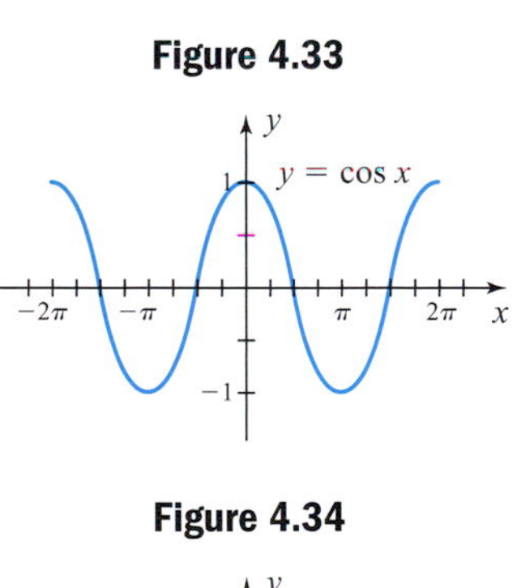

Figure 4.32

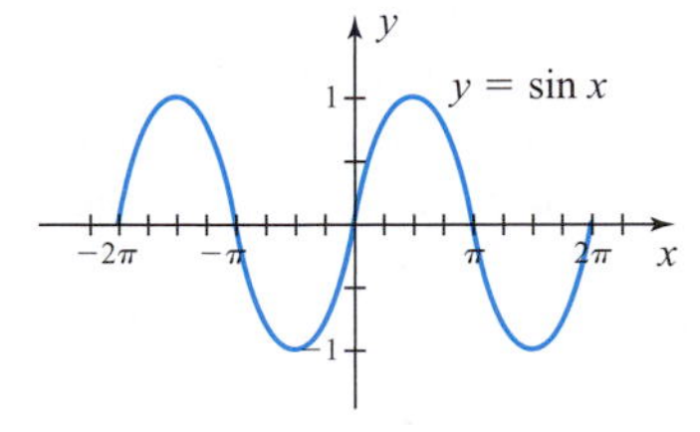

Figure 4.34

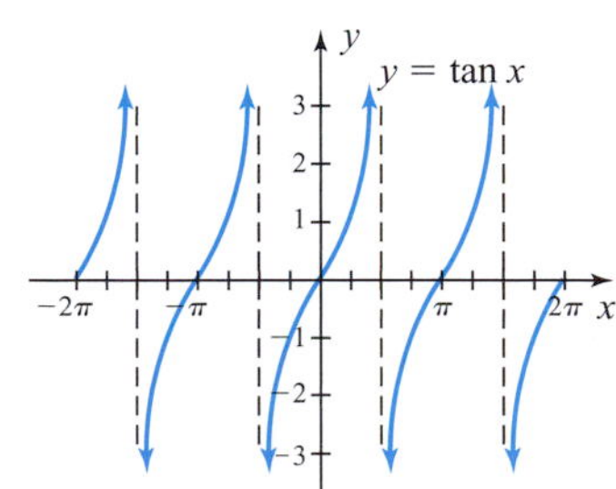

EXAMPLE 1 Consider the equation $\sin x = \frac{2}{3}$. Using a graph of $y = \sin x$ and $y = \frac{2}{3}$, (a) state the quadrant of the principal root, (b) state the number of roots in $[0, 2\pi)$ and their quadrants, and (c) comment on the number of real roots.

Solution: We begin by drawing a quick sketch of $y = \sin x$ and $y = \frac{2}{3}$, noting that solutions will occur where the graphs intersect.

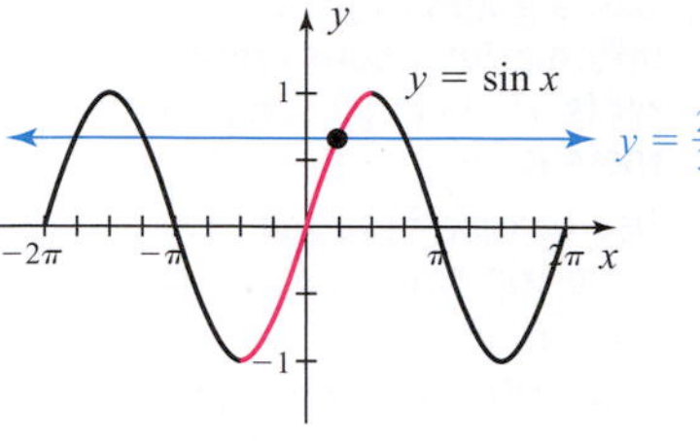

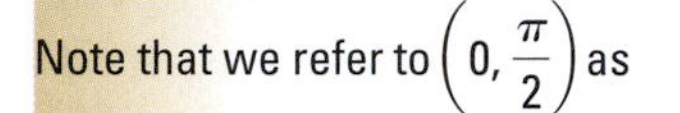

Note that we refer to $\left(0, \frac{\pi}{2}\right)$ as Quadrant I or QI, regardless of whether we're discussing the unit circle or the graph of the function.

a. The sketch shows the principal root occurs between 0 and $\frac{\pi}{2}$ (recall there is only one principal root since we required a one-to-one relationship between sine and arcsine). The solution occurs in QI.

b. For $[0, 2\pi)$ we note the graphs intersect twice. There will be two solutions in this interval, where $\sin x$ is positive (graphically—above the x-axis). Note that these solutions correspond to those found in QI and QII on the unit circle, where $\sin x$ is also positive.

c. Since the graphs of $y = \sin x$ and $y = \frac{2}{3}$ extend infinitely in both directions, they will intersect an infinite number of times—*but at regular intervals!* Once a root is found, adding integer multiples of 2π (the period of the sine function) to this root will give the location of additional roots.

NOW TRY EXERCISES 7 THROUGH 10

When this process is applied to the equation $\tan x = -2$, the graph shows the principal root occurs between $-\frac{\pi}{2}$ and 0 in QIV (see Figure 4.35). In the interval $[0, 2\pi)$ the graphs intersect twice, in QII and QIV where $\tan x$ is negative (graphically—below the x-axis). As in Example 1, the graphs continue infinitely and will intersect an infinite number of times—*but again at regular intervals!* Once a root is found, adding integer multiples of π (the period of tangent) to this root will give the location of other roots.

Figure 4.35

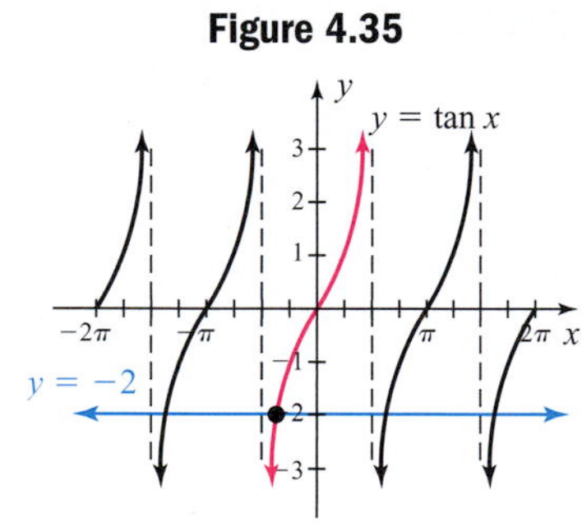

B. Inverse Functions and Principal Roots

To solve equations having a single variable term, the basic goal is to isolate the variable term and apply the inverse function or operation. This is true for algebraic equations like $2x - 1 = 0$, $2\sqrt{x} - 1 = 0$, or $2x^2 - 1 = 0$, and for trig equations like $2\sin x - 1 = 0$. In each case we would add 1 to both sides, divide by 2, then apply the appropriate inverse function. Recall that when the inverse trig functions are applied, the result is only the principal root and other solutions may exist depending on the interval under consideration.

EXAMPLE 2 Find the principal root of $\sqrt{3}\tan x - 1 = 0$.

Solution: We begin by isolating the variable term, then apply the inverse function.

$$\sqrt{3}\tan x - 1 = 0 \quad \text{given equation}$$

$$\tan x = \frac{1}{\sqrt{3}} \quad \text{add 1 and divide by } \sqrt{3}$$

$$\tan^{-1}(\tan x) = \tan^{-1}\left(\frac{1}{\sqrt{3}}\right) \quad \text{apply inverse tangent to both sides}$$

$$x = \frac{\pi}{6} \quad \text{result (exact form)}$$

NOW TRY EXERCISES 11 THROUGH 28

GRAPHICAL SUPPORT

Graphing $Y_1 = \tan x$ and $Y_2 = \frac{1}{\sqrt{3}}$ on a graphing calculator supports the solution from Example 2. Note the line intersects the principal branch of the tangent function once (using the 2nd TRACE (**CALC**) 5: intersect feature) in QI where tan x is positive. Noting $\frac{\pi}{6} \approx 0.5236$ further supports the solution indicated.

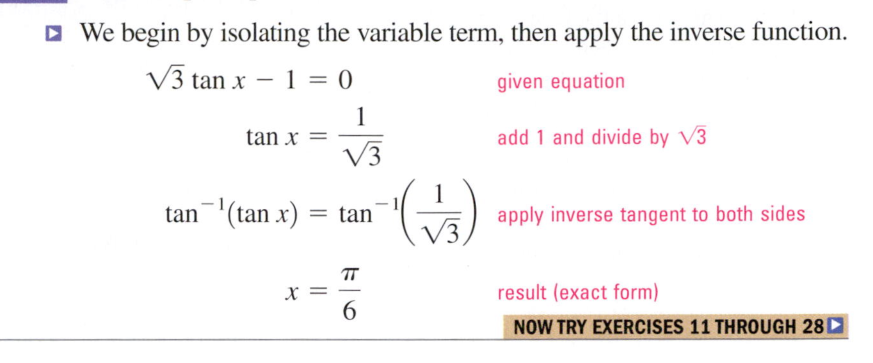

Table 4.2

θ	$\sin\theta$	$\cos\theta$
0	0	1
$\frac{\pi}{6}$	$\frac{1}{2}$	$\frac{\sqrt{3}}{2}$
$\frac{\pi}{4}$	$\frac{\sqrt{2}}{2}$	$\frac{\sqrt{2}}{2}$
$\frac{\pi}{3}$	$\frac{\sqrt{3}}{2}$	$\frac{1}{2}$
$\frac{\pi}{2}$	1	0
$\frac{2\pi}{3}$	$\frac{\sqrt{3}}{2}$	$-\frac{1}{2}$
$\frac{3\pi}{4}$	$\frac{\sqrt{2}}{2}$	$-\frac{\sqrt{2}}{2}$
$\frac{5\pi}{6}$	$\frac{1}{2}$	$-\frac{\sqrt{3}}{2}$
π	0	-1

Equations like the one in Example 2 demonstrate the need to be *very* familiar with the functions of a standard angle. They are frequently used in equations and applications to ensure results don't get so messy they obscure the main ideas. For convenience, the values of $\sin\theta$ and $\cos\theta$ are repeated in Table 4.2 for $x \in [0, \pi)$. Using symmetry and the appropriate sign, the table can easily be extended to all values in $[0, 2\pi)$. Using the reciprocal and ratio relationships, values for the other trig functions can also be found.

C. Solving Trig Equations for Roots in $[0, 2\pi)$ or $[0°, 360°)$

To find multiple solutions to a trig equation, we simply take the reference angle of the principal root, and *use this angle to find all solutions* within a specified range. A mental image of the graph still guides us, and the standard table of values (also held in memory) allows for a quick solution while helping solidify concepts.

EXAMPLE 3 For $2\cos\theta + \sqrt{2} = 0$, find all solutions in $[0, 2\pi)$.

Solution: Isolate the variable term, then apply the inverse function.

$$2\cos\theta + \sqrt{2} = 0 \quad \text{given equation}$$

$$\cos\theta = -\frac{\sqrt{2}}{2} \quad \text{subtract } \sqrt{2} \text{ and divide by 2}$$

$$\cos^{-1}(\cos\theta) = \cos^{-1}\left(-\frac{\sqrt{2}}{2}\right) \quad \text{apply inverse cosine to both sides}$$

$$\theta = \frac{3\pi}{4} \quad \text{result}$$

WORTHY OF NOTE

Note how the graph of a trig function displays the information regarding *quadrants*. From the graph of $y = \cos x$ we "read" that cosine is negative in QII and QIII [the lower "hump" of the graph is below the x-axis in $(\pi/2, 3\pi/2)$] and positive in QI and QIV [the graph is above the x-axis in the intervals $(0, \pi/2)$ and $(3\pi/2, 2\pi)$].

With $\frac{3\pi}{4}$ as the principal root, we have a reference angle of $\theta_r = \frac{\pi}{4}$. Since we are solving $\cos\theta = -\frac{\sqrt{2}}{2}$ and $\cos x$ is negative in QII and QIII, the reference angle $\frac{\pi}{4}$ gives a second solution of $\frac{5\pi}{4}$. Our (mental) graph verifies these are the only solutions in $[0, 2\pi)$, and supports $\theta = \frac{3\pi}{4}$ and $\theta = \frac{5\pi}{4}$ as solutions. The second solution could also have been found from memory, recognition, or symmetry on the unit circle.

NOW TRY EXERCISES 29 THROUGH 34

EXAMPLE 4 For $\tan^2 x - 1 = 0$, find all solutions in $[0, 2\pi)$.

Solution: As with the other equations having a single variable term, we try to isolate this term or attempt a solution by factoring.

$$\tan^2 x - 1 = 0 \quad \text{given equation}$$
$$\sqrt{\tan^2 x} = \pm\sqrt{1} \quad \text{add 1 to both sides and take square roots}$$
$$\tan x = \pm 1 \quad \text{result}$$

This time the algebra gives two solutions, which we pursue independently.

$$\tan x = 1 \qquad\qquad \tan x = -1$$
$$\tan^{-1}(\tan x) = \tan^{-1}(1) \qquad \tan^{-1}(\tan x) = \tan^{-1}(-1) \quad \text{apply inverse tangent}$$
$$x = \frac{\pi}{4} \qquad\qquad x = -\frac{\pi}{4} \quad \text{principal roots}$$

WORTHY OF NOTE

To make the point more forcefully, the standard values are used so regularly that students must know them as well as they know the multiplication tables. This will help give an automatic response to the numerous equations that involve these values.

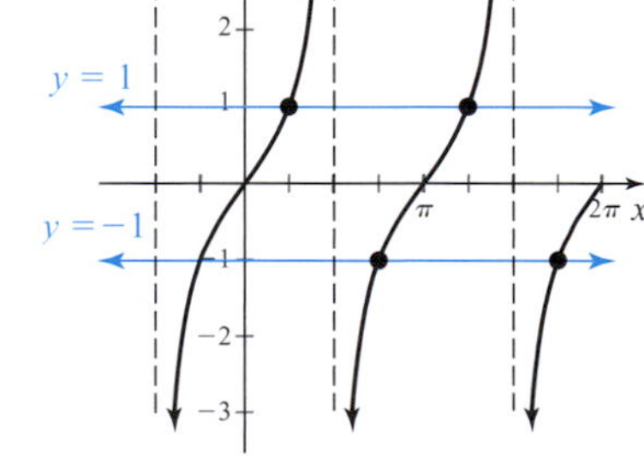

Of the principal roots, only $x = \frac{\pi}{4}$ is in the specified interval. Its reference angle is $\frac{\pi}{4}$ (itself), and with $\tan x$ positive in QI and QIII, a second solution is $\frac{5\pi}{4}$. Since $x = -\frac{\pi}{4}$ is not in the interval, we use it as a reference angle in QII and QIV (where $\tan x$ is negative, since we're solving $\tan x = -1$) and find the solutions $x = \frac{3\pi}{4}$ and

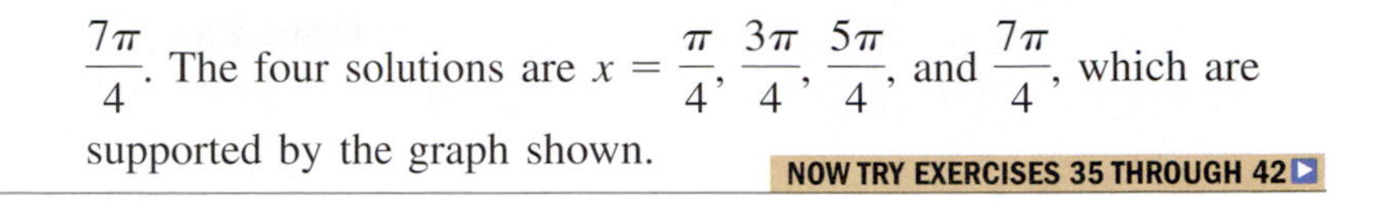

$\frac{7\pi}{4}$. The four solutions are $x = \frac{\pi}{4}, \frac{3\pi}{4}, \frac{5\pi}{4}$, and $\frac{7\pi}{4}$, which are supported by the graph shown.

NOW TRY EXERCISES 35 THROUGH 42

For any trig function that is not equal to a standard value, we use a calculator to approximate the principal root, and apply the same ideas to this root to find all solutions in the interval.

EXAMPLE 5 Find all solutions in $[0°, 360°)$ for $3\cos^2\theta + \cos\theta - 2 = 0$.

Solution: Use a u-substitution to simplify the equation and help select an appropriate strategy. For $u = \cos\theta$, the equation becomes $3u^2 + u - 2 = 0$ and factoring seems the best approach. The factored form is $(u + 1)(3u - 2) = 0$, with solutions $u = -1$ and $u = \frac{2}{3}$. Re-substituting $\cos\theta$ for u gives

$$\cos\theta = -1 \qquad \cos\theta = \frac{2}{3} \qquad \text{equations from factored form}$$

$$\cos^{-1}(\cos\theta) = \cos^{-1}(-1) \qquad \cos^{-1}(\cos\theta) = \cos^{-1}\left(\frac{2}{3}\right) \qquad \text{apply inverse cosine}$$

$$\theta = 180° \qquad \theta \approx 48.2° \qquad \text{principal roots}$$

Both principal roots are in the specified interval. The first is quadrantal, the second is found using a calculator and is approximately 48.2°. Its reference angle is 48.2° (itself), and with $\cos x$ positive in QI and QIV (from $\cos\theta = \frac{2}{3}$), a second solution is $(360 - 48.2)° = 311.8°$. The three solutions are $\theta = 48.2°$, 180°, and 311.8°.

NOW TRY EXERCISES 43 THROUGH 50

D. Solving Trig Equations for All Real Roots ($\mathbb{R}$)

As we noted, the intersections of a trig function with a horizontal line occur at regular, *predictable* intervals. This makes finding solutions from the set of real numbers a simple matter of extending the solutions we found in $[0, 2\pi)$ or $[0°, 360°)$. To illustrate, consider the solutions to Example 3. For $2\cos\theta + \sqrt{2} = 0$, we found the solutions in $[0, 2\pi)$ were $\theta = \frac{3\pi}{4}$ and $\theta = \frac{5\pi}{4}$. For solutions in $\mathbb{R}$, we note the "predictable interval" between roots *is identical to the period of the function.* This means all real solutions are represented by $\theta = \frac{3\pi}{4} + 2\pi k$ and $\theta = \frac{5\pi}{4} + 2\pi k$, $k \in \mathbb{Z}$ (k is an integer). Both are illustrated in Figures 4.36 and 4.37 with the primary solution indicated with a "*."

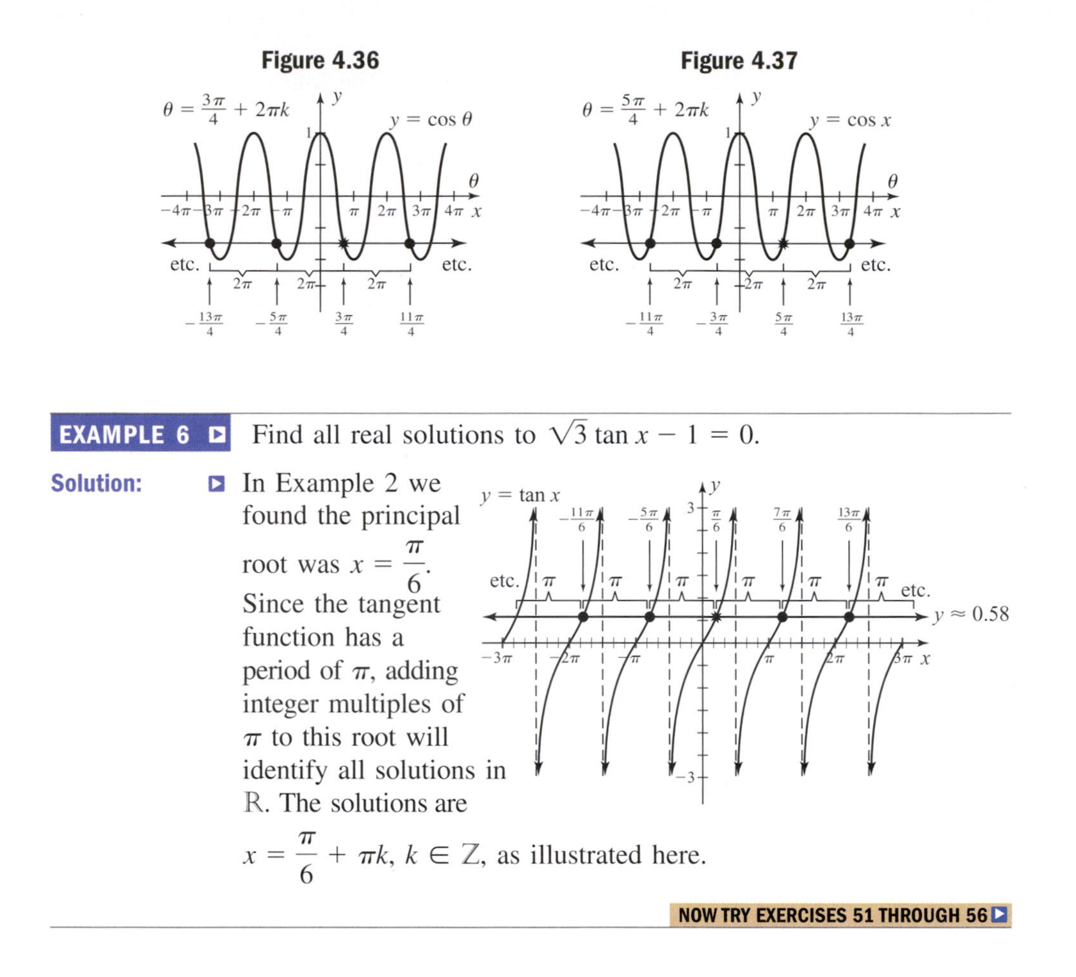

Figure 4.36 **Figure 4.37**

EXAMPLE 6 Find all real solutions to $\sqrt{3}\tan x - 1 = 0$.

Solution: In Example 2 we found the principal root was $x = \frac{\pi}{6}$. Since the tangent function has a period of π, adding integer multiples of π to this root will identify all solutions in $\mathbb{R}$. The solutions are $x = \frac{\pi}{6} + \pi k,\ k \in \mathbb{Z}$, as illustrated here.

NOW TRY EXERCISES 51 THROUGH 56

These fundamental ideas can be extended to many different situations. When asked to find *all real solutions,* be sure you find all roots in a stipulated interval before naming solutions by applying the period of the function. For instance, $\cos x = 0$ has two solutions in $[0, 2\pi)$ $\left[x = \frac{\pi}{2} \text{ and } x = \frac{3\pi}{2}\right]$, which we can quickly extend to find all real roots. But using $x = \cos^{-1}0$ or a calculator limits us to the single (principal) root $x = \frac{\pi}{2}$, and we'd miss all solutions stemming from $\frac{3\pi}{2}$. Note that solutions involving multiples of an angle (or fractional parts of an angle) should likewise be "handled with care," as in Example 7.

EXAMPLE 7 Find all real solutions to $2\sin(2x)\cos x - \cos x = 0$.

Solution: Since we have a common factor of $\cos x$, we begin by rewriting the equation as $\cos x[2\sin(2x) - 1] = 0$ and solve using the zero factor property. The resulting equations are $\cos x = 0$ and $2\sin(2x) - 1 = 0 \rightarrow \sin(2x) = \frac{1}{2}$.

$$\cos x = 0 \qquad \sin(2x) = \frac{1}{2} \qquad \text{equations from factored form}$$

In $[0, 2\pi)$, $\cos x = 0$ has solutions $x = \frac{\pi}{2}$ and $x = \frac{3\pi}{2}$, giving $x = \frac{\pi}{2} + 2\pi k$ and $x = \frac{3\pi}{2} + 2\pi k$ as solutions in $\mathbb{R}$. Note these can actually be combined and written as the single statement, $x = \frac{\pi}{2} + \pi k$, $k \in \mathbb{Z}$. In $[0, 2\pi)$, the solution process for $\sin(2x) = \frac{1}{2}$ yields two possibilities, $2x = \frac{\pi}{6}$ and $2x = \frac{5\pi}{6}$. If our only interest was solutions in $[0, 2\pi)$, we would next divide by 2 and be done. Since we seek all real roots, *we first extend each solution by $2\pi k$ before dividing by 2,* otherwise multiple solutions would be overlooked.

$$2x = \frac{\pi}{6} + 2\pi k \qquad 2x = \frac{5\pi}{6} + 2\pi k \qquad \text{solutions from } \sin(2x) = \tfrac{1}{2};\ k \in \mathbb{Z}$$

$$x = \frac{\pi}{12} + \pi k \qquad x = \frac{5\pi}{12} + \pi k \qquad \text{divide by 2}$$

NOW TRY EXERCISES 57 THROUGH 66

E. Trig Equations and Trig Identities

In the process of solving trig equations, we sometimes employ fundamental identities to help simplify an equation, or to make factoring or some other solution method possible.

EXAMPLE 8 Find all solutions in $[0°, 360°)$ for $\cos(2\theta) + \sin^2\theta - 3\cos\theta = 1$.

Solution: With a mixture of functions, exponents, and arguments, the equation is almost impossible to solve as it stands. But we can eliminate the sine function using the identity $\cos(2\theta) = \cos^2\theta - \sin^2\theta$, leaving us a quadratic equation in $\cos x$.

$$\cos(2\theta) + \sin^2\theta - 3\cos\theta = 1 \qquad \text{given equation}$$

$$\cos^2\theta - \sin^2\theta + \sin^2\theta - 3\cos\theta = 1 \qquad \text{substitute } \cos^2\theta - \sin^2\theta \text{ for } \cos(2\theta)$$

$$\cos^2\theta - 3\cos\theta = 1 \qquad \text{combine like terms}$$

$$\cos^2\theta - 3\cos\theta - 1 = 0 \qquad \text{subtract 1}$$

Let's substitute u for $\cos\theta$ to give us a simpler view of the equation. This gives $u^2 - 3u - 1 = 0$, which is clearly not factorable over the integers. Using the quadratic formula with $a = 1$, $b = -3$, and $c = -1$ gives

$$u = \frac{3 \pm \sqrt{(-3)^2 - 4(1)(-1)}}{2(1)} \qquad \text{quadratic formula in } u$$

$$= \frac{3 \pm \sqrt{13}}{2} \qquad \text{simplified}$$

To four decimal places the solutions are $u = 3.3028$ and $u = -0.3028$. To answer in terms of the original variable we re-substitute $\cos\theta$ for u, realizing that $\cos\theta \approx 3.3028$ has no solution, so solutions in $[0°, 360°)$ must be provided by $\cos\theta \approx -0.3028$. With cosine negative in QII and QIII, the solutions are $\theta = 107.6°$ and $\theta = 252.4°$ to the nearest tenth of a degree.

NOW TRY EXERCISES 67 THROUGH 82

F. Trig Equations and Graphing Technology

A majority of the trig equations you'll encounter in your studies can be solved using the ideas and methods presented here. But as mentioned in the *Point of Interest*, there are some equations that cannot be solved using standard methods because they mix polynomial functions (linear, quadratic, and so on) that can be solved using algebraic methods, with what are called **transcendental functions** (trigonometric, logarithmic, and so on). By definition, transcendental functions are those that *transcend* the reach of standard algebraic methods. These kinds of equations serve to highlight the value of graphing and calculating technology to today's problem solvers.

EXAMPLE 9 Use a graphing calculator in radian mode to find all real roots of $2\sin x + \frac{3x}{5} - 2 = 0$. Round solutions to four decimal places.

Solution: As always, when using graphing technology our initial concern is the size of the viewing window. After carefully entering the equation on the Y= screen, we note the term $2\sin x$ will never be larger than 2 or less than -2 for any real number x. On the other hand, the term $\frac{3x}{5}$ becomes larger for larger values of x, which would seem to cause $2\sin x + \frac{3x}{5}$ to "grow" as x gets larger. We conclude that the standard window is likely a good place to start, and the resulting graph is shown in Figure 4.38. From this screen it appears there are three real roots, but to be sure none are hidden to the right, we extend the Xmax value to 20 (Figure 4.39). Using 2nd TRACE CALC **2:zero,** we follow the prompts of the

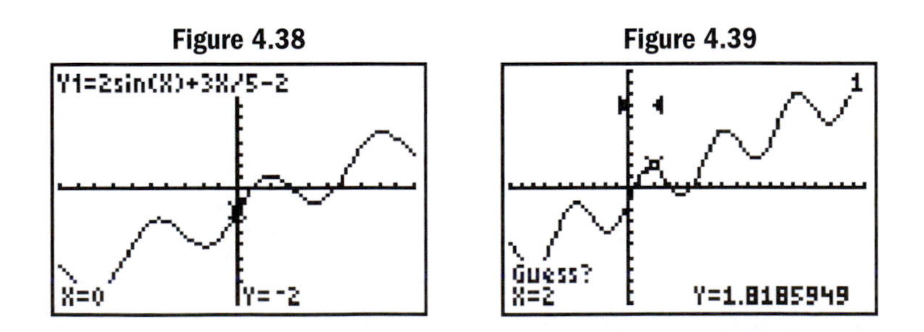

Figure 4.38

Figure 4.39

calculator and enter a left bound of 0 (a number to the left of the zero) and a right bound of 2 (a number to the right of the zero—see Figure 4.39). If you can visually approximate the root, the calculator prompts you for a GUESS, otherwise just bypass the request by pressing ENTER. The smallest root is approximately $x = 0.8435$. Repeating this sequence we find the other roots are $x \approx 3.0593$ and $x \approx 5.5541$.

NOW TRY EXERCISES 83 THROUGH 88

TECHNOLOGY HIGHLIGHT

Solving Equations Graphically Using a Graphing Calculator

The keystrokes illustrated refer to a TI-84 Plus model. For other models please consult your manual or visit our Internet site.

The periodic behavior of the trig functions is often used to form a solution set, which can be very helpful when solutions are nonstandard values. In addition, some equations are very difficult to solve analytically, and even with the use of a graphing calculator a strong combination of analytical skills with technical skills is required to state the solution set. Consider the equation $5\sin\left(\frac{1}{2}x\right) + 5 = \cot\left(\frac{1}{2}x\right)$. There appears to be no quick analytical solution, and the first attempt at a graphical solution holds some hidden surprises. Enter $Y_1 = 5\sin\left(\frac{1}{2}x\right) + 5$ and $Y_2 = \frac{1}{\tan(\frac{1}{2}x)}$ on the Y= screen. Pressing ZOOM **7:ZTrig** gives the screen shown in Figure 4.40, where we note there are at least two and possibly three solutions, depending on how the sine graph "cuts" the cotangent graph. We are also uncertain as to whether the graphs intersect again between $-\frac{\pi}{2}$ and $\frac{\pi}{2}$. Increasing the maximum Y-value to Ymax = 8 shows they do indeed. But once again, are there now three or four solutions? In situations like this it is helpful to use the **Zeroes Method** for solving graphically, where we actually compute the difference between the two functions, looking for a difference of zero (meaning they intersect), with every x-intercept indicating an intersection point. On the Y= screen, disable Y_1 and Y_2 and enter Y_3 as $Y_1 - Y_2$. Pressing ZOOM **7:ZTrig** at this point clearly shows that there are four solutions (Figure 4.41), which can easily be found using 2nd CALC **2:zero:** $x \approx -5.7543, -4.0094, -3.1416$, and 0.3390. To investigate solutions in $\mathbb{R}$, note the period of Y_1 is $P = \frac{2\pi}{\frac{1}{2}} = 4\pi$. On the 2nd WINDOW (**TBLSET**) screen, enter any one of the four solutions as the **TblStart** value, then set ΔTBL to π, put the calculator in AUTO mode, and go to 2nd GRAPH (**TABLE**). Scrolling through the TABLE shows this solution indeed repeats every 4π and we write the solution (to four decimal places) as $x \approx -5.7543 + 4\pi k$ for all integers k. The other solutions are likewise found. Use these ideas to find all real solutions to the exercises that follow.

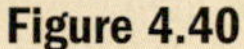

Figure 4.40

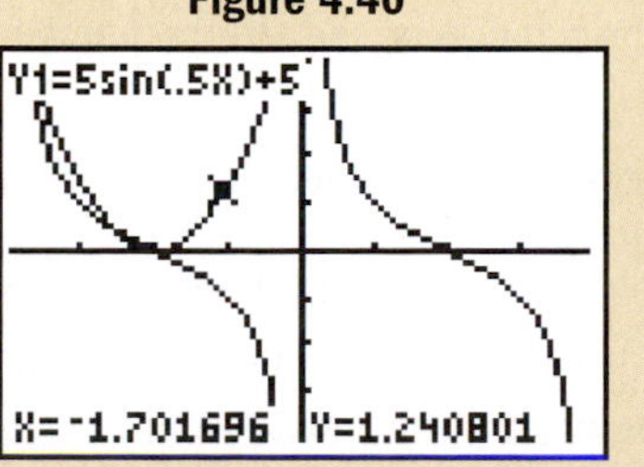

Figure 4.41

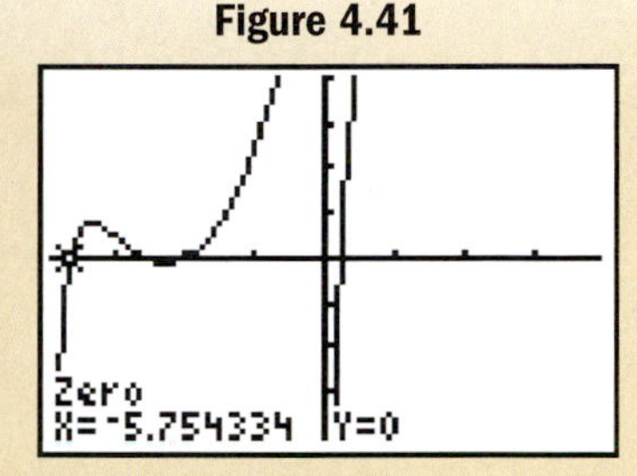

Exercise 1: $(1 + \sin x)^2 + \cos(2x) = 4\cos x(1 + \sin x)$

Exercise 2: $4\sin x = 2\cos^2\left(\frac{x}{2}\right)$

4.3 EXERCISES

CONCEPTS AND VOCABULARY

Fill in each blank with the appropriate word or phrase. Carefully reread the section if necessary.

1. For simple equations, a mental graph will tell us the quadrant of the ________ root, the number of roots in ______, and show a pattern for all ______ roots.

2. Solving trig equations is similar to solving algebraic equations, in that we first ______ the variable term, then apply the appropriate ______ function.

3. For $\sin x = \frac{\sqrt{2}}{2}$ the principal root is ______, solutions in $[0, 2\pi)$ are ______ and ______, and an expression for all real roots is ______ and ______; $k \in \mathbb{Z}$.

4. For $\tan x = -1$, the principal root is ______, solutions in $[0, 2\pi)$ are ______ and ______, and an expression for all real roots is ______.

5. Discuss/explain/illustrate why $\tan x = \frac{3}{4}$ and $y = \cos x$ have two solutions in $[0, 2\pi)$, even though the period of $y = \tan x$ is π while the period of $y = \cos x$ is 2π.

6. The equation $\sin^2 x = \frac{1}{2}$ has four solutions in $[0, 2\pi)$. Explain how these solutions can be viewed as the vertices of a square inscribed in the unit circle.

DEVELOPING YOUR SKILLS

7. For the equation $\sin x = -\frac{3}{4}$ and the graphs of $y = \sin x$ and $y = -\frac{3}{4}$ given, state (a) the quadrant of the principal root and (b) the number of roots in $[0, 2\pi)$.

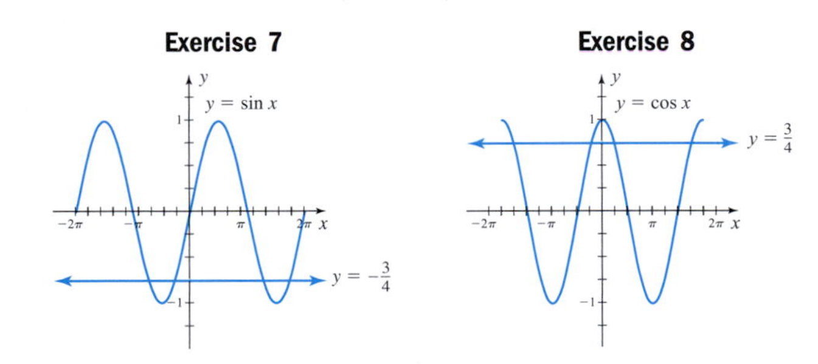

8. For the equation $\cos x = \frac{3}{4}$ and the graphs of $y = \cos x$ and $y = \frac{3}{4}$ given, state (a) the quadrant of the principal root and (b) the number of roots in $[0, 2\pi)$.

9. Given the graph $y = \tan x$ shown here, draw the horizontal line $y = -1.5$ and state (a) the quadrant of the principal root and (b) the number of roots in $[0, 2\pi)$.

Exercise 9

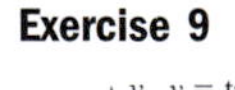
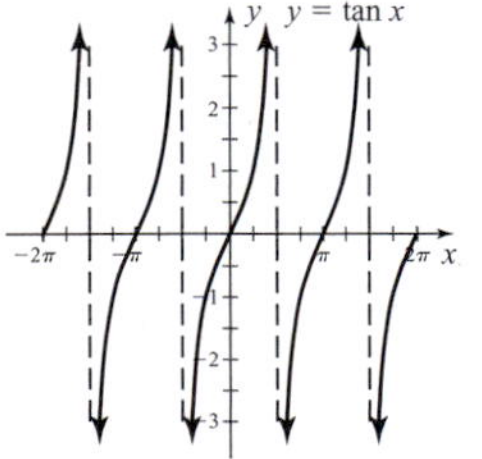

10. Given the graph of $y = \sec x$ shown, draw the horizontal line $y = \frac{5}{4}$ and state (a) the quadrant of the principal root and (b) the number of roots in $[0, 2\pi)$.

Exercise 10

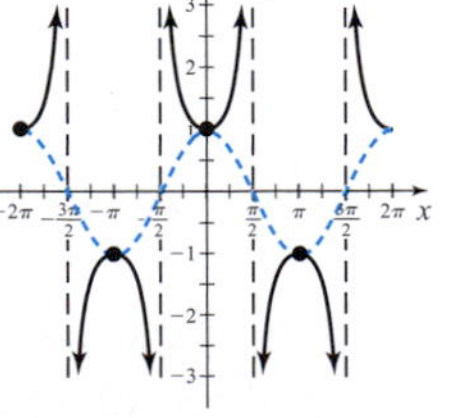

11. The table on page 257 shows θ in multiples of $\frac{\pi}{6}$ between 0 and $\frac{4\pi}{3}$, with the values for $\sin\theta$ given. Complete the table without a calculator or references using your knowledge of the unit circle, the signs of $f(\theta)$ in each quadrant, memory/recognition, $\tan\theta = \frac{\sin\theta}{\cos\theta}$, and so on.

Exercise 11

θ	$\sin\theta$	$\cos\theta$	$\tan\theta$
0	0		
$\frac{\pi}{6}$	$\frac{1}{2}$		
$\frac{\pi}{3}$	$\frac{\sqrt{3}}{2}$		
$\frac{\pi}{2}$	1		
$\frac{2\pi}{3}$	$\frac{\sqrt{3}}{2}$		
$\frac{5\pi}{6}$	$\frac{1}{2}$		
π	0		
$\frac{7\pi}{6}$	$-\frac{1}{2}$		
$\frac{4\pi}{3}$	$-\frac{\sqrt{3}}{2}$		

Exercise 12

θ	$\sin\theta$	$\cos\theta$	$\tan\theta$
0		1	
$\frac{\pi}{4}$		$\frac{\sqrt{2}}{2}$	
$\frac{\pi}{2}$		0	
$\frac{3\pi}{4}$		$-\frac{\sqrt{2}}{2}$	
π		-1	
$\frac{5\pi}{4}$		$-\frac{\sqrt{2}}{2}$	
$\frac{3\pi}{2}$		0	
$\frac{7\pi}{4}$		$\frac{\sqrt{2}}{2}$	
2π		1	

12. The table shows θ in multiples of $\frac{\pi}{4}$ between 0 and 2π, with the values for $\cos\theta$ given. Complete the table without a calculator or references using your knowledge of the unit circle, the signs of $f(\theta)$ in each quadrant, memory/recognition, $\tan\theta = \frac{\sin\theta}{\cos\theta}$, and so on.

Find the principal root of each equation.

13. $2\cos x = \sqrt{2}$

14. $2\sin x = -1$

15. $-4\sin x = 2\sqrt{2}$

16. $-4\cos x = 2\sqrt{3}$

17. $\sqrt{3}\tan x = 1$

18. $-2\sqrt{3}\tan x = 2$

19. $2\sqrt{3}\sin x = -3$

20. $-3\sqrt{2}\csc x = 6$

21. $-6\cos x = 6$

22. $4\sec x = -8$

23. $\frac{7}{8}\cos x = \frac{7}{16}$

24. $-\frac{5}{3}\sin x = \frac{5}{6}$

25. $2 = 4\sin\theta$

26. $\pi\tan x = 0$

27. $-5\sqrt{3} = 10\cos\theta$

28. $4\sqrt{3} = 4\tan\theta$

Find all solutions in $[0, 2\pi)$.

29. $9\sin x - 3.5 = 1$

30. $6.2\cos x + 4 = 7.1$

31. $8\tan x + 7\sqrt{3} = -\sqrt{3}$

32. $\frac{1}{2}\sec x - \frac{3}{4} = -\frac{7}{4}$

33. $\frac{2}{3}\cot x - \frac{5}{6} = -\frac{3}{2}$

34. $-110\sin x = -55\sqrt{3}$

35. $4\cos^2 x = 3$

36. $4\sin^2 x = 1$

37. $-7\tan^2 x = -21$

38. $3\sec^2 x = 6$

39. $-4\csc^2 x = -8$

40. $6\sqrt{3}\cos^2 x = 3\sqrt{3}$

41. $4\sqrt{2}\sin^2 x = 4\sqrt{2}$

42. $\frac{2}{3}\cos^2 x + \frac{5}{6} = \frac{4}{3}$

Solve the following equations by factoring. State all real solutions in radians using the exact form where possible and rounded to four decimal places if the result is not a standard value.

43. $3\cos^2\theta + 14\cos\theta - 5 = 0$

44. $6\tan^2\theta - 2\sqrt{3}\tan\theta = 0$

45. $2\cos x\sin x - \cos x = 0$

46. $2\sin^2 x + 7\sin x = 4$

47. $\sec^2 x - 6\sec x = 16$

48. $2\cos^3 x + \cos^2 x = 0$

49. $4\sin^2 x - 1 = 0$

50. $4\cos^2 x - 3 = 0$

Find all real solutions. Note that identities are not required to solve these exercises.

51. $-2\sin x = \sqrt{2}$

52. $2\cos x = 1$

53. $-4\cos x = 2\sqrt{2}$

54. $4\sin x = 2\sqrt{3}$

55. $\sqrt{3}\tan x = -\sqrt{3}$

56. $2\sqrt{3}\tan x = 2$

57. $6\cos(2x) = -3$

58. $2\sin(3x) = -\sqrt{2}$

59. $\sqrt{3}\tan(2x) = -\sqrt{3}$

60. $2\sqrt{3}\tan(3x) = 6$

61. $-2\sqrt{3}\cos\left(\frac{1}{3}x\right) = 2\sqrt{3}$

62. $-8\sin\left(\frac{1}{2}x\right) = -4\sqrt{3}$

63. $\sqrt{2}\cos x\sin(2x) - 3\cos x = 0$

64. $\sqrt{3}\sin x\tan(2x) - \sin x = 0$

65. $\cos(3x)\csc(2x) - 2\cos(3x) = 0$

66. $\sqrt{3}\sin(2x)\sec(2x) - 2\sin(2x) = 0$

Solve each equation using calculator and inverse trig functions to determine the principal root (not by graphing). Clearly state (a) the principal root and (b) all real roots.

67. $3\cos x = 1$

68. $5\sin x = -2$

69. $\sqrt{2}\sec x + 3 = 7$

70. $\sqrt{3}\csc x + 2 = 11$

71. $\frac{1}{2}\sin(2\theta) = \frac{1}{3}$

72. $\frac{2}{5}\cos(2\theta) = \frac{1}{4}$

73. $-5\cos(2\theta) - 1 = 0$

74. $6\sin(2\theta) - 3 = 2$

Solve the following equations using an identity. State all real solutions in radians using the exact form where possible and rounded to four decimal places if the result is not a standard value.

75. $\cos^2 x - \sin^2 x = \frac{1}{2}$

76. $4\sin^2 x - 4\cos^2 x = 2\sqrt{3}$

77. $2\cos\left(\frac{1}{2}x\right)\cos x - 2\sin\left(\frac{1}{2}x\right)\sin x = 1$

78. $\sqrt{2}\sin(2x)\cos(3x) + \sqrt{2}\sin(3x)\cos(2x) = 1$

79. $(\cos\theta + \sin\theta)^2 = 1$

80. $(\cos\theta + \sin\theta)^2 = 2$

81. $\cos(2\theta) + 2\sin^2\theta - 3\sin\theta = 0$

82. $3\sin(2\theta) - \cos^2(2\theta) - 1 = 0$

Find all roots in $[0, 2\pi)$ using a graphing calculator. State answers in radians rounded to four decimal places.

83. $5\cos x - x = 3$

84. $3\sin x + x = 4$

85. $\cos^2(2x) + x = 3$

86. $\sin^2(2x) + 2x = 1$

87. $x^2 + \sin(2x) = 1$

88. $\cos(2x) - x^2 = -5$

WORKING WITH FORMULAS

89. Range of a projectile: $R = \frac{5}{49}v^2\sin(2\theta)$

The distance a projectile travels is called its range and is modeled by the formula shown, where R is the range in meters, v is the initial velocity in meters per second, and θ is the angle of release. Two friends are standing 16 m apart playing catch. If the first throw has an initial velocity of 15 m/sec, what *two* angles will insure the ball travels the 16 m between the friends?

90. Fine tuning a golf swing: (club head to shoulder)^2 = (club length)^2 + (arm length)^2 − 2 (club length)(arm length)cos θ

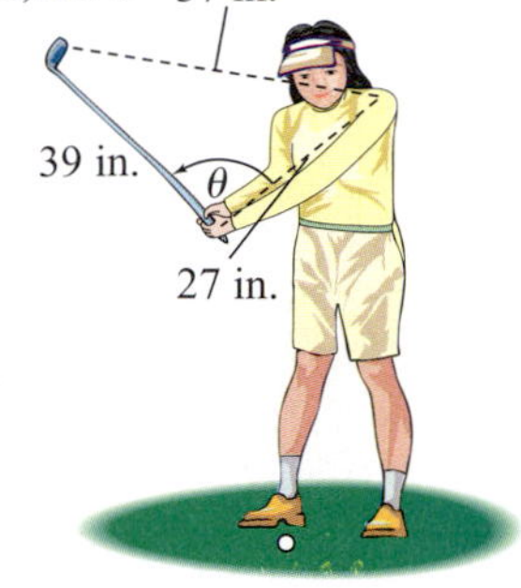

A golf pro is taking specific measurements on a client's swing to help improve her game. If the angle θ is too small, the ball is hit late and "too thin" (you *top the ball*). If θ is too large, the ball is hit early and "too fat" (you *scoop the ball*). Approximate the angle θ formed by the club and the extended (left) arm using the given measurements and formula shown.

APPLICATIONS

Acceleration due to gravity: When a steel ball is released down an inclined plane, the rate of the ball's acceleration depends on the angle of incline. The acceleration can be approximated by the formula $A(\theta) = 9.8 \sin \theta$, where θ is in degrees and the acceleration is measured in meters per second/per second. To the nearest tenth of a degree,

91. What angle produces an acceleration of 0 m/sec^2 when the ball is released? Explain why this is reasonable.

92. What angle produces an acceleration of 9.8 m/sec^2? What does this tell you about the acceleration due to gravity?

93. What angle produces an acceleration of 5 m/sec^2? Will the angle be larger or smaller for an acceleration of 4.5 m/sec^2?

94. Will an angle producing an acceleration of 2.5 m/sec^2 be one-half the angle required for an acceleration of 5 m/sec^2? Explore and discuss.

Snell's law states that when a ray of light passes from one medium into another, the sine of the angle of incidence α *varies directly with* the sine of the angle of refraction β (see the figure). This phenomenon is modeled by the formula $\sin \alpha = k \sin \beta$, where k is called the **index of refraction.** Note the angle θ is the angle at which the light strikes the surface, so that $\alpha = 90° - \theta$. Use this information to work Exercises 95 to 98.

Exercises 95 to 98

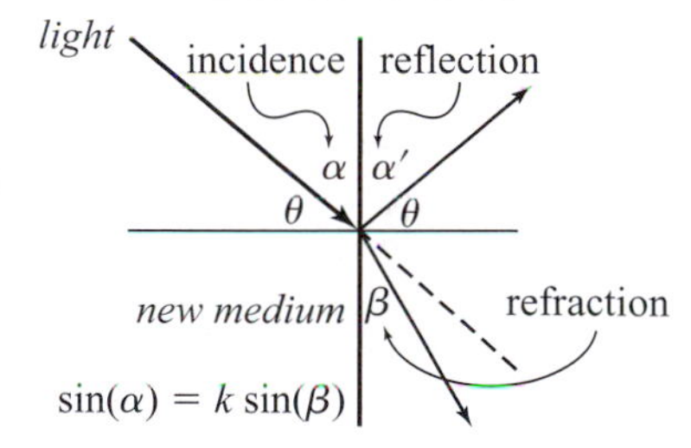

95. A ray of light passes from air into water, striking the water at an angle of 55°. Find the angle of incidence α and the angle of refraction β, if the index of refraction for water is $k = 1.33$.

96. A ray of light passes from air into a diamond, striking the surface at an angle of 75°. Find the angle of incidence α and the angle of refraction β, if the index of refraction for a diamond is $k = 2.42$.

97. Find the index of refraction for ethyl alcohol if a beam of light strikes the surface of this medium at an angle of 40° and produces an angle of refraction $\beta = 34.3°$. Use this index to find the angle of incidence if a second beam of light created an angle of refraction measuring 15°.

98. Find the index of refraction for rutile (a type of mineral) if a beam of light strikes the surface of this medium at an angle of 30° and produces an angle of refraction $\beta = 18.7°$. Use this index to find the angle of incidence if a second beam of light created an angle of refraction measuring 10°.

99. Roller coaster design: As part of a science fair project, Ronnie builds a scale model of a roller coaster using the equation $y = 5\sin\left(\frac{1}{2}x\right) + 7$, where y is the height of the model in inches and x is the distance from the "loading platform" in inches. (a) How high is the platform? (b) What distances from the platform does the model attain a height of 9.5 in.?

100. Company logo: Part of the logo for an engineering firm was modeled by a cosine function. The logo was then manufactured in steel and installed on the entrance marquee of the home office. The position and size of the logo is modeled by the function $y = 9\cos x + 15$, where y is the height of the graph above the base of the marquee in inches and x represents the distance from the edge of the marquee. Assume the graph begins flush with the edge. (a) How far above the base is the beginning of the cosine graph? (b) What distances from the edge does the graph attain a height of 19.5 in.?

Geometry applications: Solve Exercises 101 and 102 graphically using a calculator. For Exercise 101, give θ in radians rounded to four decimal places. For Exercise 102, answer in degrees to the nearest tenth of a degree.

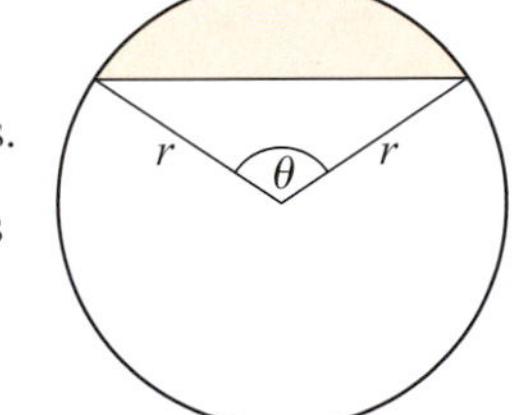

101. The area of a circular segment (the shaded portion shown) is given by the formula $A = \frac{1}{2}r^2(\theta - \sin\theta)$, where θ is in radians. If the circle has a radius of 10 cm, find the angle θ that gives an area of 12 cm^2.

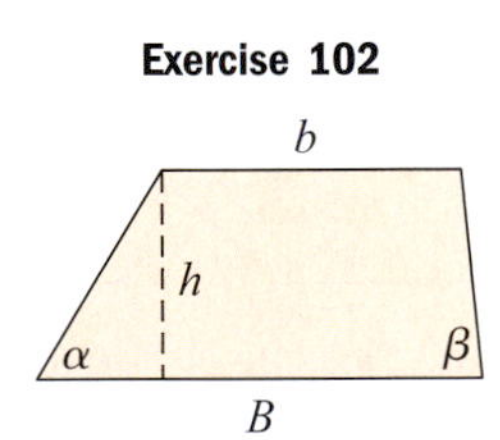

102. The perimeter of a trapezoid with parallel sides B and b, altitude h, and base angles α and β is given by the formula $P = B + b + h(\csc\alpha + \csc\beta)$. If $b = 30$ m, $B = 40$ m, $h = 10$ m, and $\alpha = 45°$, find the angle β that gives a perimeter of 105 m.

WRITING, RESEARCH, AND DECISION MAKING

103. Find all real solutions to $5\cos x - x = -x$ in two ways. First use a calculator with $Y_1 = 5\cos x - x$ and $Y_2 = -x$ to determine the regular intervals between points of intersection. Second, simplify by adding x to both sides, and draw a quick sketch of the result to locate x-intercepts. Explain why both methods give the same result, even though the first presents you with a very different graph.

104. A polyhedron is a solid whose faces are plane figures, like a cube. For each polyhedron, the **dihedral angles** α formed between plane faces differ, as does the angle β subtended by an edge from the center of the solid. The dihedral angles for a cube are $\alpha = 90°$, with $\beta = \cos^{-1}\left(\frac{1}{3}\right)$. Do some research on polyhedra and find the number of faces and the dihedral angles for a regular tetrahedron, octahedron, dodecahedron, and an icosahedron. Express the angles in exact form using radicals and inverse functions, and in approximate form rounded to the nearest tenth.

EXTENDING THE CONCEPT

105. Once the fundamental ideas of solving a given family of equations is understood and practiced, a student usually begins to generalize them—making the numbers or symbols used in the equation irrelevant. Use the inverse sine function to find the principal root of $y = A\sin(Bx - C) + D$, by solving for x in terms of y, A, B, C, and D. The result won't be very "pretty," but nevertheless represents the complete principal solution for all equations of this form.

106. Solve the following equation two ways: (a) using the techniques addressed in this section, and (b) using the "formula" developed in Exercise 105: $5 = 2\sin\left(\frac{1}{2}x + \frac{\pi}{4}\right) + 3$. Do the results agree?

MAINTAINING YOUR SKILLS

107. (1.3) Find the value of all six trig functions given $\cos\theta = -\frac{5}{13}$ and $\tan\theta > 0$.

108. (3.2) Verify that the following is an identity: $\frac{\cot x + \tan x}{\csc x} = \sec x$

109. (4.2) A road sign cautions truckers to slow down as the upcoming down hill roads have a -12% grade, meaning the ratio $\frac{\text{rise}}{\text{run}} = -0.12$. Find the angle of descent to the nearest tenth of a degree.

110. (4.2) Evaluate without using a calculator:

a. $\tan\left[\sin^{-1}\left(-\frac{1}{2}\right)\right]$

b. $\sin[\tan^{-1}(-1)]$

111. (2.1) Write the equation of the function shown in two ways—first using a secant function, and second as a cosecant function.

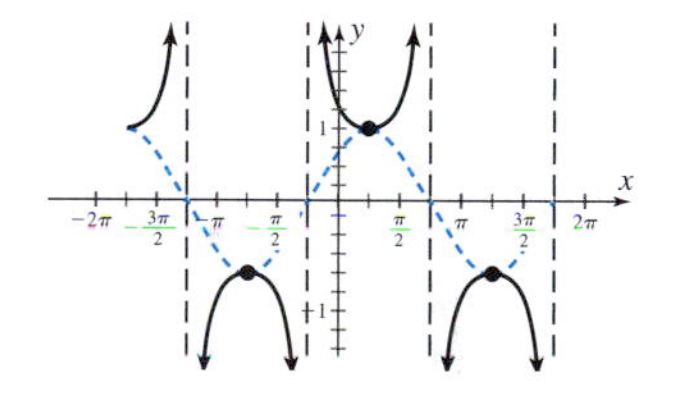

112. (1.1) The largest Ferris wheel in the world, located in Yokohama, Japan, has a radius of 50 m. To the nearest hundredth of a meter, how far does a seat on the rim travel as the wheel turns through $\theta = 292.5°$?

MID-CHAPTER CHECK

1. The algebraic method for finding the inverse function for $f(x) = \tan x$ is illustrated here. The equation in (3) is called the __________ form of the inverse equation. The equation in (4) is called the __________ form.

(1)	$\tan x = f(x)$	given function
(2)	$\tan x = y$	use y instead of $f(x)$
(3)	$\tan y = x$	interchange x and y
(4)	$y = \tan^{-1}x$	solve for y
(5)	$f^{-1}(x) = \tan^{-1}x$	substitute $f^{-1}(x)$ for y

2. Use the algebraic method to find the inverse function for $f(x) = 2\sqrt[3]{2x+1} + 3$, then verify your work using a composition.

3. Comment on the third line in the following sequence. In particular, why does the "pattern" of inverses not hold in the third line?

$$\sin(60°) = \frac{\sqrt{3}}{2} \quad \rightarrow \quad \sin^{-1}\left(\frac{\sqrt{3}}{2}\right) = 60°$$
$$\sin(90°) = 1 \quad \rightarrow \quad \sin^{-1}(1) = 90°$$
$$\sin(120°) = \frac{\sqrt{3}}{2} \quad \rightarrow \quad \sin^{-1}\left(\frac{\sqrt{3}}{2}\right) = 60°(?)$$

4. Evaluate the following expressions without a calculator.

a. $\sec^{-1}(\sqrt{2})$ **b.** $\csc^{-1}\left(\frac{2}{\sqrt{3}}\right)$

5. Evaluate the following expressions without a calculator.

a. $\cos^{-1}\left[\cos\left(\frac{\pi}{6}\right)\right]$ **b.** $\cos^{-1}\left[\cos\left(\frac{7\pi}{6}\right)\right]$

6. Evaluate the following expressions exactly by drawing the corresponding triangle.

a. $\cos\left[\tan^{-1}\left(\frac{13}{84}\right)\right]$ **b.** $\sec\left[\sin^{-1}\left(\frac{x}{\sqrt{x^2+49}}\right)\right]$

7. For $Y_1 = \frac{\sqrt{81 - x^2}}{x}$: (a) Find an equivalent expression Y_2 using the substitution $x = 9\cos\theta$; (b) express Y_2 in terms of x using $x = 9\cos\theta$ (solve for θ) and substitution.

8. Find exact solutions in $[0, 2\pi)$ without using a calculator: $2\sin x + 5\sqrt{3} = 6\sqrt{3}$.

9. Find approximate solutions in $[0, 2\pi)$ using a calculator: $-\frac{1}{2}\tan(2x) + 3 = -7$.

10. Find approximate solutions in $[-\pi, \pi]$ by solving the equation graphically: $x^2\cos(4x) = 1$. Use symmetry to cut down the number of calculations and keystrokes needed. Round to four decimal places.

REINFORCING BASIC CONCEPTS

More on Equation Solving

"An idea reaches its maximum level of usefulness only when you understand it so well, it seems as though you have always known it" (Henri Lebesque).

In this RBC we look at the fundamental concepts that will help you become a better, more agile equation solver. The word "agile" implies a *ready ability to move with ease and grace* (Webster). In the same way that knowledge of the multiplication tables helps you see patterns between numbers and leads to factoring quadratics, reducing rational expressions, and other skill areas, mastering basic concepts related to equation solving will have tremendous value in the sections that follow.

Concept (1): Solving Basic Equations

To become an agile equation solver, try to see the similarities that exist in the equation-solving process. As with other basic equations, the primary goal is to isolate the basic function so that the inverse function can be applied.

a.
$$\begin{aligned} -3x^2 + 1 &= -11 \\ -3x^2 &= -12 \\ x^2 &= 4 \\ \sqrt{x^2} &= \pm\sqrt{4} \\ x &= \pm 2 \end{aligned}$$

b.
$$\begin{aligned} -5\sqrt[3]{x} + 1 &= -9 \\ -5\sqrt[3]{x} &= -10 \\ \sqrt[3]{x} &= 2 \\ (\sqrt[3]{x})^3 &= 2^3 \\ x &= 8 \end{aligned}$$

c.
$$\begin{aligned} -6\sin x + 5\sqrt{2} &= 2\sqrt{2} && \text{given equation} \\ -6\sin x &= -3\sqrt{2} && \text{isolate variable term} \\ \sin x &= \frac{\sqrt{2}}{2} && \text{isolate basic function} \\ \sin^{-1}(\sin x) &= \sin^{-1}\left(\frac{\sqrt{2}}{2}\right) && \text{apply inverse function} \\ x &= \frac{\pi}{4} && \text{result} \end{aligned}$$

Concept (2): Standard values and standard angles in degrees or radians

To become a better "trig equation" solver, a ready knowledge of the standard values and standard angles is vital. For algebraic equations, solutions are found by performing an inverse operation on a number. While it can be argued that $\sin^{-1}\left(\frac{\sqrt{2}}{2}\right)$ is "an inverse operation on a number," solutions to trig equations are better characterized as coming from the connections and definitions we established for each trig function. In other words, $\sin^{-1}\left(\frac{\sqrt{2}}{2}\right) = \frac{\pi}{4}$ because $\frac{\sqrt{2}}{2}$ is the ratio $\frac{\text{opp}}{\text{hyp}}$ on a 45-45-90 triangle, or $\frac{\sqrt{2}}{2}$ is the y-coordinate of the point on the unit circle corresponding to $\theta = \frac{\pi}{4}$. The table shown should first of all be *understood*, then committed to memory for ready recall.

θ	$\sin\theta$	$\cos\theta$	$\tan\theta$
0	0	1	0
$\frac{\pi}{6}$	$\frac{1}{2}$	$\frac{\sqrt{3}}{2}$	$\frac{1}{\sqrt{3}}$
$\frac{\pi}{4}$	$\frac{\sqrt{2}}{2}$	$\frac{\sqrt{2}}{2}$	1
$\frac{\pi}{3}$	$\frac{\sqrt{3}}{2}$	$\frac{1}{2}$	$\sqrt{3}$
$\frac{\pi}{2}$	1	0	—

Concept (3): Periodic solutions and the basic graphs of each function

To become a more agile "trig equation" solver, becoming very familiar with the period and graph of each function is also vital. The cyclic nature of the trig functions makes them unique. In a study of systems of equations, it was possible for a system of two or more equations to have an infinite number of solutions. For trig equations, a single equation can have an infinite number of solutions occurring at regular intervals—as dictated by the period of the function. The basic graph of each function should be committed to memory, ready for instant recall, as the period can be "read" from each: $\sin\theta$, $\cos\theta$, $\csc\theta$, and $\sec\theta$ have period 2π; $\tan\theta$ and $\cot\theta$ have period π. Find all real solutions to the following equations without the use of a calculator.

Exercise 1: $2\cos x + 6\sqrt{3} = 7\sqrt{3}$

Exercise 2: $\sqrt{3}\tan x + 2.7 = 3.7$

Exercise 3: $2\sin(4x) + 7 = 8$

Exercise 4: $2\sec^2(2x) + 1 = 5$

4.4 General Trig Equations and Applications

LEARNING OBJECTIVES

In Section 4.4 you will learn how to:

A. Use additional algebraic techniques to solve trig equations

B. Solve trig equations using multiple angle, sum and difference, and sum-to-product identities

C. Solve trig equations of the form $A\sin(Bx + C) + D = k$

D. Use a combination of skills to model and solve a variety of applications

INTRODUCTION

At this point you're likely beginning to understand the true value of trigonometry to the scientific world. Essentially, any phenomenon that is cyclic or periodic is beyond the reach of polynomial (and other) functions, and may require trig for an accurate understanding. And while there is an abundance of trig applications in oceanography, astronomy, meteorology, geology, zoology, and engineering, their value is not limited to the hard sciences. There are also rich applications in business and economics, and a growing number of modern artists are creating works based on attributes of the trig functions. In this section, we try to place some of these applications within your reach, with the exercise set offering an appealing variety from many of these fields.

POINT OF INTEREST

In *Troilus and Cressida,* Shakespeare wrote, "The heavens themselves, the planets and this centre; Observe degree, priority, and place; Insisture, course, proportion, season, form; Office, and custom, in all line of order" (iii.85). The *season, form,* and *order* of the planets is many times modeled by a trigonometric equation, with temperature, daylight, tides, and many other aspects of "this center" (the Earth) being prime examples.

A. Trig Equations and Algebraic Methods

We begin this section with a follow-up to Section 4.3, by introducing trig equations that require slightly more sophisticated methods to work out a solution. The methods are likely familiar, we simply apply them in the context of trigonometric equations.

EXAMPLE 1 Find all solutions in $[0, 2\pi)$: $\sec x + \tan x = \sqrt{3}$.

Solution: Our first instinct might be to rewrite the equation in terms of sine and cosine, but that simply leads to a similar equation which still has two different functions $[\sqrt{3}\cos x - \sin x = 1]$. Instead, we *square both sides* and see if the Pythagorean identity $1 + \tan^2 x = \sec^2 x$ will be of use. Prior to squaring, we separate the functions on opposite sides to avoid the mixed term $2\tan x \sec x$.

$$\sec x + \tan x = \sqrt{3} \qquad \text{given equation}$$
$$(\sec x)^2 = (\sqrt{3} - \tan x)^2 \qquad \text{subtract } \tan x \text{ and square}$$
$$\sec^2 x = 3 - 2\sqrt{3}\tan x + \tan^2 x \qquad \text{result}$$

Since $\sec^2 x = 1 + \tan^2 x$, we substitute directly and obtain an equation in tangent alone.

$$1 + \tan^2 x = 3 - 2\sqrt{3}\tan x + \tan^2 x \qquad \text{substitute } 1 + \tan^2 x \text{ for } \sec^2 x$$
$$-2 = -2\sqrt{3}\tan x \qquad \text{simplify}$$
$$\frac{1}{\sqrt{3}} = \tan x \qquad \text{solve for } \tan x$$

$\tan x > 0$ in QI and QIII

The proposed solutions are $x = \frac{\pi}{6}$ [QI] and $\frac{7\pi}{6}$ [QIII]. Since squaring an equation sometimes introduces extraneous roots, both should be checked in the original equation. The check shows only $x = \frac{\pi}{6}$ is a solution.

NOW TRY EXERCISES 7 THROUGH 12

Here is one additional example that uses a factoring strategy commonly employed when an equation has more than three terms.

EXAMPLE 2 Find all solutions in $[0°, 360°)$:
$8\sin^2\theta\cos\theta - 2\cos\theta - 4\sin^2\theta + 1 = 0.$

Solution: The four terms in the equation share no common factors, so we attempt to factor by grouping. We could factor $2\cos\theta$ from the first two terms but instead elect to group the $\sin^2\theta$ terms and begin there.

$$8\sin^2\theta\cos\theta - 2\cos\theta - 4\sin^2\theta + 1 = 0 \quad \text{given equation}$$
$$(8\sin^2\theta\cos\theta - 4\sin^2\theta) - (2\cos\theta - 1) = 0 \quad \text{rearrange and group terms}$$
$$4\sin^2\theta(2\cos\theta - 1) - 1(2\cos\theta - 1) = 0 \quad \text{remove common factors}$$
$$(2\cos\theta - 1)(4\sin^2\theta - 1) = 0 \quad \text{remove common binomial factors}$$

Using the zero factor property, we write two equations and solve each independently.

$$2\cos\theta - 1 = 0 \qquad 4\sin^2\theta - 1 = 0 \quad \text{resulting equations}$$
$$2\cos\theta = 1 \qquad \sin^2\theta = \frac{1}{4} \quad \text{isolate variable term}$$
$$\cos\theta = \frac{1}{2} \qquad \sin\theta = \pm\frac{1}{2} \quad \text{solve}$$

$\cos\theta > 0$ in QI and QIV $\qquad \sin\theta > 0$ in QI and QII

$\theta = 60°, 300° \qquad \sin\theta < 0$ in QIII and QIV

$\theta = 30°, 150°, 210°, 330°$ solutions

Initially factoring $2\cos\theta$ from the first two terms and proceeding from there would have produced the same result.

NOW TRY EXERCISES 13 THROUGH 16

B. Trig Equations and Other Identities

To solve equations effectively, a student should strive to develop *all* of the necessary "tools." Certainly the underlying concepts and graphical connections are of primary importance, as are the related algebraic skills. But to solve *trig* equations effectively we must also have a ready command of commonly used identities. Observe how the following example combines a double-angle identity with factoring by grouping.

EXAMPLE 3 Find all solutions in $[0, 2\pi)$: $3\sin(2x) + 2\sin x - 3\cos x = 1$. Round solutions to four decimal places as necessary.

Solution: Noting that one of the terms involves a double angle, we attempt to replace that term to make factoring a possibility. Using the double identity for sine, we have

$$3(2\sin x\cos x) + 2\sin x - 3\cos x = 1$$ substitute $2\sin x\cos x$ for $\sin(2x)$

$$(6\sin x\cos x + 2\sin x) - (3\cos x + 1) = 0$$ set equal zero and group terms

$$2\sin x(3\cos x + 1) - 1(3\cos x + 1) = 0$$ factor out common factor

$$(3\cos x + 1)(2\sin x - 1) = 0$$ factor common binomial factor

We again use the zero factor property, solving each equation independently.

$$3\cos x + 1 = 0 \qquad 2\sin x - 1 = 0$$ resulting equations

$$\cos x = -\frac{1}{3} \qquad \sin x = \frac{1}{2}$$ isolate variable term

$\cos x < 0$ in QII and QIII $\quad \sin x > 0$ in QI and QII

$$x \approx 1.9106, 4.3726 \qquad x = \frac{\pi}{6}, \frac{5\pi}{6}$$ solutions

Should you prefer the exact form, the solutions from the cosine equation could be written as $x = \cos^{-1}\left(-\frac{1}{3}\right)$ and $x = 2\pi - \cos^{-1}\left(-\frac{1}{3}\right)$.

NOW TRY EXERCISES 17 THROUGH 26

C. Solving Equations of the Form $A\sin(Bx \pm C) \pm D = k$

You may remember equations of this form from Section 2.3. They actually occur quite frequently in the investigation of many natural phenomena and in the modeling of data from a periodic or seasonal context. Solving these equations requires a good combination of algebra skills with the fundamentals of trig.

EXAMPLE 4 Given $f(x) = 160\sin\left(\frac{\pi}{3}x + \frac{\pi}{3}\right) + 320$ and $x \in [0, 2\pi)$, for what real numbers x is $f(x)$ less than 240?

Solution: We reason that to find values where $f(x) < 240$, we should begin by finding values where $f(x) = 240$. The result is

$$160\sin\left(\frac{\pi}{3}x + \frac{\pi}{3}\right) + 320 = 240$$ equation

$$\sin\left(\frac{\pi}{3}x + \frac{\pi}{3}\right) = -0.5$$ subtract 320 and divide by 160; isolate variable term

At this point we elect to use a u-substitution for $\left(\frac{\pi}{3}x + \frac{\pi}{3}\right) = \frac{\pi}{3}(x + 1)$ to obtain a "clearer view."

$$\sin u = -0.5$$ substitute the place-holder u for $\frac{\pi}{3}(x + 1)$

$$\sin u < 0 \text{ in QIII and QIV}$$

$$u = \frac{7\pi}{6} \qquad u = \frac{11\pi}{6}$$ solutions in u

To complete the solution we re-substitute $\frac{\pi}{3}(x + 1)$ for u and solve.

$$\frac{\pi}{3}(x + 1) = \frac{7\pi}{6} \qquad \frac{\pi}{3}(x + 1) = \frac{11\pi}{6}$$ re-substitute $\frac{\pi}{3}(x + 1)$ for u

$$x + 1 = \frac{7}{2} \qquad x + 1 = \frac{11}{2}$$ multiply both sides by $\frac{3}{\pi}$

$$x = 2.5 \qquad x = 4.5$$ solutions

We now know $f(x) = 240$ when $x = 2.5$ and $x = 4.5$ but when will $f(x)$ be *less than* 240? By analyzing the equation, we find the function has period of $P = \frac{2\pi}{\frac{\pi}{3}} = 6$ and is shifted to the left $\frac{\pi}{3}$ units. This would indicate the graph peaks early in the interval $[0, 2\pi)$ with a "valley" in the interior. We conclude $f(x) < 240$ for $x \in (2.5, 4.5)$.

NOW TRY EXERCISES 27 THROUGH 30

GRAPHICAL SUPPORT

Support for the result in Example 4 can be obtained by graphing the equation over the specified interval. Enter $Y_1 = 160\sin\left(\frac{\pi}{3}x + \frac{\pi}{3}\right) + 320$ on the Y= screen, then $Y_2 = 240$. After locating points of intersection, we note the graphs indeed verify that in the interval $[0, 2\pi)$, $f(x) < 240$ for $x \in (2.5, 4.5)$.

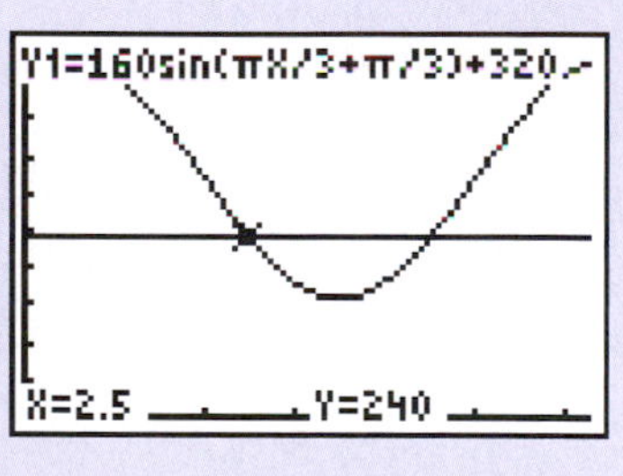

D. Applications Using Trigonometric Equations

We first introduce an application from the mathematical sciences. Using characteristics of the trig functions, we can often generalize and extend many of the formulas that are familiar to you. For example, the formulas for the volume of a right circular cylinder and a right circular cone are well known, but what about the volume of a non-right figure? Here, trigonometry provides the answer, as the most general volume formula is $V = V_0\sin\theta$, where V_0 is a "standard" volume formula and θ is the complement of angle of deflection (see Exercises 43 and 44).

As for other applications, consider the following from the environmental sciences. Natural scientists are very interested in the discharge rate of major rivers, as this gives an indication of rainfall over the inland area served by the river. In addition, the discharge rate has a large impact on the freshwater and saltwater zones found at the river's estuary (where it empties into the sea).

EXAMPLE 5 For May through December, the discharge rate of the Ganges River (Bangladesh) can be modeled by $D(t) = 16{,}580 \sin\left(\frac{\pi}{3}t - \frac{2\pi}{3}\right) + 17{,}760$ where $t = 1$ represents May 1, and $D(t)$ is the discharge rate in m^3/sec.

Source: Global River Discharge Database Project; www.rivdis.sr.unh.edu.

a. What is the discharge rate in mid-October?

b. For what months (within this interval) is the discharge rate over 26,050 m^3/sec?

Solution:

a. To find the discharge rate in mid-October we simply evaluate the function at $t = 6.5$:

$$D(t) = 16{,}580 \sin\left(\frac{\pi}{3}t - \frac{2\pi}{3}\right) + 17{,}760 \quad \text{given function}$$

$$D(6.5) = 16{,}580 \sin\left[\frac{\pi}{3}(6.5) - \frac{2\pi}{3}\right] + 17{,}760 \quad \text{substitute 6.5 for } t$$

$$= 1180 \quad \text{compute result on a calculator}$$

In mid-October the discharge rate is 1180 m^3/sec.

b. We first find when the rate is *equal* to 26,050 m^3/sec: $D(t) = 26{,}050$.

$$D(t) = 16{,}580 \sin\left(\frac{\pi}{3}t - \frac{2\pi}{3}\right) + 17{,}760 \quad \text{given function}$$

$$26{,}050 = 16{,}580 \sin\left(\frac{\pi}{3}t - \frac{2\pi}{3}\right) + 17{,}760 \quad \text{substitute 26,050 for } D(t)$$

$$0.5 = \sin\left(\frac{\pi}{3}t - \frac{2\pi}{3}\right) \quad \text{subtract 17,760; divide by 16,580}$$

Using a u-substitution for $\left(\frac{\pi}{3}t - \frac{2\pi}{3}\right)$ we obtain the equation

$$0.5 = \sin u$$

$$\sin u > 0 \text{ in QI and QII}$$

$$u = \frac{\pi}{6} \qquad u = \frac{5\pi}{6} \quad \text{solutions in } u$$

To complete the solution we re-substitute $\left(\frac{\pi}{3}t - \frac{2\pi}{3}\right) = \frac{\pi}{3}(t - 2)$ for u and solve.

$$\frac{\pi}{3}(t-2)=\frac{\pi}{6} \qquad \frac{\pi}{3}(t-2)=\frac{5\pi}{6} \qquad \text{re-substitute } \frac{\pi}{3}(t-2) \text{ for } u$$

$$t-2=0.5 \qquad t-2=2.5 \qquad \text{multiply both sides by } \frac{3}{\pi}$$

$$t=2.5 \qquad t=4.5 \qquad \text{solutions}$$

The Ganges River will have a flow rate of over 26,050 m^3/sec between mid-June (2.5) and mid-August (4.5).

NOW TRY EXERCISES 45 AND 46

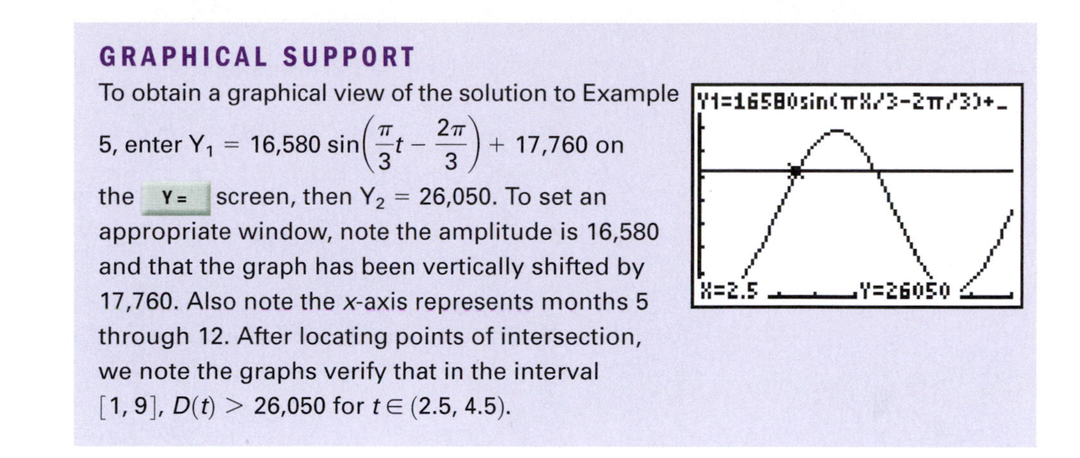

GRAPHICAL SUPPORT

To obtain a graphical view of the solution to Example 5, enter $Y_1 = 16{,}580 \sin\left(\frac{\pi}{3}t - \frac{2\pi}{3}\right) + 17{,}760$ on the Y= screen, then $Y_2 = 26{,}050$. To set an appropriate window, note the amplitude is 16,580 and that the graph has been vertically shifted by 17,760. Also note the x-axis represents months 5 through 12. After locating points of intersection, we note the graphs verify that in the interval $[1, 9]$, $D(t) > 26{,}050$ for $t \in (2.5, 4.5)$.

Our final application comes from the business/economics arena. Many businesses specialize in seasonal merchandise and run through cycles of high-sales/low-sales/high-sales/low-sales that are fairly regular from year to year. To even out the "feast or famine" effect, revenue projections need to be made and budgets set up.

EXAMPLE 6 Joe's Water Sports sells boats, equipment for water skiing, camping supplies, and other supplies for water sports. Sales are great in summer but weak in winter. Revenue projections from day to day can be modeled by the function $R(d) = 750 \sin\left(\frac{2\pi}{365}d - \frac{\pi}{2}\right) + 950$, where $R(d)$ is the daily revenue on the dth day of the year ($d = 1 \rightarrow$ Jan 1). (a) Estimate the revenue for May 1. (b) For what days of the year is revenue over \$1250?

Solution: **a.** To find the revenue for May we evaluate $R(d)$ at $d = (31 + 28 + 31 + 30 + 1) = 121$:

$$R(d) = 750 \sin\left(\frac{2\pi}{365}d - \frac{\pi}{2}\right) + 950 \qquad \text{given function}$$

$$R(121) = 750 \sin\left[\frac{2\pi}{365}(121) - \frac{\pi}{2}\right] + 950 \qquad \text{substitute 121 for } d$$

$$\approx \$1317.52 \qquad \text{result}$$

On May 1 the projected revenue is about \$1317.52.

b. We first find when the amount of revenue is *equal* to \$1250: $R(d) = 1250$.

$$R(d) = 750\sin\left(\frac{2\pi}{365}d - \frac{\pi}{2}\right) + 950 \quad \text{given function}$$

$$1250 = 750\sin\left(\frac{2\pi}{365}d - \frac{\pi}{2}\right) + 950 \quad \text{substitute 1250 for } R(d)$$

$$0.4 = \sin\left(\frac{2\pi}{365}d - \frac{\pi}{2}\right) \quad \text{subtract 950; divide by 750}$$

Using a u-substitution for $\left(\frac{2\pi}{365}d - \frac{\pi}{2}\right)$ we obtain the equation

$$0.4 = \sin u \quad \text{substitute } u \text{ for } \frac{2\pi}{365}d - \frac{\pi}{2}$$

$$\sin u > 0 \text{ in QI and QII}$$

$$u \approx 0.4115 \qquad u \approx 2.7301 \quad \text{solutions in } u\text{, rounded to four decimal places}$$

To complete the solution we re-substitute $\frac{2\pi}{365}d - \frac{\pi}{2}$ for u and solve.

$$\frac{2\pi}{365}d - \frac{\pi}{2} \approx 0.4115 \qquad \frac{2\pi}{365}d - \frac{\pi}{2} \approx 2.7301 \quad \text{re-substitute } \frac{2\pi}{365}d - \frac{\pi}{2} \text{ for } u$$

$$\frac{2\pi}{365}d \approx 1.9823 \qquad \frac{2\pi}{365}d \approx 4.3009 \quad \text{add } \frac{\pi}{2}$$

$$d \approx 115 \qquad t \approx 250 \quad \text{multiply by } \frac{365}{2\pi}$$

Joe's Water Sports will have revenue income of over \$1250 between late April (day 115) and early September (day 250), during the summer swim season.

NOW TRY EXERCISES 47 AND 48

There is a variety of additional exercises in the Exercise Set. Enjoy.

TECHNOLOGY HIGHLIGHT

Window Size and Applications of the Trig Equations

The keystrokes illustrated refer to a TI-84 Plus model. For other models please consult your manual or visit our Internet site.

As for graphical studies of other types of equations, setting an appropriate window size for trigonometric equations is a high priority. For applications involving sine and cosine functions, several things can be done to ensure a "good" window.

Ymax, Ymin, and Yscl

For equations of the form $y = A\sin(Bx - C) + D$, note the largest value that $\sin(Bx - C)$ can attain is 1, and the smallest is -1. This immediately tells us the largest value (Ymax) for y is $A(1) + D$, and the smallest value for y (Ymin) is $A(-1) + D$. However, to leave room for the calculator to display function names, TRACE values, and other information, we always add a "frame." Use this information to set Ymax, Ymin, and Yscl $= \frac{\text{Ymax} - \text{Ymin}}{10}$.

Xmax, Xmin, and Xscl

The values for Xmax, Xmin, and Xscl are always taken from the period of the function. For sine and cosine graphs, $P = \frac{2\pi}{B}$. If *any* period is sufficient we simply use Xmin = 0 and Xmax = P. If we want to see a primary interval, say where $\cos\theta$ begins at $y = A$ and ends at $y = A$, we set $(Bx - C) = 0$ and $(Bx - C) = 2\pi$ to find this interval. For Xscl, $\frac{P}{10}$ usually works well, but if the period is 12 and represents the months of the year, we would use $\frac{P}{12}$ as it places each month at a tick mark. Although scaling the axes is somewhat arbitrary, it should be used to improve the "readability" of a graph. We'll illustrate

using the function

$R(t) = 7500\sin\left[\frac{\pi}{6}(t - 4.9)\right] + 8500.$

Enter the function as $Y_1 = 7500\sin\left[\frac{\pi}{6}(X - 4.9)\right] + 8500$ on the Y= screen, then press WINDOW to begin setting the window size. For this example $A = 7500$, $B = \frac{\pi}{6}$, $C = \frac{4.9\pi}{6}$, and $D = 8500$.

$\text{Ymax} = 7500(1) + 8500 = 16{,}000$	$\text{Xmax} = \frac{2\pi}{(\pi/6)} = 12$
$\text{Ymin} = 7500(-1) + 8500 = 1000$	$\text{Xmin} = 0$
$\text{Yscl} = 1500$	$\text{Xscl} = \frac{12}{12} = 1$

The Y= screen, WINDOW screen, and GRAPH screen are shown in Figures 4.42, 4.43, and 4.44, respectively. Note that we have actually used a wider range of y-values than was necessary for our window, to ensure that the equation and displayed values do not obscure the graph.

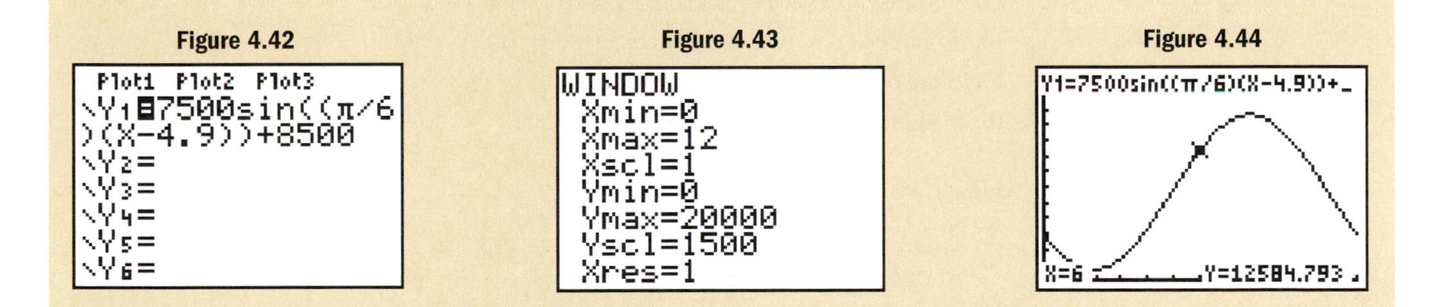

Figure 4.42 Figure 4.43 Figure 4.44

Use these ideas to complete the following exercises.

Exercise 1: Verify the results from Example 4.

$$f(x) = 160\sin\left(\frac{\pi}{3}x + \frac{\pi}{3}\right) + 320$$

Exercise 2: Verify the results from Example 5.

$$D(t) = 16{,}580\sin\left(\frac{\pi}{3}t - \frac{2\pi}{3}\right) + 17{,}760$$

4.4 EXERCISES

CONCEPTS AND VOCABULARY

Fill in each blank with the appropriate word or phrase. Carefully reread the section if needed.

1. The three Pythagorean identities are ________, ________, and ________.

2. When an equation contains two functions from a Pythagorean identity, sometimes ________ both sides will lead to a solution.

3. One strategy to solve equations with four terms and no common factors is ________ by ________.

4. To combine two sine or cosine terms with different arguments, we can use the ________ to ________ formulas.

5. Regarding Example 5, discuss/explain how to determine the months of the year the discharge rate is *under* 26,050 m^3/sec, using the solution set given.

6. Regarding Example 6, discuss/explain how to determine the months of the year the revenue projection is *under* \$1250 using the solution set given.

DEVELOPING YOUR SKILLS

Solve each equation in $[0, 2\pi)$ using the method indicated. Round nonstandard values to four decimal places.

• Squaring both sides

7. $\sin x + \cos x = \frac{\sqrt{6}}{2}$

8. $\cot x - \csc x = \sqrt{3}$

9. $\tan x - \sec x = -1$

10. $\sin x + \cos x = \sqrt{2}$

11. $\cos x + \sin x = \frac{4}{3}$

12. $\sec x + \tan x = 2$

• Factor by grouping

13. $\cot x \csc x - 2 \cot x - \csc x + 2 = 0$

14. $4 \sin x \cos x - 2\sqrt{3} \sin x - 2 \cos x + \sqrt{3} = 0$

15. $3 \tan^2 x \cos x - 3 \cos x + 2 = 2 \tan^2 x$

16. $4\sqrt{3} \sin^2 x \sec x - \sqrt{3} \sec x + 2 = 8 \sin^2 x$

• Using identities

17. $\frac{1 + \cot^2 x}{\cot^2 x} = 4$

18. $\frac{1 + \tan^2 x}{\tan^2 x} = \frac{4}{3}$

19. $3 \cos(2x) + 7 \sin x - 5 = 0$

20. $3 \cos(2x) - \cos x + 1 = 0$

21. $2 \sin^2\left(\frac{x}{2}\right) - 3 \cos\left(\frac{x}{2}\right) = 0$

22. $2 \cos^2\left(\frac{x}{3}\right) + 3 \sin\left(\frac{x}{3}\right) - 3 = 0$

23. $\cos(3x) + \cos(5x)\cos(2x) + \sin(5x)\sin(2x) - 1 = 0$

24. $\sin(7x)\cos(4x) + \sin(5x) - \cos(7x)\sin(4x) + \cos x = 0$

25. $\sec^4 x - 2 \sec^2 x \tan^2 x + \tan^4 x = \tan^2 x$

26. $\tan^4 x - 2 \sec^2 x \tan^2 x + \sec^4 x = \cot^2 x$

State the period P of each function and find all solutions in $[0, P)$. Round to four decimal places as needed.

27. $250 \sin\left(\frac{\pi}{6}x + \frac{\pi}{3}\right) - 125 = 0$

28. $-75\sqrt{2} \sec\left(\frac{\pi}{4}x + \frac{\pi}{6}\right) + 150 = 0$

29. $1235 \cos\left(\frac{\pi}{12}x - \frac{\pi}{4}\right) + 772 = 1750$

30. $-0.075 \sin\left(\frac{\pi}{2}x + \frac{\pi}{3}\right) - 0.023 = -0.068$

• Using any appropriate method

31. $\cos x - \sin x = \frac{\sqrt{2}}{2}$

32. $5 \sec^2 x - 2 \tan x - 8 = 0$

33. $\frac{1 - \cos^2 x}{\tan^2 x} = \frac{\sqrt{3}}{2}$

34. $5 \csc^2 x - 5 \cot x - 5 = 0$

35. $\csc x + \cot x = 1$

36. $\frac{1 - \sin^2 x}{\cot^2 x} = \frac{\sqrt{2}}{2}$

37. $\sec x \cos\left(\frac{\pi}{2} - x\right) = -1$

38. $\sin\left(\frac{\pi}{2} - x\right)\csc x = \sqrt{3}$

39. $\sec^2 x \tan\left(\frac{\pi}{2} - x\right) = 4$

40. $2 \tan\left(\frac{\pi}{2} - x\right)\sin^2 x = \frac{\sqrt{3}}{2}$

WORKING WITH FORMULAS

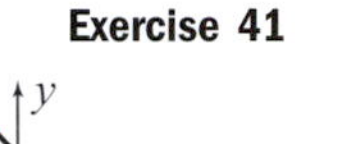

Exercise 41

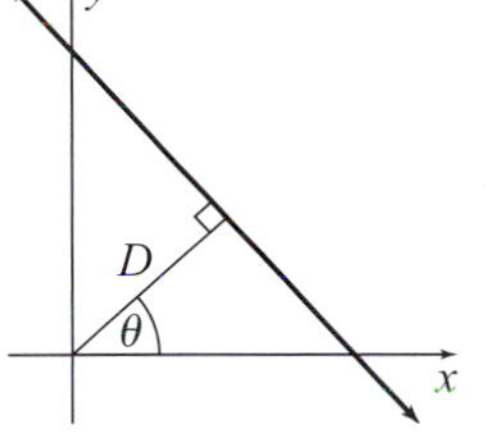

41. The equation of a line in trigonometric form: $y = \frac{D - x\cos\theta}{\sin\theta}$

The trigonometric form of a linear equation is given by the formula shown, where D is the perpendicular distance from the origin to the line and θ is the angle between the perpendicular segment and the x-axis. For each pair of perpendicular lines here, (a) find the point (a, b) of their intersection; (b) compute the distance $D = \sqrt{a^2 + b^2}$ and the angle $\theta = \tan^{-1}\left(\frac{b}{a}\right)$, and give the equation of the line in trigonometric form; and (c) use the GRAPH or the 2nd GRAPH **TABLE** feature of a graphing calculator to verify that both equations name the same line.

I. L_1: $y = -x + 5$
L_2: $y = x$

II. L_1: $y = -\frac{1}{2}x + 5$
L_2: $y = 2x$

III. L_1: $y = -\frac{\sqrt{3}}{3}x + \frac{4\sqrt{3}}{3}$
L_2: $y = \sqrt{3}x$

42. Rewriting $y = a\cos x + b\sin x$ as a single function: $y = k\sin(x + \theta)$

Linear terms of sine and cosine can be rewritten as a single function using the formula shown, where $k = \sqrt{a^2 + b^2}$ and $\theta = \sin^{-1}\left(\frac{a}{k}\right)$. Rewrite the equations given using these relationships and verify they are equivalent using the GRAPH or the 2nd GRAPH **TABLE** feature of a graphing calculator:

a. $y = 2\cos x + 2\sqrt{3}\sin x$

b. $y = 4\cos x + 3\sin x$

The ability to rewrite a trigonometric equation in simpler form has a tremendous number of applications in graphing, equation solving, working with identities, and solving applications.

APPLICATIONS

Exercise 43

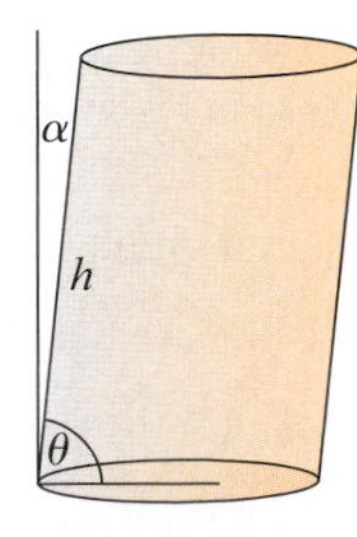

43. Volume of a cylinder: The volume of a cylinder is given by the formula $V = \pi r^2 h \sin\theta$, where r is the radius and h is the height of the cylinder, and θ is the indicated complement of the angle of deflection α. Note that when $\theta = \frac{\pi}{2}$, the formula becomes that of a right circular cylinder (if $\theta \neq \frac{\pi}{2}$, then h is called the *slant height or lateral height* of the cylinder). An old farm silo is built in the form of a right circular cylinder with a radius of 10 ft and a height of 25 ft. After an earthquake, the silo became tilted with an angle of deflection $\alpha = 5°$. (a) Find the volume of the silo before the earthquake. (b) Find the volume of the silo after the earthquake. (c) What angle θ is required to bring the original volume of the silo down 2%?

Exercise 44

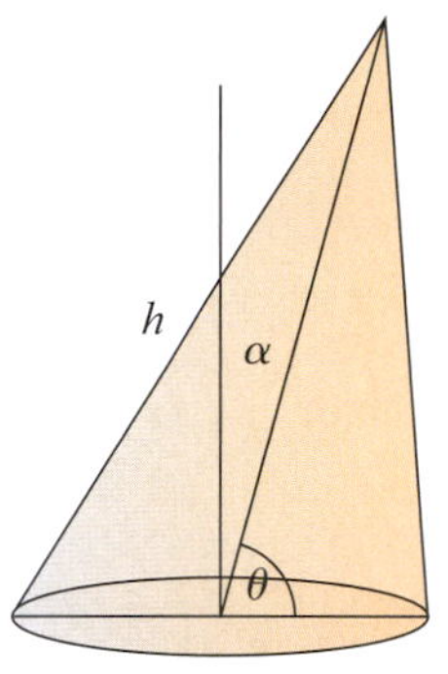

44. Volume of a cone: The volume of a cone is given by the formula $V = \frac{1}{3}\pi r^2 h \sin\theta$, where r is the radius and h is the height of the cone, and θ is the indicated complement of the angle of deflection α. Note that when $\theta = \frac{\pi}{2}$, the formula becomes that of a right circular cone (if $\theta \neq \frac{\pi}{2}$, then h is called the *slant height or lateral height* of the cone). As part of a sculpture exhibit, an artist is constructing three such structures each with a radius of 2 m and a slant height of 3 m. (a) Find the volume of the sculptures if the angle of deflection is $\alpha = 15°$. (b) What angle θ was used if the volume of each sculpture is 12 m^3?

45. River discharge rate: For June through February, the discharge rate of the La Corcovada River (Venezuela) can be modeled by the function $D(t) = 36\sin\left(\frac{\pi}{4}t - \frac{9}{4}\right) + 44$, where t represents the months of the year with $t = 1$ corresponding to June, and $D(t)$ is the discharge rate in cubic meters per second. (a) What is the discharge rate in mid-September? (b) For what months of the year is the discharge rate over 50 m^3/sec?

Source: Global River Discharge Database Project; www.rivdis.sr.unh.edu.

46. River discharge rate: For February through June, the average monthly discharge of the Point Wolfe River (Canada) can be modeled by the function $D(t) = 4.6\sin\left(\frac{\pi}{2}t + 3\right) + 7.4$, where t represents the months of the year with $t = 1$ corresponding to February, and $D(t)$ is the discharge rate in cubic meters/second. (a) What is the discharge rate in mid-March? (b) For what months of the year is the discharge rate less than 7.5 m^3/sec?

Source: Global River Discharge Database Project; www.rivdis.sr.unh.edu.

47. Seasonal sales: Hank's Heating Oil is a very seasonal enterprise, with sales in the winter far exceeding sales in the summer. Monthly sales for the company can be modeled by $S(x) = 1600\cos\left(\frac{\pi}{6}x - \frac{\pi}{12}\right) + 5100$, where $S(x)$ is the average sales in month x ($x = 1 \rightarrow$ January). (a) What is the average sales amount for July? (b) For what months of the year are sales less than \$4000?

48. Seasonal income: As a roofing company employee, Mark's income fluctuates with the seasons and the availability of work. For the past several years his average monthly income could be approximated by the function $I(m) = 2100\sin\left(\frac{\pi}{6}m - \frac{\pi}{2}\right) + 3520$, where $I(m)$ represents income in month m ($m = 1 \rightarrow$ January). (a) What is Mark's average monthly income in October? (b) For what months of the year is his average monthly income over \$4500?

49. Seasonal ice thickness: The average thickness of the ice covering an arctic lake can be modeled by the function $T(x) = 9\cos\left(\frac{\pi}{6}x\right) + 15$, where $T(x)$ is the average thickness in month x ($x = 1 \rightarrow$ January). (a) How thick is the ice in mid-March? (b) For what months of the year is the ice at most 10.5 in. thick?

50. Seasonal temperatures: The function $T(x) = 19\sin\left(\frac{\pi}{6}x - \frac{\pi}{2}\right) + 53$ models the average monthly temperature of the water in a mountain stream, where $T(x)$ is the temperature (°F) of the water in month x ($x = 1 \rightarrow$ January). (a) What is the temperature of the water in October? (b) What two months are most likely to give a temperature reading of 62°F? (c) For what months of the year is the temperature below 50°F?

51. Coffee sales: Coffee sales fluctuate with the weather, with a great deal more coffee sold in the winter than in the summer. For Joe's Diner, assume the function $G(x) = 21\cos\left(\frac{2\pi}{365}x + \frac{\pi}{2}\right) + 29$ models daily coffee sales (for non-leap years), where $G(x)$ is the number of gallons sold and x represents the days of the year ($x = 1 \rightarrow$ January 1). (a) How many gallons are projected to be sold on March 21? (b) For what days of the year are more than 40 gal of coffee sold?

52. Park attendance: Attendance at a popular state park varies with the weather, with a great deal more visitors coming in during the summer months. Assume daily attendance at the park can be modeled by the function $V(x) = 437\cos\left(\frac{2\pi}{365}x - \pi\right) + 545$ (for non-leap years), where $V(x)$ gives the number of visitors on day x ($x = 1 \rightarrow$ January 1). (a) Approximately how many people visited the park on November 1 ($11 \times 30.5 = 335.5$)? (b) For what days of the year are there more than 900 visitors?

53. Exercise routine: As part of his yearly physical, Manu Tuiosamoa's heart rate is closely monitored during a 12-min, cardiovascular exercise routine. His heart rate in beats per minute (bpm) is modeled by the function $B(x) = 58\cos\left(\frac{\pi}{6}x + \pi\right) + 126$ where x represents the duration of the workout in minutes. (a) What was his resting heart rate? (b) What was his heart rate 5 min into the workout? (c) At what times during the workout was his heart rate over 170 bpm?

54. Exercise routine: As part of her workout routine, Sara Lee programs her treadmill to begin at a slight initial grade (angle of incline), gradually increase to a maximum grade, then gradually decrease back to the original grade. For the duration of her workout, the grade is modeled by the function $G(x) = 3\cos\left(\frac{\pi}{5}x - \pi\right) + 4$, where $G(x)$ is the percent grade x minutes after the workout has begun. (a) What is the initial grade for her workout? (b) What is the grade at $x = 4$ min? (c) At $G(x) = 4.9\%$, how long has she been working out? (d) What is the duration of the treadmill workout?

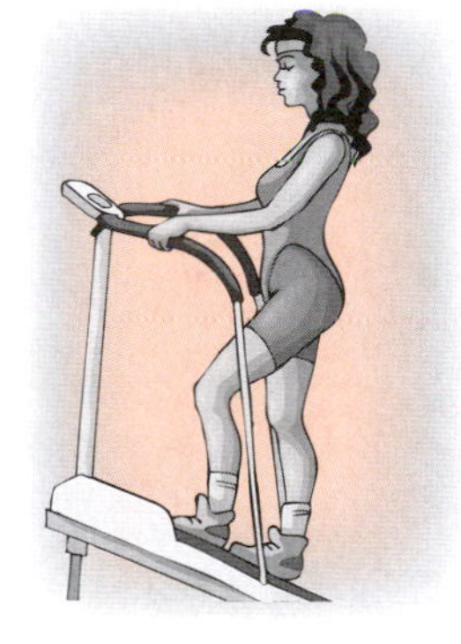

WRITING, RESEARCH, AND DECISION MAKING

55. When pottery and other ceramic items are fired in a kiln, the temperature of the kiln must be raised gradually to a maximum temperature, then decreased gradually in order to prevent cracking of the vessel. In many cases, this gradual increase/decrease can be modeled by a sine or cosine function. Contact a pottery shop and talk to the kiln technician, or visit the Internet to see if you can find an equation model for kiln temperature during firing. Is the increase/decrease programmed or done manually? At what temperature are ceramic items placed in the kiln and removed from the kiln?

56. As we saw in Chapter 3, cosine is the cofunction of sine and each can be expressed in terms of the other: $\cos\left(\frac{\pi}{2} - \theta\right) = \sin\theta$ and $\sin\left(\frac{\pi}{2} - \theta\right) = \cos\theta$. This implies that either function can be used to model the phenomenon described in Section 4.4 by adjusting the phase shift. By experimentation, (a) find a model using cosine that will produce results identical to the sine function in Exercise 50 and (b) find a model using sine that will produce results identical to the cosine function in Exercise 51.

EXTENDING THE CONCEPT

57. Use multiple identities to find all real solutions for the equation given: $\sin(5x) + \sin(2x)\cos x + \cos(2x)\sin x = 0$.

58. A rectangular parallelepiped with square ends has 12 edges and six surfaces. If the sum of all edges is 176 cm and the total surface area is 1288 cm^2, find (a) the length of the diagonal of the parallelepiped (shown in bold) and (b) the angle the diagonal makes with the base (two answers are possible).

Exercise 58

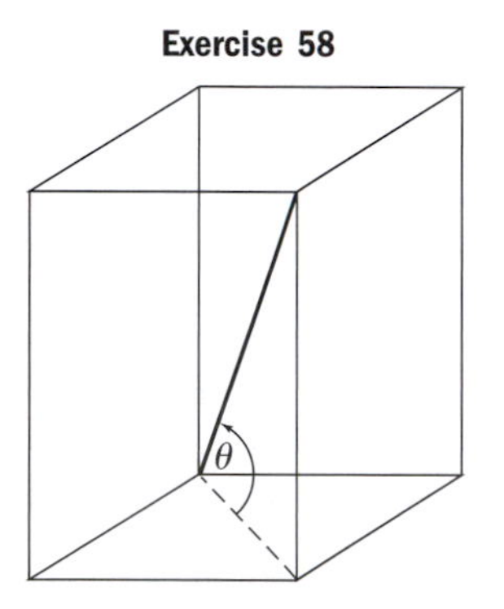

MAINTAINING YOUR SKILLS

59. (1.3) Find $f(\theta)$ for all six trig functions, given $P(-51, 68)$ is on the terminal side.

60. (4.2) Draw a corresponding triangle to help write $\cos\left[\tan^{-1}\left(\frac{x}{6}\right)\right]$ in algebraic form.

61. (3.4) Use identities to evaluate $\cos(4x)$ exactly, given $\sin x = \frac{3}{5}$ and $\cos x = \frac{4}{5}$.

62. (3.3) If $\sin\alpha = \frac{\sqrt{2}}{2}$ and $\cos\beta = \frac{\sqrt{3}}{2}$, find the value of $\sin(\alpha + \beta)$ and the value of $\cos(\alpha + \beta)$.

63. (2.1) The graph of the function shown models one of the seven primary colors. Use the wavelength chart on page 94 to identify the color. The x-axis is marked in nanometers.

64. (1.2) The Sears Tower in Chicago, Illinois, remains one of the tallest structures in the world. The top of the roof reaches 1450 ft above the street below and the antenna extends an additional 280 ft into the air. Find the viewing angle θ for the antenna from a distance of 1000 ft (the angle formed from the base of the antenna to its top).

Exercise 63 **Exercise 64**

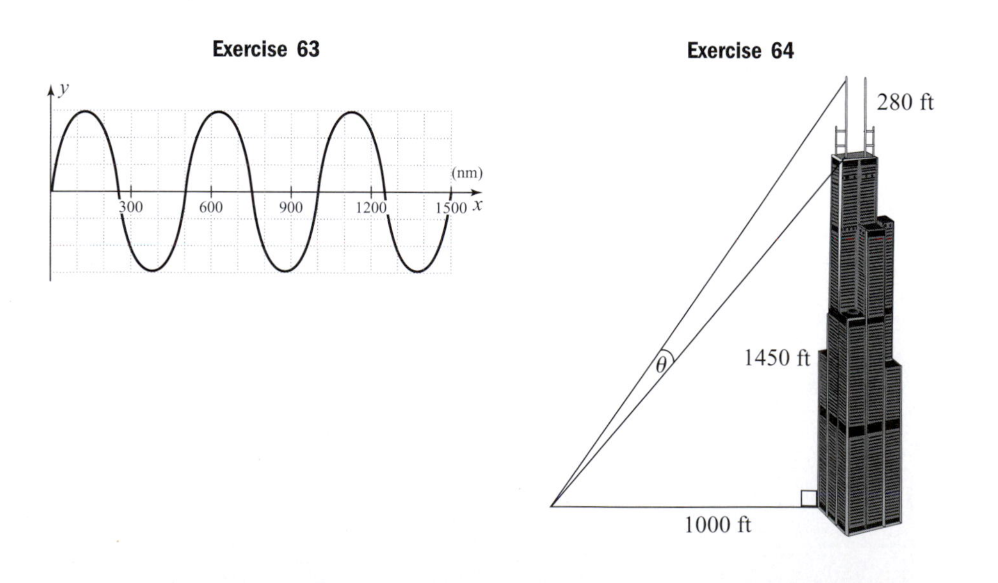

4.5 Parametric Equations and Graphs

LEARNING OBJECTIVES

In Section 4.5 you will learn how to:

- **A. Sketch the graph of a parametric equation**
- **B. Write parametric equations in rectangular form**
- **C. Graph curves from the cycloid family**
- **D. Solve applications involving parametric equations**

INTRODUCTION

A large portion of the mathematics curriculum is devoted to functions, due to their overall importance and widespread applicability. But there are a host of applications for which nonfunctions are a more natural fit. In this section, we show that many *nonfunctions* can be expressed as **parametric equations,** where each is actually a *function*. These equations can be appreciated for the diversity and versatility they bring to the mathematical spectrum.

POINT OF INTEREST

Parametric equations make it possible for us to investigate a special family of curves known as the cycloids. The most basic cycloid is formed by tracing the path of a point on the circumference of a wheel as it rolls in a straight line. In the mid-seventeenth century, this curve was enjoying a considerable amount of attention, with the discovery of many interesting and useful properties. However, this attention also prompted a large number of quarrels and disagreements among mathematicians of the day. For this reason many have called the cycloid the *Helen of Geometry,* a metaphor for Helen of Troy and the numerous battles fought over her.

A. Sketching a Curve Defined Parametrically

Suppose you were given the set of points in the table here, and asked to come up with an equation model for the data. To begin, you might plot the points to see if any patterns or clues emerge, but in this case the result seems to be a curve we've never seen before (see Figure 4.45).

Figure 4.45

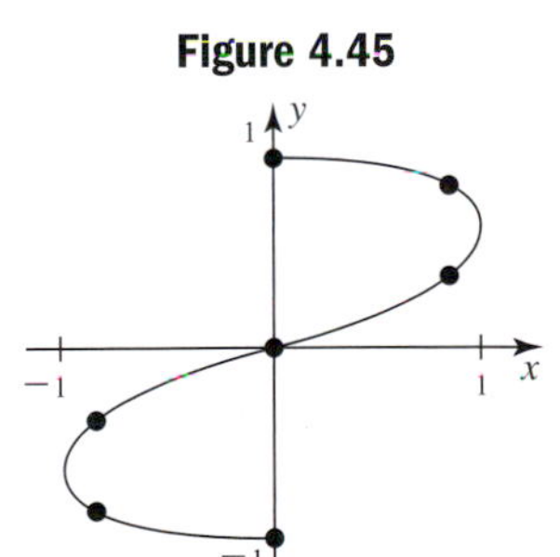

x	0	$\frac{\sqrt{3}}{2}$	$\frac{\sqrt{3}}{2}$	0	$-\frac{\sqrt{3}}{2}$	$-\frac{\sqrt{3}}{2}$	0
y	1	$\frac{\sqrt{3}}{2}$	$\frac{1}{2}$	0	$-\frac{1}{2}$	$-\frac{\sqrt{3}}{2}$	-1

You also might consider running a regression on the data, but it's not possible since the graph is obviously not a function. However, a closer look at the data reveals the y-values could be modeled *independently of the x-values* by a cosine function, $y = \cos t$ for $t \in [0, \pi]$. This observation leads to a closer look at the x-values, which we find could be modeled by a sine function over the same interval, namely, $x = \sin(2t)$ for $t \in [0, \pi]$. These two functions combine to name all points on this curve, and both use the independent variable t called a **parameter.** The functions $x = \sin(2t)$ and $y = \cos t$ are called the parametric equations for this curve. The complete curve, shown in Figure 4.46, is called a **Lissajous figure,** or a closed graph (coincident beginning and ending points) that crosses itself to form two or more

Figure 4.46

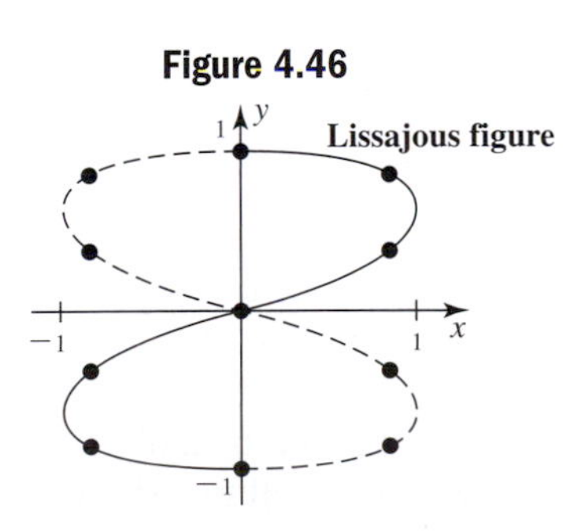

loops. Note that since the maximum value of x and y is 1 (the amplitude of each function), the entire figure will fit within a 1×1 rectangle centered at the origin. This observation can often be used to help sketch parametric graphs with trigonometric parameters. In general, parametric equations can take many forms, including polynomial, exponential, trigonometric, and other forms.

PARAMETRIC EQUATIONS

Given the set of points $P(x, y)$ such that $x = f(t)$ and $y = g(t)$, where f and g are both defined on an interval of the domain, the equations $x = f(t)$ and $y = g(t)$ are called parametric equations, with parameter t.

EXAMPLE 1 Graph the curve defined by the parametric equations $x = t^2 - 3$ and $y = 2t + 1$.

Solution: Begin by creating a table of values using $t \in [-3, 3]$. After plotting ordered pairs (x, y), the result appears to be a parabola, opening to the right.

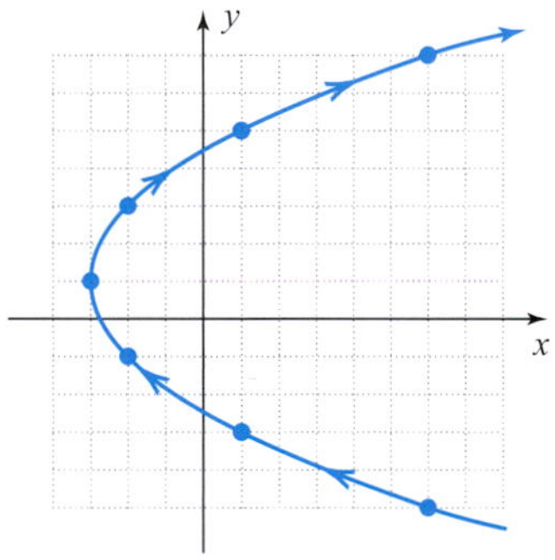

t	$x = t^2 - 3$	$y = 2t + 1$
−3	6	−5
−2	1	−3
−1	−2	−1
0	−3	1
1	−2	3
2	1	5
3	6	7

NOW TRY EXERCISES 7 THROUGH 12, PART A

If the parameter is a trig function, we'll often use standard angles as inputs to simplify calculations and the period of the function(s) to help sketch the resulting graph.

EXAMPLE 2 Graph the curve defined by the parametric equations $x = 2 \cos t$ and $y = 4 \sin t$.

Solution: Using standard angle inputs and knowing the maximum value of any x- and y-coordinate will be 2 and 4, respectively, we begin computing and graphing a few points. After going from 0 to π, we note the graph appears to be a vertical ellipse. This is verified using standard values from π to 2π. Plotting the points and connecting them with a smooth curve produces the ellipse shown in the figure.

t	$x = 2\cos t$	$y = 4\sin t$
0	2	0
$\frac{\pi}{6}$	$\sqrt{3}$	2
$\frac{\pi}{3}$	1	$2\sqrt{3}$
$\frac{\pi}{2}$	0	4
$\frac{2\pi}{3}$	-1	$2\sqrt{3}$
$\frac{5\pi}{6}$	$-\sqrt{3}$	2
π	-2	0

NOW TRY EXERCISES 13 THROUGH 18, PART A

A complete study of ellipses and the other conic sections is given in Chapter 6.

B. Writing Parametric Equations in Rectangular Form

When graphing parametric equations, there are sometimes alternatives to simply plotting points. One alternative is to try and *eliminate the parameter*, writing the parametric equations in standard, rectangular form. To accomplish this we use some connection that allows us to "rejoin" the parameterized equations, such as variable t itself, a trigonometric identity, or some other connection.

EXAMPLE 3 Eliminate the parameter from the equations in Example 1: $x = t^2 - 3$ and $y = 2t + 1$.

Solution: Solving for t in the second equation gives $t = \frac{y-1}{2}$, which we then substitute into the first. The result is $x = \left(\frac{y-1}{2}\right)^2 - 3 = \frac{1}{4}(y-1)^2 - 3$. Notice this is indeed a horizontal parabola, opening to the right, with vertex at $(-3, 1)$.

NOW TRY EXERCISES 7 THROUGH 12, PART B

EXAMPLE 4 Eliminate the parameter from the equations in Example 2: $x = 2\cos t$ and $y = 4\sin t$.

Solution: Instead of trying to solve for t, we note the parametrized equations involve sine and cosine functions with the same argument (t), and opt to use the identity $\cos^2 t + \sin^2 t = 1$. Squaring both equations and solving for $\cos^2 t$ and $\sin^2 t$ yields $\frac{x^2}{4} = \cos^2 t$ and $\frac{y^2}{16} = \sin^2 t$. This shows $\cos^2 t + \sin^2 t = \frac{x^2}{4} + \frac{y^2}{16} = 1$, and as we suspected—the result is a vertical ellipse with vertices at $(0, \pm 4)$ and endpoints of the minor axis at $(\pm 2, 0)$.

NOW TRY EXERCISES 13 THROUGH 16, PART B

It's important to realize that a given curve can be represented parametrically in infinitely many ways. This flexibility sometimes enables us to simplify the given form, or to write a given polynomial form in an equivalent nonpolynomial form. The easiest way to write the function $y = f(x)$ in parametric form is $x = t$; $y = f(t)$, which is valid *as long as t is in the domain of* $f(t)$.

EXAMPLE 5 Write the equation $y = 4(x - 3)^2 + 1$ in three different parametric forms.

Solution:

1. If we let $x = t$, we have $y = 4(t - 3)^2 + 1$.
2. Letting $x = t + 3$ simplifies the related equation for y, and we begin to see some of the advantages of using a parameter: $x = t + 3$; $y = 4t^2 + 1$.
3. As a third alternative, we can let $x = \frac{1}{2}\tan t + 3$, which gives $x = \frac{1}{2}\tan t + 3$; $y = 4\left(\frac{1}{2}\tan t\right)^2 + 1 = \tan^2 t + 1$ or $y = \sec^2 t$.

NOW TRY EXERCISES 19 THROUGH 26

C. Graphing Curves from the Cycloid Family

As you might have noticed in the *Point of Interest*, **cycloids** are an important family of curves, and are used extensively to solve what are called **brachistochrone** applications. The name comes from the Greek *brakhus*, meaning short, and *khronos*, meaning time, and deal with finding the path along which a weight will fall in the shortest time possible. Cycloids are an excellent example of why parametric equations are important, as it's very difficult to name them in rectangular form. Consider a point fixed to the circumference of a wheel as it rolls from left to right. If we trace the path of the point as the wheel rolls, the resulting curve is a cycloid. Figure 4.47 shows the location of the point every one-quarter turn.

Figure 4.47

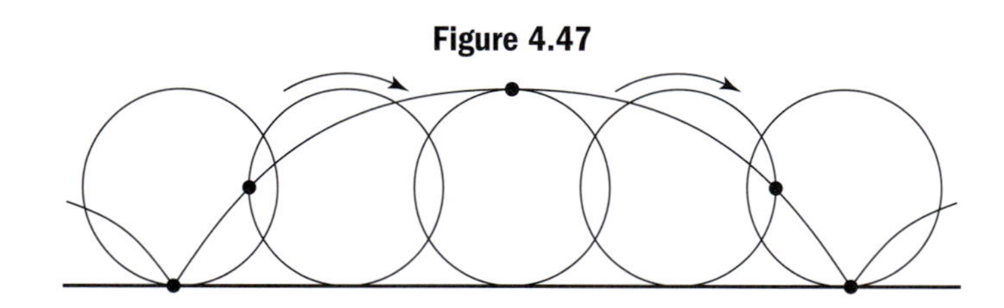

By superimposing a coordinate grid on the diagram in Figure 4.47, we can construct parametric equations that will produce the graph. This is done by developing equations for the location of a point $P(x, y)$ on the circumference of a circle with center (h, k), as the circle rotates through angle t. After a rotation of t rad, the x-coordinate of $P(x, y)$ is $x = h - a$ (Figure 4.48), and the y-coordinate is $y = k - b$. Using a right triangle with the radius as the hypotenuse, we find $\sin t = \frac{a}{r}$ and $\cos t = \frac{b}{r}$, giving $a = r\sin t$ and $b = r\cos t$. Substituting into $x = h - a$ and $y = k - b$ yields $x = h - r\sin t$ and $y = k - r\cos t$. Since the circle has radius r, we know $k = r$

Figure 4.48

(the "height" of the center is constantly $k = r$). The arc length subtended by t is the same as the distance h (see Figure 4.49), meaning $h = rt$ (t in radians) Substituting rt for h and r for k in the equations $x = h - r\sin t$ and $y = k - r\cos t$, gives the equation of the cycloid in parametric form: $x = rt - r\sin t$ and $y = r - r\cos t$, sometimes written $x = r(t - \sin t)$ and $y = r(1 - \cos t)$.

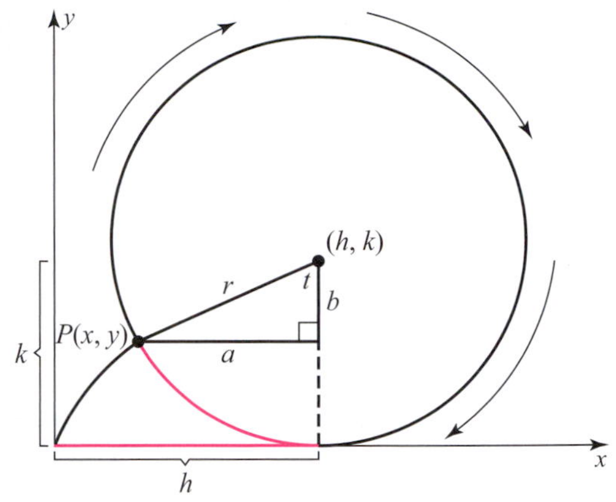

Most graphers have a parametric MODE that enables you to enter the equations for x and y separately, and graph the resulting points as a single curve. After pressing the Y= key (in parametric mode), the screen in Figure 4.49 comes into view using a TI-84 Plus, and we enter the equation of the cycloid formed by a circle of radius $r = 3$. To set the viewing window (including a frame), press WINDOW and set Ymin $= -1$ and Ymax at slightly more than 6 (since $r = 3$). Since the cycloid completes one cycle every $2\pi r$, we set Xmax at $2\pi rn$, where n is the number of cycles we'd like to see. In this case, we set it for four cycles $(2\pi)(3)(4) = 24\pi$ (Figure 4.50). With $r = 3$ we conveniently set Xscl at $3(2\pi) = 6\pi \approx 18.8$ to tick each cycle, and Xscl $= 3\pi \approx 9.4$ to tick each half cycle (Figure 4.50). For parametric equations, we must also specify a range of values for t, which we set at Tmin $= 0$, Tmax $= 8\pi \approx 25.1$ for the four cycles, and Tstep $= \frac{\pi}{6} \approx 0.52$ (Tstep controls the number of points plotted and joined to form the curve). The window settings and resulting graph are shown in Figure 4.51, which doesn't look much like a cycloid because the current settings do not produce a square viewing window. Using ZOOM **5:ZSquare** (and changing Yscl) produces the graph shown in Figure 4.52, which looks much more like the cycloid we expected.

Figure 4.49

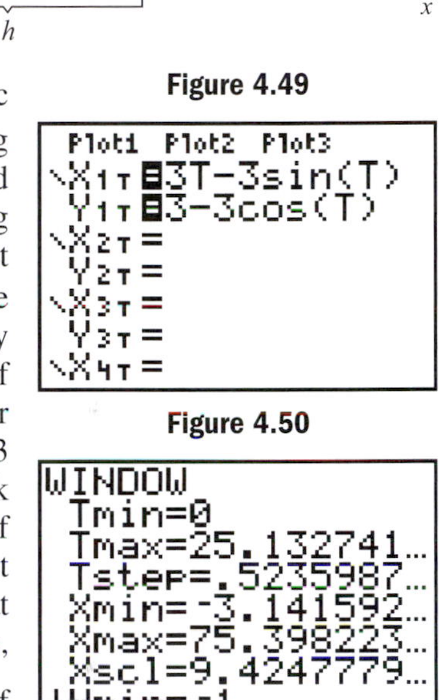

Figure 4.50

Figure 4.51

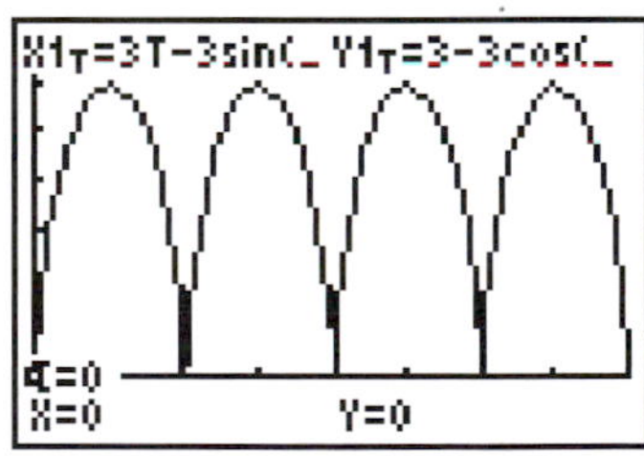

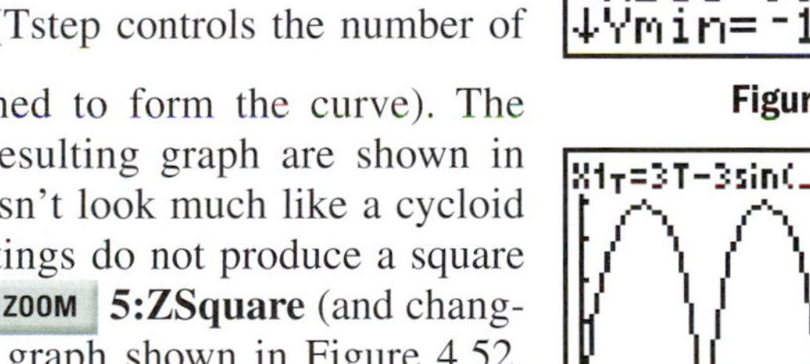

Figure 4.52

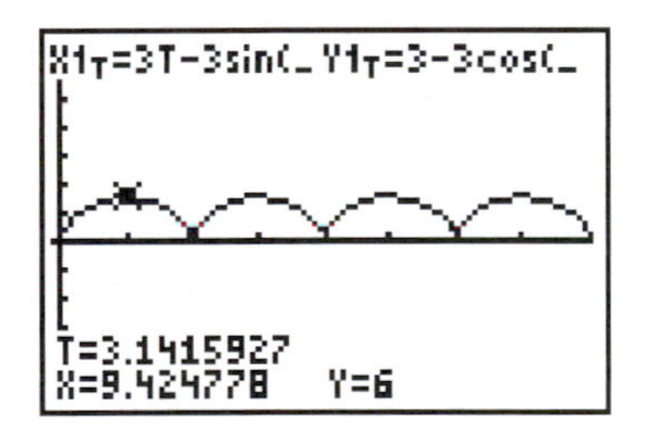

Figure 4.53

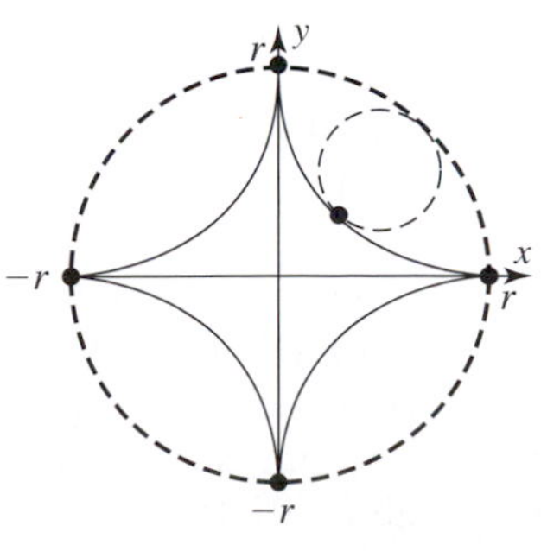

EXAMPLE 6 Use a graphing calculator to graph the curve defined by the equations $x = 3\cos^3 t$ and $y = 3\sin^3 t$, called a **hypocycloid with four cusps.**

Solution: A hypocycloid is a curve traced out by the path of a point on the circumference of a circle as it rolls *inside a larger circle* of radius r (see Figure 4.53). Here $r = 3$ and we set Xmax and Ymax accordingly. Knowing ahead of time the hypocycloid will have four cusps, we set Tmax $= 4(2\pi) \approx 25.13$ to show all four. The window settings used and the resulting graph are shown in Figures 4.54 and 4.55.

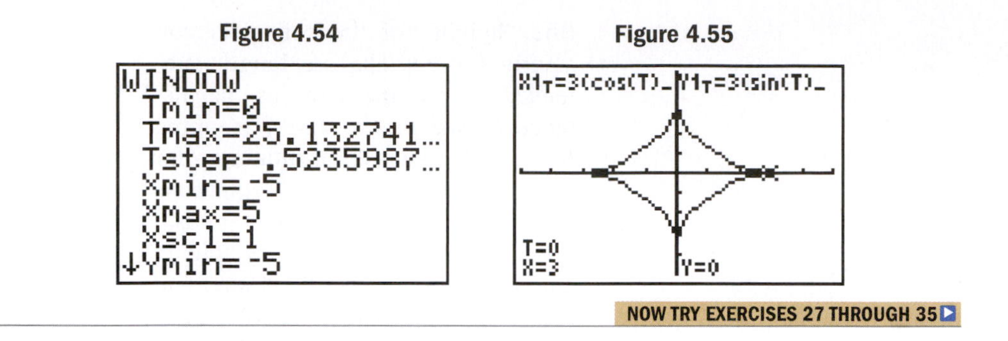

Figure 4.54 Figure 4.55

NOW TRY EXERCISES 27 THROUGH 35

D. Common Applications of Parametric Equations

In Example 1 the parameter was simply the *real number t*, which enabled us to model the x- and y-values of an ordered pair (x, y) independently. In Examples 2 and 6, the parameter t represented an *angle*. Here we introduce yet another kind of parameter, that of *time t*.

A **projectile** is any object thrown, dropped, or projected in some way with no continuing source of propulsion. The parabolic path traced out by the projectile (assuming negligible air resistance) will be fully developed in Section 5.5. It is stated here in parametric terms. For the projectile's location $P(x, y)$ and any time t in seconds, the x-coordinate (horizontal distance from point of projection) is given by $x = v_0 t \cos\theta$, where v_0 is the initial velocity in feet per second and t is the time in seconds. The y-coordinate (vertical height) is $y = v_0 t \sin\theta - 16t^2$.

EXAMPLE 7 As part of a circus act, Karl the Human Cannonball is shot out of a specially designed cannon at an angle of 40° with an initial velocity of 120 ft/sec. Use a graphing calculator to graph the resulting parametric curve. Then use the graph to determine how high the Ring Master must place a circular ring for Karl to be shot through at the maximum height of his trajectory, and how far away the net must be placed to catch Karl.

Solution: The information given leads to the equations $x = 120t\cos 40°$ and $y = 120t\sin 40° - 16t^2$. Enter these equations on the Y= screen of your calculator, remembering to reset the MODE to degrees (circus clowns may not know or understand radians). To set the window size, we can use trial and error, or estimate using $\theta = 45°$ (instead of 40°) and an estimate for t (the time that Karl will stay aloft). With $t = 6$ we get estimates of $x = 120(6)\left(\frac{\sqrt{2}}{2}\right) = 360\sqrt{2}$ for the horizontal distance. To find a range for y, use $t = 3$ since the maximum height of the parabolic path will occur halfway through the flight. This gives an estimate of $120(3)\left(\frac{\sqrt{2}}{2}\right) - 16(9) = 180\sqrt{2} - 144$ for y. The results are shown in Figures 4.56 and 4.57. Using the TRACE feature or 2nd GRAPH (**TABLE**) feature, we find the center of the net used to catch Karl should be set at a distance of about 450 ft from the cannon, and the ring should be located 220 ft from the cannon at a height of about 93 ft.

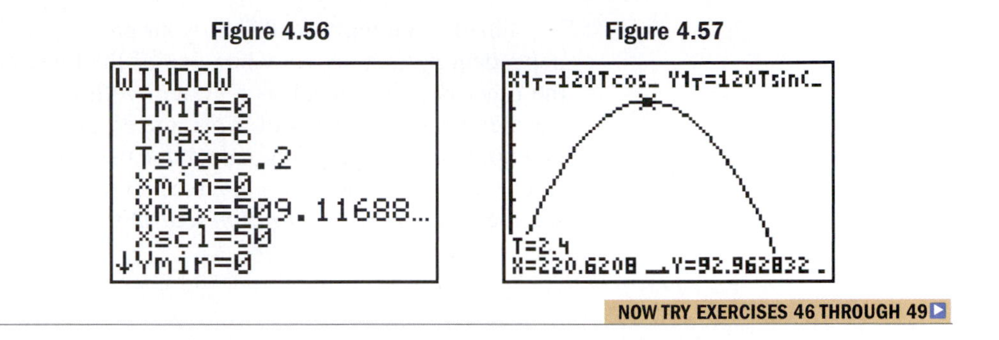

Figure 4.56

Figure 4.57

NOW TRY EXERCISES 46 THROUGH 49 ▶

It is well known that planets orbit the Sun in elliptical paths. While we're able to model their orbits in both rectangular and polar form, neither of these forms can give a true picture of the *direction they travel.* This gives parametric forms a great advantage, in that they can model the shape of the orbit, *while also indicating the direction of travel.* We illustrate in Example 8 using a "planet" with a very simple orbit.

EXAMPLE 8 ▶ The elliptical orbit of a certain planet is defined parametrically as $x = 4 \sin t$ and $y = -3 \cos t$. Graph the orbit and verify that for increasing values of t, the planet orbits in a counterclockwise direction.

Solution: ▶ Eliminating the parameter as in Example 4, we obtain the equation $\frac{x^2}{16} + \frac{y^2}{9} = 1$, or the equation of an ellipse with center at (0, 0), major axis of length 8, and minor axis of length 6. The path of the planet is traced out by the ordered pairs (x, y) generated by the parametric equations, shown in the table for $t \in [0, \pi]$. Starting at $t = 0$,

$P(x, y)$ begins at $(0, -3)$ with x and y both increasing until $t = \frac{\pi}{2}$. Then from $t = \frac{\pi}{2}$ to $t = \pi$, y continues to increase as x decreases, indicating a counterclockwise orbit in this case. The orbit is illustrated in the figure.

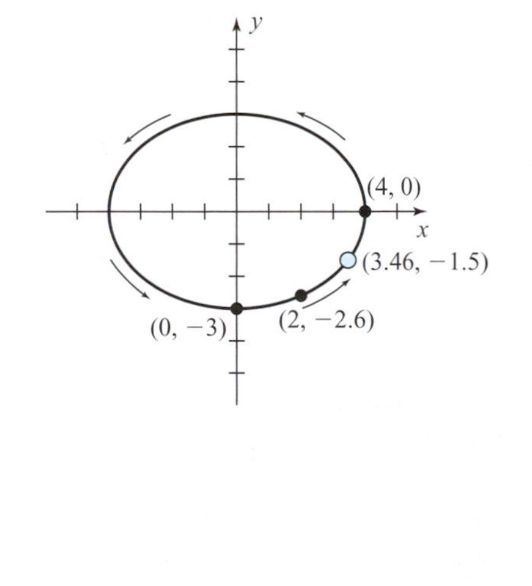

t	$x = 4 \sin t$	$y = -3 \cos t$
0	0	−3
$\frac{\pi}{6}$	2	−2.6
$\frac{\pi}{3}$	3.46	−1.5
$\frac{\pi}{2}$	4	0
$\frac{2\pi}{3}$	3.46	1.5
$\frac{5\pi}{6}$	2	2.6
π	0	3

NOW TRY EXERCISES 50 AND 51 ▶

Finally, you may recall from your previous work with linear 3×3 systems, that a dependent system occurs when one of the three equations is a linear combination of the other two. The result is a system with more variables than equations, with solutions expressed in terms of a parameter, or in *parametric form.* These solutions can be explored on a graphing calculator using ordered triples of the form $(t, f(t), g(t))$, where $Y_1 = f(t)$ and $Y_2 = g(t)$ (see Exercises 52 through 55). For more information, see the *Calculator Exploration and Discovery* feature on page 296.

TECHNOLOGY HIGHLIGHT

Exploring Parametric Graphs on a Graphing Calculator

The keystrokes shown apply to a TI-84 Plus model. Please consult our Internet site or your manual for other models.

Most graphing calculators have features that make it easy (and fun) to investigate parametric equations. For example, the TI-84 Plus can use a circular cursor to trace the path of the plotted points, as they are generated by the equations. This can be used to illustrate the path of a projectile, the distance of a runner, or the orbit of a planet. Operations can also be applied to the parameter T to give the effect of "speed" (the points from one set of equations are plotted faster than the points of a second set). To help illustrate their use, consider again the simple, elliptical orbit of a planet given in Example 8. Physics tells us that the closer a planet is to the Sun, the faster its orbit. In fact, the orbital speed of Mercury is about twice that of Mars and about 10 times as fast as the dwarf planet Pluto (29.8, 15, and 2.9 mi/sec, respectively). With this information, we can explore a number of interesting questions. On the Y= screen, let the orbits of Planet 1 and Planet 2 be modeled parametrically by the equations shown in Figure 4.58. Since the orbit of Planet 1 is "smaller" (closer to the Sun), we have T-values growing at a rate that is *four times as fast* as for Planet 2. Notice to the far left of X_{1T}, there is a symbol that looks like an old key "**–0**." By moving the cursor to the far left of the equation, you can change how the graph will look by repeatedly pressing ENTER. With this symbol in view, the calculator will trace out the curve with a circular cursor, which in this case represents the planets as they orbit (be sure you are in simultaneous MODE). Setting the window as in Figure 4.59 and pressing GRAPH produces the screen in Figure 4.60, which displays their elliptical paths as they race around the Sun. Notice the inner planet has already completed one orbit while the outer planet has just completed one-fourth of an orbit. Use your calculator to complete the following exercises.

Figure 4.58

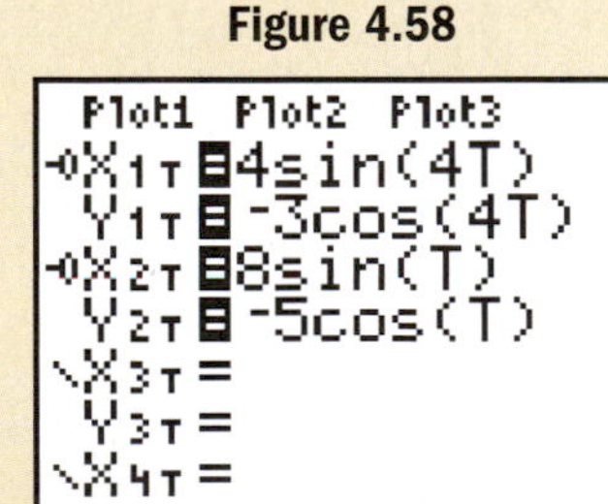

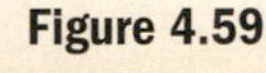
Figure 4.59

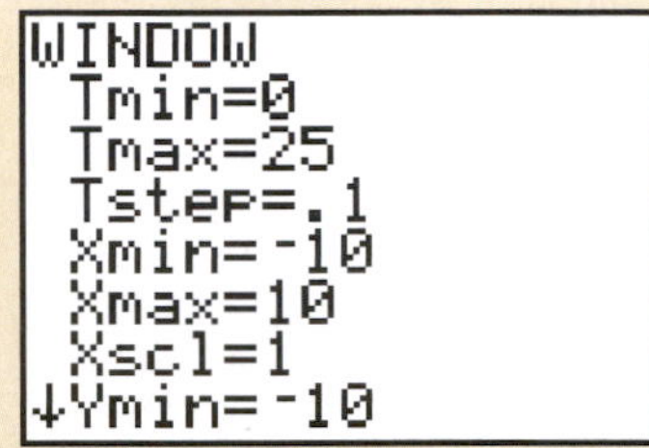

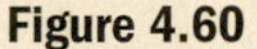
Figure 4.60

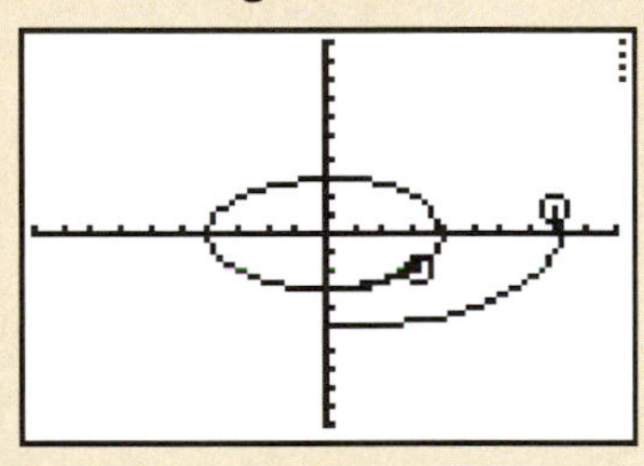

Exercise 1: Verify that the inner planet completes four orbits for every single orbit of the outer planet.

Exercise 2: Suppose that due to some cosmic interference, the orbit of the faster planet begins to decay at a rate of $T^{0.84}$ (replace T with $T^{0.84}$ in both equations for the inner planet). By observation, how many orbits did the inner planet make for the first revolution of the outer planet? What is the ratio of orbits for the next complete orbit of the outer planet?

Exercise 3: Create new planets with different orbits, and formulate some interesting questions of your own.

4.5 EXERCISES

CONCEPTS AND VOCABULARY

Fill in each blank with the appropriate word or phrase. Carefully reread the section if needed.

1. When the coordinates of a point (x, y) are generated independently using $x = f(t)$ and $y = g(t)$, t is called a(n) ________.

2. The equations $x = f(t)$ and $y = g(t)$ used to generate the ordered pairs (x, y) are called ________ equations.

3. Parametric equations can both graph a curve *and* indicate the ________ traveled by a point on the curve.

4. To write parametric equations in rectangular form, we must ________ the parameter to write a single equation.

5. Discuss the connection between solutions to dependent systems and the parametric equations studied in this section.

6. In your own words, explain and illustrate the process used to develop the equation of a cycloid. Illustrate with a specific example.

DEVELOPING YOUR SKILLS

For Exercises 7 through 18, (a) graph the curves defined by the parametric equations using the specified interval and identify the graph (if possible) and (b) eliminate the parameter (Exercises 7 to 16 only) and write the corresponding rectangular form.

7. $x = t + 2;\ t \in [-3, 3]$
$y = t^2 - 1$

8. $x = t - 3;\ t \in [-5, 5]$
$y = 2 - 0.5t^2$

9. $x = (2 - t)^2;\ t \in [0, 5]$
$y = (t - 3)^2$

10. $x = t^3 - 3;\ t \in [-2, 2.5]$
$y = t^2 + 1$

11. $x = \dfrac{5}{t},\ t \neq 0;\ t \in [-3.5, 3.5]$
$y = t^2$

12. $x = \dfrac{t^3}{10};\ t \in [-5, 5]$
$y = |t|$

13. $x = 4\cos t;\ t \in [0, 2\pi)$
$y = 3\sin t$

14. $x = 2\sin t;\ t \in [0, 2\pi)$
$y = -3\cos t$

15. $x = 4\sin(2t);\ t \in [0, 2\pi)$
$y = 6\cos t$

16. $x = 4\cos(2t);\ t \in \left[\dfrac{\pi}{2}, \dfrac{3\pi}{2}\right]$
$y = 6\sin t$

17. $x = \dfrac{3}{\tan t};\ t \in (0, \pi)$
$y = 5\sin(2t)$

18. $x = \tan^2 t;\ t \neq \dfrac{\pi}{2},\ t \in [0, \pi]$
$y = 3\cos t$

Write each function in three different parametric forms by altering the parameter. For Exercises 19–22 use at least one trigonometric form, restricting the domain as needed.

19. $y = 3x - 2$

20. $y = 0.5x + 6$

21. $y = (x + 3)^2 + 1$

22. $y = 2(x - 5)^2 - 1$

23. $y = \tan^2(x - 2) + 1$

24. $y = \sin(2x - 1)$

25. Use a graphing calculator or computer to verify that the parametric equations from Example 5 all produce the same graph.

26. Use a graphing calculator or computer to verify that your parametric equations from Exercise 21 all produce the same graph.

The curves defined by the following parametric equations are from the cycloid family. (a) Use a graphing calculator or computer to draw the graph and (b) use the graph to approximate all x- and y-intercepts, and maximum and minimum values to one decimal place.

27. $x = 8\cos t + 2\cos(4t)$, $y = 8\sin t - 2\sin(4t)$, hypocycloid (5-cusp)

28. $x = 8\cos t + 4\cos(2t)$, $y = 8\sin t - 4\sin(2t)$, hypocycloid (3-cusp)

29. $x = \dfrac{2}{\tan t}$, $y = 8\sin t\cos t$, serpentine curve

30. $x = 8\sin^2 t$, $y = \dfrac{8\sin^3 t}{\cos t}$, cissoid of Diocles

31. $x = 2(\cos t + t\sin t)$, $y = 2(\sin t - t\cos t)$, involute of a circle

32. $4x = (16 - 36)\cos^3 t$, $6y = (16 - 36)\sin^3 t$, evolute of an ellipse

33. $x = 3t - \sin t$, $y = 3 - \cos t$, curtate cycloid

34. $x = t - 3\sin t$, $y = 1 - 3\cos t$, prolate cycloid

35. $x = 2[3\cos t - \cos(3t)]$, $y = 2[3\sin t - \sin(3t)]$, nephroid

Use a graphing calculator or computer to draw the following parametrically-defined graphs, called Lissajous figures (Exercise 37 is a scaled version of the initial example from this section). Then find the dimensions of the rectangle necessary to frame the figure and state the number of times the graph crosses itself.

36. $x = 6\sin(3t)$, $y = 8\cos t$

37. $x = 6\sin(2t)$, $y = 8\cos t$

38. $x = 8\sin(4t)$, $y = 10\cos t$

39. $x = 5\sin(7t)$, $y = 7\cos(4t)$

40. $x = 8\sin(4t)$, $y = 10\cos(3t)$

41. $x = 10\sin(1.5t)$, $y = 10\cos(2.5t)$

42. Use a graphing calculator to experiment with parametric equations of the form $x = A\sin(mt)$ and $y = B\cos(nt)$. Try different values of A, B, m, and n, then discuss their effect on the Lissajous figures.

43. Use a graphing calculator to experiment with parametric equations of the form $x = \dfrac{a}{\tan t}$ and $y = b\sin t\cos t$. Try different values of a and b, then discuss their effect on the resulting graph, called a serpentine curve. Also see Exercise 29.

WORKING WITH FORMULAS

44. The Folium of Descartes: $x(t) = \dfrac{3kt}{1 + t^3}$; $y(t) = \dfrac{3kt^2}{1 + t^3}$

The Folium of Descartes is a parametric curve developed by Descartes in order to test the ability of Fermat to find its maximum and minimum values.

a. Graph the curve on a graphing calculator with $k = 1$ using a reduced window (ZOOM 4), with Tmin $= -6$, Tmax $= 6$, and Tstep $= 0.1$. Locate the coordinates of the tip of the folium (the loop).

b. This graph actually has a discontinuity (a break in the graph). At what value of t does this occur?

c. Experiment with different values of k and generalize its effect on the basic graph.

45. The Witch of Agnesi: $x(t) = 2kt$; $y(t) = \dfrac{2k}{1 + t^2}$

The Witch of Agnesi is a parametric curve named by Maria Agnesi in 1748. Some believe she confused the Italian word for *witch* (*versiera*), with a similar word that meant *free to move*. In any case, the name stuck. The curve can also be stated in trigonometric form: $x(t) = 2k\cot t$ and $y = 2k\sin^2 t$.

a. Graph the curve with $k = 1$ on a calculator or computer on a reduced window (ZOOM 4) using both of the forms shown with Tmin $= -6$, Tmax $= 6$, and Tstep $= 0.1$. Try to determine the maximum value.

b. Explain why the x-axis is a horizontal asymptote.

c. Experiment with different values of k and generalize its effect on the basic graph.

APPLICATIONS

Model each application using parametric equations, then solve using the GRAPH and TRACE features of a graphing calculator.

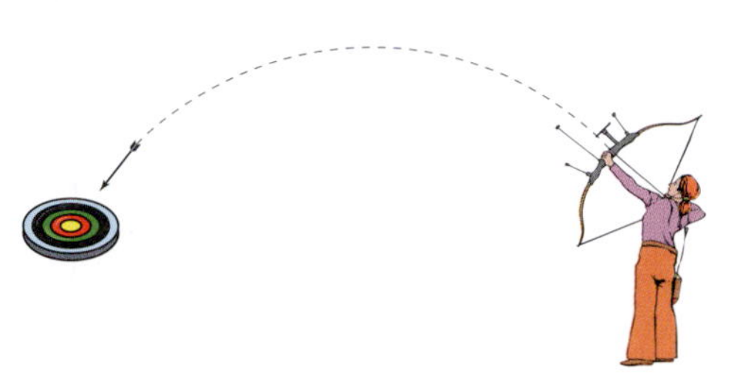

46. Archery competition: At an archery contest, a large circular target 5 ft in diameter is laid flat on the ground with the bull's-eye exactly 180 yd (540 ft) away from the archers. Marion draws her bow and shoots an arrow at an angle of 25° above horizontal with an initial velocity of 150 ft/sec (assume the archers are standing in a depression and the arrow is shot from ground level). (a) What was the maximum height of the arrow? (b) Does the arrow hit the target? (c) What is the distance between Marion's arrow and the bull's-eye after the arrow hits?

47. Football competition: As part of their contribution to charity, a group of college quarterbacks participate in a contest. The object is to throw a football through a hoop whose center is 30 ft high and 25 yd (75 ft) away, trying to hit a stationary (circular) target laid on the ground with the center 56 yd (168 ft) away. The hoop and target both have a diameter of 4 ft. On his turn, Lance throws the football at an angle of 36° with an initial velocity of 75 ft/sec. (a) Does the football make it through the hoop? (b) Does the ball hit the target? (c) What is the approximate distance between the football and the center of the target when the ball hits the ground?

48. Walk-off home run: It's the bottom of the ninth, two outs, the count is full, and the bases are loaded with the opposing team ahead 5 to 2. The home team has Charmaine, their best hitter at the plate; the opposition has Raylene the Rocket on the mound. Here's the pitch . . . it's hit . . . a long fly ball to left-center field! If the ball left the bat at an angle of 30° with an initial velocity of 112 ft/sec, will it clear the home run fence, 9 ft high and 320 ft away?

Exercise 49

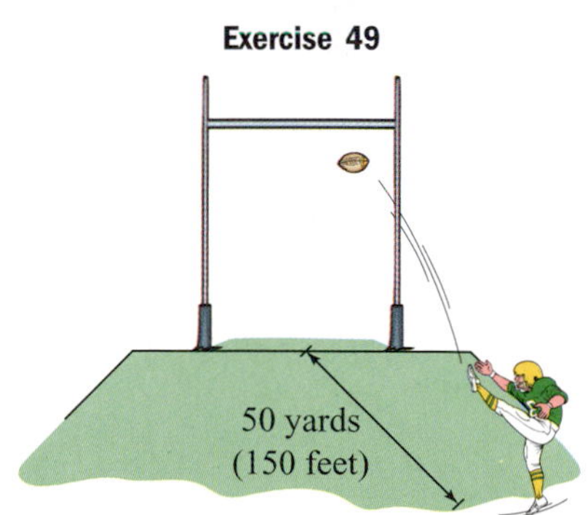

49. Last-second win: It's fourth-and-long, late in the fourth quarter of the homecoming football game, with the home team trailing 29 to 27. The coach elects to kick a field goal, even though the goal posts are 50 yd (150 ft) away from the spot of the kick. If the ball leaves the kicker's foot at an angle of 29° with an initial velocity of 80 ft/sec, will the home team win (does the ball clear the 10-ft high cross bar)?

50. Particle motion: The motion of a particle is modeled by the parametric equations $\begin{cases} x = 5t - 2t^2 \\ y = 3t - 2 \end{cases}$. Between $t = 0$ and $t = 1$, is the particle moving to the right or to the left? Is the particle moving upward or downward?

51. Electron motion: The motion of an electron as it orbits the nucleus is modeled by the parametric equations $\begin{cases} x = 6\cos t \\ y = 2\sin t \end{cases}$ with t in radians. Between $t = 2$ and $t = 3$, is the electron moving to the right or to the left? Is the electron moving upward or downward?

Systems applications: Solve the following systems using elimination. If the system is dependent, write the general solution in parametric form and use a calculator to generate several solutions.

52. $\begin{cases} 2x - y + 3z = -3 \\ 3x + 2y - z = 4 \\ 8x + 3y + z = 5 \end{cases}$

53. $\begin{cases} x - 5y + z = 3 \\ 5x + y - 7z = -9 \\ 2x + 3y - 4z = -6 \end{cases}$

54. $\begin{cases} -5x - 3z = -1 \\ x + 2y - 2z = -3 \\ -2x + 6y - 9z = -10 \end{cases}$

55. $\begin{cases} x + y - 5z = -4 \\ 2y - 3z = -1 \\ x - 3y + z = -3 \end{cases}$

56. Regressions and parameters: Draw a scatter-plot of the data given in the table. Note that connecting the points with a smooth curve will not result in a function, so a standard regression cannot be run on the data. Now consider the x-values alone—what do you notice? Find a sinusoidal model for the x-values, using T = 0, 1, 2, 3, . . . , 8. Use the same inputs to run some form of regression on the y-values, then use the results to form the "best-fit" parametric equations for this data (use L1 for T, L2 for the x-values, and L3 for the y-values). With your calculator in parametric MODE, enter the equations as X_{1T} and Y_{1T}, then graph these along with the scatterplot (L2, L3) to see the finished result. Use the TABLE feature of your calculator to comment on the accuracy of the model.

x	y
0	0
$\sqrt{2}$	0.25
2	2
$\sqrt{2}$	6.75
0	16
$-\sqrt{2}$	31.25
−2	54
$-\sqrt{2}$	85.75
0	128

57. Regressions and parameters: Draw a scatter-plot of the data given in the table, and connect the points with a smooth curve. The result is a function, but no standard regression seems to give an accurate model. The x-values alone are actually generated by an exponential function. Run a regression on these values using T = 0, 1, 2, 3, . . . , 8 as inputs to find the exponential model. Then use the same inputs to run some form of regression on the y-values and use the results to form the "best-fit" parametric equations for this data (use L1 for T, L2 for the x-values, and L3 for the y-values). With your calculator in parametric MODE, enter the equations as X_{1T} and Y_{1T}, then graph these along with the scatterplot (L2, L3) to see the finished result. Use the TABLE feature of your calculator to comment on the accuracy of the model.

x	y
1	0
1.2247	−1.75
1.5	−3
1.8371	−3.75
2.25	−4
2.7557	−3.75
3.375	−3
4.1335	−1.75
5.0625	0

WRITING, RESEARCH, AND DECISION MAKING

58. What is the difference between an *epicycloid,* a *hypercycloid,* and a *hypocycloid*? Do a word study on the prefixes *epi-*, *hyper-*, and *hypo-*, and see how their meanings match with the mathematical figures graphed in Exercises 27 to 35. Comment on what you find. To what other shapes or figures are these prefixes applied?

59. As mentioned in the *Point of Interest*, the cycloids created quite a stir when they first drew the attention of prominent mathematicians. Using any of the resources available to you, do some research on the cycloids and try to find some of the more controversial questions of the day and some of the mathematicians who participated in these quarrels. In the end, who was right?

60. The motion of a particle in a certain medium is modeled by the parametric equations $\begin{cases} x = 6\sin(4t) \\ y = 8\cos t \end{cases}$. Initially, use only the 2nd GRAPH (**TABLE**) feature of your calculator (not the graph) to name the intervals for which the particle is moving (a) to the left and upward and (b) to the left and downward. Answer to the nearest tenth (set ΔTbl = 0.1). Is it *possible* for this particle to collide with another particle in this medium whose movement is modeled by $\begin{cases} x = 3\cos t + 7 \\ y = 2\sin t + 2 \end{cases}$? Discuss why or why not.

EXTENDING THE CONCEPT

61. Consider the ellipses defined parametrically, as indicated. Find and state the location of all points where the two ellipses intersect. Round to two decimal places as needed.

$$\begin{cases} x = 6\cos t \\ y = 2\sin t \end{cases} \qquad \begin{cases} x = 4\sin t \\ y = 3\cos t \end{cases}$$

62. Write the function $y = \frac{1}{2}(x + 3)^2 - 1$ in parametric form using the substitution $x = 2\cos t - 3$ and the appropriate identity for $\cos(2t)$. Is the result equivalent to the original function? Why or why not?

MAINTAINING YOUR SKILLS

63. (3.4) Find all solutions in $[0, 2\pi)$: $\sin(2\theta) - \cos\theta = 0$.

64. (3.2) Verify that the following is an identity: $2\tan^2 x = \dfrac{\sin x}{\csc x - 1} + \dfrac{\sin x}{\csc x + 1}$.

65. (4.2) Evaluate the expression exactly by drawing the corresponding triangle: $\sin\left[\arctan\left(\frac{12}{5}\right)\right]$.

66. (3.3) Use an identity to rewrite the expression as a single term, then determine its value: $\sin 52^\circ \cos 38^\circ + \cos 52^\circ \sin 38^\circ$.

67. (1.2) When the tip of the antenna atop the Eiffel Tower is viewed at a distance of 265 ft from its base, the angle of elevation is 76°. Is the Eiffel Tower taller or shorter than the Chrysler Building (New York City) at 1046 ft?

68. (4.3) The maximum height a projectile will attain depends on the angle it is projected and its initial velocity. This phenomena is modeled by the function $H = \dfrac{v^2 \sin^2\theta}{64}$, where v is the initial velocity (in feet/sec) of the projectile and θ is the angle of projection. Find the angle of projection if the projectile attained a maximum height of 151 feet, and the initial velocity was 120 ft/sec.

SUMMARY AND CONCEPT REVIEW

SECTION 4.1 One-to-One and Inverse Functions

KEY CONCEPTS

- A function is one-to-one if each element of the range corresponds to a unique element of the domain.
- If every horizontal line intersects the graph of a function in at most one point, the function is one-to-one.

- If f is a one-to-one function with ordered pairs (a, b), then the inverse of f is that one-to-one function f^{-1} with ordered pairs of the form (b, a).
- To find f^{-1} using the algebraic method, use the following four-step process:
  1. Use y instead of $f(x)$.
  2. Interchange x and y.
  3. Solve the equation for y.
  4. Substitute $f^{-1}(x)$ for y.
- If f is a one-to-one function, the inverse of f is the function f^{-1} such that $(f \circ f^{-1})(x) = x$ and $(f^{-1} \circ f)(x) = x$.
- The graphs of $f(x)$ and $f^{-1}(x)$ are symmetric with respect to the identity function $y = x$.

EXERCISES

Determine whether the functions given are one-to-one by noting the function family to which each belongs and mentally picturing the shape of the graph.

1. $h(x) = -|x - 2| + 3$ **2.** $p(x) = 2x^2 + 7$ **3.** $s(x) = \sqrt{x - 1} + 5$

Find the inverse of each function given. Then show graphically and using composition that your inverse function is correct. State any necessary restrictions.

4. $f(x) = -3x + 2$ **5.** $f(x) = x^2 - 2, x \geq 0$ **6.** $f(x) = \sqrt{x - 1}$

Determine the domain and range for each function whose graph is given, and use this information to state the domain and range of the inverse function. Then sketch in the line $y = x$, estimate the location of three points on the graph, and use these to graph $f^{-1}(x)$ on the same grid.

7. **8.** **9.**

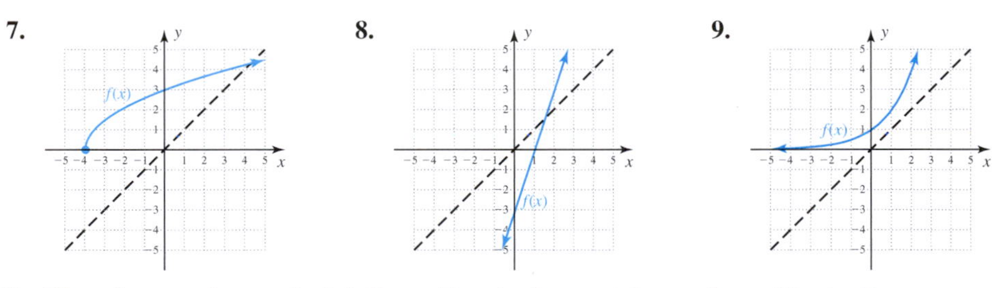

10. **Fines for overdue material:** Some libraries have set fees and penalties to discourage patrons from holding borrowed materials for an extended period. Suppose the fine for overdue DVDs is given by the function $f(t) = 0.15t + 2$, where $f(t)$ is the amount of the fine t days after it is due. (a) What is the fine for keeping a DVD seven extra days? (b) Find $f^{-1}(t)$, then input your answer from part (a) and comment on the result. (c) If a fine of \$3.80 was assessed, how many days was the DVD overdue?

SECTION 4.2 The Inverse Trig Functions and Their Applications

KEY CONCEPTS

- In order to create one-to-one functions, the domains of $y = \sin t$, $y = \cos t$, and $y = \tan t$ are restricted as follows: (a) $y = \sin t, t \in \left[-\frac{\pi}{2}, \frac{\pi}{2}\right]$; (b) $y = \cos t, t \in [0, \pi]$; and (c) $y = \tan t; t \in \left[-\frac{\pi}{2}, \frac{\pi}{2}\right]$.
- For $y = \sin x$, the inverse function is given implicitly as $x = \sin y$ and explicitly as $y = \sin^{-1} x$ or $y = \arcsin x$.

- The expression $y = \sin^{-1}x$ is read, "y is the angle or real number whose sine is x." The other inverse functions are similarly read/understood.
- For $y = \cos x$, the inverse function is given implicitly as $x = \cos y$ and explicitly as $y = \cos^{-1}x$ or $y = \arccos x$.
- For $y = \tan x$, the inverse function is given implicitly as $x = \tan y$ and explicitly as $y = \tan^{-1}x$ or $y = \arctan x$.
- The domains of $y = \sec x$, $y = \csc x$, and $y = \cot x$ are likewise restricted to create one-to-one functions: (a) $y = \sec t$; $t \in \left[0, \frac{\pi}{2}\right) \cup \left(\frac{\pi}{2}, \pi\right]$; (b) $y = \csc t$, $t \in \left[-\frac{\pi}{2}, 0\right) \cup \left(0, \frac{\pi}{2}\right]$; and (c) $y = \cot t$, $t \in (0, \pi)$.
- In some applications, inverse functions occur in a composition with other trig functions, with the expression best evaluated by drawing a diagram using the ratio definition of the trig functions.
- To evaluate $y = \sec^{-1}t$, we use $y = \cos^{-1}\left(\frac{1}{t}\right)$; for $y = \cot^{-1}t$, use $\tan^{-1}\left(\frac{1}{t}\right)$; and so on.
- Trigonometric substitutions can be used to simplify certain algebraic expressions.

EXERCISES

Evaluate without the aid of calculators or tables. State answers in both radians and degrees in exact form.

11. $y = \sin^{-1}\left(\frac{\sqrt{2}}{2}\right)$ **12.** $y = \csc^{-1}2$ **13.** $y = \arccos\left(-\frac{\sqrt{3}}{2}\right)$

Evaluate the following using a calculator, *keeping the domain and range of each function in mind.* Answer in radians to the nearest ten-thousandth *and* in degrees to the nearest tenth. Some may be undefined.

14. $y = \tan^{-1}4.3165$ **15.** $y = \sin^{-1}0.8892$ **16.** $f(x) = \arccos\left(\frac{7}{8}\right)$

Evaluate the following without the aid of a calculator. Some may be undefined.

17. $\sin\left[\sin^{-1}\left(\frac{1}{2}\right)\right]$ **18.** $\text{arcsec}\left[\sec\left(\frac{\pi}{4}\right)\right]$ **19.** $\cos(\cos^{-1}2)$

Evaluate the following using a calculator. Some may be undefined.

20. $\sin^{-1}(\sin 1.0245)$ **21.** $\arccos[\cos(-60°)]$ **22.** $\cot^{-1}\left[\cot\left(\frac{11\pi}{4}\right)\right]$

Evaluate each expression by drawing a right triangle and labeling the sides.

23. $\sin\left[\cos^{-1}\left(\frac{12}{37}\right)\right]$ **24.** $\tan\left[\text{arcsec}\left(\frac{7}{3x}\right)\right]$ **25.** $\cot\left[\sin^{-1}\left(\frac{x}{\sqrt{81+x^2}}\right)\right]$

Use an inverse function to solve the following equations for θ in terms of x.

26. $x = 5\cos\theta$ **27.** $7\sqrt{3}\sec\theta = x$ **28.** $x = 4\sin\left(\theta - \frac{\pi}{6}\right)$

29. For $Y_1 = \frac{\sqrt{169 + x^2}}{x}$: (a) find an equivalent expression Y_2 using the substitution $x = 13 \tan \theta$; (b) express Y_2 in terms of x using $x = 13 \tan \theta$ [solve for θ] and substitution; and (c) use the TABLE feature of a graphing calculator to help demonstrate that $Y_1 = Y_2$ for $x \neq 0$.

SECTION 4.3 Solving Basic Trig Equations

KEY CONCEPTS

- When solving trig equations, we often consider either the principal root, roots in $[0, 2\pi)$, or all real roots.
- Keeping the graph of each function in mind helps to determine the desired solution set.
- After isolating the trigonometric term containing the variable, we solve by applying the appropriate inverse function, realizing the result is only the principal root.
- Once the principal root is found, roots in $[0, 2\pi)$ or all real roots can be found using reference angles and the period of the function under consideration.
- Trig identities can be used to obtain an equation that can be solved by factoring or other solution methods.

EXERCISES

Solve each equation without the aid of a calculator (all solutions are standard values). Clearly state (a) the principal root; (b) all solutions in the interval $[0, 2\pi)$; and (c) all real roots.

30. $2 \sin x = \sqrt{2}$

31. $3 \sec x = -6$

32. $8 \tan x + 7\sqrt{3} = -\sqrt{3}$

Solve using a calculator and the inverse trig functions (not by graphing). Clearly state (a) the principal root; (b) solutions in $[0, 2\pi)$; and (c) all real roots. Answer in radians to the nearest ten-thousandth as needed.

33. $9 \cos x = 4$

34. $\frac{2}{5}\sin(2\theta) = \frac{1}{4}$

35. $\sqrt{2} \csc x + 3 = 7$

36. The area of a circular segment (the shaded portion shown in the diagram) is given by the formula $A = \frac{1}{2}r^2(\theta - \sin \theta)$, where θ is in radians. If the circle has a radius of 10 cm, find the angle θ that gives an area of 12 cm^2.

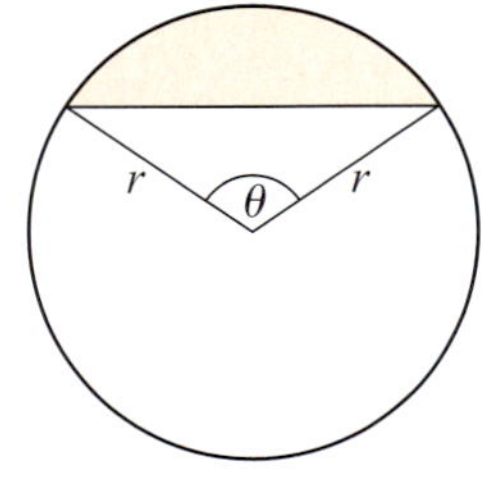

SECTION 4.4 General Trig Equations and Applications

KEY CONCEPTS

- In addition to the basic solution methods from Section 4.3, additional strategies include squaring both sides, factoring by grouping, and using the full range of identities to simplify an equation.
- Many applications result in equations of the form $A\sin(Bx + C) + D = k$. To solve, isolate the factor $\sin(Bx + C)$ (subtract D and divide by A), then apply the inverse function.
- Once the principal root is found, roots in $[0, 2\pi)$ or all real roots can be found using reference angles and the period of the function under consideration.

EXERCISES

Solve each equation in $[0, 2\pi]$ using the method indicated. Round nonstandard values to four decimal places.

37. squaring both sides

$$\sin x + \cos x = \frac{\sqrt{6}}{2}$$

38. using identities

$$3\cos(2x) + 7\sin x - 5 = 0$$

39. factor by grouping

$$4\sin x\cos x - 2\sqrt{3}\sin x - 2\cos x + \sqrt{3} = 0$$

40. using any appropriate method

$$\csc x + \cot x = 1$$

State the period P of each function and find all solutions in $[0, P)$. Round to four decimal places as needed.

41. $-750\sin\left(\frac{\pi}{6}x + \frac{\pi}{2}\right) + 120 = 0$

42. $80\cos\left(\frac{\pi}{3}x + \frac{\pi}{4}\right) - 40\sqrt{2} = 0$

43. The revenue earned by Waipahu Joe's Tanning Lotions fluctuates with the seasons, with a great deal more lotion sold in the summer than in the winter. The function $R(x) = 15\sin\left(\frac{\pi}{6}x - \frac{\pi}{2}\right) + 30$ models the monthly sales of lotion nationwide, where $R(x)$ is the revenue in thousands of dollars and x represents the months of the year ($x = 1 \rightarrow$ Jan). (a) How much revenue is projected for July? (b) For what months of the year does revenue exceed $37,000?

SECTION 4.5 Parametric Equations and Graphs

KEY CONCEPTS

- If we consider the set of points $P(x, y)$ such that the x-values are generated by $f(t)$ and the y-values are generated by $g(t)$ (assuming f and g are both defined on an interval of the domain), the equations $x = f(t)$ and $y = g(t)$ are called parametric equations, with parameter t.
- Parametric equations can be converted to rectangular form by eliminating the parameter. This can sometimes be done by solving for t in one equation and substituting in the other, or by using trigonometric forms.
- A function can be written in parametric form many different ways, by altering the parameter or using trigonometric identities.
- The cycloids are an important family of curves, with equations $x = r(t - \sin t)$ and $y = r(1 - \cos t)$.

EXERCISES

Graph the curves defined by the parametric equations over the specified intervals and identify the graph. Then eliminate the parameter and write the corresponding rectangular form.

44. $x = t - 4;\ y = -2t^2 + 3;\ t \in [-3, 3]$

45. $x = (2 - t)^2;\ y = (t - 3)^2;\ t \in [0, 5]$

46. $x = -3\sin t;\ y = 4\cos t;\ t \in [0, 2\pi)$

47. Write each function in three different parametric forms by altering the parameter. $y = 2(x - 5)^2 - 1$

48. Use a graphing calculator to graph the Lissajous figure, then state the size of the rectangle needed to frame it. $x = 4\sin(5t);\ y = 8\cos t$

MIXED REVIEW

1. Given $f(x) = \frac{1}{(x+2)^2}, x > -2$: (a) use the algebraic method to find the inverse function, (b) state the domain and range of $f^{-1}(x)$, and (c) verify your inverse is correct using a composition.

2. Determine the domain and range for the function f whose graph is given, and use this information to state the domain and range of f^{-1}. Then sketch the line $y = x$ and sketch the graph of $y = f^{-1}(x)$ on the same grid.

Exercise 2

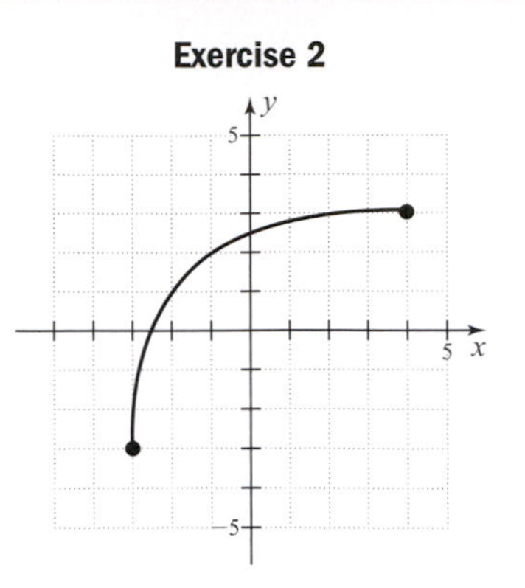

Solve the following equations for $x \in [0, 2\pi)$ without the aid of a calculator.

3. $-12\sin x + 5 = 11$

4. $\frac{3}{4}\cos^2 x - \frac{1}{3} = \frac{1}{24}$

Evaluate each expression by drawing a right triangle and labeling the sides appropriately.

5. $\tan\left[\operatorname{arccsc}\left(\frac{10}{x}\right)\right]$

6. $\sin\left[\sec^{-1}\left(\frac{\sqrt{64 + x^2}}{x}\right)\right]$

7. Solve for x in the interval $[0, 2\pi)$. Round to four decimal places as needed: $-100\sin\left(\frac{\pi}{4}x - \frac{\pi}{6}\right) + 80 = 100$

8. Without the aid of a calculator, find: (a) the principal roots, (b) all solutions in $[0, 2\pi)$, and (c) all real solutions: $(\cos x - 1)[2\cos^2(x) - 1] = 0$.

9. Graph the curve defined by the parametric equations given, using the interval $t \in [0, 10]$. Then identify the graph: $x = (t - 2)^2$; $y = (t - 4)^2$

10. A go-cart travels around an elliptical track with a 100-m major axis that is horizontal. The minor axis measures 60 m. Write an equation model for the track in parametric form.

Evaluate without the aid of a calculator or tables. Answer in both radians and degrees.

11. $y = \operatorname{arcsec}(-\sqrt{2})$

12. $y = \sin^{-1} 0$

13. $y = \arctan\sqrt{3}$

Use an inverse function to solve each equation for θ in terms of x.

14. $\frac{x}{10} = \tan\theta$

15. $2\sqrt{2}\csc\left(\theta - \frac{\pi}{4}\right) = x$

16. On a large clock, the distance from the tip of the hour hand to the base of the "12" can be approximated by the function $D(t) = \left|8\sin\left(\frac{\pi t}{12}\right)\right| + 2$, where $D(t)$ is this distance in feet at time t in hours. Use this function to approximate (a) the time of day when the hand is 6 ft from the 12 and 10 ft from the 12 and (b) the distance between the tip and the 12 at 4:00. Check your answer graphically.

17. The figure shows a smaller pentagon inscribed within a larger pentagon. Find the measure of angle θ using the diagram and equation given: $3.2^2 = 11^2 + 9.4^2 - 2(11)(9.4)\cos\theta$

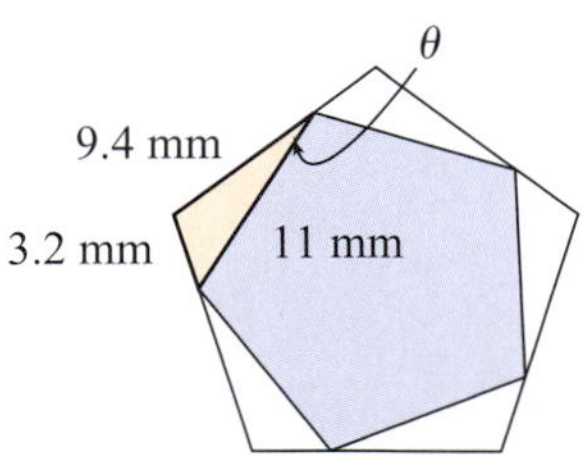

18. Given $100\sin t = 70$, use a calculator to find (a) the principal root, (b) all solutions in $[0, 2\pi)$, and (c) all real solutions. Round to the nearest ten-thousandth.

19. Use an identity to find all real solutions for $\cos(2\theta) + 2\sin^2\theta = -2\sin\theta$.

20. State the period P of the function and find all exact solutions in $x \in [0, P)$, given $124\cos\left[\frac{\pi}{12}(x-3)\right] + 82 = 144$

Practice Test

1. For $f(x) = x^3$, the inverse function is $f^{-1}(x) = \sqrt[3]{x}$, but for $f(x) = x^2$, the inverse function is *not* $f^{-1}(x) = \sqrt{x}$. Explain why.

2. For t in radians, the area of a circular sector is given by $A(t) = \frac{1}{2}r^2t$. (a) Find the area if $r = 10$ cm and $t = \frac{\pi}{4}$. (b) Use the algebraic method to find $A^{-1}(t)$, then substitute the result from part (a). What do you notice?

3. Sketch the graph of $y = \cos^{-1}x$ and state its domain and range.

4. Use a calculator to approximate the value of $y = \sec^{-1}(3)$ to four decimal places.

Evaluate without the aid of calculators or tables.

5. $y = \tan^{-1}\left(\frac{1}{\sqrt{3}}\right)$

6. $f(x) = \sin\left[\sin^{-1}\left(\frac{1}{2}\right)\right]$

7. $y = \arccos(\cos 30°)$

Evaluate the following. Answer in exact form, where possible.

8. $y = \sin^{-1}0.7528$

9. $y = \arctan(\tan 78.5°)$

10. $y = \sec^{-1}\left[\sec\left(\frac{7\pi}{24}\right)\right]$

Evaluate the expressions by drawing a right triangle and labeling the sides.

11. $\cos\left[\tan^{-1}\left(\frac{56}{33}\right)\right]$

12. $\cot\left[\cos^{-1}\left(\frac{x}{\sqrt{25 + x^2}}\right)\right]$

Solve without the aid of a calculator (all solutions are standard values). Clearly state (a) the principal root, (b) all solutions in the interval $[0, 2\pi)$, and (c) all real roots.

13. $8\cos x = -4\sqrt{2}$

14. $\sqrt{3}\sec x + 2 = 4$

Solve each equation using a calculator and inverse trig functions to find the principal root (not by graphing). Then state (a) the principal root, (b) all solutions in the interval $[0, 2\pi)$, and (c) all real roots.

15. $\frac{2}{3}\sin(2x) = \frac{1}{4}$

16. $-3\cos(2x) - 0.8 = 0$

Solve the equations graphically in the indicated interval using a graphing calculator. State answers in radians rounded to the nearest ten-thousandth.

17. $3\cos(2x - 1) = \sin x;\ x \in [-\pi, \pi]$

18. $2\sqrt{x} - 1 = 3\cos^2 x;\ x \in [0, 2\pi)$

Solve the following equations for $x \in [0, 2\pi)$ using a combination of identities and/or factoring. State solutions in radians using the exact form, where possible.

19. $2\sin x\sin(2x) + \sin(2x) = 0$

20. $(\cos x + \sin x)^2 = \frac{1}{2}$

Solve each equation in $[0, 2\pi)$ by squaring both sides, factoring, using identities or by using any appropriate method. Round nonstandard values to four decimal places.

21. $3\sin(2x) + \cos x = 0$

22. $\frac{2}{3}\sin\left(2x - \frac{\pi}{6}\right) + \frac{3}{2} = \frac{5}{6}$

For Exercises 23 and 24, sketch the graph using the parametric equations given. Then remove the parameter, convert to rectangular form, and identify the graph.

23. $x = 4\sin t$
$y = 5\cos t$

24. $x = (t - 3)^2 + 1$
$y = t + 2$

25. Use a graphing calculator to graph the cycloid, then identify the maximum and minimum values, and the period: $x = 4T - 4\sin T$, $\quad y = 4 - 4\cos T$.

Calculator Exploration and Discovery

Parametric Equations and 3 × 3 Systems

The keystrokes shown apply to a TI-84 Plus model. Please consult your manual or our Internet site for other models.

Parametric Equations and Dependent Systems

As mentioned in Section 4.5, solutions to a linearly dependent system of equations can be represented parametrically. Consider the following illustration.

ILLUSTRATION 1 Solve using elimination: $\begin{cases} 3x - 2y + z = -1 \\ 2x + y - z = 5 \\ 10x - 2y = 8 \end{cases}$

Solution: Using Rl as the *SOURCE*, we'll *TARGET* z in R2, since R3 has no z-term.

$$\begin{array}{r} SOURCE: \\ TARGET\ z: \\ \\ \end{array} \begin{cases} 3x - 2y + z = -1 & \text{R1} \\ 2x + y - z = 5 & \text{R2} \\ 10x - 2y = 8 & \text{R3} \end{cases}$$

R1 + R2 eliminates the z-term from R2, yielding a new R2: $5x - y = 4$, and we obtain the new system shown.

$$\begin{cases} 3x - 2y + z = -1 \\ 2x + y - z = 5 \\ 10x - 2y = 8 \end{cases} \xrightarrow[\text{R3} \to \text{R3}]{\text{R1} + \text{R2} \to \text{R2}} \begin{cases} 3x - 2y + z = -1 \\ 5x - y = 4 \\ 10x - 2y = 8 \end{cases}$$

To finish, we use $-2\text{R2} + \text{R3}$ to eliminate the y-term in R3, but this *also eliminates all other terms*:

$$\begin{array}{r} -2\text{R2} \\ + \\ \text{R3} \end{array} \left\{ \begin{array}{rcr} -10x + 2y &=& -8 \\ \\ 10x - 2y &=& 8 \\ \hline 0 &=& 0 \end{array} \right. \text{ result}$$

Since R3 = 2R2, the system is linearly dependent and equivalent to $\begin{cases} 3x - 2y + z = -1 \\ 5x - y = 4 \end{cases}$. For a solution, we express the remaining equations in terms of a common variable, called the parameter. For R2 we have $y = 5x - 4$, and substituting $5x - 4$ for y in R1 we can also write z in terms of x:

$$\begin{aligned} 3x - \quad 2y \quad + z &= -1 && \text{R1} \\ 3x - 2(5x - 4) + z &= -1 && \text{substitute } 5x - 4 \text{ for } y \\ z &= 7x - 9 && \text{simplify and solve for } z \end{aligned}$$

The general solution is $(x, 5x - 4, 7x - 9)$. In cases of linear dependence, it is common practice to express the solution in terms of a new parameter, with t being a common choice. The solution to the original system is any triple satisfying $(t, 5t - 4, 7t - 9)$. We can actually enter this solution in parametric form using our calculator, with $X_{1T} = 5T - 4$, $Y_{1T} = 7T - 9$, and T as the parameter (Figure 4.61). Since the slope of both parameterized lines is very steep, window settings are a concern. Using $T \in [-10, 10]$, with $x \in [-50, 50]$ and $y \in [-50, 50]$ will display a large number of the possible solutions (Figure 4.62). The graph is shown in Figure 4.63, with the TRACE feature being used to view various possibilities. The solution illustrated in Figure 4.63 is the ordered triple (3, 11, 12).

Figure 4.61

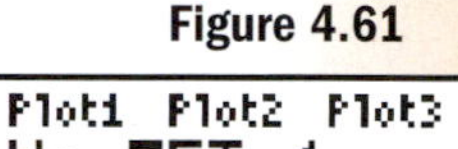

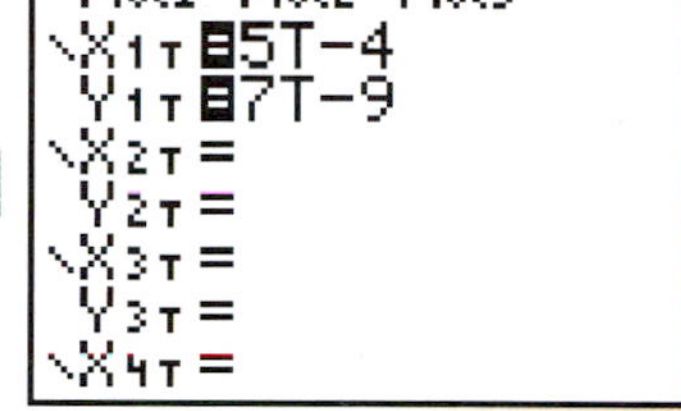

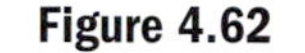

Figure 4.62

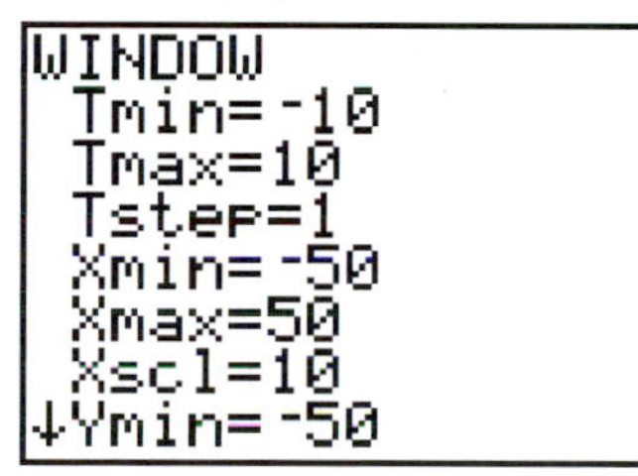

Figure 4.63

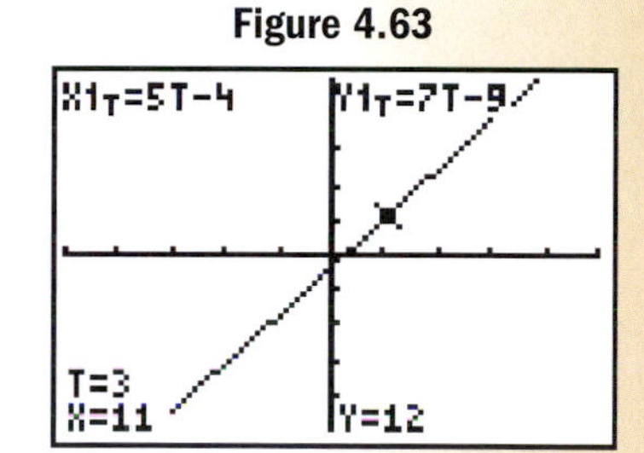

Solve each system using elimination. If the system is linearly dependent, write the general solution in parametric form and use a calculator to generate several solutions. Many representations are possible.

Exercise 1: $\begin{cases} 2x - y - z = -1 \\ -3x + 2y + z = 4 \\ 3x - 3y = -9 \end{cases}$

Exercise 2: $\begin{cases} -x + y + 2z = 1 \\ 2x - 3y - 3z = 1 \\ 4x - 7y - 5z = 5 \end{cases}$

STRENGTHENING CORE SKILLS

Trigonometric Equations and Inequalities

The ability to draw a quick graph of the trigonometric functions is a tremendous help in understanding equations and inequalities. A basic sketch can help reveal the number of solutions in $[0, 2\pi)$ and the quadrant of each solution. For nonstandard angles, the value given by the inverse function can then be used as a basis for stating the solution set for all real numbers. We'll illustrate the process using a few simple examples, then generalize our observations to solve more realistic applications. Consider the function $f(x) = 2\sin x + 1$, a sine wave with amplitude 2, and a vertical translation of +1. To find intervals in $[0, 2\pi)$ where

$f(x) > 2.5$, we reason that f has a maximum of $2(1) + 1 = 3$ and a minimum of $2(-1) + 1 = -1$, since $-1 \leq \sin x \leq 1$. With no phase shift and a standard period of 2π, we can easily draw a quick sketch of f by vertically translating x-intercepts and max/min points 1 unit up. After drawing the line $y = 2.5$ (see Figure 4.64), it appears there are two intersections in the interval, one in QI and one in QII. More importantly, it is clear that $f(x) > 2.5$ *between these two solutions.* Substituting 2.5 for $f(x)$ in $f(x) = 2 \sin x + 1$, we solve for $\sin x$ to obtain $\sin x = 0.75$, which we use to state the solution in exact form: $f(x) > 2.5$ for $x \in (\sin^{-1}0.75, \pi - \sin^{-1}0.75)$. In approximate form the solution interval is $x \in (0.85, 2.29)$. If the function involves a horizontal shift, the graphical analysis will reveal which intervals should be chosen to satisfy the given inequality.

Figure 4.64

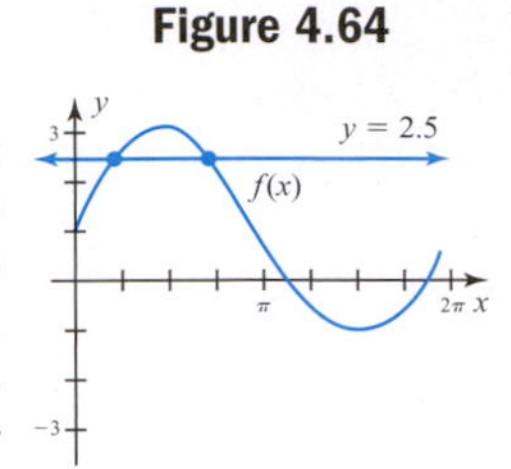

The basic ideas remain the same regardless of the complexity of the equation, and we illustrate by studying Example 6 from Section 4.4. Given the function $R(d) = 750 \sin\left(\frac{2\pi}{365}d - \frac{\pi}{2}\right) + 950$, we were asked to solve the inequality $R(d) > 1250$. Remember—our current goal is not a supremely accurate graph, just a sketch that will guide us to the solution using the inverse functions and the correct quadrants. Perhaps that greatest challenge is recalling that when $B \neq 1$, the horizontal shift is $-\frac{C}{B}$, but other than this a fairly accurate sketch can quickly be obtained.

ILLUSTRATION 1 Given $R(d) = 750 \sin\left(\frac{2\pi}{365}d - \frac{\pi}{2}\right) + 950$, find intervals in $[0, 365]$ where $R(d) > 1250$.

Solution: This is a sine wave with a period of 365 days, an amplitude of 750, shifted $-\frac{C}{B} = 91.25$ units to the right and 950 units up. The maximum value will be 1700 and the minimum value will be 200. For convenience, scale the axes from 0 to 360 (as though the period were 360 days), and plot the x-intercepts and maximum/minimum values for a standard sine wave with amplitude 750 (by scaling the axes). Then shift these points about 90 units in the positive direction (to the right), and 950 units up, again using a scale that makes this convenient (see Figure 4.65). This sketch along with the graph of $y = 1250$ is sufficient to show that solutions to $R(d) = 1250$ occur early in the second quarter and late in the third quarter, with solutions to $R(d) > 1250$ occurring *between* these solutions. For $R(d) = 750 \sin\left(\frac{2\pi}{365}d - \frac{\pi}{2}\right) + 950$, we substitute 1250 for $R(d)$ and isolate the sine function, obtaining $\sin\left(\frac{2\pi}{365}d - \frac{\pi}{2}\right) = 0.4$, which leads to exact form solutions of $d = \left(\sin^{-1}0.4 + \frac{\pi}{2}\right)\left(\frac{365}{2\pi}\right)$ and $d = \left(\pi - \sin^{-1}0.4 + \frac{\pi}{2}\right)\left(\frac{365}{2\pi}\right)$. In approximate form the solution interval is $x \in [115, 250]$.

Figure 4.65

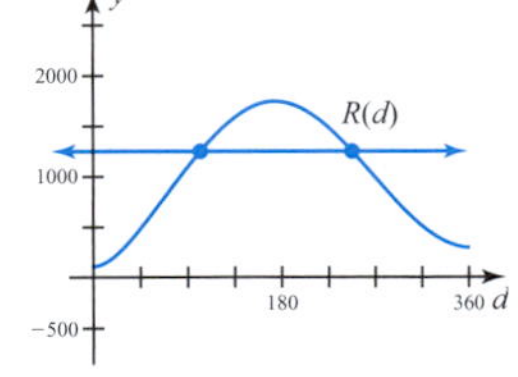

Practice with these ideas by finding solutions to the following inequalities in the intervals specified.

Exercise 1: $f(x) = 3\sin x + 2; f(x) > 3.7;\ x \in [0, 2\pi)$

Exercise 2: $g(x) = 4\sin\left(x - \frac{\pi}{3}\right) - 1;\ g(x) \le -2;\ x \in [0, 2\pi)$

Exercise 3: $h(x) = 125\sin\left(\frac{\pi}{6}x - \frac{\pi}{2}\right) + 175;\ h(x) \le 150;\ x \in [0, 12)$

Exercise 4: $f(x) = 15{,}750\sin\left(\frac{2\pi}{360}x - \frac{\pi}{4}\right) + 19{,}250;\ f(x) > 25{,}250;\ x \in [0, 360)$

CUMULATIVE REVIEW CHAPTERS 1-4

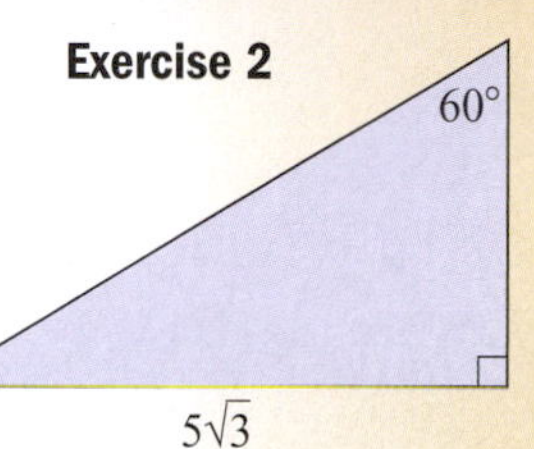

1. Find $f(\theta)$ for all six trig functions, given $P(-13, 84)$ is on the terminal side with θ in QII.
2. Find the lengths of the missing sides.
3. Convert 56° 20′ 06″ to (a) decimal degrees, and (b) radians (round to four places).
4. Verify $\left(-\frac{5}{6}, \frac{\sqrt{11}}{6}\right)$ is on the unit circle and use it to state the values $\sin t$, $\cos t$, and $\tan t$.
5. Standing 5 mi (26,400 ft) from the base of Mount Logan (Yukon), the angle of elevation to the summit is 36° 56′. How much taller is Mount McKinley (Alaska), which stands at 20,320 ft high?
6. The largest clock face in the world, located in Toi, Japan, has a diameter of 100 ft. To the nearest inch, how far does the tip of the minute hand travel from 1:05 P.M. to 1:37 P.M.? Assume the minute hand has a length of 50 ft.
7. Find all solutions in $[0, 2\pi)$. Round to four decimal places as needed: $3\sin(2x) + \cos x = 0$.
8. The Petronas Towers in Malaysia are two of the tallest structures in the world. The top of the roof reaches 1483 ft above the street below and the stainless steel pinnacles extend an additional 241 ft into the air (see the figure). Find the viewing angle θ for the pinnacles from a distance of 1000 ft (the angle formed from the base of the antennae to its top).

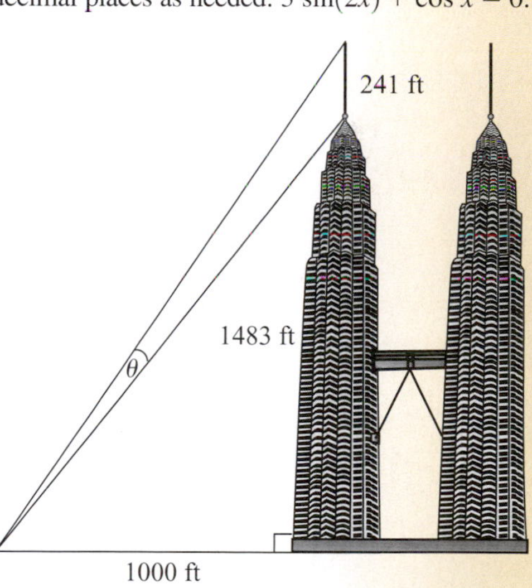

9. A wheel with radius 45 cm is turning at 5 revolutions per second. Find the linear velocity of a point on the rim in kilometers per hour, rounded to the nearest tenth of a kilometer.
10. Find all real solutions to the equation $-2\sin\left(\frac{\pi}{6}x\right) = \sqrt{3}$
11. Evaluate without using a calculator: $\cos\left[\sin^{-1}\left(\frac{1}{2}\right)\right]$.
12. The Earth has a radius of 3960 mi. Tokyo, Japan, is located at 35.4° N latitude, very near the 139° E latitude line. Adelaide, Australia, is at 34.6° S latitude, and also very near 139° E latitude. How many miles separate the two cities?

13. The table shown gives the percentage of the Moon that is illuminated for the days of a particular month, at a given latitude. (a) Use a graphing calculator to find a sinusoidal regression model. (b) If this data applies to May, use the regression model to estimate the percent of the Moon that will be illuminated on June 7.

Day	% Illum.	Day	% Illum.
1	15	19	63
4	41	22	29
7	69	25	8
10	91	28	2
13	100	31	18
16	90	—	—

14. List the three Pythagorean identities and three identities equivalent to $\cos(2\theta)$.

15. For $f(x) = 325\cos\left(\frac{\pi}{6}x - \frac{\pi}{2}\right) + 168$, what values of x in $[0, 2\pi)$ satisfy $f(x) > 330.5$?

16. The revenue for Otake's Mower Repair is seasonal, with business in the summer months far exceeding business in the winter months. Monthly revenue for the company can be modeled by the function $R(x) = 7.5\cos\left(\frac{\pi}{6}x + \pi\right) + 12.5$, where $R(x)$ is the average revenue (in thousands of dollars) for month $x (x = 1 \rightarrow \text{Jan})$. (a) What is the average revenue for October? (b) For what months of the year is revenue more than $16,250?

17. Use the triangle given to find the exact value of $\sin(2\theta)$.

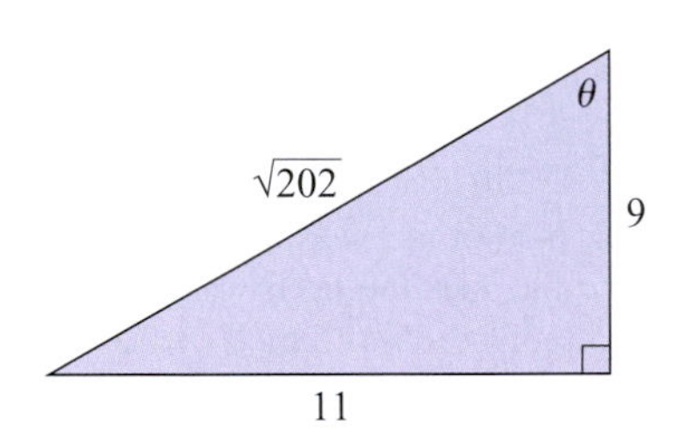

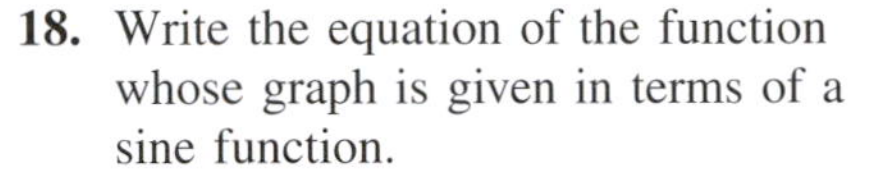

18. Write the equation of the function whose graph is given in terms of a sine function.

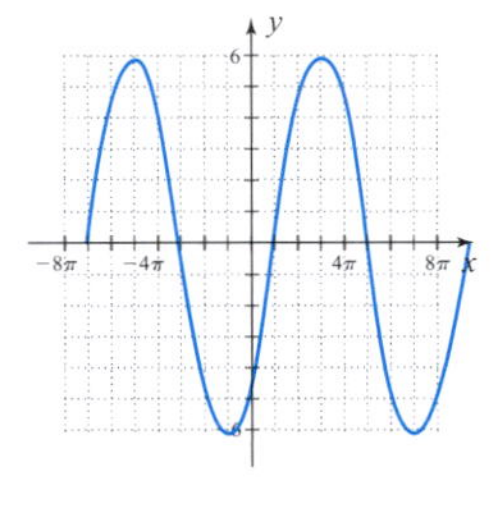

19. Verify that the following is an identity: $\frac{\cos x + 1}{\tan^2 x} = \frac{\cos x}{\sec x - 1}$.

20. The amount of waste product released by a manufacturing company varies according to its production schedule, which is much heavier during the summer months and lighter in the winter. Waste products reach a maximum of 32.5 tons in the month of July, and falls to a minimum of 21.7 tons in January ($t = 1$). (a) Use this information to build a sinusoidal equation that models the amount of waste produced each month. (b) During what months of the year does output exceed 30 tons?

Chapter 5

Applications of Trigonometry

Chapter Outline

Preview

This chapter introduces applications that will further extend fundamental ideas from trigonometry, while involving a combination of algebraic and trigonometric concepts. For example, to find the height of a projectile thrown vertically upward, we need only use the standard "projectile equation" seen in all college algebra texts: $h(t) = -16t^2 + v_0t + k$, where h is the height in feet, v_0 is the initial velocity, k is the initial height, and t is the time in seconds (air resistance is ignored). But to find the height of a projectile thrown nonvertically, say at an angle of $\theta°$, requires both algebra and trigonometry as evidenced by the formula $h(t) = -16t^2 + v_0 \sin \theta t + k$. See Section 5.5 for more details.

5.1 Oblique Triangles and the Law of Sines

LEARNING OBJECTIVES

In Section 5.1 you will learn how to:

A. **Develop the law of sines using right triangles**

B. **Solve ASA and AAS triangles**

C. **Use the law of sines to solve applications**

INTRODUCTION

Many applications of trigonometry involve *oblique triangles*, or triangles that do not have a 90° angle. For example, suppose a trolley carries passengers from ground level up to a mountain chateau, as shown in Figure 5.1. Assuming the cable could be held taut, what is its approximate length? Can we also determine the slant height of the mountain? To answer questions like these, we'll develop techniques that enable us to solve acute and obtuse triangles using fundamental trigonometric relationships.

Figure 5.1

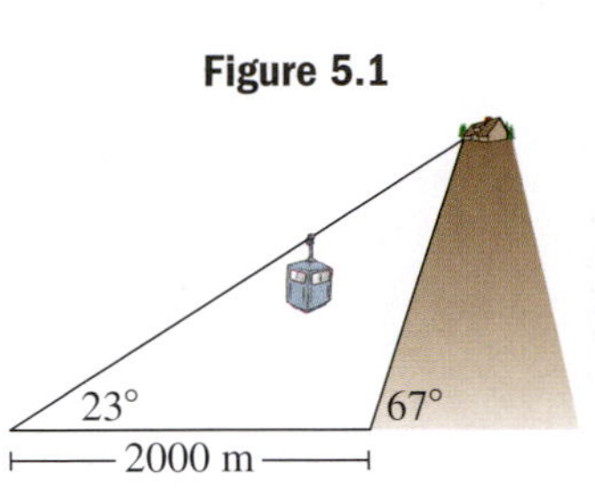

POINT OF INTEREST

One of the more famous ratios involving a trig function is the ratio $\frac{\sin\theta}{\theta}$, with θ in radians. Although it seems counterintuitive, when we evaluate this rational expression for smaller and smaller $\theta > 0$, its value actually becomes very close to 1 (see the table). This relationship provides fundamental links between various topics in advanced mathematics and is another example of the intriguing ways trigonometry and trigonometric relationships can be used.

θ (radians)	$\frac{\sin\theta}{\theta}$
0.5	0.95885
0.3	0.98507
0.1	0.99833
0.01	0.99998

A. The Law of Sines

Figure 5.2

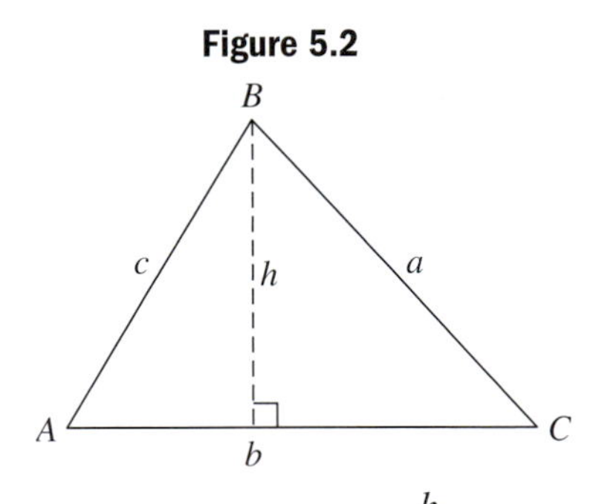

Consider the oblique triangle ABC pictured in Figure 5.2. Since it is not a right triangle, it seems the trig ratios studied earlier cannot be applied. But if we draw the altitude h (from vertex B), two right triangles are formed that *share a common side*. By applying the sine ratio to angles A and C, we can develop a relationship that will help us "solve the triangle."

WORTHY OF NOTE

As with right triangles, solving an oblique triangle involves determining the lengths of all three sides and the measures of all three angles.

For $\angle A$ we have $\sin A = \frac{h}{c}$ or $h = c\sin A$. For $\angle C$ we have $\sin C = \frac{h}{a}$ or $h = a\sin C$. Since both products are equal to h, the transitive property gives $c\sin A = a\sin C$, which leads to

$$c\sin A = a\sin C \quad \text{since } h = h$$

$$\frac{c\sin A}{ac} = \frac{a\sin C}{ac} \quad \text{divide by } ac$$

$$\frac{\sin A}{a} = \frac{\sin C}{c} \quad \text{simplify}$$

Using the same triangle and the altitude drawn from C (Figure 5.3), we note a similar relationship involving angles A and B: $\sin A = \frac{h}{b}$ or $h = b \sin A$, and $\sin B = \frac{h}{a}$ or $h = a \sin B$. As before, we can then write $\frac{\sin A}{a} = \frac{\sin B}{b}$, and the relationship can now be stated using any two angles and their corresponding sides. The result is called the **law of sines:** $\frac{\sin A}{a} = \frac{\sin B}{b} = \frac{\sin C}{c}$. See Exercise 44 for a development of the proof in the case of an obtuse triangle.

Figure 5.3

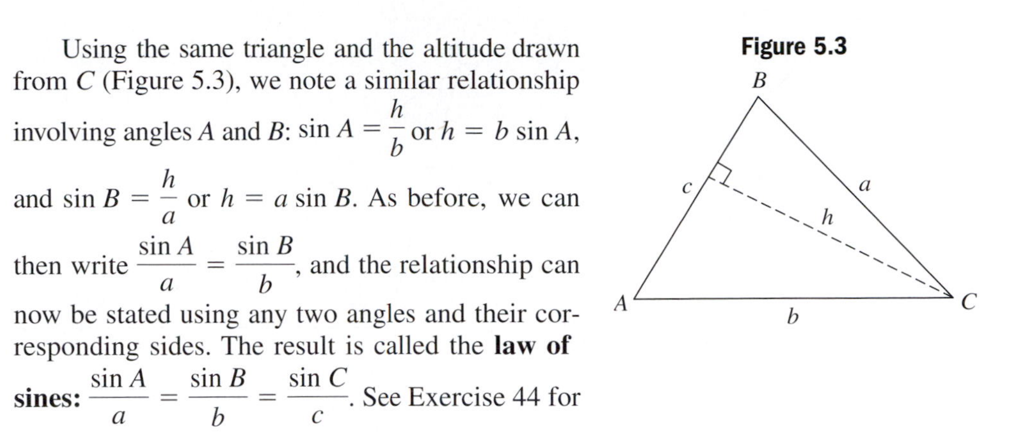

> **THE LAW OF SINES**
>
> For any triangle ABC, the ratio of the sine of an angle to the side opposite that angle is constant.
>
> $$\frac{\sin A}{a} = \frac{\sin B}{b} = \frac{\sin C}{c}$$

As with any rule or formula, it helps to know what is required to use it effectively. For the law of sines, we need (1) an angle, (2) a side opposite this angle, and (3) an additional side or angle. If two angles and a side are known, the situation is described as angle-side-angle (ASA) or angle-angle-side (AAS), depending on whether the side is between two angles or opposite one of them. If one angle and two sides are known, the situation is described as side-side-angle (SSA). See Figures 5.4, 5.5, and 5.6.

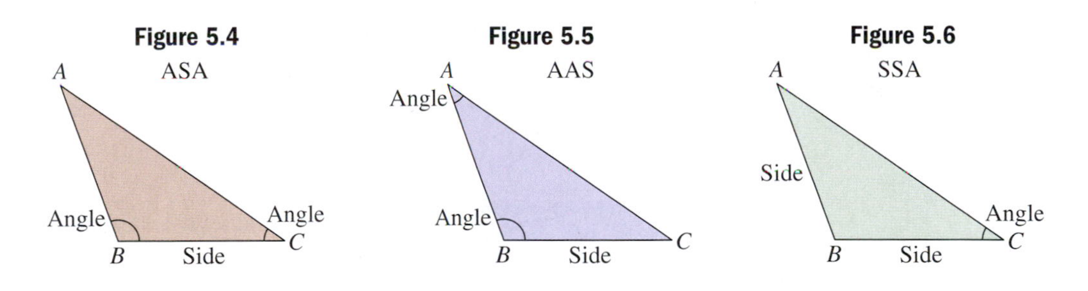

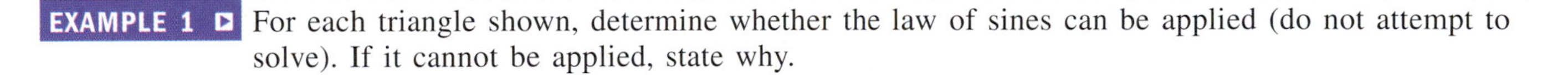

EXAMPLE 1 For each triangle shown, determine whether the law of sines can be applied (do not attempt to solve). If it cannot be applied, state why.

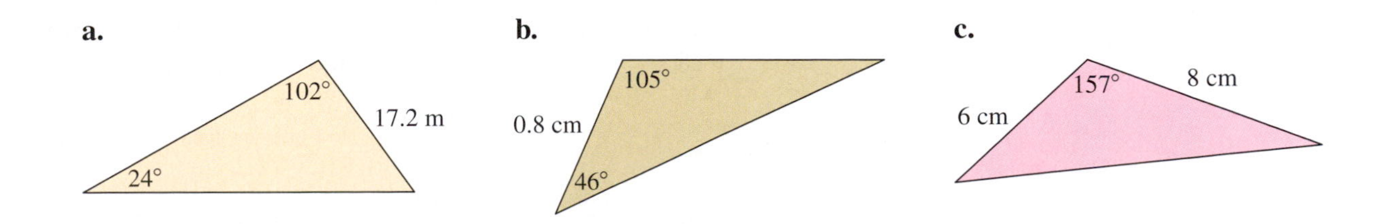

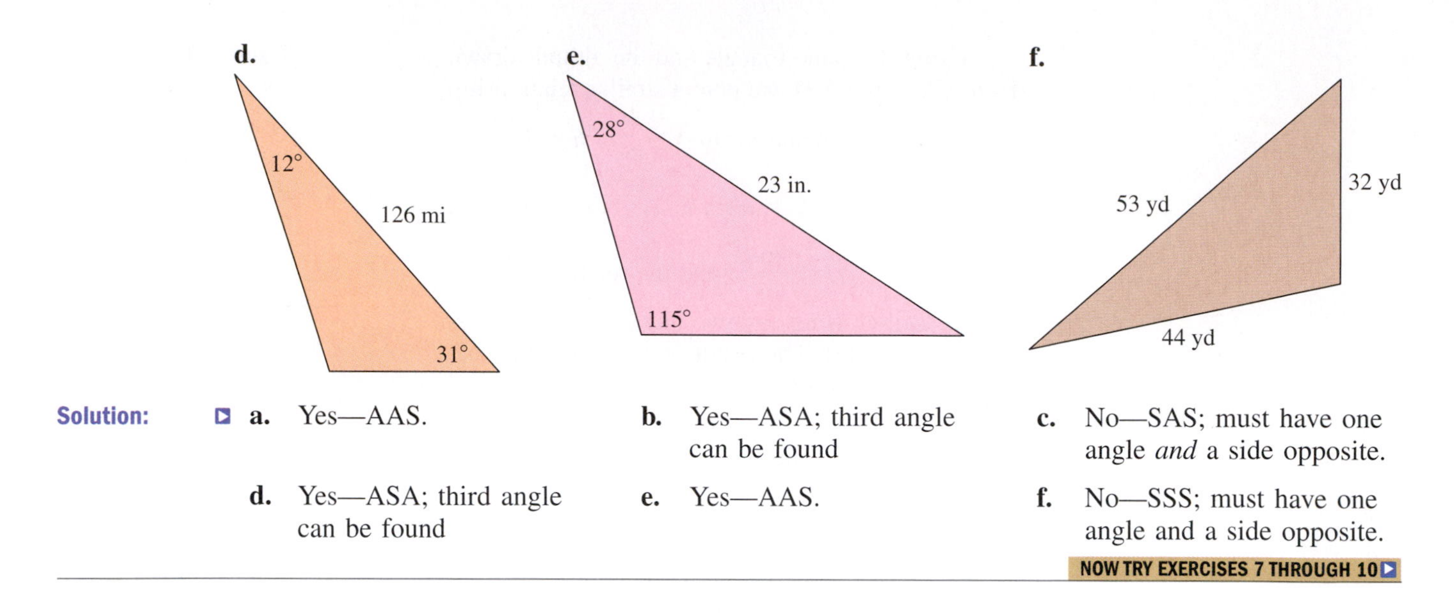

Solution:

a. Yes—AAS.

b. Yes—ASA; third angle can be found

c. No—SAS; must have one angle *and* a side opposite.

d. Yes—ASA; third angle can be found

e. Yes—AAS.

f. No—SSS; must have one angle and a side opposite.

NOW TRY EXERCISES 7 THROUGH 10

B. Unique Triangles and the Law of Sines: AAS or ASA

Figure 5.7

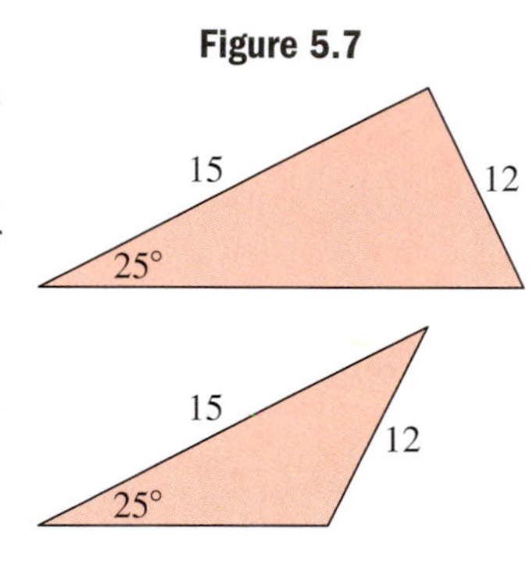

To understand the concept of unique or nonunique solutions regarding the law of sines, consider a case where students are asked to draw a triangle containing a 25° angle and sides of 15 and 12 units. Unavoidably, more than one correct solution will be offered, with two of the possibilities shown in Figure 5.7. For the SSA case, there is some doubt as to the number of solutions possible, or whether a solution even exists. For this reason SSA is called *the ambiguous case*, which we study in greater detail in Section 5.2. Given any two angles and the length of one side (AAS or ASA), *a unique triangle is formed*. This is because the third angle is determined by the first two since $\angle A + \angle B + \angle C = 180°$, and the remaining sides must be of fixed length. Here the law of sines can be applied with the confidence that a solution exists, and only one solution exists.

EXAMPLE 2 Solve the triangle shown and state your answer using a table.

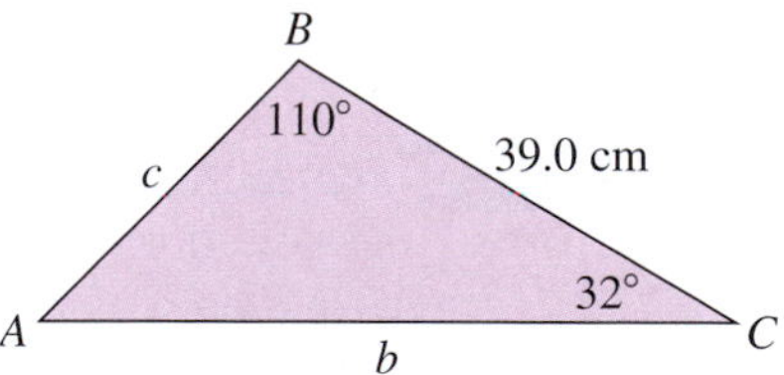

Solution: Once again, this is *not* a right triangle, so the standard ratios cannot be used. The information given is ASA but at first glance, it appears the law of sines cannot be applied (no length of side is given opposite $\angle B$ or $\angle C$). However, we know $\angle A = 180° - (110° + 32°) = 38°$ and with $\angle A$ and side a, we have

$$\frac{\sin A}{a} = \frac{\sin B}{b} \quad \text{law of sines applied to } \angle A \text{ and } \angle B$$

$$\frac{\sin 38°}{39} = \frac{\sin 110°}{b} \quad \text{substitute given values}$$

$$\begin{aligned} b \sin 38° &= 39 \sin 110° && \text{multiply by } 39b \\ b &= \frac{39 \sin 110°}{\sin 38°} && \text{divide by } \sin 38° \\ b &\approx 59.5 && \text{result} \end{aligned}$$

Repeating this procedure using $\frac{\sin A}{a} = \frac{\sin C}{c}$ shows side $c \approx 33.6$ cm.

In table form we have

Angles	Sides (cm)
$A = 38°$	$a = \mathbf{39.0}$
$B = \mathbf{110°}$	$b \approx 59.5$
$C = \mathbf{32°}$	$c \approx 33.6$

NOW TRY EXERCISES 11 THROUGH 28

WORTHY OF NOTE

Although not a definitive check, always review the solution table to ensure the smallest side is opposite the smallest angle, the largest side is opposite the largest angle, and so on. If this is not the case, you should go back and check your work.

C. Applications Involving the Law of Sines

The applications of the law of sines are as diverse as they are plentiful. In Example 3 we solve the trolley car example cited in the introduction.

EXAMPLE 3 A trolley car carries passengers from ground level up to a mountain chateau, as shown in Figure 5.8. (a) How long is the trolley cable? (b) How tall is the mountain?

Figure 5.8

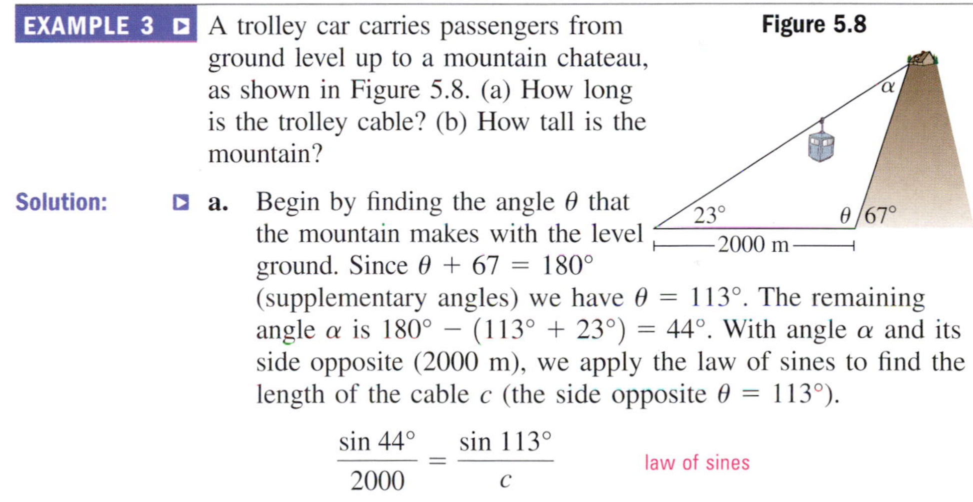

Solution: **a.** Begin by finding the angle θ that the mountain makes with the level ground. Since $\theta + 67 = 180°$ (supplementary angles) we have $\theta = 113°$. The remaining angle α is $180° - (113° + 23°) = 44°$. With angle α and its side opposite (2000 m), we apply the law of sines to find the length of the cable c (the side opposite $\theta = 113°$).

$$\begin{aligned} \frac{\sin 44°}{2000} &= \frac{\sin 113°}{c} && \text{law of sines} \\ c \sin 44° &= 2000 \sin 113° && \text{multiply by } 2000c \\ c &= \frac{2000 \sin 113°}{\sin 44°} && \text{divide by } \sin 44° \\ c &\approx 2650 \text{ m} && \text{result} \end{aligned}$$

The trolley cable is about 2650 m long.

b. To find the mountain's height h, note the trolley cable forms the hypotenuse of a right triangle, with the height of the mountain as one leg (see Figure 5.9). Using the standard relationship for sine, we have

Figure 5.9

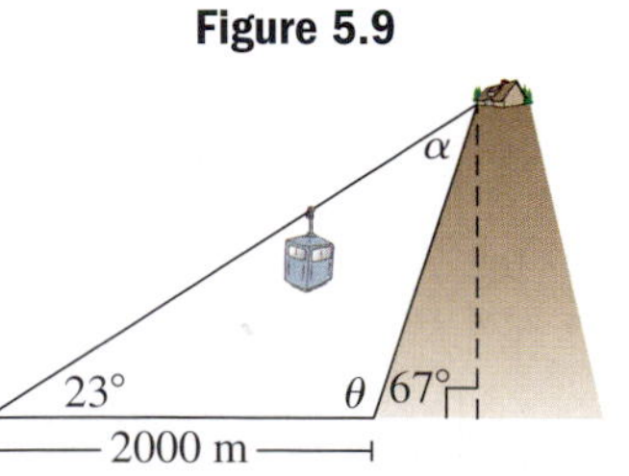

$$\sin 23° = \frac{\text{height of mountain}}{\text{length of cable}}.$$

After the appropriate substitutions,

$$\sin 23° \approx \frac{h}{2650} \qquad \sin\theta = \frac{\text{opp}}{\text{hyp}}$$

$$h \approx 2650 \sin 23° \qquad \text{solve for } h$$

$$\approx 1035 \qquad \text{result}$$

The mountain is about 1035 m high.

NOW TRY EXERCISES 31 THROUGH 38

Here is one additional example showing the versatility of the law of sines.

EXAMPLE 4 The orbit of a communications satellite carries it directly over the cities of San Antonio and Houston, Texas. If the cities are approximately 180 mi apart and the observation angles from each city are as shown, what is the orbiting altitude of the satellite?

Solution: A careful analysis shows we need the length of either side c or side a, which we can then use to solve for h. We choose to solve for side a, and after finding $\angle B = 180° - (80° + 55°)$ or $45°$, we apply the law of sines:

$$\frac{\sin 45°}{180} = \frac{\sin 80°}{a} \qquad \text{law of sines}$$

$$a \sin 45° = 180 \sin 80° \qquad \text{multiply by } 180a$$

$$a = \frac{180 \sin 80°}{\sin 45°} \qquad \text{divide by } \sin 45°$$

$$\approx 251 \text{ mi} \qquad \text{result}$$

Since the altitude forms one side of right triangle BDC, we can use the standard sine function to solve for h:

$$\sin 55° = \frac{h}{a} \qquad \sin\theta = \frac{\text{opp}}{\text{hyp}}$$

$$\sin 55° \approx \frac{h}{251} \qquad \text{substitute 251 for } a$$

$$h \approx 251 \sin 55° \qquad \text{solve for } h$$

$$\approx 206 \text{ mi} \qquad \text{result}$$

The satellite is orbiting at an altitude of approximately 206 mi.

NOW TRY EXERCISES 39 THROUGH 42

TECHNOLOGY HIGHLIGHT

Graphing Calculators and Approximate Solutions

The keystrokes shown apply to a TI-84 Plus model. Please consult our Internet site or your manual for other models.

Generally speaking, users of graphing technology should be very familiar with the storage and recall functions of their calculators, and while working through a problem, use them for any related or intermediate calculations (see *Technology Highlight* in Section 1.2). Only the final answer should be rounded or stated to the desired number of decimal places. In special situations where only a stated degree of accuracy is required, you can designate the number of decimal places your calculator "rounds to" automatically. Most graphing calculators have the ability to display a total of 10 digits, plus the decimal point. On the TI-84 Plus, press the MODE key, which will display the screen shown in Figure 5.10, then press the down arrow ▼ to the second line (FLOAT). Leaving the calculator in "FLOAT" mode will display all ten digits, so evaluating sin(25°) on the home screen would give .4226182617. Suppose only four-decimal-place accuracy is required. Move the cursor to the right ► until it overlays the "4" and press ENTER. Then press 2nd MODE (QUIT) to get back to the home screen. Recall sin 25° using 2nd ENTER then press enter once again. The calculator automatically rounds the value of sin 25° to four decimal places: 0.4226 (Figure 5.11). Pressing the MODE key and changing the number of decimal places to "5" produces a value of 0.42262. Changing the number of decimal places you wish the calculator to display will affect results on the home screen, the coordinates while TRACE ing a graph, results of any CALC operations, and the results of a regression.

Figure 5.10

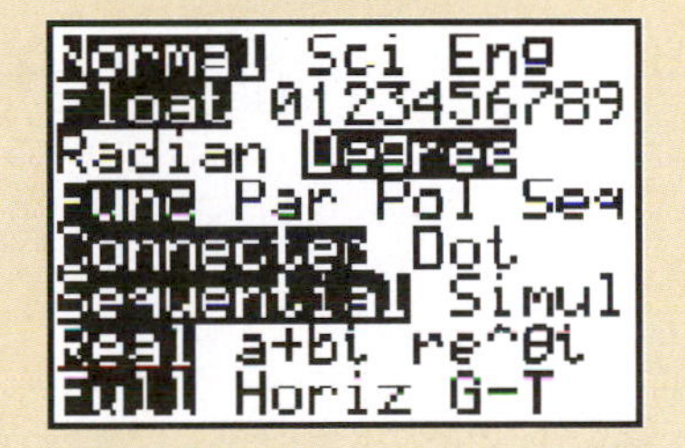

Figure 5.11

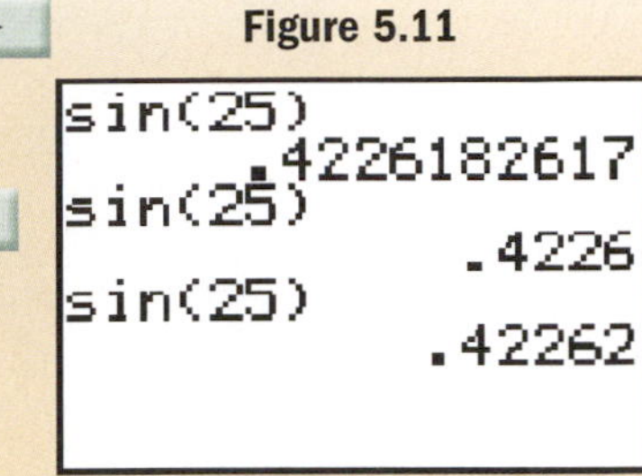

Exercise 1: The value of sin 34° is approximately 0.5591929035 (to 10 decimal places). Round this value to four-, five-, six-, seven-, and eight-decimal-place accuracy, then verify that your calculator gives the identical results using the "FLOAT" settings described above.

Exercise 2: In FLOAT mode, find the value of cos 45°34.34′. Then reset your calculator so that it displays two decimal places, then three, four, and five decimal places. What do you notice?

5.1 EXERCISES

CONCEPTS AND VOCABULARY

Fill in each blank with the appropriate word or phrase. Carefully reread the section if needed.

1. The law of sines states that the ratio of the ______ of an angle to the side ______ the angle is constant.

2. To use the law of sines, we need an ______, a ______ opposite the angle, and an additional ______ or ______.

3. An important part of solving triangles using the law of sines is the triangle property $\angle A + \angle B + \angle C =$ ______.

4. After a triangle is solved, you should always check to ensure that the ______ side is opposite the ______ angle.

5. A property of reciprocals states, "If $\frac{a}{b} = \frac{c}{d}$, then $\frac{b}{a} = \frac{d}{c}$." Show/discuss how this also applies to the law of sines.

6. A property of reciprocals states, "If $\frac{a}{b} = \frac{c}{d}$, then $ad = bc$." Show/discuss how this also applies to the law of sines.

DEVELOPING YOUR SKILLS

Determine if the law of sines can be applied. If it cannot, state why.

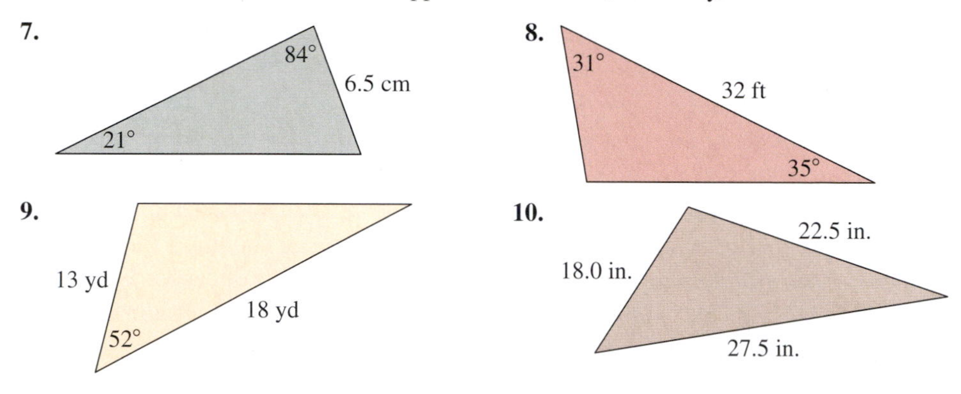

Solve each of the following equations for the unknown part (if possible). Round sides to the nearest hundredth and degrees to the nearest tenth.

11. $\frac{\sin 32°}{15} = \frac{\sin 18.5°}{a}$

12. $\frac{\sin 52°}{b} = \frac{\sin 30°}{12}$

13. $\frac{\sin 63°}{21.9} = \frac{\sin C}{18.6}$

14. $\frac{\sin B}{3.14} = \frac{\sin 105°}{6.28}$

15. $\frac{\sin C}{48.5} = \frac{\sin 19°}{43.2}$

16. $\frac{\sin 38°}{125} = \frac{\sin B}{190}$

Solve each triangle using the law of sines. If the law of sines cannot be used, state why. Draw and label a triangle or label the triangle given before you begin.

17. side $a = 75$ cm
$\angle A = 38°$
$\angle B = 64°$

18. side $b = 385$ m
$\angle B = 47°$
$\angle A = 108°$

19. side $b = 10\sqrt{3}$ in.
$\angle A = 30°$
$\angle B = 60°$

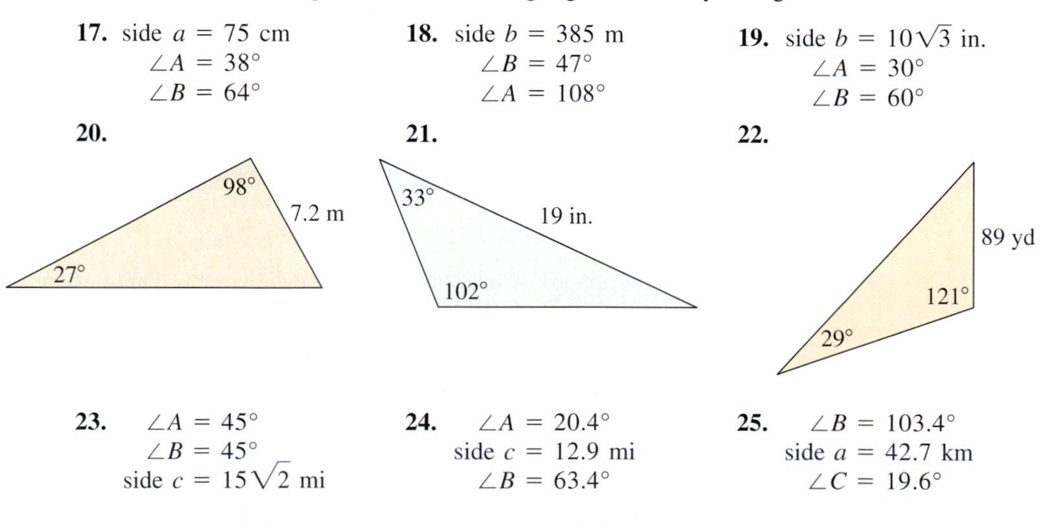

23. $\angle A = 45°$
$\angle B = 45°$
side $c = 15\sqrt{2}$ mi

24. $\angle A = 20.4°$
side $c = 12.9$ mi
$\angle B = 63.4°$

25. $\angle B = 103.4°$
side $a = 42.7$ km
$\angle C = 19.6°$

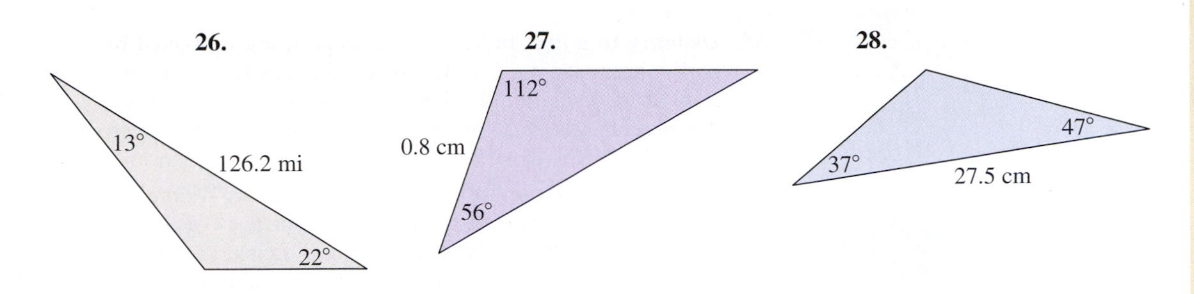

WORKING WITH FORMULAS

29. Area of a triangle: $A = \frac{1}{2}bc \sin\theta$

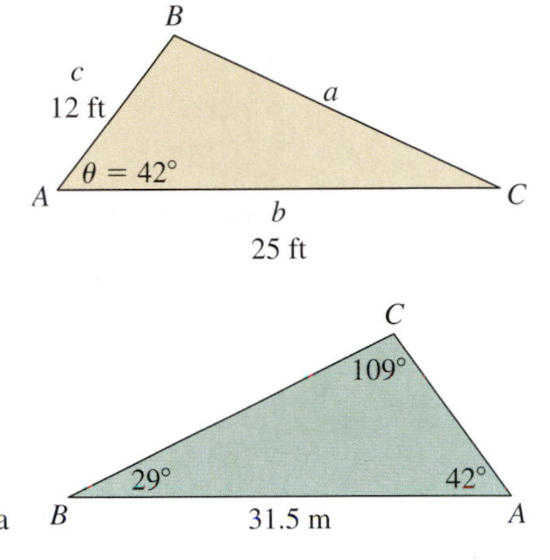

If two sides and an included angle are known, the area of a triangle can be found using the formula shown, where b and c are the known sides and θ is the included angle. Find the area of the triangle given.

30. Area of a triangle: $A = \frac{c^2 \sin A \sin B}{2 \sin C}$

If one side and three angles are known, the area of a triangle can be found using the formula shown, where c is the known side and A, B, and C are the known angles. Find the area of the triangle given.

APPLICATIONS

Length of a rafter: Determine the length of both roof rafters in the diagrams given.

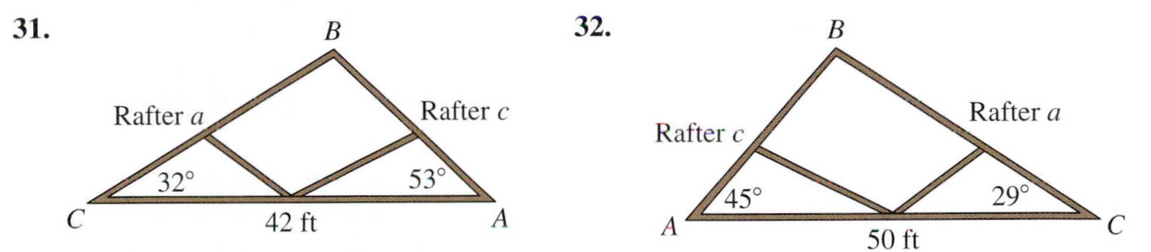

Exercise 33

33. Map distance: A cartographer is using aerial photographs to prepare a map for publication. The distance from Sexton to Rhymes is known to be 27.2 km. Using a protractor, the map maker measures an angle of 96° from Sexton to Tarryson (a newly developed area) and an angle of 58° from Rhymes to Tarryson. Compute each unknown distance.

Exercise 34

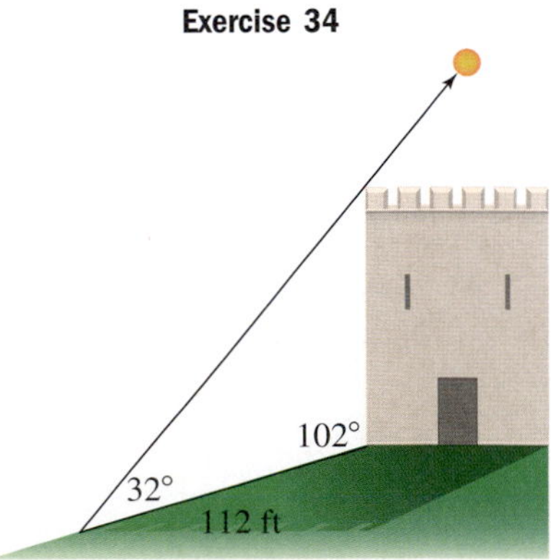

34. Height of a fortress: An ancient fortress is built on a steep hillside, with the base of the fortress walls making a 102° angle with the hill. At the moment the fortress casts a 112-ft shadow, the angle of elevation from the tip of the shadow to the top of the wall is 32°. What is the distance from the base of the fortress to the top of the tower?

35. Distance to a fire: In Yellowstone Park, a fire is spotted by park rangers stationed in two towers that are known to be 5 mi apart. Using the line between them as a baseline, tower A reports the fire is at an angle of 39°, while tower B reports an angle of 58°. How far is the fire from the closer tower?

Exercise 35

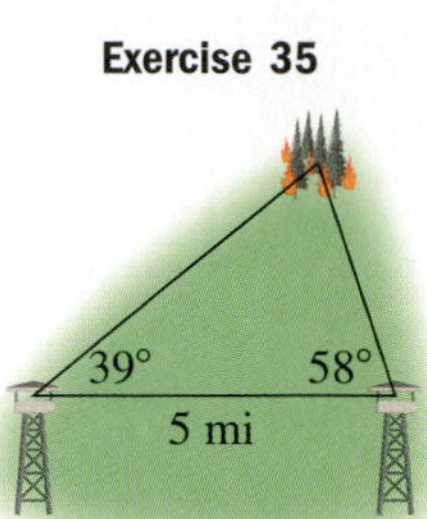

36. Width of a canyon: To find the distance across Waimea Canyon (on the island of Kauai), a surveyor marks a 1000-m baseline along the southern rim. Using a transit, she sights on a large rock formation on the north rim, and finds the angles indicated. How wide is the canyon from point B to point C?

Exercise 36

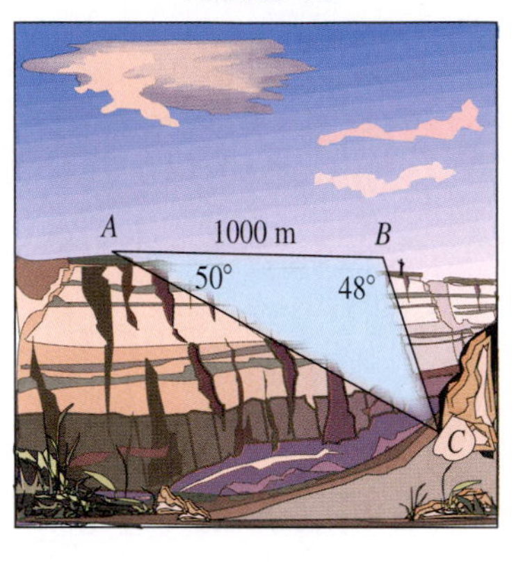

37. Height of a blimp: When the Good-Year Blimp is viewed from the field-level bleachers near the southern end-zone of a football stadium, the angle of elevation is 62°. From the field-level bleachers near the northern end-zone, the angle of elevation is 70°. Find the height of the blimp if the distance from the southern bleachers to the northern bleachers is 145 yd.

Exercise 37

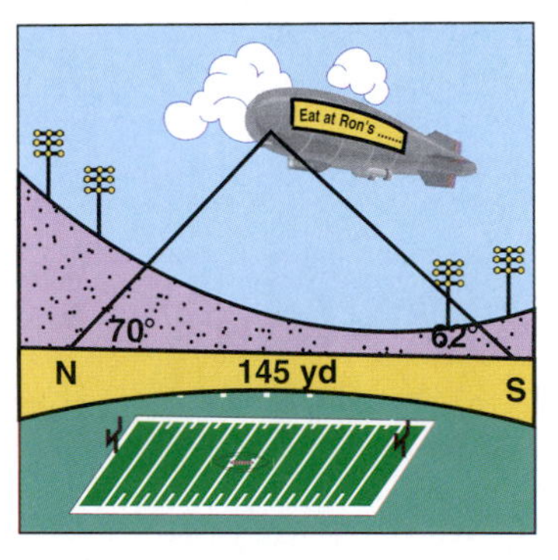

38. Height of a blimp: The rock-n-roll group *Pink Floyd* just finished their most recent tour and has moored their touring blimp at a hangar near the airport in Indianapolis, Indiana. From an unknown distance away, the angle of elevation is measured at 26.5°. After moving 110 yd closer, the angle of elevation has become 48.3°. At what height is the blimp moored?

39. Circumscribed triangles: A triangle is circumscribed within the upper semicircle drawn in the figure. Use the law of sines to solve the triangle given the measures shown. What is the diameter of the circle? What do you notice about the triangle?

Exercises 39 and 40

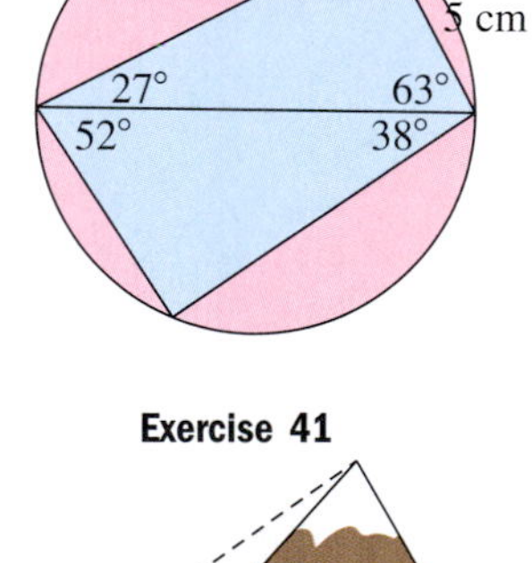

40. Circumscribed triangles: A triangle is circumscribed within the lower semicircle drawn to the right. Use the law of sines to solve the triangle given the measures shown. How long is the longer chord? What do you notice about the triangle?

41. Approaching from the west, a group of hikers notes the angle of elevation to the summit of a steep mountain is 35° at a distance of 1250 meters. Arriving at the base of the mountain, they estimate this side of the mountain has an average slope of 48°. (a) Find the slant height of the mountain's west side. (b) Find the slant height of the east side of the mountain, if the east side has an average slope of 65°. (c) How tall is the mountain?

Exercise 41

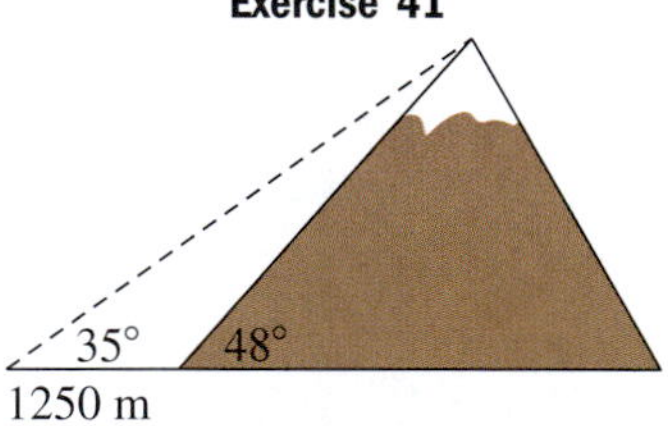

42. Coffeyville and Liberal, Kansas, lie along the state's southern border and are roughly 298 miles apart. Olathe, Kansas, is very near the state's eastern border at an angle of 23° with Liberal and 72° with Coffeyville (using the southern border as one side of the angle). (a) Compute the distance between these cities. (b) What is the shortest (straight line) distance from Olathe to the southern border of Kansas?

Exercise 42

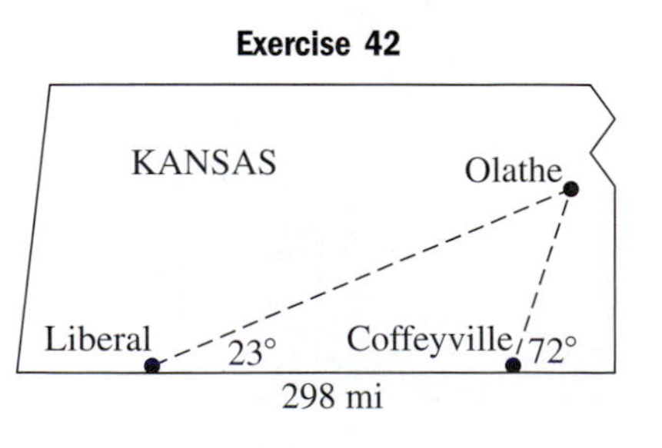

WRITING, RESEARCH, AND DECISION MAKING

43. Solve the triangle shown in three ways—first by using the law of sines, second using right triangle trigonometry, and third using the standard 30-60-90 triangle. Was one method "easier" than the others? Use these connections to express the irrational number $\sqrt{3}$ as a quotient of two trigonometric functions of an angle. Can you find a similar expression for $\sqrt{2}$?

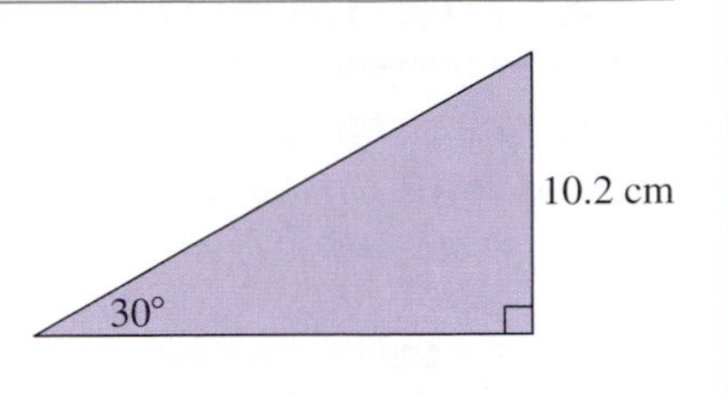

44. For obtuse angle β shown, the altitude h from $\angle A$ actually falls outside the triangle, as seen in the figure. In this case, consider that $\sin(180° - \beta) = \sin\beta$ from the difference formula for sines (Exercise 59, Section 3.3).

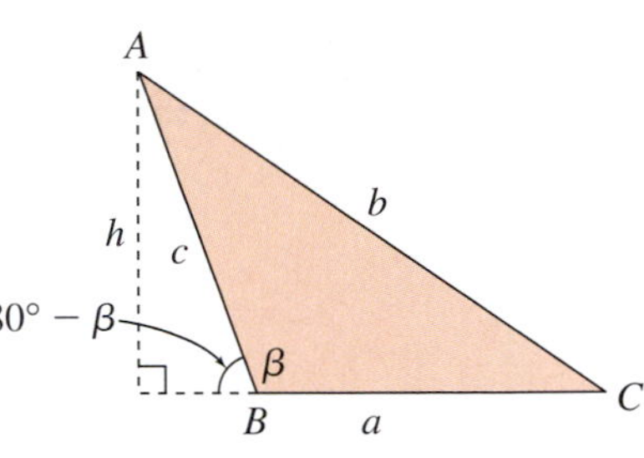

In the figure we note $\sin(180° - \beta) = \frac{h}{c} = \sin\beta$, yielding $h = c\sin\beta$. (a) Write the ratio for $\sin C$ and verify that $\frac{\sin B}{b} = \frac{\sin C}{c}$. (b) Draw the altitude from $\angle C$ to side $\overline{AB}$ and use the result to verify that $\frac{\sin B}{b} = \frac{\sin A}{a}$. (c) What can you conclude?

EXTENDING THE CONCEPT

45. Lines L_1 and L_2 shown are parallel. The three triangles between these lines all share the same base (in bold). Explain why all three triangles must have the same area.

Exercise 45

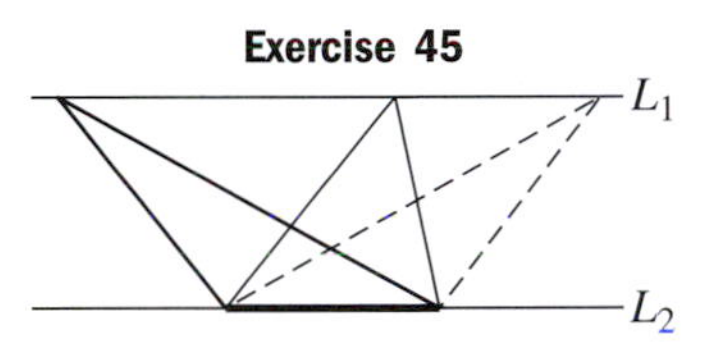

46. A UFO is sighted on a direct line between the towns of Batesville and Cave City, sitting stationary in the sky. The towns are 13 mi apart as the crow flies. A student in Batesville calls a friend in Cave City and both take measurements of the angle of elevation: 35° from Batesville and 42° from Cave City. Suddenly the UFO zips across the sky at a level altitude heading directly for Cave City, then stops and hovers long enough for an additional measurement from Batesville: 24°. If the UFO was in motion for 1.2 sec, at what average speed (in mph) did it travel?

MAINTAINING YOUR SKILLS

47. (4.4) Find all solutions to the equation $2\sin x = \cos(2x)$

48. (3.2) Prove the given identity: $\tan^2 x - \sin^2 x = \tan^2 x \sin^2 x$

49. (3.4) Use sum-to-product formulas to verify that the following is an identity:

$$\cot x = \frac{\cos(3x) + \cos x}{\sin(3x) - \sin x}$$

50. (4.4) Find all solutions in $[0, 2\pi)$: $2\sin^2 x - 7\sin x = -3$

51. (1.1) A Ferris wheel has a diameter of 50 ft. What is the linear distance a seat on the ferris wheel moves as the wheel rotates through an angle of 245°?

52. (2.2) Determine the equation of the graph shown, given it is of the form $y = A\tan(Bx \pm C)$.

Exercise 52

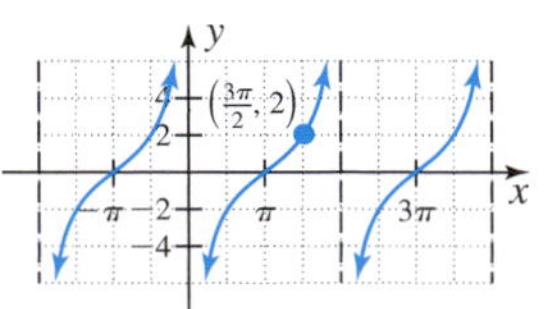

5.2 The Law of Sines and the Ambiguous Case

LEARNING OBJECTIVES

In Section 5.2 you will learn how to:

A. Identify the "ambiguous case" (SSA)

B. Solve the ambiguous case with the aid of scaled drawings and triangle properties

C. Solve applications that involve the ambiguous case of the law of sines

INTRODUCTION

As noted in Section 5.1, when the information given for a triangle includes two sides and an angle opposite one side (SSA case), there is some ambiguity as to the number of solutions. The word *ambiguous* derives from the Latin *ambiguus*, which means "shifting" or "doubtful." The ambiguous case is appropriately named, because without further analysis, even the *existence* of a triangle is uncertain.

POINT OF INTEREST

One of the more remarkable advances from Renaissance mathematics was the realization that an infinite series could be used to estimate the value of the trigonometric functions to any desired degree of accuracy. For any real number x, the value of $\sin x$ can be computed as $x - \frac{x^3}{3!} + \frac{x^5}{5!} - \frac{x^7}{7!} + \frac{x^9}{9!} \ldots$. If greater accuracy is desired, more terms can be used, but even with only five terms we have an estimate for sin 2 (in radians) of $2 - \frac{2^3}{3!} + \frac{2^5}{5!} - \frac{2^7}{7!} + \frac{2^9}{9!} \approx 0.90935$, which is remarkably close to the calculated value of $\sin 2 \approx 0.90930$.

A. The Ambiguous Case of the Law of Sines: Two Sides and an Angle (SSA)

In the ASA and AAS cases from Section 5.1, a unique triangle is formed since the size of the third angle is fixed by the two given (they must sum to 180°). In the SSA case where two *sides* and an angle opposite one side are given, any possible triangle depends on the length of the given side. To see why, consider a triangle with side $c = 30$ cm, $\angle A = 30°$, and side a opposite the 30° angle (Figure 5.12—note the length of side b is yet to be determined). From our work with 30-60-90 triangles, we know if $a = 15$ cm, it is exactly the length needed to form a right triangle (Figure 5.13).

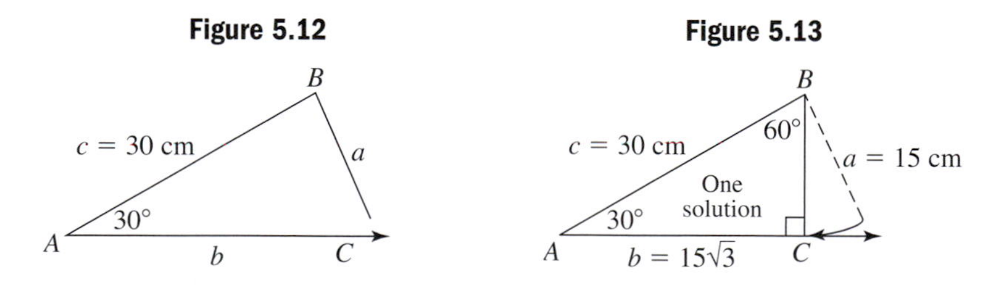

By varying the length of side a, we note three other possibilities. If side $a < 15$ cm, no triangle is possible since a is too short to contact side b (Figure 5.14), while if $15 \text{ cm} < \text{side } a < 30$ cm, two triangles are possible since side a will then intersect side b at two points, C_1 and C_2 (Figure 5.15).

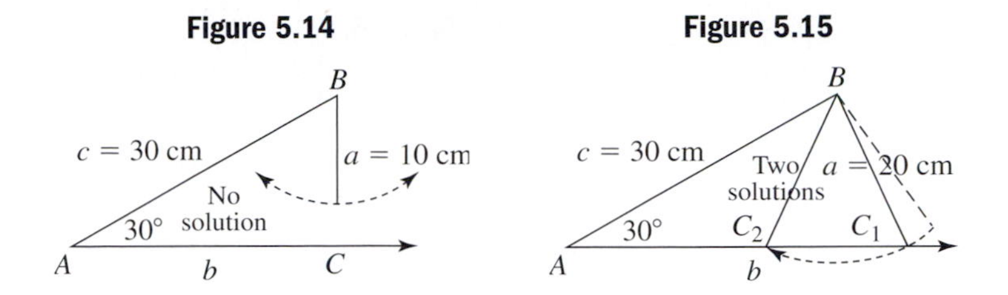

Figure 5.14 Figure 5.15

For future use, note that when two triangles are possible, angles C_1 and C_2 must be supplements (see Exercise 43). Finally, if side $a > 30$ cm, it will intersect side b only once, forming the obtuse triangle shown in Figure 5.16, where we've assumed $a = 35$ cm.

Figure 5.16

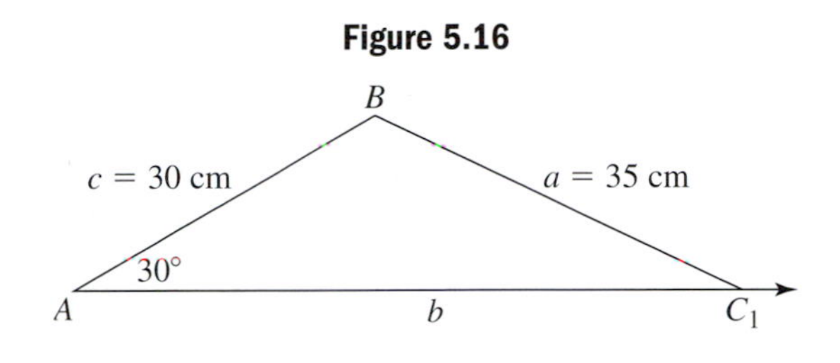

Since the final solution is in doubt until we do further work, the SSA case is called the **ambiguous case** of the law of sines.

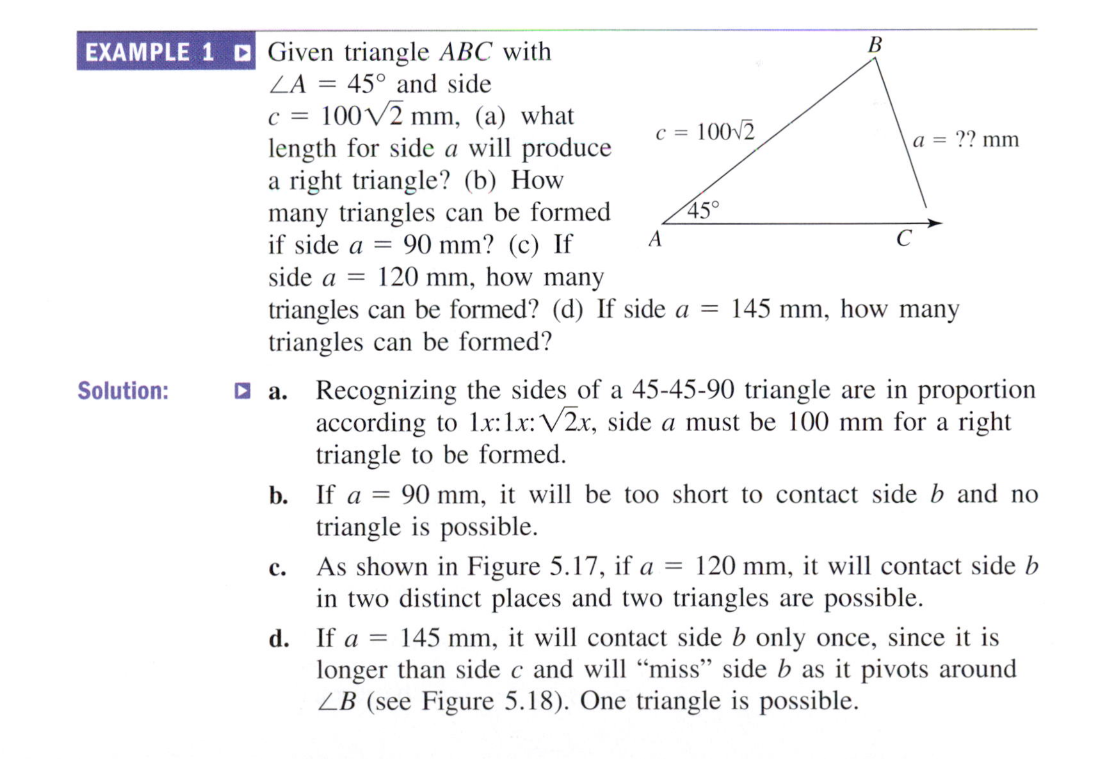

EXAMPLE 1 Given triangle ABC with $\angle A = 45°$ and side $c = 100\sqrt{2}$ mm, (a) what length for side a will produce a right triangle? (b) How many triangles can be formed if side $a = 90$ mm? (c) If side $a = 120$ mm, how many triangles can be formed? (d) If side $a = 145$ mm, how many triangles can be formed?

Solution:

a. Recognizing the sides of a 45-45-90 triangle are in proportion according to $1x{:}1x{:}\sqrt{2}x$, side a must be 100 mm for a right triangle to be formed.

b. If $a = 90$ mm, it will be too short to contact side b and no triangle is possible.

c. As shown in Figure 5.17, if $a = 120$ mm, it will contact side b in two distinct places and two triangles are possible.

d. If $a = 145$ mm, it will contact side b only once, since it is longer than side c and will "miss" side b as it pivots around $\angle B$ (see Figure 5.18). One triangle is possible.

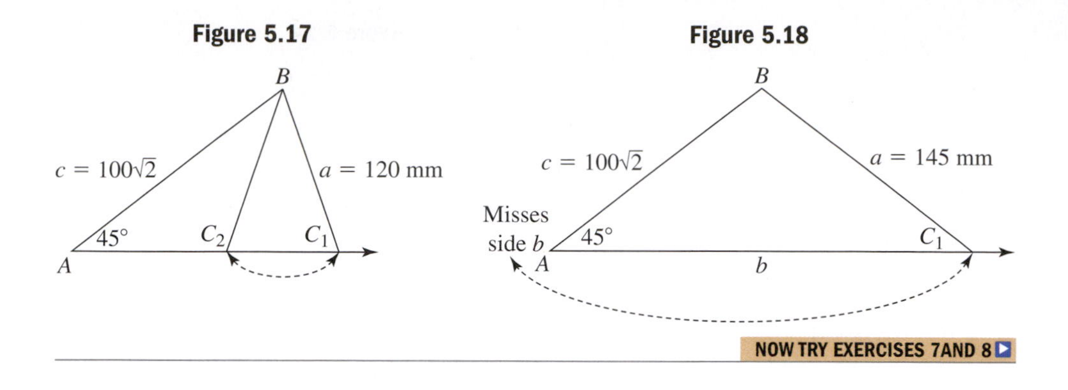

Figure 5.17 Figure 5.18

NOW TRY EXERCISES 7AND 8

B. The Ambiguous Case and Scaled Drawings

For a better understanding of the SSA (ambiguous) case, scaled drawings can initially be used along with a metric ruler and protractor. Begin with a horizontal line segment of undetermined length to represent the third (unknown) side, and use the protractor to draw the given angle on either the left or right side of this segment (we chose the left). Then use the metric ruler to draw an adjacent side of appropriate length, *choosing a scale that enables a complete diagram*. For instance, if the given sides are 3 ft and 5 ft, use 3 cm and 5 cm instead (1 cm = 1 ft). If the sides are 80 mi and 120 mi, use 8 cm and 12 cm (1 cm = 10 mi), and so on. Once the adjacent side is drawn, start at the free endpoint and draw a vertical segment to represent the remaining side. A careful sketch will often indicate if the vertical segment will intersect the horizontal side at 0, 1, or 2 points, indicating that none, one, or two triangles may be possible.

EXAMPLE 2 Solve the triangle with side $c = 12$ cm, side $a = 8$ cm, and $\angle A = 45°$. State the complete solution in table form.

Solution: Two sides and an angle opposite are given (SSA), indicating there may be no triangle, one triangle, or two triangles possible. Draw a horizontal segment of some length first, and use a protractor to mark the 45° angle A. Then measure and draw a segment 12 cm in length for side c (see the figure). Note that since $\angle A$ is given, side c is the adjacent side. Finally, draw side $a = 8$ cm as a vertical segment from the free end of side c. As the resulting diagram shows, no triangles are possible. If we attempt to apply the law of sines, we obtain $\frac{3\sqrt{2}}{4} = \sin C$. Since $\frac{3\sqrt{2}}{4} \approx 1.06 > 1$, no such triangle can be formed.

c = 12 cm
a = 8 cm
45°
A

NOW TRY EXERCISES 9 AND 10

EXAMPLE 3 Solve the triangle with side $b = 100$ ft, side $c = 60$ ft, and $\angle C = 28.0°$.

Solution: Once again, two sides and an angle opposite are given (SSA), and we draw a diagram to help determine the possibilities. Draw the horizontal segment and use a protractor to mark $\angle C = 28°$. Then draw a segment 10 cm long (to represent $b = 100$ ft) as the adjacent side of the angle, and a vertical segment 6 cm long from the free end of b (to represent $c = 60$ ft). It seems apparent that side c will intersect the horizontal side in two places, and two triangles are possible. We apply the law of sines to solve the first triangle, whose features we'll note with a subscript of 1.

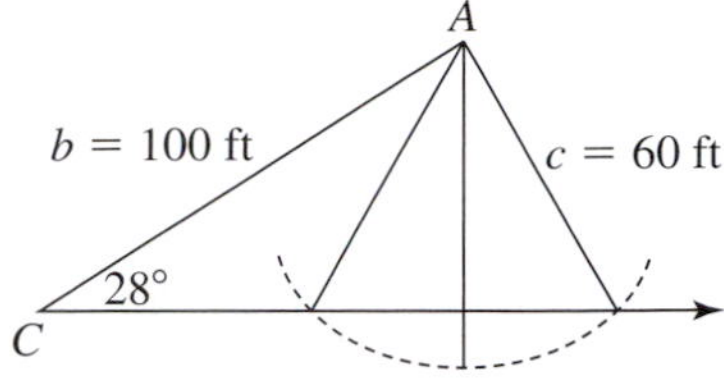

$$\frac{\sin B_1}{b} = \frac{\sin C}{c} \quad \text{law of sines}$$

$$\frac{\sin B_1}{100} = \frac{\sin 28°}{60} \quad \text{substitute}$$

$$\sin B_1 = \frac{5}{3}\sin 28° \quad \text{solve for } \sin B_1$$

$$B_1 \approx 51.5° \quad \text{apply arcsine}$$

Since $\angle B_1 + \angle B_2 = 180°$, we know $\angle B_2 = 128.5°$. These values give 100.5° and 23.5° as the measures of $\angle A_1$ and $\angle A_2$, respectively. By once again applying the law of sines, we find side $a_1 \approx 125.7$ ft and $a_2 \approx 51.0$ ft. See Figure 5.19.

WORTHY OF NOTE

In Example 3, we found $\angle B_2$ using the property that states the angles in a triangle must sum to 180°. We could also view B_1 as a QI reference angle, which also gives a QII solution of $(180 - 51.5)° = 128.5°$.

Angles	Sides (ft)	Angles	Sides (ft)
$A_1 \approx 100.5°$	$a_1 \approx 125.7$	$A_2 \approx 23.5°$	$a_2 \approx 51.0$
$B_1 \approx 51.5°$	**b = 100**	$B_2 \approx 128.5°$	**b = 100**
C = 28°	**c = 60**	**C = 28°**	**c = 60**

Figure 5.19

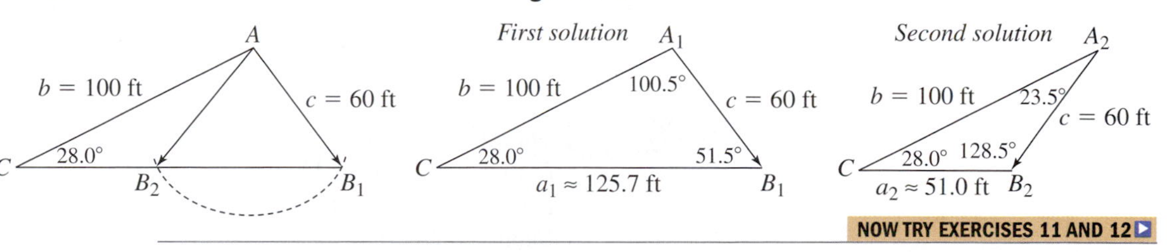

NOW TRY EXERCISES 11 AND 12

Admittedly, the scaled drawing approach has some drawbacks—it takes time to draw the diagrams and is of little use if the situation is a close call. It does, however, offer a deeper understanding of the subtleties involved in solving the SSA case. Instead of a scaled drawing, we can use a simple sketch *as a guide*, while keeping needed properties in mind: (1) the angles of a triangle must sum to 180°, (2) a sine ratio can never be greater than 1, and (3) for $0 < \theta < 180°$, $\sin\theta > 0$ has two solutions that are supplements.

EXAMPLE 4 Solve the triangle with side $a = 220$ ft, side $b = 200$ ft, and $\angle A = 40°$.

Solution: The information given is again SSA, and we apply the law of sines with this in mind.

$$\frac{\sin A}{a} = \frac{\sin B}{b} \quad \text{law of sines}$$

$$\frac{\sin 40°}{220} = \frac{\sin B}{200} \quad \text{substitute}$$

$$\sin B = \frac{200 \sin 40°}{220} \quad \text{solve for } \sin B$$

$$B_1 \approx 36° \quad \text{apply arcsine}$$

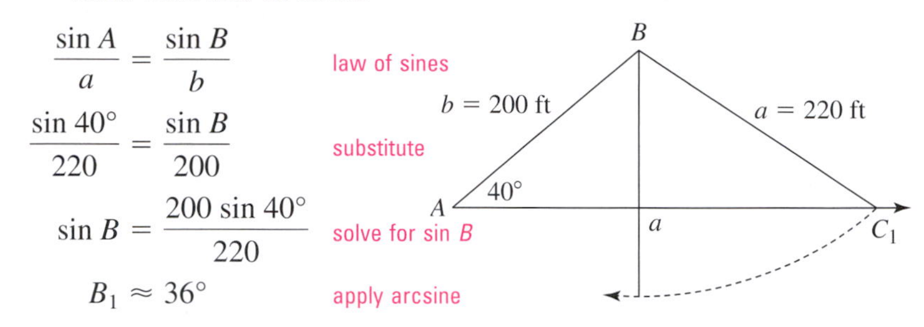

This is the solution from Quadrant I. The QII solution is about $(180 - 36)° = 144°$. At this point our solution tables have this form:

Angles	Sides (ft)
$A = 40°$	$a = 220$
$B_1 \approx 36°$	$b = 200$
$C_1 =$	$c_1 =$

Angles	Sides (ft)
$A = 40°$	$a = 220$
$B_2 \approx 144°$	$b = 200$
$C_2 =$	$c_2 =$

It seems reasonable to once again find the remaining angles and finish by reapplying the law of sines, but observe that the sum of the two angles from the second solution *already exceeds 180°*: $40° + 144° = 188°$! This means no second solution is possible (side a is too long). We find that $C_1 \approx 104°$, and applying the law of sines gives a value of $c_1 \approx 332$ ft.

NOW TRY EXERCISES 13 THROUGH 20

C. Applications of the Ambiguous Case

As "ambiguous" as it is, the ambiguous case has a number of applications in engineering, astronomy, physics, and other areas. Here is an example from astronomy.

EXAMPLE 5 The planet Venus can be seen from Earth with the naked eye, but as the diagram indicates, the position of Venus is uncertain (we are unable to tell if Venus is in the near position or the far position). Given the Earth is 93 million miles from the Sun and Venus is 67 million miles from the Sun, determine the closest and farthest possible distances that separate the

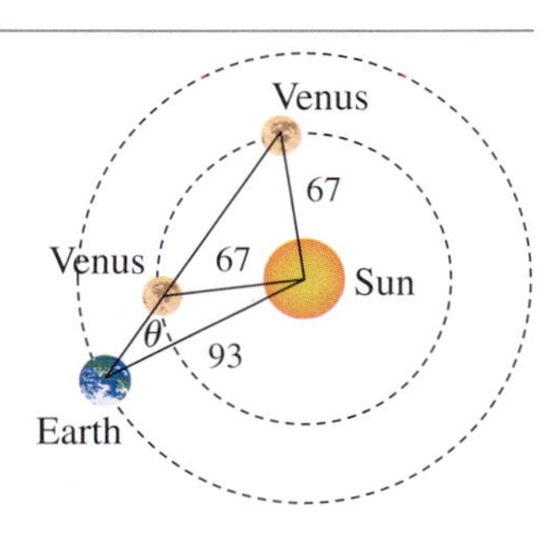

planets in this alignment. Assume a viewing angle of $\theta \approx 18°$ and that the orbits of both planets are roughly circular.

Solution: A close look at the information and diagram shows a SSA case. Begin by applying the law of sines where $E \rightarrow$ Earth, $V \rightarrow$ Venus, and $S \rightarrow$ Sun.

$$\frac{\sin E}{e} = \frac{\sin V}{v} \quad \text{law of sines}$$

$$\frac{\sin 18°}{67} = \frac{\sin V}{93} \quad \text{substitute given values}$$

$$\sin V = \frac{93 \sin 18°}{67} \quad \text{solve for } \sin V$$

$$V \approx 25.4° \quad \text{apply arcsine}$$

This is the angle V_1 formed when Venus is farthest away. The angle V_2 at the closer distance is $180° - 25.4° = 154.6°$. At this point our solution tables have this form:

Angles	Sides (10^6 mi)
$E = 18°$	$e = 67$
$V_1 \approx 25.4°$	$v = 93$
$S_1 =$	$s_1 =$

Angles	Sides (10^6 mi)
$E = 18°$	$e = 67$
$V_2 = 154.6°$	$v = 93$
$S_2 =$	$s_2 =$

For S_1 and S_2 we have $S_1 \approx 180 - (18 + 25.4°) = 136.6°$ (larger angle) and $S_2 \approx 180 - (18 + 154.6°) = 7.4°$ (smaller angle). Re-applying the law of sines for s_1 shows the farthest distance between the planets is about 149 million miles. Solving for s_2 shows that the closest distance is approximately 28 million miles.

NOW TRY EXERCISES 29 AND 30

EXAMPLE 6 As shown in Figure 5.20, a radar ship is 30.0 mi off shore when a large fleet of ships leaves port at an angle of 43.0°. (a) If the maximum range of the ship's radar is 20.0 mi, will the departing fleet be detected? (b) If the maximum range of the ship's radar is 25.0 mi, how far from port is the fleet when it is first detected?

Figure 5.20

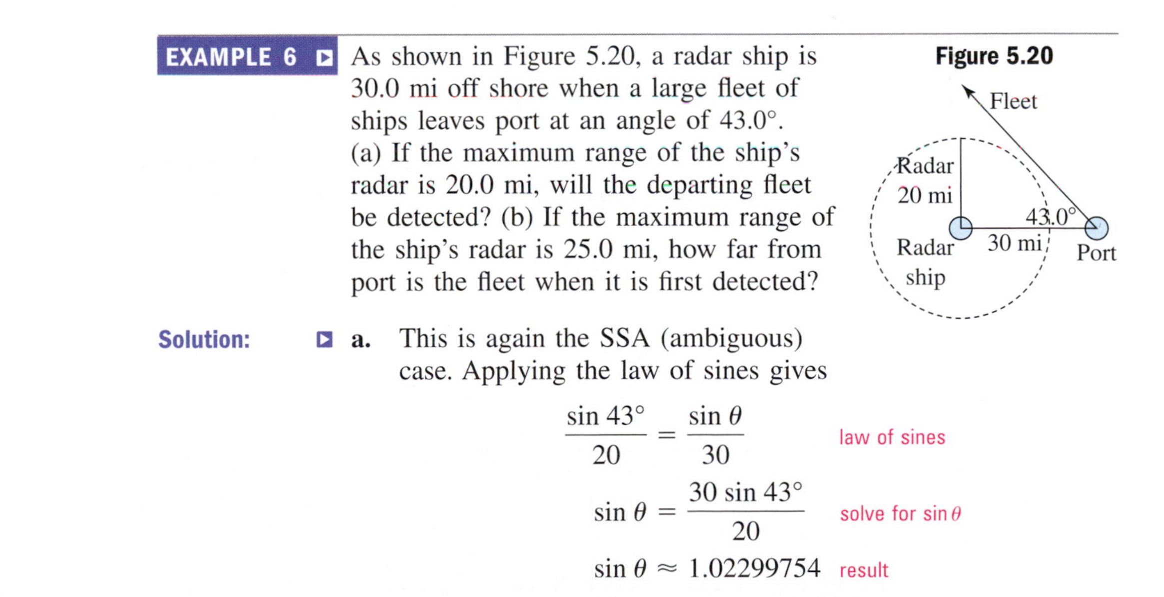

Solution: **a.** This is again the SSA (ambiguous) case. Applying the law of sines gives

$$\frac{\sin 43°}{20} = \frac{\sin \theta}{30} \quad \text{law of sines}$$

$$\sin \theta = \frac{30 \sin 43°}{20} \quad \text{solve for } \sin\theta$$

$$\sin \theta \approx 1.02299754 \quad \text{result}$$

No triangle is possible and the departing fleet will not be detected.

b. If the radar has a range of 25.0 mi, the radar beam will intersect the projected course of the fleet in two places.

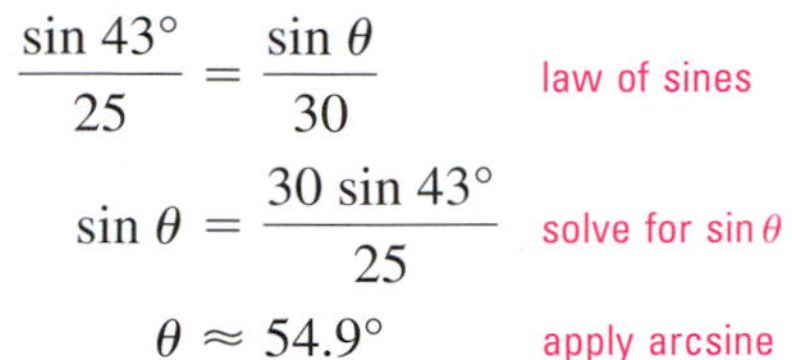

$$\frac{\sin 43°}{25} = \frac{\sin \theta}{30} \quad \text{law of sines}$$

$$\sin \theta = \frac{30 \sin 43°}{25} \quad \text{solve for } \sin \theta$$

$$\theta \approx 54.9° \quad \text{apply arcsine}$$

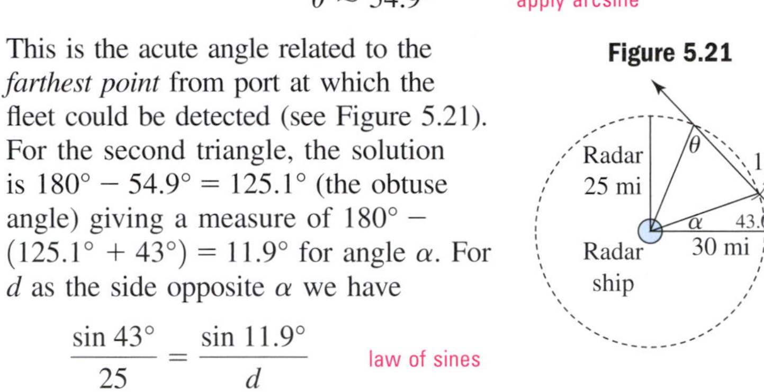

Figure 5.21

This is the acute angle related to the *farthest point* from port at which the fleet could be detected (see Figure 5.21). For the second triangle, the solution is $180° - 54.9° = 125.1°$ (the obtuse angle) giving a measure of $180° - (125.1° + 43°) = 11.9°$ for angle α. For d as the side opposite α we have

$$\frac{\sin 43°}{25} = \frac{\sin 11.9°}{d} \quad \text{law of sines}$$

$$d = \frac{25 \sin 11.9°}{\sin 43°} \quad \text{solve for } d$$

$$\approx 7.6 \quad \text{simplify}$$

This shows the fleet is first detected about 7.6 mi from port.

NOW TRY EXERCISES 31 AND 32

There are a number of additional, interesting applications in the exercise set (see Exercises 33 through 40). Enjoy.

TECHNOLOGY HIGHLIGHT

Using TABLES to Understand the Law of Sines

The keystrokes shown apply to a TI-84 Plus model. Please consult our Internet site or your manual for other models.

Using the TABLE feature of a graphing calculator can increase our understanding of the ambiguous case. Consider the "unfinished" $\triangle ABC$ shown in Figure 5.22, and the following questions:

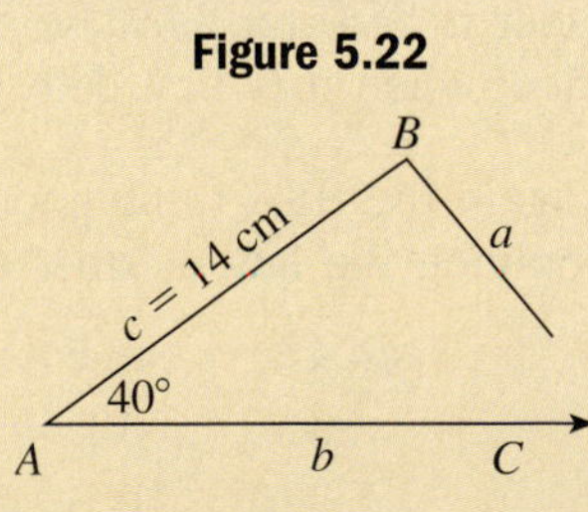

Figure 5.22

1. What length(s) of side a will result in one obtuse triangle?
2. What length(s) of side a will result in two triangles?
3. What length(s) of side a will result in one right triangle?

We can easily answer all three questions with the TABLE feature of a graphing calculator. Using the law of sines we have the following sequence:

$$\frac{\sin C}{14} = \frac{\sin 40°}{a}$$

$$\sin C = \frac{14 \sin 40°}{a}$$

$$C = \sin^{-1}\left(\frac{14 \sin 40°}{a}\right)$$

We can enter this expression for $\angle C$ as Y_1 on the Y= screen of the calculator (see Figure 5.23), then use the TABLE feature to evaluate the expression for *different lengths of side a*. The expression for Y_2 gives the second value for $\angle C$ if two triangles are possible. A casual observation shows that side a must be 14 cm or longer to answer question 1. For $a > 14$ cm, side a will intersect side b one time (to the right) and miss side c entirely (to the left) as it pivots at vertex B. To investigate other possibilities for side a, set up the TBLSET screen (2nd WINDOW), as shown in Figure 5.24, which has us start at $a = 14$ and counting backward by 1's (ΔTbl $= -1$). Pressing 2nd GRAPH (TABLE) produces the TABLE shown in Figure 5.25. Note when $a = 14$, angle $C = 40°$ and an isosceles triangle is formed (base angles are equal). For $9 < x < 14$, two triangles are formed. Figure 5.26 shows the case where side $a \approx 10$ cm. When side $a \approx 9$ cm, a single right triangle is formed. When side $a < 9$ cm, the table returns an error message because the value of $\left[\frac{14 \sin 40°}{a}\right]$ is greater than one and no triangle can be formed. Use a diagram similar to the one shown and these ideas/methods to investigate triangles given the following conditions:

Figure 5.23

```
Plot1 Plot2 Plot3
\Y1=sin⁻¹(14sin(4
0)/X)
\Y2=180-Y1
\Y3=
\Y4=
\Y5=
\Y6=
```

Figure 5.24

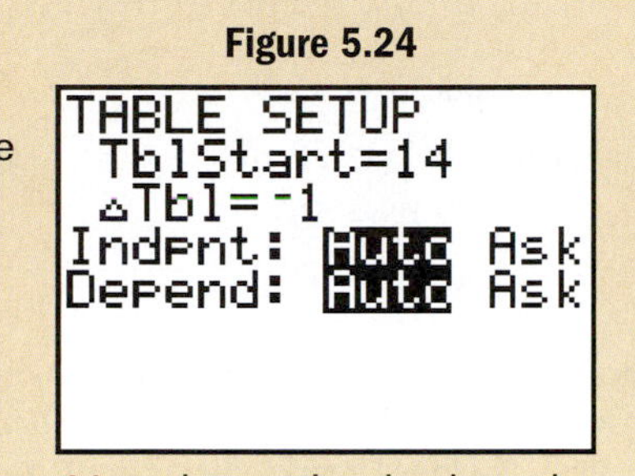

Figure 5.25

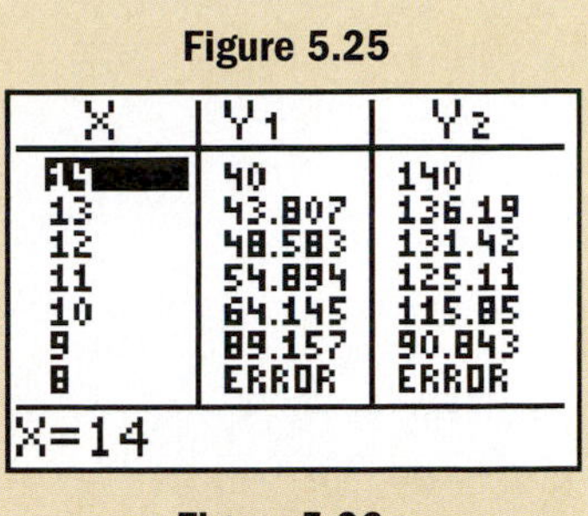

X	Y1	Y2
14	40	140
13	43.807	136.19
12	48.583	131.42
11	54.894	125.11
10	64.145	115.85
9	89.157	90.843
8	ERROR	ERROR

X=14

Figure 5.26

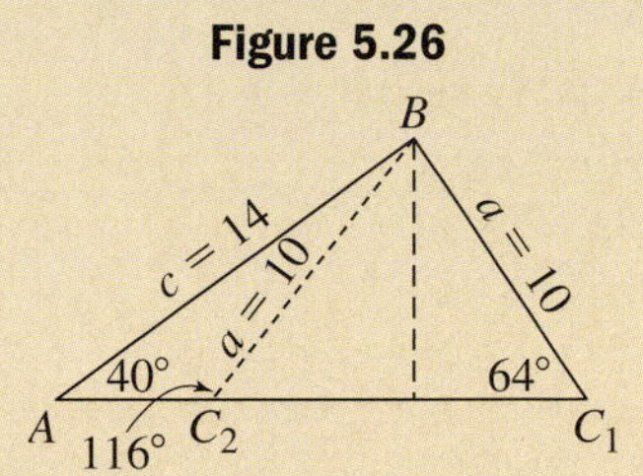

Exercise 1: $\angle A = 35°$, side $c = 25$ mm, side $a =$ ____.

Exercise 2: $\angle A = 30°$, side $c = 8$ in., side $a =$ ____.

Exercise 3: $\angle C = 52°$, side $a = 27.5$ cm, side $c =$ ____.

5.2 EXERCISES

CONCEPTS AND VOCABULARY

Fill in each blank with the appropriate word or phrase. Carefully reread the section if needed.

1. For the law of sines, if two sides and an angle opposite one side are given, this is referred to as the ________ case, since the solution is in doubt until further work.

2. Two inviolate properties of a triangle that can be used to help solve the ambiguous case are: (a) the angles must sum to ____ and (b) no sine ratio can exceed ____.

3. For positive k, the equation $\sin\theta = k$ has two solutions, one in Quadrant ____ and the other in Quadrant ____.

4. With graph paper, a protractor, and a compass, ________ drawings can be used to help see the number of solutions in the ambiguous case.

5. In your own words, explain why the AAS case results in a unique solution while the SSA case does not. Give supporting diagrams.

6. Explain why no triangle is possible in each case:

a. $A = 34°, B = 73°, C = 52°,$
$a = 14', b = 22', c = 18'$

b. $A = 42°, B = 57°, C = 81°,$
$a = 7'', b = 9'', c = 22''$

DEVELOPING YOUR SKILLS

Answer each question and justify your response using a diagram, but do not solve the triangle(s).

7. Given $\triangle ABC$ with $\angle A = 30°$ and side $c = 20$ cm, (a) what length for side a will produce a right triangle? (b) How many triangles can be formed if side $a = 8$ cm? (c) If side $a = 12$ cm, how many triangles can be formed? (d) If side $a = 25$ cm, how many triangles can be formed?

8. Given $\triangle ABC$ with $\angle A = 60°$ and side $c = 6\sqrt{3}$ m, (a) what length for side a will produce a right triangle? (b) How many triangles can be formed if side $a = 8$ m? (c) If side $a = 10$ m, how many triangles can be formed? (d) If side $a = 15$ m, how many triangles can be formed?

Solve using the law of sines and a scaled drawing. If two triangles exist, solve both completely.

9. side $b = 385$ m
$\angle B = 67°$
side $a = 490$ m

10. side $a = 36.5$ yd
$\angle B = 67°$
side $b = 12.9$ yd

11. side $c = 25.8$ mi
$\angle A = 30°$
side $a = 12.9$ mi

12. side $c = 10\sqrt{3}$ in.
$\angle A = 60°$
side $a = 15$ in.

13. side $c = 58$ mi
$\angle C = 59°$
side $b = 67$ mi

14. side $b = 24.9$ km
$\angle B = 45°$
side $a = 32.8$ km

Use the law of sines to determine if no triangle, one triangle, or two triangles can be formed from the diagrams given (diagrams *may not be to scale*). If two solutions exist, solve both completely. Note the arrowhead marks the side of undetermined length.

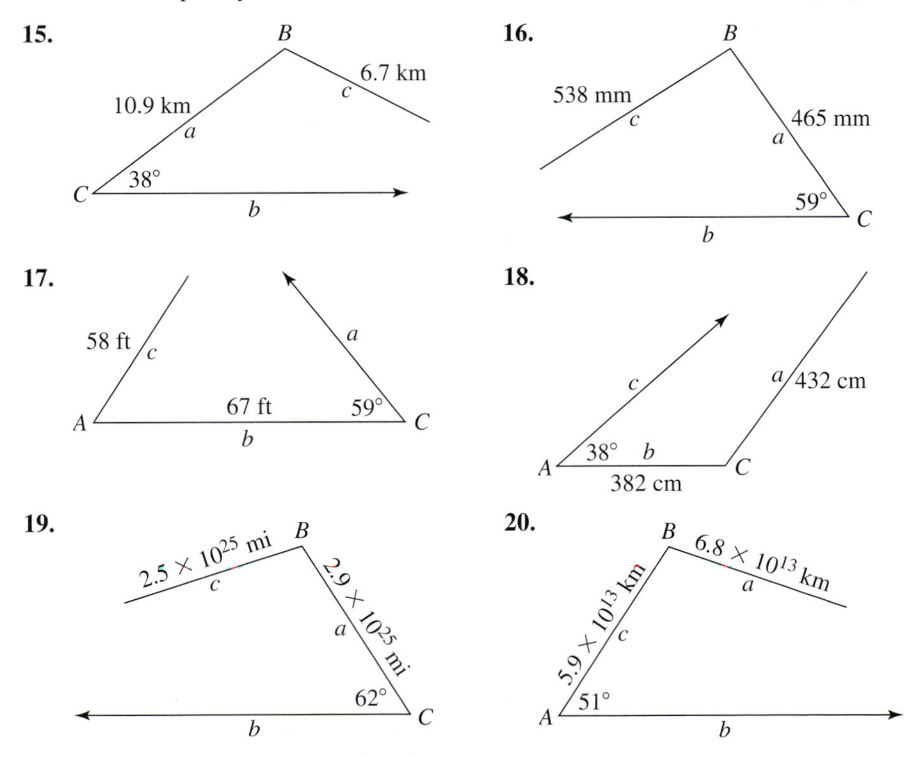

For Exercises 21 to 26, assume the law of sines is being applied to solve a triangle. Solve for the unknown angle (if possible), then determine if a second angle ($0° < \theta < 180°$) exists that also satisfies the proportion.

21. $\dfrac{\sin A}{12} = \dfrac{\sin 48°}{27}$

22. $\dfrac{\sin 60°}{32} = \dfrac{\sin B}{9}$

23. $\dfrac{\sin 57°}{35.6} = \dfrac{\sin C}{40.2}$

24. $\dfrac{\sin B}{5.2} = \dfrac{\sin 65°}{4.9}$

25. $\dfrac{\sin A}{280} = \dfrac{\sin 15°}{52}$

26. $\dfrac{\sin 29°}{121} = \dfrac{\sin B}{321}$

WORKING WITH FORMULAS

27. Triple angle formula for sine: $\sin(3\theta) = 3\sin\theta - 4\sin^3\theta$

Most students are familiar with the double angle formula for sine: $\sin(2\theta) = 2\sin\theta\cos\theta$. The triple angle formula for sine is given here. Use the formula to find an exact value for sin 135°, then verify the result using a reference angle.

28. Radius of a circumscribed circle: $R = \dfrac{b}{2\sin B}$

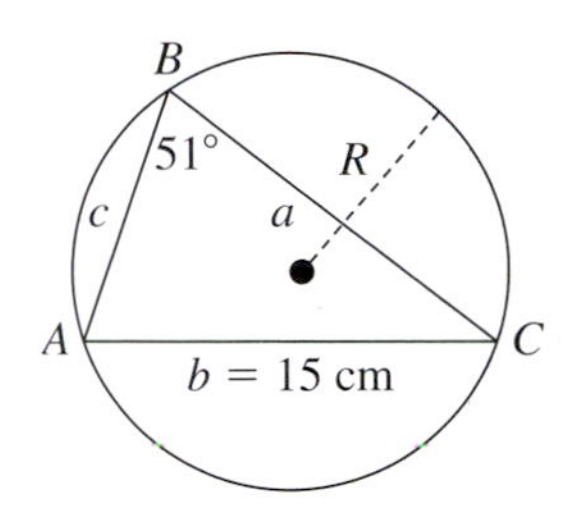

Given $\triangle ABC$ is circumscribed by a circle of radius R, the radius of the circle can be found using the formula shown, where side b is opposite angle B. Find the radius of the circle shown.

APPLICATIONS

29. Planetary distances: In a solar system that parallels our own, the planet Sorus can be seen from a Class M planet with the naked eye, but as the diagram indicates, the position of Sorus is uncertain. Assume the orbits of both planets are roughly circular and that the viewing angle θ is about 20°. If the Class M planet is 82 million miles from the Sun and Sorus is 51 million miles from the Sun, determine the closest and farthest possible distances that separate the planets in this alignment.

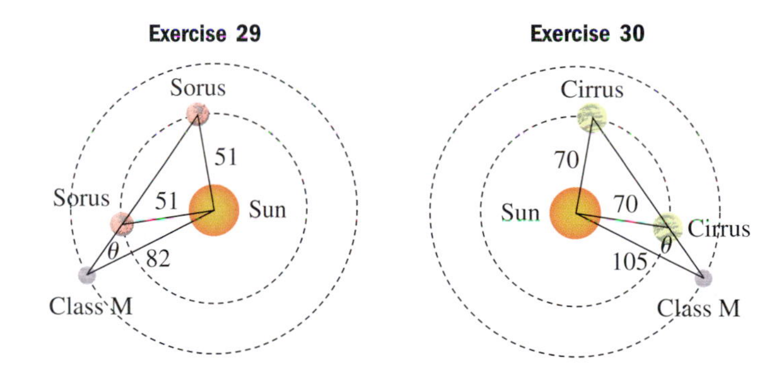

30. Planetary distances: In a solar system that parallels our own, the planet Cirrus can be seen from a Class M planet with the naked eye, but as the diagram indicates, the position of Cirrus is uncertain. Assume the orbits of both planets are roughly circular and that the viewing angle θ is about 15°. If the Class M planet is 105 million miles from the Sun and Cirrus is 70 million miles from the Sun, determine the closest and farthest possible distances that separate the planets in this alignment.

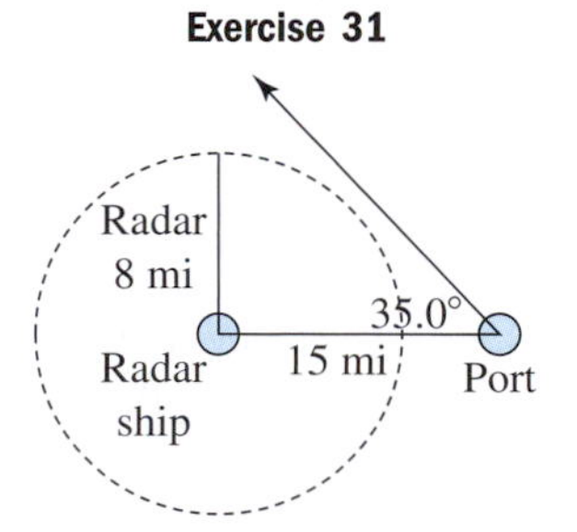

31. Radar detection: A radar ship is 15.0 mi off shore from a major port when a large fleet of ships leaves the port at the 35.0° angle shown. (a) If the maximum range of the ship's radar is 8.0 mi, will the departing fleet be detected? (b) If the maximum range of the ship's radar is 12 mi, how far from port is the fleet when it is first detected?

32. Radar detection: To notify environmentalists of the presence of big game, motion detectors are installed 200 yd from a watering hole. A pride of lions has just visited the hole and is leaving the area at the 29.0° angle shown. (a) If the maximum range of the motion detector is 90 yd, will the pride be detected? (b) If the maximum range of the motion detector is 120 yd, how far from the watering hole is the pride when first detected?

Exercise 32

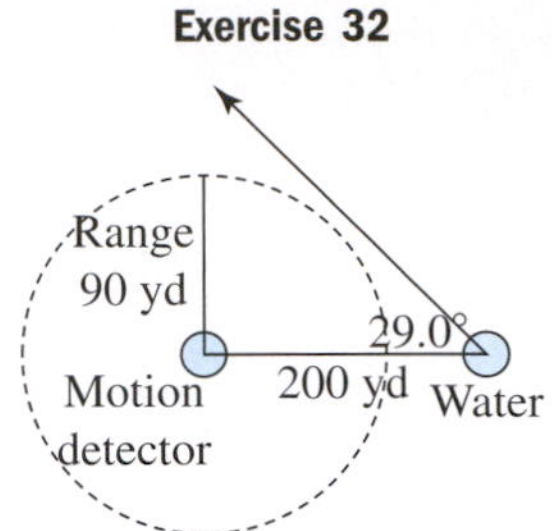

33. Distance between cities: The cities of ***V***an Gogh, ***R***embrandt, ***P***issarro, and ***S***eurat are situated as shown in the diagram. Assume that triangle *RSP* is isosceles and use the law of sines to find the distance between *V*an Gogh and *S*eurat, and between *V*an Gogh and *P*issarro.

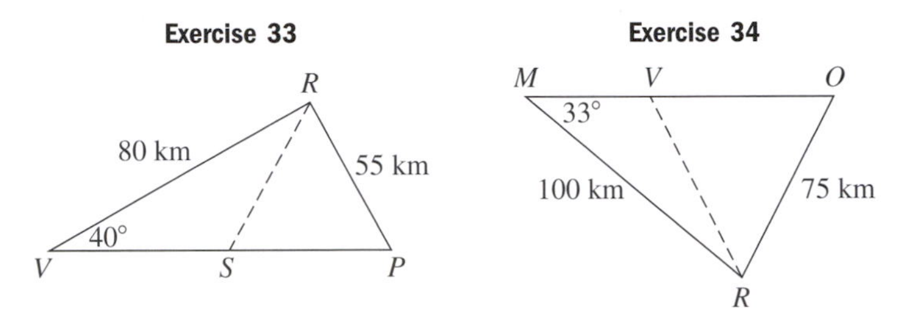

34. Distance between cities: The cities of ***M***ozart, ***R***ossini, ***O***ffenbach, and ***V***erdi are situated as shown in the diagram. Assume that triangle *ROV* is isosceles and use the law of sines to find the distance between *M*ozart and *V*erdi, and between *M*ozart and *O*ffenbach.

35. Distance to target: To practice for a competition, an archer stands as shown in the diagram and attempts to hit a moving target. (a) If the archer has a maximum effective range of about 180 ft, can the target be hit? (b) If the archer's range is 202 ft, how many "effective" shots can be taken? (c) If the archer's range is 215 ft and the target is moving at 10 ft/sec, how many seconds is the target within range?

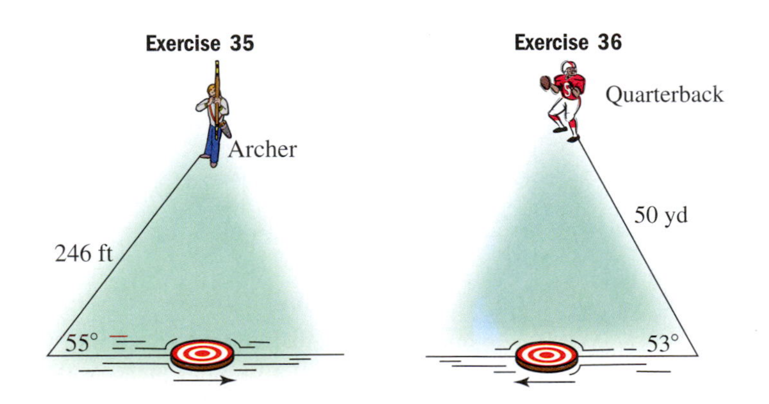

36. Distance to target: As part of an All-Star competition, a quarterback stands as shown in the diagram and attempts to hit a moving target with a football. (a) If the quarterback has a maximum effective range of about 35 yd, can the target be hit? (b) If the quarterback's range is 40 yd, how many "effective" throws can be made? (c) If the quarterback's range is 45 yd and the target is moving at 5 yd/sec, how many seconds is the target within range?

In Exercises 37 and 38, three rods are attached via pivot joints so the rods can be manipulated to form a triangle. How many triangles can be formed if angle B must measure 26°? If one triangle, solve it. If two, solve both. Diagrams are drawn to scale.

37. **38.**

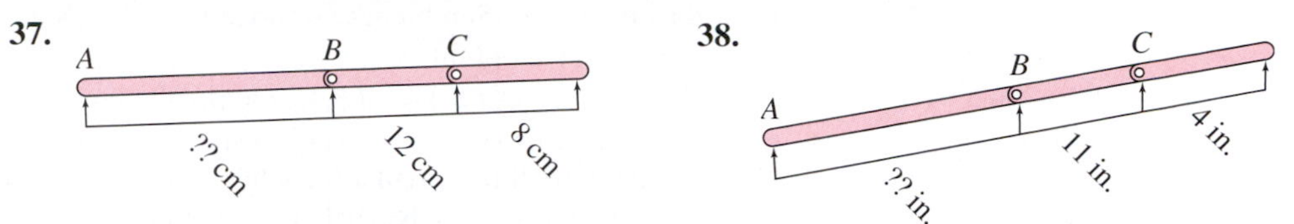

In the diagrams given, the measure of angle C and the length of sides a and c are fixed. Side c can be rotated at pivot point B. Solve any triangles that can be formed. (*Hint:* Begin by using the grid to find lengths a and c, then find angle C.)

39. **40.**

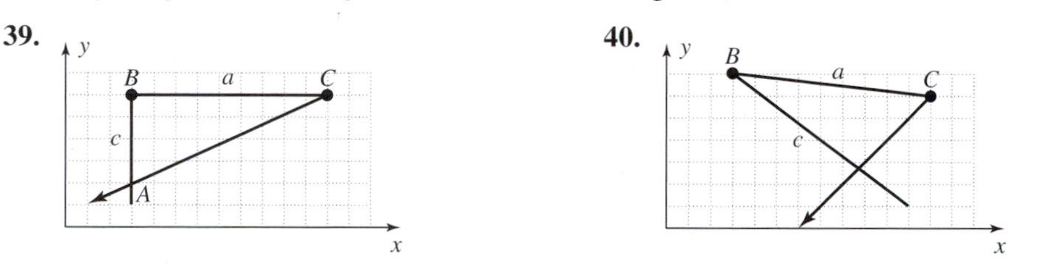

WRITING, RESEARCH, AND DECISION MAKING

41. Consider the triangle shown. Use a *table of values* or a *guess-and-check approach* to determine (a) the length of side a (rounded to the nearest whole) that yields one right triangle, (b) the lengths of side a that will yield two triangles, and (c) the lengths of side a that will yield one obtuse triangle. Use the information to create two new exercises that involve the ambiguous case.

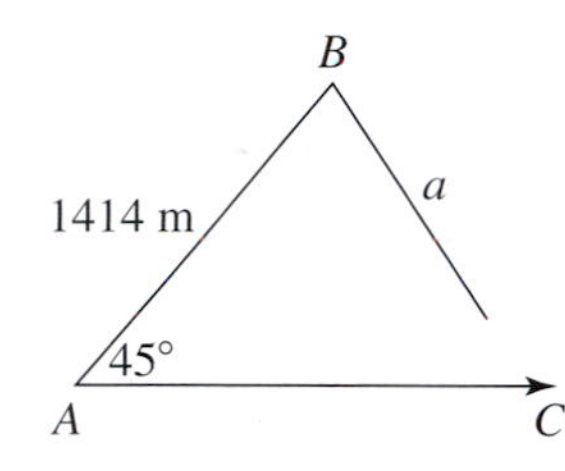

42. Hipparchus of Nicea (~135 B.C.), an ancient astronomer, is often credited with laying the foundations of modern trigonometry. By thorough research, determine the methods and ideas Hipparchus used to advance the study of trig, and include references and details to other early mathematicians who based their discoveries on Hipparchus' earlier work.

43. In studying case III of the ambiguous case, the statement is made that, "when two triangles are possible, the angles C_1 and C_2 are supplements." Discuss why the statement is true, noting that an isosceles triangle (base angles equal) is formed.

EXTENDING THE CONCEPT

44. Use the law of sines and any needed identities to solve the triangle shown.

45. Similar to the law of sines, there is a *law of tangents*. The law says for any triangle ABC, $\dfrac{a+b}{a-b} = \dfrac{\tan\left[\frac{1}{2}(A+B)\right]}{\tan\left[\frac{1}{2}(A-B)\right]}$. Use the law of tangents to solve the triangle shown.

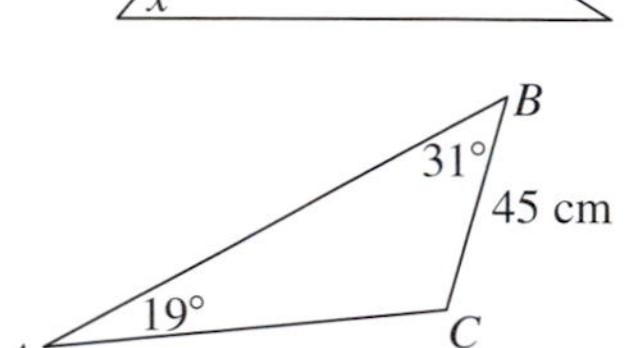

MAINTAINING YOUR SKILLS

46. (4.4) Solve the equation for $x \in \mathbb{R}$: $\sin(2x) = \cos x$

47. (5.1) If you connect the cities of Honolulu, Hawaii, San Francisco, California, and Tokyo, Japan, on a map, a triangle is formed. Find the distance from Honolulu to Tokyo, and from Honolulu to San Francisco, given that the angle at the Honolulu vertex is 108.4°, the angle at the San Francisco vertex is 45.5°, and the distance between Tokyo and San Francisco is 5142 mi.

48. (1.1) At the local hobby shop, a toy train is traveling around a circular track when to everyone's amazement a beautiful speckled butterfly finds its way indoors and alights on the caboose. If the circular track has a radius of 8 ft and the train makes one circuit every 4 sec, how fast is the butterfly traveling in miles per hour?

49. (2.3) Use a reference rectangle and the *rule of fourths* to graph $y = -2\sin\left(x + \frac{\pi}{4}\right)$ in $[0, 2\pi)$.

50. (3.2) Verify the following is an identity: $\sin^2\theta = \tan^2\theta - \frac{\sin^4\theta}{\cos^2\theta}$

51. (1.4) A central angle is drawn through a unit circle and a larger circle of unknown radius. Assume that (33, 56) is a point on the larger circle. What is the corresponding point on the unit circle?

5.3 The Law of Cosines

LEARNING OBJECTIVES

In Section 5.3 you will learn how to:

A. **Establish the law of cosines**

B. **Apply the law of cosines when two sides and an included angle are known (SAS)**

C. **Apply the law of cosines when three sides are known (SSS)**

D. **Solve applications using the law of cosines**

INTRODUCTION

The distance formula $d = \sqrt{(x_2 - x_1)^2 + (y_2 - y_1)^2}$ is traditionally developed by placing two arbitrary points on a coordinate grid and using the Pythagorean theorem. The relationship known as the *law of cosines* is developed in much the same way, but this time by placing *three* arbitrary points (the vertices of a triangle) on a coordinate grid. After giving the location of one vertex in trigonometric form, we obtain a formula that enables us to solve SSS and SAS triangles, which cannot be solved using the law of sines alone.

POINT OF INTEREST

While a study of trigonometry seems preoccupied with solving triangles, its beginnings and maturation are the result of an intense interest in a single applied science—astronomy. In fact, the two fields were not considered as separate entities until nearly the fourteenth century.

A. The Law of Cosines

In situations where all three sides are known (but no angles), the law of sines cannot be applied. The same is true when two sides and the angle between them are known, since we must have an angle opposite one of the sides in order to apply it. In these two cases (Figure 5.27), side-side-side (**SSS**) and side-angle-side (**SAS**), we use the **law of cosines.**

WORTHY OF NOTE

Keep in mind that the sum of any two sides of a triangle must be greater than the remaining side. For example, if $a = 7$, $B = 20$, and $C = 12$, no triangle is possible (see the figure).

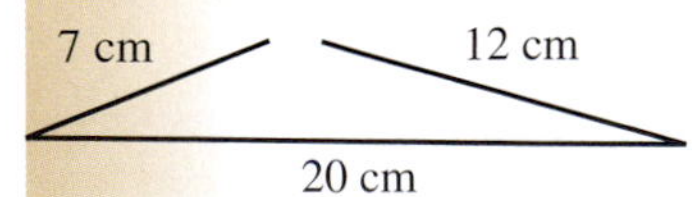

Figure 5.27
Law of Sines cannot be applied.

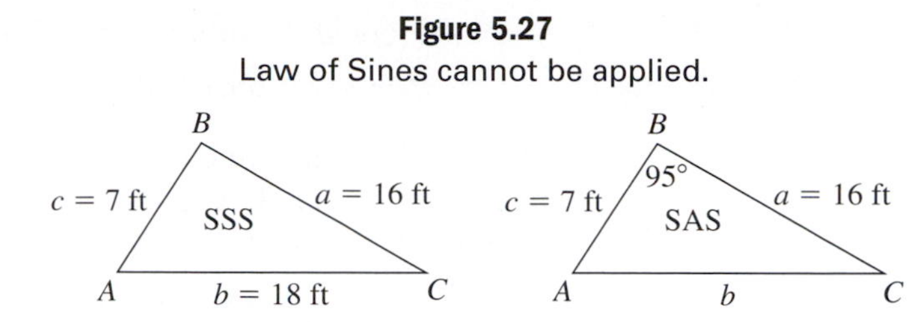

For each triangle, determine whether the law of sines could be used to begin the solution process. Justify your answers. Do not solve.

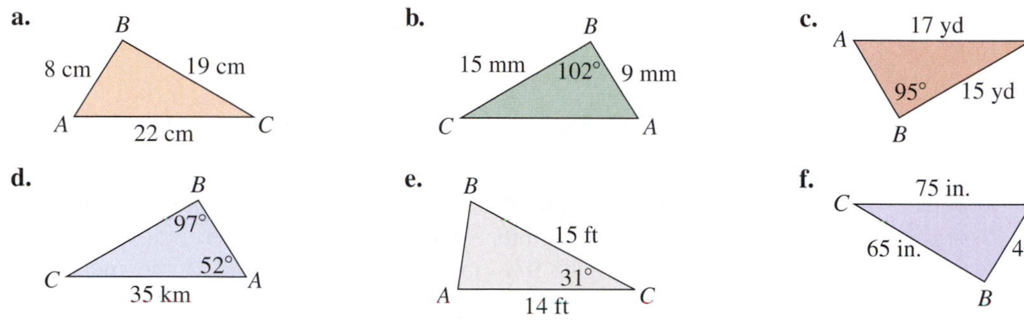

Solution:

a. The law of sines cannot be used since no angles are known.

b. The law of sines cannot be used since no angle is opposite a known side.

c. The law of sines can be used:

$$\frac{\sin 95°}{17} = \frac{\sin A}{15}$$

d. The law of sines can be used:

$$\frac{\sin 97°}{35} = \frac{\sin 52°}{a}$$

e. The law of sines cannot be used since no angle is opposite a known side.

f. The law of sines cannot be used since no angles are known.

NOW TRY EXERCISES 7 THROUGH 12

To solve the SSS and SAS cases, it is evident that we need additional insight on the unknown angles. For this purpose, we consider a general triangle ABC on the coordinate grid, conveniently placed with vertex A at the origin, side c along the x-axis, and the vertex C at some point (x, y) in QI (Figure 5.28). Note $\cos\theta = \frac{x}{b}$

Figure 5.28

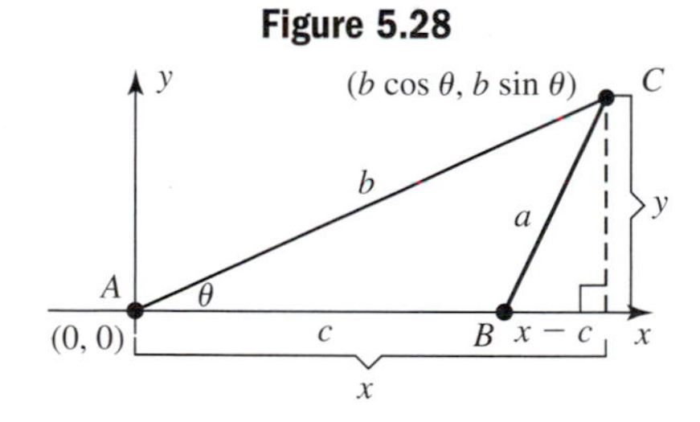

giving $x = b\cos\theta$, and $\sin\theta = \frac{y}{b}$ or $y = b\sin\theta$. This means we can write the point (x, y) as $(b\cos\theta, b\sin\theta)$ as shown, and use the Pythagorean theorem with the side $x - c$ to find the length of side a of the exterior, right triangle. It follows that

$$\begin{aligned}
a^2 &= (x - c)^2 + y^2 && \text{Pythagorean theorem} \\
&= (b\cos\theta - c)^2 + (b\sin\theta)^2 && \text{substitute } b\cos\theta \text{ for } x \text{ and } b\sin\theta \text{ for } y \\
&= b^2\cos^2\theta - 2bc\cos\theta + c^2 + b^2\sin^2\theta && \text{square binomial, square term} \\
&= b^2\cos^2\theta + b^2\sin^2\theta + c^2 - 2bc\cos\theta && \text{rearrange terms}
\end{aligned}$$

$$= b^2(\cos^2\theta + \sin^2\theta) + c^2 - 2bc\cos\theta \quad \text{factor out } b^2$$
$$= b^2 + c^2 - 2bc\cos\theta \quad \text{substitute 1 for } \cos^2\theta + \sin^2\theta$$

We now have a formula relating all three sides and an included angle. Since the naming of the angle is purely arbitrary, the formula can be used in any of the three forms shown here. For the derivation of the formula where $\angle B$ is acute, see Exercise 56.

THE LAW OF COSINES

For any triangle ABC and corresponding sides a, b, and c,

$$a^2 = b^2 + c^2 - 2bc\cos A$$
$$b^2 = a^2 + c^2 - 2ac\cos B$$
$$c^2 = a^2 + b^2 - 2ab\cos C$$

Note the relationship between the indicated angle and the squared term.

In words, the law of cosines says that the square of any side is equal to the sums of the squares of the other two sides, minus twice their product times the cosine of the included angle. It is interesting to note that if the included angle is 90°, the formula reduces to the Pythagorean theorem since $\cos 90° = 0$.

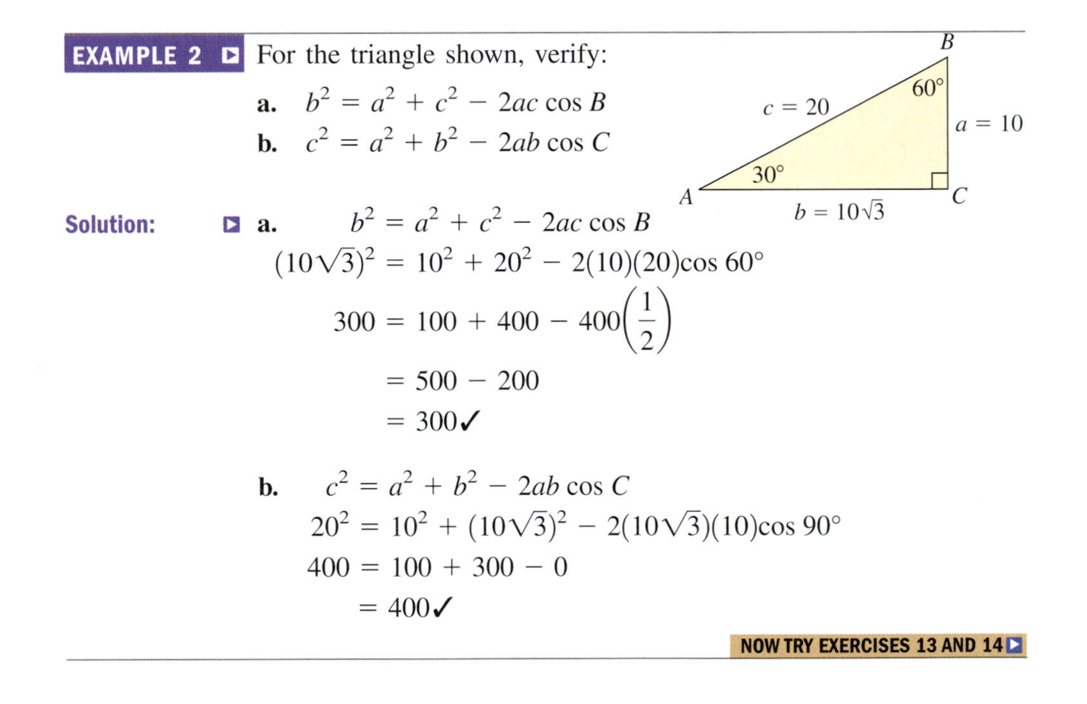

EXAMPLE 2 For the triangle shown, verify:

a. $b^2 = a^2 + c^2 - 2ac\cos B$

b. $c^2 = a^2 + b^2 - 2ab\cos C$

Solution:

a.
$$b^2 = a^2 + c^2 - 2ac\cos B$$
$$(10\sqrt{3})^2 = 10^2 + 20^2 - 2(10)(20)\cos 60°$$
$$300 = 100 + 400 - 400\left(\frac{1}{2}\right)$$
$$= 500 - 200$$
$$= 300 ✓$$

b.
$$c^2 = a^2 + b^2 - 2ab\cos C$$
$$20^2 = 10^2 + (10\sqrt{3})^2 - 2(10\sqrt{3})(10)\cos 90°$$
$$400 = 100 + 300 - 0$$
$$= 400 ✓$$

NOW TRY EXERCISES 13 AND 14

CAUTION

When evaluating the law of cosines, a common error is to combine the coefficient of $\cos\theta$ with the squared terms (the terms shown in blue): $a^2 = b^2 + c^2 - 2bc\cos A$. Be sure to use the correct order of operations when simplifying the expression.

B. The Law of Cosines and SAS

Once again, the law of cosines must be used in the SAS case, since we have no angle opposite a known side. However, once we gain any additional information about the triangle and have an angle and a side opposite, the law of sines can be used to complete the solution.

EXAMPLE 3 Solve the triangle shown. Write the solution in table form.

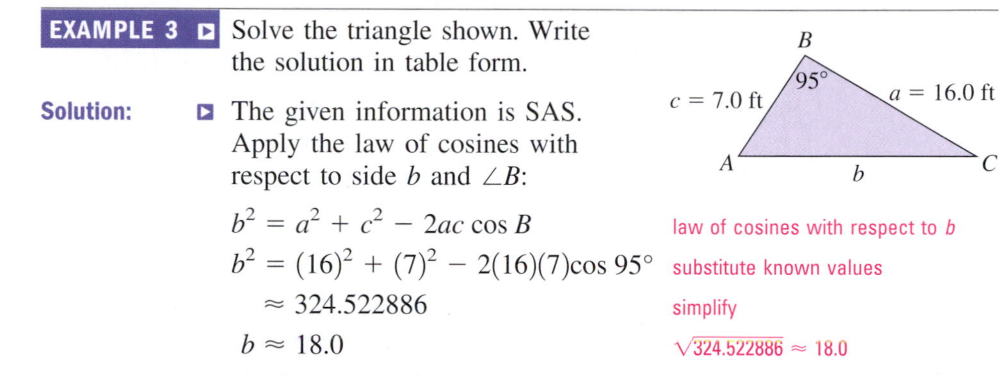

Solution: The given information is SAS. Apply the law of cosines with respect to side b and $\angle B$:

$$b^2 = a^2 + c^2 - 2ac\cos B \quad \text{law of cosines with respect to } b$$

$$b^2 = (16)^2 + (7)^2 - 2(16)(7)\cos 95° \quad \text{substitute known values}$$

$$\approx 324.522886 \quad \text{simplify}$$

$$b \approx 18.0 \quad \sqrt{324.522886} \approx 18.0$$

We now have side b opposite $\angle B$, and complete the solution using the law of sines, selecting the smaller angle to avoid the ambiguous case (we *could* apply the law of cosines again, if we chose).

$$\frac{\sin C}{c} = \frac{\sin B}{b} \quad \text{law of sines applied to } \angle C \text{ and } \angle B$$

$$\frac{\sin C}{7} \approx \frac{\sin 95°}{18} \quad \text{substitute given values}$$

$$\sin C \approx 7 \cdot \frac{\sin 95°}{18} \quad \text{solve for } \sin C$$

$$C \approx \sin^{-1}\left(\frac{7\sin 95°}{18}\right) \quad \text{apply } \sin^{-1}$$

$$\approx 22.8° \quad \text{result}$$

Last of all, find the remaining angle, $\angle C$: $180° - (95° + 22.8°) = 62.2°$. The finished solution is shown in the table (given information is in bold).

Angles	Sides (ft)
$A \approx 62.2°$	$\boldsymbol{a = 16.0}$
$\boldsymbol{B = 95.0°}$	$b \approx 18.0$
$C \approx 22.8°$	$\boldsymbol{c = 7.0}$

NOW TRY EXERCISES 15 THROUGH 26

WORTHY OF NOTE

After using the law of cosines, we often use the law of sines to complete a solution. With a little foresight, we can avoid the ambiguous case—since the ambiguous case occurs only if θ *could be* obtuse (the largest angle of the triangle). After calculating the third side of a SAS triangle using the law of cosines, use the law of sines to find the smallest angle, since it cannot be obtuse. For SSS triangles, using the law of cosines to find the largest angle will ensure that when the second angle is found using the law of sines, it cannot be obtuse.

C. The Law of Cosines and SSS

When three sides of a triangle are given, the law of cosines can be used to find any one of the three angles. It is a good practice to first find the *largest* angle, or the angle opposite the largest side. This will ensure that the remaining two angles are acute, avoiding the analysis involved if the law of sines is used to complete the solution (the ambiguous case).

EXAMPLE 4 Solve the triangle shown. Write the solution in table form, with angles rounded to tenths of a degree.

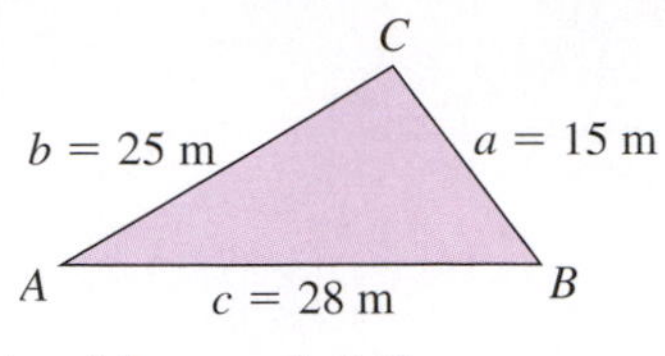

Solution: The information is given as SSS. Since side c is the longest side, we apply the law of cosines with respect to side c and $\angle C$:

$$c^2 = a^2 + b^2 - 2ab\cos C$$ law of cosines with respect to c

$$28^2 = (15)^2 + (25)^2 - 2(15)(25)\cos C$$ substitute known values

$$784 = 850 - 750\cos C$$ simplify

$$-66 = -750\cos C$$ isolate variable term

$$0.088 = \cos C$$ divide

$$\cos^{-1}0.088 = C$$ solve for C

$$85.0 \approx C$$ result

We now have side c opposite $\angle C$ and finish up using the law of sines, which yields $A \approx 32.3°$. Since the remaining angle must be acute, we compute it directly. $\angle B$: $180° - (85° + 32.3°) = 62.7°$. The finished solution is shown in the table, with the information originally given shown in bold.

Angles	Sides (m)
$A \approx 32.3°$	**a = 15**
$B \approx 62.7°$	**b = 25**
$C \approx 85°$	**c = 28**

NOW TRY EXERCISES 27 THROUGH 34

D. Applications Using the Law of Cosines

As with the law of sines, the law of cosines has a large number of applications from very diverse fields including geometry, navigation, surveying, and astronomy, as well as being put to use in solving recreational problems (see Exercises 37 through 40).

EXAMPLE 5 A volcanologist needs to measure the distance across the base of an active volcano. Distance AB is measured at 1.5 km, while distance AC is 3.2 km. Using a theodolite (a sighting instrument used by surveyors), angle BAC is found to be 95.7°. What is the distance across the base?

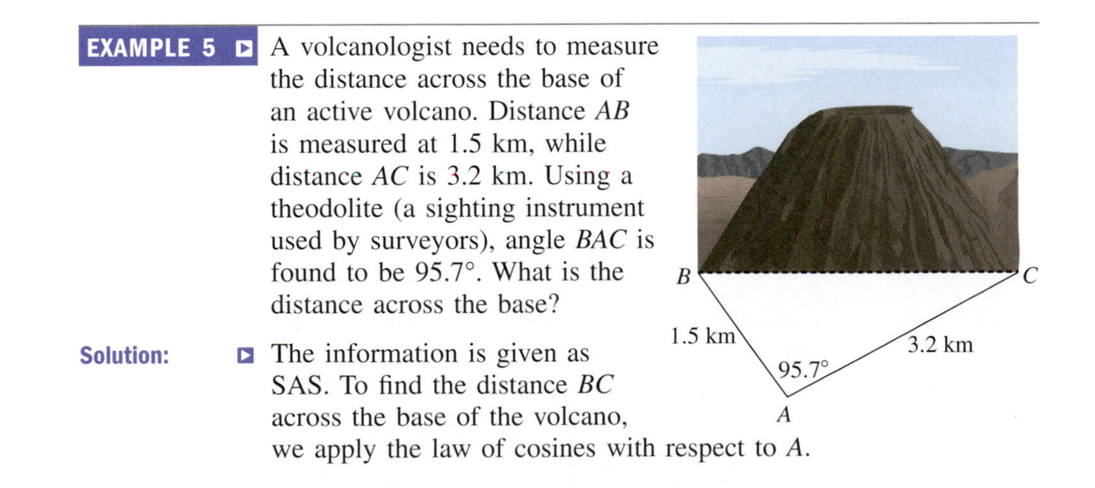

Solution: The information is given as SAS. To find the distance BC across the base of the volcano, we apply the law of cosines with respect to A.

$$a^2 = b^2 + c^2 - 2bc\cos A$$ law of cosines with respect to a

$$= (1.5)^2 + (3.2)^2 - 2(1.5)(3.2)\cos 95.7°$$ substitute known values

$$\approx 13.44347$$ simplify

$$a \approx 3.7$$ solve for a

The volcano is approximately 3.7 km wide at its base.

NOW TRY EXERCISES 41 THROUGH 44

A variety of additional applications can be found in the exercise set (see Exercises 45 through 52).

TECHNOLOGY HIGHLIGHT

A Simple Program for the Law of Cosines (SSS)

The keystrokes shown apply to a TI-84 Plus model. Please consult our Internet site or your manual for other models.

Our website at www.mhhe.com/coburn has a number of interesting and useful features. In particular, there are some *Technology Extensions*, which are simple programs that may be of use in your study of Algebra and Trig. Writing programs tends to offer great insight as to how the various parts of a formula are related, and how the formula itself operates. You may notice that we've given this program a *descriptive* name, but the TI-84 Plus limits us to eight characters: COSBYSSS (Cosines by SSS). The blank lines indicate that you should write the purpose and meaning of certain lines in the program, based on what is given elsewhere, or what you see as the output.

PROGRAM:COSBYSSS

Program line	Description
:ClrHome	Clears the home screen, places cursor in upper left position
:Disp "THIS PRGM SOLVES"	Displays *THIS PRGM SOLVES*
:Disp "SSS TRIANGLES"	Displays *SSS TRIANGLES*
:Disp "USING THE LAW"	Displays *USING THE LAW*
:Disp "OF COSINES"	Displays *OF COSINES*
:Pause:ClrHome	Pauses execution, allows user to view results until ENTER is pressed
:Disp "PLEASE ENTER"	______
:Disp "THE LONGEST"	______
:Disp "SIDE AS SIDE A"	______
:Pause	______
:Disp " "	Displays a blank line for formatting purposes
:Disp "THE 3 SIDES ARE:"	Displays *THE 3 SIDES ARE*
:Prompt A, B, C	Prompts user to input 3 lengths, stores them in A, B, and C
:ClrHome	______
:$\cos^{-1}((A^2-(B^2+C^2))/(-2BC)\to D$	Computes the value of $\angle A$ and stores it in memory location D
:Disp "ANGLE A IS:"	Displays *ANGLE A IS:*
:round(D,1)→D	Rounds the value stored in D to one decimal, stores the new value
:Disp "D"	Displays the value now stored in memory location D
:Pause	______

```
:sin⁻¹((B*sin(D))/20)→E
:Disp "ANGLE B IS:"
:round (E,1)→E
:Disp "E"
:Pause
:(180 – (A + B))→F
:Disp "ANGLE C IS:"
:round (F,1)→F
:Disp "F"
```

Exercise 1: Using this program as a guide, write a program that will solve SAS triangles using the law of cosines.

5.3 EXERCISES

CONCEPTS AND VOCABULARY

Fill in each blank with the appropriate word or phrase. Carefully reread the section if needed.

1. When the information given is SSS or SAS, the law of ________ is used to solve the triangle.

2. Fill in the blank so that the law of cosines is complete: $c^2 = a^2 + b^2 - \_\_\_\_ \cos C$

3. If the law of cosines is applied to a right triangle, the result is the same as the ______________ theorem, since $\cos 90° = 0$.

4. Write out which version of the law of cosines you would use to begin solving the triangle shown: ______________

5. Solve the triangle in Exercise 4 using only the law of cosines, then by using the law of cosines followed by the law of sines. Which method was more efficient?

6. Begin with $a^2 = b^2 + c^2 - 2bc \cos A$ and write $\cos A$ in terms of a, b, and c (solve for $\cos A$). Why must $c^2 + b^2 - a^2 < 2bc$ hold in order for a solution to exist?

DEVELOPING YOUR SKILLS

Determine whether the law of cosines can be used to begin the solution process for each triangle.

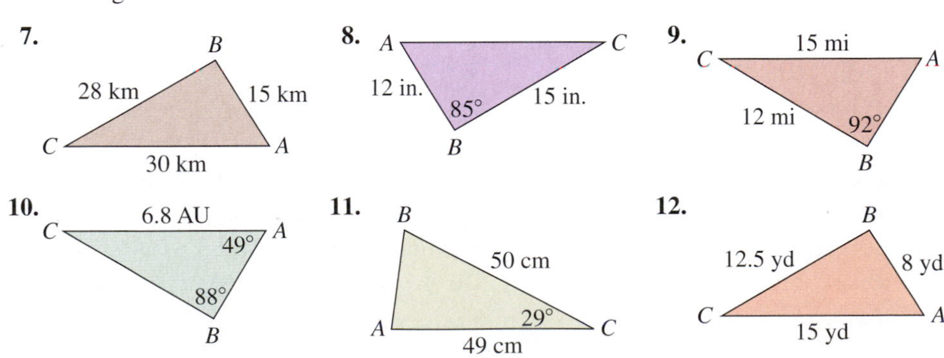

For each triangle, verify that all three forms of the law of cosines result in an equality.

13. **14.**

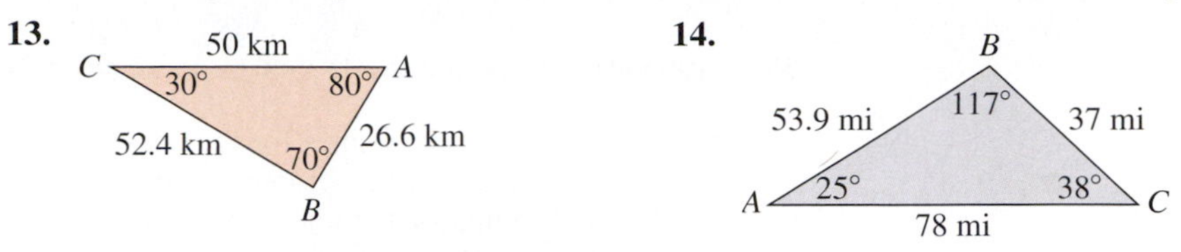

Solve each of the following equations for the unknown part.

15. $4^2 = 5^2 + 6^2 - 2(5)(6)\cos B$

16. $12.9^2 = 15.2^2 + 9.8^2 - 2(15.2)(9.8)\cos C$

17. $a^2 = 9^2 + 7^2 - 2(9)(7)\cos 52°$

18. $b^2 = 3.9^2 + 9.5^2 - 2(3.9)(9.5)\cos 30°$

19. $10^2 = 12^2 + 15^2 - 2(12)(15)\cos A$

20. $202^2 = 182^2 + 98^2 - 2(182)(98)\cos B$

Solve each triangle using the law of cosines.

21. side $a = 75$ cm
$\angle C = 38°$
side $b = 32$ cm

22. side $b = 385$ m
$\angle C = 67°$
side $a = 490$ m

23. side $c = 25.8$ mi
$\angle B = 30°$
side $a = 12.9$ mi

Solve using the law of cosines (if possible). Label each triangle appropriately before you begin.

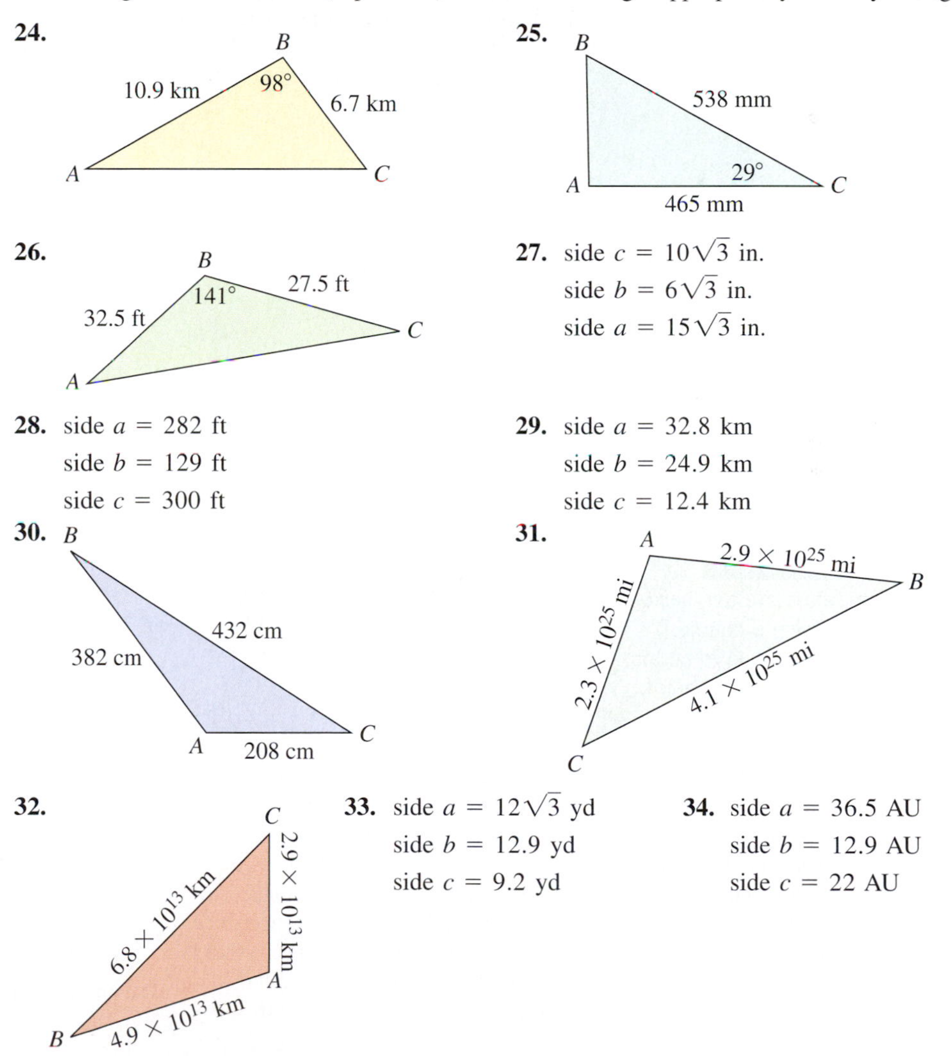

27. side $c = 10\sqrt{3}$ in.
side $b = 6\sqrt{3}$ in.
side $a = 15\sqrt{3}$ in.

28. side $a = 282$ ft
side $b = 129$ ft
side $c = 300$ ft

29. side $a = 32.8$ km
side $b = 24.9$ km
side $c = 12.4$ km

33. side $a = 12\sqrt{3}$ yd
side $b = 12.9$ yd
side $c = 9.2$ yd

34. side $a = 36.5$ AU
side $b = 12.9$ AU
side $c = 22$ AU

WORKING WITH FORMULAS

35. Alternative form for the law of cosines: $\cos A = \dfrac{b^2 + c^2 - a^2}{2bc}$

By solving the law of cosines for the cosine of the angle, the formula can be written as shown. Derive this formula (solve for $\cos\theta$), beginning from $a^2 = b^2 + c^2 - 2bc\cos A$, then use this form to begin the solution of the triangle given.

B, 52 m, 39 m, A, C, 37 m

36. Heron's formula for the area of a triangle: $A = \sqrt{p(p-a)(p-b)(p-c)}$

Heron of Alexandria was an ancient engineer and offered the first known proof that the area of a general triangle could be expressed solely in terms of its sides. The area is given by the formula shown above, where p represents the triangle's semiperimeter: $p = \dfrac{a+b+c}{2}$. Use Heron's formula to find the area of the triangle given in Exercise 35.

APPLICATIONS

37. Distance between cities: The satellite Mercury II measures its distance from Portland and from Green Bay using radio waves as shown. Using an on-board sighting device, the satellite determines that $\angle M$ is 99°. How many miles is it from Portland to Green Bay?

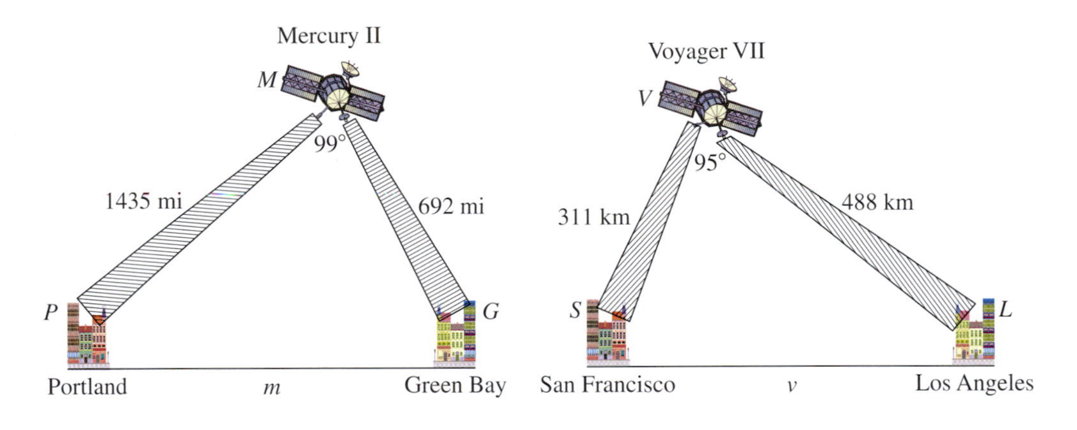

WORTHY OF NOTE

In navigation, there are two basic methods for defining a course. *Headings* are understood to be the amount of rotation from due north in the clockwise direction ($0 \le \theta < 360°$). *Bearings* give the number of degrees East or West from a due North or due South orientation, hence the angle indicated is always less than 90°. For instance, the bearing N 25° W and a heading of 335° would have the ship or plane traveling in the same direction.

38. Distance between cities: Voyager VII measures its distance from Los Angeles and from San Francisco using radio waves as shown. Using an on-board sighting device, the satellite determines $\angle V$ is 95°. Approximately how many kilometers separate Los Angeles and San Francisco?

39. Trip planning: A business executive is going to fly the corporate jet from Providence to College Cove. She calculates the distances shown using a map, with Mannerly Main for reference since it is due east of Providence. What is the measure of angle P? What heading should she set for this trip?

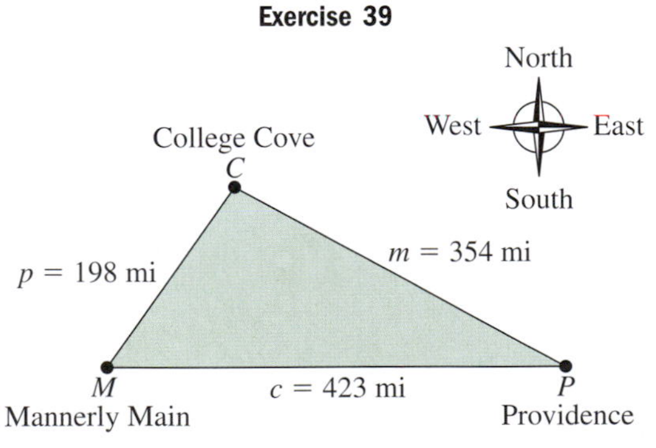

40. Trip planning: A troop of Scouts is planning a hike from ***M***ontgomery to ***P***attonville. They calculate the distances shown using a map, using ***B***radleyton for reference since it is due east of ***M***ontgomery. What is the measure of angle M? What heading should they set for this trip?

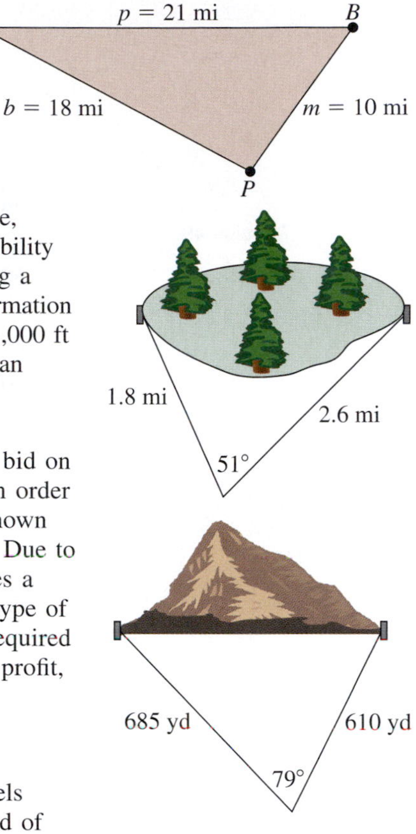

41. Runway length: Surveyors are measuring a large, marshy area outside of the city as part of a feasibility study for the construction of a new airport. Using a theodolite and the markers shown gives the information indicated. If the main runway must be at least 11,000 ft long, and environmental concerns are satisfied, can the airport be constructed at this site (recall that 1 mi = 5280 ft)?

42. Tunnel length: An engineering firm decides to bid on a proposed tunnel through Harvest Mountain. In order to find the tunnel's length, the measurements shown are taken. (a) How long will the tunnel be? (b) Due to previous tunneling experience, the firm estimates a cost of $5000 per foot for boring through this type of rock and constructing the tunnel according to required specifications. If management insists on a 25% profit, what will be their minimum bid to the nearest hundred?

43. Aerial distance: Two planes leave Los Angeles International Airport at the same time. One travels due west (at heading 270°) with a cruising speed of 450 mph, going to Tokyo, Japan, with a group that seeks tranquility at the foot of Mount Fuji. The other travels at heading 225° with a cruising speed of 425 mph, going to Brisbane, Australia, with a group seeking adventure in the Great Outback. Approximate the distance between the planes after 5 hr of flight.

44. Nautical distance: Two ships leave Honolulu Harbor at the same time. One travels 15 knots (nautical miles per hour) at heading 150°, and is going to the Marquesas Islands (*Crosby, Stills, and Nash*). The other travels 12 knots at heading 200°, and is going to the Samoan Islands (*Samoa, le galu a tu*). How far apart are the two ships after 10 hr?

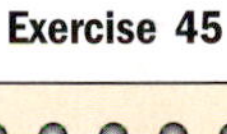

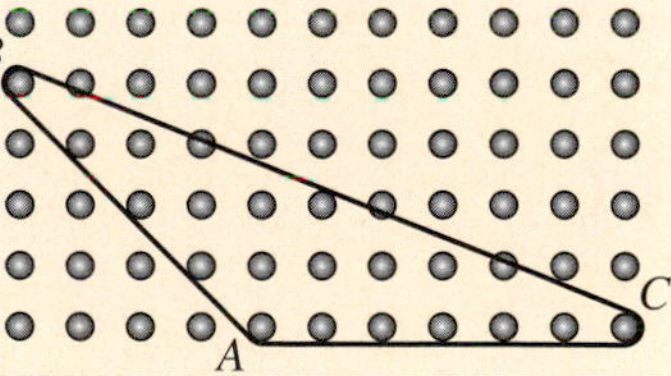

45. Geoboard geometry: A rubber band is placed on a geoboard (a board with all pegs 1 cm apart) as shown. Approximate the perimeter of the triangle formed by the rubber band *and* the angle formed at each vertex. (*Hint:* Use a standard triangle to find $\angle A$ and length $\overline{AB}$.)

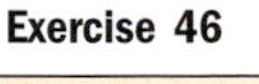

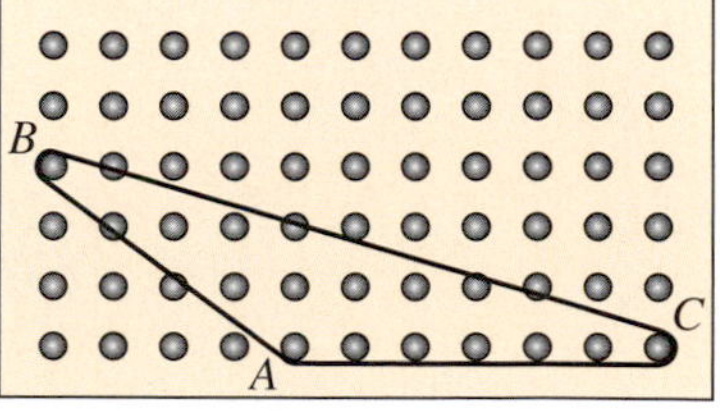

46. Geoboard geometry: A rubber band is placed on a geoboard as shown. Approximate the perimeter of the triangle formed by the rubber band *and* the angle formed at each vertex. (*Hint:* Use a Pythagorean triple, then find angle A.)

In Exercises 47 and 48, three rods are attached via pivot joints so the rods can be manipulated to form a triangle. Find the three angles of the triangle formed.

47. **48.**

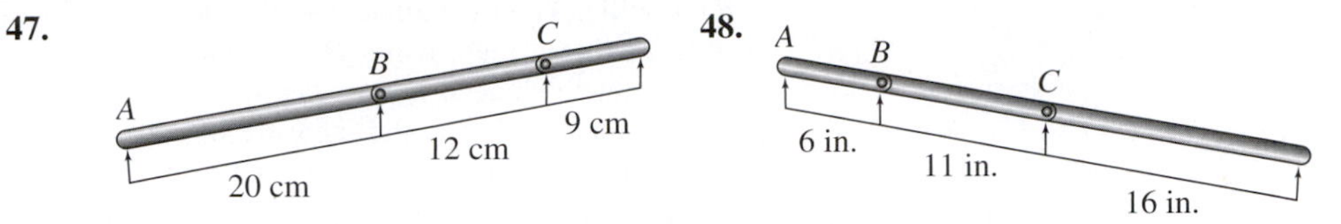

Exercise 49

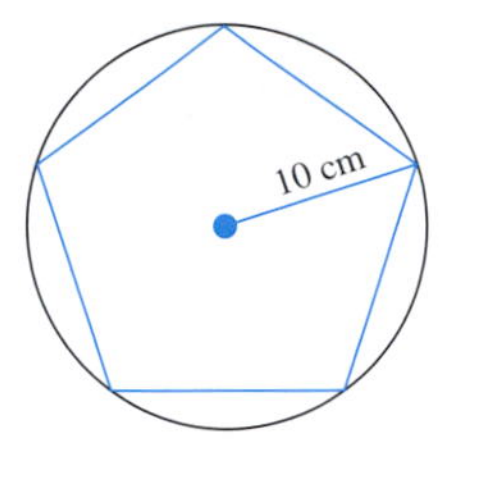

49. Pentagon perimeter: Find the perimeter of a regular *pentagon* that is circumscribed by a circle with radius $r = 10$ cm.

50. Hexagon perimeter: Find the perimeter of a regular *hexagon* that is circumscribed by a circle with radius $r = 15$ cm.

Solve the following triangles. Round sides and angles to the nearest tenth. (*Hint:* Use Pythagorean triples.)

51.

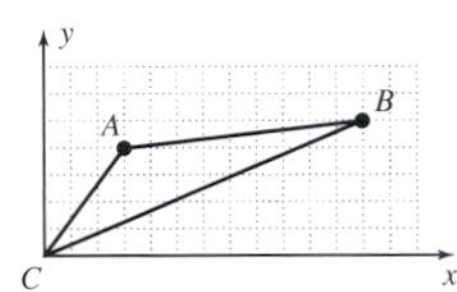

52.

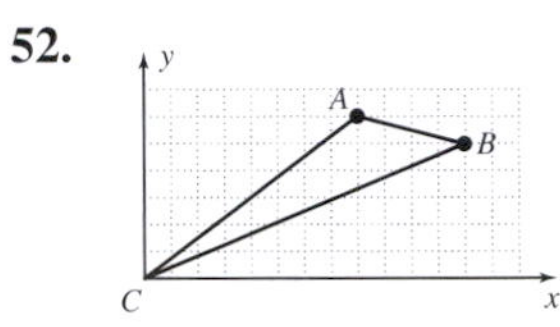

WRITING, RESEARCH, AND DECISION MAKING

53. Nicolas Copernicus (1473–1543), best known for insisting that the solar system was heliocentric (planets revolve around the Sun), also made substantial contributions to trigonometry in his work *De revolutionibus orbium coelestium* (1543). Do some research on Copernicus, determine what these contributions were, and discuss how they advanced the study of trigonometry.

54. Mathematics is viewed by many as a huge playground—full of beautiful wonders, intrigue, curiosities, and connections. Let's join the fun. Find the area of the right triangle shown in the figure using the standard geometric formula, the formula given in Section 5.1 $\left(A = \frac{1}{2}bc \sin \theta\right)$ and Heron's formula (Exercise 36). Verify that you obtain the same answer regardless of the method used, and comment as to which was most efficient.

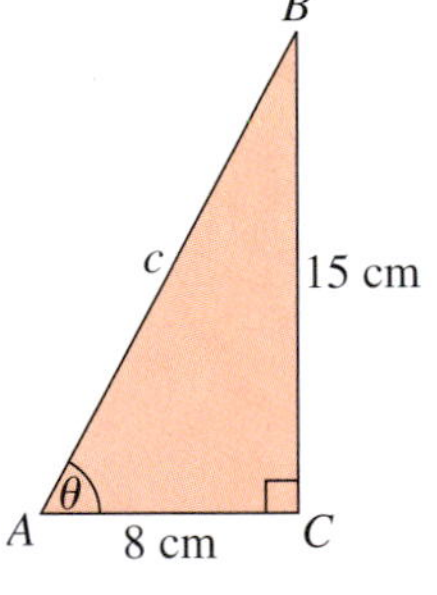

55. Find the area of the circumscribed pentagon and circumscribed hexagon given in Exercises 49 and 50.

56. For the diagram in Figure 5.28 (page 325), note that if the x-coordinate of vertex B is greater than the x-coordinate of vertex C, $\angle B$ becomes acute, and $\angle C$ obtuse. How does this change the relationship between x and c? Verify the law of cosines remains unchanged.

EXTENDING THE CONCEPT

57. No matter how hard I try, I cannot solve the following triangle. Why?

Exercise 57

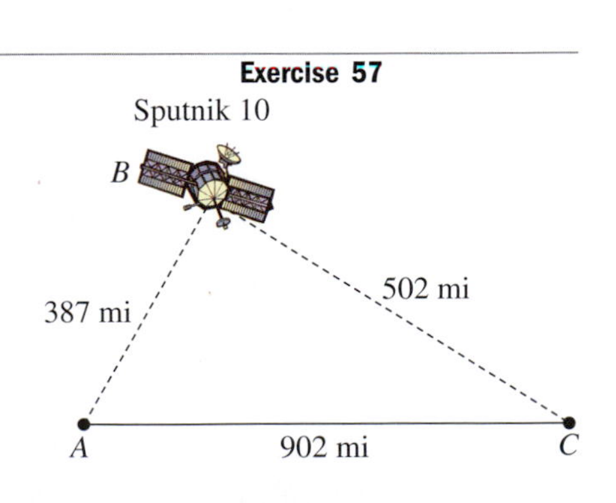

58. Most students are familiar with this double-angle formula for cosine: $\cos(2\theta) = \cos^2\theta - \sin^2\theta$. The *triple* angle formula for cosine is $\cos(3\theta) = 4\cos^3\theta - 3\cos\theta$. Use the formula to find an exact value for $\cos 135°$. Show that you get the same result as when using a reference angle.

Exercise 59

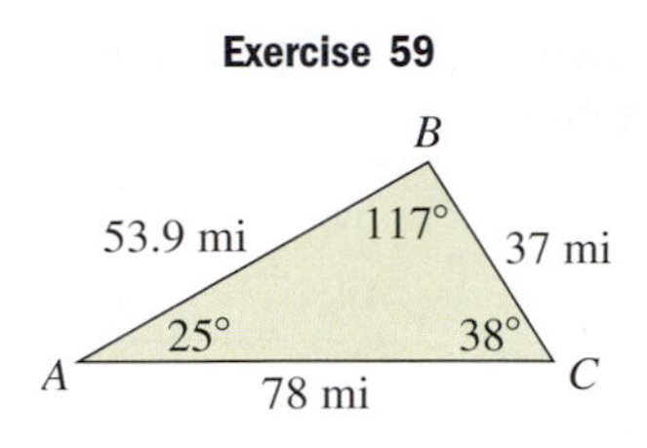

59. For the triangle shown, verify that $c = b\cos A + a\cos B$, then use two different forms of the law of cosines to show this relationship holds for *any* triangle ABC.

MAINTAINING YOUR SKILLS

60. (4.3) Find the primary solution to: $-3\sec\theta + 7\sqrt{3} = 5\sqrt{3}$.

61. (2.1) State exact forms for each of the following: $\sin\left(\frac{\pi}{6}\right)$, $\cos\left(\frac{7\pi}{6}\right)$, and $\tan\left(\frac{\pi}{3}\right)$.

62. (3.1) Derive the other two common versions of the Pythagorean identities, given $\sin^2x + \cos^2x = 1$.

63. (3.1) Use fundamental identities to find the values of all six trig functions that satisfy the conditions $\sin x = -\frac{5}{13}$ and $\cos x > 0$.

64. (3.3) Use a sum identity to find the value of $\sin 75°$ in exact form.

65. (4.2) Evaluate the expression by drawing a representative triangle: $\csc\left[\tan^{-1}\left(\frac{55}{48}\right)\right]$.

5.4 Vectors and Vector Diagrams

LEARNING OBJECTIVES

In Section 5.4 you will learn how to:

A. Represent a vector quantity geometrically

B. Represent a vector quantity graphically

C. Perform defined operations on vectors

D. Represent a vector quantity algebraically and find unit vectors

E. Use vector diagrams to solve applications

INTRODUCTION

The study of vectors is closely connected to the study of force, motion, velocity, and other related phenomena. Vectors enable us to quantify certain characteristics of these phenomena and to physically represent them with a simple model. To quantify something means we assign it a relative numeric value for purposes of study and comparison. While very uncomplicated, this model turns out to be a powerful mathematical tool.

POINT OF INTEREST

The word *vector* comes from the Latin verb *vehere,* which means "to carry." Related modern words include *vehicles,* which carry people or goods, and *convection,* which carries heat or current. Mathematically, a vector models a displacement from one point to another while carrying (illustrating) information regarding the magnitude and direction of the displacement.

A. The Notation and Geometry of Vectors

Measurements involving time, area, volume, energy, and temperature are called **scalar measurements** or **scalar quantities** because each can be adequately described by their magnitude alone and the appropriate unit or "scale." The related real number is simply called a **scalar.** Concepts that require more than a single quantity to describe their attributes are called **vector quantities.** Examples might include force, velocity, and displacement, which require knowing a magnitude *and* direction to describe them completely.

To begin our study, consider two identical airplanes flying at 300 mph, on a parallel course and in the same direction. Although we don't know how far apart they are, what direction they're flying, or if one is "ahead" of the other, we can still model, "300 mph on a parallel course," using **directed line segments** (Figure 5.29). Drawing these segments parallel with the arrowheads pointing the same way models the direction of flight, while drawing segments the *same length* indicates the velocities are equal. The directed segment used to represent a vector quantity is simply called a **vector.** In this case the length of the vector models the **magnitude** of the velocity, while the arrowhead indicates the **direction** of travel. The origin of the segment is called the **initial point,** with the arrowhead pointing to the **terminal point.** Both are labeled using capital letters as shown in Figure 5.30.

Figure 5.29

Figure 5.30

Figure 5.31

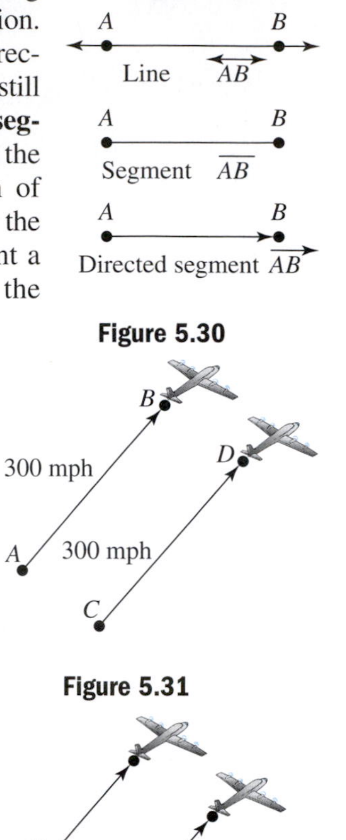

Vectors can be named using the initial and terminal points that define them (initial point first) as in $\overrightarrow{AB}$ and $\overrightarrow{CD}$, or using a bold, small case letter with the favorites being **v** (first letter of the word vector) and **u**. Other small case, bold letters can be used and subscripted vector names ($\mathbf{v}_1, \mathbf{v}_2, \mathbf{v}_3, \ldots$) are also common. Two **vectors are equal** if they have the same magnitude and direction. For $\mathbf{u} = \overrightarrow{AB}$ and $\mathbf{v} = \overrightarrow{CD}$, we can say $\mathbf{u} = \mathbf{v}$ or $\overrightarrow{AB} = \overrightarrow{CD}$ since both airplanes are flying at the same speed and in the same direction.

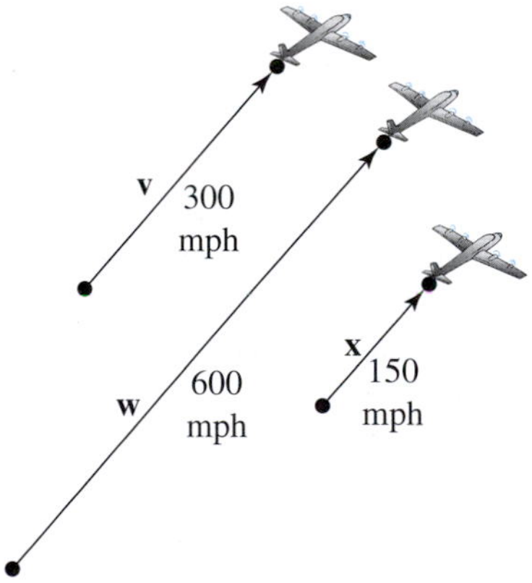

Based on these conventions, it seems reasonable to represent an airplane flying at 600 mph with a vector that is twice as long as **u** and **v,** and one flying at 150 mph with a vector that is half as long. If all planes are flying in the same direction on a parallel course, we can represent them geometrically as shown in Figure 5.31, and state that $\mathbf{w} = 2\mathbf{v}$, $\mathbf{x} = \frac{1}{2}\mathbf{v}$, and $\mathbf{w} = 4\mathbf{x}$. The multiplication of a vector by a constant is called **scalar multiplication,** since the product changes only the scale or size of the vector and not its direction.

Figure 5.32

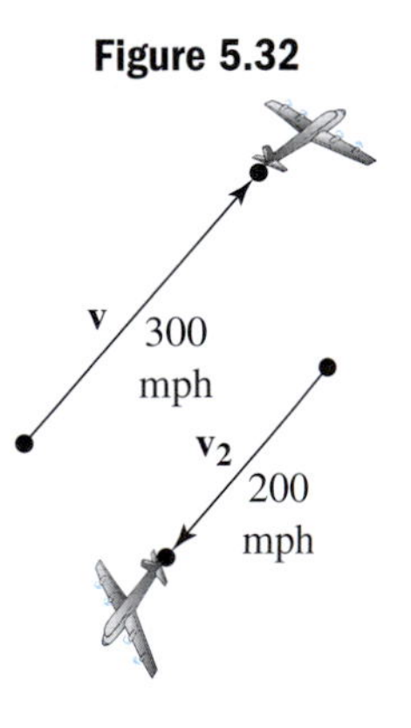

Finally, consider the airplane represented by vector $\mathbf{v}_2$, flying at 200 mph on a parallel course *but in the opposite direction* (see Figure 5.32). In this case, the directed segment will be $\frac{200}{300} = \frac{2}{3}$ as long as **v** and point in the opposite or "negative" direction. In perspective we can now state: $\mathbf{v}_2 = -\frac{2}{3}\mathbf{v}$, $\mathbf{v}_2 = -\frac{1}{3}\mathbf{w}$, $\mathbf{v}_2 = -\frac{4}{3}\mathbf{x}$, or any equivalent form of these equations.

EXAMPLE 1 Two tugboats are attempting to free a barge that is stuck on a sand bar. One is pulling with a force of 2000 newtons (N) in a certain direction, the other is pulling with a force of 1500 N in a direction that is *perpendicular to the first.* Represent the situation geometrically using vectors.

Solution: We could once again draw a vector of arbitrary length and let it represent the 2000-N force applied by the first tugboat. For better perspective, we can actually use a ruler and choose a convenient length, say 6 cm. We then represent the pulling force of the second tug with a vector that is $\frac{1500}{2000} = \frac{3}{4}$ as long (4.5 cm), drawn at a 90° angle with relation to the first. Note that many correct solutions are possible, depending on the direction of the first vector drawn.

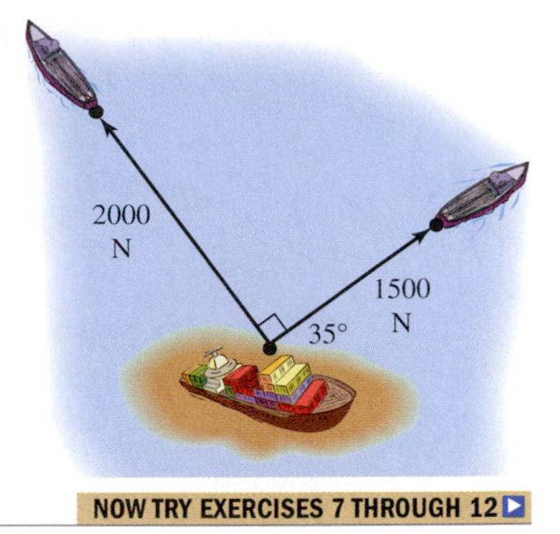

NOW TRY EXERCISES 7 THROUGH 12

B. Vectors and the Coordinate Grid

Representing vectors geometrically (with a directed line segment) is fine for simple comparisons, but many applications involve numerous vectors acting on a single point or changes in a vector quantity over time. For these situations, a graphical representation in the coordinate plane helps to analyze this interaction. The only question is *where* to place the vector on the grid, and the answer is—it really doesn't matter. Consider the three vectors shown in Figure 5.33. From the initial point of each, counting four units in the vertical direction, then three units in the horizontal direction, puts us at the terminal point. This shows the vectors are all 5 units long (since a 3-4-5 triangle is formed) and are all parallel (since slopes are equal: $\frac{\Delta y}{\Delta x} = \frac{4}{3}$). In other words, they are **equivalent vectors.**

Figure 5.33

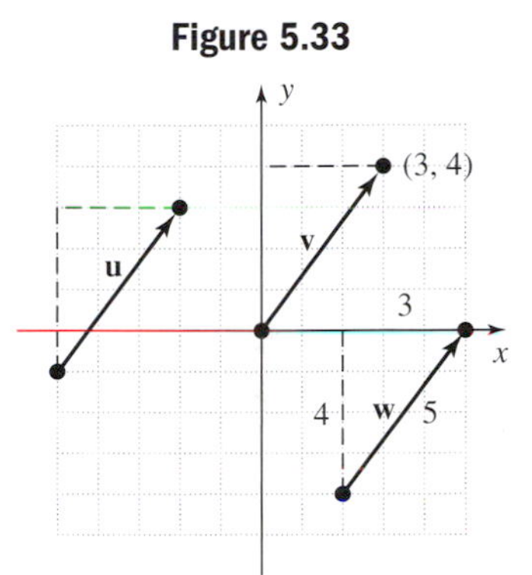

Since a vector's location is unimportant, we can replace any given vector with a unique and equivalent vector whose initial point is (0, 0), called the **position vector.**

WORTHY OF NOTE

For vector **u**, the initial and terminal points are $(-5, -1)$ and $(-2, 3)$, respectively, yielding the position vector $\langle -2 - (-5), 3 - (-1) \rangle = \langle 3, 4 \rangle$ as before.

POSITION VECTORS

For a vector **v** with initial point (x_1, y_1) and terminal point (x_2, y_2), the unique position vector for **v** is $\mathbf{v} = \langle x_2 - x_1, y_2 - y_1 \rangle$, an equivalent vector with initial point (0, 0) and terminal point $(x_2 - x_1, y_2, -y_1)$.

For instance, the initial and terminal points of vector **w** in Figure 5.33 have coordinates $(2, -4)$ and $(5, 0)$, respectively, with $(5 - 2, 0 - (-4)) = (3, 4)$. Since (3, 4) is also the terminal point of **v** (whose initial point is at the origin), **v** is the position vector for **u** and **w**. This observation also indicates that every geometric vector in the xy-plane corresponds to a unique ordered pair of real numbers (a, b), with a as the **horizontal component** and b as the **vertical component** of the vector. We actually denote the vector in **component form** as $\langle \boldsymbol{a}, \boldsymbol{b} \rangle$, using the new notation to prevent confusing vector $\langle a, b \rangle$ with the ordered pair (a, b). Finally, while each of the vectors in Figure 5.33 has a component form of $\langle 3, 4 \rangle$, the horizontal and vertical components can be read directly only from $\mathbf{v} = \langle 3, 4 \rangle$, giving it a distinct advantage.

EXAMPLE 2 Vector $\mathbf{v}_2 = \langle 12, -5 \rangle$ has initial point $(-4, 3)$. (a) Find the coordinates of the terminal point. (b) Verify the position vector for $\mathbf{v}_2$ is also $\langle 12, -5 \rangle$ and find its length.

Solution:

a. Since $\mathbf{v}_2$ has a horizontal component of 12 and a vertical component of -5, we add 12 to the x-coordinate and -5 to the y-coordinate of the initial point. This gives a terminal point of $(12 + (-4), -5 + 3) = (8, -2)$.

b. To verify we use the initial and terminal points to compute $\langle x_2 - x_1, y_2, -y_1 \rangle$, giving $\langle 8 - (-4), -2 - 3 \rangle = \langle 12, -5 \rangle$. To find its length we can use either the Pythagorean theorem or simply note that a 5-12-13 Pythagorean triple is formed. Vector $\mathbf{v}_1$ has length 13 units.

NOW TRY EXERCISES 13 THROUGH 20

For the remainder of this section, vector $\mathbf{v} = \langle a, b \rangle$ will refer to the unique position vector for all those equivalent to $\mathbf{v}$. Upon considering the graph of $\langle a, b \rangle$ (shown in QI for convenience in Figure 5.34), several things are immediately evident. The length or **magnitude** of the vector, which is denoted $|\mathbf{v}|$, can be determined using the Pythagorean theorem: $|\mathbf{v}| = \sqrt{a^2 + b^2}$. In addition, basic trigonometry shows the horizontal component can be found using $\cos\theta = \frac{a}{|\mathbf{v}|}$ or $a = |\mathbf{v}|\cos\theta$, with the vertical component being $\sin\theta = \frac{b}{|\mathbf{v}|}$ or $b = |\mathbf{v}|\sin\theta$. Finally, we note the angle θ can be determined using $\tan\theta = \left(\frac{b}{a}\right)$, or $\theta_r = \tan^{-1}\left(\frac{b}{a}\right)$ and the quadrant of $\boldsymbol{v}$.

Figure 5.34

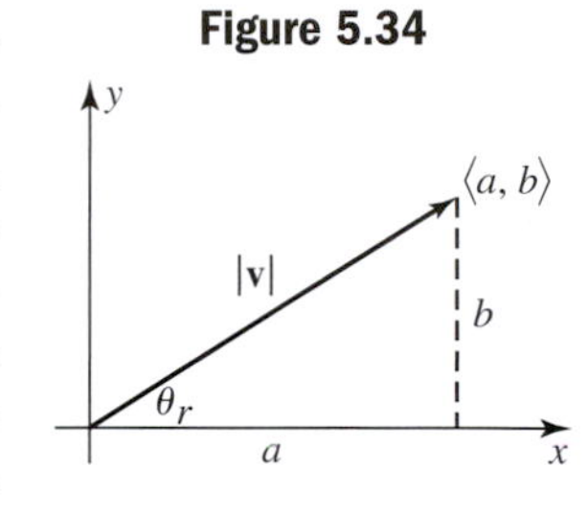

VECTOR COMPONENTS IN TRIG FORM

For a position vector $\mathbf{v} = \langle a, b \rangle$ and angle θ,

$$a = |\mathbf{v}|\cos\theta \qquad b = |\mathbf{v}|\sin\theta$$

where

$$\theta_r = \tan^{-1}\left(\frac{b}{a}\right) \text{ and}$$

$$|\mathbf{v}| = \sqrt{a^2 + b^2}$$

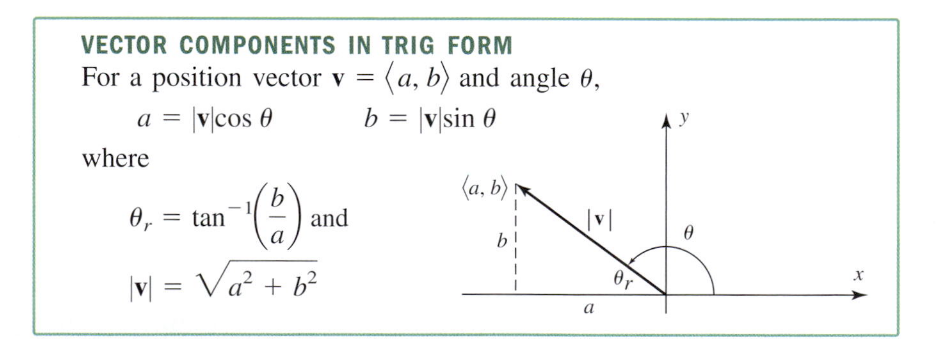

The ability to model characteristics of a vector using these equations is a huge benefit to solving applications, since we must often work out solutions using only the partial information given.

EXAMPLE 3 For $\mathbf{v}_1 = \langle -2.5, -6 \rangle$ and $\mathbf{v}_2 = \langle 3\sqrt{3}, 3 \rangle$, (a) graph each vector and name the quadrant, (b) find their magnitudes, and (c) find the angle θ (round to tenths of a degree as needed).

Solution: **a.** The graphs of $\mathbf{v}_1$ and $\mathbf{v}_2$ are shown in the figure. Using the signs of each coordinate, we note that $\mathbf{v}_1$ is in QIII, and $\mathbf{v}_2$ is in QI.

b.
$$|\mathbf{v}_1| = \sqrt{(-2.5)^2 + (-6)^2} = \sqrt{6.25 + 36} = \sqrt{42.25} = 6.5$$
$$|\mathbf{v}_2| = \sqrt{(3\sqrt{3})^2 + (3)^2} = \sqrt{27 + 9} = \sqrt{36} = 6$$

c. For $\mathbf{v}_1$: $\theta_r = \tan^{-1}\left(\frac{-6}{-2.5}\right) = \tan^{-1}(2.4) \approx 67.4°$

In QIII, $\theta \approx 247.4°$.

For $\mathbf{v}_2$: $\theta_r = \tan^{-1}\left(\frac{3}{3\sqrt{3}}\right) = \tan^{-1}\left(\frac{\sqrt{3}}{3}\right) = 30°$

In QI, $\theta = 30°$.

NOW TRY EXERCISES 21 THROUGH 24

EXAMPLE 4 The vector $\mathbf{v} = \langle a, b \rangle$ is in QIII, has a magnitude of $|\mathbf{v}| = 21$, and forms an angle of 25° with the negative x-axis. Find the horizontal and vertical components of the vector, rounded to one decimal place.

Solution: Begin by graphing the vector and setting up the equations for its components.

For the horizontal component:
$$a = |\mathbf{v}|\cos\theta = 21\cos 205° \approx -19$$

For the vertical component:
$$b = |\mathbf{v}|\sin\theta = 21\sin 205° \approx -8.9$$

With $\mathbf{v}$ in QIII, its component form is approximately $\langle -19, -8.9 \rangle$. As a check, we apply the Pythagorean theorem: $\sqrt{(-19)^2 + (-8.9)^2} \approx 21$ ✓.

NOW TRY EXERCISES 25 THROUGH 30

C. Operations on Vectors and Vector Properties

The operations defined for vectors have a close knit graphical representation. Consider a local park having a large pond with pathways around both sides, so that a park visitor can enjoy the view from either side. Suppose $\mathbf{v} = \langle 8, 2 \rangle$ is the position vector representing a person who decides to turn to the right at the pond, while $\mathbf{u} = \langle 2, 6 \rangle$ represents a person who decides to first turn left. At (8, 2) the first person changes direction and walks to (10, 8) on the other side of the pond, while the second person arrives at (2, 6) and turns to head for (10, 8) as well. This is shown graphically in Figure 5.35 and demonstrates that (1) a parallelogram is formed (opposite sides equal and parallel), (2) the path taken is unimportant relative to the destination, and (3) the coordinates of the destination represent the *sum of corresponding coordinates* from the terminal points of **u** and **v**: $(2, 6) + (8, 2) = (2 + 8, 6 + 2) = (10, 8)$. In other words, the result of adding **u** and **v** gives the new position vector $\mathbf{u} + \mathbf{v} = \mathbf{w}$, called the **resultant** or the **resultant vector.** Geometrically or graphically, the addition of vectors can be viewed as a "tail-to-tip" combination of one with another, by shifting one vector (without changing its direction) so that its tail is at the tip of the other vector. This is illustrated in Figures 5.36 through 5.38.

Figure 5.35

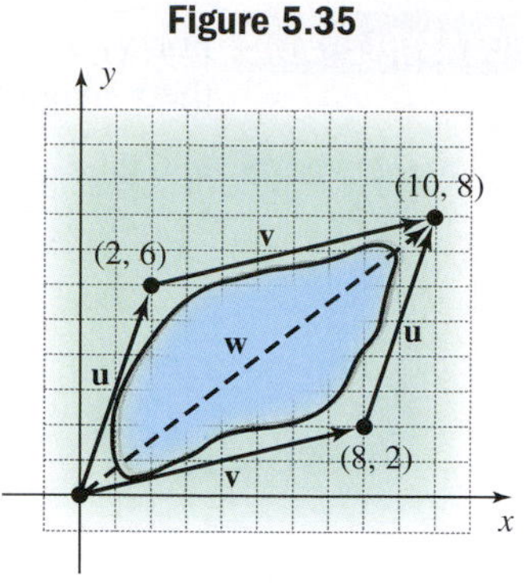

Figure 5.36 Given vectors **u** and **v**

Figure 5.37 Shift vector **v**

Figure 5.38 Shift vector **u**

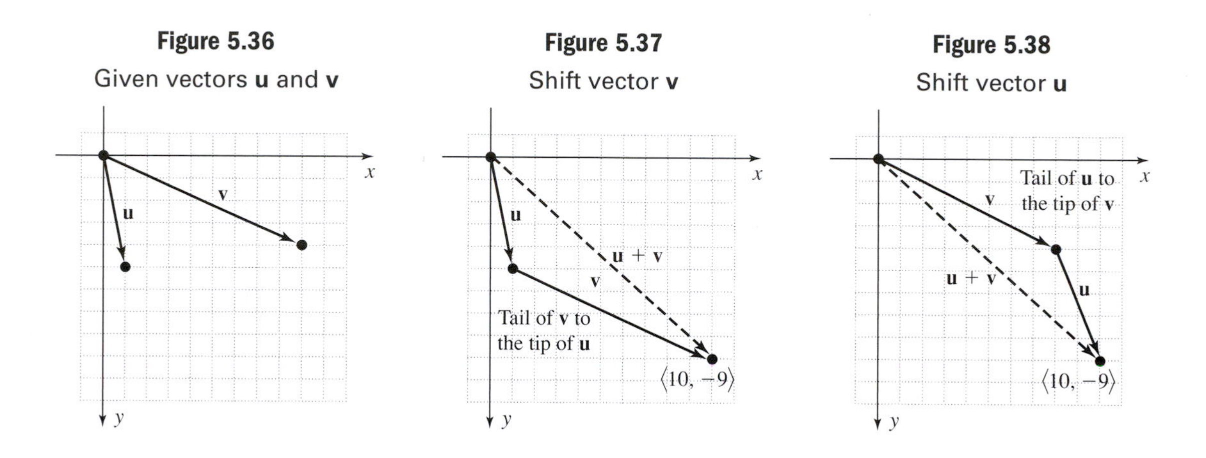

The subtraction of vectors is likewise defined. Scalar multiplication of vectors also has a graphical representation that corresponds to the geometric description given earlier.

OPERATIONS ON VECTORS

For vectors $\mathbf{u} = \langle a, b\rangle$, $\mathbf{v} = \langle c, d\rangle$, and a scalar k,

$$\mathbf{u} + \mathbf{v} = \langle a + c, b + d\rangle \qquad \mathbf{u} - \mathbf{v} = \langle a - c, b - d\rangle$$

$$k\mathbf{u} = \langle ka, kb\rangle \text{ for } k \in \mathbb{R}$$

If $k > 0$, the new vector points in the same direction as **u**.
If $k < 0$, the new vector points in the opposite direction as **u**.

EXAMPLE 5 Given $\mathbf{u} = \langle -3, -2\rangle$ and $\mathbf{v} = \langle 4, -6\rangle$ compute each of the following and represent the result graphically: (a) $-2\mathbf{u}$, (b) $\frac{1}{2}\mathbf{v}$, and (c) $-2\mathbf{u} + \frac{1}{2}\mathbf{v}$. Note the relationship between part (c) and parts (a) and (b).

Solution: a. $-2\mathbf{u} = -2\langle -3, -2\rangle$
$= \langle 6, 4\rangle$

b. $\frac{1}{2}\mathbf{v} = \frac{1}{2}\langle 4, -6\rangle$
$= \langle 2, -3\rangle$

c. $-2\mathbf{u} + \frac{1}{2}\mathbf{v} = \langle 6, 4\rangle + \langle 2, -3\rangle$
$= \langle 8, 1\rangle$

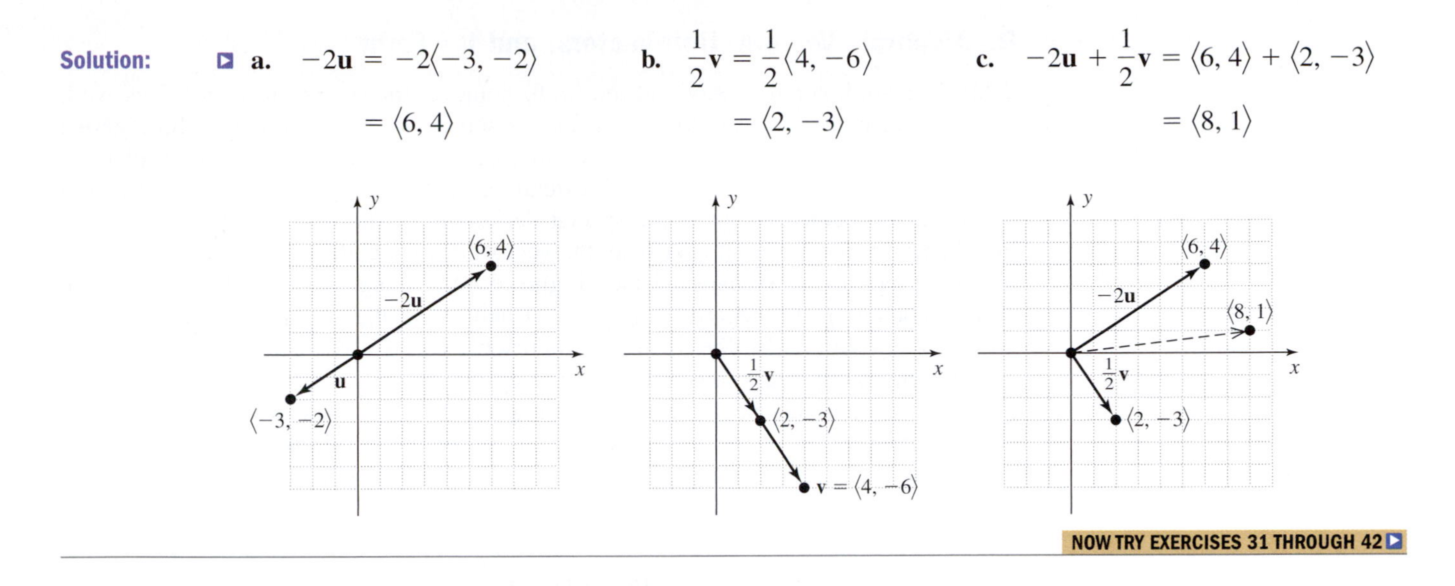

NOW TRY EXERCISES 31 THROUGH 42

The properties that guide operations on vectors closely resemble the familiar properties of real numbers. Note we define the zero vector $\mathbf{0} = \langle 0, 0\rangle$ as one having no magnitude or direction.

PROPERTIES OF VECTORS

For vector quantities $\mathbf{u}$, $\mathbf{v}$, and $\mathbf{w}$ and real numbers c and k,

1. $1\mathbf{u} = \mathbf{u}$	2. $0\mathbf{u} = \mathbf{0} = k\mathbf{0}$
3. $\mathbf{u} + \mathbf{v} = \mathbf{v} + \mathbf{u}$	4. $\mathbf{u} - \mathbf{v} = \mathbf{u} + (-\mathbf{v})$
5. $(\mathbf{u} + \mathbf{v}) + \mathbf{w} = \mathbf{u} + (\mathbf{v} + \mathbf{w})$	6. $(ck)\mathbf{u} = c(k\mathbf{u}) = k(c\mathbf{u})$
7. $\mathbf{u} + \mathbf{0} = \mathbf{u}$	8. $\mathbf{u} + (-\mathbf{u}) = \mathbf{0}$
9. $k(\mathbf{u} + \mathbf{v}) = k\mathbf{u} + k\mathbf{v}$	10. $(c + k)\mathbf{u} = c\mathbf{u} + k\mathbf{u}$

PROOF OF PROPERTY 3

For $\mathbf{u} = \langle a, b\rangle$ and $\mathbf{v} = \langle c, d\rangle$, we have

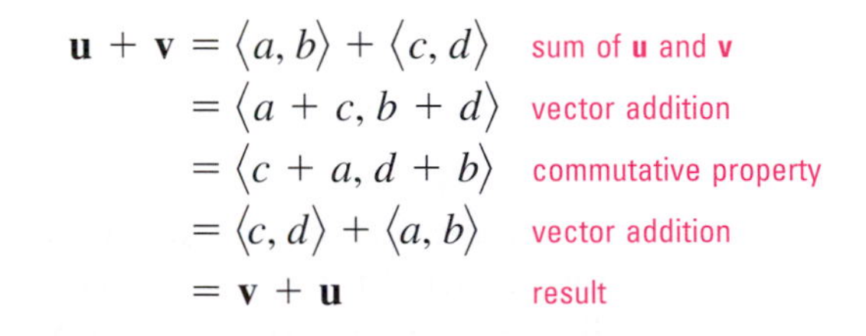

$\mathbf{u} + \mathbf{v} = \langle a, b\rangle + \langle c, d\rangle$ — sum of **u** and **v**
$= \langle a + c, b + d\rangle$ — vector addition
$= \langle c + a, d + b\rangle$ — commutative property
$= \langle c, d\rangle + \langle a, b\rangle$ — vector addition
$= \mathbf{v} + \mathbf{u}$ — result

Proofs of the other properties are similarly derived (see Exercises 89 through 97). Property 4 shows the subtraction of vectors is viewed as adding the opposite of the second vector. Graphically, this indicates we can use the "tail-to-tip" addition seen earlier, with the shifted vector pointing in the opposite direction. See Exercises 43 through 48.

D. Algebraic Vectors, Unit Vectors, and i, j Form

While the bold, small case **v** and the $\langle a, b\rangle$ notation for vectors has served us well, we now introduce an alternative form that is somewhat better suited to the **algebra of vectors,** and is used extensively in some of the physical sciences. Consider the vector $\langle 1, 0\rangle$, a vector 1 unit in length extending along the x-axis. It is called the **horizontal unit vector** and given the special designation **i** (not to be confused with the imaginary unit $i = \sqrt{-1}$). Likewise, the vector $\langle 0, 1\rangle$ is called the **vertical unit vector** and given the designation **j** (see Figure 5.39). Using scalar multiplication, the unit vector along the negative x-axis is $-\mathbf{i}$ and along the negative y-axis is $-\mathbf{j}$. Similarly, the vector $4\mathbf{i}$ represents a position vector 4 units long along the x-axis, and $-5\mathbf{j}$ represents a position vector 5 units long along the negative y-axis. Using these conventions, any nonquadrantal vector $\langle a, b\rangle$ can be written as a **linear combination** of **i** and **j**, with a and b expressed as multiples of **i** and **j**, respectively. These ideas can easily be generalized and applied to any vector.

Figure 5.39

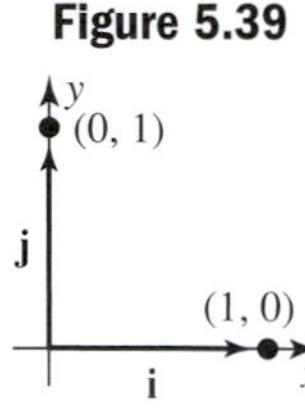

ALGEBRAIC VECTORS AND i, j FORM

For the unit vectors $\mathbf{i} = \langle 1, 0\rangle$ and $\mathbf{j} = \langle 0, 1\rangle$, any arbitrary vector $\mathbf{v} = \langle a, b\rangle$ can be written as a linear combination of **i** and **j**:

$$\mathbf{v} = a\mathbf{i} + b\mathbf{j}$$

Graphically, **v** is being expressed as the resultant of a vector sum.

WORTHY OF NOTE

Earlier we stated, "Two vectors were equal if they have the same magnitude and direction." Note that this means two vectors are equal if *their components are equal.*

EXAMPLE 6 Vector **u** is in QII, has a magnitude of 15, and makes an angle of 20° with the negative x-axis. (a) Graph the vector, (b) find the horizontal and vertical components (round to one decimal place) then write **u** in component form, and (c) write **u** in terms of **i** and **j**.

Solution:

a. The vector is graphed in the figure.

b.

Horizontal Component	Vertical Component
$a = \lvert\mathbf{v}\rvert\cos\theta$	$b = \lvert\mathbf{v}\rvert\sin\theta$
$= 15\cos 160°$	$= 15\sin 160°$
≈ -14.1	≈ 5.1

With the vector in QII, $\mathbf{u} = \langle -14.1, 5.1\rangle$ in component form.

c. In terms of **i** and **j** we have $\mathbf{u} = -14.1\mathbf{i} + 5.1\mathbf{j}$.

NOW TRY EXERCISES 49 THROUGH 62

Figure 5.40

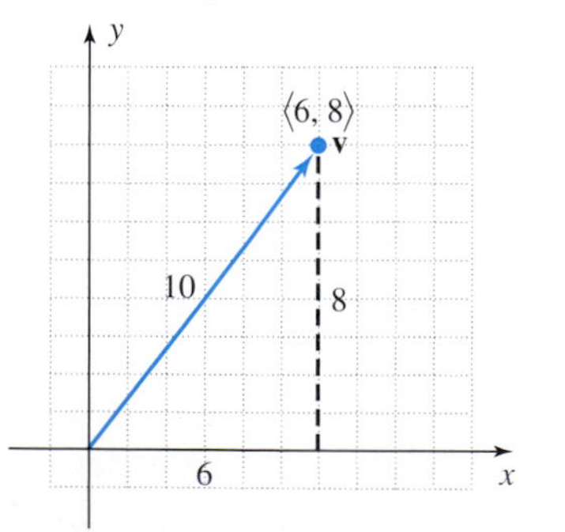

Some applications require that we find a nonhorizontal, nonvertical vector one unit in length, having the same direction as a given vector **v**. To understand how this is done, consider vector $\mathbf{v} = \langle 6, 8\rangle$. Using the Pythagorean theorem we find $\lvert\mathbf{v}\rvert = 10$, and can form a 6-8-10 triangle using the horizontal and vertical components (Figure 5.40). Knowing that similar triangles have sides that are proportional, we can find a unit vector in the same direction as **v** by dividing all three sides by 10, giving a triangle with sides $\frac{3}{5}$, $\frac{4}{5}$, and 1. The new vector "**u**" (along the hypotenuse) indeed points in the same direction since we have merely shortened **v**, and is a unit vector since $\left(\frac{3}{5}\right)^2 + \left(\frac{4}{5}\right)^2 \rightarrow \frac{9}{25} + \frac{16}{25} = 1$. In retrospect, we have divided the components of vector **v** by its magnitude $|\mathbf{v}|$ (or multiplied components by the reciprocal of $|\mathbf{v}|$) to obtain the desired unit vector: $\frac{\mathbf{v}}{|\mathbf{v}|} = \frac{\langle 6, 8\rangle}{10} = \left\langle \frac{6}{10}, \frac{8}{10}\right\rangle = \left\langle \frac{3}{5}, \frac{4}{5}\right\rangle$. In general we have the following:

UNIT VECTORS

For any nonzero vector $\mathbf{v} = \langle a, b\rangle = a\mathbf{i} + b\mathbf{j}$, the vector $\mathbf{u} = \frac{\mathbf{v}}{|\mathbf{v}|}$ is a unit vector in the same direction as **v**.

You are asked to verify this relationship in Exercise 101. For vector $\mathbf{w} = 8\mathbf{i} + 15\mathbf{j}$, we find $|\mathbf{w}| = \sqrt{8^2 + 15^2} = 17$, so the unit vector pointing in the same direction is $\frac{\mathbf{w}}{|\mathbf{w}|} = \frac{8}{17}\mathbf{i} + \frac{15}{17}\mathbf{j}$. See Exercises 63 through 74.

EXAMPLE 7 Vectors **u** and **v** form the 37° angle illustrated in the figure. Find the vector **w** (in red), which points in the same direction as **v** and forms the base of the right triangle shown.

Solution: Using the Pythagorean theorem we find $|\mathbf{u}| \approx 7.3$ and $|\mathbf{v}| = 10$. Using the cosine of 37° the magnitude of **w** is then $|\mathbf{w}| \approx 7.3\cos 37°$ or about 5.8. To ensure that **w** will point in the same direction as **v**, we simply multiply the 5.8 magnitude by the unit vector for **v**: $|\mathbf{w}|\frac{\mathbf{v}}{|\mathbf{v}|} \approx (5.8)\frac{\langle 8, 6\rangle}{10} = (5.8)\langle 0.8, 0.6\rangle \approx \langle 4.6, 3.5\rangle$. The components of **w** are approximately $\langle 4.6, 3.5\rangle$, which can again be checked using the Pythagorean theorem: $\sqrt{4.6^2 + 3.5^2} = \sqrt{33.41} \approx 5.8$.

NOW TRY EXERCISES 75 THROUGH 78

E. Vector Diagrams and Vector Applications

The applications of vectors are unlimited and are used extensively in all of the applied sciences. Here we'll look at two applications that are an extension of our work in this section. In Section 5.5 we'll see how vectors can be applied in a number of other creative and useful ways.

In Example 1, two tugboats were pulling on a barge to dislodge it from a sand bar, with the pulling force exerted by each represented by a vector. Using our knowledge of vector components, vector addition, and **resultant forces** (a force exerted along the resultant), we can now determine the direction and magnitude of the resultant force if we know the angle formed by one of the vector forces and the barge.

EXAMPLE 8 Two tugboats are attempting to free a barge that is stuck on a sand bar, and are exerting the forces shown in Figure 5.41. Find the direction and magnitude of the resultant force.

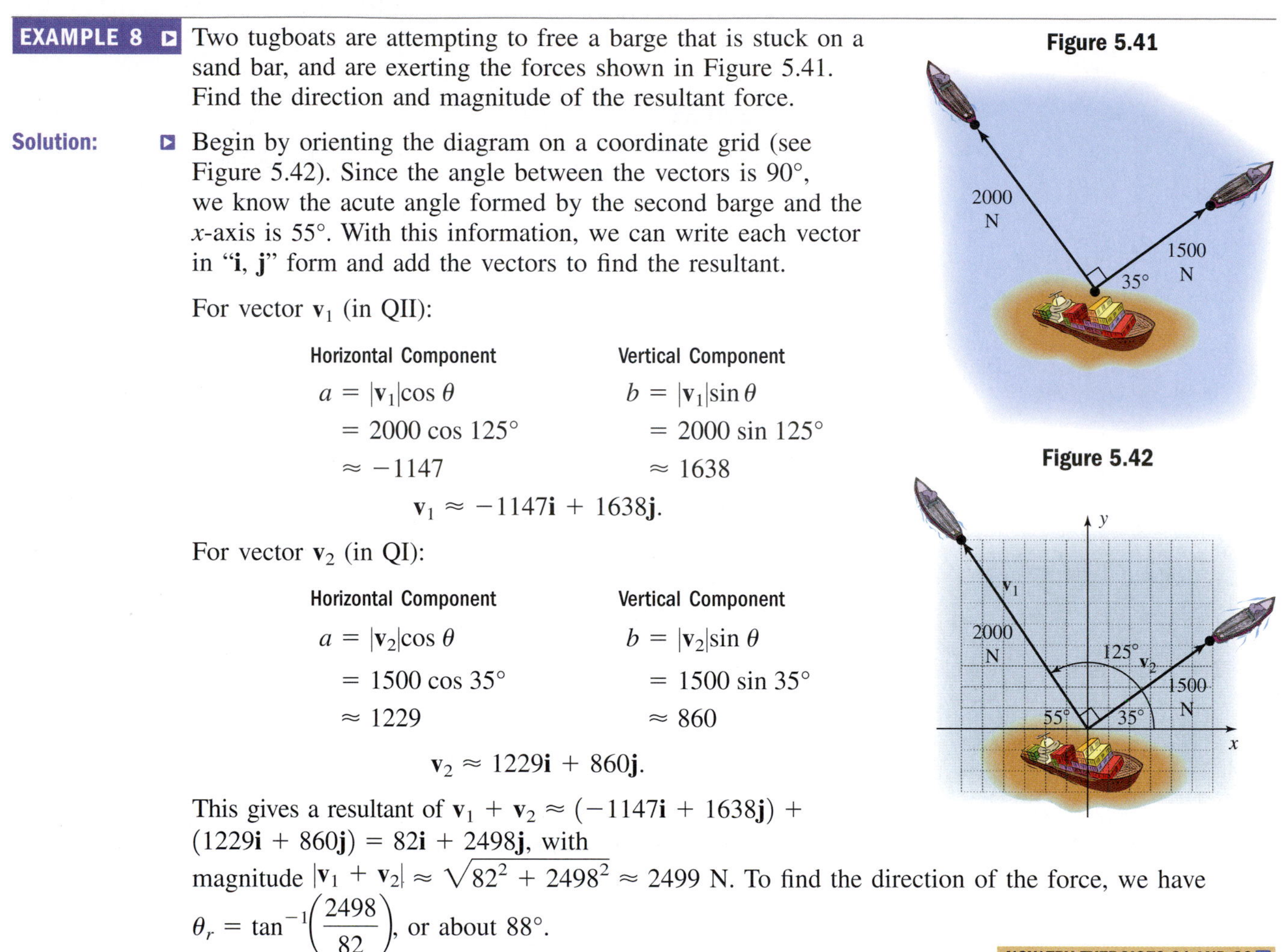

Figure 5.41

Figure 5.42

Solution: Begin by orienting the diagram on a coordinate grid (see Figure 5.42). Since the angle between the vectors is 90°, we know the acute angle formed by the second barge and the x-axis is 55°. With this information, we can write each vector in "**i**, **j**" form and add the vectors to find the resultant.

For vector $\mathbf{v}_1$ (in QII):

Horizontal Component	Vertical Component
$a = \lvert\mathbf{v}_1\rvert\cos\theta$	$b = \lvert\mathbf{v}_1\rvert\sin\theta$
$= 2000\cos 125°$	$= 2000\sin 125°$
≈ -1147	≈ 1638

$$\mathbf{v}_1 \approx -1147\mathbf{i} + 1638\mathbf{j}.$$

For vector $\mathbf{v}_2$ (in QI):

Horizontal Component	Vertical Component
$a = \lvert\mathbf{v}_2\rvert\cos\theta$	$b = \lvert\mathbf{v}_2\rvert\sin\theta$
$= 1500\cos 35°$	$= 1500\sin 35°$
≈ 1229	≈ 860

$$\mathbf{v}_2 \approx 1229\mathbf{i} + 860\mathbf{j}.$$

This gives a resultant of $\mathbf{v}_1 + \mathbf{v}_2 \approx (-1147\mathbf{i} + 1638\mathbf{j}) + (1229\mathbf{i} + 860\mathbf{j}) = 82\mathbf{i} + 2498\mathbf{j}$, with magnitude $\lvert\mathbf{v}_1 + \mathbf{v}_2\rvert \approx \sqrt{82^2 + 2498^2} \approx 2499$ N. To find the direction of the force, we have $\theta_r = \tan^{-1}\left(\frac{2498}{82}\right)$, or about 88°.

NOW TRY EXERCISES 81 AND 82

It's worth noting that a single tugboat pulling at 88° with a force of 2499 N would have the same effect as the two tugs in the original diagram. In other words, a resultant vector of $82\mathbf{i} + 2498\mathbf{j}$ truly represents the "result" of the two original forces.

Knowing that the location of a vector is unimportant enables us to model and solve a great number of seemingly unrelated applications. Although the final example concerns aviation, **headings,** and crosswinds, the solution process has a striking similarity to the "tugboat" example just discussed. As mentioned earlier in the text, "headings" involve a single angle, which is understood to be the amount of rotation from due north in the clockwise direction. For review, several headings are illustrated in Figures 5.43 through 5.46.

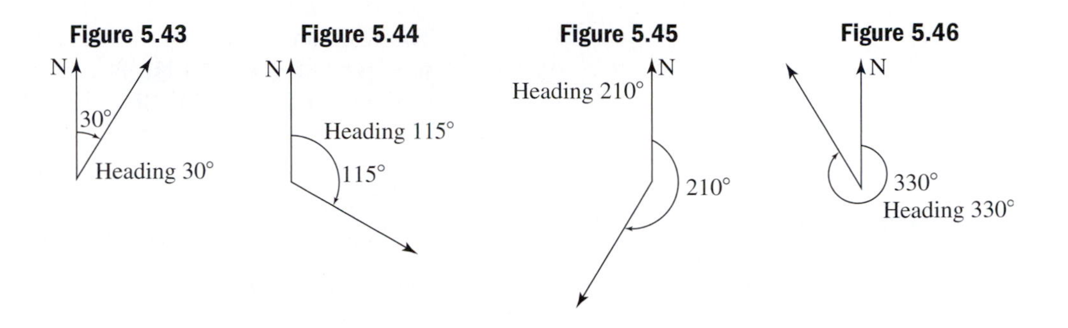

In order to keep an airplane on course, the captain must consider the direction and speed of any wind currents, since the plane's true course (relative to the ground) will be affected. Both the plane and the wind can be represented by vectors, with the plane's true course being the resultant vector.

EXAMPLE 9 An airplane is flying at 240 mph, heading 75°, when it suddenly encounters a strong, 60 mph wind blowing *from* the southwest, heading 10°. What is the actual course and speed of the plane (relative to the ground) as it flies through this wind?

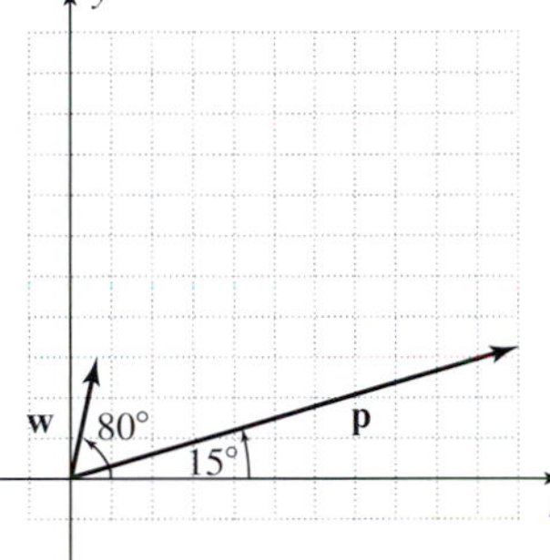

Figure 5.47

Solution: Begin by drawing a vector **p** to represent the speed and direction of the airplane (Figure 5.47). Since the heading is 75°, the angle between the vector and the x-axis must be 15°. For convenience (and because location is unimportant) we draw it as a position vector. Note the vector **w** representing the wind will be $\frac{60}{240} = \frac{1}{4}$ as long, and can also be drawn as a position vector—with an acute 80° angle. To find the resultant, we first find the components of each vector, then add. For vector **w** (in QI):

Horizontal Component	Vertical Component
$a = \lvert\mathbf{w}\rvert\cos\theta$	$b = \lvert\mathbf{w}\rvert\sin\theta$
$= 60\cos 80°$	$= 60\sin 80°$
≈ 10.4	≈ 59.1

$$\mathbf{w} \approx 10.4\mathbf{i} + 59.1\mathbf{j}.$$

For vector **p** (in QI):

Horizontal Component	Vertical Component
$a = \lvert\mathbf{p}\rvert\cos\theta$	$b = \lvert\mathbf{p}\rvert\sin\theta$
$= 240\cos 15°$	$= 240\sin 15°$
≈ 231.8	≈ 62.1

$$\mathbf{p} \approx 231.8\mathbf{i} + 62.1\mathbf{j}.$$

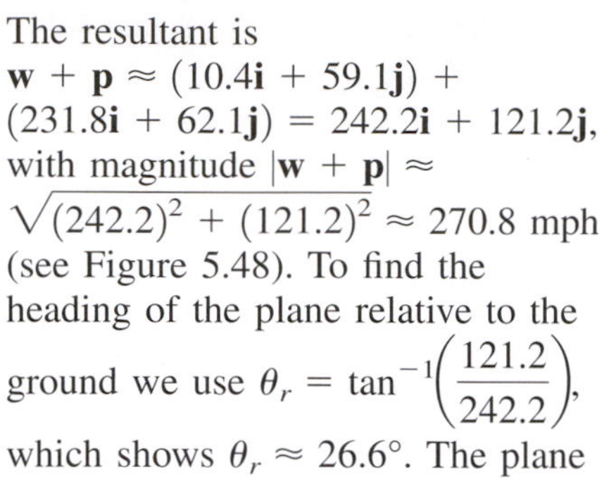

The resultant is $\mathbf{w} + \mathbf{p} \approx (10.4\mathbf{i} + 59.1\mathbf{j}) + (231.8\mathbf{i} + 62.1\mathbf{j}) = 242.2\mathbf{i} + 121.2\mathbf{j}$, with magnitude $|\mathbf{w} + \mathbf{p}| \approx \sqrt{(242.2)^2 + (121.2)^2} \approx 270.8$ mph (see Figure 5.48). To find the heading of the plane relative to the ground we use $\theta_r = \tan^{-1}\left(\frac{121.2}{242.2}\right)$, which shows $\theta_r \approx 26.6°$. The plane is flying on a course heading of $90° - 26.6° = 63.4°$ at a speed of about 270.8 mph relative to the ground. Note the airplane has actually "increased speed" due to the wind.

Figure 5.48

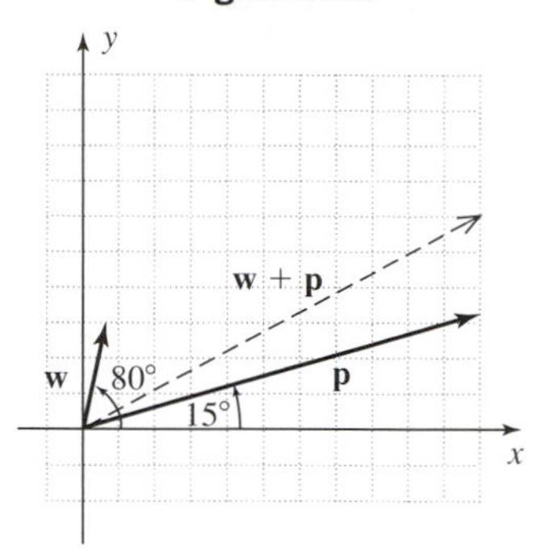

NOW TRY EXERCISES 83 THROUGH 86

WORTHY OF NOTE

Be aware that using the rounding values of intermediate calculations may cause slight variations in the final result. If we calculate $\mathbf{w} + \mathbf{p} = (60 \cos 80° + 240 \cos 15°)\mathbf{i} + (60 \sin 80° + 240 \sin 15°)\mathbf{j}$, then find $|\mathbf{w} + \mathbf{p}|$, the result is actually closer to 270.9 mph.

TECHNOLOGY HIGHLIGHT

Vector Components Given the Magnitude and the Angle θ

The keystrokes shown apply to a T1-84 Plus model. Please consult our Internet site or your manual for other models.

The TABLE feature of a graphing calculator can help us find the horizontal and vertical components of any vector with ease. Consider the vector **v** shown in Figure 5.49, which has a magnitude of 9.5 and $\theta = 15°$. Knowing this magnitude is used in both computations, first store 9.5 in storage location A: 9.5 STO▸ ALPHA MATH. Next, enter the expressions for the horizontal and vertical components as Y_1 and Y_2 on the Y= screen (see Figure 5.50). Note that storing the magnitude 9.5 in memory will prevent our having to alter Y_1 and Y_2 as we apply these ideas to other vectors. As an additional check, note that Y_3 recomputes the magnitude of the vector using the components generated in Y_1 and Y_2. Recall that to access the function variables we press: VARS ▶ ENTER and select the desired function. Although our primary interest is the components for an angle of $\theta = 15°$, we use the TBLSET screen to begin at TblStart = 0°, ΔTbl = 5, and have it count **AUTO**matically, which will enable us to make additional observations. Once these features have been set, pressing 2nd GRAPH (**TABLE**) brings up the screen shown in Figure 5.51. As expected, at $\theta = 0°$ the horizontal component is the same as the magnitude and the vertical component is zero. At $\theta = 15°$ we have the

Figure 5.49

Figure 5.50

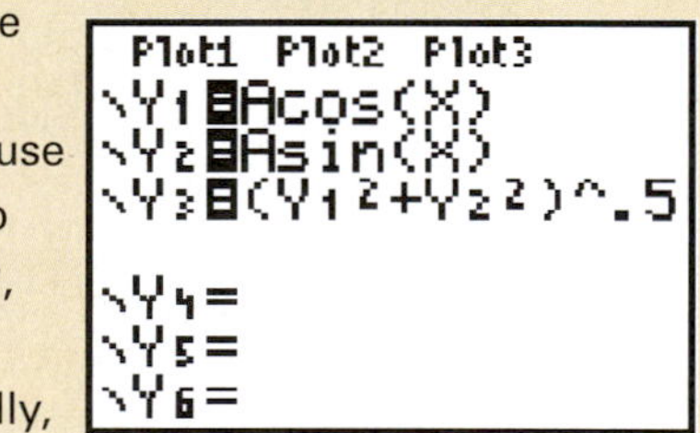

Figure 5.51

X	Y1	Y2
0	9.5	0
5	9.4638	.82798
10	9.3557	1.6497
15	9.1763	2.4588
20	8.9271	3.2492
25	8.6099	4.0149
30	8.2272	4.75

X=0

components of the vector pictured in Figure 5.49, approximately $\langle 9.18, 2.46\rangle$. If the angle were increased to $\theta = 30°$, a 30-60-90 triangle could be formed and one component should be $\sqrt{3}$ times the other. Sure enough, $\sqrt{3}(4.75) \approx 8.2272$.

Exercise 1: If $\theta = 45°$, what would you know about the lengths of the horizontal and vertical components? Scroll down to $\theta = 45°$ to verify.

Exercise 2: If $\theta = 60°$, what would you know about the lengths of the horizontal and vertical components? Scroll down to $\theta = 60°$ to verify.

Exercise 3: We used column Y_3 as a double check on the magnitude of **v** for any given θ. What would this value be for $\theta = 45°$ and $\theta = 60°$? Press the right arrow ▶ to verify. What do you notice?

5.4 EXERCISES

CONCEPTS AND VOCABULARY

Fill in each blank with the appropriate word or phrase. Carefully reread the section if needed.

1. Measurements that can be described using a single number are called ________ quantities.

2. ________ quantities require more than a single number to describe their attributes. Examples are force, velocity, and displacement.

3. To represent a vector quantity geometrically we use a ________ ______ segment.

4. Two vectors are equal if they have the same ________ and ________.

5. Discuss/explain the geometric interpretation of vector addition. Give several examples and illustrations.

6. Describe the process of finding a resultant vector given the magnitude and direction of two arbitrary vectors **u** and **v**. Follow-up with an example.

DEVELOPING YOUR SKILLS

Draw the comparative geometric vectors indicated.

7. Three oceanic research vessels are traveling on a parallel course in the same direction, mapping the ocean floor. One ship is traveling at 12 knots (nautical miles per hour), one at 9 knots, and the third at 6 knots.

8. As part of family reunion activities, the Williams Clan is at a bowling alley and using three lanes. Being amateurs they all roll the ball straight on, aiming for the 1 pin. Grand Dad in Lane 1 rolls his ball at 50 ft/sec. Papa in Lane 2 lets it rip at 60 ft/sec, while Junior in Lane 3 can muster only 30 ft/sec.

9. Vector $\mathbf{v}_1$ is a geometric vector representing a boat traveling at 20 knots. Vectors $\mathbf{v}_2$, $\mathbf{v}_3$, and $\mathbf{v}_4$ are geometric vectors representing boats traveling at 10 knots, 15 knots, and 25 knots, respectively. Draw these vectors given that $\mathbf{v}_2$ and $\mathbf{v}_3$ are traveling the same direction and parallel to $\mathbf{v}_1$, while $\mathbf{v}_4$ is traveling in the opposite direction and parallel to $\mathbf{v}_1$.

10. Vector $\mathbf{F}_1$ is a geometric vector representing a force of 50 N. Vectors $\mathbf{F}_2$, $\mathbf{F}_3$, and $\mathbf{F}_4$ are geometric vectors representing forces of 25 N, 35 N, and 65 N, respectively. Draw these vectors given that $\mathbf{F}_2$ and $\mathbf{F}_3$ are applied in the same direction and parallel to $\mathbf{F}_1$, while $\mathbf{F}_4$ is applied in the opposite direction and parallel to $\mathbf{F}_1$.

Represent each situation described using geometric vectors.

11. Two tractors are pulling at a stump in an effort to clear land for more crops. The Massey-Ferguson is pulling with a force of 250 N, while the John Deere is pulling with a force of 210 N. The chains attached to the stump and each tractor form a 25° angle.

12. In an effort to get their mule up and plowing again, Jackson and Rupert are pulling on ropes attached to the mule's harness. Jackson pulls with 200 lb of force, while Rupert, who is really upset, pulls with 220 lb of force. The angle between their ropes is 16°.

Draw the vector **v** indicated, then graph the equivalent position vector.

13. initial point $(-3, 2)$; terminal point $(4, 5)$

14. initial point $(-4, -4)$; terminal point $(2, 3)$

15. initial point $(5, -3)$; terminal point $(-1, 2)$

16. initial point $(1, 4)$; terminal point $(-2, -2)$

For each vector $\mathbf{v} = \langle a, b \rangle$ and initial point (x, y) given, find the coordinates of the terminal point and the magnitude $|\mathbf{v}|$ of the vector.

17. $\mathbf{v} = \langle 7, 2 \rangle$; initial point $(-2, -3)$

18. $\mathbf{v} = \langle -6, 1 \rangle$; initial point $(5, -2)$

19. $\mathbf{v} = \langle -3, -5 \rangle$; initial point $(2, 6)$

20. $\mathbf{v} = \langle 8, -2 \rangle$; initial point $(-3, -5)$

For each position vector given, (a) graph the vector and name the quadrant, (b) compute its magnitude, and (c) find the acute angle θ formed by the vector and the nearest x-axis.

21. $\langle 8, 3 \rangle$

22. $\langle -7, 6 \rangle$

23. $\langle -2, -5 \rangle$

24. $\langle 8, -6 \rangle$

For Exercises 25 through 30, the magnitude of a vector is given, along with the quadrant of the terminal point and the angle it makes with the nearest x-axis. Find the horizontal and vertical components of each vector and write the result in component form.

25. $|\mathbf{v}| = 12$; $\theta = 25°$; QII

26. $|\mathbf{u}| = 25$; $\theta = 32°$; QIII

27. $|\mathbf{w}| = 140.5$; $\theta = 41°$; QIV

28. $|\mathbf{p}| = 15$; $\theta = 65°$; QI

29. $|\mathbf{q}| = 10$; $\theta = 15°$; QIII

30. $|\mathbf{r}| = 4.75$; $\theta = 62°$; QII

For each pair of vectors **u** and **v** given, compute (a) through (d) and illustrate the indicated operations graphically.

a. $\mathbf{u} + \mathbf{v}$ **b.** $\mathbf{u} - \mathbf{v}$ **c.** $2\mathbf{u} + 1.5\mathbf{v}$ **d.** $\mathbf{u} - 2\mathbf{v}$

31. $\mathbf{u} = \langle 2, 3 \rangle$; $\mathbf{v} = \langle -3, 6 \rangle$

32. $\mathbf{u} = \langle -3, -4 \rangle$; $\mathbf{v} = \langle 0, 5 \rangle$

33. $\mathbf{u} = \langle 7, -2 \rangle$; $\mathbf{v} = \langle 1, 6 \rangle$

34. $\mathbf{u} = \langle -5, -3 \rangle$; $\mathbf{v} = \langle 6, -4 \rangle$

35. $\mathbf{u} = \langle -4, 2 \rangle$; $\mathbf{v} = \langle 1, 4 \rangle$

36. $\mathbf{u} = \langle 7, 3 \rangle$; $\mathbf{v} = \langle -7, 3 \rangle$

Use the graphs of vectors **a**, **b**, **c**, **d**, **e**, **f**, **g**, and **h** given to determine if the following statements are true or false.

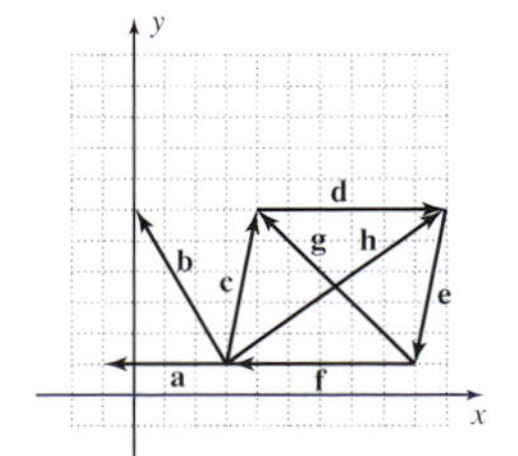

37. $\mathbf{a} + \mathbf{c} = \mathbf{b}$

38. $\mathbf{f} - \mathbf{e} = \mathbf{g}$

39. $\mathbf{c} + \mathbf{f} = \mathbf{h}$

40. $\mathbf{b} + \mathbf{h} = \mathbf{c}$

41. $\mathbf{d} - \mathbf{e} = \mathbf{h}$

42. $\mathbf{d} + \mathbf{f} = \mathbf{0}$

For the vectors **u** and **v** shown, compute **u** + **v** and **u** − **v** and represent each result graphically.

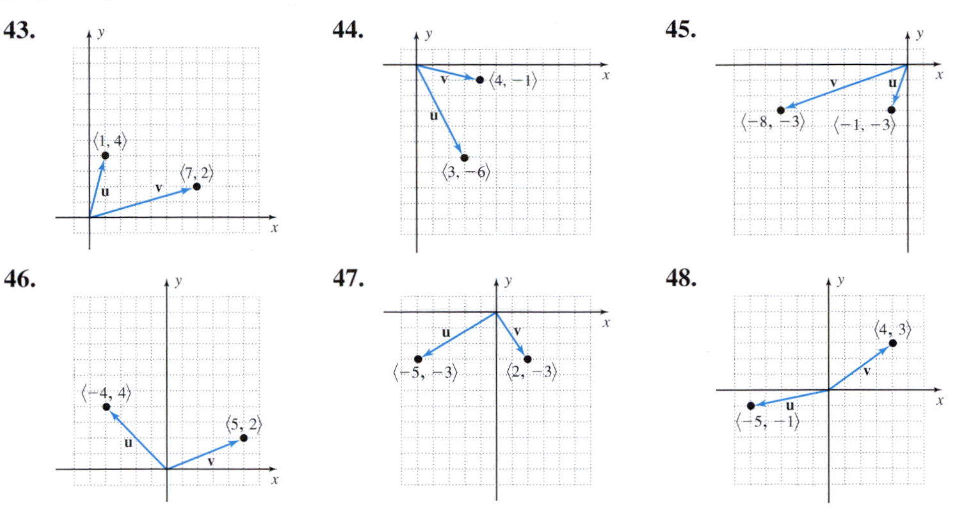

Graph each vector and write it as a linear combination of **i** and **j**. Then compute its magnitude.

49. $\mathbf{u} = \langle 8, 15 \rangle$

50. $\mathbf{v} = \langle -5, 12 \rangle$

51. $\mathbf{p} = \langle -3.2, -5.7 \rangle$

52. $\mathbf{q} = \langle 7.5, -3.4 \rangle$

For each vector here, θ represents the acute angle formed by the vector and the x-axis. (a) Graph each vector, (b) find the horizontal and vertical components and write the vector in component form, and (c) write the vector in **i**, **j** form. Round to the nearest tenth.

53. $\mathbf{v}$ in QIII, $|\mathbf{v}| = 12$, $\theta = 16°$

54. $\mathbf{u}$ in QII, $|\mathbf{u}| = 10.5$, $\theta = 25°$

55. $\mathbf{w}$ in QI, $|\mathbf{w}| = 9.5$, $\theta = 74.5°$

56. $\mathbf{v}$ in QIV, $|\mathbf{v}| = 20$, $\theta = 32.6°$

For vectors $\mathbf{v}_1$ and $\mathbf{v}_2$ given, compute the vector sums (a) through (d) and find the magnitude and direction of each resultant.

a. $\mathbf{v}_1 + \mathbf{v}_2 = \mathbf{p}$

b. $\mathbf{v}_1 - \mathbf{v}_2 = \mathbf{q}$

c. $2\mathbf{v}_1 + 1.5\mathbf{v}_2 = \mathbf{r}$

d. $\mathbf{v}_1 - 2\mathbf{v}_2 = \mathbf{s}$

57. $\mathbf{v}_1 = 2\mathbf{i} - 3\mathbf{j}$; $\mathbf{v}_2 = -4\mathbf{i} + 5\mathbf{j}$

58. $\mathbf{v}_1 = 7.8\mathbf{i} + 4.2\mathbf{j}$; $\mathbf{v}_2 = 5\mathbf{j}$

59. $\mathbf{v}_1 = 5\sqrt{2}\mathbf{i} + 7\mathbf{j}$; $\mathbf{v}_2 = -3\sqrt{2}\mathbf{i} - 5\mathbf{j}$

60. $\mathbf{v}_1 = 6.8\mathbf{i} - 9\mathbf{j}$; $\mathbf{v}_2 = -4\mathbf{i} + 9\mathbf{j}$

61. $\mathbf{v}_1 = 12\mathbf{i} + 4\mathbf{j}$; $\mathbf{v}_2 = -4\mathbf{i}$

62. $\mathbf{v}_1 = 2\sqrt{3}\mathbf{i} - 6\mathbf{j}$; $\mathbf{v}_2 = -4\sqrt{3}\mathbf{i} + 2\mathbf{j}$

Find a unit vector pointing in the same direction as the vector given. Verify that a unit vector was found.

63. $\mathbf{u} = \langle 7, 24 \rangle$

64. $\mathbf{v} = \langle -15, 36 \rangle$

65. $\mathbf{p} = \langle -20, 21 \rangle$

66. $\mathbf{q} = \langle 12, -35 \rangle$

67. $20\mathbf{i} - 21\mathbf{j}$

68. $-4\mathbf{i} - 7.5\mathbf{j}$

69. $3.5\mathbf{i} + 12\mathbf{j}$

70. $-9.6\mathbf{i} + 18\mathbf{j}$

71. $\mathbf{v}_1 = \langle 13, 3 \rangle$

72. $\mathbf{v}_2 = \langle -4, 7 \rangle$

73. $6\mathbf{i} + 11\mathbf{j}$

74. $-2.5\mathbf{i} + 7.2\mathbf{j}$

Vectors **p** and **q** form the angle indicated in each diagram. Find the vector **r** that points in the same direction as **q** and forms the base of the right triangle shown.

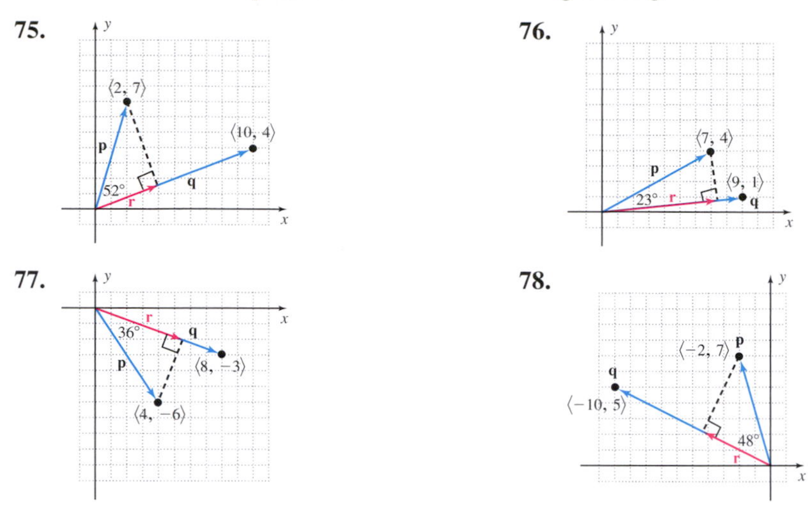

WORKING WITH FORMULAS

79. The magnitude of a vector in three dimensions: $|\mathbf{v}| = \sqrt{a^2 + b^2 + c^2}$

The magnitude of a vector in three dimensional space is given by the formula shown, where the components of the position vector **v** are $\langle a, b, c\rangle$. Find the magnitude of **v** if $\mathbf{v} = \langle 5, 9, 10\rangle$.

80. Find a cardboard box of any size and carefully measure its length, width, and height. Then use the given formula to find the magnitude of the box's diagonal. Verify your calculation by direct measurement.

APPLICATIONS

Exercise 81

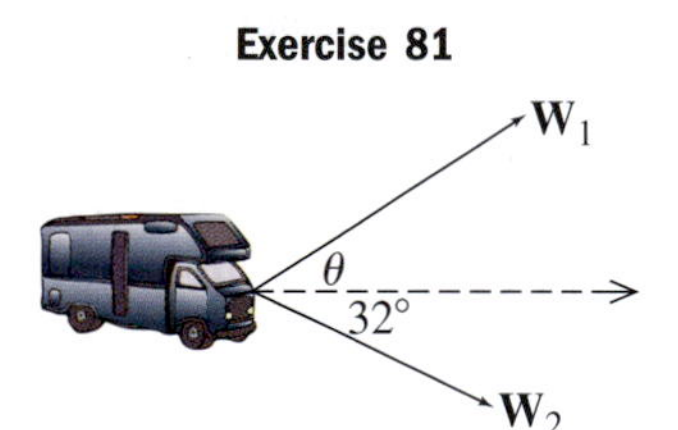

81. Tow forces: A large van has careened off of the road into a ditch, and two tow trucks are attempting to winch it out. The cable from the first winch exerts a force of 900 lb, while the cable from the second exerts a force of 700 lb. Determine the angle θ for the first tow truck that will bring the van directly out of the ditch and along the line indicated.

82. Tow forces: Two tugboats are pulling a large ship into dry dock. The first is pulling with a force of 1250 N and the second with a force of 1750 N. Determine the angle θ for the second tugboat that will keep the ship moving straight forward and into the dock.

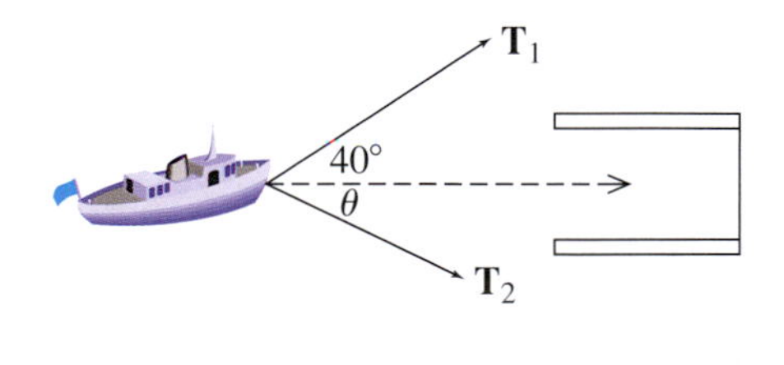

83. Projectile components: An arrow is shot into the air at an angle of 37° with an initial velocity of 100 ft/sec. Compute the horizontal and vertical components of the representative vector.

84. **Projectile components:** A football is punted (kicked) into the air at an angle of 42° with an initial velocity of 20 m/sec. Compute the horizontal and vertical components of the representative vector.

85. **Headings and cross-winds:** An airplane is flying at 390 mph on a heading of 45°. There is a strong, 50 mph wind blowing from the southeast on a heading of 345°. What is the true course and speed of the plane (relative to the ground)?

86. **Headings and currents:** A cruise ship is traveling at 16 knots on a heading of 300°. There is a strong water current flowing at 6 knots from the northwest on a heading of 120°. What is the true course and speed of the cruise ship?

The lights used in a dentist's office are multijointed so they can be configured in multiple ways to accommodate various needs. As a simple model, consider such a light that has the three joints, as illustrated. The first segment has a length of 45 cm, the second is 40 cm in length, and the third is 35 cm.

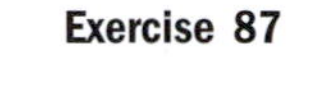
Exercise 87

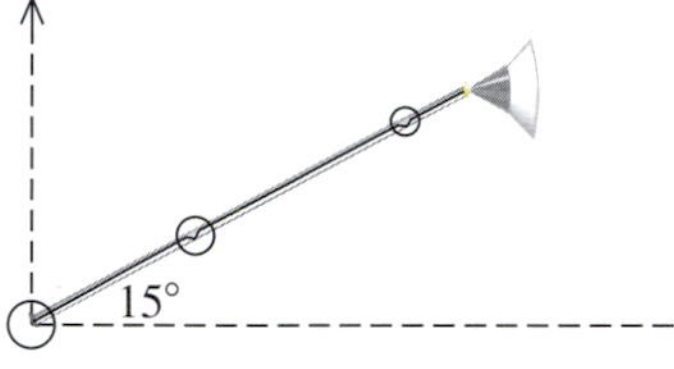

87. If the joints of the light are positioned so a straight line is formed and the angle made with the horizontal is 15°, determine the approximate coordinates of the joint nearest the light.

88. If the first segment is rotated 75° above horizontal, the second segment −30° (below the horizontal), and the third segment is parallel to the horizontal, determine the approximate coordinates of the joint nearest the light.

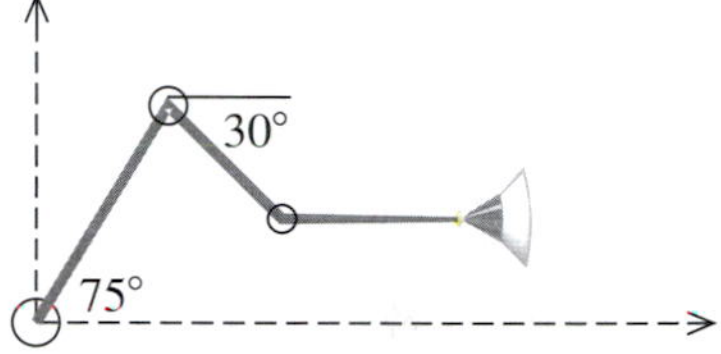

For the arbitrary vectors $\mathbf{u} = \langle a, b\rangle$, $\mathbf{v} = \langle c, d\rangle$, and $\mathbf{w} = \langle e, f\rangle$ and the scalars c and k, prove the following vector properties using the properties of real numbers.

89. $1\mathbf{u} = \mathbf{u}$

90. $0\mathbf{u} = \mathbf{0} = k\mathbf{0}$

91. $\mathbf{u} - \mathbf{v} = \mathbf{u} + (-\mathbf{v})$

92. $(\mathbf{u} + \mathbf{v}) + \mathbf{w} = \mathbf{u} + (\mathbf{v} + \mathbf{w})$

93. $(ck)\mathbf{u} = c(k\mathbf{u}) = k(c\mathbf{u})$

94. $\mathbf{u} + \mathbf{0} = \mathbf{u}$

95. $\mathbf{u} + (-\mathbf{u}) = \mathbf{0}$

96. $k(\mathbf{u} + \mathbf{v}) = k\mathbf{u} + k\mathbf{v}$

97. $(c + k)\mathbf{u} = c\mathbf{u} + k\mathbf{u}$

WRITING, RESEARCH, AND DECISION MAKING

98. The origin of the vector concept and vector algebra is somewhat uncertain, but it is well known that the concepts we view today as common and natural are the result of some very advanced work. Do some research on the history of vectors, and include some notes on the people who aided their development.

99. Consider an airplane flying at 200 mph at a heading of 45°. Compute the groundspeed of the plane under the following conditions. A strong, 40-mph wind is blowing (a) in the same direction; (b) in the direction of due north (0°); (c) in the direction heading 315°; (d) in the direction heading 270°; and (e) in the direction heading 225°. What did you notice about the groundspeed for (a) and (b)? Explain why the plane's speed is greater than 200 mph for (a) and (b), but less than 200 mph for the others.

Exercise 100

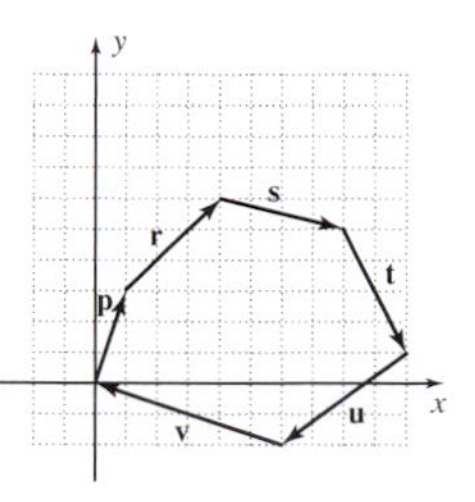

EXTENDING THE CONCEPT

100. Show that the sum of the vectors given, which form the sides of a closed polygon, is the zero vector. Assume all vectors have integer coordinates and each tick mark is 1 unit.

101. Verify that for $\mathbf{v} = a\mathbf{i} + b\mathbf{j}$ and $|\mathbf{v}| = \sqrt{a^2 + b^2}$, $= \dfrac{\mathbf{v}}{|\mathbf{v}|} = 1$.

(*Hint:* Create the vector $\mathbf{u} = \dfrac{\mathbf{v}}{|\mathbf{v}|}$ and find its magnitude.)

102. Referring to Exercises 87 and 88, suppose the dentist needed the pivot joint at the light (the furthest joint from the wall) to be at (80, 20) for a certain patient or procedure. Find at least one set of "joint angles" that will make this possible.

MAINTAINING YOUR SKILLS

103. (1.2) Mt. Tortolas lies on the Argentine-Chilean border. When viewed from a distance of 5 mi, the angle of elevation to the top of the peak is 38°. How tall is Mount Tortolas? State your answer in feet.

104. (3.2) Prove the following is an identity: $2\sec^2\theta = \dfrac{1}{1+\sin\theta} + \dfrac{1}{1-\sin\theta}$

105. (2.3) Graph the function $y = 3\sin\left(2x - \dfrac{\pi}{4}\right)$ using a reference rectangle and the *rule of fourths*.

106. (1.3) Use fundamental identities to find the values of all six trig functions that satisfy the conditions, $\sec x = \dfrac{13}{12}$ and $\sin x > 0$.

107. (5.1) The satellite Jupiter VII measures its distance from Denver and from Louisville using radio waves as shown. With an onboard sighting device, the satellite is able to determine that $\angle\mathbf{J}$ is 98°. How many miles is it from Denver to Louisville?

Exercise 107

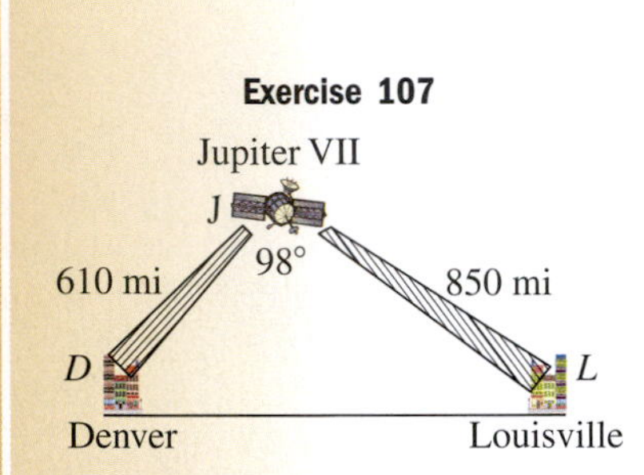

108. (2.1) Find the equation of the graph shown, (a) given it is of the form $y = A\sin(Bx \pm C)$ and (b) given it is of the form $y = A\cos(Bx \pm C)$.

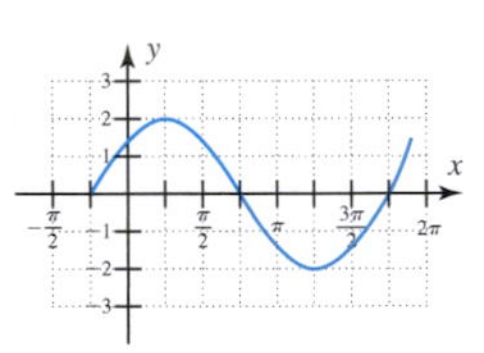

MID-CHAPTER CHECK

1. Given $\dfrac{\sin A}{a} = \dfrac{\sin B}{b}$, solve for $\sin B$.

2. Given $b^2 = a^2 + c^2 - 2ac\cos B$, solve for $\cos B$.

Solve the triangles shown below using any appropriate method.

3.

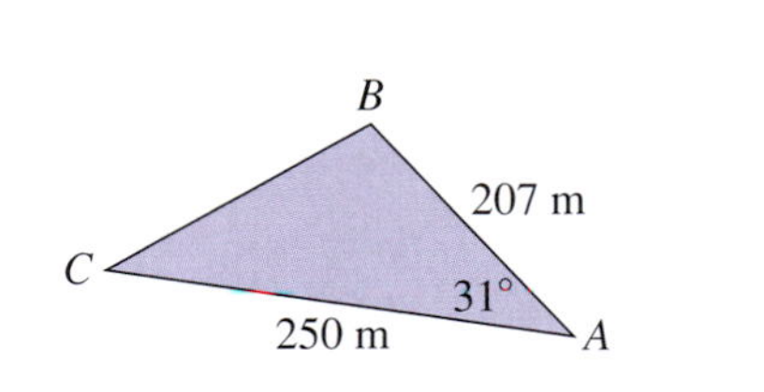

4.

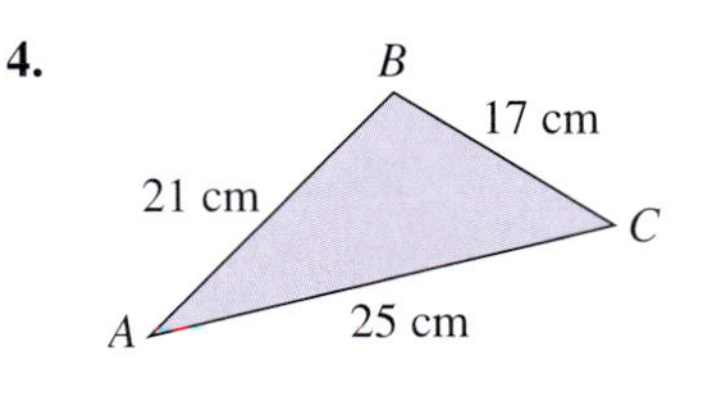

Solve the triangles described below using the law of sines. If more than one triangle exists, solve both.

5. $A = 44°$, $a = 2.1$ km, $c = 2.8$ km

6. $C = 27°$, $a = 70$ yd, $c = 100$ yd

7. A large highway sign is erected on a steep hillside that is inclined 28° from the horizontal. At 9:00 A.M. the sign casts a 75 ft shadow. Find the height of the sign if the angle of elevation (measured from a horizontal line) from the tip of the shadow to the top of the sign is 48°.

Exercise 7

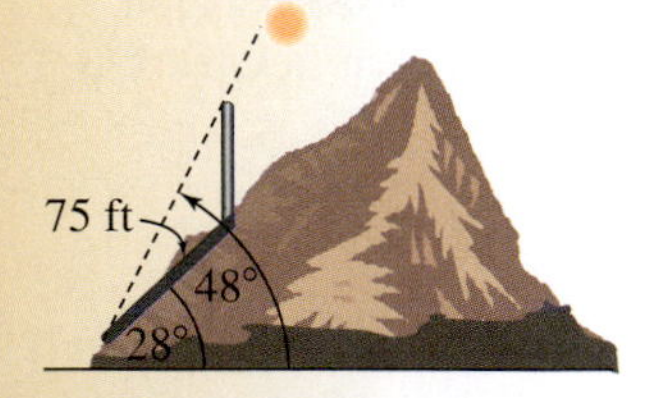

Exercise 8

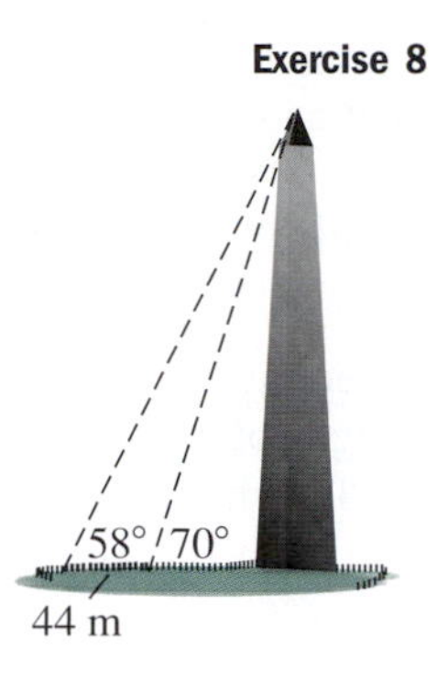

8. Modeled after an Egyptian obelisk, the Washington Monument (Washington, D.C.) is one of the tallest masonry buildings in the world. Find the height of the monument given the measurements shown (see the figure).

9. The circles shown here have radii of 4 cm, 9 cm, and 12 cm, and are tangent to each other. Find the angles formed by the line segments joining their centers.

Exercise 9

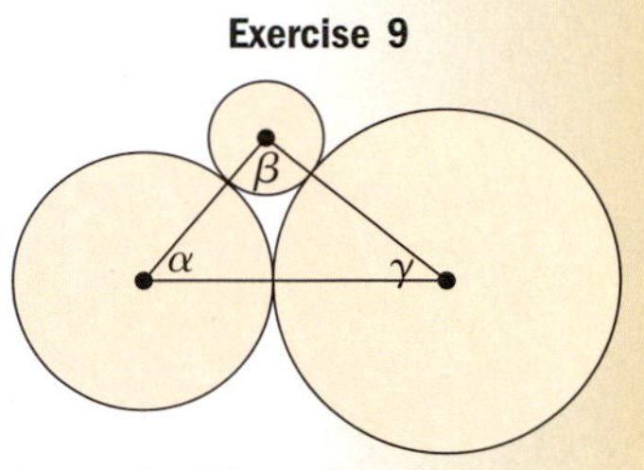

10. (5.4) A submarine leaves its base at Diego Garcia (Indian Ocean) to make a port call in Sri Lanka, traveling at 12 knots, at heading 50°. Enroute, it has to travel through the Indian Equatorial Current, which moves at 0.9 knots, at heading 95°. Assuming these factors were to remain constant, find the true direction and speed of the submarine on this trip.

Reinforcing Basic Concepts

Scaled Drawings and the Laws of Sine and Cosine

In mathematics, there are few things as satisfying as the tactile verification of a concept or computation. In this *Reinforcing Basic Concepts*, we'll use scaled drawings to verify the relationships stated by the law of sines and the law of cosines. First, gather a blank sheet of paper, a ruler marked in centimeters/millimeters, and a protractor. When working with scale models, always measure and mark as carefully as possible. The greater the care, the better the results. For the first illustration (see Figure 5.52), we'll draw a 20-cm horizontal line segment near the bottom of the paper, then use the left endpoint to mark off a 35° angle. Draw the second side a length of 18 cm. Our first goal is to compute the length of the side needed to complete the triangle, then verify our computation by measurement. Since the current "triangle" is SAS, we use the law of cosines. Label the 35° as $\angle A$, the top vertex as $\angle B$, and the right endpoint as $\angle C$.

Figure 5.52

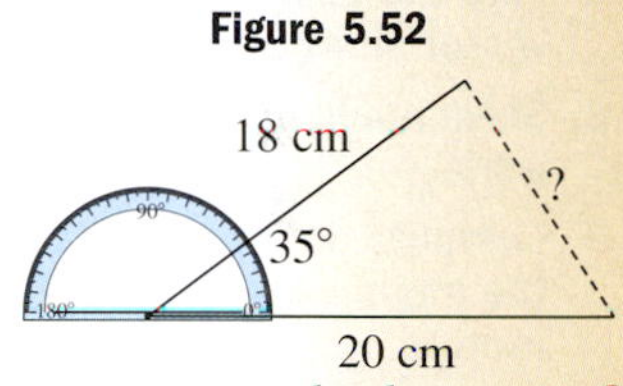

$$a^2 = b^2 + c^2 - 2bc\cos A \quad \text{law of cosines with respect to } a$$
$$= (20)^2 + (18)^2 - 2(20)(18)\cos 35 \quad \text{substitute known values}$$
$$\approx 724 - 589.8 \quad \text{simplify (round to tenths)}$$
$$= 134.2 \quad \text{combine terms}$$
$$a \approx 11.6 \quad \text{solve for } a$$

The computed length of side a is 11.6 cm, and if you took great care in drawing your diagram, you'll find the missing side is indeed very close to this length.

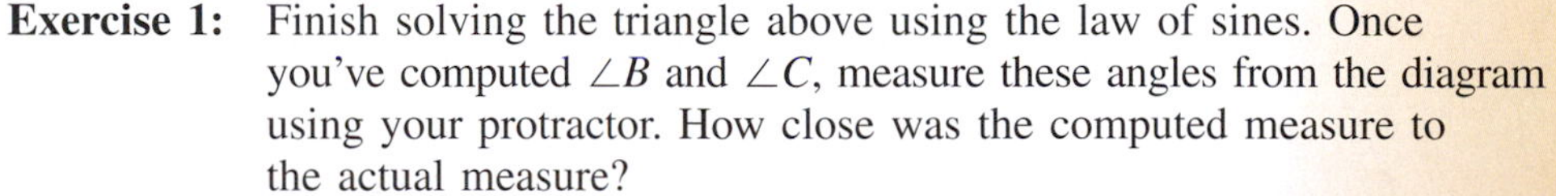
Exercise 1: Finish solving the triangle above using the law of sines. Once you've computed $\angle B$ and $\angle C$, measure these angles from the diagram using your protractor. How close was the computed measure to the actual measure?

For the second illustration (see Figure 5.53), draw *any arbitrary triangle* on a separate blank sheet, noting that the larger the triangle, the easier it is to measure the angles. After you've drawn it, measure the length of each side to the nearest millimeter (our triangle turned out to be 21.2 cm × 13.3 cm × 15.3 cm). Now use the

Figure 5.53

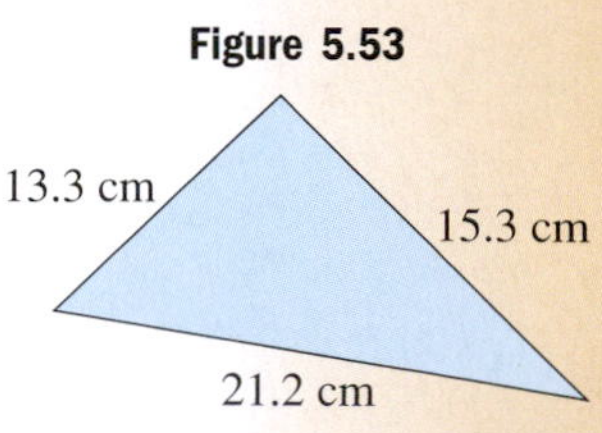

law of cosines to find one angle, then the law of sines to solve the triangle. The computations for our triangle gave angles of 95.4°, 45.9°, and 38.7°. What angles did your computations give? Finally, use your protractor to measure the angles of the triangle you drew. With careful drawings, the measured results are often remarkably accurate!

Exercise 2: Using sides of 18 cm and 15 cm, draw a 35° angle, a 50° angle, and a 70° angle, then complete each triangle by connecting the endpoints. Use the law of cosines to compute the length of this third side, then actually measure each one. Was the actual length close to the computed length?

5.5 Vector Applications and the Dot Product

LEARNING OBJECTIVES

In Section 5.5 you will learn how to:

A. **Use vectors to investigate forces in equilibrium**

B. **Find the components of one vector along another**

C. **Solve applications involving work**

D. **Compute dot products and the angle between two vectors**

E. **Find the projection of one vector along another and resolve a vector into orthogonal components**

F. **Use vectors to develop an equation for nonvertical, projectile motion and solve related applications**

INTRODUCTION

In Section 5.4 we introduced the concept of a vector along with its geometric, graphical, and algebraic representations. We also looked at operations on vectors and employed vector diagrams to solve basic applications. In this section we introduce additional ideas that enable us to solve a variety of new applications and lay a strong foundation for future studies.

POINT OF INTEREST

In every day usage, a projection is an image shown on a screen using light or shadow, like a movie or a shadow play. In mathematics, the word carries much the same meaning, as we look at how one vector is projected onto another. The word *projection* stems from the Latin "*pro*" and "*iactus*," which imply that something has been "thrown forward" onto something else. Vectors can be used to describe the projected image.

A. Vectors and Equilibrium

Much like the intuitive meaning of the word, vector forces are in **equilibrium** when they "counterbalance" each other. The simplest example is two vector forces of equal magnitude acting on the same point but in opposite directions. Similar to a tug-of-war with both sides equally matched, no one wins. If vector $\mathbf{F}_1$ has a magnitude of 500 lb in the positive direction, $\mathbf{F}_1 = \langle 500, 0\rangle$ would need vector $\mathbf{F}_2 = \langle -500, 0\rangle$ to counter it. If the forces are nonquadrantal, we intuitively sense the components must still sum to zero, and that $\mathbf{F}_3 = \langle 600, -200\rangle$ would need $\mathbf{F}_4 = \langle -600, 200\rangle$ for equilibrium to occur (see Figure 5.54). In other words, two vectors are in equilibrium when their sum is the zero vector $\mathbf{0}$. If the forces have unequal magnitudes or do not pull in opposite directions, recall a resultant vector $\mathbf{F} = \mathbf{F}_a + \mathbf{F}_b$ can be found and represents the combined force. Equilibrium will then occur by adding the vector $-1(\mathbf{F})$.

Figure 5.54

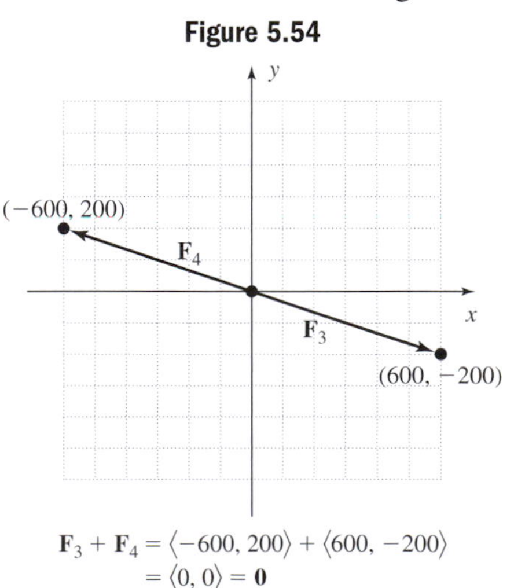

$\mathbf{F}_3 + \mathbf{F}_4 = \langle -600, 200\rangle + \langle 600, -200\rangle = \langle 0, 0\rangle = \mathbf{0}$

These ideas can be extended to include any number of vector forces acting on the same point. In general, we have the following:

VECTORS AND EQUILIBRIUM

Given vectors $\mathbf{v}_1, \mathbf{v}_2, \ldots, \mathbf{v}_n$ acting on a point P,

1. The resultant vector is $\mathbf{V} = \mathbf{v}_1 + \mathbf{v}_2 + \cdots + \mathbf{v}_n$.
2. Equilibrium for these forces requires the vector $-1\mathbf{V}$, where $\mathbf{V} + (-1)\mathbf{V} = \mathbf{0}$ (the zero vector).

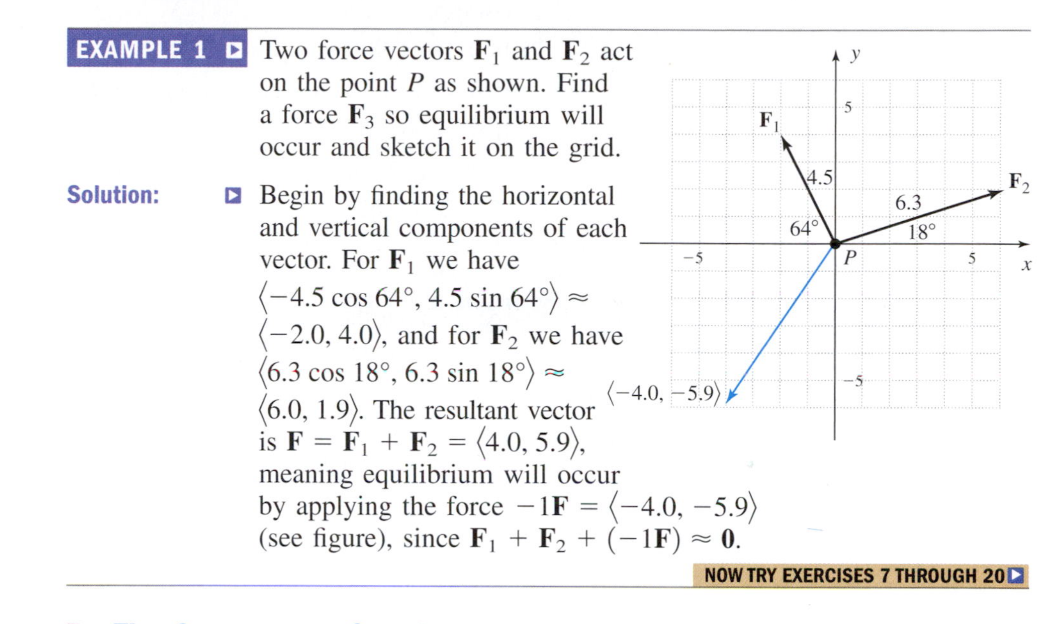

EXAMPLE 1 Two force vectors $\mathbf{F}_1$ and $\mathbf{F}_2$ act on the point P as shown. Find a force $\mathbf{F}_3$ so equilibrium will occur and sketch it on the grid.

Solution: Begin by finding the horizontal and vertical components of each vector. For $\mathbf{F}_1$ we have $\langle -4.5\cos 64°, 4.5\sin 64°\rangle \approx \langle -2.0, 4.0\rangle$, and for $\mathbf{F}_2$ we have $\langle 6.3\cos 18°, 6.3\sin 18°\rangle \approx \langle 6.0, 1.9\rangle$. The resultant vector is $\mathbf{F} = \mathbf{F}_1 + \mathbf{F}_2 = \langle 4.0, 5.9\rangle$, meaning equilibrium will occur by applying the force $-1\mathbf{F} = \langle -4.0, -5.9\rangle$ (see figure), since $\mathbf{F}_1 + \mathbf{F}_2 + (-1\mathbf{F}) \approx \mathbf{0}$.

NOW TRY EXERCISES 7 THROUGH 20

B. The Component of u along v: $\text{comp}_\mathbf{v}\mathbf{u}$

As in Example 1, many simple applications involve position vectors where the angle and horizontal/vertical components are known or can easily be found. In these situations, the components are often quadrantal, that is, they lie along the x- and y-axes and meet at a right angle. Many other applications require us to find components of a vector that are nonquadrantal, with one of the components parallel to, or lying along a second vector. Given vectors **u** and **v**, as shown in Figure 5.55, we symbolize the component of **u** that lies along **v** as $\text{comp}_\mathbf{v}\mathbf{u}$, noting its value is simply $|\mathbf{u}|\cos\theta$ since $\cos\theta = \frac{\text{comp}_\mathbf{v}\mathbf{u}}{|\mathbf{u}|}$. As the diagrams further indicate, $\text{comp}_\mathbf{v}\mathbf{u} = |\mathbf{u}|\cos\theta$ regardless of how the vectors are oriented. Note that even when the components of a vector do not lie along the x- or y-axes, they are still **orthogonal** (meet at a 90° angle).

Figure 5.55

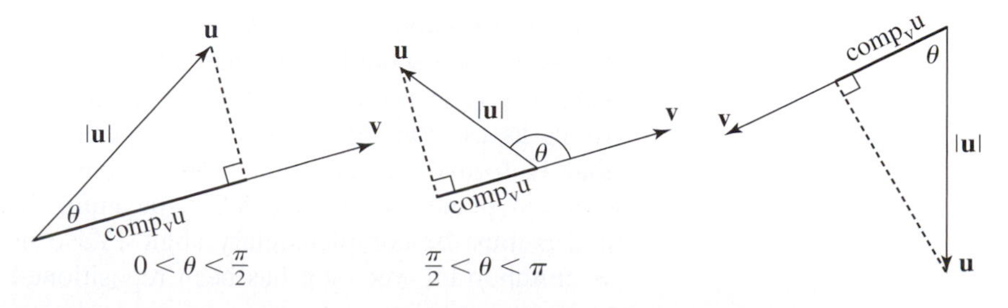

It is important to note that $\text{comp}_{\mathbf{v}}\mathbf{u}$ is a *scalar quantity* (not a vector), giving only the magnitude of this component (the **vector projection** of **u** along **v** is studied later in this section). From these developments we make the following observations regarding angle θ at which vectors **u** and **v** meet:

VECTORS AND THE COMPONENT OF u ALONG v

Given vectors **u** and **v** which met an angle θ,

1. $\text{comp}_{\mathbf{v}}\mathbf{u} = |\mathbf{u}|\cos\theta$.
2. If $0 < \theta < 90°$, $\text{comp}_{\mathbf{v}}\mathbf{u} > 0$; if $90° < \theta < 180°$, $\text{comp}_{\mathbf{v}}\mathbf{u} < 0$.
3. If $\theta = 0$, **u** and **v** have the same direction and $\text{comp}_{\mathbf{v}}\mathbf{u} = |\mathbf{u}|$.
4. If $\theta = 90°$, **u** and **v** are orthogonal and $\text{comp}_{\mathbf{v}}\mathbf{u} = 0$.
5. If $\theta = 180°$, **u** and **v** have opposite directions and $\text{comp}_{\mathbf{v}}\mathbf{u} = -|\mathbf{u}|$.

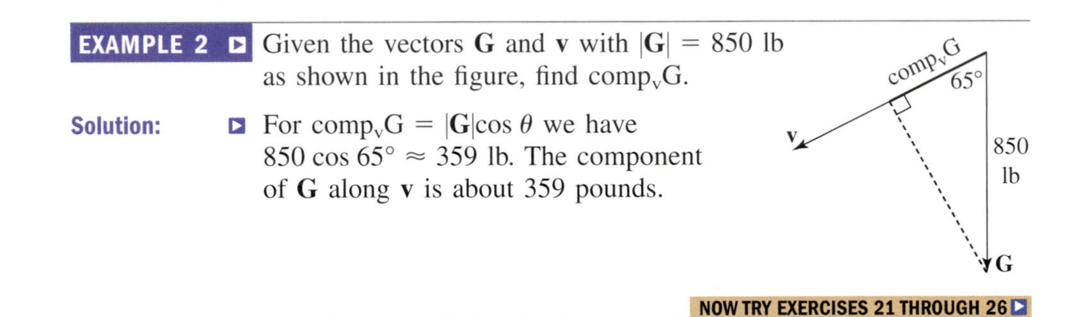

EXAMPLE 2 Given the vectors **G** and **v** with $|\mathbf{G}| = 850$ lb as shown in the figure, find $\text{comp}_{\mathbf{v}}\mathbf{G}$.

Solution: For $\text{comp}_{\mathbf{v}}\mathbf{G} = |\mathbf{G}|\cos\theta$ we have $850\cos 65° \approx 359$ lb. The component of **G** along **v** is about 359 pounds.

NOW TRY EXERCISES 21 THROUGH 26

One interesting application of equilibrium and $\text{comp}_{\mathbf{v}}\mathbf{u}$ involves the force of gravity acting on an object placed on an inclined plane (for this study, we assume there is no friction between the object and the inclined plane). The greater the incline, the greater the tendency of the object to slide down the plane. While the force of gravity continues to pull straight downward (represented by the vector **G** in Figure 5.56), **G** is now the resultant of a force acting parallel to the plane along vector **v** (causing the object to slide) and a force acting perpendicular to the plane along vector **p** (causing the object to press against the plane). If we knew the component of **G** along **v** (indicated by the shorter, bold segment), we would know the force required to keep the object stationary (in a state of equilibrium) as the two forces must be opposites. Note that **G** forms a right angle with the base of the inclined plane (see Figure 5.57), meaning that α and β must be complementary angles. Also note that since the location of a vector is unimportant, vector **p** has been repositioned for clarity.

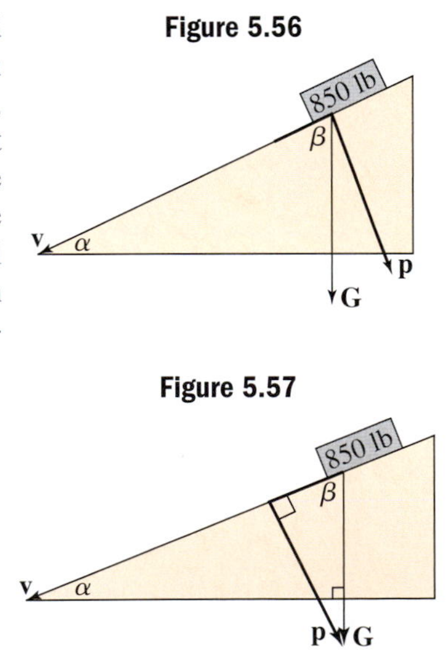

Figure 5.56

Figure 5.57

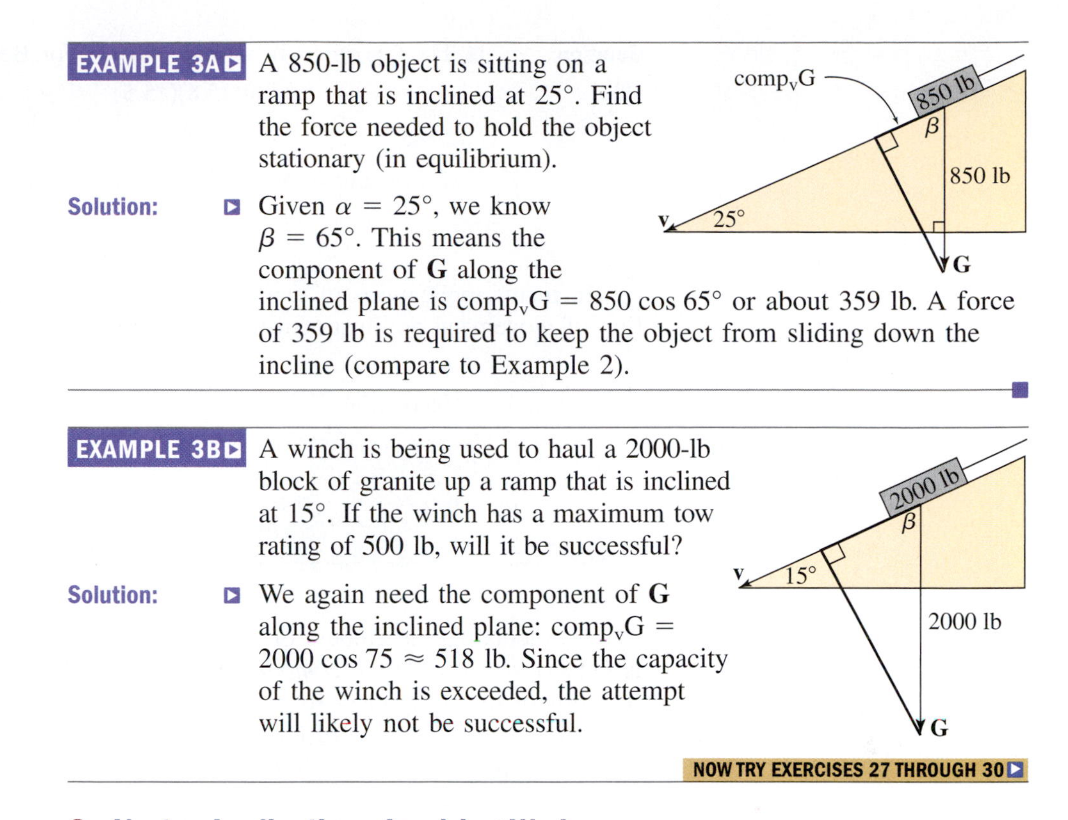

EXAMPLE 3A A 850-lb object is sitting on a ramp that is inclined at 25°. Find the force needed to hold the object stationary (in equilibrium).

Solution: Given $\alpha = 25°$, we know $\beta = 65°$. This means the component of **G** along the inclined plane is $\text{comp}_{\mathbf{v}}\text{G} = 850 \cos 65°$ or about 359 lb. A force of 359 lb is required to keep the object from sliding down the incline (compare to Example 2).

EXAMPLE 3B A winch is being used to haul a 2000-lb block of granite up a ramp that is inclined at 15°. If the winch has a maximum tow rating of 500 lb, will it be successful?

Solution: We again need the component of **G** along the inclined plane: $\text{comp}_{\mathbf{v}}\text{G} = 2000 \cos 75 \approx 518$ lb. Since the capacity of the winch is exceeded, the attempt will likely not be successful.

NOW TRY EXERCISES 27 THROUGH 30

C. Vector Applications Involving Work

In common, everyday usage, **work** is understood to involve the exertion of energy or force to move an object a certain distance. For example, digging a ditch is hard work and involves moving dirt (exerting a force) from the trench to the bankside (over a certain distance). In an office, moving a filing cabinet likewise involves work. If the filing cabinet is heavier, or the distance it needs to be moved is greater, more work is required to move it (Figures 5.58 and 5.59).

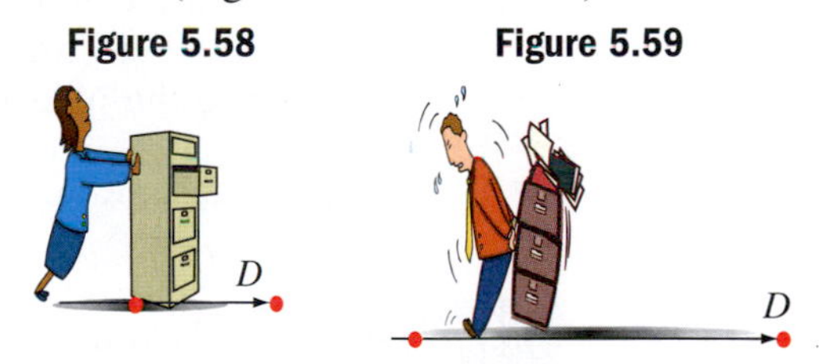

Figure 5.58 **Figure 5.59**

To determine how much work was done by each person, we need to quantify the concept. Consider a constant force **F**, applied to move an object a distance D *in the same direction as the force*. In this case work is defined as the product of the force applied with the distance the object is moved: Work = Force × Distance or $W = |\mathbf{F}|D$. If the force is given in pounds and the distance in feet, the amount of work is measured in a unit called **foot-pounds** (ft-lb). If the force is in newtons and the distance in meters, the amount of work is measured in **newton-meters** (N-m), also known as **joules** (J).

EXAMPLE 4 While rearranging the office, Carrie must apply a force of 5.8 N to relocate a filing cabinet 15.5 m, while Bernard applies a 7.5-N force to move a second cabinet 11.6 m. Who did the most work? Express your answer in joules.

Solution: For Carrie: $W = |\mathbf{F}|D = (5.8)(15.5) = 89.9$ J For Bernard: $W = |\mathbf{F}|D = (7.5)(11.6) = 87.0$ J

Carrie did $89.9 - 87 = 2.9$ J more work than Bernard.

NOW TRY EXERCISES 31 AND 32

In many applications of work, the force **F** is not applied parallel to the direction of movement, as illustrated in Figures 5.60 and 5.61.

Figure 5.60 **Figure 5.61**

In calculating the amount of work done, the general concept of force × distance is preserved, *but only the component of force in the direction of movement is used.* In terms of the component forces discussed earlier, if **F** is a constant force applied at angle θ to the direction of movement, the amount of work done is *the component of force along D times the distance the object is moved.*

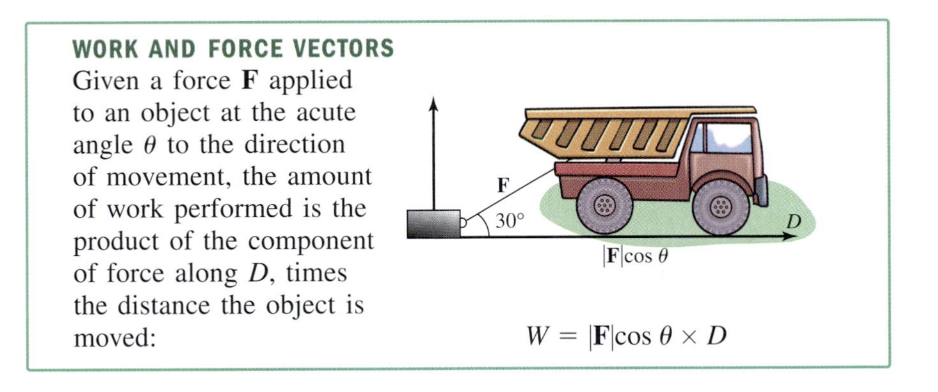

WORK AND FORCE VECTORS

Given a force **F** applied to an object at the acute angle θ to the direction of movement, the amount of work performed is the product of the component of force along D, times the distance the object is moved:

$$W = |\mathbf{F}|\cos\theta \times D$$

EXAMPLE 5 To help move heavy pieces of furniture across the floor, movers sometime employ a body harness similar to that used for a plow horse. A mover applies a constant 200-lb force to drag a piano 100 ft down a long hallway and into another room. If the straps make a 40° angle with the direction of movement, find the amount of work performed.

Solution: The component of force in the direction of movement is $200 \cos 40°$ or about 153 lb. The amount of work done is $W \approx 153(100) = 15{,}300$ ft-lb.

NOW TRY EXERCISES 35 THROUGH 40

These ideas can be generalized to include work problems where the component of force in the direction of motion is along a *nonhorizontal* vector **v**. Consider Example 6.

EXAMPLE 6 The force vector $\mathbf{F} = \langle 5, 12 \rangle$ moves an object along the vector $\mathbf{v} = \langle 15.44, 2 \rangle$ as shown. If the angle between the vectors is 60°, find the amount of work required to move the object along the entire length of **v**. Assume force is in pounds and distance in feet.

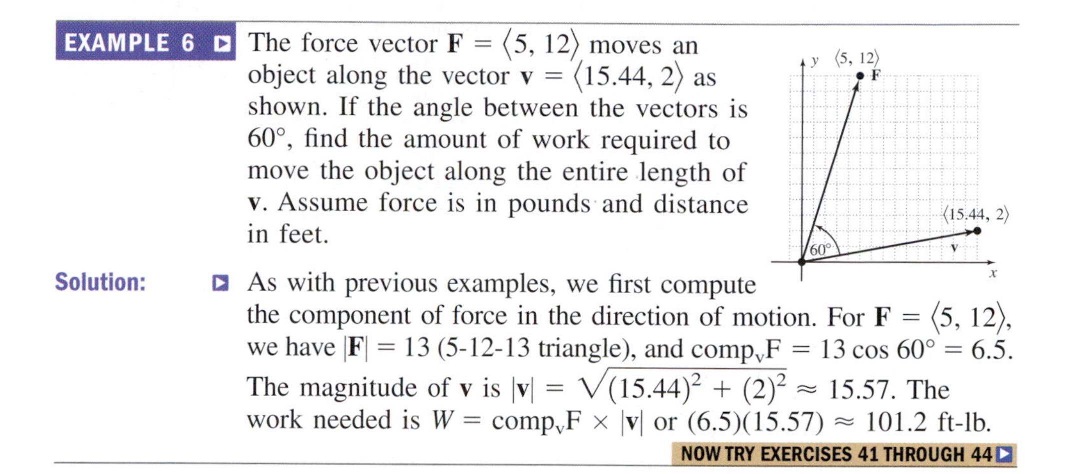

Solution: As with previous examples, we first compute the component of force in the direction of motion. For $\mathbf{F} = \langle 5, 12 \rangle$, we have $|\mathbf{F}| = 13$ (5-12-13 triangle), and $\text{comp}_{\mathbf{v}}\text{F} = 13 \cos 60° = 6.5$. The magnitude of **v** is $|\mathbf{v}| = \sqrt{(15.44)^2 + (2)^2} \approx 15.57$. The work needed is $W = \text{comp}_{\mathbf{v}}\text{F} \times |\mathbf{v}|$ or $(6.5)(15.57) \approx 101.2$ ft-lb.

NOW TRY EXERCISES 41 THROUGH 44

D. Dot Products and the Angle Between Two Vectors

When the component of force in the direction of motion lies along a *nonhorizontal* vector, the amount of work performed can be computed more efficiently using an operation called the **dot product.** For any two vectors **u** and **v**, the dot product $\mathbf{u} \cdot \mathbf{v}$ is equivalent to $\text{comp}_{\mathbf{v}}\text{u} \times |\mathbf{v}|$, yet is much easier to compute. The operation is defined as follows:

THE DOT PRODUCT

Given vectors $\mathbf{u} = \langle a, b \rangle$ and $\mathbf{v} = \langle c, d \rangle$, their dot product is denoted $\mathbf{u} \cdot \mathbf{v}$ and is defined as $\mathbf{u} \cdot \mathbf{v} = \langle a, b \rangle \cdot \langle c, d \rangle = ac + bd$. In words, it is the *real number* found by taking the sum of corresponding component products.

EXAMPLE 7 Verify the answer to Example 6 using the dot product $\mathbf{u} \cdot \mathbf{v}$.

Solution: For $\mathbf{u} = \langle 5, 12 \rangle$ and $\mathbf{v} = \langle 15.44, 2 \rangle$, we have $\mathbf{u} \cdot \mathbf{v} = \langle 5, 12 \rangle \cdot \langle 15.44, 2 \rangle$ giving $5(15.44) + 12(2) = 101.2$. The result is 101.2, as in Example 6.

NOW TRY EXERCISES 45 THROUGH 48

Note that dot products can also be used in the simpler case where the direction of motion is along a horizontal distance (Examples 4 and 5). While the dot product offers a powerful and efficient way to compute the amount of work performed, it has many other applications; for example, to find the angle between two vectors. Consider that for any two vectors **u** and **v**, $\mathbf{u} \cdot \mathbf{v} = |\mathbf{u}|\cos\theta \times |\mathbf{v}|$, leading directly to $\cos\theta = \frac{\mathbf{u}}{|\mathbf{u}|} \cdot \frac{\mathbf{v}}{|\mathbf{v}|}$ (solve for $\cos\theta$).

In summary,

THE ANGLE θ BETWEEN TWO VECTORS

Given the nonzero vectors **u** and **v**:

$$\cos\theta = \frac{\mathbf{u}}{|\mathbf{u}|}\cdot\frac{\mathbf{v}}{|\mathbf{v}|} \text{ and } \theta_r = \cos^{-1}\left(\frac{\mathbf{u}}{|\mathbf{u}|}\cdot\frac{\mathbf{v}}{|\mathbf{v}|}\right)$$

Figure 5.62

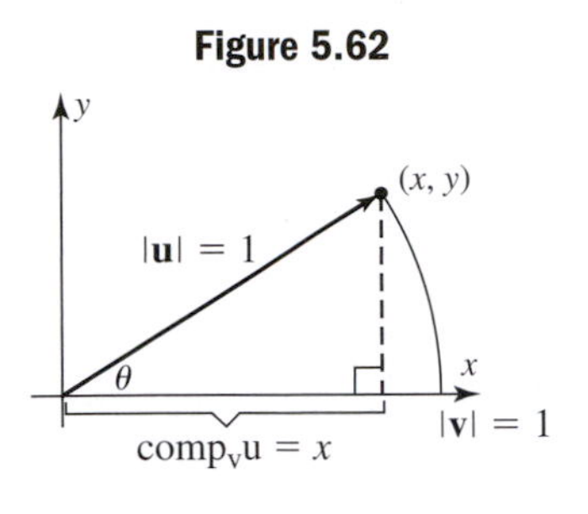

In the special case where **u** *and* **v** *are unit vectors*, this simplifies to $\cos\theta = \mathbf{u}\cdot\mathbf{v}$ since $|\mathbf{u}| = |\mathbf{v}| = 1$. This relationship is shown in Figure 5.62. The dot product $\mathbf{u}\cdot\mathbf{v}$ gives $\text{comp}_{\mathbf{v}}\mathbf{u} \times |\mathbf{v}|$, but $|\mathbf{v}| = 1$ and the component of **u** along **v** is simply the adjacent side of a right triangle whose hypotenuse is 1. Hence $\mathbf{u}\cdot\mathbf{v} = \cos\theta$.

EXAMPLE 8 Find the angle between the vectors given.

a. $\mathbf{u} = \langle -3, 4\rangle$; $\mathbf{v} = \langle 5, 12\rangle$

b. $\mathbf{v}_1 = 2\mathbf{i} - 3\mathbf{j}$; $\mathbf{v}_2 = 6\mathbf{i} + 4\mathbf{j}$

Solution:

a.
$$\begin{aligned}\cos\theta &= \frac{\mathbf{u}}{|\mathbf{u}|}\cdot\frac{\mathbf{v}}{|\mathbf{v}|}\\ &= \left\langle\frac{-3}{5}, \frac{4}{5}\right\rangle\cdot\left\langle\frac{5}{13}, \frac{12}{13}\right\rangle\\ &= \frac{-15}{65} + \frac{48}{65}\\ &= \frac{33}{65}\\ \theta &= \cos^{-1}\left(\frac{33}{65}\right)\\ &\approx 59.5°\end{aligned}$$

b.
$$\begin{aligned}\cos\theta &= \frac{\mathbf{v}_1}{|\mathbf{v}_1|}\cdot\frac{\mathbf{v}_2}{|\mathbf{v}_2|}\\ &= \left\langle\frac{2}{\sqrt{13}}, \frac{-3}{\sqrt{13}}\right\rangle\cdot\left\langle\frac{6}{\sqrt{52}}, \frac{4}{\sqrt{52}}\right\rangle\\ &= \frac{12}{\sqrt{676}} + \frac{-12}{\sqrt{676}}\\ &= \frac{0}{26} = 0\\ \theta &= \cos^{-1}0\\ &= 90°\end{aligned}$$

NOW TRY EXERCISES 49 THROUGH 66

Note we have implicitly shown that if $\mathbf{u}\cdot\mathbf{v} = 0$, then $\mathbf{u}\perp\mathbf{v}$. As with other vector operations, recognizing certain properties of the dot product will enable us to work with them more efficiently.

PROPERTIES OF THE DOT PRODUCT

Given vectors **u**, **v**, and **w** and a constant k,

1. $\mathbf{u}\cdot\mathbf{v} = \mathbf{v}\cdot\mathbf{u}$
2. $\mathbf{u}\cdot\mathbf{u} = |\mathbf{u}|^2$
3. $\mathbf{w}\cdot(\mathbf{u}+\mathbf{v}) = \mathbf{w}\cdot\mathbf{u} + \mathbf{w}\cdot\mathbf{v}$
4. $k(\mathbf{u}\cdot\mathbf{v}) = k\mathbf{u}\cdot\mathbf{v} = \mathbf{u}\cdot k\mathbf{v}$
5. $\mathbf{0}\cdot\mathbf{u} = \mathbf{u}\cdot\mathbf{0} = 0$
6. $\frac{\mathbf{u}}{|\mathbf{u}|}\cdot\frac{\mathbf{v}}{|\mathbf{v}|} = \frac{\mathbf{u}\cdot\mathbf{v}}{|\mathbf{u}||\mathbf{v}|}$

Property 6 offers an alternative to unit vectors when finding $\cos\theta$—the dot product of the vectors can be computed first, and the result divided by the product of their magnitudes: $\cos\theta = \frac{\mathbf{u}\cdot\mathbf{v}}{|\mathbf{u}||\mathbf{v}|}$. Proofs of the first two properties are given here. Proofs of the others have a similar development (see Exercises 79 through 82). For any two nonzero vectors $\mathbf{u} = \langle a, b\rangle$ and $\mathbf{v} = \langle c, d\rangle$:

Property 1:
$$\begin{aligned}\mathbf{u}\cdot\mathbf{v} &= \langle a, b\rangle\cdot\langle c, d\rangle\\ &= ac + bd\\ &= ca + db\\ &= \langle c, d\rangle\cdot\langle a, b\rangle\\ &= \mathbf{v}\cdot\mathbf{u}\end{aligned}$$

Property 2:
$$\begin{aligned}\mathbf{u}\cdot\mathbf{u} &= \langle a, b\rangle\cdot\langle a, b\rangle\\ &= a^2 + b^2\\ &= |\mathbf{u}|^2\\ &(\text{since } |\mathbf{u}| = \sqrt{a^2 + b^2})\end{aligned}$$

Using $\text{comp}_{\mathbf{v}}\mathbf{u} = |\mathbf{u}|\cos\theta$ and $\mathbf{u}\cdot\mathbf{v} = \text{comp}_{\mathbf{v}}\mathbf{u} \times |\mathbf{v}|$, we can also state the following relationships, which give us some flexibility on how we approach applications of the dot product.

For any two vectors $\mathbf{u} = \langle a, b\rangle$ and $\mathbf{v} = \langle c, d\rangle$:

(1) $\mathbf{u}\cdot\mathbf{v} = ac + bd$ — standard computation of the dot product

(2) $\mathbf{u}\cdot\mathbf{v} = |\mathbf{u}|\cos\theta \times |\mathbf{v}|$ — alternative computation of the dot product

(3) $\mathbf{u}\cdot\mathbf{v} = \text{comp}_{\mathbf{v}}\mathbf{u} \times |\mathbf{v}|$ — replace $|\mathbf{u}|\cos\theta$ in (2) with $\text{comp}_{\mathbf{v}}\mathbf{u}$

(4) $\dfrac{\mathbf{u}\cdot\mathbf{v}}{|\mathbf{u}||\mathbf{v}|} = \cos\theta$ — divide (2) by scalars $|\mathbf{u}|$ and $|\mathbf{v}|$

(5) $\dfrac{\mathbf{u}\cdot\mathbf{v}}{|\mathbf{v}|} = \text{comp}_{\mathbf{v}}\mathbf{u}$ — divide (3) by $|\mathbf{v}|$

Figure 5.63

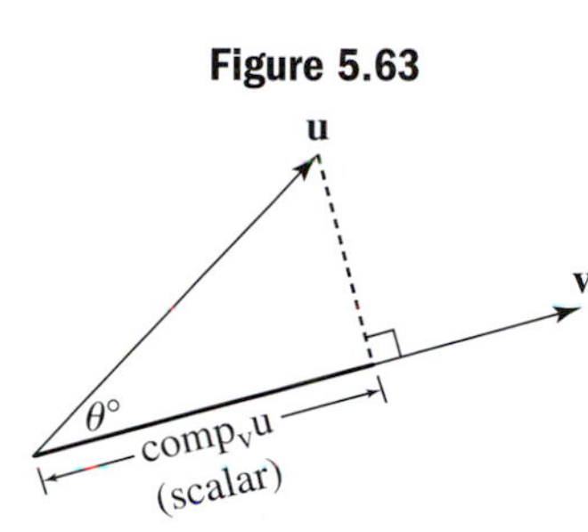

E. Vector Projections and Orthogonal Components

In work problems and other simple applications, it is enough to find and apply $\text{comp}_{\mathbf{v}}\mathbf{u}$ (Figure 5.63). However, applications involving thrust and drag forces, tension and stress limits in a cable, electronic circuits, and cartoon animations often require that we also find the *vector form of* $\text{comp}_{\mathbf{v}}\mathbf{u}$. This is called the **projection** of **u** along **v** or $\mathbf{proj_v u}$, and is a vector in the same direction of **v** with magnitude $\text{comp}_{\mathbf{v}}\mathbf{u}$ (Figures 5.64 and 5.65). By its design, the unit vector $\dfrac{\mathbf{v}}{|\mathbf{v}|}$ has a length of one and points in the same direction as **v**, so $\mathbf{proj_v u}$ can be computed as $\text{comp}_{\mathbf{v}}\mathbf{u} \times \dfrac{\mathbf{v}}{|\mathbf{v}|}$ (see Example 7, Section 5.4). Using equation (5) above and the properties shown earlier, a more convenient formula for $\mathbf{proj_v u}$ can be found:

Figure 5.64

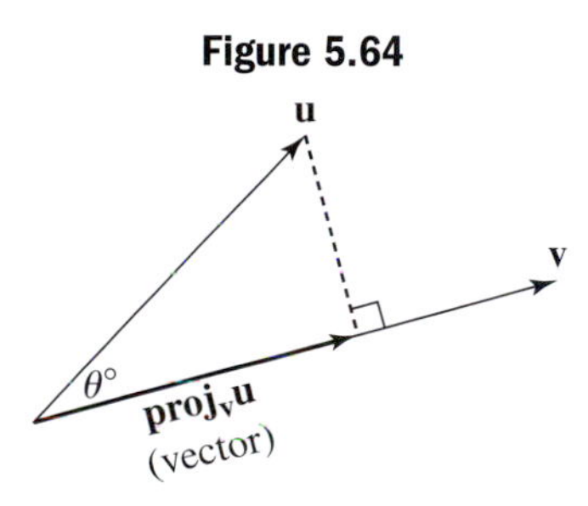

Figure 5.65

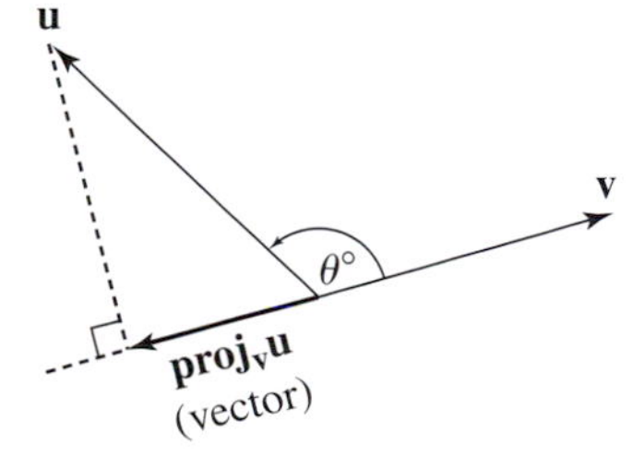

$$\begin{aligned}\mathbf{proj_v u} &= \text{comp}_{\mathbf{v}}\mathbf{u} \times \frac{\mathbf{v}}{|\mathbf{v}|} && \text{definition of a projection}\\ &= \frac{\mathbf{u}\cdot\mathbf{v}}{|\mathbf{v}|} \times \frac{\mathbf{v}}{|\mathbf{v}|} && \text{substitute } \frac{\mathbf{u}\cdot\mathbf{v}}{|\mathbf{v}|} \text{ for comp}_{\mathbf{v}}\mathbf{u}\\ &= \frac{\mathbf{u}\cdot\mathbf{v}}{|\mathbf{v}|^2} \times \mathbf{v} && \text{rewrite factors}\end{aligned}$$

VECTOR PROJECTIONS

Given vectors **u** and **v**, the projection of **u** along **v** is the *vector* $\mathbf{proj_v u}$ defined by $\mathbf{proj_v u} = \left(\dfrac{\mathbf{u}\cdot\mathbf{v}}{|\mathbf{v}|^2}\right)\mathbf{v}$.

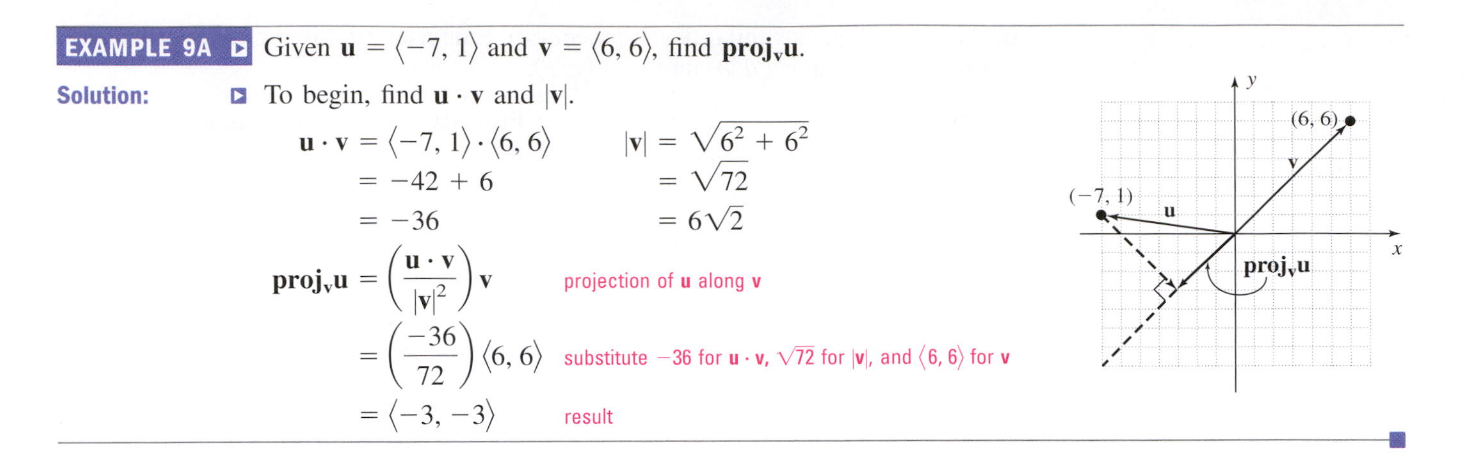

EXAMPLE 9A Given $\mathbf{u} = \langle -7, 1 \rangle$ and $\mathbf{v} = \langle 6, 6 \rangle$, find $\mathbf{proj_v u}$.

Solution: To begin, find $\mathbf{u} \cdot \mathbf{v}$ and $|\mathbf{v}|$.

$$\begin{aligned} \mathbf{u} \cdot \mathbf{v} &= \langle -7, 1 \rangle \cdot \langle 6, 6 \rangle & |\mathbf{v}| &= \sqrt{6^2 + 6^2} \\ &= -42 + 6 & &= \sqrt{72} \\ &= -36 & &= 6\sqrt{2} \end{aligned}$$

$$\begin{aligned} \mathbf{proj_v u} &= \left(\frac{\mathbf{u} \cdot \mathbf{v}}{|\mathbf{v}|^2}\right)\mathbf{v} && \text{projection of } \mathbf{u} \text{ along } \mathbf{v} \\ &= \left(\frac{-36}{72}\right)\langle 6, 6 \rangle && \text{substitute } -36 \text{ for } \mathbf{u}\cdot\mathbf{v},\ \sqrt{72} \text{ for } |\mathbf{v}|, \text{ and } \langle 6, 6\rangle \text{ for } \mathbf{v} \\ &= \langle -3, -3 \rangle && \text{result} \end{aligned}$$

A useful consequence of computing $\mathbf{proj_v u}$ is we can then **resolve** an arbitrary vector $\mathbf{u}$ into **orthogonal components,** with one component parallel to $\mathbf{v}$ and the other perpendicular to $\mathbf{v}$ (the dashed line in the diagram in Example 9A). In general terms, this means we can write $\mathbf{u}$ as the vector sum $\mathbf{u}_1 + \mathbf{u}_2$, where $\mathbf{u}_1 = \mathbf{proj_v u}$ and $\mathbf{u}_2 = \mathbf{u} - \mathbf{u}_1$ (note $\mathbf{u}_1 \parallel \mathbf{v}$).

RESOLVING A VECTOR INTO ORTHOGONAL COMPONENTS

Given vectors $\mathbf{u}$, $\mathbf{v}$, and $\mathbf{proj_v u}$, $\mathbf{u}$ can be resolved into the orthogonal components $\mathbf{u}_1$ and $\mathbf{u}_2$, where $\mathbf{u} = \mathbf{u}_1 + \mathbf{u}_2$, $\mathbf{u}_1 = \mathbf{proj_v u}$, and $\mathbf{u}_2 = \mathbf{u} - \mathbf{u}_1$.

EXAMPLE 9B Given $\mathbf{u} = \langle 2, 8 \rangle$ and $\mathbf{v} = \langle 8, 6 \rangle$, resolve $\mathbf{u}$ into orthogonal components $\mathbf{u}_1$ and $\mathbf{u}_2$, where $\mathbf{u}_1 \parallel \mathbf{v}$ and $\mathbf{u}_2 \perp \mathbf{v}$. Also verify $\mathbf{u}_1 \perp \mathbf{u}_2$.

Solution: Once again, begin by finding $\mathbf{u} \cdot \mathbf{v}$ and $|\mathbf{v}|$.

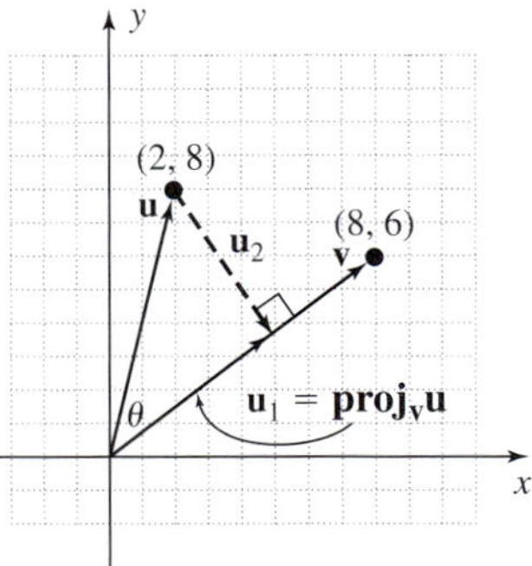

$$\begin{aligned} \mathbf{u} \cdot \mathbf{v} &= \langle 2, 8 \rangle \cdot \langle 8, 6 \rangle & |\mathbf{v}| &= \sqrt{8^2 + 6^2} \\ &= 16 + 48 & &= \sqrt{100} \\ &= 64 & &= 10 \end{aligned}$$

$$\begin{aligned} \mathbf{proj_v u} &= \left(\frac{\mathbf{u} \cdot \mathbf{v}}{|\mathbf{v}|^2}\right)\mathbf{v} && \text{projection of } \mathbf{u} \text{ along } \mathbf{v} \\ &= \left(\frac{64}{100}\right)\langle 8, 6 \rangle && \text{substitute 64 for } \mathbf{u}\cdot\mathbf{v},\ 10 \text{ for } |\mathbf{v}|, \text{ and } \langle 8, 6\rangle \text{ for } \mathbf{v} \\ &= \langle 5.12, 3.84 \rangle && \text{result} \end{aligned}$$

For $\mathbf{proj_v u} = \mathbf{u}_1 = \langle 5.12, 3.84 \rangle$, we have $\mathbf{u}_2 = \mathbf{u} - \mathbf{u}_1 = \langle 2, 8 \rangle - \langle 5.12, 3.84 \rangle = \langle -3.12, 4.16 \rangle$. To verify $\mathbf{u}_1 \perp \mathbf{u}_2$, we need only show $\mathbf{u}_1 \cdot \mathbf{u}_2 = 0$:

$$\begin{aligned} \mathbf{u}_1 \cdot \mathbf{u}_2 &= \langle 5.12, 3.84 \rangle \cdot \langle -3.12, 4.16 \rangle \\ &= (5.12)(-3.12) + (3.84)(4.16) \\ &= 0 \checkmark \end{aligned}$$

NOW TRY EXERCISES 67 THROUGH 72

F. Vectors and the Height of a Projectile

Our final application of vectors involves **projectile motion.** A projectile is any object that is thrown or projected upward, with no source of propulsion to sustain its motion. In this case, the only force acting on the projectile is gravity (air resistance is neglected), so the maximum height and the range of the projectile depend solely on its initial velocity and the angle θ at which it is projected. In a college algebra course, the equation $y = v_0t - 16t^2$ is developed to model the height in feet (at time t) of a projectile thrown vertically upward with initial velocity v_0 in feet per second. Here, we'll modify the equation slightly to take into account that the object is now moving horizontally as well as vertically. As you can see in Figure 5.66, the vector $\mathbf{v}$ representing the initial velocity, as well as the velocity vector at other times, can easily be decomposed into horizontal and vertical components. This will enable us to find a more general relationship for the position of the projectile. For now, we'll let $\mathbf{v}_y$ represent the component of velocity in the vertical (y) direction, and $\mathbf{v}_x$ represent the component of velocity in the horizontal (x) direction. Since gravity acts only in the vertical (and negative) direction, the horizontal component of the velocity remains constant at $\mathbf{v}_x = |\mathbf{v}|\cos\theta$. Using $D = RT$, the x-coordinate of the projectile at time t is $x = (|\mathbf{v}|\cos\theta)t$. For the vertical component $\mathbf{v}_y$ we use the projectile equation developed earlier, *substituting* $|\mathbf{v}|\sin\theta$ *for* $\mathbf{v}_0$, since the angle of projection is no longer 90°. This gives the y-coordinate at time t as $y = v_0t - 16t^2 \rightarrow (|\mathbf{v}|\sin\theta)t - 16t^2$.

Figure 5.66

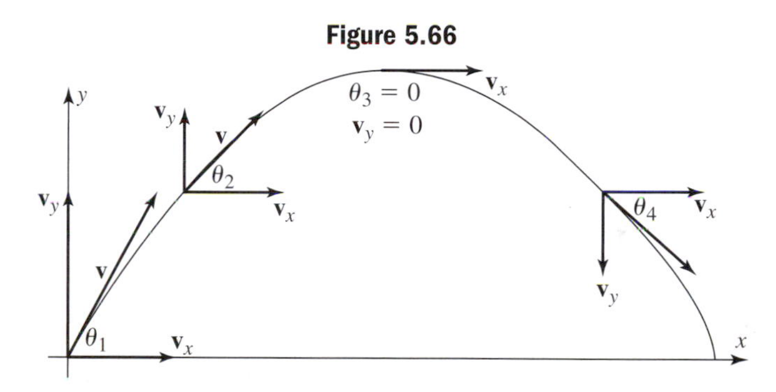

PROJECTILE MOTION

Given an object is projected upward from the origin with initial velocity $|\mathbf{v}|$ at an angle of $\theta°$.

The x-coordinate of its position at time t is $x = (|\mathbf{v}|\cos\theta)t$.

The y-coordinate of its position at time t is $y = (|\mathbf{v}|\sin\theta)t - 16t^2$.

EXAMPLE 10 An arrow is shot upward with an initial velocity of 150 ft/sec at an angle of 50°. (a) Find the position of the arrow after 2 sec. (b) How many seconds does it take to reach a height of 190 ft?

Solution: **a.** Using the preceding equations yields these coordinates for its position at $t = 2$:

$$x = (|\mathbf{v}|\cos\theta)t \qquad y = (|\mathbf{v}|\sin\theta)t - 16t^2$$
$$= (150\cos 50°)(2) \qquad = (150\sin 50°)(2) - 16(2)^2$$
$$\approx 193 \qquad \approx 166$$

The arrow has traveled a horizontal distance of about 193 ft and is 166 ft high.

b. To find the time required to reach 190 ft in height, set the equation for the y coordinate equal to 190, which yields a quadratic equation in t:

$$y = (|\mathbf{v}|\sin\theta)t - 16t^2 \quad \text{equation for } y$$
$$190 = (150\sin 50°)t - 16t^2 \quad \text{substitute 150 for } |\mathbf{v}| \text{ and } 50° \text{ for } \theta$$
$$0 \approx -16(t)^2 + 115t - 190 \quad 150\sin 50 \approx 115$$

Using the quadratic formula we find that $t \approx 2.6$ sec and $t \approx 4.6$ sec are solutions. This makes sense, since the arrow reaches a given height once on the way up and again on the way down, as long as it hasn't reached its maximum height.

NOW TRY EXERCISES 73 THROUGH 78

For more on projectile motion, see the *Calculator Exploration and Discovery* feature at the end of this chapter.

TECHNOLOGY HIGHLIGHT

The Magnitude and Reference Angle of a Vector

The keystrokes shown apply to a TI-84 Plus model. Please consult our Internet site or your manual for other models.

As you've seen in this chapter, a vector can be written or named in the component form $\langle a, b\rangle$ or in terms of its magnitude and direction. The vector $\langle 5\sqrt{3}, 5\rangle$ is equal to the position vector with magnitude 10 and angle $\theta = 30°$, since $10\cos 30° = 5\sqrt{3}$ and $10\sin 30° = 5$. Many calculators use "r" (instead of $|\mathbf{v}|$) for the magnitude of a vector. When the magnitude and angle of a vector are written in the form (r, θ) we say the vector is written in **polar form.** Most graphing calculators offer features that enable a quick conversion from the component form $\langle a, b\rangle$ of a vector to its polar form (r, θ) and vice versa. On the TI-84 Plus, these features are accessed using 2nd APPS **(ANGLE)**. As you can see from the screen in Figure 5.67, we are interested in options 5 through 8 (press the up arrow ▲ to bring option 8 into view). Option **5:R ▶ Pr(** takes the **R**ectangular coordinates and returns the magnitude **r** of the **P**olar Coordinates. Option **6:R ▶ Pθ(** returns the angle of the **P**olar form. Option **7:P ▶ Rx(** takes the **P**olar form (r, θ) of the vector and returns the horizontal component of the **R**ectangular form. As you might expect, option **8:P ▶ Ry(** takes

Figure 5.67

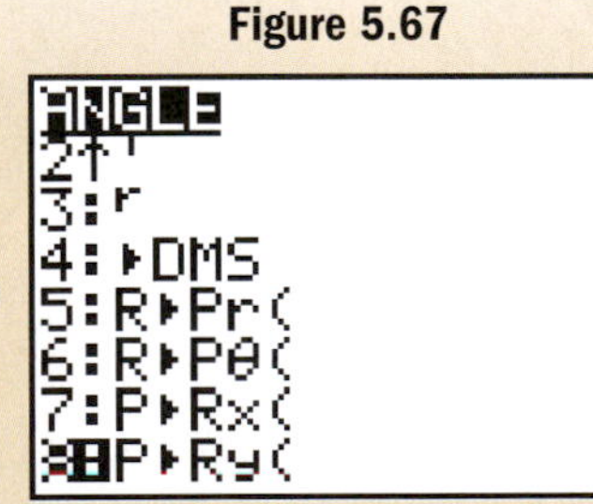

the **P**olar form (r, θ) of the vector and returns the vertical component of the **R**ectangular form. To verify the previous illustration for $\langle 5\sqrt{3}, 5\rangle$, we CLEAR the home screen and press 2nd APPS 5 to access the desired option and enter the components as **5:R ▶ Pr(**$5\sqrt{3}$, 5). Pressing ENTER gives a magnitude 10, as expected (Figure 5.68).

Exercise 1: Use your calculator to write the vector $\langle 5\sqrt{2}, 5\sqrt{2}\rangle$ in polar form. Check the result by converting manually.

Exercise 2: Use your calculator to write the vector $(r, \theta) = \langle 20\sqrt{2}, 45\rangle$ in rectangular form. Check the result by converting manually.

Figure 5.68

```
R▸Pr(5√(3),5)
                10
R▸Pθ(5√(3),5)
                30
P▸Rx(10,30)
       8.660254038
5√(3)
```

5.5 EXERCISES

CONCEPTS AND VOCABULARY

Fill in each blank with the appropriate word or phrase. Carefully reread the section if needed.

1. Vector forces are in ________ when they counterbalance each other. Such vectors have a sum of ________.

2. The component of a vector **u** along another vector **v** is written notationally as ______, and is computed as ______.

3. Two vectors that meet at a right angle are said to be ________.

4. The component of **u** along **v** is a ______ quantity. The projection of **u** along **v** is a ______.

5. Explain/discuss exactly what information the dot product of two vectors gives us. Illustrate with a few examples.

6. Compare and contrast the projectile equations $y = v_0 t - 16t^2$ and $y = (v_0 \sin\theta)t - 16t^2$. Discuss similarities/differences using illustrative examples.

DEVELOPING YOUR SKILLS

The force vectors given are acting on a common point P. Find an additional force vector so that equilibrium takes place.

7. $\mathbf{F}_1 = \langle -8, -3\rangle$; $\mathbf{F}_2 = \langle 2, -5\rangle$

8. $\mathbf{F}_1 = \langle -2, 7\rangle$; $\mathbf{F}_2 = \langle 5, 3\rangle$

9. $\mathbf{F}_1 = \langle -2, -7\rangle$; $\mathbf{F}_2 = \langle 2, -7\rangle$; $\mathbf{F}_3 = \langle 5, 4\rangle$

10. $\mathbf{F}_1 = \langle -3, 10\rangle$; $\mathbf{F}_2 = \langle -10, 3\rangle$; $\mathbf{F}_3 = \langle -9, -2\rangle$

11. $\mathbf{F}_1 = 5\mathbf{i} - 2\mathbf{j}$; $\mathbf{F}_2 = \mathbf{i} + 10\mathbf{j}$

12. $\mathbf{F}_1 = -7\mathbf{i} + 6\mathbf{j}$; $\mathbf{F}_2 = -8\mathbf{i} - 3\mathbf{j}$

13. $\mathbf{F}_1 = 2.5\mathbf{i} + 4.7\mathbf{j}$; $\mathbf{F}_2 = -0.3\mathbf{i} + 6.9\mathbf{j}$; $\mathbf{F}_3 = -12\mathbf{j}$

14. $\mathbf{F}_1 = 3\sqrt{2}\mathbf{i} - 2\sqrt{3}\mathbf{j}$; $\mathbf{F}_2 = -2\mathbf{i} + 7\mathbf{j}$; $\mathbf{F}_3 = 5\mathbf{i} + 2\sqrt{3}\mathbf{j}$

15.

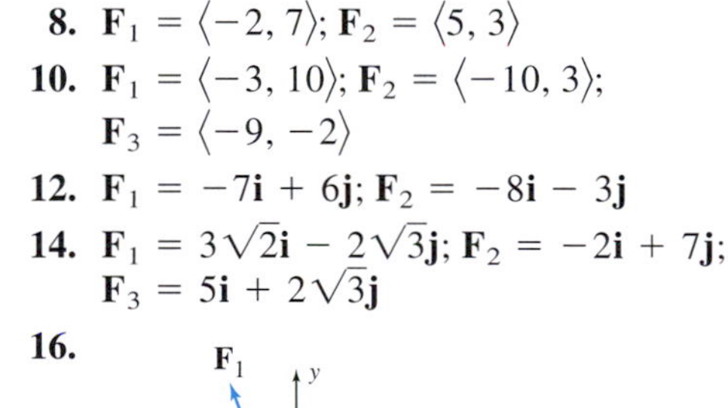

$\mathbf{F}_1$, y, 10, 104°, 6, $\mathbf{F}_2$, 25°, 20°, x, 9, $\mathbf{F}_3$

16.

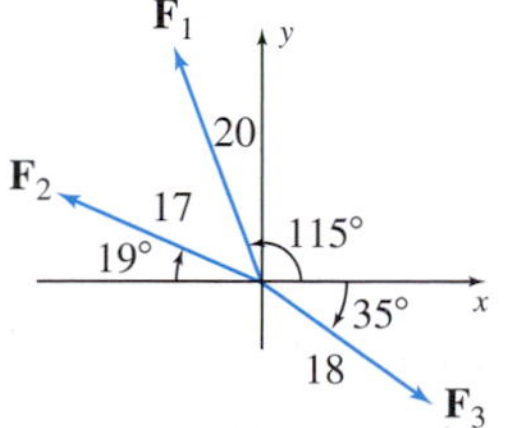

17. The force vectors $\mathbf{F}_1$ and $\mathbf{F}_2$ are simultaneously acting on a point P. Find a third vector $\mathbf{F}_3$ so that equilibrium takes place if $\mathbf{F}_1 = \langle 19, 10\rangle$ and $\mathbf{F}_2 = \langle 5, 17\rangle$.

18. The force vectors $\mathbf{F}_1$, $\mathbf{F}_2$, and $\mathbf{F}_3$ are simultaneously acting on a point P. Find a fourth vector $\mathbf{F}_4$ so that equilibrium takes place if $\mathbf{F}_1 = \langle -12, 2\rangle$, $\mathbf{F}_2 = \langle -6, 17\rangle$, and $\mathbf{F}_3 = \langle 3, 15\rangle$.

19. A new "Survivor" game involves a three-team tug-of-war. Teams 1 and 2 are pulling with the magnitude and at the angles indicated in the diagram. If the teams are currently in a stalemate, find the magnitude and angle of the rope held by team 3.

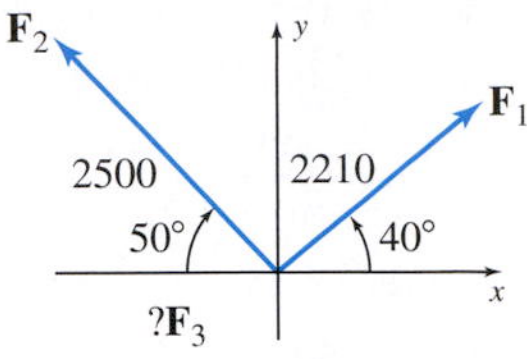

20. Three cowhands have roped a wild stallion and are attempting to hold him steady. The first and second cowhands are pulling with the magnitude and at the angles indicated in the diagram. If the stallion is held fast by the three cowhands, find the magnitude and angle of the rope from the third cowhand.

Exercise 20

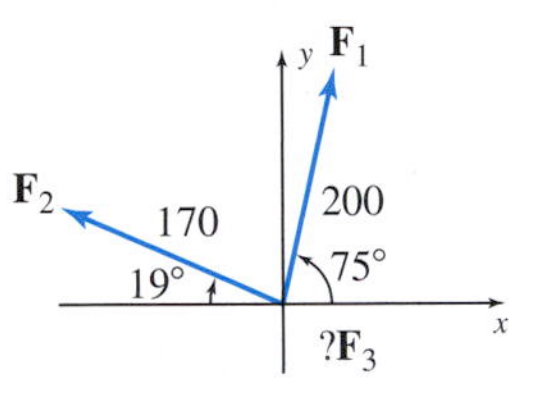

Find the component of **u** along **v** (compute $\text{comp}_{\mathbf{v}}\mathbf{u}$) for the vectors **u** and **v** given.

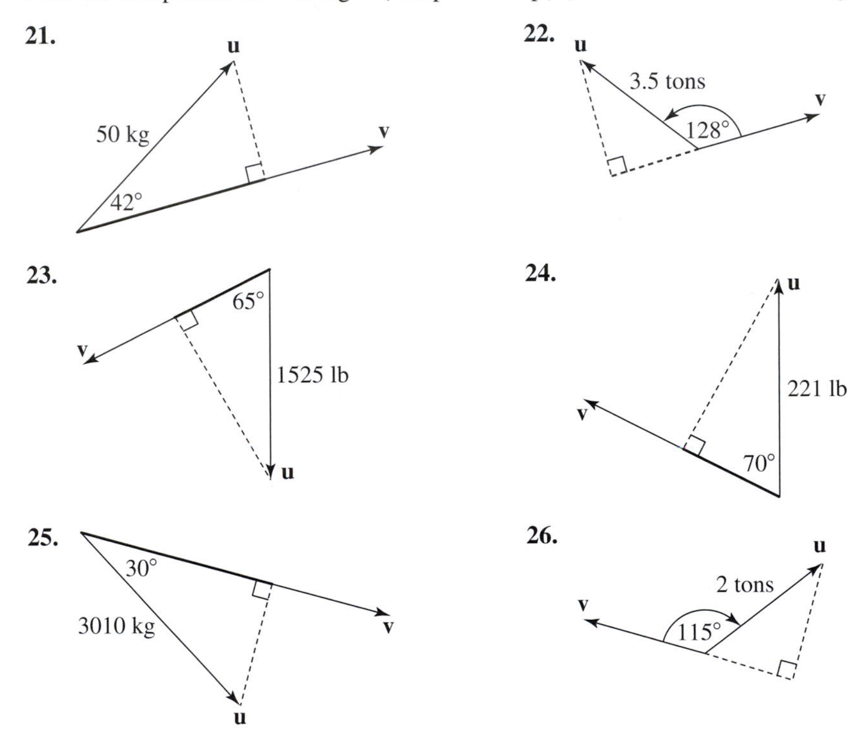

27. **Static equilibrium:** A 500-lb crate is sitting on a ramp that is inclined at 35°. Find the force needed to hold the object stationary.

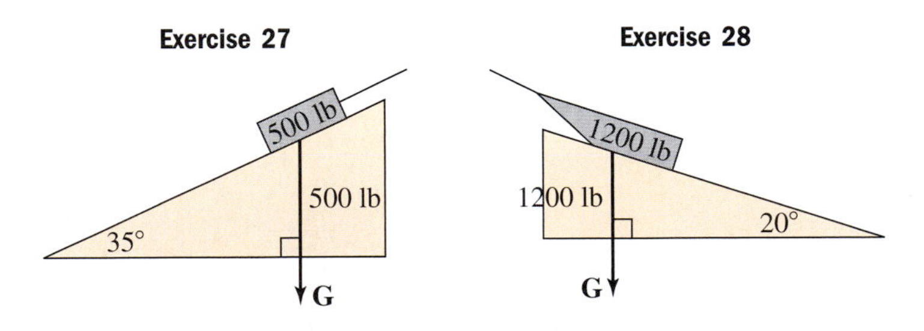

28. **Static equilibrium:** A 1200-lb skiff is being pulled from a lake, using a boat ramp inclined at 20°. Find the minimum force needed to dock the skiff.

29. Static equilibrium: A 325-kg carton is sitting on a ramp, held stationary by 225 kg of tension in a restraining rope. Find the ramps's angle of incline.

30. Static equilibrium: A heavy dump truck is being winched up a ramp with an 18° incline. Find the weight of the truck if the winch is working at its maximum capacity of 1.75 tons.

31. While rearranging the patio furniture, Rick has to push the weighted base of the umbrella stand 15 ft. If he uses a constant force of 75 lb, how much work did he do? Did he break a sweat?

32. Vinny's car just broke down in the middle of the road. Luckily, a buddy is with him and offers to steer if Vinny will get out and push. If he pushes with a constant force of 185 lb to move the car 30 ft onto the shoulder, how much work did he do?

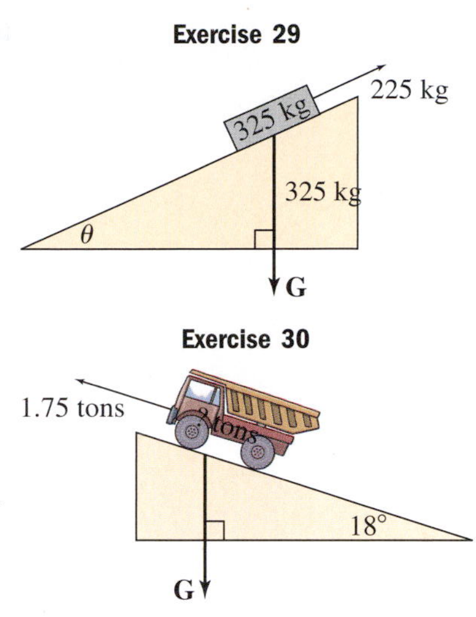

WORKING WITH FORMULAS

The range of a projectile: $R = \dfrac{v^2 \sin\theta \cos\theta}{16}$

33. The range of a projected object (total horizontal distance traveled) is given by the formula shown, where v is the initial velocity and θ is the angle at which it is projected. If an arrow leaves the bow traveling 175 ft/sec at an angle of 45°, what horizontal distance will it travel?

34. A collegiate javelin thrower releases the javelin at a 40° angle, with an initial velocity of about 95 ft/sec. If the NCAA record is 280 ft, will this throw break the record? What is the smallest angle of release that will break this record? If the javelin were released at the optimum 45°, by how many feet would the record be broken?

APPLICATIONS

35. Plowing a field: An old-time farmer is plowing his field with a mule. How much work does the mule do in plowing one length of a field 300 ft long, if it pulls the plow with a constant force of 250 lb and the straps make a 30° angle with the horizontal.

36. Pulling a sled: To enjoy a beautiful snowy day, a mother is pulling her three children on a sled along a level street. How much work (play) is done if the street is 100 ft long and she pulls with a constant force of 55 lb with the tow-rope making an angle of 32° with the street?

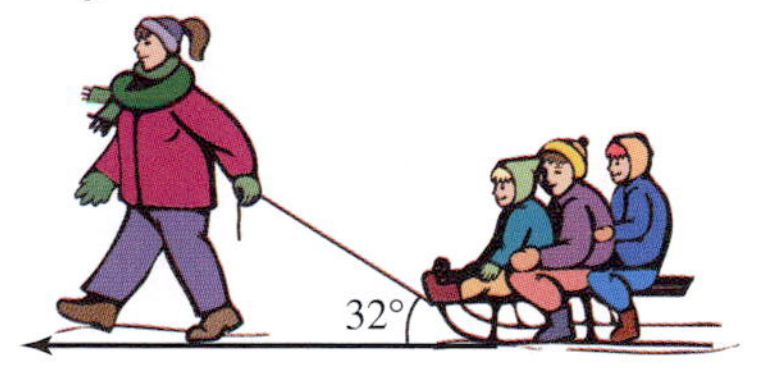

Exercise 37

37. Tough-man contest: As part of a "tough-man" contest, participants are required to pull a bus along a level street for 100 ft. If one contestant did 45,000 ft-lb of work to accomplish the task and the straps used made an angle of 5° with the street, find the tension in the strap during the pull.

38. Moving supplies: An arctic explorer is hauling supplies from the supply hut to her tent, a distance of 150 ft, in a sled she is dragging behind her. If 9000 ft-lb of work was done and the straps used made an angle of 25° with the snow-covered ground, find the tension in the strap during the task.

39. Wheelbarrow rides: To break up the monotony of a long, hot, boring Saturday, a father decides to (carefully) give his kids a ride in a wheelbarrow. He applies a force of 30 N to move the "load" 100 m, then stops to rest. Find the amount of work done if the wheelbarrow makes an angle of 20° with level ground while in motion.

40. Mowing the lawn: A home owner applies a force of 7.2 N to push her lawn mower back and forth across the back yard. Find the amount of work done if the yard is 50 m long, requires 24 passes to get the lawn mowed, and the mower arm makes an angle of 39° with the level ground.

Force vectors: For the force vector **F** and vector **v** given, find the amount of work required to move an object along the entire length of **v**. Assume force is in pounds and distance in feet.

41. $\mathbf{F} = \langle 15, 10\rangle$; $\mathbf{v} = \langle 50, 5\rangle$

42. $\mathbf{F} = \langle -5, 12\rangle$; $\mathbf{v} = \langle -25, 10\rangle$

43. $\mathbf{F} = \langle 8, 2\rangle$; $\mathbf{v} = \langle 15, -1\rangle$

44. $\mathbf{F} = \langle 15, -3\rangle$; $\mathbf{v} = \langle 24, -20\rangle$

45. Use the dot product to verify the solution to Exercise 41.

46. Use the dot product to verify the solution to Exercise 42.

47. Use the dot product to verify the solution to Exercise 43.

48. Use the dot product to verify the solution to Exercise 44.

For each pair of vectors given, (a) compute the dot product $\mathbf{p} \cdot \mathbf{q}$ and (b) find the angle between the vectors to the nearest tenth of a degree.

49. $\mathbf{p} = \langle 5, 2\rangle$; $\mathbf{q} = \langle 3, 7\rangle$

50. $\mathbf{p} = \langle -3, 6\rangle$; $\mathbf{q} = \langle 2, -5\rangle$

51. $\mathbf{p} = -2\mathbf{i} + 3\mathbf{j}$; $\mathbf{q} = -6\mathbf{i} - 4\mathbf{j}$

52. $\mathbf{p} = -4\mathbf{i} + 3\mathbf{j}$; $\mathbf{q} = -6\mathbf{i} - 8\mathbf{j}$

53. $\mathbf{p} = 7\sqrt{2}\mathbf{i} - 3\mathbf{j}$; $\mathbf{q} = 2\sqrt{2}\mathbf{i} + 9\mathbf{j}$

54. $\mathbf{p} = \sqrt{2}\mathbf{i} - 3\mathbf{j}$; $\mathbf{q} = 3\sqrt{2}\mathbf{i} + 5\mathbf{j}$

Determine if the pair of vectors given are orthogonal.

55. $\mathbf{u} = \langle 7, -2\rangle$; $\mathbf{v} = \langle 4, 14\rangle$

56. $\mathbf{u} = \langle -3.5, 2.1\rangle$; $\mathbf{v} = \langle -6, -10\rangle$

57. $\mathbf{u} = \langle -6, -3\rangle$; $\mathbf{v} = \langle -8, 15\rangle$

58. $\mathbf{u} = \langle -5, 4\rangle$; $\mathbf{v} = \langle -9, -11\rangle$

59. $\mathbf{u} = -2\mathbf{i} - 6\mathbf{j}$; $\mathbf{v} = 9\mathbf{i} - 3\mathbf{j}$

60. $\mathbf{u} = 3\sqrt{2}\mathbf{i} - 2\mathbf{j}$; $\mathbf{v} = 2\sqrt{2}\mathbf{i} + 6\mathbf{j}$

Find $\text{comp}_{\mathbf{v}}\mathbf{u}$ for the vectors **u** and **v** given.

61. $\mathbf{u} = \langle 3, 5\rangle$; $\mathbf{v} = \langle 7, 1\rangle$

62. $\mathbf{u} = \langle 3, 5\rangle$; $\mathbf{v} = \langle -7, 1\rangle$

63. $\mathbf{u} = -7\mathbf{i} + 4\mathbf{j}$; $\mathbf{v} = -10\mathbf{j}$

64. $\mathbf{u} = 8\mathbf{i}$; $\mathbf{v} = 10\mathbf{i} + 3\mathbf{j}$

65. $\mathbf{u} = 7\sqrt{2}\mathbf{i} - 3\mathbf{j}$; $\mathbf{v} = 6\mathbf{i} + 5\sqrt{3}\mathbf{j}$

66. $\mathbf{u} = -3\sqrt{2}\mathbf{i} + 6\mathbf{j}$; $\mathbf{v} = 2\mathbf{i} + 5\sqrt{5}\mathbf{j}$

For each pair of vectors given, (a) find the projection of **u** along **v** (compute $\mathbf{proj_v u}$) and (b) resolve **u** into vectors $\mathbf{u}_1$ and $\mathbf{u}_2$, where $\mathbf{u}_1 \parallel \mathbf{v}$ and $\mathbf{u}_2 \perp \mathbf{v}$.

67. $\mathbf{u} = \langle 2, 6\rangle$; $\mathbf{v} = \langle 8, 3\rangle$

68. $\mathbf{u} = \langle -3, 8\rangle$; $\mathbf{v} = \langle -12, 3\rangle$

69. $\mathbf{u} = \langle -2, -8\rangle$; $\mathbf{v} = \langle -6, 1\rangle$

70. $\mathbf{u} = \langle -4.2, 3\rangle$; $\mathbf{v} = \langle -5, -8.3\rangle$

71. $\mathbf{u} = 10\mathbf{i} + 5\mathbf{j}$; $\mathbf{v} = 12\mathbf{i} + 2\mathbf{j}$

72. $\mathbf{u} = -3\mathbf{i} - 9\mathbf{j}$; $\mathbf{v} = 5\mathbf{i} - 3\mathbf{j}$

Projectile motion: A projectile is launched from a catapult with the initial velocity v_0 and angle θ indicated. Find (a) the position of the object after 3 sec and (b) the time required to reach a height of 250 ft.

73. $v_0 = 250$ ft/sec; $\theta = 60°$

74. $v_0 = 300$ ft/sec; $\theta = 55°$

75. $v_0 = 200$ ft/sec; $\theta = 45°$

76. $v_0 = 500$ ft/sec; $\theta = 70°$

77. At the circus, a "human cannon ball" is shot from a large cannon with an initial velocity of 90 ft/sec at an angle of 65° from the horizontal. How high is the acrobat after 1.2 sec? How long until the acrobat is again at this same height?

78. A center fielder runs down a long hit by an opposing batter and whirls to throw the ball to the infield to keep the hitter to a double. If the initial velocity of the throw is 130 ft/sec and the ball is released at an angle of 30° with level ground, how high is the ball after 1.5 sec? How long until the ball again reaches this same height?

For the arbitrary vectors $\mathbf{u} = \langle a, b \rangle$, $\mathbf{v} = \langle c, d \rangle$, and $\mathbf{w} = \langle e, f \rangle$ and the scalars c and k, prove the following vector properties using the properties of real numbers.

79. $\mathbf{w} \cdot (\mathbf{u} + \mathbf{v}) = \mathbf{w} \cdot \mathbf{u} + \mathbf{w} \cdot \mathbf{v}$

80. $k(\mathbf{u} \cdot \mathbf{v}) = k\mathbf{u} \cdot \mathbf{v} = \mathbf{u} \cdot k\mathbf{v}$

81. $\mathbf{0} \cdot \mathbf{u} = \mathbf{u} \cdot \mathbf{0} = 0$

82. $\dfrac{\mathbf{u}}{|\mathbf{u}|} \cdot \dfrac{\mathbf{v}}{|\mathbf{v}|} = \dfrac{\mathbf{u} \cdot \mathbf{v}}{|\mathbf{u}||\mathbf{v}|}$

WRITING, RESEARCH, AND DECISION MAKING

83. As an alternative to $\cos\theta = \dfrac{\mathbf{u} \cdot \mathbf{v}}{|\mathbf{u}||\mathbf{v}|}$ for finding the angle between two vectors, the equation $\tan\theta = \dfrac{m_2 - m_1}{1 + m_2 m_1}$ can be used, where m_1 and m_2 represent the slopes of the vectors. Find the angle between the vectors $1\mathbf{i} + 5\mathbf{j}$ and $5\mathbf{i} + 2\mathbf{j}$ using each equation and comment on which you found more efficient. Then see if you can find a geometric connection between the two equations.

84. As a curiosity, research the number of countries that are members of the United Nations, and how many of these countries use the metric system. The answer may surprise you. In fact, the equation for projectile motion is most often stated with v_0 in m/s (rather than ft/sec), changing the equation slightly. Do some research to find what these changes are (find the equation stated in meters per second). Use the fact that 50 m/s ≈ 164 ft/sec in an example of your own making that will verify both forms give like results.

EXTENDING THE CONCEPT

85. The range of an object placed in projectile motion is the horizontal distance the object travels before reaching a height equal to its initial height, and is given by $R = \dfrac{v^2 \sin\theta\cos\theta}{16}$. Rewrite the expression *in terms of sine only,* using a well-known identity, then find the angle θ that will produce the maximum range. Discuss the implications of this θ in terms of various athletic skills.

86. Use the equations for the horizontal and vertical components of the projected object's position to obtain the equation of trajectory $y = (\tan\theta)x - \dfrac{16}{v^2\cos^2\theta}x^2$. This is a quadratic equation in x. What can you say about its graph? Include comments about the concavity, x-intercepts, maximum height, and so on.

MAINTAINING YOUR SKILLS

87. (5.1) Solve using the law of sines.

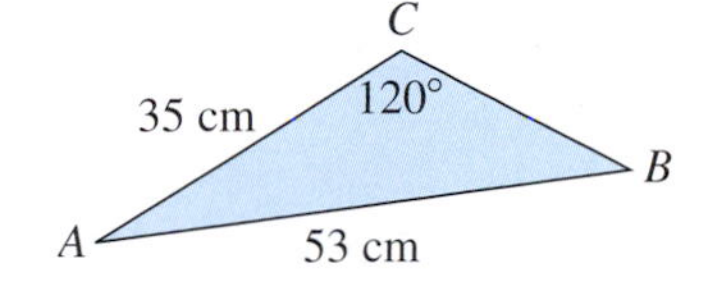

88. (5.3) Solve using the law of cosines.

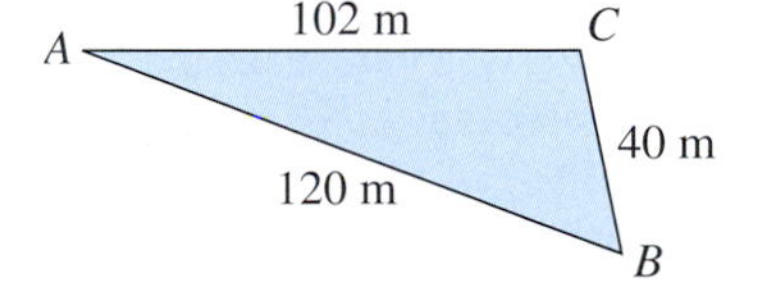

89. (4.2) Draw the graph of $y = \sin^{-1}x$ and state the domain and range of the function.

90. (3.3) Use a sum identity to find all solutions in $[0, 2\pi)$. Answer in exact form. $\sin(2x)\cos x + \cos(2x)\sin x = 0.5$.

91. (3.4) Use a half-angle identity to find the value of sin 15° and cos 75° in exact form. What do you notice?

92. (1.2) From the top of a 1000-ft cliff to a hut in the valley below, the angle of depression is 27°. How far is the hut from the foot of the cliff?

5.6 Complex Numbers

LEARNING OBJECTIVES

In Section 5.6 you will learn how to:

A. **Identify and simplify imaginary and complex numbers**

B. **Add and subtract complex numbers**

C. **Multiply complex numbers and find powers of *i***

D. **Divide complex numbers**

INTRODUCTION

Methods for solving quadratic equations were developed by almost every ancient civilization and a general formula has been known for millennia. But when early mathematicians encountered numbers like $2 + \sqrt{-121}$ in the solution of $x^2 - 4x + 125 = 0$, they found them useless and strange since there is no real number whose square is -121. In the sixteenth century, this attitude began to change. The equation $x^3 - 15x + 4 = 0$ was known to have three real roots, but produced solutions like $x = \sqrt[3]{2 + \sqrt{-121}} + \sqrt[3]{2 - \sqrt{-121}}$, where $\sqrt{-121}$ could no longer be dismissed—it was somehow a part of the real solutions known to exist. Today such numbers are much better understood, and known to be an important part of mathematics. In fact, the numbers we know today as *complex* and *imaginary*, are actually neither.

POINT OF INTEREST

Some of the most celebrated names in mathematics can be associated with the history of imaginary numbers and the complex number system. François Viéte (1540–1603) realized their existence, but did not accept them. Girolamo Cardano (1501–1576) found them puzzling, but actually produced real number solutions to cubic equations that required complex numbers to represent them. Albert Girard (1595–1632) was apparently the first to advocate their acceptance, suggesting that this would establish that a polynomial equation has exactly as many roots as its degree. Then in 1799, German mathematician Carl F. Gauss (1777–1855) proved the fundamental theorem of algebra, which states that every polynomial with degree $n \geq 1$ has at least one complex solution. For more information, see Exercise 80.

A. Identifying and Simplifying Imaginary and Complex Numbers

The equation $x^2 = -1$ has no real solutions, since the square of any real number must be positive. But if we apply the principle of square roots we get $x = \sqrt{-1}$ and $x = -\sqrt{-1}$, which seem to check when substituted into the original equation:

$$x^2 + 1 = 0 \quad \text{original equation}$$

$$(1) \quad (\sqrt{-1})^2 + 1 = 0 \quad \text{substitute } x = \sqrt{-1}$$

$$-1 + 1 = 0 \checkmark \quad \text{answer "checks"}$$

$$(2) \quad (-\sqrt{-1})^2 + 1 = 0 \quad \text{substitute } x = -\sqrt{-1}$$

$$-1 + 1 = 0 \checkmark \quad \text{answer "checks"}$$

This observation may have prompted the renaissance mathematicians to study these numbers in greater depth, reasoning that although they were not *real* number solutions—*they must be solutions of a different kind.* Their study eventually resulted in the introduction of the set of **imaginary numbers** and the **imaginary unit** *i*.

WORTHY OF NOTE

It was René Descartes (in 1637) who first used the term *imaginary* to describe these numbers, Leonhard Euler (in 1777) who introduced the letter *i* to represent $\sqrt{-1}$, and Carl F. Gauss (in 1831) who first used the phrase *complex number* to describe solutions that had both a real number part and an imaginary part. For more on complex numbers and their story, see www.mhhe.com/coburn.

IMAGINARY NUMBERS AND THE IMAGINARY UNIT

- Imaginary numbers are those of the form $\sqrt{k}$, where k is a negative real number.
- The imaginary unit i represents the number whose square is -1, meaning $i^2 = -1$ and $i = \sqrt{-1}$.

To simplify an imaginary number such as $\sqrt{-3} = \sqrt{-1 \cdot 3}$, we allow that the product property of radicals ($\sqrt{AB} = \sqrt{A}\sqrt{B}$) still applies if at most one of the factors is negative. In other words, $\sqrt{-1 \cdot 3} = \sqrt{-1}\sqrt{3} = i\sqrt{3}$. However, the property leads to a contradiction if *both* A and B are negative *and it cannot be applied in such cases* (see Exercise 81). For $\sqrt{-12}$ we have $\sqrt{-12} = \sqrt{-1 \cdot 4 \cdot 3} = i \cdot 2\sqrt{3} = 2i\sqrt{3}$, and we say the expression has been *simplified and written in terms of i*. It's best to write imaginary numbers with the unit "i" *in front of the radical* to prevent it being interpreted as being *under the radical*. In other words, $2i\sqrt{3}$ is preferred over $2\sqrt{3i}$.

EXAMPLE 1 Rewrite the imaginary numbers in terms of i and simplify.

a. $\sqrt{-121}$ **b.** $\sqrt{-7}$ **c.** $\sqrt{-24}$ **d.** $-3\sqrt{-16}$

Solution:

a. $\sqrt{-121} = \sqrt{-1 \cdot 121} = \sqrt{-1} \cdot \sqrt{121} = i \cdot 11$ or $11i$

b. $\sqrt{-7} = \sqrt{-1 \cdot 7} = \sqrt{-1} \cdot \sqrt{7} = i\sqrt{7}$

c. $\sqrt{-24} = \sqrt{-1 \cdot 24} = \sqrt{-1} \cdot \sqrt{4} \cdot \sqrt{6} = 2i\sqrt{6}$

d. $-3\sqrt{-16} = -3 \cdot \sqrt{-1 \cdot 16} = -3 \cdot \sqrt{-1} \cdot \sqrt{16} = -12i$

NOW TRY EXERCISES 7 THROUGH 12

EXAMPLE 2 The numbers $x = \frac{-6 + \sqrt{-16}}{2}$ and $x = \frac{6 - \sqrt{-16}}{2}$ are not real, but are known to be solutions of $x^2 + 6x + 13 = 0$. Simplify $\frac{-6 + \sqrt{-16}}{2}$.

Solution: Using the i notation and properties of radicals we have:

$$x = \frac{-6 + \sqrt{-1}\sqrt{16}}{2} \quad \text{write in } i \text{ notation}$$

$$x = \frac{-6 + 4i}{2} \quad \text{simplify}$$

$$x = \frac{2(-3 + 2i)}{2} = -3 + 2i \quad \text{factor numerator and reduce}$$

NOW TRY EXERCISES 13 THROUGH 16

The solutions to Example 2 contained both a **real number part** (-3) and an **imaginary part** $(2i)$. Numbers of this type are called **complex numbers.**

COMPLEX NUMBERS

Complex numbers are those that can be written in the form $a + bi$, where a and b are real numbers and $i = \sqrt{-1}$. The expression $a + bi$ is called the **standard form** of a complex number.

From this definition we note that all real numbers are also complex numbers, since $a + 0i$ is complex with $b = 0$. In addition, *all imaginary numbers are complex numbers*, since $0 + bi$ is a complex number with $a = 0$.

EXAMPLE 3 Write each complex number in the form $a + bi$ and identify the values of a and b.

a. $2 + \sqrt{-49}$ **b.** $\sqrt{-18}$ **c.** 7 **d.** $\dfrac{4 + 3\sqrt{-25}}{20}$

Solution:

a. $2 + \sqrt{-49}$
$= 2 + \sqrt{-1}\sqrt{49}$
$= 2 + 7i$
$a = 2, b = 7$

b. $\sqrt{-18}$
$= 0 + \sqrt{-1}\sqrt{18}$
$= 0 + 3i\sqrt{2}$
$a = 0, b = 3\sqrt{2}$

c. 7
$= 7 + 0i$
$a = 7, b = 0$

d. $\dfrac{4}{20} + \dfrac{3\sqrt{-1 \cdot 25}}{20}$
$= \dfrac{1}{5} + \dfrac{3 \cdot 5i}{20}$
$= \dfrac{1}{5} + \dfrac{3}{4}i$
$a = \dfrac{1}{5}, b = \dfrac{3}{4}$

NOW TRY EXERCISES 17 THROUGH 24

Complex numbers complete the "numerical landscape" of mathematics. The various sets of numbers and their relationship to each other can be seen in Figure 5.69, which also shows how some sets of numbers are nested within larger sets.

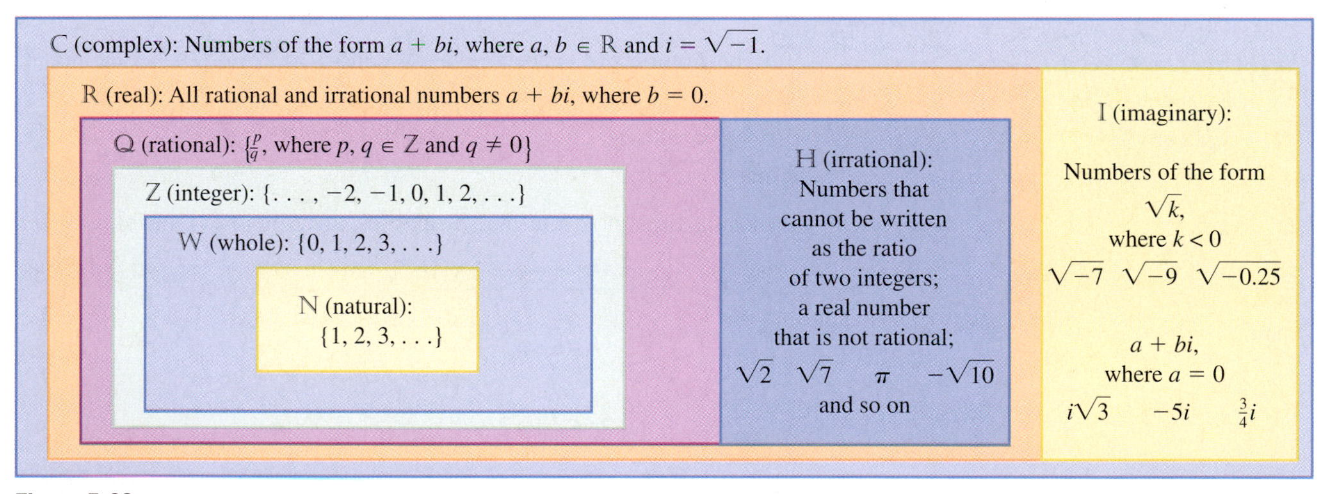

Figure 5.69

B. Adding and Subtracting Complex Numbers

The sum and difference of two polynomials is computed by identifying and combining like terms. The sum or difference of two complex numbers is computed in a similar way, by adding the real number parts from each, and the imaginary parts from each. Notice from Example 4 that the commutative, associative, and distributive properties also apply to the terms of a complex number.

EXAMPLE 4 Perform the indicated operation and write the result in the form $a + bi$.

a. $(2 + 3i) + (-5 + 2i)$ **b.** $(-5 - 4i) - (-2 - \sqrt{2}i)$

Solution:

a.
$$\begin{aligned} &(2 + 3i) + (-5 + 2i) && \text{original sum} \\ &= 2 + 3i + (-5) + 2i && \text{distribute} \\ &= 2 + (-5) + 3i + 2i && \text{commute terms} \\ &= [2 + (-5)] + (3i + 2i) && \text{associate terms} \\ &= -3 + 5i && \text{result} \end{aligned}$$

b.
$$\begin{aligned} &(-5 - 4i) - (-2 - \sqrt{2}i) && \text{original difference} \\ &= -5 + (-4i) + 2 + \sqrt{2}i && \text{distribute} \\ &= -5 + 2 + (-4i) + \sqrt{2}i && \text{commute terms} \\ &= (-5 + 2) + [(-4i) + \sqrt{2}i] && \text{associate terms} \\ &= -3 + (-4 + \sqrt{2})i && \text{result} \end{aligned}$$

NOW TRY EXERCISES 25 THROUGH 30

C. Multiplying Complex Numbers; Powers of *i*

The numbers $a + bi$ and $a - bi$ are called **complex conjugates.** The product of two complex numbers is computed using the distributive property and the F-O-I-L process in the same way we apply these to binomials. If any result gives a factor of i^2, remember that $i^2 = -1$ (meaning any term containing a factor of i^2 will simply change sign).

EXAMPLE 5 Find the indicated product and write the answer in $a + bi$ form.

a. $2i(-1 + 3i)$ **b.** $(6 - 5i)(4 + i)$ **c.** $(1 - 2i)^2$ **d.** $(2 + 3i)(2 - 3i)$

Solution:

a.
$$\begin{aligned} &2i(-1 + 3i) && \text{monomial} \cdot \text{binomial} \\ &= 2i(-1) + 2i(3i) && \text{distribute} \\ &= -2i + 6i^2 && i \cdot i = i^2 \\ &= -2i + 6(-1) && i^2 = -1 \\ &= -6 - 2i && \text{result} \end{aligned}$$

b.
$$\begin{aligned} &(6 - 5i)(4 + i) && \text{binomial} \cdot \text{binomial} \\ &= (6)(4) + 6i + (-5i)(4) + (-5i)(i) && \text{F-O-I-L} \\ &= 24 + 6i + (-20i) + (-5)i^2 && i \cdot i = i^2 \\ &= 24 + 6i + (-20i) + (-5)(-1) && i^2 = -1 \\ &= 29 - 14i && \text{result} \end{aligned}$$

c.
$$\begin{aligned} &(1 - 2i)^2 && \text{binomial square} \\ &= (1)^2 - 2(1)(2i) + (-2i)^2 && A^2 \quad 2AB + B^2 \\ &= 1 - 4i + 4i^2 && i \cdot i = i^2 \\ &= 1 - 4i + 4(-1) && i^2 = -1 \\ &= -3 - 4i && \text{result} \end{aligned}$$

d.
$$\begin{aligned} &(2 + 3i)(2 - 3i) && \text{binomial} \cdot \text{conjugate} \\ &= (2)^2 - (3i)^2 && (A + B)(A - B) = A^2 - B^2 \\ &= 4 - 9i^2 && i \cdot i = i^2 \\ &= 4 - 9(-1) && i^2 = -1 \\ &= 13 + 0i = 13 && \text{result} \end{aligned}$$

NOW TRY EXERCISES 31 THROUGH 48

Note from Example 5(d) that the *product* of the complex number $a + bi$ with its complex conjugate $a - bi$ *is a real number.* This relationship is useful when rationalizing expressions with a complex number in the denominator, and we generalize the result as follows:

> **PRODUCT OF COMPLEX CONJUGATES**
> Given the complex number $a + bi$ and its complex conjugate $a - bi$, their product is a real number given by $(a + bi)(a - bi) = a^2 + b^2$.

Showing that $(a + bi)(a - bi) = a^2 + b^2$ is left as an exercise (see Exercise 79), but from here on, when asked to compute the product of complex conjugates, simply refer to the formula as illustrated here: $(-3 + 5i)(-3 - 5i) = (-3)^2 + 5^2$ or 34.

These operations on complex numbers enable us to verify complex solutions by substitution, in the same way we verify solutions for real numbers. In addition to offering contextual practice with these skills, it is fascinating to observe how the complex roots balance or "cancel each other out" to arrive at the solution. In Example 2 we stated that $x = -3 + 2i$ was one solution to $x^2 + 6x + 13 = 0$. This is verified in Example 6.

EXAMPLE 6 Verify that $x = -3 + 2i$ is a solution to the equation $x^2 + 6x + 13 = 0$.

Solution:

$$x^2 + 6x + 13 = 0 \quad \text{original equation}$$
$$(-3 + 2i)^2 + 6(-3 + 2i) + 13 = 0 \quad \text{substitute } 3 + 2i \text{ for } x$$
$$(-3)^2 + 2(-3)(2i) + (2i)^2 - 18 + 12i + 13 = 0 \quad \text{square binomial and distribute}$$
$$9 - 12i + 4i^2 + 12i - 5 = 0 \quad \text{simplify}$$
$$9 + (-4) - 5 = 0 \quad \text{combine like terms } (12i - 12i = 0;\ i^2 = -1)$$
$$0 = 0 \checkmark$$

NOW TRY EXERCISES 49 THROUGH 60

The imaginary unit i has another interesting and useful property. Since $i = \sqrt{-1}$ and $i^2 = -1$, we know that $i^3 = i^2 \cdot i = (-1)i = -i$ and $i^4 = (i^2)^2 = 1$. We can now simplify any **power of *i*** by rewriting the expression in terms of i^4.

$$i^1 = i \qquad i^5 = i^4 \cdot i = 1 \cdot i = i$$
$$i^2 = -1 \qquad i^6 = i^4 \cdot i^2 = 1 \cdot (-1) = -1$$
$$i^3 = i^2 \cdot i = (-1)i = -i \qquad i^7 = i^4 \cdot i^3 = 1 \cdot (-i) = -i$$
$$i^4 = (i^2)^2 = (-1)^2 = 1 \qquad i^8 = (i^4)^2 = 1^2 = 1$$

Notice the powers of i "cycle through" the four values i, -1, $-i$ and 1. Here we learn to reduce higher powers using the power property of exponents and $i^4 = 1$.

EXAMPLE 7 Simplify: **a.** i^{22} **b.** i^{28} **c.** i^{57} **d.** i^{75}

Solution: **a.** $i^{22} = (i^4)^5 \cdot (i^2) = -1$ **b.** $i^{28} = (i^4)^7 = 1^7 = 1$

c. $i^{57} = (i^4)^{14} \cdot (i) = i$ **d.** $i^{75} = (i^4)^{18} \cdot (i^3) = 1^{18} \cdot -i = -i$

NOW TRY EXERCISES 61 AND 62

D. Division of Complex Numbers

With $i = \sqrt{-1}$, expressions like $\frac{3 - i}{2 + i}$ actually have a radical in the denominator, which leads to our method for complex number division. We simply apply the method of rationalizing denominators using a complex conjugate. The goal is to write the quotient in standard $a + bi$ form.

EXAMPLE 8 Divide and write each result in the form $a + bi$. Clearly state the values of a and b.

a. $\dfrac{2}{5 - i}$ **b.** $\dfrac{3 - i}{2 + i}$

Solution:

a.
$$\frac{2}{5 - i} = \frac{2}{5 - 1i} \cdot \frac{5 + 1i}{5 + 1i}$$
$$= \frac{2(5 + i)}{5^2 + 1^2}$$
$$= \frac{10 + 2i}{26}$$
$$= \frac{10}{26} + \frac{2}{26}i$$
$$= \frac{5}{13} + \frac{1}{13}i$$
$$a = \frac{5}{13}, b = \frac{1}{13}$$

b.
$$\frac{3 - i}{2 + i} = \frac{3 - 1i}{2 + 1i} \cdot \frac{2 - 1i}{2 - 1i}$$
$$= \frac{6 - 3i - 2i + 1i^2}{2^2 + 1^2}$$
$$= \frac{6 - 5i + (-1)}{5}$$
$$= \frac{5 - 5i}{5} = \frac{5}{5} - \frac{5i}{5}$$
$$= 1 - i$$
$$a = 1, b = -1$$

NOW TRY EXERCISES 63 THROUGH 68

It is important to note that results from operations in the complex number system can be checked using inverse operations, just as we do for real number operations. In Example 8(b) we found that $(3 - i) \div (2 + i) = 1 - i$. The related multiplication would be $(1 - i)(2 + i)$ giving $2 + 1i - 2i - i^2 = 2 - 1i - (-1) = 3 - i$.✓ Several checks are asked for in the exercises.

TECHNOLOGY HIGHLIGHT

Graphing Calculators and Operations on Complex Numbers

The keystrokes shown apply to a TI-84 Plus model. Please consult our Internet site or your manual for other models.

Virtually all graphing calculators have the ability to find imaginary and complex roots, as well as perform operations on complex numbers.

To use this capability on the TI-84 Plus, we first put the calculator in $a + bi$ mode. Press the MODE key (next to the yellow 2nd key) and the screen shown in Figure 5.70 appears. On the second line from the bottom, note the calculator may be in "Real" mode. To change to "$a + bi$" mode, simply navigate the cursor down to this line using the down arrow, then overlay the "$a + bi$" selection using the right arrow and press the ENTER key. The calculator is now in complex number mode. Press 2nd MODE (QUIT) to return to the home screen. To compute the product $(-2 - 3i)(5 + 4i)$, enter the expression on the home screen exactly as it is written. The number "i" is located above the decimal point on the bottom row. After pressing ENTER the result $2 - 23i$ immediately appears. Compute the product by hand to see if results match.

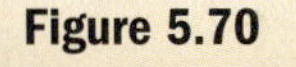

Figure 5.70

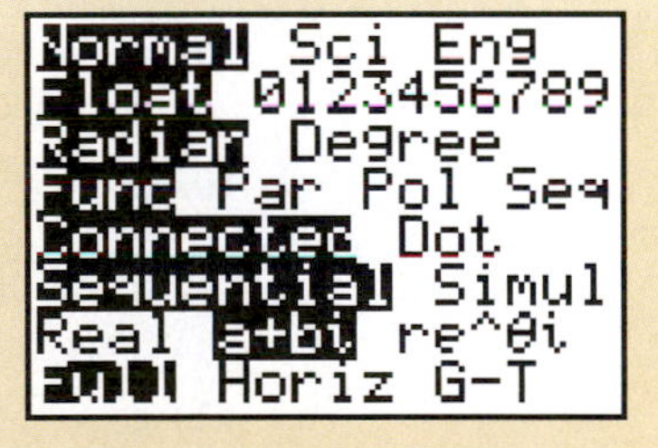

Exercise 1: Use a graphing calculator to compute the sum $(-2 + \sqrt{-108}) + (5 - \sqrt{-192})$. Note the result is in approximate form. Compute the sum by hand in exact form and compare the results.

Exercise 2: Use a graphing calculator to compute the product $(-3 + 7i)(4 - 5i)$. Then compute the product by hand and compare results. Check your answer using complex number division.

Exercise 3: Use a graphing calculator to compute the quotient $(2i)/(3 + i)$. Then compute the quotient by hand and compare results. Check your answer using multiplication.

5.6 EXERCISES

CONCEPTS AND VOCABULARY

Fill in each blank with the appropriate word or phrase. Carefully reread the section if needed.

1. Given the complex number $3 + 2i$, its complex conjugate is ______.

2. The product $(3 + 2i)(3 - 2i)$ gives the real number ______.

3. If the expression $\dfrac{4 + 6i\sqrt{2}}{2}$ is written in the standard form $a + bi$, then $a =$ ______ and $b =$ ______.

4. For $i = \sqrt{-1}$, $i^2 =$ ______, $i^4 =$ ______, $i^6 =$ ______, and $i^8 =$ ______; $i^3 =$ ______, $i^5 =$ ______, $i^7 =$ ______; and $i^9 =$ ______.

5. Discuss/explain which is correct:

a. $\sqrt{-4} \cdot \sqrt{-9} = \sqrt{(-4)(-9)} = \sqrt{36} = 6$

b. $\sqrt{-4} \cdot \sqrt{-9} = 2i \cdot 3i = 6i^2 = -6$

6. Compare/contrast the product $(1 + \sqrt{2})(1 - \sqrt{3})$ with the product $(1 + i\sqrt{2})(1 - i\sqrt{3})$. What is the same? What is different?

DEVELOPING YOUR SKILLS

Simplify each radical (if possible). If imaginary, rewrite in terms of i and simplify.

7. **a.** $\sqrt{-16}$ **b.** $\sqrt{-49}$ **c.** $\sqrt{27}$ **d.** $\sqrt{72}$

8. **a.** $\sqrt{-81}$ **b.** $\sqrt{-169}$ **c.** $\sqrt{64}$ **d.** $\sqrt{98}$

9. **a.** $-\sqrt{-18}$ **b.** $-\sqrt{-50}$ **c.** $3\sqrt{-25}$ **d.** $2\sqrt{-9}$

10. **a.** $-\sqrt{-32}$ **b.** $-\sqrt{-75}$ **c.** $3\sqrt{-144}$ **d.** $2\sqrt{-81}$

11. **a.** $\sqrt{-19}$ **b.** $\sqrt{-31}$ **c.** $\sqrt{\dfrac{-12}{25}}$ **d.** $\sqrt{\dfrac{-9}{32}}$

12. **a.** $\sqrt{-17}$ **b.** $\sqrt{-53}$ **c.** $\sqrt{\dfrac{-45}{36}}$ **d.** $\sqrt{\dfrac{-49}{75}}$

Write each complex number in the standard form $a + bi$ and clearly identify the values of a and b.

13. **a.** $\dfrac{2 + \sqrt{-4}}{2}$ **b.** $\dfrac{6 + \sqrt{-27}}{3}$

14. **a.** $\dfrac{16 - \sqrt{-8}}{2}$ **b.** $\dfrac{4 + 3\sqrt{-20}}{2}$

15. **a.** $\dfrac{8 + \sqrt{-16}}{2}$ **b.** $\dfrac{10 - \sqrt{-50}}{5}$

16. **a.** $\dfrac{6 - \sqrt{-72}}{4}$ **b.** $\dfrac{12 + \sqrt{-200}}{8}$

17. **a.** 5 **b.** $3i$

18. **a.** -2 **b.** $-4i$

19. **a.** $2\sqrt{-81}$ **b.** $\dfrac{\sqrt{-32}}{8}$

20. **a.** $-3\sqrt{-36}$ **b.** $\dfrac{\sqrt{-75}}{15}$

21. **a.** $4 + \sqrt{-50}$ **b.** $-5 + \sqrt{-27}$

22. **a.** $-2 + \sqrt{-48}$ **b.** $7 + \sqrt{-75}$

23. **a.** $\dfrac{14 + \sqrt{-98}}{8}$ **b.** $\dfrac{5 + \sqrt{-250}}{10}$

24. **a.** $\dfrac{21 + \sqrt{-63}}{12}$ **b.** $\dfrac{8 + \sqrt{-27}}{6}$

Perform the addition or subtraction. Write each result in $a + bi$ form.

25. **a.** $(12 - \sqrt{-4}) + (7 + \sqrt{-9})$
b. $(3 + \sqrt{-25}) + (-1 - \sqrt{-81})$
c. $(11 + \sqrt{-108}) - (2 - \sqrt{-48})$

26. **a.** $(-7 - \sqrt{-72}) + (8 + \sqrt{-50})$
b. $(\sqrt{3} + \sqrt{-2}) - (\sqrt{12} + \sqrt{-8})$
c. $(\sqrt{20} - \sqrt{-3}) + (\sqrt{5} - \sqrt{-12})$

27. **a.** $(2 + 3i) + (-5 - i)$
b. $(5 - 2i) + (3 + 2i)$
c. $(6 - 5i) - (4 + 3i)$

28. **a.** $(-2 + 5i) + (3 - i)$
b. $(7 - 4i) - (2 - 3i)$
c. $(2.5 - 3.1i) + (4.3 + 2.4i)$

29. **a.** $(3.7 + 6.1i) - (1 + 5.9i)$
b. $\left(8 + \frac{3}{4}i\right) - \left(-7 + \frac{2}{3}i\right)$
c. $\left(-6 - \frac{5}{8}i\right) + \left(4 + \frac{1}{2}i\right)$

30. **a.** $(9.4 - 8.7i) - (6.5 + 4.1i)$
b. $\left(3 + \frac{3}{5}i\right) - \left(-11 + \frac{7}{15}i\right)$
c. $\left(-4 - \frac{5}{6}i\right) + \left(13 + \frac{3}{8}i\right)$

Multiply and write each answer in $a + bi$ form.

31. **a.** $5i \cdot (-3i)$
b. $(4i)(-4i)$

32. **a.** $3(2 - 3i)$
b. $-7(3 + 5i)$

33. **a** $-7i(5 - 3i)$
b. $6i(-3 + 7i)$

34. **a.** $(-4 - 2i)(3 + 2i)$
b. $(2 - 3i)(-5 + i)$

35. **a.** $(-3 + 2i)(2 + 3i)$
b. $(3 + 2i)(1 + i)$

36. **a.** $(5 + 2i)(-7 + 3i)$
b. $(4 - i)(7 + 2i)$

For each complex number, name the complex conjugate. Then find the product.

37. **a.** $4 + 5i$ **b.** $3 - i\sqrt{2}$

38. **a.** $2 - i$ **b.** $-1 + i\sqrt{5}$

39. **a.** $7i$ **b.** $\frac{1}{2} - \frac{2}{3}i$

40. **a.** $-5i$ **b.** $\frac{3}{4} + \frac{1}{5}i$

Compute the special products and write each answer in $a + bi$ form.

41. **a.** $(4 - 5i)(4 + 5i)$
b. $(7 - 5i)(7 + 5i)$

42. **a.** $(-2 - 7i)(-2 + 7i)$
b. $(2 + i)(2 - i)$

43. **a.** $(3 - i\sqrt{2})(3 + i\sqrt{2})$
b. $(\frac{1}{6} + \frac{2}{3}i)(\frac{1}{6} - \frac{2}{3}i)$

44. **a.** $(5 + i\sqrt{3})(5 - i\sqrt{3})$
b. $(\frac{1}{2} + \frac{3}{4}i)(\frac{1}{2} - \frac{3}{4}i)$

45. **a.** $(2 + 3i)^2$ **b.** $(3 - 4i)^2$

46. **a.** $(2 - i)^2$ **b.** $(3 - i)^2$

47. **a.** $(-2 + 5i)^2$ **b.** $(3 + i\sqrt{2})^2$

48. **a.** $(-2 - 5i)^2$ **b.** $(2 - i\sqrt{3})^2$

Use substitution to determine if the value shown is a solution to the given equation.

49. $x^2 + 36 = 0; x = -6$

50. $x^2 + 16 = 0; x = -4$

51. $x^2 + 49 = 0; x = -7i$

52. $x^2 + 25 = 0; x = -5i$

53. $(x - 3)^2 = -9; x = 3 - 3i$

54. $(x + 1)^2 = -4; x = -1 + 2i$

55. $x^2 - 2x + 5 = 0; x = 1 - 2i$

56. $x^2 + 6x + 13 = 0; x = -3 + 2i$

57. $x^2 - 4x + 9 = 0; x = 2 + i\sqrt{5}$

58. $x^2 - 2x + 4 = 0; x = 1 - \sqrt{3}i$

59. Verify that $x = 1 + 4i$ is a solution to $x^2 - 2x + 17 = 0$. Then verify its complex conjugate $1 - 4i$ is also a solution.

60. Verify that $x = 2 - 3\sqrt{2}i$ is a solution to $x^2 - 4x + 22 = 0$. Then verify its complex conjugate $2 + 3\sqrt{2}i$ is also a solution.

Simplify using powers of i.

61. **a.** i^{48} **b.** i^{26} **c.** i^{39} **d.** i^{53}

62. **a.** i^{36} **b.** i^{50} **c.** i^{19} **d.** i^{65}

Divide and write each answer in $a + bi$ form. Check answers using multiplication.

63. **a.** $\frac{-2}{\sqrt{-49}}$ **b.** $\frac{4}{\sqrt{-25}}$

64. **a.** $\frac{2}{1 - \sqrt{-4}}$ **b.** $\frac{3}{2 + \sqrt{-9}}$

65. **a.** $\frac{7}{3 + 2i}$ **b.** $\frac{-5}{2 - 3i}$

66. **a.** $\frac{6}{1 + 3i}$ **b.** $\frac{7}{7 - 2i}$

67. **a.** $\frac{3 + 4i}{4i}$ **b.** $\frac{2 - 3i}{3i}$

68. **a.** $\frac{-4 + 8i}{2 - 4i}$ **b.** $\frac{3 - 2i}{-6 + 4i}$

WORKING WITH FORMULAS

69. Absolute value of a complex number: $|a + bi| = \sqrt{a^2 + b^2}$
The absolute value of any complex number $a + bi$ (sometimes called the *modulus* of the number) is computed by taking the square root of the sums of the squares of a and b. Find the absolute value of the given complex numbers.

a. $|2 + 3i|$ **b.** $|4 - 5i|$ **c.** $|3 + \sqrt{2}i|$

70. Binomial cubes: $(A + B)^3 = A^3 + 3A^2B + 3AB^2 + B^3$
The cube of any binomial can be found using this formula, where A and B are the terms of the binomial. Use the formula to compute the cube of $1 - 2i$ (note $A = 1$ and $B = -2i$).

APPLICATIONS

71. In a day when imaginary numbers were imperfectly understood, Girolamo Cardano (1501–1576) once posed the problem, "Find two numbers that have a sum of 10 and whose product is 40." In other words, $A + B = 10$ and $AB = 40$. Although the solution is routine today, at the time the problem posed an enormous challenge. Verify that $A = 5 + \sqrt{15}i$ and $B = 5 - \sqrt{15}i$ satisfy these conditions.

72. Suppose Cardano had said, "Find two numbers that have a sum of 4 and a product of 7" (see Exercise 71). Verify that $A = 2 + \sqrt{3}i$ and $B = 2 - \sqrt{3}i$ satisfy these conditions.

Although it may seem odd, imaginary numbers have several applications in the real world. Many of these involve a study of electrical circuits, in particular *alternating current* or AC circuits. Briefly, the components of an AC circuit are current I (in amperes), voltage V (in volts), and the impedance Z (in ohms). The impedance of an electrical circuit is a measure of the total opposition to the flow of current through the circuit and is calculated as $Z = R + iX_L - iX_C$ where R represents a pure resistance, X_C represents the capacitance, and X_L represents the inductance. Each of these is also measured in ohms (symbolized by Ω).

73. Find the impedance Z if $R = 7\ \Omega$, $X_L = 6\ \Omega$, and $X_C = 11\ \Omega$.

74. Find the impedance Z if $R = 9.2\ \Omega$, $X_L = 5.6\ \Omega$, and $X_C = 8.3\ \Omega$.

The voltage V (in volts) across any element in an AC circuit is calculated as a product of the current I and the impedance Z: $V = IZ$.

75. Find the voltage in a circuit with a current of $3 - 2i$ amperes and an impedance of $Z = 5 + 5i\ \Omega$.

76. Find the voltage in a circuit with a current of $2 - 3i$ amperes and an impedance of $Z = 4 + 2i\ \Omega$.

In an AC circuit, the total impedance (in ohms) is given by the formula $Z = \frac{Z_1 Z_2}{Z_1 + Z_2}$, where Z represents the total impedance of a circuit that has Z_1 and Z_2 wired in parallel.

77. Find the total impedance Z if $Z_1 = 1 + 2i$ and $Z_2 = 3 - 2i$.

78. Find the total impedance Z if $Z_1 = 3 - i$ and $Z_2 = 2 + i$.

WRITING, RESEARCH, AND DECISION MAKING

79. (a) Up to this point, we have said that $x^2 - 9$ is factorable and $x^2 + 9$ is prime. Actually we mean that $x^2 + 9$ is nonfactorable *using real numbers*, but it can be factored using complex numbers. Do some research and exploration and see if you can accomplish the task. (b) Verify that $(a + bi)(a - bi) = a^2 + b^2$.

80. Locate and read the following article. Then turn in a one page summary. "Thinking the Unthinkable: The Story of Complex Numbers," Israel Kleiner, *Mathematics Teacher*, Volume 81, Number 7, October 1988: pages 583–592.

81. In this section, we noted that the product property of radicals, $\sqrt{AB} = \sqrt{A}\sqrt{B}$, can still be applied when at most one of the factors is negative. So what happens if *both* are negative? First consider the expression $\sqrt{-4 \cdot -9}$. What happens if you first simplify the radicand, then compute the square root? Next consider the product $\sqrt{-4} \cdot \sqrt{-9}$. Rewrite each factor using the i notation, then compute the product. Do you get the same result as before? What can you say about $\sqrt{-4 \cdot -9}$ and $\sqrt{-4} \cdot \sqrt{-9}$?

EXTENDING THE CONCEPT

82. Use the formula from Exercise 70 to compute the cube of $-\frac{1}{2} + \frac{\sqrt{3}}{2}i$.

83. If $a = 1$ and $b = 4$, the expression $\sqrt{b^2 - 4ac}$ represents an imaginary number for what values of c?

84. While it is a simple concept for real numbers, the square root of a complex number is much more involved due to the interplay between its real and imaginary parts. For $z = a + bi$, the square root of z can be found using the formula:

$\sqrt{z} = \frac{\sqrt{2}}{2}\left(\sqrt{|z| + a} \pm i\sqrt{|z| - a}\right)$, where the sign is chosen to match the sign of b (see Exercise 69). Use the formula to find the square root of each complex number, then check by squaring.

a. $z = -7 + 24i$ **b.** $z = 5 - 12i$ **c.** $z = 4 + 3i$

MAINTAINING YOUR SKILLS

85. (5.5) Find the angle θ between the vectors $\mathbf{u} = \langle -2, 9 \rangle$ and $\mathbf{v} = \langle 6, 5 \rangle$.

86. (4.4) Find all real solutions to $2\sin(2\theta) + \sqrt{3} = 0$.

87. (3.2) Prove the following identity: $1 + \sin\theta = \frac{\cos^2\theta}{1 - \sin\theta}$.

88. (2.1) Graph the function using a reference rectangle and the *rule of fourths*:

$$y = 3\cos\left(2\theta - \frac{\pi}{4}\right)$$

89. (5.3) Solve the triangle shown, then compute its perimeter and area.

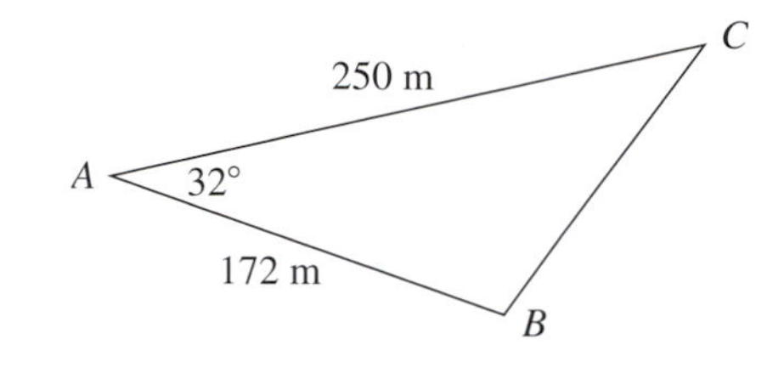

90. (5.4) A plane is flying 200 mph at heading 30°, with a 40 mph wind blowing from the west. Find the true course and speed of the plane.

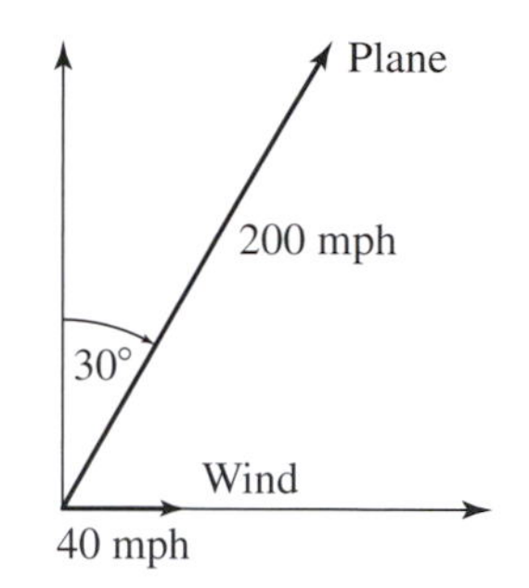

5.7 Complex Numbers in Trigonometric Form

LEARNING OBJECTIVES

In Section 5.7 you will learn how to:

A. Graph a complex number

B. Write a complex number in trigonometric form

C. Convert from trigonometric form to rectangular form

D. Interpret products and quotients geometrically

E. Compute products and quotients in trigonometric form

F. Solve applications involving complex numbers

INTRODUCTION

Once the set of complex numbers became recognized and defined, the related basic operations matured very quickly. With little modification—sums, differences, products, quotients, and powers all lent themselves fairly well to the algebraic techniques used for real numbers. But roots of complex numbers did not yield so easily and additional tools and techniques were needed. Writing complex numbers in trigonometric form allows us to find complex roots (Section 5.8) and in some cases, makes computing products, quotients, and powers more efficient.

POINT OF INTEREST

In 1629, Albert Girard (1595–1632) began suggesting that if one were to accept imaginary numbers, it could be claimed that a polynomial of degree *n* has exactly *n* roots. He further noted several relationships between the roots of a polynomial (both real and complex) and its coefficients. Many other mathematicians contributed to the wider acceptance of imaginary numbers, including Descartes, Hudde, Wessel, Argand, and Gauss. Even so, Gauss himself was heard to comment that, "the true metaphysics of $\sqrt{-1}$ remains very difficult."

A. Graphing Complex Numbers

In previous sections we defined a vector quantity as one that required more than a single component to describe its attributes. The complex number $z = a + bi$ certainly fits this description, since both a real number "component" and an imaginary

WORTHY OF NOTE

Surprisingly, the study of complex numbers matured much earlier than the study of vectors, and representing complex numbers as directed line segments actually preceded their application to a vector quantity.

"component" are needed to define it. In many respects, we can treat complex numbers in the same way we treated vectors and in fact, there is much we can learn from this connection.

Since both axes in the xy-plane have real number values, it's not possible to graph a complex number in $\mathbb{R}$ (the real plane). However, in the same way we used the x-axis for the horizontal component of a vector and the y-axis for the vertical, we can let the x-axis represent the real valued part of a complex number and the y-axis the imaginary part. The result is called the **complex plane** $\mathbb{C}$. Every point (a, b) in $\mathbb{C}$ can be associated with a complex number $a + bi$, and any complex number $a + bi$ can be associated with a point (a, b) in $\mathbb{C}$ (Figure 5.71). The point (a, b) can also be regarded as the terminal point of a position vector representing the complex number, generally named using the letter z.

Figure 5.71

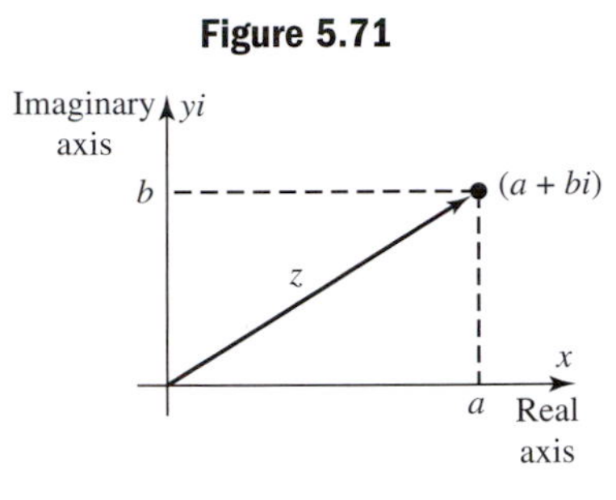

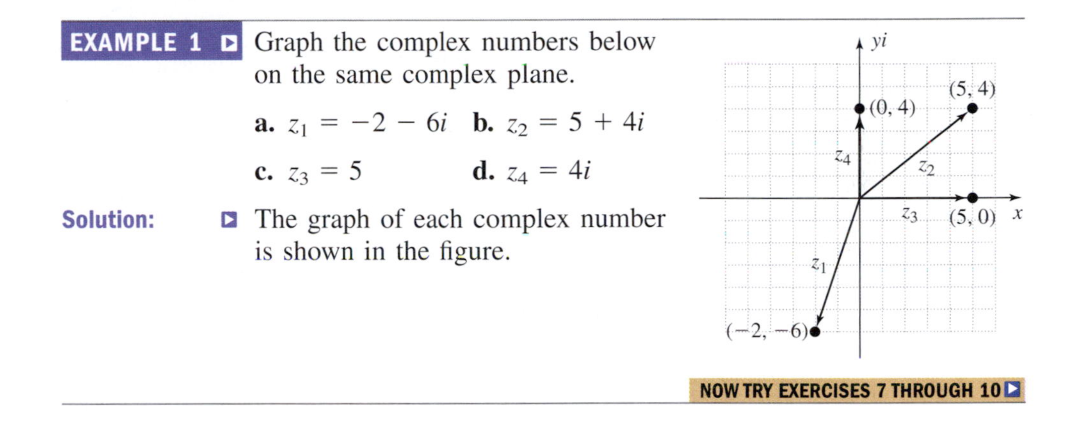

EXAMPLE 1 Graph the complex numbers below on the same complex plane.

a. $z_1 = -2 - 6i$ **b.** $z_2 = 5 + 4i$

c. $z_3 = 5$ **d.** $z_4 = 4i$

Solution: The graph of each complex number is shown in the figure.

NOW TRY EXERCISES 7 THROUGH 10

Figure 5.72

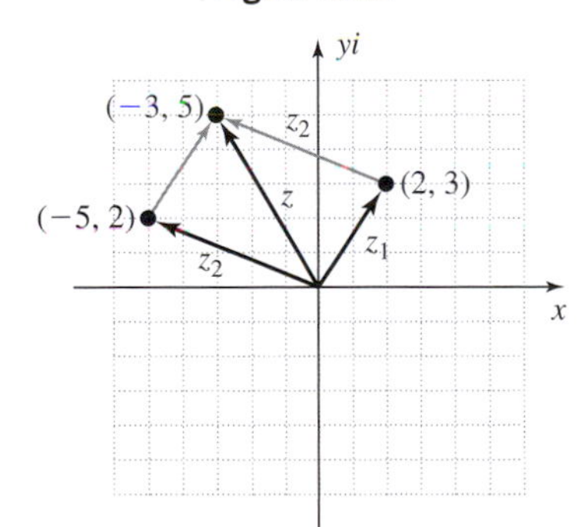

In Example 1, you likely noticed that from a vector perspective, z_2 is the "resultant vector" for the sum $z_3 + z_4$. To investigate further, consider $z_1 = (2 + 3i)$, $z_2 = (-5 + 2i)$, and the sum $z_1 + z_2 = z$ shown in Figure 5.72. The figure helps to confirm that the sum of complex numbers can be illustrated geometrically using the parallelogram (tail-to-tip) method employed for vectors in previous sections.

B. Complex Numbers in Trigonometric Form

The complex number $z = a + bi$ is said to be in **rectangular form** since it can be graphed using the rectangular coordinates of the complex plane. Complex numbers can also be written in **trigonometric form.** Similar to how $|x|$ represents the distance between the real number x and zero, $|z|$ represents the distance between (a, b) and the origin, and is computed as $|z| = \sqrt{a^2 + b^2}$. With any nonzero z, we can also associate an angle θ, which is the angle in standard position whose terminal side coincides with the graph of z. If we let r represent $|z|$, Figure 5.73 shows $\cos\theta = \frac{a}{r}$ and $\sin\theta = \frac{b}{r}$, yielding

Figure 5.73

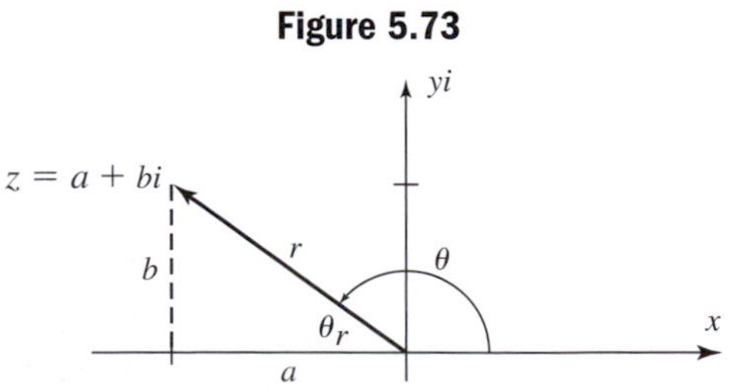

$r\cos\theta = a$ and $r\sin\theta = b$. The appropriate substitutions into $a + bi$ give the trigonometric form:

$$z = a + bi$$
$$= r\cos\theta + r\sin\theta \cdot i$$

Factoring out r and writing the imaginary unit as the lead factor of $\sin\theta$ gives the relationship in its more common form, $z = r(\cos\theta + i\sin\theta)$, where $\tan\theta = \frac{b}{a}$.

WORTHY OF NOTE

While it is true the trigonometric form can more generally be written as $z = r[\cos(\theta + 2\pi k) + i\sin(\theta + 2\pi k)]$ for $k \in \mathbb{Z}$, the result is identical for any integer k and we will select θ so that $0 \le \theta < 2\pi$ or $0° \le \theta < 360°$, depending on whether we are working in radians or degrees.

THE TRIGONOMETRIC FORM OF A COMPLEX NUMBER

For the complex number $z = a + bi$ and angle θ shown, $z = r(\cos\theta + i\sin\theta)$ is the trigonometric form of z, where $r = \sqrt{a^2 + b^2}$, and $\tan\theta = \frac{b}{a}$; $a \ne 0$.

- $r = |z|$ represents the magnitude of z (also called the **modulus** of z).
- θ is often refered to as the **argument** of z.

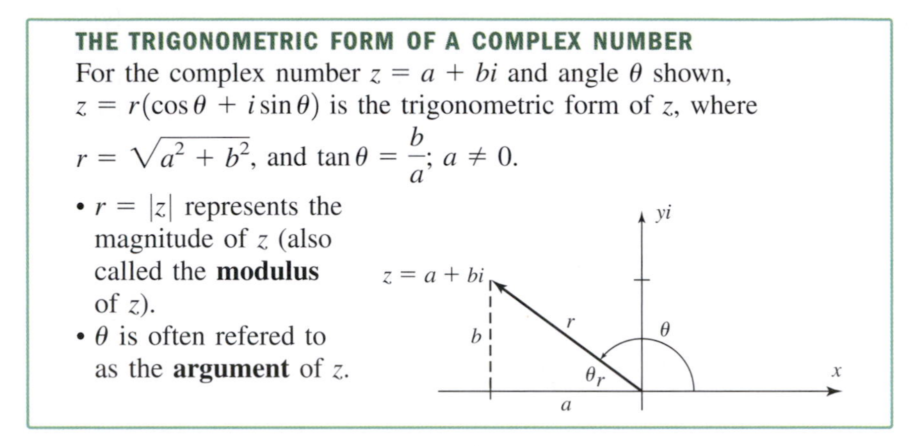

Be sure to note that for $\tan\theta = \frac{b}{a}$, $\tan^{-1}\left(\frac{b}{a}\right)$ is equal to θ_r (the reference angle for θ) and the value of θ will ultimately *depend on the quadrant of z.*

EXAMPLE 2 State the quadrant of the complex number, then write each in trigonometric form.

a. $z_1 = -2 - 2i$ **b.** $z_2 = 6 + 2i$

Solution: Knowing that modulus r and angle θ are needed for the trigonometric form, we first determine these values. Once again, to find the correct value of θ, *it's important to note the quadrant of the complex number.*

a. $z_1 = -2 - 2i$; QIII

$$r = \sqrt{(-2)^2 + (-2)^2} = \sqrt{8} = 2\sqrt{2}$$

$$\theta_r = \tan^{-1}\left(\frac{-2}{-2}\right) = \tan^{-1}(1) = \frac{\pi}{4}$$

with z_1 in QIII, $\theta = \frac{5\pi}{4}$.

$$z_1 = 2\sqrt{2}\left[\cos\left(\frac{5\pi}{4}\right) + i\sin\left(\frac{5\pi}{4}\right)\right]$$

See Figure 5.74.

b. $z = 6 + 2i$; QI

$$r = \sqrt{(6)^2 + (2)^2} = \sqrt{40} = 2\sqrt{10}$$

$$\theta_r = \tan^{-1}\left(\frac{2}{6}\right) = \tan^{-1}\left(\frac{1}{3}\right)$$

z is in QI, so $\theta = \tan^{-1}\left(\frac{1}{3}\right)$

$$z = 2\sqrt{10}\left(\cos\left[\tan^{-1}\left(\frac{1}{3}\right)\right] + i\sin\left[\tan^{-1}\left(\frac{1}{3}\right)\right]\right)$$

Figure 5.74

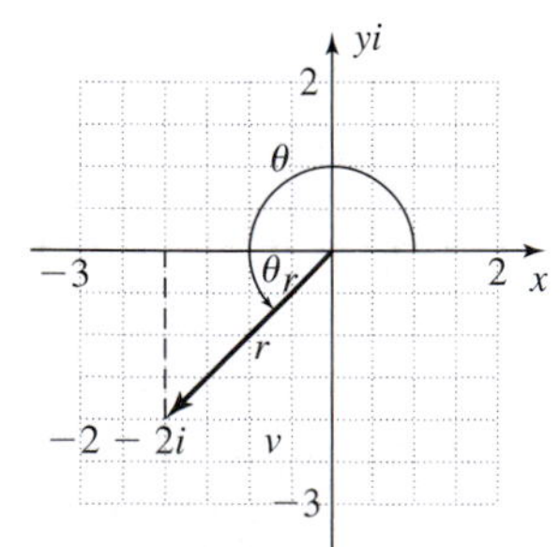

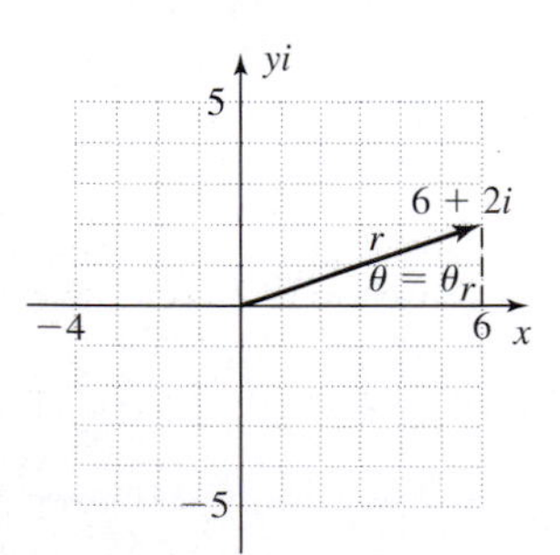

NOW TRY EXERCISES 11 THROUGH 26

Since the angle θ is repeated for both cosine and sine, we often use an abbreviated notation for the trigonometric form, called "cis" (sis) notation: $z = r(\cos\theta + i\sin\theta) = r\operatorname{cis}\theta$. The results of Example 2(a) and 2(b) would then be written $z = 2\sqrt{2}\operatorname{cis}\left(\frac{5\pi}{4}\right)$ and $z = 2\sqrt{10}\operatorname{cis}\left[\tan^{-1}\left(\frac{1}{3}\right)\right]$, respectively.

WORTHY OF NOTE

Using the triangle diagrams from Section 4.2, $\cos\left[\tan^{-1}\left(\frac{1}{3}\right)\right]$ and $\sin\left[\tan^{-1}\left(\frac{1}{3}\right)\right]$ can easily be evaluated and used to verify $2\sqrt{10}\operatorname{cis}\left[\tan^{-1}\left(\frac{1}{3}\right)\right] = 6 + 2i$.

As in Example 2b, when $\theta_r = \tan^{-1}\left(\frac{b}{a}\right)$ is not a standard angle, we either answer in exact form as shown, or use a four-decimal-place approximation: $2\sqrt{10}\operatorname{cis}(0.3218)$.

C. Converting from Trigonometric Form to Rectangular Form

Converting from trigonometric form back to rectangular form is simply a matter of evaluating $r\operatorname{cis}\theta$. This can be done regardless of whether θ is a standard angle or in the form $\tan^{-1}\left(\frac{b}{a}\right)$, since in the latter case we can construct a right triangle with side b opposite θ and side a adjacent θ, and find the needed values as in Section 4.2.

It's interesting to note that the "cis θ" in $z = r\operatorname{cis}\theta$ can be regarded as a unit vector in the direction of z, with multiplication by r correctly naming the complex number. In addition, we could also note that $\sqrt{a^2 + b^2} = \sqrt{\cos^2\theta + \sin^2\theta} = 1$.

EXAMPLE 3 Graph the following complex numbers, then write them in rectangular form.

a. $z = 12\operatorname{cis}\left(\frac{\pi}{6}\right)$ **b.** $z = 13\operatorname{cis}\left[\tan^{-1}\left(\frac{5}{12}\right)\right]$

Solution: **a.** We have $r = 12$ and $\theta = \frac{\pi}{6}$, which yields the graph in Figure 5.75.

In the nonabbreviated form we have $z = 12\left[\cos\left(\frac{\pi}{6}\right) + i\sin\left(\frac{\pi}{6}\right)\right]$. Evaluating within the brackets gives

$$z = 12\left[\frac{\sqrt{3}}{2} + \frac{1}{2}i\right] = 6\sqrt{3} + 6i.$$

Figure 5.75

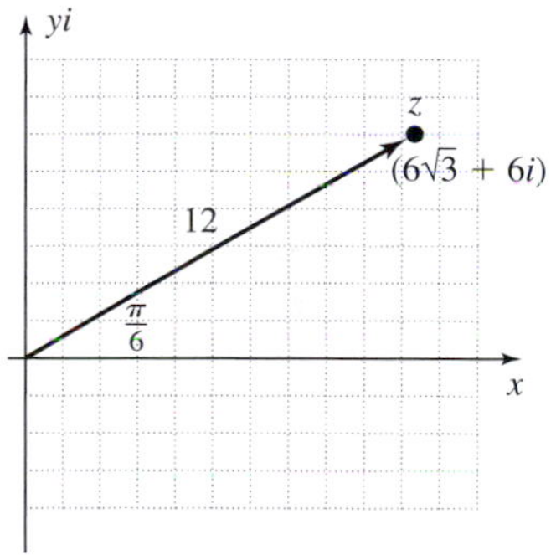

b. For $r = 13$ and $\theta = \tan^{-1}\left(\frac{5}{12}\right)$, we have the graph shown in Figure 5.76. Here we obtain the rectangular form directly from the diagram with $z = 12 + 5i$. Verify by noting that for $\theta = \tan^{-1}\left(\frac{5}{12}\right)$, $\cos\theta = \frac{12}{13}$, and $\sin\theta = \frac{5}{13}$, meaning $z = 13(\cos\theta + i\sin\theta) = 13\left[\frac{12}{13} + \frac{5}{13}i\right] = 12 + 5i$.

Figure 5.76

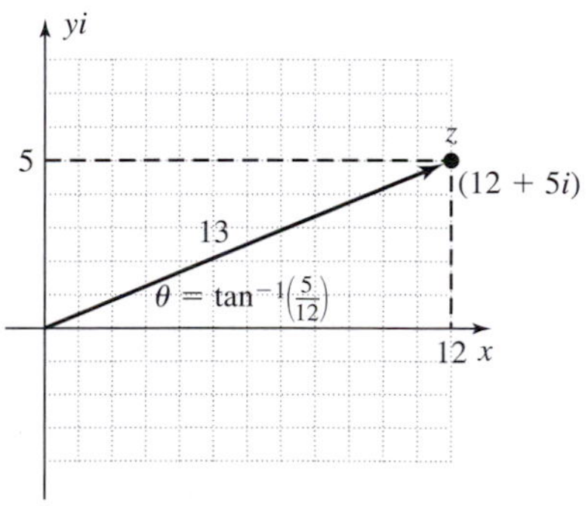

NOW TRY EXERCISES 27 THROUGH 34

D. Interpreting Products and Quotients Geometrically

The multiplication and division of complex numbers has some geometric connections that can help us understand their computation in trigonometric form. Note the relationship between the modulus and argument of the following product, with the moduli (plural of modulus) and arguments from each factor.

EXAMPLE 4 For $z_1 = 3 + 3i$ and $z_2 = 0 + 2i$, (a) graph the complex numbers and compute their moduli and arguments and (b) compute and graph the product z_1z_2 and find its modulus and argument. Discuss any connections you see between the factors and the resulting product.

Solution:

a. The graphs of z_1 and z_2 are shown in the figure. For the modulus and argument we have:

$z_1 = 3 + 3i$; QI	$z_2 = 0 + 2i$;
$r = \sqrt{(3)^2 + (3)^2}$	(quadrantal)
$= \sqrt{18} = 3\sqrt{2}$	$r = 2$ directly
$\theta = \tan^{-1}1$	$\theta = 90°$ directly
$\Rightarrow \theta = 45°$	

b. The product z_1z_2 is $(3 + 3i)(2i) = -6 + 6i$, which is in QII. The modulus is $\sqrt{(-6)^2 + (6)^2} = \sqrt{72} = 6\sqrt{2}$, with an argument of $\theta_r = \tan^{-1}(-1)$ or 135° (QII). Note the product of the two moduli is *equal to the modulus of the final product:* $2 \cdot 3\sqrt{2} = 6\sqrt{2}$. Also note that the sum of the arguments for z_1 and z_2 *is equal to the argument of the product:* $45° + 90° = 135°$!

NOW TRY EXERCISES 35 THROUGH 38

E. Products and Quotients in Trigonometric Form

The connections illustrated in Example 4 are much more than a coincidence, and can be proven to hold for all complex numbers. Consider any two nonzero complex numbers $z_1 = r_1(\cos\alpha + i\sin\alpha)$ and $z_2 = r_2(\cos\beta + i\sin\beta)$. For the product z_1z_2 we have

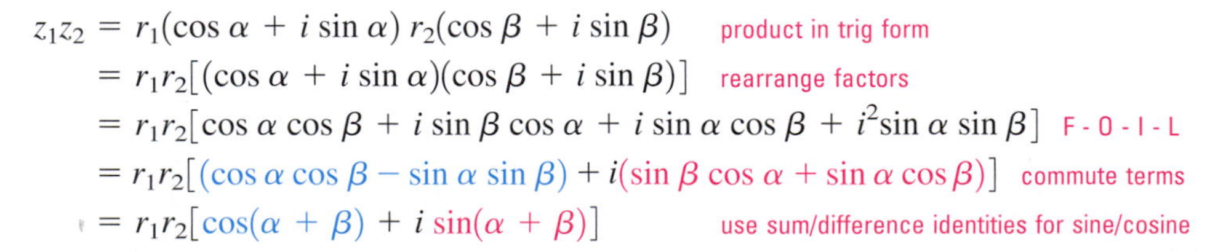

$$z_1z_2 = r_1(\cos\alpha + i\sin\alpha)\,r_2(\cos\beta + i\sin\beta) \quad \text{product in trig form}$$
$$= r_1r_2[(\cos\alpha + i\sin\alpha)(\cos\beta + i\sin\beta)] \quad \text{rearrange factors}$$
$$= r_1r_2[\cos\alpha\cos\beta + i\sin\beta\cos\alpha + i\sin\alpha\cos\beta + i^2\sin\alpha\sin\beta] \quad \text{F-O-I-L}$$
$$= r_1r_2[(\cos\alpha\cos\beta - \sin\alpha\sin\beta) + i(\sin\beta\cos\alpha + \sin\alpha\cos\beta)] \quad \text{commute terms}$$
$$= r_1r_2[\cos(\alpha + \beta) + i\sin(\alpha + \beta)] \quad \text{use sum/difference identities for sine/cosine}$$

In words, the proof says that to *multiply* complex numbers in trigonometric form, we *multiply* the moduli and *add* the arguments. For *division*, we *divide* the moduli and *subtract* the arguments. The proof for division resembles that for multiplication and is asked for in Exercise 71.

PRODUCTS AND QUOTIENTS OF COMPLEX NUMBERS IN TRIGONOMETRIC FORM

For the complex numbers $z_1 = r_1(\cos\alpha + i\sin\alpha)$ and

$$z_2 = r_2(\cos\beta + i\sin\beta),$$

$$z_1z_2 = r_1r_2[\cos(\alpha+\beta) + i\sin(\alpha+\beta)]$$

and

$$\frac{z_1}{z_2} = \frac{r_1}{r_2}[\cos(\alpha-\beta) + i\sin(\alpha-\beta)],\ z_2 \neq 0.$$

EXAMPLE 5 For $z_1 = -3 + \sqrt{3}i$ and $z_2 = \sqrt{3} + 1i$,

a. write z_1 and z_2 in trigonometric form and compute z_1z_2,

b. compute the quotient $\frac{z_1}{z_2}$ in trigonometric form, and

c. verify the product using the rectangular form.

Solution: **a.** For z_1 in QII we find $r = 2\sqrt{3}$ and $\theta = 150°$, for z_2 in QI, $r = 2$ and $\theta = 30°$. In trigonometric form,

$z_1 = 2\sqrt{3}(\cos 150° + i\sin 150°)$ and

$z_2 = 2(\cos 30° + i\sin 30°)$:

$$\begin{aligned} z_1z_2 &= 2\sqrt{3}(\cos 150° + i\sin 150°)\cdot 2(\cos 30° + i\sin 30°) \\ &= 2\sqrt{3}\cdot 2[\cos(150°+30°) + i\sin(150°+30°)] \quad \text{multiply moduli, add arguments} \\ &= 4\sqrt{3}(\cos 180° + i\sin 180°) \\ &= 4\sqrt{3}(-1 + 0i) \\ &= -4\sqrt{3} + 0i \end{aligned}$$

b.
$$\begin{aligned} \frac{z_1}{z_2} &= \frac{2\sqrt{3}(\cos 150° + i\sin 150°)}{2(\cos 30° + i\sin 30°)} \\ &= \sqrt{3}[\cos(150°-30°) + i\sin(150°-30°)] \quad \text{divide moduli, subtract arguments} \\ &= \sqrt{3}(\cos 120° + i\sin 120°) \\ &= \sqrt{3}\left(-\frac{1}{2} + \frac{\sqrt{3}}{2}i\right) \\ &= -\frac{\sqrt{3}}{2} + \frac{3}{2}i \end{aligned}$$

c.
$$\begin{aligned} z_1z_2 &= (-3 + \sqrt{3}i)(\sqrt{3} + 1i) \\ &= -3\sqrt{3} - 3i + 3i + \sqrt{3}i^2 \\ &= -4\sqrt{3} + 0i \checkmark \end{aligned}$$

NOW TRY EXERCISES 39 THROUGH 46

Converting to trigonometric form for multiplication and division seems too clumsy for practical use, as we can often compute these results more efficiently in rectangular form. However, this approach leads to powers and roots of complex numbers, *an indispensable part of advanced equation solving*, and these are not easily found in rectangular form. In any case, note that the power and simplicity of computing products/quotients in trigonometric form is highly magnified when the complex numbers are *given in trig form*:

$$(12 \text{ cis } 50°)(3 \text{ cis } 20°) = 36 \text{ cis } 70° \qquad \frac{12 \text{ cis } 50°}{3 \text{ cis } 20°} = 4 \text{ cis } 30°.$$

See Exercises 47 through 50.

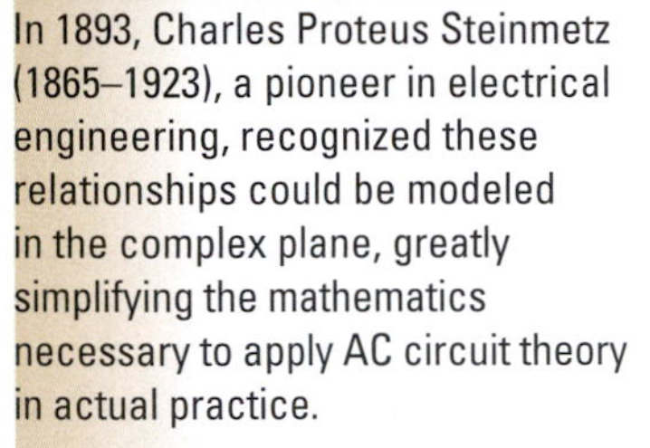

WORTHY OF NOTE

In 1893, Charles Proteus Steinmetz (1865–1923), a pioneer in electrical engineering, recognized these relationships could be modeled in the complex plane, greatly simplifying the mathematics necessary to apply AC circuit theory in actual practice.

F. Applications of Complex Numbers

Complex numbers have several applications in the real world. Many of these involve a study of electricity, and in particular **AC (alternating current) circuits.** The chief components of AC circuits are **voltage** (V) and **current** (I). Due to the nature of how the current is generated, V and I can be modeled by sine functions. Other characteristics of electricity include pure **resistance** (R), **inductive reactance** (X_L), and **capacitive reactance** (X_C) (see Figure 5.77). Each of these is measured in a unit called ohms (Ω), while current I is measured in amperes (A), and voltages are measured in volts (V). These components of electricity *are related by fixed and inherent traits*, which include the following: (1) voltage across a resistor is always *in phase* with the current, meaning the phase shift or **phase angle** between them is 0° (Figure 5.78); (2) voltage across an inductor *leads the current* by 90° (Figure 5.79); (3) voltage across a capacitor *lags the current* by 90° (Figure 5.80); and (4) voltage is equal to the product of the current times the resistance or reactance: $V = IR$, $V = IX_L$, and $V = IX_C$.

Figure 5.77

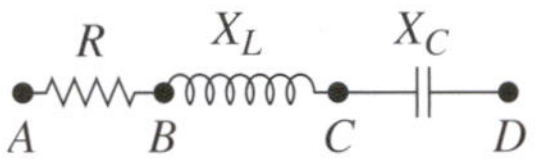

Figure 5.78

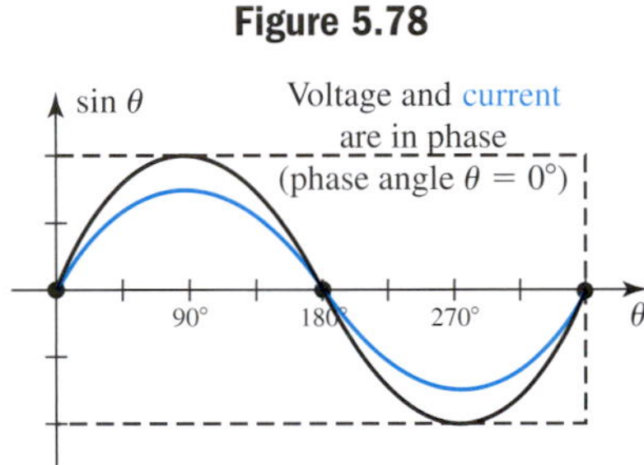

Figure 5.79

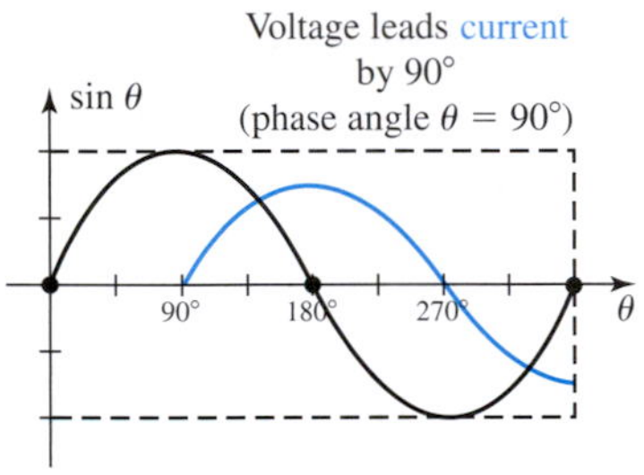

Voltage leads current by 90° (phase angle θ = 90°)

Figure 5.80

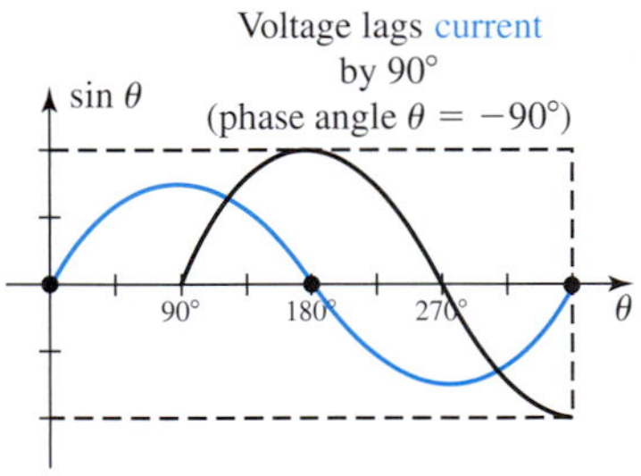

Voltage lags current by 90° (phase angle θ = −90°)

Different combinations of R, X_L, and X_C in a combined (series) circuit alter the phase angle and the resulting voltage. Since voltage across a resistance is always in phase with the current (trait 1), we can model the resistance as a vector along the positive real axis (since the phase angle is 0°). For traits (2) and (3), X_L is modeled on the positive imaginary axis since voltage leads current by 90°, and X_C on the negative imaginary axis since voltage lags current by 90° (see Figure 5.81). These natural characteristics make the complex plane *a perfect fit for describing the characteristics of the circuit.*

Figure 5.81

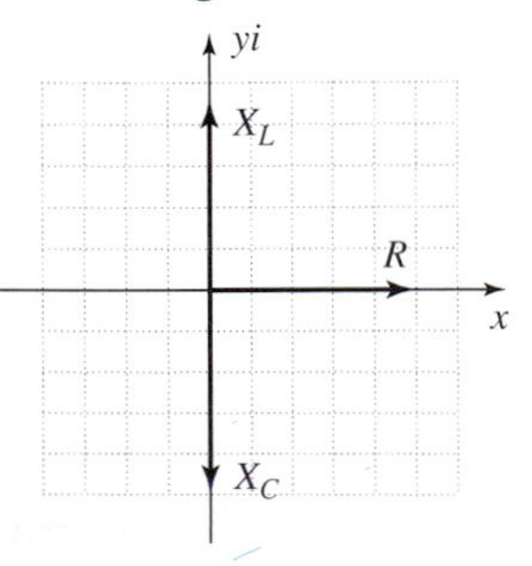

Consider a series circuit (Figure 5.77), where $R = 12\ \Omega$, $X_L = 9\ \Omega$, and $X_C = 4\ \Omega$. For a current of $I = 2$ A through this circuit, the voltage across each individual element would be $V_R = (2)(12) = 24$ V (A to B), $V_L = (2)(9) = 18$ V (B to C), and $V_C = (2)(4) = 8$ V (C to D). However, the resulting voltage across this circuit *cannot be an arithmetic sum*, since R is real while X_L and X_C are represented by imaginary numbers. The joint effect of resistance (R) and reactance (X_L, X_C) in a circuit is called the **impedance,** denoted by the letter Z, and is a measure of the total resistance to the flow of electrons. It is computed $Z = R + X_L j - X_C j$ (see *Worthy of Note*, page 387), due to the phase angle relationship of the voltage in each element (X_L and X_C point in opposite directions, hence the subtraction). The expression for Z is

WORTHY OF NOTE

While mathematicians generally use the symbol i to represent $\sqrt{-1}$, the "i" is used in other fields to represent an electric current and the symbol $j = \sqrt{-1}$ is used instead. In conformance with this convention, we will temporarily use the j for $\sqrt{-1}$ as well.

more commonly written $R + (X_L - X_C)j$, where we more clearly note *Z is a complex number* whose magnitude and angle with the x-axis can be found as before: $|Z| = \sqrt{R^2 + (X_L - X_C)^2}$ and $\theta_r = \tan^{-1}\left(\frac{X_L - X_C}{R}\right)$. The angle θ represents the phase angle between the voltage and current brought about by this combination of elements. The resulting voltage of the circuit is then calculated as the product of the current with the magnitude of the impedance, or $V_{RLC} = I|Z|$ (Z is also measured in ohms, Ω).

EXAMPLE 6 For the circuit diagrammed in the figure, (a) find the magnitude of Z, the phase angle between current and voltage, and write the result in trigonometric form; and (b) find the total voltage across this circuit.

R X_L X_C
A 12 Ω B 9 Ω C 4 Ω D

Solution:

a. Using the values given, we find $Z = R + (X_L - X_C)j = 12 + (9 - 4)j = 12 + 5j$ (QI). This gives a magnitude of $|Z| = \sqrt{(12)^2 + (5)^2} = \sqrt{169} = 13\ \Omega$, with a phase angle of $\theta = \tan^{-1}\left(\frac{5}{12}\right) \approx 22.6°$ (voltage leads the current by about 22.6°). In trigonometric form we have $Z \approx 13 \text{ cis } 22.6°$.

b. With $I = 2$ A, the total voltage across this circuit is $V_{RLC} = I|Z| = 2(13) = 26$ V.

NOW TRY EXERCISES 53 THROUGH 66

TECHNOLOGY HIGHLIGHT

Graphing Calculators and Complex Numbers

The Keystrokes shown apply to a TI-84 Plus model. Please consult our Internet site or your manual for other models.

With a graphing calculator in "$a + bi$" mode, we have the ability to confirm and extend many of the results obtained in this section. To find the argument of a complex number using the calculator, press MATH then ▶ ▶ to get to the **CPX** (complex) submenu (Figure 5.82). Option **"4:angle("** displays the angle (argument) of the complex number entered. This option can also find the argument for a product or quotient if entered in that form. Let's use the calculator to verify that for the product of complex numbers, we add the arguments of each factor. We'll use $z_1 = 2 + 3i$ and $z_2 = 8 + 2i$ for this purpose. Results are displayed on the screen in Figure 5.83, where we note that arguments do indeed sum to ≈70.35°. Note that option **"5:abs("** can be used to find the modulus of a complex number.

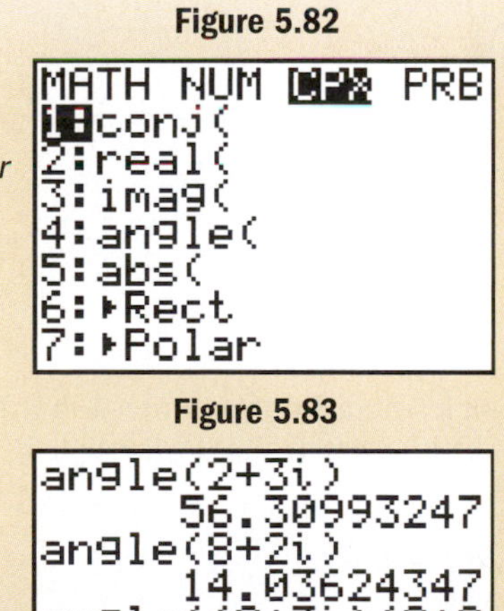

Exercise 1: Use $z_1 = 2 + 3i$ and $z_2 = 8 + 2i$ and a graphing calculator to confirm that for the quotient of

complex numbers, we subtract the arguments of each factor.

Exercise 2: Create two complex numbers on your own, one with $\alpha = 45°$ and another with $\beta = 30°$. $\left(\textit{Hint:} \text{ Use } \tan^{-1}\frac{b}{a}\right)$. Use these numbers and your graphing calculator to verify the moduli and argument relationships for products and quotients.

Exercise 3: Compute the product $(87 - 87i)(-187.5 + 62.5\sqrt{3}i)$ by hand, then once again using the trigonometric form. Comment on what you notice.

5.7 EXERCISES

CONCEPTS AND VOCABULARY

Fill in each blank with the appropriate word or phrase. Carefully reread the section if needed.

1. For a complex number written in the form $z = r(\cos\theta + i\sin\theta)$, r is called the ______ and θ is called the ______.

2. The complex number $z = 2\left[\cos\left(\frac{\pi}{4}\right) + i\sin\left(\frac{\pi}{4}\right)\right]$ can be written as the abbreviated "cis" notation as ______.

3. To multiply complex numbers in trigonometric form, we ______ the moduli and ____ the arguments.

4. To divide complex numbers in trigonometric form, we ______ the moduli and ______ the arguments.

5. Write $z = -1 - \sqrt{3}i$ in trigonometric form and explain why the argument is $\theta = 240°$ instead of $60°$ as indicated by your calculator.

6. Discuss the similarities between finding the components of a vector and writing a complex number in trigonometric form.

DEVELOPING YOUR SKILLS

Graph the complex numbers z_1, z_2, and z_3 given, then express one as the sum of the other two.

7. $z_1 = 7 + 2i$, $z_2 = 8 + 6i$, $z_3 = 1 + 4i$

8. $z_1 = 2 + 7i$, $z_2 = 3 + 4i$, $z_3 = -1 + 3i$

9. $z_1 = -2 - 5i$, $z_2 = 1 - 7i$, $z_3 = 3 - 2i$

10. $z_1 = -2 + 6i$, $z_2 = 7 - 2i$, $z_3 = 5 + 4i$

State the quadrant of each complex number, then write it in trigonometric form. For Exercises 11 through 14, answer in degrees. For 15 through 18, answer in radians.

11. $-2 - 2i$ **12.** $7 - 7i$ **13.** $-5\sqrt{3} - 5i$ **14.** $2 - 2\sqrt{3}i$

15. $-3\sqrt{2} + 3\sqrt{2}i$ **16.** $5\sqrt{7} - 5\sqrt{7}i$ **17.** $4\sqrt{3} - 4i$ **18.** $-6 + 6\sqrt{3}i$

Write each complex number in trigonometric form. For Exercises 19 through 22, answer in degrees using both an exact form and an approximate form, rounding to tenths. For 23 through 26, answer in radians using both an exact form and an approximate form, rounding to four decimal places.

19. $8 + 6i$ **20.** $-9 + 12i$ **21.** $-5 - 12i$ **22.** $-8 + 15i$

23. $6 + 17.5i$ **24.** $30 - 5.5i$ **25.** $-6 + 10i$ **26.** $12 - 4i$

Graph each complex number using its trigonometric form, then convert each to rectangular form.

27. $2 \text{ cis}\left(\frac{\pi}{4}\right)$

28. $12 \text{ cis}\left(\frac{\pi}{6}\right)$

29. $4\sqrt{3} \text{ cis}\left(\frac{\pi}{3}\right)$

30. $5\sqrt{3} \text{ cis}\left(\frac{7\pi}{6}\right)$

31. $17 \text{ cis}\left[\tan^{-1}\left(\frac{15}{8}\right)\right]$

32. $10 \text{ cis}\left[\tan^{-1}\left(\frac{3}{4}\right)\right]$

33. $6 \text{ cis}\left[\pi - \tan^{-1}\left(\frac{5}{\sqrt{11}}\right)\right]$

34. $4 \text{ cis}\left[\pi + \tan^{-1}\left(\frac{\sqrt{7}}{3}\right)\right]$

For the complex numbers z_1 and z_2 given, find their moduli r_1 and r_2 and arguments θ_1 and θ_2. Then compute their *product* in rectangular form. For modulus r and argument θ of the product, verify that $r_1 r_2 = r$ and $\theta_1 + \theta_2 = \theta$.

35. $z_1 = -2 + 2i;\quad z_2 = 3 + 3i$

36. $z_1 = 1 + \sqrt{3}i;\quad z_2 = 3 + \sqrt{3}i$

For the complex numbers z_1 and z_2 given, find their moduli r_1 and r_2 and arguments θ_1 and θ_2. Then compute their *quotient* in rectangular form. For modulus r and argument θ of the quotient, verify that $\frac{r_1}{r_2} = r$ and $\theta_1 - \theta_2 = \theta$.

37. $z_1 = \sqrt{3} + i;\quad z_2 = 1 + \sqrt{3}i$

38. $z_1 = -\sqrt{3} + i;\quad z_2 = 3 + 0i$

Compute the product $z_1 z_2$ and quotient $\frac{z_1}{z_2}$ using the trigonometric form. Answer in exact form where possible, otherwise round all values to two decimal places.

39. $z_1 = -4\sqrt{3} + 4i$
$z_2 = \frac{3\sqrt{3}}{2} + \frac{3}{2}i$

40. $z_1 = \frac{5\sqrt{3}}{2} + \frac{5}{2}i$
$z_2 = 0 + 6i$

41. $z_1 = -2\sqrt{3} + 0i$
$z_2 = -\frac{21}{2} + \frac{7i\sqrt{3}}{2}$

42. $z_1 = 0 - 6i\sqrt{2}$
$z_2 = \frac{3\sqrt{2}}{2} + \frac{3i\sqrt{6}}{2}$

43. $z_1 = 9\left[\cos\left(\frac{\pi}{15}\right) + i\sin\left(\frac{\pi}{15}\right)\right]$
$z_2 = 1.8\left[\cos\left(\frac{2\pi}{3}\right) + i\sin\left(\frac{2\pi}{3}\right)\right]$

44. $z_1 = 2\left[\cos\left(\frac{3\pi}{5}\right) + i\sin\left(\frac{3\pi}{5}\right)\right]$
$z_2 = 8.4\left[\cos\left(\frac{\pi}{5}\right) + i\sin\left(\frac{\pi}{5}\right)\right]$

45. $z_1 = 10(\cos 60° + i\sin 60°)$
$z_2 = 4(\cos 30° + i\sin 30°)$

46. $z_1 = 7(\cos 120° + i\sin 120°)$
$z_2 = 2(\cos 300° + i\sin 300°)$

47. $z_1 = 5\sqrt{2} \text{ cis } 210°$
$z_2 = 2\sqrt{2} \text{ cis } 30°$

48. $z_1 = 5\sqrt{3} \text{ cis } 240°$
$z_2 = \sqrt{3} \text{ cis } 90°$

49. $z_1 = 6 \text{ cis } 82°$
$z_2 = 1.5 \text{ cis } 27°$

50. $z_1 = 1.6 \text{ cis } 59°$
$z_2 = 8 \text{ cis } 275°$

WORKING WITH FORMULAS

51. Equilateral triangles in the complex plane: $u^2 + v^2 + w^2 = uv + uw + vw$

If the line segments connecting the complex numbers u, v, and w form the vertices of an equilateral triangle, the formula shown above holds true. Verify that $u = 2 + \sqrt{3}i$, $v = 10 + \sqrt{3}i$, and $w = 6 + 5\sqrt{3}i$ form the vertices of an equilateral triangle using the distance formula, then verify the formula given.

52. The cube of a complex number: $(A + B)^3 = A^3 + 3A^2B + 3AB^2 + B^3$

The cube of any binomial can be found using the formula here, where A and B are the terms of the binomial. Use the formula to compute the cube of $1 - 2i$ (note $A = 1$ and $B = -2i$).

APPLICATIONS

AC circuits: For the circuits indicated in Exercises 53 through 58, (a) find the magnitude of Z, the phase angle between current and voltage, and write the result in trigonometric form; and (b) find the total voltage across this circuit. Recall $Z = R + (X_L - X_C)j$ and $|Z| = \sqrt{R^2 + (X_L - X_C)^2}$.

Exercises 53 through 56

53. $R = 15\ \Omega$, $X_L = 12\ \Omega$, and $X_C = 4\ \Omega$, with $I = 3$ A.

54. $R = 24\ \Omega$, $X_L = 12\ \Omega$, and $X_C = 5\ \Omega$, with $I = 2.5$ A.

55. $R = 7\ \Omega$, $X_L = 6\ \Omega$, and $X_C = 11\ \Omega$, with $I = 1.8$ A.

56. $R = 9.2\ \Omega$, $X_L = 5.6\ \Omega$, and $X_C = 8.3\ \Omega$, with $I = 2.0$ A.

Exercises 57 and 58

57. $R = 12\ \Omega$ and $X_L = 5\ \Omega$, with $I = 1.7$ A.

58. $R = 35\ \Omega$ and $X_L = 12\ \Omega$, with $I = 4$ A.

AC circuits—voltage: The current I and the impedance Z for certain AC circuits are given. Write I and Z in trigonometric form and find the voltage in each circuit. Recall $V = IZ$.

59. $I = \sqrt{3} + 1j$ A and $Z = 5 + 5j\ \Omega$

60. $I = \sqrt{3} - 1j$ A and $Z = 2 + 2j\ \Omega$

61. $I = 3 - 2j$ A and $Z = 2 + 3.75j\ \Omega$

62. $I = 4 + 3j$ A and $Z = 2 - 4j\ \Omega$

AC circuits—current: If the voltage and impedance are known, the current I in the circuit is calculated as the quotient $I = \frac{V}{Z}$. Write V and Z in trigonometric form to find the current in each circuit.

63. $V = 2 + 2\sqrt{3}j$ and $Z = 4 - 4j\ \Omega$

64. $V = 4\sqrt{3} - 4j$ and $Z = 1 - 1j\ \Omega$

65. $V = 3 - 4j$ and $Z = 4 + 7.5j\ \Omega$

66. $V = 2.8 + 9.6j$ and $Z = 1.4 - 4.8j\ \Omega$

Parallel circuits: For AC circuits *wired in parallel*, the total impedance is given by $Z = \frac{Z_1Z_2}{Z_1 + Z_2}$, where Z_1 and Z_2 represent the impedance in each branch. Find the total impedance for the values given. Compute the product in the numerator using trigonometric form, and the sum in the denominator in rectangular form.

67. $Z_1 = 1 + 2j$ and $Z_2 = 3 - 2j$

68. $Z_1 = 3 - j$ and $Z_2 = 2 + j$

WRITING, RESEARCH, AND DECISION MAKING

69. Using the Internet or your local library, do some research on the concept of **resonance** in AC circuits, and how resonance helps us to selectively tune in to a particular television or radio station. In particular, try to determine how resonance is related to the capacitance and inductance of a circuit. Describe what you find in a paragraph or two, and include an illustrative example.

70. Some of the units used in a study of electricity have an interesting history. In particular, volts (voltage), ohms (resistance), farads (capacitance), and henrys (inductance) were all named after people. Do some research on each person and give a short biographical sketch.

71. Verify/prove that for the complex numbers $z_1 = r_1(\cos\alpha + i\sin\alpha)$ and $z_2 = r_2(\cos\beta + i\sin\beta)$, $\dfrac{z_1}{z_2} = \dfrac{r_1}{r_2}[\cos(\alpha - \beta) + i\sin(\alpha - \beta)]$.

EXTENDING THE CONCEPT

72. Recall that two lines are perpendicular if their slopes have a product of -1. For the directed line segment representing the complex number $z_1 = 7 + 24i$, find complex numbers z_2 and z_3 whose directed line segments are perpendicular to z_1 and have a magnitude one-fifth as large.

73. The magnitude of the impedance is $|Z| = \sqrt{R^2 + (X_L - X_C)^2}$. If R, X_L, and X_C are all nonzero, what conditions would make the magnitude of Z as small as possible?

MAINTAINING YOUR SKILLS

74. (4.4) Solve for $x \in [0, 2\pi)$:
$$350 = 750\sin\left(2x - \frac{\pi}{4}\right) - 25$$

75. (3.4) Write cos 15° in exact form using (a) a half angle identity and (b) a difference identity. Verify the results obtained are equivalent.

76. (3.2) Verify the following is an identity:
$$\frac{1 + \cos\alpha}{1 - \cos\alpha} = \frac{\sec\alpha + 1}{\sec\alpha - 1}$$

77. (1.1) For 24°12′36″, (a) determine its complement, (b) convert the complement to decimal degrees, and (c) convert the complement to radians.

78. (5.2) A ship is spotted by two observation posts that are 4 mi apart. Using the line between them for reference, the first post reports the ship is at an angle of 41°, while the second reports an angle of 63°, as shown. How far is the ship from the closest post?

79. (5.4) Two tugboats are pulling a large ship into dry dock. The first is pulling with a force of 1500 N and the second with a force of 1800 N. Determine the angle θ for the second tugboat that will keep the ship moving straight forward and into the dock.

5.8 De Moivre's Theorem and the Theorem on *n*th Roots

LEARNING OBJECTIVES

In Section 5.8 you will learn how to:

A. Use De Moivre's theorem to raise complex numbers to any power

B. Use De Moivre's theorem to check solutions to polynomial equations

C. Use the *n*th roots theorem to find the *n*th roots of a complex number

D. Use the *n*th roots theorem to solve a polynomial equation

INTRODUCTION

The material in this section represents some of the most significant developments in the history of mathematics. After hundreds of years of struggle, mathematical scientists had not only come to recognize the existence of complex numbers, but were able to make operations on them commonplace and routine. This allowed for the unification of many ideas related to the study of polynomial equations, and answered questions that had puzzled scientists from many different fields for centuries. In this section we will look at two fairly simple theorems, that actually represent over 1000 years in the evolution of mathematical thought.

POINT OF INTEREST

In addition to the theorem that bears his name, Abraham De Moivre (dé-muah) (1667–1754) made many contributions to the field of probability and much of his research was published in the *Transactions of the Royal Society,* to which he was elected at the age of 30. He was a close friend of Sir Isaac Newton and often acted in his behalf during the dispute between Newton and Leibniz over the invention of the calculus.

A. De Moivre's Theorem

Having found acceptable means for applying the four basic operations to complex numbers, our attention naturally shifts to the computation of powers and roots. Without them, we'd remain wholly unable to offer complete solutions to polynomial equations and find solutions for many applications. The computation of powers, squares, and cubes offer little challenge, as they can be computed easily using the formula for binomial squares $[(A + B)^2 = A^2 + 2AB + B^2]$ or by applying the **binomial theorem.** For larger powers, the binomial theorem becomes too time consuming and a more efficient method is desired. The key here is to use the trigonometric form of the complex number. In Section 5.7 we noted the product of two complex numbers involved multiplying their moduli and adding their arguments:

For $z_1 = r_1(\cos\theta_1 + i\sin\theta_1)$ and $z_2 = r_2(\cos\theta_2 + i\sin\theta_2)$ we have

$$r_1(\cos\theta_1 + i\sin\theta_1)r_2(\cos\theta_2 + i\sin\theta_2) = r_1r_2[\cos(\theta_1 + \theta_2) + i\sin(\theta_1 + \theta_2)]$$

For the square of a complex number, $r_1 = r_2$ and $\theta_1 = \theta_2$. Using θ itself yields

$$\begin{aligned} r(\cos\theta + i\sin\theta)r(\cos\theta + i\sin\theta) &= r^2[\cos(\theta + \theta) + i\sin(\theta + \theta)] \\ &= r^2[\cos(2\theta) + i\sin(2\theta)] \end{aligned}$$

Applying the techniques once again for the cube of the same complex number yields

$$\begin{aligned} r^2[\cos(2\theta) + i\sin(2\theta)]\, r(\cos\theta + i\sin\theta) &= r^3[\cos(2\theta + \theta) + i\sin(2\theta + \theta)] \\ &= r^3[\cos(3\theta) + i\sin(3\theta)] \end{aligned}$$

The result can be extended further and generalized into **De Moivre's theorem.**

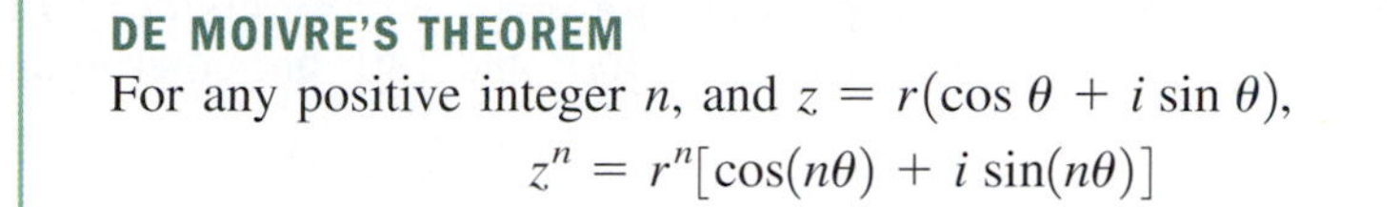

DE MOIVRE'S THEOREM
For any positive integer n, and $z = r(\cos\theta + i\sin\theta)$,
$$z^n = r^n[\cos(n\theta) + i\sin(n\theta)]$$

For a proof of the theorem where n is an integer and $n \geq 1$, interested students can investigate a **proof by induction** by going to www.mhhe.com/coburn.

WORTHY OF NOTE

Sometimes the argument of cosine and sine becomes very large after applying De Moivre's theorem. In these cases, we use the fact that $\theta = \theta \pm 360°k$ and $\theta = \theta \pm 2\pi k$ represent coterminal angles for integers k, and use the coterminal angle θ where $0 \leq \theta < 360°$ or $0 \leq \theta < 2\pi$.

EXAMPLE 1 Use De Moivre's theorem to compute z^9, given $z = -\frac{1}{2} - \frac{1}{2}i$.

Solution: Here we have $r = \sqrt{\left(-\frac{1}{2}\right)^2 + \left(-\frac{1}{2}\right)^2} = \frac{\sqrt{2}}{2}$. With z in QIII, $\tan\theta = 1$ yields $\theta = \frac{5\pi}{4}$. The trigonometric form is $z = \frac{\sqrt{2}}{2}\left[\cos\left(\frac{5\pi}{4}\right) + i\sin\left(\frac{5\pi}{4}\right)\right]$ and applying the theorem with $n = 9$ gives

$$z^9 = \left(\frac{\sqrt{2}}{2}\right)^9\left[\cos\left(9\cdot\frac{5\pi}{4}\right) + i\sin\left(9\cdot\frac{5\pi}{4}\right)\right] \quad \text{De Moivre's theorem}$$
$$= \frac{\sqrt{2}}{32}\left[\cos\left(\frac{45\pi}{4}\right) + i\sin\left(\frac{45\pi}{4}\right)\right] \quad \text{simplify}$$
$$= \frac{\sqrt{2}}{32}\left[\cos\left(\frac{5\pi}{4}\right) + i\sin\left(\frac{5\pi}{4}\right)\right] \quad \text{coterminal angles}$$
$$= \frac{\sqrt{2}}{32}\left(-\frac{\sqrt{2}}{2} - \frac{\sqrt{2}}{2}i\right) \quad \text{evaluate functions}$$
$$= -\frac{1}{32} - \frac{1}{32}i \quad \text{result}$$

NOW TRY EXERCISES 7 THROUGH 14

As with products and quotients of complex numbers in trig form, computing powers in trig form is both elegant and efficient: $(2 \text{ cis } 40°)^6 = 64 \text{ cis } 240°$. See Exercises 15 through 18.

For cases where θ is not a standard angle, De Moivre's theorem requires an intriguing application of the skills developed in Chapter 3, including the use of multiple angle identities and working from a right triangle drawn relative to $\theta_r = \tan^{-1}\left(\frac{b}{a}\right)$. See Exercises 57 and 58.

B. Checking Solutions to Polynomial Equations

One application of De Moivre's theorem is checking the complex roots of a polynomial, as in Example 2.

EXAMPLE 2 Use De Moivre's theorem to show that $z = -2 - 2i$ is a solution to $z^4 - 3z^3 - 38z^2 - 128z - 144 = 0$.

Solution: We will apply the theorem to the third and fourth degree terms, and compute the square directly. Since z is in QIII, the trigonometric form is $z = 2\sqrt{2}\,\text{cis}\,225°$. In the following illustration, note that 900° and 180° are coterminal, as are 675° and 315°.

$$\begin{array}{lll}
(-2-2i)^4 & (-2-2i)^3 & (-2-2i)^2 \\
= (2\sqrt{2})^4\text{cis}(4 \cdot 225°) & = (2\sqrt{2})^3\text{cis}(3 \cdot 225°) & = 4 + 8i + (2i)^2 \\
= (2\sqrt{2})^4\text{cis}\,900° & = (2\sqrt{2})^3\text{cis}\,675° & = 4 + 8i + 4i^2 \\
= 64\,\text{cis}\,180° & = (2\sqrt{2})^3\text{cis}\,315° & = 4 + 8i - 4 \\
= 64(-1 + 0i) & = 16\sqrt{2}\left(\frac{\sqrt{2}}{2} - \frac{\sqrt{2}}{2}i\right) & = 0 + 8i \\
= -64 & = 16 - 16i & = 8i
\end{array}$$

Substituting back into the original equation gives

$$\begin{aligned}
1z^4 - 3z^3 - 38z^2 - 128z - 144 &= 0 \\
1(-64) - 3(16 - 16i) - 38(8i) - 128(-2 - 2i) - 144 &= 0 \\
-64 - 48 + 48i - 304i + 256 + 256i - 144 &= 0 \\
(-64 - 48 + 256 - 144) + (48 - 304 + 256)i &= 0 \\
0 &= 0 \checkmark
\end{aligned}$$

NOW TRY EXERCISES 19 THROUGH 26

Regarding Example 2, we know from a study of algebra that complex roots must occur in conjugate pairs, meaning $-2 + 2i$ is also a root. This equation actually has two real and two complex roots, with $z = 9$ and $z = -2$ being the two real roots.

C. The *n*th Roots Theorem

Having looked at De Moivre's theorem, which raises a complex number to any power, we now consider the ***n*th roots theorem,** which will compute the nth roots of a complex number. If we allow that De Moivre's theorem also holds for rational values $\frac{1}{n}$, instead of only the integers n illustrated previously, the formula for computing an nth root would be a direct result:

$$\begin{aligned}
z^{\frac{1}{n}} &= r^{\frac{1}{n}}\left[\cos\left(\frac{1}{n}\theta\right) + i\sin\left(\frac{1}{n}\theta\right)\right] && \text{De Moivre's theorem} \\
&= \sqrt[n]{r}\left[\cos\left(\frac{\theta}{n}\right) + i\sin\left(\frac{\theta}{n}\right)\right] && \text{simplify}
\end{aligned}$$

However, this formula would *find only the principal nth root*! In other words, periodic solutions would be ignored. As in Section 5.7, it's worth noting the most general form of a complex number is $z = r[\cos(\theta + 360°k) + i\sin(\theta + 360°k)]$, for $k \in \mathbb{Z}$. When De Moivre's theorem is applied to this form for *integers n*, we obtain $z^n = r^n[\cos(n\theta + 360°kn) + i\sin(n\theta + 360°kn)]$, which returns a result identical to $r^n[\cos(n\theta) + i\sin(n\theta)]$. However, for the rational exponent $\frac{1}{n}$, the general form takes additional solutions into account and will return all n, nth roots.

$$\begin{aligned}
z^{\frac{1}{n}} &= r^{\frac{1}{n}}\left\{\cos\left[\frac{1}{n}(\theta + 360°k)\right] + i\sin\left[\frac{1}{n}(\theta + 360°k)\right]\right\} && \text{De Moivre's theorem for rational exponents} \\
&= \sqrt[n]{r}\left[\cos\left(\frac{\theta}{n} + \frac{360°k}{n}\right) + i\sin\left(\frac{\theta}{n} + \frac{360°k}{n}\right)\right] && \text{simplify}
\end{aligned}$$

THE nTH ROOTS THEOREM

For $z = r(\cos\theta + i\sin\theta)$, a positive integer n, and $r \in \mathbb{R}$, z has exactly n distinct nth roots determined by

$$\sqrt[n]{z} = \sqrt[n]{r}\left[\cos\left(\frac{\theta}{n} + \frac{360°k}{n}\right) + i\sin\left(\frac{\theta}{n} + \frac{360°k}{n}\right)\right]$$

where $k = 0, 1, 2, \ldots, n - 1$.

For ease of computation, it helps to note that once the argument for the principal root is found using $k = 0$, $\frac{\theta}{n} + \frac{360°k}{n}$ simply adds $\frac{360}{n}\left(\text{or } \frac{2\pi}{n}\right)$ to the previous argument for $k = 1, 2, 3, \ldots, n$.

EXAMPLE 3 Use the nth roots theorem to find the five fifth roots of $z = 16\sqrt{3} + 16i$.

Solution: In trigonometric form, $16\sqrt{3} + 16i = 32(\cos 30° + i\sin 30°)$. With $n = 5$, $r = 32$, and $\theta = 30°$, we have $\sqrt[5]{r} = \sqrt[5]{32} = 2$, and $\frac{30°}{5} + \frac{360°k}{5} = 6° + 72°k$. The principal root is $z_0 = 2(\cos 6° + i\sin 6°)$. Adding 72° to each previous argument, we find the other four roots are

$$z_1 = 2(\cos 78° + i\sin 78°) \qquad z_2 = 2(\cos 150° + i\sin 150°)$$
$$z_3 = 2(\cos 222° + i\sin 222°) \qquad z_4 = 2(\cos 294° + i\sin 294°)$$

NOW TRY EXERCISES 27 THROUGH 30

Figure 5.84

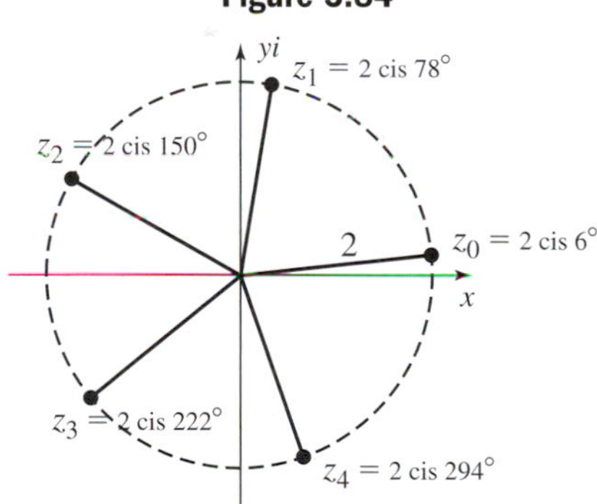

Of the five roots in Example 3, only $z_2 = 2(\cos 150° + i\sin 150°)$ uses a standard angle. Applying De Moivre's theorem with $n = 5$ gives $(2 \text{ cis } 150°)^5 = 32 \text{ cis } 720° = 32 \text{ cis } 30°$ or $16\sqrt{3} + 16i$.✓ See Exercise 54.

As a consequence of the arguments in a solution being separated uniformly by $\frac{360°}{n}$, the graphs of complex roots are equally spaced about a circle of radius r. The five fifth roots from Example 3 are shown in Figure 5.84 (note each argument differs by 72°).

For additional insight, we reason that the nth roots of a complex number must also be complex. To find the four fourth roots of $z = 8 + 8\sqrt{3}i = 16(\cos 60° + i\sin 60°)$, we seek a number of the form $r(\cos\alpha + i\sin\alpha)$ such that $[r(\cos\alpha + i\sin\alpha)]^4 = 16(\cos 60° + i\sin 60°)$. Applying De Moivre's theorem to the left-hand side and equating equivalent parts we obtain

$$r^4[\cos(4\alpha) + i\sin(4\alpha)] = 16(\cos 60° + i\sin 60°), \text{ which leads to}$$
$$r^4 = 16 \text{ and}$$
$$4\alpha = 60°$$

From this it is obvious that $r = 2$, but as with similar equations solved in Chapter 4, the equation $4\alpha = 60°$ has multiple solutions. To find them, we first add $360°k$ to $60°$, *then* solve for α.

$$4\alpha = 60° + 360°k; \text{ for any integer } k$$
$$\alpha = \frac{60° + 360°k}{4}$$
$$= 15° + 90°k$$

For convenience, we start with $k = 0, 1, 2$, and so on, which leads to

For $k = 0$: $\alpha = 15° + 90°(0) = 15°$ For $k = 1$: $\alpha = 15° + 90°(1) = 105°$

For $k = 2$: $\alpha = 15° + 90°(2) = 195°$ For $k = 3$: $\alpha = 15° + 90°(3) = 285°$

At this point it should strike us that we have four roots—exactly the number required. Indeed, using $k = 4$ gives $\alpha = 15° + 90°(4) = 375°$, which is coterminal with the $15°$ obtained when $k = 0$. Hence the four fourth roots are

$$z_0 = 2(\cos 15° + i \sin 15°) \qquad z_1 = 2(\cos 105° + i \sin 105°)$$
$$z_2 = 2(\cos 195° + i \sin 195°) \qquad z_3 = 2(\cos 285° + i \sin 285°).$$

The check for these solutions is asked for in Exercise 53.

In Example 4 you're asked to find the three cube roots of 1, also called the **cube roots of unity,** and graph the results. The nth roots of unity play a significant role in the solution of many polynomial equations. For an in-depth study of this connection, visit www.mhhe.com/coburn and go to Section 5.10: Trigonometry, Complex Numbers and Cubic Equations.

EXAMPLE 4 Use the nth roots theorem to solve the equation $x^3 - 1 = 0$. Write the results in rectangular form and graph.

Solution: From $x^3 - 1 = 0$, we have $x^3 = 1$ and must find the three cube roots of unity. As before, we begin in trigonometric form: $1 + 0i = 1(\cos 0° + i \sin 0°)$. With $n = 3$, $r = 1$, and $\theta = 0°$, we have $\sqrt[3]{r} = \sqrt[3]{1} = 1$, and $\frac{0°}{3} + \frac{360°k}{3} = 0° + 120°k$.

The principal root ($k = 0$) is $z_0 = 1(\cos 0° + i \sin 0°)$. Adding $120°$ to each previous argument, we find the other roots are

$$z_1 = 1(\cos 120° + i \sin 120°)$$
$$z_2 = 1(\cos 240° + i \sin 240°)$$

In rectangular form these are $1, -\frac{1}{2} + \frac{\sqrt{3}}{2}i$, and $-\frac{1}{2} - \frac{\sqrt{3}}{2}i$, as shown in Figure 5.85.

Figure 5.85

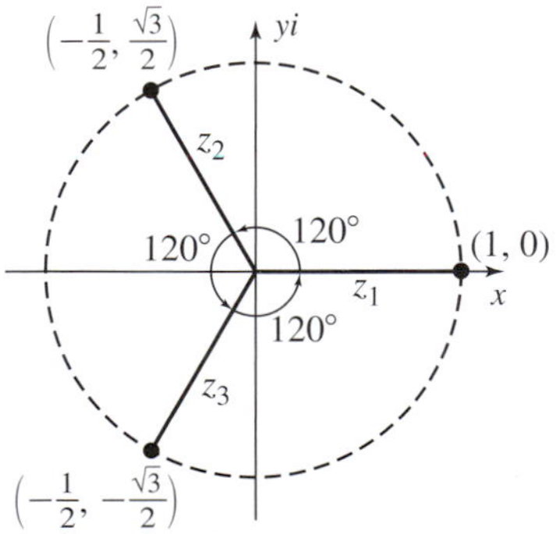

NOW TRY EXERCISES 31 THROUGH 44

D. Polynomial Equations and the *n*th Roots Theorem

In addition to simply computing the *n*th roots of a complex number, the *n*th roots theorem provides the ability to solve certain polynomial equations, which up to this time have been "unsolvable." Consider the equation $z^4 - 6z^2 + 25 = 0$, noting an equation of degree 4 will have 4 roots, which may be a combination of real roots, complex roots, and/or roots of multiplicity (repeated roots).

EXAMPLE 5 Find all four solutions to $z^4 - 6z^2 + 25 = 0$.

Solution: The equation is not factorable using real numbers, so we'll attempt a solution by writing the equation in quadratic form. Using the substitution $u = z^2$ we obtain $u^2 - 6u + 25 = 0$ and apply the quadratic formula to solve for u:

$$u = \frac{-b \pm \sqrt{b^2 - 4ac}}{2a} \quad \text{quadratic formula}$$

$$= \frac{-(-6) \pm \sqrt{(-6)^2 - 4(1)(25)}}{2(1)} \quad \text{substitute 1 for } a, -6 \text{ for } b, \text{ and 25 for } c$$

$$= \frac{6 \pm \sqrt{-64}}{2} \quad \text{simplify}$$

$$= \frac{6 \pm 8i}{2} = 3 \pm 4i \quad \text{result}$$

Replacing u with the original z^2, we find the solutions to the original equation must satisfy $z^2 = 3 + 4i$ and $z^2 = 3 - 4i$. Writing these in trigonometric form gives

$$z^2 = 3 + 4i \qquad\qquad z^2 = 3 - 4i$$

$$= 5\,\text{cis}\left[\tan^{-1}\left(\frac{4}{3}\right)\right] \qquad\qquad = 5\,\text{cis}\left[\tan^{-1}\left(-\frac{4}{3}\right)\right]$$

At this point we'll store $\theta_r = \tan^{-1}\left(\frac{4}{3}\right)$ in memory location "A" of a calculator (since it is not a standard angle) and recall this stored value of θ as needed.

For $z^2 = 5\,\text{cis}(A°)$ we have $z = \sqrt{5}\,\text{cis}\left(\frac{A°}{2} + \frac{360°k}{2}\right) =$ $\sqrt{5}\,\text{cis}\left(\frac{A°}{2} + 180°k\right)$, which for $k = 0$ gives

$$z_0 = \sqrt{5}\,\text{cis}\left(\frac{A}{2}\right) \quad n\text{th roots theorem } (k = 0)$$

$$= 2 + i \quad \text{compute using a calculator and the stored value of } A$$

$$z_1 = \sqrt{5}\,\text{cis}\left(\frac{A}{2} + 180°\right) \quad n\text{th roots theorem } (k = 1)$$

$$= -2 - i \quad \text{compute using a calculator}$$

From algebra we know that complex roots must occur in conjugate pairs, so $z_2 = 2 - i$ (from $k = 0$) and $z_3 = -2 + i$ (from $k = 1$)

WORTHY OF NOTE

On the TI-84 Plus, we store θ in memory location "A" using the keystrokes $\tan^{-1}\left(\frac{4}{3}\right)$ STO➡ ALPHA MATH. Consult our Internet site or your calculator's manual for information on other calculator models.

are the remaining two solutions and there is no need to extract them from $z^2 = 3 - 4i$. The four roots are $2 + i$, $2 - i$, $-2 - i$, and $-2 + i$.

NOW TRY EXERCISES 47 THROUGH 52

As a final note, it must have struck the mathematicians who pioneered these discoveries with some amazement, that complex numbers and the trigonometric functions should be so closely related. The amazement must have been all the more profound upon discovering an additional connection between complex numbers and *exponential functions*. For more on these connections, visit www.mhhe.com/coburn and review Section 5.9: Complex Numbers in Exponential Form.

TECHNOLOGY HIGHLIGHT

Templates, Graphing Calculators, and Complex Numbers

The keystrokes shown apply to a TI-84 Plus model. Please consult our Internet site or your manual for other models.

To work efficiently with larger expressions on a graphing calculator, it helps to set up a general template that can repeatedly be recalled and reused. For De Moivre's theorem, different complex numbers will give different values for n, r, and θ. Generally n and r will consist of a single digit, but θ will often have one, two, or three digits so we set the template to accommodate this fact. After clearing the home screen, we enter the generic template R^N*(cos(N*θ) + *i* sin(N*θ)) as in Figure 5.86, store the value of θ, then recall the expression each time we wish to use it.

Figure 5.86

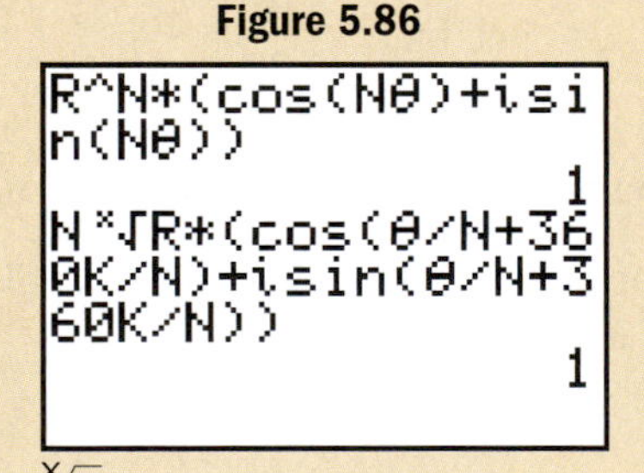

For nth roots we enter N$\sqrt[x]{\text{R}}$*(cos(θ/N + 360K/N) + *i* sin(θ/N + 360K/N)). In either case we can recall and overwrite the values to use the formula (as in Example 1 here), or store values in N, R, θ, and K to use the formula as it stands (as in Example 2).

EXAMPLE 1 Use the template above for De Moivre's theorem and your graphing calculator to find the value of $(2\sqrt{3} - 2i)^6$.

Solution: In this case the complex number is $z = 2\sqrt{3} - 2i$ (in QIV) with $n = 6$, $r = 4$, and $\theta = -30°$. Storing -30 in location θ (-30 STO➡ ALPHA 3), we recall the expression for De Moivre's theorem and enter/overwrite N and R. Here, $z^6 = -4096$. See Figure 5.87.

Figure 5.87

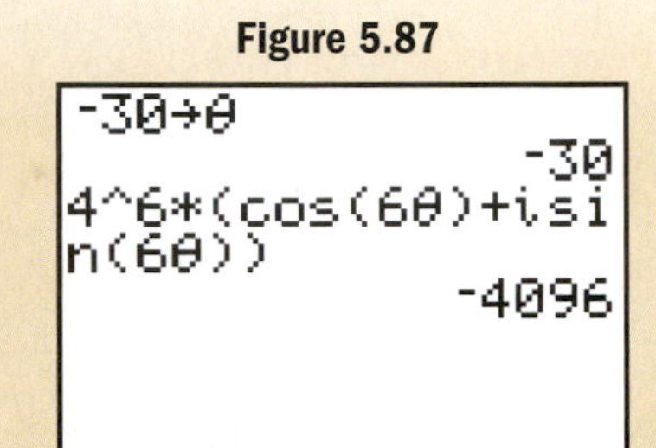

EXAMPLE 2 Use the template above for the nth roots theorem and the storage/recall abilities of your graphing calculator to find the three cube roots of $z = -2 - 2i$.

Solution: The complex number $z = -2 - 2i$ (in QIII) yields $r = \sqrt{8}$ and $\theta = 225°$. With $n = 3$ and $k = 0$ to begin, we store these values in their respective locations: $\sqrt{8}$ STO➡ ALPHA X for R, 3 STO➡ ALPHA LOG for N, 225 STO➡ ALPHA 3 for θ, and 0 STO➡ ALPHA (for K (see Figure 5.88). Recalling the template for nth roots and pressing ENTER gives $z_0 = 0.3660254038 + 1.366025404i$. Changing the value of k to $k = 1$ and $k = 2$ gives $z_1 = -1.366025404 - 0.3660254038i$, with $z_2 = 1 - i$.

Figure 5.88

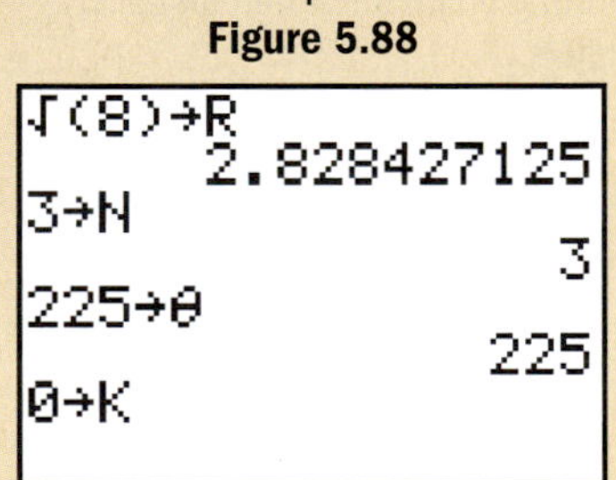

Exercise 1: Use the appropriate template to find the three cube roots of unity, $z = 1 + 0i$ $\left(\text{note } \frac{\sqrt{3}}{2} \approx 0.8660\right)$.

Exercise 2: Use the appropriate template to find z^3, given $z = 1 - \sqrt{3}i$.

5.8 EXERCISES

CONCEPTS AND VOCABULARY

Fill in each blank with the appropriate word or phrase. Carefully reread the section if needed.

1. For $z = r(\cos\theta + i\sin\theta)$, z^5 is computed as ________ according to ________ theorem.

2. If $z = 6i$, then z raised to an ____ power will be real and z raised to an ____ power will be ________ since $\theta =$ ____.

3. One application of De Moivre's theorem is to check ________ solutions to a polynomial equation.

4. The nth roots of a complex number are equally spaced on a circle of radius r, since their arguments all differ by ________ degrees or ________ radians.

5. From Example 3, go ahead and compute the value of z_5, z_6, and z_7. What do you notice? Discuss how this reaffirms that there are exactly n, nth roots.

6. Use a calculator to find $(1 - 3i)^4$. Then use it again to find the fourth root of the result. What do you notice? Explain the discrepancy and then resolve it using the nth roots theorem to find all four roots.

DEVELOPING YOUR SKILLS

Use De Moivre's theorem to compute the following. Clearly state the value of r, n, and θ before you begin.

7. $(3 + 3i)^4$

8. $(-2 + 2i)^6$

9. $(-1 + \sqrt{3}i)^3$

10. $(\sqrt{3} - i)^3$

11. $\left(\frac{1}{2} - \frac{\sqrt{3}}{2}i\right)^5$

12. $\left(-\frac{\sqrt{3}}{2} + \frac{1}{2}i\right)^6$

13. $\left(\frac{\sqrt{2}}{2} - \frac{\sqrt{2}}{2}i\right)^6$

14. $\left(-\frac{\sqrt{2}}{2} + \frac{\sqrt{2}}{2}i\right)^5$

15. $(4 \text{ cis } 330°)^3$

16. $(4 \text{ cis } 300°)^3$

17. $\left(\frac{\sqrt{2}}{2} \text{ cis } 135°\right)^5$

18. $\left(\frac{\sqrt{2}}{2} \text{ cis } 135°\right)^8$

Use De Moivre's theorem to verify the solution given for each polynomial equation.

19. $z^4 + 3z^3 - 6z^2 + 12z - 40 = 0$; $z = 2i$

20. $z^4 - z^3 + 7z^2 - 9z - 18 = 0$; $z = -3i$

21. $z^4 + 6z^3 + 19z^2 + 6z + 18 = 0$; $z = -3 - 3i$

22. $2z^4 + 3z^3 - 4z^2 + 2z + 12 = 0$; $z = 1 - i$

23. $z^5 + z^4 - 4z^3 - 4z^2 + 16z + 16 = 0$; $z = \sqrt{3} - i$

24. $z^5 + z^4 - 16z^3 - 16z^2 + 256z + 256 = 0$; $z = 2\sqrt{3} + 2i$

25. $z^4 - 4z^3 + 7z^2 - 6z - 10 = 0$; $z = 1 + 2i$

26. $z^4 - 2z^3 - 7z^2 + 28z + 52 = 0$; $z = 3 - 2i$

Use the nth roots theorem to find the nth roots. Clearly state r, n, and θ (from the trigonometric form of z) as you begin. Answer in exact form when possible, otherwise use a four decimal place approximation.

27. four fourth roots of $-8 + 8\sqrt{3}i$

28. five fifth roots of $16 - 16\sqrt{3}i$

29. four fourth roots of $-7 - 7i$

30. three cube roots of $9 + 9i$

31. five fifth roots of unity

32. six sixth roots of unity

33. five fifth roots of 243

34. three cube roots of 8

35. three cube roots of $-27i$

36. five fifth roots of $32i$

Solve each equation using the nth roots theorem.

37. $x^5 - 32 = 0$

38. $x^5 - 243 = 0$

39. $x^3 - 27i = 0$

40. $x^3 + 64i = 0$

41. $x^5 - \sqrt{2} - \sqrt{2}i = 0$

42. $x^5 - 1 + \sqrt{3}i = 0$

43. Solve the equation $x^3 - 1 = 0$ by factoring it as the difference of cubes and applying the quadratic formula. Compare results to those obtained in Example 4.

44. Use the nth roots theorem to find the four fourth roots of unity, then find all solutions to $x^4 - 1 = 0$ by factoring it as a difference of squares. What do you notice?

WORKING WITH FORMULAS

The discriminant of a cubic equation: $D = \dfrac{4p^3 + 27q^2}{108}$

For cubic equations of the form $z^3 + pz + q = 0$, where p and q are real numbers, one solution has the form $z = \sqrt[3]{-\frac{q}{2} + \sqrt{D}} + \sqrt[3]{-\frac{q}{2} - \sqrt{D}}$, where D is called the discriminant. Compute the value of D for the cubic equations given, then use the nth roots theorem to find the three cube roots of $-\frac{q}{2} + \sqrt{D}$ and $-\frac{q}{2} - \sqrt{D}$ in trigonometric form (also see Exercises 61 and 62).

45. $z^3 - 6z + 4 = 0$

46. $z^3 - 12z - 8 = 0$

APPLICATIONS

Polynomial equations: Solve the following equations using a u-substitution, the quadratic formula, and the nth roots theorem. Verify solutions using a calculator.

47. $z^4 - 8z^2 + 20 = 0$

48. $z^4 + 2z^2 + 26 = 0$

49. $z^4 + 12z^2 + 45 = 0$

50. $z^4 - 10z^2 + 41 = 0$

51. $z^4 - 10z^2 + 34 = 0$

52. $z^4 + 4z^2 + 53 = 0$

53. Powers and roots: Just prior to Example 4, the four fourth roots of $z = 8 + 8\sqrt{3}i$ were given as

$z_0 = 2(\cos 15° + i \sin 15°)$ $\quad z_1 = 2(\cos 105° + i \sin 105°)$

$z_2 = 2(\cos 195° + i \sin 195°)$ $\quad z_3 = 2(\cos 285° + i \sin 285°)$.

Verify these are the four fourth roots of $z = 8 + 8\sqrt{3}i$ using a calculator and De Moivre's theorem.

54. Powers and roots: In Example 3 we found the five fifth roots of $z = 16\sqrt{3} + 16i$ were

$z_0 = 2(\cos 6° + i \sin 6°)$ $\quad z_1 = 2(\cos 78° + i \sin 78°)$ $\quad z_2 = 2(\cos 150° + i \sin 150°)$

$z_3 = 2(\cos 222° + i \sin 222°)$ $\quad z_4 = 2(\cos 294° + i \sin 294°)$

Verify these are the five fifth roots of $16\sqrt{3} + 16i$ using a calculator and De Moivre's theorem.

Electrical circuits: For an AC circuit with three branches wired in parallel, the total impedance is given by $Z_T = \dfrac{Z_1Z_2Z_3}{Z_1Z_2 + Z_1Z_3 + Z_2Z_3}$, where Z_1, Z_2, and Z_3 represent the impedance in each branch of the circuit. If the impedance in each branch is identical, $Z_1 = Z_2 = Z_3 = Z$, and the numerator becomes Z^3 and the denominator becomes $3Z^2$, (a) use De Moivre's theorem to calculate the numerator and denominator for each value of Z given, (b) find the total impedance by computing the quotient $\dfrac{Z^3}{3Z^2}$, and (c) verify your result is identical to $\dfrac{Z}{3}$.

55. $Z = 3 + 4j$ in all three branches

56. $Z = 5\sqrt{3} + 5j$ in all three branches

WRITING, RESEARCH, AND DECISION MAKING

In Chapter 3, you were asked to verify that $\sin(3\theta) = 3\sin\theta - 4\sin^3\theta$ and $\cos(4\theta) = 8\cos^2\theta - 8\cos^2\theta + 1$ were identities (Section 3.4, Exercises 21 and 22). For $z = 3 + \sqrt{7}i$, verify $|z| = 4$ and $\theta = \tan^{-1}\left(\frac{\sqrt{7}}{3}\right)$, then draw a right triangle with $\sqrt{7}$ opposite θ and 3 adjacent to θ. Discuss how this right triangle and the identities given can be used in conjunction with De Moivre's theorem to find the exact value of the powers given (also see Exercises 59 and 60).

57. $(3 + \sqrt{7}i)^3$

58. $(3 + \sqrt{7}i)^4$

EXTENDING THE CONCEPT

For cases where θ is not a standard angle, working toward an exact answer using De Moivre's theorem requires the use of multiple angle identities and drawing the right triangle related to

$\theta = \tan^{-1}\left(\frac{b}{a}\right)$. When $n = 4$, the general result gives $z^4 = r^4[\cos(4\theta) + i\sin(4\theta)]$, where $\theta = \tan^{-1}\left(\frac{b}{a}\right)$. For Exercises 59 and 60, use De Moivre's theorem to compute the complex powers by (a) constructing the related right triangle for θ, (b) evaluating $\sin(4\theta)$ using two applications of double-angle identities, and (c) evaluating $\cos(4\theta)$ using a Pythagorean identity and the computed value of $\sin(4\theta)$.

59. $z = (1 + 2i)^4$

60. $(2 + \sqrt{5}i)^4$

The solutions to the cubic equations in Exercises 45 and 46 (repeated in Exercises 61 and 62) can be found by adding the cube roots of $-\frac{q}{2} + \sqrt{D}$ and $-\frac{q}{2} - \sqrt{D}$ that have arguments summing to 360°.

61. Find the roots of $z^3 - 6z + 4 = 0$

62. Find the roots of $z^3 - 12z - 8 = 0$

MAINTAINING YOUR SKILLS

63. (5.6) In AC (alternating current) circuits, the current I is computed as the quotient of the voltage V and the impedance Z: $I = \frac{V}{Z}$. (a) Find the amount of current given $V = 14 - 5i$ and $Z = 3 - 2i$. (b) Verify your answer by showing $ZI = V$.

64. (4.2) Using her calculator Julia finds that $\sin 120° = \frac{\sqrt{3}}{2}$ but $\sin^{-1}\left(\frac{\sqrt{3}}{2}\right) \neq 120°$. Explain why.

65. (3.2) Prove the following is an identity:

$$\frac{\tan^2 x}{\sec x + 1} = \frac{1 - \cos x}{\cos x}$$

66. (5.7) Convert $z = 4 - 4i$ to trigonometric form.

67. (1.2) Solve the triangle shown using a 30-60-90 standard triangle.

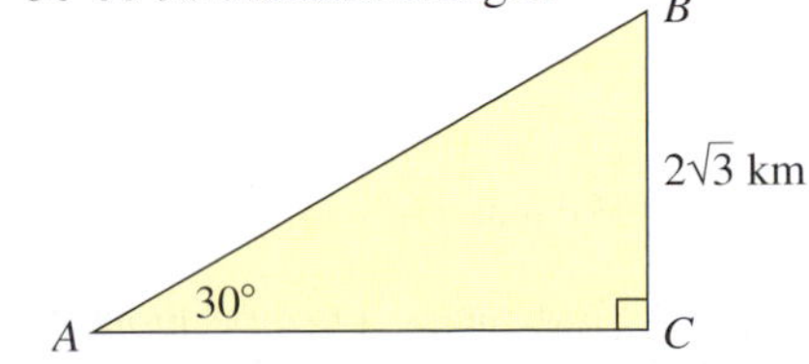

68. (1.2) Solve the triangle given. Round lengths to hundredths of a meter.

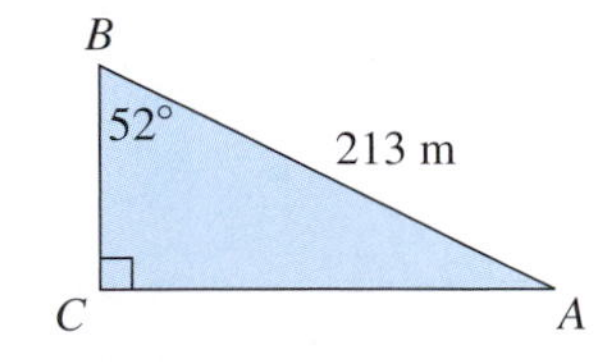

SUMMARY AND CONCEPT REVIEW

SECTION 5.1 Oblique Triangles and the Law of Sines

KEY CONCEPTS

- The law of sines states that in a triangle, the ratio of the sine of an angle to its opposite side is constant: $\frac{\sin A}{a} = \frac{\sin B}{b} = \frac{\sin C}{c}$.
- The law of sines requires a known angle and a side opposite this angle, hence cannot be applied for SSS and SAS triangles.
- For AAS and ASA triangles, the law of sines yields a unique solution.

EXERCISES

Solve the following triangles.

1. 2.

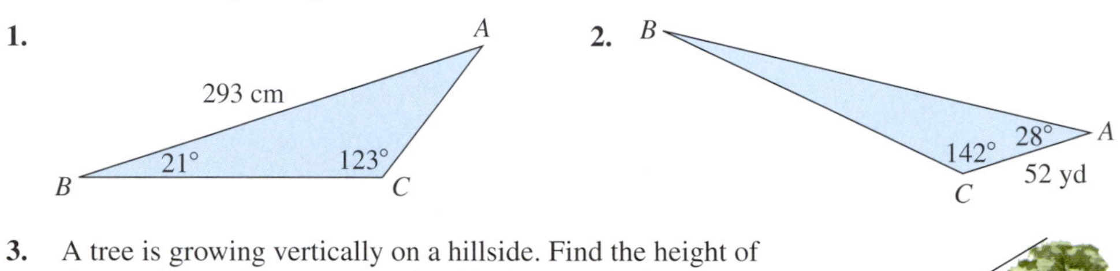

3. A tree is growing vertically on a hillside. Find the height of the tree if it makes an angle of 110° with the hillside and the angle of elevation from the base of the hill to the top of the tree is 25° at a distance of 70 ft.

SECTION 5.2 The Law of Sines and the Ambiguous Case

KEY CONCEPTS

- When given two sides of a triangle and an angle opposite one of these sides (SSA), the number of solutions is *in doubt,* giving rise to the designation, "the ambiguous case."
- SSA triangles may have no solution, one solution, or two solutions, depending on the values given.
- When solving triangles, always remember:
  - the sum of all angles must be 180°: $\angle A + \angle B + \angle C = 180°$
  - the sum of any two sides must exceed the length of the remaining side: $a + b > c$
  - $\sin^{-1}\theta = k$ has no solution for $k > 1$
  - $\sin^{-1}\theta = k$ has two solutions for $0 < |k| < 1$
- In many cases *scaled drawings*, using a metric ruler and a protractor, can be very helpful in "seeing" the number of solutions possible.
- Although it is indeed called the ambiguous case, SSA triangles have a number of useful applications.

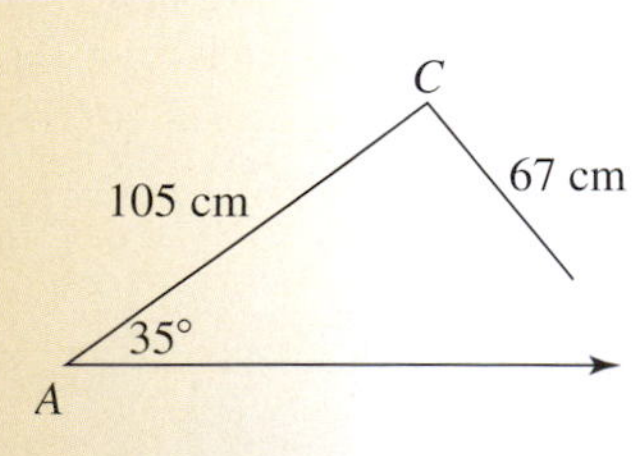

EXERCISES

4. Find two values of θ that will make the equation true: $\frac{\sin\theta}{14} = \frac{\sin 50}{31}$.
5. Solve using the law of sines. If two solutions exist, find both (figure not drawn to scale).

6. Jasmine is flying her tethered, gas-powered airplane at a local park, where a group of bystanders is watching from a distance of 60 ft, as shown. If the tether has a radius of 35 ft and one of the bystanders walks away at an angle of 40°, will he get hit by the plane? What is the smallest angle of exit he could take (to the nearest whole) without being struck by Jasmine's plane?

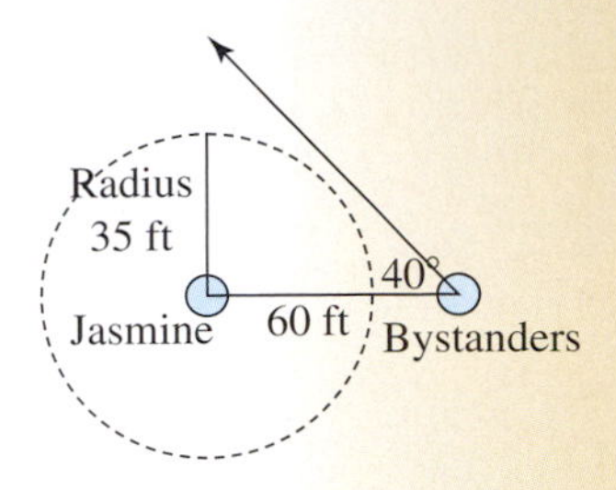

SECTION 5.3 The Law of Cosines

KEY CONCEPTS

- The law of cosines is used to solve SSS and SAS triangles (the law of sines cannot be applied).
- The law of cosines states that in a triangle, the square of any side is equal to the sums of the squares of the other two sides, minus twice their product times the cosine of the included angle:

$$a^2 = b^2 + c^2 - 2bc \cos A$$

- When solving triangles using the law of cosines, always begin with the largest angle or the angle opposite the largest side.

EXERCISES

7. Solve for B: $9^2 = 12^2 + 15^2 - 2(12)(15) \cos B$
8. Use the law of cosines to find the missing side.
9. While preparing for the day's orienteering meet, Rick finds that the distances between the first three markers he wants to pick up are 1250 yd, 1820 yd, and 720 yd. Find the measure of each angle in the triangle formed so that Rick is sure to find all three markers.

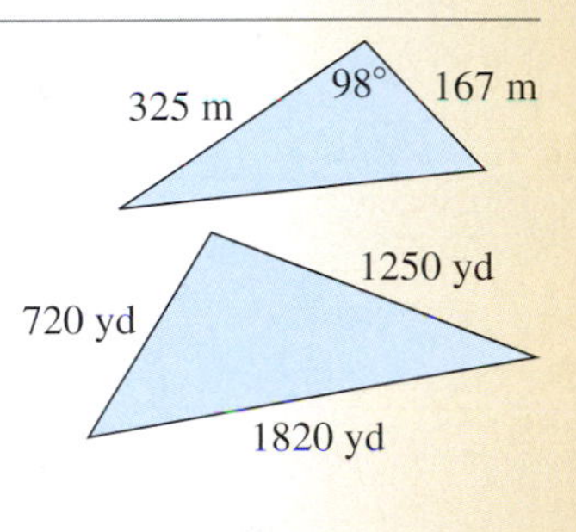

SECTION 5.4 Vectors and Vector Diagrams

KEY CONCEPTS

- Quantities/concepts that can be described using a single number are called scalar quantities or simply, scalars. Examples are time, perimeter, area, volume, energy, temperature, weight, and so on.
- Quantities and concepts that require more than a single number to describe their attributes are called vector quantities or simply, vectors. Examples are force, velocity, displacement, pressure, and so on.
- Vectors can be represented using directed line segments p to indicate magnitude and direction of a vector quantity. The origin of the segment is called the initial point, with the arrowhead pointing to the terminal point. When used solely for comparative analysis, they are called geometric vectors.
- Two vectors are equal if they have the same magnitude and direction.
- Vectors can be represented graphically in the xy-plane by naming the initial and terminal points of the vector or by giving the magnitude and angle of the related position vector [initial point at (0, 0)].
- For a vector with initial point (x_1, y_1) and terminal point (x_2, y_2), the related position vector can be written in the component form $\langle a, b\rangle$ where $a = x_2 - x_1$ and $b = y_2 - y_1$.

- For a vector written in the component form $\langle a, b\rangle$, a is called the horizontal component and b is called the vertical component of the vector.
- For vector $\mathbf{v} = \langle a, b\rangle$, the length or magnitude of $\mathbf{v}$ is denoted $|\mathbf{v}|$ and is computed as $\sqrt{a^2 + b^2}$.
- Vector components can also be written in trigonometric form. See page 338.
- For $\mathbf{u} = \langle a, b\rangle$, $\mathbf{v} = \langle c, d\rangle$, and any scalar k, we have the following operations defined:
  $\mathbf{u} + \mathbf{v} = \langle a + c, b + d\rangle$ $\quad \mathbf{u} - \mathbf{v} = \langle a - c, b - d\rangle$ $\quad k\mathbf{u} = \langle ka, kb\rangle$ for $k \in \mathbb{R}$
  If $k > 0$, the new vector has the same direction as $\mathbf{u}$; $k < 0$, the opposite direction.
- Vectors can also be written in algebraic form using $\mathbf{i}$, $\mathbf{j}$ notation, where $\mathbf{i}$ is a unit vector along the x-axis and $\mathbf{j}$ is a unit vector along the y-axis. The vector $\langle a, b\rangle$ is written as a linear combination of $\mathbf{i}$ and $\mathbf{j}$, meaning the terminal point (a, b) is expressed in multiples of $\mathbf{i}$ and $\mathbf{j}$, respectively: $\langle a, b\rangle = a\mathbf{i} + b\mathbf{j}$.
- For any nonzero vector $\mathbf{v} = \langle a, b\rangle$, vector $\mathbf{u} = \frac{\mathbf{v}}{|\mathbf{v}|}$ is a unit vector in the same direction as $\mathbf{v}$.
- In aviation and shipping, the heading of a ship or plane is understood to be the amount of rotation from due north in the clockwise direction.

EXERCISES

10. Graph the vector $\mathbf{v} = \langle 9, 5\rangle$, then compute its magnitude and direction angle.

11. Write the vector $\mathbf{u} = \langle -8, 3\rangle$ in $\mathbf{i}$, $\mathbf{j}$ form and compute its magnitude and direction angle.

12. Approximate the horizontal and vertical components of the vector $\mathbf{u}$, where $|\mathbf{u}| = 18$ and $\theta = 52°$.

13. Compute $2\mathbf{u} + \mathbf{v}$, then find the magnitude and direction of the resultant: $\mathbf{u} = \langle -3, -5\rangle$ and $\mathbf{v} = \langle 2, 8\rangle$.

14. Find a unit vector that points in the same direction as $\mathbf{u} = 7\mathbf{i} + 12\mathbf{j}$.

15. Without computing, if $\mathbf{u} = \langle -9, 2\rangle$ and $\mathbf{v} = \langle 2, 8\rangle$, will the resultant sum lie in Quadrant I or II? Why?

16. It's once again time for the Great River Race, a $\frac{1}{2}$-mi swim across the Panache River. If Karl fails to take the river's 1-mph current into account and he swims the race at 3 mph, how far from the finish marker does he end up when he makes it to the other side?

17. Two Coast Guard vessels are towing a large yacht into port. The first is pulling with a force of 928 N and the second with a force of 850 N. Determine the angle θ for the second Coast Guard vessel that will keep the ship moving safely in a straight line.

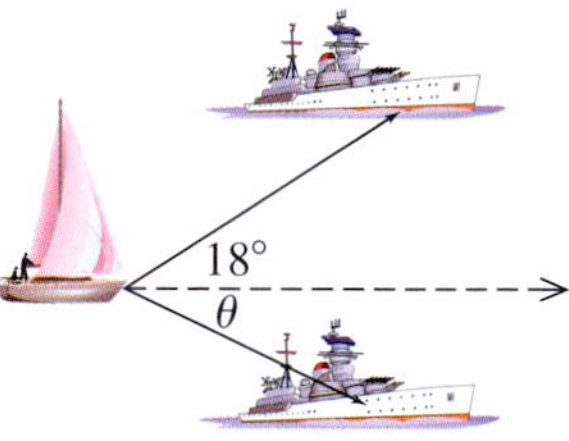

SECTION 5.5 Vector Applications and the Dot Product

KEY CONCEPTS

- Two or more vector forces are in equilibrium when the sum of their components is the zero vector.
- When the components of a vector $\mathbf{u}$ are nonquadrantal, with one of its components lying along another vector $\mathbf{v}$, we call this component the "component of $\mathbf{u}$ along $\mathbf{v}$" and abbreviate it $\text{comp}_{\mathbf{v}}\mathbf{u}$.
- For vectors $\mathbf{u}$ and $\mathbf{v}$, $\text{comp}_{\mathbf{v}}\mathbf{u} = |\mathbf{u}|\cos\theta$, where θ is the angle between $\mathbf{u}$ and $\mathbf{v}$.
- Work done is computed as the product of the constant force $\mathbf{F}$ applied times the distance $\mathbf{D}$ that the force is applied: $W = \mathbf{F} \cdot \mathbf{D}$.

- If force is not applied parallel to the direction of movement, only the component of the force in the direction of movement is used. If $\mathbf{u}$ is a force vector but not applied in the direction of a vector $\mathbf{v}$, the equation becomes $W = \text{comp}_{\mathbf{v}}\mathbf{u} \cdot |\mathbf{v}|$.
- For vectors $\mathbf{u} = \langle a, b\rangle$ and $\mathbf{v} = \langle c, d\rangle$, the dot product $\mathbf{u} \cdot \mathbf{v}$ is defined as the scalar $ac + bd$.
- The dot product $\mathbf{u} \cdot \mathbf{v}$ is equivalent to $\text{comp}_{\mathbf{u}}\mathbf{v} \cdot |\mathbf{v}|$ and to $|\mathbf{u}|\cos\theta$.
- The angle between two vectors can be computed using $\cos\theta = \frac{\mathbf{u}}{|\mathbf{u}|} \cdot \frac{\mathbf{v}}{|\mathbf{v}|} = \frac{\mathbf{u} \cdot \mathbf{v}}{|\mathbf{u}||\mathbf{v}|}$.
- Given vectors $\mathbf{u}$ and $\mathbf{v}$, the projection of $\mathbf{u}$ along $\mathbf{v}$ is the *vector* $\mathbf{proj_v u}$ defined by $\mathbf{proj_v u} = \left(\frac{\mathbf{u} \cdot \mathbf{v}}{|\mathbf{v}|^2}\right)\mathbf{v}$.
- Given vectors $\mathbf{u}$ and $\mathbf{v}$ and $\mathbf{proj_v u}$, $\mathbf{u}$ can be resolved into the orthogonal components $\mathbf{u}_1$ and $\mathbf{u}_2$ where $\mathbf{u} = \mathbf{u}_1 + \mathbf{u}_2$, $\mathbf{u}_1 = \mathbf{proj_v u}$, and $\mathbf{u}_2 = \mathbf{u} - \mathbf{u}_1$. In this case we have $\mathbf{u}_1 \| \mathbf{v}$ and $\mathbf{u}_2 \perp \mathbf{v}$.
- The horizontal distance x that a projectile has traveled at time t can be found using $x = (|\mathbf{v}|\cos\theta)t$.
- The vertical height y of a projectile at time t can be found using the equation $y = (|\mathbf{v}|\sin\theta)t - 16t^2$, where $|\mathbf{v}|$ is the magnitude of the initial velocity, θ is the angle at which the object is projected, and t is the time after release in seconds.

EXERCISES

18. For the force vectors $\mathbf{F}_1$ and $\mathbf{F}_2$ given, find the resultant and an additional force vector so that equilibrium takes place: $\mathbf{F}_1 = \langle -20, 70\rangle$; $\mathbf{F}_2 = \langle 45, 53\rangle$.

19. Find the component of $\mathbf{u}$ along $\mathbf{v}$ (compute $\text{comp}_{\mathbf{v}}\mathbf{u}$) for the vectors $\mathbf{u} = -12\mathbf{i} - 16\mathbf{j}$ and $\mathbf{v} = 19\mathbf{i} - 13\mathbf{j}$.

20. Find the component d so that the vectors $\mathbf{u}$ and $\mathbf{v}$ will be orthogonal: $\mathbf{u} = \langle 2, 9\rangle$ and $\mathbf{v} = \langle -18, d\rangle$.

21. Compute the dot product $\mathbf{p} \cdot \mathbf{q}$ and find the angle between them: $\mathbf{p} = \langle -5, -2\rangle$; $\mathbf{q} = \langle 4, -7\rangle$.

22. For the force vector $\mathbf{F} = \langle 50, 15\rangle$ and $\mathbf{v} = \langle 85, 6\rangle$, find the amount of work required to move an object along the entire length of $\mathbf{v}$. Assume force is in pounds and distance in feet.

23. A 650-lb crate is sitting on a ramp which is inclined at 40°. Find the force needed to hold the crate stationary.

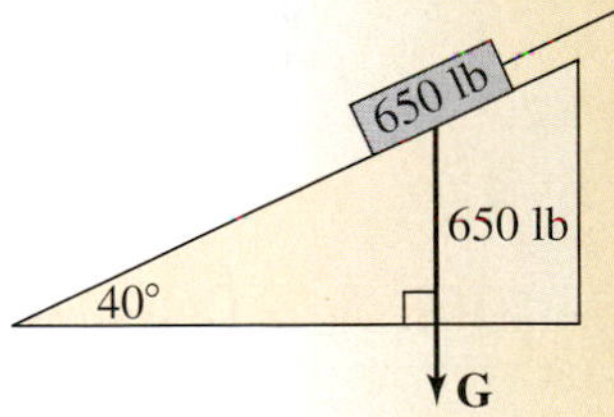

24. An arctic explorer is hauling supplies from the supply hut to her tent, a distance of 120 ft, in a sled she is dragging behind her. If the straps used make an angle of 25° with the snow-covered ground and she pulls with a constant force of 75 lb, find the amount of work done.

25. A projectile is launched from a sling-shot with an initial velocity of $v_0 = 280$ ft/sec at an angle of $\theta = 50°$. Find (a) the position of the object after 1.5 sec and (b) the time required to reach a height of 150 ft.

SECTION 5.6 Complex Numbers

KEY CONCEPTS

- The italicized i represents the number whose square is -1. This means $i^2 = -1$ and $i = \sqrt{-1}$.
- Since $i^4 = i^2 \cdot i^2 = (-1)\cdot(-1)$ or 1, larger powers of i can be simplified by writing them in terms of i^4.

- The square root of a negative number can be rewritten using "i" notation: $\sqrt{-4} = \sqrt{-1 \cdot 4} = \sqrt{-1}\sqrt{4}$ or $2i$. We say the expression has been *written in terms of i and simplified.*
- The standard form of a *complex number* is $a + bi$, where a is the *real number part* and bi is the *imaginary number part.*
- The commutative, associative, and distributive properties also apply to complex numbers and are used to perform basic operations.
- To add or subtract complex numbers, combine the like terms.
- For any complex number $a + bi$, its *complex conjugate* is $a - bi$.
- The *product* of a complex number and its conjugate is a real number.
- The *discriminant* of the quadratic formula $b^2 - 4ac$ gives the number and nature of the roots.
- To multiply complex numbers, use the F-O-I-L method and combine like terms.
- The *sum* of a complex number and its complex conjugate is a real number.
- To find a *quotient* of complex numbers, multiply the numerator and denominator by the conjugate of the denominator.
- If $a + bi$ is one solution to a polynomial equation, then its complex conjugate $a - bi$ is also a solution.

EXERCISES

Simplify each expression and write the result in standard form.

26. $\sqrt{-72}$ **27.** $6\sqrt{-48}$ **28.** $\dfrac{-10 + \sqrt{-50}}{5}$

29. $\sqrt{3}\sqrt{-6}$ **30.** i^{57}

Perform the operation indicated and write the result in standard form.

31. $(5 + 2i)^2$ **32.** $\dfrac{5i}{1 - 2i}$ **33.** $(-3 + 5i) - (2 - 2i)$

34. $(2 + 3i)(2 - 3i)$ **35.** $4i(-3 + 5i)$

Use substitution to show the given complex number and its conjugate are solutions to the equation shown.

36. $x^2 - 9 = -34; x = 5i$ **37.** $x^2 - 4x + 9 = 0; x = 2 + i\sqrt{5}$

SECTION 5.7 Complex Numbers in Trigonometric Form

KEY CONCEPTS

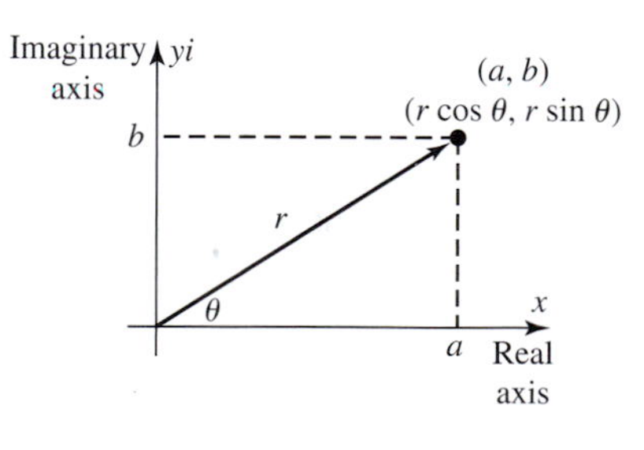

- A complex number $a + bi = (a, b)$ can be written in trigonometric form by noting (from its graph) that $a = r\cos\theta$ and $b = r\sin\theta$: $a + bi = r(\cos\theta + i\sin\theta)$.
- The angle θ is called the argument of z and r is called the modulus of z.
- The argument of a complex number z is not unique, since any rotation of $\theta + 2\pi k$ (k an integer) will yield a coterminal angle.
- To convert from trigonometric to rectangular form, evaluate $\cos\theta$ and $\sin\theta$ and multiply by the modulus.

- To multiply complex numbers in trig form, multiply the moduli and add the arguments. To divide complex numbers in trig form, divide the moduli and subtract the arguments.
- Complex numbers have numerous real-world applications, particularly in a study of AC electrical circuits.
- The impedance of an AC circuit is given as $Z = R + j(X_L - X_C)$, where R represents a pure resistance, X_C represents the capacitive reactance, X_L represents the inductive reactance, and $j = \sqrt{-1}$.
- Z is a complex number whose magnitude is given by $|Z| = \sqrt{(R)^2 + (X_L - X_C)^2}$ and whose phase angle is $\theta = \tan^{-1}\left(\frac{X_L - X_C}{R}\right)$. The angle θ represents the angle between the voltage and current.
- The voltage V across any element in the circuit is $V = IZ$. The current I in the circuit is $I = \frac{V}{Z}$.

EXERCISES

38. Write in trigonometric form: $z = -1 - \sqrt{3}i$

39. Write in rectangular form: $z = 3\sqrt{2}\left[\text{cis}\left(\frac{\pi}{4}\right)\right]$

40. Graph in the complex plane: $z = 5(\cos 30° + i \sin 30°)$

41. For $z_1 = 8\,\text{cis}\left(\frac{\pi}{4}\right)$ and $z_2 = 2\,\text{cis}\left(\frac{\pi}{6}\right)$, compute z_1z_2 and $\frac{z_1}{z_2}$.

42. Find the current I in a circuit where $V = 4\sqrt{3} - 4j$ and $Z = 1 - \sqrt{3}j\ \Omega$.

43. In the V_{RLC} series circuit shown, $R = 10\ \Omega$, $X_L = 8\ \Omega$, and $X_C = 5\ \Omega$. Find the magnitude of Z and the phase angle between current and voltage. Express the result in trigonometric form.

SECTION 5.8 De Moivre's Theorem and the Theorem on *n*th Roots

KEY CONCEPTS

- While powers of a complex number can be found by expanding the binomial $(a + bi)^n$, this becomes impractical for larger values of n and De Moivre's theorem will compute the result more efficiently.
- For a complex number $z = r(\cos\theta + i\sin\theta)$, De Moivre's theorem states that: $z^n = r^n[\cos(n\theta) + i\sin(n\theta)]$.
- One application of De Moivre's theorem is to check complex solutions of polynomial equations.
- The nth roots theorem can be viewed as an extension of De Moivre's theorem, where a complex number is raised to the power $\frac{1}{n}$ ($n \rightarrow$ a natural number). In this case, the most general form of the complex number $z = r[\cos(\theta + 2\pi k) + i\sin(\theta + 2\pi k)]$ is used so as to include all n nth roots. This gives $z^{\frac{1}{n}} = r^{\frac{1}{n}}\left\{\cos\left[\frac{1}{n}(\theta + 2\pi k)\right] + i\sin\left[\frac{1}{n}(\theta + 2\pi k)\right]\right\}$, which is more commonly written $\sqrt[n]{z} = \sqrt[n]{r}\left[\cos\left(\frac{\theta}{n} + \frac{2\pi k}{n}\right) + i\sin\left(\frac{\theta}{n} + \frac{2\pi k}{n}\right)\right]$, where $k = 1, 2, 3, \ldots, n - 1$.
- The nth roots of a complex number are equally spaced around a circle of radius r in the complex plane.

EXERCISES

44. Use De Moivre's theorem to compute the value of $(-1 + i\sqrt{3})^5$.

45. Use De Moivre's theorem to verify that $z = 1 - i$ is a solution of $z^4 + z^3 - 2z^2 + 2z + 4 = 0$.

46. Use the nth roots theorem to find the three cube roots of $125i$.

47. Solve the equation using the nth roots theorem: $x^3 - 216 = 0$.

48. Given that $z = 2 + 2i$ is a fourth root of -64, state the other three roots.

49. Solve using the quadratic formula and the nth roots theorem: $z^4 + 6z^2 + 25 = 0$.

50. Use De Moivre's theorem to verify the three roots of $125i$ found in Exercise 46.

MIXED REVIEW

Solve each triangle using either the law of sines or law of cosines, whichever is appropriate.

1.

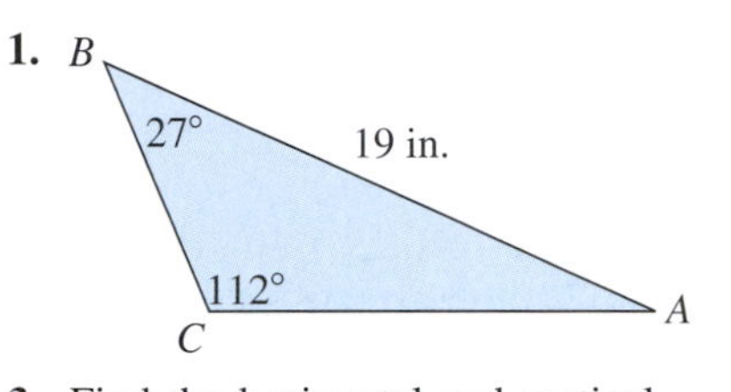

2.

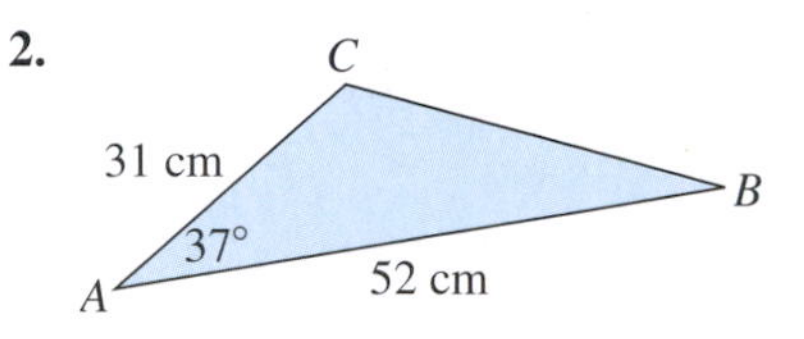

3. Find the horizontal and vertical components of the vector **u**, where $|\mathbf{u}| = 21$ and $\theta = 40°$.

4. Compute $2\mathbf{u} + \mathbf{v}$, then find the magnitude and direction of the resultant: $\mathbf{u} = \langle 6, -3\rangle$, $\mathbf{v} = \langle -2, 8\rangle$.

5. Find the height of a flagpole that sits atop a hill, if it makes an angle of 122° with the hillside, and the angle of elevation between the side of the hill to the top of the flagpole is 35° at a distance of 120 ft.

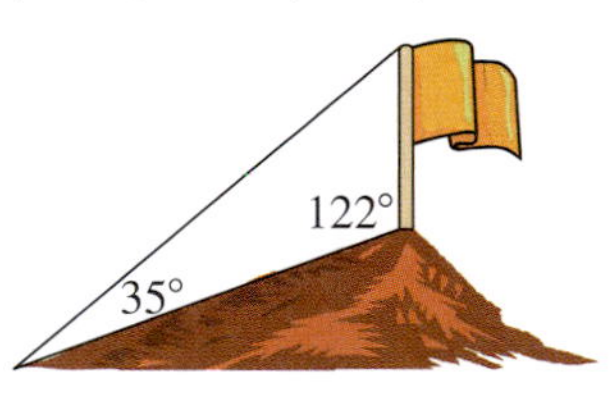

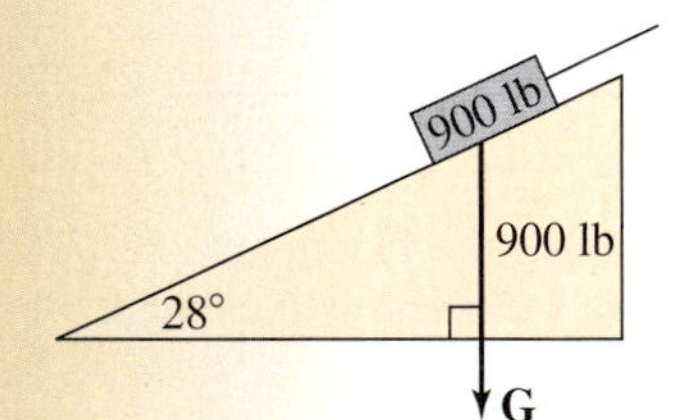

6. A 900-lb crate is sitting on a ramp that is inclined at 28°. Find the force needed to hold the object stationary.

7. A jet plane is flying at 750 mph on a heading of 30°. There is a strong, 50-mph wind blowing from due south (heading 0°). What is the true course and speed of the plane (relative to the ground)?

8. Graph the vector $\mathbf{v} = \langle -8, 5\rangle$, then compute its magnitude and direction.

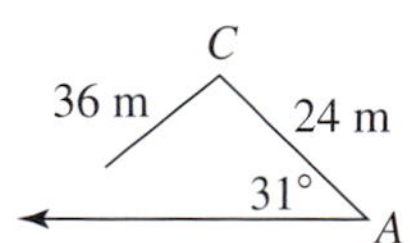

9. Solve using the law of sines. If two solutions exist, find both.

Exercise 10

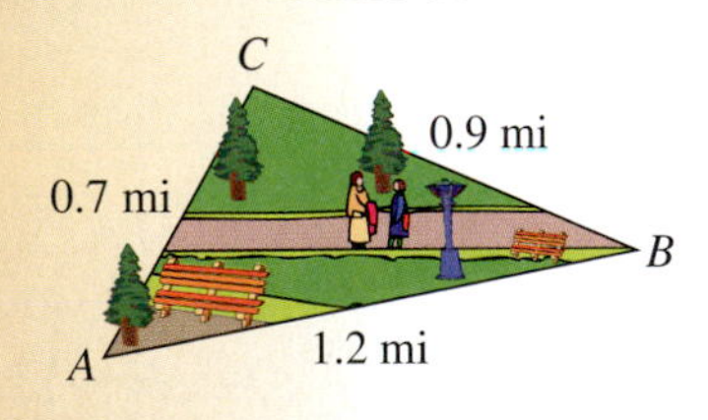

10. A local Outdoors Club sponsors a treasure hunt activity for its members, and has placed surprise packages at the corners of the triangular park shown. Find the measure of each angle to help club members find their way to the treasure.

11. Given the vectors $\mathbf{p} = \langle -5, 2\rangle$ and $\mathbf{q} = \langle 4, 7\rangle$, use the dot product $\mathbf{p} \cdot \mathbf{q}$ to find the angle between them.

12. As part of a lab demonstrating centrifugal and centripetal forces, a physics teacher is whirling a tethered weight above her head while a group of students looks on from a distance of 20 ft as shown. If the tether has a radius of 10 ft and a student departs at the 35° angle shown, will the student be struck by the weight? What is the smallest angle of exit the student could take (to the nearest whole) without being struck by the whirling weight?

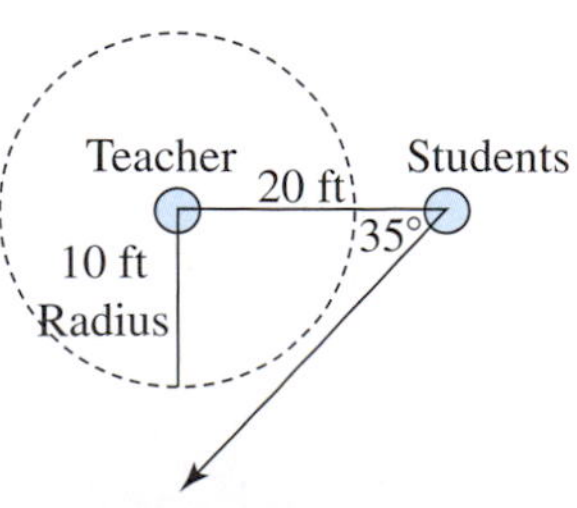

13a. Graph the complex number using the rectangular form, then convert to trigonometric form: $z = 4 - 4i$.

b. Graph the complex number using the trigonometric form, then convert to rectangular form: $z = 6(\cos 120° + i \sin 120°)$.

14a. Verify that $z = 4 - 5i$ *and* its conjugate are solutions to $z^2 - 8z + 41 = 0$.

b. Solve using the quadratic formula: $z^2 - 6iz + 7 = 0$

15. Two tractors are dragging a large, fallen tree into the brush pile that's being prepared for a large Fourth of July bonfire. The first is pulling with a force of 418 N and the second with a force of 320 N. Determine the angle θ for the second tractor that will keep the tree headed straight for the brush pile.

16. Given $z_1 = 8(\cos 45° + i \sin 45°)$ and $z_2 = 4(\cos 15° + i \sin 15°)$ compute:

a. the product $z_1 z_2$ **b.** the quotient $\frac{z_1}{z_2}$

17. Given the vectors $\mathbf{u} = -12\mathbf{i} - 16\mathbf{j}$ and $\mathbf{v} = 19\mathbf{i} - 13\mathbf{j}$, find $\text{comp}_\mathbf{v}\mathbf{u}$ and $\mathbf{proj_v u}$.

18. Find the result using De Moivre's theorem: $(2\sqrt{3} - 2i)^6$.

19. Use the nth roots theorem to find the four fourth roots of $-2 + 2i\sqrt{3}$.

20. The impedance of an AC circuit is $Z = R + j(X_L - X_C)$. The voltage across the circuit is $V_{RLC} = I|Z|$. Given $R = 12\ \Omega$, $X_L = 15.2\ \Omega$, and $X_C = 9.4\ \Omega$, write Z in trigonometric form and find the voltage in the circuit if the current is $I = 6.5$ A.

PRACTICE TEST

1. Within the Kilimanjaro Game Reserve, a fire is spotted by park rangers stationed in two towers known to be 10 mi apart. Using the line between them as a baseline, tower A reports the fire is at an angle of 39°, while tower B reports an angle of 68°. How far is the fire from the closer tower?

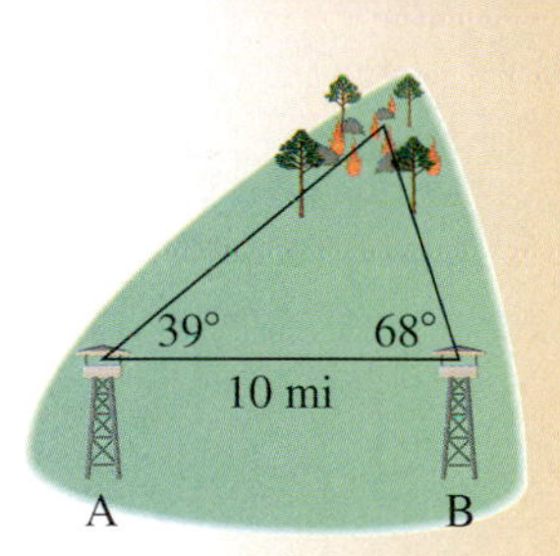

Exercise 2

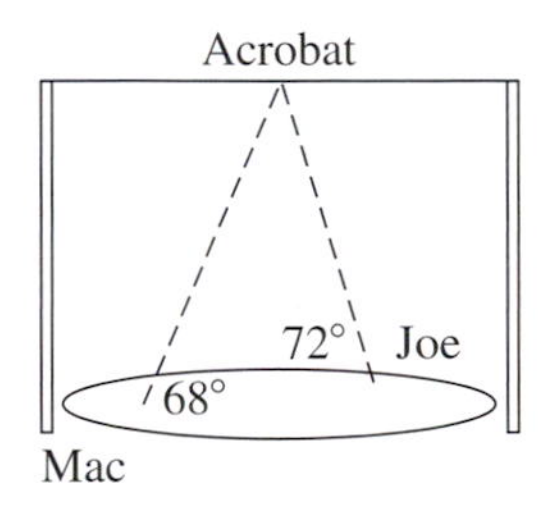

2. At the circus, Mac and Joe are watching a high-wire act from first-row seats on opposite sides of the center ring. Find the height of the performing acrobats at the instant Mac measures an angle of elevation of 68° while Joe measures an angle of 72°. Assume Mac and Joe are sitting 100 ft apart.

3. Three rods are attached via two joints and shaped into a triangle. How many triangles can be formed if the angle at the joint B must measure 20°? If two triangles can be formed, solve both.

4. Jackie and Sam are rounding up cattle in the brush country, and are communicating via walkie-talkie. Jackie is at the water hole and Sam is at Dead Oak, which are 6 mi apart. Sam finds some strays and heads them home at the 32° indicated. (a) If the maximum range of Jackie's unit is 3 mi, will she be able to communicate with Sam as he heads home? (b) If the maximum range were 4 mi, how far from Dead Oak is Sam when he is first contacted by Jackie?

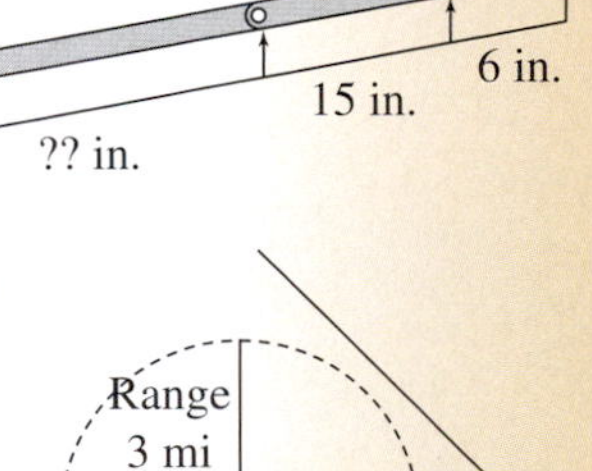

Exercise 5

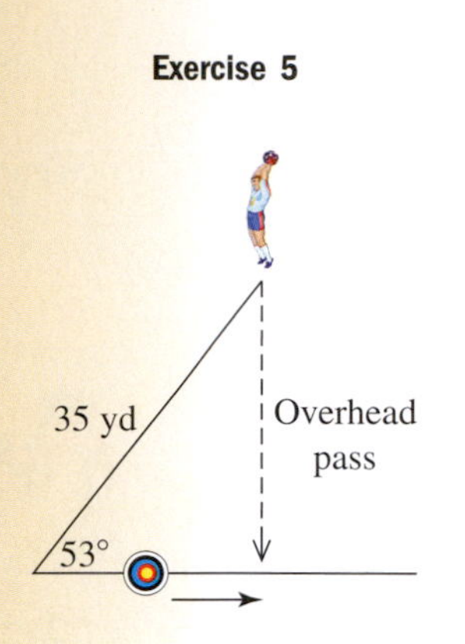

5. As part of an All-Star competition, a group of soccer players (forwards) stand where shown in the diagram and attempt to hit a moving target with a two-handed overhead pass. If a player has a maximum effective range of approximately (a) 25 yd, can the target be hit? (b) about 28 yd, how many "effective" throws can be made? (c) 35 yd and the target is moving at 5 yd/sec, how many seconds is the target within range?

6. The summit of Triangle Peak can only be reached from one side, using a trail straight up the side that is approximately 3.5 mi long. If the peak is 5 mi wide at its base and the trail makes a 24° angle with the horizontal, (a) what is the approximate length of the opposing side? (b) How tall is the peak (in feet)?

7. The Bermuda Triangle is generally thought to be the triangle formed by Miami, Florida, San Juan, Puerto Rico, and Bermuda itself. If the distances between these locations are the 1025 mi, 1020 mi, and 977 mi indicated, find the measure of each angle in the Bermuda Triangle.

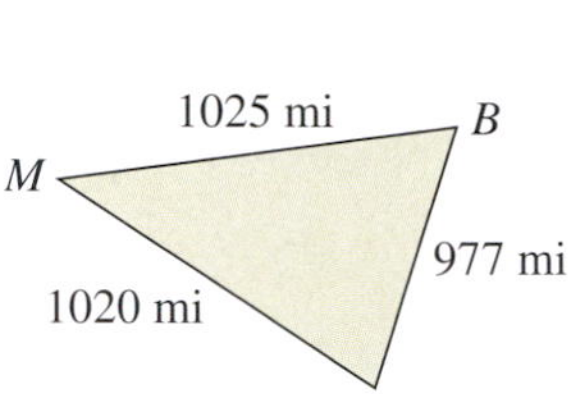

8. A helicopter is flying at 90 mph on a heading of 40°. A 20-mph wind is blowing from the NE on a heading of 190°. What is the true course and speed of the helicopter relative to the ground? Draw a diagram as part of your solution.

9. Two mules walking along a river bank are pulling a heavy barge up river. The first is pulling with a force of 250 N and the second with a force of 210 N. Determine the angle θ for the second mule that will ensure the barge stays midriver and does not collide with the shore.

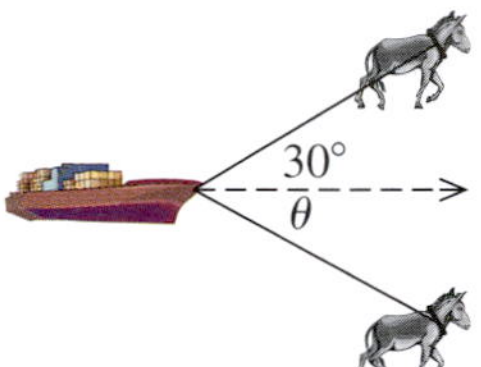

Exercise 10

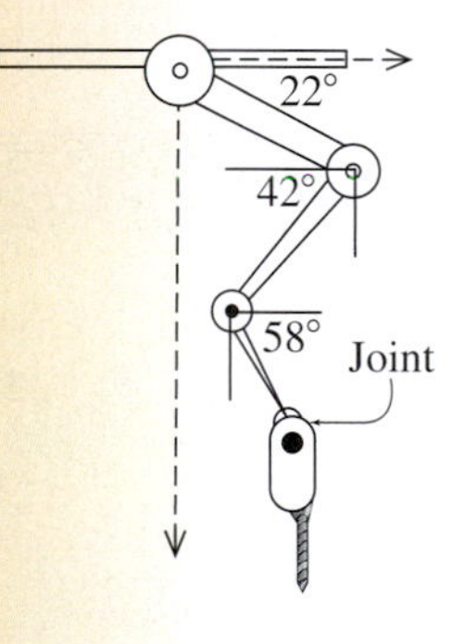

10. Along a production line, various tools are attached to the ceiling with a multijointed arm so that workers can draw one down, position it for use, then move it up out of the way for the next tool (see the diagram). If the first segment is 100 cm, the second is 75 cm, and the third is 50 cm, determine the approximate coordinates of the last joint.

11. Three ranch hands have roped a run-away steer and are attempting to hold him steady. The first and second ranch hands are pulling with the magnitude and at the angles indicated in the diagram. If the steer is held fast by the efforts of all three, find the magnitude of the tension and angle of the rope from the third cowhand.

12. For $\mathbf{u} = \langle -9, 5\rangle$ and $\mathbf{v} = \langle -2, 6\rangle$, (a) compute the angle between $\mathbf{u}$ and $\mathbf{v}$; (b) find the projection of $\mathbf{u}$ along $\mathbf{v}$ (find $\mathbf{proj_v u}$); and (c) resolve $\mathbf{u}$ into vectors $\mathbf{u}_1$ and $\mathbf{u}_2$, where $\mathbf{u}_1 \parallel \mathbf{v}$ and $\mathbf{u}_2 \perp \mathbf{v}$.

13. A lacrosse player flips a long pass to a teammate way down field who is near the opponent's goal. If the initial velocity of the pass is 110 ft/sec and the ball is released at an angle of 50° with level ground, how high is the ball after 2 sec? How long until the ball again reaches this same height?

14. Compute the quotient $\frac{z_1}{z_2}$, given

$$z_1 = 6\sqrt{5}\operatorname{cis}\left(\frac{\pi}{8}\right) \text{ and}$$

$$z_2 = 3\sqrt{5}\operatorname{cis}\left(\frac{\pi}{12}\right).$$

15. Compute the product $z = z_1z_2$ in trigonometric form, then verify $|z_1||z_2| = |z|$ and $\theta_1 + \theta_2 = \theta$: $z_1 = -6 + 6i$; $z_2 = 4 - 4\sqrt{3}i$

16. Use De Moivre's theorem to compute the value of $(\sqrt{3} - i)^4$.

17. Use De Moivre's theorem to verify $2 + 2\sqrt{3}i$ is a solution to $z^5 + 3z^3 + 64z^2 + 192 = 0$.

18. Use the nth roots theorem to solve $x^3 - 125i = 0$.

19. Solve using u-substitution, the quadratic formula, and the nth roots theorem: $z^4 - 6z^2 + 58 = 0$.

20. In an AC circuit, the impedance in the circuit is given by $Z = R + (X_L - X_C)j$. Given $R = 8\ \Omega$, $X_L = 9.4\ \Omega$, $X_C = 3.4\ \Omega$, with a current of $I = 5$ A, (a) find the magnitude and phase angle of the impedance and write the result in trigonometric form, and (b) find the total voltage in the circuit (recall $V = I|Z|$).

CALCULATOR EXPLORATION AND DISCOVERY

Investigating Projectile Motion

The keystrokes shown apply to a TI-84 Plus model. Please consult our Internet site or your manual for other models.

There are two important aspects of projectile motion that were not discussed in Section 5.5, the **range** of the projectile and the **optimum angle** θ that will maximize this range (these were introduced in Section 3.4). Both can be explored using the equations for the horizontal and vertical components of the projectile's position: horizontal $\rightarrow (|\mathbf{v}|\cos\theta)t$ and vertical $\rightarrow (|\mathbf{v}|\sin\theta)t - 16t^2$. In Example 10 of Section 5.5, an arrow was shot from a bow with an initial velocity of $|\mathbf{v}| = 150$ ft/sec at an angle of $\theta = 50°$. Enter the equations above on the Y= screen as Y_1 and Y_2, using these values (Figure 5.89). Then set up the TABLE using TblStart = 0, ΔTbl = 0.5 and the AUTO mode. The resulting table is shown in Figure 5.90, where Y_1 represents the horizontal distance the arrow has traveled, and Y_2 represents the height of the arrow. To find the *range* of the arrow, scroll downward ▼ until the height (Y_2) shows a value that is less than or equal to zero (the arrow has hit the ground). As Figure 5.91 shows, this happens somewhere between $t = 7$ and $t = 7.5$ sec. We could now change the TBLSET settings to TblStart = 0 and ΔTbl = 0.1 to get a better approximation of the time the arrow is in flight (it's just less than 7.2 sec) and the horizontal range of the arrow (about 692.4 ft), but our main interest is how to *compute these values exactly.* We begin with the equation for the arrow's vertical position $y = (|\mathbf{v}|\sin\theta)t - 16t^2$. Since the object returns to earth when $y = 0$, we substitute 0 for y and factor out t: $0 = t(|\mathbf{v}|\sin\theta - 16t)$. Solving for t gives $t = 0$ or $t = \frac{|\mathbf{v}|\sin\theta}{16}$. Since the component of velocity in the horizontal direction is $|\mathbf{v}|\cos\theta$, the basic distance relationship $D = \mathbf{r} \cdot \mathbf{t}$ gives the horizontal range of $R = |\mathbf{v}|\cos\theta \cdot \frac{|\mathbf{v}|\sin\theta}{16}$ or $\frac{|\mathbf{v}|^2\sin\theta\cos\theta}{16}$! Checking the values given for the arrow ($|\mathbf{v}| = 150$ ft/sec and $\theta = 50°$) verifies the range is $R \approx 692.4$. But what about the *maximum possible range* for the arrow? Using $|\mathbf{v}| = 150$ for R results in an equation in theta only: $R(\theta) = \frac{150^2\sin\theta\cos\theta}{16}$, which we can enter as Y_3 and investigate for various θ. After carefully entering $R(\theta)$ as Y_3 and resetting TBLSET to TblStart = 30 and ΔTbl = 5, the TABLE in Figure 5.92 shows a maximum range of about 703 ft at 45°. Resetting TBLSET to TblStart = 40 and ΔTbl = 1 verifies this fact.

Figure 5.89

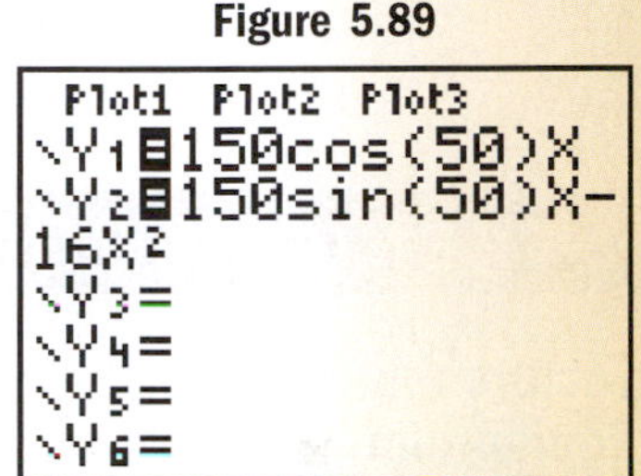

Figure 5.90

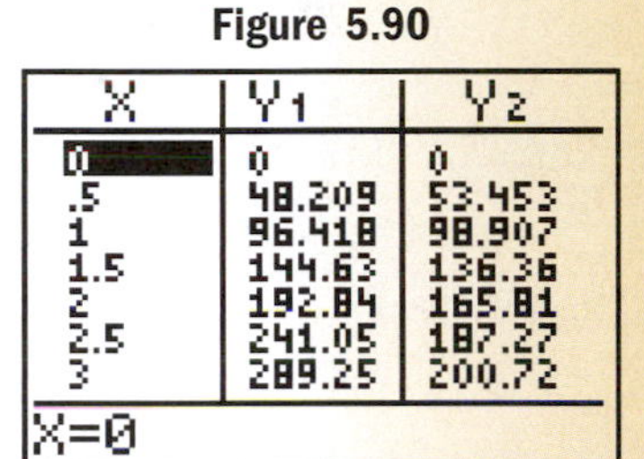

X	Y1	Y2
0	0	0
.5	48.209	53.453
1	96.418	98.907
1.5	144.63	136.36
2	192.84	165.81
2.5	241.05	187.27
3	289.25	200.72

X=0

Figure 5.91

X	Y1	Y2
5	482.09	174.53
5.5	530.3	147.99
6	578.51	113.44
6.5	626.72	70.893
7	674.93	20.347
7.5	723.14	-38.2
8	771.35	-104.7

X=7.5

For each of the following exercises, find (a) the height of the projectile after 1.75 sec, (b) the maximum height of the projectile, (c) the range of the projectile, and (d) the number of seconds the projectile is airborne.

Exercise 1: A javelin is thrown with an initial velocity of 85 ft/sec at an angle of 42°.

Exercise 2: A cannon ball is shot with an initial velocity of 1120 ft/sec at an angle of 30°.

Exercise 3: A baseball is hit with an initial velocity of 120 ft/sec at an angle of 50°. Will it clear the center field fence, 10 ft high and 375 ft away?

Exercise 4: A field goal (American football) is kicked with an initial velocity of 65 ft/sec at an angle of 35°. Will it clear the crossbar, 10 ft high and 40 yd away?

Figure 5.92

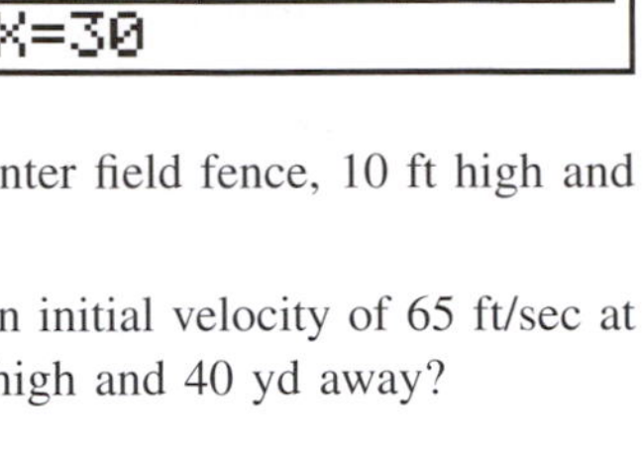

X	Y_3
30	608.92
35	660.72
40	692.44
45	703.13
50	692.44
55	660.72
60	608.92

X=30

STRENGTHENING CORE SKILLS

Vectors and Static Equilibrium

In Sections 5.4 and 5.5, the concepts of vector forces, resultant forces, and equilibrium were studied extensively. A nice extension of these concepts involves what is called **static equilibrium.** Assuming that only coplanar forces are acting on an object, the object is said to be in static equilibrium if *the sum of all vector forces acting on it is* ***0***. This implies that the object is stationary, since the forces all counterbalance each other. The methods involved are simple and direct, with a wonderful connection to the systems of equations you've likely seen previously. Consider the following example.

ILLUSTRATION 1 As part of their training, prospective FBI agents must move hand-over-hand across a rope strung between two towers. An agent-in-training weighing 180 lb is two-thirds of the way across, causing the rope to deflect from the horizontal at the angles shown. What is the tension in each part of the rope at this point?

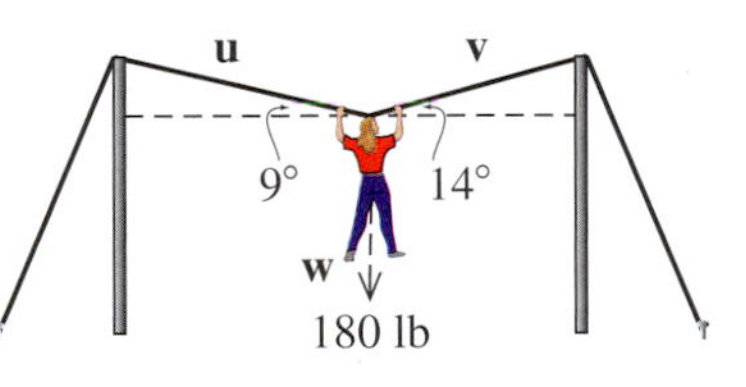

Solution: We have three concurrent forces acting on the point where the agent grasps the rope. Begin by drawing a vector diagram and computing the components of each force, using the **i**, **j** notation. Note that $\mathbf{w} = -180\mathbf{j}$.

$$\begin{aligned}\mathbf{u} &= -|\mathbf{u}|\cos(9°)\mathbf{i} + |\mathbf{u}|\sin(9°)\mathbf{j}\\ &\approx -0.9877|\mathbf{u}|\mathbf{i} + 0.1564|\mathbf{u}|\mathbf{j}\\ \mathbf{v} &= |\mathbf{v}|\cos(14°)\mathbf{i} + |\mathbf{v}|\sin(14°)\mathbf{j}\\ &\approx 0.9703|\mathbf{v}|\mathbf{i} + 0.2419|\mathbf{v}|\mathbf{j}\end{aligned}$$

For equilibrium, all vector forces must sum to the zero vector: $\mathbf{u} + \mathbf{v} + \mathbf{w} = \mathbf{0}$, which results in the following equation:
$-0.9877|\mathbf{u}|\mathbf{i} + 0.1564|\mathbf{u}|\mathbf{j} + 0.9703|\mathbf{v}|\mathbf{i} + 0.2419|\mathbf{v}|\mathbf{j} - 180\mathbf{j} = 0\mathbf{i} + 0\mathbf{j}$.
Factoring out **i** and **j** from the left-hand side yields
$(-0.9877|\mathbf{u}| + 0.9703|\mathbf{v}|)\mathbf{i} + (0.1564|\mathbf{u}| + 0.2419|\mathbf{v}| - 180)\mathbf{j} = 0\mathbf{i} + 0\mathbf{j}$.
Since any two vectors are equal only when corresponding components

are equal, we obtain a system in the two variables $|\mathbf{u}|$ and $|\mathbf{v}|$: $\begin{cases} -0.9877|\mathbf{u}| + 0.9703|\mathbf{v}|) = 0 \\ 0.1564|\mathbf{u}| + 0.2419|\mathbf{v}| - 180 = 0 \end{cases}$. Solving the system using matrix equations and a calculator (or any desired method), gives $|\mathbf{u}| \approx 447$ lb and $|\mathbf{v}| \approx 455$ lb.

At first it may seem surprising that the vector forces (tension) in each part of the rope are so much greater than the 180-lb the agent weighs. But with a 180-lb object hanging from the middle of the rope, the tension required to keep the rope taut (with small angles of deflection) must be very great. This should become more obvious to you after you work Exercise 2.

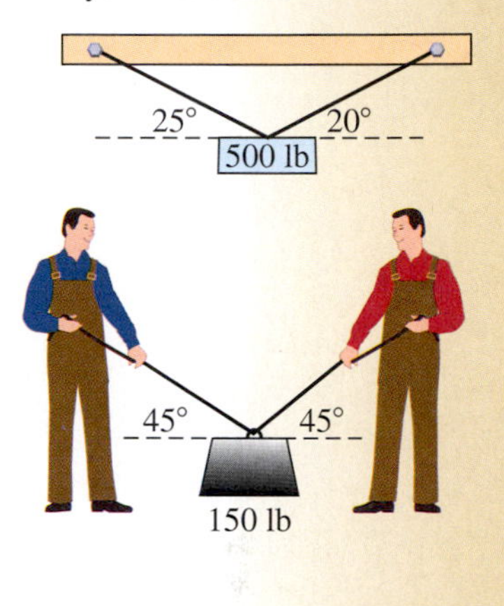

Exercise 1: A 500-lb crate is suspended by two ropes attached to the ceiling rafters. Find the tension in each rope.

Exercise 2: Two people team up to carry a 150-lb weight by passing a rope through an eyelet in the object. Find the tension in each rope.

Exercise 3: Referring to Illustration 1, if the rope has a tension limit of 600 lb (before it snaps), can a 200-lb agent make it across?

Cumulative Review Chapters 1–5

1. Solve using a standard triangle.

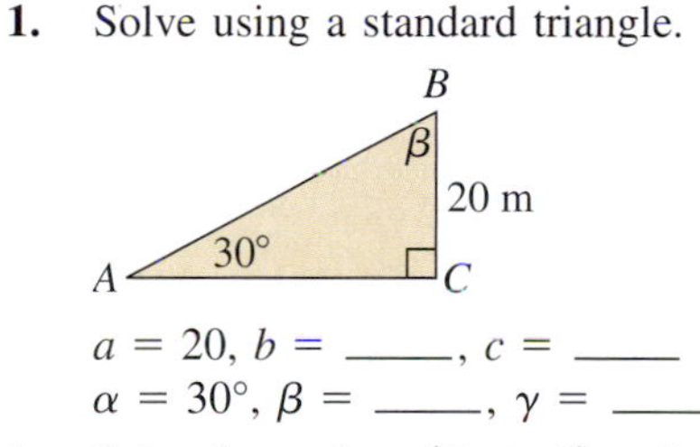

$a = 20$, $b =$ _____, $c =$ _____
$\alpha = 30°$, $\beta =$ _____, $\gamma =$ _____

2. Solve using trigonometric ratios.

B, 82 ft, 63°, A, α, C

$a \approx$ _____, $b \approx$ _____, $c = 82$
$\alpha =$ _____, $\beta = 63°$, $\gamma =$ _____

3. Solve for t: $A\cos(Bt + C) - D = 0$

4. For a complex number $a + bi$, (a) verify the sum of a complex number and its conjugate is a real number, and (b) verify the product of a complex number and its conjugate is a real number.

5. State the value of all six trig functions given $\tan\alpha = -\frac{3}{4}$ with $\cos\alpha > 0$.

6. Sketch the graph of $y = 3\cos\left(\frac{\pi}{6}x - \frac{\pi}{3}\right)$ using a reference rectangle and the *rule of fourths.*

7. State each related identity:

a. $\sin(2\alpha)$ **b.** $\sin\left(\frac{\alpha}{2}\right)$ **c.** $\sin(\alpha + \beta)$

8. State the domain and range of each function:

a. $y = \sin^{-1}t$ **b.** $y = \cos^{-1}t$ **c.** $y = \tan^{-1}t$

9. Given $\cos 53° \approx 0.6$ and $\cos 72° \approx 0.3$, approximate the value of $\cos 19°$ and $\cos 125°$ without using a calculator.

10. Find all real values of x that satisfy the equation $\sqrt{3} + 2\sin(2x) = 2\sqrt{3}$. State the answer in degrees.

11. The approximate number of daylight hours for Juneau, Alaska (58° N latitude), is given in the table. Use the data to find an appropriate regression equation, then answer the following:

a. Approximately how many daylight hours were there on April 15 ($t = 4.5$)?

b. Approximate the dates between which there are over 15 hr of daylight.

Source: Alaskan Alternative Energy @ www.absak.com/design/sunhours.

Month (Jan → 1)	Daylight (hours)	Month (Jan → 1)	Daylight (hours)
1	6.5	7	18.1
2	8.3	8	16.5
3	10.5	9	14.0
4	13.2	10	11.5
5	15.7	11	9.0
6	17.8	12	6.9

12. Verify that $\csc^2 t - \sec^2 t = \csc^2 t \sec^2 t$ is an identity.

13. Verify that the following is an identity: $\dfrac{2\cos^2\theta}{\csc^2\theta} + \dfrac{2\sin^2\theta}{\sec^2\theta} = \sin^2(2\theta)$

Solve each triangle using the law of sines or the law of cosines, whichever is appropriate.

14. **15.**

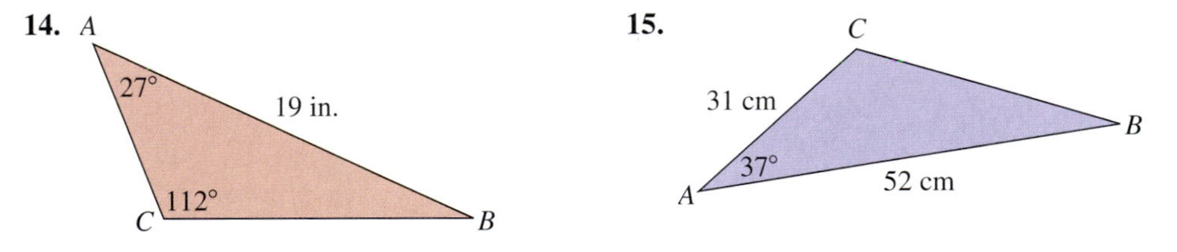

16. A mountain climber has slipped off the edge of a cliff, and his partner is attempting to pull him 30 ft up the face of the cliff and back onto the plateau. If the rope makes an angle of 10° with the plateau and she must pull with a force of 200 lb to make headway, how much work is done to save the fallen climber?

17. A 900-lb crate is sitting on a ramp that is inclined at 28°. Find the force needed to hold the object stationary.

18. A jet plane is flying at 750 mph on a heading of 30°. There is a strong, 50-mph wind blowing from due south (heading of 0°). What is the true course and speed of the plane (relative to the ground)?

Exercise 17

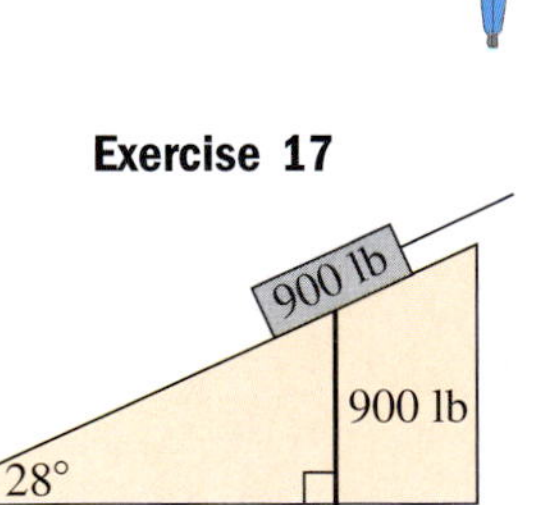

19. Use the dot product to find the angle θ between the vectors $\langle -3, 8\rangle$ and $\langle 7, 6\rangle$.

20. Find the three cube roots of unity (find all solutions to $x^3 - 1 = 0$).

21. Find $(1 - \sqrt{3}i)^8$ using De Moivre's theorem.

For Exercises 22 and 23, $z_1 = 8(\cos 45° + i \sin 45°)$ and $z_2 = 4(\cos 15° + i \sin 15°)$.

22. Compute the product $z_1 \cdot z_2$.

23. Compute the quotient $\frac{z_1}{z_2}$.

24. Mount Tortolas lies on the Argentine-Chilean border. When viewed from a distance of 5 mi, the angle of elevation to the top of the peak is 38°. How tall is Mount Tortolas? State the answer in feet.

25. The graph given is of the form $y = A \sin(Bx \pm C)$. Find the values of A, B, and C.

Exercise 25

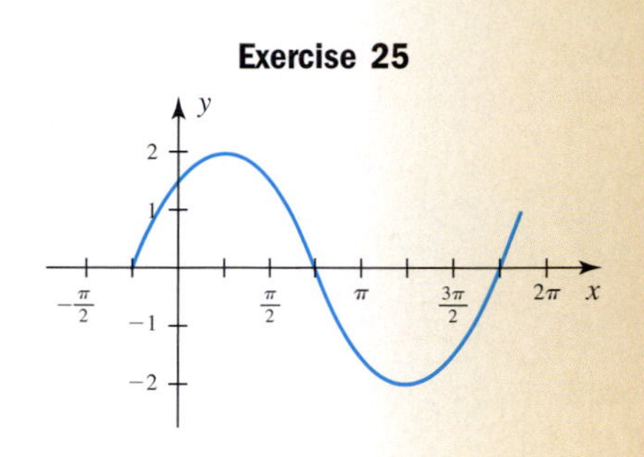

Chapter 6

Conic Sections and Polar Coordinates

Chapter Outline

Preview

In this chapter we study a family of curves called the **conic sections.** In common use, a cone brings to mind the cone-shaped paper cups at a water cooler (see photo). The point of the cone is called a **vertex** and the sheet of paper forming the sides is called a **nappe.** Mathematically speaking, a cone can have two nappes that meet at a vertex and extend infinitely in both directions (Figure 6.1). The conic sections are so named because all curves in the family can be formed by looking at a *section* of the *cone*, or more precisely—the intersection

Figure 6.1

Nappe
Vertex
Nappe

of a plane with a **right circular cone.*** The intersection can be manipulated to form a point or a line, but generally the term *conic* refers to a **circle, ellipse, hyperbola,** or **parabola,** which can also be formed. Each conic section can be represented by a second-degree equation in two variables.

6.1 The Circle and the Ellipse

LEARNING OBJECTIVES

In Section 6.1 you will learn how to:

A. Identify the factored form, polynomial form, and standard form of the equation of a circle and graph central and noncentral circles

B. Identify the factored form, polynomial form, and standard form of the equation of an ellipse and graph central and noncentral ellipses

INTRODUCTION

Consider the different ways a plane can intersect a right circular cone. If the plane is perpendicular to the axis of the cone and does not contain the vertex, a circle is formed (Figure 6.2). The size of the circle depends on the distance of the plane from the vertex. If the plane is slightly tilted from perpendicular, an ellipse is formed (Figure 6.3). In this section we introduce the equation of each conic.

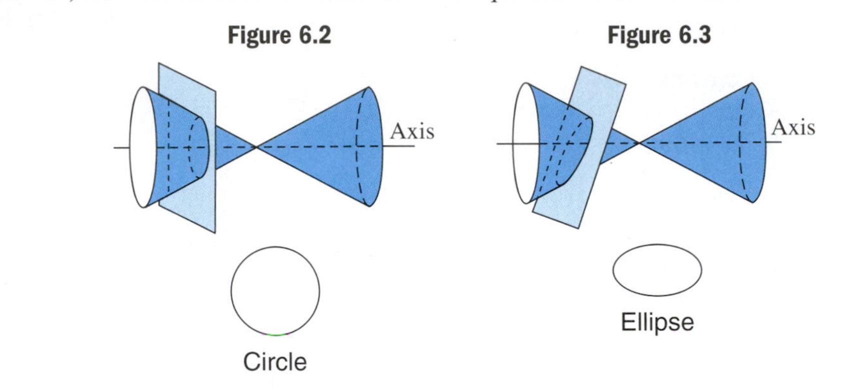

POINT OF INTEREST

Suppose a satellite is orbiting the Earth at an altitude of 200 mi. If the satellite maintains a velocity of approximately 4.8 mi/sec, the orbit will be circular. If the velocity of the satellite is greater or less than 4.8 mi/sec, the orbit becomes elliptical—unless velocity is so small that the satellite returns to Earth, or the velocity is so great that the satellite escapes Earth's gravity.

A. The Equation of a Circle

A circle can be defined as the set of all points in a plane that are a *fixed distance* called the **radius,** from a *fixed point* called the **center.** Since the definition involves *distance*, we can construct the general equation of a circle using the distance formula. Assume the center has coordinates (h, k), and let (x, y) represent any point on the graph. Since the distance between these points must be r, the distance formula yields: $\sqrt{(x-h)^2 + (y-k)^2} = r$.

*A line through the vertex (called the axis) is perpendicular to a circular base.

Squaring both sides gives the general equation of a circle in factored form: $(x - h)^2 + (y - k)^2 = r^2$.

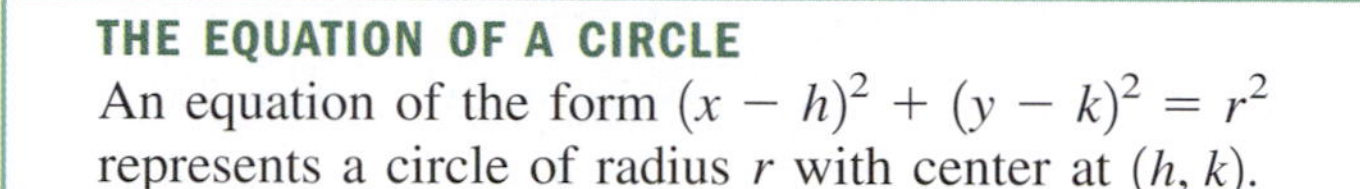

THE EQUATION OF A CIRCLE

An equation of the form $(x - h)^2 + (y - k)^2 = r^2$ represents a circle of radius r with center at (h, k).

If $h = 0$ and $k = 0$, the circle is centered at $(0, 0)$ and the graph is a **central circle** with equation $x^2 + y^2 = r^2$. At other values for h or k, the circle shifts horizontally h units and vertically k units with no change in the radius. Note this implies that shifts will be "opposite the sign."

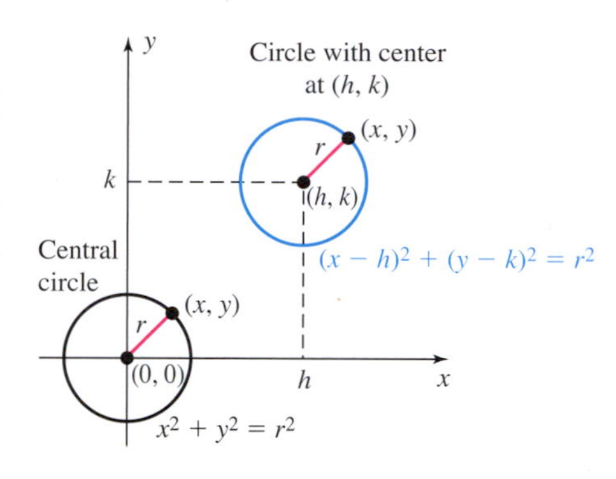

EXAMPLE 1 Find the equation of a circle with center $(0, -1)$ and radius 4, then sketch the graph.

Solution: Since the center is at $(0, -1)$ we have $h = 0$, $k = -1$, and $r = 4$. Using the factored form $(x - h)^2 + (y - k)^2 = r^2$ we obtain

$$(x - 0)^2 + [y - (-1)]^2 = 4^2 \quad \text{substitute 0 for } h, -1 \text{ for } k, \text{ and 4 for } r$$

$$x^2 + (y + 1)^2 = 16 \quad \text{simplify}$$

To graph the circle, it's easiest to begin at $(0, -1)$ and count $r = 4$ units in each horizontal direction, and $r = 4$ units in each vertical direction, knowing the radius is four in *any* direction. Neatly complete the circle by freehand drawing or using a compass. The graph shown is obtained.

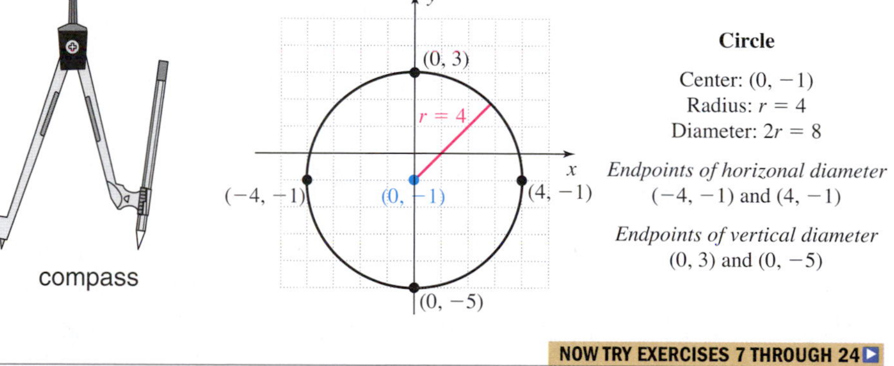

NOW TRY EXERCISES 7 THROUGH 24

EXAMPLE 2 Graph the circle represented by $(x - 2)^2 + (y + 3)^2 = 12$. Clearly label the center, radius, diameter, and the coordinates of the endpoints of the horizontal and vertical diameters.

Solution: Comparing the given equation with the factored form, we find the center is at $(2, -3)$ and the radius is $r = 2\sqrt{3} \approx 3.5$.

$$(x - h)^2 + (y - k)^2 = r^2 \quad \text{factored form}$$
$$\downarrow \qquad \downarrow \qquad \downarrow$$
$$(x - 2)^2 + (y + 3)^2 = 12 \quad \text{given equation}$$
$$-h = -2 \qquad -k = 3 \qquad r^2 = 12$$
$$h = 2 \qquad k = -3 \qquad r = \sqrt{12} = 2\sqrt{3} \quad \text{radius must be positive}$$

Plot the center $(2, -3)$ and count approximately 3.5 units in the horizontal and vertical directions. Complete the circle by freehand drawing or using a compass. The graph shown is obtained.

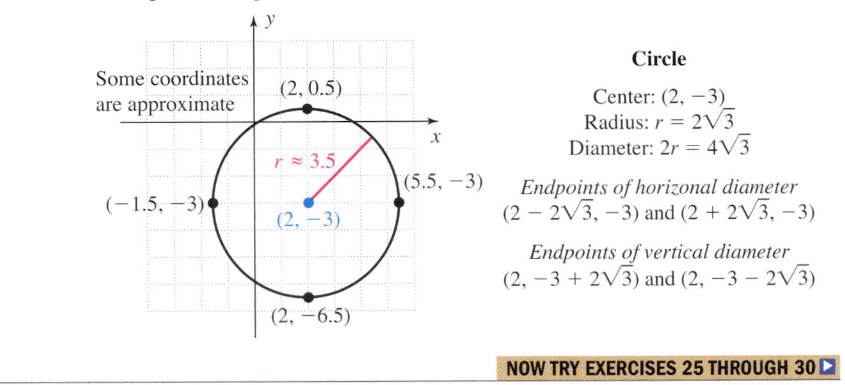

NOW TRY EXERCISES 25 THROUGH 30

In Example 2, note the equation is composed of binomial squares in both x and y. Expanding the binomials and collecting like terms places the equation of the circle in **polynomial form:**

$$(x - 2)^2 + (y + 3)^2 = 12 \quad \text{equation in factored form}$$
$$x^2 - 4x + 4 + y^2 + 6y + 9 = 12 \quad \text{expand binomials}$$
$$x^2 + y^2 - 4x + 6y + 1 = 0 \quad \text{combine like terms—polynomial form}$$

For future reference, observe the polynomial form contains a *sum* of second degree terms in x and y, and that *both terms have the same coefficient* (in this case, "1").

Since this form of the equation was derived by squaring binomials, it seems reasonable to assume we can go back to the factored form by creating binomial squares in x and y. This is accomplished by *completing the square*.

WORTHY OF NOTE

After writing the equation in factored form, it is possible to end up with a constant that is zero or negative. In the first case, the graph is a single point. In the second case, no graph is possible since roots of the equation will be complex numbers. These are called *degenerate cases*.

EXAMPLE 3 Find the center and radius of the circle whose equation is given, then sketch its graph: $x^2 + y^2 + 2x - 4y - 4 = 0$. Clearly label the center, radius, diameter, and the coordinates of the endpoints of the horizontal and vertical diameters.

Solution: To find the center and radius, we complete the square in both x and y.

$$x^2 + y^2 + 2x - 4y - 4 = 0 \quad \text{given equation}$$
$$(x^2 + 2x + \underline{\quad}) + (y^2 - 4y + \underline{\quad}) = 4 \quad \text{group } x\text{-terms and } y\text{-terms; add 4}$$
$$(x^2 + 2x + 1) + (y^2 - 4y + 4) = 4 + 1 + 4$$
adds 1 to left side — adds 4 to left side — add 1 + 4 to right side
$$(x + 1)^2 + (y - 2)^2 = 9 \quad \text{factor and simplify}$$

The center is at $(-1, 2)$ and the radius is $r = \sqrt{9} = 3$.

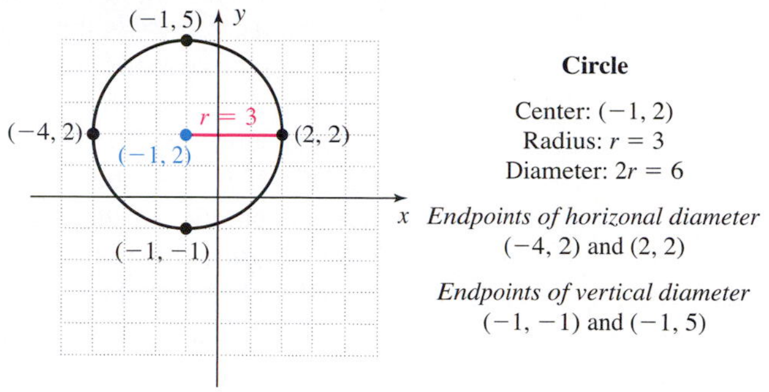

Circle

Center: $(-1, 2)$
Radius: $r = 3$
Diameter: $2r = 6$

Endpoints of horizonal diameter
$(-4, 2)$ and $(2, 2)$

Endpoints of vertical diameter
$(-1, -1)$ and $(-1, 5)$

NOW TRY EXERCISES 31 THROUGH 42

B. The Equation of an Ellipse

The equation of a circle in **standard form** provides a useful link to some of the other conic sections, and is obtained by *setting the equation equal to 1*. In the case of a circle, this means we simply divide by r^2.

$$(x - h)^2 + (y - k)^2 = r^2 \quad \text{factored form}$$

$$\frac{(x - h)^2}{r^2} + \frac{(y - k)^2}{r^2} = 1 \quad \text{divide by } r^2 \rightarrow \textit{standard form}$$

In this form, the value of r in each denominator gives the *horizontal* and *vertical* distances, respectively, from the center to the graph. This is not so important in the case of a circle, since this distance is the same in *any* direction. But for other conics, these horizontal and vertical distances are *not* the same, making the standard form a valuable tool for graphing. To distinguish the horizontal from the vertical distance, r^2 is replaced by a^2 in the "x-term" (horizontal distance), and by b^2 in the "y-term" (vertical distance).

It then seems reasonable to ask, "What happens to the graph when $a \neq b$?" To answer, consider the equation from Example 3. In standard form we have $\frac{(x + 1)^2}{3^2} + \frac{(y - 2)^2}{3^2} = 1$ (after dividing by 9), which we now compare to $\frac{(x + 1)^2}{4^2} + \frac{(y - 2)^2}{3^2} = 1$, where $a = 4$ and $b = 3$. The center of the curve is still at $(-1, 2)$, since $h = -1$ and $k = 2$ remain unchanged. For $y = 2$, $(x + 1)^2 = 16$, which gives $x = 3$ and $x = -5$ by inspection: $(3 + 1)^2 = 16$✓ and $(-5 + 1)^2 = 16$✓. This shows the horizontal distance from the center to the graph is $a = 4$, and the points $(3, 2)$ and $(-5, 2)$ are on the graph (see Figure 6.4). For $x = -1$ we have $(y - 2)^2 = 9$, giving $y = 5$ and $y = -1$ by inspection, and showing the vertical distance from the center to the graph is $b = 3$, with points $(-1, 5)$ and $(-1, -1)$ also on the graph. Using this information to sketch the curve reveals the "circle" is elongated and has become an **ellipse.**

Figure 6.4

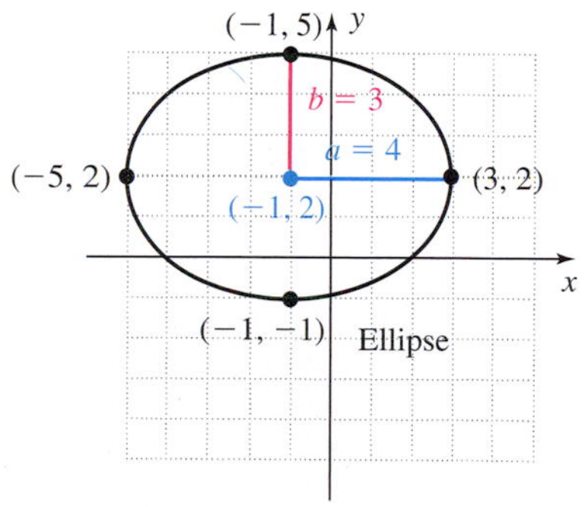

For an ellipse, the longest distance across the graph is called the **major axis,** with the endpoints of the major axis called **vertices.** The segment perpendicular to and bisecting the major axis with its endpoints on the ellipse is called the **minor axis.** If $a > b$, the major axis is horizontal (parallel to the x-axis) with length $2a$, and the minor axis is vertical with length $2b$ (see Figure 6.5). If $b > a$, the major axis is vertical (parallel to the y-axis) with length $2b$, and the minor axis is horizontal with length $2a$.

Figure 6.5

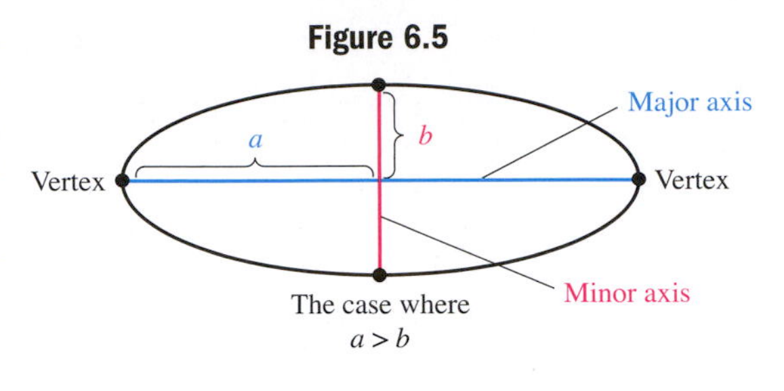

Generalizing this observation we obtain the equation of an ellipse in standard form.

THE EQUATION OF AN ELLIPSE IN STANDARD FORM

Given $\dfrac{(x-h)^2}{a^2} + \dfrac{(y-k)^2}{b^2} = 1$.

If $a \neq b$ the equation represents the graph of an ellipse with center at (h, k).

- $|a|$ gives the horizontal distance from center to graph.
- $|b|$ gives the vertical distance from center to graph.

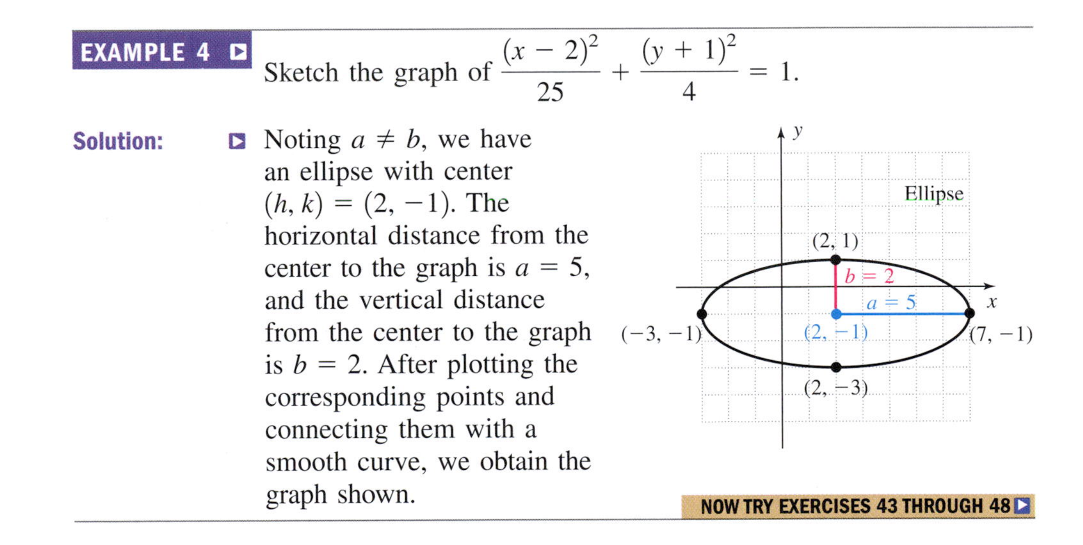

EXAMPLE 4 Sketch the graph of $\dfrac{(x-2)^2}{25} + \dfrac{(y+1)^2}{4} = 1$.

Solution: Noting $a \neq b$, we have an ellipse with center $(h, k) = (2, -1)$. The horizontal distance from the center to the graph is $a = 5$, and the vertical distance from the center to the graph is $b = 2$. After plotting the corresponding points and connecting them with a smooth curve, we obtain the graph shown.

NOW TRY EXERCISES 43 THROUGH 48

To obtain the factored form of the ellipse from Example 4, we clear the denominators using the least common multiple of 25 and 4, which is 100. This gives

$$(100)\frac{(x-2)^2}{25} + (100)\frac{(y+1)^2}{4} = 1(100)$$
$$4(x-2)^2 + 25(y+1)^2 = 100$$

In this form, we can still identify the center as $(2, -1)$, but now the distinguishing characteristic is that the coefficients of the binomial squares are not equal. The binomials can be expanded to obtain the polynomial form, yielding

$4x^2 + 25y^2 - 16x + 50y - 59 = 0$, where we note a sum of second degree terms with unequal coefficients. In general, we have

THE EQUATION OF AN ELLIPSE
An equation of the form $A(x - h)^2 + B(y - k)^2 = F$, with $A, B, F > 0$, represents an ellipse with center at (h, k).

If $A = B$, the equation becomes that of a circle. See Exercises 49 through 54.

EXAMPLE 5 For $25x^2 + 4y^2 = 100$, (a) write the equation in standard form and identify the center and the value of a and b, (b) identify the major and minor axes and name the vertices, and (c) sketch the graph.

Solution: The coefficients of x^2 and y^2 are unequal, and 25, 4, and 100 have like signs. The equation represents an ellipse with center at (0, 0). To obtain standard form:

a.

$$25x^2 + 4y^2 = 100 \quad \text{given equation}$$

$$\frac{25x^2}{100} + \frac{4y^2}{100} = 1 \quad \text{divide by 100}$$

$$\frac{x^2}{4} + \frac{y^2}{25} = 1 \quad \text{standard form}$$

$$\frac{x^2}{2^2} + \frac{y^2}{5^2} = 1 \quad \text{write denominators in squared form; } a = 2, b = 5$$

b. The result shows $a = 2$ and $b = 5$, indicating the major axis will be vertical and the minor axis will be horizontal. With the center at the origin, the x-intercepts will be (2, 0) and (−2, 0), with the vertices (and y-intercepts) at (0, 5) and (0, −5).

c. Plotting these intercepts and sketching the ellipse results in the graph shown.

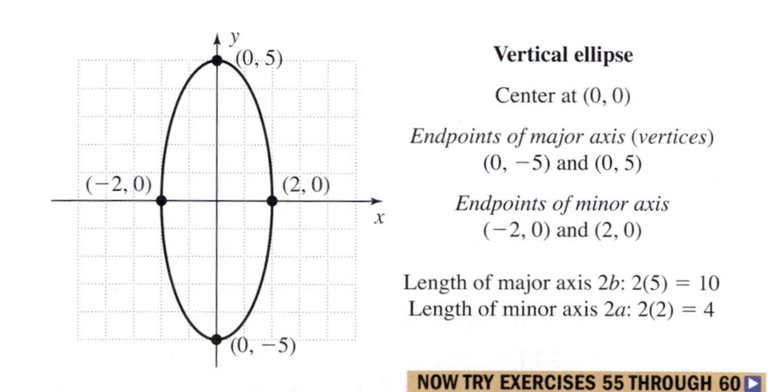

NOW TRY EXERCISES 55 THROUGH 60

If the equation is given in polynomial form, we complete the square in both x and y, then write the equation in standard form to sketch the graph. Figure 6.6 illustrates how the central ellipse and the shifted ellipse are related.

While the factored form is helpful for *identifying* ellipses and provides an intermediate link between the polynomial form and standard form, it is not very helpful when it comes to *graphing* an ellipse. For graphing, the standard form is more useful since it immediately tells us the *distance from the center to the graph in the horizontal and vertical directions.*

Figure 6.6

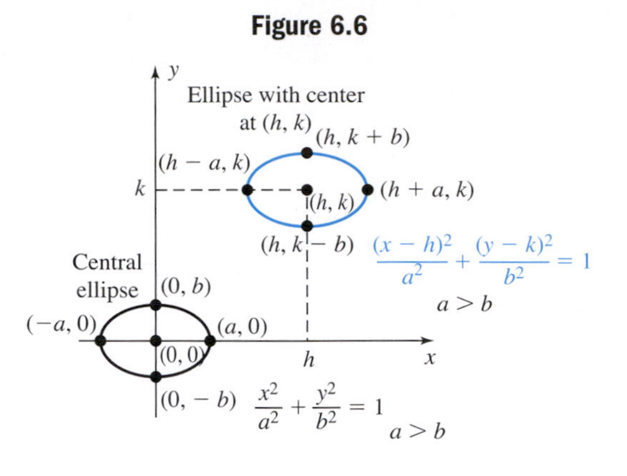

EXAMPLE 6 Sketch the graph of $25x^2 + 4y^2 + 150x - 16y + 141 = 0$.

Solution: The coefficients of x^2 and y^2 are unequal and have like signs, and we assume the equation represents an ellipse but wait until we have the factored form to be certain.

$$25x^2 + 4y^2 + 150x - 16y + 141 = 0 \quad \text{given equation (polynomial form)}$$

$$25x^2 + 150x + 4y^2 - 16y = -141 \quad \text{group like terms; subtract 141}$$

$$25(x^2 + 6x + \_\_\_) + 4(y^2 - 4y + \_\_\_) = -141 \quad \text{factor out leading coefficient from each group}$$

$$25(x^2 + 6x + 9) + 4(y^2 - 4y + 4) = -141 + 225 + 16 \quad \text{complete the square}$$

adds $25(9) = 225$; adds $4(4) = 16$; add $225 + 16$ to right

$$25(x + 3)^2 + 4(y - 2)^2 = 100 \quad \text{factored form (an ellipse)}$$

$$\frac{25(x + 3)^2}{100} + \frac{4(y - 2)^2}{100} = \frac{100}{100} \quad \text{divide both sides by 100}$$

$$\frac{(x + 3)^2}{4} + \frac{(y - 2)^2}{25} = 1 \quad \text{simplify (standard form)}$$

$$\frac{(x + 3)^2}{2^2} + \frac{(y - 2)^2}{5^2} = 1 \quad \text{write denominators in squared form}$$

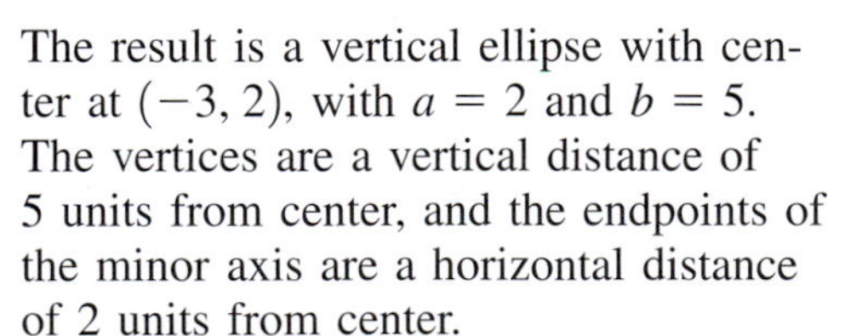

The result is a vertical ellipse with center at $(-3, 2)$, with $a = 2$ and $b = 5$. The vertices are a vertical distance of 5 units from center, and the endpoints of the minor axis are a horizontal distance of 2 units from center.

Note this is the same ellipse as in Example 5, but shifted 3 units left and 2 up.

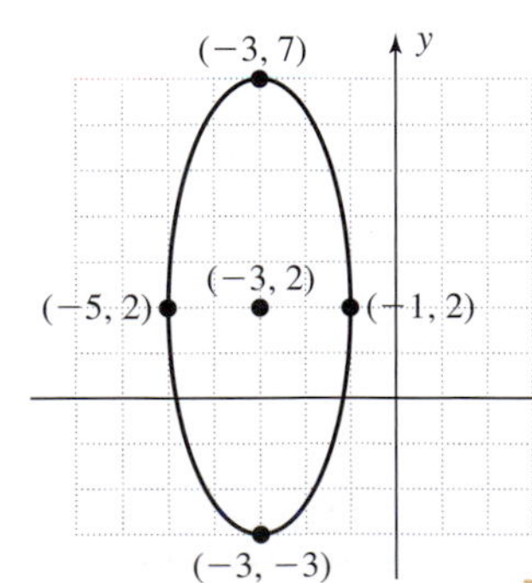

Vertical ellipse

Center at $(-3, 2)$

Endpoints of major axis (vertices) $(-3, -3)$ and $(-3, 7)$

Endpoints of minor axis $(-5, 2)$ and $(-1, 2)$

Length of major axis $2b$: $2(5) = 10$
Length of minor axis $2a$: $2(2) = 4$

NOW TRY EXERCISES 61 THROUGH 68

TECHNOLOGY HIGHLIGHT

Using a Graphing Calculator to Study Circles

The keystrokes shown apply to a TI-84 Plus model. Please consult our Internet site or your manual for other models.

When using a graphing calculator to study the conic sections, it is important to keep two things in mind. First, most graphing calculators are only capable of graphing *functions,* which means we must modify the equations of those conic sections that are relations (the circle, ellipse, hyperbola, and horizontal parabola) before they can be graphed using this technology. Second, most standard viewing windows have the x- and y-values preset at $[-10, 10]$ even though the calculator screen may be 94 pixels wide and 64 pixels high. This tends to compress the y-values and give a skewed image of the graph. Consider the *relation* $x^2 + y^2 = 25$. From our work in this section, we know this is the equation of a circle centered at (0, 0) with radius $r = 5$. To enable the calculator to graph this relation, we must define it in two pieces, each of which is a *function,* by solving for y:

$$x^2 + y^2 = 25 \quad \text{original equation}$$
$$y^2 = 25 - x^2 \quad \text{isolate } y^2$$
$$y = \pm\sqrt{25 - x^2} \quad \text{solve for } y$$

Note that we can separate this result into two parts, each of which is a function, enabling the calculator to draw the graph: $Y_1 = \sqrt{25 - x^2}$ gives the "upper half" of the circle, and $Y_2 = -\sqrt{25 - x^2}$, which gives the "lower half." Enter these on the Y= screen (note that $Y_2 = -Y_1$ can be used instead of reentering the entire expression: VARS ▶ ENTER). But if we graph Y_1 and Y_2 on the standard screen, the result appears more elliptical than circular (Figure 6.7). One way to fix this (there are other ways), is to use the ZOOM **5:ZSquare** option, which places the tick marks equally spaced on both axes, instead of trying to force both to display points from -10 to 10. Accessing this option by using the keystrokes ZOOM **5** gives the final result shown (Figure 6.8). Although it is a much improved graph, the circle does not appear "closed" as the calculator lacks sufficient pixels to show the proper curvature.

Figure 6.7

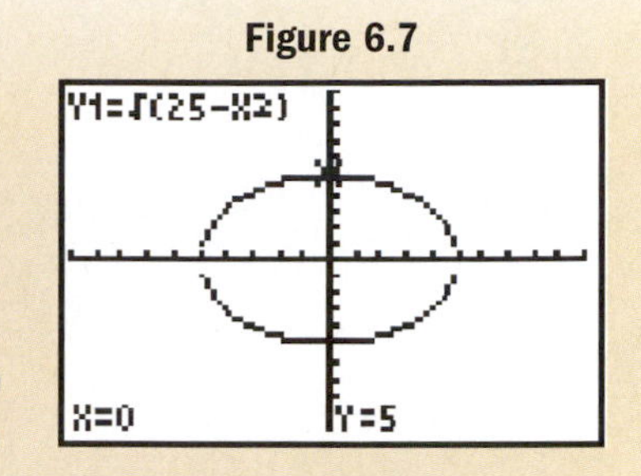

A second alternative is to manually set a "friendly" window. Using Xmin = −9.4, Xmax = 9.4, Ymin = −6.2, and Ymax = 6.2 will generate a better graph, which we can use to study the relation more closely. Note that we can jump between the upper and lower halves of the circle using the up ▲ or down ▼ arrows.

Figure 6.8

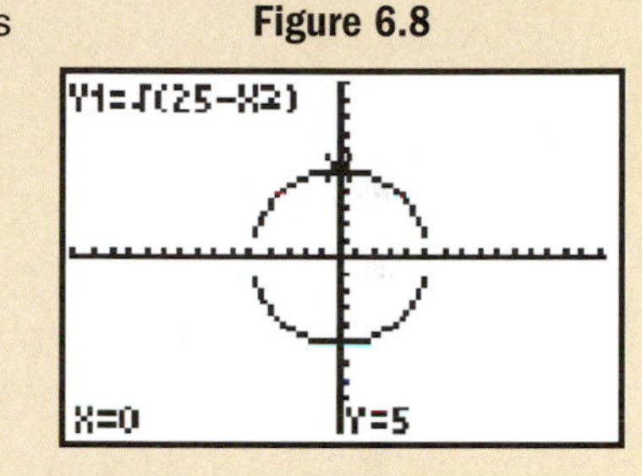

Exercise 1: Use these ideas to graph the circle defined by $x^2 + y^2 = 36$ using a friendly window. Then use the TRACE feature to find the value of y when $x = 3.6$. Now find the value of y when $x = 4.8$. Explain why the values seem "interchangeable."

Exercise 2: Use these ideas to graph the ellipse defined by $4x^2 + y^2 = 36$ using a friendly window. Then use the TRACE feature to find the value of the x- and y-intercepts.

6.1 EXERCISES

CONCEPTS AND VOCABULARY

Fill in the blank with the appropriate word or phrase. Carefully reread the section if needed.

1. A circle is defined to be the set of all points an equal distance, called the ________, from a given point, called the ________.

2. For $x^2 + y^2 = r^2$, the center of the circle is at ____ and the length of the radius is ___. The graph is called a(n) ________ circle.

3. To write the equation $x^2 + y^2 - 6x = 7$ in standard form, ________ the ________ in x, then set the equation equal to __.

4. The longest distance across an ellipse is called the ________ ________ and the endpoints are called ________.

5. Explain/discuss how the relations $a > b$, $a = b$, and $a < b$ affect the graph of a conic section with equation $\frac{(x-h)^2}{a^2} + \frac{(y-k)^2}{b^2} = 1$. Include several illustrative examples.

6. Compare/contrast the factored, polynomial, and standard forms of the equations discussed in this section. Discuss times when one form may be more useful than another (include examples).

DEVELOPING YOUR SKILLS

Find the equation of a circle satisfying the conditions given, then sketch its graph.

7. center (0, 0), radius 3
8. center (0, 0), radius 6
9. center (5, 0), radius $\sqrt{3}$
10. center (0, 4), radius $\sqrt{5}$
11. center $(4, -3)$, radius 2
12. center $(3, -8)$, radius 9
13. center $(-7, -4)$, radius $\sqrt{7}$
14. center $(-2, -5)$, radius $\sqrt{6}$
15. center $(1, -2)$, radius $2\sqrt{3}$
16. center $(-2, 3)$, radius $3\sqrt{2}$
17. center (4, 5), diameter $4\sqrt{3}$
18. center (5, 1), diameter $4\sqrt{5}$
19. center at (7, 1), graph contains the point $(1, -7)$
20. center at $(-8, 3)$, graph contains the point $(-3, 15)$
21. center at (3, 4), graph contains the point (7, 9)
22. center at $(-5, 2)$, graph contains the point $(-1, 3)$
23. diameter has endpoints (5, 1) and (5, 7)
24. diameter has endpoints (2, 3) and (8, 3)

Identify the center and radius of each circle, then graph. Also state the domain and range of the relation.

25. $(x - 2)^2 + (y - 3)^2 = 4$
26. $(x - 5)^2 + (y - 1)^2 = 9$
27. $(x + 1)^2 + (y - 2)^2 = 12$
28. $(x - 7)^2 + (y + 4)^2 = 20$
29. $(x + 4)^2 + y^2 = 81$
30. $x^2 + (y - 3)^2 = 49$

Write each equation in factored form to find the center and radius of the circle. Then sketch the graph.

31. $x^2 + y^2 - 10x - 12y + 4 = 0$
32. $x^2 + y^2 + 6x - 8y - 6 = 0$
33. $x^2 + y^2 - 10x + 4y + 4 = 0$
34. $x^2 + y^2 + 6x + 4y + 12 = 0$
35. $x^2 + y^2 + 6y - 5 = 0$
36. $x^2 + y^2 - 8x + 12 = 0$
37. $x^2 + y^2 + 4x + 10y + 18 = 0$
38. $x^2 + y^2 - 8x - 14y - 47 = 0$
39. $x^2 + y^2 + 14x + 12 = 0$
40. $x^2 + y^2 - 22y - 5 = 0$
41. $2x^2 + 2y^2 - 12x + 20y + 4 = 0$
42. $3x^2 + 3y^2 - 24x + 18y + 3 = 0$

Sketch the graph of each ellipse.

43. $\frac{(x-1)^2}{9} + \frac{(y-2)^2}{16} = 1$
44. $\frac{(x-3)^2}{4} + \frac{(y-1)^2}{25} = 1$
45. $\frac{(x-2)^2}{25} + \frac{(y+3)^2}{4} = 1$
46. $\frac{(x+5)^2}{1} + \frac{(y-2)^2}{16} = 1$
47. $\frac{(x+1)^2}{16} + \frac{(y+2)^2}{9} = 1$
48. $\frac{(x+1)^2}{36} + \frac{(y+3)^2}{9} = 1$

Identify each equation as that of an ellipse or circle, then sketch its graph.

49. $(x + 1)^2 + 4(y - 2)^2 = 16$

50. $9(x - 2)^2 + (y + 3)^2 = 36$

51. $2(x - 2)^2 + 2(y + 4)^2 = 18$

52. $(x - 6)^2 + y^2 = 49$

53. $4(x - 1)^2 + 9(y - 4)^2 = 36$

54. $25(x - 3)^2 + 4(y + 2)^2 = 100$

For each exercise, (a) write the equation in standard form then identify the center and the values of a and b, (b) state the coordinates of the vertices and the coordinates of the endpoints of the minor axis, and (c) sketch the graph.

55. $x^2 + 4y^2 = 16$

56. $9x^2 + y^2 = 36$

57. $16x^2 + 9y^2 = 144$

58. $25x^2 + 9y^2 = 225$

59. $2x^2 + 5y^2 = 10$

60. $3x^2 + 7y^2 = 21$

Complete the square in both x and y to write each equation in standard form. Then draw a complete graph of the relation and identify all important features.

61. $4x^2 + y^2 + 6y + 5 = 0$

62. $x^2 + 3y^2 + 8x + 7 = 0$

63. $x^2 + 4y^2 - 8y + 4x - 8 = 0$

64. $3x^2 + y^2 - 8y + 12x - 8 = 0$

65. $5x^2 + 2y^2 + 20y - 30x + 75 = 0$

66. $4x^2 + 9y^2 - 16x + 18y - 11 = 0$

67. $2x^2 + 5y^2 - 12x + 20y - 12 = 0$

68. $6x^2 + 3y^2 - 24x + 18y - 3 = 0$

WORKING WITH FORMULAS

69. Area of an inscribed square: $A = 2r^2$

The area of a square inscribed in a circle is found by using the formula given where r is the radius of the circle. Find the area of the inscribed square shown.

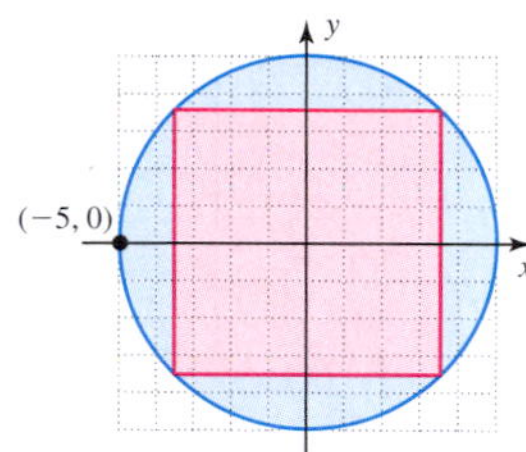

70. Area of an ellipse: $A = \pi ab$

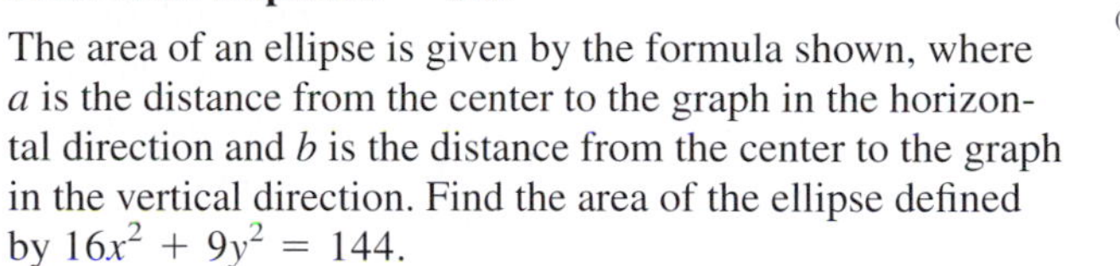

The area of an ellipse is given by the formula shown, where a is the distance from the center to the graph in the horizontal direction and b is the distance from the center to the graph in the vertical direction. Find the area of the ellipse defined by $16x^2 + 9y^2 = 144$.

APPLICATIONS

71. Radar detection: The radar on a luxury liner has a range of 25 nautical miles in any direction. If this ship is located at coordinates (5, 12), can the radar pick up its sister ship located at coordinates (15, 36)? Assume coordinates indicate a location in nautical miles from (0, 0).

72. Earthquake movement: If the epicenter (point of origin) of an earthquake is located at map coordinates (3, 7) and the quake could be felt up to 12 mi away, would a person located at (13, 1) have felt the quake? Assume coordinates indicate a location in statute miles from (0, 0).

73. Inscribed circle: Find the equation for both the red and blue circles, then find the area of the region shaded in blue.

Exercise 73

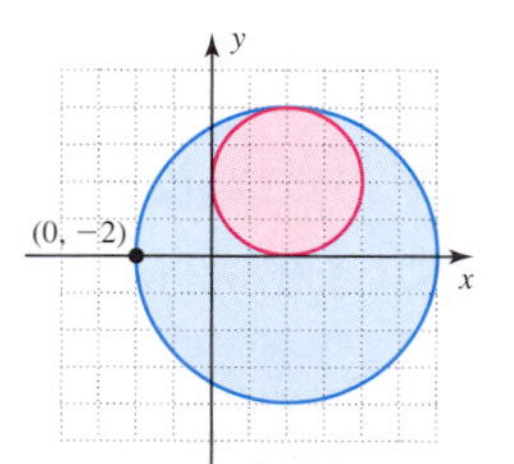

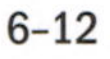

Exercise 74

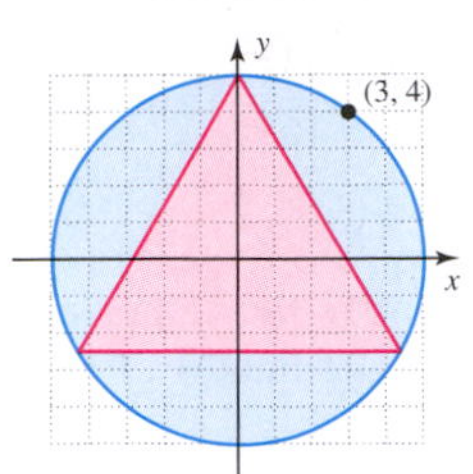

74. Inscribed triangle: The area of an equilateral triangle inscribed in a circle is given by the formula $A = \frac{3\sqrt{3}}{4}r^2$, where r is the radius of the circle. Find the area of the equilateral triangle shown.

75. Radio broadcast range: Two radio stations may not use the same frequency if their broadcast areas *overlap.* Suppose station KXRQ has a broadcast area bounded by $x^2 + y^2 + 8x - 6y = 0$ and WLRT has a broadcast area bounded by $x^2 + y^2 - 10x + 4y = 0$. Graph the circle representing each broadcast area on the same grid to determine if both stations may broadcast on the same frequency.

76. Radio broadcast range: The emergency radio broadcast system is designed to alert the population by relaying an emergency signal to all points of the country. A signal is sent from a station whose broadcast area is bounded by $x^2 + y^2 = 2500$ (x and y in miles) and the signal is picked up and relayed by a transmitter with range $(x - 20)^2 + (y - 30)^2 = 900$. Graph the circle representing each broadcast area on the same grid to determine the greatest distance from the original station that this signal can be received. Be sure to scale the axes appropriately.

As a planet orbits around the Sun, it traces out an ellipse. If the center of the ellipse were placed at (0, 0) on a coordinate grid, the Sun would be actually off-centered (located at a point called the *focus* of the ellipse). Use this information and the graphs provided to complete Exercises 77 and 78.

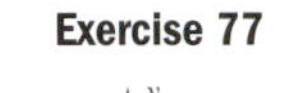
Exercise 77

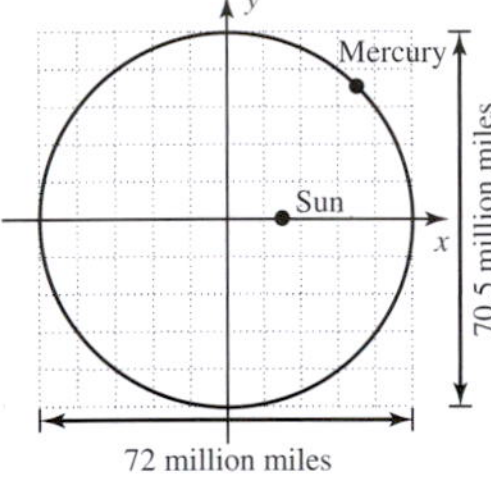

Exercise 78

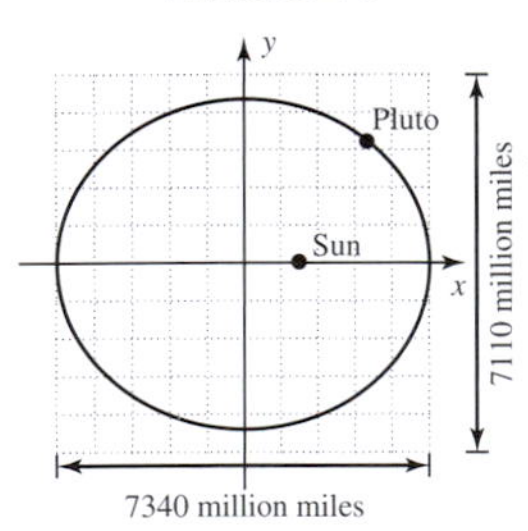

77. Orbit of Mercury: The approximate orbit of the planet Mercury is shown in the figure given. Find an equation that models this orbit.

78. Orbit of Pluto: The approximate orbit of the dwarf planet Pluto is shown in the figure given. Find an equation that models this orbit.

79. Race track area: Suppose the *Coronado 500* is a car race that is run on an elliptical track. The track is bounded by two ellipses with equations of $4x^2 + 9y^2 = 900$ and $9x^2 + 25y^2 = 900$, where x and y are in hundreds of yards. Use the formula given in Exercise 70 to find the area of the race track.

80. Area of a border: The tablecloth for a large oval table is elliptical in shape. It is designed with two concentric ellipses (one within the other), as shown in the figure. The equation of the outer ellipse is $9x^2 + 25y^2 = 225$, and the equation of the inner ellipse is $4x^2 + 16y^2 = 64$ with x and y in feet. Use the formula given in Exercise 70 to find the area of the border of the tablecloth.

81. Elliptical arches: In some situations, bridges are built using uniform elliptical archways, as shown in the figure. Find the equation of the ellipse forming each arch if it has a total width of 30 ft and a maximum center height (above level ground) of 8 ft. What is the height of a point 9 ft to the right of the center of the ellipse?

8 ft

60 ft

82. Elliptical arches: An elliptical arch bridge is built across a one-lane highway. The arch is 20 ft across and has a maximum center height of 12 ft. Will a farm truck hauling a load 10 ft wide with a clearance height of 11 ft be able to go through the bridge without damage? (*Hint:* See Exercise 81.)

WRITING, RESEARCH, AND DECISION MAKING

83. In Exercises 77 and 78, we saw that the orbits of the planets around the Sun are elliptical in shape. The maximum distance from a planet to the Sun is called the *aphelion*, and the maximum distance from the center of the orbit is called the *semimajor axis.* Use a reference book to find the aphelion of the planet Mars and the length of its semimajor axis.

84. You have likely heard that Pluto (a dwarf planet) is the farthest planet from the Sun, but may be surprised to hear that this is not always true. Do some reading and research on the orbits of the outer planets and try to determine why. How often does the phenomenon occur? How long does it last?

85. Attempt to find the x- and y-intercepts for the relation defined by $-9x^2 + 16y^2 = 144$. What happens? Why is this *not* the equation of an ellipse?

EXTENDING THE CONCEPT

86. A circle centered at (3, 4) is tangent to the y-axis. Find all values of y that satisfy $(1, y)$ for this circle.

87. Find the equation of the ellipse that passes through the four points $(-1, 1)$, $(5, 1)$, $(2, 3)$ and $(2, -1)$.

MAINTAINING YOUR SKILLS

88. (2.2) Find the value of all six trig functions given $\sin\theta = -\frac{5}{13}$ and $\sec\theta > 0$.

89. (3.2) Verify that the following is an identity: $\frac{\cos x}{1 - \sin x} = \frac{1 + \sin x}{\cos x}$

90. (5.5) Find the angle θ between the vectors $\mathbf{u} = -12\mathbf{i} + 5\mathbf{j}$ and $\mathbf{v} = -2\mathbf{i} + 10\mathbf{j}$.

91. (5.7) Which complex number has the greater absolute value: $z_1 = 15 + 1i$ or $z_2 = 13 + 9i$?

92. (2.3) The graph of $f(x) = A\sin(Bx + C)$ is shown here. Find the value of A, B, and C given that $\sin\frac{3\pi}{8} = -3$.

93. (1.2) The blueprint for a ride at Wet Willy's Water Park is pictured here. Find the entire length L of the slide given the dimensions and angles indicated.

Exercise 92

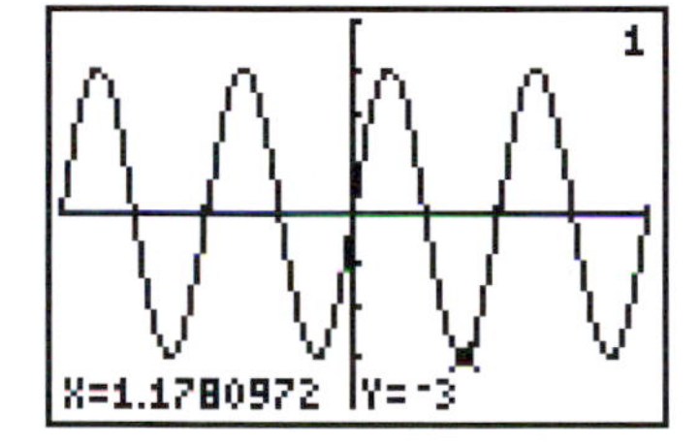

Exercise 93

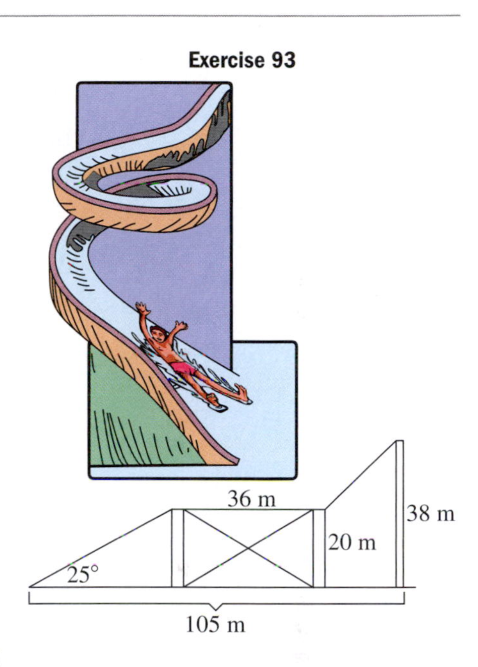

6.2 The Hyperbola

LEARNING OBJECTIVES

In Section 6.2 you will learn how to:

A. Identify the factored form, polynomial form, and standard form of the equation of a hyperbola and graph central and noncentral hyperbolas

B. Distinguish between the equations of a circle, ellipse, and hyperbola

INTRODUCTION

As shown in Figure 6.9, a hyperbola is a conic section formed by a plane that cuts both nappes of a right circular cone (the plane need not be parallel to the axis). A hyperbola has two symmetric parts called **branches,** which open in opposite directions. Although the branches appear to resemble parabolas, we will soon discover they are actually a very different curve.

Figure 6.9

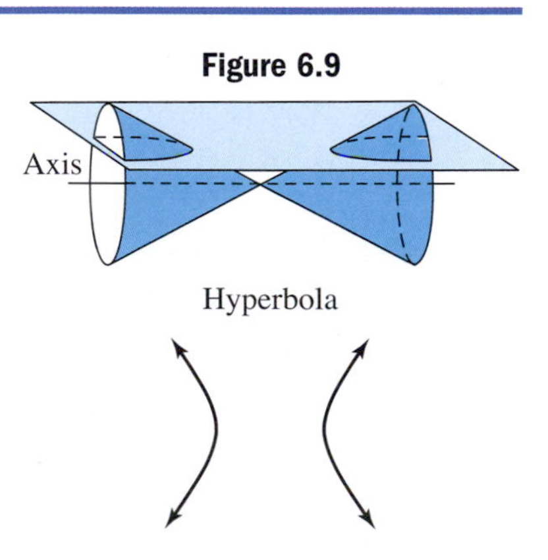

POINT OF INTEREST

By comparing the orbits of a number of earlier comets, the British astronomer Edmond Halley (1656–1742) showed the great comet of 1682 to be the same as those that had appeared in 1607 and 1531, and successfully predicted the return of the comet in 1759. Earlier appearances of Halley's comet have now been identified from records dating as early as 240 B.C. Most comets have elliptical orbits, and the periods (the time they take to orbit the Sun) of about 200 comets have been calculated. Comets that have a very high velocity and a large mass cannot be captured by the Sun's gravitational field, and the path of these comets trace out one branch of a hyperbola.

A. The Equation of a Hyperbola in Polynomial and Standard Form

In Section 6.1, we noted the equation $A(x - h)^2 + B(y - k)^2 = F\ (A, B, F > 0)$, could be used to describe the equation of both a circle and an ellipse. If $A = B$, the equation is that of a circle; if $A \neq B$, the equation represents an ellipse. Both cases contain a *sum* of second-degree terms. Perhaps driven by curiosity, we might wonder what happens if the equation has a *difference* of second-degree terms. As you'll see, the result is noteworthy.

Consider the equation $9x^2 - 16y^2 = 144$. It appears the graph will be centered at (0, 0) since no shifts are applied (h and k are both zero). Using the intercept method to graph this equation reveals an entirely new curve, called a *hyperbola.*

EXAMPLE 1 Graph the equation $9x^2 - 16y^2 = 144$ using intercepts and additional points as needed.

Solution:

$$9x^2 - 16y^2 = 144 \quad \text{given}$$
$$9(0)^2 - 16y^2 = 144 \quad \text{substitute 0 for } x$$
$$-16y^2 = 144 \quad \text{simplify}$$
$$y^2 = -9 \quad \text{divide by } -16$$

Since y^2 can never be negative, we conclude that the graph has *no y-intercepts.* Substituting $y = 0$ to find the x-intercepts gives

$$9x^2 - 16y^2 = 144 \quad \text{given}$$
$$9x^2 - 16(0)^2 = 144 \quad \text{substitute 0 for } y$$
$$9x^2 = 144 \quad \text{simplify}$$
$$x^2 = 16 \quad \text{divide by 9}$$
$$x = \sqrt{16} \text{ and } x = -\sqrt{16} \quad \text{square root property}$$
$$x = 4 \quad \text{and} \quad x = -4 \quad \text{simplify}$$
$$(4, 0) \quad \text{and} \quad (-4, 0) \quad x\text{-intercepts}$$

Knowing the graph has no y-intercepts, we select inputs greater than 4 and less than -4 to help sketch the graph. Using $x = -5$ and $x = 5$ yields

$9x^2 - 16y^2 = 144$		given	$9x^2 - 16y^2 = 144$	
$9(5)^2 - 16y^2 = 144$		substitute for x	$9(-5)^2 - 16y^2 = 144$	
$9(25) - 16y^2 = 144$		$5^2 = (-5)^2 = 25$	$9(25) - 16y^2 = 144$	
$225 - 16y^2 = 144$		simplify	$225 - 16y^2 = 144$	
$-16y^2 = -81$		subtract 225	$-16y^2 = -81$	
$y^2 = \frac{81}{16}$		divide by -16	$y^2 = \frac{81}{16}$	
$y = \frac{9}{4}$	$y = -\frac{9}{4}$	square root property	$y = \frac{9}{4}$	$y = -\frac{9}{4}$
$y = 2.25$	$y = -2.25$	decimal form	$y = 2.25$	$y = -2.25$
$(5, 2.25)$	$(5, -2.25)$	ordered pairs	$(-5, 2.25)$	$(-5, -2.25)$

Plotting these points and connecting them with a smooth curve, while *knowing there are no y-intercepts*, produces the graph in the figure. The point at the origin (in blue) is not a part of the graph, and is given only to indicate the "center" of the hyperbola.

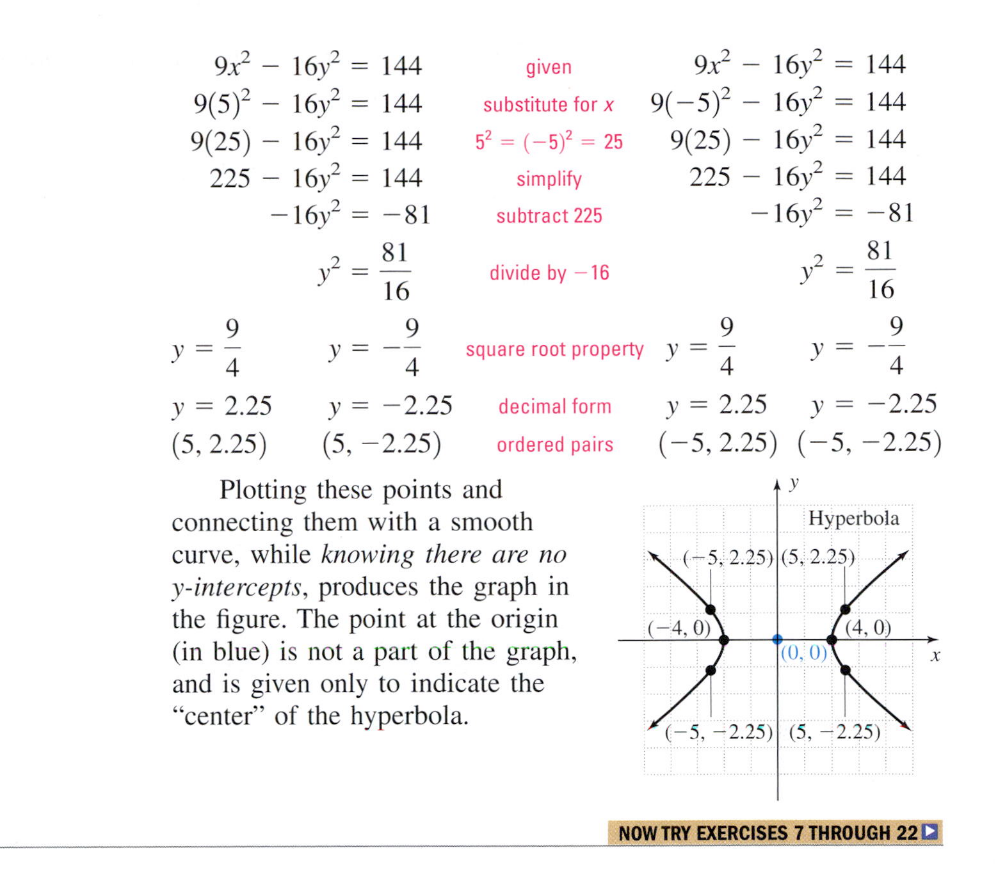

NOW TRY EXERCISES 7 THROUGH 22

Since the hyperbola crosses a horizontal axis, it is referred to as a **horizontal hyperbola.** The points $(-4, 0)$ and $(4, 0)$ are called **vertices,** and the **center** of the hyperbola is always the point halfway between them. If the center is at the origin, we have a **central hyperbola.** The line passing through the center and both vertices is called the **transverse axis** (vertices are always on the transverse axis), and the line passing through the center and perpendicular to this axis is called the **conjugate axis** (see Figure 6.10).

In Example 1, the coefficient of the term containing x^2 was positive and we were subtracting the term containing y^2. If the y^2-term is positive and we subtract the term containing x^2, the result is a vertical hyperbola (Figure 6.11).

Figure 6.10 **Figure 6.11**

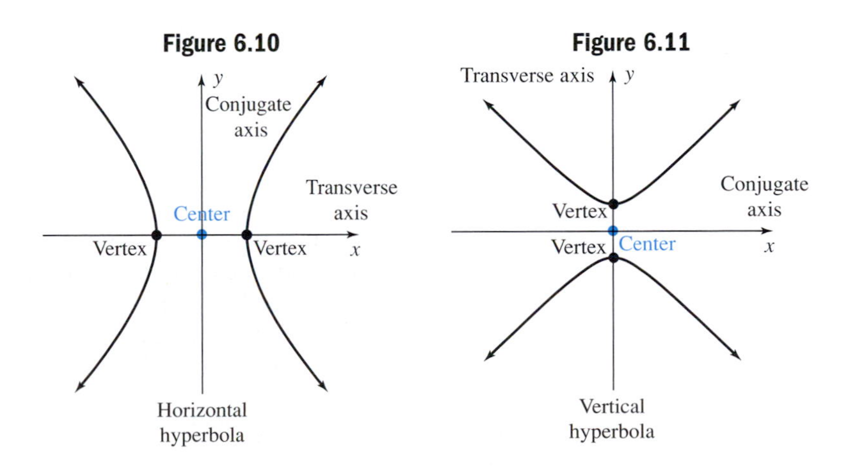

These observations lead us to the equation of a hyperbola in factored form.

THE EQUATION OF A HYPERBOLA (A, B, $F > 0$)

The equation	The equation
$A(x - h)^2 - B(y - k)^2 = F$	$B(y - k)^2 - A(x - h)^2 = F$
represents a *horizontal hyperbola* with center at (h, k):	represents a *vertical hyperbola* with center at (h, k):
transverse axis $y = k$, conjugate axis $x = h$.	transverse axis $x = h$, conjugate axis $y = k$.

Note each equation contains a *difference* of squared terms in x and y ($A \neq B$ is not a requirement for hyperbolas).

EXAMPLE 2 For the hyperbola shown, state the location of the vertices and the equation of the transverse axis. Then identify the location of the center and the equation of the conjugate axis.

Solution: By inspection we locate the vertices at (0, 0) and (0, 4). The equation of the transverse axis is $x = 0$. The center is halfway between the vertices at (0, 2), meaning the equation of the conjugate axis is $y = 2$.

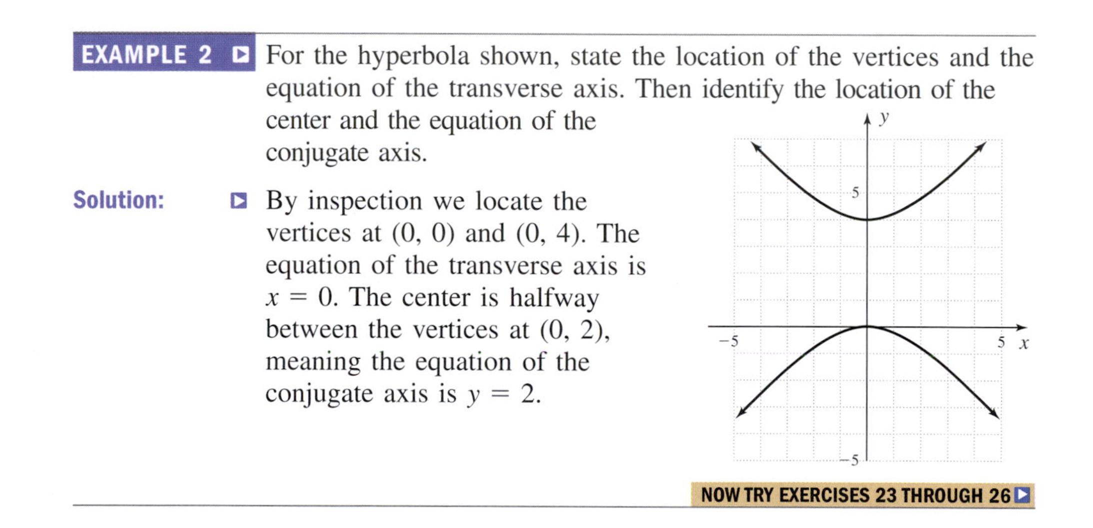

NOW TRY EXERCISES 23 THROUGH 26

Standard Form

As with the ellipse, the factored form of the equation is helpful for *identifying* hyperbolas, but not very helpful when it comes to *graphing* a hyperbola (since we still must go through the laborious process of finding additional points). For graphing, standard form is once again preferred, with the equation being set equal to 1. Consider the hyperbola $9x^2 - 16y^2 = 144$ from Example 1. To write the equation in standard form, we divide by 144 and obtain $\frac{x^2}{4^2} - \frac{y^2}{3^2} = 1$. By comparing the standard form to the graph, we note $a = 4$ represents the distance from center to vertices, similar to the way we used a previously. But since the graph has no y-intercepts, what could $b = 3$ represent? The answer lies in the fact that branches of a hyperbola are **asymptotic,** meaning they will approach and become very close to imaginary lines that can be used to sketch the graph. The slope of the asymptotic lines are given by the ratios $\frac{b}{a}$ and $-\frac{b}{a}$, with the related equations being $y = \frac{b}{a}x$ and $y = -\frac{b}{a}x$ since this is a central hyperbola. The graph from Example 1 is repeated in Figure 6.12, with the asymptotes drawn. For a clearer understanding of how the equations for the asymptotes were determined, see Exercise 70.

Figure 6.12 **Figure 6.13**

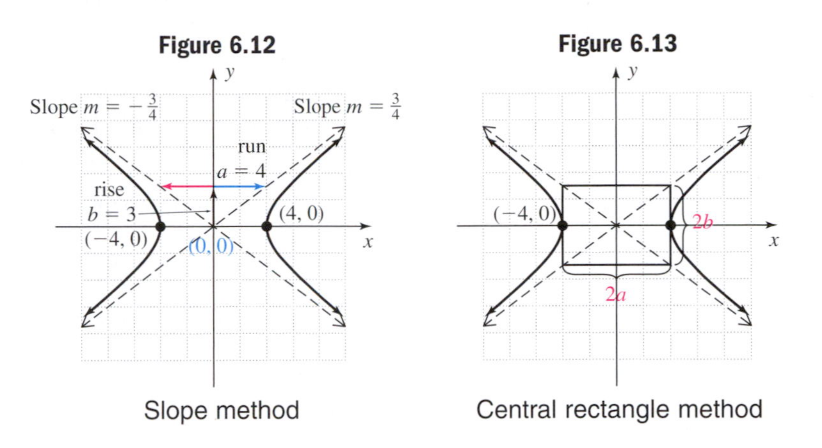

Slope method Central rectangle method

A second method of drawing the asymptotes involves drawing a **central rectangle** with dimensions $2a$ by $2b$, as shown in Figure 6.13. The asymptotes will be the *extended diagonals* of this rectangle. This brings us to the equation of a hyperbola in standard form.

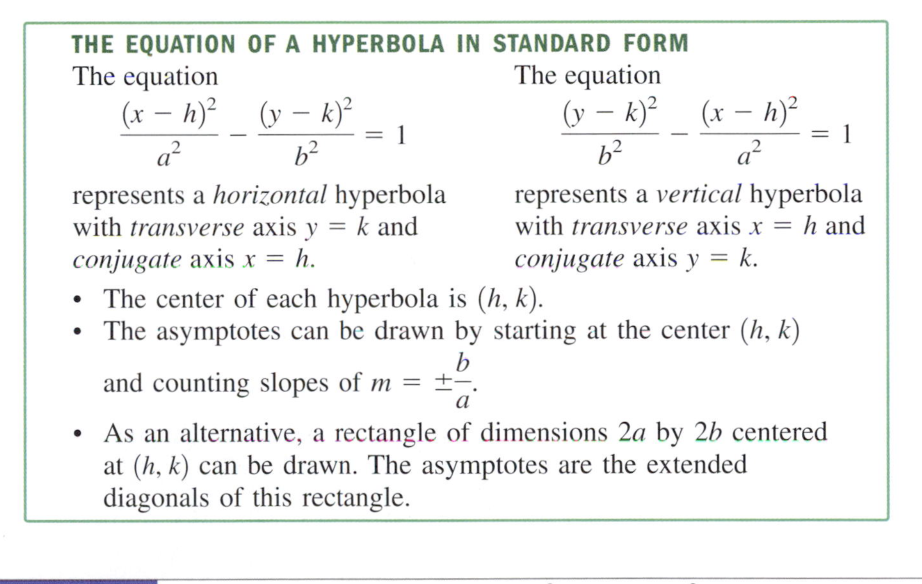

THE EQUATION OF A HYPERBOLA IN STANDARD FORM

The equation

$$\frac{(x-h)^2}{a^2} - \frac{(y-k)^2}{b^2} = 1$$

represents a *horizontal* hyperbola with *transverse* axis $y = k$ and *conjugate* axis $x = h$.

The equation

$$\frac{(y-k)^2}{b^2} - \frac{(x-h)^2}{a^2} = 1$$

represents a *vertical* hyperbola with *transverse* axis $x = h$ and *conjugate* axis $y = k$.

- The center of each hyperbola is (h, k).
- The asymptotes can be drawn by starting at the center (h, k) and counting slopes of $m = \pm\frac{b}{a}$.
- As an alternative, a rectangle of dimensions $2a$ by $2b$ centered at (h, k) can be drawn. The asymptotes are the extended diagonals of this rectangle.

EXAMPLE 3 Sketch the graph of $16(x-2)^2 - 9(y-1)^2 = 144$. Include the center, vertices, and asymptotes.

Solution: Begin by noting a difference of the second-degree terms, with the x^2-term occurring first. This means we'll be graphing a horizontal hyperbola whose center is at (2, 1). Continue by writing the equation in standard form.

$$16(x-2)^2 - 9(y-1)^2 = 144 \quad \text{given equation}$$

$$\frac{16(x-2)^2}{144} - \frac{9(y-1)^2}{144} = \frac{144}{144} \quad \text{divide by 144}$$

$$\frac{(x-2)^2}{9} - \frac{(y-1)^2}{16} = 1 \quad \text{simplify}$$

$$\frac{(x-2)^2}{3^2} - \frac{(y-1)^2}{4^2} = 1 \quad \text{write denominators in squared form}$$

Since $a = 3$, the vertices are a horizontal distance of 3 units from the center (2, 1), giving $(2 + 3, 1) \rightarrow (5, 1)$ and $(2 - 3, 1) \rightarrow (-1, 1)$. After plotting the center and vertices, we can begin at the center and count off slopes of $m = \pm\frac{b}{a} = \pm\frac{4}{3}$, or draw a rectangle with dimensions $2(3) = 6$ (horizontal dimension) by $2(4) = 8$ (vertical dimension) centered at (2, 1) to sketch the asymptotes. The complete graph is shown here.

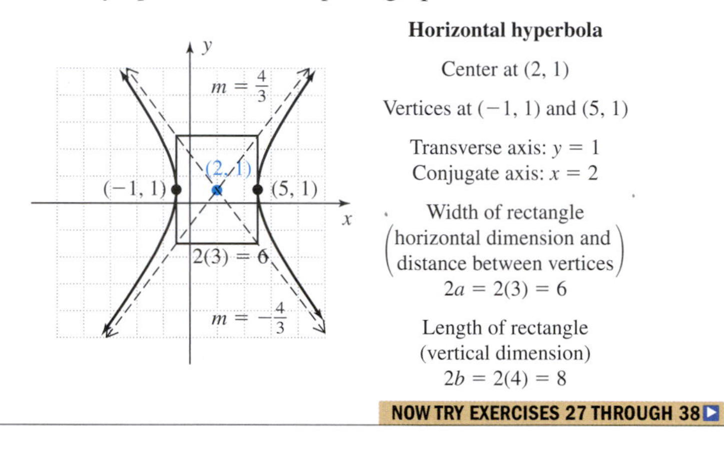

NOW TRY EXERCISES 27 THROUGH 38

Polynomial Form

If the equation is given in polynomial form, complete the square in x and y to write the equation in standard form.

EXAMPLE 4 Graph the equation $9y^2 - x^2 + 54y + 4x + 68 = 0$.

Solution: Since the y^2-term occurs first, we assume the equation represents a vertical hyperbola, but wait for the factored form to be sure (see Exercise 68).

$$9y^2 - x^2 + 54y + 4x + 68 = 0 \quad \text{given}$$

$$9y^2 + 54y - x^2 + 4x = -68 \quad \text{collect like-variable terms; subtract 68}$$

$$9(y^2 + 6y + \underline{\quad}) - 1(x^2 - 4x + \underline{\quad}) = -68 \quad \text{factor out 9 from } y\text{-terms and } -1 \text{ from } x\text{-terms}$$

$$9(y^2 + 6y + 9) - 1(x^2 - 4x + 4) = -68 + 81 + (-4) \quad \text{complete the square}$$

adds $9(9) = 81$ adds $-1(4) = -4$ add $81 + (-4)$ to right

$$9(y+3)^2 - 1(x-2)^2 = 9 \quad \text{factored form} \rightarrow \text{vertical hyperbola}$$

$$\frac{(y+3)^2}{1} - \frac{(x-2)^2}{9} = 1 \quad \text{divide by 9 (standard form)}$$

$$\frac{(y+3)^2}{1^2} - \frac{(x-2)^2}{3^2} = 1 \quad \text{write denominators in squared form}$$

The center of the hyperbola is $(2, -3)$, with $a = 3$, $b = 1$, and a transverse axis of $x = 2$. The vertices are at $(2, -3 + 1)$ and $(2, -3 - 1) \rightarrow (2, -2)$ and $(2, -4)$. After plotting the center and vertices, we draw a rectangle centered at $(2, -3)$ with a horizontal "width" of $2(3) = 6$ and a vertical "length" of $2(1) = 2$ to sketch the asymptotes. The completed graph is given in the figure.

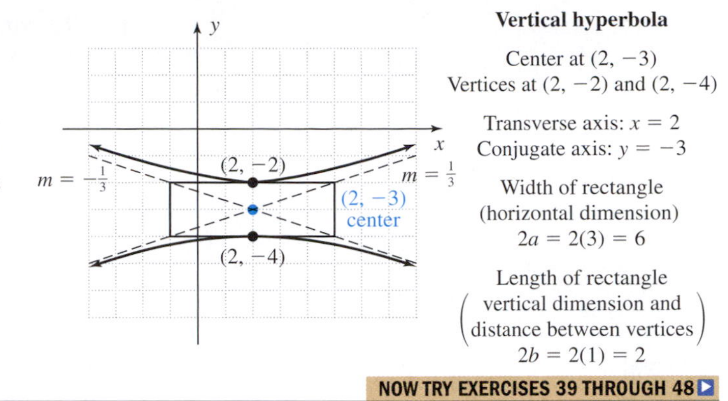

NOW TRY EXERCISES 39 THROUGH 48

B. Distinguishing between the Equations of a Circle, Ellipse, and Hyperbola

So far we've explored numerous graphs of circles, ellipses, and hyperbolas. In Example 5 we'll attempt to identify a given conic section from its equation alone (without graphing the equation). As you've seen, the corresponding equations have unique characteristics that can help distinguish one from the other.

EXAMPLE 5 Identify each equation as that of a circle, ellipse, or hyperbola. Justify your choice and name the center, but do not draw the graphs.

a. $y^2 = 36 + 9x^2$ **b.** $4x^2 = 16 - 4y^2$

c. $x^2 = 225 - 25y^2$ **d.** $25x^2 = 100 + 4y^2$

e. $3(x - 2)^2 + 4(y + 3)^2 = 12$

f. $4(x + 5)^2 = 36 + 9(y - 4)^2$

Solution:

a. Writing the equation in factored form gives $y^2 - 9x^2 = 36$ $(h = 0, k = 0)$. Since the equation contains a difference of second-degree terms, it is the equation of a (vertical) hyperbola. The center is at $(0, 0)$.

b. Rewriting the equation as $4x^2 + 4y^2 = 16$ and dividing by 4 gives $x^2 + y^2 = 4$. The equation represents a circle of radius 2, with the center at $(0, 0)$.

c. Writing the equation as $x^2 + 25y^2 = 225$ we note a sum of second-degree terms with unequal coefficients. The equation is that of an ellipse, with the center at $(0, 0)$.

d. Rewriting the equation as $25x^2 - 4y^2 = 100$ we note the equation contains a difference of second-degree terms. The equation represents a central (horizontal) hyperbola, whose center is at $(0, 0)$.

e. The equation is in factored form and contains a sum of second-degree terms with unequal coefficients. This is the equation of an ellipse with the center at $(2, -3)$.

f. Rewriting the equation as $4(x + 5)^2 - 9(y - 4)^2 = 36$, we note a difference of second-degree terms. The equation represents a horizontal hyperbola with center $(-5, 4)$.

NOW TRY EXERCISES 49 THROUGH 60

TECHNOLOGY HIGHLIGHT

Using a Graphing Calculator to Study Hyperbolas

The keystrokes shown apply to a TI-84 Plus model. Please consult our Internet site or your manual for other models.

As with the circle and ellipse, the hyperbola must also be defined in two pieces in order to use a graphing calculator to study its graph. Consider the *relation* $4x^2 - 9y^2 = 36$. From our work in this section, we know this is the equation of a horizontal hyperbola centered at (0, 0). Solving for y gives:

$$4x^2 - 9y^2 = 36 \quad \text{original equation}$$
$$-9y^2 = 36 - 4x^2 \quad \text{isolate } y^2\text{-term}$$
$$y^2 = \frac{36 - 4x^2}{-9} \quad \text{divide by } -9$$
$$y = \pm\sqrt{\frac{36 - 4x^2}{-9}} \quad \text{solve for } y$$

We can again separate this result into two parts: $Y_1 = \sqrt{\frac{36 - 4x^2}{-9}}$ gives the "upper half" of the hyperbola, and $Y_2 = -\sqrt{\frac{36 - 4x^2}{-9}}$ gives the "lower half."

Entering these on the Y= screen, graphing them on the standard screen, and pressing the TRACE key gives the graph shown in Figure 6.14 (this time the standard screen gives a fairly nice graph of the function, even though the y-values are still compressed). Note the location of the cursor at $x = 0$, but no y-value is displayed.

Figure 6.14

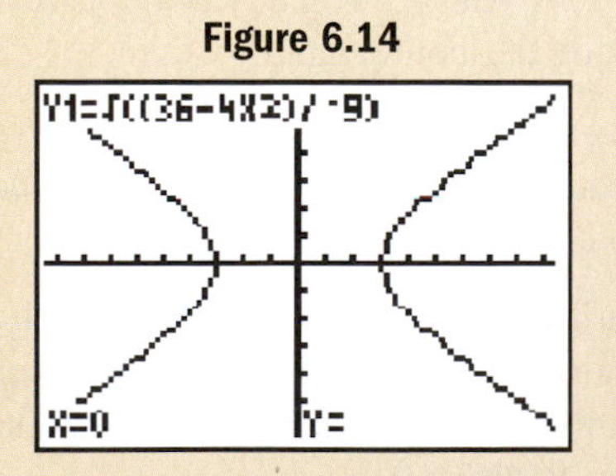

This is because the hyperbola is not defined at $x = 0$. Press the right arrow key ▶ and walk the cursor to the right until the y-values begin appearing. In fact, they begin to appear at (3, 0), which is one of the vertices of the hyperbola. We could also graph the asymptotes for this hyperbola ($y = \pm\frac{2}{3}x$) by entering the lines as Y_3 and Y_4 on the Y= screen. The resulting graph is shown in Figure 6.15 (the TRACE key has been pushed and the down arrow used to highlight Y_2). Use these ideas and the features of your graphing calculator to complete the following exercises.

Figure 6.15

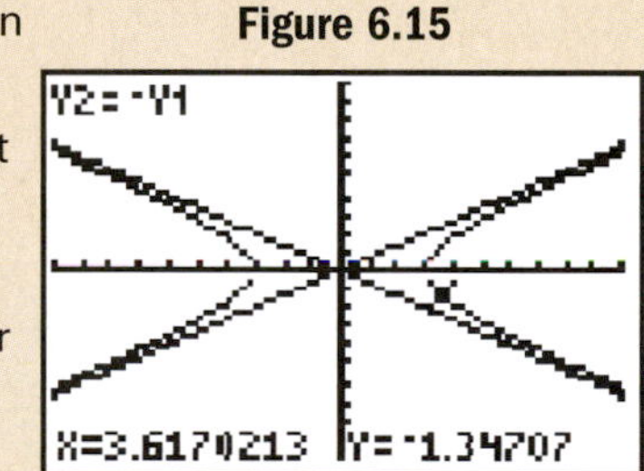

Exercise 1: Graph the hyperbola defined by $25y^2 - 4x^2 = 100$ using a friendly window. What are the coordinates of the vertices of this hyperbola? Use the TRACE feature to find the value(s) of y when $x = 4$. Determine (from the graph) the value(s) of y when $x = -4$, then verify your response using the TABLE feature of your calculator.

Exercise 2: Graph the hyperbola defined by $9x^2 - 16y^2 = 144$ using the standard window. Then determine the equations of the asymptotes and graph these as well. Why do the asymptotes of this hyperbola intersect at the origin? When will the asymptotes of a hyperbola *not* intersect at the origin?

6.2 EXERCISES

CONCEPTS AND VOCABULARY

Fill in the blank with the appropriate word or phrase. Carefully reread the section if needed.

1. The line that passes through the vertices of a hyperbola is called the ________ axis.

2. The conjugate axis is ________ to the ________ axis and contains the ______ of the hyperbola.

3. The center of a hyperbola is located ______ between the vertices.

4. The center of the hyperbola defined by $\frac{(x-2)^2}{4^2} - \frac{(y-3)^2}{5^2} = 1$ is at _____.

5. Compare/contrast the two methods used to find the asymptotes of a hyperbola. Include an example illustrating both methods.

6. Explore/explain why $A(x-h)^2 - B(y-k)^2 = F$ results in a hyperbola regardless of whether $A = B$ or $A \neq B$. Illustrate with an example.

DEVELOPING YOUR SKILLS

Graph each hyperbola. Label the center, vertices, and any additional points used.

7. $\frac{x^2}{4} - \frac{y^2}{9} = 1$ **8.** $\frac{x^2}{16} - \frac{y^2}{9} = 1$ **9.** $\frac{x^2}{4} - \frac{y^2}{9} = 1$ **10.** $\frac{x^2}{25} - \frac{y^2}{16} = 1$

11. $\frac{x^2}{49} - \frac{y^2}{16} = 1$ **12.** $\frac{x^2}{25} - \frac{y^2}{9} = 1$ **13.** $\frac{x^2}{36} - \frac{y^2}{16} = 1$ **14.** $\frac{x^2}{81} - \frac{y^2}{16} = 1$

15. $\frac{y^2}{9} - \frac{x^2}{1} = 1$ **16.** $\frac{y^2}{1} - \frac{x^2}{4} = 1$ **17.** $\frac{y^2}{12} - \frac{x^2}{4} = 1$ **18.** $\frac{y^2}{9} - \frac{x^2}{18} = 1$

19. $\frac{y^2}{9} - \frac{x^2}{9} = 1$ **20.** $\frac{y^2}{4} - \frac{x^2}{4} = 1$ **21.** $\frac{y^2}{36} - \frac{x^2}{25} = 1$ **22.** $\frac{y^2}{16} - \frac{x^2}{4} = 1$

For the graphs given, state the location of the vertices and the equation of the transverse axis. Then identify the location of the center and the equation of the conjugate axis. Assume all coordinates are lattice points. Note the scale used on each axis.

23. **24.** **25.** **26.**

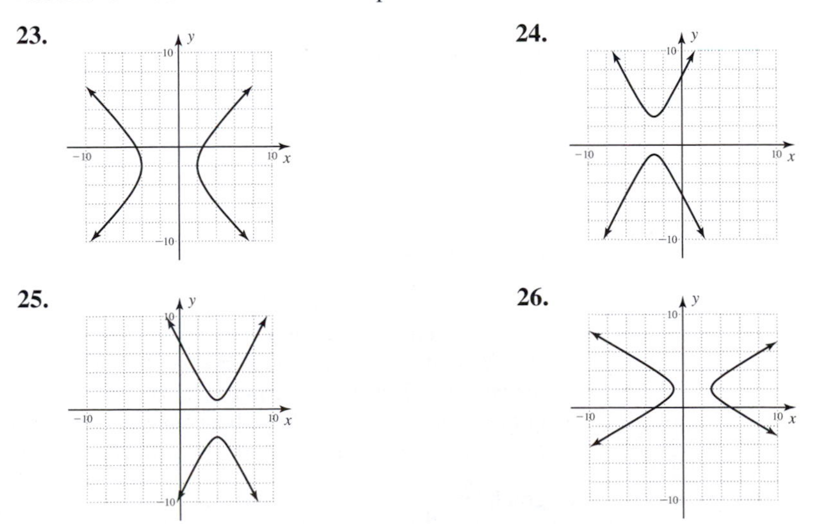

Sketch a complete graph of each equation, including the asymptotes. Be sure to identify the center and vertices.

27. $\frac{(y+1)^2}{4} - \frac{x^2}{25} = 1$

28. $\frac{y^2}{4} - \frac{(x-2)^2}{9} = 1$

29. $\frac{(x-3)^2}{36} - \frac{(y+2)^2}{49} = 1$

30. $\frac{(x-2)^2}{9} - \frac{(y-1)^2}{4} = 1$

31. $\frac{(y+1)^2}{7} - \frac{(x+5)^2}{9} = 1$

32. $\frac{(y-3)^2}{16} - \frac{(x+2)^2}{5} = 1$

33. $(x-2)^2 - 4(y+1)^2 = 16$

34. $9(x+1)^2 - (y-3)^2 = 81$

35. $2(y+3)^2 - 5(x-1)^2 = 50$

36. $9(y-4)^2 - 5(x-3)^2 = 45$

37. $12(x-4)^2 - 5(y-3)^2 = 60$

38. $8(x-4)^2 - 3(y-3)^2 = 24$

39. $16x^2 - 9y^2 = 144$

40. $16x^2 - 25y^2 = 400$

41. $9y^2 - 4x^2 = 36$

42. $25y^2 - 4x^2 = 100$

43. $12x^2 - 9y^2 = 72$

44. $36x^2 - 20y^2 = 180$

45. $4x^2 - y^2 + 40x - 4y + 60 = 0$

46. $x^2 - 4y^2 - 12x - 16y + 16 = 0$

47. $x^2 - 4y^2 - 24y - 4x - 36 = 0$

48. $-9x^2 + 4y^2 - 18x - 24y - 9 = 0$

Classify each equation as that of a circle, ellipse, or hyperbola. Justify your response.

49. $-4x^2 - 4y^2 = -24$

50. $9y^2 = -4x^2 + 36$

51. $x^2 + y^2 = 2x + 4y + 4$

52. $x^2 = y^2 + 6y - 7$

53. $2x^2 - 4y^2 = 8$

54. $36x^2 + 25y^2 = 900$

55. $x^2 + 5 = 2y^2$

56. $x + y^2 = 3x^2 + 9$

57. $2x^2 = -2y^2 + x + 20$

58. $2y^2 + 3 = 6x^2 + 8$

59. $16x^2 + 5y^2 - 3x + 4y = 538$

60. $9x^2 + 9y^2 - 9x + 12y + 4 = 0$

WORKING WITH FORMULAS

61. Equation of a semi-hyperbola: $y = \sqrt{\frac{36 - 4x^2}{-9}}$

The "upper half" of a certain hyperbola is given by the equation shown. (a) Simplify the radicand, (b) state the domain of the expression, and (c) enter the expression as Y_1 on a graphing calculator and graph. What is the equation for the "lower half" of this hyperbola?

62. Focal chord of a hyperbola: $L = \frac{2b^2}{a}$

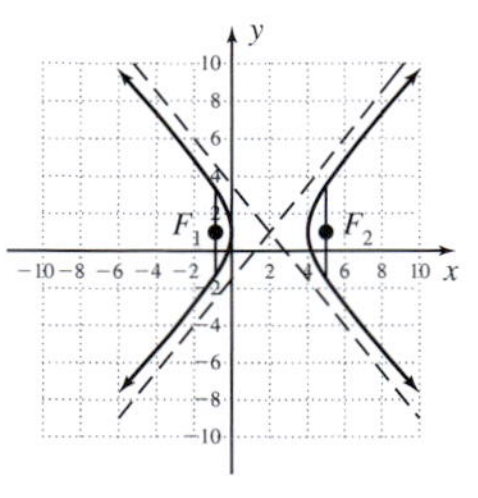

The focal chords of a hyperbola are line segments parallel to the conjugate axis with endpoints on the hyperbola, and containing certain points F_1 and F_2 called the *foci* (see grid). The length of the chord is given by the formula shown. Use it to find the length of the focal chord for the hyperbola indicated, then compare the calculated value with the length estimated from the given graph: $\frac{(x-2)^2}{4} - \frac{(y-1)^2}{5} = 1$.

APPLICATIONS

63. **Stunt pilots:** At an air show, a stunt plane dives along a hyperbolic path whose vertex is directly over the grandstands. If the plane's flight path can be modeled by the hyperbola $25y^2 - 1600x^2 = 40{,}000$, what is the minimum altitude of the plane as it passes over the stands? Assume x and y are in yards.

64. **Flying clubs:** To test their skill as pilots, the members of a flight club attempt to drop sandbags on a target placed in an open field, by diving along a hyperbolic path whose vertex is directly over the target area. If the flight path of the plane flown by the club's president is modeled by $9y^2 - 16x^2 = 14{,}400$, what is the minimum altitude of her plane as it passes over the target? Assume x and y are in feet.

65. **Nuclear cooling towers:** The natural draft cooling towers for nuclear power stations are called *hyperboloids of one sheet.* The perpendicular cross sections of these hyperboloids form two branches of a hyperbola. Suppose the central cross section of one such tower is modeled by the hyperbola $1600x^2 - 400(y - 50)^2 = 640{,}000$. What is the minimum distance between the sides of the tower? Assume x and y are in feet.

66. **Charged particles:** It has been shown that when like particles with a common charge are hurled at each other, they deflect and travel along paths that are hyperbolic. Suppose the path of two such particles is modeled by the hyperbola $x^2 - 9y^2 = 36$. What is the minimum distance between the particles as they approach each other? Assume x and y are in microns.

WRITING, RESEARCH, AND DECISION MAKING

67. Hyperbolas such as $x^2 - y^2 = 9$ and $y^2 - x^2 = 9$ are referred to as **conjugate hyperbolas.** Graph both on the same grid and discuss why the name is appropriate. Compare and contrast the equations and graphs, then give the equations of two other conjugate hyperbolas.

68. Referring to the introduction and Figure 6.9, it is possible for the plane to intersect only the vertex of the cone or to be tangent to the sides. These are called **degenerate cases** of a conic section. Many times we're unable to tell if the equation represents a degenerate case until it's written in factored or standard form. For example, write the following equations in standard form and comment.

    a. $4x^2 - 32x - y^2 + 4y + 60 = 0$ **b.** $x^2 - 4x + 5y^2 - 40y + 84 = 0$

69. LORAN is a long distance radio-navigation system for ships and aircraft, developed and deployed extensively during World War II. Using any of the resources available to you, determine how this system uses hyperbolas to pinpoint the location of a ship or aircraft. Submit a detailed report that includes diagrams and examples of how LORAN works.

70. For a greater understanding as to *why* the branches of a hyperbola are asymptotic, solve the basic equation $\frac{x^2}{a^2} - \frac{y^2}{b^2} = 1$ for y, then consider what happens as $x \to \infty$ (note that $x^2 - k \approx x^2$ for large x).

EXTENDING THE CONCEPT

71. Find the equation of the circle that shares the same center as the hyperbola given, if the vertices of the hyperbola are on the circle: $9(x - 2)^2 - 25(y - 3)^2 = 225$.

72. Find the equation of the ellipse that shares the same center as the hyperbola given, if the length of the minor axis is equal to the height of the central rectangle, and the hyperbola and ellipse share the same vertices: $9(x - 2)^2 - 25(y - 3)^2 = 225$.

73. Which has a greater area: (a) The central rectangle of the hyperbola given by $(x - 5)^2 - (y + 4)^2 = 57$, (b) the circle given by $(x - 5)^2 + (y + 4)^2 = 57$, or (c) the ellipse given by $9(x - 5)^2 + 10(y + 4)^2 = 570$?

MAINTAINING YOUR SKILLS

74. (5.3) At 2000 m from the base of a mountain, the angle of elevation to its peak is 40°. If the slope of the mountain itself is 72°, how tall is the mountain?

40° 72° h 2000 m

75. (2.4) The table shown gives the average rainfall on a tropical island for selected months. Use your calculator to find a sinusoidal equation model for the data, then use the equation to determine the number of months in a year that rainfall is less than 2 in.

Month (Jan = 1)	Rainfall (inches)
1	4.3
3	5.5
5	4.0
7	1.7
9	0.5
11	1.9

76. (5.8) Use De Moivre's theorem to find the value of $(-2 + 2i)^5$.

77. (4.3) Solve the equation for $x \in [0, 360°)$: $|\tan x| = \sqrt{3}$.

78. (3.4) Find $\sin(2t)$ and $\cos(2t)$ given $\cos t = \frac{12}{13}$ with t in Quadrant IV.

79. (5.8) Use the nth roots theorem to find all solutions (real and complex) to $x^3 - 729 = 0$.

6.3 Foci and the Analytic Ellipse and Hyperbola

LEARNING OBJECTIVES

In Section 6.3 you will learn how to:

A. Locate the foci of an ellipse and use the foci and other features to construct the equation of an ellipse

B. Locate the foci of a hyperbola and use the foci and other features to construct the equation of a hyperbola

C. Solve applications involving foci

INTRODUCTION

Previously, we developed equations for the ellipse and hyperbola by looking at changes in the factored form of the equation of a circle. While this development sheds some light on their equations and related graphs, it limited our ability to use these conics in some significant ways. In this section we develop the equation of the ellipse and hyperbola from their analytic definition.

POINT OF INTEREST

Until the time of Johannes Kepler (1571–1630) astronomers assumed, for philosophical and aesthetic reasons, that all heavenly bodies moved in circular orbits. However, Kepler noted that the careful planetary observations of Tycho Brahe (1546–1601) could not be explained or predicted by such orbits. After years of careful study and searching, Kepler discovered that planetary orbits are actually elliptical, a result he published in his book *New Astronomy* in 1609. This is now referred to as Kepler's first law. Additional discoveries followed soon after. Kepler's second law states that a line joining the planet and the Sun sweeps out equal areas in equal intervals of time, meaning a planet moves slower near its aphelion, and very fast near its perihelion.

A. The Foci of an Ellipse

The Museum of Science and Industry in Chicago, Illinois (http://www.msichicago.org), has a permanent exhibit called the *Whispering Gallery.* The construction of the room is based on some of the reflective properties of an ellipse. If two people stand at designated points in the room and one of them whispers very softly, the other person can hear the whisper quite clearly—even though they are over 40 ft apart! The point at which each person stands is called a **focus** of the ellipse (together they are called the **foci**). This reflective property also applies to light and radiation, giving the ellipse some powerful applications in science, medicine, acoustics, and other areas. To understand and appreciate these applications, we introduce the analytic definition of an ellipse.

WORTHY OF NOTE

You can easily draw an ellipse that satisfies the definition. Press two pushpins (these form the foci of the ellipse) halfway down into a piece of heavy cardboard about 6 in. apart. Take an 8-in. piece of string and loop each end around the pins. Use a pencil to draw the string taut and keep it taut as you move the pencil in a circular motion—and the result is an ellipse! A different length of string or a different distance between the foci will produce a different ellipse.

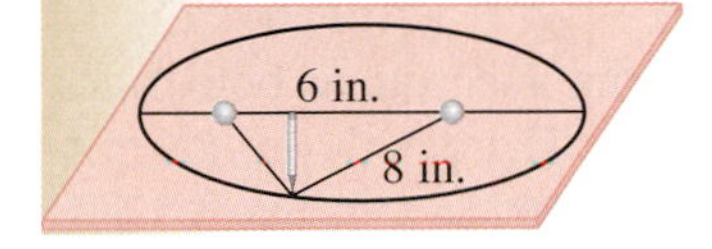

DEFINITION OF AN ELLIPSE

Given any two fixed points F_1 and F_2 in a plane, an ellipse is defined to be the set of all points $P(x, y)$ such that the distance $|F_1P|$ added to the distance $|F_2P|$ remains constant. In symbols,

$$|F_1P| + |F_2P| = k$$

The fixed points F_1 and F_2 are called the *foci* of the ellipse, and the points $P(x, y)$ are points on the graph of the ellipse.

The equation of an ellipse is obtained by combining the definition just given with the distance formula. Consider the general ellipse shown in Figure 6.16 (for calculating ease we use a central ellipse). Note the vertices have coordinates $(-a, 0)$ and $(a, 0)$, and the endpoints of the minor axis have coordinates $(0, -b)$ and $(0, b)$ as before. It is customary to assign foci the coordinates $F_1 \rightarrow (-c, 0)$ and $F_2 \rightarrow (c, 0)$. We can calculate the distance between $(c, 0)$ and any point $P(x, y)$ on the ellipse using the distance formula: $\sqrt{(x - c)^2 + (y - 0)^2}$. Likewise the distance between $(-c, 0)$ and any point (x, y) is $\sqrt{(x + c)^2 + (y - 0)^2}$. According to the definition, the sum must be constant: $\sqrt{(x - c)^2 + (y - 0)^2} + \sqrt{(x + c)^2 + (y - 0)^2} = k$.

Figure 6.16

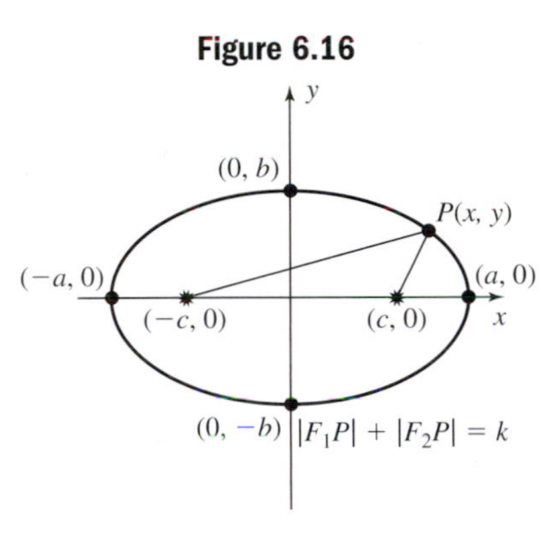

EXAMPLE 1 Use the definition of an ellipse and the diagram given to determine the "length of the string" used to form this ellipse (also see the *Worthy of Note*). Note that $a = 5$, $b = 3$, and $c = 4$.

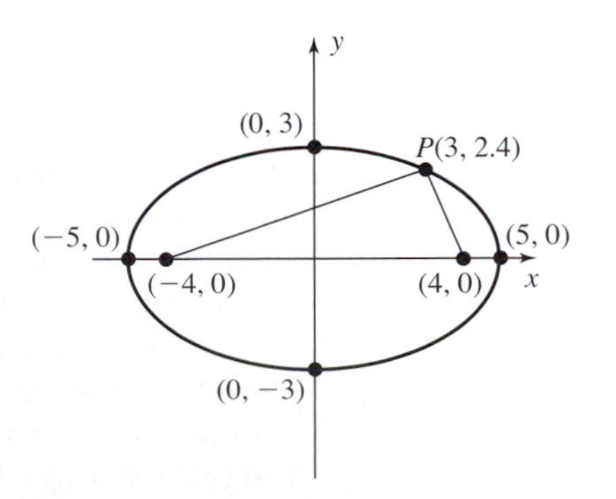

WORTHY OF NOTE

Note that if the foci are coincident (both at the origin) the "ellipse" will actually be a circle with radius $\frac{k}{2}$; $\sqrt{x^2 + y^2} + \sqrt{x^2 + y^2} = k$ leads to $x^2 + y^2 = \frac{k^2}{4}$. In Example 1 we found $k = 10$, giving $\frac{10}{2} = 5$, and if we used the "string" to draw the circle, the pencil would be 5 units from the center, creating a circle of radius 5.

Solution:

$$\sqrt{(x - c)^2 + (y - 0)^2} + \sqrt{(x + c)^2 + (y - 0)^2} = k \quad \text{given}$$
$$\sqrt{(3 - 4)^2 + (2.4 - 0)^2} + \sqrt{(3 + 4)^2 + (2.4 - 0)^2} = k \quad \text{substitute}$$
$$\sqrt{(-1)^2 + 2.4^2} + \sqrt{7^2 + 2.4^2} = k \quad \text{add}$$
$$\sqrt{6.76} + \sqrt{54.76} = k \quad \text{simplify radicals}$$
$$2.6 + 7.4 = k \quad \text{compute square roots}$$
$$10 = k \quad \text{result}$$

The "length of string" used to form this ellipse is 10 units long.

NOW TRY EXERCISES 7 THROUGH 10

In Example 1, the length of the string could also be found by moving the point P to the location of a vertex, then using the symmetry of the ellipse. This helps to show the constant k is equal to $2a$ *regardless of the distance between foci.* When $P(x, y)$ is coincident with vertex $(a, 0)$, the length of the "string" is identical to the length of the major axis, since the overlapping part of the string from $(c, 0)$ to $(a, 0)$ is the same length as from $(-a, 0)$ to $(-c, 0)$.

The result is an equation with two radicals:

$$\sqrt{(x - c)^2 + (y - 0)^2} + \sqrt{(x + c)^2 + (y - 0)^2} = 2a$$

To simplify this equation, we isolate one of the radicals and square both sides, then isolate the resulting radical expression and square again. The details are given in Appendix V, and the result is very close to the standard form we saw in Section 6.1:

$$\frac{x^2}{a^2} + \frac{y^2}{a^2 - c^2} = 1$$

By comparing the standard form $\frac{x^2}{a^2} + \frac{y^2}{b^2} = 1$ with $\frac{x^2}{a^2} + \frac{y^2}{a^2 - c^2} = 1$, we might suspect that $b^2 = a^2 - c^2$, and this is indeed the case. Note from Example 1 the relationship yields

$$b^2 = a^2 - c^2$$
$$3^2 = 5^2 - 4^2$$
$$9 = 25 - 16 \checkmark$$

Additionally, when we consider that $(0, b)$ is a point on the ellipse, the distance from $(0, b)$ to $(c, 0)$ *must be equal to a* due to symmetry (the "constant distance" used to form the ellipse is always $2a$). See Figure 6.17. The Pythagorean theorem (with a as the hypotenuse) gives $b^2 + c^2 = a^2$ or $b^2 = c^2 - a^2$.

While the equation $\frac{x^2}{a^2} + \frac{y^2}{b^2} = 1$ is identical to the one obtained in Section 6.1, we now have the ability to *locate the foci of any ellipse*—an important step toward using the ellipse in practical applications. Because we're often asked to find the location of the foci, it is best to remember the relationship as $c^2 = |a^2 - b^2|$, with the absolute value bars used to allow for a vertical major axis. Also note that for an ellipse,

Figure 6.17

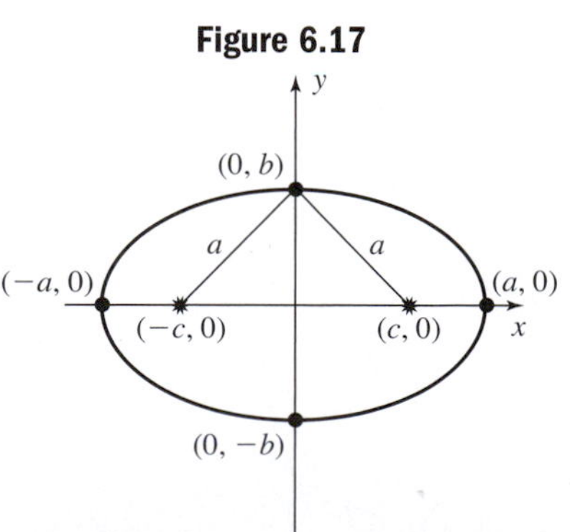

$c < a$ (*major axis horizontal*) or $c < b$ (*major axis vertical*). Note that a noncentral ellipse with center at (h, k) will still have an equation of $\frac{(x-h)^2}{a^2} + \frac{(y-k)^2}{b^2} = 1$.

EXAMPLE 2 For the ellipse defined by $25x^2 + 9y^2 - 100x - 54y - 44 = 0$, find the coordinates of the center, vertices, foci, and endpoints of the minor axis. Then sketch the graph.

Solution:

$$25x^2 + 9y^2 - 100x - 54y - 44 = 0 \quad \text{given}$$

$$25x^2 - 100x + 9y^2 - 54y = 44 \quad \text{group terms; add 44}$$

$$25(x^2 - 4x + \_\_) + 9(y^2 - 6y + \_\_) = 44 \quad \text{factor out lead coefficients}$$

$$25(x^2 - 4x + 4) + 9(y^2 - 6y + 9) = 44 + 100 + 81 \quad \text{add } 100 + 81 \text{ to right-hand side}$$

adds $25(4) = 100$ adds $9(9) = 81$

$$25(x-2)^2 + 9(y-3)^2 = 225 \quad \text{factored form}$$

$$\frac{25(x-2)^2}{225} + \frac{9(y-3)^2}{225} = \frac{225}{225} \quad \text{divide by 225}$$

$$\frac{(x-2)^2}{9} + \frac{(y-3)^2}{25} = 1 \quad \text{simplify (standard form)}$$

$$\frac{(x-2)^2}{3^2} + \frac{(y-3)^2}{5^2} = 1 \quad \text{write denominators in squared form}$$

The result shows a vertical ellipse with $a = 3$ and $b = 5$. The center of the ellipse is at $(2, 3)$. The vertices are a vertical distance of 5 units from center at $(2, 8)$ and $(2, -2)$. The endpoints of the minor axis are a horizontal distance of 3 units from center at $(-1, 3)$ and $(5, 3)$. To locate the foci, we use the foci formula for an ellipse: $c^2 = |a^2 - b^2|$, giving $c^2 = |3^2 - 5^2| = 16$. The result indicates the foci "✸" are located a vertical distance of 4 units from center at $(2, 7)$ and $(2, -1)$.

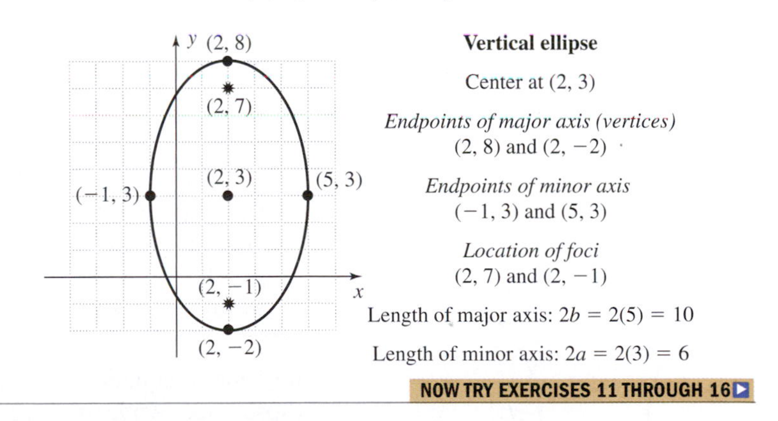

NOW TRY EXERCISES 11 THROUGH 16

For future reference, remember the foci of an ellipse always occur on the major axis, with $a > c$ and $a^2 > c^2$ for a horizontal ellipse. This makes it easier to remember the **foci formula** for ellipses: $c^2 = |a^2 - b^2|$. Since a^2 is larger, it must be decreased by b^2 to equal c^2.

If any two of the values for a, b, and c are known, the relationship between them can be used to construct the equation of the ellipse.

LOOKING AHEAD

For the hyperbola, we'll find that $c > a$, and the formula for the foci of a hyperbola will be $c^2 = a^2 + b^2$.

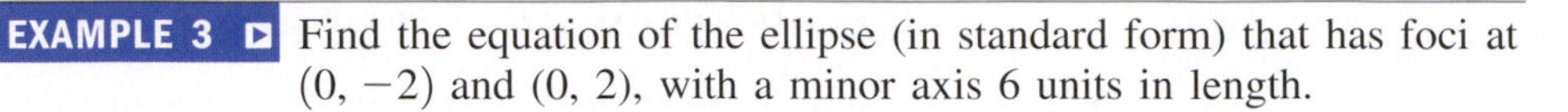

EXAMPLE 3 Find the equation of the ellipse (in standard form) that has foci at $(0, -2)$ and $(0, 2)$, with a minor axis 6 units in length.

Solution: Since the foci must be on the major axis, we know this is a vertical and central ellipse with $c = 2$ and $c^2 = 4$. The minor axis has a length of $2a = 6$ units, meaning $a = 3$ and $a^2 = 9$. To find b^2, use the foci equation and solve.

$$c^2 = |a^2 - b^2| \quad \text{foci equation (ellipse)}$$

$$4 = |9 - b^2| \quad \text{substitute}$$

$$-4 = 9 - b^2 \qquad 4 = 9 - b^2 \quad \text{solve}$$

$$b^2 = 13 \qquad b^2 = 5 \quad \text{result}$$

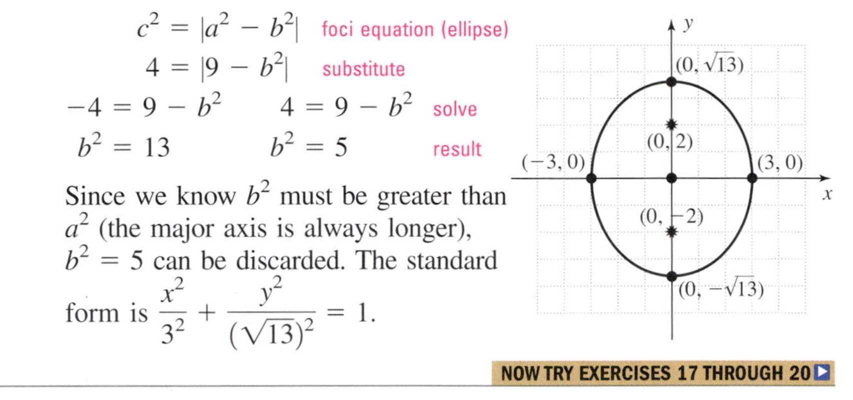

Since we know b^2 must be greater than a^2 (the major axis is always longer), $b^2 = 5$ can be discarded. The standard form is $\frac{x^2}{3^2} + \frac{y^2}{(\sqrt{13})^2} = 1$.

NOW TRY EXERCISES 17 THROUGH 20

B. The Foci of a Hyperbola

Like the ellipse, the foci of a hyperbola play an important part in their application. A long distance radio navigation system (called LORAN for short), can be used to determine the location of ships and airplanes and is based on the characteristics of a hyperbola (see Exercises 55 and 56). Hyperbolic mirrors are also used in some telescopes, and have the property that a beam of light directed at one focus will be reflected to the second focus. To understand and appreciate these applications, we use the analytic definition of a hyperbola:

DEFINITION OF A HYPERBOLA

Given any two fixed points F_1 and F_2 in a plane, a hyperbola is defined to be the set of all points $P(x, y)$ such that the distance $|F_2P|$ subtracted from the distance $|F_1P|$ is a positive constant. In symbols,

$$|F_1P| - |F_2P| = k, \; k > 0$$

The fixed points F_1 and F_2 are called the foci of the hyperbola, and the points $P(x, y)$ are points on the graph of the hyperbola.

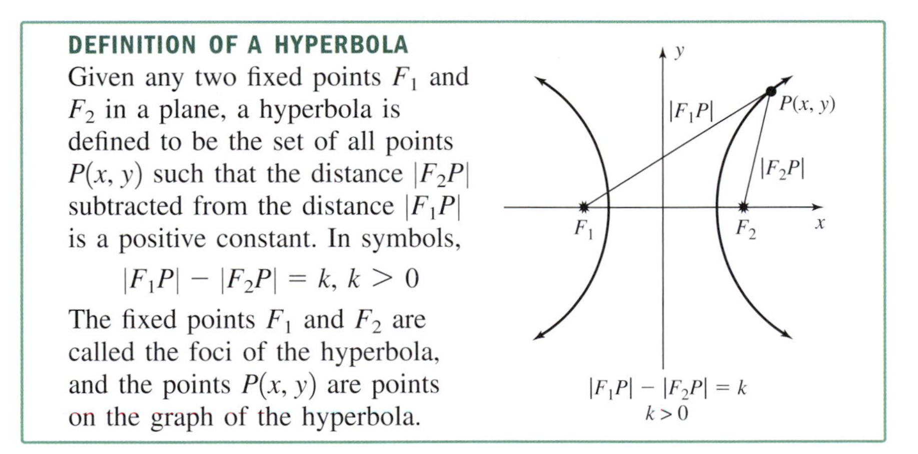

The general equation of a hyperbola is obtained in a manner similar to that of the ellipse (see Appendix V), and is the same as that given in Section 6.2: $\frac{x^2}{a^2} - \frac{y^2}{b^2} = 1$ (see Figure 6.18).

We now have the ability to *find the foci of any hyperbola*—and can use this information in many significant applications. Since the location of the foci play such an important role, it is best to remember the relationship as $c^2 = a^2 + b^2$ (called the **foci**

Figure 6.18

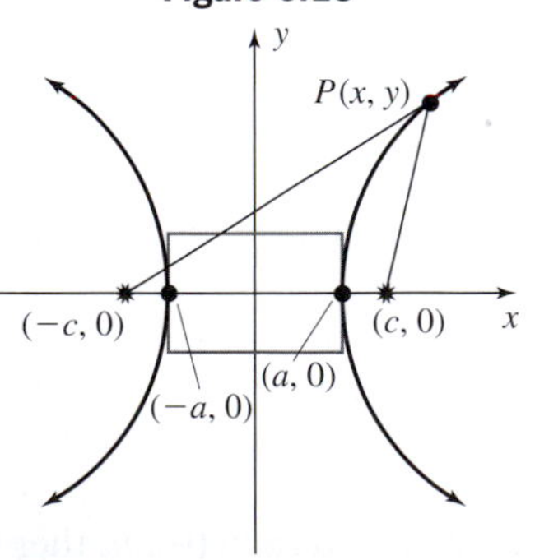

formula for hyperbolas), noting that for a hyperbola, $c > a$ and $c^2 > a^2$ (also $c > b$ and $c^2 > b^2$). Recall the asymptotes of any hyperbola can be found by counting off the slope ratio $\frac{\text{rise}}{\text{run}} = \pm\frac{b}{a}$ beginning at the center, or by drawing a central rectangle of dimensions $2a$ by $2b$ and using the extended diagonals of the rectangle. Again note that a noncentral (*horizontal*) hyperbola with center at (h, k) will still have an equation of

$$\frac{(x-h)^2}{a^2} - \frac{(y-k)^2}{b^2} = 1.$$

EXAMPLE 4 For the hyperbola defined by $7x^2 - 9y^2 - 14x + 72y - 200 = 0$, find the coordinates of the center, vertices, foci, and the dimensions of the central rectangle. Then sketch the graph.

Solution:

$$7x^2 - 9y^2 - 14x + 72y - 200 = 0 \quad \text{given}$$

$$7x^2 - 14x - 9y^2 + 72y = 200 \quad \text{group terms; add 200}$$

$$7(x^2 - 2x + \_\_) - 9(y^2 - 8y + \_\_) = 200 \quad \text{factor out lead coefficients}$$

$$7(x^2 - 2x + 1) - 9(y^2 - 8y + 16) = 200 + 7 + (-144) \quad \text{complete the square}$$

adds $7(1) = 7$; adds $-9(16) = -144$; $\rightarrow$ add $7 + (-144)$ to right-hand side

$$7(x-1)^2 - 9(y-4)^2 = 63 \quad \text{factored form}$$

$$\frac{(x-1)^2}{9} - \frac{(y-4)^2}{7} = 1 \quad \text{divide by 63 and simplify}$$

$$\frac{(x-1)^2}{3^2} - \frac{(y-4)^2}{(\sqrt{7})^2} = 1 \quad \text{write denominators in squared form}$$

From the result we find this is a horizontal hyperbola with $a = 3$ and $a^2 = 9$ and $b = \sqrt{7}$ and $b^2 = 7$. The center of the hyperbola is at $(1, 4)$. The vertices are a horizontal distance of 3 units from center at $(-2, 4)$ and $(4, 4)$. To locate the foci, we use the foci formula for a hyperbola: $c^2 = a^2 + b^2$. This yields $c^2 = 16$, showing the foci are located a horizontal distance of 4 units from center at $(-3, 4)$ and $(5, 4)$. The central rectangle is $2\sqrt{7} \approx 5.29$ by $2(3) = 6$. Draw the rectangle and sketch the asymptotes using the extended diagonals. The completed graph is shown in the figure.

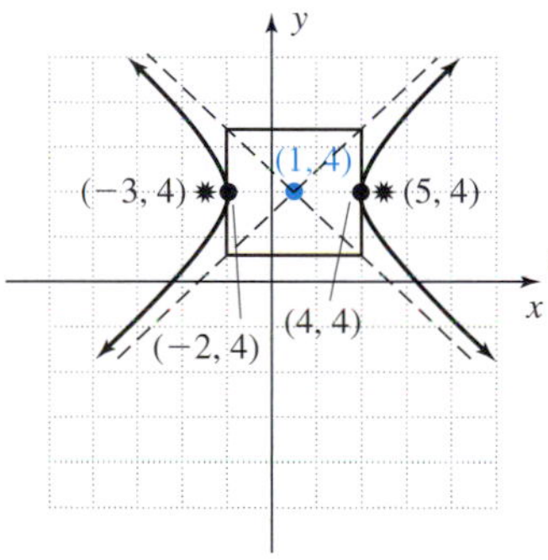

Horizontal hyperbola

Center at $(1, 4)$
Vertices at $(-2, 4)$ and $(4, 4)$

Transverse axis: $y = 4$
Conjugate axis: $x = 1$
Location of foci: $(-3, 4)$ and $(5, 4)$

Width of rectangle (horizontal dimension and distance between vertices)
$2a = 2(3) = 6$

Length of rectangle (vertical dimension)
$2b = 2(\sqrt{7}) \approx 5.29$

NOW TRY EXERCISES 21 THROUGH 30

As with the ellipse, if any two of the values for a, b, and c are known, the relationship between them can be used to construct the equation of the hyperbola. See Exercises 31 through 34.

C. Applications Involving Foci

Applications involving the foci of a conic section can take various forms. In many cases, only partial information about the ellipse or hyperbola is available and the ideas from Example 3 must be used to "fill in the gaps." In other applications, we must rewrite a known or given equation to find information related to the values of a, b, and c.

EXAMPLE 5 In Washington, D.C., there is a park called the *Ellipse* located between the White House and the Washington Monument. The park is surrounded by a path that forms an ellipse with the length of the major axis being about 1502 ft and the minor axis having a length of 1280 ft. Suppose the park manager wants to install water fountains at the location of the foci. Find the distance between the fountains rounded to the nearest foot.

Solution: Assume the center of the park has the coordinates (0, 0) and that the ellipse is horizontal. Since the major axis has length $2a = 1502$, we know $a = 751$ and $a^2 = 564{,}001$. The minor axis has length $2b = 1280$, meaning $b = 640$ and $b^2 = 409{,}600$. To find c, use the foci equation:

$$\begin{aligned} c^2 &= a^2 - b^2 \\ &= 564{,}001 - 409{,}600 \\ &= 154{,}401 \\ c \approx -393 \text{ and } c &\approx 393 \end{aligned}$$

The distance between the water fountains would be $2(393) = 786$ ft.

NOW TRY EXERCISES 49 THROUGH 54

TECHNOLOGY HIGHLIGHT

Graphing Calculators and the Definition of a Conic

The keystrokes shown apply to a TI-84 Plus model. Please consult our Internet site or your manual for other models.

Recall that if F_1 and F_2 are the foci of an ellipse and $P(x, y)$ is a point on the graph of the ellipse, then the distance from P to F_1 plus the distance from P to F_2 must be equal to some constant k regardless of the point chosen.

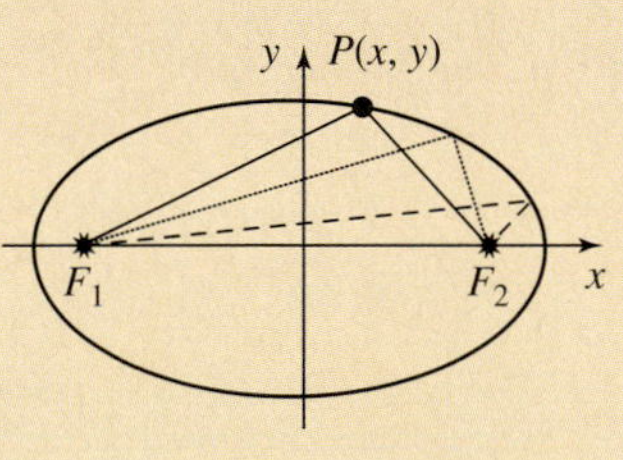

In this *Technology Highlight* we'll use lists L1 through L5, a horizontal ellipse, and the distance formula to check this definition for a select number of points on the ellipse. To begin, write the equation in standard form, clearly identify the values of a, b, and c, then solve for y and enter the positive root as Y_1 on the Y= screen (only the upper half of the ellipse is used for this exercise). For $4x^2 + 9y^2 = 36$, this leads to $\frac{x^2}{9} + \frac{y^2}{4} = 1$, giving $a = 3$, $b = 2$, and $c = \sqrt{9 - 4} = \sqrt{5}$. Solving for y and simplifying yields $y = \frac{2}{3}\sqrt{9 - x^2}$, which we enter as Y_1. Since the domain of this relation is $x \in [-3, 3]$,

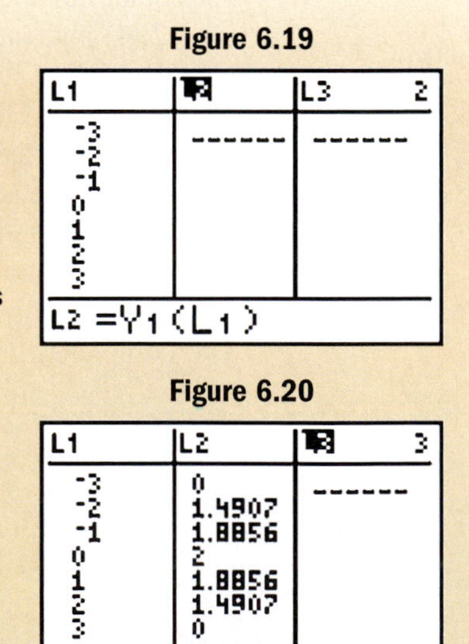

we enter the integers from this interval in L1, and the related y-values in L2, using L2 = Y_1(L1). Be sure the cursor is in the header of L2 as you begin (see Figure 6.19). Next we'll calculate the distance between the points on the ellipse stored as $(x, y) \rightarrow$ (L1, L2), and the foci located at $F_1 = (-\sqrt{5}, 0)$ and $F_2 = (\sqrt{5}, 0)$. For the distance between $(\sqrt{5}, 0)$ and (L1, L2) we'll use L3 = $\sqrt{(\text{L1} - \sqrt{5})^2 + (\text{L2} - 0)^2}$ (see Figure 6.20). For the distance between $(-\sqrt{5}, 0)$ and (L1, L2) we'll use L4 = $\sqrt{(\text{L1} + \sqrt{5})^2 + (\text{L2} - 0)^2}$. Figure 6.21 shows the results of these calculations. Finally, we compute the sum of these two distances using L5 = L3 + L4, noting that for all points the sum is equal to 6 (Figure 6.22), which is identical to $2a = 2(3)$.

Figure 6.21

L2	L3	L4
0	5.2361	.76393
1.4907	4.4907	1.5093
1.8856	3.7454	2.2546
2	3	3
1.8856	2.2546	3.7454
1.4907	1.5093	4.4907
0	.76393	5.2361

L2(1)=0

Figure 6.22

L3	L4	L5
5.2361	.76393	6
4.4907	1.5093	6
3.7454	2.2546	6
3	3	6
2.2546	3.7454	6
1.5093	4.4907	6
.76393	5.2361	6

L5(1)=6

Exercise 1: Rework the exercise using the ellipse $4x^2 + 25y^2 = 100$. What do you notice?

Exercise 2: Modify the exercise so that it checks the definition of a hyperbola. Use the hyperbola $9x^2 - 16y^2 = 144$ for verification.

6.3 EXERCISES

CONCEPTS AND VOCABULARY

Fill in the blank with the appropriate word or phrase. Carefully reread the section if needed.

1. For an ellipse, the relationship between a, b, and c is given by the foci equation __________, since $c < a$ or $c < b$.

2. For a hyperbola, the relationship between a, b, and c is given by the foci equation __________, since $c > a$ and $c > b$.

3. For a horizontal hyperbola, the length of the transverse axis is _____ and the length of the conjugate axis is _____.

4. For a vertical ellipse, the length of the minor axis is _____ and the length of the major axis is _____.

5. Suppose foci are located at $(-2, 5)$ and $(-2, -3)$. Discuss/explain the conditions necessary for the graph to be a hyperbola.

6. Suppose foci are located at $(-3, 2)$ and $(5, 2)$. Discuss/explain the conditions necessary for the graph to be an ellipse.

DEVELOPING YOUR SKILLS

Use the definition of an ellipse to find the length of the major axes (figures are not drawn to scale).

7. 8.

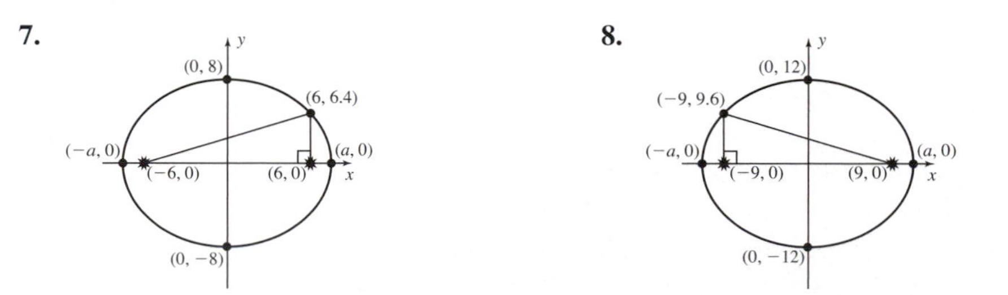

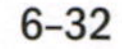

9.

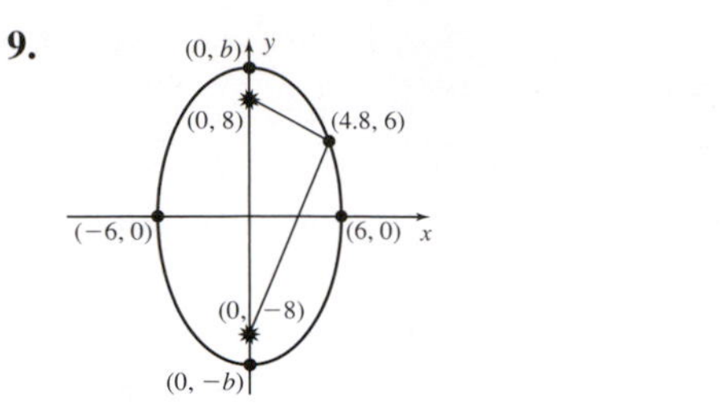

10.

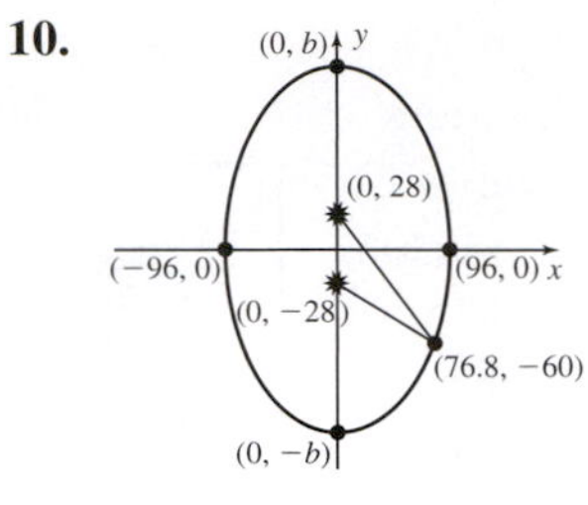

Find the coordinates of the (a) center, (b) vertices, (c) foci, and (d) endpoints of the minor axis. Then (e) sketch the graph.

11. $4x^2 + 25y^2 - 16x - 50y - 59 = 0$

12. $9x^2 + 16y^2 - 54x - 64y + 1 = 0$

13. $25x^2 + 16y^2 - 200x + 96y + 144 = 0$

14. $49x^2 + 4y^2 + 196x - 40y + 100 = 0$

15. $6x^2 + 24x + 9y^2 + 36y + 6 = 0$

16. $5x^2 - 50x + 2y^2 - 12y + 93 = 0$

Find the equation of an ellipse (in standard form) that satisfies the following conditions:

17. vertices at $(-6, 0)$ and $(6, 0)$;
foci at $(-4, 0)$ and $(4, 0)$

18. vertices at $(-8, 0)$ and $(8, 0)$;
foci at $(-5, 0)$ and $(5, 0)$

19. foci at $(0, -4)$ and $(0, 4)$;
length of minor axis: 6 units

20. foci at $(-6, 0)$ and $(6, 0)$;
length of minor axis: 8 units

Use the definition of a hyperbola to find the distance between the vertices and the dimensions of the rectangle centered at (h, k). Figures are not drawn to scale. Note that Exercises 23 and 24 are *vertical hyperbolas*.

21. **22.** **23.** **24.**

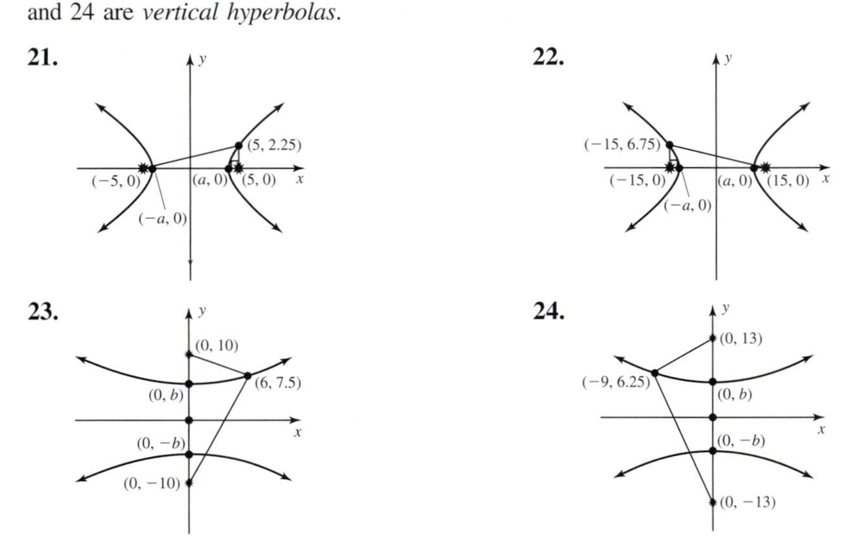

Find and list the coordinates of the (a) center, (b) vertices, (c) foci, and (d) dimensions of the central rectangle. Then (e) sketch the graph, including the asymptotes.

25. $4x^2 - 9y^2 - 24x + 72y - 144 = 0$

26. $4x^2 - 36y^2 - 40x + 144y - 188 = 0$

27. $16x^2 - 4y^2 + 24y - 100 = 0$

28. $81x^2 - 162x - 4y^2 - 243 = 0$

29. $9x^2 - 3y^2 - 54x - 12y + 33 = 0$

30. $10x^2 + 60x - 5y^2 + 20y - 20 = 0$

Find the equation of the hyperbola (in standard form) that satisfies the following conditions:

31. vertices at $(-6, 0)$ and $(6, 0)$;
foci at $(-8, 0)$ and $(8, 0)$

32. vertices at $(-4, 0)$ and $(4, 0)$;
foci at $(-6, 0)$ and $(6, 0)$

33. foci at $(0, -3\sqrt{2})$ and $(0, 3\sqrt{2})$;
length of conjugate axis: 6 units

34. foci at $(-6, 0)$ and $(6, 0)$;
length of conjugate axis: 8 units

Find the coordinates of the foci for the conic sections defined by the equations given. Note that both ellipses and hyperbolas are represented.

35. $\frac{x^2}{49} + \frac{y^2}{4} = 1$ **36.** $\frac{x^2}{9} + \frac{y^2}{4} = 1$ **37.** $\frac{x^2}{9} + \frac{y^2}{25} = 1$ **38.** $\frac{x^2}{16} + \frac{y^2}{36} = 1$

39. $\frac{x^2}{18} + \frac{y^2}{12} = 1$ **40.** $\frac{x^2}{20} + \frac{y^2}{8} = 1$ **41.** $\frac{x^2}{4} - \frac{y^2}{9} = 1$ **42.** $\frac{x^2}{25} - \frac{y^2}{16} = 1$

43. $\frac{y^2}{36} - \frac{x^2}{25} = 1$ **44.** $\frac{y^2}{16} - \frac{x^2}{4} = 1$ **45.** $\frac{x^2}{28} - \frac{y^2}{32} = 1$ **46.** $\frac{x^2}{40} - \frac{y^2}{20} = 1$

WORKING WITH FORMULAS

47. The eccentricity of a conic: $e = \frac{c}{a}$

In lay terms, the eccentricity of a conic section is a measure of its "roundness," or more exactly, how much the conic section deviates from being "round." A circle has an eccentricity of $e = 0$, since it is perfectly round. Ellipses have an eccentricity between zero and one, or $0 < e < 1$. An ellipse with $e = 0.16$ is closer to circular than one where $e = 0.72$. Here, c represents the distance from the center of the ellipse to its focus, and a represents the length of the semimajor axis (the distance from center to either vertex). Determine which of the following ellipses is closest to being circular: $\frac{x^2}{9} + \frac{y^2}{25} = 1$ or $\frac{x^2}{16} + \frac{y^2}{36} = 1$.

48. The perimeter of an ellipse: $P \approx 2\pi\sqrt{\frac{a^2 + b^2}{2}}$

The perimeter of an ellipse can be *approximated* by the formula shown, where a represents the length of the semimajor axis and b represents the length of the semiminor axis. Estimate the perimeter of the orbit of the planet Mercury, defined by the equation $\frac{x^2}{1296} + \frac{y^2}{1243} = 1$ (answer will be in millions of miles).

APPLICATIONS

Exercise 49

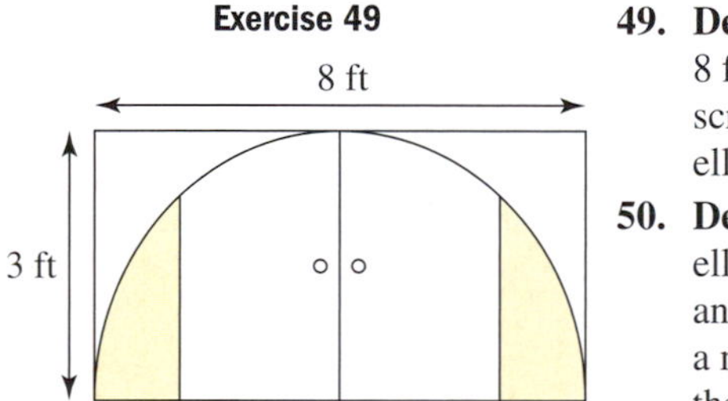

49. Decorative fireplaces: A bricklayer intends to build an elliptical fireplace 3 ft high and 8 ft wide, with two glass doors that open at the middle. The hinges to these doors are to be screwed onto a spine that is perpendicular to the hearth and goes through the foci of the ellipse. How far from center will the spines be located? What is the height of the spine?

50. Decorative gardens: A retired math teacher decides to present her husband with a beautiful elliptical garden to help celebrate their 50th anniversary. The ellipse is to be 8 m long and 5 m across, with decorative fountains located at the foci. To the nearest hundredth of a meter, how far from the center of the ellipse should the fountains be? How far apart are the fountains?

51. Attracting attention to art: As part of an art show, a gallery owner asks a student from the local university to design a unique exhibit that will highlight one of the more significant pieces in the collection, an ancient sculpture. The student decides to create an elliptical showroom with reflective walls, with a rotating laser light on a stand at one foci, and the sculpture placed at the other foci on a stand of equal height. The laser light then points continually at the sculpture as it rotates. If the elliptical room is 24 ft long and 16 ft wide, how far from the center of the ellipse should the stands be located (round to the nearest tenth of a foot)? How far apart are the stands?

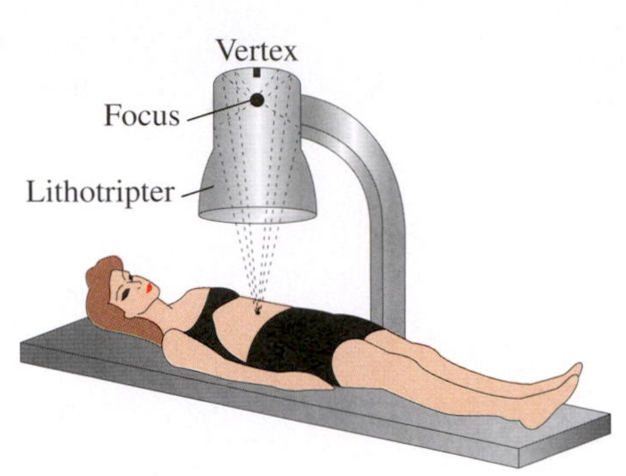

52. **Medical procedures:** The medical procedure called *lithotripsy* is a noninvasive medical procedure that is used to break up kidney and bladder stones in the body. A machine called a *lithotripter* uses its three-dimensional semielliptical shape and the foci properties of an ellipse to concentrate shock waves generated at one focus on a kidney stone located at the other focus (see diagram—not drawn to scale). If the lithotripter has a length (semimajor axis) of 16 cm and a radius (semiminor axis) of 10 cm, how far from the vertex should a kidney stone be located for the best result? Round to the nearest hundredth.

53. **Planetary orbits:** Except for small variations, a planet's orbit around the Sun is elliptical with the Sun at one foci. The aphelion (maximum distance from the Sun) of the planet Mars is approximately 156 million miles, while the perihelion (minimum distance from the sun) of Mars is about 128 million miles. Use this information to find the lengths of the semimajor and semiminor axes, rounded to the nearest million. If Mars has an orbital velocity of 54,000 mph (1.296 million miles per day), how many days does it take Mars to orbit the Sun? (*Hint:* Use the formula from Exercise 48.)

54. **Planetary orbits:** The aphelion (maximum distance from the Sun) of the planet Saturn is approximately 940 million miles, while the perihelion (minimum distance from the Sun) of Saturn is about 840 million miles. Use this information to find the lengths of the semimajor and semiminor axes, rounded to the nearest million. If Saturn has an orbital velocity of 21,650 mph (about 0.52 million miles per day), how many days does it take Saturn to orbit the Sun? How many years?

55. **Locating a ship using radar:** Under certain conditions, the properties of a hyperbola can be used to help locate the position of a ship. Suppose two radio stations are located 100 km apart along a straight shoreline. A ship is sailing parallel to the shore and is 60 km out to sea. The ship sends out a distress call that is picked up by the closer station in 0.4 milliseconds (msec—one-thousandth of a second), while it takes 0.5 msec to reach the station that is farther away. Radio waves travel at a speed of approximately 300 km/msec. Use this information to find the equation of a hyperbola that will help you find the location of the ship, then find the coordinates of the ship. (*Hint:* Draw the hyperbola on a coordinate system with the radio stations on the x-axis at the foci, then use the definition of a hyperbola.)

56. **Locating a plane using radar:** Two radio stations are located 80 km apart along a straight shoreline, when a "mayday" call (a plea for immediate help) is received from a plane that is about to ditch in the ocean (attempt a water landing). The plane was flying at low altitude, parallel to the shoreline, and 20 km out when it ran into trouble. The plane's distress call is picked up by the closer station in 0.1 msec, while it takes 0.3 msec to reach the other. Use this information to construct the equation of a hyperbola that will help you find the location of the ditched plane, then find the coordinates of the plane. Also see Exercise 55.

WRITING, RESEARCH, AND DECISION MAKING

Exercise 57

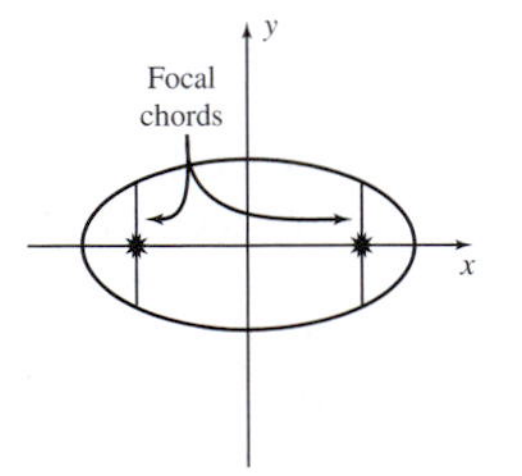

57. When graphing the conic sections, it is often helpful to use what is called a **focal chord,** as it gives additional points on the graph with very little effort. A focal chord is a line segment through a focus (perpendicular to the major or transverse axis), with the endpoints on the graph. For ellipses and hyperbolas, the length of the focal chord is given by $L = \frac{2b^2}{a}$, where a is a vertex. The focus will always be the midpoint of this line segment. Find the length of the focal chord for the hyperbola $\frac{x^2}{4} - \frac{y^2}{9} = 1$ and the coordinates of the endpoints.

Verify (by substituting into the equation) that these endpoints are indeed points on the graph, then use them to help complete the graph.

58. Using graph paper, draw a complete and careful graph of the hyperbola $9x^2 - 16y^2 = 144$. Be particularly sure that the central rectangle is carefully drawn and the foci are accurately located. What are the coordinates of the upper-right corner of this central rectangle? Use the Pythagorean theorem to find the distance from (0, 0) to this corner point. How does this distance compare to the distance from the center to the focus? What does this relationship have to do with the formula for finding the focus? Explain and discuss what you find.

EXTENDING THE CONCEPT

Exercise 59

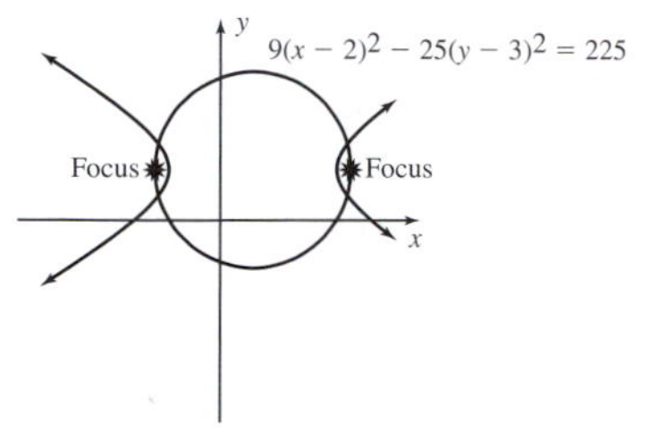

Exercise 60

$9(x-2)^2 - 25(y-3)^2 = 225$

59. Find the equation of the circle shown, given the equation of the hyperbola.

60. Find the equation of the ellipse shown, given the equation of the hyperbola and (2, 0) is on the graph of the ellipse. The hyperbola and ellipse share the same foci.

61. Verify that for the horizontal ellipse $\frac{x^2}{a^2} + \frac{y^2}{b^2} = 1$, the length of the focal chord is $\frac{2b^2}{a}$. Also see Exercise 57.

62. Verify that for a central hyperbola, a circle that circumscribes the central rectangle must also go through both foci.

MAINTAINING YOUR SKILLS

63. (5.4) A ship is traveling at 12 mph on a heading of 315°, when it encounters a 3-mph current with a heading of 30°. Find the true course and speed of the ship as it travels through the current.

64. (2.4) Use the graph shown to write an equation of the form $y = A\sec(Bx + C)$.

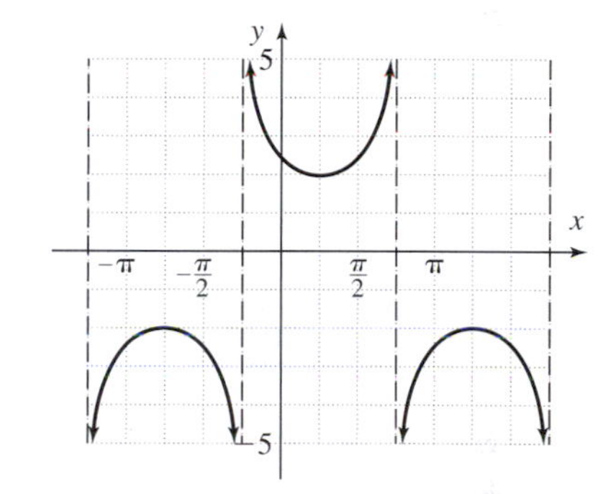

65. (3.3) Use a sum or difference identity to write the value of cos 105° in exact form.

66. (4.4) Find all values of $x \in [0, 2\pi)$ that make the equation true:
$1560 = 250\sin\left(2x - \frac{\pi}{6}\right) + 1735$. Round to four decimal places as needed.

67. (5.2) Determine the number of triangles that can be formed from the dimensions given here. If two triangles exist, solve both.

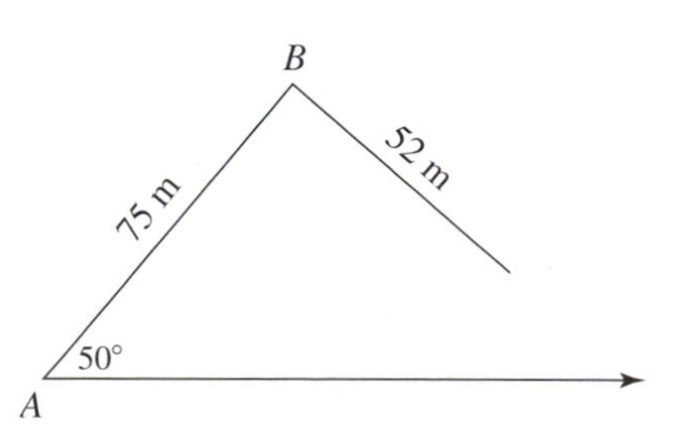

68. (1.1) The wheels on a motorcycle are rotating at 1000 rpms. If they have a 12 in. radius, how fast is the motorcycle travelling in miles per hour?

MID-CHAPTER CHECK

Sketch the graph of each conic section.

1. $(x - 4)^2 + (y + 3)^2 = 9$
2. $x^2 + y^2 - 10x + 4y + 4 = 0$
3. $\dfrac{(x - 2)^2}{16} + \dfrac{(y + 3)^2}{1} = 1$
4. $9x^2 + 4y^2 + 18x - 24y + 9 = 0$
5. $\dfrac{(x + 3)^2}{9} - \dfrac{(y - 4)^2}{4} = 1$
6. $9x^2 - 4y^2 + 18x - 24y - 63 = 0$
7. Find the equation of each relation and state its domain and range.

a.

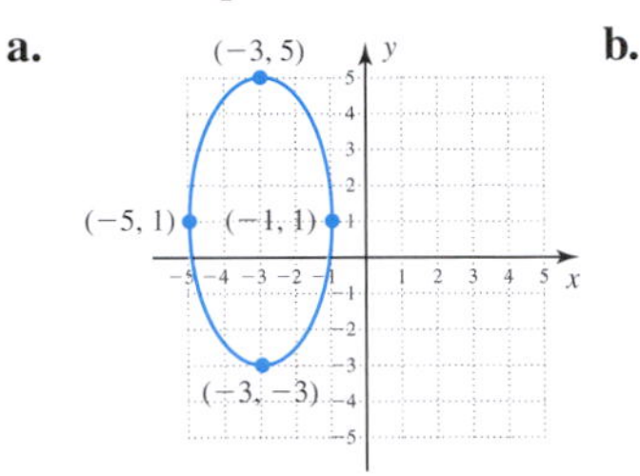

b.

(3, 6)
(−1, 2)
(7, 2)
(3, −2)

c.

(−√13, 0)
(√13, 0)
(−2, 0)
(2, 0)

8. Find the equation of the hyperbola having foci at (0, 13) and (0, −13), with a conjugate axis of length 10 units.
9. Find the equation of the ellipse (in standard form) if the vertices are (−4, 0) and (4, 0) and the distance between the foci is $4\sqrt{3}$ units.
10. The radio signal emanating from a tall radio tower spreads evenly in all directions with a range of 50 mi. If the tower is located at coordinates (20, 30) and my home is at coordinates (10, 78), will I be able to pick up this station on my home radio? Assume coordinates are in miles.

REINFORCING BASIC CONCEPTS

Ellipses and Hyperbolas with Rational/Irrational Values of *a* and *b*

Using the process known as completing the square, we were able to convert from the polynomial form of a conic section to the standard form. However, for some equations, values of a and b are somewhat difficult to identify, since the coefficients are not factors. Consider the equation $20x^2 + 120x + 27y^2 - 54y + 192 = 0$, the equation of an ellipse.

$$20x^2 + 120x + 27y^2 - 54y + 192 = 0 \qquad \text{original equation}$$
$$20(x^2 + 6x + \underline{\quad\quad}) + 27(y^2 - 2y + \underline{\quad\quad}) = -192 \qquad \text{subtract 192, begin process}$$
$$20(x^2 + 6x + 9) + 27(y^2 - 2y + 1) = -192 + 27 + 180 \qquad \text{complete the square in } x \text{ and } y$$
$$20(x + 3)^2 + 27(y - 1)^2 = 15 \qquad \text{factor and simplify}$$
$$\frac{4(x + 3)^2}{3} + \frac{9(y - 1)^2}{5} = 1 \qquad \text{standard form}$$

Unfortunately, we cannot easily identify the values of a and b, since the coefficients of each binomial square were not "1." In these cases, we can write the equation in standard form by using a simple property of fractions—the numerator and denominator of any fraction can be divided by the same quantity to obtain an equivalent fraction. Although the result may look odd, it can nevertheless be applied here, giving a result of $\dfrac{(x + 3)^2}{\frac{3}{4}} + \dfrac{(y - 1)^2}{\frac{5}{9}} = 1$. We can now identify a and b by writing these denominators in squared form, which gives the following expression: $\dfrac{(x+3)^2}{\left(\frac{\sqrt{3}}{2}\right)^2} + \dfrac{(y-1)^2}{\left(\frac{\sqrt{5}}{3}\right)^2} = 1$. The values of a and b are now seen to be $a = \frac{\sqrt{3}}{2} \approx 0.866$ and $b = \frac{\sqrt{5}}{3} \approx 0.745$. Use this idea to complete the following exercises.

Exercise 1: Identify the values of a and b by writing the equation $100x^2 - 400x - 18y^2 - 108y + 230 = 0$ in standard form.

Exercise 2: Identify the values of a and b by writing the equation $28x^2 - 56x + 48y^2 + 192y + 195 = 0$ in standard form.

Exercise 3: Write the equation in standard form, then identify the values of a and b and use them to graph the ellipse.

$$\frac{4(x+3)^2}{49} + \frac{25(y-1)^2}{36} = 1$$

Exercise 4: Write the equation in standard form, then identify the values of a and b and use them to graph the hyperbola.

$$\frac{9(x+3)^2}{80} - \frac{4(y-1)^2}{81} = 1$$

6.4 The Analytic Parabola

LEARNING OBJECTIVES

In Section 6.4 you will learn how to:

A. Graph parabolas with a horizontal axis of symmetry

B. Identify and use the focus-directrix form of the equation of a parabola

INTRODUCTION

In previous coursework, you likely noted that the graph of a quadratic function was a parabola. Parabolas are actually the fourth and final member of the family of conic sections, and like the others, the graph can be obtained by observing the intersection of a plane and a cone. If the plane is parallel to one **element** of the cone (shown as a dark line in Figure 6.23), the intersection of the plane with one nappe forms a parabola. In this section we develop the equation of a parabola from its analytic definition, opening a new realm of applications that extends far beyond those involving only zeroes and extreme values.

Figure 6.23

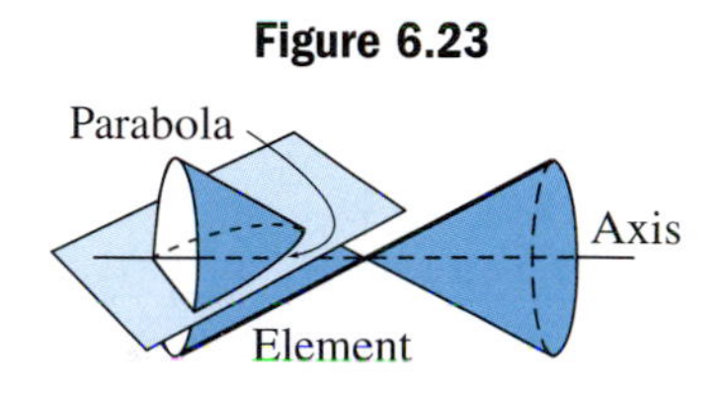

POINT OF INTEREST

Parabolas have a reflective property similar to that of ellipses. For an ellipse, light or sound emanating at one focus is reflected to the second. For a parabola, light or sound emanating from the focus reflects in a path parallel to the parabola's axis. This makes parabolas singularly valuable for lighting—where rays of light from a light source at the focus can be directed—and for telescopes—where parallel rays of light from objects far out in space are brought together and observed at the focus of the telescope's parabolic mirror.

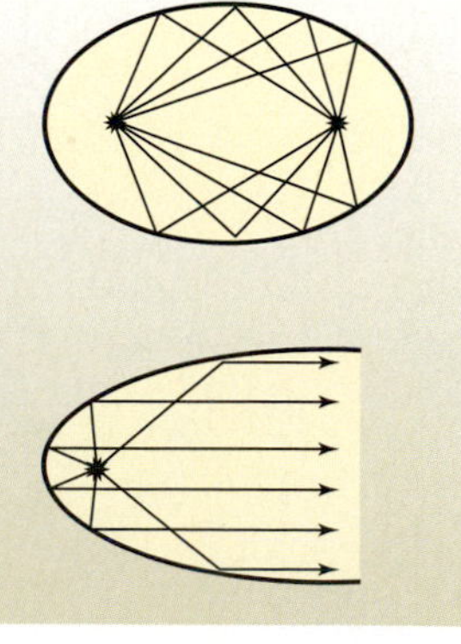

A. Parabolas with a Horizontal Axis

An introductory study of parabolas generally involves those with a vertical axis, defined by the equation $f(x) = ax^2 + bx + c$. Unlike the previous conic sections, this equation has *only one second-degree (squared) term in x* and defines a function rather than a relation. As a review, recall that to graph this function, a five-step method can be used, as outlined next and shown in Figure 6.24.

Figure 6.24

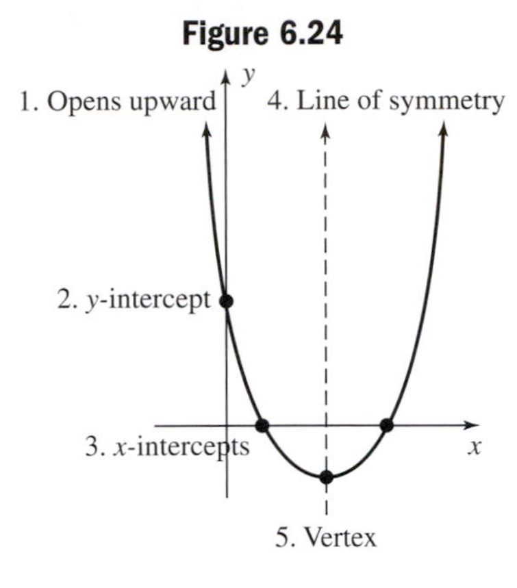

Vertical Parabolas

1. *Determine the concavity:* Concave up if $a > 0$, down if $a < 0$.
2. *Find the y-intercept:* The y-intercept is the ordered pair $(0, c)$.
3. *Find the x-intercepts (if they exist):* Solve the related equation $0 = ax^2 + bx + c$ by factoring or using the quadratic formula.
4. *Graph the line of symmetry:* The line of symmetry is the vertical line $x = \frac{-b}{2a}$.
5. *Find the coordinates of the vertex and determine the maximum/minimum value:*

   Since the axis of symmetry will contain the vertex, it has coordinates $\left(\frac{-b}{2a}, f\left(\frac{-b}{2a}\right)\right)$. If $a > 0$, the y-coordinate of the vertex is the minimum value of the function. If $a < 0$, the y-coordinate of the vertex is the maximum value. See Exercises 7 through 12.

Horizontal Parabolas

Similar to our study of horizontal and vertical hyperbolas, the graph of a parabola can open *to the right or left*, as well as up or down. After interchanging the variables x and y in the standard equation, we obtain the parabola $x = ay^2 + by + c$, which opens to the right if $a > 0$ and to the left if $a < 0$. By completing the square, this equation can also be written in shifted form as $x = a(y - k)^2 + h$, where (h, k) is the vertex of the parabola. However, this time the axis of symmetry is the horizontal line $y = k$ and factoring or the quadratic formula is used to find the *y-intercepts* (if they exist). It is important to note that although the graph is still a parabola—*it is not the graph of a function*!

EXAMPLE 1 Graph the relation whose equation is $x = y^2 + 3y - 4$, then state the domain and range of the relation.

Solution: Since the equation has a single squared term in y, the graph will be a horizontal parabola. With $a > 0$ $(a = 1)$, the parabola opens to the right. The x-intercept is $(-4, 0)$. Factoring shows the y-intercepts are $y = -4$ and $y = 1$. The axis of symmetry is

$y = \frac{-3}{2} = -1.5$, and substituting this value into the original equation gives $x = -6.25$. The coordinates of the vertex are $(-6.25, -1.5)$. Using horizontal and vertical boundary lines we find the domain for this relation is $x \in [-6.25, \infty)$ and the range is $y \in (-\infty, \infty)$. The graph is shown.

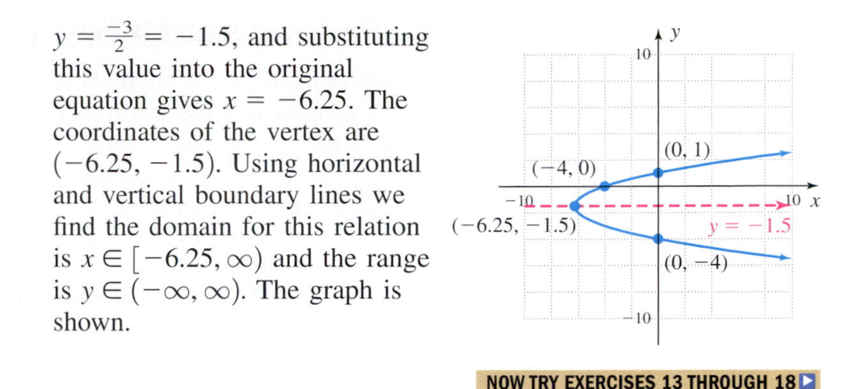

NOW TRY EXERCISES 13 THROUGH 18

The characteristics of a horizontal parabola are summarized here:

HORIZONTAL PARABOLAS

For a second-degree equation of the form $x = ay^2 + by + c$,

1. The graph is a parabola that opens right if $a > 0$, left if $a < 0$.
2. The x-intercept is $(c, 0)$.
3. The y-intercept(s) can be found by substituting $x = 0$, then solving by factoring or using the quadratic formula.
4. The axis of symmetry is $y = \frac{-b}{2a}$.
5. The vertex (h, k) can be found by completing the square and writing the equation in *shifted form* as $x = a(y - k)^2 + h$.

EXAMPLE 2 Graph by completing the square: $x = -2y^2 - 8y - 9$.

Solution: Using the original equation, we note the graph will be a horizontal parabola opening to the left $(a = -2)$ and have an x-intercept of $(-9, 0)$. Completing the square gives $x = -2(y^2 + 4y + 4) - 9 + 8$, so $x = -2(y + 2)^2 - 1$. The vertex is at $(-1, -2)$ and $y = -2$ is the axis of symmetry. This means there are no y-intercepts, a fact that comes to light when we attempt to solve the equation after substituting 0 for x:

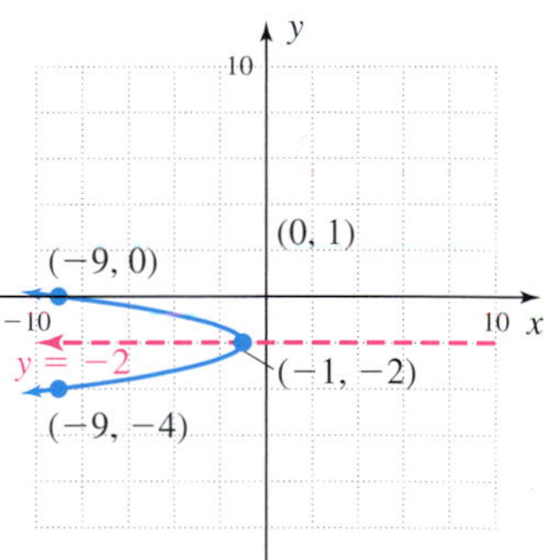

$$-2(y + 2)^2 - 1 = 0 \quad \text{substitute 0 for } x$$

$$(y + 2)^2 = -\frac{1}{2} \quad \text{isolate squared term}$$

The equation has no real roots and there are no y-intercepts. Using symmetry, the point $(-9, -4)$ is also on the graph. After plotting these points we obtain the graph shown.

NOW TRY EXERCISES 19 THROUGH 36

B. The Focus-Directrix Form of the Equation of a Parabola

As with the ellipse and hyperbola, many significant applications of the parabola rely on its analytical definition rather than its algebraic form. From the construction of radio telescopes to the manufacture of flashlights, the location of the focus of a parabola is critical. To understand these and other applications, we introduce the analytic definition of a parabola.

DEFINITION OF A PARABOLA

Given a fixed point F and fixed line D in the plane, a parabola is defined as the set of all points $P(x, y)$ such that the distance from F to P is equal to the perpendicular distance from line D to P. In symbols, $|FP| = |PP_d|$, where P_d is a point on line D. The fixed point F is called the **focus** of the parabola, and the fixed line D is called the **directrix.**

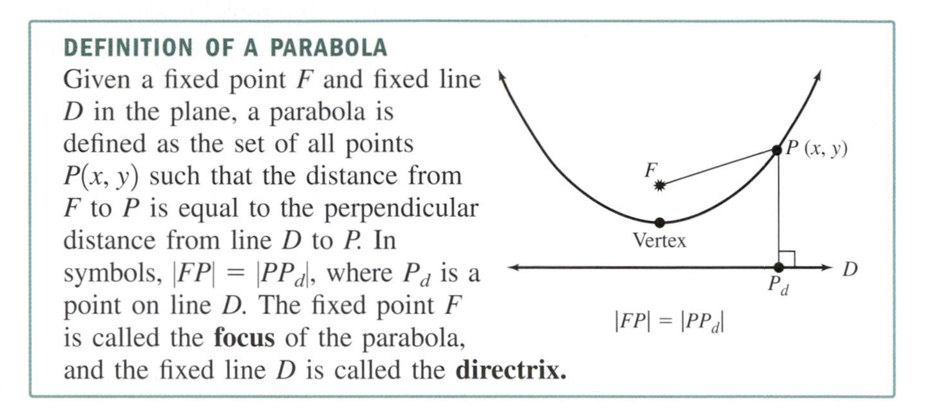

The equation of a parabola can be obtained by combining this definition with the distance formula. With no loss of generality, we can assume the parabola shown in the definition box is oriented in the plane with the vertex at (0, 0) and the focus at $(0, p)$. For the parabola, we use the coordinates $(0, p)$ for the focus, since we already designated $(0, c)$ as the y-intercept. As the diagram in Figure 6.25 indicates, this gives the directrix an equation of $y = -p$ and the point P_d coordinates of $(x, -p)$.

Figure 6.25

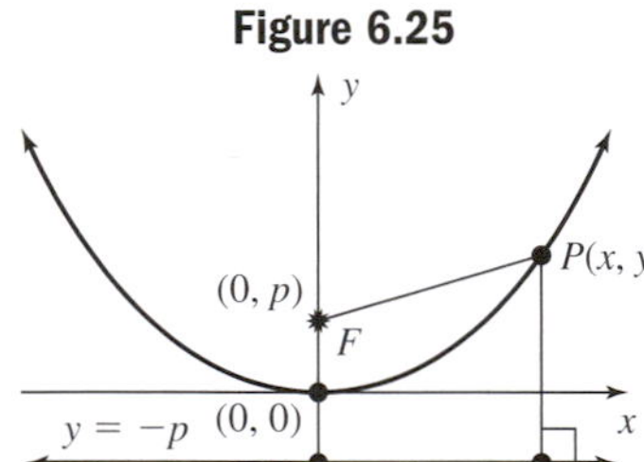

Using $|FP| = |PP_d|$, the distance formula yields

$$\sqrt{(x-0)^2 + (y-p)^2} = \sqrt{(x-x)^2 + (y+p)^2} \quad \text{from definition}$$
$$(x-0)^2 + (y-p)^2 = (x-x)^2 + (y+p)^2 \quad \text{square both sides}$$
$$x^2 + y^2 - 2py + p^2 = 0 + y^2 + 2py + p^2 \quad \text{simplify; expand binomials}$$
$$x^2 - 2py = 2py \quad \text{subtract } p^2 \text{ and } y^2$$
$$x^2 = 4py \quad \text{isolate } x^2$$

The resulting equation is called the **focus-directrix form** of a *vertical parabola* with center at (0, 0). If we had begun by orienting the parabola so it opened to the right, we would have obtained the equation of a *horizontal parabola* with center (0, 0): $y^2 = 4px$.

THE EQUATION OF A PARABOLA IN FOCUS-DIRECTRIX FORM WITH VERTEX (0, 0)

Vertical Parabola	Horizontal Parabola
$x^2 = 4py$	$y^2 = 4px$
focus $(0, p)$, directrix: $y = -p$	focus at $(p, 0)$, directrix: $x = -p$
If $p > 0$, concave up.	If $p > 0$, parabola opens right.
If $p < 0$, concave down.	If $p < 0$, parabola opens left.

For a parabola, note there is only one second-degree term.

EXAMPLE 3 Find the vertex, focus, and directrix for the parabola defined by the equation $x^2 = -12y$. Then sketch the graph, including the focus and directrix.

Solution: Since the x-term is squared and no shifts have been applied, the graph will be a vertical parabola with a vertex of (0, 0). Use a direct comparison between the given equation and the focus-directrix form to determine the value of p:

$$x^2 = -12y \quad \text{given equation}$$
$$\downarrow$$
$$x^2 = 4py \quad \text{focus-directrix form}$$

This shows:

$$4p = -12$$
$$p = -3$$

Since $p = -3$ ($p < 0$), the parabola is concave down, with the focus at $(0, -3)$ and directrix $y = 3$. To complete the graph we need a few additional points. Since 36 (6^2) is divisible by 12, we can use inputs of $x = 6$ and $x = -6$, giving the points $(6, -3)$ and $(-6, -3)$. Note the axis of symmetry is $x = 0$. The graph is shown.

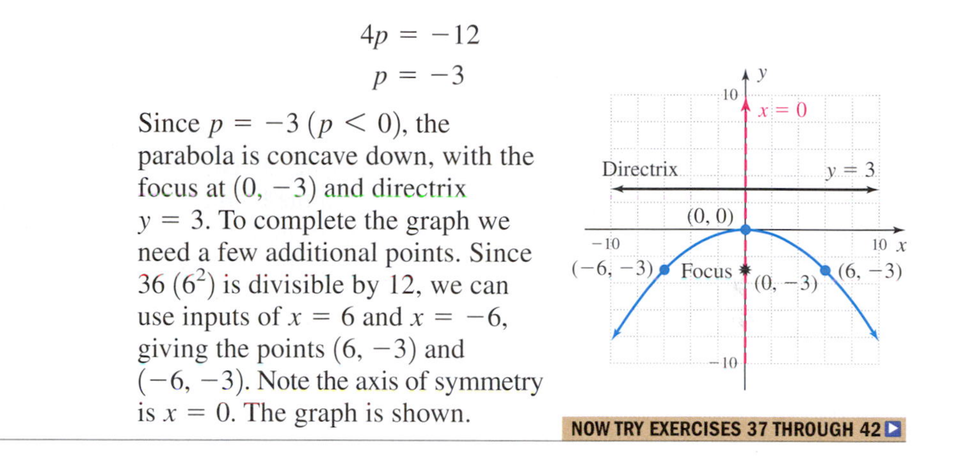

NOW TRY EXERCISES 37 THROUGH 42

Figure 6.26

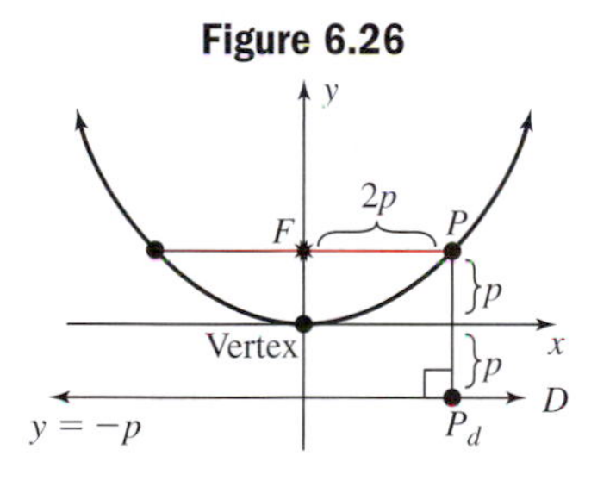

As an alternative to calculating additional points to sketch the graph, we can use what is called the **focal chord** of the parabola. Similar to the ellipse and hyperbola, the focal chord is a line segment that contains the focus, is parallel to the directrix, and has its endpoints on the graph. Using the definition of a parabola and the diagram in Figure 6.26, we see the distance $|PP_d|$ is $2p$. Since $|PP_d| = |FP|$, a line segment parallel to the directrix from the focus to the graph will also have a length of $|2p|$, and the focal chord of any parabola has a total length of $|4p|$. Note that in Example 3, the points we happened to choose were actually the endpoints of the focal chord. Also recall that when we graph a relation using transformations and shifts of a basic graph, all features of the graph are likewise shifted. These shifts apply to the focus and directrix of a parabola, as well as to the vertex and axis of symmetry.

EXAMPLE 4 Find the vertex, focus, and directrix for the parabola whose equation is given, then sketch the graph, including the focus and directrix: $x^2 - 6x + 12y - 15 = 0$.

Solution: Since only the x-term is squared, the graph will be a vertical parabola. To find the concavity, vertex, focus, and directrix, we complete the square in x and use a direct comparison between the shifted form and the focus-directrix form:

$$x^2 - 6x + 12y - 15 = 0 \quad \text{given equation}$$
$$x^2 - 6x + \_\_\_ = -12y + 15 \quad \text{complete the square in } x$$
$$x^2 - 6x + 9 = -12y + 24 \quad \text{add 9}$$
$$(x - 3)^2 = -12(y - 2) \quad \text{factor}$$

Notice the parabola has been shifted 3 right and 2 up, so *all features of the parabola will likewise be shifted.* Since we have $4p = -12$ (the coefficient of the linear term), we know $p = -3$ ($p < 0$) and the parabola is concave down. If the parabola were in standard position, the vertex would be at (0, 0), the focus at (0, −3) and the directrix a horizontal line at $y = 3$. But since the parabola is shifted 3 right and 2 up, we add 3 to all x-values and 2 to all y-values to locate the features of the shifted parabola. The vertex is at $(0 + 3, 0 + 2) = (3, 2)$. The focus is $(0 + 3, -3 + 2) = (3, -1)$ and the directrix is $y = 3 + 2 = 5$. Finally, the horizontal distance from the focus to the graph is $|2p| = 6$ units (since $|4p| = 12$), giving us the additional points (−3, −1) and (9, −1). See the figure.

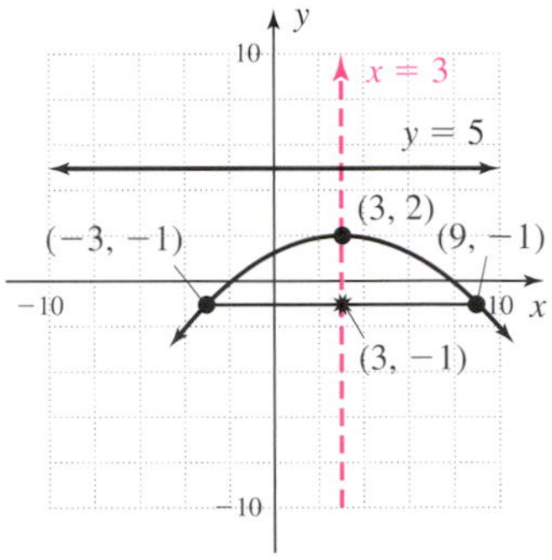

NOW TRY EXERCISES 43 THROUGH 60

Here is just one of the many ways the analytic definition of a parabola can be applied. There are several others in the exercise set.

EXAMPLE 5 The diagram shows the cross section of a radio antenna dish. Engineers have located a point on the cross section that is 0.75 m above and 6 m to the right of the vertex. At what coordinates should the engineers build the focus of the antenna?

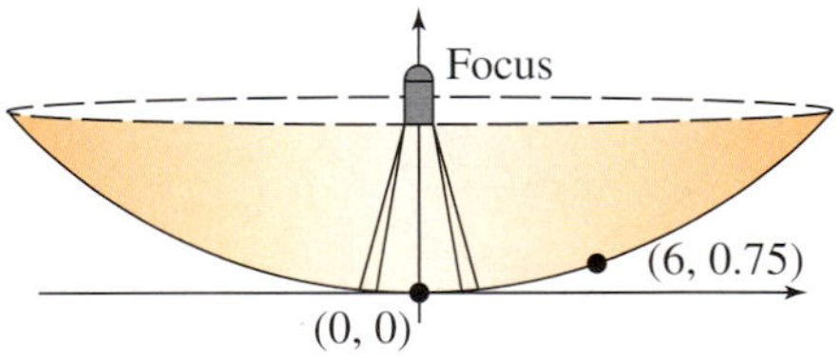

Solution: By inspection we see this is a vertical parabola with center at (0, 0). This means its equation must be of the form $x^2 = 4py$. Because we know (6, 0.75) is a point on this graph, we can substitute (6, 0.75) in this equation and solve for p:

$$x^2 = 4py \quad \text{equation for vertical parabola, vertex at (0, 0)}$$
$$(6)^2 = 4p(0.75) \quad \text{substitute 6 for } x \text{ and 0.75 for } y$$
$$36 = 3p \quad \text{simplify}$$
$$p = 12 \quad \text{result}$$

With $p = 12$, we see that the focus must be located at (0, 12), or 12 m directly above the vertex.

NOW TRY EXERCISES 63 THROUGH 68

Note that in many cases, the focus of a parabolic dish may be taller than the rim of the dish.

TECHNOLOGY HIGHLIGHT

The Focus of a Parabola Given in the Form $y = ax^2 + bx + c$

The keystrokes shown apply to a TI-84 Plus model. Please consult our Internet site or your manual for other models.

In this *Technology Highlight,* we attempt to verify that for *any* parabola, the distance from the focus to a point on the graph is equal to the distance from this point to the directrix. While our website contains a TI-84 Plus program that accomplishes this very nicely, we'll use a more deliberate and rudimentary approach here. The quadratic function $y = ax^2 + bx + c$ and its graphs are studied extensively in developmental courses, but usually no mention is made of the parabola's focus and directrix. Generally, when $a \geq 1$, the focus of a parabola is very near its vertex. To see why, write this function in focus-directrix form by completing the square. For convenience, assume $c = 0$:

$$y = ax^2 + bx \qquad \text{quadratic function; } c = 0$$

$$y = a\left(x^2 + \frac{b}{a}x + \_\_\right) \qquad \text{factor out } a$$

$$y = a\left(x^2 + \frac{b}{a}x + \frac{b^2}{4a^2}\right) - \frac{b^2}{4a} \qquad \text{complete the square; } a\left(\frac{b^2}{4a^2}\right) = \frac{b^2}{4a}$$

$$\frac{1}{a}\left(y + \frac{b^2}{4a}\right) = \left(x + \frac{b}{2a}\right)^2 \qquad \text{focus-directrix form}$$

Regardless of the coefficients chosen, this development shows that $4p = \frac{1}{a}$ so $p = \frac{1}{4a}$, so the larger the lead coefficient, the smaller p becomes. Consider the function $y = (x - 2)^2 + 1$, which is a parabola, concave up, with a vertex of (2, 1). Enter this function as Y_1 on the Y= screen of your graphing calculator. Since $a = 1$, the focus of the parabola *in standard position* would be $\left(0, \frac{1}{4}\right)$, but this parabola is shifted 2 right and 1 up, so the focus is actually at (2, 1.25). The directrix is $y = 0.75$ ($y = 1 - 0.25$). Store the coordinates of the focus (2, 1.25) in locations A and B, respectively, and the value of the directrix in location D. To find the distance between the focus (A, B) and a point $(x, f(x))$ on the graph we use $\sqrt{(X - A)^2 + (Y_1(X) - B)^2}$ (the distance formula), entering this expression as Y_2. To find the distance between $(x, f(x))$ and (x, D) we enter $\sqrt{(Y_1(X) - D)^2}$ as Y_3 (only one addend under the radical since $(X - X)^2 = 0$). Deactivate Y_1, leaving Y_2 and Y_3 active (see Figure 6.27).

Figure 6.27

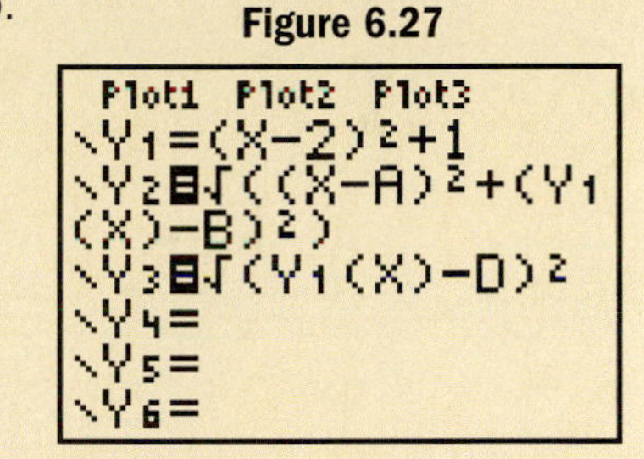

To verify that these distances are identical, go to the TABLE (use 2nd GRAPH) and compare the entries in Y_2 with those in Y_3. For the following functions, locate the focus and use the ideas from this *Technology Highlight* to verify the definition of a parabola.

Exercise 1: $y = (x + 3)^2 - 2$

Exercise 2: $y = -2(x - 1)^2 + 5$

Exercise 3: $y = \frac{1}{2}(x - 4)^2 + 2$

Exercise 4: $y = x^2 - 4x + 7$

6.4 EXERCISES

CONCEPTS AND VOCABULARY

Fill in the blank with the appropriate word or phrase. Carefully reread the section if needed.

1. The equation $x = ay^2 + by + c$ is that of a(n) __________ parabola, opening to the ______ if $a > 0$ and to the left if ______.

2. If point P is on the graph of a parabola with directrix D, the distance from P to line D is equal to the distance between P and the ______ of the parabola.

3. Given $y^2 = 4px$, the focus is at ______ and the equation of the directrix is ______.

4. Given $x^2 = -16y$, the value of p is ______ and the coordinates of the focus are ________.

5. Discuss/explain how to find the vertex, directrix, and focus from the equation $(x - h)^2 = 4p(y - k)$.

6. If a horizontal parabola has a vertex of $(2, -3)$ with $a > 0$, what can you say about the y-intercepts? Will the graph always have an x-intercept? Explain.

DEVELOPING YOUR SKILLS

Find the x- and y-intercepts (if they exist) and the vertex of the parabola. Then sketch the graph by using symmetry and a few additional points or completing the square and shifting a parent function. Scale the axes as needed to comfortably fit the graph and state the domain and range.

7. $y = x^2 - 2x - 3$
8. $y = x^2 + 6x + 5$
9. $y = 2x^2 - 8x - 10$
10. $y = 3x^2 + 12x - 15$
11. $y = 2x^2 + 5x - 7$
12. $y = 2x^2 - 7x + 3$

Find the x- and y-intercepts (if they exist) and the vertex of the graph. Then sketch the graph using symmetry and a few additional points (scale the axes as needed). Finally, state the domain and range of the relation.

13. $x = y^2 - 2y - 3$
14. $x = y^2 - 4y - 12$
15. $x = -y^2 + 6y + 7$
16. $x = -y^2 + 8y - 12$
17. $x = -y^2 + 8y - 16$
18. $x = -y^2 + 6y - 9$

Sketch using symmetry and shifts of a basic function. Be sure to find the x- and y-intercepts (if they exist) and the vertex of the graph, then state the domain and range of the relation.

19. $x = y^2 - 6y$
20. $x = y^2 - 8y$
21. $x = y^2 - 4$
22. $x = y^2 - 9$
23. $x = -y^2 + 2y - 1$
24. $x = -y^2 + 4y - 4$
25. $x = y^2 + y - 6$
26. $x = y^2 + 4y - 5$
27. $x = y^2 - 10y + 4$
28. $x = y^2 + 12y - 5$
29. $x = 3 - 8y - 2y^2$
30. $x = 2 - 12y + 3y^2$
31. $y = (x - 2)^2 + 3$
32. $y = (x + 2)^2 - 4$
33. $x = (y - 3)^2 + 2$
34. $x = (y + 1)^2 - 4$
35. $x = 2(y - 3)^2 + 1$
36. $x = -2(y + 3)^2 - 5$

Find the vertex, focus, and directrix for the parabolas defined by the equations given, then use this information to sketch a complete graph (illustrate and name these features). For Exercises 43 to 60, also include the focal chord.

37. $x^2 = 8y$
38. $x^2 = 16y$
39. $x^2 = -24y$
40. $x^2 = -20y$
41. $x^2 = 6y$
42. $x^2 = 18y$
43. $y^2 = -4x$
44. $y^2 = -12x$
45. $y^2 = 18x$
46. $y^2 = 20x$
47. $y^2 = -10x$
48. $y^2 = -14x$
49. $x^2 - 8x - 8y + 16 = 0$
50. $x^2 - 10x - 12y + 25 = 0$
51. $x^2 - 14x - 24y + 1 = 0$
52. $x^2 - 10x - 12y + 1 = 0$
53. $3x^2 - 24x - 12y + 12 = 0$
54. $2x^2 - 8x - 16y - 24 = 0$
55. $y^2 - 12y - 20x + 36 = 0$
56. $y^2 - 6y - 16x + 9 = 0$
57. $y^2 - 6y + 4x + 1 = 0$
58. $y^2 - 2y + 8x + 9 = 0$
59. $2y^2 - 20y + 8x + 2 = 0$
60. $3y^2 - 18y + 12x + 3 = 0$

WORKING WITH FORMULAS

61. The area of a right parabolic segment: $A = \frac{2}{3}ab$

A *right parabolic segment* is that part of a parabola formed by a line perpendicular to its axis, which cuts the parabola. The area of this segment is given by the formula shown, where b is the length of the chord cutting the parabola and a is the perpendicular distance from the vertex to this chord. What is the area of the parabolic segment shown in the figure?

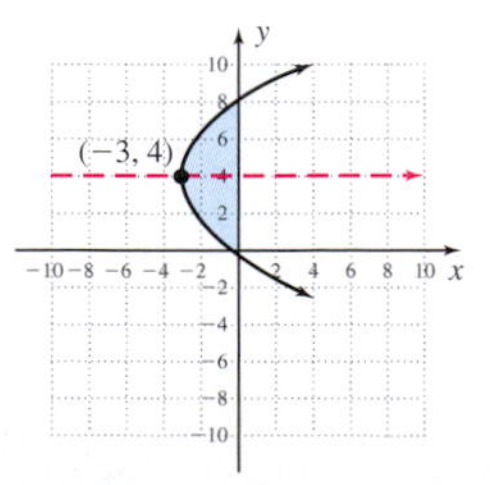

62. The arc length of a right parabolic segment:

$$\frac{1}{2}\sqrt{b^2 + 16a^2} + \frac{b^2}{8a}\ln\left(\frac{4a + \sqrt{b^2 + 16a^2}}{b}\right)$$

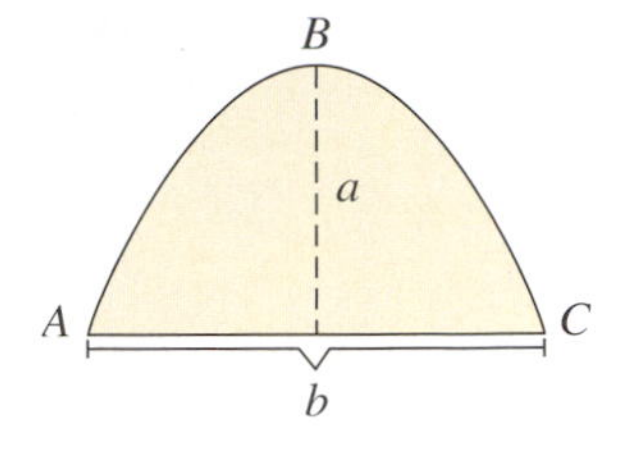

Although a fairly simple concept, finding the length of the parabolic arc traversed by a projectile requires a good deal of computation. To find the length of the arc *ABC* shown, we use the formula given where *a* is the maximum height attained by the projectile, *b* is the horizontal distance it traveled, and "ln" represents the natural log function. Suppose a baseball thrown from centerfield reaches a maximum height of 20 ft and traverses an arc length of 340 ft. Will the ball reach the catcher 310 ft away without bouncing?

APPLICATIONS

63. **Parabolic car headlights:** The cross section of a typical car headlight can be modeled by an equation similar to $25x = 16y^2$, where x and y are in inches and $x \in [0, 4]$. Use this information to graph the relation for the indicated domain.

64. **Parabolic flashlights:** The cross section of a typical flashlight reflector can be modeled by an equation similar to $4x = y^2$, where x and y are in centimeters and $x \in [0, 2.25]$. Use this information to graph the relation for the indicated domain.

Exercise 65

65. **Parabolic sound receivers:** Sound technicians at professional sports events often use parabolic receivers (see the diagram) as they move along the sidelines. If a two-dimensional cross section of the receiver is modeled by the equation $y^2 = 54x$, and is 36 in. in *diameter*, how deep is the parabolic receiver? What is the location of the focus? [*Hint:* Graph the parabola on the coordinate grid (scale the axes).]

Exercise 65

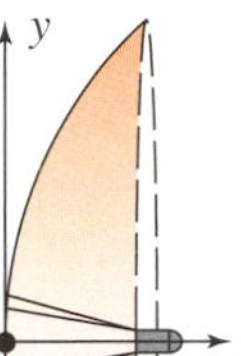

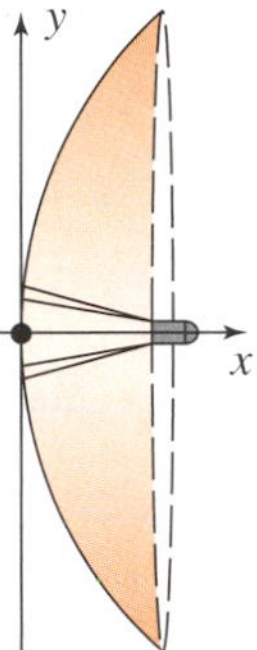

66. **Parabolic sound receivers:** Private investigators will often use a smaller and less expensive parabolic receiver (see Exercise 65) to gather information for their clients. If a two-dimensional cross section of the receiver is modeled by the equation $y^2 = 24x$, and the receiver is 12 in. in *diameter*, how deep is the parabolic dish? What is the location of the focus?

67. **Parabolic radio wave receivers:** The program known as S.E.T.I. (Search for Extra-Terrestrial Intelligence) identifies a group of scientists using radio telescopes to look for radio signals from possible intelligent species in outer space. The radio telescopes are actually parabolic dishes that vary in size from a few feet to hundreds of feet in diameter. If a particular radio telescope is 100 ft in diameter and has a cross section modeled by the equation $x^2 = 167y$, how deep is the parabolic dish? What is the location of the focus? [*Hint:* Graph the parabola on the coordinate grid (scale the axes).]

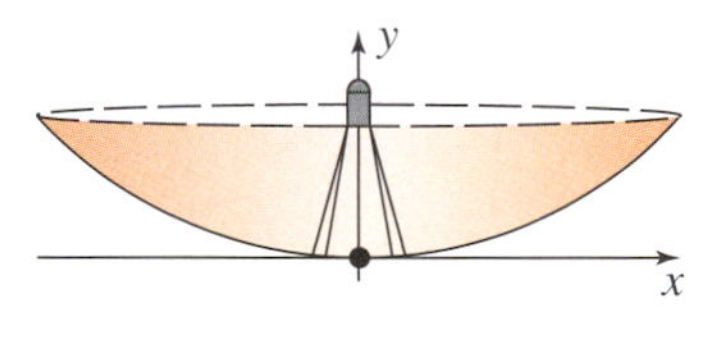

68. **Solar furnace:** Another form of technology that uses a parabolic dish is called a solar furnace. In general, the rays of the Sun are reflected by the dish and concentrated at the focus, producing extremely high temperatures. Suppose the dish of one of these parabolic reflectors had a 30-ft diameter and a cross section modeled by the equation $x^2 = 50y$. How deep is the parabolic dish? What is the location of the focus?

WRITING, RESEARCH, AND DECISION MAKING

69. In a study of quadratic graphs from the equation $y = ax^2 + bx + c$, no mention is made of a parabola's focus and directrix. Generally, when $a \geq 1$, the focus of a parabola is very near its vertex. Complete the square of the function $y = 2x^2 - 8x$ and write the

result in the form $(x - h)^2 = 4p(y - k)$. What is the value of p? What are the coordinates of the vertex?

70. Have someone in your class bring an inexpensive flashlight to class, one where the bulb assembly can easily be removed from the body of the flashlight. As carefully as you can, measure the diameter and depth of the parabolic reflector in millimeters. Use these measurements to draw a cross section of the parabolic reflector on a coordinate grid, with the vertex of the parabola at (0, 0). The parabola will have an equation of the form $y = ax^2$, where a point (x, y) can be determined from the graph (the endpoints of the diameter). Use these values of x and y to find the value of a, then use the equation to locate the focus of this parabolic reflector. How closely does your answer seem to fit the location of the filament of the lightbulb when it is held in place?

71. Match the graph with the correct equation. Then write a short paragraph explaining how you made your choice.

a. $y = (x - 2)^2 - 3$

b. $x = -(y + 3)^2 + 2$

c. $x = (y - 3)^2 - 2$

d. $y = -(x - 3)^2 + 2$

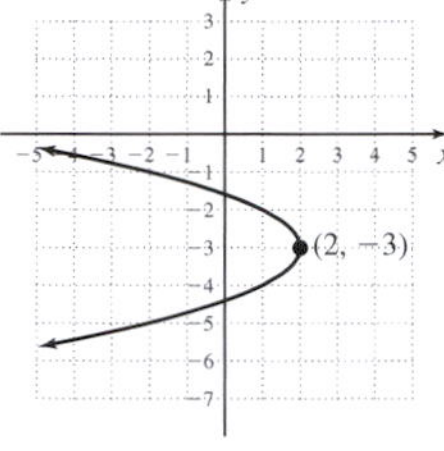

72. With a diameter of 1000 ft, the radio telescope located at Arecibo, Puerto Rico, is the largest in the world. Do some research on this remarkable telescope, and be sure to include information on the unique way the telescope was manufactured and the ingenious way the focus is held in place.

EXTENDING THE CONCEPT

73. Find the equation and area of the circle whose center is the vertex of the parabola $y = -x^2 - 6x - 5$ and that intersects the parabola at the two points $(-1, 0)$ and $(-5, 0)$.

Exercise 74

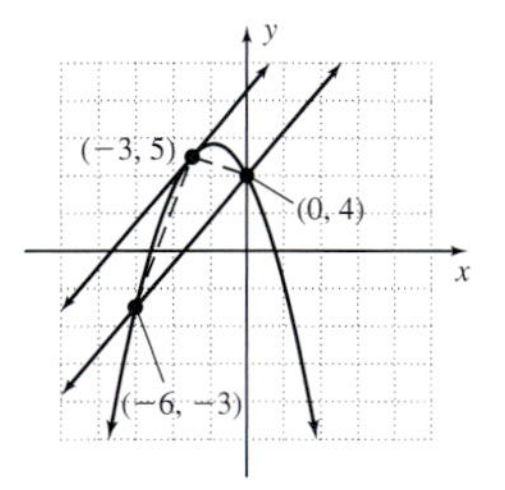

74. In Exercise 61, a formula was given for the area of a right parabolic segment. The area of an *oblique* parabolic segment (the line segment cutting the parabola is *not* perpendicular to the axis) is more complex, as it involves locating the point where a line parallel to this segment is tangent (touches at only one point) to the parabola. The formula is $A = \frac{4}{3}T$, where T represents the area of the triangle formed by the endpoints of the segment and this point of tangency. What is the area of the parabolic segment shown (assuming the lines are parallel)? See Section 5.1, Exercises 29 and 30 and Section 5.3, Exercise 36.

MAINTAINING YOUR SKILLS

75. (2.3) The graph shown displays the variation in daylight from an average of 12 hr per day (i.e., the maximum is 15 hr and the minimum is 9 hr). Use the graph to *approximate* the number of days in a year there are 10.5 hr or less of daylight.

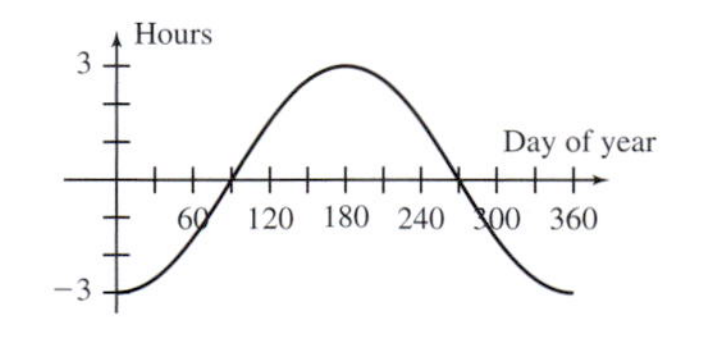

76. (5.5) In tough-man competitions, Bobby the Bouncer has shown he can lift up to 500 lb. Use a vector analysis to determine whether he will be able to pull the crate up the ramp shown. Discuss your response.

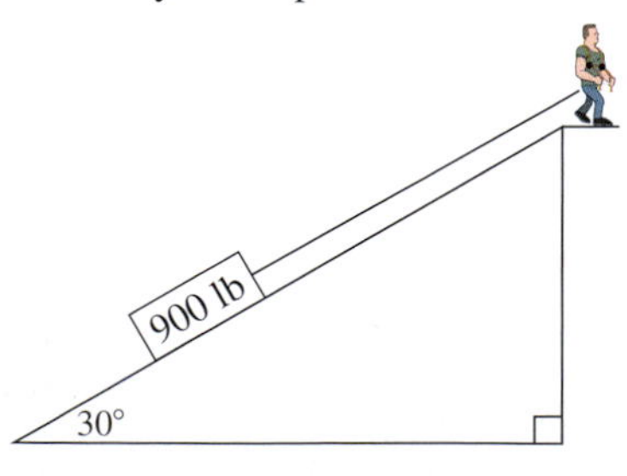

77. (4.4) Solve the equation for $x \in [0, 2\pi)$, rounding to four decimal places as needed:
$$-225 = 600 + 875\sin\left(x + \frac{\pi}{6}\right)$$

78. (3.2) Prove the following is an identity:
$$\frac{\cos(2x) + \sin^2 x}{1 - \cos^2 x} = \cot^2 x$$

79. (1.2) Estimate the value of $\tan 12°$, given $\csc 12° \approx 4.8097$ and $\sec 12° \approx 1.0223$.

80. (1.3) What is the measure of central angle α whose terminal side contains the point (25, 19)?

6.5 Polar Coordinates, Equations, and Graphs

LEARNING OBJECTIVES

In Section 6.5 you will learn how to:

A. **Plot points given in polar form**

B. **Convert from rectangular form to polar form**

C. **Convert from polar form to rectangular form**

D. **Sketch basic polar graphs using an *r*-value analysis**

E. **Use symmetry and families of curves to write a polar equation given a polar graph or information about the graph**

INTRODUCTION

One of the most enduring goals of mathematics is to express relations with the greatest possible simplicity and ease of use. For $\frac{\tan\theta - \cot\theta}{\tan^2\theta - \cot^2\theta} = \sin\theta\cos\theta$, we would definitely prefer working with $\sin\theta\cos\theta$, although the expressions are equivalent. Similarly, we would prefer computing $(3 + \sqrt{3}i)^6$ in trigonometric form rather than algebraic form—and would quickly find the result is -1728. In just this way, many equations and graphs are easier to work with in **polar form** rather than rectangular form. In rectangular form, a circle of radius 2 centered at (0, 2) has the equation $x^2 + (y - 2)^2 = 4$. In polar form, the equation of the same circle is simply $r = 4\sin\theta$. As you'll see, polar coordinates offer an alternative method for plotting points and graphing relations.

POINT OF INTEREST

Although there were vague references to similar systems before his time, Sir Isaac Newton (1642–1727) is generally credited with the development of polar coordinates. In fact, his publication *Method of Fluxions* (written in 1671 but not published until 1736) offered 10 different types of coordinate systems that one could use in the study of analytical geometry. Independently of Newton, Jacob Bernoulli (1654–1705) published a paper in the *Acta Eruditorum* in 1691, which also made extensive use of polar coordinates.

Figure 6.28

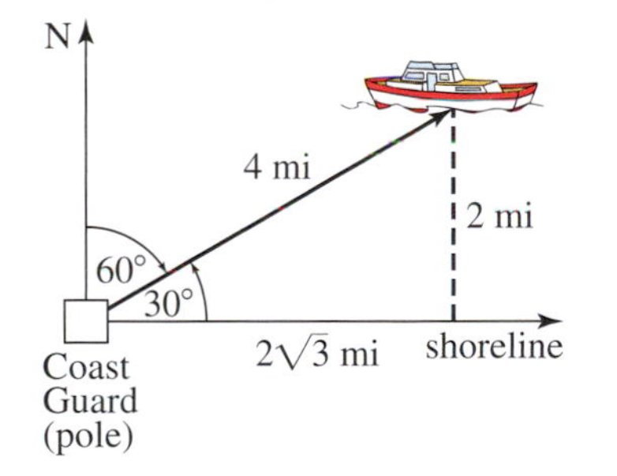

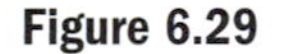

Figure 6.29

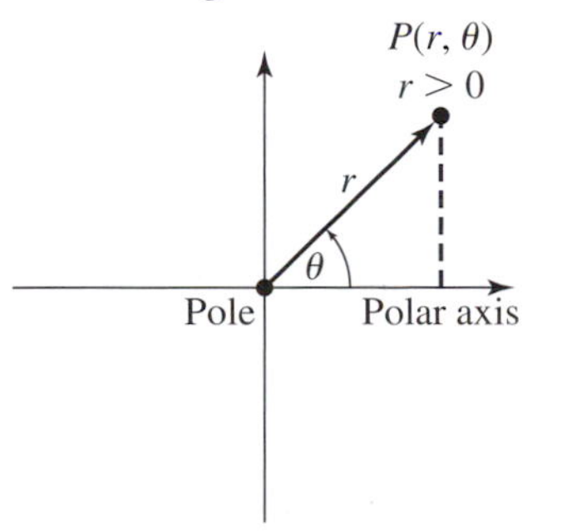

A. Plotting Points Using Polar Coordinates

Suppose a Coast Guard station receives a distress call from a stranded boat. The boater could attempt to give the location in rectangular form, but this might require imposing an arbitrary coordinate grid on an uneven shoreline, using uncertain points of reference. However, if the radio message said, "We're stranded 4 miles out, bearing 60°," the Coast Guard could immediately locate the boat and send help. In **polar coordinates,** "4 miles out, bearing 60°" would simply be written $(r, \theta) = (4, 30°)$, with r representing the distance from the station and $\theta > 0$ measured from a horizontal axis in the counterclockwise direction as before (see Figure 6.28). If we placed the scenario on a rectangular grid (assuming a straight shoreline), the coordinates of the boat would be $(2\sqrt{3}, 2)$ using basic trigonometry. As you see, the **polar coordinate system** uses angles and distances to locate a point in the plane. In this example, the Coast Guard station would be considered the **pole** or origin, with the x-axis as the **polar axis** or axis of reference (Figure 6.29). A distinctive feature of polar

Figure 6.30

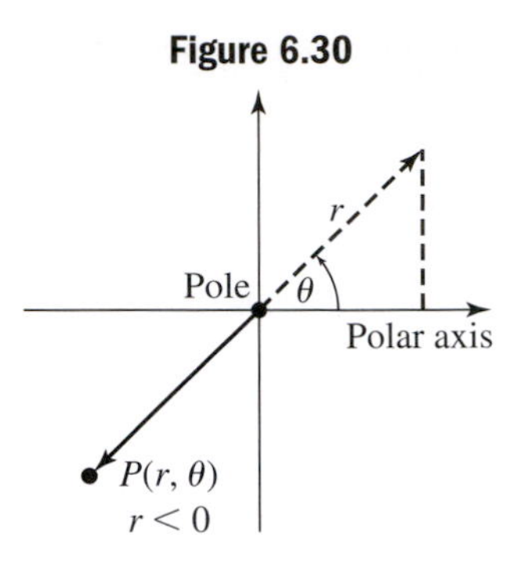

coordinates is that *we allow r to be negative,* in which case $P(r, \theta)$ is the point $|r|$ units from the pole in a direction opposite (180°) to that of θ (Figure 6.30). For convenience, polar graph paper is often used when working with polar coordinates. It consists of a series of concentric circles that share the same center and have integer radii. The standard angles are marked off in multiples of $\frac{\pi}{12} = 15°$ depending on whether you're working in radians or degrees (Figure 6.31). To plot the point $P(r, \theta)$, go a distance of $|r|$ at 0° then move $\theta°$ counterclockwise along a circle of radius r. If $r > 0$, plot a point at that location (you're finished). If $r < 0$, the point is plotted on a circle of the same radius, but 180° in the opposite direction.

Figure 6.31

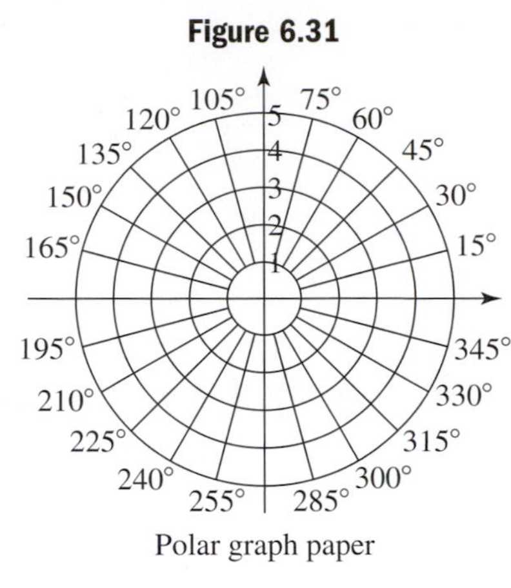

Polar graph paper

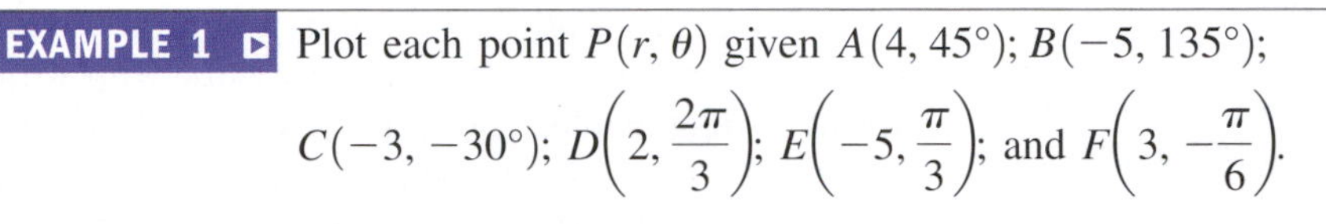

EXAMPLE 1 Plot each point $P(r, \theta)$ given $A(4, 45°)$; $B(-5, 135°)$; $C(-3, -30°)$; $D\left(2, \frac{2\pi}{3}\right)$; $E\left(-5, \frac{\pi}{3}\right)$; and $F\left(3, -\frac{\pi}{6}\right)$.

Solution: For $A(4, 45°)$ go 4 units at 0°, then rotate 45° counterclockwise and plot point A. For $B(-5, 135°)$, move $|-5| = 5$ units at 0°, rotate 135°, then actually plot point B 180° in the opposite direction, as shown. Point $C(-3, -30°)$ is plotted by moving $|-3| = 3$ units at 0°, rotating $-30°$, then plotting point C 180° in the opposite direction (since $r < 0$). See Figure 6.32. The points $D\left(2, \frac{2\pi}{3}\right)$, $E\left(-5, \frac{\pi}{3}\right)$, and $F\left(3, -\frac{\pi}{6}\right)$ are plotted on the grid in Figure 6.33.

Figure 6.32 Figure 6.33

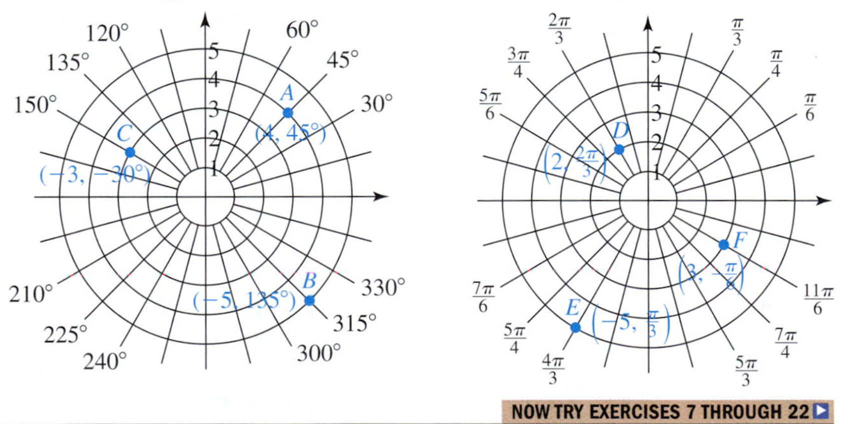

NOW TRY EXERCISES 7 THROUGH 22

While plotting the points $B(-5, 135°)$ and $F\left(3, -\frac{\pi}{6}\right)$, you likely noticed that the coordinates of a point in polar coordinates are not unique. For $B(-5, 135°)$ it appears more natural to name the location $(5, 315°)$; while for $F\left(3, -\frac{\pi}{6}\right)$, the expression $\left(3, \frac{11\pi}{6}\right)$ is just as reasonable. In fact, for any point $P(r, \theta)$ in polar coordinates, $P(r, \theta \pm 2\pi)$ and $P(-r, \theta \pm \pi)$ name the same location. See Exercises 23 through 36.

B. Converting from Rectangular Coordinates to Polar Coordinates

Figure 6.34

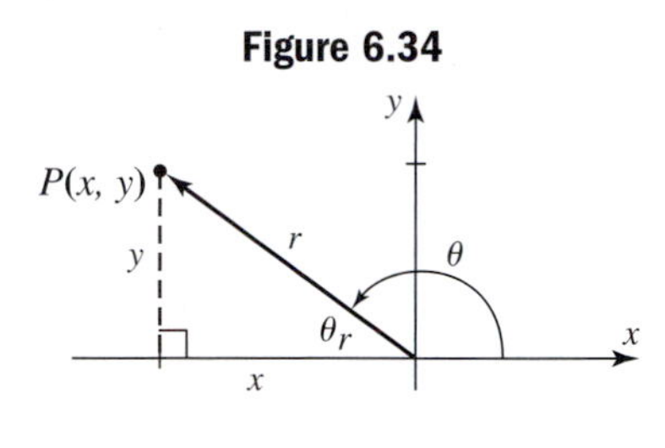

Conversions between rectangular and polar coordinates is a simple application of skills from previous sections, and closely resembles the conversion from the rectangular form to the trigonometric form of a complex number. To make the connection, we first assume $r > 0$ with θ in Quadrant II (see Figure 6.34). In rectangular form, the coordinates of the point are simply (x, y), with the lengths of x and y forming the sides of a right triangle. The distance r from the origin to point P resembles the modulus of a complex number and is computed in the same way: $r = \sqrt{x^2 + y^2}$. As long as $x \neq 0$, we have $\theta_r = \tan^{-1}\left(\frac{y}{x}\right)$, noting θ_r is a reference angle if the terminal side is not in Quadrant I. If needed, refer to Section 1.4 for a review of reference arcs and reference angles.

CONVERTING FROM RECTANGULAR TO POLAR COORDINATES

Any point $P(x, y)$ in rectangular coordinates can be represented as $P(r, \theta)$ in polar coordinates, where $r = \sqrt{x^2 + y^2}$ and $\theta_r = \tan^{-1}\left(\frac{y}{x}\right)$, $x \neq 0$.

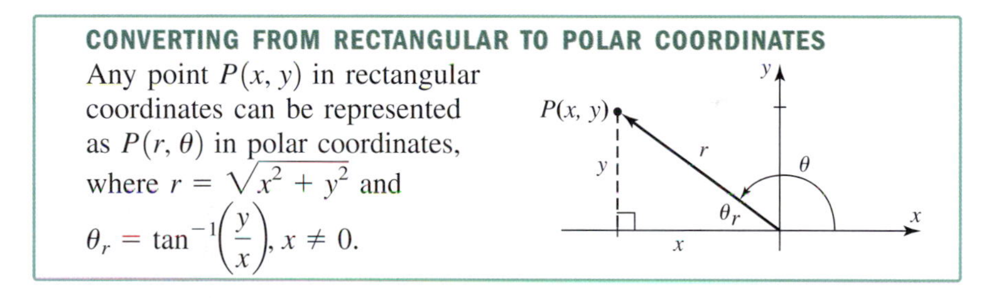

EXAMPLE 2 Convert from rectangular to polar form, with $r > 0$ and $0 \leq \theta \leq 360°$ (round values to one decimal place as needed).

a. $P(-5, 12)$ **b.** $P(3\sqrt{2}, -3\sqrt{2})$

Solution: **a.** Point $P(-5, 12)$ is in Quadrant II.

$$r = \sqrt{(-5)^2 + 12^2} \qquad \theta = \tan^{-1}\left(\frac{12}{-5}\right)$$
$$= \sqrt{169} \qquad \theta_r \approx -67.4°$$
$$= 13 \qquad \theta \approx 112.6°$$
$$P(-5, 12) \rightarrow P(13, 112.6°)$$

b. Point $P(3\sqrt{2}, -3\sqrt{2})$ is in Quadrant IV.

$$r = \sqrt{(3\sqrt{2})^2 + (-3\sqrt{2})^2} \qquad \theta = \tan^{-1}\left(\frac{-3\sqrt{2}}{3\sqrt{2}}\right)$$
$$= \sqrt{36} \qquad \theta_r = -45°$$
$$= 6 \qquad \theta = 315°$$
$$P(3\sqrt{2}, -3\sqrt{2}) \rightarrow P(6, 315°)$$

NOW TRY EXERCISES 37 THROUGH 44

C. Converting from Polar Coordinates to Rectangular Coordinates

The conversion from polar form to rectangular form is likewise straightforward. From Figure 6.35 we again note $\cos\theta = \frac{x}{r}$ and $\sin\theta = \frac{y}{r}$, giving $x = r\cos\theta$ and $y = r\sin\theta$. The conversion simply consists of making these substitutions and simplifying.

Figure 6.35

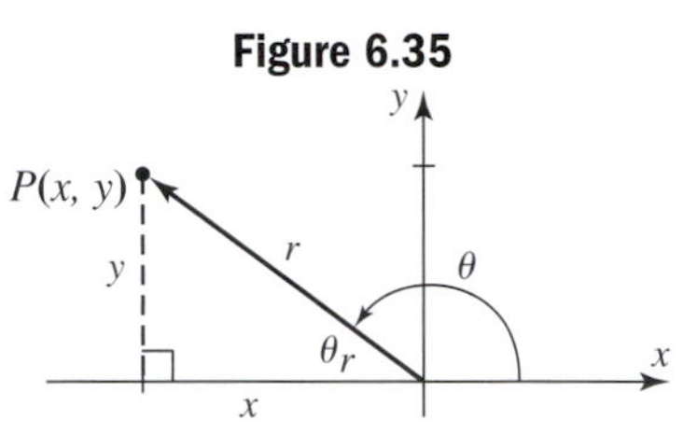

CONVERTING FROM POLAR TO RECTANGULAR COORDINATES

Any point $P(r, \theta)$ in polar coordinates can be represented as $P(x, y)$ in rectangular coordinates, where $x = r\cos\theta$ and $y = r\sin\theta$.

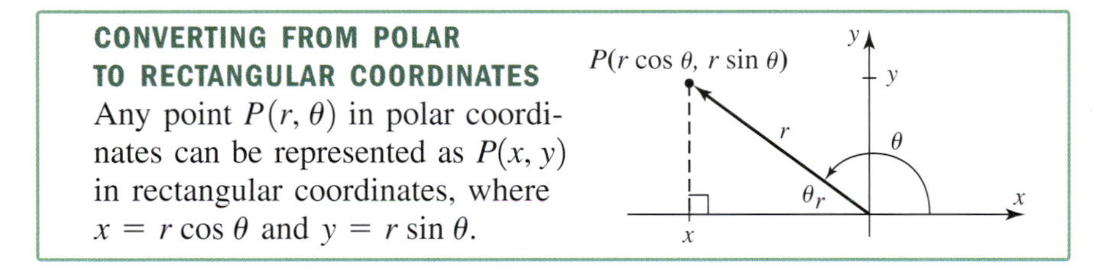

EXAMPLE 3 Convert from polar to rectangular form (round values to one decimal place as needed).

a. $P\left(12, \frac{5\pi}{3}\right)$ **b.** $P(6, 240°)$

Solution: **a.** Point $P\left(12, \frac{5\pi}{3}\right)$ is in Quadrant IV.

$$x = r\cos\theta \qquad y = r\sin\theta$$
$$= 12\cos\left(\frac{5\pi}{3}\right) \qquad = 12\sin\left(\frac{5\pi}{3}\right)$$
$$= 12\left(\frac{1}{2}\right) \qquad = 12\left(\frac{-\sqrt{3}}{2}\right)$$
$$= 6 \qquad = -6\sqrt{3}$$
$$P\left(12, \frac{5\pi}{3}\right) \rightarrow P(6, -6\sqrt{3}) \approx P(6, -10.4)$$

b. Point $P(6, 240°)$ is in Quadrant III.

$$x = 6\cos 240° \qquad y = 6\sin 240°$$
$$= 6\left(-\frac{1}{2}\right) \qquad = 6\left(\frac{-\sqrt{3}}{2}\right)$$
$$= -3 \qquad = -3\sqrt{3}$$
$$P(6, 240°) \rightarrow P(-3, -3\sqrt{3}) \approx P(-3, -5.2)$$

NOW TRY EXERCISES 45 THROUGH 52

D. Basic Polar Graphs and r-Value Analysis

To really understand polar graphs, an intuitive sense of how they're developed is needed. Polar equations are generally stated in terms of r and trigonometric functions of θ, with θ being the input value and r being the output value. First, it helps to view the length r as the long second hand of a clock, but extending an equal distance in both directions from center (Figure 6.36). This "second hand" ticks around the face of the clock in the counterclockwise direction, with the angular measure of each tick being $\frac{\pi}{12}$ radians $= 15°$. As each angle "ticks by," we locate a point somewhere along the radius, depending on whether r is positive or negative, and plot it on the face of the clock before going on to the next tick. For the purposes of this study, we will allow that all polar graphs are continuous and smooth curves, without presenting a formal proof.

Figure 6.36

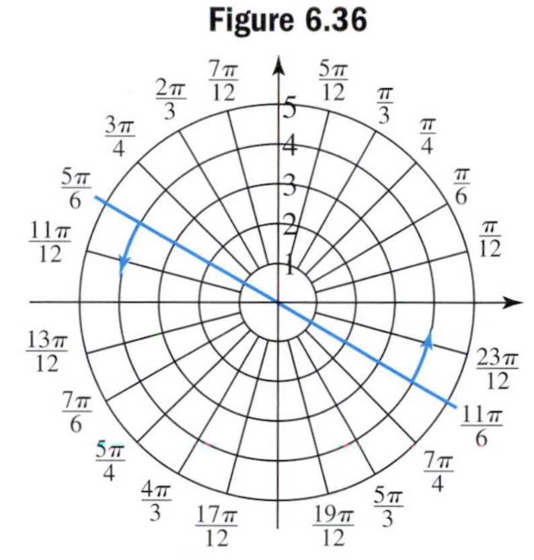

EXAMPLE 4 Graph the polar equations (a) $r = 4$ and (b) $\theta = \frac{\pi}{4}$.

Solution: **a.** For $r = 4$, we're plotting all points of the form $(4, \theta)$, where r has a constant value and θ varies. As the second hand "ticks around the polar grid," we plot all points a distance of 4 units from the pole. As you might imagine, the graph is a circle with radius 4.

b. For $\theta = \frac{\pi}{4}$, all points have the form $\left(r, \frac{\pi}{4}\right)$ with $\frac{\pi}{4}$ constant and r varying. In this case, the "second hand" is frozen at $\frac{\pi}{4}$, and we plot any selection of r-values, producing the straight line shown in the figure.

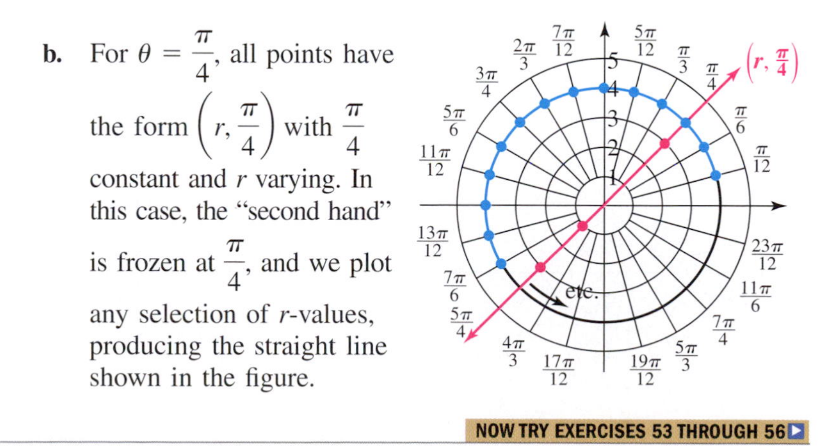

NOW TRY EXERCISES 53 THROUGH 56

To develop an "intuitive sense" that allows for the efficient graphing of more sophisticated equations, we use a technique called ***r*-value analysis.** This technique basically takes advantage of the predictable patterns in $r = \sin\theta$ and $r = \cos\theta$ taken from their graphs, including the zeros and maximum/minimum values.

We begin with the r-value analysis for $r = \sin\theta$, using the graph shown in Figure 6.37. Note the analysis occurs in the four colored parts corresponding to Quadrants I, II, III, and IV, and that the maximum value of $|\sin\theta| = 1$.

Figure 6.37

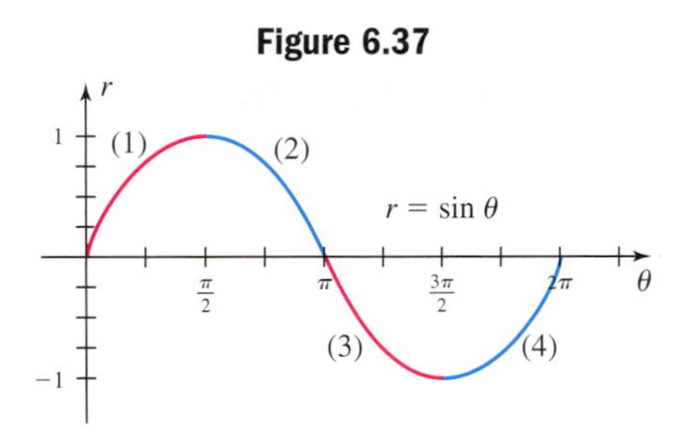

1. As θ moves from 0 to $\frac{\pi}{2}$, sin θ is positive and $|\sin\theta|$ increases from 0 to 1.
   $\Rightarrow$ for $r = \sin\theta$, r is increasing
2. As θ moves from $\frac{\pi}{2}$ to π, sin θ is positive and $|\sin\theta|$ decreases from 1 to 0.
   $\Rightarrow$ for $r = \sin\theta$, r is decreasing
3. As θ moves from π to $\frac{3\pi}{2}$, sin θ is negative and $|\sin\theta|$ increases from 0 to 1.
   $\Rightarrow$ for $r = \sin\theta$, r is increasing
4. As θ moves from $\frac{3\pi}{2}$ to 2π, sin θ is negative and $|\sin\theta|$ decreases from 1 to 0.
   $\Rightarrow$ for $r = \sin\theta$, r is decreasing

In summary, note that the value of $|r|$ goes through four cycles, two where it is increasing from 0 to 1 (in red), and two where it is decreasing from 1 to 0 (in blue).

WORTHY OF NOTE

It is important to remember that if $r < 0$, the related point on the graph is $|r|$ units from center, *180° in the opposite direction*: $(-r, \theta) \rightarrow (r, \theta + 180°)$. In addition, students are encouraged not to use a table of values, a conversion to rectangular coordinates, or a graphing calculator until after the r-value analysis.

EXAMPLE 5 Sketch the graph of $r = 4\sin\theta$ using an r-value analysis.

Solution: Begin by noting that $r = 0$ at $\theta = 0$, and will increase from 0 to 4 as the clock "ticks" from 0 to $\frac{\pi}{2}$, since sin θ is increasing from 0 to 1. (1) For $\theta = \frac{\pi}{6}, \frac{\pi}{4}$, and $\frac{\pi}{3}$, $r = 2$, $r \approx 2.8$, and $r \approx 3.5$, respectively (at $\theta = \frac{\pi}{2}$, $r = 4$). See Figure 6.38. (2) As θ continues "ticking" from $\frac{\pi}{2}$ to π, $|r|$ decreases from 4 to 0, since sin θ is decreasing from 1 to 0. For $\theta = \frac{2\pi}{3}, \frac{3\pi}{4}$, and $\frac{5\pi}{6}$, $r \approx 3.5$, $r \approx 2.8$, and $r = 2$ respectively

Figure 6.38

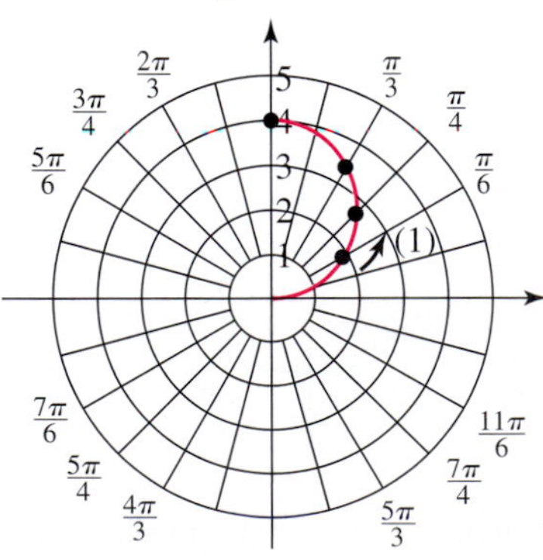

Figure 6.39

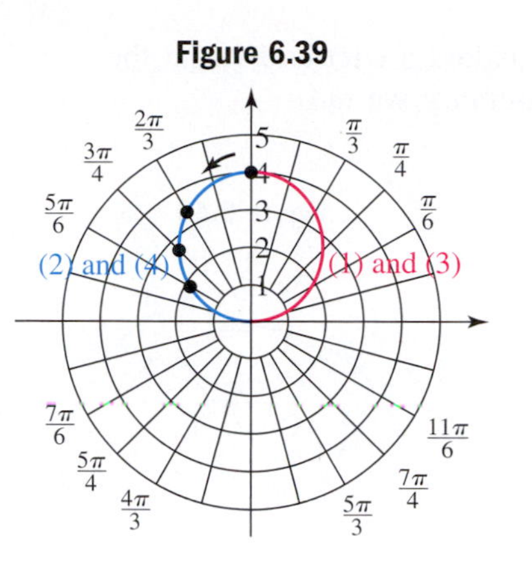

(at $\theta = \pi$, $r = 0$). See Figure 6.39. (3) From π to $\frac{3\pi}{2}$, $|r|$ increases from 0 to 4, but since $r < 0$, this portion of the graph is reflected back into Quadrant I, overlapping the portion already drawn from 0 to $\frac{\pi}{2}$. (4) From $\frac{3\pi}{2}$ to 2π, $|r|$ decreases from 4 to 0, overlapping the portion drawn from $\frac{\pi}{2}$ to π. We conclude the graph is a closed figure limited to Quadrants I and II as shown in Figure 6.39. This is a circle with radius 2, centered at (0, 2). In summary:

$r = 4 \sin \theta$

θ	0 to $\frac{\pi}{2}$	$\frac{\pi}{2}$ to π	π to $\frac{3\pi}{2}$	$\frac{3\pi}{2}$ to 2π
$\lvert r \rvert$	0 to 4	4 to 0	0 to 4	4 to 0

NOW TRY EXERCISES 57 AND 58

Although it takes some effort, r-value analysis offers an efficient way to graph polar equations, and gives a better understanding of graphing in polar coordinates. In addition, it often enables you to sketch the graph with a minimum number of calculations and plotted points. As you continue using the technique, it will help to have Figure 6.37 in plain view for quick reference, as well as the corresponding analysis of $y = \cos \theta$ for polar graphs involving cosine (see Exercise 98).

EXAMPLE 6 Sketch the graph of $r = 2 + 2 \sin \theta$ using an r-value analysis.

Solution: Since the minimum value of $\sin \theta$ is -1, we note that r will always be greater than or equal to zero. At $\theta = 0$, r has a value of 2 ($\sin 0 = 0$), and will increase from 2 to 4 as the clock "ticks" from 0 to $\frac{\pi}{2}$ ($\sin \theta$ is positive and $|\sin \theta|$ is increasing). From $\frac{\pi}{2}$ to π, r decreases from 4 to 2 ($\sin \theta$ is positive and $|\sin \theta|$ is decreasing). From π to $\frac{3\pi}{2}$, r decreases from 2 to 0 ($\sin \theta$ is negative and $|\sin \theta|$ is increasing); and from $\frac{3\pi}{2}$ to 2π, r increases from 0 to 2 ($\sin \theta$ is negative and $|\sin \theta|$ is decreasing). We conclude the graph is a closed figure containing the points (2, 0), $\left(4, \frac{\pi}{2}\right)$, $(2, \pi)$, and $\left(0, \frac{3\pi}{2}\right)$. Noting that $\theta = \frac{\pi}{6}$ and $\theta = \frac{5\pi}{6}$ will produce integer values, we evaluate $r = 2 + 2 \sin \theta$ and obtain the additional points $\left(3, \frac{\pi}{6}\right)$ and $\left(3, \frac{5\pi}{6}\right)$. Using these points and the r-value analysis produces the graph shown in Figure 6.40, called a **cardioid** (from the limaçon family of curves). In summary we have:

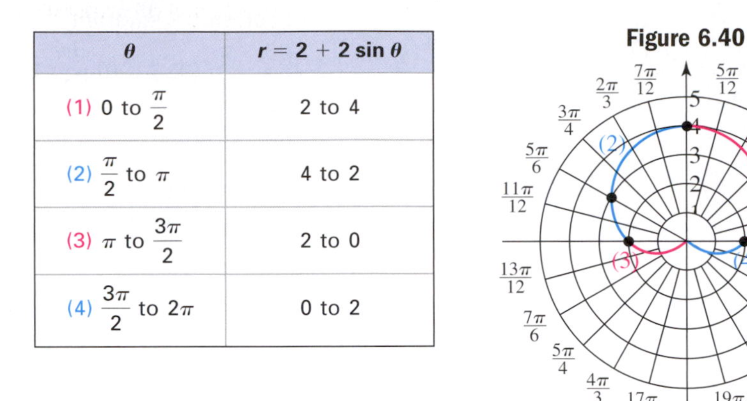

θ	$r = 2 + 2\sin\theta$
(1) 0 to $\frac{\pi}{2}$	2 to 4
(2) $\frac{\pi}{2}$ to π	4 to 2
(3) π to $\frac{3\pi}{2}$	2 to 0
(4) $\frac{3\pi}{2}$ to 2π	0 to 2

Figure 6.40

E. Symmetry and Families of Polar Graphs

Even with a careful r-value analysis, some polar graphs require a good deal of effort to produce. In many cases, symmetry can be a big help, as can recognizing certain families of equations and their related graphs. As with other forms of graphing, gathering this information beforehand will enable you to graph relations with a smaller number of plotted points. Figures 6.41 to 6.44 offer some examples of symmetry for polar graphs.

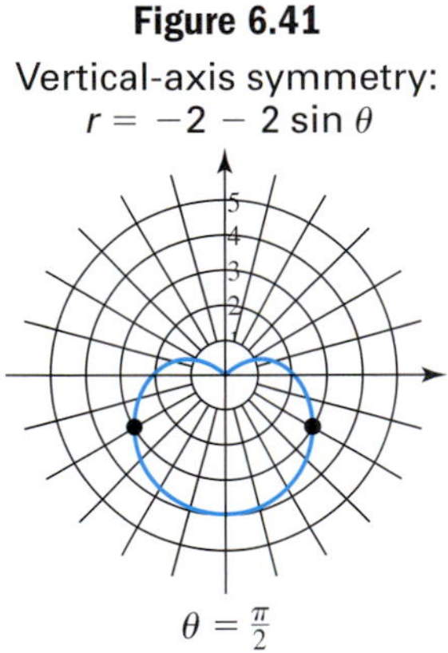

Figure 6.41
Vertical-axis symmetry: $r = -2 - 2\sin\theta$

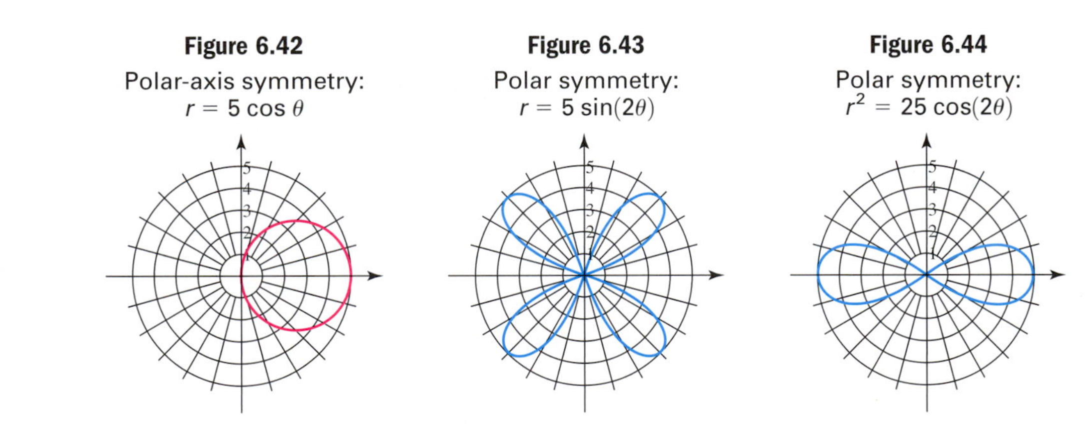

Figure 6.42
Polar-axis symmetry: $r = 5\cos\theta$

Figure 6.43
Polar symmetry: $r = 5\sin(2\theta)$

Figure 6.44
Polar symmetry: $r^2 = 25\cos(2\theta)$

WORTHY OF NOTE

In mathematics we refer to the tests for polar symmetry as *sufficient but not necessary conditions.* The tests are *sufficient* to show symmetry (if the test is satisfied, the graph must be symmetric), but the tests are *not necessary* to show symmetry (the graph may be symmetric even if the test is not satisfied).

The tests for symmetry in polar coordinates bear a strong resemblance to those for rectangular coordinates, but there is a major difference. Since there are many different ways to name a point in polar coordinates, a polar graph may actually exhibit a form of symmetry without satisfying the related test. In other words, the tests are *sufficient* to establish symmetry, but not *necessary.*

The formal tests for symmetry are explored in Exercises 100 to 102. For our purposes, we'll rely on a somewhat narrower view, one that is actually a synthesis of our observations here and our previous experience with the sine and cosine.

SYMMETRY FOR GRAPHS OF CERTAIN POLAR EQUATIONS

Given the polar equation $r = f(\theta)$,

1. If $f(\theta)$ represents an expression in terms of sine(s), the graph will be symmetric to $\theta = \frac{\pi}{2}$: (r, θ) and $(r, \pi - \theta)$ are on the graph.
2. If $f(\theta)$ represents an expression in terms of cosine(s), the graph will be symmetric to $\theta = 0$: (r, θ) and $(r, -\theta)$ are on the graph.

While the fundamental ideas from Examples 5 and 6 go a long way toward graphing other polar equations, our discussion would not be complete without a review of the *period* of sine and cosine. Many polar equations have factors of $\sin(n\theta)$ or $\cos(n\theta)$ in them, and it helps to recall the period formula $P = \frac{2\pi}{n}$. Comparing $r = 4\sin\theta$ from Example 5 with $r = 4\sin(2\theta)$, we note the period of sine changes from $P = 2\pi$ to $P = \frac{2\pi}{2} = \pi$, *meaning there will be twice as many cycles* and $|r|$ will now go through *eight* cycles—four where $|\sin(2\theta)|$ is increasing from 0 to 1 (in red), and four where it is decreasing from 1 to 0 (in blue). See Figure 6.45.

Figure 6.45

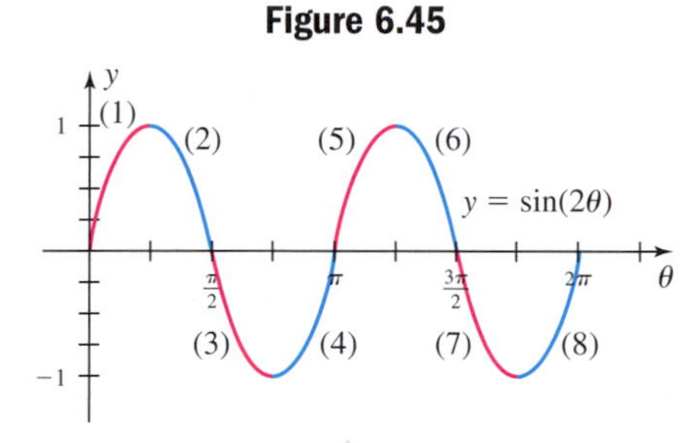

EXAMPLE 7 Sketch the graph of $r = 4\sin(2\theta)$ using symmetry and an r-value analysis.

Solution: Since r is expressed in terms of sine, the graph will be symmetric to $\theta = \frac{\pi}{2}$. We note that $r = 0$ at $\theta = \frac{n\pi}{2}$, where n is even, and the graph will go through the pole at these points. This also tells us the graph will be a closed figure. From the graph of $\sin(2\theta)$ in Figure 6.45, we see $|\sin(2\theta)| = 1$ at $\theta = \frac{\pi}{4}, \frac{3\pi}{4}, \frac{5\pi}{4}$, and $\frac{7\pi}{4}$, so the graph will include the points $\left(4, \frac{\pi}{4}\right), \left(4, \frac{3\pi}{4}\right), \left(4, \frac{5\pi}{4}\right)$, and $\left(4, \frac{7\pi}{4}\right)$. Only the analysis of the first four cycles is given next, since the remainder of the graph can be drawn using symmetry.

	Cycle	r-Value Analysis	Location of Graph
(1)	0 to $\frac{\pi}{4}$	$\|r\|$ increases from 0 to 4	QI $(r > 0)$
(2)	$\frac{\pi}{4}$ to $\frac{\pi}{2}$	$\|r\|$ decreases from 4 to 0	QI $(r > 0)$
(3)	$\frac{\pi}{2}$ to $\frac{3\pi}{4}$	$\|r\|$ increases from 0 to 4	QIV $(r < 0)$
(4)	$\frac{3\pi}{4}$ to π	$\|r\|$ decreases from 4 to 0	QIV $(r < 0)$

Plotting the points and applying the r-value analysis with the symmetry involved produces the graph in Figure 6.46, called a **four-leaf rose.** At any time during this process, additional points can be calculated to "round-out" the graph.

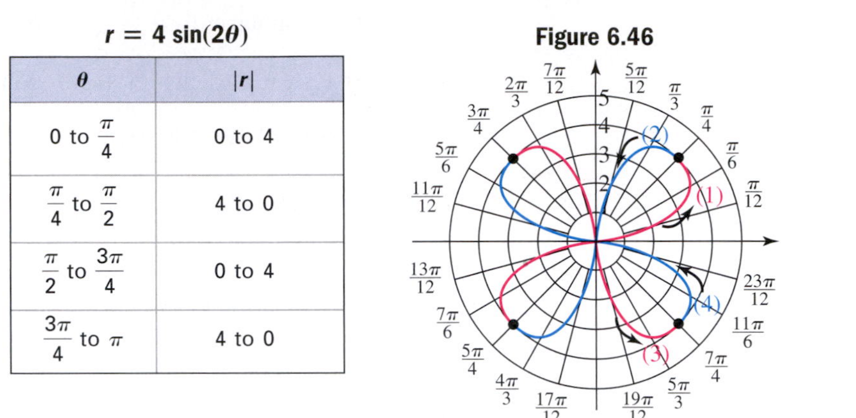

$r = 4\sin(2\theta)$

θ	$\lvert r \rvert$
0 to $\frac{\pi}{4}$	0 to 4
$\frac{\pi}{4}$ to $\frac{\pi}{2}$	4 to 0
$\frac{\pi}{2}$ to $\frac{3\pi}{4}$	0 to 4
$\frac{3\pi}{4}$ to π	4 to 0

Figure 6.46

NOW TRY EXERCISES 63 THROUGH 70

GRAPHING POLAR EQUATIONS

To assist the process of graphing polar equations:

1. Carefully note any symmetries you can use.
2. Have graphs of $y = \sin(n\theta)$ and $y = \cos(n\theta)$ in view for quick reference.
3. Use these graphs to analyze the value of r as the "clock ticks" around the polar grid: (a) determine the max/min r-values and write them in polar form, and (b) determine the polar-axis intercepts and write them in polar form.

Plot the points, then use the r-value analysis and any symmetries to complete the graph.

Similar to polynomial graphs, polar graphs come in numerous shapes and varieties, yet many of them share common characteristics and can be organized into certain families. Some of the more common families are illustrated in Appendix VI, and give the general equation and related graph for common family members. Also included are characteristics of certain graphs that will enable you to develop the polar equation given its graph or information about its graph.

EXAMPLE 8 Find the equation of the polar curve satisfying the given conditions, then sketch the graph: limaçon, symmetric to $\theta = 90°$, with $a = 2$ and $b = -3$ (see Appendix VI if needed).

Solution: The general equation of a limaçon symmetric to $\theta = 90°$ is $r = a + b\sin\theta$, so our desired equation is $r = 2 - 3\sin\theta$. Since $|a| < |b|$, the limaçon has an inner loop of length $3 - 2 = 1$ and a maximum distance from the origin of $2 + 3 = 5$. The

(2, 180°)
(2, 0°)
(−1, 90°)
(5, 270°)

WORTHY OF NOTE

You've likely been wondering how the different families of polar graphs were named. The roses are easy to figure as each graph has a flower-like appearance. The limaçon (pronounced li-ma-sawn) family takes its name from the Latin words *limax* or *lamacis*, meaning "snail." With some imagination, these graphs do have the appearance of a snail shell. The cardioids are a subset of the limaçon family and are so named due to their obvious resemblance to the human heart. In fact, the name stems from the Greek *kardia* meaning heart, and many derivative words are still in common use (a cardiologist is one who specializes in a study of the heart). Finally, there is the lemniscate family, a name derived from the Latin *lemniscus*, which describes a certain kind of ribbon. Once again, a little creativity enables us to make the connection between ribbons, bows, and the shape of this graph.

polar-axis intercepts are (2, 0) and (2, 180°). With $b < 0$, the graph is reflected across the polar axis (facing "downward"). The complete graph is shown in the figure.

NOW TRY EXERCISES 79 THROUGH 94

EXAMPLE 9 Scavenger birds sometimes fly over dead or dying animals (called carrion) in a "figure-eight" formation, closely resembling the graph of a lemniscate. Suppose the flight path of one of these birds was plotted and found to contain the polar coordinates (81, 0°) and (0, 45°). Find the equation of the lemniscate. If the bird lands at the point (r, 136°), how far is it from the carrion? Assume r is in yards.

Solution: Since (81, 0°) is a point on the graph, the lemniscate is symmetric to the polar axis and the general equation is $r^2 = a^2\cos(2\theta)$. The point (81, 0°) indicates $a = 81$, hence the equation is $r^2 = 6561\cos(2\theta)$. At $\theta = 136°$ we have $r^2 = 6561\cos 272°$, and the bird has landed $r \approx 15$ yd away.

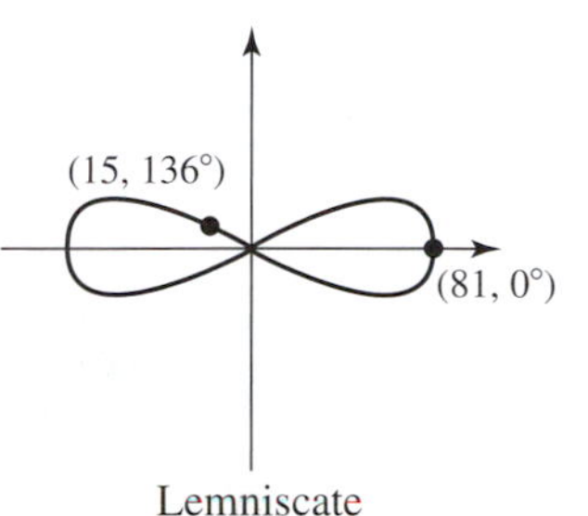

Lemniscate

NOW TRY EXERCISES 95 THROUGH 97

TECHNOLOGY HIGHLIGHT

Polar and Rectangular Coordinates (and Symmetry) on a Graphing Calculator

The keystrokes shown apply to a TI-84 Plus model. Please consult our Internet site or your manual for other models.

Most graphing calculators are programmed to evaluate and graph four kinds of relations—functions, sequences, polar equations, and parametric equations. This is the reason the "default" variable key X,T,θ,n is embossed with four different input variables. The X is used for functions, T for parametric equations, θ for polar equations, and n for relations defined by a sequence. To change the operating mode of the calculator, press the MODE key, highlight the desired operating mode, and press ENTER (Figure 6.47). In this case, our main interest is polar equations. Note the Y= screen obtained while in **Pol** (polar mode) is similar to that obtained in **Func** (function mode), except the dependent variable is now r_i instead of $\mathbf{Y}_i$ (Figure 6.48). In addition, a θ now appears when the variable key X,T,θ,n is pressed, rather than an X as when in function mode. Verify by entering the equation from Example 5, $r_1 = 4\sin\theta$, as shown. Since polar graphs are by nature different from polynomial graphs, the WINDOW screen also takes on a different appearance (Figure 6.49). We can still set a

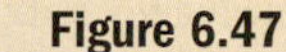

Figure 6.47

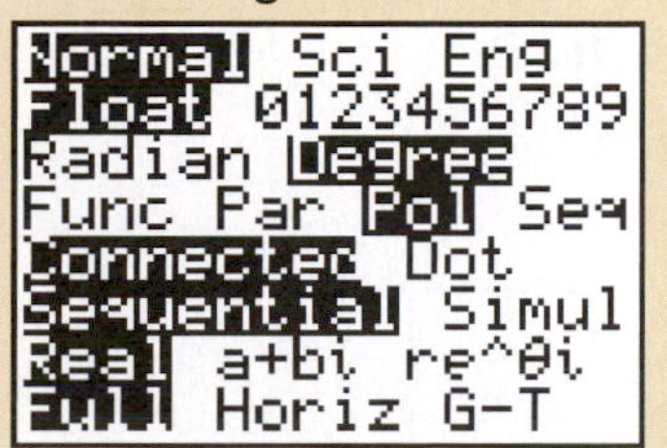

Figure 6.48

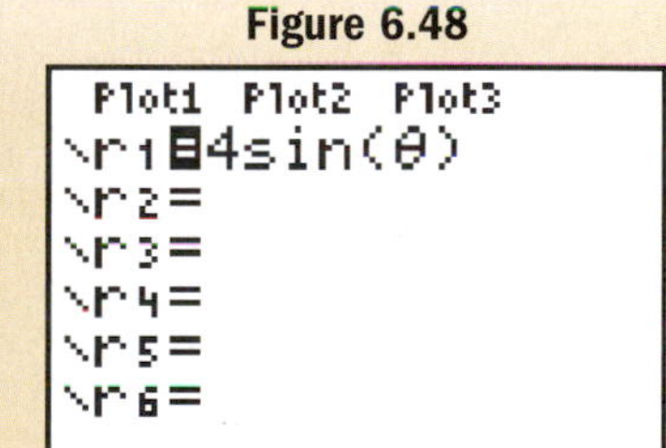

Figure 6.49

WINDOW
θmin=0
θmax=360
θstep=7.5
Xmin=-10
Xmax=10
Xscl=1
↓Ymin=-10

desired window size or use the ZOOM options as before, and we can also state the desired interval for θ, and indicate the interval between plotted points using θstep. The standard settings (ZOOM 6) are shown in Figure 6.49. Knowing that the graph of r_1 is a circle of radius 4 centered at (0, 2), we set the window appropriately to obtain the GRAPH shown in Figure 6.50.

Figure 6.50

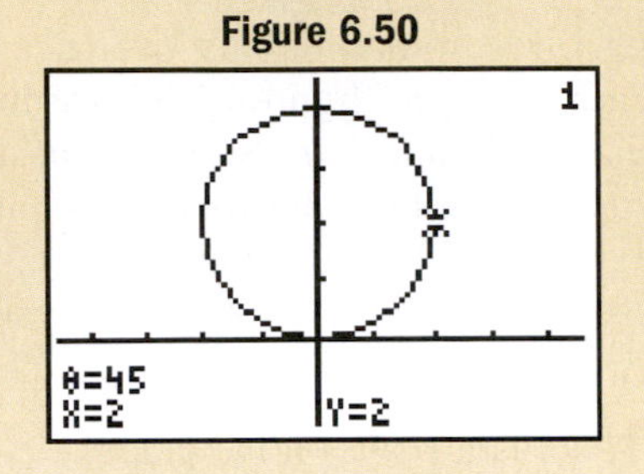

Exercise 1: Graph $r_1 = 4 \sin \theta$ three more times, using θstep = 15, θstep = 45, and θstep = 60. What do you notice?

Exercise 2: Graph $r_1 = 2 + 4 \cos\left(\frac{\theta}{2}\right)$ on the standard screen (ZOOM 6). It appears that only one-half of the graph is drawn. Can you determine why? Graph this equation three more times using θmax = 540, θmax = 630, and θmax = 720. What do you notice?

Exercise 3: Graph $r_1 = 4 \sin(n\theta)$ for $n = 3, 4, 5$, and 6. What pattern do you see? Verify using $r_1 = 4 \cos(n\theta)$.

6.5 EXERCISES

CONCEPTS AND VOCABULARY

Fill in each blank with the appropriate word or phrase. Carefully reread the section if needed.

1. The point (r, θ) is said to be written in ______ coordinates.
2. In polar coordinates, the origin is called the ______ and the horizontal axis is called the ______ axis.
3. The point $(4, 135°)$ is located in Q ___, while $(-4, 135°)$ is located in Q ___.
4. If a polar equation is given in terms of cosine, the graph will be symmetric to __________.
5. Write out the procedure for plotting points in polar coordinates, as though you were explaining the process to a friend.
6. Discuss the graph of $r = 6 \cos \theta$ in terms of an r-value analysis, using $y = \cos \theta$ and a color-coded graph.

DEVELOPING YOUR SKILLS

Plot the following points using polar graph paper.

7. $\left(4, \frac{\pi}{2}\right)$
8. $\left(3, \frac{3\pi}{2}\right)$
9. $\left(2, \frac{5\pi}{4}\right)$
10. $\left(4.5, -\frac{\pi}{3}\right)$
11. $\left(-5, \frac{5\pi}{6}\right)$
12. $\left(-4, \frac{7\pi}{4}\right)$
13. $\left(-3, -\frac{2\pi}{3}\right)$
14. $\left(-4, -\frac{\pi}{4}\right)$

Express the points shown using polar coordinates with θ in radians, $0 \le \theta < 2\pi$ and $r > 0$.

15.
16.
17.

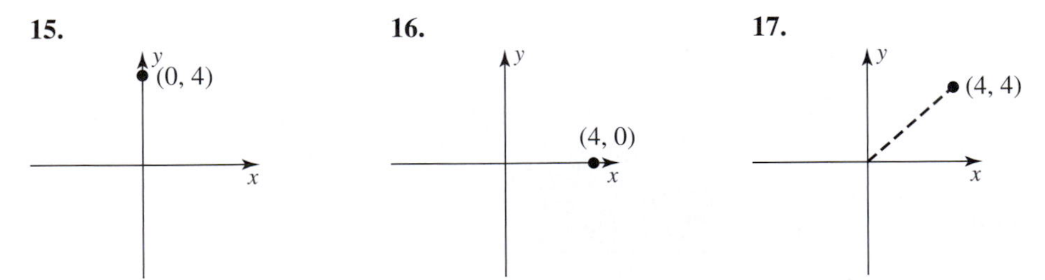

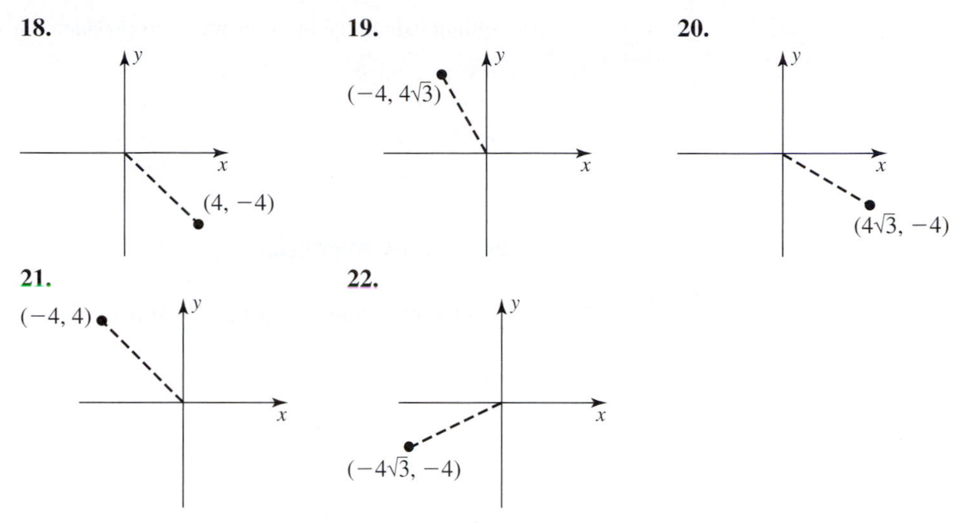

Exercises 27–36

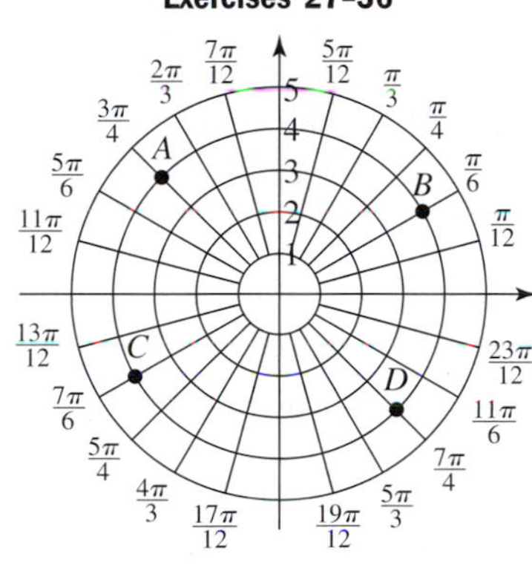

List three alternative ways the given points can be expressed in polar coordinates using $r > 0$, $r < 0$, and $\theta \in [-2\pi, 2\pi)$.

23. $\left(3\sqrt{2}, \frac{3\pi}{4}\right)$ **24.** $\left(4\sqrt{3}, -\frac{5\pi}{3}\right)$ **25.** $\left(-2, \frac{11\pi}{6}\right)$ **26.** $\left(-3, -\frac{7\pi}{6}\right)$

Match each (r, θ) given to one of the points A, B, C, or D shown.

27. $\left(4, -\frac{5\pi}{6}\right)$ **28.** $\left(4, -\frac{5\pi}{4}\right)$ **29.** $\left(-4, \frac{\pi}{6}\right)$ **30.** $\left(-4, \frac{3\pi}{4}\right)$ **31.** $\left(-4, -\frac{5\pi}{4}\right)$

32. $\left(-4, -\frac{\pi}{4}\right)$ **33.** $\left(4, \frac{13\pi}{6}\right)$ **34.** $\left(4, \frac{19\pi}{6}\right)$ **35.** $\left(-4, -\frac{21\pi}{4}\right)$ **36.** $\left(4, -\frac{35\pi}{6}\right)$

Convert from rectangular coordinates to polar coordinates. A diagram may help.

37. $(-8, 0)$ **38.** $(0, -7)$ **39.** $(4, 4)$ **40.** $(4\sqrt{3}, 4)$

41. $(5\sqrt{2}, 5\sqrt{2})$ **42.** $(6, -6\sqrt{3})$ **43.** $(-5, -12)$ **44.** $(-3.5, 12)$

Convert from polar coordinates to rectangular coordinates. A diagram may help.

45. $(8, 45°)$ **46.** $(6, 60°)$ **47.** $\left(4, \frac{3\pi}{4}\right)$ **48.** $\left(5, \frac{5\pi}{6}\right)$

49. $\left(-2, \frac{7\pi}{6}\right)$ **50.** $\left(-10, \frac{4\pi}{3}\right)$ **51.** $(-5, -135°)$ **52.** $(-4, -30°)$

Sketch each polar graph using an r-value analysis (a table may help), symmetry, and any convenient points.

53. $r = 5$ **54.** $r = 6$ **55.** $\theta = \frac{\pi}{6}$

56. $\theta = -\frac{3\pi}{4}$ **57.** $r = 4\cos\theta$ **58.** $r = 2\sin\theta$

59. $r = 3 + 3\sin\theta$ **60.** $r = 2 + 2\cos\theta$ **61.** $r = 2 - 4\sin\theta$

62. $r = 1 - 2\cos\theta$ **63.** $r = 5\cos(2\theta)$ **64.** $r = 3\sin(4\theta)$

65. $r = 4\sin(3\theta)$ **66.** $r = 6\cos(5\theta)$ **67.** $r^2 = 9\sin(2\theta)$

68. $r^2 = 16\cos(2\theta)$ **69.** $r = 4\sin\left(\frac{\theta}{2}\right)$ **70.** $r = 6\cos\left(\frac{\theta}{2}\right)$

Use a graphing calculator in polar mode to produce the following polar graphs.

71. $r = 4\sqrt{1 - \sin^2\theta}$, *a hippopede*

72. $r = 3 + \csc\theta$, *a conchoid*

73. $r = 2\cos\theta\cot\theta$, *a cissoid*

74. $r = \cot\theta$, *a kappa curve*

75. $r = 8\sin\theta\cos^2\theta$, *a bifoliate*

76. $r = 8\cos\theta(4\sin^2\theta - 2)$, *a folium*

WORKING WITH FORMULAS

77. The midpoint formula in polar coordinates: $M = \left(\frac{r\cos\alpha + R\cos\beta}{2}, \frac{r\sin\alpha + R\sin\beta}{2}\right)$

The midpoint of a line segment connecting the points (r, α) and (R, β) in polar coordinates can be found using the formula shown. Find the midpoint of the line segment between $(R, \alpha) = (6, 45°)$ and $(r, \beta) = (8, 30°)$, then convert these points to rectangular coordinates and find the midpoint using the "standard" formula. Do the results match?

78. The distance formula in polar coordinates: $d = \sqrt{R^2 + r^2 - 2Rr\cos(\alpha - \beta)}$

Using the law of cosines, it can be shown that the distance between the points (R, α) and (r, β) in polar coordinates is given by the formula indicated. Use the formula to find the distance between $(R, \alpha) = (6, 45°)$ and $(r, \beta) = (8, 30°)$, then convert these to rectangular coordinates and compute the distance between them using the "standard" formula. Do the results match?

APPLICATIONS

Polar graphs: Find the equation of a polar graph satisfying the given conditions, then sketch the graph.

79. limaçon, symmetric to polar axis, $a = 4$ and $b = 4$

80. rose, four petals, two petals symmetric to the polar axis, $a = 6$

81. rose, five petals, one petal symmetric to the polar axis, $a = 4$

82. limaçon, symmetric to $\theta = \frac{\pi}{2}$, $a = 2$ and $b = 4$

83. lemniscate, $a = 4$ through $(\pi, 4)$

84. lemniscate, $a = 8$ through $\left(8, \frac{\pi}{4}\right)$

85. circle, symmetric to $\theta = \frac{\pi}{2}$, center at $\left(2, \frac{\pi}{2}\right)$, containing $\left(2, \frac{\pi}{6}\right)$

86. circle, symmetric to polar axis, through $(6, \pi)$

Matching: Match each graph to its equation a through h, which follow. Justify your answers.

87. **88.** **89.**

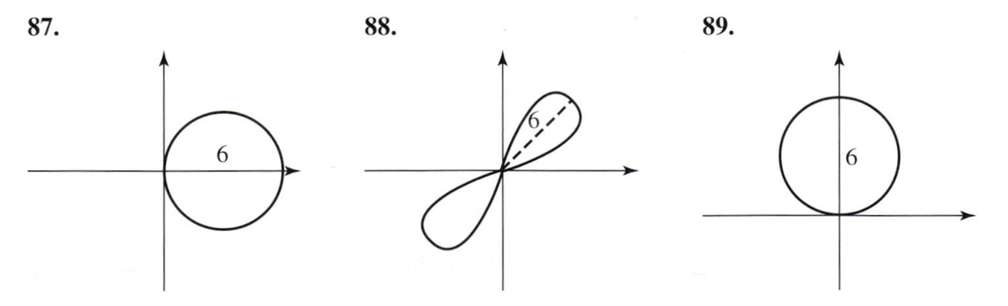

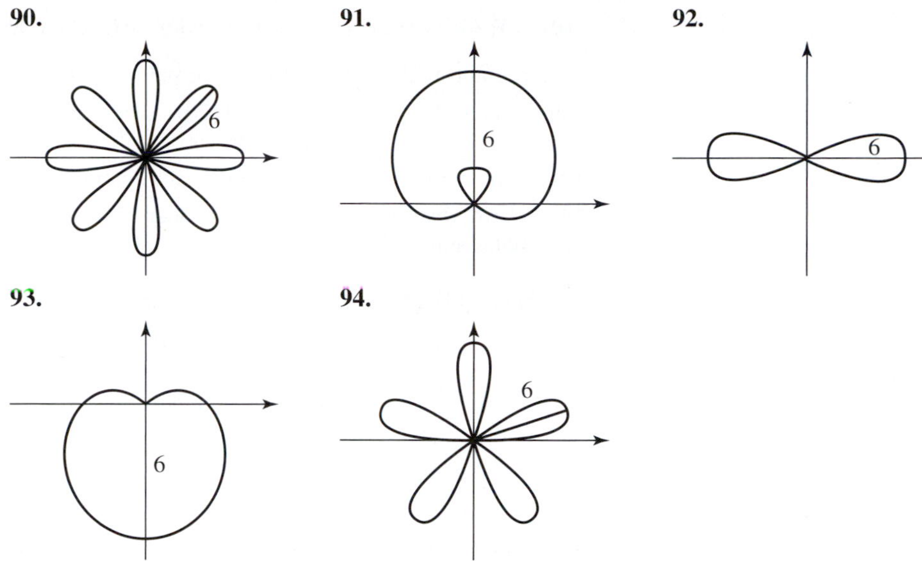

a. $r = 6\cos\theta$ **b.** $r = 3 - 3\sin\theta$ **c.** $r = 6\cos(4\theta)$ **d.** $r^2 = 36\cos(2\theta)$

e. $r^2 = 36\sin(2\theta)$ **f.** $r = 2 + 4\sin\theta$ **g.** $r = 6\sin\theta$ **h.** $r = 6\sin(5\theta)$

Exercise 95

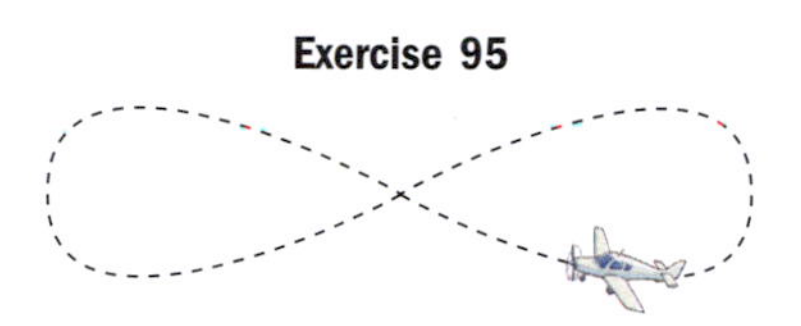

95. Figure eights: Waiting for help to arrive on foot, a light plane is circling over some stranded hikers using a "figure eight" formation, closely resembling the graph of a lemniscate. Suppose the flight path of the plane was plotted (using the hikers as the origin) and found to contain the polar coordinates (7200, 45°) and (0, 90°) with r in meters. Find the equation of the lemniscate.

96. Animal territories: Territorial animals often prowl the borders of their territory, marking the boundaries with various bodily excretions. Suppose the territory of one such animal was limaçon shaped, with the pole representing the den of the animal. Find the polar equation defining the animal's territory if markings are left at (750, 0°), (1000, 90°), and (750, 180°). Assume r is in meters.

Exercise 97

97. Prop manufacturing: The propellers for a toy boat are manufactured by stamping out a rose with n petals and then bending each blade. If the manufacturer wants propellers with five blades and a radius of 15 mm, what two polar equations will satisfy these specifications?

98. Polar curves and cosine: Do a complete r-value analysis for graphing polar curves involving cosine. Include a color-coded graph showing the relationship between r and θ, similar to the analysis for sines that preceded Example 6.

WRITING, RESEARCH, AND DECISION MAKING

99. The polar graph $r = a\theta$ is called the *Spiral of Archimedes.* Consider the spiral $r = \frac{1}{2}\theta$. As this graph spirals around the origin, what is the distance between each positive, polar intercept? In QI, what is the distance between consecutive branches of the spiral each time it intersects $\theta = \frac{\pi}{4}$? What is the distance between consecutive branches of the spiral at $\theta = \frac{\pi}{2}$? What can you conclude?

As mentioned in the exposition, tests for symmetry of polar graphs are sufficient to show symmetry (if the test is satisfied, the graph must be symmetric), but the tests are not necessary to show symmetry (the graph may be symmetric even if the test is not satisfied). For $r = f(\theta)$, the formal tests for the symmetry are: (1) the graph will be symmetric to the polar axis if

$f(\theta) = f(-\theta)$; (2) the graph will be symmetric to the line $\theta = \frac{\pi}{2}$ if $f(\pi - \theta) = f(\theta)$; and (3) the graph will be symmetric to the pole if $f(\theta) = -f(\theta)$.

100. Sketch the graph of $r = 4\sin(2\theta)$. Show the equation fails the first test, yet the graph is still symmetric to the polar axis.

101. Why is the graph of every lemniscate symmetric to the pole?

102. Verify that the graph of every limaçon of the form $r = a + b\cos\theta$ is symmetric to the polar axis.

EXTENDING THE CONCEPT

103. Although the equations are very different, systems of polar equations are solved in much the same way as systems of polynomial equations (substitution and elimination). Begin by graphing the system and estimating the intersection points, then solve the system algebraically.

a. $\begin{cases} r = \sin\theta \\ r = 1 - \sin\theta \end{cases}$ **b.** $\begin{cases} r = 2\sin(2\theta) \\ r = 2\sin\theta \end{cases}$

104. The graphs of $r = a\sin(n\theta)$ and $r = a\cos(n\theta)$ are from the rose family of polar graphs. If n is odd, there are n petals in the rose, and if n is even, there are $2n$ petals. An interesting extension of this fact is that the n petals enclose exactly 25% of the area of the circumscribed circle, and the $2n$ petals enclose exactly 50%. Find the area within the boundaries of the rose defined by $r = 6\sin(5\theta)$.

Exercise 108

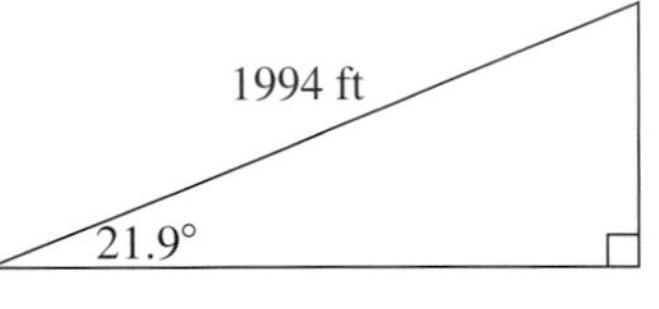

Exercise 110

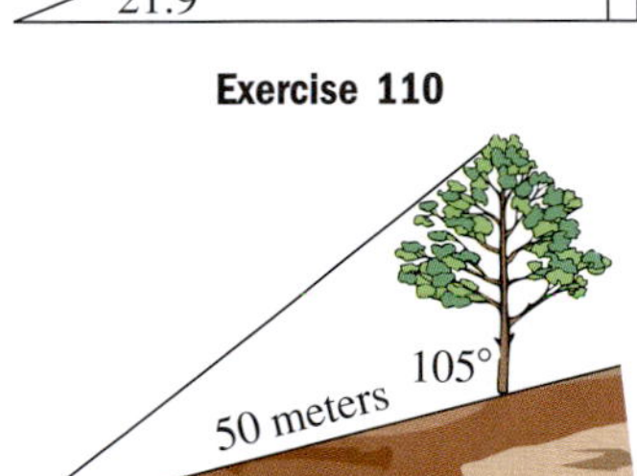

MAINTAINING YOUR SKILLS

105. (6.3) Identify the conic section and sketch its graph: $x^2 + 4y^2 - 6x + 8y - 3 = 0$.

106. (3.2) Verify the following is an identity: $\cos^2 x - \sin^2 x = 1 - \sin(2x)\tan x$.

107. (5.5) Find the angle between the vectors $\mathbf{u} = \langle -3, 8\rangle$ and $\mathbf{v} = \langle 5, 2\rangle$.

108. (1.2) Given that 1 acre = 43,560 ft^2, how many acres (to the nearest tenth) are covered by the triangular lot shown?

109. (4.4) Solve for $t \in [0, 2\pi)$:
$$20 = 5 - 30\sin\left(2t - \frac{\pi}{6}\right).$$

110. (1.3) Find the height of the tree, assuming the angle of elevation (from horizontal) is 41°, and the slope of the hillside is 14°.

6.6 More on the Conic Sections: Rotation of Axes and Polar Form

LEARNING OBJECTIVES

In Section 6.6 you will learn how to:

A. Graph conic sections that have nonvertical and nonhorizontal axes (rotated conics)

B. Identify conics using the discriminant of the polynomial form—the invariant $B^2 - 4AC$

C. Write the equation of a conic section in polar form

D. Solve applications involving the conic sections in polar form

INTRODUCTION

Our study of conic sections would not be complete without considering conic sections whose graphs are not symmetric to a vertical or horizontal axis. The axis of symmetry still exists, but is rotated by some angle. We'll first study these **rotated conics** using the equation in its polynomial form, then investigate some interesting applications of the polar form.

POINT OF INTEREST

In mathematics, an *invariant* is a quantity, expression, or relationship that remains unchanged when a given transformation is applied. For example, the relationship $V - E + F = 2$ applies to any polyhedron regardless of how it is oriented, where V represents the number of vertices, E represents the number of edges, and F

represents the number of faces. Felix Klein (1849–1925) used invariants extensively in his organization and description of hyperbolic, parabolic, and elliptic geometry. His work (along with that of many others) helped open the branch of mathematics known as topology. In this section, we introduce the invariant $B^2 - 4AC$, used to identify the conic sections.

A. Rotated Conics and the Rotation of Axes

It's always easier to understand a new idea in terms of a known idea, so we begin our study with a review of the reciprocal function $y = \frac{1}{x}$. From the equation we note:

1. The denominator is zero when $x = 0$, and the y-axis is a vertical asymptote (the vertical line $x = 0$).
2. Since the degree of the numerator is less than the degree of the denominator, the x-axis is a horizontal asymptote (the horizontal line $y = 0$).
3. Since $x < 0$ implies $y < 0$ and $x > 0$ implies $y > 0$, the graph will have two branches—one in the first quadrant and one in the third.

Note the polynomial form of this equation is $xy = 1$. The resulting graph is shown in Figure 6.51, *and is actually the graph of a hyperbola with a transverse axis of* $y = x$. Using the 45-45-90 triangle indicated, we find the distance from the origin to each vertex is $\sqrt{2}$. If we rotated the hyperbola 45° clockwise, we would obtain a more "standard" graph with a horizontal transverse axis and vertices at $(\pm a, 0) \rightarrow (\pm\sqrt{2}, 0)$. The asymptotes would be $y = \pm 1x$, and since $y = \pm\frac{b}{a}x$ is the general form we know $b = \pm\sqrt{2}$. This information can be used to find the equation of the rotated hyperbola.

Figure 6.51

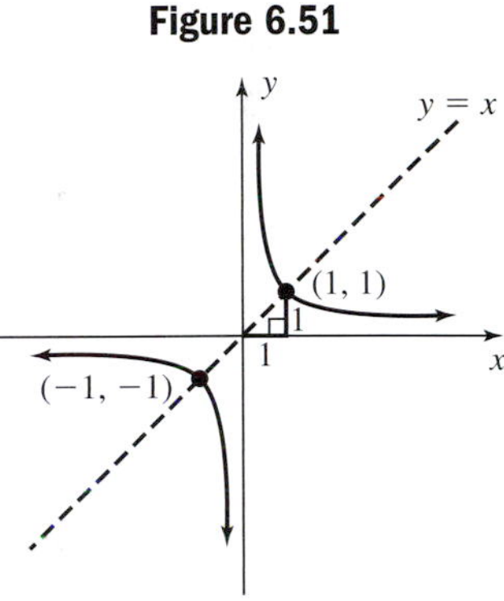

EXAMPLE 1 The hyperbola $xy = 1$ is rotated clockwise 45°, with new vertices at $(\pm\sqrt{2}, 0)$, asymptotes at $y = \pm 1x$ and $b = \pm\sqrt{2}$. Find the equation and graph the hyperbola.

Solution: Using the standard form $\frac{x^2}{a^2} - \frac{y^2}{b^2} = 1$ and substituting $\pm\sqrt{2}$ for a and b, the equation of the rotated hyperbola is $\frac{x^2}{2} - \frac{y^2}{2} = 1$ or $x^2 - y^2 = 2$ in polynomial form. The resulting graph is the central hyperbola shown in Figure 6.52.

Figure 6.52

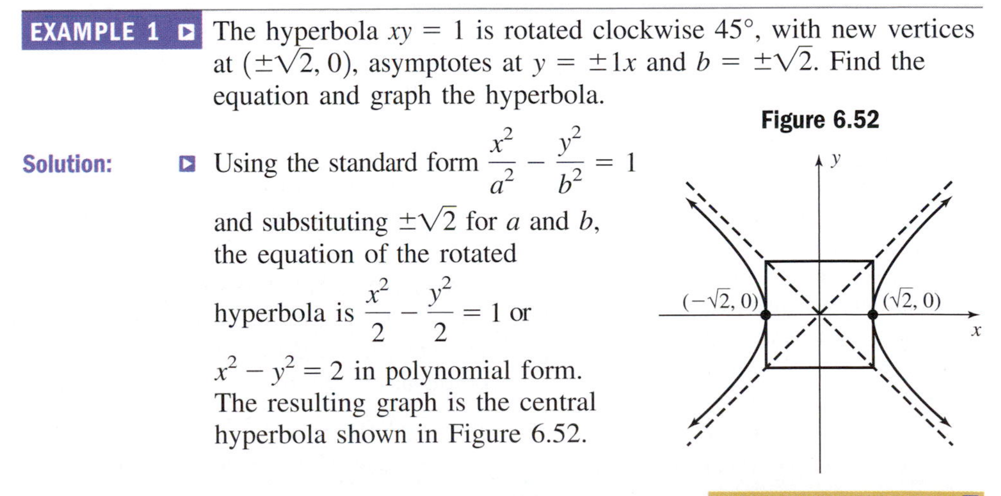

NOW TRY EXERCISES 7 AND 8

It's important to note the equation of the rotated hyperbola *is devoid of the mixed "xy" term.* In nondegenerate cases, the equation $Ax^2 + Cy^2 + Dx + Ey + F = 0$ is the polynomial form of a conic with axes that are vertical/horizontal. However, the most general form of the equation is $Ax^2 + Bxy + Cy^2 + Dx + Ey + F = 0$, and includes this Bxy term. As noted in Example 1, the inclusion of this term will rotate the graph through some angle β. Based on these observations, we reason that one approach to graphing these conics is to find the angle of rotation β with respect to the xy-axes. We can then use β to rewrite the equation so that it corresponds to a new set of XY-axes, *which are parallel to the axes of the conic*. The mixed xy-term will be absent from the new equation and we can graph the conic on the new axes using the same ideas as before (identifying a, b, foci, and so on). To find β, recall that a point (x, y) in the xy-plane can be written $x = r\cos\alpha$, $y = r\sin\alpha$, as in Figure 6.53. The diagram in Figure 6.54 shows the axes of a new XY-plane, rotated counterclockwise by angle β. In this new plane, the coordinates of the point (x, y) become $X = r\cos(\alpha - \beta)$ and $Y = r\sin(\alpha - \beta)$ as shown. Using the difference identities for sine and cosine and substituting $x = r\cos\alpha$ and $y = r\sin\alpha$ leads to

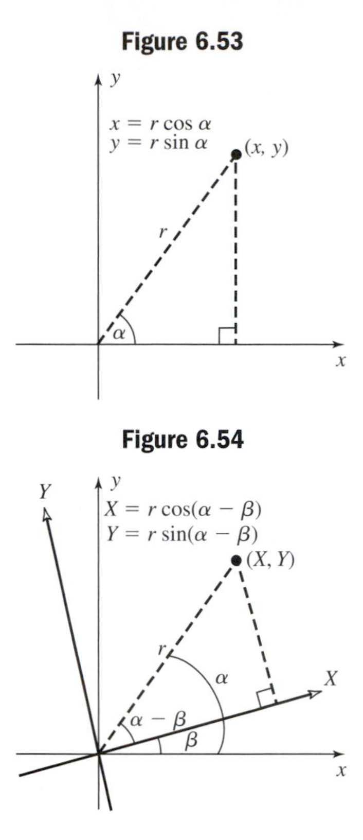

$$\begin{aligned} X &= r\cos(\alpha - \beta) \\ &= r(\cos\alpha\cos\beta + \sin\alpha\sin\beta) \\ &= r\cos\alpha\cos\beta + r\sin\alpha\sin\beta \\ &= x\cos\beta + y\sin\beta \end{aligned} \qquad \begin{aligned} Y &= r\sin(\alpha - \beta) \\ &= r(\sin\alpha\cos\beta - \cos\alpha\sin\beta) \\ &= r\sin\alpha\cos\beta - r\cos\alpha\sin\beta \\ &= y\cos\beta - x\sin\beta \end{aligned}$$

The last two equations can be written as a system, which we will use to solve for x and y in terms of X and Y.

$$\begin{cases} X = x\cos\beta + y\sin\beta \\ Y = y\cos\beta - x\sin\beta \end{cases} \quad \text{original system}$$

$$\begin{cases} X\cos\beta = x\cos^2\beta + y\sin\beta\cos\beta \\ Y\sin\beta = y\sin\beta\cos\beta - x\sin^2\beta \end{cases} \quad \begin{matrix}\text{multiply first equation by } \cos\beta \\ \text{multiply second equation by } \sin\beta\end{matrix}$$

$$X\cos\beta - Y\sin\beta = x\cos^2\beta + x\sin^2\beta \quad \text{first equation} - \text{second equation}$$

$$X\cos\beta - Y\sin\beta = x \quad \text{factor out } x\,(\cos^2\beta + \sin^2\beta = 1)$$

Re-solving the system for y results in $y = X\sin\beta + Y\cos\beta$, yielding what are called the **rotation of axes formulas** (see Exercise 79).

WORTHY OF NOTE

If you are familiar with matrices, it may be easier to remember the rotation formulas in their matrix form, since the pattern of functions is the same, with only a difference in sign:

$$\begin{bmatrix} x \\ y \end{bmatrix} = \begin{bmatrix} \cos\beta & -\sin\beta \\ \sin\beta & \cos\beta \end{bmatrix}\begin{bmatrix} X \\ Y \end{bmatrix}$$

$$\begin{bmatrix} X \\ Y \end{bmatrix} = \begin{bmatrix} \cos\beta & \sin\beta \\ -\sin\beta & \cos\beta \end{bmatrix}\begin{bmatrix} x \\ y \end{bmatrix}$$

ROTATION OF AXES FORMULAS

If the x- and y-axes of the xy-plane are rotated counterclockwise by the (acute) angle β to form the X- and Y-axes of an XY-plane, the coordinates of the points (x, y) and (X, Y) are related by the formulas

$$x = X\cos\beta - Y\sin\beta \qquad X = x\cos\beta + y\sin\beta$$
$$y = X\sin\beta + Y\cos\beta \qquad Y = -x\sin\beta + y\cos\beta$$

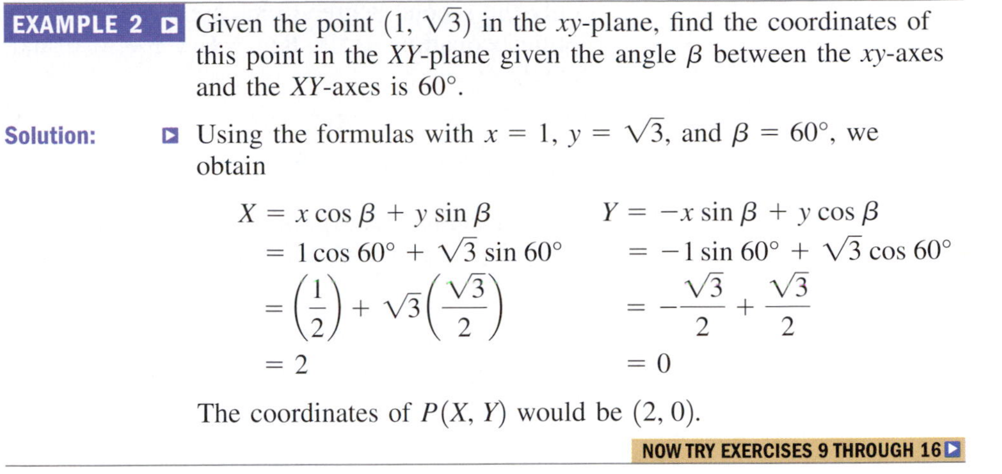

EXAMPLE 2 Given the point $(1, \sqrt{3})$ in the xy-plane, find the coordinates of this point in the XY-plane given the angle β between the xy-axes and the XY-axes is 60°.

Solution: Using the formulas with $x = 1$, $y = \sqrt{3}$, and $\beta = 60°$, we obtain

$$\begin{aligned} X &= x\cos\beta + y\sin\beta \\ &= 1\cos 60° + \sqrt{3}\sin 60° \\ &= \left(\frac{1}{2}\right) + \sqrt{3}\left(\frac{\sqrt{3}}{2}\right) \\ &= 2 \end{aligned} \qquad \begin{aligned} Y &= -x\sin\beta + y\cos\beta \\ &= -1\sin 60° + \sqrt{3}\cos 60° \\ &= -\frac{\sqrt{3}}{2} + \frac{\sqrt{3}}{2} \\ &= 0 \end{aligned}$$

The coordinates of $P(X, Y)$ would be $(2, 0)$.

NOW TRY EXERCISES 9 THROUGH 16

Figure 6.55

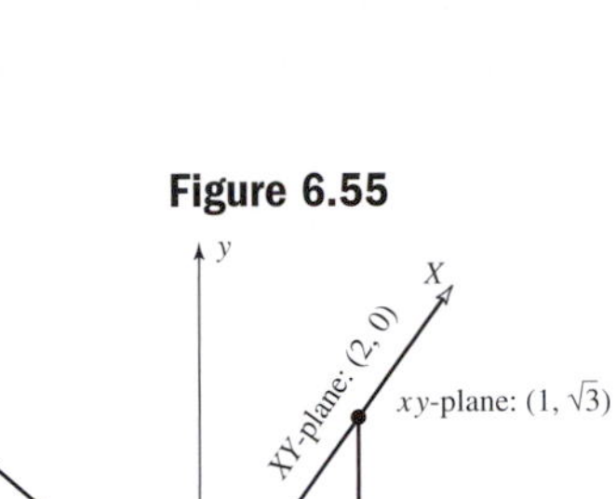

The diagram in Figure 6.55 provides a more intuitive look at the rotation from Example 2. As you can see, a 30-60-90 triangle is formed with a hypotenuse of 2, giving coordinates (2, 0) in the XY-plane.

EXAMPLE 3 The ellipse $X^2 + 4Y^2 = 16$ is rotated clockwise 45°. What is the corresponding equation in the xy-plane?

Solution: We proceed as before, using the rotation formulas $X = x\cos\beta + y\sin\beta$ and $Y = y\cos\beta - x\sin\beta$. With $\beta = 45°$ we have $\cos\beta = \sin\beta = \frac{\sqrt{2}}{2}$, yielding

$$\begin{aligned} X^2 + 4Y^2 &= 16 \\ (x\cos\beta + y\sin\beta)^2 + 4(y\cos\beta - x\sin\beta)^2 &= 16 && \text{use rotation formulas} \\ \left(\frac{\sqrt{2}}{2}x + \frac{\sqrt{2}}{2}y\right)^2 + 4\left(\frac{\sqrt{2}}{2}y - \frac{\sqrt{2}}{2}x\right)^2 &= 16 && \text{substitute } \tfrac{\sqrt{2}}{2} \text{ for } \sin\beta \text{ and } \cos\beta \\ \left(\frac{1}{2}x^2 + xy + \frac{1}{2}y^2\right) + 4\left(\frac{1}{2}x^2 - xy + \frac{1}{2}y^2\right) &= 16 && \text{square binomials} \\ \frac{1}{2}x^2 + xy + \frac{1}{2}y^2 + 2x^2 - 4xy + 2y^2 &= 16 && \text{distribute} \\ \frac{5}{2}x^2 - 3xy + \frac{5}{2}y^2 &= 16 && \text{result} \end{aligned}$$

NOW TRY EXERCISES 17 THROUGH 20

Note the equation of the conic in the standard xy-plane contains the "mixed" Bxy-term. In practice, we seek to reverse this procedure by starting in the xy-plane, and finding the angle β needed to *eliminate* the Bxy-term. Using the rotation formulas and the appropriate angle β, the equation $Ax^2 + Bxy + Cy^2 + Dx + Ey + F = 0$ becomes $aX^2 + cY^2 + dX + eY + f = 0$, where the xy-term is absent. To find the angle β, note that without loss of generality, we can assume $D = E = 0$ since only the second-degree terms are used to identify a conic. Starting with the simplified equation $Ax^2 + Bxy + Cy^2 + F = 0$ and using the rotation formulas we obtain

$$Ax^2 \quad + \quad Bx \quad \cdot \quad y \quad + \quad Cy^2 \quad + F = 0$$
$$A(X\cos\beta - Y\sin\beta)^2 + B(X\cos\beta - Y\sin\beta)(X\sin\beta + Y\cos\beta) + C(X\sin\beta + Y\cos\beta)^2 + F = 0$$

Expanding this expression and collecting like terms (see Exercise 80), gives the following expressions for coefficients a, b, and c of the corresponding equation $aX^2 + bXY + cY^2 + f = 0$:

$a \rightarrow A\cos^2\beta + B\sin\beta\cos\beta + C\sin^2\beta$ — a is the coefficient of X^2

$b \rightarrow -2A\sin\beta\cos\beta + B(\cos^2\beta - \sin^2\beta) + 2C\sin\beta\cos\beta$ — b is the coefficient of XY

$c \rightarrow A\sin^2\beta - B\sin\beta\cos\beta + C\cos^2\beta$ — c is the coefficient of Y^2

$f \rightarrow F$ — $f = F$ (the constant remains unchanged)

To accomplish our purpose, we require the coefficient b to be zero. While this expression looks daunting, the double-angle identities for sine and cosine simplify it very nicely:

$$b \rightarrow -A(2\sin\beta\cos\beta) + B(\cos^2\beta - \sin^2\beta) + C(2\sin\beta\cos\beta) = 0 \quad (1)$$
$$-A\sin(2\beta) + B\cos(2\beta) + C\sin(2\beta) = 0 \quad (2)$$
$$(C - A)\sin(2\beta) = -B\cos(2\beta) \quad (3)$$
$$\tan(2\beta) = \frac{-B}{C - A} \quad (4)$$
$$\tan(2\beta) = \frac{B}{A - C};\ A \neq C \quad (5)$$

Note from line (3) that $A = C$ would imply $\cos(2\beta) = 0$, giving $2\beta = 90°$ or $-90°$, with $\beta = 45°$ or $-45°$ (for the sake of convenience, we select the angle in QI). This fact can many times be used to great advantage. If $A \neq C$, $\tan(2\beta) = \frac{B}{A - C}$ and we choose 2β between 0 and 180° so that β will be in the first quadrant $[0 < \beta < 90°]$.

THE EQUATION OF A CONIC AFTER ROTATING THE AXES

Given (1) $Ax^2 + Bxy + Cy^2 + Dx + Ey + F = 0$ and its related graph in the xy-plane, an angle β can be found using $\tan(2\beta) = \frac{B}{A - C}$, with β used in the rotation formulas to determine the coefficients of polynomial (2) $aX^2 + cY^2 + dX + eY + f = 0$. The orientation of the graphs from (1) and (2) are identical, but the graph of (2) is a vertical or horizontal conic in the XY-plane. Recall that if $A = C$, $\beta = 45°$.

EXAMPLE 4 For $x^2 - 2\sqrt{3}xy + 3y^2 - \sqrt{3}x - y - 16 = 0$, eliminate the xy-term using a rotation of axes and identify the conic associated with the resulting equation. Then sketch the graph of the rotated conic in the XY-plane.

Solution: Since $A \neq C$, we find β using $\tan(2\beta) = \frac{B}{A - C}$, giving $\tan(2\beta) = \frac{-2\sqrt{3}}{1 - 3} = \sqrt{3}$. This shows $2\beta = \tan^{-1}\sqrt{3}$, yielding $2\beta = 60°$ so $\beta = 30°$. Using $\cos 30° = \frac{\sqrt{3}}{2}$ and $\sin 30° = \frac{1}{2}$ along with the rotation formulas we obtain the following XY-equation, with corresponding terms shown side-by-side for clarity:

Given term in xy-plane	Corresponding term in XY-plane
x^2	$\left(\frac{\sqrt{3}}{2}X - \frac{1}{2}Y\right)^2 = \frac{3}{4}X^2 - \frac{\sqrt{3}}{2}XY + \frac{1}{4}Y^2$
$-2\sqrt{3}xy$	$-2\sqrt{3}\left(\frac{\sqrt{3}}{2}X - \frac{1}{2}Y\right)\left(\frac{1}{2}X + \frac{\sqrt{3}}{2}Y\right) = -\frac{3}{2}X^2 - \sqrt{3}XY + \frac{3}{2}Y^2$
$3y^2$	$3\left(\frac{1}{2}X + \frac{\sqrt{3}}{2}Y\right)^2 = \frac{3}{4}X^2 + 3\frac{\sqrt{3}}{2}XY + \frac{9}{4}Y^2$
$-\sqrt{3}x$	$-\sqrt{3}\left(\frac{\sqrt{3}}{2}X - \frac{1}{2}Y\right) = -\frac{3}{2}X + \frac{\sqrt{3}}{2}Y$
$-y$	$-\left(\frac{1}{2}X + \frac{\sqrt{3}}{2}Y\right) = -\frac{1}{2}X - \frac{\sqrt{3}}{2}Y$
-16	-16

Adding the like terms to the far right, the X^2-terms (in red), the Y-terms (**in bold**), and the mixed XY-terms (in blue) sum to zero, leaving the equation $-2X + 4Y^2 - 16 = 0$, which is the parabola defined by $Y^2 = \frac{1}{2}(X + 8)$. This parabola is symmetric to the X-axis, concave up, with a vertex at $(-8, 0)$, Y-intercepts at $(0, -2)$ and $(0, 2)$, focus at $(-\frac{63}{8}, 0)$ and directrix through $(-\frac{65}{8}, 0)$. The graph is shown in the figure.

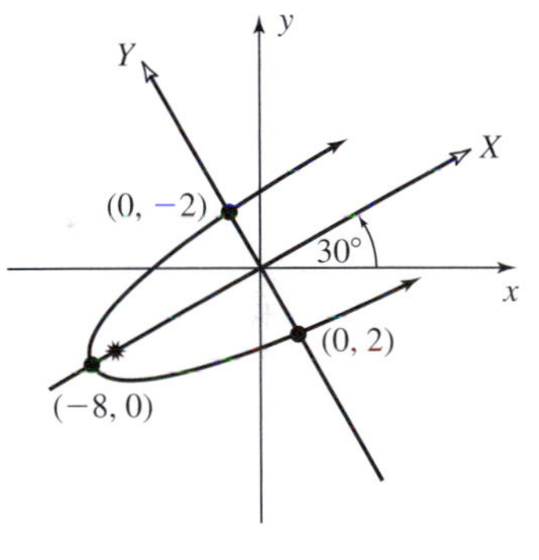

NOW TRY EXERCISES 21 THROUGH 30

In Example 4, the angle β was a **standard angle** and easily found. In general, this is not the case and finding exact values of $\cos\beta$ and $\sin\beta$ for use in the rotation formulas requires using $\tan(2\beta) = \frac{\sin(2\beta)}{\cos(2\beta)}$, the corresponding (triangle) diagram, and the identities $\cos\beta = \sqrt{\frac{1 + \cos(2\beta)}{2}}$ and $\sin\beta = \sqrt{\frac{1 - \cos(2\beta)}{2}}$. See Exercises 31, 32, 87, and 88 for further study.

B. Identifying Conics Using the Discriminant

In addition to rotating the axes, the inclusion of the "xy-term" makes it impossible to identify the conic section using the tests seen earlier. For example, having $A = C$ no longer guarantees a circle, and $A = 0$ or $C = 0$ does not guarantee a parabola. Rather than continuing to look at what the mixed term and the resulting rotation *changes,* we now look at what the rotation *does not change*, called **invariants** of the transformation. These invariants can be used to double-check the algebra involved and to identify the conic using the **discriminant.** These are given here without proof.

INVARIANTS OF A ROTATION AND CLASSIFICATION USING THE DISCRIMINANT

By rotating the coordinate axes through a predetermined angle β,

the equation $Ax^2 + Bxy + Cy^2 + Dx + Ey + F = 0$
can be transformed into $aX^2 + cY^2 + dX + eY + f = 0$

in which the xy-term is absent. This rotation has the following invariants:

(1) $F = f$ (2) $A + C = a + c$ (3) $B^2 - 4AC = b^2 - 4ac$.

The discriminant of a conic equation in polynomial form is $B^2 - 4AC$. Except in degenerate cases, the graph of the equation can be classified as follows:

If $B^2 - 4AC = 0$, the graph will be a parabola.

If $B^2 - 4AC < 0$, the graph will be a circle or an ellipse.

If $B^2 - 4AC > 0$, the graph will be a hyperbola.

EXAMPLE 5A Verify the invariants just given using the equations from Example 4. Also verify the discriminant test.

Solution: From the equation $x^2 - 2\sqrt{3}xy + 3y^2 - \sqrt{3}x - y - 16 = 0$, we have $A = 1$, $B = -2\sqrt{3}$, $C = 3$, $D = -\sqrt{3}$, $E = -1$, and $F = -16$. After applying the rotation the equation became $-2X + 4Y^2 - 16 = 0$, with $a = 0$, $b = 0$, $c = 4$, $d = -2$, $e = 0$, and $f = -16$. Checking each invariant gives (1) $-16 = -16$✓, (2) $1 + 3 = 0 + 4$✓, and (3) $(-2\sqrt{3})^2 - 4(1)(3) = (0)^2 - 4(0)(4)$✓. With $B^2 - 4AC = 0$, the discriminant test indicates the conic is a parabola✓.

EXAMPLE 5B Use the discriminant to identify each equation as that of a circle, ellipse, parabola, or hyperbola, but do not graph the equation.

a. $3x^2 - 4xy + 3y^2 + 6x + 12y - 2 = 0$

b. $4x^2 + 9xy + 4y^2 - 8x + 24y + 9 = 0$

c. $6x^2 - 7xy + y^2 - 5 = 0$

d. $x^2 - 6xy + 9y^2 + 6x = 0$

Solution:

a. $A = 3;\ B = -4;\ C = 3$

$B^2 - 4AC = (-4)^2 - 4(3)(3)$
$= -20$

circle or ellipse

b. $A = 4;\ B = 9;\ C = 4$

$B^2 - 4AC = (9)^2 - 4(4)(4)$
$= 17$

hyperbola

c. $A = 6;\ B = -7;\ C = 1$

$B^2 - 4AC = (-7)^2 - 4(6)(1)$
$= 25$

hyperbola

d. $A = 1;\ B = -6;\ C = 9$

$B^2 - 4AC = (-6)^2 - 4(1)(9)$
$= 0$

parabola

NOW TRY EXERCISES 33 THROUGH 36

C. Conic Equations in Polar Form

You might recall that earlier in this chapter we defined ellipses and hyperbolas in terms of a distance between two points, but a parabola in terms of a distance between a point and a line (the focus and directrix). Actually, all conic sections can be defined using a focus/directrix development *and written in polar form.* This serves to unify and greatly simplify their study. We begin by revisiting the focus/directrix development of a parabola, using a directrix $\mathscr{L}$ and placing the focus at the origin. With the polar axis as the axis of symmetry and the point $P(r, \theta)$ in polar coordinates, we obtain the graph shown in Figure 6.56. Given D and A are points on $\mathscr{L}$ (with A on the polar axis), we note the following:

Figure 6.56

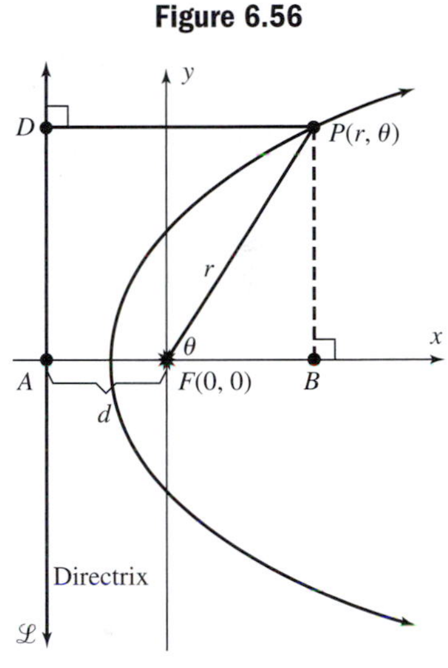

$$\overline{DP} = \overline{FP} \quad \text{definition of a parabola}$$
$$\overline{DP} = \overline{AB} \quad \text{equal line segments}$$
$$\overline{FB} = r\cos\theta \quad \cos\theta = \frac{\overline{FB}}{r}$$
$$\overline{AB} = \overline{AF} + \overline{FB} \quad \text{sum of line segments}$$

Using the preceding and representing the distance $\overline{AF}$ by the constant d, we obtain this sequence:

$$\overline{AB} = d + r\cos\theta \quad \text{substitute } d \text{ for } \overline{AF} \text{ and } r\cos\theta \text{ for } \overline{FB}$$
$$\overline{FP} = d + r\cos\theta \quad \text{substitute } \overline{FP} \text{ for } \overline{AB} \text{ since } \overline{FP} = \overline{DP} = \overline{AB}$$
$$r = d + r\cos\theta \quad \text{substitute } r \text{ for } \overline{FP}$$

Solving the last equation for r we have $r - r\cos\theta = d$, then $r = \dfrac{d}{1 - \cos\theta}$, which is the equation of a parabola in polar form with its focus at the origin, vertex at $\left(\dfrac{d}{2}, \pi\right)$, and y-intercepts at $\left(d, \dfrac{\pi}{2}\right)$ and $\left(d, \dfrac{3\pi}{2}\right)$. Note the constant "1" in the denominator is a key characteristic of polar equations, and helps define the standard form.

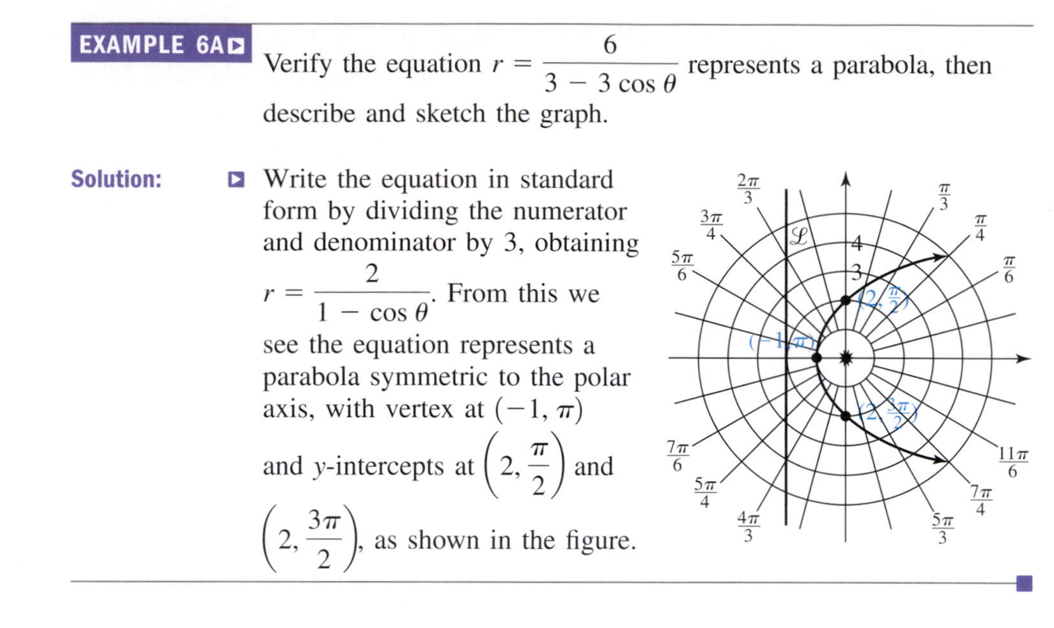

EXAMPLE 6A Verify the equation $r = \dfrac{6}{3 - 3\cos\theta}$ represents a parabola, then describe and sketch the graph.

Solution: Write the equation in standard form by dividing the numerator and denominator by 3, obtaining $r = \dfrac{2}{1 - \cos\theta}$. From this we see the equation represents a parabola symmetric to the polar axis, with vertex at $(-1, \pi)$ and y-intercepts at $\left(2, \dfrac{\pi}{2}\right)$ and $\left(2, \dfrac{3\pi}{2}\right)$, as shown in the figure.

The polar equation for a parabola depended on $\overline{DP}$ and $\overline{FP}$ being equal in length, with ratio $\dfrac{\overline{FP}}{\overline{DP}} = 1$. But what if this ratio is not equal to 1? To explore this question, assume $\dfrac{\overline{FP}}{\overline{DP}} = \dfrac{1}{2}$ and investigate the graph that results. Cross-multiplying gives $2\overline{FP} = \overline{DP}$, which states that the distance from D to P is twice the distance from F to P. Note that we are able to locate *two points* P_1 *and* P_2 *on the polar axis* that satisfy this relation, rather than only one as in the case of the parabola.

Figure 6.57

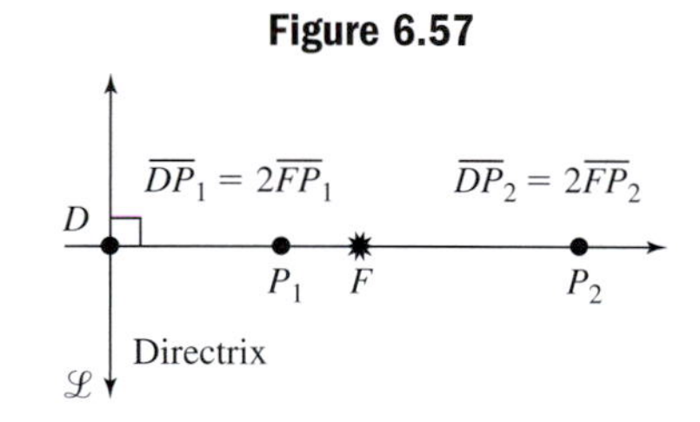

Figure 6.57 illustrates the location of these points. Using the focal chord for convenience, two additional points P_3 and P_4 can be located that also satisfy the stated condition (see Figure 6.58). In fact, we can locate an infinite number of these points using $\dfrac{\overline{FP}}{\overline{DP}} = \dfrac{1}{2}$, and the resulting graph appears to be an ellipse (and is definitely *not* a parabola). These illustrations provide the basis for stating a general focus/directrix definition of the conic sections. The ratio $\dfrac{\overline{FP}}{\overline{DP}}$ is often represented by the letter e, and represents the **eccentricity** of the conic. Using $\overline{FP} = r$ and $\overline{DP} = d + r\cos\theta$ from our initial development, $\dfrac{\overline{FP}}{\overline{DP}} = \dfrac{r}{d + r\cos\theta} = e$, which enables us to state the general equation of a conic in polar form. Solving for r leads to the equation $r = \dfrac{de}{1 - e\cos\theta}$, where the type of conic depends solely on e. Depending on the orientation of the conic, the general form may involve sine instead of cosine, and have a sum of terms in the denominator rather than a difference. Note once again that if $e = 1$, the relation simplifies into the parabolic equation seen earlier.

Figure 6.58

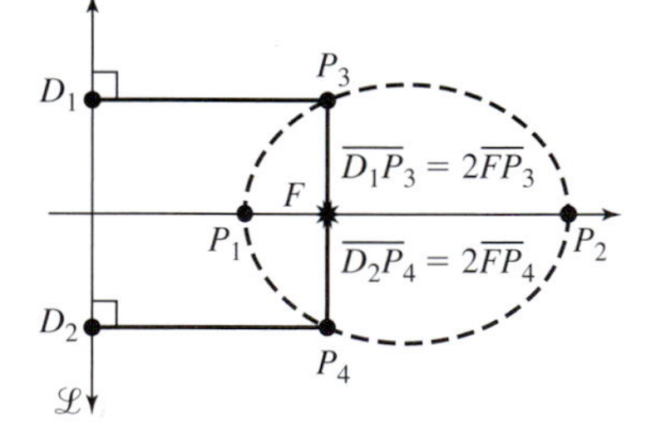

THE STANDARD EQUATION OF A CONIC IN POLAR FORM

Given a conic section with eccentricity e, one foci at the pole of the $r\theta$-plane, and directrix $\mathscr{L}$ located d units from this focus. Then the polar equations

$$r = \frac{de}{1 \pm e\cos\theta} \quad \text{and} \quad r = \frac{de}{1 \pm e\sin\theta}$$

represent one of the conic sections as determined by the value of e.

If $e = 1$, the graph is a parabola.
If $0 < e < 1$, the graph is an ellipse.
If $e > 1$, the graph is a hyperbola.

As in our previous study of polar equations, if the equation involves cosine the graph will be symmetric to the polar axis. If the graph involves sine, the line $\theta = \frac{\pi}{2}$ is the axis of symmetry. In addition, if the denominator contains a difference of terms (as in Example 6A), the graph will be above or to the right of the directrix (depending on whether the equation involves sine or cosine). If the denominator contains a sum of terms, the graph will be below or to the left of the directrix.

EXAMPLE 6B Determine if the equation $r = \frac{10}{5 - 3\sin\theta}$ represents a parabola, ellipse, or hyperbola. Then describe and sketch the graph.

Solution: To write the equation in standard form, we divide both numerator and denominator by 5, obtaining the equation $r = \frac{2}{1 - \frac{3}{5}\sin\theta}$.

From the standard form we note $e = \frac{3}{5}$ so the equation represents an ellipse. With a difference of terms and the sine function involved, the graph is symmetric to $\theta = \frac{\pi}{2}$ and is above the directrix. Given so much information by the equation, we require very few points to sketch the graph and settle for those generated by $\theta = 0, \frac{\pi}{2}, \pi$, and $\frac{3\pi}{2}$, yielding the points $(2, 0)$, $\left(5, \frac{\pi}{2}\right)$, $(2, \pi)$, and $\left(\frac{5}{4}, \frac{3\pi}{2}\right)$. The graph is shown in the figure.

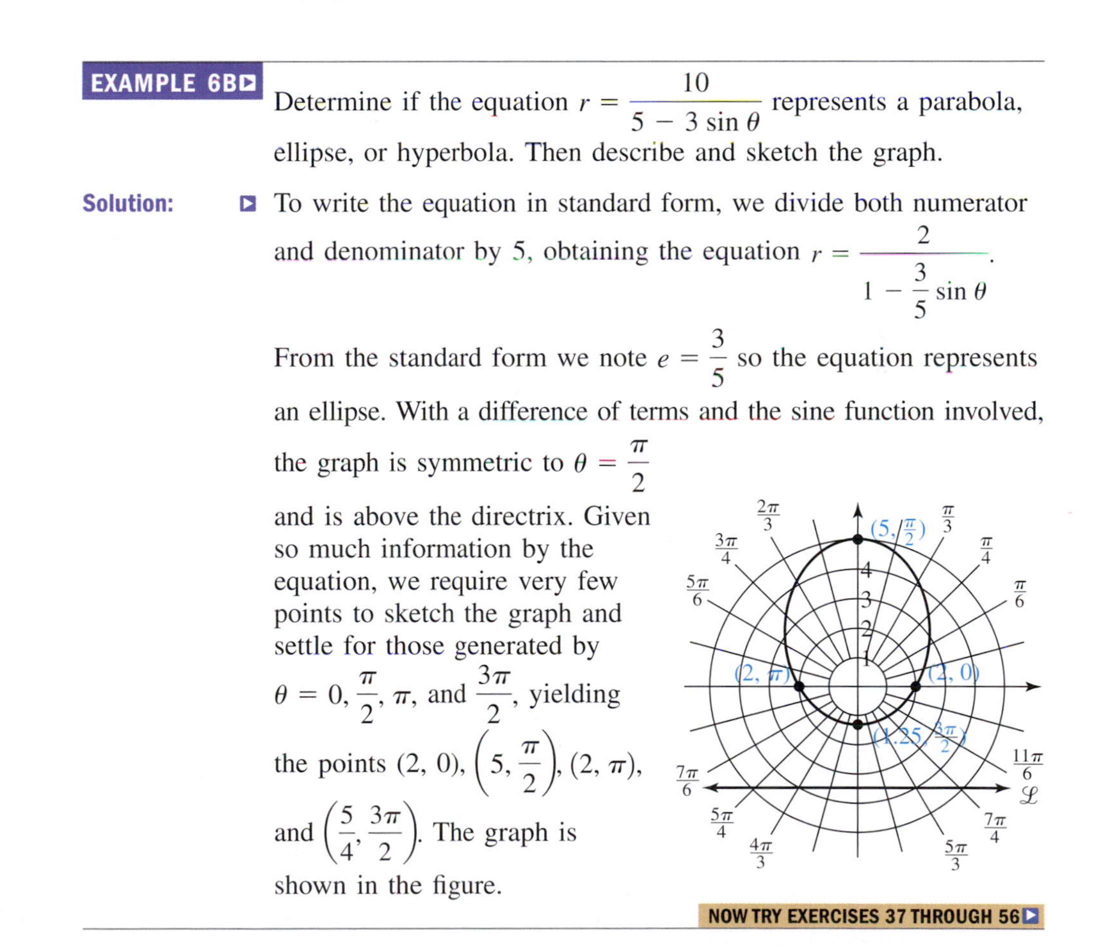

NOW TRY EXERCISES 37 THROUGH 56

D. Applications of Conics in Polar Form

For centuries it has been known that the orbits of the planets around the Sun are elliptical, with the Sun at one focus. In addition, comets may approach our Sun in an elliptical, hyperbolic, or parabolic path with the Sun again at the foci. This makes planetary studies a very natural application of the conic sections in polar form. To aid this study, it helps to know that in an elliptical orbit, the maximum distance of a planet from the Sun is called its **aphelion,** and the shortest distance is the **perihelion** (Figure 6.59). This means the length of the major axis is "aphelion + perihelion," enabling us to find the value of c if the aphelion and perihelion are known (Figure 6.60). Using our earlier, alternate definition of eccentricity $e = \frac{c}{a}$ (Exercise 47, Section 6.3), we can then find the eccentricity of the planet's orbit.

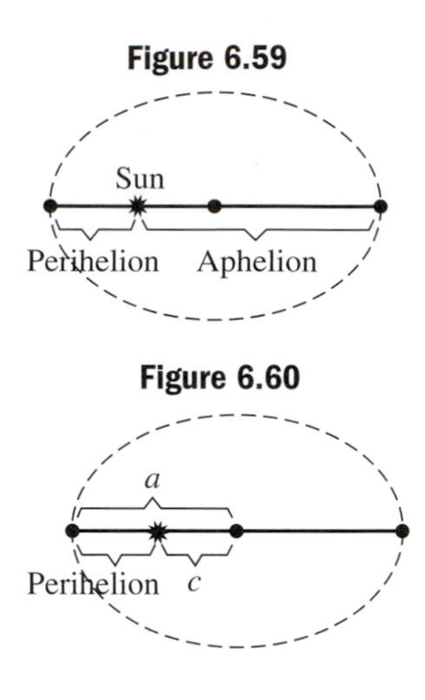

EXAMPLE 7 In its elliptical orbit around the Sun, Mars has an aphelion of 154.9 million miles and a perihelion of 128.4 million miles. What is the eccentricity of its orbit?

Solution: The length of the major axes would be $2a = (154.9 + 128.4)$ mi, yielding a semimajor axis of $a = 141.65$ million miles. Since $a = c +$ perihelion (Figure 6.60), we have $141.65 = c + 128.4$ so $c = 13.25$. The eccentricity of the orbit is $e = \frac{c}{a} = \frac{13.25}{141.65}$ or about 0.0935.

NOW TRY EXERCISES 59 AND 60

We can also find the perihelion and aphelion directly in terms of a (semimajor axis) and e (eccentricity) if these quantities are known. Using $a = c +$ perihelion, we obtain: perihelion $= a - c$. For $e = \frac{c}{a}$, we have $ea = c$ and by direct substitution we obtain: perihelion $= a - ea = a(1 - e)$. You are asked for the aphelion equation in Exercise 84. For Example 8, recall that "AU" designates an *astronomical unit*, and represents the mean distance from the Earth to the Sun, approximately 92.96 million miles.

EXAMPLE 8 The orbit of the planet Jupiter has a semimajor axis of 5.2 AU (1 AU $\approx$ 92.96 million miles) and an eccentricity of 0.0489. What is the closest distance from Jupiter to the Sun?

Solution: With perihelion $= a(1 - e)$, we have $5.2(1 - 0.0489) \approx 4.946$. At its closest approach, Jupiter is 4.946 AU from the Sun (about 460 million miles).

NOW TRY EXERCISES 61 THROUGH 64

Figure 6.61

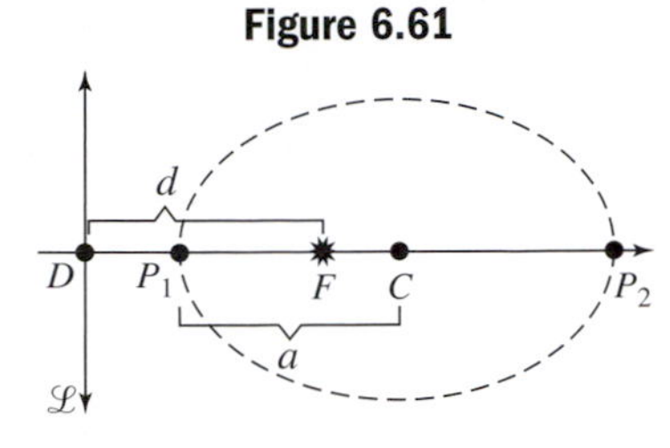

To find the polar equation of a planetary orbit, it's helpful to write the general polar equation in terms of the semimajor axis a, which is often known or easily found, rather than in terms of the distance d from directrix to focus, which is often unknown. Consider the diagram in Figure 6.61, which shows an elliptical orbit with the Sun at one focus, vertices P_1 and P_2 (perihelion and aphelion), and the center C of the ellipse. Assume the point P used to define the conic sections is at position P_1, giving $\dfrac{\overline{FP_1}}{\overline{DP_1}} = e$. From Example 8 we have $\overline{FP_1} = a(1 - e)$. Substituting $a(1 - e)$ for $\overline{FP_1}$ and solving for $\overline{DP_1}$ gives $\overline{DP_1} = \dfrac{a(1 - e)}{e}$. Using $d = \overline{DP_1} + \overline{FP_1}$, we obtain the following sequence:

$$d = \overline{DP_1} + \overline{FP_1}$$

$$= \frac{a(1 - e)}{e} + a(1 - e) \qquad \text{substitute } \frac{a(1 - e)}{e} \text{ for } \overline{DP_1} \text{ and } a(1 - e) \text{ for } \overline{FP_1}$$

$$= \frac{a(1 - e)}{e} + \frac{ae(1 - e)}{e} \qquad \text{common denominator}$$

$$= \frac{a(1 - e)(1 + e)}{e} \qquad \text{combine terms, factor out } a(1 - e)$$

$$= \frac{a(1 - e^2)}{e} \qquad (1 - e)(1 + e) = 1 - e^2$$

$$de = a(1 - e^2) \qquad \text{multiply by } e$$

Substituting $a(1 - e^2)$ for de in the standard equation $r = \dfrac{de}{1 - e\cos\theta}$ gives the equation of the orbit entirely in terms of a and e: $r = \dfrac{a(1 - e^2)}{1 - e\cos\theta}$.

EXAMPLE 9 At its aphelion, the dwarf planet Pluto is the most distant from the Sun at 4538 million miles. It has a perihelion of 2756 million miles. Use this information to find the polar equation that models the orbit of Pluto, then find the length of the focal chord for this ellipse.

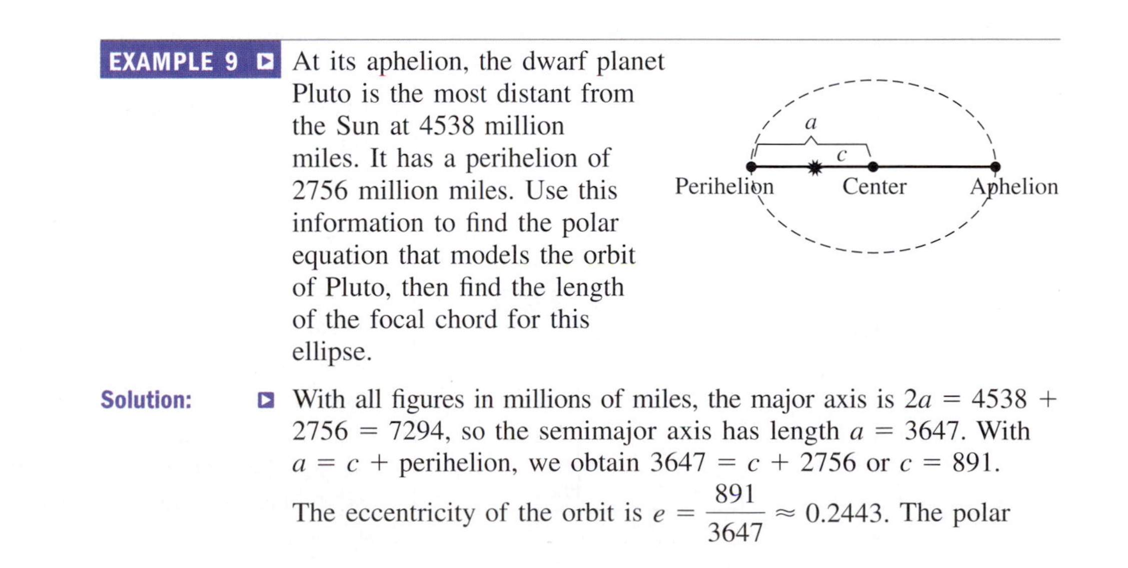

Solution: With all figures in millions of miles, the major axis is $2a = 4538 + 2756 = 7294$, so the semimajor axis has length $a = 3647$. With $a = c +$ perihelion, we obtain $3647 = c + 2756$ or $c = 891$. The eccentricity of the orbit is $e = \dfrac{891}{3647} \approx 0.2443$. The polar equation for the orbit of Pluto is $r \approx \dfrac{(3647)(1 - [0.244]^2)}{1 - [0.2443]\cos\theta}$ or $r \approx \dfrac{3430}{1 - 0.2443\cos\theta}$. Substituting $\theta = \dfrac{\pi}{2}$ (since the left-most focus is at the pole), we obtain $r = 3430$, so the length of the focal chord is $2(3430) = 6860$ million miles.

NOW TRY EXERCISES 65 THROUGH 70

TECHNOLOGY HIGHLIGHT

Using a Graphing Calculator to Investigate the Eccentricity e

The keystrokes shown apply to a TI-84 Plus model, Please consult our Internet site or your manual for other models.

One dictionary meaning of the word eccentric is "to deviate from a circular pattern." In a very real sense, this is the role that eccentricity plays as it helps to describe the conic sections. For an ellipse we've learned that $0 < e < 1$. If the eccentricity is near zero, there is little deviation and the ellipse appears nearly circular. If e is near 1, the ellipse is very elongated. To explore the eccentricity of the ellipse, enter the equation $r = \dfrac{a(1 - e^2)}{1 - e\cos\theta}$ on the Y= screen, using $a = 2$ (arbitrarily chosen) and ALPHA SIN "E" for the eccentricity. The resulting screen is shown in Figure 6.62. We will enter and store various values for E on the home screen and graph the resulting ellipse (see Exercise 2 for an alternative method). Return to the home screen and enter 0.1 STO▸ ALPHA SIN and graph the result on the ZOOM 4:ZDecimal screen. This results in a graph that is not distorted by the rectangular **4:Zstandard** screen, giving a "true" picture of the eccentricity. Repeat the procedures using $e = 0.25$, 0.5, 0.75, and 0.9. The graphs for $e = 0.1$ and $e = 0.9$ are shown in Figures 6.63 and 6.64. As you can see, when $e = 0.1$ the ellipse is very nearly a circle, while $e = 0.9$ produces a graph that is almost cigar shaped.

Figure 6.62

```
Plot1 Plot2 Plot3
\r1■2(1-E²)/(1-E
cos(θ))
\r2=
\r3=
\r4=
\r5=
\r6=
```

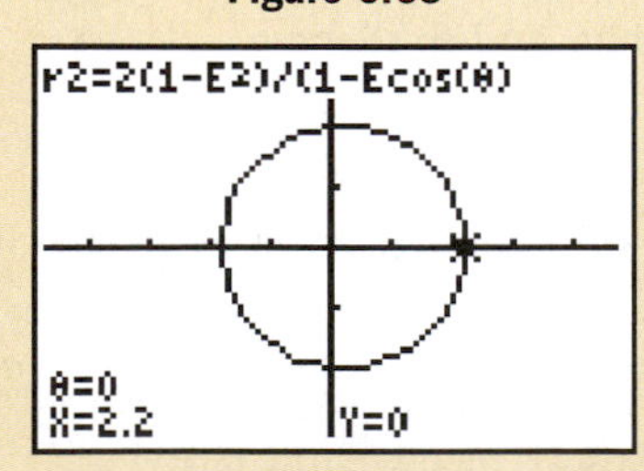

Figure 6.63

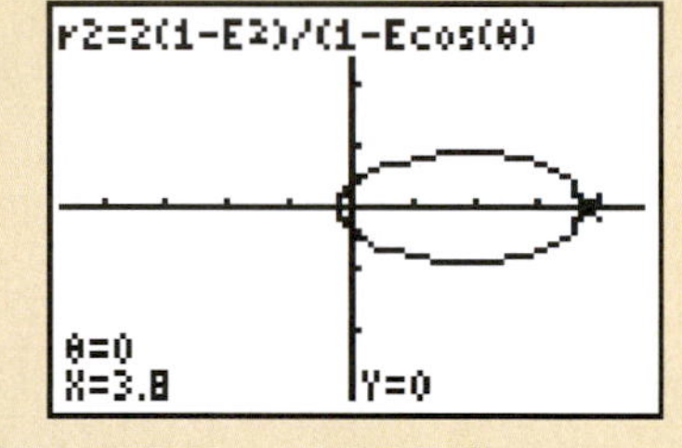

Figure 6.64

Exercise 1: Try entering a value of $e = 0$, then use your graphing calculator and basic knowledge to verify the resulting graph is a circle.

Exercise 2: Try the same exercise using the set/list option of the TI-84 Plus. In other words, enter the equation as shown here, with the values of e in braces { }:

$r_1 = \dfrac{2(1 - \{0.1, 0.25, 0.5, 0.75, 0.9\}^2)}{(1 - \{0.1, 0.25, 0.5, 0.75, 0.9\}\cos(\theta))}$. This will enable you to view all five ellipses on the same screen. Discuss the similarities and differences of this family of graphs.

Exercise 3: Use your graphing calculator to verify that changing the eccentricity does not change the length "$2a$" of the major axis. [*Hint:* Use $r_1(0) + r_1(\pi)$ on the home screen for various values of e.]

6.6 EXERCISES

CONCEPTS AND VOCABULARY

Fill in each blank with the appropriate word or phrase. Carefully reread the section if needed.

1. The set of points (x, y) in the xy-plane are related to points (X, Y) in the XY-plane by the ____________ ____ ________ formulas. To find the angle β between the original axes and the rotated axes, we use $\tan(2\beta) =$ ________.
2. For a point P on the graph of a conic with focus F and D a point on the directrix, the ratio $\frac{\overline{FP}}{\overline{DP}}$ gives the ____________ of the graph. For the eccentricity e, if $e = 1$ the graph is a ____________, if $e > 1$ the graph is a ____________, and if $0 < e < 1$ the graph will be an ellipse.
3. Features or relationships that do not change when certain transformations are applied are called ____________ of the transformation.
4. The ____________ form of the equation of a conic is $r = \frac{de}{1 \pm e\cos\theta}$ if the graph is symmetric to the ____________ axis, and $r = \frac{de}{1 \pm e\sin\theta}$ if symmetric to the line ____________.
5. Discuss the advantages of graphing a rotated conic using the rotation of axes, over graphing by simply plotting points.
6. Discuss the primary advantages of using $r = \frac{a(1 - e^2)}{1 - e\cos\theta}$ rather than $r = \frac{de}{1 - e\cos\theta}$ to develop the equation of planetary orbit.

DEVELOPING YOUR SKILLS

The graph of a conic rotated in the xy-plane is given. Use the graph (not the rotation of axes formulas) to find the equation of the conic in the XY-plane.

7. 8.

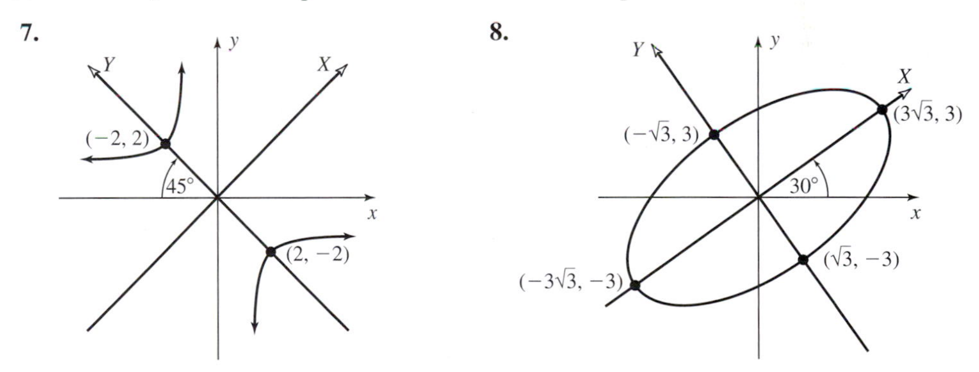

Given the point (x, y) in the xy-plane, find the coordinates of this point in the XY-plane given the angle β between the xy-axes and the XY-axes is 45°.

9. $(6\sqrt{2}, 6)$
10. $(4, 3\sqrt{2})$
11. $(0, 5)$
12. $(8, 0)$

Given the point (X, Y) in the XY-plane, find the coordinates of this point in the xy-plane given the angle β between the xy-axes and the XY-axes is 30°.

13. $(2, 2\sqrt{3})$
14. $(\sqrt{3}, 3)$
15. $(3, 4)$
16. $(12, 5)$

The conic sections whose equations are given in the XY-plane are rotated clockwise by the indicated angle. Find the corresponding equation in the xy-plane.

17. $X^2 - Y^2 = 9$; $60°$

18. $X^2 + Y = 4$; $60°$

The conic sections whose equations are given in the xy-plane are rotated by the indicated angle. What is the corresponding equation in the XY-plane?

19. $3x^2 + 2xy + 3y^2 = 9$; $45°$

20. $x^2 + \sqrt{3}xy + 2y^2 = 8$; $60°$

For the given conics in the xy-plane, (a) use a rotation of axes to find the corresponding equation in the XY-plane (clearly state the angle of rotation β), and (b) sketch its graph. Be sure to indicate the characteristic features of each conic in the XY-plane.

21. $x^2 + 4xy + y^2 - 2 = 0$

22. $x^2 + 2xy + y^2 - 12 = 0$

23. $5x^2 + 6xy + 5y^2 = 16$

24. $5x^2 - 26xy + 5y^2 = -72$

25. $x^2 + 10\sqrt{3}xy + 11y^2 = -64$

26. $37x^2 + 42\sqrt{3}xy + 79y^2 - 400 = 0$

27. $3x^2 - 2\sqrt{3}xy + y^2 - 8x - 8\sqrt{3}y = 0$

28. $6x^2 - 4\sqrt{3}xy + 2y^2 + 2x + 2\sqrt{3}y = 0$

29. $13x^2 - 6\sqrt{3}xy + 7y^2 - 100 = 0$

30. $x^2 + 4xy + y^2 + \sqrt{2}x + \sqrt{2}y = -11$

Identify the graph of each equation using the discriminant, then find the value of $\cos(2\beta)$ using $\tan(2\beta) = \dfrac{\sin(2\beta)}{\cos(2\beta)}$ and the related triangle diagram. Finally, find $\sin\beta$ and $\cos\beta$ using the half-angle identities $\cos\beta = \sqrt{\dfrac{1 + \cos(2\beta)}{2}}$ and $\sin\beta = \sqrt{\dfrac{1 - \cos(2\beta)}{2}}$.

31. $12x^2 + 24xy + 5y^2 - 40x - 30y = 25$

32. $25x^2 + 840xy - 16y^2 - 400 = 0$

For the following equations, (a) use the discriminant to identify the equation as that of a circle, ellipse, parabola, or hyperbola; (b) find the angle of rotation β and use it to find the corresponding equation in the XY-plane; and (c) verify all invariants of the transformation.

33. $x^2 - 2xy + y^2 - 5 = 0$

34. $2x^2 - 3xy + 2y^2 = 0$

35. $3x^2 + \sqrt{3}xy + 4y^2 + 4x = 1$

36. $3x^2 + 8\sqrt{3}xy - 5y^2 + 12y = -2$

Match each graph to its corresponding equation. Justify your answers (two equations have no match).

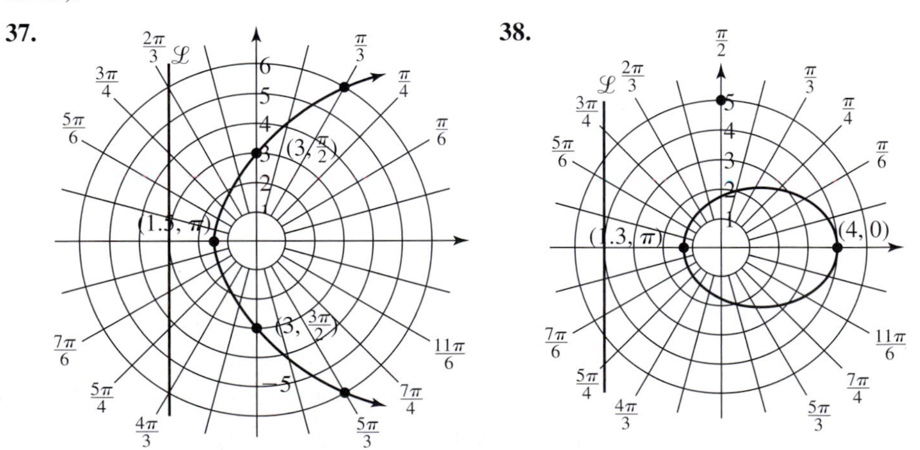

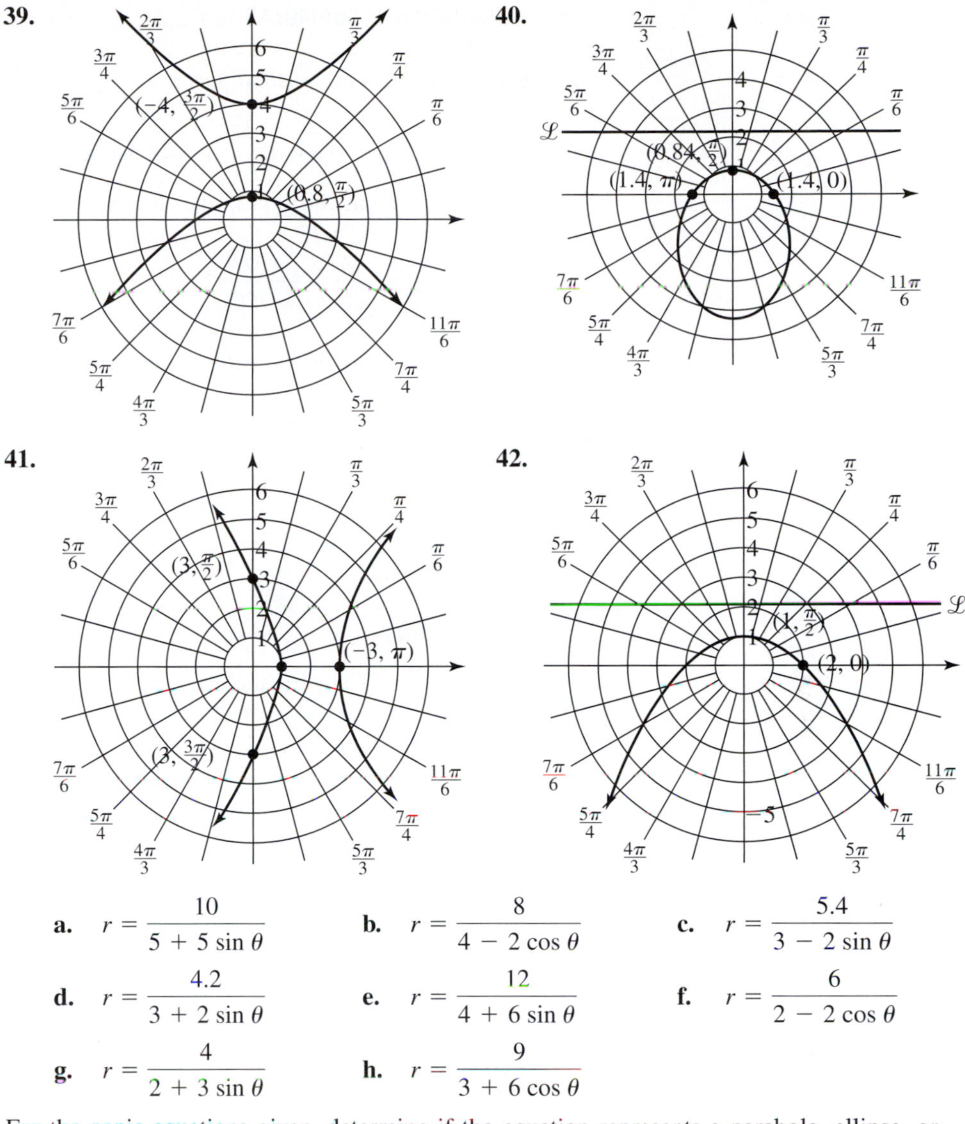

a. $r = \dfrac{10}{5 + 5\sin\theta}$ **b.** $r = \dfrac{8}{4 - 2\cos\theta}$ **c.** $r = \dfrac{5.4}{3 - 2\sin\theta}$

d. $r = \dfrac{4.2}{3 + 2\sin\theta}$ **e.** $r = \dfrac{12}{4 + 6\sin\theta}$ **f.** $r = \dfrac{6}{2 - 2\cos\theta}$

g. $r = \dfrac{4}{2 + 3\sin\theta}$ **h.** $r = \dfrac{9}{3 + 6\cos\theta}$

For the conic equations given, determine if the equation represents a parabola, ellipse, or hyperbola. Then describe and sketch the graphs using polar graph paper.

43. $r = \dfrac{4}{2 + 2\sin\theta}$ **44.** $r = \dfrac{10}{5 - 5\sin\theta}$ **45.** $r = \dfrac{12}{6 - 3\sin\theta}$

46. $r = \dfrac{6}{4 + 3\cos\theta}$ **47.** $r = \dfrac{6}{2 + 4\cos\theta}$ **48.** $r = \dfrac{2}{2 - 3\sin\theta}$

49. $r = \dfrac{5}{5 + 4\cos\theta}$ **50.** $r = \dfrac{2}{4 - 5\sin\theta}$

Write the equation of a conic that satisfies the conditions given. Assume each has one focus at the pole.

51. ellipse, $e = 0.8$, directrix to focus: $d = 4$

52. hyperbola, $e = 1.25$, directrix to focus: $d = 6$

53. parabola, vertex at $(2, \pi)$

54. ellipse, $e = 0.35$, vertex at $(4, 0)$

55. hyperbola, $e = 1.5$, vertex at $\left(3, \dfrac{\pi}{2}\right)$

56. parabola, directrix to focus: $d = 5.4$

WORKING WITH FORMULAS

57. Equation of a line in polar form: $r = \dfrac{C}{A\cos\theta + B\sin\theta}$

For the line $Ax + By = C$ in the xy-plane with slope $m = -\frac{A}{B}$ and y-intercept $\left(0, \frac{C}{B}\right)$, the corresponding equation in the $r\theta$-plane is given by the formula shown. (a) Given the line $2x + 3y = 12$ in the xy-plane, find the corresponding polar equation and (b) verify that $-\frac{A}{B} = -\frac{r(\pi/2)}{r(0)}$.

58. Polar form of an ellipse with center at the pole: $r^2 = \dfrac{a^2b^2}{a^2\sin^2\theta + b^2\cos^2\theta}$

If an ellipse in the $r\theta$-plane has its center at the pole (with major axis parallel to the x-axis), its equation is given by the formula here, where $2a$ and $2b$ are the lengths of the major and minor axes, respectively. (a) Given an ellipse with center at the pole has a major axis of length 8 and a minor axis of length 4, find the equation of the ellipse in polar form and (b) graph the result on a calculator and verify that $2a = 8$ and $2b = 4$.

APPLICATIONS

Planetary motion: The perihelion, aphelion, and orbital period of the planets Jupiter, Saturn, Uranus, and Neptune are shown in the table. Use the information to answer or complete the following exercises. The formula $L = 2\pi\sqrt{0.5(a^2 + b^2)}$ can be used to estimate the length of the orbital path. Recall for an ellipse, $c^2 = a^2 - b^2$.

Planet	Perihelion (10^6 mi)	Aphelion (10^6 mi)	Period (yr)
Jupiter	460	507	11.9
Saturn	840	941	29.5
Uranus	1703	1866	84
Neptune	2762	2824	164.8

59. Find the eccentricity of the planets Jupiter and Saturn.

60. Find the eccentricity of the planets Uranus and Neptune.

61. The orbit of Pluto (a dwarf planet) has a semimajor axis of 3647 million miles and an eccentricity of $e = 0.2443$. Find the perihelion of Pluto.

62. The orbit of Ceres (a large asteroid) has a semimajor axis 257 million miles and an eccentricity of $e = 0.097$. Find the perihelion of Ceres.

63. Which of the four planets in the table given has the greatest orbital eccentricity?

64. Which of these four planets has the greatest orbital velocity?

65. Find the polar equation modeling the orbit of Jupiter.

66. Find the polar equation modeling the orbit of Saturn.

67. Find the polar equation modeling the orbit of Uranus.

68. Find the polar equation modeling the orbit of Neptune.

69. Suppose all four major planets arrived at the focal chord of their orbit $\left(\theta = \frac{\pi}{2}\right)$ simultaneously. Use the equations in Exercises 65 to 68 to determine the distance between each of the planets at this moment.

70. The polar equation for the orbit of Pluto (a dwarf planet) was developed in Example 9. From an earlier exercise, the polar equation for the orbit of Neptune is $r \approx \dfrac{2793}{1 - 0.0111\cos\theta}$. Using the TABLE of your graphing calculator, determine if Pluto is *always* the farthest planet from the Sun. If not, how much further from the Sun is Neptune than Pluto at their perihelion?

Mirror manufacturing: A modern manufacturer of oval (elliptical) mirrors for consumer use has programmed the equipment to automatically cut the glass for each mirror (major axis horizontal). The most popular mirrors are those that fit within a golden rectangle (ratio of L to W is approximately 1 to 0.618). Find the polar equation the manufacturer should use to program the equipment for mirror orders of the following lengths. Recall that $c^2 = a^2 - b^2$ and $e = \dfrac{c}{a}$ and assume one focus is at the pole.

71. $L = 4$ ft **72.** $L = 3.5$ ft **73.** $L = 1.5$ m **74.** $L = 0.5$ m

75. Referring to Exercises 71 to 74, find the total cost of each mirror (to the consumer) if they sell for \$75 per square foot (\$807 per square meter). The area of an ellipse is given by $A = \pi ab$.

76. Referring to Exercises 71 to 74, find the total cost of an elliptical frame for each mirror (to the consumer) if the frame sells for \$12.50 per linear foot (\$41.01 per meter). The circumference of an ellipse is approximated by $C = \pi\sqrt{2(a^2 + b^2)}$.

Exercise 77

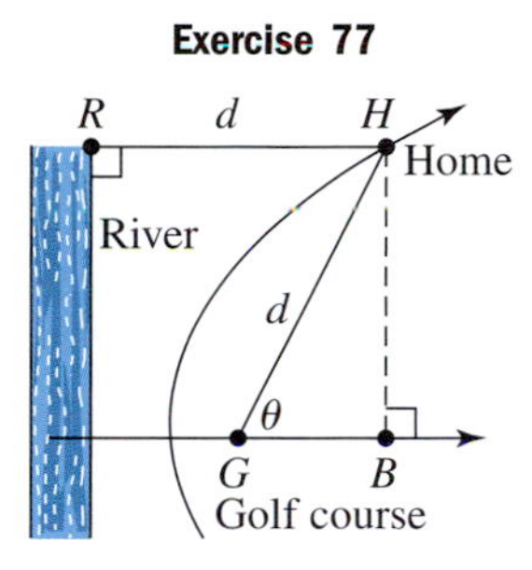

77. Home location: Candice is an enthusiastic golfer and an avid swimmer. After being transferred to a new city, she decides to buy a house that is an equal distance from the local golf course and the river running through the city. If the distance between the river and the golf course at the closest point is 3 mi, find the polar equation of the parabola that will trace through the possible locations for her new home. Assume the golf course is at the focus of the parabola.

78. Home location: Referring to Exercise 77, assume Candice finds the perfect dream house in a subdivision located at $\left(6, \dfrac{\pi}{3}\right)$. Does this home fit the criteria (is it an equal distance from the river and golf course)?

79. Solve the system below for y to verify the rotation formula for y given on page 480.

$$\begin{cases} X = x\cos\beta + y\sin\beta \\ Y = y\cos\beta - x\sin\beta \end{cases} \quad \text{original system}$$

80. Rotation of a conic section: Expand the following, collect like terms, and simplify. Show the result is the equation $aX^2 + bXY + cY^2 + f = 0$, where the coefficients a, b, c, and f are as given on page 482. $A(X\cos\beta - Y\sin\beta)^2 + B(X\cos\beta - Y\sin\beta)(X\sin\beta + Y\cos\beta) + C(X\sin\beta + Y\cos\beta)^2 + F = 0$

WRITING, RESEARCH, AND DECISION MAKING

81. Using the rotation of axes formulas in the general equation $Ax^2 + Bxy + Cy^2 + F = 0$ $(D = E = 0)$, we were able to obtain the equation $aX^2 + bXY + cY^2 + f = 0$ (see page 482), where

$$a \rightarrow A\cos^2\beta + B\sin\beta\cos\beta + C\sin^2\beta$$
$$b \rightarrow -2A\sin\beta\cos\beta + B(\cos^2\beta - \sin^2\beta) + 2C\sin\beta\cos\beta$$
$$c \rightarrow A\sin^2\beta - B\sin\beta\cos\beta + C\cos^2\beta \text{ and } f \rightarrow F$$

a. Use these to verify $b^2 - 4ac = B^2 - 4AC$.

b. Use these to verify $a + c = A + C$.

c. Explain why the invariant $f = F$ must always hold.

82. If we take degenerate cases into consideration, the discriminant test gives us the following information about the "special cases," shown in italics. If $B^2 - 4AC = 0$, the graph will be a parabola, *two parallel lines, one line, or no graph is possible.* If $B^2 - 4AC < 0$, the graph will be an ellipse, circle, *single point or no graph is possible*. If $B^2 - 4AC > 0$, the graph will be a hyperbola or *two intersecting lines*. Do some research and/or experimentation to find equations that generate each of the special cases.

83. A short-period comet is one that orbits the Sun in 200 yr or less. Two of the best known are Halley's Comet and Encke's Comet. Using any of the resources available to you, find the perihelion and aphelion of each comet and use the information to find the lengths of the semimajor and semiminor axes. Also find the period of each comet. If the length of an elliptical (orbital) path is approximated by $L = 2\pi\sqrt{0.5(a^2 + b^2)}$, find the approximate average speed of each comet in miles per hour. Finally, determine the polar equation of each orbit.

84. Use the equation $e = \frac{c}{a}$ and the relationship between the semimajor axis a, the focus c, and the aphelion of an orbit to find an equation that expresses the aphelion directly in terms of a and e.

EXTENDING THE CONCEPT

85. The polar form of the equation of an ellipse whose center is at the pole was given in Exercise 58. (a) Experiment with this equation to find the polar form of a hyperbola whose center is at the pole (with major axis parallel to the x-axis). (b) Then use it to find the equation of a hyperbola with center at the pole having vertices at $(0, \sqrt{6})$ and $(\pi, \sqrt{6})$, and a minor axis of length $2b = 4\sqrt{3}$.

86. In the $r\theta$-plane, the equation of a circle having radius R, center at (R, β), and going through the pole is given by $r = 2R\cos(\theta - \beta)$. Consider the circle defined by $x^2 + y^2 - 6\sqrt{2}x - 6\sqrt{2}y = 0$ in the xy-plane. Verify this circle goes through the origin, then find the equation of the circle in polar form.

For the given conics in the xy-plane, use a rotation of axes to find the corresponding equation in the XY-plane. See Exercises 31 and 32.

87. $12x^2 + 24xy + 5y^2 - 40x - 30y = 25$

88. $25x^2 + 840xy - 16y^2 - 400 = 0$

MAINTAINING YOUR SKILLS

89. (2.4) The revenue for Lizzie's Yard Service is very seasonal, peaking at \$4000/month in July, and dropping to \$1200 per month in January. Assuming a 12-month weather cycle, (a) find a sinusoidal equation model for her revenue and (b) use the model to estimate the number of months that revenue is below \$2500 dollars.

Exercise 90

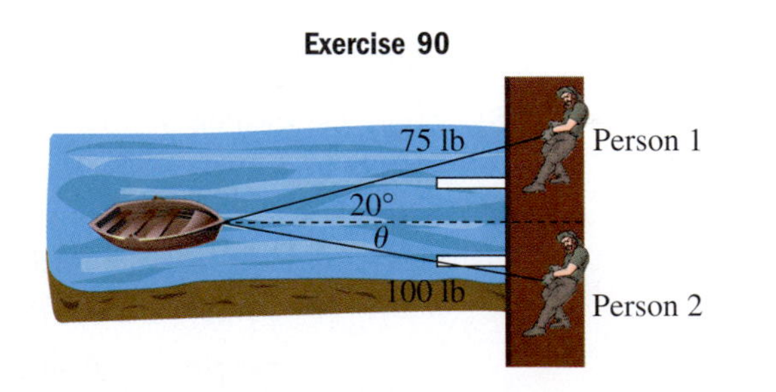

90. (5.4) Use the diagram given to determine the angle θ at which the second person should pull in order to bring the stranded boat directly into its berth.

91. (4.4) Given $24\sin^2 t - 1 = 5$, find all solutions in $[0, 2\pi)$. Answer in exact form.

92. (5.7) Find the quotient in trigonometric form:
$z_1 = 3 + \sqrt{3}i$ and $z_2 = \sqrt{3} + 3i$

93. (3.2) State the three identities for $\cos(2\theta)$ and the three Pythagorean identities.

94. (4.5) Write the parametric equations in rectangular form:
$x = 3\cos T$
$y = -4\sin T$

Summary and Concept Review

SECTION 6.1 The Circle and the Ellipse

KEY CONCEPTS

- The equation of a circle centered at (h, k) with radius r is $(x-h)^2 + (y-k)^2 = r^2$.
- After dividing both sides by r^2, we obtain the standard form $\frac{(x-h)^2}{r^2} + \frac{(y-k)^2}{r^2} = 1$, showing the horizontal and vertical distance from center to graph is r.
- The equation of an ellipse in factored form is $A(x-h)^2 + B(y-k)^2 = F$, where $A \neq B$.
- After dividing both sides by F and simplifying, we obtain the standard form $\frac{(x-h)^2}{a^2} + \frac{(y-k)^2}{b^2} = 1$. The center of the ellipse is (h, k) with horizontal distance a and vertical distance b from center to graph.
- For an ellipse, note the *sum* of second-degree terms with $a \neq b$.

EXERCISES

Sketch the graph of each equation in Exercises 1 through 5.

1. $x^2 + y^2 = 16$

2. $x^2 + 4y^2 = 36$

3. $9x^2 + y^2 - 18x - 27 = 0$

4. $x^2 + y^2 + 6x + 4y + 12 = 0$

5. $\frac{(x+3)^2}{16} + \frac{(y-2)^2}{9} = 1$

6. Find the equation of a circle if $(-4, 5)$ and $(2, -3)$ are endpoints of the diameter.

SECTION 6.2 The Hyperbola

KEY CONCEPTS

- The equation of a *horizontal* hyperbola in factored form is $A(x-h)^2 - B(y-k)^2 = F$.
- After dividing both sides by F and simplifying, we obtain the standard form $\frac{(x-h)^2}{a^2} - \frac{(y-k)^2}{b^2} = 1$. The center of the hyperbola is (h, k), with horizontal distance a from

center to vertices and vertical distance b from center to midpoint of one side of the central rectangle.

- The equation of a vertical hyperbola in factored form is $B(y - k)^2 - A(x - h)^2 = F$.
- For a hyperbola, note the *difference* of second-degree terms.

EXERCISES

Sketch the graph of each equation, indicating the center, vertices, and asymptotes. For Exercise 12, also give the equation of the hyperbola in standard form.

7. $4y^2 - 25x^2 = 100$

8. $\frac{(y - 3)^2}{16} - \frac{(x + 2)^2}{9} = 1$

9. $\frac{(x + 2)^2}{9} - \frac{(y - 1)^2}{4} = 1$

10. $9y^2 - x^2 - 18y - 72 = 0$

11. $x^2 - 4y^2 - 12x - 8y + 16 = 0$

12. vertices at $(-3, 0)$ and $(3, 0)$, asymptotes of $y = \pm\frac{4}{3}x$

SECTION 6.3 Foci and the Analytic Ellipse and Hyperbola

KEY CONCEPTS

- Given any two fixed points F_1 and F_2 in a plane, an ellipse is defined to be the set of all points $P(x, y)$ such that the distance from the first focus to point P, plus the distance from the second focus to P, remains constant.
- For an ellipse, the distance from the center to one of the vertices is greater than the distance from the center to one of the foci.
- To find the foci of an ellipse: $c^2 = |a^2 - b^2|$ (since $a > c$ or $b > c$).
- Given any two fixed points F_1 and F_2 in a plane, a hyperbola is defined to be the set of all points $P(x, y)$ such that the distance from the first focus to point P, less the distance from the second focus to point P, remains constant.
- For a hyperbola, the distance from center to one of the vertices is less than the distance from center to one of the foci.
- To find the foci of a hyperbola: $c^2 = a^2 + b^2$ (since $c > a$ or $c > b$).

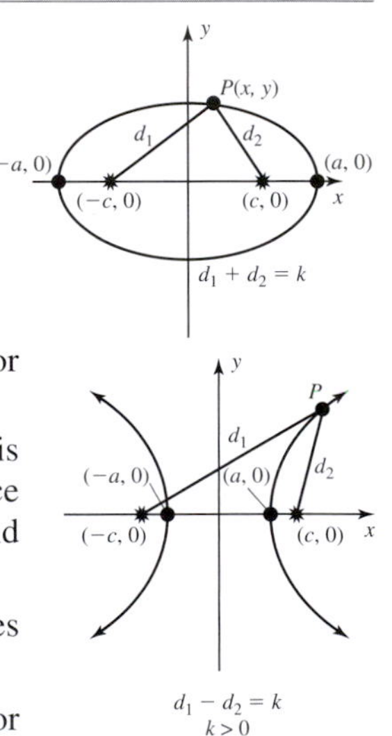

EXERCISES

Sketch each graph, noting all special features.

13. $4x^2 + 25y^2 - 16x - 50y - 59 = 0$

14. $4x^2 - 36y^2 - 40x + 144y - 188 = 0$

15. Find the equation of the ellipse given

a. vertices at $(-13, 0)$ and $(13, 0)$; foci at $(-12, 0)$ and $(12, 0)$

b. foci at $(0, -16)$ and $(0, 16)$; length of major axis: 40 units

16. Find the equation of the hyperbola given

a. vertices at $(-15, 0)$ and $(15, 0)$; foci at $(-17, 0)$ and $(17, 0)$

b. foci at $(0, -5)$ and $(0, 5)$; vertical length of central rectangle: 8 units

SECTION 6.4 The Analytic Parabola

KEY CONCEPTS

- Horizontal parabolas have equations of the form $x = ay^2 + by + c$.
- A horizontal parabola will open to the right if $a > 0$ and to the left if $a < 0$. The axis of symmetry is $y = \frac{-b}{2a}$, with the vertex (h, k) found by evaluating at $y = \frac{-b}{2a}$ or by completing the square and writing the equation in shifted form: $x = a(y - k)^2 + h$.
- Given a fixed point F and fixed line D in the plane, a parabola is defined to be the set of all points $P(x, y)$ such that the distance from point F to point P is equal to the perpendicular distance from point P to line D.

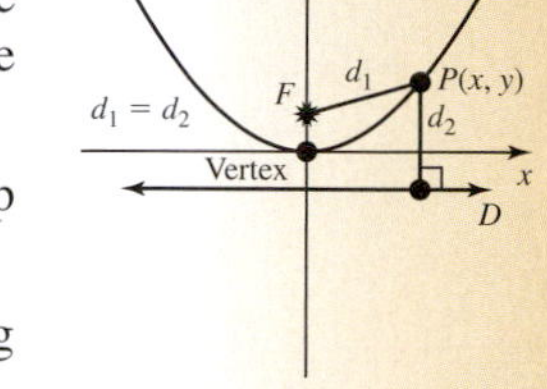

- The equation $x^2 = 4py$ describes a vertical parabola, concave up if $p > 0$ and concave down if $p < 0$.
- The equation $y^2 = 4px$ describes a horizontal parabola, opening to the right if $p > 0$ and opening to the left if $p < 0$.
- The focal chord of a parabola is a line segment that contains the focus and is parallel to the directrix, with its endpoints on the graph. It has a total length of $|4p|$, meaning the distance from the focus to a point of the graph is $|2p|$. It is commonly used to assist in drawing a graph of the parabola.

EXERCISES

For Exercises 17 and 18, find the vertex and x- and y-intercepts, if they exist. Then sketch the graphs using symmetry and a few points, or by completing the square and shifting a parent function.

17. $x = y^2 - 4$

18. $x = y^2 + y - 6$

For Exercises 19 and 20, find the vertex, focus, and directrix for each parabola. Then sketch the graphs using the vertex, focus, and focal chord. Also graph the directrix.

19. $x^2 = -20y$

20. $x^2 - 8x - 8y + 16 = 0$

SECTION 6.5 Polar Coordinates, Equations, and Graphs

KEY CONCEPTS

- In polar coordinates, the location of a point in the plane is denoted (r, θ), where r is the distance to the point from the origin or *pole*, and θ is the angle between a stipulated polar axis and a ray containing P.
- In the polar coordinate system, the location (r, θ) of a point is not unique for two reasons: (1) the angles θ and $\theta + 2\theta n$ are coterminal (n an integer), and (2) r may be negative.
- The point $P(r, \theta)$ can be converted to $P(x, y)$ in rectangular coordinates where $x = r\cos\theta$ and $y = r\sin\theta$.
- The point $P(x, y)$ in rectangular coordinates can be converted to $P(r, \theta)$ in polar coordinates, where $r = \sqrt{x^2 + y^2}$ and $\theta_r = \tan^{-1}\left(\frac{y}{x}\right)$.

- To sketch a polar graph, we view the length r as being along the second hand of a clock, ticking in a counterclockwise direction. Each "tick" is $\frac{\pi}{12}$ rad or 15°. For each tick we locate a point on the radius and plot it on the face of the clock before going on.
- For graphing, we also apply an "r-value" analysis, which looks where r is increasing, decreasing, zero, maximized, and/or minimized.
- If the polar equation is given in terms of sines, the graph will be symmetric to $\theta = \frac{\pi}{2}$.
- If the polar equation is given in terms of cosines, the graph will be symmetric to the polar axis.
- The graphs of several common polar equations are given in Appendix VI.

EXERCISES

Sketch using an r-value analysis (include a table), symmetry, and any convenient points.

21. $r = 5\sin\theta$ **22.** $r = 4 + 4\cos\theta$ **23.** $r = 2 + 4\cos\theta$ **24.** $r = 8\sin(2\theta)$

SECTION 6.6 More on the Conic Sections: Rotation of Axes and Polar Form

KEY CONCEPTS

- Using a rotation, the conic equation $Ax^2 + Bxy + Cy^2 + Dx + Ey + F = 0$ in the xy-plane can be transformed into $aX^2 + cY^2 + dX + eY + f = 0$ in the XY-plane, in which the mixed xy-term is absent.
- The required angle of rotation β is found using $\tan(2\beta) = \frac{B}{A - C}$; $0 < 2\beta < 180°$.
- The change in coordinates from the xy-plane to the XY-plane is accomplished using the rotation formulas:

$$x = X\cos\beta - Y\sin\beta \qquad y = X\sin\beta + Y\cos\beta$$

- In the process of this conversion, certain quantities, called invariants, remain unchanged and can be used to check that the conversion was correctly performed. These invariants are (1) $F = f$, (2) $A + C = a + c$, and (3) $B^2 - 4AC = b^2 - 4ac$.
- The invariants $B^2 - 4AC = b^2 - 4ac$ are called discriminants and can be used to classify the type of graph the equation will give, except in degenerate cases:

  If $B^2 - 4AC = 0$, the equation is that of a parabola.

  If $B^2 - 4AC < 0$, the equation is that of a circle or an ellipse.

  If $B^2 - 4AC > 0$, the equation is that of a hyperbola.
- All conics (not only the parabola) can be stated in terms of a focus/directrix definition. This is done using the concept of eccentricity, symbolized by the letter e.
- If F is a fixed point and $\mathscr{L}$ a fixed line in the plane with the point D on $\mathscr{L}$, the set of all points P such that $\frac{\overline{FP}}{\overline{DP}} = e$ (e a constant) is the graph of a conic section. If $e = 1$, the graph is a parabola. If $0 < e < 1$, the graph is an ellipse. If $e > 1$, the graph is a hyperbola.
- Given a conic section with eccentricity e, one focus at the pole of the $r\theta$-plane, and directrix $\mathscr{L}$ located d units from this focus, then the polar equations $r = \frac{de}{1 \pm e\cos\theta}$ and $r = \frac{de}{1 \pm e\sin\theta}$ represent one of the conic sections as determined by the value of e.

EXERCISES

For the given conics in the xy-plane, use a rotation of axes to find the corresponding equation in the XY-plane, then sketch its graph.

25. $2x^2 - 4xy + 2y^2 - 8\sqrt{2}y - 24 = 0$

26. $x^2 + 6\sqrt{3}xy + 7y^2 - 160 = 0$

For the conic equations given, determine if the equation represents a parabola, ellipse, or hyperbola. Then describe and sketch the graphs using polar graph paper.

27. $r = \dfrac{9}{3 - 2\cos\theta}$

28. $r = \dfrac{8}{4 - 6\cos\theta}$

29. $r = \dfrac{4}{3 + 3\sin\theta}$

30. Mars has a perihelion of 128.4 million miles and an aphelion of 154.9 million miles. Use this information to find a polar equation that models the elliptical orbit, then find the length of the focal chord.

MIXED REVIEW

For Exercises 1 through 14, graph the conic section and locate the center, vertices, directrix, foci, focal chords, asymptotes, and other important features as these apply to a particular equation and conic.

1. $9x^2 + 9y^2 = 54$

2. $16x^2 + 25y^2 = 400$

3. $9y^2 - 25x^2 = 225$

4. $\dfrac{(x - 3)^2}{9} + \dfrac{(y + 1)^2}{25} = 1$

5. $4(x - 1)^2 - 36(y + 2)^2 = 144$

6. $16(x + 2)^2 + 4(y - 1)^2 = 64$

7. $x = y^2 + 2y + 3$

8. $x = (y + 2)^2 - 3$

9. $x^2 - 8x - 8y + 16 = 0$

10. $x^2 = -24y$

11. $4x^2 - 25y^2 - 24x + 150y - 289 = 0$

12. $4x^2 + 16y^2 - 12x - 48y - 19 = 0$

13. $49(x + 2)^2 + (y - 3)^2 = 49$

14. $x^2 + y^2 - 8x + 12y + 16 = 0$

15. Plot the polar coordinates given, then convert to rectangular coordinates.

a. $\left(3.5, \dfrac{2\pi}{3}\right)$

b. $\left(-4, \dfrac{5\pi}{4}\right)$

16. Match each equation to its corresponding graph. Justify each response.

(i) $r = 3.5 + \cos\theta$ **(ii)** $r^2 = 20.25\sin(-2\theta)$ **(iii)** $r = 4.5\cos\theta$

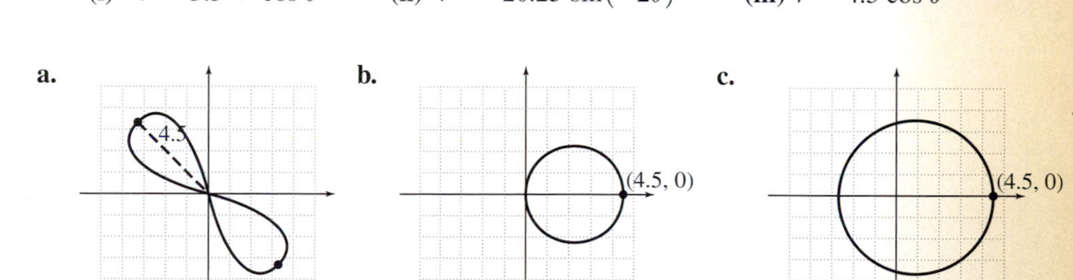

17. Except for small variations, a planet's orbit around the Sun is elliptical, with the Sun at one focus. The *perihelion* or minimum distance from the planet Mercury to the Sun is about 46 million kilometers. Its *aphelion* or maximum distance from the Sun is approximately 70 million kilometers. Use this information to find the length of the major and minor axes, then determine the equation model for the orbit of Mercury in the standard form $\frac{x^2}{a^2} + \frac{y^2}{b^2} = 1$.

18. The orbit of a comet can also be modeled by one of the conic sections, with the Sun at one focus. Assuming the equations given model a comet's path, (1) determine if the path is circular, elliptic, hyperbolic, or parabolic; and (2) determine the closest distance the comet will come to the Sun (in millions of miles).

a. $r = \frac{84}{100 + 70\cos\theta}$ **b.** $r = \frac{31}{5 - 5\sin\theta}$

19. In the design of their corporate headquarters, Centurion Computing includes a seven-leaf rose in a large foyer, with a fountain in the center. Each of the leaves is 5 m long (when measured from the center of the fountain), and will hold flower beds for carefully chosen perennials. The rose is to be symmetric to a vertical axis, with the leaf bisected by $\theta = \frac{\pi}{2}$ pointing directly to the elevators. Find the equation of the rose in polar form.

20. The hyperbola defined by $\frac{X^2}{80^2} - \frac{Y^2}{400^2} = 1$ in the XY-plane is rotated clockwise by 45°. What is the corresponding equation in the xy-plane?

PRACTICE TEST

By inspection only (no graphing), match each equation to its correct description.

1. $x^2 + y^2 - 6x + 4y + 9 = 0$ ______ **2.** $4y^2 + x^2 - 4x + 8y + 20 = 0$ ______

3. $x^2 - 4y^2 - 4x + 12y + 20 = 0$ ______ **4.** $y - x^2 - 4x + 20 = 0$ ______

a. Parabola **b.** Hyperbola **c.** Circle **d.** Ellipse

Identify and then graph each of the following conic sections. State the center, vertices, foci, asymptotes, and other important points when applicable.

5. $x^2 + y^2 - 4x + 10y + 20 = 0$ **6.** $25(x + 2)^2 + 4(y - 1)^2 = 100$

7. $r = \frac{10}{5 - 4\cos\theta}$ **8.** $r = \frac{12}{5 - 5\cos\theta}$

9. $\frac{(y + 3)^2}{9} - \frac{(x - 2)^2}{16} = 1$ **10.** $4(x - 1)^2 - 25(y + 2)^2 = 100$

Use the equation $80x^2 + 120xy + 45y^2 - 100y - 44 = 0$ to complete Exercises 11 and 12.

11. Use the discriminant $B^2 - 4AC$ to identify the graph, and $\tan(2\beta) = \frac{B}{A - C}$ to find $\cos\beta$ and $\sin\beta$.

12. Find the equation in the XY-plane and use a rotation of axes to draw a neat sketch of the graph. Include the X- and Y-intercepts.

Graph each polar equation.

13. $r = 3 + 3\cos\theta$ **14.** $r = 4 + 8\cos\theta$ **15.** $r = 6\sin(2\theta)$

16. Halley's comet has a perihelion of 54.5 million miles and an aphelion of 3253 million miles. Use this information to find a polar equation that models its elliptical orbit. How does its eccentricity compare with that of the planets in our solar system?

17. The soccer match is tied, with time running out. In a desperate attempt to win, the opposing coach pulls his goalie and substitutes a forward. Suddenly, Marques gets a break-away and has an open shot at the empty net, 165 ft away. If the kick is on-line and leaves his foot at an angle of 28° with an initial velocity of 80 ft/sec, is the ball likely to go in the net and score the winning goal?

Determine the equation of each relation and state its domain and range. For the parabola and the ellipse, also give the location of the foci.

18. **19.** **20.**

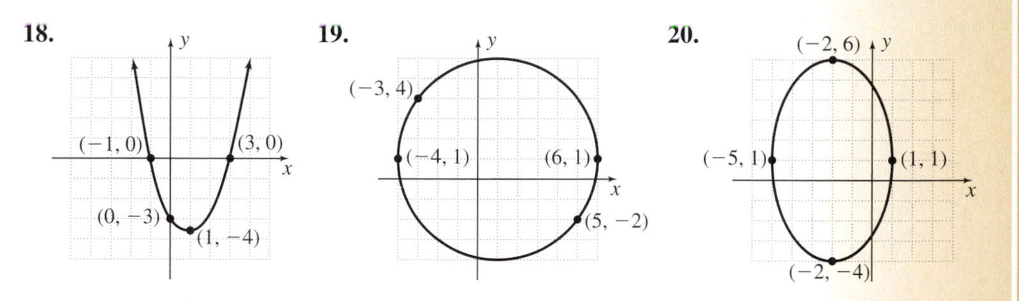

Calculator Exploration and Discovery

Conic Rotations in Polar Form

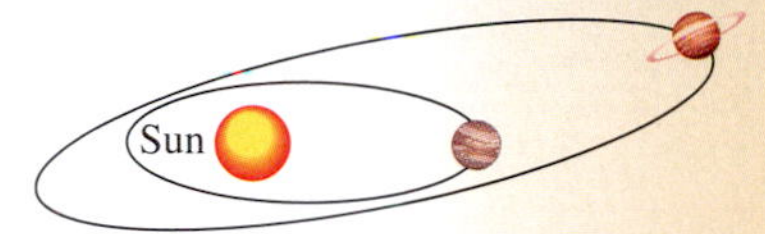

While all planets orbit around the sun in an elliptical path, their **ecliptic planes,** or the planes containing the orbits, differ considerably. For example, using the ecliptic plane of the Earth for reference, the plane containing Mercury's orbit is inclined by 7° and the plane of the dwarf planet Pluto by 17°! In addition, if we use the major axis of Earth's orbit for reference, the major axes of the other planets, assuming they are transformed to the ecliptic plane, are rotated by some angle θ. We can gain a basic understanding of the rotations of an elliptical path (relative to some point of reference) using skills developed in this chapter. Here we've seen that the equation of a conic can be given in rectangular form, polar form, and parametric form. Each form seems to have its advantages. When it comes to the rotations of a conic section, it's hard to match the ease and versatility of the polar form. To illustrate, recall that in polar form the general equation of a horizontal ellipse with one focus (the Sun) at the origin is $r = \dfrac{a(1 - e^2)}{1 - e\cos\theta}$. The constant a gives the length of the semimajor axis and e represents the eccentricity of the orbit. With the exception of Mercury and Pluto (a dwarf planet), the orbits of most planets are close to circular (e is very near zero). This makes the rotations difficult to see. Instead we will explore the concept of axes rotation using "planets" with higher eccentricities. Consider the following planets and their orbital equations. The planet Agnesi has an eccentricity of $e = 0.5$, while the planet Erdös is the most eccentric at $e = 0.75$.

$$\text{Agnesi: } \frac{2.9}{1 - 0.5\cos\theta} \qquad \text{Galois: } \frac{5.75}{1 - 0.7\cos\theta} \qquad \text{Erdös: } \frac{7.875}{1 - 0.75\cos\theta}$$

Figure 6.65

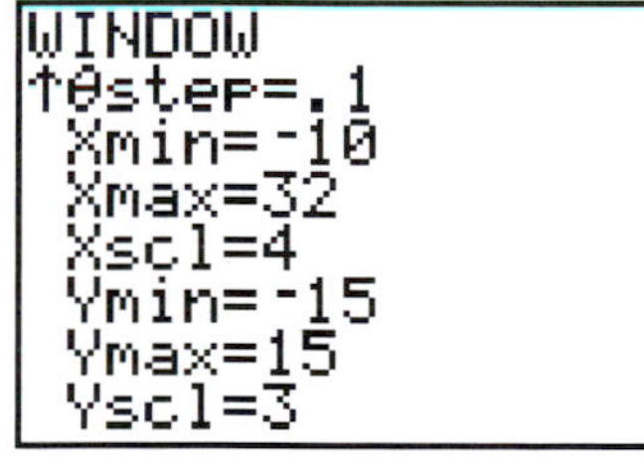

Figure 6.66

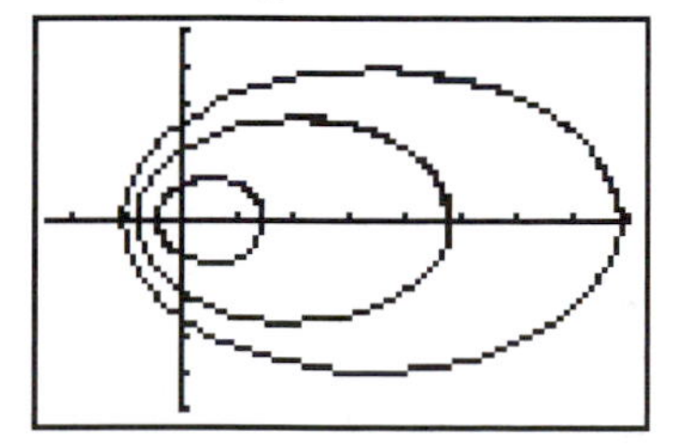

We'll investigate the concept of conic rotations in polar form by rotating these ellipses. With your calculator in polar MODE, enter these three equations on the Y= screen and use the settings shown in Figure 6.65 to set the window size (use θmax = 7).

The resulting graph is displayed in Figure 6.66, showing the very hypothetical case where all planets share the same major axis. To show a more realistic case where the planets approach the Sun along orbits with differing major axes, we'll use Galois as a reference and rotate Agnesi

$\frac{\pi}{4}$ rad clockwise and Erdös $\frac{\pi}{12}$ rad counterclockwise. This is done by *simply adjusting the argument of cosine* in each equation, using $\cos\left(\theta - \frac{\pi}{4}\right)$ for Agnesi and $\cos\left(\theta + \frac{\pi}{12}\right)$ for Erdös. The adjusted Y= screen is shown in Figure 6.67, and new graphs in Figure 6.68.

Figure 6.67 **Figure 6.68**

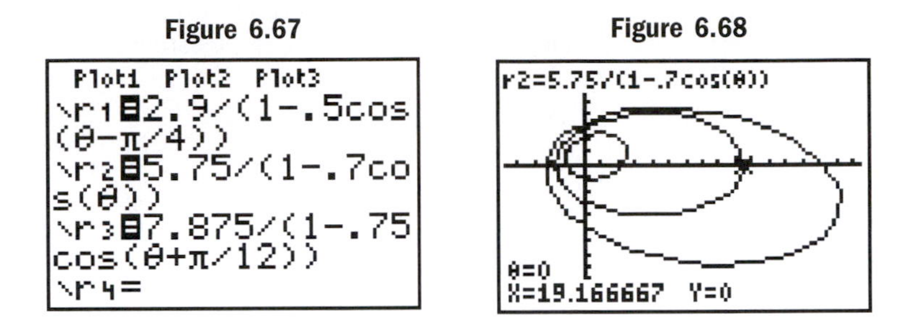

Use these ideas to explore and investigate other rotations by completing the following exercises.

Exercise 1: What happens if the angle of rotation is π? Is the orbit identical if you rotate by $-\pi$?

Exercise 2: If the denominator in the equation is changed to a sum, what effect does it have on the graph?

Exercise 3: If the sign in the numerator is changed, what effect does it have on how the graph is generated?

Exercise 4: After resetting the orbits as originally given, use trial and error to approximate the smallest angle of rotation required for the orbit of Galois to intersect the orbit of Erdös.

Exercise 5: What minimum rotation is required for the orbit of Galois to intersect the orbit of *both* Agnesi and Erdös?

Exercise 6: What is the minimum rotation required for the orbit of Agnesi to intersect the orbit of Galois?

STRENGTHENING CORE SKILLS

Simplifying and Streamlining Computations for the Rotation of Axes

While the calculations involved for eliminating the mixed xy-term require a good deal of concentration, there are a few things we can do to simplify the overall process. Basically this involves two things. First, in Figure 6.69 we've organized the process in flowchart form to help you "see" the sequence involved in finding $\cos\beta$ and $\sin\beta$ (for use in the rotation formulas). Second, calculating x^2, y^2, and xy (from the equations $x = X\cos\beta - Y\sin\beta$ and $y = X\sin\beta + Y\cos\beta$) *as single terms and apart from their actual substitution* is somewhat less restrictive and seems to help to streamline the algebra.

ILLUSTRATION 1 For $2x^2 + 12xy - 3y^2 - 42 = 0$, use a rotation of axes to eliminate the xy-term, then identify the conic and its characteristic features.

Solution: Since $A \neq C$, we find β using $\tan(2\beta) = \frac{B}{A - C}$, giving $\tan(2\beta) = \frac{12}{5}$.

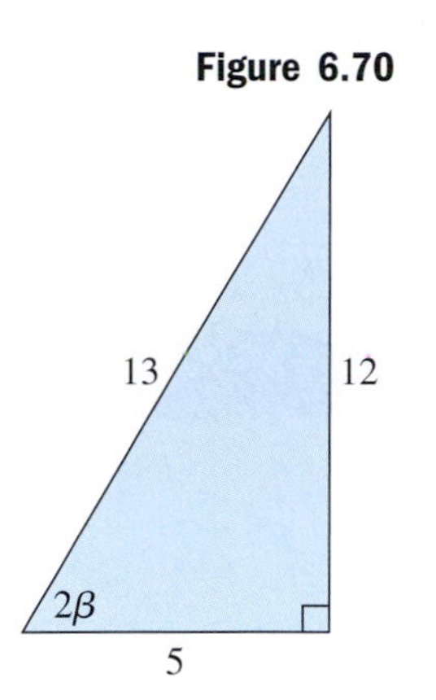

Figure 6.70

Using the triangle shown in Figure 6.70 we find $\cos(2\beta) = \frac{5}{13}$. We then find the values of $\cos\beta$ and $\sin\beta$ (choosing 2β in QII), using the double-angle identities as follows:

$$\cos\beta = \sqrt{\frac{1+\cos(2\beta)}{2}}$$

$$= \sqrt{\frac{1+\frac{5}{13}}{2}} \rightarrow \sqrt{\frac{\frac{18}{13}}{2}}$$

$$= \frac{3}{\sqrt{13}}$$

$$\sin\beta = \sqrt{\frac{1-\cos(2\beta)}{2}}$$

$$= \sqrt{\frac{1-\frac{5}{13}}{2}} \rightarrow \sqrt{\frac{\frac{8}{13}}{2}}$$

$$= \frac{2}{\sqrt{13}}$$

Figure 6.69

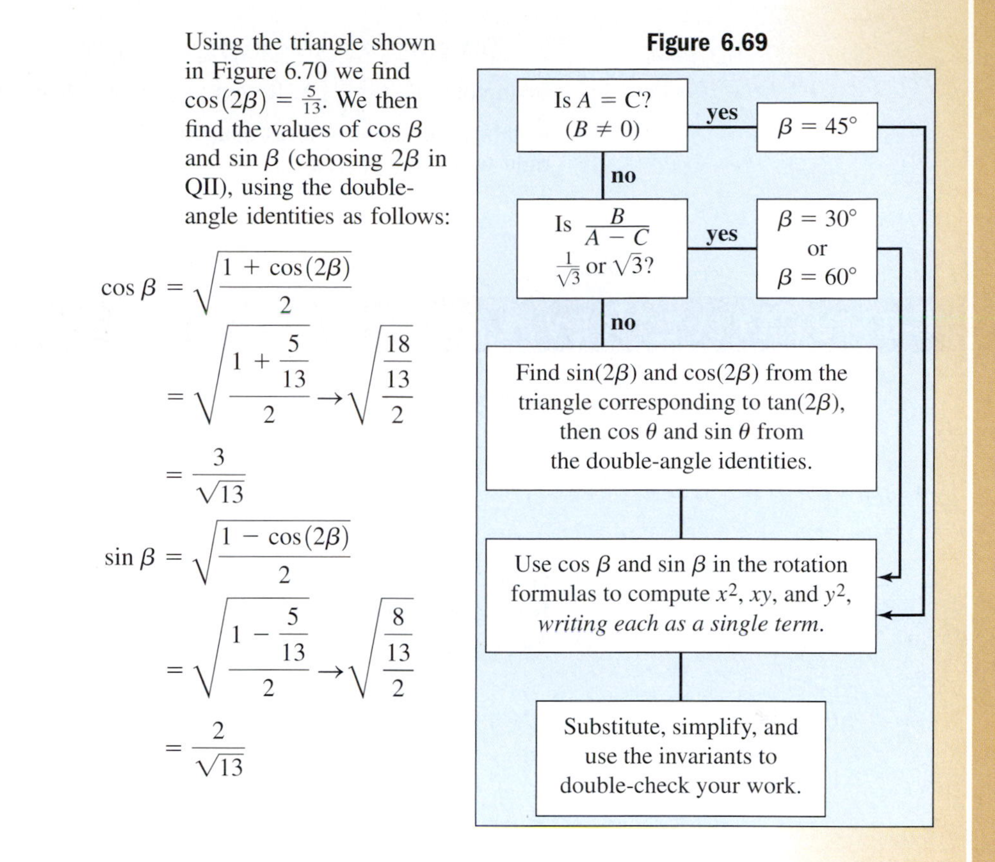

We now compute x^2, xy, and y^2 prior to substitution in the original equation, *writing each as a single term*:

- $x = \frac{3}{\sqrt{13}}X - \frac{2}{\sqrt{13}}Y \rightarrow \frac{3X-2Y}{\sqrt{13}}$
- $y = \frac{2}{\sqrt{13}}X + \frac{3}{\sqrt{13}}Y \rightarrow \frac{2X+3Y}{\sqrt{13}}$

$$x^2 = \left(\frac{3X-2Y}{\sqrt{13}}\right)^2 = \frac{9X^2-12XY+4Y^2}{13}$$

$$y^2 = \left(\frac{2X+3Y}{\sqrt{13}}\right)^2 = \frac{4X^2+12XY+9Y^2}{13}$$

$$xy = \frac{(3X-2Y)(2X+3Y)}{\sqrt{13}} = \frac{6X^2+5XY-6Y^2}{13}$$

Next, we substitute into the original equation, clearing denominators *prior* to using the distributive property.

$$42 = 2x^2 + 12xy - 3y^2$$

$$42 = 2\left(\frac{9X^2-12XY+4Y^2}{13}\right) + 12\left(\frac{6X^2+5XY-6Y^2}{13}\right) - 3\left(\frac{4x^2+12XY+9Y^2}{13}\right)$$

multiply both sides by 13, *then* distribute

$$546 = 18X^2 - 24XY + 8Y^2 + 72X^2 + 60XY - 72Y^2 - 12X^2 - 36XY - 27Y^2$$

$$546 = 78X^2 - 91Y^2 \quad \text{combine like terms}$$

$$42 = 6X^2 - 7Y^2 \quad \text{simplify and check invariants: } F = f✓ \quad A + C = a + c✓ \quad B^2 - 4AC = b^2 - 4ac✓$$

$$1 = \frac{X^2}{(\sqrt{7})^2} - \frac{Y^2}{(\sqrt{6})^2} \quad \text{standard form}$$

The graph is a central hyperbola along the X-axis, with vertices at $(\pm\sqrt{7}, 0)$ and asymptotes $Y = \pm\sqrt{\frac{6}{7}}X$. Return to Section 6.6 and resolve Exercises 31 and 32 using these methods. Do the new ideas make a difference? The calculations are much cleaner with the right triangle definitions.

Cumulative Review Chapters 1–6

Solve each equation.

1. $-6\tan x = 2\sqrt{3}$
2. $\dfrac{\sin 27°}{18} = \dfrac{\sin x}{35}$
3. $25\sin\left(\dfrac{\pi}{3}x - \dfrac{\pi}{6}\right) + 3 = 15.5$
4. $619^2 = 450^2 + 325^2 - 2(450)(325)\cos\theta$

Graph each relation. Include and label vertices, x- and y-intercepts, asymptotes, and other features as applicable.

5. $x = y^2 + 4y + 5$
6. $x^2 + y^2 + 10x - 4y + 20 = 0$
7. $4(x - 1)^2 - 36(y + 2)^2 = 144$
8. $y = -2\cos\left(x - \dfrac{\pi}{4}\right) + 1$
9. $r = 4\cos(2\theta)$
10. $x = 2\sin t$
$y = \tan t$
11. Use the dot product to find the angle between the vectors $\mathbf{u} = \langle -4, 5\rangle$ and $\mathbf{v} = \langle 3, 7\rangle$.
12. Solve using a standard triangle (no trig).

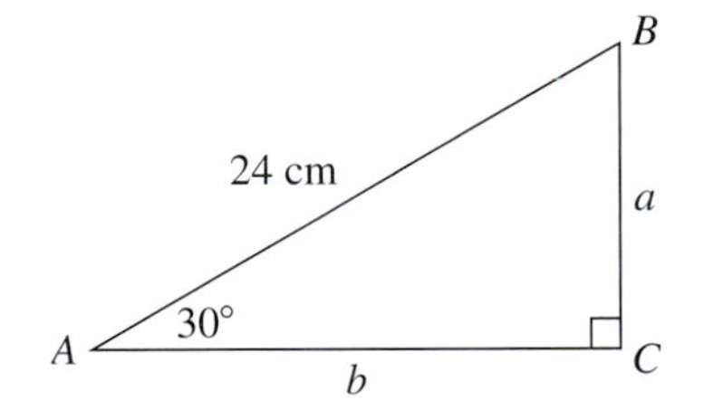

13. Find the area of the triangle shown.

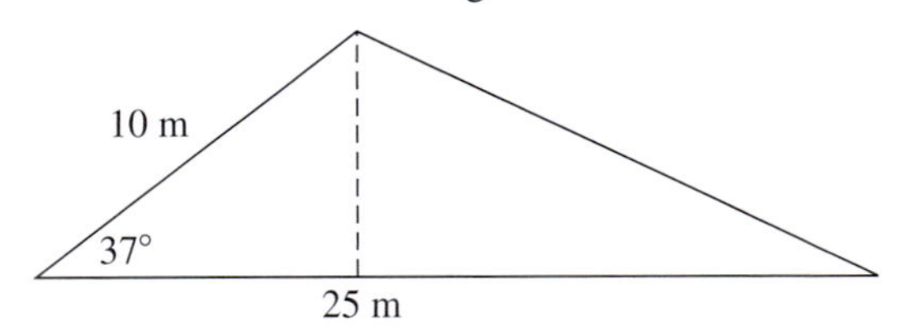

14. Verify the following is an identity: $(\cos\theta + \sin\theta)^2 = 1 + \sin(2\theta)$
15. For what values of $\theta \in [0, 2\pi)$ is the following relation not defined?
$$\sec\theta - \cot\theta = \frac{\cos\theta}{1 + \sin\theta}$$
16. Convert $\gamma = 2.426$ to degrees, rounded to the nearest tenth.
17. Name an angle coterminal with $\alpha = 32°$ and one coterminal with $\beta = 0.7169$.
18. An arc is subtended by $\theta = 125°$. If the circle has a radius of 6 ft, how long is the arc?
19. What is the area of the circular sector described in Exercise 18?

20. A child's bicycle has wheels with an 8-in. radius that are rotating at 300 rpm (revolutions per minute). (a) Find the angular velocity of the wheels and, (b) the linear velocity of the bicycle in miles per hour.

21. When viewed from a distance of 100 yd, the angle of elevation from the canyon floor to the top of the canyon is 52°. How tall are the canyon walls?

22. Starting with $\sin^2\theta + \cos^2\theta = 1$, derive the Pythagorean identities involving sec/tan and csc/cot.

23. Starting with $\cos(2\theta) = \cos^2\theta - \sin^2\theta$, use a Pythagorean identity to derive two other forms of the double angle formula for cosine.

24. Use a sum/difference identity to find the value of sin 15° and sin 105° in exact form.

25. Write in terms of a single sine function: $\sin\left(\frac{\pi}{8}\right)\cos\left(\frac{\pi}{12}\right) - \cos\left(\frac{\pi}{8}\right)\sin\left(\frac{\pi}{12}\right)$

26. A mover applies a constant force of $18\sqrt{3}$ N at 30° to move a piano 15 m. How much work was done?

To notify nature enthusiasts of the presence of nocturnal animals, a motion detector is installed 20 m from a watering hole. A rare brown hyaena (Hyanea Brunnea) just made a midnight visit to the hole and is leaving the area at the 32° angle shown.

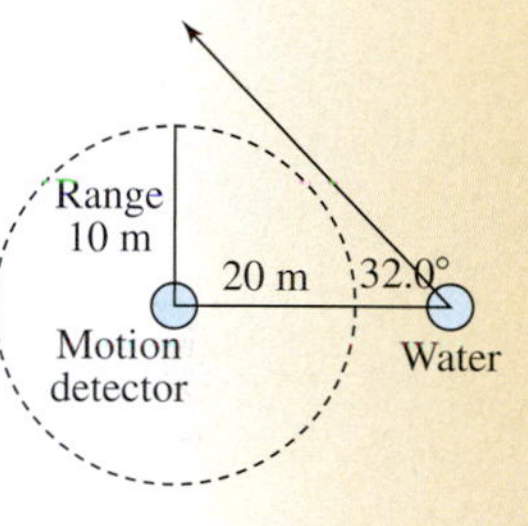

27. If the maximum range of the motion detector is 10 m, will the hyaena be detected?

28. If the maximum range of the motion detector is 12 m, how far from the watering hole is the hyaena when it is first detected?

29. Given $z = 1 - \sqrt{3}i$, use De Moivre's theorem to calculate the value of z^4.

30. Use the nth roots theorem to find the four fourth roots of $-8 + 8\sqrt{3}i$.

Appendix I

Transformations of a Basic Graph

In previous coursework you've likely noted the graph of any function from a given family maintains the same general shape. The graphs of $y = -2x^2 - 5x + 3$ and $y = x^2$ are both parabolas, the graphs of $y = \sqrt[3]{x}$ and $y = -\sqrt[3]{x-2} + 1$ are both "horizontal propellers," and so on for other functions. Once you're aware of the main features of a basic function, you can graph any function from that family using far fewer points, and analyze the graph more efficiently. As we study specific transformations of a graph, it's important to develop a *global view of the transformations*, as they can be applied to virtually any function (see Illustration 7).

A. Vertical and Horizontal Shifts

We'll begin our review using the absolute value function family.

Vertical Translations

If a constant k is added to the output of a basic function $y = f(x)$, the result is a vertical shift since the output values are altered uniformly.

ILLUSTRATION 1 Construct a table of values for $f(x) = |x|$, $g(x) = |x| + 1$, and $h(x) = |x| - 3$ and graph the functions on the same coordinate grid. Then discuss what you observe.

Solution: A table of values for all three functions is shown here, with the corresponding graphs shown in the figure.

x	$f(x) = \|x\|$	$g(x) = \|x\| + 1$	$h(x) = \|x\| - 3$
−3	3	4	0
−2	2	3	−1
−1	1	2	−2
0	0	1	−3
1	1	2	−2
2	2	3	−1
3	3	4	0

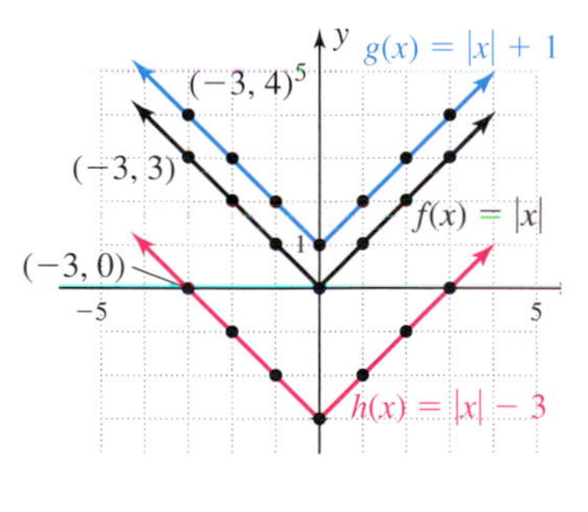

From the table we note that outputs of $g(x)$ are one more than the outputs for $f(x)$, and that each point on the graph of f has been shifted *upward 1 unit* to form the graph of g. Similarly, each point on the graph of f has been shifted *downward 3 units* to form the graph of h: $h(x) = f(x) - 3$.

We describe the transformations in Illustration 1 as **vertical shifts** or the **vertical translation** of a basic graph. The graph of g is the same as the graph of f but *shifted up 1 unit,* and the graph of h is the same as f but *shifted down 3 units*. In general, we have the following:

VERTICAL TRANSLATIONS OF A BASIC GRAPH

Given any function whose graph is determined by $y = f(x)$ and a constant $k > 0$,

1. The graph of $y = f(x) + k$ is the graph of $f(x)$ shifted *upward* k units.
2. The graph of $y = f(x) - k$ is the graph of $f(x)$ shifted *downward* k units.

Horizontal Translations

The graph of a parent function can also be shifted left or right. This happens when we *alter the inputs to the basic function*, as opposed to adding or subtracting something to the basic function. For $Y_1 = x^2 + 2$, it's clear that we first square inputs, then add 2, which results in a vertical shift. For $Y_2 = (x + 2)^2$, we add 2 to x *prior to squaring* and since the input values are affected, we might anticipate the graph will shift along the x-axis—horizontally.

ILLUSTRATION 2 Construct a table of values for $f(x) = x^2$ and $g(x) = (x + 2)^2$, then graph the functions on the same grid and discuss what you observe.

Solution: Both f and g belong to the quadratic family and their graphs will be parabolas. A table of values is shown along with the corresponding graphs.

x	$f(x) = x^2$	$g(x) = (x + 2)^2$
−3	9	1
−2	4	0
−1	1	1
0	0	4
1	1	9
2	4	16
3	9	25

It is apparent the graphs of g and f are identical, but the graph of g has been shifted horizontally 2 units left (the left branch of g can be completed using additional inputs or by simply completing the parabola).

We describe the transformation in Illustration 2 as a **horizontal shift** or **horizontal translation** of a basic graph. The graph of g is the same as that of f, but *shifted 2 units to the left*. Once again it seems reasonable that since *input* values were altered, the shift must be horizontal rather than vertical. From this example, we also learn the direction of the shift is **opposite the sign:** $y = (x + 2)^2$ is 2 units *to the left* of $y = x^2$. Although it may seem counterintuitive, the shift *opposite the sign* can be "seen" by locating the new x-intercept, which in this case is also the vertex. Substituting 0 for y gives $0 = (x + 2)^2$ with $x = -2$, as shown in the graph in Illustration 2. In general, we have

HORIZONTAL TRANSLATIONS OF A BASIC GRAPH
Given any function whose graph is determined by $y = f(x)$ and a constant $h > 0$,

1. The graph of $y = f(x + h)$ is the graph of $f(x)$ shifted *to the left h* units.
2. The graph of $y = f(x - h)$ is the graph of $f(x)$ shifted *to the right h* units.

B. Vertical and Horizontal Reflections

The next transformation we investigate is called a **vertical reflection,** in which we compare the function $Y_1 = f(x)$ with the negative of the function: $Y_2 = -f(x)$.

Vertical Reflections

ILLUSTRATION 3 Construct a table of values for $Y_1 = x^2$ and $Y_2 = -x^2$, then graph the functions on the same grid and discuss what you observe.

Solution: A table of values is given for both functions, along with the corresponding graphs.

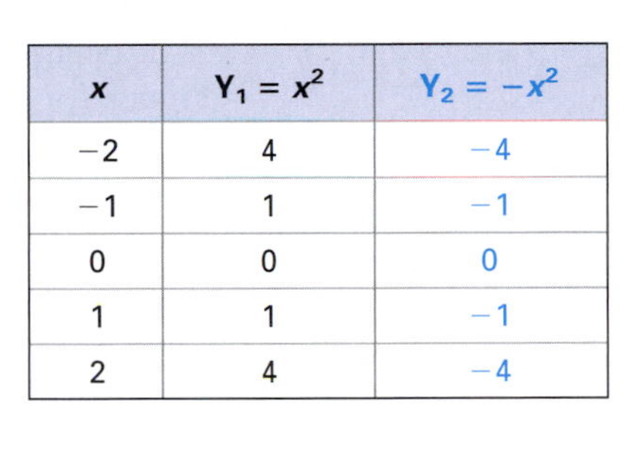

x	$Y_1 = x^2$	$Y_2 = -x^2$
−2	4	−4
−1	1	−1
0	0	0
1	1	−1
2	4	−4

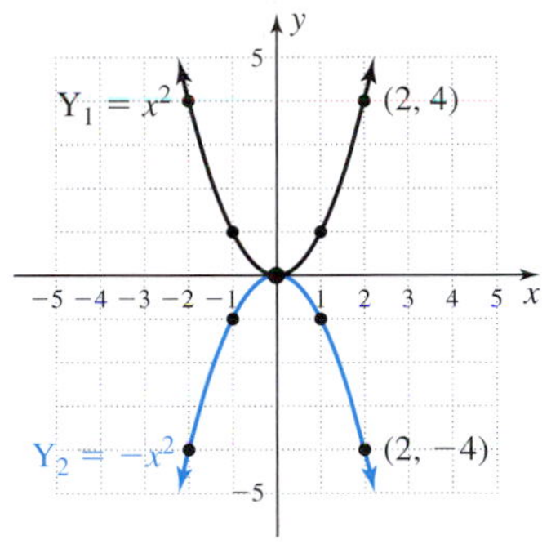

As you might have anticipated, the outputs for f and g differ only in sign. Each output is a **reflection** of the other, being an equal distance from the x-axis but on opposite sides.

The vertical reflection in Illustration 3 is sometimes called a **reflection across the x-axis** or a **north/south reflection.** In general,

VERTICAL REFLECTIONS OF A BASIC GRAPH
Given any function whose graph is determined by $y = f(x)$, the graph of $y = -f(x)$ is the graph of $f(x)$ reflected across the x-axis.

Horizontal Reflections

It's also possible for a graph to be reflected horizontally *across the y-axis.* Just as we noted that $f(x)$ versus $-f(x)$ resulted in a vertical reflection, $f(x)$ versus $f(-x)$ results in a horizontal reflection.

ILLUSTRATION 4 Construct a table of values for $f(x) = \sqrt{x}$ and $g(x) = \sqrt{-x}$, then graph the functions on the same coordinate grid and discuss what you observe.

Solution: A table of values is given here, along with the corresponding graphs.

x	$f(x) = \sqrt{x}$	$g(x) = \sqrt{-x}$
-4	not real	2
-2	not real	$\sqrt{2} \approx 1.41$
-1	not real	1
0	0	0
1	1	not real
2	$\sqrt{2} \approx 1.41$	not real
4	2	not real

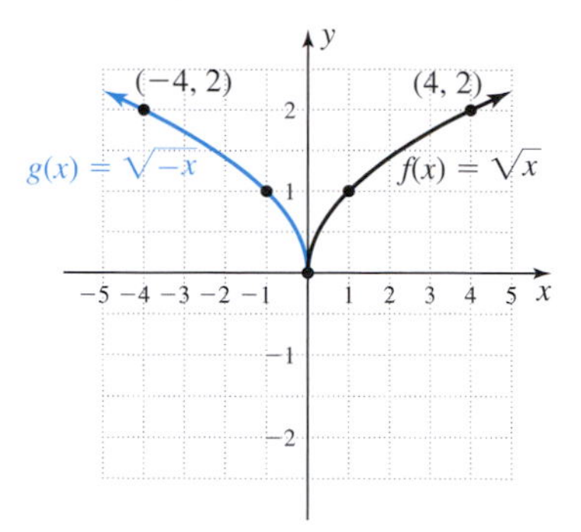

The graph of g is the same as the graph of f, but has been reflected across the y-axis. A study of the domain shows why—f represents a real number only for nonnegative inputs, so its graph occurs to the right of the y-axis, while g represents a real number for nonpositive inputs, so its graph occurs to the left.

The transformation in Illustration 4 is called a **horizontal reflection** (or an **east/west reflection**) of a basic graph. In general,

HORIZONTAL REFLECTIONS OF A BASIC GRAPH
Given any function whose graph is determined by $y = f(x)$, the graph of $y = f(-x)$ is the graph of $f(x)$ reflected across the y-axis.

Since the actual shape of a graph remains unchanged after the previous transformations are applied, they are often referred to as **rigid transformations.**

C. Stretching/Compressing a Basic Graph

Stretches and compressions of a basic graph are called **nonrigid transformations.** As the name implies, the shape of a graph is changed or transformed when these are applied. However, the transformation doesn't actually "deform" the graph, and we can still identify the function family as well as all important characteristics.

ILLUSTRATION 5 Construct a table of values for $f(x) = x^2$, $g(x) = 3x^2$, and $h(x) = \frac{1}{3}x^2$, then graph the functions on the same grid and discuss what you observe.

Solution: A table of values is given for all three functions with the corresponding graphs.

x	$f(x) = x^2$	$g(x) = 3x^2$	$h(x) = \frac{1}{3}x^2$
−3	9	27	3
−2	4	12	$\frac{4}{3}$
−1	1	3	$\frac{1}{3}$
0	0	0	0
1	1	3	$\frac{1}{3}$
2	4	12	$\frac{4}{3}$
3	9	27	3

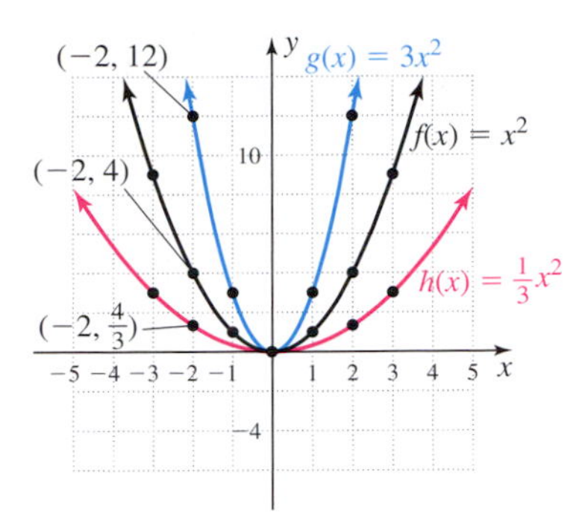

The outputs of g are triple those of f, *stretching* g upward and causing its branches to hug the vertical axis (making it more narrow). The outputs of h are one-third those of f and the graph of h is *compressed* downward, with its branches farther away from the vertical axis (making it wider).

The transformations in Illustration 5 are called **vertical stretches** or **compressions.** In general, we have,

> **STRETCHES AND COMPRESSIONS OF A BASIC GRAPH**
> Given $a > 0$ and any function whose graph is determined by $y = f(x)$, the graph of $y = af(x)$ is the graph of $f(x)$ stretched vertically if $|a| > 1$ and compressed vertically if $0 < |a| < 1$.

D. Transformations of a General Function $y = f(x)$

Often more than one transformation acts on the same function at the same time. Although the transformations can be applied in almost any sequence, it's helpful to use an organized sequence when graphing them.

> **GENERAL TRANSFORMATIONS OF A BASIC GRAPH**
> Given any transformation of a function whose graph is defined by $y = f(x)$, the graph of the transformed function can be found by
>
> 1. Applying the stretch or compression.
> 2. Reflecting the result.
> 3. Applying the horizontal and/or vertical shifts.
>
> These are usually applied to a few characteristic points, with the new graph drawn through the shifted points.

ILLUSTRATION 6 Sketch the graphs of $g(x) = -(x + 2)^2 + 3$ and $h(x) = 4\sqrt[3]{x - 2} - 1$ using transformations of a parent function and a few characteristic points.

Solution: The graph of $g(x) = -(x + 2)^2 + 3$ is a basic parabola reflected across the x-axis, shifted left 2 and up 3. This sequence of transformations in shown in Figures A.1 through A.3. The graph of $h(x) = 4\sqrt[3]{x - 2} - 1$ is a horizontal propeller, stretched by a factor of 4, then shifted right 2 and down 1. This sequence is shown in Figures A.4 through A.6.

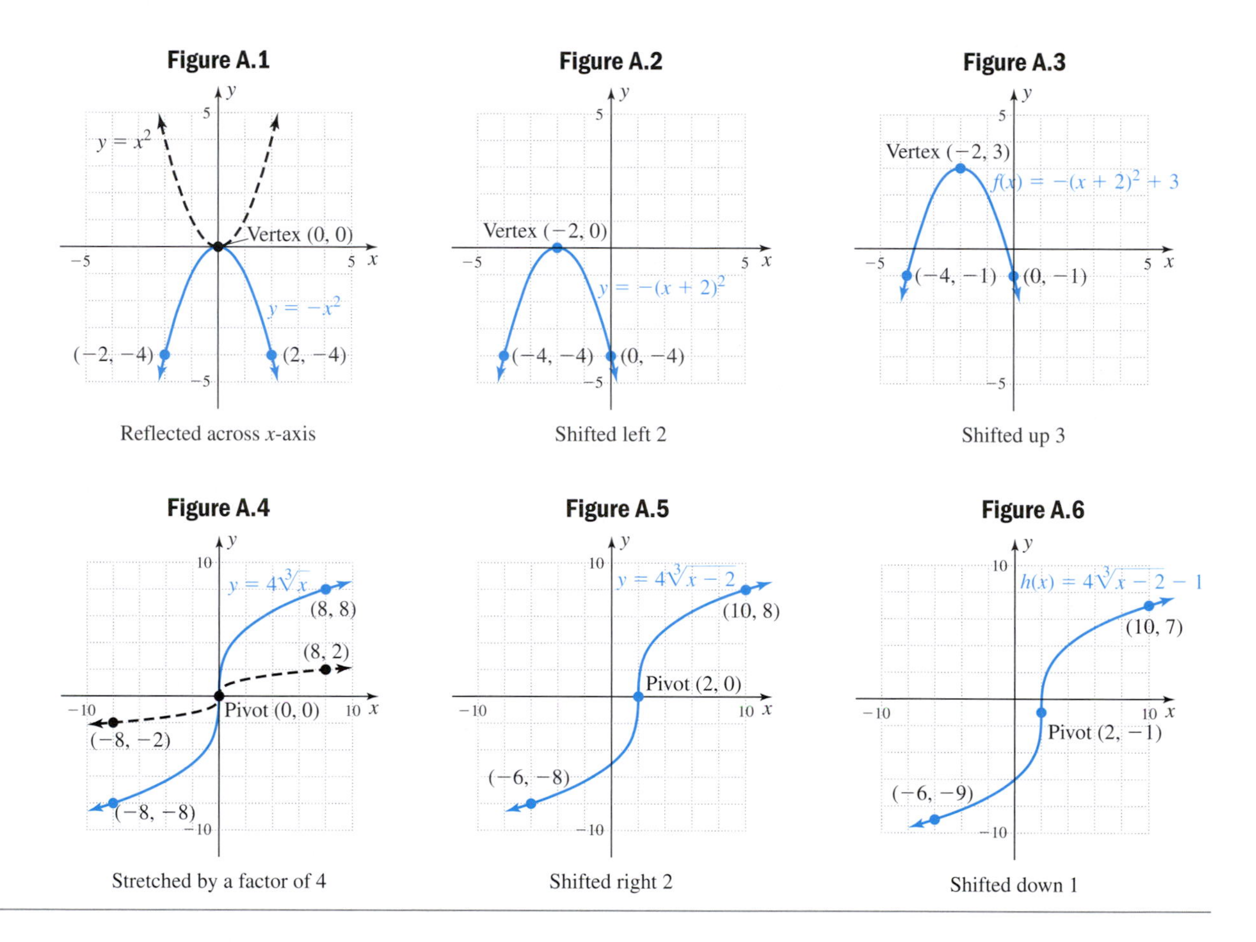

As mentioned, it's important to realize that these transformations can actually be applied to *any function*, even those that are new and unfamiliar. Consider the following pattern:

Parent Function		Transformation of Parent Function
quadratic:	$y = x^2$	$y = -2(x - 3)^2 + 1$
cubic:	$y = x^3$	$y = -2(x - 3)^3 + 1$
absolute value:	$y = \|x\|$	$y = -2\|x - 3\| + 1$
square root:	$y = \sqrt{x}$	$y = -2\sqrt{x - 3} + 1$
cube root:	$y = \sqrt[3]{x}$	$y = -2\sqrt[3]{x - 3} + 1$
general:	$y = f(x)$	$y = -2f(x - 3) + 1$

In each case, the transformation involves a vertical stretch, then a vertical reflection with the result shifted right 3 and up 1. Since the shifts are the same regardless of the initial function, we can extend and globalize these results to a general function $y = f(x)$.

Parent Function	**Transformation of Parent Function**
$y = f(x)$	$y = af(x \pm h) \pm k$

- a: north/south reflections, vertical stretches and compressions
- h: horizontal shift *h* units, opposite direction of sign
- k: vertical shift *k* units, same direction as sign

Remember—if the graph of a function is shifted, the *individual points* on the graph are likewise shifted.

ILLUSTRATION 7 Given the graph of $f(x)$ shown in Figure A.7, graph $g(x) = -f(x + 1) - 2$.

Solution: For g, the graph of f is reflected across the x-axis, then shifted horizontally 1 unit left and vertically 2 units down. The result is shown in Figure A.8.

Figure A.7 **Figure A.8**

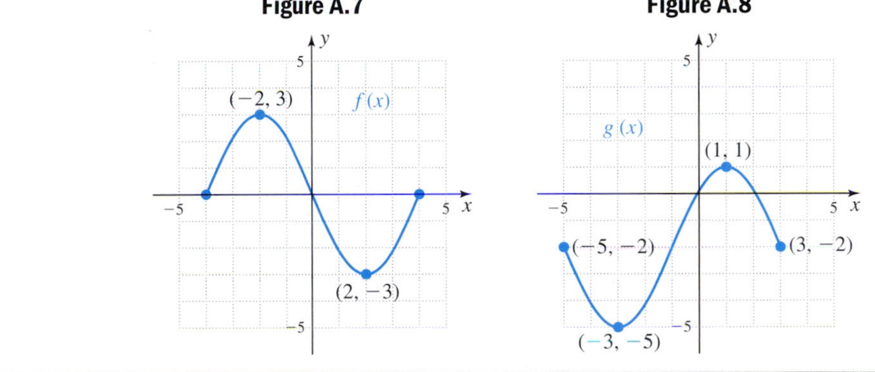

Appendix II

Solving Equations Graphically

At the heart of understanding graphical solutions to an equation is this basic definition: *An equation is a statement that two expressions are equal for a certain input value(s).* The linear equation $3(x - 2) = x - 8$ literally says that for some unknown input x, the expression $3(x - 2)$ is equal to the expression $x - 8$. While overly simplistic, if we were to solve this equation by trial and error, our method might involve systematically using various inputs, in a search for one that resulted in a like output for both expressions. Table A.1 shows the solution to $3(x - 2) = x - 8$ is $x = -1$, since the left-hand expression is equal to the right-hand expression for this input. Note that we're actually treating each expression as the independent functions $Y_1 = 3(x - 2)$ and $Y_2 = x - 8$, and can actually view the solution method as *an attempt to find where the graphs of these two lines intersect.* This basic idea is very powerful and can be extended to expressions and equations of all kinds. With the help of graphing and calculating technology, we have the ability to solve some very sophisticated equations. The keystrokes illustrated here apply to a TI-84 Plus calculator model. Please consult our Internet site or your manual for other models.

Table A.1

Inputs	$3(x - 2)$	$x - 8$
−4	−18	−12
−3	−15	−11
−2	−12	−10
−1	**−9**	**−9**
0	−6	−8
1	−3	−7
2	0	−6

To solve the equation $2\sin\left(x - \frac{\pi}{6}\right) + 1 = 1.5$ graphically, we begin by assigning Y_1 to the left-hand expression, Y_2 to the right-hand expression, and carefully entering them on the **Y=** screen (Figure A.9). The TI-84 Plus is programmed with a "standard window" for trigonometric functions, which is preset as shown in Figure A.10 (with the calculator in radian **MODE**), and accessed by pressing **ZOOM** **7:ZTrig**. Note the range is $y \in [-4, 4]$, the x-axis is scaled in units of $\frac{\pi}{2} \approx 1.57$, and the independent variable has a domain of roughly -2π to 2π. This should be adequate for the equation at hand, and pressing **ENTER** and **TRACE** produces the graph shown in

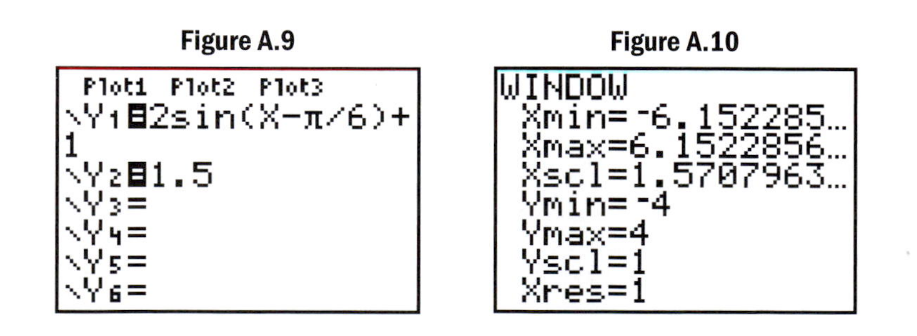

Figure A.9

Figure A.10

Figure A.11. To have the calculator locate a point of intersection, we press 2nd CALC and select option **5:intersect** by pressing the number 5 or using the down arrow ▼ to access this option. We then press ENTER *three* times: The first " ENTER " selects the graph of Y_1, the second " ENTER " selects the graph of Y_2, and the third " ENTER " (the "**GUESS**" option) uses the x-value of the current cursor location to begin its search for a point of intersection (this means we can help the calculator find a specific point of intersection if there is more than one showing). The calculator will "think" for a moment or two, then display x- and y-coordinates of the point of intersection at the bottom of the screen (Figure A.12).

Figure A.11

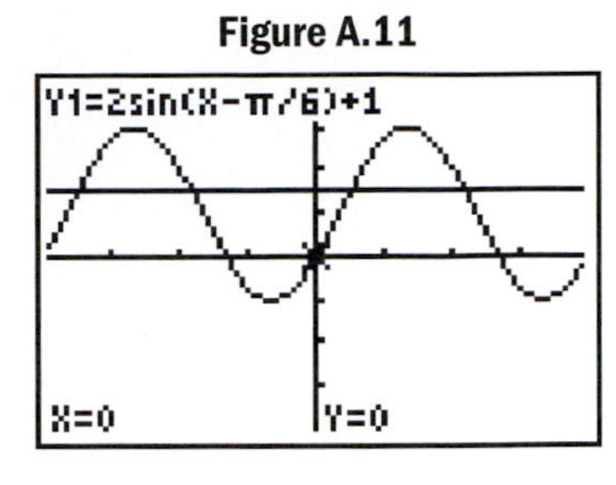

Figure A.12

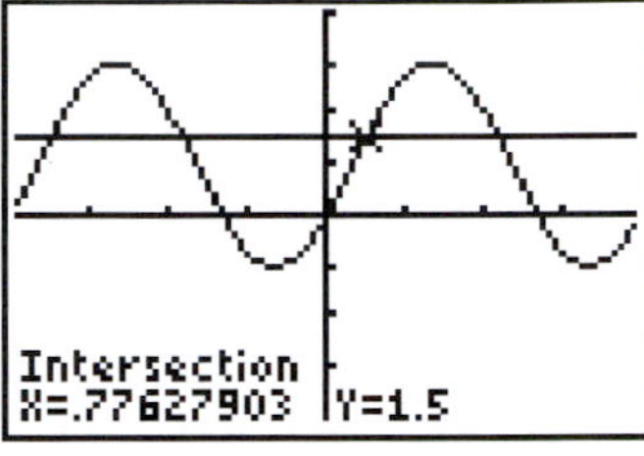

Figure A.13

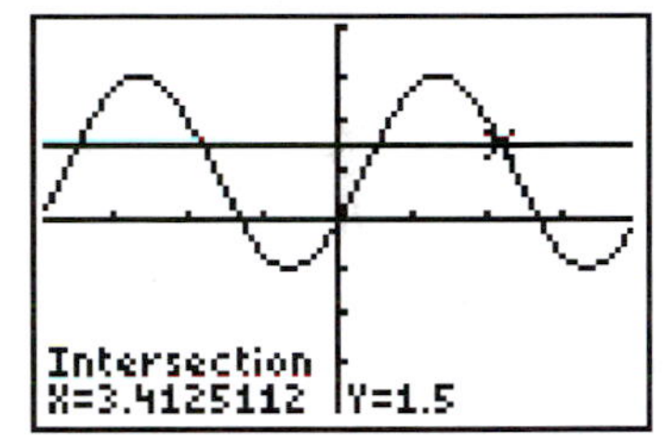

For $2\sin\left(x - \frac{\pi}{6}\right) + 1 = 1.5$, the calculator finds a solution of $x \approx 0.7763$ (Figure A.12). To find the next point of intersection to the right, recall that the x-axis is scaled in units of $\frac{\pi}{2}$, so we can repeat the preceding sequence of keystrokes, except that we'll enter $x = \pi$ (the point of intersection occurs at roughly the second tick-mark) at the "**GUESS**" option. The calculator then finds a second solution at $x \approx 3.4125$ (Figure A.13). Use these ideas to find points of intersection where $x < 0$.

Regression and Calculator Use

Here we'll review/introduce the basic fundamentals of regression and calculator use, using data collected from various sources or from observed real-world relationships. You can hardly pick up a newspaper or magazine without noticing it contains a large volume of data—graphs, charts, and tables seem to appear throughout the pages.

A. Scatter-Plots and Positive/Negative Association

There are many simple experiments or activities that enable you to collect your own data. After it's been collected, we can begin analyzing the data using a **scatter-plot,** which is simply a graph of all of the ordered pairs in a data set. Much of the time, real data (sometimes called **raw data**) is not very "well behaved" and the points may be somewhat scattered—which is the reason for the name.

Positive and Negative Associations

You have likely noted that lines with positive slope rise from left to right, while lines with negative slope fall from left to right. We can extend this idea to the data from a scatter-plot. The data points in Illustration 1 seem to *rise* as you move from left to right, with larger input values resulting in larger outputs. In this case, we say there is a **positive association** between the variables. If the data seems to decrease or fall as you move left to right, we say there is a **negative association.**

ILLUSTRATION 1 ▸ The ratio of the federal debt to the total population is known as the *per capita debt*. The per capita debt of the United States is shown in the table for the odd-numbered years from 1995 to 2003. Draw a scatter-plot of the data and state whether the association is positive or negative.

Data from the Bureau of Public Debt at www.publicdebt.treas.gov

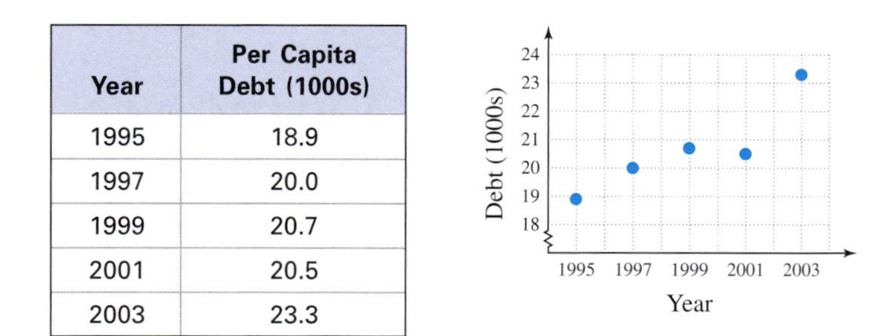

Year	Per Capita Debt (1000s)
1995	18.9
1997	20.0
1999	20.7
2001	20.5
2003	23.3

Solution: ▸ Since the amount of debt depends on the year, *year* is the input x and *per capita debt* is the output y. Scale the x-axis from 1995 to 2003 and the y-axis from 18 to 23 to comfortably fit the data (the "squiggly line" near the 18 in the graph is used to show that some initial values have been skipped). The graph indicates there is a positive association between the variables, meaning the debt is generally *increasing* as time goes on.

ILLUSTRATION 2 ▷ A cup of coffee is placed on a table and allowed to cool. The temperature of the coffee is measured every 10 min and the data are shown in the table. Draw the scatter-plot and state whether the association is positive or negative.

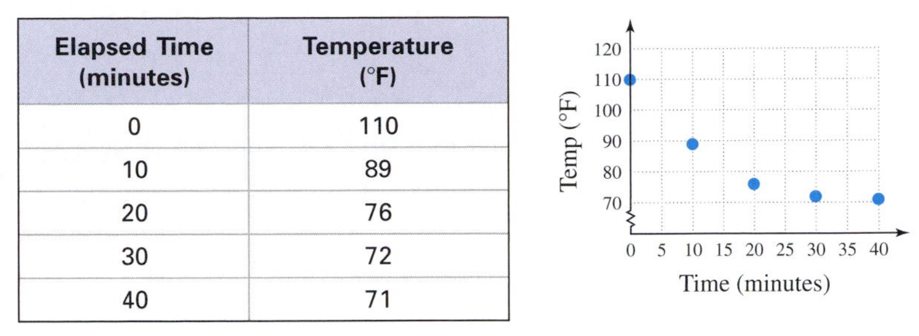

Elapsed Time (minutes)	Temperature (°F)
0	110
10	89
20	76
30	72
40	71

Solution: ▷ Since temperature depends on cooling time, *time* is the input x and *temperature* is the output y. Scale the x-axis from 0 to 40 and the y-axis from 70 to 110 to comfortably fit the data. As you see in the figure, there is a negative association between the variables, meaning the temperature *decreases* over time.

B. Scatter-Plots and Linear/Nonlinear Associations

The data in Illustration 1 had a positive association, while the association in Illustration 2 was negative. But the data from these examples differ in another important way. In Illustration 1, the data seem to cluster about an imaginary line. This indicates a linear equation model might be a good approximation for the data, and we say there is a **linear association** between the variables. The data in Illustration 2 could not accurately be modeled using a straight line, and we say the variables *time* and *cooling temperature* exhibit a **nonlinear association.**

ILLUSTRATION 3 ▷ A college professor tracked her annual salary for 1997 to 2004 and the data are shown in the table. Draw the scatter-plot and determine if there is a linear or nonlinear association between the variables. Also state whether the association is positive, negative, or cannot be determined.

Year	Salary (1000s)
1997	30.5
1998	31
1999	32
2000	33.2
2001	35.5
2002	39.5
2003	45.5
2004	52

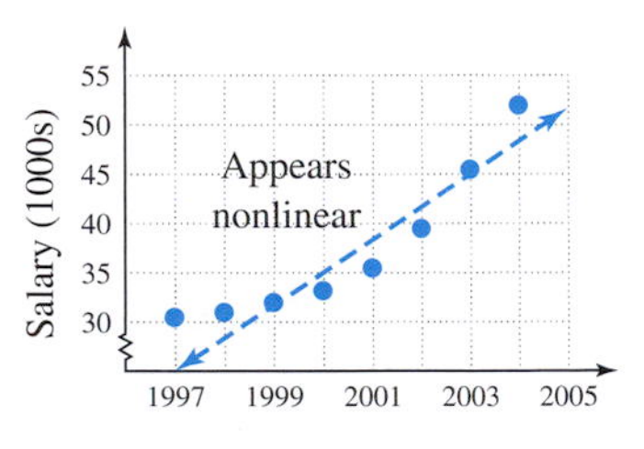

Solution: Since salary earned depends on a given year, *year* is the input x and *salary* is the output y. Scale the x-axis from 1996 to 2005, and the y-axis from 30 to 55 to comfortably fit the data. A line doesn't seem to model the data very well, and the association appears to be nonlinear. The data rises from left to right, indicating a positive association between the variables. This makes good sense, since we expect our salaries to increase over time.

C. Strong and Weak Associations

Using Figures A.14 and A.15, we can make one additional observation regarding the data in a scatter-plot. While both associations appear linear, the data in Figure A.14 seems to cluster more tightly about an imaginary straight line than the data in Figure A.15.

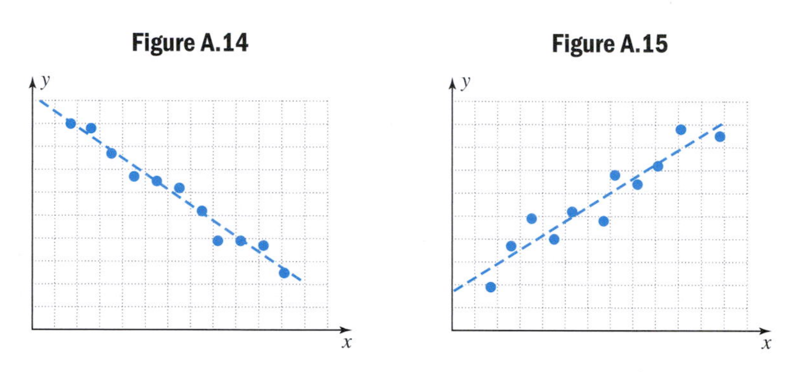

We refer to this "clustering" as the **tightness of fit** or in statistical terms, the **strength of the correlation.** To quantify this fit we use a measure called the **correlation coefficient *r*,** which tells whether the association is positive or negative—$r > 0$ or $r < 0$, and quantifies the strength of the association: $|r| \leq 100\%$. Actually, the coefficient is given in decimal form, making it a number from -1.0 to $+1.0$, depending on the association. If the data points form a perfectly straight line, we say the strength of the correlation is either -1 or 1. If the data points appear clustered about the line, but are scattered on either side of it, the strength of the correlation falls somewhere between -1 and 1, depending on how tightly or loosely they're scattered. This is summarized in Figure A.16.

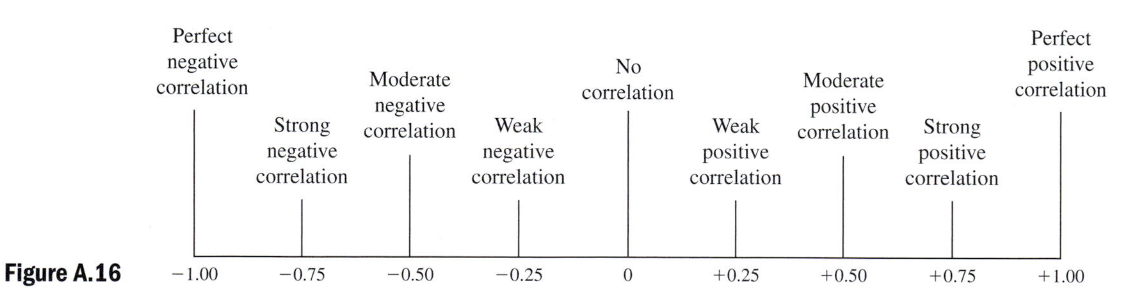

Figure A.16

The following scatter-plots help to further illustrate this idea. Figure A.17 shows a linear and negative association between the value of a car and the age of a car, with a strong correlation. Figure A.18 shows there is no apparent association between family income and the number of children, and Figure A.19 shows a linear and positive association between a man's height and weight, with a moderate correlation.

Figure A.17

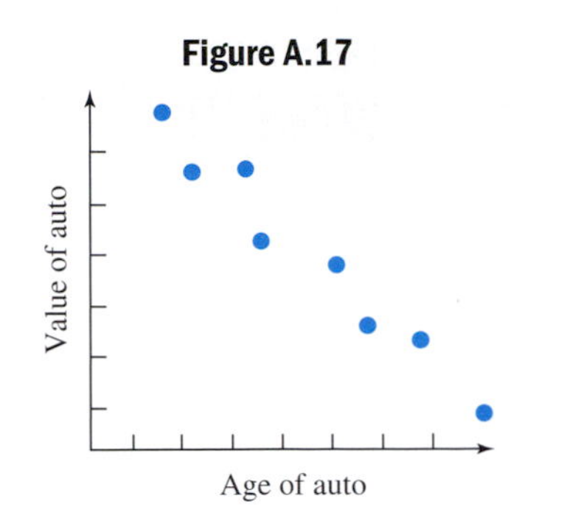

Figure A.18

Figure A.19

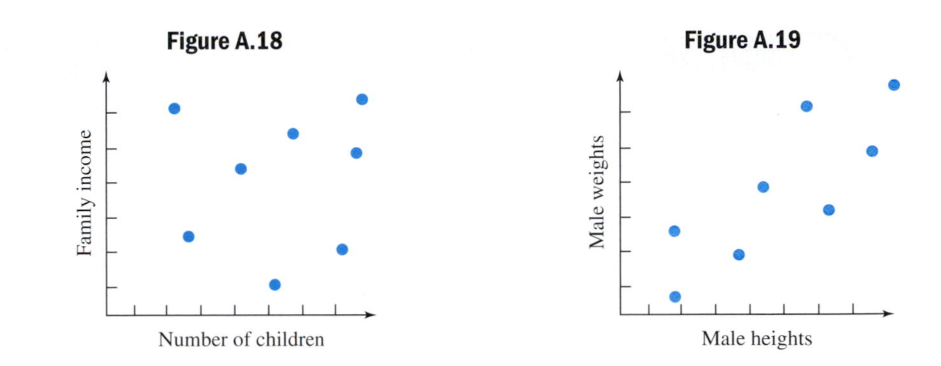

D. Linear Regression and the Line of Best Fit

Calculating a **linear equation model** (by hand) for a set of data involves visually estimating and sketching a line that appears to fit the data. This means answers will vary slightly, but a good, usable equation model can often be obtained.

There is actually a sophisticated method for calculating the equation of a line that best fits a data set, called the **regression line.** The method minimizes the vertical distance between all data points and the line itself, making it the unique **line of best fit.** Most graphing calculators have the ability to perform this calculation quickly, and we'll illustrate using the TI-84 Plus. The process involves these steps: (1) clearing old data, (2) entering new data; (3) displaying the data; (4) calculating the regression line; and (5) displaying and using the regression line.

ILLUSTRATION 4 The men's 400-m freestyle times (to the nearest second) for the 1960 through 2000 Olympics are given in the table shown. Let the year be the input x, and race time be the output y.

Year (x) (1900 → 0)	Time (y) (sec)
60	258
64	252
68	249
72	240
76	232
80	231
84	231
88	227
92	225
96	228
100	221

Step 1: Clear Old Data
While on the home screen, press the STAT key and select option **4:ClrList,** then tell the calculator which list(s) to clear. Press 2nd 1 to indicate List 1 (L1), then any other list we wish to clear (2nd 2 for L2, and so on), separating the list names with a comma. Pressing ENTER then clears all lists indicated.

Figure A.20

L1	L2	L3 3
60	258	
64	252	
68	249	
72	240	
76	232	
80	231	
84	231	

L3(1)=

Step 2: Enter New Data
We can now enter the data for the Year (x) in L1. Press the STAT key and select option **1:Edit.** This places the cursor at the first position in L1, where we simply begin listing the x-values in order: 60 ENTER , 64 ENTER , 68 ENTER , and so on. Next, enter the y-values (Time) in L2. When finished, you should obtain the screen shown in Figure A.20.

Figure A.21a

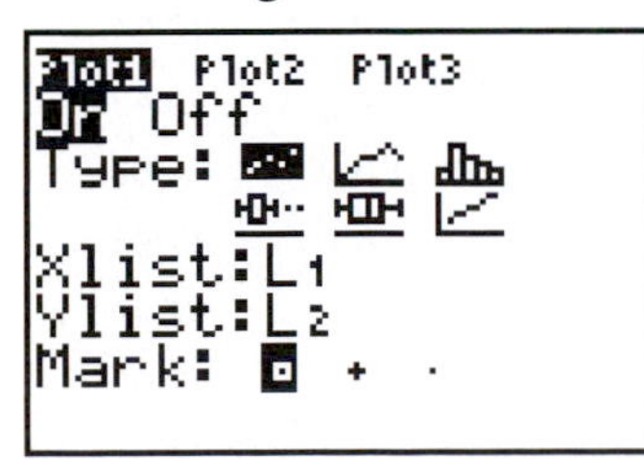

Step 3: Display the Data
To display the data in these lists as plotted points, press 2nd Y= to access the **STATPLOTS** screen. With the cursor over option 1, press ENTER and be sure the options shown in Figure A.21a are highlighted. Then go to the WINDOW screen

WORTHY OF NOTE

As a rule of thumb, the tic marks for Xscl can be set by mentally calculating $\frac{|Xmax| + |Xmin|}{10}$ and using a convenient number in the neighborhood of the result. The same goes for Yscl.

to set up a good viewing window. The data in L1 (the Xlist) ranges from 60 to 100, and the data in L2 (the Ylist) ranges from 221 to 258, so we set the display window on the calculator accordingly, allowing for a **frame around the window** to comfortably display all points. For instance, we'll use [50, 110] and [210, 270] for the Xlist and Ylist, respectively.

Figure A.21b

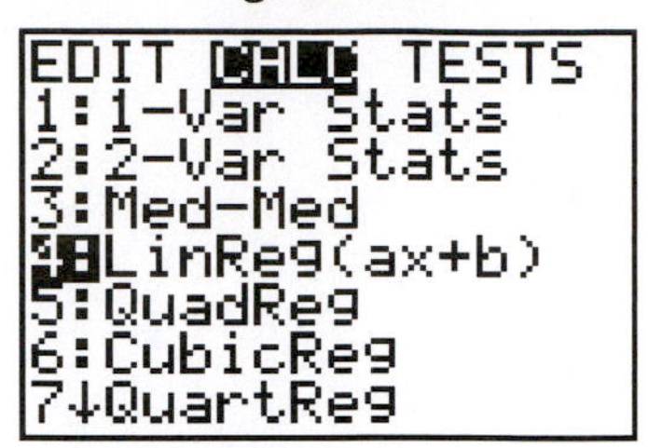

Step 4: Calculate the Regression Equation

To have the calculator compute the regression equation, press the STAT and ▶ keys to move the cursor over to the **CALC** options (see Figure A.21b). Note that the fourth option reads **4:LinReg (ax + b).** Pressing the number 4 places **LinReg(ax + b)** on the home screen, and pressing ENTER computes the values of a, b, and the correlation coefficient r (the calculator automatically uses the data in L1 and L2 unless instructed otherwise). Rounded to hundredths, the equation is $y = -0.86x + 304.91$ (Figure A.22). An ***r*-value** (correlation coefficient) of -0.94 tells us the association is *negative* and *very strong*.

Figure A.22

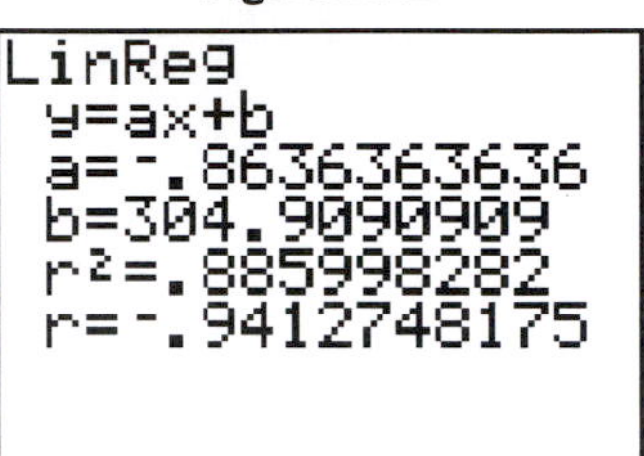

WORTHY OF NOTE

The correlation coefficient is a *diagnostic feature* of the regression calculation and can sometimes be misleading. Other indications of a good or poor correlation include a study of **residuals.** To turn the diagnostic feature on or off, use 2nd 0 **(CATALOG)** and scroll down to the "D's" (all options offered on the TI-84 Plus are listed in the CATALOG) until you reach **DiagnosticOff** and **DiagnosticOn.** Highlight your choice and press ENTER.

Step 5: Display and Use the Results

Although the TI-84 Plus can paste the regression equation directly into Y_1 on the Y= screen, for now we'll enter $y = -0.86x + 304.91$ by hand. Afterward, pressing the GRAPH key will plot the data points (if Plot1 is still active) and graph the line. Your display screen should now look like the one in Figure A.23. The regression line is the best estimator for the set of data as a whole, but there will still be some difference between the values it generates and values from the set of raw data.

Figure A.23

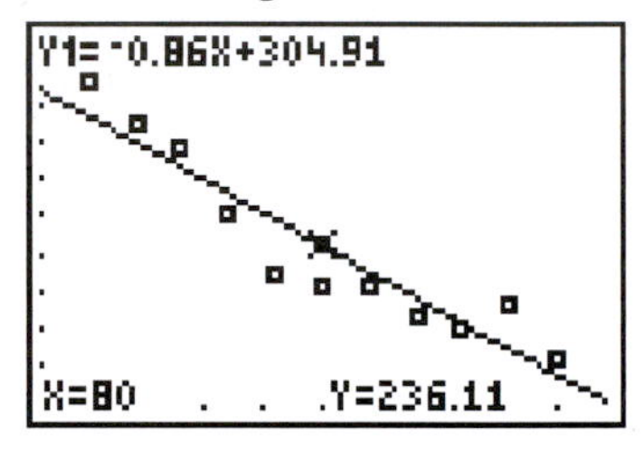

ILLUSTRATION 5 Riverside Electronics reviews employee performance semiannually, and awards increases in their hourly rate of pay based on the review. The table shows Thomas's hourly wage for the last 4 yr (eight reviews). Find the regression equation for the data and use it to project his hourly wage for the year 2007, after his fourteenth review.

Year (x)	Wage (y)
(2001) 1	9.58
2	9.75
(2002) 3	10.54
4	11.41
(2003) 5	11.60
6	11.91
(2004) 7	12.11
8	13.02

Solution: Following the prescribed sequence produces the equation $y \approx 0.48x + 9.09$. For $x = 14$ we obtain $y = 0.48(14) + 9.09$, or a wage of \$15.81. According to this model, Thomas will be earning \$15.81 per hour in 2007.

If the input variable is a unit of time, particularly the time in years, we often **scale the data** to avoid working with large numbers. For instance, if the data involved the cost of attending a major sporting event for the years 1980 to 2000, we would say 1980 corresponds to 0 and use input values of 0 to 20 (subtracting the smallest value from itself and all other values has the effect of scaling down the data). This is easily done on a graphing calculator. Simply enter the four-digit years in L1, then with the cursor in the header of L1—use the keystrokes 2nd 1 **(L1)** – 1980 ENTER and the data in this list automatically adjusts.

E. Quadratic Regression and the Parabola of Best Fit

Once the data have been entered, graphing calculators have the ability to find many different regression equations. The choice of regression depends on the context of the data, patterns formed by the scatter-plot, and/or some foreknowledge of how the data are related. We now turn our attention to quadratic regression equations.

ILLUSTRATION 6A Since 1990, the number of *new* books published each year has been growing at a rate that can be approximated by a quadratic function. The table shows the number of books published in the United States for selected years. Draw a scatter-plot and sketch an estimated parabola of best fit by hand.

Source: 1998, 2000, 2002, and 2004 *Statistical Abstract of the United States.*

Year (1990→0)	Books Published (1000s)
0	46.7
2	49.2
3	49.7
4	51.7
5	62.0
6	68.2
7	65.8
9	102.0
10	122.1

Solution: Begin by drawing the scatter-plot, being sure to scale the axes appropriately. The data appears to form a quadratic pattern, and we sketch a parabola that seems to best fit the data (see graph).

The regression abilities of a graphing calculator can be used to find a **parabola of best fit** and the steps are identical to those for linear regression.

WORTHY OF NOTE

The TI-84 Plus can round all coefficients and the correlation coefficient to any desired number of decimal places. For three decimal places, press MODE and change the **Float** setting to "3." Also, be aware that there are additional methods for pasting the equation in Y_1.

ILLUSTRATION 6B Use the data from Illustration 6A to calculate a quadratic regression equation, then display the data and graph. How well does the equation match the data?

Solution: Begin by entering the data in L1 and L2 as shown in Figure A.24. Press 2nd Y= to be sure that Plot 1 is still active and is using L1 and L2 with the desired point type. Set the window size to comfortably fit the data. Finally, press STAT and the right arrow ▶ to overlay the **CALC** option. The quadratic regression option is number **5:QuadReg.** Pressing 5 places this option directly on the home screen. Lists L1 and L2 are the default lists, so pressing ENTER will have the calculator compute the regression equation for the data in L1 and L2. After "chewing on the data" for a short while, the calculator returns the regression equation in the form shown in Figure A.25. To maintain a higher degree of accuracy, we can actually paste the entire regression equation in Y_1. Recall the last operation using 2nd ENTER, and **QuadReg** should (re)appear. Then enter the function Y_1 after the QuadReg option by pressing VARS ▶ (**Y-Vars**) and ENTER (**1:Function**) and ENTER (Y_1). After pressing ENTER once again, the full equation is automatically pasted in Y_1. To compare this equation model with the data, simply press GRAPH and both the graph and plotted data will appear. The graph and data seem to match very well (Figure A.26).

Figure A.24

L1	L2	L3 2
3	49.7	
4	51.7	
5	62	
6	68.2	
7	65.8	
9	102	
10		

L2(9) =122.1

Figure A.25

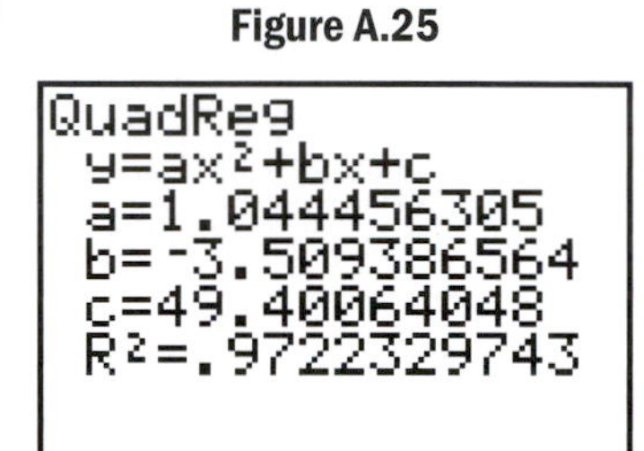

Figure A.26

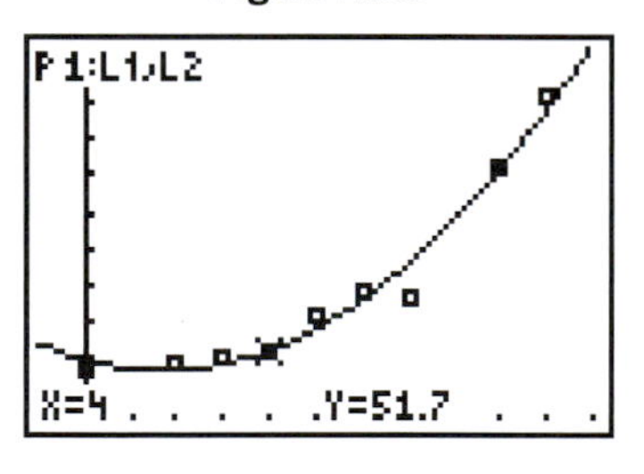

ILLUSTRATION 6C Use the equation from Illustration 6B to answer the following questions: According to the function model, how many new books were published in 1991? If this trend continues, how many new books will be published in 2009?

Solution: Since the year 1990 corresponds to 0 in this data set, we use an input value of 1 for 1991, and an input of 19 for 2009. Accessing the table (2nd GRAPH) feature and inputting 1 and 19 gives the screen shown.

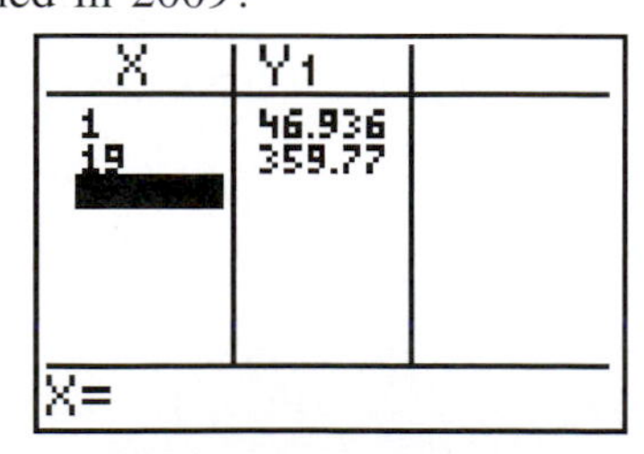

Approximately 47,000 new books were published in 1991, and just under 360,000 will be published in the year 2009.

The sequence of steps used is the same regardless of the form of regression, so the fundamentals illustrated previously can still be applied to data that can be modeled by a sine graph.

ILLUSTRATION 7 The data shown give the record high temperature for selected months for Bismarck, North Dakota. (a) Use the data to draw a scatter-plot, then find a sinusoidal regression model and graph both on the same screen. (b) Use the equation model to estimate the record high temperatures for months 2, 6, and 8. (c) Determine what month gives the largest difference between the actual data and the computed results.

Month (Jan → 1)	Temp. (°F)	Month (Jan → 1)	Temp. (°F)
1	63	9	105
3	81	11	79
5	98	12	65
7	109		

Source: NOAA Comparative Climate Data 2004.

Solution:

a. Entering the data and running the regression results in the coefficients shown in Figure A.27. After pasting the equation into Y_1 and pressing ZOOM **9:Zoom Stat** (used to comfortably display all data points) we obtain the graph shown in Figure A.28.

Figure A.27

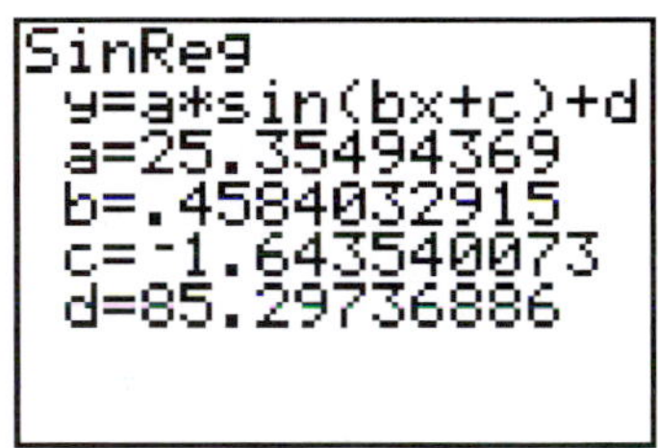

b. Using $x = 2$, $x = 6$, and $x = 8$ as inputs, projects record high temperatures of 68.5°, 108.0°, and 108.1° respectively, for these months.

Figure A.28

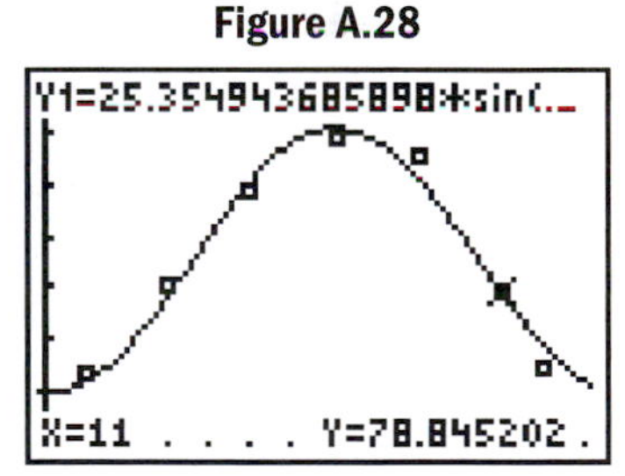

c. In the header of L3, use Y_1(L1) ENTER to evaluate the regression model using the inputs from L1, and place the results in L3. Entering L2 − L3 in the header of L4 gives the result shown in Figure A.29, and we note the largest difference occurs in September—about 4°.

Figure A.29

L2	L3	L4 4
63	61.805	1.1953
81	78.575	2.4248
98	100.61	-2.611
109	110.65	-1.652
105	100.83	4.167
79	78.845	.1548
65	68.661	-3.661

L4(1)=1.195267717...

Miscellaneous Algebra Review

Figure A.30

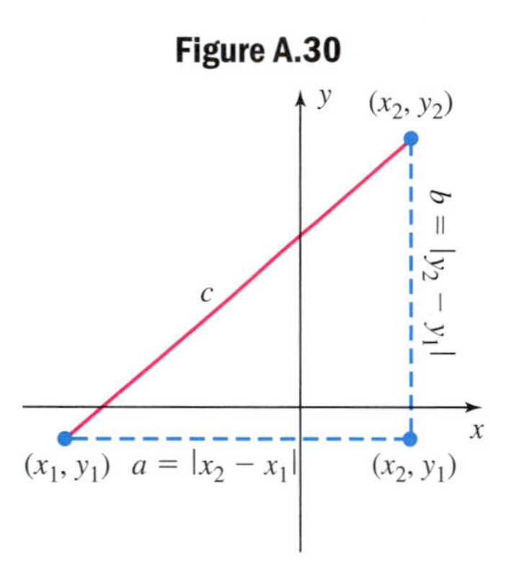

The Distance Formula

Consider the "slope triangle" for (x_1, y_1) and (x_2, y_2) as shown in Figure A.30 and note the base of the triangle is $a = |x_2 - x_1|$ units long and the height (vertical distance) is $b = |y_2 - y_1|$ units. From the Pythagorean theorem we see that $c^2 = a^2 + b^2$ corresponds to $c^2 = (|x_2 - x_1|)^2 + (|y_2 - y_1|)^2$, and taking the square root of both sides yields the **distance formula:** $c = \sqrt{(x_2 - x_1)^2 + (y_2 - y_1)^2}$, although it is most often written using d for distance, rather than c. Note the absolute value bars are dropped since the square of any quantity is always positive.

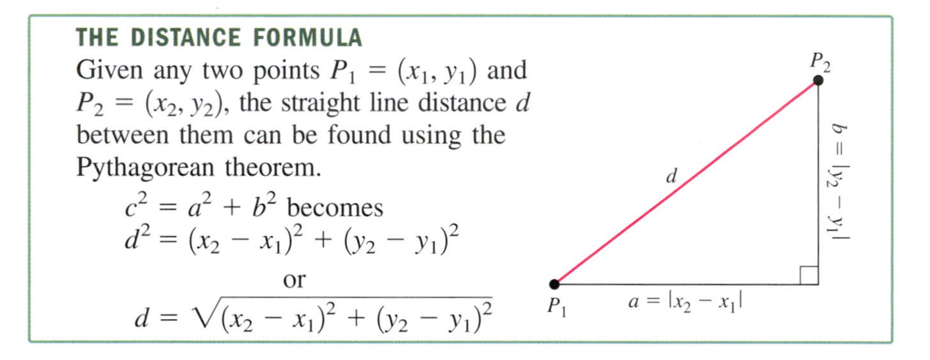

THE DISTANCE FORMULA

Given any two points $P_1 = (x_1, y_1)$ and $P_2 = (x_2, y_2)$, the straight line distance d between them can be found using the Pythagorean theorem.

$$c^2 = a^2 + b^2 \text{ becomes}$$
$$d^2 = (x_2 - x_1)^2 + (y_2 - y_1)^2$$
or
$$d = \sqrt{(x_2 - x_1)^2 + (y_2 - y_1)^2}$$

ILLUSTRATION 1 Use the distance formula to find the diameter of a circle if the endpoints are $(-3, -2)$ and $(5, 4)$.

Solution: For $(x_1, y_1) = (-3, -2)$ and $(x_2, y_2) = (5, 4)$, the distance formula gives

$$\begin{aligned} d &= \sqrt{(x_2 - x_1)^2 + (y_2 - y_1)^2} \\ &= \sqrt{[5 - (-3)]^2 + [4 - (-2)]^2} \\ &= \sqrt{8^2 + 6^2} \text{ or } \sqrt{100} = 10 \end{aligned}$$

The diameter of the circle is 10 units long.

Composition of Functions

The composition of functions is best understood by studying the "input/output" nature of a function. Consider $g(x) = x^2 - 3$. To describe how this function operates on input values, we might say, "inputs are squared, then decreased by three." Using a function box, we could "program" the box to perform these operations and in diagram form we have:

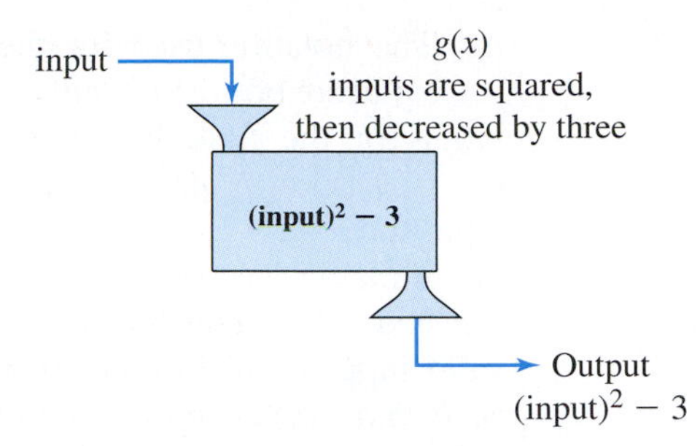

In many respects, a function box can be regarded as a very simple machine, running a simple program. It doesn't matter what the input is, this machine is going to *square the input, then subtract three.*

ILLUSTRATION 2 For $g(x) = x^2 - 3$, find (a) $g(-5)$, (b) $g(t)$, and (c) $g(t - 4)$.

Solution:

a. $g(x) = x^2 - 3$ original function

input −5

$g(-5) = (-5)^2 - 3$ square input, then subtract 3

$= 25 - 3$ simplify

$= 22$ result

b. $g(x) = x^2 - 3$ original function

input t

$g(t) = (t)^2 - 3$ square input, then subtract 3

$= t^2 - 3$ result

c. $g(x) = x^2 - 3$ original function

input $t - 4$

$g(t - 4) = (t - 4)^2 - 3$ square input, then subtract 3

$= t^2 - 8t + 16 - 3$ expand binomial

$= t^2 - 8t + 13$ result

When the input value is itself a function (rather than a single number or variable), this process is called the **composition of functions.** The evaluation method is exactly the same; we are simply using a function input. Using a general function $g(x)$ and a function box as before, we show the process in Figure A.31.

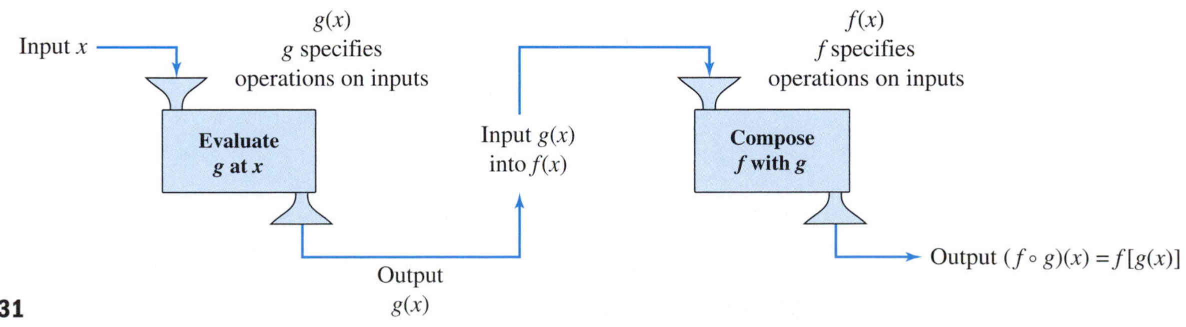

Figure A.31

The notation used for the composition of functions f and g is an open circle "$\circ$" placed between them, and indicates we will use the second function as an input for the first. In other words, $(f \circ g)(x)$ indicates that $g(x)$ is an input for f: $(f \circ g)(x) = f[g(x)]$. If the order is reversed as in $(g \circ f)(x)$, $f(x)$ becomes the input for g: $(g \circ f)(x) = g[f(x)]$. The diagram in Figure A.31 also helps us determine the domain of a composite function, in that the first function box can operate only if x is a valid input for g, and the second function box can operate only if $g(x)$ is a valid input for f. In other words, $(f \circ g)(x)$ is defined for *all x in the domain of g, such that $g(x)$ is in the domain of f.*

THE COMPOSITION OF FUNCTIONS

Given two functions f and g, the composition of f with g is defined by

$$(f \circ g)(x) = f[g(x)],$$

for all x in the domain of g such that $g(x)$ is in the domain of f.

In Figure A.32 the ideas are displayed using a "mapping notation," which can sometimes help clarify concepts related to the domain. The diagram shows that not all elements in the domain of g are automatically in the domain of $(f \circ g)$, since $g(x)$ may represent inputs unsuitable for f. This means the range of g and the domain of f will intersect, while the domain of $(f \circ g)$ is a subset of the domain of g.

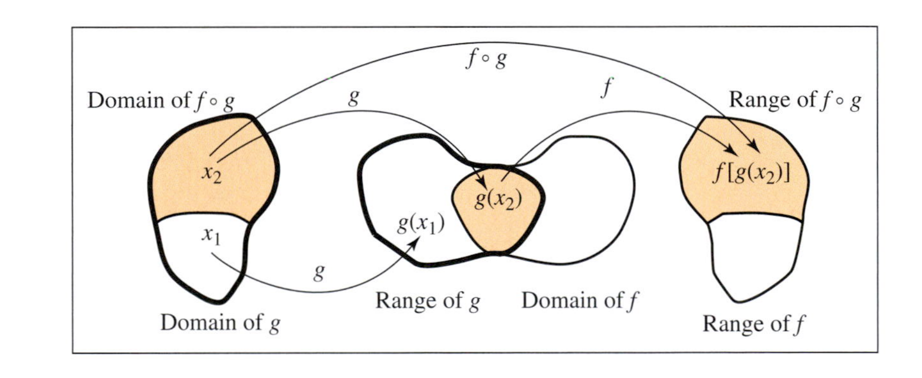

Figure A.32

ILLUSTRATION 3 Given $f(x) = \sqrt{x - 4}$ and $g(x) = 3x + 2$, find (a) $(f \circ g)(x)$ and (b) $(g \circ f)(x)$. Also determine the domain for each.

Solution: **a.** Begin by describing what the function f does to inputs: $f(x) = \sqrt{x - 4}$ says "decrease inputs by 4, and take the square root of the result."

$$(f \circ g)(x) = f[g(x)] \quad g(x) \text{ is an input for } f$$
$$= \sqrt{g(x) - 4} \quad \text{decrease input by 4, and take the square root of the result}$$
$$= \sqrt{(3x + 2) - 4} \quad \text{substitute } 3x + 2 \text{ for } g(x)$$
$$= \sqrt{3x - 2} \quad \text{result}$$

For the domain of $f[g(x)]$, we note g is defined for all real numbers x, but we must have $g(x) \geq 4$ (in blue) from part (a) or $f[g(x)]$ will not be real. This gives $3x + 2 \geq 4$ so $x \geq \frac{2}{3}$. In interval notation, the domain of $(f \circ g)(x)$ is $x \in [\frac{2}{3}, \infty)$.

b. The function g says "inputs are multiplied by 3, then increased by 2."

$$\begin{aligned}(g \circ f)(x) &= g[f(x)] && f(x) \text{ is an input for } g \\ &= 3f(x) + 2 && \text{multiply input by 3, then increase by 2} \\ &= 3\sqrt{x-4} + 2 && \text{substitute } \sqrt{x-4} \text{ for } f(x)\end{aligned}$$

For the domain of $g[f(x)]$, although g can accept any real number input, f can supply only those where $x \geq 4$. The domain of $(g \circ f)(x)$ is $x \in [4, \infty)$.

Illustration 3 shows $(f \circ g)(x)$ is generally not equal to $(g \circ f)(x)$. On those occasions when they *are* equal, the functions have the unique relationship studied in Section 4.1.

Deriving the Equation of a Conic

The Equation of an Ellipse

In Section 6.1, the equation $\sqrt{(x + c)^2 + y^2} + \sqrt{(x - c)^2 + y^2} = 2a$ was developed using the distance formula and the definition of an ellipse. To find the standard form of the equation, we treat this result as a radical equation, isolating one of the radicals and squaring both sides.

$$\sqrt{(x + c)^2 + y^2} = 2a - \sqrt{(x - c)^2 + y^2} \quad \text{isolate one radical}$$
$$(x + c)^2 + y^2 = 4a^2 - 4a\sqrt{(x - c)^2 + y^2} + (x - c)^2 + y^2 \quad \text{square both sides}$$

We continue by simplifying the equation, isolating the remaining radical, and squaring again.

$$x^2 + 2xc + c^2 + y^2 = 4a^2 - 4a\sqrt{(x - c)^2 + y^2} + x^2 - 2xc + c^2 + y^2 \quad \text{expand binomials}$$
$$4xc = 4a^2 - 4a\sqrt{(x - c)^2 + y^2} \quad \text{simplify}$$
$$a\sqrt{(x - c)^2 + y^2} = a^2 - xc \quad \text{isolate radical; divide by 4}$$
$$a^2[(x - c)^2 + y^2] = a^4 - 2a^2xc + x^2c^2 \quad \text{square both sides}$$
$$a^2x^2 - 2a^2xc + a^2c^2 + a^2y^2 = a^4 - 2a^2xc + x^2c^2 \quad \text{expand and distribute } a^2 \text{ on the left}$$
$$a^2x^2 - x^2c^2 + a^2y^2 = a^4 - a^2c^2 \quad \text{simplify}$$
$$x^2(a^2 - c^2) + a^2y^2 = a^2(a^2 - c^2) \quad \text{factor}$$
$$\frac{x^2}{a^2} + \frac{y^2}{a^2 - c^2} = 1 \quad \text{divide by } a^2(a^2 - c^2)$$

Since $a > c$, we know $a^2 > c^2$ and $a^2 - c^2 > 0$. For convenience, we let $b^2 = a^2 - c^2$ and it also follows that $a^2 > b^2$ and $a > b$ (since $c > 0$). Substituting b^2 for $a^2 - c^2$ we obtain the standard form of the equation of an ellipse (major axis horizontal, since we stipulated $a > b$): $\frac{x^2}{a^2} + \frac{y^2}{b^2} = 1$. Note once again the x-intercepts are $(\pm a, 0)$, while the y-intercepts are $(0, \pm b)$.

The Equation of a Hyperbola

In Section 6.2, the equation $\sqrt{(x + c)^2 + y^2} - \sqrt{(x - c)^2 + y^2} = 2a$ was developed using the distance formula and the definition of a hyperbola. To find the standard form of this equation, we apply the same procedures as before.

$$
\begin{aligned}
\sqrt{(x + c)^2 + y^2} &= 2a + \sqrt{(x - c)^2 + y^2} && \text{isolate one radical} \\
(x + c)^2 + y^2 &= 4a^2 + 4a\sqrt{(x - c)^2 + y^2} + (x - c)^2 + y^2 && \text{square both sides} \\
x^2 + 2xc + c^2 + y^2 &= 4a^2 + 4a\sqrt{(x - c)^2 + y^2} + x^2 - 2xc + c^2 + y^2 && \text{expand binomials} \\
4xc &= 4a^2 + 4a\sqrt{(x - c)^2 + y^2} && \text{simplify} \\
xc - a^2 &= a\sqrt{(x - c)^2 + y^2} && \text{isolate radical; divide by 4} \\
x^2c^2 - 2a^2xc + a^4 &= a^2[(x - c)^2 + y^2] && \text{square both sides} \\
x^2c^2 - 2a^2xc + a^4 &= a^2x^2 - 2a^2xc + a^2c^2 + a^2y^2 && \text{expand binomial then distribute } a^2 \\
x^2c^2 - a^2x^2 - a^2y^2 &= a^2c^2 - a^4 && \text{simplify} \\
x^2(c^2 - a^2) - a^2y^2 &= a^2(c^2 - a^2) && \text{factor} \\
\frac{x^2}{a^2} - \frac{y^2}{c^2 - a^2} &= 1 && \text{divide by } a^2(c^2 - a^2)
\end{aligned}
$$

From the definition of a hyperbola we have $0 < a < c$, showing $c^2 > a^2$ and $c^2 - a^2 > 0$. For convenience, we let $b^2 = c^2 - a^2$ and substitute to obtain the standard form of the equation of a hyperbola (transverse axis horizontal): $\frac{x^2}{a^2} - \frac{y^2}{b^2} = 1$. Note the x-intercepts are $(0, \pm a)$ and there are no y-intercepts.

Appendix VI

Families of Polar Graphs

Circles and Spiral Curves

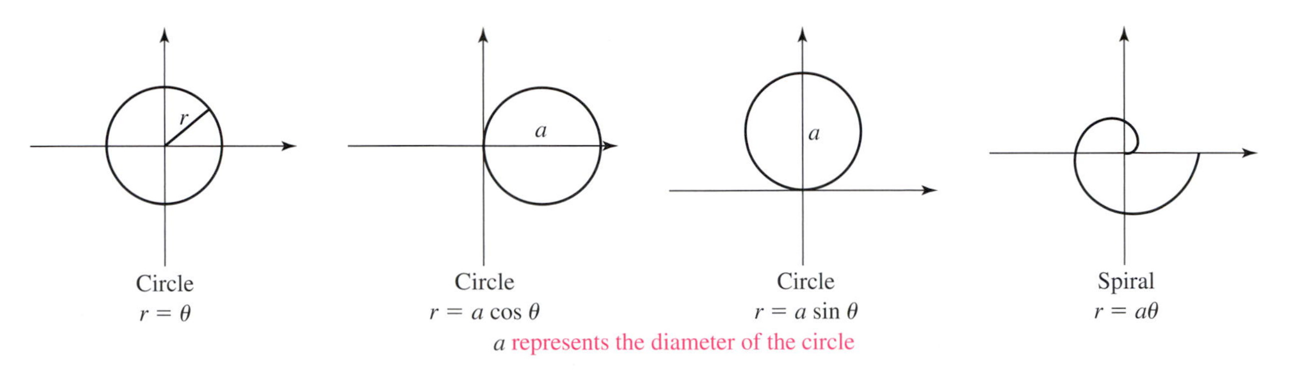

Circle $r = \theta$ | Circle $r = a \cos \theta$ | Circle $r = a \sin \theta$ | Spiral $r = a\theta$

a represents the diameter of the circle

Roses: $r = a \sin(n\theta)$ (illustrated here) and $r = a \cos(n\theta)$

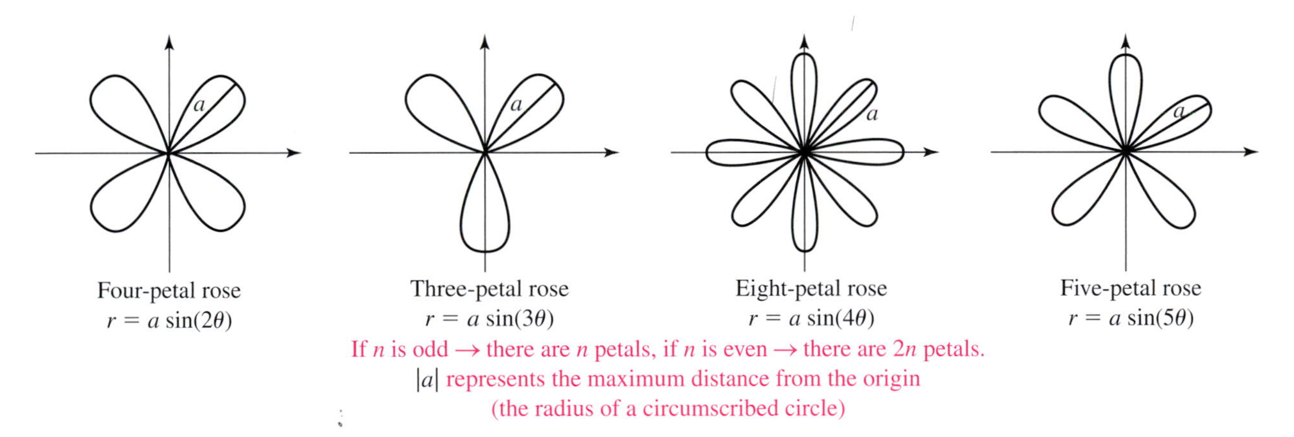

Four-petal rose $r = a \sin(2\theta)$ | Three-petal rose $r = a \sin(3\theta)$ | Eight-petal rose $r = a \sin(4\theta)$ | Five-petal rose $r = a \sin(5\theta)$

If n is odd $\rightarrow$ there are n petals, if n is even $\rightarrow$ there are $2n$ petals.
$|a|$ represents the maximum distance from the origin
(the radius of a circumscribed circle)

Limaçons: $r = a + b \sin\theta$ (illustrated here) and $r = a + b \cos\theta$

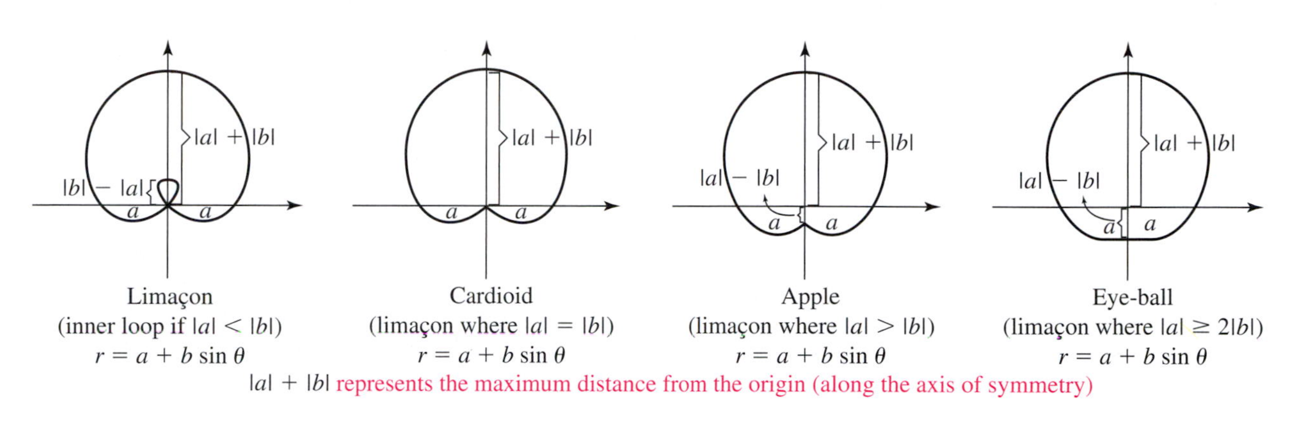

Limaçon (inner loop if $|a| < |b|$) $r = a + b \sin\theta$

Cardioid (limaçon where $|a| = |b|$) $r = a + b \sin\theta$

Apple (limaçon where $|a| > |b|$) $r = a + b \sin\theta$

Eye-ball (limaçon where $|a| \geq 2|b|$) $r = a + b \sin\theta$

$|a| + |b|$ represents the maximum distance from the origin (along the axis of symmetry)

Lemniscates: $r^2 = a^2\sin(2\theta)$ and $r^2 = a^2\cos(2\theta)$

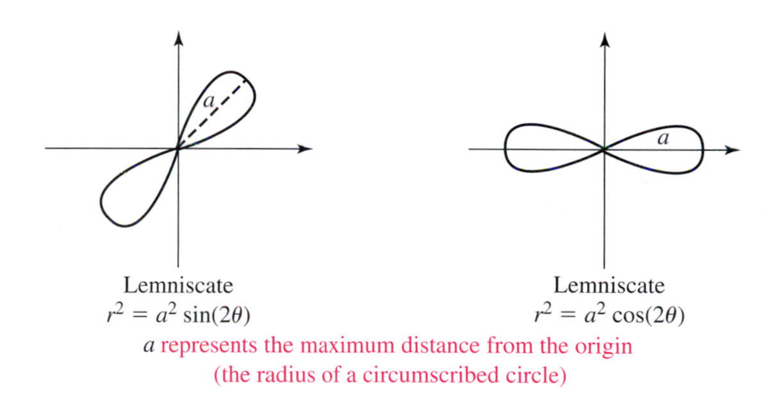

Lemniscate $r^2 = a^2 \sin(2\theta)$

Lemniscate $r^2 = a^2 \cos(2\theta)$

a represents the maximum distance from the origin (the radius of a circumscribed circle)

Student Answer Appendix

CHAPTER 1

Exercises 1.1, pp. 12–18

1. Complementary; 180; less; greater **3.** $r\theta$; $\frac{1}{2}r^2\theta$; radians **5.** Answers will vary. **7. a.** 77.5° **b.** 30.8° **9.** 53° **11.** 42.5° **13.** 67.38° **15.** 285.01° **17.** 45.76° **19.** 20°15′00″ **21.** 67°18′00″ **23.** 275°19′48″ **25.** 5°27′9″ **27.** No, $19 + 16 < 40$ **29.** 69° **31.** 25° **33.** 62.5 m **35.** $41\sqrt{2}$ ft ≈ 58 ft + 10 ft = 68 ft **37.** −645°, −285°, 435°, 795° **39.** −765°, −405°, 315°, 675° **41.** $s = 980$ m **43.** $\theta = 0.75$ rad **45.** $r \approx 1760$ yd **47.** $s = \frac{8\pi}{3}$ mi **49.** $\theta = 0.2575$ rad **51.** $r \approx 9.4$ km **53.** $A = 115.6\ \text{km}^2$ **55.** $\theta = 0.6$ rad **57.** $r \approx 3$ m **59.** $\theta = 1.5$ rad; $s = 7.5$ cm; $r = 5$ cm; $A = 18.75\ \text{cm}^2$ **61.** $\theta = 4.3$ rad; $s = 43$ m; $r = 10$ m; $A = 215\ \text{m}^2$ **63.** $\theta = 3$ rad; $A = 864\ \text{mm}^2$; $s = 72$ mm; $r = 24$ mm **65.** 2π rad **67.** $\frac{\pi}{4}$ rad **69.** $\frac{7\pi}{6}$ rad **71.** $\frac{-2\pi}{3}$ rad **73.** 0.4712 rad **75.** 3.9776 rad **77.** 60° **79.** 30° **81.** 120° **83.** 720° **85.** 165° **87.** 186.4° **89.** 171.9° **91.** −143.2° **93.** 960.7 mi apart **95. a.** ≈50.3 m² **b.** 80° **c.** ≈17 m **97.** $h \approx 7.06$ cm; $m \approx 3.76$ cm; $n \approx 13.24$ cm **99. a.** 1.5π rad/sec **b.** about 15 mi/hr **101. a.** 40π rad/min **b.** $\frac{\pi}{6}$ ft/sec ≈ 0.52 ft/sec **c.** about 11.5 sec **103. a.** 1000 m **b.** 1000 m **c.**

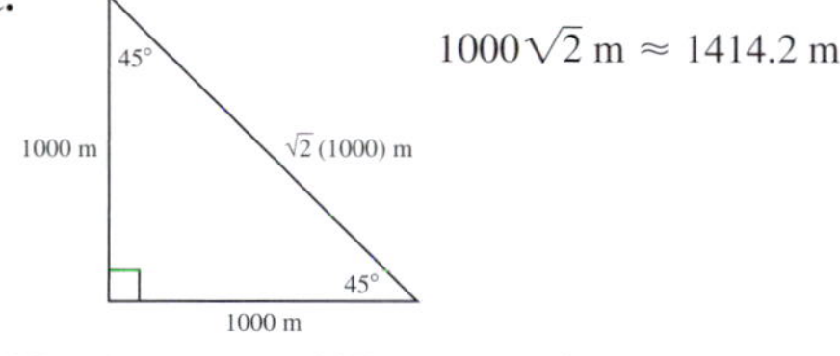

105. 50 mi apart **107. a.** ≈50.3°/day; ≈0.8788 rad/day **b.** ≈0.0366 rad/hr **c.** ≈6.67 mi/sec **109.** Answers will vary. **111. a.** ≈192 yd **b.** ≈86.6 rpm

Exercises 1.2, pp. 27–34

1. $\theta = \tan^{-1}x$ **3.** opposite; hypotenuse **5.** To find the measure of all three angles and all three sides. **7.** $\sin\theta = \frac{12}{13}$, $\csc\theta = \frac{13}{12}$, $\sec\theta = \frac{13}{5}$, $\tan\theta = \frac{12}{5}$, $\cot\theta = \frac{5}{12}$ **9.** $\cos\theta = \frac{13}{85}$, $\sec\theta = \frac{85}{13}$, $\cot\theta = \frac{13}{84}$, $\sin\theta = \frac{84}{85}$, $\csc\theta = \frac{85}{84}$ **11.** $\sin\theta = \frac{11}{5\sqrt{5}}$, $\tan\theta = \frac{11}{2}$, $\csc\theta = \frac{5\sqrt{5}}{11}$, $\cos\theta = \frac{2}{5\sqrt{5}}$, $\sec\theta = \frac{5\sqrt{5}}{2}$

13.

Angles	Sides
A = 30°	$a = 98$ cm
$B = 60°$	$b = 98\sqrt{3}$ cm
C = 90°	**c = 196 cm**

15.

Angles	Sides
$A = 45°$	$a = 9.9$ mm
B = 45°	**b = 9.9 mm**
C = 90°	$c = 9.9\sqrt{2}$ mm

17.

Angles	Sides
A = 22°	**a = 14 m**
$B = 68°$	$b \approx 34.65$ m
C = 90°	$c \approx 37.37$ m

19.

Angles	Sides
$A = 32°$	**a = 5.6 mi**
B = 58°	$b \approx 8.96$ mi
C = 90°	$c \approx 10.57$ mi

21.

Angles	Sides
A = 65°	**a = 625 mm**
$B = 25°$	$b \approx 291.44$ mm
C = 90°	$c \approx 689.61$ mm

23. 0.4540 **25.** 0.8391 **27.** 1.3230 **29.** 0.9063 **31.** 27° **33.** 40° **35.** 40.9° **37.** 65° **39.** 44.7° **41.** 82.0° **43.** 18.4° **45.** 46.2° **47.** 61.6° **49.** 21.98 mm **51.** 3.04 mi **53.** 177.48 furlongs **55.** They have like values. **57.** They have like values. **59.** 43° **61.** 21° **63.** $\frac{1}{2}, \frac{\sqrt{3}}{2}, \frac{1}{\sqrt{3}}, \frac{\sqrt{3}}{2}, \frac{1}{2}, \sqrt{3}, 2, \frac{2}{\sqrt{3}}, \sqrt{3}$ **65.** $\sqrt{5} - 1$ **67.** $\frac{1}{2}\sqrt{2 - \sqrt{2}}$ **69.** $\theta \approx 11.0°, \beta \approx 23.9°, \gamma \approx 145.1°$ **71.** approx. 300.6 m **73.** approx. 1483.8 ft **75.** approx. 118.1 mph **77. a.** approx. 250.0 yd **b.** approx. 351.0 yd **c.** approx. 23.1 yd **79.** approx. 1815.2 ft; approx. 665.3 ft **81.** approx. 386.0 Ω **83. a.** 875 m **b.** 1200 m **c.** 1485 m; 36.1°

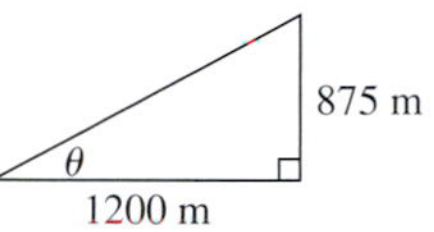

85. approx. 254.8 ft **87. a.** approx. 20.2 cm for each side **b.** approx. 35.3° **89.** Answers will vary.

91.
$$\cot u = \frac{x}{h}$$
$$x = h\cot u$$
$$\cot v = \frac{x - d}{h}$$
$$\cot v = \frac{h\cot u - d}{h}$$
$$h\cot v = h\cot u - d$$
$$d = h\cot u - h\cot v$$
$$h = \frac{d}{\cot u - \cot v}$$

93. a. approx. 4,585,757 mi **b.** approx. 91,673,352 mi **c.** Answers will vary. **95.** 132.715° **97.** 240° **99. a.** $\theta = 3.7$ **b.** $\mathcal{A} = 46.25\ \text{cm}^2$

Mid-Chapter Check, pp. 34–35

1. a. 36.11°, 115.08° **b.** 0.6302, 2.0085 **2.** about 2496 mi **3.** 4.3, 860 cm² **4. a.** 1.3270 **b.** 0.7837 **c.** 0.7071 **d.** 3.0963

5.

Angles	Sides
$A = 35°$	$a \approx 16.80$
B = 55°	**b = 24**
C = 90°	$c \approx 29.30$

6. $\cot 15° = 2 + \sqrt{3}$, since $\cot 15° = \tan(90 - 15)° = \tan 75°$ **7. a.** 378°, −342° **b.** $\frac{17\pi}{6}, -\frac{7\pi}{6}$

8.

θ	$\sin\theta$	$\cos\theta$	$\tan\theta$	$\csc\theta$	$\sec\theta$	$\cot\theta$
30°	$\frac{1}{2}$	$\frac{\sqrt{3}}{2}$	$\frac{\sqrt{3}}{3}$	2	$\frac{2\sqrt{3}}{3}$	$\sqrt{3}$
45°	$\frac{\sqrt{2}}{2}$	$\frac{\sqrt{2}}{2}$	1	$\sqrt{2}$	$\sqrt{2}$	1
60°	$\frac{\sqrt{3}}{2}$	$\frac{1}{2}$	$\sqrt{3}$	$\frac{2\sqrt{3}}{3}$	2	$\frac{\sqrt{3}}{3}$

9. a. 20π rad/min **b.** about 10.7 mph **10.** about 3 ft 5.6 in.

Reinforcing Basic Concepts, pp. 35–37

1. 70.4 cm^2 **2.** 441.3 mm^2 **3.** 2161.7 in.^{2} **4.** 248.3 ft^2

Exercises 1.3, pp. 46–49

1. origin; x-axis **3.** positive; clockwise **5.** Answers will vary.
7. slope $= \sqrt{3}$, equation: $y = \sqrt{3}x$, $\sin 60° = \frac{\sqrt{3}}{2}$, $\cos 60° = \frac{1}{2}$, $\tan 60° = \sqrt{3}$ **9.** $\sin\theta = \frac{15}{17}$, $\csc\theta = \frac{17}{15}$, $\cos\theta = \frac{8}{17}$, $\sec\theta = \frac{17}{8}$, $\tan\theta = \frac{15}{8}$, $\cot\theta = \frac{8}{15}$ **11.** $\sin\theta = \frac{21}{29}$, $\csc\theta = \frac{29}{21}$, $\cos\theta = \frac{-20}{29}$, $\sec\theta = \frac{-29}{20}$, $\tan\theta = \frac{-21}{20}$, $\cot\theta = \frac{-20}{21}$
13. $\sin\theta = \frac{-\sqrt{2}}{2}$, $\csc\theta = \frac{-2}{\sqrt{2}}$, $\cos\theta = \frac{\sqrt{2}}{2}$, $\sec\theta = \frac{2}{\sqrt{2}}$, $\tan\theta = -1$, $\cot\theta = -1$ **15.** $\sin\theta = \frac{1}{2}$, $\csc\theta = 2$, $\cos\theta = \frac{\sqrt{3}}{2}$, $\sec\theta = \frac{2}{\sqrt{3}}$, $\tan\theta = \frac{1}{\sqrt{3}}$, $\cot\theta = \sqrt{3}$ **17.** $\sin\theta = \frac{4}{\sqrt{17}}$, $\csc\theta = \frac{\sqrt{17}}{4}$, $\cos\theta = \frac{1}{\sqrt{17}}$, $\sec\theta = \sqrt{17}$, $\tan\theta = 4$, $\cot\theta = \frac{1}{4}$ **19.** $\sin\theta = \frac{-2}{\sqrt{13}}$, $\csc\theta = \frac{-\sqrt{13}}{2}$, $\cos\theta = \frac{-3}{\sqrt{13}}$, $\sec\theta = \frac{-\sqrt{13}}{3}$, $\tan\theta = \frac{2}{3}$, $\cot\theta = \frac{3}{2}$
21. $\sin\theta = \frac{6}{\sqrt{61}}$, $\csc\theta = \frac{\sqrt{61}}{6}$, $\cos\theta = \frac{-5}{\sqrt{61}}$, $\sec\theta = \frac{-\sqrt{61}}{5}$, $\tan\theta = \frac{-6}{5}$, $\cot\theta = \frac{-5}{6}$ **23.** $\sin\theta = \frac{-2\sqrt{5}}{\sqrt{21}}$, $\csc\theta = -\frac{\sqrt{21}}{2\sqrt{5}}$, $\cos\theta = \frac{1}{\sqrt{21}}$, $\sec\theta = \sqrt{21}$, $\tan\theta = -2\sqrt{5}$, $\cot\theta = \frac{-1}{2\sqrt{5}}$

25. QI/QIII;

$(4, 3)$: $\sin\theta = \frac{3}{5}$; $(-4, -3)$: $\sin\theta = -\frac{3}{5}$
$\cos\theta = \frac{4}{5}$ $\cos\theta = -\frac{4}{5}$
$\tan\theta = \frac{3}{4}$ $\tan\theta = \frac{3}{4}$

27. QII/QIV;

$(-3, \sqrt{3})$: $\sin\theta = \frac{1}{2}$; $(3, -\sqrt{3})$: $\sin\theta = -\frac{1}{2}$
$\cos\theta = -\frac{\sqrt{3}}{2}$ $\cos\theta = \frac{\sqrt{3}}{2}$
$\tan\theta = -\frac{1}{\sqrt{3}}$ $\tan\theta = -\frac{1}{\sqrt{3}}$

29. $x = 0$, $y = k$; $k > 0$; $r = |k|$;
$\sin 90° = \frac{k}{k}$, $\cos 90° = \frac{0}{k}$, $\tan 90° = \frac{k}{0}$,
$\sin 90° = 1$, $\cos 90° = 0$, $\tan 90°$ undefined
$\csc 90° = 1$, $\sec 90°$ undefined
$\cot 90° = 0$
31. 60° **33.** 45° **35.** 45° **37.** 68° **39.** 40° **41.** 11.6° **43.** QII **45.** QII **47.** $\sin\theta = -\frac{1}{2}$; $\cos\theta = \frac{\sqrt{3}}{2}$; $\tan\theta = -\frac{1}{\sqrt{3}}$
49. $\sin\theta = \frac{-\sqrt{2}}{2}$; $\cos\theta = \frac{\sqrt{2}}{2}$; $\tan\theta = -1$ **51.** $\sin\theta = \frac{-\sqrt{3}}{2}$; $\cos\theta = \frac{-1}{2}$; $\tan\theta = \sqrt{3}$ **53.** $\sin\theta = -\frac{1}{2}$; $\cos\theta = \frac{-\sqrt{3}}{2}$; $\tan\theta = \frac{1}{\sqrt{3}}$
55. $x = 4$, $y = -3$, $r = 5$; QIV; $\sin\theta = \frac{-3}{5}$, $\csc\theta = \frac{-5}{3}$, $\cos\theta = \frac{4}{5}$, $\sec\theta = \frac{5}{4}$, $\tan\theta = \frac{-3}{4}$, $\cot\theta = \frac{-4}{3}$ **57.** $x = -12$, $y = -35$, $r = 37$; QIII; $\sin\theta = \frac{-35}{37}$, $\csc\theta = \frac{-37}{35}$, $\cos\theta = \frac{-12}{37}$, $\sec\theta = \frac{-37}{12}$, $\tan\theta = \frac{35}{12}$, $\cot\theta = \frac{12}{35}$ **59.** $x = 2\sqrt{2}$, $y = 1$, $r = 3$; QI; $\sin\theta = \frac{1}{3}$, $\csc\theta = 3$, $\cos\theta = \frac{2\sqrt{2}}{3}$, $\sec\theta = \frac{3}{2\sqrt{2}}$, $\tan\theta = \frac{1}{2\sqrt{2}}$, $\cot\theta = 2\sqrt{2}$ **61.** $x = -\sqrt{15}$, $y = -7$, $r = 8$; QIII; $\sin\theta = \frac{-7}{8}$, $\csc\theta = \frac{-8}{7}$, $\cos\theta = \frac{-\sqrt{15}}{8}$, $\sec\theta = -\frac{8}{\sqrt{15}}$, $\tan\theta = \frac{7}{\sqrt{15}}$, $\cot\theta = \frac{\sqrt{15}}{7}$ **63.** $52° + 360°k$ **65.** $87.5° + 360°k$ **67.** $225° + 360°k$
69. $-107° + 360°k$ **71.** $\frac{\sqrt{3}}{2}, \frac{-1}{2}, -\sqrt{3}$ **73.** $-\frac{1}{2}, \frac{\sqrt{3}}{2}, -\frac{1}{\sqrt{3}}$
75. $\sin\theta = \frac{-\sqrt{3}}{2}$, $\cos\theta = \frac{-1}{2}$, $\tan\theta = \sqrt{3}$
77. $\sin\theta = \frac{-\sqrt{3}}{2}$, $\cos\theta = -\frac{1}{2}$, $\tan\theta = \sqrt{3}$
79. $\sin\theta = \frac{-1}{2}$, $\cos\theta = \frac{-\sqrt{3}}{2}$, $\tan\theta = \frac{1}{\sqrt{3}}$
81. $\sin\theta = \frac{-1}{2}$, $\cos\theta = \frac{-\sqrt{3}}{2}$, $\tan\theta = \frac{1}{\sqrt{3}}$ **83.** QIV, neg., -0.0175
85. QIV, neg., -1.6643 **87.** QIV, neg., -1.5890 **89.** QI, pos., 0.0872
91. a. approx. 144.78 units^2 **b.** 53° **c.** The parallelogram is a rectangle whose area is $A = ab$. **d.** $A = \frac{ab}{2}\sin\theta$ **93.** $\theta = 60° + 360°k$ and $\theta = 300° + 360°k$ **95.** $\theta = 240° + 360°k$ and $\theta = 300° + 360°k$
97. $\theta = 61.1° + 360°k$ and $\theta = 118.9° + 360°k$
99. $\theta = 113.0° + 360°k$ and $\theta = 293.0° + 360°k$

101. $1890°$; $90° + 360°k$ **103.** head first; $900°$ **105.** approx. $701.6°$ **107.** 343.12 in^2 **109.** Answers will vary. **111. a.** approx. $34.5°$ **b.** about 5.4 units **113.** $\frac{\pi}{6}, \frac{\pi}{4}, \frac{\pi}{3}, \frac{\pi}{2}, \pi, \frac{3\pi}{2}, 2\pi$

115.

Angles	Sides
$A \approx 28.7°$	$a = 25$ **cm**
$B \approx 61.3°$	$b \approx 45.6$ cm
$C = 90°$	$c = 52$ **cm**

117. about 32 mph

Exercises 1.4, pp. 59–64

1. x; y; origin **3.** x; y; $\frac{y}{x}$; sec t; csc t; cot t **5.** Answers will vary.

7. $(-0.6, -0.8)$ **9.** $\left(\frac{5}{13}, \frac{-12}{13}\right)$ **11.** $\left(\frac{\sqrt{11}}{6}, \frac{5}{6}\right)$ **13.** $\left(\frac{-\sqrt{11}}{4}, \frac{\sqrt{5}}{4}\right)$

15. $(-0.9769, -0.2137)$ **17.** $(-0.9928, 0.1198)$

19. $\left(\frac{-\sqrt{3}}{2}, \frac{-1}{2}\right), \left(\frac{\sqrt{3}}{2}, \frac{1}{2}\right), \left(\frac{\sqrt{3}}{2}, \frac{-1}{2}\right)$

21. $\left(\frac{-\sqrt{11}}{6}, \frac{-5}{6}\right), \left(\frac{-\sqrt{11}}{6}, \frac{5}{6}\right), \left(\frac{\sqrt{11}}{6}, \frac{5}{6}\right)$

23. $(-0.3325, 0.9431), (-0.3325, -0.9431), (0.3325, -0.9431)$

25. $(0.9937, 0.1121), (-0.9937, 0.1121), (-0.9937, -0.1121)$

27. $\left(\frac{1}{2}, \frac{\sqrt{3}}{2}\right)$ is on unit circle **29.** $\frac{\pi}{4}; \left(\frac{-\sqrt{2}}{2}, \frac{-\sqrt{2}}{2}\right)$

31. $\frac{\pi}{6}; \left(\frac{-\sqrt{3}}{2}, \frac{-1}{2}\right)$ **33.** $\frac{\pi}{4}; \left(\frac{-\sqrt{2}}{2}, \frac{\sqrt{2}}{2}\right)$ **35.** $\frac{\pi}{6}; \left(\frac{\sqrt{3}}{2}, \frac{1}{2}\right)$

37. a. $\frac{\sqrt{2}}{2}$ **b.** $\frac{\sqrt{2}}{2}$ **c.** $\frac{-\sqrt{2}}{2}$ **d.** $\frac{-\sqrt{2}}{2}$ **e.** $\frac{\sqrt{2}}{2}$ **f.** $\frac{-\sqrt{2}}{2}$ **g.** $\frac{\sqrt{2}}{2}$ **h.** $\frac{-\sqrt{2}}{2}$ **39. a.** -1 **b.** 1 **c.** 0 **d.** 0 **41. a.** $\frac{\sqrt{3}}{2}$ **b.** $\frac{-\sqrt{3}}{2}$ **c.** $\frac{-\sqrt{3}}{2}$ **d.** $\frac{\sqrt{3}}{2}$ **e.** $\frac{\sqrt{3}}{2}$ **f.** $\frac{\sqrt{3}}{2}$ **g.** $\frac{-\sqrt{3}}{2}$ **h.** $\frac{\sqrt{3}}{2}$ **43. a.** 0 **b.** 0 **c.** undefined **d.** undefined **45.** $\sin t = 0.6$, $\cos t = -0.8$, $\tan t = -0.75$, $\csc t = 1.\overline{6}$, $\sec t = -1.25$, $\cot t = -1.\overline{3}$ **47.** $\sin t = -\frac{12}{13}$, $\cos t = -\frac{5}{13}$, $\tan t = \frac{12}{5}$, $\csc t = -\frac{13}{12}$, $\sec t = -\frac{13}{5}$, $\cot t = \frac{5}{12}$ **49.** $\sin t = \frac{\sqrt{11}}{6}$, $\cos t = \frac{5}{6}$, $\tan t = \frac{\sqrt{11}}{5}$, $\csc t = \frac{6\sqrt{11}}{11}$, $\sec t = \frac{6}{5}$, $\cot t = \frac{5\sqrt{11}}{11}$

51. $\sin t = \frac{\sqrt{21}}{5}$, $\cos t = \frac{-2}{5}$, $\tan t = \frac{-\sqrt{21}}{2}$, $\csc t = \frac{5\sqrt{21}}{21}$, $\sec t = \frac{-5}{2}$, $\cot t = \frac{-2\sqrt{21}}{21}$ **53.** $\sin t = \frac{-2\sqrt{2}}{3}$, $\cos t = \frac{-1}{3}$, $\tan t = 2\sqrt{2}$, $\csc t = \frac{-3\sqrt{2}}{4}$, $\sec t = -3$, $\cot t = \frac{\sqrt{2}}{4}$ **55.** $\sin t = \frac{\sqrt{3}}{2}$, $\cos t = \frac{1}{2}$, $\tan t = \sqrt{3}$, $\csc t = \frac{2\sqrt{3}}{3}$, $\sec t = 2$, $\cot t = \frac{\sqrt{3}}{3}$

57. $\sin t = \frac{\sqrt{2}}{2}$, $\cos t = \frac{-\sqrt{2}}{2}$, $\tan t = -1$, $\csc t = \sqrt{2}$, $\sec t = -\sqrt{2}$, $\cot t = -1$ **59.** QI, 0.7 **61.** QIV, 0.7 **63.** QI, 1 **65.** QII, 1.1 **67.** QII, -0.4 **69.** QIV, -3.1 **71.** $\frac{2\pi}{3}$ **73.** $\frac{7\pi}{6}$ **75.** $\frac{2\pi}{3}$ **77.** $\frac{\pi}{2}$

79. $\frac{3\pi}{4}, \frac{5\pi}{4}$ **81.** $\frac{\pi}{2}, \frac{3\pi}{2}$ **83.** $\frac{3\pi}{4}, \frac{5\pi}{4}$ **85.** $0, \pi$ **87.** 5.3056

89. 5.5644 **91.** 4.5522 **93.** 0.1538 **95.** 2.3416 **97.** 1.7832 **99.** 3.5416 **101. a.** $\left(\frac{3}{4}, \frac{4}{5}\right)$ **b.** $\left(\frac{-3}{4}, \frac{4}{5}\right)$

103. a. $\left(\frac{5}{13}, \frac{12}{13}, 1\right), \left(\frac{5}{13}\right)^2 + \left(\frac{12}{13}\right)^2 = \frac{25}{169} + \frac{144}{169} = \frac{169}{169} = 1$; $\sin t = \frac{12}{13}$, $\cos t = \frac{5}{13}$, $\tan t = \frac{12}{5}$, $\csc t = \frac{13}{12}$, $\sec t = \frac{13}{5}$, $\cot t = \frac{5}{12}$ **b.** $\left(\frac{7}{25}, \frac{24}{25}, 1\right), \left(\frac{7}{25}\right)^2 + \left(\frac{24}{25}\right)^2 = \frac{49}{625} + \frac{576}{625} = \frac{625}{625} = 1$; $\sin t = \frac{24}{25}$, $\cos t = \frac{7}{25}$, $\tan t = \frac{24}{7}$, $\csc t = \frac{25}{24}$, $\sec t = \frac{25}{7}$, $\cot t = \frac{7}{24}$ **c.** $\left(\frac{12}{37}, \frac{35}{37}, 1\right)$, $\left(\frac{12}{37}\right)^2 + \left(\frac{35}{37}\right)^2 = \frac{144}{1369} + \frac{1225}{1369} = \frac{1369}{1369} = 1$; $\sin t = \frac{35}{37}$, $\cos t = \frac{12}{37}$, $\tan t = \frac{35}{12}$, $\csc t = \frac{37}{35}$, $\sec t = \frac{37}{12}$, $\cot t = \frac{12}{35}$ **d.** $\left(\frac{9}{41}, \frac{40}{41}, 1\right)$, $\left(\frac{9}{41}\right)^2 + \left(\frac{40}{41}\right)^2 = \frac{81}{1681} + \frac{1600}{1681} = \frac{1681}{1681} = 1$; $\sin t = \frac{40}{41}$, $\cos t = \frac{9}{41}$, $\tan t = \frac{40}{9}$, $\csc t = \frac{41}{40}$, $\sec t = \frac{41}{9}$, $\cot t = \frac{9}{40}$

105. a. 5 rad **b.** 30 rad **107. a.** 5 dm **b.** ≈ 6.28 dm **109. a.** 2.5 AU **b.** ≈ 6.28 AU **111.** yes **113.** range of $\sin t$ and $\cos t$ is $[-1, 1]$ **115. a.** $2t \approx 2.2$ **b.** QI **c.** $\cos t \approx 0.5$ **d.** No **117. a.** They are similar triangles. **b.** the smaller one **c.** $t \approx 0.5566$ **119.** $d \approx 53.74$ in., $D \approx 65.82$ in. **121.** 18 ft **123.** $12^2 + 35^2 = 37^2$; $A = 18.9°$, $B = 71.1°$, $C = 90°$

Summary and Concept Review, pp. 65–69

1. $147.61\overline{3}$ **2.** $32° 52' 12''$ **3.** $10.125 \times 13.5 \times 16.875$

4. approx. 692.82 yd **5.** $120°$ **6.** $\frac{7\pi}{6}$ **7.** approx 4.97 units

8. $-\frac{1}{2}$ **9.** $s = 25.5$ cm, $A = 191.25 \text{ cm}^2$

10. $r \approx 41.74$ in., $A \approx 2003.48 \text{ in}^2$ **11.** $\theta = 4.75$ rad, $s = 38$ m

12. a. approx. 9.4248 rad/sec **b.** approx. 3.9 ft/sec **c.** about 15.4 sec

13. a. $A \approx 0.80$ **b.** $A \approx 64.3°$

14. a. $\cot 32.6°$ **b.** $\cos(70° 29' 45'')$

15.

Angles	Sides
$A = 49°$	$a = 89$ **in.**
$B = 41°$	$b \approx 77.37$ in.
$C = 90°$	$c \approx 117.93$ in.

16.

Angles	Sides
$A \approx 43.6°$	$a = 20$ **m**
$B \approx 46.4°$	$b = 21$ **m**
$C = 90°$	$c = 29$ m

17. approx. 5.18 m **18. a.** approx. 239.32 m **b.** approx. 240.68 m apart

19. approx. $54.5°$ and $35.5°$ **20.** $207° + 360°k$; answers will vary.

21. $28°$, $19°$, $30°$ **22. a.** $\sin \theta = \frac{35}{37}$, $\csc \theta = \frac{37}{35}$, $\cos \theta = \frac{-12}{37}$, $\sec \theta = \frac{-37}{12}$, $\tan \theta = \frac{-35}{12}$, $\cot \theta = \frac{-12}{35}$

b. $\sin \theta = \frac{-3}{\sqrt{13}}$, $\csc \theta = \frac{-\sqrt{13}}{3}$, $\cos \theta = \frac{2}{\sqrt{13}}$, $\sec \theta = \frac{\sqrt{13}}{2}$, $\tan \theta = \frac{-3}{2}$, $\cot \theta = \frac{-2}{3}$

23. a. $x = 4$, $y = -3$, $r = 5$; QIV; $\sin \theta = -\frac{3}{5}$, $\csc \theta = -\frac{5}{3}$, $\cos \theta = \frac{4}{5}$, $\sec \theta = \frac{5}{4}$, $\tan \theta = \frac{-3}{4}$, $\cot \theta = \frac{-4}{3}$

b. $x = 5$, $y = -12$, $r = 13$; QIV; $\sin \theta = \frac{-12}{13}$, $\csc \theta = \frac{-13}{12}$, $\cos \theta = \frac{5}{13}$, $\sec \theta = \frac{13}{5}$, $\tan \theta = \frac{-12}{5}$, $\cot \theta = \frac{-5}{12}$

24. a. $\theta = 135° + 180°k$ **b.** $\theta = 30° + 360°k$ or $\theta = 330° + 360°k$ **c.** $\theta \approx 76.0° + 180°k$ **d.** $\theta \approx -27.0° + 360°k$ or $\theta = 207.0° + 360°k$

25. $y = -\frac{6}{7}, \left(-\frac{\sqrt{13}}{7}, \frac{6}{7}\right), \left(-\frac{\sqrt{13}}{7}, -\frac{6}{7}\right)$, and $\left(\frac{\sqrt{13}}{7}, \frac{6}{7}\right)$

26. $\sin t = -\frac{\sqrt{7}}{4}$, $\csc t = -\frac{4}{\sqrt{7}}$, $\cos t = \frac{3}{4}$, $\sec t = \frac{4}{3}$, $\tan t = -\frac{\sqrt{7}}{3}$, $\cot t = -\frac{3}{\sqrt{7}}$ **27.** $\frac{\pi}{3}$ and $\frac{2\pi}{3}$ **28.** $t \approx 2.44$

29. a. approx. 19.6667 rad **b.** 25 rad

Mixed Review, pp. 69–70

1. a. 15 cm **b.** $10\sqrt{3}$ cm **3.** $t = \dfrac{2\pi}{3}$ and $t = \dfrac{4\pi}{3}$ **5.** $220°\,48'\,50''$ **7.** $12\sqrt{2}$ in.; $60\sqrt{2} \approx 84.9$ in. **9.** arc length: $\dfrac{28}{3}\pi \approx 29.3$ units; area: $\dfrac{112\pi}{3} \approx 117.3$ units^2 **11.** 86.915° **13.** $\sin\theta = \dfrac{-8}{17}$, $\sec\theta = \dfrac{17}{15}$, $\cos\theta = \dfrac{15}{17}$, $\csc\theta = \dfrac{-17}{8}$, $\tan\theta = \dfrac{-8}{15}$, $\cot\theta = \dfrac{-15}{8}$ **15.** 60° **17. a.** 6π rad/sec **b.** $20(6\pi)$ cm/sec ≈ 377 cm/sec **19.** about 19.5 ft

Practice Test, pp. 71–72

1. complement: 55°; supplement: 145° **2.** $30° + 360°k$; $k \in \mathbb{Z}$ **3. a.** 45° **b.** 30° **c.** $\dfrac{\pi}{6}$ **d.** $\dfrac{\pi}{3}$ **4. a.** 100.755° **b.** $48°12'45''$ **5. a.** 430 mi **b.** $215\sqrt{3} \approx 372$ mi

6.

t	$\sin t$	$\cos t$	$\tan t$	$\csc t$	$\sec t$	$\cot t$
0	0	1	0	undefined	1	undefined
$\frac{2\pi}{3}$	$\frac{\sqrt{3}}{2}$	$-\frac{1}{2}$	$-\sqrt{3}$	$\frac{2\sqrt{3}}{3}$	-2	$-\frac{\sqrt{3}}{3}$
$\frac{7\pi}{6}$	$-\frac{1}{2}$	$-\frac{\sqrt{3}}{2}$	$\frac{\sqrt{3}}{3}$	-2	$\frac{-2\sqrt{3}}{3}$	$\sqrt{3}$
$\frac{5\pi}{4}$	$-\frac{\sqrt{2}}{2}$	$-\frac{\sqrt{2}}{2}$	1	$-\sqrt{2}$	$-\sqrt{2}$	1
$\frac{5\pi}{3}$	$-\frac{\sqrt{3}}{2}$	$\frac{1}{2}$	$-\sqrt{3}$	$-\frac{2\sqrt{3}}{3}$	2	$-\frac{\sqrt{3}}{3}$
$\frac{7\pi}{4}$	$\frac{-\sqrt{2}}{2}$	$\frac{\sqrt{2}}{2}$	-1	$-\sqrt{2}$	$\sqrt{2}$	-1
$\frac{13\pi}{6}$	$\frac{1}{2}$	$\frac{\sqrt{3}}{2}$	$\frac{\sqrt{3}}{3}$	2	$\frac{2\sqrt{3}}{3}$	$\sqrt{3}$

7. $\sec\theta = \dfrac{5}{2}$, $\sin\theta = \dfrac{-\sqrt{21}}{5}$, $\tan\theta = \dfrac{-\sqrt{21}}{2}$, $\csc\theta = \dfrac{-5}{\sqrt{21}}$, $\cot\theta = \dfrac{-2}{\sqrt{21}}$ **8.** $\left(\dfrac{1}{3}\right)^2 + \left(-\dfrac{2\sqrt{2}}{3}\right)^2 = \dfrac{1}{9} + \dfrac{8}{9} = 1$; $\sin\theta = \dfrac{-2\sqrt{2}}{3}$, $\cos\theta = \dfrac{1}{3}$, $\tan\theta = -2\sqrt{2}$, $\csc\theta = \dfrac{-3\sqrt{2}}{4}$, $\sec\theta = 3$, $\cot\theta = \dfrac{-\sqrt{2}}{4}$ **9. a.** ≈ 225.8 ft or 225 ft 9.6 in. **b.** $\dfrac{23\pi}{480} \approx 0.1505$ rad/sec **c.** 11.29 ft/sec ≈ 7.7 mph **10.**

Angles	Sides
$A = 33°$	$a \approx 8.2$ cm
$B = 57°$	$b \approx 12.6$ cm
$C = 90°$	**$c \approx 15.0$ cm**

11. about 67 cm, 49.6° **12.** 57.9 m **13. a.** $\dfrac{7\pi}{6}$ **b.** $\dfrac{11\pi}{6}$ **c.** $\dfrac{3\pi}{4}$ **14.** about 600.6 km **15.** $\left(-\dfrac{20}{29}, \dfrac{21}{29}\right)$, $\left(-\dfrac{20}{29}, -\dfrac{21}{29}\right)$, $\left(\dfrac{20}{29}, -\dfrac{21}{29}\right)$ **16.** about 26.3 cm **17.** 1260° **18. a.** about 39 ft **b.** about 74 ft **c.** about 4.8 mph **19. a.** ≈ 0.53 **b.** ≈ -0.53 **c.** ≈ -0.53 **20. a.** ≈ 4 **b.** ≈ 2.3

Strengthening Core Skills, pp. 74–75

1. a. $(0.97, 0.24)$, $(0.97)^2 + (0.24)^2 \approx 1$, $\cos t = 0.9689124217$, $\sin t = 0.2474039593$ **b.** $(0.88, 0.48)$, $(0.88)^2 + (0.48)^2 \approx 1$, $\cos t = 0.8775825619$, $\sin t = 0.4794255386$ **c.** $(0.73, 0.68)$, $(0.73)^2 + (0.68)^2 \approx 1$, $\cos t = 0.73168886889$, $\sin t = 0.68163876$ **d.** $(0.54, 0.84)$, $(0.54)^2 + (0.84)^2 \approx 1$, $\cos t = 0.5403023059$, $\sin t = 0.8414709848$ **2. a.** $(-0.5, 0.87)$, $(-0.5)^2 + (0.87)^2 \approx 1$, $\cos t = -\dfrac{1}{2}$, $\sin t = \dfrac{\sqrt{3}}{2}$ **b.** $(-0.71, 0.71)$, $(-0.71)^2 + (0.71)^2 \approx 1$, $\cos t = -\dfrac{\sqrt{2}}{2}$, $\sin t = \dfrac{\sqrt{2}}{2}$ **c.** $(-0.87, 0.5)$, $(-0.87)^2 + (0.5)^2 \approx 1$, $\cos t = -\dfrac{\sqrt{3}}{2}$, $\sin t = \dfrac{1}{2}$ **d.** $(-1, 0)$, $(-1)^2 + (0)^2 = 1$, $\cos t = -1$, $\sin t = 0$

CHAPTER 2

Exercises 2.1, pp. 90–96

1. increasing **3.** $(-\infty, \infty)$; $[-1, 1]$ **5.** Answers will vary.

7.

t	$y = \cos t$
π	-1
$\frac{7\pi}{6}$	$-\frac{\sqrt{3}}{2}$
$\frac{5\pi}{4}$	$-\frac{\sqrt{2}}{2}$
$\frac{4\pi}{3}$	$-\frac{1}{2}$
$\frac{3\pi}{2}$	0
$\frac{5\pi}{3}$	$\frac{1}{2}$
$\frac{7\pi}{4}$	$\frac{\sqrt{2}}{2}$
$\frac{11\pi}{6}$	$\frac{\sqrt{3}}{2}$
2π	1

9. **11.**

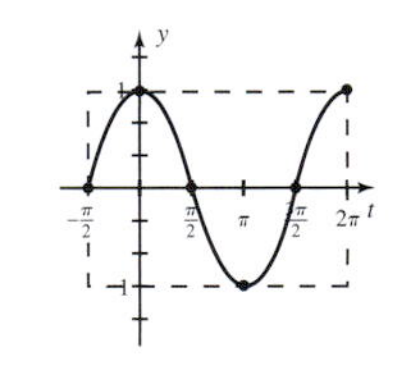

13. $|A| = 3$, $P = 2\pi$

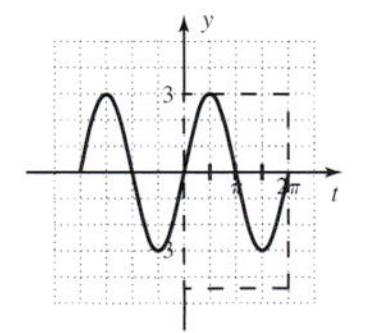

15. $|A| = 2$, $P = 2\pi$

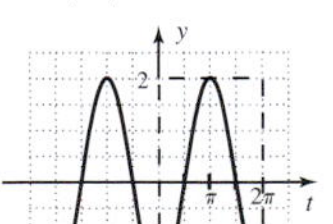

17. $|A| = \frac{1}{2}$, $P = 2\pi$

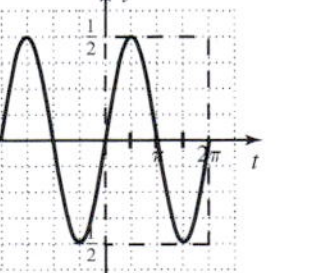

19. $|A| = 1$, $P = \pi$

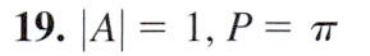

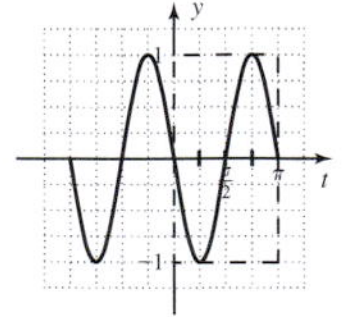

21. $|A| = 0.8, P = \pi$ **23.** $|A| = 4, P = 4\pi$ **25.** $|A| = 3, P = \frac{1}{2}$

27. $|A| = 4, P = \frac{6}{5}$ **29.** $|A| = 2, P = \frac{1}{128}$ **31.**

33.

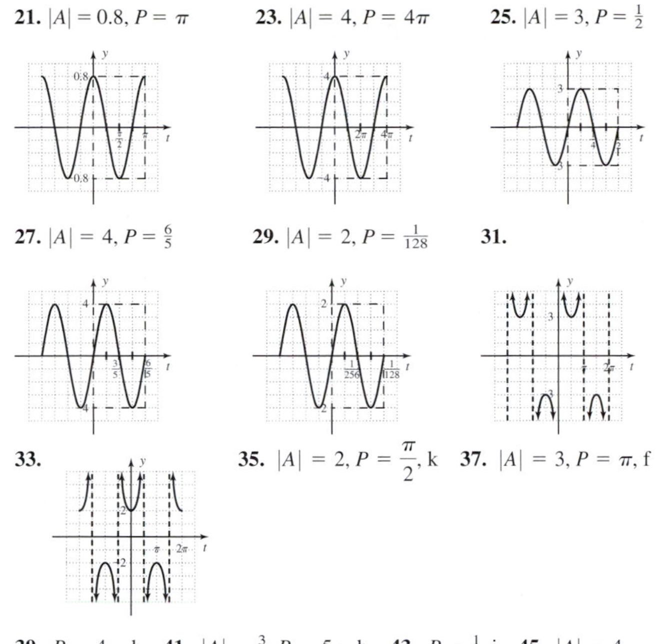

35. $|A| = 2, P = \frac{\pi}{2}$, k **37.** $|A| = 3, P = \pi$, f

39. $P = 4\pi$, h **41.** $|A| = \frac{3}{4}, P = 5\pi$, b **43.** $P = \frac{1}{4}$, j **45.** $|A| = 4$, $P = \frac{1}{72}$, d **47.** $y = -\frac{3}{4}\cos(8t)$ **49.** $y = -0.2\csc\left(\frac{1}{2}t\right)$

51. $y = 6\cos\left(\frac{2\pi}{3}t\right)$ **53.** red: $y = -\cos x$; blue: $y = \sin x$; $x = \frac{3\pi}{4}, \frac{7\pi}{4}$

55. red: $y = -2\cos x$; blue: $y = 2\sin(3x)$; $x = \frac{3\pi}{8}, \frac{3\pi}{4}, \frac{7\pi}{8}, \frac{11\pi}{8}, \frac{7\pi}{4}, \frac{15\pi}{8}$

57. $\cos t = \frac{112}{113}$, (15, 112, 113) **59. a.** 3 ft **b.** 80 mi **c.** $h = 1.5\cos\left(\frac{\pi}{40}x\right)$ **61. a.** $D = -4\cos\left(\frac{\pi}{12}t\right)$ **b.** $D \approx 3.86$ **c.** 72°

63. a. $D = 15\cos(\pi t)$ **b.** at center **c.** Swimming leisurely. One complete cycle in 2 sec! **65. a.** Graph a **b.** 76 days **c.** 96 days **67. a.** 480 nm → blue **b.** 620 nm → orange **69.** $I = 30\sin(50\pi t)$, $I \approx 21.2$ amps

71. Since $m = -M$, 0;

t	y
0	3
$\frac{\pi}{2}$	5
π	3
$\frac{3\pi}{2}$	1
2π	3

avg. value = 3; shifted up 3 units; avg. value = 1; amplitude is "centered" on average value.

73. Answers will vary.

75. $g(t)$ has the shortest period;

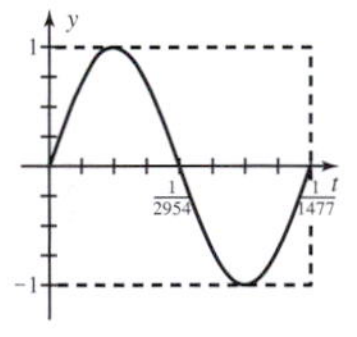

77. $t = \pi - 1.12 \approx 2.02$ **79. a.** $x = 28, y = -45, r = 53$ **b.** QIV **c.** $\sin t = -\frac{45}{53}$, $\tan t = -\frac{45}{28}$, $\sec t = \frac{53}{28}$, $\csc t = -\frac{53}{45}$, $\cot t = -\frac{28}{45}$ **81. a.** $\omega = 28\pi$ rad/min **b.** $s \approx 98.2$ ft **c.** about 25 mph

Exercises 2.2, pp. 105–111

1. π; $P = \frac{\pi}{B}$ **3.** odd; $-f(t)$; -0.268 **5. a.** use $\frac{\cos t}{\sin t}$ **b.** use reciprocals of tan t **7.** $0, \frac{1}{\sqrt{3}}, 1, \sqrt{3}$, und. **9.** 1.6, 0.8, 0.5, 1.4, 0.7, 1.2 **11. a.** -1 **b.** $\sqrt{3}$ **c.** -1 **d.** $\sqrt{3}$ **13. a.** $\frac{7\pi}{4}$ **b.** $\frac{7\pi}{6}$ **c.** $\frac{5\pi}{3}$ **d.** $\frac{3\pi}{4}$

15. und., $\sqrt{3}, 1, \frac{1}{\sqrt{3}}, 0$ **17.** $\frac{-13\pi}{24}, \frac{35\pi}{24}, \frac{59\pi}{24}$ **19.** $-1.6, 4.6, 7.8$

21. $\frac{\pi}{10} + \pi k, k \in \mathbb{Z}$ **23.** $\frac{\pi}{12} + \pi k, k \in \mathbb{Z}$

25. **27.** **29.**

31. **33.** **35.**

37. **39.**

41. $y = 3\tan\left(\frac{1}{2}t\right)$ **43.** $y = 2\cot\left(\frac{2\pi}{3}t\right)$ **45.** $\frac{\pi}{8}, \frac{3\pi}{8}$ **47.** about 1.2 mm

49. $y = 5.2\tan\left(\frac{\pi}{12}x\right)$; $P = 12$; asymptotes at $x = 6 + 12k, k \in \mathbb{Z}$; using (3, 5.2), $|A| = 5.2$; at $x = 2$, model gives $y \approx 3.002$; at $x = -2$, model gives $y \approx -3.002$; answers will vary. **51.** Answers will vary; $y = 11.95\tan\theta$; $P = 180°$; asymptotes at $\theta = 90° + 180°k$; $|A| = 11.95$ from (30°, 6.9 cm); pen is ≈12 cm long **53. a.** 20π cm ≈ 62.8 cm **b.** 80 cm; it is a square

c.

n	P
10	64.984
20	63.354
30	63.063
100	62.853

getting close to 20π

55.

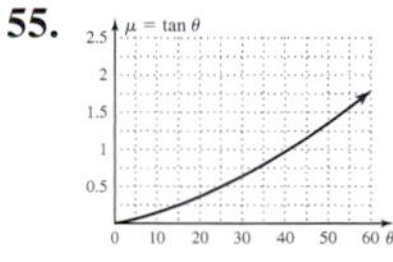

a. no; ≈35° **b.** 1.05 **c.** Angles will be greater than 68.2°; soft rubber on sandstone **57. a.** 5.67 units **b.** 86.5° **c.** Yes. Range of tan θ is $(-\infty, \infty)$. **d.** The closer θ gets to 90°, the longer the line segment gets.

59. $\sin 0.6662394325 = 0.6180339887 \approx \frac{-1 + \sqrt{5}}{2}$; $\cos x = \tan x$ can be rewritten as $\sin^2 x = 1 - \sin x$, which can in turn be converted to $\sin^2(-x) = 1 + \sin(-x)$, which is the basis of the golden ratio. **61.** slope of secant line is 2; slope of tangent line is 2; tangent of α and β is 2; tangent gives slope of line; $y - y_1 = \tan\theta\,(x - x_1)$ **63.** 21,266,032 km²

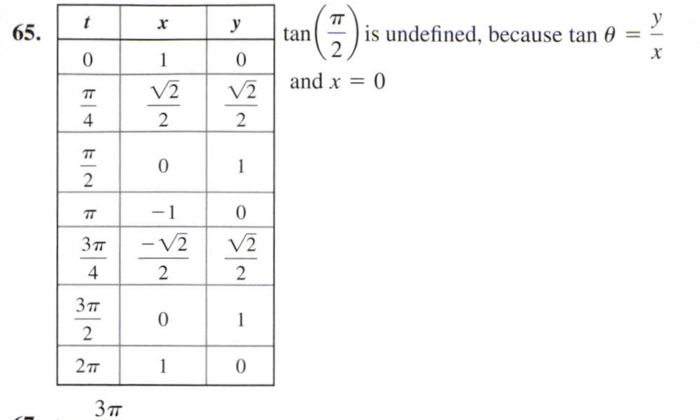

65.

t	x	y
0	1	0
$\frac{\pi}{4}$	$\frac{\sqrt{2}}{2}$	$\frac{\sqrt{2}}{2}$
$\frac{\pi}{2}$	0	1
π	-1	0
$\frac{3\pi}{4}$	$\frac{-\sqrt{2}}{2}$	$\frac{\sqrt{2}}{2}$
$\frac{3\pi}{2}$	0	1
2π	1	0

$\tan\left(\frac{\pi}{2}\right)$ is undefined, because $\tan\theta = \frac{y}{x}$ and $x = 0$

67. $t = \frac{3\pi}{2}$

Mid-Chapter Check, p. 112

1. $y = \cot t$; $y = \cos t$ **2.** $P = 4$, at $t = 1$ and $t = 3$ **3. a.** $\frac{1}{\sqrt{3}}$ **b.** $\frac{-\sqrt{2}}{2}$ **4. a.** ≈ 1.0353 **b.** ≈ 2.2858 **5.** $y = \cos t$ and $y = \sec t$

6. $t \in (-\infty, \infty)$; $t \neq \frac{\pi}{2}k, k \in \mathbb{Z}$

7. asymptotes: $x = -5, -3, -1, 1, 3, 5$; **8.** $|A| = 3, P = 4$;

9. a. QIV **b.** $2\pi - 5.94 \approx 0.343$ **c.** $\sin t, \tan t$

10. a. $|A| = 6, P = \frac{3\pi}{4}$ **b.** $f(t) = -6\cos\left(\frac{8}{3}t\right)$ **c.** $f(\pi) = 3$

Exercises 2.3, pp. 124–130

1. $y = A\sin(Bt + C) + D$; $y = A\cos(Bt + C) + D$

3. $0 \leq Bt + C < 2\pi$ **5.** Answers will vary. **7. a.** $|A| = 50, P = 24$ **b.** ≈ -25 **c.** $[1.6, 10.4]$ **9. a.** $|A| = 200, P = 3$ **b.** -175 **c.** $[1.75, 2.75]$

11. $y = 40\sin\left(\frac{\pi}{15}t\right) + 60$

13. $y = 8\sin\left(\frac{\pi}{180}t\right) + 12$

15. a. $y = 5\sin\left(\frac{\pi}{12}t\right) + 34$ **b.**

c. $\approx$1:30 A.M., 10:30 A.M.

17. a. $y = -6.4\cos\left(\frac{\pi}{6}t\right) + 12.4$ **b.** **c.** $\approx$134 days

19. a. $P = 11$ yr **b.** **c.** max = 1200, min = 700 **d.** about 2 yr.

21. $P(t) = 250\cos\left[\frac{2\pi}{11}(t - 2.75)\right] + 950$; $P(t) = 250\sin\left(\frac{2\pi}{11}t\right) + 950$

23. $|A| = 120$; $P = 24$; HS: 6 units right; VS: (none); PI: $6 \leq t < 30$

25. $|A| = 1$; $P = 12$; HS: 2 units right; VS: (none); PI: $2 \leq t < 14$

27. $|A| = 1$; $P = 8$; HS: $\frac{2}{3}$ unit right; VS: (none); PI: $\frac{2}{3} \leq t < \frac{26}{3}$

29. $|A| = 24.5$; $P = 20$; HS: 2.5 units right; VS: 15.5 units up; PI: $2.5 \leq t < 22.5$ **31.** $|A| = 28$; $P = 12$; HS: $\frac{5}{2}$ units right; VS: 92 units up; PI: $\frac{5}{2} \leq t < \frac{29}{2}$

33. $|A| = 2500$; $P = 8$; HS: $\frac{1}{3}$ unit left; VS: 3150 units up; PI: $-\frac{1}{3} \leq t < \frac{23}{3}$

35. $y = 250\sin\left(\frac{\pi}{12}t\right) + 350$ **37.** $y = 5\sin\left(\frac{\pi}{50}t + \frac{\pi}{2}\right) + 13$

39. $y = 4\sin\left(\frac{\pi}{180}t + \frac{\pi}{4}\right) + 7$

41. **43.**

45. $P = \frac{2\pi}{B}$, $B = \frac{2\pi}{P}$; $f = \frac{1}{P}$, $P = \frac{1}{f}$; $B = \frac{2\pi}{1/f} = 2\pi f$. $A\sin(Bt) = A\sin[(2\pi f)t]$ **47. a.** $P = 4$ sec, $f = \frac{1}{4}$ cycle/sec

b. -4.24 cm, moving away **c.** -4.24 cm, moving toward **d.** about 1.76 cm, avg. vel. = 3.52 cm/sec, greater, still gaining speed

49. $d(t) = 15\cos\left(\frac{5\pi}{4}t\right)$ **51.** red $\rightarrow$ D_3; blue $\rightarrow$ $A\#_3$

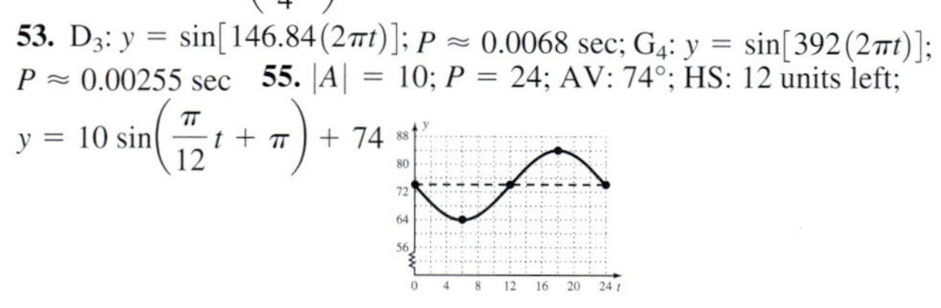

53. D_3: $y = \sin[146.84(2\pi t)]$; $P \approx 0.0068$ sec; G_4: $y = \sin[392(2\pi t)]$; $P \approx 0.00255$ sec **55.** $|A| = 10$; $P = 24$; AV: 74°; HS: 12 units left; $y = 10\sin\left(\frac{\pi}{12}t + \pi\right) + 74$

57. a. Caracas: $\approx$11.4 hr, Tokyo: $\approx$9.9 hr **b.** (i) Same # of hours on 79th day & 261st day (ii) Caracas: $\approx$81 days, Tokyo: $\approx$158 days

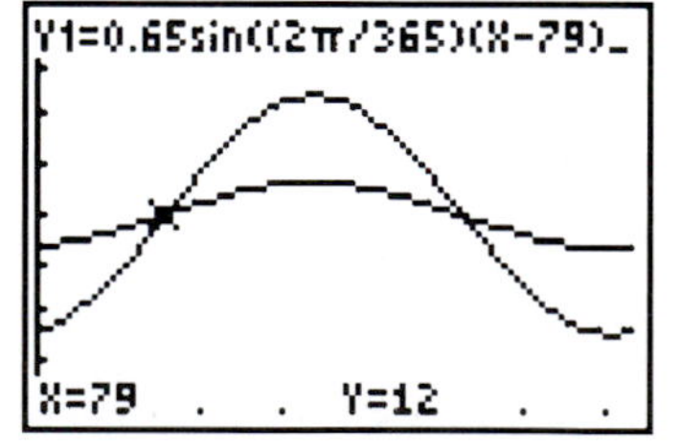

59. Answers will vary. **61. a.** Adds 12 hr. The sinusoidal behavior is actually based on hours more/less than an average of 12 hr of light.
b. Means 12 hours of light and dark on March 20th, day 79 (Solstice!).
c. Additional hours of deviation from average. In the north, the planet is tilted closer toward the Sun or farther from Sun, depending on date. Variations will be greater!

63. QIII; $3.7 - \pi \approx 0.5584$

65. $\sin\theta = -\frac{5}{12}$, $\cos\theta = \frac{\sqrt{119}}{12}$, $\tan\theta = -\frac{5}{\sqrt{119}}$, $\csc\theta = \frac{-12}{5}$, $\sec\theta = \frac{12}{\sqrt{119}}$, $\cot\theta = \frac{-\sqrt{119}}{5}$ **67.** verified; the acute angles are both 45°.

Exercises 2.4, pp. 139–144

1. $\sin(Bx + C); A + D$ **3.** critical **5.** Answers will vary.

7. $y = 25\sin\left(\frac{\pi}{6}x\right) + 50$ **9.** $y = 2.25\sin\left(\frac{\pi}{12}x + \frac{\pi}{4}\right) + 5.25$

11. $y = 503\sin\left(\frac{\pi}{6}x + \frac{2\pi}{3}\right) + 782$ **13. a.** $D(t) = 2000\cos\left(\frac{\pi}{60}t\right)$

b. 30 min **c.** north, 1258.6 mi **15. a.** $T(x) = 19.6\sin\left(\frac{\pi}{6}x + \frac{4\pi}{3}\right) + 84.6$

b. about 94.4°F **c.** beginning of May ($x \approx 5.1$) to end of August ($x \approx 8.9$)

17. a. $T(x) = 0.4\sin\left(\frac{\pi}{12}x + \frac{13\pi}{12}\right) + 98.6$ **b.** at 11 A.M. and 11 P.M.

c. from $x = 1$ to $x = 9$, about 8 hr **19.** $P = 12$, $B = \frac{\pi}{12}$, $C = \frac{\pi}{2}$; using (4, 3) gives $A = -3\sqrt{3}$, so $f(x) = -3\sqrt{3}\tan\left(\frac{\pi}{12}x + \frac{\pi}{2}\right)$.

a. $f(2.5) \approx 6.77$ **b.** $f(x) = 16$ for $x \approx 1.20$ **21. a.** using (18; 10) gives $A \approx 4.14$; $H(d) = 4.14\tan\left(\frac{\pi}{48}d\right)$ **b.** ≈ 12.2 cm **c.** ≈ 21.9 mi

23. a. $y \approx 49.26\sin(0.213x - 1.104) + 51.43$
b. $y \approx 49\sin(0.203x - 0.963) + 51$ **c.** at day 31 ≈ 5.6

25. a. $y \approx 5.88\sin(0.523x - 0.521) + 16.00$
b. $y \approx 6\sin(0.524x - 0.524) + 16$ **c.** at month 9 ≈ 0.12

27. a. $T(m) \approx 15.328\sin(0.461m - 1.610) + 85.244$

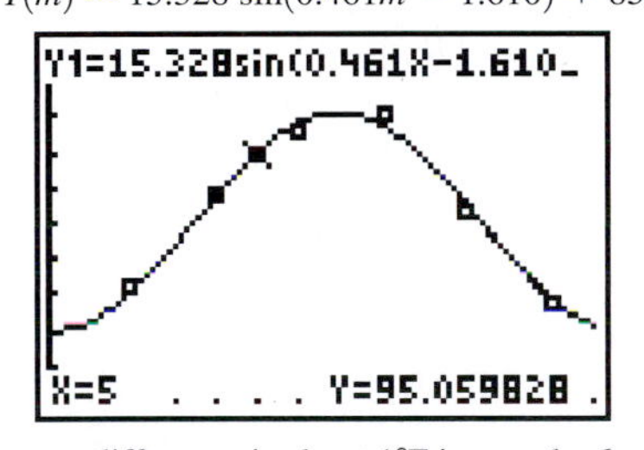

b.

Month	Temp. (°F)
1	71
3	82
5	95
7	101
9	94
11	80

c. max difference is about 1°F in months 6 and 8

29. a. Reno: $R(t) \approx 0.452\sin(0.396t + 1.831) + 0.750$
b. The graphs intersect at $t \approx 2.6$ and $t \approx 10.5$.
Reno gets more rainfall than Cheyenne for about 4 months of the year: $2.6 + (12 - 10.5) = 4.1$

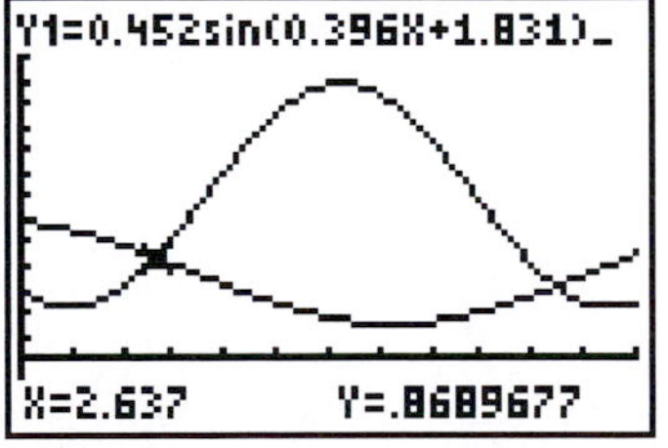

31. a. $f(x) \approx 49.659\sin(0.214x - 0.689) + 48.328$ **b.** about 26.8%
c. $g(x) = 49.5\sin\left(\frac{2\pi}{31}x - \frac{7\pi}{62}\right) + 49.5$
values for A, B, and D are very close; some variation in C.

33. $$\frac{m - D}{A} = \frac{m - \left(\frac{M + m}{2}\right)}{\frac{M - m}{2}} = \frac{2m - M - m}{M - m} = \frac{m - M}{M - m} = -1$$

35. the dampening factor is quadratic, $f(x) \approx 0.02x^2 - 0.32x + 2.28$, $Y_1 = \sin(3x)$, $Y_2 = 0.02x^2 - 0.32x + 2.28$, $x \in [-2\pi, 7\pi]$, $y \in [-5, 5]$

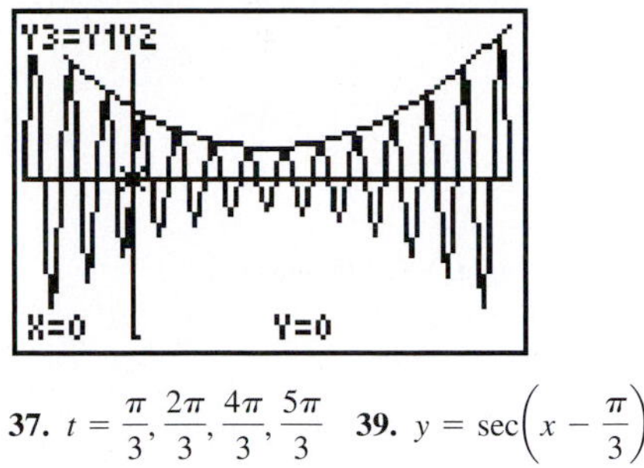

37. $t = \frac{\pi}{3}, \frac{2\pi}{3}, \frac{4\pi}{3}, \frac{5\pi}{3}$ **39.** $y = \sec\left(x - \frac{\pi}{3}\right)$

41. $2 \times \frac{1}{2}(24^2)\left(\frac{7\pi}{36}\right) = 112\pi$ in.^{2}

Summary and Concept Review, pp. 145–150

1. $|A| = 3$, $P = 2\pi$ **2.** $P = 2\pi$

3. $|A| = 1$, $P = \pi$ **4.** $|A| = 1.7$, $P = \frac{\pi}{2}$

5. $|A| = 2$, $P = \frac{1}{2}$ **6.** $|A| = 3$, $P = \frac{1}{199}$

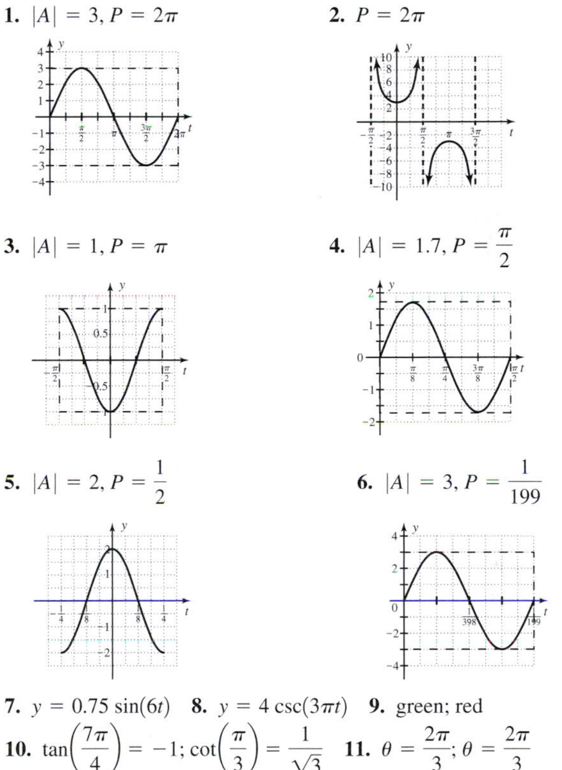

7. $y = 0.75\sin(6t)$ **8.** $y = 4\csc(3\pi t)$ **9.** green; red

10. $\tan\left(\frac{7\pi}{4}\right) = -1$; $\cot\left(\frac{\pi}{3}\right) = \frac{1}{\sqrt{3}}$ **11.** $\theta = \frac{2\pi}{3}$; $\theta = \frac{2\pi}{3}$

12. **13.**

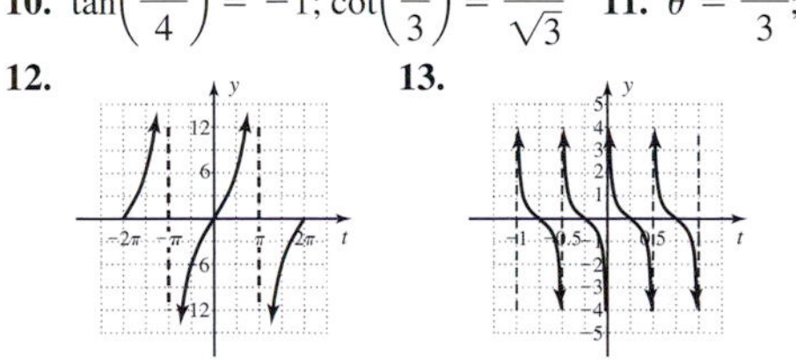

14. $1.55 + k\pi$ radians; $k \in Z$ **15.** 3.5860 **16.** ≈ 151.14 m

17. $y = 5.2\tan\left(\frac{\pi}{12}x\right)$; period = 12; $A = 5.2$; asymptotes $x = -6$, $x = 6$

18. a. $|A| = 240$, $P = 12$, HS: 3 units right, VS: 520 units up **b.**

19. a. $|A| = 3.2$, $P = 8$, HS: 6 units left, VS: 6.4 units up

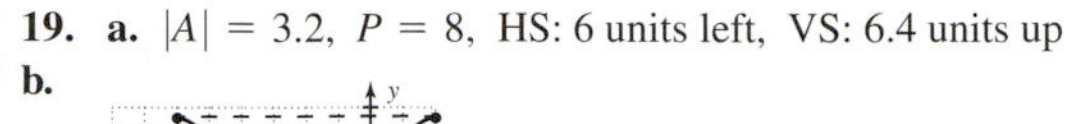

b.

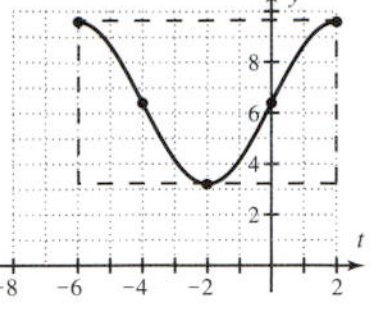

20. $|A| = 125$, $P = 24$, HS: 3 units right, VS: 175 units up,

$$y = 125\cos\left[\frac{\pi}{12}(t-3)\right] + 175$$

21. $A = 75$, $P = \frac{3\pi}{8}$, HS: (none), VS: 105 units up,

$$y = 75\sin\left(\frac{16}{3}t\right) + 105$$

22. a. $P(t) = 0.91\sin\left(\frac{\pi}{6}t\right) + 1.35$ **b.** August: 1.81 in., Dec: 0.44 in.

23. a. $y = 2187.723\sin(0.017x + 1.751) + 2307.437$

b. $A = \frac{4450 - 90}{2} = 2180;$

$D = 90 + 2180 = 2270;$

$\frac{2\pi}{B} = 365, \frac{2\pi}{365} = B; C = \frac{3\pi}{2} - \frac{2\pi}{365}(184) = \frac{359}{730}\pi$

$y = 2180\sin\left(\frac{2\pi}{365}x + \frac{359}{730}\pi\right) + 2270$

c. The largest difference from $0 \le x \le 365$ is about 372 at $x = 85$.

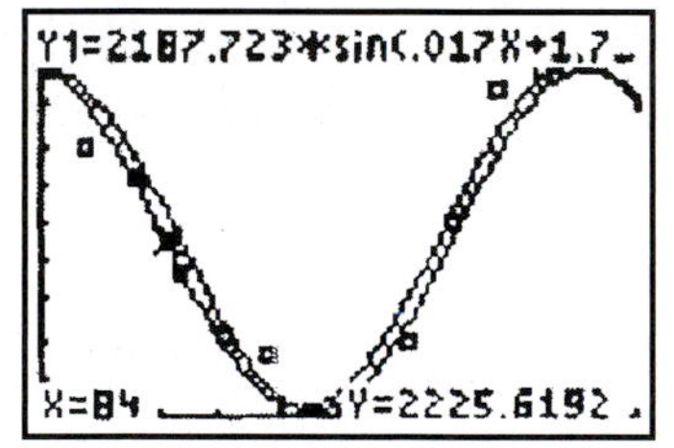

24. a. $y = 19.424\sin(0.145x - 0.748) + 79.581$

b. $A = \frac{100 - 67}{2} = 16.5;$

$D = 67 + 16.5 = 83.5;$

$\frac{2\pi}{B} = 31,$

$\frac{2\pi}{31} = B$

$C = \frac{3\pi}{2} - \frac{2\pi}{31}(31) = -\frac{\pi}{2}$

$y = 16.5\sin\left(\frac{2\pi}{31}x - \frac{\pi}{2}\right) + 83.5$

c. The largest difference is about 7, at $x = 26$.

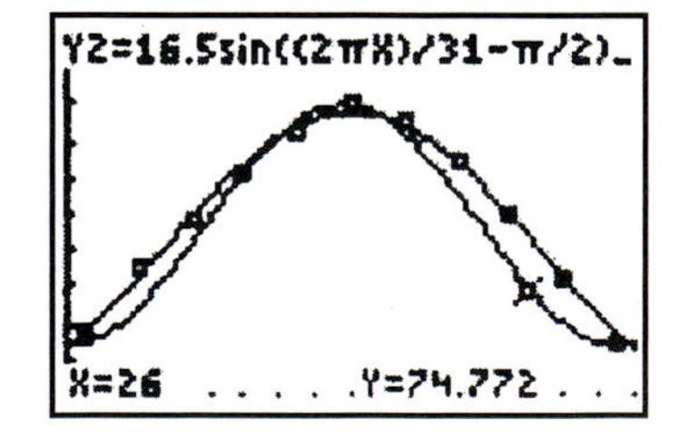

25. a. $y = 16.800\sin(0.602x - 2.341) + 70.968$

b. $x = 7 \quad y = 87.01°F$

c. The model alternates between slightly overpredicting and underpredicting output values, and appears to be a fairly accurate model.

Mixed Review pp. 150–151

1. $f(t) = 10\sin(2t)$

3. a. $y = 30.249\sin(0.018x - 1.813) + 54.917$

b. about 54.7°

c. average is about 52°; close to value of D from the regression model.

5. $y = -2\cot\left[\frac{3}{2}\left(x + \frac{\pi}{3}\right)\right]$, other solutions are possible.

7. **9.** **11.**

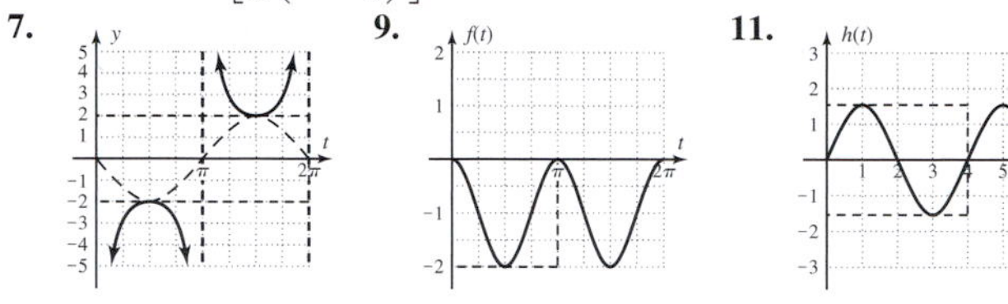

13. a. $P(t) = 1.63\sin\left(\frac{\pi}{6}t\right) + 2.42$ **b.** July: 4.05 in., Dec: 1.01 in.

15. $t \approx 5.99$

17. $|A| = 5$; $P = \pi$;
HS: (none);
VS: 8 units down;
PI: $0 \le t < \pi$

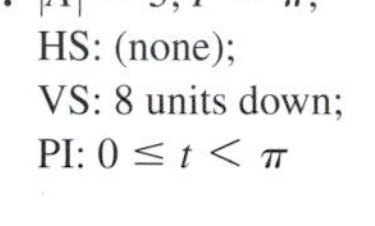

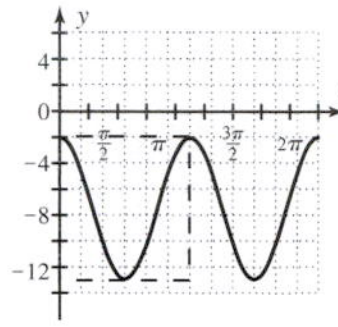

19. $P = 4\pi$; no horizontal shift

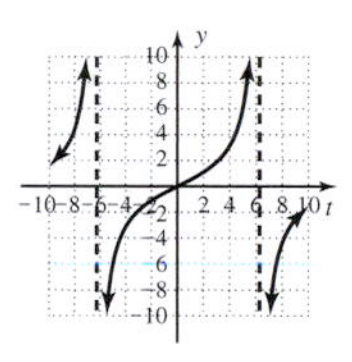

Practice Test pp. 151–153

1.

t	$\sin t$	$\cos t$	$\tan t$	$\csc t$	$\sec t$	$\cot t$	$P(x, y)$
0	0	1	0	—	1	—	(1, 0)
$\frac{\pi}{6}$	$\frac{1}{2}$	$\frac{\sqrt{3}}{2}$	$\frac{\sqrt{3}}{3}$	2	$\frac{2\sqrt{3}}{3}$	$\sqrt{3}$	$\left(\frac{\sqrt{3}}{2}, \frac{1}{2}\right)$
$\frac{\pi}{4}$	$\frac{\sqrt{2}}{2}$	$\frac{\sqrt{2}}{2}$	1	$\sqrt{2}$	$\sqrt{2}$	1	$\left(\frac{\sqrt{2}}{2}, \frac{\sqrt{2}}{2}\right)$
$\frac{\pi}{2}$	1	0	—	1	—	0	(0, 1)
$\frac{2\pi}{3}$	$\frac{\sqrt{3}}{2}$	$-\frac{1}{2}$	$-\sqrt{3}$	$\frac{2\sqrt{3}}{3}$	-2	$-\frac{\sqrt{3}}{3}$	$\left(-\frac{1}{2}, \frac{\sqrt{3}}{2}\right)$
$\frac{5\pi}{6}$	$\frac{1}{2}$	$-\frac{\sqrt{3}}{2}$	$-\frac{\sqrt{3}}{3}$	2	$-\frac{2\sqrt{3}}{3}$	$-\sqrt{3}$	$\left(-\frac{\sqrt{3}}{2}, \frac{1}{2}\right)$
$\frac{5\pi}{4}$	$-\frac{\sqrt{2}}{2}$	$-\frac{\sqrt{2}}{2}$	1	$-\sqrt{2}$	$-\sqrt{2}$	1	$\left(-\frac{\sqrt{2}}{2}, -\frac{\sqrt{2}}{2}\right)$
$\frac{4\pi}{3}$	$-\frac{\sqrt{3}}{2}$	$-\frac{1}{2}$	$\sqrt{3}$	$-\frac{2\sqrt{3}}{3}$	-2	$\frac{\sqrt{3}}{3}$	$\left(-\frac{1}{2}, -\frac{\sqrt{3}}{2}\right)$
$\frac{3\pi}{2}$	-1	0	—	-1	—	0	(0, −1)

2. a. 0 **b.** 0 **c.** 1 **d.** $\frac{2\sqrt{3}}{3}$

3. a. $\frac{\pi}{3}$ **b.** $\frac{2\pi}{3}$ **c.** $\frac{4\pi}{3}$ **d.** $\frac{7\pi}{4}$

4. $|A| = 3, P = 10$

5. no amplitude, $P = \pi$

6. no amplitude, $P = \frac{\pi}{3}$

7. $|A| = 12, P = \frac{2\pi}{3}$, HS: $\frac{\pi}{6}$ right, VS: 19 units up

8. $|A| = \frac{3}{4}$, $P = 4$, HS: 0, VS: $\frac{1}{2}$ unit down

9. no amplitude, $P = 3$, HS: $\frac{3}{2}$ right, VS: none

10. 6 or $6,000 **b.** January through July

11. $t \approx 1.11, t \approx 7.39$

12. $t \approx 4.6018$

13. a. $r(t) = 0.41 \sin\left[\frac{\pi}{6}(x + 2)\right] + 0.65$, **b.** January

14. a. $y = 35.223 \sin(0.576x - 2.589) + 6.120$

b.

Month (Jan → 1)	Low Temp. (°F)
1	−26
3	−21
5	16
7	41
9	25
11	−14

15. $y = 7.5 \sin\left(\frac{\pi}{6}t - \frac{\pi}{2}\right) + 12.5$

16. $y = 2 \csc\left(\frac{\pi}{2}t\right) - 1$

17. a **18.** c **19.** b **20.** d

Strengthening Core Skills, p. 156

Exercise 1.

t	0	$\frac{\pi}{6}$	$\frac{\pi}{4}$	$\frac{\pi}{3}$	$\frac{\pi}{2}$	$\frac{2\pi}{3}$	$\frac{3\pi}{4}$
$\sin t = y$	0	$\frac{1}{2}$	$\frac{\sqrt{2}}{2}$	$\frac{\sqrt{3}}{2}$	1	$\frac{\sqrt{3}}{2}$	$\frac{\sqrt{2}}{2}$
$\cos t = x$	1	$\frac{\sqrt{3}}{2}$	$\frac{\sqrt{2}}{2}$	$\frac{1}{2}$	0	$\frac{-1}{2}$	$\frac{-\sqrt{2}}{2}$
$\tan t = \frac{y}{x}$	0	$\frac{\sqrt{3}}{3}$	1	$\sqrt{3}$	–	$-\sqrt{3}$	−1

t	$\frac{5\pi}{6}$	π	$\frac{7\pi}{6}$	$\frac{5\pi}{4}$
$\sin t = y$	$\frac{1}{2}$	0	$\frac{-1}{2}$	$\frac{-\sqrt{2}}{2}$
$\cos t = x$	$\frac{-\sqrt{3}}{2}$	−1	$\frac{-\sqrt{3}}{2}$	$\frac{-\sqrt{2}}{2}$
$\tan t = \frac{y}{x}$	$\frac{-\sqrt{3}}{3}$	0	$\frac{\sqrt{3}}{3}$	1

Exercise 2. a. $t = \frac{4\pi}{3}, \frac{5\pi}{3}$ **b.** $t = \frac{\pi}{4}, \frac{7\pi}{4}$ **c.** $t = \frac{\pi}{6}, \frac{7\pi}{6}$ **d.** $t = \frac{\pi}{6}$

Exercise 3. a. no solution **b.** $t \approx 1.2310, t \approx 5.0522$ **c.** $t \approx 6.0382, t \approx 2.8966$ **d.** $t \approx 1.9823, t \approx 4.3009$

Cumulative Review, pp. 156–158

1.

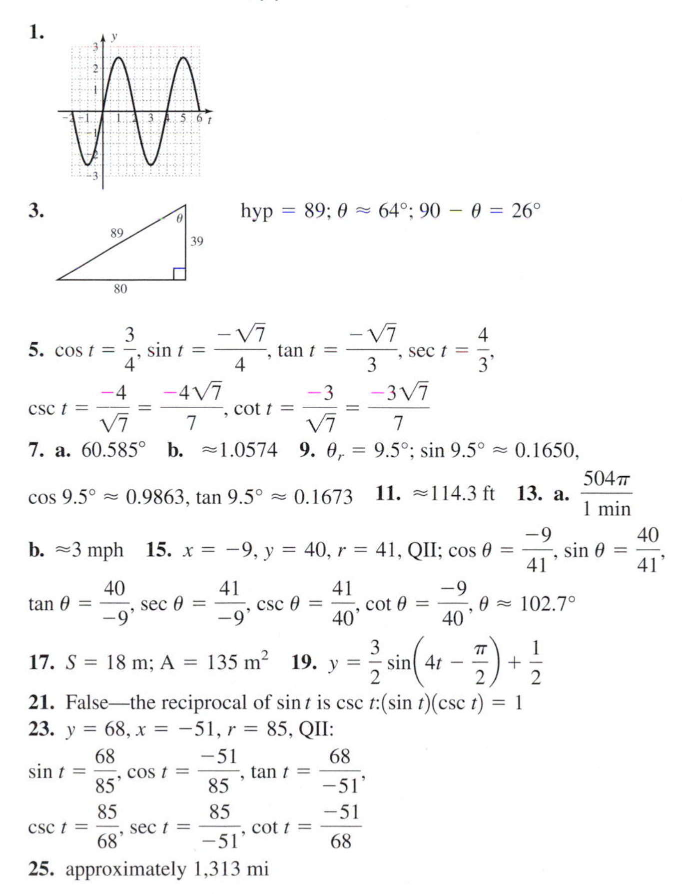

3. hyp = 89; $\theta \approx 64°$; $90 - \theta = 26°$

5. $\cos t = \frac{3}{4}$, $\sin t = \frac{-\sqrt{7}}{4}$, $\tan t = \frac{-\sqrt{7}}{3}$, $\sec t = \frac{4}{3}$, $\csc t = \frac{-4}{\sqrt{7}} = \frac{-4\sqrt{7}}{7}$, $\cot t = \frac{-3}{\sqrt{7}} = \frac{-3\sqrt{7}}{7}$

7. a. 60.585° **b.** ≈1.0574 **9.** $\theta_r = 9.5°$; $\sin 9.5° \approx 0.1650$, $\cos 9.5° \approx 0.9863$, $\tan 9.5° \approx 0.1673$ **11.** ≈114.3 ft **13. a.** $\frac{504\pi}{1 \text{ min}}$ **b.** ≈3 mph **15.** $x = -9, y = 40, r = 41$, QII; $\cos\theta = \frac{-9}{41}$, $\sin\theta = \frac{40}{41}$, $\tan\theta = \frac{40}{-9}$, $\sec\theta = \frac{41}{-9}$, $\csc\theta = \frac{41}{40}$, $\cot\theta = \frac{-9}{40}$, $\theta \approx 102.7°$

17. $S = 18$ m; $A = 135 \text{ m}^2$ **19.** $y = \frac{3}{2}\sin\left(4t - \frac{\pi}{2}\right) + \frac{1}{2}$

21. False—the reciprocal of $\sin t$ is $\csc t$: $(\sin t)(\csc t) = 1$

23. $y = 68, x = -51, r = 85$, QII: $\sin t = \frac{68}{85}$, $\cos t = \frac{-51}{85}$, $\tan t = \frac{68}{-51}$, $\csc t = \frac{85}{68}$, $\sec t = \frac{85}{-51}$, $\cot t = \frac{-51}{68}$

25. approximately 1,313 mi

CHAPTER 3

Exercises 3.1, pp. 165–169

1. $\sin\theta$; $\sec\theta$; $\cos\theta$ **3.** one; false **5.** $\dfrac{1-\sin^2 x}{\sin x \sec x}$; Answers will vary.

7. Answers may vary; $\tan x = \dfrac{\sec x}{\csc x}$; $\dfrac{\sin x}{\cos x} = \dfrac{\sec x}{\csc x}$; $\dfrac{1}{\cot x} = \dfrac{\sec x}{\csc x}$; $\dfrac{1}{\cot x} = \dfrac{\sin x}{\cos x}$ **9.** $1 = \sec^2 x - \tan^2 x$; $\tan^2 x = \sec^2 x - 1$; $1 = (\sec x + \tan x)(\sec x - \tan x)$; $\tan x = \pm\sqrt{\sec^2 x - 1}$

11. $\sin x \cot x = \cos x$; $\cancel{\sin x}\dfrac{\cos x}{\cancel{\sin x}} = \cos x$

13. $\sec^2 x \cot^2 x = \csc^2 x$; $\dfrac{1}{\cancel{\cos^2 x}}\dfrac{\cancel{\cos^2 x}}{\sin^2 x} = \dfrac{1}{\sin^2 x} = \csc^2 x$

15. $\cos x(\sec x - \cos x) = \sin^2 x$; $\cos x \sec x - \cos^2 x = \cos x\dfrac{1}{\cos x} - \cos^2 x = 1 - \cos^2 x = \sin^2 x$

17. $\sin x(\csc x - \sin x) = \cos^2 x$; $1 - \sin^2 x = \cos^2 x$

19. $\tan x(\csc x + \cot x) = \sec x + 1$; $\tan x \csc x + \tan x \cot x = \dfrac{\cancel{\sin x}}{\cos x}\dfrac{1}{\cancel{\sin x}} + \dfrac{\cancel{\sin x}}{\cancel{\cos x}}\dfrac{\cancel{\cos x}}{\cancel{\sin x}} = \dfrac{1}{\cos x} + 1 = \sec x + 1$

21. $\tan^2 x \csc^2 x - \tan^2 x = 1$; $\tan^2 x(\csc^2 x - 1) = 1$; $\tan^2 x(\cot^2 x) = 1$; $1 = 1$ **23.** $\dfrac{\sin x \cos x + \sin x}{\cos x + \cos^2 x} = \tan x$; $\dfrac{\sin x\cancel{(\cos x + 1)}}{\cos x\cancel{(1 + \cos x)}} = \tan x$; $\dfrac{\sin x}{\cos x} = \tan x$ **25.** $\dfrac{1 + \sin x}{\cos x + \cos x \sin x} = \dfrac{(1)\cancel{(1 + \sin x)}}{(\cos x)\cancel{(1 + \sin x)}} = \dfrac{1}{\cos x} = \sec x$ **27.** $\dfrac{\sin x \tan x + \sin x}{\tan x + \tan^2 x} = \dfrac{\sin x\cancel{(\tan x + 1)}}{\tan x\cancel{(1 + \tan x)}} = \dfrac{\sin x}{\tan x} = \dfrac{\sin x}{\sin x/\cos x} = \dfrac{\cancel{\sin x}\cos x}{\cancel{\sin x}} = \cos x$

29. $\dfrac{(\sin x + \cos x)^2}{\cos x} = \dfrac{\sin^2 x + 2\sin x\cos x + \cos^2 x}{\cos x} = \dfrac{\sin^2 x + \cos^2 x + 2\sin x \cos x}{\cos x} = \dfrac{1 + 2\sin x\cos x}{\cos x} = \dfrac{1}{\cos x} + \dfrac{2\sin x \cos x}{\cos x} = \sec x + 2\sin x$

31. $(1 + \sin x)[1 + \sin(-x)] = (1 + \sin x)(1 - \sin x) = 1 - \sin^2 x = \cos^2 x$

33. $\dfrac{(\csc x - \cot x)(\csc x + \cot x)}{\tan x} = \dfrac{\csc^2 x - \cot^2 x}{\tan x} = \dfrac{1}{\tan x} = \cot x$

35. $\dfrac{\cos^2 x}{\sin x} + \dfrac{\sin x}{1} = \dfrac{\cos^2 x + \sin^2 x}{\sin x} = \dfrac{1}{\sin x} = \csc x$

37. $\dfrac{\tan x}{\csc x} - \dfrac{\sin x}{\cos x} = \dfrac{\tan x\cos x - \sin x\csc x}{\csc x \cos x} = \dfrac{\dfrac{\sin x}{\cancel{\cos x}}\cancel{\cos x} - 1}{\dfrac{1}{\sin x}\cos x} = \dfrac{\sin x - 1}{\cot x}$

39. $\dfrac{\sec x}{\sin x} - \dfrac{\csc x}{\sec x} = \dfrac{\sec^2 x - \sin x\csc x}{\sin x \sec x} = \dfrac{\sec^2 x - 1}{\sin x\dfrac{1}{\cos x}} = \dfrac{\tan^{\cancel{2}} x}{\cancel{\tan x}} = \tan x$

41. $\dfrac{\sin x}{\pm\sqrt{1 - \sin^2 x}}$ **43.** $\pm\sqrt{\dfrac{1}{\cot^2 x} + 1}$ **45.** $\dfrac{\pm\sqrt{1 - \sin^2 x}}{\sin x}$

47. Answers will vary. **49.** Answers will vary. **51.** Answers will vary. **53.** Answers will vary.

55. $\cos\theta = -\dfrac{35}{37}$, $\tan\theta = -\dfrac{12}{35}$, $\sec\theta = -\dfrac{37}{35}$, $\csc\theta = \dfrac{37}{12}$, $\cot\theta = -\dfrac{35}{12}$

57. $\cos\theta = \dfrac{27}{45} = \dfrac{3}{5}$, $\sin\theta = -\dfrac{36}{45} = -\dfrac{4}{5}$, $\tan\theta = -\dfrac{36}{27} = -\dfrac{4}{3}$, $\csc\theta = -\dfrac{45}{36} = \dfrac{-5}{4}$, $\cot\theta = -\dfrac{27}{36} = -\dfrac{3}{4}$

59. $\cos\theta = -\sqrt{1 - \dfrac{x^2}{49}} = \dfrac{-\sqrt{49 - x^2}}{7}$, $\sin\theta = \dfrac{x}{7}$, $\tan\theta = -\dfrac{x}{\sqrt{49 - x^2}}$, $\sec\theta = -\dfrac{7}{\sqrt{49 - x^2}}$, $\cot\theta = -\dfrac{\sqrt{49 - x^2}}{x}$ **61.** $\sin\theta = -\dfrac{4\sqrt{6}}{25}$, $\tan\theta = -\dfrac{4\sqrt{6}}{23}$, $\sec\theta = \dfrac{25}{23}$, $\csc\theta = -\dfrac{25}{4\sqrt{6}}$, $\cot\theta = -\dfrac{23}{4\sqrt{6}}$

63. a. $A = \dfrac{nx^2}{4}\cot\left(\dfrac{\pi}{n}\right)$

b. $A = \dfrac{\cancel{4}(8\text{ m})^2}{\cancel{4}}\cot\left(\dfrac{\pi}{4}\right) = 64\text{ m}^2 \cdot 1 = 64\text{ m}^2$

c. $\approx 1119.62\text{ in}^2$

65. $\cos^3 x = (\cos x)(\cos^2 x) = (\cos x)(1 - \sin^2 x)$

67. $\tan x + \tan^3 x = (\tan x)(1 + \tan^2 x) = (\tan x)(\sec^2 x)$

69. $\tan^2 x \sec x - 4\tan^2 x = (\tan^2 x)(\sec x - 4)$
$= (\sec x - 4)(\tan^2 x) = (\sec x - 4)(\sec^2 - 1)$
$= (\sec x - 4)(\sec x - 1)(\sec x + 1)$

71. $\cos^2 x \sin x - \cos^2 x = (\cos^2 x)(\sin x - 1)$
$= (1 - \sin^2 x)(\sin x - 1)$
$= (1 + \sin x)(1 - \sin x)(\sin x - 1)$
$= (1 + \sin x)(1 - \sin x)(-1)(1 - \sin x)$
$= (-1)(1 + \sin x)(1 - \sin x)^2$

73. a. $A = nr^2\tan\left(\dfrac{\pi}{n}\right)$ **b.** $A = 4 \cdot 4^2\tan\left(\dfrac{\pi}{4}\right) = 64\text{ m}^2$ **c.** $A \approx 51.45\text{ m}^2$ **75.** $\tan\theta = \dfrac{1 + m_1 m_2}{m_2 - m_1}$ **77.** $\theta = 45°$

79. Answers will vary. **81. a.** 9 sides **b.** 13 sides **83.** about 1148 ft

85. $\sin\theta = \frac{-63}{65}$, $\csc\theta = \frac{65}{-63}$, $\cos\theta = \frac{-16}{65}$, $\sec\theta = \frac{65}{-16}$, $\tan\theta = \frac{63}{16}$, $\cot\theta = \frac{16}{63}$ **87.**

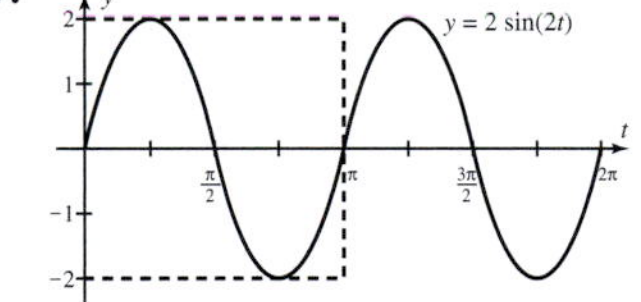

Exercises 3.2, pp. 174–177

1. reversible **3.** complicated; simplify; build

5. Because we don't know if the equation is true.

7. $\dfrac{1 + \sin x}{\cos x}$ **9.** $\cos x$ **11.** $\dfrac{1 - \cos x}{\sin x}$

13. $\cos^2 x\tan^2 x = 1 - \cos^2 x$
$\cancel{\cos^2 x}\dfrac{\sin^2 x}{\cancel{\cos^2 x}} =$
$\sin^2 x =$
$1 - \cos^2 x =$

15. $\tan x + \cot x = \sec x\csc x$
$\dfrac{\sin x}{\cos x} + \dfrac{\cos x}{\sin x} =$
$\dfrac{\sin^2 x + \cos^2 x}{\cos x \sin x} =$
$\dfrac{1}{\cos x\sin x} =$
$\dfrac{1}{\cos x}\dfrac{1}{\sin x} =$
$\sec x\csc x =$

17. $\dfrac{\cos x}{\tan x} = \csc x - \sin x$

$= \dfrac{1}{\sin x} - \sin x$

$= \dfrac{1 - \sin^2 x}{\sin x}$

$= \dfrac{\cos^2 x}{\sin x}$

$= \dfrac{\cos x}{\sin x / \cos x}$

$= \dfrac{\cos x}{\tan x}$

19. $\dfrac{\cos\theta}{1 - \sin\theta} = \sec\theta + \tan\theta$

$= \dfrac{1}{\cos\theta} + \dfrac{\sin\theta}{\cos\theta}$

$= \dfrac{1 + \sin\theta}{\cos\theta}$

$= \dfrac{(1 + \sin\theta)(1 - \sin\theta)}{\cos\theta(1 - \sin\theta)}$

$= \dfrac{1 - \sin^2\theta}{\cos\theta(1 - \sin\theta)}$

$= \dfrac{\cos^2\theta}{\cancel{\cos\theta}(1 - \sin\theta)}$

$= \dfrac{\cos\theta}{1 - \sin\theta}$

21. $\dfrac{1 - \sin x}{\cos x} = \dfrac{\cos x}{1 + \sin x}$

$\dfrac{(1 - \sin x)(1 + \sin x)}{\cos x(1 + \sin x)} =$

$\dfrac{1 - \sin^2 x}{\cos x\,(1 + \sin x)} =$

$\dfrac{\cos^2 x}{\cancel{\cos x}(1 + \sin x)} =$

$\dfrac{\cos x}{1 + \sin x} =$

23. $\dfrac{\csc x}{\cos x} - \dfrac{\cos x}{\csc x} = \dfrac{\cot^2 x + \sin^2 x}{\cot x}$

$\dfrac{\csc^2 x - \cos^2 x}{\cos x \csc x} =$

$\dfrac{\csc^2 x - (1 - \sin^2 x)}{\cos x \dfrac{1}{\sin x}} =$

$\dfrac{\csc^2 x - 1 + \sin^2 x}{\cot x} =$

$\dfrac{\cot^2 x + \sin^2 x}{\cot x} =$

25. $\dfrac{\sin x}{1 + \sin x} - \dfrac{\sin x}{1 - \sin x} = -2\tan^2 x$

$\dfrac{\sin x(1 - \sin x) - \sin x(1 + \sin x)}{(1 + \sin x)(1 - \sin x)} =$

$\dfrac{\cancel{\sin x} - \sin^2 x - \cancel{\sin x} - \sin^2 x}{1 - \sin^2 x} =$

$\dfrac{-2\sin^2 x}{\cos^2 x} =$

$-2\tan^2 x =$

27. $\dfrac{\cot x}{1 + \csc x} - \dfrac{\cot x}{1 - \csc x} = 2\sec x$

$\dfrac{\cot x(1 - \csc x) - \cot x(1 + \csc x)}{(1 + \csc x)(1 - \csc x)} =$

$\dfrac{\cancel{\cot x} - \cot x \csc x - \cancel{\cot x} - \cot x \csc x}{1 - \csc^2 x} =$

$\dfrac{2\cot x \csc x}{\cot^2 x} =$

$\dfrac{2\csc x}{\cot x} =$

$\dfrac{2\dfrac{1}{\cancel{\sin x}}}{\dfrac{\cos x}{\cancel{\sin x}}} =$

$\dfrac{2}{\cos x} =$

$2\sec x =$

29. $\dfrac{\sec^2 x}{1 + \cot^2 x} = \tan^2 x$

$\dfrac{\sec^2 x}{\csc^2 x} =$

$\dfrac{\dfrac{1}{\cos^2 x}}{\dfrac{1}{\sin^2 x}} =$

$\dfrac{\sin^2 x}{\cos^2 x} =$

$\tan^2 x =$

31. $\sin^2 x(\cot^2 x - \csc^2 x) = -\sin^2 x$

$\sin^2 x \cot^2 x - \sin^2 x \csc^2 x =$

$\cancel{\sin^2 x}\dfrac{\cos^2 x}{\cancel{\sin^2 x}} - \cancel{\sin^2 x}\dfrac{1}{\cancel{\sin^2 x}} =$

$\cos^2 x - 1 =$

$-\sin^2 x =$

33. $\cos x \cot x + \sin x = \csc x$

$\cos x \dfrac{\cos x}{\sin x} + \sin x =$

$\dfrac{\cos^2 x}{\sin x} + \sin x =$

$\dfrac{\cos^2 x + \sin^2 x}{\sin x} =$

$\dfrac{1}{\sin x} =$

$\csc x =$

35. $\dfrac{\sec x}{\cot x + \tan x} = \sin x$

$\dfrac{\dfrac{1}{\cos x}(\sin x)(\cos x)}{\left(\dfrac{\cos x}{\sin x} + \dfrac{\sin x}{\cos x}\right)(\sin x)(\cos x)} =$

$\dfrac{\sin x}{\cos^2 x + \sin^2 x} =$

$\dfrac{\sin x}{1} =$

$\sin x =$

37. $$\frac{\sin x - \csc x}{\csc x} = -\cos^2 x$$
$$\frac{\sin x}{\csc x} - \frac{\csc x}{\csc x} =$$
$$\sin^2 x - 1 =$$
$$-\cos^2 x =$$

39. $$\frac{1}{\csc x - \sin x} = \tan x \sec x$$
$$\frac{1}{(\csc x - \sin x)}\,\frac{\sin x}{\sin x} =$$
$$\frac{\sin x}{1 - \sin^2 x} =$$
$$\frac{\sin x}{\cos^2 x} =$$
$$\frac{\sin x}{\cos x}\,\frac{1}{\cos x} =$$
$$\tan x \sec x =$$

41. $$\frac{1 + \sin x}{1 - \sin x} = (\tan x + \sec x)^2$$
$$\frac{(1 + \sin x)}{(1 - \sin x)}\,\frac{(1 + \sin x)}{(1 + \sin x)} =$$
$$\frac{1 + 2\sin x + \sin^2 x}{1 - \sin^2 x} =$$
$$\frac{1 + 2\sin x + \sin^2 x}{\cos^2 x} =$$
$$\frac{1}{\cos^2 x} + 2\frac{\sin x}{\cos x}\,\frac{1}{\cos x} + \frac{\sin^2 x}{\cos^2 x} =$$
$$\sec^2 x + 2\tan x \sec x + \tan^2 x =$$
$$(\sec x + \tan x)^2 =$$
$$(\tan x + \sec x)^2 =$$

43. $$\frac{\cos x - \sin x}{1 - \tan x} = \frac{\cos x + \sin x}{1 + \tan x}$$
$$\frac{(\cos x - \sin x)}{(1 - \tan x)}\,\frac{(\cos x + \sin x)}{(\cos x + \sin x)} =$$
$$\frac{(\cos x - \sin x)(\cos x + \sin x)}{\cos x + \sin x - \sin x - \dfrac{\sin^2 x}{\cos x}} =$$
$$\frac{(\cos x - \sin x)(\cos x + \sin x)}{\cos x\left(1 - \dfrac{\sin^2 x}{\cos^2 x}\right)} =$$
$$\frac{(\cos x - \sin x)(\cos x + \sin x)}{\cos x(1 - \tan^2 x)} =$$
$$\frac{(\cos x - \sin x)(\cos x + \sin x)}{\cos x(1 - \tan x)(1 + \tan x)} =$$
$$\frac{\cancel{(\cos x - \sin x)}(\cos x + \sin x)}{\cancel{(\cos x - \sin x)}(1 + \tan x)} =$$
$$\frac{\cos x + \sin x}{1 + \tan x} =$$

45. $$\frac{\tan^2 x - \cot^2 x}{\tan x - \cot x} = \csc x \sec x$$
$$\frac{(\tan x + \cot x)(\tan x - \cot x)}{(\tan x - \cot x)} =$$
$$\tan x + \cot x =$$
$$\frac{\sin x}{\cos x} + \frac{\cos x}{\sin x} =$$
$$\frac{\sin^2 x + \cos^2 x}{\cos x \sin x} =$$
$$\frac{1}{\cos x \sin x} =$$
$$\frac{1}{\cos x}\,\frac{1}{\sin x} =$$
$$\sec x \csc x =$$

47. $$\frac{\cot x}{\cot x + \tan x} = 1 - \sin^2 x$$
$$\frac{\dfrac{\cos x}{\sin x}}{\dfrac{\cos x}{\sin x} + \dfrac{\sin x}{\cos x}}\,\frac{(\cos x)(\sin x)}{(\cos x)(\sin x)} =$$
$$\frac{\cos^2 x}{\cos^2 x + \sin^2 x} =$$
$$\frac{\cos^2 x}{1} =$$
$$1 - \sin^2 x =$$

49. $$\frac{\sec^4 x - \tan^4 x}{\sec^2 x + \tan^2 x} = 1$$
$$\frac{(\sec^2 x - \tan^2 x)\cancel{(\sec^2 x + \tan^2 x)}}{\cancel{(\sec^2 x + \tan^2 x)}} =$$
$$\sec^2 x - \tan^2 x =$$
$$1 =$$

51. $$\frac{\cos^4 x - \sin^4 x}{\cos^2 x} = 2 - \sec^2 x$$
$$\frac{(\cos^2 x - \sin^2 x)(\cos^2 x + \sin^2 x)}{\cos^2 x} =$$
$$\frac{(\cos^2 x - \sin^2 x)(1)}{\cos^2 x} =$$
$$\frac{\cos^2 x}{\cos^2 x} - \frac{\sin^2 x}{\cos^2 x} =$$
$$1 - \tan^2 x =$$
$$1 - (\sec^2 x - 1) =$$
$$1 - \sec^2 x + 1 =$$
$$2 - \sec^2 x =$$

53. $$(\sec x + \tan x)^2 = \frac{(\sin x + 1)^2}{\cos^2 x}$$
$$\sec^2 x + 2\sec x \tan x + \tan^2 x =$$
$$\frac{1}{\cos^2 x} + \frac{2\sin x}{\cos^2 x} + \frac{\sin^2 x}{\cos^2 x} =$$
$$\frac{1 + 2\sin x + \sin^2 x}{\cos^2 x} =$$
$$\frac{(1 + \sin x)^2}{\cos^2 x} =$$
$$\frac{(\sin x + 1)^2}{\cos^2 x} =$$

55. $$\frac{\cos x}{\sin x} + \frac{\sin x}{\cos x} + \frac{\csc x}{\sec x} = \frac{\sec x + \cos x}{\sin x}$$
$$\frac{\cos^2 x \sec x + \sin^2 x \sec x + \csc x \sin x \cos x}{\sin x \cos x \sec x} =$$
$$\frac{\sec x(\cos^2 x + \sin^2 x) + (1)\cos x}{\sin x(1)} =$$
$$\frac{\sec x + \cos x}{\sin x} =$$

57. $\dfrac{\sin^4 x - \cos^4 x}{\sin^3 x + \cos^3 x} = \dfrac{\sin x - \cos x}{1 - \sin x \cos x}$

$\dfrac{(\sin^2 x + \cos^2 x)(\sin^2 x - \cos^2 x)}{(\sin x + \cos x)(\sin^2 x - \sin x \cos x + \cos^2 x)} =$

$\dfrac{(1)(\sin x + \cos x)(\sin x - \cos x)}{(\sin x + \cos x)(\sin^2 x + \cos^2 x - \sin x \cos x)} =$

$\dfrac{\sin x - \cos x}{1 - \sin x \cos x} =$

59. a. $d^2 = (20 + x\cos\theta)^2 + (20 - x\sin\theta)^2$
$= 400 + 40x\cos\theta + x^2\cos^2\theta + 400 - 40x\sin\theta + x^2\sin^2\theta$
$= 800 + 40x(\cos\theta - \sin\theta) + x^2(\cos^2\theta + \sin^2\theta)$
$= 800 + 40x(\cos\theta - \sin\theta) + x^2$

b. ≈ 42.2 ft

61. a. $h = \sqrt{\cot x + \tan x}$;
$h \approx 3.76$

b. $\cot x + \tan x = \dfrac{\cos x}{\sin x} + \dfrac{\sin x}{\cos x}$
$= \dfrac{\cos^2 x + \sin^2 x}{\sin x \cos x}$
$= \dfrac{1}{\sin x \cos x}$
$= \csc x \sec x$;
$h = \sqrt{\csc x \sec x}$
$h \approx 3.76$; yes

63. $D^2 = 400 + 40x\cos\theta + x^2$
$D \approx 40.5$ ft

65. $\sin\alpha = \cos\theta$ **67.** Answers will vary.

69. $\dfrac{\sin^6 x - \cos^6 x}{\sin^4 x - \cos^4 x} = 1 - \sin^2 x \cos^2 x$

$\dfrac{(\sin^3 x - \cos^3 x)(\sin^3 x + \cos^3 x)}{(\sin^2 x + \cos^2 x)(\sin^2 x - \cos^2 x)} =$

$\dfrac{(\sin x - \cos x)(\sin^2 x + \sin x \cos x + \cos^2 x)(\sin x + \cos x)(\sin^2 x - \sin x \cos x + \cos^2 x)}{(1)(\sin x + \cos x)(\sin x - \cos x)} =$

$(1 + \sin x \cos x)(1 - \sin x \cos x) = 1 - \sin^2 x \cos^2 x$

71. $\sin t = \dfrac{3}{4}, \cos t = \dfrac{\sqrt{7}}{4}, \tan t = \dfrac{3}{\sqrt{7}}$

73.

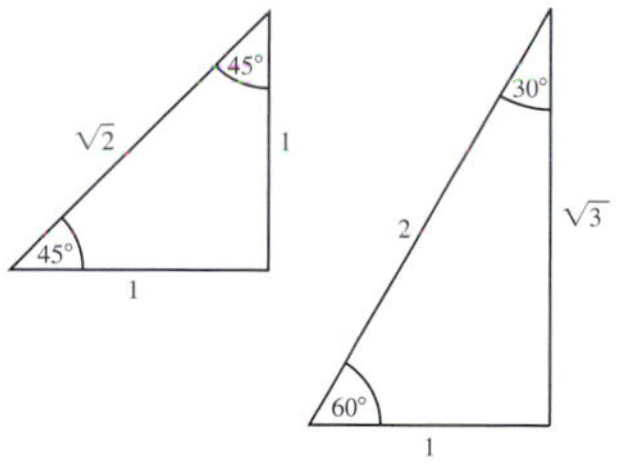

75.

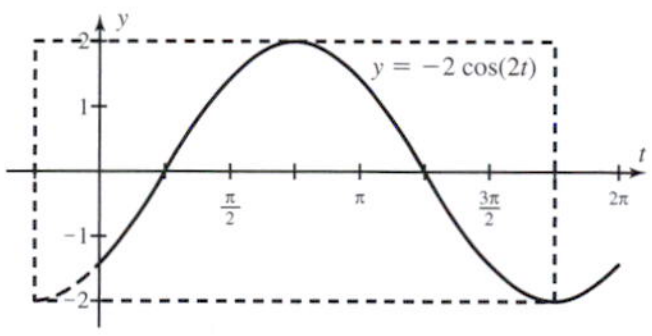

Mid-Chapter Check, p. 178

1. $\sin x[\csc x - \sin x] = \cos^2 x$
$\sin x \csc x - \sin^2 x =$
$\sin x \dfrac{1}{\sin x} - \sin^2 x = 1 - \sin^2 x = \cos^2 x$

2. $(1 + \sec t)(1 - \sec t) = \tan^2 t$
$1 - \sec^2 t =$
$\tan^2 t =$

3. $\cos^2 x - \cot^2 x = -\cos^2 x \cot^2 x$
$\cos^2 x - \dfrac{\cos^2 x}{\sin^2 x} =$
$\cos^2 x\left(1 - \dfrac{1}{\sin^2 x}\right) =$
$\cos^2 x(1 - \csc^2 x) =$
$\cos^2 x(-\cot^2 x) =$
$-\cos^2 x \cot^2 x =$

4. $(1 - \sin^4 t) = (1 + \sin^2 t)\cos^2 t$
$(1 + \sin^2 t)(1 - \sin^2 t) =$
$(1 + \sin^2 t)\cos^2 t =$

5. $\dfrac{2\sin x}{\sec x} - \dfrac{\cos x}{\csc x} = \cos x \sin x$

$\dfrac{2\sin x \csc x - \cos x \sec x}{\sec x \csc x} =$

$\dfrac{2(1) - 1}{\sec x \csc x} =$

$\dfrac{1}{\sec x \csc x} =$

$\cos x \sin x =$

6. $\dfrac{1 - \cos t}{\cos t} + \dfrac{\sec t - 1}{\sec t} = \sec t - \cos t$

$\dfrac{(\sec t - 1) + (1 - \cos t)}{\cos t \sec t} =$

$\sec t - \cos t =$

7. $\csc^2 x - \cot^2 x = 1$

$\dfrac{1}{\sin^2 x} - \dfrac{\cos^2 x}{\sin^2 x} =$

$\dfrac{1 - \cos^2 x}{\sin^2 x} =$

$\dfrac{\sin^2 x}{\sin^2 x} =$

8. $\dfrac{\tan x}{\sec x} = \sin x$

$\dfrac{\frac{\sin x}{\cos x}}{\frac{1}{\cos x}} =$

$\dfrac{\sin x}{\cos x} \cdot \dfrac{\cos x}{1} =$

9. $1 + \sec^2 x = \tan^2 x$
$1 + \sec^2 0 = \tan^2 0$
$1 + 1^2 = 0^2$
$1 + 1 = 0$
$2 = 0$ False

10. $\cos^2\frac{\pi}{6} = \sin^2\frac{\pi}{6} - 1$

$\left(\frac{\sqrt{3}}{2}\right)^2 = \left(\frac{1}{2}\right)^2 - 1$

$\frac{3}{4} = \frac{1}{4} - 1$

$\frac{3}{4} = -\frac{3}{4}$; false

Reinforcing Basic Concepts, pp. 178–179

Exercise 1. For α,

$a \approx 34$ mm

$b \approx 18.5$ mm

$34^2 + 18.5^2 = h^2$

$\sqrt{1498.25} = h$

$38.7 \approx h$

38.5 vs 38.7; 0.5%

For β

$a \approx 34$ mm

$b \approx 23.5$ mm

$34^2 + 23.5^2 = h^2$

$\sqrt{1708.25} = h$

$41.3 \approx h$

41.5 vs 41.3; 0.4%

Exercise 2. $\cos\alpha = \frac{34}{38.5}$; $\cos\beta = \frac{34}{41.5}$;

$\sin\alpha = \frac{18.5}{38.5}$; $\sin\beta = \frac{23.5}{41.5}$;

$\left(\frac{18.5}{38.5}\right)^2 + \left(\frac{34}{38.5}\right)^2 \approx 1$; $\left(\frac{34}{41.5}\right)^2 + \left(\frac{23.5}{41.5}\right)^2 \approx 1$;

Their sum is very close to 1.

Exercise 3. yes, yes **Exercise 4.** verified

Exercises 3.3, pp. 186–191

1. false **3.** repeat; alternate **5.** Answers will vary.

7. $\frac{\sqrt{2} - \sqrt{6}}{4}$ **9.** $\frac{\sqrt{2} - \sqrt{6}}{4}$

11. **a.** $\cos(45° + 30°) = \cos 45° \cos 30° - \sin 45° \sin 30° = \frac{\sqrt{6} - \sqrt{2}}{4}$;

b. $\cos(120° - 45°) = \cos 120° \cos 45° + \sin 120° \sin 45° = \frac{-\sqrt{2} + \sqrt{6}}{4} = \frac{\sqrt{6} - \sqrt{2}}{4}$

13. $\cos(5\theta)$ **15.** $\frac{\sqrt{3}}{2}$ **17.** $\frac{-16}{65}$ **19.** $\sin 33°$ **21.** $\cot\left(\frac{\pi}{12}\right)$

23. $\cos\left(\frac{\pi}{3} + \theta\right)$ **25.** $\frac{-\sqrt{2}}{2}$ **27.** $\frac{1}{2}$ **29.** $\sin(8x)$ **31.** $\tan(3\theta)$

33. 1 **35.** $\sqrt{3}$ **37.** **a.** $\frac{-304}{425}$ **b.** $\frac{-304}{297}$ **39.** $\frac{\sqrt{6} + \sqrt{2}}{4}$

41. $\frac{\sqrt{6} + \sqrt{2}}{4}$ **43.** $-\frac{1}{\sqrt{3}} = -\frac{\sqrt{3}}{3}$ **45.** $-\sqrt{3}$

47. **a.** $\sin(45° - 30°) = \sin 45° \cos 30° - \cos 45° \sin 30° = \frac{\sqrt{6} - \sqrt{2}}{4}$;

b. $\sin(135° - 120°) = \sin 135° \cos 120° - \cos 135° \sin 120°$

$= \left(\frac{\sqrt{2}}{2}\right)\left(-\frac{1}{2}\right) - \left(\frac{-\sqrt{2}}{2}\right)\left(\frac{\sqrt{3}}{2}\right)$

$= \frac{-\sqrt{2}}{4} + \frac{\sqrt{6}}{4}$

$= \frac{\sqrt{6} - \sqrt{2}}{4}$

49. $\frac{-\sqrt{2} - \sqrt{6}}{4}$ **51.** **a.** $\frac{319}{481}$ **b.** $\frac{480}{481}$ **c.** $-\frac{319}{360}$

53. **a.** $\frac{3416}{4505}$ **b.** $\frac{-1767}{4505}$ **c.** $\frac{3416}{2937}$ **55.** **a.** $\frac{12 + 5\sqrt{3}}{26}$

b. $\frac{12\sqrt{3} - 5}{26}$ **c.** $\frac{12 + 5\sqrt{3}}{12\sqrt{3} - 5}$

57. $(90° - \alpha) + \theta + (90° - \beta) = 180°$; **a.** $\frac{247}{265}$ **b.** $\frac{96}{265}$ **c.** $\frac{247}{96}$

59. $\sin(\pi - \alpha) = \sin\pi\cos\alpha - \cos\pi\sin\alpha$

$= 0 - (-1)\sin\alpha$

$= \sin\alpha$

61. $\cos\left(x + \frac{\pi}{4}\right) = \cos x\cos\left(\frac{\pi}{4}\right) - \sin x\sin\left(\frac{\pi}{4}\right) =$

$\cos x\left(\frac{\sqrt{2}}{2}\right) - \sin x\left(\frac{\sqrt{2}}{2}\right) = \frac{\sqrt{2}}{2}(\cos x - \sin x)$

63. $\tan\left(x + \frac{\pi}{4}\right) = \frac{\tan x + \tan\left(\frac{\pi}{4}\right)}{1 - \tan x\tan\left(\frac{\pi}{4}\right)} = \frac{\tan x + 1}{1 - \tan x} = \frac{1 + \tan x}{1 - \tan x}$

65. $\cos(\alpha + \beta) + \cos(\alpha - \beta) =$

$\cos\alpha\cos\beta - \cancel{\sin\alpha\sin\beta} + \cos\alpha\cos\beta + \cancel{\sin\alpha\sin\beta} =$

$2\cos\alpha\cos\beta$

67. $\cos(2t) = \cos^2 t - \sin^2 t$

$\cos(t + t) =$

$\cos t\cos t - \sin t\sin t =$

$\cos^2 t - \sin^2 t =$

69. $\sin(3t) = -4\sin^3 t + 3\sin t$

$\sin(2t + t) =$

$\sin(2t)\cos t + \cos(2t)\sin t =$

$2\sin t\cos t\cos t + (\cos^2 t - \sin^2 t)\sin t =$

$2\sin t\cos^2 t + \sin t\cos^2 t - \sin^3 t =$

$3\sin t\cos^2 t - \sin^3 t =$

$3\sin t(1 - \sin^2 t) - \sin^3 t =$

$3\sin t - 3\sin^3 t - \sin^3 t =$

$-4\sin^3 t + 3\sin t =$

71. $\cos\left(x - \frac{\pi}{4}\right) = \cos x\cos\left(\frac{\pi}{4}\right) + \sin x\sin\left(\frac{\pi}{4}\right)$

$= \cos x\left(\frac{\sqrt{2}}{2}\right) + \sin x\left(\frac{\sqrt{2}}{2}\right)$

$= \frac{\sqrt{2}}{2}(\cos x + \sin x)$

73. $F = \frac{Wk}{c}\,\frac{1 - \sqrt{3}}{1 + \sqrt{3}}$ **75.** $\frac{\cos h - 1}{h}$ **77.** $\frac{2\tan h}{h(1 - \tan h)}$

79. $\frac{f(x + h) - f(x)}{h} = \frac{\sin(x + h) - \sin x}{h}$

$= \frac{\sin x\cos h + \cos x\sin h - \sin x}{h} = \frac{\sin x\cos h - \sin x + \cos x\sin h}{h}$

$= \frac{\sin x(\cos h - 1) + \cos x\sin h}{h} = \sin x\frac{\cos h - 1}{h} + \cos x\frac{\sin h}{h}$

81. $R = \dfrac{\cos s \cos t}{\varpi C \sin(s+t)}$

$= \dfrac{\cos s \cos t}{\varpi C(\sin s \cos t + \cos s \sin t)}$

$= \dfrac{\cos s \cos t \dfrac{1}{\cos s \cos t}}{\varpi C(\sin s \cos t + \cos s \sin t)\dfrac{1}{\cos s \cos t}}$

$= \dfrac{1}{\varpi C\left(\dfrac{\sin s \cancel{\cos t}}{\cos s \cancel{\cos t}} + \dfrac{\cancel{\cos s} \sin t}{\cancel{\cos s} \cos t}\right)}$

$= \dfrac{1}{\varpi C(\tan s + \tan t)}$

83. $\dfrac{A}{B} = \dfrac{\sin\theta\cos(90° - \theta)}{\cos\theta\sin(90° - \theta)}$

$\dfrac{A}{B} = \dfrac{\sin\theta(\cos 90°\cos\theta + \sin 90°\sin\theta)}{\cos\theta(\sin 90°\cos\theta - \cos 90°\sin\theta)}$

$= \dfrac{\sin\theta(0 + \sin\theta)}{\cos\theta(\cos\theta - 0)}$

$= \dfrac{\sin^2\theta}{\cos^2\theta}$

$= \tan^2\theta$

85. $\beta = \dfrac{\pi}{2} - \alpha$

$\sin^2\alpha + \sin^2\beta = \sin^2\alpha + \sin^2\left(\dfrac{\pi}{2} - \alpha\right)$

$= \sin^2\alpha + \cos^2\alpha$

$= 1;$

answers will vary.

87. Answers will vary. **89. a.** $P = 16$ **b.** $P = \dfrac{\pi}{2}$

91.

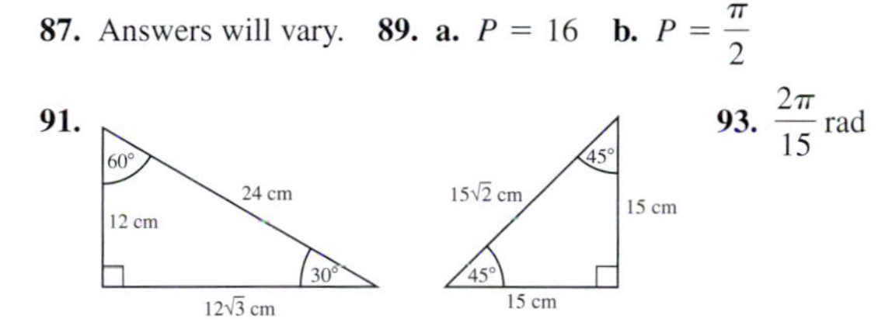

93. $\dfrac{2\pi}{15}$ rad

Exercises 3.4, pp. 201–206

1. sum; $\alpha = \beta$ **3.** $2x$; x **5.** Answers will vary.

7. $\sin(2\theta) = \dfrac{-120}{169}$, $\cos(2\theta) = \dfrac{119}{169}$, $\tan(2\theta) = \dfrac{-120}{119}$

9. $\sin(2\theta) = \dfrac{-720}{1681}$, $\cos(2\theta) = \dfrac{-1519}{1681}$, $\tan(2\theta) = \dfrac{720}{1519}$

11. $\sin(2\theta) = \dfrac{2184}{7225}$, $\cos(2\theta) = \dfrac{6887}{7225}$, $\tan(2\theta) = \dfrac{2184}{6887}$

13. $\sin(2\theta) = \dfrac{-5280}{5329}$, $\cos(2\theta) = \dfrac{721}{5329}$, $\tan(2\theta) = \dfrac{-5280}{721}$

15. $\sin(2\theta) = \dfrac{-24}{25}$, $\cos(2\theta) = \dfrac{7}{25}$, $\tan(2\theta) = \dfrac{-24}{7}$

17. $\sin\theta = \dfrac{4}{5}$, $\cos\theta = \dfrac{3}{5}$, $\tan\theta = \dfrac{4}{3}$

19. $\sin\theta = \dfrac{21}{29}$, $\cos\theta = \dfrac{20}{29}$, $\tan\theta = \dfrac{21}{20}$

21. $\sin(3\theta) = 3\sin\theta - 4\sin^3\theta$

$\sin(2\theta + \theta) = \sin(2\theta)\cos\theta + \cos(2\theta)\sin\theta$

$= (2\sin\theta\cos\theta)\cos\theta + (1 - 2\sin^2\theta)\sin\theta$

$= 2\sin\theta\cos^2\theta + \sin\theta - 2\sin^3\theta$

$= 2\sin\theta(1 - \sin^2\theta) + \sin\theta - 2\sin^3\theta$

$= 2\sin\theta - 2\sin^3\theta + \sin\theta - 2\sin^3\theta$

$= 3\sin\theta - 4\sin^3\theta$

23. $\dfrac{1}{4}$ **25.** $\dfrac{\sqrt{2}}{2}$ **27.** 1 **29.** $4.5\sin(6x)$ **31.** $\dfrac{1}{8} - \dfrac{1}{8}\cos(4x)$

33. $\dfrac{9}{8} + \dfrac{3}{2}\cos(2x) + \dfrac{3}{8}\cos(4x)$

35. $\dfrac{5}{8} - \dfrac{7}{8}\cos(2x) + \dfrac{3}{8}\cos(4x) - \dfrac{1}{8}\cos(2x)\cos(4x)$

37. $\sin\theta = \dfrac{\sqrt{2 - \sqrt{2}}}{2}$, $\cos\theta = \dfrac{\sqrt{2 + \sqrt{2}}}{2}$, $\tan\theta = \sqrt{2} - 1$

39. $\sin\theta = \dfrac{\sqrt{2 - \sqrt{3}}}{2}$, $\cos\theta = \dfrac{\sqrt{2 + \sqrt{3}}}{2}$, $\tan\theta = 2 - \sqrt{3}$

41. $\sin\theta = \dfrac{\sqrt{2 + \sqrt{2}}}{2}$, $\cos\theta = \dfrac{\sqrt{2 - \sqrt{2}}}{2}$, $\tan\theta = \sqrt{2} + 1$

43. $\sin\theta = \dfrac{\sqrt{2 + \sqrt{2}}}{2}$, $\cos\theta = \dfrac{\sqrt{2 - \sqrt{2}}}{2}$, $\tan\theta = \sqrt{2} + 1$

45. $\dfrac{\sqrt{2 - \sqrt{2 + \sqrt{2}}}}{2}$ **47.** $\dfrac{\sqrt{2 - \sqrt{2 + \sqrt{3}}}}{2}$

49. $\cos 15°$ **51.** $\tan 2\theta$ **53.** $\tan x$

55. $\sin\left(\dfrac{\theta}{2}\right) = \dfrac{3}{\sqrt{13}}$, $\cos\left(\dfrac{\theta}{2}\right) = \dfrac{2}{\sqrt{13}}$, $\tan\left(\dfrac{\theta}{2}\right) = \dfrac{3}{2}$

57. $\sin\left(\dfrac{\theta}{2}\right) = \dfrac{3}{\sqrt{10}}$, $\cos\left(\dfrac{\theta}{2}\right) = \dfrac{1}{\sqrt{10}}$, $\tan\left(\dfrac{\theta}{2}\right) = 3$

59. $\sin\left(\dfrac{\theta}{2}\right) = \dfrac{7}{\sqrt{74}}$, $\cos\left(\dfrac{\theta}{2}\right) = \dfrac{5}{\sqrt{74}}$, $\tan\left(\dfrac{\theta}{2}\right) = \dfrac{7}{5}$

61. $\sin\left(\dfrac{\theta}{2}\right) = \dfrac{1}{\sqrt{226}}$, $\cos\left(\dfrac{\theta}{2}\right) = \dfrac{15}{\sqrt{226}}$, $\tan\left(\dfrac{\theta}{2}\right) = \dfrac{1}{15}$

63. $\sin\left(\dfrac{\theta}{2}\right) = \dfrac{5}{\sqrt{29}}$, $\cos\left(\dfrac{\theta}{2}\right) = \dfrac{-2}{\sqrt{29}}$, $\tan\left(\dfrac{\theta}{2}\right) = -\dfrac{5}{2}$

65. $-\dfrac{1}{2}[\cos(4\theta) - \cos(12\theta)]$ **67.** $\cos(2t) + \cos(5t)$

69. $\cos(1540\pi t) + \cos(2418\pi t)$ **71.** $\dfrac{1 + \sqrt{3}}{2}$ **73.** $\dfrac{-1}{4}$

75. $2\sin\left(\dfrac{55}{2}k\right)\cos\left(\dfrac{27}{2}k\right)$ **77.** $-2\sin x\sin\left(\dfrac{x}{6}\right)$

79. $2\cos\left(\dfrac{2061}{2}\pi t\right)\cos\left(\dfrac{357}{2}\pi t\right)$ **81.** $\dfrac{-\sqrt{2}}{2}$

83. $\dfrac{2\sin x\cos x}{\cos^2 x - \sin^2 x} = \tan(2x)$

$\dfrac{\sin(2x)}{\cos(2x)} =$

$\tan(2x) =$

85. $(\sin x + \cos x)^2 = 1 + \sin(2x)$

$\sin^2 x + 2\sin x\cos x + \cos^2 x =$

$\sin^2 x + \cos^2 x + 2\sin x\cos x =$

$1 + 2\sin x\cos x =$

$1 + \sin(2x) =$

87.
$$\cos(8\theta) = \cos^2(4\theta) - \sin^2(4\theta)$$
$$\cos(2 \cdot 4\theta) =$$
$$\cos^2(4\theta) - \sin^2(4\theta) =$$

89.
$$\frac{\cos(2\theta)}{\sin^2\theta} = \cot^2\theta - 1$$
$$\frac{\cos^2\theta - \sin^2\theta}{\sin^2\theta} =$$
$$\frac{\cos^2\theta}{\sin^2\theta} - \frac{\sin^2\theta}{\sin^2\theta} =$$
$$\cot^2\theta - 1 =$$

91.
$$\tan(2\theta) = \frac{2}{\cot\theta - \tan\theta}$$
$$\frac{2\tan\theta}{1 - \tan^2\theta} =$$
$$\frac{(2\tan\theta)\frac{1}{\tan\theta}}{(1 - \tan^2\theta)\frac{1}{\tan\theta}} =$$
$$\frac{2}{\frac{1}{\tan\theta} - \tan\theta} =$$
$$\frac{2}{\cot\theta - \tan\theta} =$$

93. $\tan x + \cot x = 2\csc(2x)$
$$= \frac{2}{\sin(2x)}$$
$$= \frac{2}{2\sin x\cos x}$$
$$= \frac{1}{\sin x\cos x}$$
$$= \frac{\sin^2 x + \cos^2 x}{\sin x\cos x}$$
$$= \frac{\sin^2 x}{\cancel{\sin x}\cos x} + \frac{\cos^2 x}{\sin x\cancel{\cos x}}$$
$$= \frac{\sin x}{\cos x} + \frac{\cos x}{\sin x}$$
$$= \tan x + \cot x$$

95. $\cos^2\left(\frac{x}{2}\right) - \sin^2\left(\frac{x}{2}\right) = \cos x$
$$\cos\left(2 \cdot \frac{x}{2}\right) =$$
$$\cos x =$$

97. $1 - \sin^2(2\theta) = 1 - 4\sin^2\theta + 4\sin^4\theta$
$$= (1 - 2\sin^2\theta)^2$$
$$= [\cos(2\theta)]^2$$
$$= \cos^2(2\theta)$$
$$= 1 - \sin^2(2\theta)$$

99.
$$\frac{\sin(120\pi t) + \sin(80\pi t)}{\cos(120\pi t) - \cos(80\pi t)} = -\cot(20\pi t)$$
$$\frac{2\cancel{\sin(100\pi t)}\cos(20\pi t)}{-2\cancel{\sin(100\pi t)}\sin(20\pi t)} =$$
$$-\frac{\cos(20\pi t)}{\sin(20\pi t)} =$$
$$-\cot(20\pi t) =$$

101. a. $\mathcal{M} = \frac{2}{\sqrt{2 - \sqrt{3}}}$, $\mathcal{M} \approx 3.9$ **b.** $\mathcal{M} = \frac{2}{\sqrt{2 - \sqrt{2}}}$, $\mathcal{M} \approx 2.6$

c. $\theta = 60°$ **103. a.** $288 - 144\sqrt{2}$ ft ≈ 84.3 ft

b. $288 - 144\sqrt{2}$ ft ≈ 84.3 ft

105. $\cos[2\pi(1209)t] + \cos[2\pi(941)t]$; the ⊞ key

107. $\sin^2\alpha + (1 - \cos\alpha)^2 = \sin^2\alpha + 1 - 2\cos\alpha + \cos^2\alpha$
$= \sin^2\alpha + \cos^2\alpha + 1 - 2\cos\alpha = 1 + 1 - 2\cos\alpha = 2 - 2\cos\alpha$
$= 2(1 - \cos\alpha) = 4\left(\frac{1 - \cos\alpha}{2}\right) = 4\sin^2\left(\frac{\alpha}{2}\right) = \left[2\sin\left(\frac{\alpha}{2}\right)\right]^2$

109. $\sin(2\alpha) = \sin(\alpha + \alpha)$
$$= \sin\alpha\cos\alpha + \cos\alpha\sin\alpha$$
$$= \sin\alpha\cos\alpha + \sin\alpha\cos\alpha$$
$$= 2\sin\alpha\cos\alpha$$
$$\tan(\alpha + \beta) = \frac{\tan\alpha + \tan\beta}{1 - \tan\alpha\tan\beta}$$
$$\tan(\alpha + \alpha) = \frac{\tan\alpha + \tan\alpha}{1 - \tan\alpha\tan\alpha}$$
$$\tan(2\alpha) = \frac{2\tan\alpha}{1 - \tan^2\alpha}$$

111. $\frac{1}{2}[\cos(\alpha - \beta) - \cos(\alpha + \beta)] = \sin\alpha\sin\beta$

113.
$$d(t) = \left|6\sin\left(\frac{\pi t}{60}\right)\right|$$
$$= \left|6\sin\left(\frac{1}{2} \cdot \frac{\pi t}{30}\right)\right|$$
$$= \left|6\left(\pm\sqrt{\frac{1 - \cos\left(\frac{\pi t}{30}\right)}{2}}\right)\right|$$
$$= 6\sqrt{\frac{1 - \cos\left(\frac{\pi t}{30}\right)}{2}}$$
$$= \sqrt{36\frac{1 - \cos\left(\frac{\pi t}{30}\right)}{2}}$$
$$= \sqrt{18\left[1 - \cos\left(\frac{\pi t}{30}\right)\right]}$$

115. a. $\sin(2\theta - 90°) + 1$
$= \sin(2\theta)\cos 90° - \cos(2\theta)\sin 90° + 1$
$= 0 - \cos(2\theta) + 1$
$= 1 - \cos(2\theta)$

b. $2\sin^2\theta$
$= \sin^2\theta + \sin^2\theta$
$= 1 - \cos^2\theta + \sin^2\theta$
$= 1 - (\cos^2\theta - \sin^2\theta)$
$= 1 - \cos(2\theta)$

c. $1 + \sin^2\theta - \cos^2\theta$
$= 1 - (\cos^2\theta - \sin^2\theta)$
$= 1 - \cos(2\theta)$

d. $1 - \cos(2\theta) = 1 - \cos(2\theta)$

117. a. ≈ 0.9659; ≈ 0.9659

b.
$$\left(\frac{\sqrt{2 + \sqrt{3}}}{2}\right)^2 \stackrel{?}{=} \left(\frac{\sqrt{6} + \sqrt{2}}{4}\right)^2$$
$$\frac{2 + \sqrt{3}}{4} \stackrel{?}{=} \frac{6 + 2\sqrt{12} + 2}{16}$$
$$\frac{2 + \sqrt{3}}{4} \stackrel{?}{=} \frac{8 + 4\sqrt{3}}{16}$$
$$\frac{2 + \sqrt{3}}{4} = \frac{2 + \sqrt{3}}{4}$$

119. Answers will vary, one example is $\frac{\sqrt{3}}{3}$, 1, $-2 - \sqrt{3}$.

121. $\cos 15° = \dfrac{\sqrt{2+\sqrt{3}}}{2}$

$\cos 7.5° = \dfrac{\sqrt{2+\sqrt{2+\sqrt{3}}}}{2}$

$\cos 3.75° = \dfrac{\sqrt{2+\sqrt{2+\sqrt{2+\sqrt{3}}}}}{2} \approx 0.9979$

$\cos 1.875° = \dfrac{\sqrt{2+\sqrt{2+\sqrt{2+\sqrt{2+\sqrt{3}}}}}}{2} \approx 0.9995,$

they are getting close to 1.

123. Y_1 is increasing; Y_2 is defined on $(0, \pi)$

125. $\left(\dfrac{16}{65}\right)^2 + \left(\dfrac{63}{65}\right)^2 = \dfrac{256}{4225} + \dfrac{3969}{4225} = \dfrac{4225}{4225} = 1,$

$\tan\theta = \dfrac{63}{16}$; $\sec\theta = \dfrac{65}{16}$

$$1 + \left(\frac{63}{16}\right)^2 = \left(\frac{65}{16}\right)^2$$
$$1 + \frac{3969}{256} = \frac{4225}{256}$$
$$\frac{256}{256} + \frac{3969}{256} = \frac{4225}{256}$$

127. $\dfrac{\sqrt{2}-\sqrt{6}}{4}$

Summary and Concept Review, pp. 206–210

1. $\sin x(\csc x - \sin x) = \cos^2 x$
$$\sin x \csc x - \sin x \sin x =$$
$$\sin x \frac{1}{\sin x} - \sin^2 x =$$
$$1 - \sin^2 x =$$
$$\cos^2 x =$$

2. $\dfrac{\tan^2 x \csc x + \csc x}{\sec^2 x} = \csc x$
$$\frac{\csc x(\tan^2 x + 1)}{\sec^2 x} =$$
$$\frac{\csc x \sec^2 x}{\sec^2 x} =$$
$$\csc x =$$

3. $\dfrac{(\sec x - \tan x)(\sec x + \tan x)}{\csc x} = \sin x$
$$\frac{\sec^2 x + \sec x \tan x - \sec x \tan x - \tan^2 x}{\csc x} =$$
$$\frac{\sec^2 x - \tan^2 x}{\csc x} =$$
$$\frac{1 + \tan^2 x - \tan^2 x}{\csc x} =$$
$$\frac{1}{\csc x} =$$
$$\sin x =$$

4. $\dfrac{\sec^2 x}{\csc x} - \sin x = \dfrac{\tan^2 x}{\csc x}$
$$\frac{\sec^2 x - \sin x \csc x}{\csc x} =$$
$$\frac{\sec^2 x - 1}{\csc x} =$$
$$\frac{\tan^2 x}{\csc x} =$$

5. $\sin\theta = \dfrac{-35}{37}$, $\csc\theta = \dfrac{-37}{35}$, $\cot\theta = \dfrac{12}{35}$, $\tan\theta = \dfrac{35}{12}$, $\sec\theta = \dfrac{-37}{12}$

6. $\sin\theta = \dfrac{-4\sqrt{6}}{25}$, $\csc\theta = \dfrac{-25}{4\sqrt{6}}$, $\cot\theta = \dfrac{-23}{4\sqrt{6}}$, $\tan\theta = -\dfrac{4\sqrt{6}}{23}$, $\cos\theta = \dfrac{23}{25}$

7. $\dfrac{1+\cos x}{\sin x}$; answers will vary **8.** $\sec x - \tan x$; answers will vary

9. $\dfrac{\csc^2 x(1-\cos^2 x)}{\tan^2 x} = \cot^2 x$
$$\frac{\csc^2 x \sin^2 x}{\tan^2 x} =$$
$$\frac{1}{\tan^2 x} =$$
$$\cot^2 x =$$

10. $\dfrac{\cot x}{\sec x} - \dfrac{\csc x}{\tan x} = \cot x(\cos x - \csc x)$
$$\cot x \frac{1}{\sec x} - \cot x \csc x =$$
$$\cot x \cos x - \cot x \csc x =$$
$$\cot x(\cos x - \csc x) =$$

11. $\dfrac{\sin^4 x - \cos^4 x}{\sin x \cos x} = \tan x - \cot x$
$$= \frac{(\sin^2 x - \cos^2 x)(\sin^2 x + \cos^2 x)}{\sin x \cos x}$$
$$= \frac{(\sin^2 x - \cos^2 x)(1)}{\sin x \cos x}$$
$$= \frac{\sin x \cancel{\sin x}}{\cancel{\sin x}\cos x} - \frac{\cancel{\cos x}\cos x}{\sin x \cancel{\cos x}}$$
$$= \frac{\sin x}{\cos x} - \frac{\cos x}{\sin x}$$
$$= \tan x - \cot x$$

12. $\dfrac{(\sin x + \cos x)^2}{\sin x \cos x} = \csc x \sec x + 2$
$$\frac{\sin^2 x + 2\sin x\cos x + \cos^2 x}{\sin x \cos x} =$$
$$\frac{\sin^2 x + \cos^2 x}{\sin x \cos x} + \frac{2\sin x \cos x}{\sin x \cos x} =$$
$$\frac{1}{\sin x \cos x} + 2 =$$
$$\csc x \sec x + 2 =$$

13. a. $\cos 75° = \dfrac{\sqrt{6}-\sqrt{2}}{4}$

b. $\tan\left(\dfrac{\pi}{12}\right) = \dfrac{\sqrt{3}-1}{1+\sqrt{3}} = \dfrac{(\sqrt{3}-1)^2}{2} = 2 - \sqrt{3}$

14. a. $\tan 15° = \dfrac{\sqrt{3}-1}{1+\sqrt{3}} = \dfrac{(\sqrt{3}-1)^2}{2} = 2-\sqrt{3}$

b. $\sin\left(\dfrac{-\pi}{12}\right) = \dfrac{\sqrt{2}-\sqrt{6}}{4}$ **15. a.** $\cos 180° = -1$ **b.** $\sin 120° = \dfrac{\sqrt{3}}{2}$

16. a. $\cos x$ **b.** $\sin\left(\dfrac{5x}{8}\right)$ **17. a.** $\cos 1170° = \cos 90° = 0$

b. $\sin\left(\dfrac{57\pi}{4}\right) = \sin\left(\dfrac{\pi}{4}\right) = \dfrac{\sqrt{2}}{2}$ **18. a.** $\cos\left(\dfrac{x}{8}\right) = \sin\left(\dfrac{\pi}{2} - \dfrac{x}{8}\right)$

b. $\sin\left(x - \dfrac{\pi}{12}\right) = \cos\left(\dfrac{7\pi}{12} - x\right)$

19. $\tan(45° - 30°) = \dfrac{\tan 45° - \tan 30°}{1 + \tan 45° \tan 30°}$

$= \dfrac{1 - \frac{\sqrt{3}}{3}}{1 + 1\cdot\frac{\sqrt{3}}{3}} = \dfrac{1 - \frac{\sqrt{3}}{3}}{1 + \frac{\sqrt{3}}{3}} = \dfrac{\frac{3-\sqrt{3}}{3}}{\frac{3+\sqrt{3}}{3}}$

$= \dfrac{3-\sqrt{3}}{3}\cdot\dfrac{3}{3+\sqrt{3}} = \dfrac{3-\sqrt{3}}{3+\sqrt{3}} = \dfrac{\sqrt{3}(\sqrt{3}-1)}{\sqrt{3}(\sqrt{3}+1)} = \dfrac{\sqrt{3}-1}{\sqrt{3}+1}$

$\tan(135° - 120°) = \dfrac{\tan 135° - \tan 120°}{1 + \tan 135° \tan 120°}$

$= \dfrac{-1 + \sqrt{3}}{1 + (-1)(-\sqrt{3})} = \dfrac{\sqrt{3}-1}{1+\sqrt{3}} = \dfrac{\sqrt{3}-1}{\sqrt{3}+1}$

20. $\cos\left(x + \frac{\pi}{6}\right) + \cos\left(x - \frac{\pi}{6}\right) = \sqrt{3}\cos x$

$= \cos x\cos\left(\frac{\pi}{6}\right) - \sin x\sin\left(\frac{\pi}{6}\right) + \cos x\cos\left(\frac{\pi}{6}\right) + \sin x\sin\left(\frac{\pi}{6}\right)$

$= 2\cos x\cos\left(\frac{\pi}{6}\right) + 0 = 2\cos x\left(\frac{\sqrt{3}}{2}\right) = \sqrt{3}\cos x$

21. a. $\sin(2\theta) = 2\left(\frac{-84}{85}\right)\left(\frac{13}{85}\right) = \frac{-2184}{7225}$

$\cos(2\theta) = \left(\frac{13}{85}\right)^2 - \left(\frac{84}{85}\right)^2 = \frac{-6887}{7225}$

$\tan(2\theta) = \frac{2184}{-7225}\left(\frac{7225}{-6887}\right) = \frac{2184}{6887}$

b. $\sin(2\theta) = 2\left(\frac{-20}{29}\right)\left(\frac{-21}{29}\right) = \frac{840}{841}$

$\cos(2\theta) = \left(\frac{-21}{29}\right)^2 - \left(\frac{-20}{29}\right)^2 = \frac{441-400}{841} = \frac{41}{841}$

$\tan(2\theta) = \dfrac{2\left(\frac{20}{21}\right)}{1 - \left(\frac{20}{21}\right)^2} = \dfrac{840}{41}$

22. a. $\sin\theta = \frac{21}{29}, \cos\theta = \frac{-20}{29},$ $\tan\theta = -\frac{21}{20},$ **b.** $\sin\theta = \frac{7}{25}$ or $\sin\theta = \frac{24}{25}, \cos\theta = \frac{-24}{25}$ or $\cos\theta = \frac{-7}{25},$ $\tan\theta = \frac{-7}{24}$ or $\tan\theta = \frac{-24}{7}$ **23. a.** $\cos 45° = \frac{\sqrt{2}}{2}$ **b.** $\cos\left(\frac{\pi}{6}\right) = \frac{\sqrt{3}}{2}$

24. a. $\sin 67.5 = \sqrt{\dfrac{1 - \cos 135°}{2}} = \sqrt{\dfrac{1 + \frac{\sqrt{2}}{2}}{2}} = \sqrt{\dfrac{2+\sqrt{2}}{4}}$

$= \dfrac{\sqrt{2+\sqrt{2}}}{2}$

$\cos 67.5 = \sqrt{\dfrac{1 + \cos 135°}{2}} = \sqrt{\dfrac{1 - \frac{\sqrt{2}}{2}}{2}} = \sqrt{\dfrac{2-\sqrt{2}}{4}}$

$= \dfrac{\sqrt{2-\sqrt{2}}}{2}$

b. $\sin\left(\frac{5\pi}{8}\right) = \sqrt{\dfrac{1 - \cos\left(\frac{5\pi}{4}\right)}{2}} = \sqrt{\dfrac{1 + \frac{\sqrt{2}}{2}}{2}} = \sqrt{\dfrac{2+\sqrt{2}}{4}}$

$= \dfrac{\sqrt{2+\sqrt{2}}}{2}$

$\cos\left(\frac{5\pi}{8}\right) = -\sqrt{\dfrac{1 + \cos\left(\frac{5\pi}{4}\right)}{2}} = -\sqrt{\dfrac{1 - \frac{\sqrt{2}}{2}}{2}} = -\sqrt{\dfrac{2-\sqrt{2}}{4}}$

$= -\dfrac{\sqrt{2-\sqrt{2}}}{2}$

25. a. $\sin\left(\frac{\theta}{2}\right) = \sqrt{\dfrac{1 - 24/25}{2}} = \sqrt{\dfrac{25-24}{50}} = +\dfrac{1}{5\sqrt{2}}, \dfrac{\theta}{2}$ in QII

$\cos\left(\frac{\theta}{2}\right) = -\sqrt{\dfrac{1 + 24/25}{2}} = -\sqrt{\dfrac{25+24}{50}}$

$= -\sqrt{\dfrac{49}{50}} = \dfrac{-7}{5\sqrt{2}}, \dfrac{\theta}{2}$ in QII

b. $\sin\left(\frac{\theta}{2}\right) = -\sqrt{\dfrac{1 - 56/65}{2}} = -\sqrt{\dfrac{65-56}{130}}$

$= -\sqrt{\dfrac{9}{130}} = \dfrac{-3}{\sqrt{130}}, \dfrac{\theta}{2}$ in QIV

$\cos\left(\frac{\theta}{2}\right) = \sqrt{\dfrac{1 + 56/65}{2}} = \sqrt{\dfrac{65+56}{130}} = \sqrt{\dfrac{121}{130}} = +\dfrac{11}{\sqrt{130}}, \dfrac{\theta}{2}$ in QIV

26. $\dfrac{2\tan^2\alpha}{\sec^2\alpha - 2} = \dfrac{\cos(3\alpha) - \cos\alpha}{\cos(3\alpha) + \cos\alpha} = \dfrac{-2\sin(2\alpha)\sin\alpha}{2\cos(2\alpha)\cos\alpha}$

$= \dfrac{-2\sin^2\alpha}{\cos^2\alpha - \sin^2\alpha} = \dfrac{2\sin^2\alpha}{\sin^2\alpha - \cos^2\alpha} = \dfrac{2\sin^2\alpha}{1 - 2\cos^2\alpha}$

$= \dfrac{2\tan^2\alpha}{\sec^2\alpha - 2}$

27. $\cos(3x) + \cos x = 0 \to 2\cos(2x)\cos x = 0$

$\cos(2x) = 0: x = \frac{\pi}{4} + \frac{\pi}{2}k; k \in \mathbb{Z}$

$\cos x = 0: x = \frac{\pi}{2} + \pi k; k \in \mathbb{Z}$

28. a. $A = 12^2\sin\left(\frac{30°}{2}\right)\cos\left(\frac{30°}{2}\right) = 144\sqrt{\dfrac{1 - \cos 30°}{2}}\sqrt{\dfrac{1 + \cos 30°}{2}}$

$= 144\sqrt{\dfrac{1 - \frac{\sqrt{3}}{2}}{2}}\sqrt{\dfrac{1 + \frac{\sqrt{3}}{2}}{2}} = 144\sqrt{\dfrac{2-\sqrt{3}}{4}}\sqrt{\dfrac{2+\sqrt{3}}{4}}$

$= \dfrac{144\sqrt{4-3}}{4} = 36\text{ cm}^2$ **b.** $x^2\sin\left(\frac{\theta}{2}\right)\cos\left(\frac{\theta}{2}\right)$

Let $u = \frac{\theta}{2}$, then $= x^2\sin u\cos u = \frac{1}{2}x^2(2\sin u\cos u) = \frac{1}{2}x^2\sin(2u)$

$= \frac{1}{2}x^2\sin\theta;\ A = \frac{1}{2}(12)^2\sin(30°) = 72\left(\frac{1}{2}\right) = 36\text{ cm}^2$

Mixed Review, pp. 210–211

1. $\sin\theta = \frac{6}{\sqrt{117}}, \sec\theta = \frac{-\sqrt{117}}{9}, \tan\theta = \frac{-6}{9} = \frac{-2}{3};$

$\cos\theta = \frac{-9}{\sqrt{117}}, \csc\theta = \frac{\sqrt{117}}{6}, \cot\theta = \frac{-3}{2}$

3. $\sqrt{3} + 2$ **5.** $\cos\left[2\left(\frac{\pi}{12}\right)\right] = \cos\left(\frac{\pi}{6}\right) = \frac{\sqrt{3}}{2}$

7. $\dfrac{1 - (\cos^2\theta - \sin^2\theta)}{\tan^2\theta} = \dfrac{1 - \cos(2\theta)}{\frac{1 - \cos(2\theta)}{1 + \cos(2\theta)}} = \cancel{1 - \cos(2\theta)}\,\dfrac{1 + \cos(2\theta)}{\cancel{1 - \cos(2\theta)}}$

$= 1 + \cos(2\theta)$

9. $\sin\left(\frac{x}{2}\right) = \frac{-2}{\sqrt{5}}; \cos\left(\frac{x}{2}\right) = \frac{1}{\sqrt{5}}$

11. $\sin(2\alpha) = \sin(\alpha + \alpha) = \sin\alpha\cos\alpha + \sin\alpha\cos\alpha$

$= 2\sin\alpha\cos\alpha$

13. $\dfrac{-\sqrt{3}\sqrt{2-\sqrt{2}}}{2}$ **15.** $\dfrac{1+\sqrt{2}}{4}$

17. $\sin(\alpha+\beta)\sin(\alpha-\beta) = (\sin\alpha\cos\beta+\cos\alpha\sin\beta)(\sin\alpha\cos\beta-\cos\alpha\sin\beta)$
$= \sin^2\alpha\cos^2\beta - \cos^2\alpha\sin^2\beta$
$= \sin^2\alpha(1-\sin^2\beta) - (1-\sin^2\alpha)\sin^2\beta$
$= \sin^2\alpha - \sin^2\alpha\sin^2\beta - \sin^2\beta + \sin^2\alpha\sin^2\beta$
$= \sin^2\alpha - \sin^2\beta$

19. $R = \dfrac{1}{16}v^2\sin\theta\cos\theta = \dfrac{1}{2}\cdot\dfrac{1}{16}v^2(2)\sin\theta\cos\theta$
$= \dfrac{1}{32}v^2\sin(2\theta)$

Practice Test, pp. 211–212

1. $\dfrac{(\csc x-\cot x)(\csc x+\cot x)}{\sec x} = \cos x$

$\dfrac{\csc^2 x+\csc x\cot x-\csc x\cot x-\cot^2 x}{\sec x} =$

$\dfrac{\csc^2 x-\cot^2 x}{\sec x} = \dfrac{(1+\cot^2 x)-\cot^2 x}{\sec x} =$

$\dfrac{1}{\sec x} =$

$\cos x =$

2. $\dfrac{\sin^3 x-\cos^3 x}{1+\cos x\sin x} = \sin x-\cos x$

$\dfrac{(\sin x-\cos x)(\sin^2 x+\sin x\cos x+\cos^2 x)}{1+\cos x\sin x} =$

$\dfrac{(\sin x-\cos x)(1+\sin x\cos x)}{1+\cos x\sin x} =$

$\sin x-\cos x =$

3. $\sin\theta = \dfrac{-55}{73}$, $\sec\theta = \dfrac{73}{48}$, $\cot\theta = \dfrac{-48}{55}$, $\tan\theta = \dfrac{-55}{48}$, $\csc\theta = \dfrac{-73}{55}$

4. $\dfrac{\sqrt{3}-1}{\sqrt{3}+1}$ **5.** $\dfrac{\sqrt{2}}{2}$ **6.** $\dfrac{-\sqrt{2}}{2}$

7. $\sin\left(x+\dfrac{\pi}{4}\right) - \sin\left(x-\dfrac{\pi}{4}\right) = \sqrt{2}\cos x$

$\sin x\cos\left(\dfrac{\pi}{4}\right) + \cos x\sin\left(\dfrac{\pi}{4}\right) - \sin x\cos\left(\dfrac{\pi}{4}\right) + \cos x\sin\left(\dfrac{\pi}{4}\right) =$

$\sin\left(\dfrac{\pi}{4}\right)\cos x + \sin\left(\dfrac{\pi}{4}\right)\cos x =$

$2\sin\left(\dfrac{\pi}{4}\right)\cos x =$

$2\dfrac{\sqrt{2}}{2}\cos x =$

$\sqrt{2}\cos x =$

8. $\sin\theta = \dfrac{15}{17}$, $\cos\theta = \dfrac{8}{17}$, $\tan\theta = \dfrac{15}{8}$ **9.** $\dfrac{-\sqrt{3}}{2}$ **10.** $\dfrac{1}{\sqrt{37}}; \dfrac{6}{\sqrt{37}}$

11. $20\sqrt{2-\sqrt{2}}$ **12.** $\dfrac{\sqrt{6}-\sqrt{2}}{4} \approx 0.2588$; $\dfrac{\sqrt{6}+\sqrt{2}}{4} \approx 0.9659$

13. $\dfrac{\tan\theta+\cot\theta}{\sin\theta\cos\theta} = \csc^2\theta\sec^2\theta$

$\dfrac{\dfrac{\sin\theta}{\cos\theta}+\dfrac{\cos\theta}{\sin\theta}}{\sin\theta\cos\theta} =$

$\dfrac{\dfrac{\sin^2\theta+\cos^2\theta}{\sin\theta\cos\theta}}{\sin\theta\cos\theta} =$

$\dfrac{1}{\sin\theta\cos\theta}\,\dfrac{1}{\sin\theta\cos\theta} =$

$\dfrac{1}{\sin^2\theta}\,\dfrac{1}{\cos^2\theta} =$

$\csc^2\theta\sec^2\theta =$

14. $-1(2\cos^4\theta - 3\cos^2\theta + 1) =$
$-1(\cos^2\theta - 1)(2\cos^2\theta - 1) =$
$(1-\cos^2\theta)(2\cos^2\theta - 1) =$
$\sin^2\theta\cos^2(2\theta) =$

15. $\dfrac{3-\sqrt{3}}{3+\sqrt{3}}$ or $\dfrac{\sqrt{3}-1}{\sqrt{3}+1}$ **16.** $\dfrac{\sqrt{2}}{2}$

17. $x = -35$, $y = -12$, $r = 37$

$\sin(2\theta) = \dfrac{840}{1369}$, $\cos(2\theta) = \dfrac{1081}{1369}$, $\tan(2\theta) = \dfrac{840}{1081}$

18. $\dfrac{\csc^2 x - 2}{2\cot^2 x - \csc^2 x} = 1$

$\dfrac{1+\cot^2 x - 2}{2\cot^2 x - (1+\cot^2 x)} =$

$\dfrac{\cot^2 x - 1}{2\cot^2 x - 1 - \cot^2 x} =$

$\dfrac{\cot^2 x - 1}{\cot^2 x - 1} =$

$1 =$

19. a. The "0" (zero) was pressed **b.** $y(t) = 2\cos(2277\pi t)\cos(395\pi t)$

20. $\cos(2418\pi t) + \cos(1540\pi t)$

Cumulative Review Chapters 1–3, pp. 214–215

1.

Angles	Sides
$\alpha = 30°$	$a = 20$ m
$\beta = 60°$	$b = 20\sqrt{3}$ m
$\gamma = 90°$	$c = 40$ m

3. $\sin^2 x + \cos^2 x = 1$; $1 + \cot^2 x = \csc^2 x$; $\tan^2 x + 1 = \sec^2 x$

5. about 15.7 ft/sec

7.

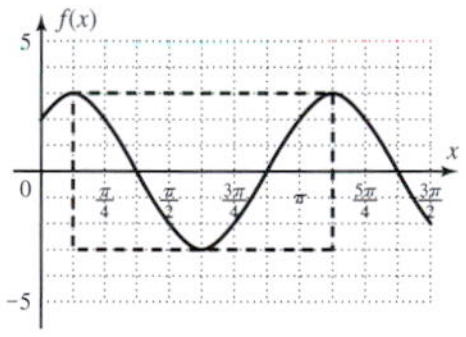

9. $\cos^2\left(\dfrac{\alpha}{2}\right) = \dfrac{\sec\alpha + 2 + \cos\alpha}{2\sec\alpha + 2}$

$= \dfrac{\sec\alpha + 2 + \cos\alpha}{2\sec\alpha + 2}\,\dfrac{\cos\alpha}{\cos\alpha}$

$= \dfrac{1 + 2\cos\alpha + \cos^2\alpha}{2 + 2\cos\alpha}$

$= \dfrac{(1+\cos\alpha)(1+\cos\alpha)}{2(1+\cos\alpha)}$

$= \dfrac{(1+\cos\alpha)}{2}$; let $\alpha = 2\theta$

11. $\sin(2x) = 2\sin x\cos x$
$\cos(2x) = \cos^2 x - \sin^2 x = 2\cos^2 x - 1 = 1 - 2\sin^2 x$

13. $\sin 195° = \frac{\sqrt{2} - \sqrt{6}}{4}$

$\cos 195° = \frac{-\sqrt{2} - \sqrt{6}}{4}$

15. a. just over 15 hr **b.** April 15 to September 1 or 2

17. $\cot x\left(\tan x - \frac{\sin x}{\cos^3 x}\right) = \tan^2 x$

$1 - \frac{\cos x}{\sin x}\frac{\sin x}{\cos^3 x} =$

$1 - \frac{1}{\cos^2 x} =$

$1 - \sec^2 x =$

$\tan^2 x =$

19. about 1034 km

CHAPTER 4

Exercises 4.1, pp. 225–230

1. second; one **3.** $(-11, -2), (-5, 0), (1, 2), (19, 4)$ **5.** False, answers will vary. **7.** one-to-one **9.** one-to-one **11.** not a function (cannot be a one-to-one *function*) **13.** one-to-one **15.** one-to-one **17.** not one-to-one, $y = 7$ is paired with $x = -2$ and $x = 2$ **19.** one-to-one **21.** one-to-one **23.** not one-to-one, $p(t) > 5$ corresponds to two x-values **25.** one-to-one **27.** one-to-one **29.** $f^{-1}(x) = \{(1, -2), (4, -1), (5, 0), (9, 2), (15, 5)\}$ **31.** $v^{-1}(x) = \{(3, -4), (2, -3), (1, 0), (0, 5), (-1, 12), (-2, 21), (-3, 32)\}$ **33.** $f^{-1}(x) = x - 5$ **35.** $p^{-1}(x) = \frac{-5}{4}x$

37. $f^{-1}(x) = \frac{x-3}{4}$ **39.** $Y_1^{-1} = x^3 + 4$ **41.** $f^{-1}(x) = \frac{x-7}{2}$

43. $f^{-1}(x) = x^2 + 2; x \geq 0$ **45.** $f^{-1}(x) = \sqrt{x-3}; x \geq 3$

47. $f^{-1}(x) = \sqrt[3]{x-1}$ **49.** $(f \circ g)(x) = x, (g \circ f)(x) = x$

51. $(f \circ g)(x) = x, (g \circ f)(x) = x$ **53.** $(f \circ g)(x) = x, (g \circ f)(x) = x$

55. $(f \circ g)(x) = x, (g \circ f)(x) = x$ **57.** $f^{-1}(x) = \frac{x+5}{3}$

59. $f^{-1}(x) = 2x + 5$ **61.** $f^{-1}(x) = 2x + 6$ **63.** $f^{-1}(x) = \sqrt[3]{x-3}$

65. $f^{-1}(x) = \frac{x^3 - 1}{2}$ **67.** $f^{-1}(x) = 2\sqrt[3]{x} + 1$

69. $f^{-1}(x) = \frac{x^2 - 2}{3}, x \geq 0$ **71.** $p^{-1}(x) = \frac{x^2}{4} + 3; x \geq 0$

73. $v^{-1}(x) = \sqrt{x-3}$

75. **77.**

79. **81.**

83. $D: x \in [0, \infty), R: y \in [-2, \infty)$;
$D: x \in [-2, \infty), R: y \in [0, \infty)$

85. $D: x \in (0, \infty), R: y \in (-\infty, \infty)$;
$D: x \in (-\infty, \infty), R: y \in (0, \infty)$

87. $D: x \in (-\infty, 4], R: y \in (-\infty, 4]$;
$D: x \in (-\infty, 4], R: y \in (-\infty, 4]$

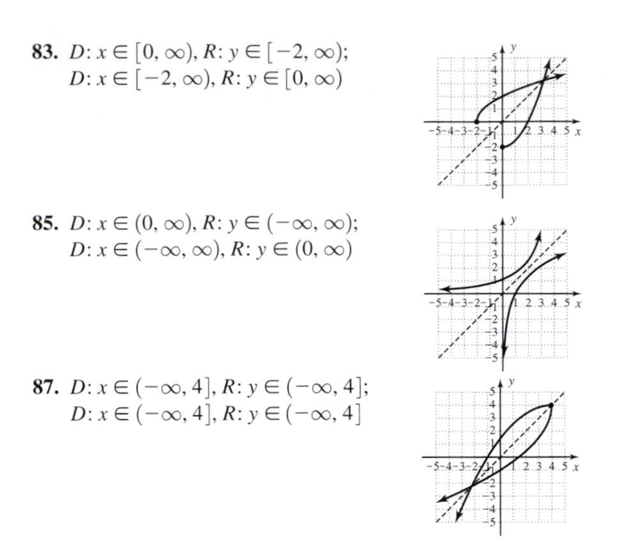

89. a. 31.5 cm **b.** The result is 80 cm. It gives the distance of the projector from the screen. **91. a.** $-63.5°$F **b.** $f^{-1}(x) = \frac{-2}{7}(x - 59)$; it is 35 **c.** 22,000 ft **93. a.** 144 ft **b.** $f^{-1}(x) = \frac{\sqrt{x}}{4}$, 3 sec, the original input for $f(x)$ **c.** 7 sec **95. a.** 28,260 ft^3 **b.** $f^{-1}(h) = \sqrt[3]{\frac{3h}{\pi}}$, 30 ft, the original input for $f(h)$ **c.** 9 ft **97. a.** 5 cm **b.** $f^{-1}(x) = \sqrt[3]{\frac{x}{\pi}}$, $f^{-1}(392.5) \approx 5$; same as original input for $f(x)$ **c.** $f^{-1}(x)$

99. a. verified **b.**

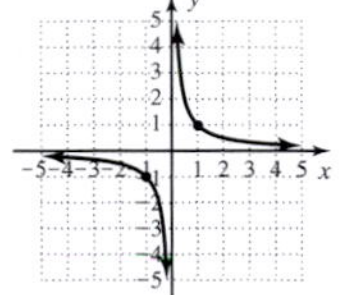

c. $(1, 1)$ and $(-1, -1)$; x and y coordinates are identical on $f(x) = x$

101. a. 0 **b.** 3 **c.** 81 **d.** 3 **103.** $\cos\beta = \frac{7\sqrt{2}}{10}, \sin\beta = \frac{\sqrt{2}}{10}$

105. $\sin\theta = \frac{15}{17}, \cos\theta = \frac{-8}{17}, \tan\theta = \frac{15}{-8}$,
$\csc\theta = \frac{17}{15}, \sec\theta = \frac{17}{-8}, \cot\theta = \frac{-8}{15}$ **107.** $\theta = \frac{5\pi}{6}$

Exercises 4.2, pp. 241–246

1. horizontal line; one; one **3.** $[-1, 1]; \left[-\frac{\pi}{2}, \frac{\pi}{2}\right]$ **5.** $\cos^{-1}(\frac{1}{5})$

7. $0; \frac{1}{2}; -\frac{\pi}{6}; -\frac{\pi}{2}$ **9.** $\frac{\pi}{4}$ **11.** $\frac{\pi}{2}$ **13.** 1.0956, 62.8° **15.** 0.3876, 22.2°

17. $\frac{\sqrt{2}}{2}$ **19.** $\frac{\pi}{3}$ **21.** 45° **23.** 0.8205 **25.** $0; \frac{\sqrt{3}}{2}$; 120°; π **27.** $\frac{\pi}{3}$

29. π **31.** 1.4352; 82.2° **33.** 0.7297; 41.8° **35.** $\frac{\pi}{4}$ **37.** 0.5560

39. $-\frac{\sqrt{2}}{2}$ **41.** $\frac{3\pi}{4}$ **43.** $0; -\sqrt{3}$; 30°; $\sqrt{3}; \frac{\pi}{3}$ **45.** $-\frac{\pi}{6}$ **47.** $\frac{\pi}{3}$

49. $-1.1170, -64.0°$ **51.** 0.9441, 54.1° **53.** $-\frac{\pi}{6}$ **55.** $\frac{\sqrt{3}}{3}$ **57.** $\sqrt{2}$

59. 120° **61.** cannot evaluate $\tan\left(\frac{\pi}{2}\right)$ **63.** $\csc\frac{\pi}{4} = \sqrt{2} > 1$, not in domain of $\sin^{-1}x$.

65. $\sin\theta = \frac{3}{5}, \cos\theta = \frac{4}{5}, \tan\theta = \frac{3}{4}$

67. $\sin\theta = \frac{\sqrt{x^2-36}}{x}, \cos\theta = \frac{6}{x}, \tan\theta = \frac{\sqrt{x^2-36}}{6}$

69. $\frac{24}{25}$ **71.** $\frac{\sqrt{5}}{3}$

73. $\frac{\sqrt{25-9x^2}}{3x}$ **75.** $\sqrt{\frac{12}{12+x^2}}$

77. $\csc\theta; D = \left(-\frac{\pi}{2}, 0\right)\cup\left(0, \frac{\pi}{2}\right)$ **79.** $\tan\theta; D = \left[0, \frac{\pi}{2}\right)\cup\left(\frac{\pi}{2}, \pi\right]$

81. 0; 2; 30°; −1; π **83.** $\frac{\pi}{6}$ **85.** $\frac{\pi}{6}$ **87.** 80.1° **89.** 67.8°

91. a. $F_N \approx 2.13$ N; $F_N \approx 1.56$ N **b.** $\theta \approx 63°$ for $F_N = 1$ N, $\theta \approx 24.9°$ for $F_N = 2$ N **93.** ≈30° **95.** $\theta \approx 72.3°$; straight line distance ≈ 157.5 yd

97. a. $Y_2 = \tan\theta; -\frac{\pi}{2} < \theta < \frac{\pi}{2}$ **b.** $Y_2 = \tan\left[\sin^{-1}\left(\frac{x}{10}\right)\right]$

c. verified **99.** $d^2 + r^2 = (x+r)^2, d^2 + r^2 = x^2 + 2rx + r^2, d^2 = x^2 + 2rx, d = \sqrt{2rx + x^2}$ **101. a.** $\theta \approx 15.5°; \theta \approx 0.2705$ rad **b.** ≈29 mi **103. a.** 413.6 ft away **b.** −503 ft **c.** ≈651.2 ft

105. $\sin(2\theta) = \frac{84}{85}$ **107.** about 5 miles per hour

109. $\sin\theta = \frac{12}{13}, \cos\theta = \frac{5}{13}, \tan\theta = \frac{12}{5}$

Exercises 4.3, pp. 256–261

1. principal; $[0, 2\pi)$; real **3.** $\frac{\pi}{4}; \frac{\pi}{4}; \frac{3\pi}{4}; \frac{\pi}{4} + 2\pi k; \frac{3\pi}{4} + 2\pi k$

5. Answers will vary. **7. a.** QIV **b.** 2 roots **9. a.** QIV **b.** 2 roots

11.

θ	$\sin\theta$	$\cos\theta$	$\tan\theta$
0	0	1	0
$\frac{\pi}{6}$	$\frac{1}{2}$	$\frac{\sqrt{3}}{2}$	$\frac{\sqrt{3}}{3}$
$\frac{\pi}{3}$	$\frac{\sqrt{3}}{2}$	$\frac{1}{2}$	$\sqrt{3}$
$\frac{\pi}{2}$	1	0	und.
$\frac{2\pi}{3}$	$\frac{\sqrt{3}}{2}$	$-\frac{1}{2}$	$-\sqrt{3}$
$\frac{5\pi}{6}$	$\frac{1}{2}$	$-\frac{\sqrt{3}}{2}$	$-\frac{\sqrt{3}}{3}$
π	0	−1	0
$\frac{7\pi}{6}$	$-\frac{1}{2}$	$-\frac{\sqrt{3}}{2}$	$\frac{\sqrt{3}}{3}$
$\frac{4\pi}{3}$	$-\frac{\sqrt{3}}{2}$	$-\frac{1}{2}$	$\sqrt{3}$

13. $\frac{\pi}{4}$ **15.** $-\frac{\pi}{4}$ **17.** $\frac{\pi}{6}$ **19.** $-\frac{\pi}{3}$ **21.** π **23.** $\frac{\pi}{3}$ **25.** $\frac{\pi}{6}$ **27.** $\frac{5\pi}{6}$

29. $\frac{\pi}{6}, \frac{5\pi}{6}$ **31.** $\frac{2\pi}{3}, \frac{5\pi}{3}$ **33.** $\frac{3\pi}{4}, \frac{7\pi}{4}$ **35.** $\frac{\pi}{6}, \frac{5\pi}{6}, \frac{7\pi}{6}, \frac{11\pi}{6}$

37. $\frac{\pi}{3}, \frac{2\pi}{3}, \frac{4\pi}{3}, \frac{5\pi}{3}$ **39.** $\frac{\pi}{4}, \frac{3\pi}{4}, \frac{5\pi}{4}, \frac{7\pi}{4}$ **41.** $\frac{\pi}{2}, \frac{3\pi}{2}$

43. $\theta = 1.2310 + 2\pi k$ or $5.0522 + 2\pi k$ **45.** $x = \frac{\pi}{2} + \pi k$ or $\frac{\pi}{6} + 2\pi k$ or $\frac{5\pi}{6} + 2\pi k$ **47.** $x = \frac{2\pi}{3} + 2\pi k$ or $\frac{4\pi}{3} + 2\pi k$ or $1.4455 + 2\pi k$ or $4.8377 + 2\pi k$ **49.** $x = \frac{\pi}{6} + \pi k$ or $\frac{5\pi}{6} + \pi k$

51. $x = \frac{5\pi}{4} + 2\pi k$ or $\frac{7\pi}{4} + 2\pi k$ **53.** $x = \frac{3\pi}{4} + 2\pi k$ or $\frac{5\pi}{4} + 2\pi k$

55. $x = \frac{3\pi}{4} + \pi k$ **57.** $x = \frac{\pi}{3} + \pi k$ or $\frac{2\pi}{3} + \pi k$ **59.** $x = \frac{3\pi}{8} + \frac{\pi}{2}k$

61. $x = 3\pi + 6\pi k$ **63.** $x = \frac{\pi}{2} + \pi k$ **65.** $x = \frac{\pi}{6} + \frac{\pi}{3}k$ or $\frac{\pi}{12} + \pi k$ or $\frac{5\pi}{12} + \pi k$ **67. a.** $x \approx 1.2310$ **b.** $x \approx 1.2310 + 2\pi k, 5.0522 + 2\pi k$

69. a. $x \approx 1.2094$ **b.** $x \approx 1.2094 + 2\pi k, 5.0738 + 2\pi k$
71. a. $\theta \approx 0.3649$ **b.** $\theta \approx 0.3649 + \pi k, 1.2059 + \pi k$
73. a. $\theta \approx 0.8861$ **b.** $\theta \approx 0.8861 + \pi k, 2.2555 + \pi k$

75. $x = \frac{\pi}{6} + \pi k$ or $\frac{5\pi}{6} + \pi k$ **77.** $x = \frac{2\pi}{9} + \frac{4\pi}{3}k$ or $\frac{10\pi}{9} + \frac{4\pi}{3}k$

79. $\theta = \frac{\pi}{2}k$ **81.** $\theta \approx 0.3398 + 2\pi k$ or $2.8018 + 2\pi k$

83. $x \approx 0.7290$ **85.** $x \approx 2.6649$ **87.** $x \approx 0.4566$ **89.** 22.1° and 67.9°
91. 0°; the ramp is horizontal. **93.** 30.7°; smaller
95. $\alpha = 35°, \beta \approx 25.5°$ **97.** $k \approx 1.36, \alpha \approx 20.6°$ **99. a.** 7 in.
b. ≈1.05 in. and ≈5.24 in. **101.** 1.1547

103. $\frac{\pi}{2} + \pi k$, explanations will vary. **105.** $x = \frac{\sin^{-1}\left(\frac{y-D}{A}\right) + C}{B}$

107. QIII: $\cos\theta = -\frac{5}{13}, \sec\theta = -\frac{13}{5}, \sin\theta = -\frac{12}{13}, \csc\theta = -\frac{13}{12}, \tan\theta = \frac{5}{12}, \cot\theta = \frac{12}{5}$ **109.** $\theta = \tan^{-1}\left(\frac{-12}{100}\right) \approx -6.8°$

111. $y = \sec\left(x - \frac{\pi}{4}\right), y = \csc\left(x + \frac{\pi}{4}\right)$

Mid-Chapter Check, pp. 261–262

1. implicit, explicit

2.

$$f(x) = 2\sqrt[3]{x+1} + 3$$
$$y = 2\sqrt[3]{x+1} + 3$$
$$x = 2\sqrt[3]{y+1} + 3$$
$$\frac{x-3}{2} = \sqrt[3]{y+1}$$
$$\left(\frac{x-3}{2}\right)^3 - 1 = y = f^{-1}(x)$$

$$(f^{-1}\circ f)(x) = f^{-1}[f(x)] = \left(\frac{2\sqrt[3]{x+1} + 3 - 3}{2}\right)^3 - 1 = (\sqrt[3]{x+1})^3 - 1 = x + 1 - 1 = x ✓$$

$$(f\circ f^{-1})(x) = f[f^{-1}(x)] = 2\sqrt[3]{\left(\frac{x-3}{2}\right)^3 - 1 + 1} + 3 = 2\left(\frac{x-3}{2}\right) + 3 = x - 3 + 3 = x ✓$$

3. The domain of $y = \sin x$ is restricted to $[-90°, 90]$ to create a one-to-one function.
4. a. $\sec^{-1}(\sqrt{2}) = \cos^{-1}\left(\frac{\sqrt{2}}{2}\right) = \frac{\pi}{4}$ **b.** $\csc^{-1}\left(\frac{2}{\sqrt{3}}\right) = \sin^{-1}\left(\frac{\sqrt{3}}{2}\right) = \frac{\pi}{3}$
5. a. $\frac{\pi}{6}$ **b.** $\cos\left(\frac{7\pi}{6}\right) = \cos\left(\frac{5\pi}{6}\right)$; $\cos^{-1}\left[\cos\left(\frac{5\pi}{6}\right)\right] = \left(\frac{5\pi}{6}\right)$
6. a. $\theta = \tan^{-1}\left(\frac{13}{84}\right)$, $\cos\theta = \frac{84}{85}$

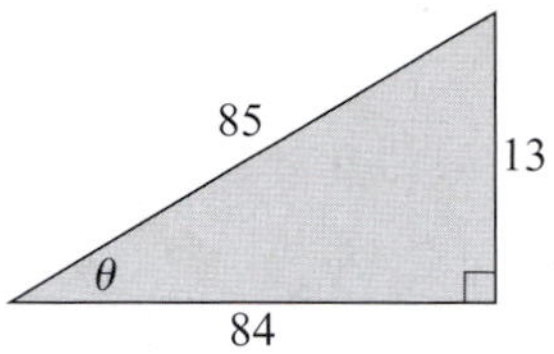

b. $\theta = \sin^{-1}\left(\frac{x}{\sqrt{x^2+49}}\right)$, $\sec\theta = \frac{\sqrt{x^2+49}}{7}$

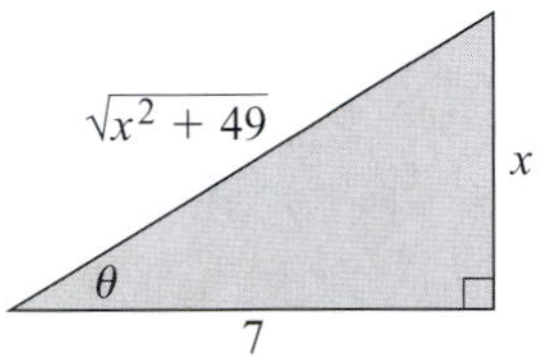

7. a. $Y_2 = \tan\theta$ **b.** $Y_2 = \tan\left[\cos^{-1}\left(\frac{x}{9}\right)\right]$ **8.** $x = \frac{\pi}{3}, \frac{2\pi}{3}$
9. $x \approx 0.7604, 2.3312, 5.4728$ **10.** $x \approx \pm 1.3287, \pm 1.8928, \pm 2.7813$

Reinforcing Basic Concepts, pp. 262–263

1. $x = \frac{\pi}{6} \pm 2\pi k, x = \frac{11\pi}{6} \pm 2\pi k; k \in \mathbb{Z}$
2. $x = \frac{\pi}{6} \pm \pi k, \frac{7\pi}{6} \pm \pi k; k \in \mathbb{Z}$
3. $x = \frac{\pi}{24} \pm \frac{\pi}{2}k, \frac{5\pi}{24} \pm \frac{\pi}{2}k; k \in \mathbb{Z}$ **4.** $x = \frac{\pi}{8} \pm \pi k, \frac{7\pi}{8} \pm \pi k; k \in \mathbb{Z}$

Exercises 4.4, pp. 272–276

1. $\sin^2 x + \cos^2 x = 1$; $1 + \tan^2 x = \sec^2 x$; $1 + \cot^2 x = \csc^2 x$
3. factor; grouping **5.** Answers will vary. **7.** $\frac{\pi}{12}, \frac{5\pi}{12}$ **9.** $0, 2\pi$
11. 0.4456, 1.1252 **13.** $\frac{\pi}{4}, \frac{5\pi}{4}, \frac{\pi}{6}, \frac{5\pi}{6}$
15. $\frac{\pi}{4}, \frac{5\pi}{4}, \frac{3\pi}{4}, \frac{7\pi}{4}$, 0.8411, 5.4421 **17.** $\frac{\pi}{6}, \frac{5\pi}{6}, \frac{7\pi}{6}, \frac{11\pi}{6}$
19. $\frac{\pi}{6}, \frac{5\pi}{6}$, 0.7297, 2.4119 **21.** $\frac{2\pi}{3}$ **23.** $\frac{\pi}{9} + \frac{2\pi}{3}k, \frac{5\pi}{9} + \frac{2\pi}{3}k$; $k = 0, 1, 2$ **25.** $\frac{\pi}{4}, \frac{5\pi}{4}, \frac{3\pi}{4}, \frac{7\pi}{4}$ **27.** $P = 12; x = 3; x = 11$
29. $P = 24; x \approx 0.4909, x \approx 5.5091$ **31.** $\frac{\pi}{12}, \frac{17\pi}{12}$ **33.** 0.3747, 5.9085, 2.7669, 3.5163 **35.** $\frac{\pi}{2}, \frac{3\pi}{2}$ **37.** $\frac{3\pi}{4}, \frac{7\pi}{4}$ **39.** $\frac{\pi}{12}, \frac{5\pi}{12}, \frac{13\pi}{12}, \frac{17\pi}{12}$
41. I. a. $\left(\frac{5}{2}, \frac{5}{2}\right)$ **b.** $D = \sqrt{12.5}, \theta = \frac{\pi}{4}, y = \frac{\sqrt{12.5} - x\cos\left(\frac{\pi}{4}\right)}{\sin\left(\frac{\pi}{4}\right)}$

c. verified **II. a.** $(2, 4)$ **b.** $D = 2\sqrt{5}, \theta \approx 1.1071$, $y = \frac{2\sqrt{5} - x\cos 1.1071}{\sin 1.1071}$ **c.** verified **III. a.** $(1, \sqrt{3})$
b. $D = 2, \theta = \frac{\pi}{3}, y = \frac{2 - x\cos\left(\frac{\pi}{3}\right)}{\sin\left(\frac{\pi}{3}\right)}$ **c.** verified
43. a. 2500π ft$^3 \approx 7853.98$ ft^3 **b.** ≈ 7824.09 ft^3 **c.** $\theta \approx 78.5°$
45. a. ≈ 78.53 m^3/sec **b.** during the months of August, September, October, and November **47. a.** ≈$3554.52 **b.** during the months of May, June, July, and August **49. a.** ≈ 12.67 in. **b.** during the months of April, May, June, July, and August **51. a.** ≈8.39 gal
b. approx. day 214 to day 333 **53. a.** 68 bpm **b.** ≈176.2 bpm
c. from about 4.6 min to 7.4 min **55.** Answers will vary.
57. $x = \frac{\pi}{4}k; k \in \mathbb{Z}$ **59.** $\sin\theta = \frac{68}{85}, \csc\theta = \frac{85}{68}, \cos\theta = \frac{-51}{85}$,
$\sec\theta = \frac{-85}{51}, \tan\theta = \frac{-68}{51}, \cot\theta = \frac{-51}{68}$
61.
$$\begin{aligned}\cos(4x) &= \cos[2(2x)] \\ &= 1 - 2\sin^2(2x) \\ &= 1 - 2(2\sin x\cos x)^2 \\ &= 1 - 8\sin^2 x\cos^2 x \\ &= 1 - 8\left(\frac{3}{5}\right)^2\left(\frac{4}{5}\right)^2 \\ &= -\frac{527}{625}\end{aligned}$$
63. wavelength 500 nm, likely green

Exercises 4.5, pp. 285–289

1. parameter **3.** direction **5.** Answers will vary.
7. a. parabola with vertex at $(2, -1)$
b. $y = x^2 - 4x + 3$

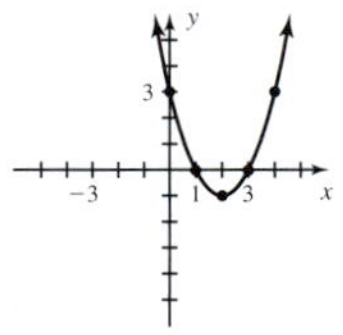

9. a. parabola
b. $y = x \pm 2\sqrt{x} + 1$

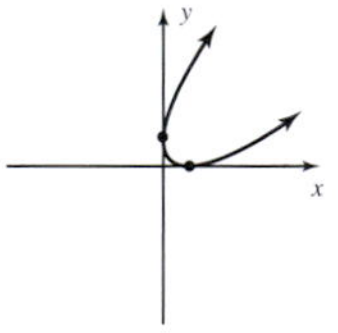

11. a. power function with $p = -2$
b. $y = \frac{25}{x^2}, x \neq 0$

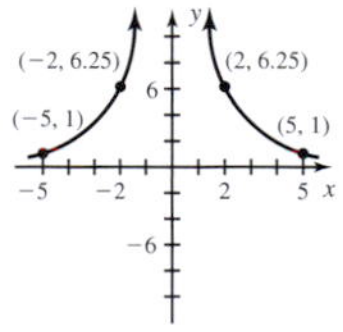

13. a. ellipse
b. $\frac{x^2}{16} + \frac{y^2}{9} = 1$

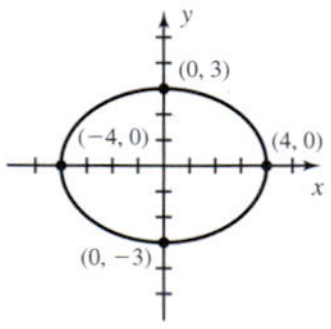

15. a. Lissajous figure 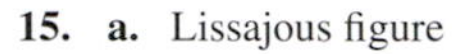

b. $y = 6\cos\left[\frac{1}{2}\sin^{-1}\left(\frac{x}{4}\right)\right]$

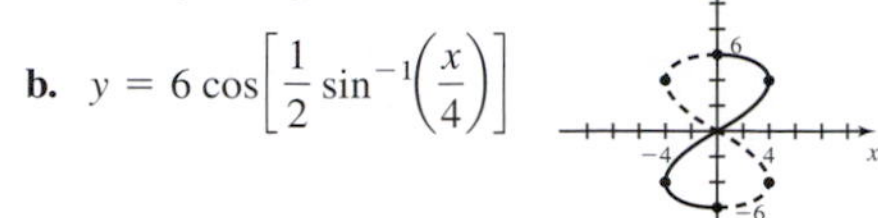

17.

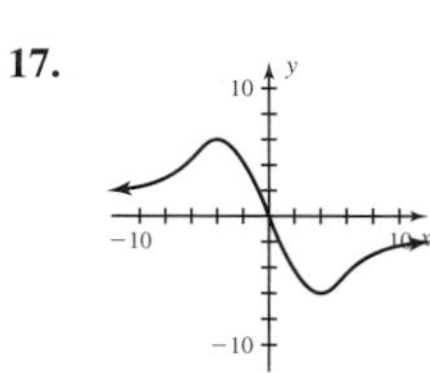

19. $x = t, y = 3t - 2; x = \frac{1}{3}t, y = t - 2; x = \cos t,\ y = 3\cos t - 2$
21. $x = t, y = (t + 3)^2 + 1; x = t - 3, y = t^2 + 1; x = \tan t - 3,$
$y = \sec^2 t, t \neq \frac{(2k + 1)\pi}{2}, k \in \mathbb{Z}$ **23.** $x = t, y = \tan^2(t - 2) + 1,$
$t \neq \pi k + \frac{\pi}{2} + 2, k \in \mathbb{Z}; x = t + 2, y = \sec^2 t, t \neq \left(k + \frac{1}{2}\right)\pi, k \in \mathbb{Z};$
$x = \tan^{-1} t + 2, y = t^2 + 1$ **25.** verified

27. a.

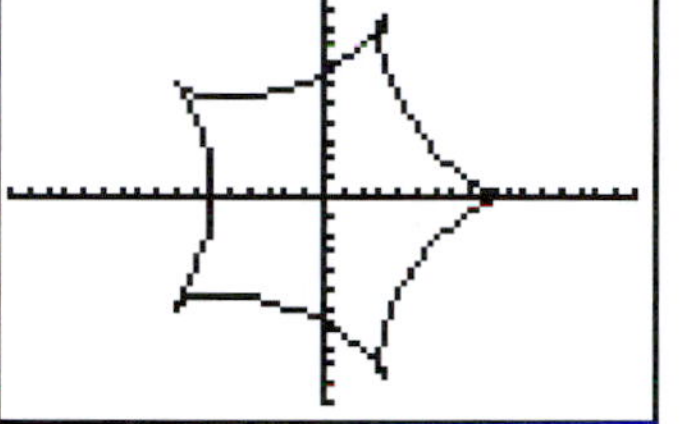

b. x-intercepts: $t = 0, x = 10, y = 0$ and $t = \pi, x = -6, y = 0$;
y-intercepts: $t \approx 1.757, x = 0, y \approx 6.5$ and
$t \approx 4.527, x = 0, y \approx -6.5$;
minimum x-value is -8.1; maximum x-value is 10;
minimum y-value is -9.5; the maximum y-value is 9.5

29. a. 

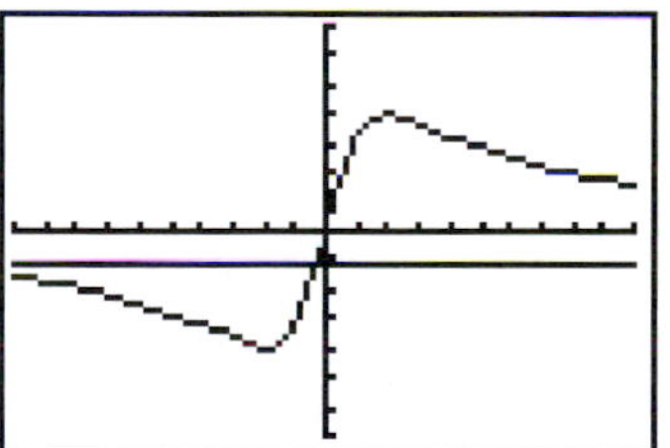

b. x-intercepts none, y-intercepts none;
no minimum or maximum x-values;
minimum y-value is -4 and maximum y-value is 4

31. a.

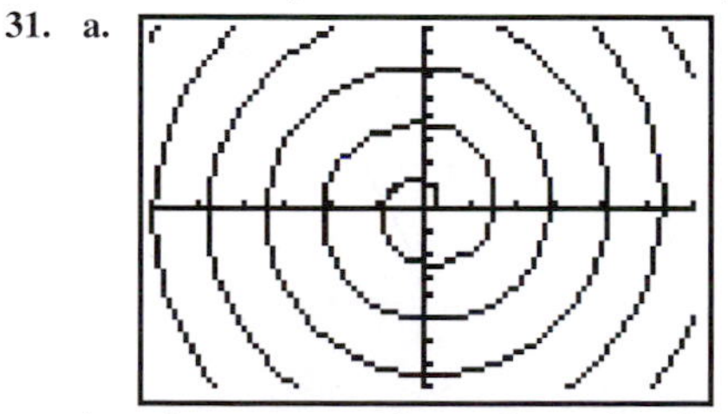

b. x-intercepts: $t = 0, x = 2, y = 0$ and $t \approx 4.493, x \approx -9.2, y = 0$;
infinitely many others;
y-intercepts: $t \approx 2.798, x = 0, y \approx 5.9$ and $t \approx 6.121$ and $x = 0$,
$y \approx -12.4$; infinitely many others; no minimum or maximum values for x or y

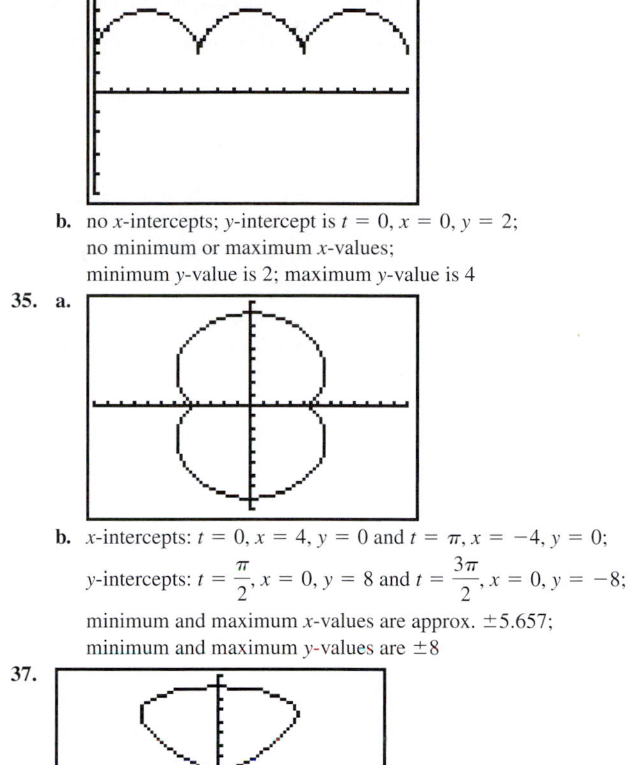

33. a.

b. no x-intercepts; y-intercept is $t = 0, x = 0, y = 2$;
no minimum or maximum x-values;
minimum y-value is 2; maximum y-value is 4

35. a.

b. x-intercepts: $t = 0, x = 4, y = 0$ and $t = \pi, x = -4, y = 0$;
y-intercepts: $t = \frac{\pi}{2}, x = 0, y = 8$ and $t = \frac{3\pi}{2}, x = 0, y = -8$;
minimum and maximum x-values are approx. ± 5.657;
minimum and maximum y-values are ± 8

37.

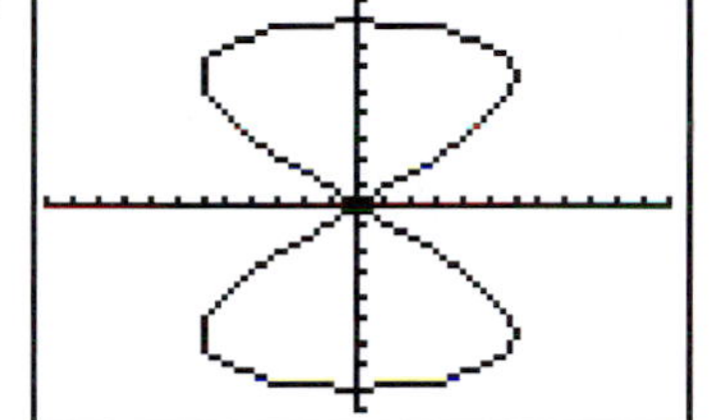

width 12 and length 16; including the endpoint $t = 2\pi$, the graph crosses itself two times from 0 to 2π.

39.

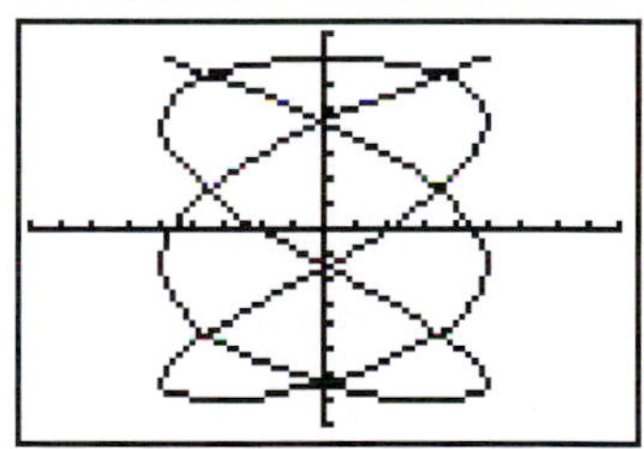

width 10 and length 14; including the endpoint $t = 2\pi$, the graph crosses itself nine times from 0 to 2π.

41.

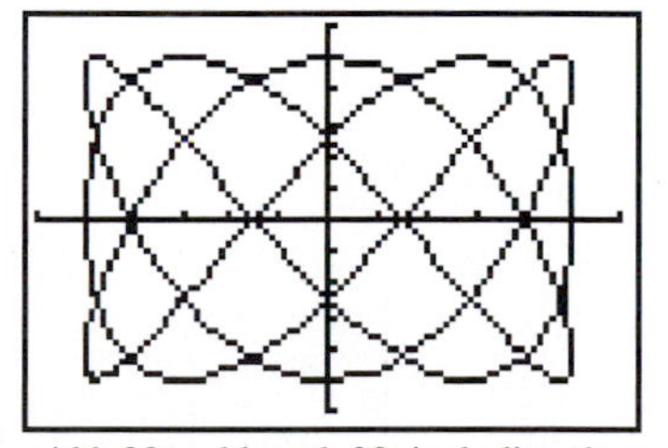

width 20 and length 20; including the endpoint $t = 4\pi$, the graph crosses itself 23 times from 0 to 4π.

43. The maximum value (as the graph swells to a peak) is at $(x, y) = (a, \frac{b}{2})$. The minimum value (as the graph dips to the valley) is at $(x, y) = (-a, \frac{-b}{2})$.

45. a. The curve is approaching $y = 2$ as t approaches $\frac{3\pi}{2}$, but $\cot\left(\frac{3\pi}{2}\right)$ is undefined, and the trig form seems to indicate a hole at $t = \frac{3\pi}{2}$, $x = 0$, $y = 2$. The algebraic form does not have this problem and shows a maximum defined at $t = 0$, $x = 0$, $y = 2$.
b. As $|t| \to \infty$, $y(t) \to 0$ **c.** The maximum value occurs at $(0, 2k)$.
47. a. Yes **b.** Yes **c.** ≈ 0.82 ft **49.** No, the kick is short.
51. The electron is moving left and downward.
53. $\left(t, \frac{6t}{17} - \frac{6}{17}, \frac{13t}{17} + \frac{21}{17}\right)$ **55.** Inconsistent, no solutions
57. $x = 1.22475^t$
$y = 0.25t^2 - 2t$

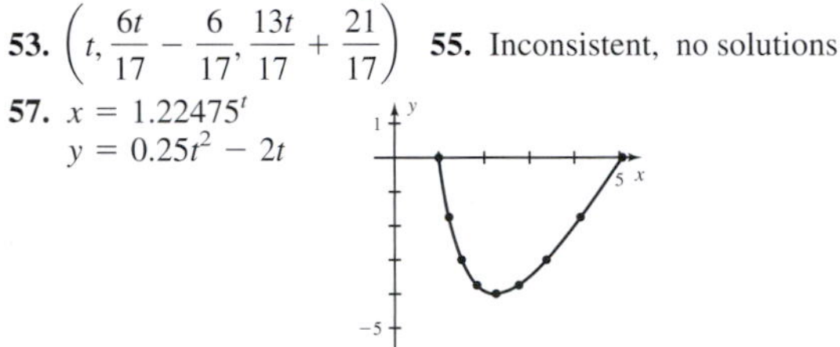

The parametric equations fit the data very well.
59. Answers will vary. **61.** $(3.3282, 1.6641)$ $(-3.3282, 1.6641)$ $(-3.3282, -1.6641)$ $(3.3282, -1.6641)$ **63.** $\theta = \frac{\pi}{6}, \frac{\pi}{2}, \frac{5\pi}{6}, \frac{3\pi}{2}$
65. $\sin\theta = \frac{12}{13}$ **67.** The Eiffel Tower is about 1063 ft tall. Taller.

Summary and Concept Review, pp. 289–293

1. no **2.** no **3.** yes **4.** $f^{-1}(x) = \frac{x-2}{-3}$ **5.** $f^{-1}(x) = \sqrt{x+2}$
6. $f^{-1}(x) = x^2 + 1; x \geq 0$
7. $f(x)$: $D: x \in [-4, \infty)$, $R: y \in [0, \infty)$; $f^{-1}(x)$: $D: x \in [0, \infty)$, $R: y \in [-4, \infty)$
8. $f(x)$: $D: x \in (-\infty, \infty)$, $R: y \in (-\infty, \infty)$; $f^{-1}(x)$: $D: (-\infty, \infty)$, $R: y \in (-\infty, \infty)$
9. $f(x)$: $D: x \in (-\infty, \infty)$, $R: y \in (0, \infty)$; $f^{-1}(x)$: $D: x \in (0, \infty)$, $R: y \in (-\infty, \infty)$
10. a. \$3.05 **b.** $f^{-1}(t) = \frac{t-2}{0.15}$; $f^{-1}(3.05) = 7$ **c.** 12 days
11. $\frac{\pi}{4}$ or 45° **12.** $\frac{\pi}{6}$ or 30° **13.** $\frac{5\pi}{6}$ or 150° **14.** 1.3431 or 77.0°
15. 1.0956 or 62.8° **16.** 0.5054 or 29.0° **17.** $\frac{1}{2}$ **18.** $\frac{\pi}{4}$
19. undefined **20.** 1.0245 **21.** 60° **22.** $\frac{3\pi}{4}$
23. $\sin\theta = \frac{35}{37}$
24. $\tan\theta = \frac{\sqrt{49 - 9x^2}}{3x}$
25. $\cot\theta = \frac{9}{x}$
26. $\theta = \cos^{-1}\left(\frac{x}{5}\right)$ **27.** $\theta = \sec^{-1}\left(\frac{x}{7\sqrt{3}}\right)$ **28.** $\theta = \sin^{-1}\left(\frac{x}{4}\right) + \frac{\pi}{6}$
29. a. $Y_2 = \frac{13\sqrt{1 + \tan^2\theta}}{13\tan\theta} = \frac{13\sec\theta}{13\tan\theta} = \csc\theta$
b. $Y_2 = \csc\left[\tan^{-1}\left(\frac{x}{13}\right)\right]$
c.

x	Y_1	Y_2
0	error	error
1	13.038	13.038
2	6.5765	6.5765
3	4.4472	4.4472
4	3.4004	3.4004
5	2.7857	2.7857
6	2.3863	2.3863

30. a. $\frac{\pi}{4}$ **b.** $\frac{\pi}{4}, \frac{3\pi}{4}$ **c.** $x = \frac{\pi}{4} + 2\pi k$ or $\frac{3\pi}{4} + 2\pi k, k \in Z$
31. a. $\frac{2\pi}{3}$ **b.** $\frac{2\pi}{3}, \frac{4\pi}{3}$ **c.** $\frac{2\pi}{3} + 2\pi k$ or $\frac{4\pi}{3} + 2\pi k, k \in Z$
32. a. $-\frac{\pi}{3}$ **b.** $\frac{2\pi}{3}, \frac{5\pi}{3}$ **c.** $\frac{2\pi}{3} + \pi k, k \in Z$
33. a. ≈ 1.1102 **b.** $\approx 1.1102, 5.1729$ **c.** $\approx 1.1102 + 2\pi k$ or $5.1729 + 2\pi k, k \in Z$
34. a. ≈ 0.3376 **b.** $\approx 0.3376, 1.2332, 3.4792, 4.3748$ **c.** $\approx 0.3376 + \pi k$ or $1.2332 + \pi k, k \in Z$
35. a. ≈ 0.3614 **b.** $\approx 0.3614, 2.7802$ **c.** $\approx 0.3614 + 2\pi k$ or $2.7802 + 2\pi k, k \in Z$
36. $\theta \approx 1.1547$ **37.** $x = \frac{\pi}{12}, \frac{5\pi}{12}$
38. $x \approx 0.7297, 2.4119$; $x = \frac{\pi}{6}, \frac{5\pi}{6}$ **39.** $x = \frac{\pi}{6}, \frac{5\pi}{6}, \frac{11\pi}{6}$
40. $x = \frac{\pi}{2}$ **41.** $P = 12$; $x \approx 2.6931$, $x \approx 9.3069$
42. $P = 6$; $x = 0$, $x = \frac{9}{2}$ **43. a.** $\approx 43(1000) = \$43{,}000$
b. April through August
44. $y = -2(x + 4)^2 + 3$ **45.** $y = (-1 \pm \sqrt{x})^2$

46. $\frac{x^2}{9} + \frac{y^2}{16} = 1$

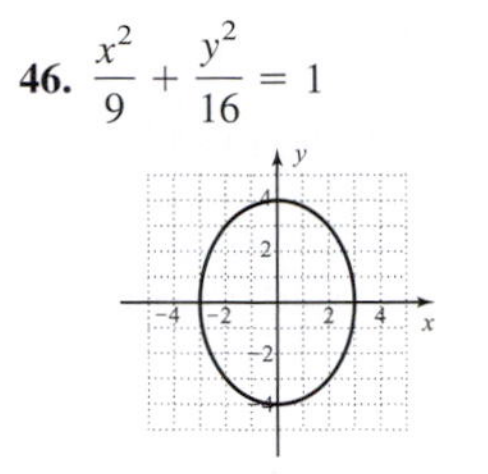

47. Answers will vary.

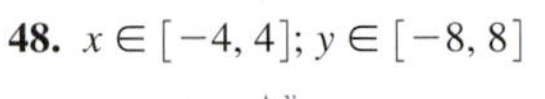

48. $x \in [-4, 4]$; $y \in [-8, 8]$

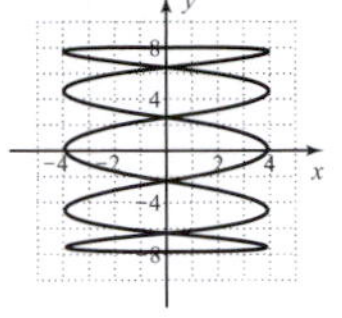

Mixed Review, pp. 294–295

1. a. $f^{-1}(x) = \sqrt{\frac{1}{x} - 2}$

b. $D: x > 0, R: y > -2$

c. verified

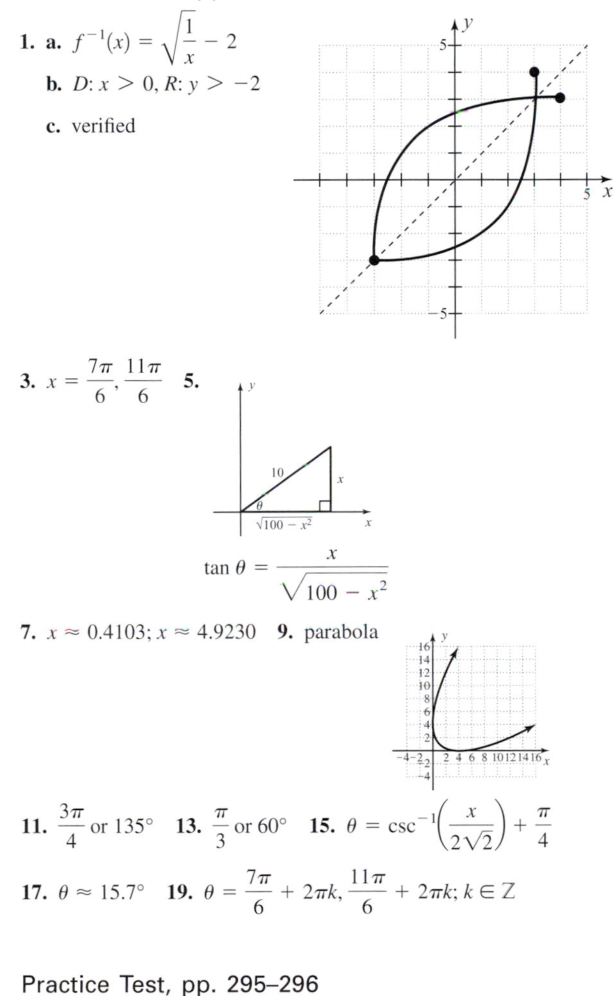

3. $x = \frac{7\pi}{6}, \frac{11\pi}{6}$ **5.**

$\tan\theta = \frac{x}{\sqrt{100 - x^2}}$

7. $x \approx 0.4103; x \approx 4.9230$ **9.** parabola

11. $\frac{3\pi}{4}$ or 135° **13.** $\frac{\pi}{3}$ or 60° **15.** $\theta = \csc^{-1}\left(\frac{x}{2\sqrt{2}}\right) + \frac{\pi}{4}$

17. $\theta \approx 15.7°$ **19.** $\theta = \frac{7\pi}{6} + 2\pi k, \frac{11\pi}{6} + 2\pi k; k \in \mathbb{Z}$

Practice Test, pp. 295–296

1. $f(x) = x^3$ is a one-to-one function, $f(x) = x^2$ is not.

2. a. $A \approx 39.27\text{ cm}^2$ **b.** $A^{-1}(t) = \frac{2t}{r^2}$; we obtain $t = \frac{\pi}{4}$

3.

$D: x \in [-1, 1], R: y \in [0, \pi]$

4. $y \approx 1.2310$

5. $y = 30°$ **6.** $f(x) = \frac{1}{2}$ **7.** $y = 30°$

8. $y = 0.8523$ rad or $y = 48.8°$ **9.** $y = 78.5°$ or $\frac{157\pi}{360}$ rad

10. $y = \frac{7\pi}{24}$ rad or 52.5°

11. $\cos\theta = \frac{33}{65}$ **12.** $\cot\theta = \frac{x}{5}$

13. a. $\cos^{-1}\left(\frac{-\sqrt{2}}{2}\right) = \frac{3\pi}{4}$ **b.** $x = \frac{3\pi}{4}, \frac{5\pi}{4}$

c. $x = \frac{3\pi}{4} + 2\pi k$ or $\frac{5\pi}{4} + 2\pi k, k \in \mathbb{Z}$

14. a. $\frac{\pi}{6}$ **b.** $x = \frac{\pi}{6}, \frac{11\pi}{6}$

c. $x = \frac{\pi}{6} + 2\pi k$ or $\frac{11\pi}{6} + 2\pi k, k \in \mathbb{Z}$

15. a. $x \approx 0.1922$ **b.** $x \approx 0.1922, 1.3786, 3.3338, 4.5202$

c. $x \approx 0.1922 + \pi k$ or $1.3786 + \pi k, k \in \mathbb{Z}$

16. a. $x \approx 0.9204$ **b.** $x \approx 0.9204, 2.2212, 4.0620, 5.3628$

c. $x \approx 0.9204 + \pi k$ or $2.2212 + \pi k, k \in \mathbb{Z}$

17. $x \approx -1.6875, -0.3413, 1.1321, 2.8967$

18. $x \approx 0.9671, 2.6110, 3.4538$ **19.** $x = 0, \pi, \frac{7\pi}{6}, \frac{11\pi}{6}$

20. $x = \frac{7\pi}{12}, \frac{11\pi}{12}, \frac{19\pi}{12}, \frac{23\pi}{12}$ **21.** $x = \frac{\pi}{2}, \frac{3\pi}{2}; x \approx 3.3090, 6.1157$

22. $x = \frac{5\pi}{6}, \frac{11\pi}{6}$

23. ellipse; $\frac{x^2}{16} + \frac{y^2}{25} = 1$ **24.** parabola; $x = (y - 5)^2 + 1$

25. max: $y = 8$; min: $y = 0$; $P = 8\pi$

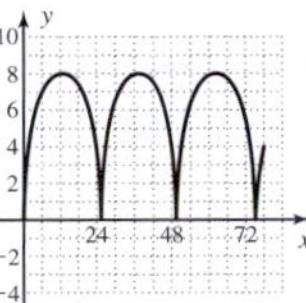

Strengthening Core Skills, pp. 297–299

Exercise 1. $x \in (0.6025, 2.5391)$
Exercise 2. $x \in [0, 0.7945] \cup [4.4415, 2\pi]$
Exercise 3. $x \in [0, 2.6154] \cup [9.3847, 12]$
Exercise 4. $x \in (67.3927, 202.6073)$

Cumulative Review Chapters 1–4, pp. 299–300

1. $\sin\theta = \frac{84}{85}$, $\csc\theta = \frac{85}{84}$, $\cos\theta = \frac{-13}{85}$, $\sec\theta = \frac{-85}{13}$, $\tan\theta = \frac{-84}{13}$, $\cot\theta = \frac{-13}{84}$ **3. a.** $56.335°, 0.9832$ **5.** about 474 ft

7. $x = \frac{\pi}{2}, \frac{3\pi}{2}, 2\pi + \sin^{-1}\left(-\frac{1}{6}\right) \approx 6.1157, \pi - \sin^{-1}\left(-\frac{1}{6}\right) \approx 3.3090$

9. 50.89 km/hr **11.** $\frac{\sqrt{3}}{2}$

13. a. $Y_1 = 48.778\sin(0.213x - 1.106) + 51.642$ **b.** about 83.2%

15. $x \in (1, 5)$ **17.** $\frac{99}{101}$

19.
$$\frac{\cos x + 1}{\tan^2 x} = \frac{\cos x}{\sec x - 1} = \frac{\cos x(\sec x + 1)}{(\sec x - 1)(\sec x + 1)} = \frac{1 + \cos x}{\sec^2 x - 1} = \frac{1 + \cos x}{\tan^2 x}$$

CHAPTER 5

Exercises 5.1, pp. 307–311

1. sine; opposite **3.** 180° **5.** $\frac{a}{\sin A} = \frac{b}{\sin B} = \frac{c}{\sin C}$ **7.** yes
9. no, we don't know the measure of any angle opposite a side
11. $a \approx 8.98$ **13.** $C \approx 49.2°$ **15.** $C \approx 21.4°$ **17.** $\angle C = 78°$, $b \approx 109.5$ cm, $c \approx 119.2$ cm **19.** $\angle C = 90°$, $a = 10$ in., $c = 20$ in.
21.

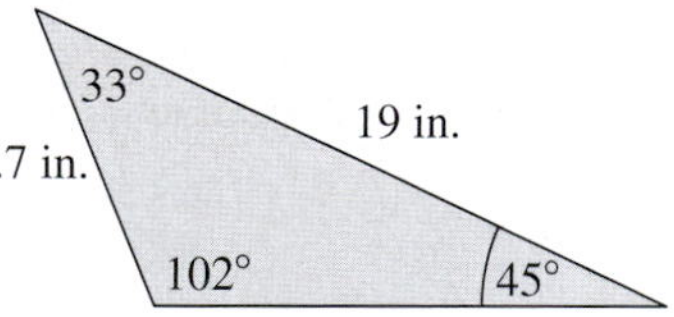

23. $\angle C = 90°$, $a = 15$ mi, $b = 15$ mi **25.** $\angle A = 57°$, $b \approx 49.5$ km, $c \approx 17.1$ km
27.

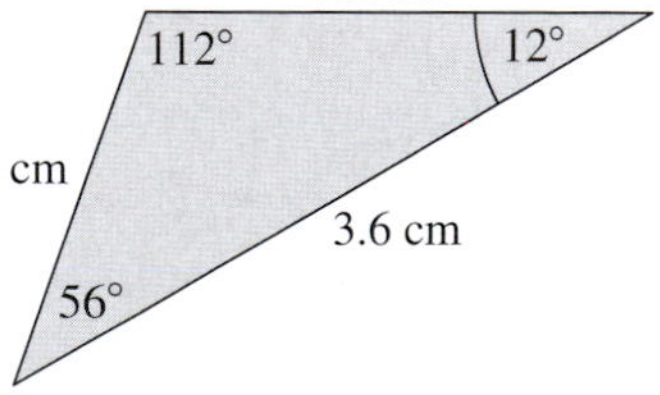

29. $A \approx 100.4$ ft^2 **31.** $a \approx 33.7$ ft, $c \approx 22.3$ ft **33.** Rhymes to Tarryson: 61.7 km, Sexton to Tarryson: 52.6 km **35.** ≈ 3.2 mi **37.** $h \approx 161.9$ yd **39.** angle = 90°; sides ≈ 9.8 cm, 11 cm; diameter ≈ 11 cm; it is a right triangle. **41. a.** about 3187 m **b.** about 2613 m **c.** about 2368 m

43.

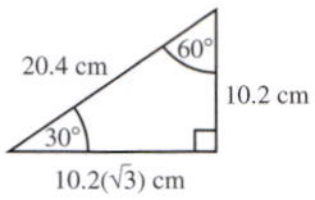

$\sqrt{3} = \frac{\sin 60°}{\sin 30°}$; $\sqrt{2} = \frac{\sin 90°}{\sin 45°}$

45. Area of triangle $= \frac{1}{2}bh$, and all three triangles have same base and height. **47.** $x \approx 0.3747 + 2\pi k$ or $2.7669 + 2\pi k$, $k \in \mathbb{Z}$

49.
$$\cot x = \frac{\cos(3x) + \cos x}{\sin(3x) - \sin x} = \frac{2\cos\left(\frac{3x + x}{2}\right)\cos\left(\frac{3x - x}{2}\right)}{2\cos\left(\frac{3x + x}{2}\right)\sin\left(\frac{3x - x}{2}\right)} = \frac{2\cos(2x)\cos x}{2\cos(2x)\sin x} = \cot x$$

51. 106.9 ft

Exercises 5.2, pp. 319–324

1. ambiguous **3.** I, II **5.** Answers will vary. **7. a.** 10 cm **b.** 0 **c.** 2 **d.** 1 **9.** not possible **11.** $B = 60°, C = 90°, b = 12.9\sqrt{3}$ mi
13. $A \approx 39°, B \approx 82°, a \approx 42.6$ mi or $A \approx 23°, B \approx 98°, a \approx 26.4$ mi
15. not possible **17.** $A \approx 39°, B \approx 82°, a \approx 42.6$ ft or $A \approx 23°$, $B \approx 98°, a \approx 26.4$ ft **19.** not possible **21.** $A_1 \approx 19.3°, A_2 \approx 160.7°$, $48° + 160.7° > 180°$; no second solution possible **23.** $A_1 \approx 71.3°$, $A_2 \approx 108.7°$, $57° + 108.7° < 180°$; two solutions possible

25. not possible, $\sin A > 1$ **27.** $\frac{\sqrt{2}}{2}$ **29.** 34.5 million miles or 119.7 million miles **31. a.** No **b.** ≈ 3.9 mi **33.** $V \longleftrightarrow S = 41.7$ km, $V \longleftrightarrow P = 80.8$ km **35. a.** No **b.** 1 **c.** ≈ 15 sec

37. Two triangles

Angles	Sides
$A_1 \approx 41.1°$	$a = 12$ cm
$B = 26°$	$b = 8$ cm
$C_1 \approx 112.9°$	$c_1 \approx 16.8$ cm

Angles	Sides
$A_2 \approx 138.9°$	$a = 12$ cm
$B = 26°$	$b = 8$ cm
$C_2 = 15.1°$	$c_2 \approx 4.8$ cm

39.

Angles	Sides
$A_1 \approx 47.0°$	$a = 9$
$B_1 \approx 109.0°$	$b_1 \approx 11.6$
$C \approx 24°$	$c = 5$

Angles	Sides
$A_2 \approx 133.0°$	$a = 9$
$B_2 = 23.0$	$b_2 = 4.8$
$C \approx 24°$	$c = 5$

41. a. $a = 1000$ or $a = 1414$ **b.** $1000 < a < 1414$ **c.** $a > 1414$ Answers will vary. **43.** Answers will vary.
45.

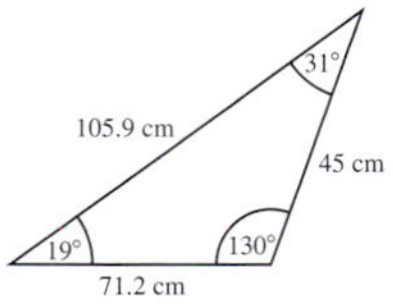

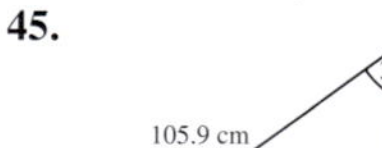

47. Honolulu to Tokyo: 3865 mi; Honolulu to San Francisco 2384 mi

49.

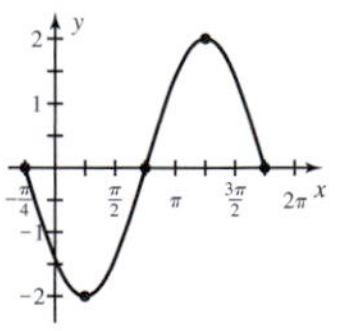

51. $\left(\frac{33}{65}, \frac{56}{65}\right)$

Exercises 5.3, pp. 330–335

1. cosines **3.** Pythagorean **5.** $B \approx 33.1°$, $C \approx 129.9°$, $a \approx 19.8$ m; law of sines **7.** yes **9.** no **11.** yes **13.** They do (with rounding). **15.** $B \approx 41.4°$ **17.** $a \approx 7.24$ **19.** $A \approx 41.6°$ **21.** $A \approx 120.4°$, $B \approx 21.6°$, $c \approx 53.5$ cm **23.** $A \approx 23.8°$, $C \approx 126.2°$, $b \approx 16$ mi

25.

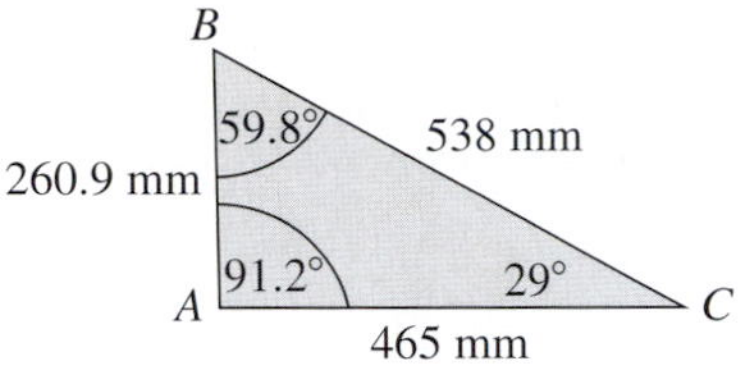

27. $A \approx 137.9°$, $B \approx 15.6°$, $C \approx 26.5°$ **29.** $A \approx 119.3°$, $B \approx 41.5°$, $C \approx 19.2°$

31.

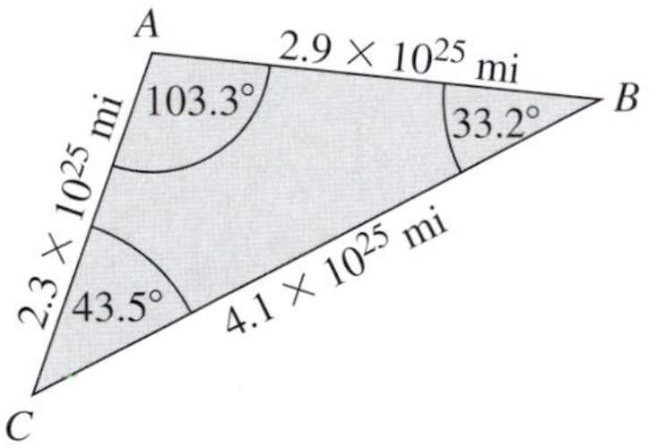

33. $A \approx 139.7°$, $B \approx 23.7°$, $C \approx 16.6°$ **35.** $C \approx 86.3°$ **37.** 1688 mi **39.** 27.7° north of west or a heading of 297.7° **41.** It cannot be constructed (available length ≈ 10,703.6 ft) **43.** 1678.2 mi **45.** $P \approx 22.4$, $A = 135°$, $B \approx 23.2°$, $C \approx 21.8°$ **47.** $A \approx 20.6°$, $B \approx 15.3°$, $C \approx 144.1°$ **49.** 58.78 cm

51. $a = 13$ $A \approx 133.2°$
$b = 5$ $B \approx 16.3°$
$c = \sqrt{82}$ $C \approx 30.5°$

53. Answers will vary. **55.** Pentagon: $A = 237.8\ \text{cm}^2$, Hexagon: $A = 150\sqrt{3} \approx 259.8\ \text{cm}^2$ **57.** $387 + 502 = 889 < 902$

59. (1) $a^2 = b^2 + c^2 - 2bc\cos A$ (2) $b^2 = a^2 + c^2 - 2ac\cos B$, use substitution for a^2 and (2) becomes $b^2 = (b^2 + c^2 - 2bc\cos A) + c^2 - 2ac\cos B$. Then $0 = 2c^2 - 2bc\cos A - 2ac\cos B$, $2bc\cos A + 2ac\cos B = 2c^2$, $b\cos A + a\cos B = c$

61. $\sin\left(\frac{\pi}{6}\right) = \frac{1}{2}$; $\cos\left(\frac{7\pi}{6}\right) = \frac{-\sqrt{3}}{2}$; $\tan\left(\frac{\pi}{3}\right) = \sqrt{3}$

63. $\sin x = \frac{-5}{13}$, $\csc x = \frac{-13}{5}$, $\cos x = \frac{12}{13}$, $\sec x = \frac{13}{12}$, $\tan x = -\frac{5}{12}$, $\cot x = \frac{-12}{5}$ **65.** $\frac{73}{55}$

Exercises 5.4, pp. 347–352

1. scalar **3.** directed, line **5.** Answers will vary.

7.

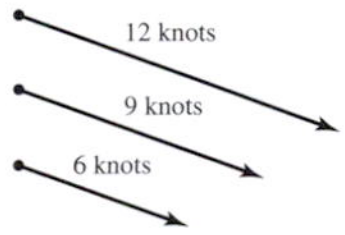

9.

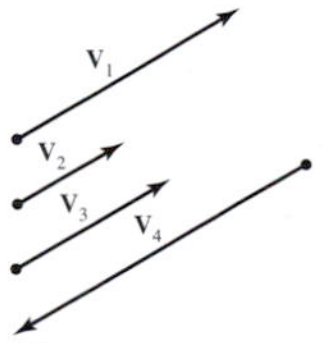

11.

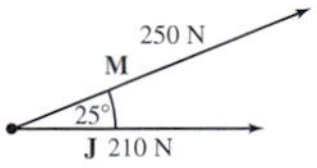

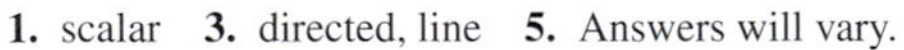

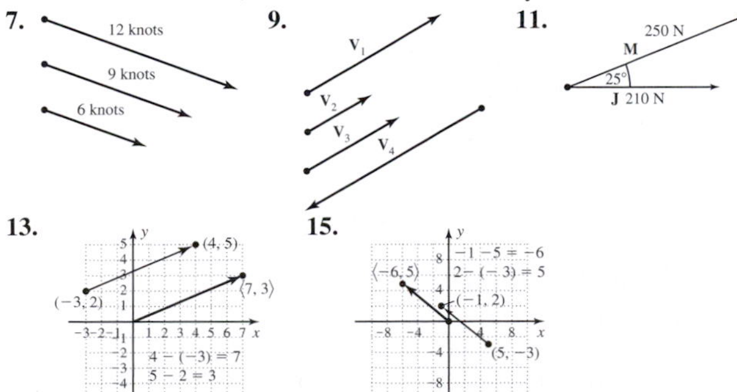

13.

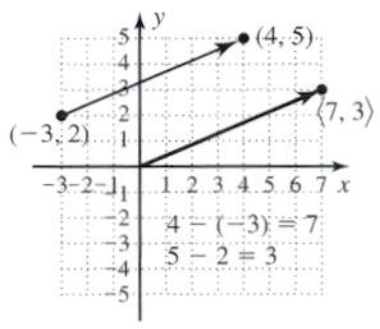

15.

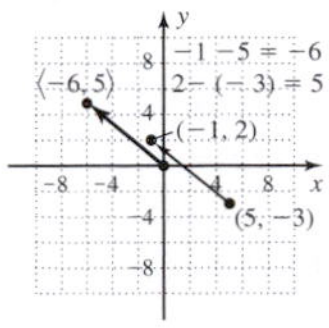

17. Terminal point: $(5, -1)$, magnitude: $\sqrt{53}$
19. Terminal point: $(-1, 1)$, magnitude: $\sqrt{34}$

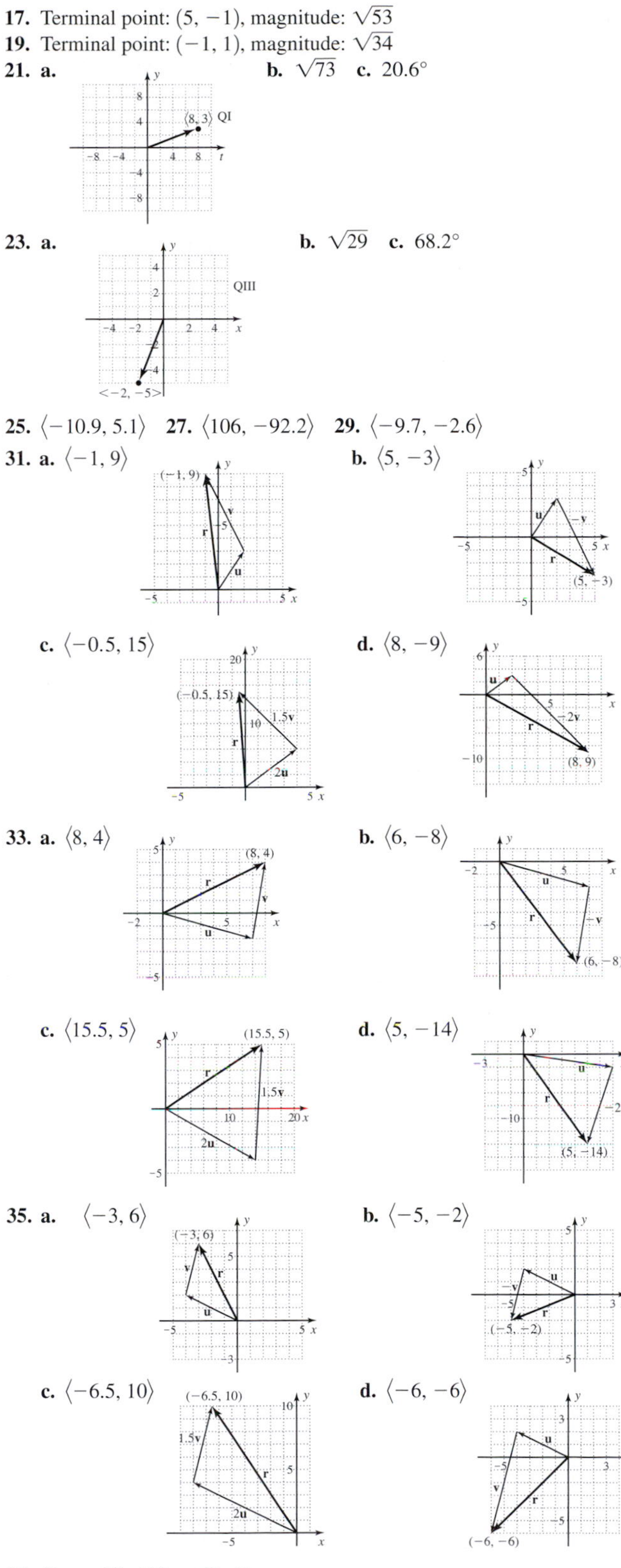

21. a.

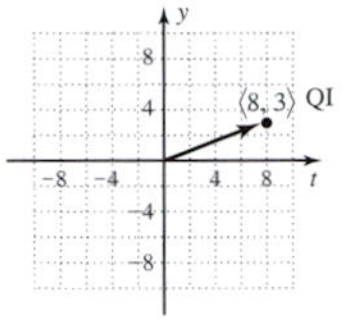

b. $\sqrt{73}$ **c.** 20.6°

23. a.

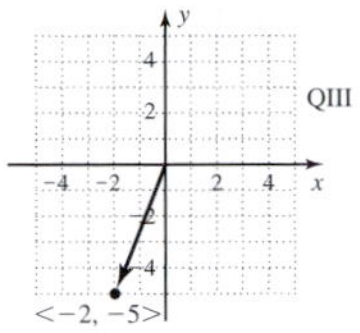

b. $\sqrt{29}$ **c.** 68.2°

25. $\langle -10.9, 5.1\rangle$ **27.** $\langle 106, -92.2\rangle$ **29.** $\langle -9.7, -2.6\rangle$

31. a. $\langle -1, 9\rangle$

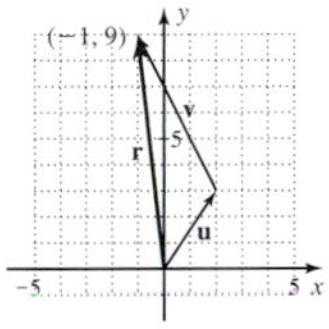

b. $\langle 5, -3\rangle$

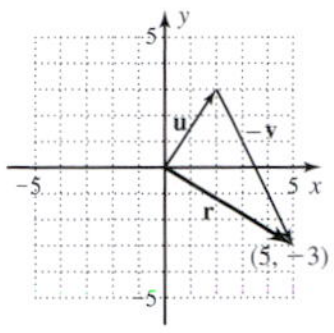

c. $\langle -0.5, 15\rangle$

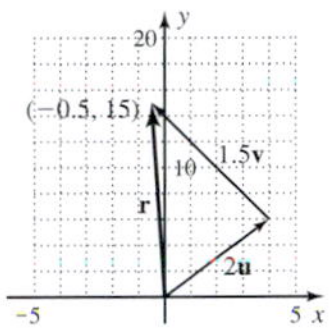

d. $\langle 8, -9\rangle$

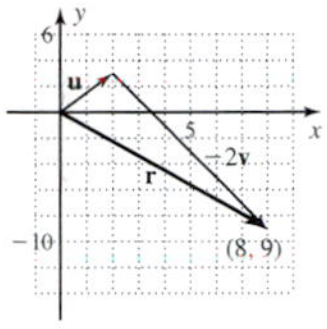

33. a. $\langle 8, 4\rangle$

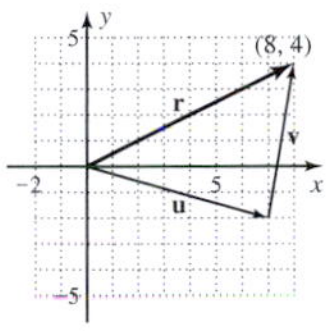

b. $\langle 6, -8\rangle$

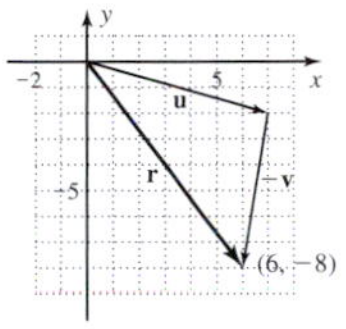

c. $\langle 15.5, 5\rangle$

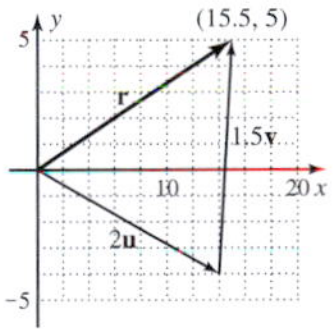

d. $\langle 5, -14\rangle$

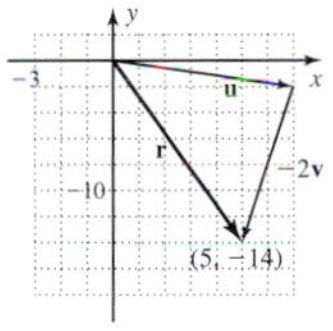

35. a. $\langle -3, 6\rangle$

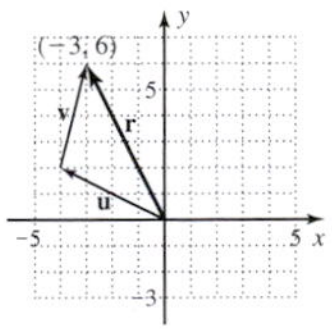

b. $\langle -5, -2\rangle$

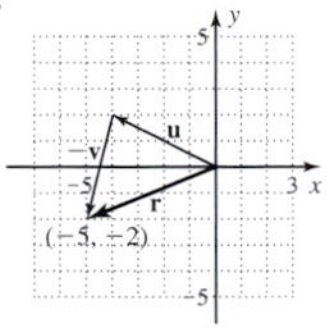

c. $\langle -6.5, 10\rangle$

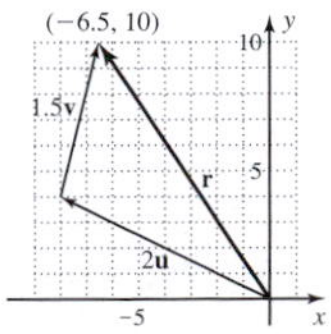

d. $\langle -6, -6\rangle$

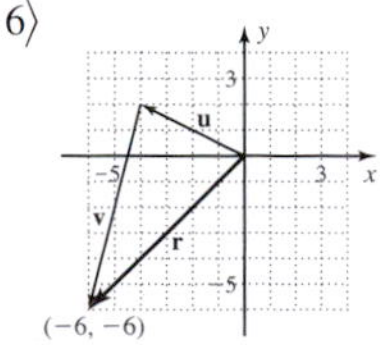

37. True **39.** False **41.** True

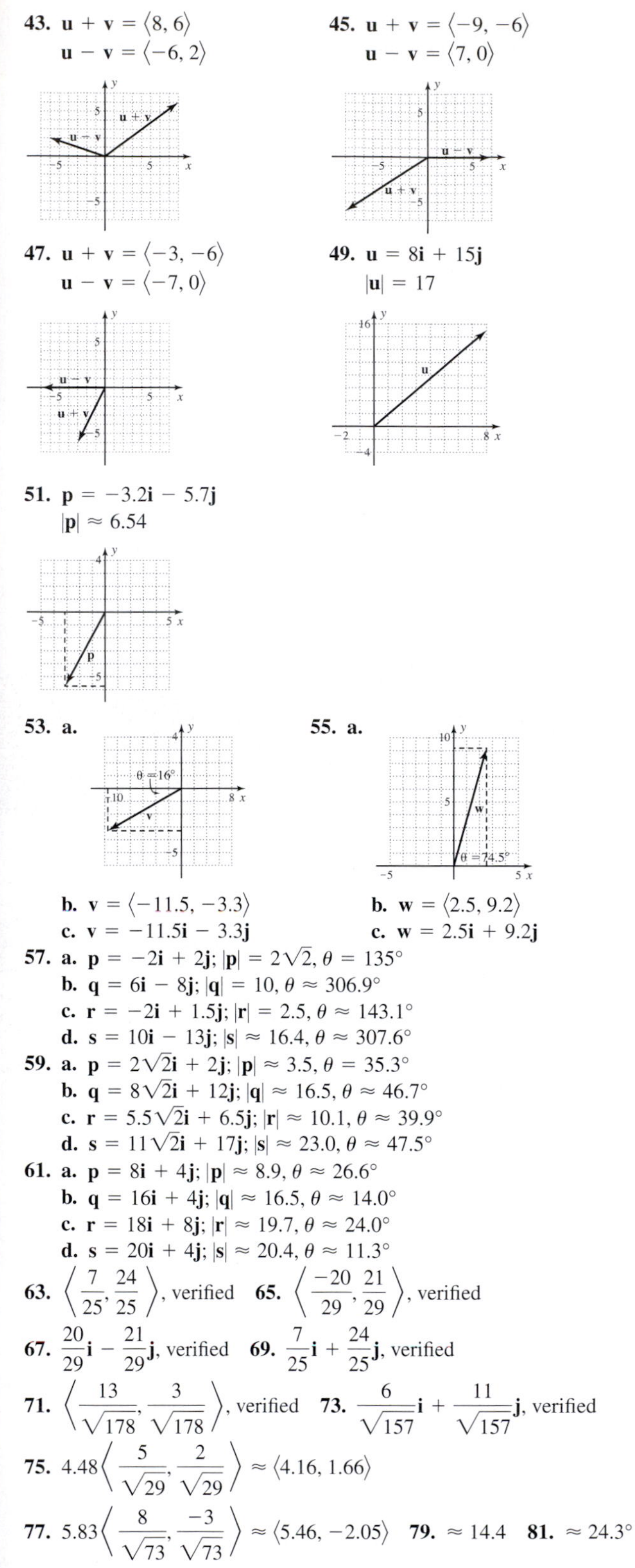

43. $\mathbf{u} + \mathbf{v} = \langle 8, 6\rangle$
$\mathbf{u} - \mathbf{v} = \langle -6, 2\rangle$

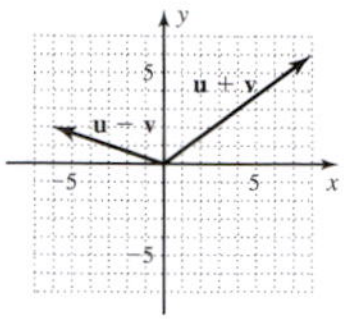

45. $\mathbf{u} + \mathbf{v} = \langle -9, -6\rangle$
$\mathbf{u} - \mathbf{v} = \langle 7, 0\rangle$

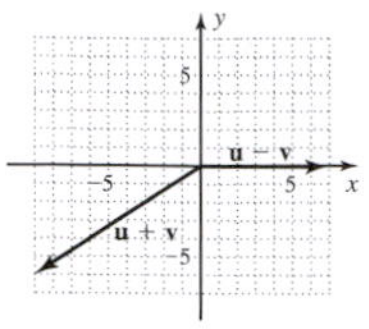

47. $\mathbf{u} + \mathbf{v} = \langle -3, -6\rangle$
$\mathbf{u} - \mathbf{v} = \langle -7, 0\rangle$

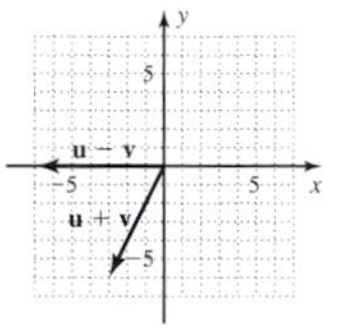

49. $\mathbf{u} = 8\mathbf{i} + 15\mathbf{j}$
$|\mathbf{u}| = 17$

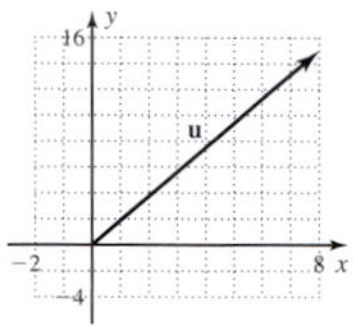

51. $\mathbf{p} = -3.2\mathbf{i} - 5.7\mathbf{j}$
$|\mathbf{p}| \approx 6.54$

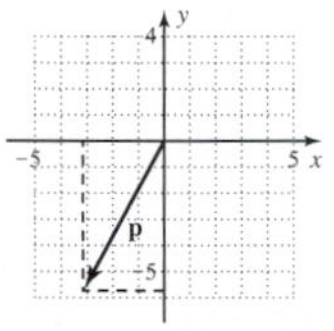

53. a.

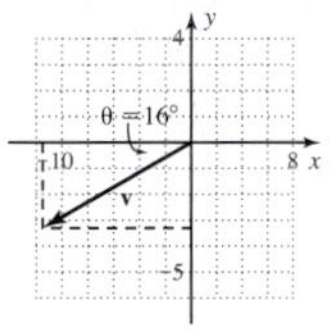

b. $\mathbf{v} = \langle -11.5, -3.3\rangle$
c. $\mathbf{v} = -11.5\mathbf{i} - 3.3\mathbf{j}$

55. a.

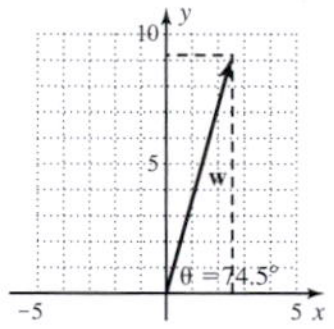

b. $\mathbf{w} = \langle 2.5, 9.2\rangle$
c. $\mathbf{w} = 2.5\mathbf{i} + 9.2\mathbf{j}$

57. a. $\mathbf{p} = -2\mathbf{i} + 2\mathbf{j}$; $|\mathbf{p}| = 2\sqrt{2}$, $\theta = 135°$
b. $\mathbf{q} = 6\mathbf{i} - 8\mathbf{j}$; $|\mathbf{q}| = 10$, $\theta \approx 306.9°$
c. $\mathbf{r} = -2\mathbf{i} + 1.5\mathbf{j}$; $|\mathbf{r}| = 2.5$, $\theta \approx 143.1°$
d. $\mathbf{s} = 10\mathbf{i} - 13\mathbf{j}$; $|\mathbf{s}| \approx 16.4$, $\theta \approx 307.6°$

59. a. $\mathbf{p} = 2\sqrt{2}\mathbf{i} + 2\mathbf{j}$; $|\mathbf{p}| \approx 3.5$, $\theta = 35.3°$
b. $\mathbf{q} = 8\sqrt{2}\mathbf{i} + 12\mathbf{j}$; $|\mathbf{q}| \approx 16.5$, $\theta \approx 46.7°$
c. $\mathbf{r} = 5.5\sqrt{2}\mathbf{i} + 6.5\mathbf{j}$; $|\mathbf{r}| \approx 10.1$, $\theta \approx 39.9°$
d. $\mathbf{s} = 11\sqrt{2}\mathbf{i} + 17\mathbf{j}$; $|\mathbf{s}| \approx 23.0$, $\theta \approx 47.5°$

61. a. $\mathbf{p} = 8\mathbf{i} + 4\mathbf{j}$; $|\mathbf{p}| \approx 8.9$, $\theta \approx 26.6°$
b. $\mathbf{q} = 16\mathbf{i} + 4\mathbf{j}$; $|\mathbf{q}| \approx 16.5$, $\theta \approx 14.0°$
c. $\mathbf{r} = 18\mathbf{i} + 8\mathbf{j}$; $|\mathbf{r}| \approx 19.7$, $\theta \approx 24.0°$
d. $\mathbf{s} = 20\mathbf{i} + 4\mathbf{j}$; $|\mathbf{s}| \approx 20.4$, $\theta \approx 11.3°$

63. $\left\langle \frac{7}{25}, \frac{24}{25}\right\rangle$, verified **65.** $\left\langle \frac{-20}{29}, \frac{21}{29}\right\rangle$, verified

67. $\frac{20}{29}\mathbf{i} - \frac{21}{29}\mathbf{j}$, verified **69.** $\frac{7}{25}\mathbf{i} + \frac{24}{25}\mathbf{j}$, verified

71. $\left\langle \frac{13}{\sqrt{178}}, \frac{3}{\sqrt{178}}\right\rangle$, verified **73.** $\frac{6}{\sqrt{157}}\mathbf{i} + \frac{11}{\sqrt{157}}\mathbf{j}$, verified

75. $4.48\left\langle \frac{5}{\sqrt{29}}, \frac{2}{\sqrt{29}}\right\rangle \approx \langle 4.16, 1.66\rangle$

77. $5.83\left\langle \frac{8}{\sqrt{73}}, \frac{-3}{\sqrt{73}}\right\rangle \approx \langle 5.46, -2.05\rangle$ **79.** ≈ 14.4 **81.** $\approx 24.3°$

83. hor. comp. $\approx$ 79.9 ft/sec; vert. comp. $\approx$ 60.2 ft/sec **85.** heading 39° at a speed of 417.3 mph, relative to the ground **87.** $\approx$ (82.10 cm, 22.00 cm)

89. $1\langle a, b\rangle = \langle 1a, 1b\rangle = \langle a, b\rangle$

91. $\langle a, b\rangle - \langle c, d\rangle = \langle a - c, b - d\rangle = \langle a + (-c), b + (-d)\rangle$
$= \langle a, b\rangle + \langle -c, -d\rangle = \langle a, b\rangle + -1\langle c, d\rangle = \mathbf{u} + (-1\mathbf{v})$

93. $(ck)\mathbf{u} = \langle cka, ckb\rangle = c\langle ka, kb\rangle = c(k\mathbf{u})$
$c(k\mathbf{u}) = \langle cka, ckb\rangle = \langle kca, kcb\rangle = k\langle ca, cb\rangle = k(c\mathbf{u})$

95. $\mathbf{u} + (-\mathbf{u}) = \langle a, b\rangle + \langle -a, -b\rangle = \langle a - a, b - b\rangle = \langle 0, 0\rangle$

97. $(c + k)\mathbf{u} = (c + k)\langle a, b\rangle = \langle (c + k)a, (c + k)b\rangle =$
$\langle ca + ka, cb + kb\rangle = \langle ca, cb\rangle + \langle ka, kb\rangle = c\mathbf{u} + k\mathbf{u}$

99. a. 240 mph **b.** 230.0 mph **c.** 204.0 mph **d.** 174.0 mph
e. 160 mph; answers will vary.

101.
$$\mathbf{u} = \frac{a\mathbf{i} + b\mathbf{j}}{\sqrt{a^2 + b^2}} = \frac{a}{\sqrt{a^2 + b^2}}\mathbf{i} + \frac{b}{\sqrt{a^2 + b^2}}\mathbf{j}$$
$$|\mathbf{u}| = \sqrt{\frac{a^2}{a^2 + b^2} + \frac{b^2}{a^2 + b^2}} = \sqrt{\frac{a^2 + b^2}{a^2 + b^2}} = 1$$

103. about 20,626 ft **105.**

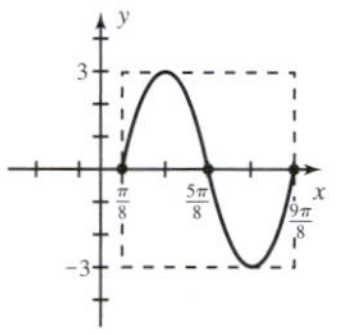

107. about 1113 mi

Mid-Chapter Check, pp. 352–353

1. $\sin B = \frac{b \sin A}{a}$ **2.** $\cos B = \frac{a^2 + c^2 - b^2}{2ac}$ **3.** $a \approx 129$ m, $B \approx 86.8°$, $C \approx 62.2°$ **4.** $A \approx 42.3°$, $B \approx 81.5°$, $C \approx 56.2°$

5. $A = 44°$ $a = 2.1$ km
$B \approx 68.1°$ $b \approx 2.8$ km
$C \approx 67.9°$ $c = 2.8$ km
or
$A = 44°$ $a = 2.1$ km
$B \approx 23.9°$ $b \approx 1.2$ km
$C \approx 112.1°$ $c = 2.8$ km

6. $A \approx 18.5°$ $a = 70$ yd
$B \approx 134.5°$ $b \approx 157.2$ yd
$C = 27°$ $c = 100$ yd

7. $\approx$ 38.3 ft **8.** $\approx$169 m **9.** $\alpha \approx 49.6°$, $\beta \approx 92.2°$, $\gamma \approx 38.2°$
10. about 12.7 knots, heading 52.9°

Reinforcing Basic Concepts, pp. 353–354

1.

Angles	Sides
$A = 35°$	$a = 11.6$ cm
$B \approx 82.0°$	$b = 20$ cm
$C \approx 63.0°$	$c = 18$ cm

Very close.

2. For $\angle A = 35°$, $a \approx 10.3$
For $\angle A = 50°$, $a \approx 14.2$
For $\angle A = 70°$, $a \approx 19.1$; yes, very close

Exercises 5.5, pp. 365–369

1. equilibrium; zero **3.** orthogonal **5.** Answers will vary.
7. $\langle 6, 8\rangle$ **9.** $\langle -5, 10\rangle$ **11.** $-6\mathbf{i} - 8\mathbf{j}$ **13.** $-2.2\mathbf{i} + 0.4\mathbf{j}$
15. $\langle -11.48, -9.16\rangle$ **17.** $\langle -24, -27\rangle$ **19.** $|\mathbf{F}_3| \approx 3336.8$; $\theta \approx 268.5°$

21. 37.16 kg **23.** 644.49 lb **25.** 2606.74 kg **27.** approx. 286.79 lb **29.** approx. 43.8° **31.** 1125 ft-lb **33.** approx. 957.0 ft **35.** approx. 64,951.90 ft-lb **37.** approx. 451.72 lb **39.** approx. 2819.08 J **41.** 800 ft-lb **43.** 118 ft-lb **45.** verified **47.** verified **49. a.** 29 **b.** 45° **51. a.** 0 **b.** 90° **53. a.** 1 **b.** 89.4° **55.** yes **57.** no **59.** yes **61.** 3.68 **63.** −4 **65.** 3.17 **67. a.** $\langle 3.73, 1.40\rangle$ **b.** $\mathbf{u}_1 = \langle 3.73, 1.40\rangle, \mathbf{u}_2 = \langle -1.73, 4.60\rangle$ **69. a.** $\langle -0.65, 0.11\rangle$ **b.** $\mathbf{u}_1 = \langle -0.65, 0.11\rangle, \mathbf{u}_2 = \langle -1.35, -8.11\rangle$ **71. a.** $10.54\mathbf{i} + 1.76\mathbf{j}$ **b.** $\mathbf{u}_1 = 10.54\mathbf{i} + 1.76\mathbf{j}, \mathbf{u}_2 = -0.54\mathbf{i} + 3.24\mathbf{j}$ **73. a.** projectile is about 375 ft away, and 505.52 ft high **b.** approx. 1.27 sec and 12.26 sec **75. a.** projectile is about 424.26 ft away, and 280.26 ft high **b.** approx. 2.44 sec and 6.40 sec **77.** about 74.84 ft; $t \approx 3.9 - 1.2 = 2.7$ sec

79.
$$\begin{aligned}\mathbf{w}\cdot\langle\mathbf{u}+\mathbf{v}\rangle &= \langle e,f\rangle\cdot\langle a+c, b+d\rangle\\ &= ea + ec + fb + fd\\ &= ea + fb + ec + fd\\ &= \langle e,f\rangle\cdot\langle a,b\rangle + \langle e,f\rangle\cdot\langle c,d\rangle\\ &= \mathbf{w}\cdot\mathbf{u} + \mathbf{w}\cdot\mathbf{v}\end{aligned}$$

81.
$$\begin{aligned}\mathbf{0}\cdot\mathbf{u} &= \langle 0,0\rangle\cdot\langle a,b\rangle\\ &= 0\cdot a + 0\cdot b\\ &= a\cdot 0 + b\cdot 0\\ &= \langle a,b\rangle\cdot\langle 0,0\rangle\\ &= \mathbf{u}\cdot\mathbf{0}\end{aligned} \qquad \begin{aligned}\mathbf{0}\cdot\mathbf{u} &= \langle 0,0\rangle\cdot\langle a,b\rangle\\ &= 0\cdot a + 0\cdot b\\ &= 0 + 0\\ &= 0\end{aligned}$$

83. $\theta \approx 56.9°$; Answers will vary. **85.** $R = \frac{v^2\sin(2\theta)}{32}$; $\theta = 45°$

87.

Angles	Sides
$A \approx 25.1°$	$a \approx 25.98$ cm
$B \approx 34.9°$	$b = 53$ cm
$C = 120°$	$c = 35$ cm

89. Domain: $[-1, 1]$ Range: $\left[\frac{-\pi}{2}, \frac{\pi}{2}\right]$

91. $\sin 15° = \frac{1}{2}\sqrt{2-\sqrt{3}}$, $\cos 75° = \frac{1}{2}\sqrt{2-\sqrt{3}}$; they are identical.

Exercises 5.6, pp. 376–380

1. $3 - 2i$ **3.** $2, 3\sqrt{2}$ **5.** (b) is correct. **7. a.** $4i$ **b.** $7i$ **c.** $3\sqrt{3}$ **d.** $6\sqrt{2}$ **9. a.** $-3i\sqrt{2}$ **b.** $-5i\sqrt{2}$ **c.** $15i$ **d.** $6i$

11. a. $i\sqrt{19}$ **b.** $i\sqrt{31}$ **c.** $\frac{2\sqrt{3}}{5}i$ **d.** $\frac{3\sqrt{2}}{8}i$

13. a. $1 + i; a = 1, b = 1$ **b.** $2 + \sqrt{3}i; a = 2, b = \sqrt{3}$

15. a. $4 + 2i; a = 4, b = 2$ **b.** $2 - \sqrt{2}i; a = 2, b = -\sqrt{2}$

17. a. $5 + 0i; a = 5, b = 0$ **b.** $0 + 3i; a = 0, b = 3$

19. a. $0 + 18i; a = 0, b = 18$ **b.** $0 + \frac{\sqrt{2}}{2}i; a = 0, b = \frac{\sqrt{2}}{2}$

21. a. $4 + 5\sqrt{2}i; a = 4, b = 5\sqrt{2}$ **b.** $-5 + 3\sqrt{3}i; a = -5, b = 3\sqrt{3}$

23. a. $\frac{7}{4} + \frac{7\sqrt{2}}{8}i; a = \frac{7}{4}, b = \frac{7\sqrt{2}}{8}$ **b.** $\frac{1}{2} + \frac{\sqrt{10}}{2}i; a = \frac{1}{2}, b = \frac{\sqrt{10}}{2}$

25. a. $19 + i$ **b.** $2 - 4i$ **c.** $9 + 10\sqrt{3}i$ **27. a.** $-3 + 2i$ **b.** 8 **c.** $2 - 8i$ **29. a.** $2.7 + 0.2i$ **b.** $15 + \frac{1}{12}i$ **c.** $-2 - \frac{1}{8}i$ **31. a.** 15 **b.** 16 **33. a.** $-21 - 35i$ **b.** $-42 - 18i$ **35. a.** $-12 - 5i$ **b.** $1 + 5i$ **37. a.** $4 - 5i$; 41 **b.** $3 + i\sqrt{2}$; 11 **39. a.** $-7i$; 49 **b.** $\frac{1}{2} + \frac{2}{3}i$; $\frac{25}{36}$ **41. a.** 41 **b.** 74 **43. a.** 11 **b.** $\frac{17}{36}$ **45. a.** $-5 + 12i$ **b.** $-7 - 24i$ **47. a.** $-21 - 20i$ **b.** $7 + 6i\sqrt{2}$ **49.** no **51.** yes **53.** yes **55.** yes **57.** yes **59.** verified **61. a.** 1 **b.** −1 **c.** $-i$ **d.** i

63. a. $\frac{2}{7}i$ **b.** $\frac{-4}{5}i$ **65. a.** $\frac{21}{13} - \frac{14}{13}i$ **b.** $\frac{-10}{13} - \frac{15}{13}i$ **67. a.** $1 - \frac{3}{4}i$ **b.** $-1 - \frac{2}{3}i$ **69. a.** $\sqrt{13}$ **b.** $\sqrt{41}$ **c.** $\sqrt{11}$ **71.** $A + B = 10$, $AB = 40$ **73.** $7 - 5i\ \Omega$ **75.** $25 + 5i$ V **77.** $\frac{7}{4} + i\ \Omega$ **79.** $(x + 3i)(x - 3i)$; verified **81.** You get a result of 6; you get a result of −6; no, $\sqrt{-4\cdot -9} \neq \sqrt{-4}\cdot\sqrt{-9}$ **83.** $c > 4$ **85.** $\theta = 62.7°$

87.
$$\begin{aligned}1 + \sin\theta &= \frac{\cos^2\theta}{1 - \sin\theta}\\ &= \frac{1 - \sin^2\theta}{1 - \sin\theta}\\ &= \frac{(1 + \sin\theta)(1 - \sin\theta)}{1 - \sin\theta}\\ &= 1 + \sin\theta\end{aligned}$$

89.

Angles	Sides
$A = 32°$	$a \approx 138.4$ m
$B \approx 106.8°$	$b = 250$ m
$C \approx 41.2°$	$c = 172$ m

$P \approx 560.4$ m, $A \approx 11{,}393.3$ m^2

Exercises 5.7, pp. 388–391

1. modulus; argument **3.** multiply; add
5. $2(\cos 240° + i\sin 240°)$, z is in QIII

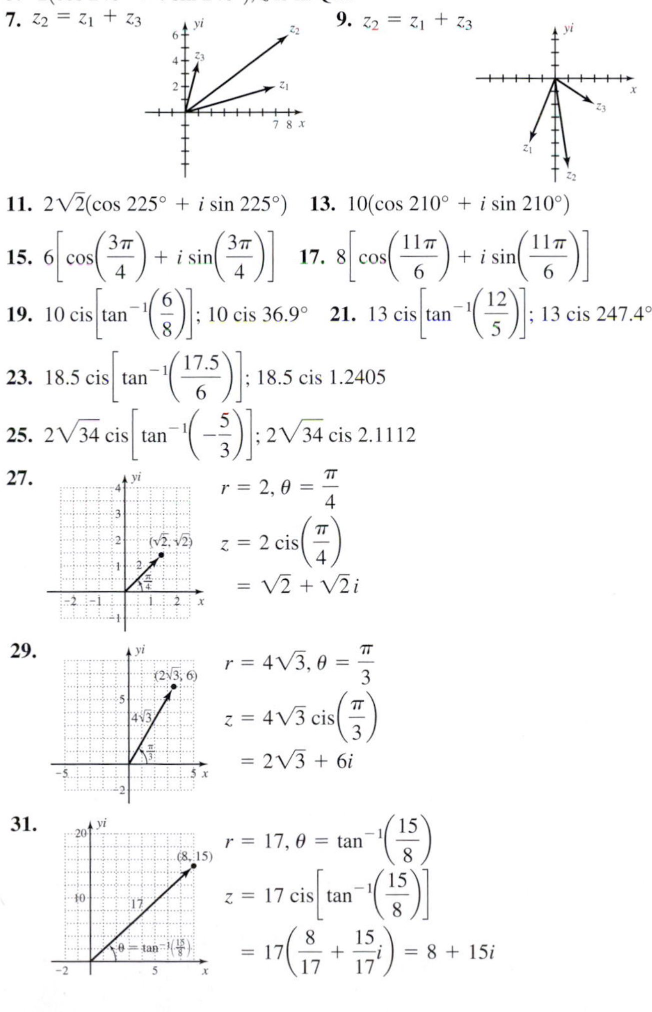

7. $z_2 = z_1 + z_3$ **9.** $z_2 = z_1 + z_3$

11. $2\sqrt{2}(\cos 225° + i\sin 225°)$ **13.** $10(\cos 210° + i\sin 210°)$

15. $6\left[\cos\left(\frac{3\pi}{4}\right) + i\sin\left(\frac{3\pi}{4}\right)\right]$ **17.** $8\left[\cos\left(\frac{11\pi}{6}\right) + i\sin\left(\frac{11\pi}{6}\right)\right]$

19. $10\ \text{cis}\left[\tan^{-1}\left(\frac{6}{8}\right)\right]$; $10\ \text{cis}\ 36.9°$ **21.** $13\ \text{cis}\left[\tan^{-1}\left(\frac{12}{5}\right)\right]$; $13\ \text{cis}\ 247.4°$

23. $18.5\ \text{cis}\left[\tan^{-1}\left(\frac{17.5}{6}\right)\right]$; $18.5\ \text{cis}\ 1.2405$

25. $2\sqrt{34}\ \text{cis}\left[\tan^{-1}\left(-\frac{5}{3}\right)\right]$; $2\sqrt{34}\ \text{cis}\ 2.1112$

27. $r = 2, \theta = \frac{\pi}{4}$; $z = 2\ \text{cis}\left(\frac{\pi}{4}\right) = \sqrt{2} + \sqrt{2}i$

29. $r = 4\sqrt{3}, \theta = \frac{\pi}{3}$; $z = 4\sqrt{3}\ \text{cis}\left(\frac{\pi}{3}\right) = 2\sqrt{3} + 6i$

31. $r = 17, \theta = \tan^{-1}\left(\frac{15}{8}\right)$; $z = 17\ \text{cis}\left[\tan^{-1}\left(\frac{15}{8}\right)\right] = 17\left(\frac{8}{17} + \frac{15}{17}i\right) = 8 + 15i$

33. $r = 6, \theta = \pi - \tan^{-1}\left(\frac{5}{\sqrt{11}}\right)$

$z = 6\operatorname{cis}\left[\pi - \tan^{-1}\frac{5}{\sqrt{11}}\right]$

$= 6\left(-\frac{\sqrt{11}}{6} + \frac{5}{6}i\right) = -\sqrt{11} + 5i$

35. $r_1 = 2\sqrt{2}, \quad r_2 = 3\sqrt{2}, \quad \theta_1 = 135°, \quad \theta_2 = 45°;$
$z = z_1z_2 = -12 + 0i \Rightarrow r = 12, \quad \theta = 180°;$
$r_1r_2 = 2\sqrt{2}\,(3\sqrt{2}) = 12$✓
$\theta_1 + \theta_2 = 135° + 45° = 180°$✓
37. $r_1 = 2, \quad r_2 = 2, \quad \theta_1 = 30°, \quad \theta_2 = 60°;$
$z = \frac{z_1}{z_2} = \frac{\sqrt{3}}{2} - \frac{1}{2}i \Rightarrow r = 1, \quad \theta = -30°;$
$\frac{r_1}{r_2} = \frac{2}{2} = 1$✓
$\theta_1 - \theta_2 = 30° - 60° = -30°$✓
39. $z_1z_2 = -24 + 0i, \frac{z_1}{z_2} = -\frac{4}{3} + \frac{4\sqrt{3}}{3}i$ **41.** $z_1z_2 = 21\sqrt{3} - 21i,$ $\frac{z_1}{z_2} = \frac{\sqrt{3}}{7} + \frac{1}{7}i$ **43.** $z_1z_2 = -10.84 + 12.04i, \frac{z_1}{z_2} = -1.55 - 4.76i$
45. $z_1z_2 = 0 + 40i, \frac{z_1}{z_2} = \frac{5\sqrt{3}}{4} + \frac{5}{4}i$ **47.** $z_1z_2 = -10 - 10\sqrt{3}i,$ $\frac{z_1}{z_2} = \frac{-5}{2} + 0i$ **49.** $z_1z_2 = -2.93 + 8.5i, \frac{z_1}{z_2} = 2.29 + 3.28i$
51. verified; verified, $u^2 + v^2 + w^2 = uv + uw + vw$
$(1 + 4\sqrt{3}i) + (97 + 20\sqrt{3}i + (-39 + 60\sqrt{3}i) = (17 + 12\sqrt{3}i)$
$+ (-3 + 16\sqrt{3}i) + (45 + 56\sqrt{3}i), 59 + 84\sqrt{3}i = 59 + 84\sqrt{3}i$
53. a. 17 cis 28.1° **b.** 51 V **55. a.** 8.60 cis 324.5° **b.** 15.48 V
57. a. 13 cis 22.6° **b.** 22.1 V **59.** $I = 2\operatorname{cis} 30°; Z = 5\sqrt{2}\operatorname{cis} 45°;$ $V = 10\sqrt{2}\operatorname{cis} 75°$
61. $I = \sqrt{13}\operatorname{cis} 326.3°; Z = \frac{17}{4}\operatorname{cis} 61.9°; V = \frac{17\sqrt{3}}{4}\operatorname{cis} 28.2°$
63. $V = 4\operatorname{cis} 60°; \; Z = 4\sqrt{2}\operatorname{cis} 315°; I = \frac{\sqrt{2}}{2}\operatorname{cis} 105°$
65. $V = 5\operatorname{cis} 306.9°; Z = 8.5\operatorname{cis} 61.9°; I = \frac{10}{17}\operatorname{cis} 245°$
67. $\frac{\sqrt{65}\operatorname{cis} 29.7°}{4}$ **69.** Answers will vary. **71.** verified
73. $X_L = X_C$ and R is as small as possible **75. a.** $\frac{\sqrt{2 + \sqrt{3}}}{2}$
b. $\frac{\sqrt{6} + \sqrt{2}}{4}$, verified **77. a.** 65°47′24″ **b.** 65.79° **c.** 1.1483
79. approx. 29.3°

Exercises 5.8, pp. 399–401

1. $r^5[\cos(5\theta) + i\sin(5\theta)]$; De Moivre's **3.** complex
5. $z_5 = 2\operatorname{cis} 366° = 2\operatorname{cis} 6°, z_6 = 2\operatorname{cis} 438° = 2\operatorname{cis} 78°, z_7 = 2\operatorname{cis} 510°$ $= 2\operatorname{cis} 150°$; answers will vary. **7.** $r = 3\sqrt{2}; n = 4; \theta = 45°; -324$
9. $r = 2; n = 3; \theta = 120°; 8$ **11.** $r = 1; n = 5; \theta = -60°; \frac{1}{2} + \frac{\sqrt{3}}{2}i$
13. $r = 1; n = 6; \theta = -45°; i$ **15.** $r = 4; n = 3; \theta = -30°; -64i$
17. $r = \frac{\sqrt{2}}{2}; n = 5; \theta = 135°; \frac{1}{8} - \frac{1}{8}i$ **19.** verified **21.** verified
23. verified **25.** verified **27.** $r = 16; n = 4; \theta = 120°;$ roots: $\sqrt{3} + i, -1 + \sqrt{3}i, -\sqrt{3} - i, 1 - \sqrt{3}i$
29. $r = 7\sqrt{2}; n = 4; \theta = 225°;$ roots: $0.9855 + 1.4749i,$ $-1.4749 + 0.9855i, -0.9855 - 1.4749i, 1.4749i - 0.9855i$
31. $r = 1; n = 5; \theta = 0°;$ roots: $1, 0.3090 \pm 0.9511i, -0.8090 \pm 0.5878i$
33. $r = 243; n = 5; \theta = 0°;$ roots: $3, 0.9271 \pm 2.8532i,$ $-2.4271 \pm 1.7634i$ **35.** $r = 27; n = 3; \theta = 270°;$ roots: $3i, \frac{-3\sqrt{3}}{2} - \frac{3}{2}i, \frac{3\sqrt{3}}{2} - \frac{3}{2}i$ **37.** $2, 0.6180 \pm 1.9021i,$ $-1.6180 \pm 1.1756i$ **39.** $\frac{3\sqrt{3}}{2} + \frac{3}{2}i, -\frac{3\sqrt{3}}{2} + \frac{3}{2}i, -3i$
41. $1.1346 + 0.1797i, 0.1797 + 1.1346i, -1.0235 + 0.5215i,$ $-0.8123 - 0.8123i, 0.5215 - 1.0235i$ **43.** $x = 1, -\frac{1}{2} \pm \frac{\sqrt{3}}{2}i.$
These are the same results as in Example 4. **45.** $D = -4, z_0 = 8^{\frac{1}{6}}\operatorname{cis} 45°,$ $z_1 = 8^{\frac{1}{6}}\operatorname{cis} 165°, z_2 = 8^{\frac{1}{6}}\operatorname{cis} 285°, z_0 = 8^{\frac{1}{6}}\operatorname{cis} 75°, z_1 = 8^{\frac{1}{6}}\operatorname{cis} 195°,$ $z_2 = 8^{\frac{1}{6}}\operatorname{cis} 315°$ **47.** $2.0582 \pm 0.4859i, -2.0582 \pm 0.4859i$
49. $0.5951 \pm 2.5207i, -0.5951 \pm 2.5207i$ **51.** $2.3271 \pm 0.6446i,$ $-2.3271 \pm 0.6446i$ **53.** verified **55. a.** numerator: $-117 + 44j,$ denominator: $-21 + 72j$ **b.** $1 + \frac{4}{3}j$ **c.** verified **57.** Answers will vary.
59. $-7 - 24i$ **61.** $z \approx -2.7320, z \approx 0.7320, z = 2$
Note: Using sum and difference identities, all three solutions can actually be given in exact form: $-1 - \sqrt{3}, -1 + \sqrt{3}, 2.$
63. a. $I = 4 + i,$ **b.** $(3 - 2i)(4 + i) = 12 + 3i - 8i - 2i^2 =$ $12 - 5i + 2 = 14 - 5i$✓ **65.** $\frac{\tan^2 x}{\sec x + 1} = \frac{1 - \cos x}{\cos x}, \frac{\sec^2 x - 1}{\sec x + 1} =$ $\frac{(\sec x + 1)(\sec x - 1)}{\sec x + 1} = \sec x - 1 = \frac{1}{\cos x} - \frac{\cos x}{\cos x} = \frac{1 - \cos x}{\cos x}$
67.

Angles	Sides
$A = 30°$	$a = 2\sqrt{3}$
$B = 60°$	$b = 6$
$C = 90°$	$c = 4\sqrt{3}$

Summary and Concept Review, pp. 402–408

1.

Angles	Sides
$A = 36°$	$a \approx 205.35$ cm
$B = 21°$	$b \approx 125.20$ cm
$C = 123°$	$c = 293$ cm

2.

Angles	Sides
$A = 28°$	$a \approx 140.59$ yd
$B = 10°$	$b = 52$ yd
$C = 142°$	$c \approx 184.36$ yd

3. approx. 41.84 ft **4.** approx. 20.2° and 159.8°
5.

Angles	Sides
$A = 35°$	$a = 67$ cm
$B_1 \approx 64.0°$	$b = 105$ cm
$C_1 \approx 81.0°$	$c_1 \approx 115.37$ cm

Angles	Sides
$A = 35°$	$a = 67$ cm
$B_2 \approx 116.0°$	$b = 105$ cm
$C_2 \approx 29.0°$	$c_2 \approx 56.65$ cm

6. no; 36° **7.** approx. 36.9° **8.** approx. 385.5 m **9.** 133.2°, 30.1°, and 16.8° **10.** $|\mathbf{v}| \approx 10.30; \theta \approx 29.1°$

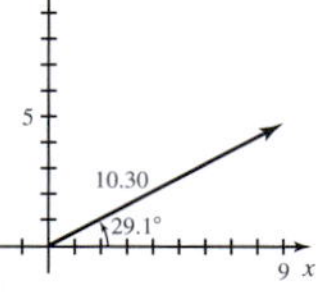

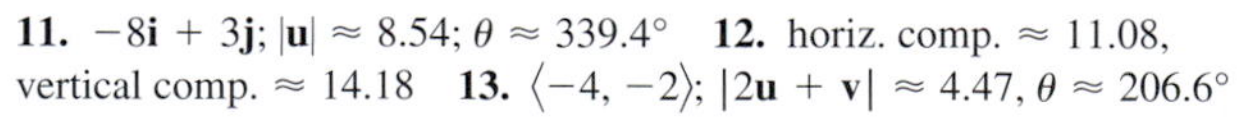
11. $-8\mathbf{i} + 3\mathbf{j}; |\mathbf{u}| \approx 8.54; \theta \approx 339.4°$ **12.** horiz. comp. ≈ 11.08, vertical comp. ≈ 14.18 **13.** $\langle -4, -2\rangle; |2\mathbf{u} + \mathbf{v}| \approx 4.47, \theta \approx 206.6°$

14. $\frac{7}{\sqrt{193}}\mathbf{i} + \frac{12}{\sqrt{193}}\mathbf{j}$ **15.** QII; since the x-component is negative and the y-component is positive. **16.** $\frac{1}{6}$ mi. **17.** approx. 19.7°
18. $\langle -25, -123 \rangle$ **19.** approx. −0.87 **20.** 4 **21.** $\mathbf{p} \cdot \mathbf{q} = -6; \theta \approx 97.9°$
22. 4340 ft-lb **23.** approx. 417.81 lb **24.** approx. 8156.77 ft-lb
25. a. $x \approx 269.97$ ft; $y \approx 285.74$ ft **b.** approx. 0.74 sec
26. $6i\sqrt{2}$ **27.** $24i\sqrt{3}$ **28.** $-2 + i\sqrt{2}$ **29.** $3\sqrt{2}\,i$ **30.** i
31. $21 + 20i$ **32.** $-2 + i$ **33.** $-5 + 7i$ **34.** 13 **35.** $-20 - 12i$
36. $(5i)^2 - 9 = -34$ $\quad (-5i)^2 - 9 = -34$
$25i^2 - 9 = -34$ $\quad 25i^2 - 9 = -34$
$-25 - 9 = -34$✓ $\quad -25 - 9 = -34$✓
37. $(2 + i\sqrt{5})^2 - 4(2 + i\sqrt{5}) + 9 = 0$
$4 + 4i\sqrt{5} + 5i^2 - 8 - 4i\sqrt{5} + 9 = 0$
$5 + (-5) = 0$✓
$(2 - i\sqrt{5})^2 - 4(2 - i\sqrt{5}) + 9 = 0$
$4 - 4i\sqrt{5} + 5i^2 - 8 + 4i\sqrt{5} + 9 = 0$
$5 + (-5) = 0$✓
38. $2(\cos 240° + i \sin 240°)$ **39.** $3 + 3i$
40.

41. $z_1 z_2 = 16 \operatorname{cis}\left(\frac{5\pi}{12}\right); \frac{z_1}{z_2} = 4 \operatorname{cis}\left(\frac{\pi}{12}\right)$ **42.** $2\sqrt{3} + 2j$
43. $|Z| \approx 10.44; \theta \approx 16.7°$, 10.44 cis 16.7° **44.** $-16 - 16\sqrt{3}i$
45. verified **46.** $\frac{5\sqrt{3}}{2} + \frac{5}{2}i, -\frac{5\sqrt{3}}{2} + \frac{5}{2}i, -5i$ **47.** $6, -3 \pm 3i\sqrt{3}$
48. $2 - 2i, -2 \pm 2i$ **49.** $1 \pm 2i, -1 \pm 2i$ **50.** verified

Mixed Review, pp. 408–409

1.

Angles	Sides
$A = 41°$	$a \approx 13.44$ in.
$B = 27°$	$b \approx 9.30$ in.
$C = 112°$	$c = 19$ in.

3. $x \approx 16.09$, $y \approx 13.50$ **5.** approx. 176.15 ft **7.** approx. 793.70 mph; heading 28.2°
9. One solution possible since side $a >$ side b

Angles	Sides
$A = 31°$	$a = 36$ m
$B \approx 20.1°$	$b \approx 24$ m
$C \approx 128.9°$	$c \approx 54.4$ m

11. approx. 97.9°
13a. $4\sqrt{2}(\cos 315° + i \sin 315°)$

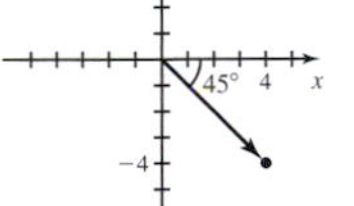

b. $-3 + 3\sqrt{3}i$

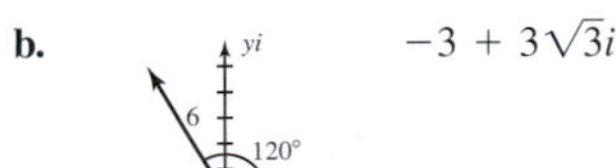

15. ≈ 13.1° **17.** $\text{comp}_{\mathbf{v}}\mathbf{u} \approx -0.87$,
$\text{proj}_{\mathbf{v}}\mathbf{u} \approx \frac{-38}{53}\mathbf{i} + \frac{26}{53}\mathbf{j}$ **19.** $z_0 = \frac{\sqrt{6}}{2} + \frac{\sqrt{2}}{2}i$,
$z_1 = \frac{-\sqrt{2}}{2} + \frac{\sqrt{6}}{2}i, z_2 = \frac{-\sqrt{6}}{2} - \frac{\sqrt{2}}{2}i, z_3 = \frac{\sqrt{2}}{2} - \frac{\sqrt{6}}{2}i$

Practice Test, pp. 409–411

1. 6.58 mi **2.** 137.18 ft
3.

Angles	Sides (in.)
$A_1 \approx 58.8°$	$a = 15$
$B = 20°$	$b = 6$
$C_1 \approx 101.2°$	$c_1 \approx 17.21$

Angles	Sides (in.)
$A_2 \approx 121.2°$	$a = 15$
$B = 20°$	$b = 6$
$C_2 \approx 38.8°$	$c_2 \approx 11.0$

4. a. No **b.** 2.66 mi **5. a.** No **b.** 1 **c.** 8.43 sec
6. a. 2.30 mi **b.** 7516.5 ft **7.** $\angle M \approx 57.1°, \angle B \approx 61.2°, \angle P \approx 61.7°$
8. speed ≈ 73.36 mph, heading ≈ 47.8° **9.** $\theta \approx 36.5°$ **10.** 63.48 cm to the right and 130.05 cm down from the initial point on the ceiling
11. $|\mathbf{F}_3| \approx 212.94$ N, $\theta \approx 251.2°$ **12. a.** $\theta \approx 42.5°$
b. $\text{proj}_{\mathbf{v}}\mathbf{u} = \langle -2.4, 7.2 \rangle$ **c.** $\mathbf{u}_1 = \langle -2.4, 7.2 \rangle, \mathbf{u}_2 = \langle -6.6, -2.2 \rangle$
13. 104.53 ft; 3.27 sec **14.** $2 \operatorname{cis}\left(\frac{\pi}{24}\right)$ **15.** $48\sqrt{2}$ cis 75°; verified
16. $-8 - 8\sqrt{3}i$ **17.** verified **18.** $\frac{5\sqrt{3}}{2} + \frac{5}{2}i, -\frac{5\sqrt{3}}{2} + \frac{5}{2}i, -5i$
19. $2.3039 \pm 1.5192i, -2.3039 \pm 1.5192i$ **20.** $|\mathbf{v}| = 10$; phase angle ≈ 36.9°; $10(\cos 36.9° + i \sin 36.9°)$; $V = 50$ V

Strengthening Core Skills, pp. 412–413

1. 664.46 lb and 640.86 lb **2.** 106.07 lb in each rope **3.** yes

Cumulative Review Chapters 1–5, pp. 413–415

1. $20\sqrt{3}$, 40, 60°, 90° **3.** $t = \frac{\cos^{-1}\left(\frac{D}{A}\right) - C}{B}$ **5.** QIV, $\sin\theta = \frac{-3}{5}$;
$\cos\theta = \frac{4}{5}$; $\tan\theta = \frac{-3}{4}$; $\csc\theta = -\frac{5}{3}$; $\sec\theta = \frac{5}{4}$; $\cot\theta = -\frac{4}{3}$
7. a. $\sin(2\alpha) = 2\sin\alpha\cos\alpha$ **b.** $\sin\left(\frac{\alpha}{2}\right) = \pm\sqrt{\frac{1-\cos\alpha}{2}}$
c. $\sin(\alpha + \beta) = \sin\alpha\cos\beta + \cos\alpha\sin\beta$ **9.** $\cos 19° \approx 0.94$, $\cos 125° \approx -0.58$ **11.** $D(t) = 5.704\sin(0.511t - 1.835) + 12.189$
a. about 14.7 hr **b.** $t \approx 4.6$ to $t \approx 8.7$, approx. April 18 to August 22
13. $\frac{2\cos^2\theta}{\csc^2\theta} + \frac{2\sin^2\theta}{\sec^2\theta} = \sin^2(2\theta)$
$\frac{2\cos^2\theta\sec^2\theta + 2\sin^2\theta\csc^2\theta}{\csc^2\theta\sec^2\theta} =$
$4\sin^2\theta\cos^2\theta =$
$(2\sin\theta\cos\theta)^2 =$
$\sin^2(2\theta) =$
15. $\angle A = 37°, a \approx 33$ cm; $\angle B \approx 34.4°, b = 31$ cm; $\angle C \approx 108.6°$, $c = 52$ cm **17.** about 422.5 lb **19.** $\theta \approx 70°$ **21.** $-128 - 128i\sqrt{3}$
23. $\sqrt{3} + 1i$ **25.** $A = 2, B = 1, C = \frac{\pi}{4}$

CHAPTER 6

Exercises 6.1, pp. 425–429

1. radius, center **3.** complete, square, 1 **5.** Answers will vary.

7. $x^2 + y^2 = 9$ **9.** $(x - 5)^2 + y^2 = 3$

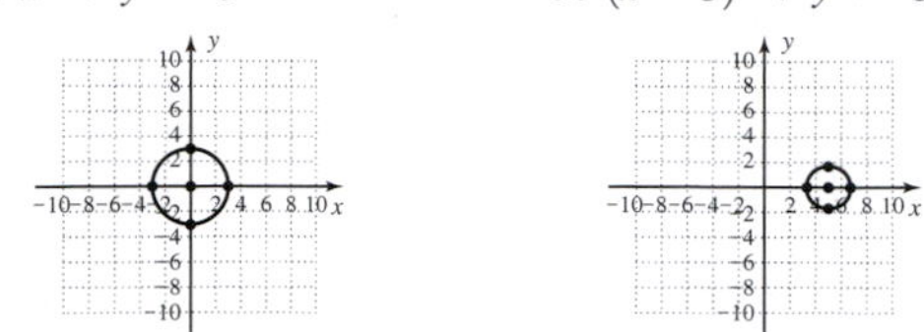

11. $(x - 4)^2 + (y + 3)^2 = 4$ **13.** $(x + 7)^2 + (y + 4)^2 = 7$

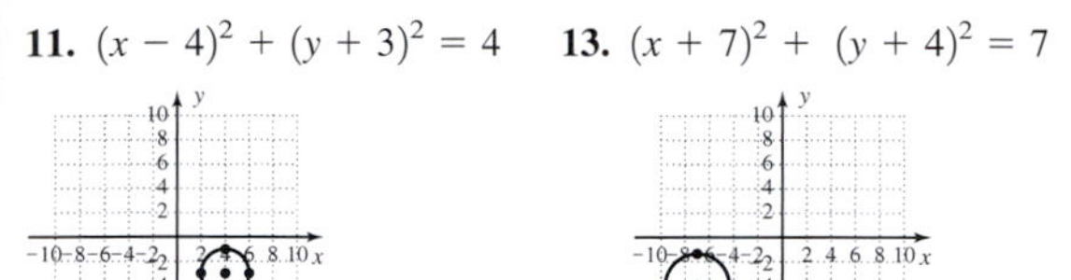

15. $(x - 1)^2 + (y + 2)^2 = 12$ **17.** $(x - 4)^2 + (y - 5)^2 = 12$

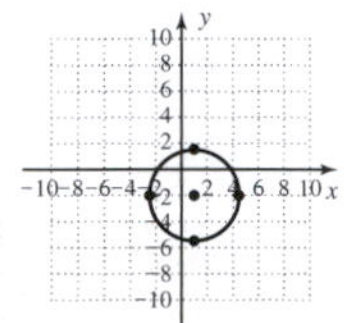

19. $(x - 7)^2 + (y - 1)^2 = 100$ **21.** $(x - 3)^2 + (y - 4)^2 = 41$

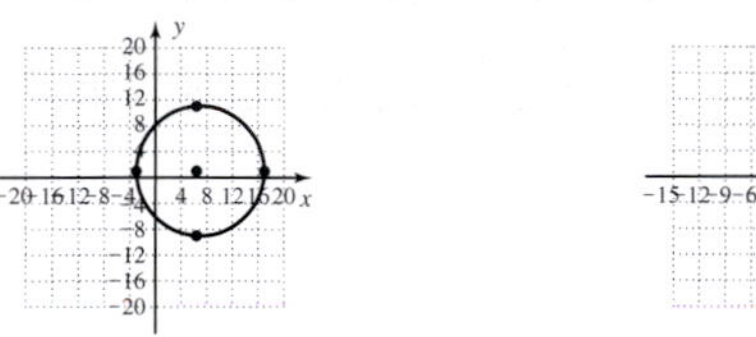

23. $(x - 5)^2 + (y - 4)^2 = 9$ **25.** $(2, 3)$, $r = 2$, $x \in [0, 4]$, $y \in [1, 5]$

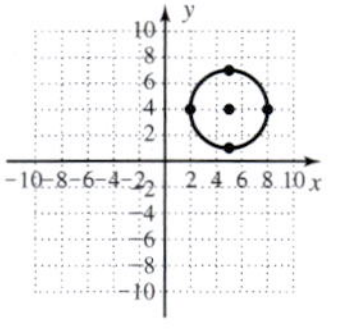

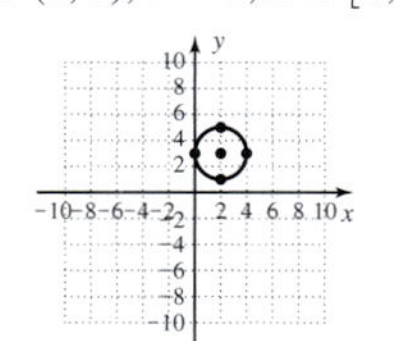

27. $(-1, 2)$, $r = 2\sqrt{3}$, $x \in [-1 - 2\sqrt{3}, -1 + 2\sqrt{3}]$, $y \in [2 - 2\sqrt{3}, 2 + 2\sqrt{3}]$

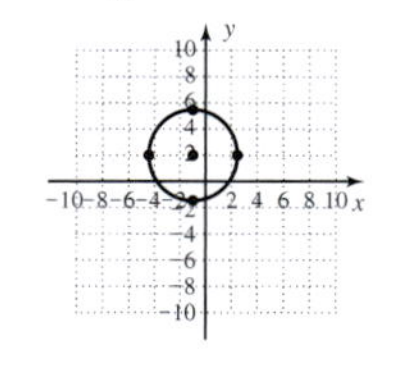

29. $(-4, 0)$, $r = 9$, $x \in [-13, 5]$, $y \in [-9, 9]$

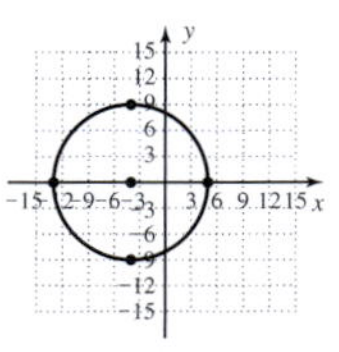

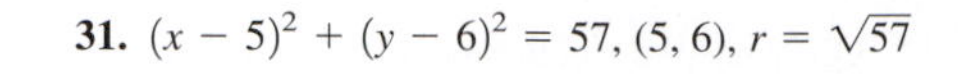

31. $(x - 5)^2 + (y - 6)^2 = 57$, $(5, 6)$, $r = \sqrt{57}$

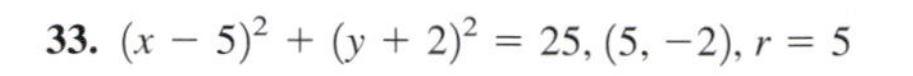

33. $(x - 5)^2 + (y + 2)^2 = 25$, $(5, -2)$, $r = 5$

35. $x^2 + (y + 3)^2 = 14$, $(0, -3)$, $r = \sqrt{14}$

37. $(x + 2)^2 + (y + 5)^2 = 11$, $(-2, -5)$, $r = \sqrt{11}$

39. $(x + 7)^2 + y^2 = 37$, $(-7, 0)$, $r = \sqrt{37}$

41. $(x - 3)^2 + (y + 5)^2 = 32$, $(3, -5)$, $r = 4\sqrt{2}$

43.

45.

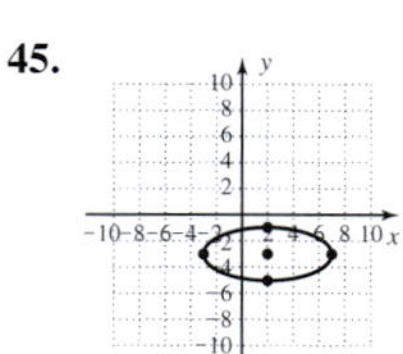

47.

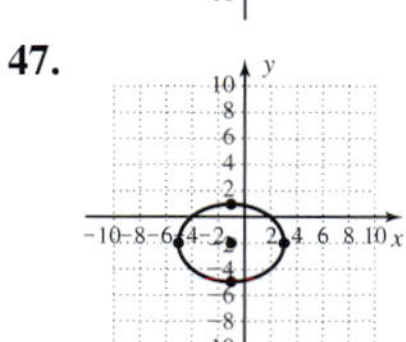

49. ellipse

51. circle

53. ellipse

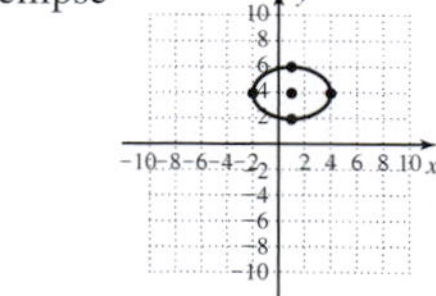

55. a. $\frac{x^2}{16} + \frac{y^2}{4} = 1$, $(0, 0)$, $a = 4$, $b = 2$
b. $(-4, 0)$, $(4, 0)$, $(0, -2)$, $(0, 2)$ **c.**

57. a. $\frac{x^2}{9} + \frac{y^2}{16} = 1$, $(0, 0)$, $a = 3$, $b = 4$
b. $(0, -4)$, $(0, 4)$ $(-3, 0)$ $(3, 0)$ **c.**

59. a. $\frac{x^2}{5} + \frac{y^2}{2} = 1$, $(0, 0)$, $a = \sqrt{5}$, $b = \sqrt{2}$
b. $(-\sqrt{5}, 0)$, $(\sqrt{5}, 0)$, $(0, -\sqrt{2})$, $(0, \sqrt{2})$ **c.**

61. $x^2 + \frac{(y+3)^2}{4} = 1$ **63.** $\frac{(x+2)^2}{16} + \frac{(y-1)^2}{4} = 1$

65. $\frac{(x-3)^2}{4} + \frac{(y+5)^2}{10} = 1$ **67.** $\frac{(x-3)^2}{25} + \frac{(y+2)^2}{10} = 1$

69. $A = 50$ units^2 **71.** No **73.** Red: $(x-2)^2 + (y-2)^2 = 4$; Blue: $(x-2)^2 + y^2 = 16$; Area blue $= 12\pi$ units^2

75.

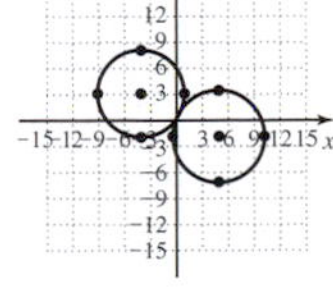

No, distance between centers is less than sum of radii.

77. $\frac{x^2}{36^2} + \frac{y^2}{(35.25)^2} = 1$ **79.** 9000π yd^2 **81.** $\frac{x^2}{15^2} + \frac{y^2}{8^2} = 1$, 6.4 ft

83. Answers will vary; aphelion: 155 million miles; semi-major axis: 142 million miles **85.** x^2 and y^2 must have the same sign.

87. $\frac{(x-2)^2}{9} + \frac{(y-1)^2}{4} = 1$

89.
$$\begin{aligned}\frac{\cos x}{1-\sin x} &= \frac{1+\sin x}{\cos x}\\ &= \frac{(1+\sin x)(1-\sin x)}{\cos x(1-\sin x)}\\ &= \frac{1-\sin^2 x}{\cos x(1-\sin x)}\\ &= \frac{\cos^2 x}{\cos x(1-\sin x)}\\ &= \frac{\cos x}{1-\sin x}\end{aligned}$$

91. $|z_1| = \sqrt{226} < \sqrt{250} = |z_2|$ **93.** $L \approx 115.0$ m

Exercises 6.2, pp. 437–440

1. transverse **3.** midway **5.** Answers will vary.

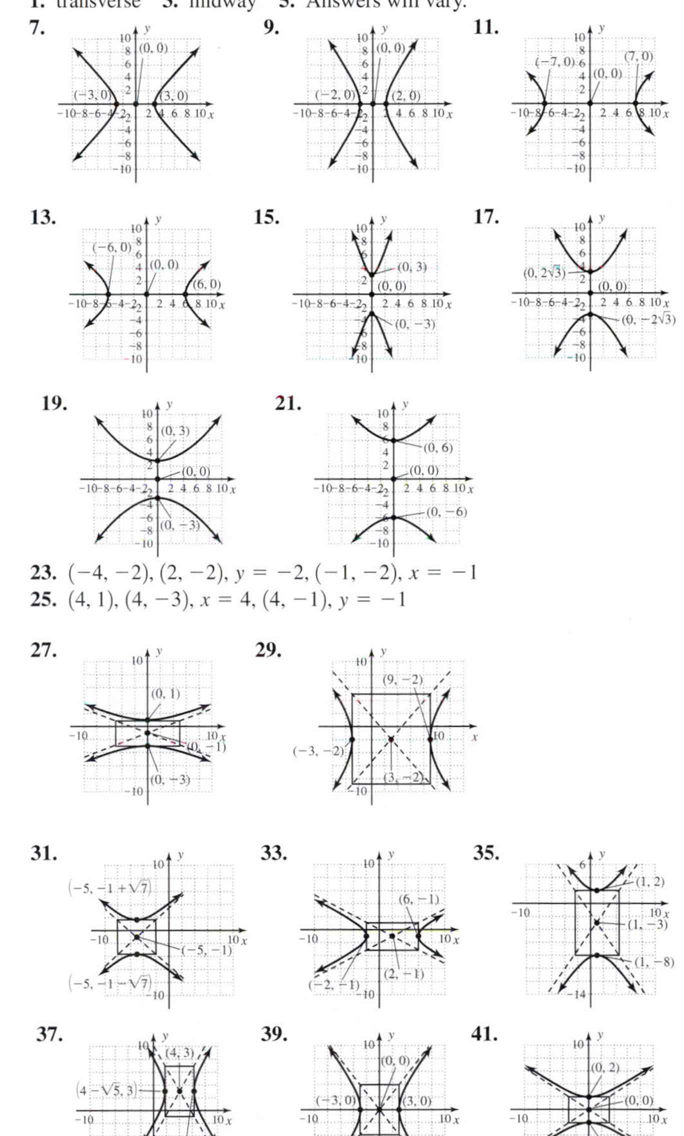

7. **9.** **11.** **13.** **15.** **17.** **19.** **21.**

23. $(-4, -2)$, $(2, -2)$, $y = -2$, $(-1, -2)$, $x = -1$
25. $(4, 1)$, $(4, -3)$, $x = 4$, $(4, -1)$, $y = -1$

27. **29.** **31.** **33.** **35.** **37.** **39.** **41.**

43. **45.** **47.**

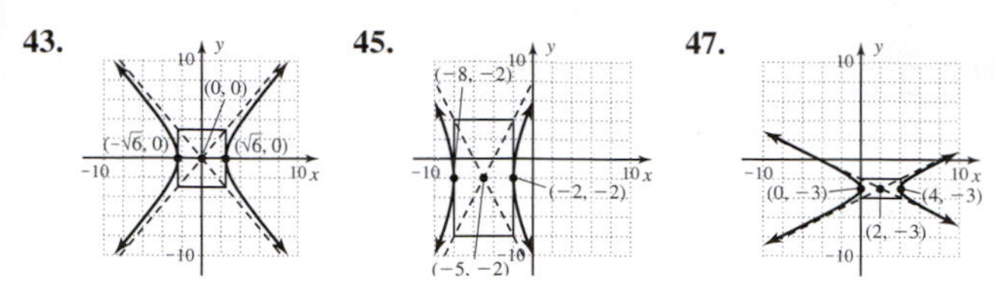

49. circle **51.** circle **53.** hyperbola **55.** hyperbola **57.** circle **59.** ellipse **61. a.** $y = \frac{2}{3}\sqrt{x^2 - 9}$ **b.** $x \in (-\infty, -3] \cup [3, \infty)$ **c.** $y = \frac{-2}{3}\sqrt{x^2 - 9}$ **63.** 40 yd **65.** 40 ft **67.** Answers will vary. **69.** Answers will vary. **71.** $(x - 2)^2 + (y - 3)^2 = 25$ **73.** a **75.** $R(t) = 2.457\sin(0.525t + 0.052) + 2.983$, $R(t) < 2$ for $t \in (6.67, 11.08)$, or from mid-June to early November. **77.** $x = 60°, 120°, 240°, 300°$ **79.** $z_0 = 9$, $z_1 = -\frac{9}{2} + \frac{9\sqrt{3}}{2}i$, $z_2 = -\frac{9}{2} - \frac{9\sqrt{3}}{2}i$

Exercises 6.3, pp. 447–451

1. $c^2 = |a^2 - b^2|$ **3.** $2a$; $2b$ **5.** Answers will vary. **7.** 20 **9.** 20 **11. a.** (2, 1) **b.** (−3, 1) and (7, 1) **c.** $(2 - \sqrt{21}, 1)$ and $(2 + \sqrt{21}, 1)$ **d.** (2, 3) and (2, −1) **e.**

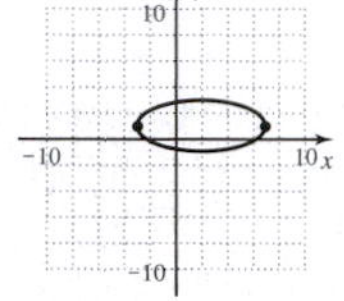

13. a. (4, −3) **b.** (4, 2) and (4, −8) **c.** (4, 0) and (4, −6) **d.** (0, −3) and (8, −3) **e.**

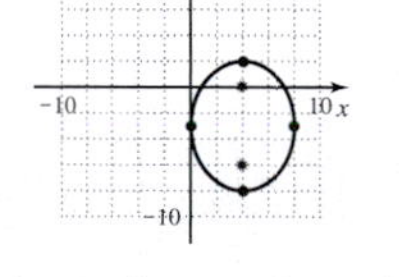

15. a. (−2, −2) **b.** (−5, −2) and (1, −2) **c.** $(-2 + \sqrt{3}, -2)$ and $(-2 - \sqrt{3}, -2)$ **d.** $(-2, -2 + \sqrt{6})$ and $(-2, -2 - \sqrt{6})$ **e.**

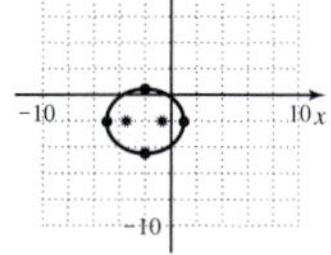

17. $\frac{x^2}{36} + \frac{y^2}{20} = 1$ **19.** $\frac{x^2}{9} + \frac{y^2}{25} = 1$ **21.** 8, $2a = 8$, $2b = 6$ **23.** 12, $2a = 16$, $2b = 12$ **25. a.** (3, 4) **b.** (0, 4) and (6, 4) **c.** $(3 - \sqrt{13}, 4)$ and $(3 + \sqrt{13}, 4)$ **d.** $2a = 6$, $2b = 4$ **e.**

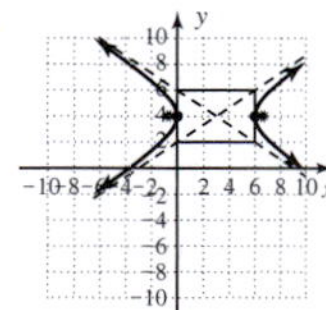

27. a. (0, 3) **b.** (−2, 3) and (2, 3) **c.** $(-2\sqrt{5}, 3)$ and $(2\sqrt{5}, 3)$ **d.** $2a = 4$, $2b = 8$ **e.**

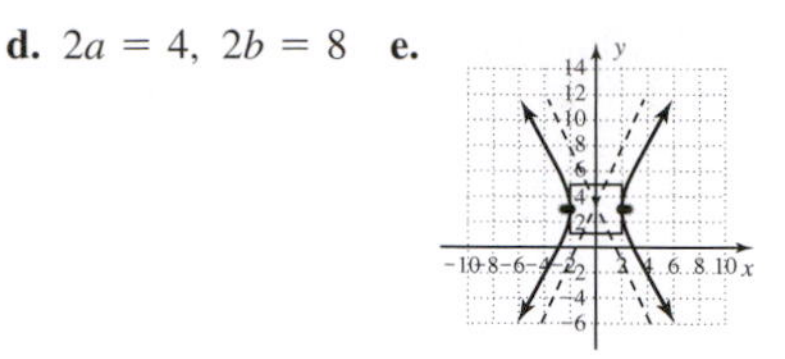

29. a. (3, −2) **b.** (1, −2) and (5, −2) **c.** (−1, −2) and (7, −2) **d.** $2a = 4$, $2b = 4\sqrt{3}$ **e.**

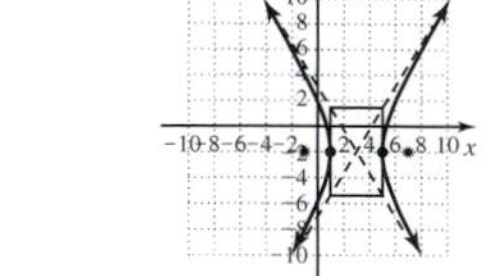

31. $\frac{x^2}{36} - \frac{y^2}{28} = 1$ **33.** $\frac{y^2}{9} - \frac{x^2}{9} = 1$ **35.** $(-3\sqrt{5}, 0)$, $(3\sqrt{5}, 0)$ **37.** (0, 4), (0, −4) **39.** $(-\sqrt{6}, 0)$, $(\sqrt{6}, 0)$ **41.** $(-\sqrt{13}, 0)$, $(\sqrt{13}, 0)$ **43.** $(0, \sqrt{61})$, $(0, -\sqrt{61})$ **45.** $(-2\sqrt{15}, 0)$, $(2\sqrt{15}, 0)$ **47.** $\frac{x^2}{16} + \frac{y^2}{36} = 1$ **49.** $\sqrt{7} \approx 2.65$ ft; 2.25 ft **51.** 8.9 ft; 17.9 ft **53.** $a \approx 142$ million miles, $b \approx 141$ million miles, orbit time ≈ 686 days **55.** $\frac{x^2}{225} - \frac{y^2}{2275} = 1$, about (24.1, 60) or (−24.1, 60) **57.** $L = 9$ units; $\left(\sqrt{13}, \frac{9}{2}\right), \left(\sqrt{13}, -\frac{9}{2}\right), \left(-\sqrt{13}, \frac{9}{2}\right), \left(-\sqrt{13}, -\frac{9}{2}\right)$; verified **59.** $(x - 2)^2 + (y - 3)^2 = 34$ **61.** verified **63.** 13.1 mph, at heading 327.8° **65.** $\cos 105° = \frac{\sqrt{2} - \sqrt{6}}{4}$ **67.** No triangles are possible.

Mid-Chapter Check, p. 452

1. **2.** **3.** **4.** **5.** **6.**

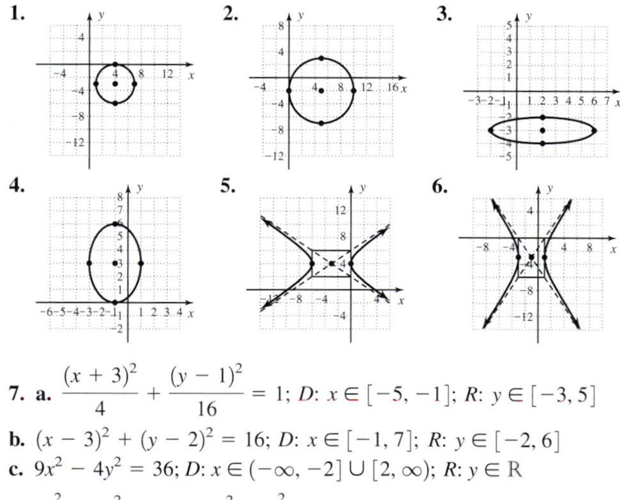

7. a. $\frac{(x + 3)^2}{4} + \frac{(y - 1)^2}{16} = 1$; D: $x \in [-5, -1]$; R: $y \in [-3, 5]$ **b.** $(x - 3)^2 + (y - 2)^2 = 16$; D: $x \in [-1, 7]$; R: $y \in [-2, 6]$ **c.** $9x^2 - 4y^2 = 36$; D: $x \in (-\infty, -2] \cup [2, \infty)$; R: $y \in \mathbb{R}$ **8.** $\frac{y^2}{144} - \frac{x^2}{25} = 1$ **9.** $\frac{x^2}{16} + \frac{y^2}{4} = 1$ **10.** yes, distance $d \approx 49$ mi

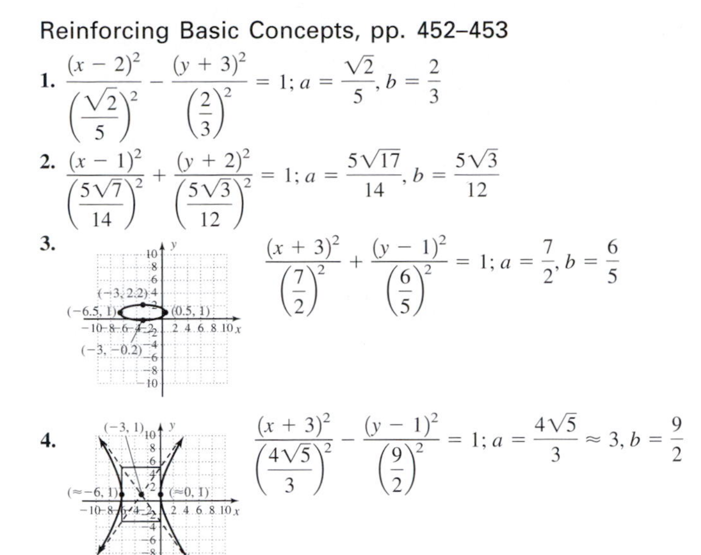

Reinforcing Basic Concepts, pp. 452–453

1. $\dfrac{(x-2)^2}{\left(\frac{\sqrt{2}}{5}\right)^2} - \dfrac{(y+3)^2}{\left(\frac{2}{3}\right)^2} = 1;\ a = \dfrac{\sqrt{2}}{5},\ b = \dfrac{2}{3}$

2. $\dfrac{(x-1)^2}{\left(\frac{5\sqrt{7}}{14}\right)^2} + \dfrac{(y+2)^2}{\left(\frac{5\sqrt{3}}{12}\right)^2} = 1;\ a = \dfrac{5\sqrt{17}}{14},\ b = \dfrac{5\sqrt{3}}{12}$

3. $\dfrac{(x+3)^2}{\left(\frac{7}{2}\right)^2} + \dfrac{(y-1)^2}{\left(\frac{6}{5}\right)^2} = 1;\ a = \dfrac{7}{2},\ b = \dfrac{6}{5}$

4. $\dfrac{(x+3)^2}{\left(\frac{4\sqrt{5}}{3}\right)^2} - \dfrac{(y-1)^2}{\left(\frac{9}{2}\right)^2} = 1;\ a = \dfrac{4\sqrt{5}}{3} \approx 3,\ b = \dfrac{9}{2}$

Exercises 6.4, pp. 459–463

1. horizontal; right; $a < 0$ **3.** $(p, 0);\ x = -p$ **5.** Answers will vary.
7. $x \in (-\infty, \infty),\ y \in [-4, \infty)$ **9.** $x \in (-\infty, \infty),\ y \in [-18, \infty)$

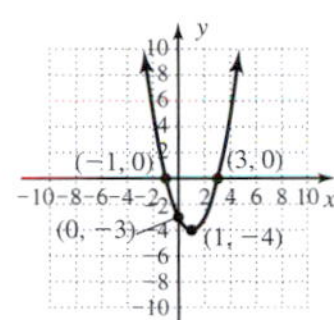

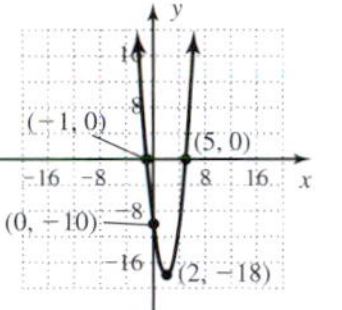

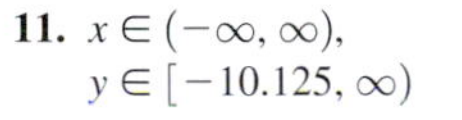

11. $x \in (-\infty, \infty),$
$y \in [-10.125, \infty)$ **13.** $x \in [-4, \infty),$
$y \in (-\infty, \infty)$

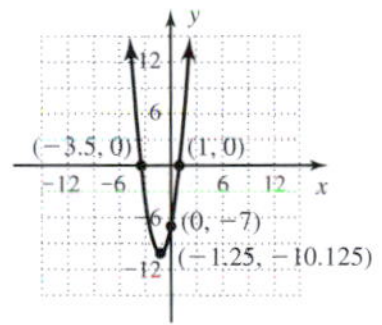

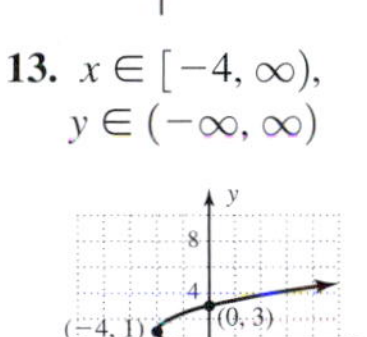

15. $x \in (-\infty, 16],\ y \in (-\infty, \infty)$ **17.** $x \in (-\infty, 0],\ y \in (-\infty, \infty)$

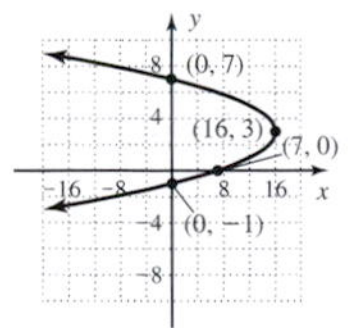

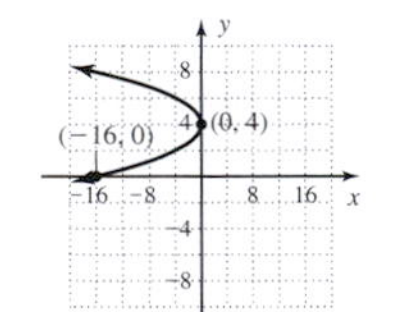

19. $x \in [-9, \infty),\ y \in (-\infty, \infty)$ **21.** $x \in [-4, \infty),\ y \in (-\infty, \infty)$

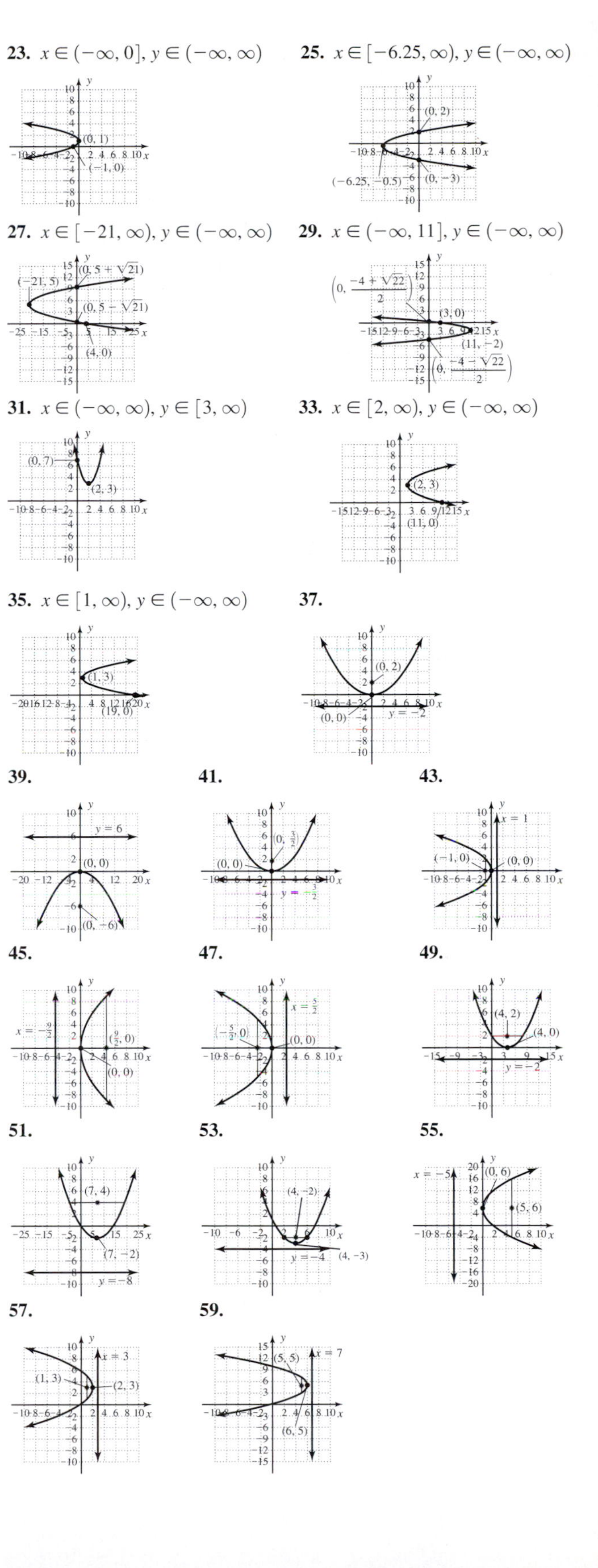

23. $x \in (-\infty, 0],\ y \in (-\infty, \infty)$ **25.** $x \in [-6.25, \infty),\ y \in (-\infty, \infty)$

27. $x \in [-21, \infty),\ y \in (-\infty, \infty)$ **29.** $x \in (-\infty, 11],\ y \in (-\infty, \infty)$

31. $x \in (-\infty, \infty),\ y \in [3, \infty)$ **33.** $x \in [2, \infty),\ y \in (-\infty, \infty)$

35. $x \in [1, \infty),\ y \in (-\infty, \infty)$ **37.**

39. **41.** **43.**

45. **47.** **49.**

51. **53.** **55.**

57. **59.**

61. 16 units^2 **63.**

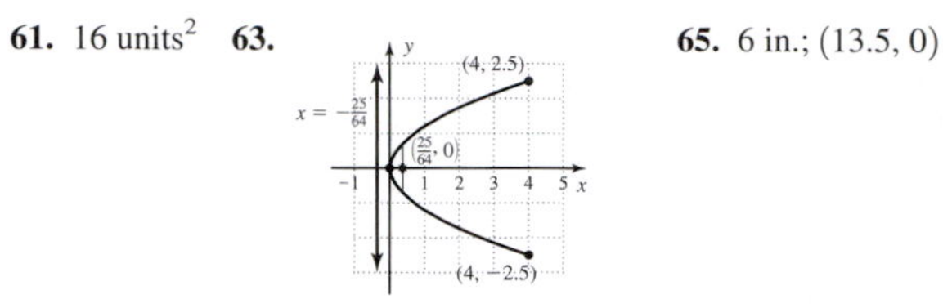

65. 6 in.; (13.5, 0)

67. 14.97 ft, (0, 41.75) **69.** $(x-2)^2 = \frac{1}{2}(y+8)$; $p = \frac{1}{8}$; $(2, -8)$
71. b, answers will vary. **73.** $(x+3)^2 + (y-4)^2 = 20$; 20π units^2
75. 120 days **77.** $x \approx 3.8490, 4.5285$ **79.** $\tan 12° \approx 0.2125$

Exercises 6.5, pp. 474–478

1. polar **3.** II; IV **5.** To plot the point (r, θ) start at the origin or pole and move $|r|$ units out along the polar axis. Then move counterclockwise an angle measure of θ. You should be r units straight out from the pole in a direction of θ from the positive polar axis. If r is negative, final resting place for the point (r, θ) will be 180° from θ.

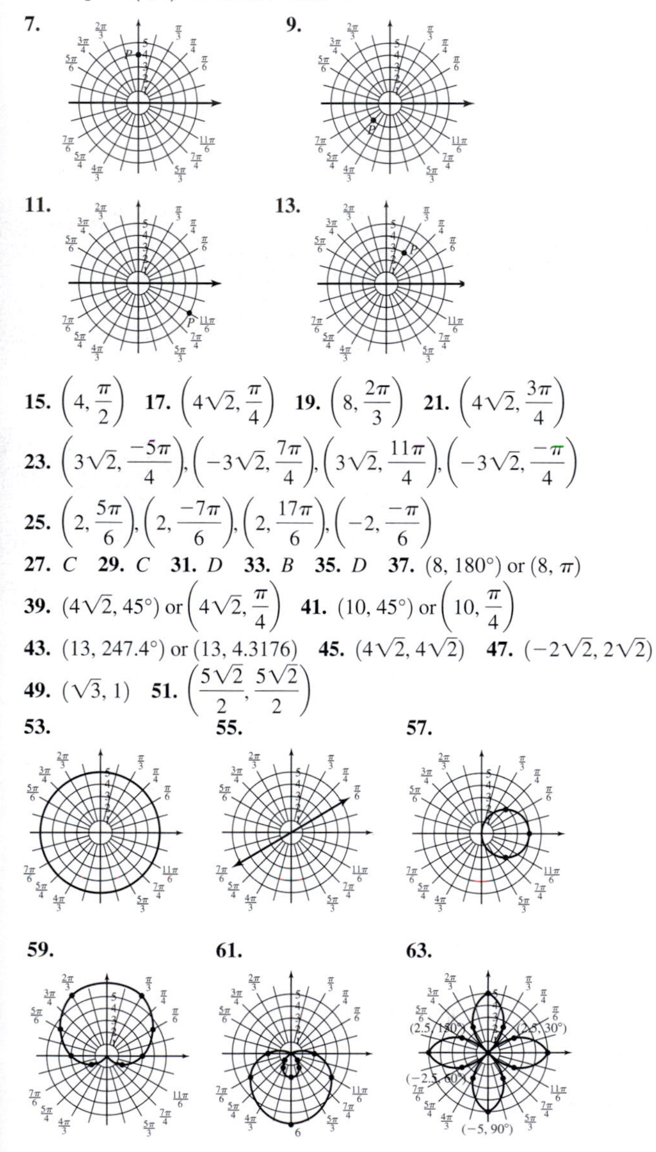

7. **9.** **11.** **13.**

15. $\left(4, \frac{\pi}{2}\right)$ **17.** $\left(4\sqrt{2}, \frac{\pi}{4}\right)$ **19.** $\left(8, \frac{2\pi}{3}\right)$ **21.** $\left(4\sqrt{2}, \frac{3\pi}{4}\right)$

23. $\left(3\sqrt{2}, \frac{-5\pi}{4}\right), \left(-3\sqrt{2}, \frac{7\pi}{4}\right), \left(3\sqrt{2}, \frac{11\pi}{4}\right), \left(-3\sqrt{2}, \frac{-\pi}{4}\right)$

25. $\left(2, \frac{5\pi}{6}\right), \left(2, \frac{-7\pi}{6}\right), \left(2, \frac{17\pi}{6}\right), \left(-2, \frac{-\pi}{6}\right)$

27. *C* **29.** *C* **31.** *D* **33.** *B* **35.** *D* **37.** (8, 180°) or $(8, \pi)$

39. $(4\sqrt{2}, 45°)$ or $\left(4\sqrt{2}, \frac{\pi}{4}\right)$ **41.** (10, 45°) or $\left(10, \frac{\pi}{4}\right)$

43. (13, 247.4°) or (13, 4.3176) **45.** $(4\sqrt{2}, 4\sqrt{2})$ **47.** $(-2\sqrt{2}, 2\sqrt{2})$

49. $(\sqrt{3}, 1)$ **51.** $\left(\frac{5\sqrt{2}}{2}, \frac{5\sqrt{2}}{2}\right)$

53. **55.** **57.**

59. **61.** **63.**

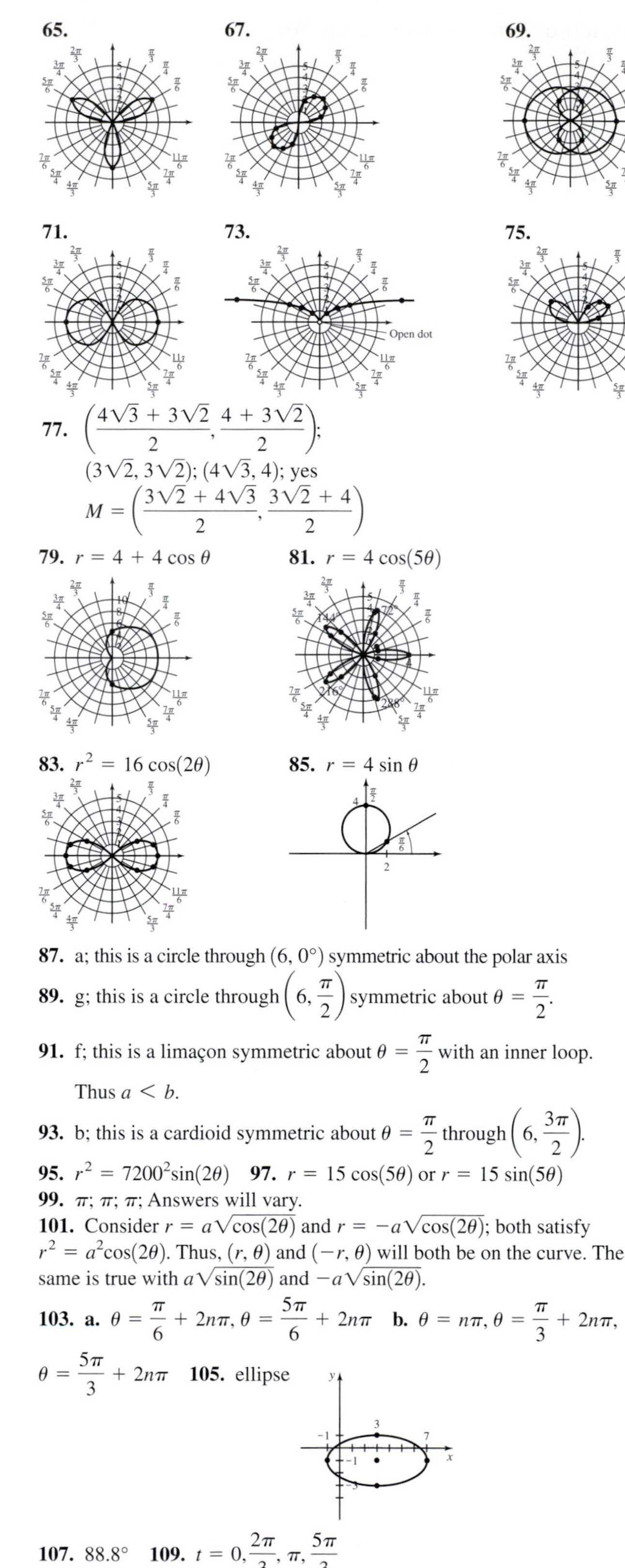

65. **67.** **69.**

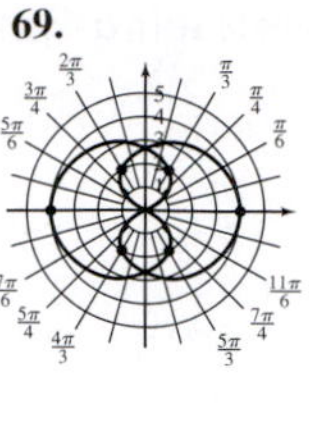

71. **73.** **75.**

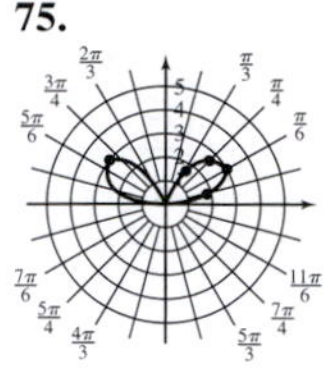

77. $\left(\frac{4\sqrt{3} + 3\sqrt{2}}{2}, \frac{4 + 3\sqrt{2}}{2}\right)$;
$(3\sqrt{2}, 3\sqrt{2})$; $(4\sqrt{3}, 4)$; yes
$M = \left(\frac{3\sqrt{2} + 4\sqrt{3}}{2}, \frac{3\sqrt{2} + 4}{2}\right)$

79. $r = 4 + 4\cos\theta$ **81.** $r = 4\cos(5\theta)$

83. $r^2 = 16\cos(2\theta)$ **85.** $r = 4\sin\theta$

87. a; this is a circle through (6, 0°) symmetric about the polar axis

89. g; this is a circle through $\left(6, \frac{\pi}{2}\right)$ symmetric about $\theta = \frac{\pi}{2}$.

91. f; this is a limaçon symmetric about $\theta = \frac{\pi}{2}$ with an inner loop. Thus $a < b$.

93. b; this is a cardioid symmetric about $\theta = \frac{\pi}{2}$ through $\left(6, \frac{3\pi}{2}\right)$.

95. $r^2 = 7200^2\sin(2\theta)$ **97.** $r = 15\cos(5\theta)$ or $r = 15\sin(5\theta)$

99. π; π; π; Answers will vary.

101. Consider $r = a\sqrt{\cos(2\theta)}$ and $r = -a\sqrt{\cos(2\theta)}$; both satisfy $r^2 = a^2\cos(2\theta)$. Thus, (r, θ) and $(-r, \theta)$ will both be on the curve. The same is true with $a\sqrt{\sin(2\theta)}$ and $-a\sqrt{\sin(2\theta)}$.

103. a. $\theta = \frac{\pi}{6} + 2n\pi, \theta = \frac{5\pi}{6} + 2n\pi$ **b.** $\theta = n\pi, \theta = \frac{\pi}{3} + 2n\pi,$
$\theta = \frac{5\pi}{3} + 2n\pi$ **105.** ellipse

107. 88.8° **109.** $t = 0, \frac{2\pi}{3}, \pi, \frac{5\pi}{3}$

Exercises 6.6, pp. 491–497

1. rotation of axes; $\frac{B}{A-C}$; first; second **3.** invariants

5. Answers will vary. **7.** $\frac{Y^2}{8} - \frac{X^2}{8} = 1$

9. $6 + 3\sqrt{2} = X, -6 + 3\sqrt{2} = Y$ **11.** $\frac{5\sqrt{2}}{2} = X, \frac{5\sqrt{2}}{2} = Y$

13. $0 = x, 4 = y$ **15.** $\frac{3\sqrt{3}}{2} - 2 = x; \frac{3}{2} + 2\sqrt{3} = y$

17. $\frac{-x^2}{2} + xy\sqrt{3} + \frac{y^2}{2} = 9$ **19.** $4X^2 + 2Y^2 = 9$

21. a. $3X^2 - Y^2 = 2$ **b.**

vertices: $(\pm\frac{\sqrt{6}}{3}, 0)$
foci: $(\pm\frac{2\sqrt{10}}{3}, 0)$
asymptotes: $Y = \pm\sqrt{3}X$

23. a. $4X^2 + Y^2 = 8$ **b.**

vertices: $(0, \pm 2\sqrt{2})$
foci: $(0, \pm\sqrt{6})$
minor axis endpoints: $(\pm\sqrt{6}, 0)$

25. a. $Y^2 - 4X^2 = 16$ **b.**

vertices: $(\pm 4, 0)$
foci: $(\pm 2\sqrt{5}, 0)$
asymptotes: $Y = \pm 2X$

27. a. $Y^2 - 4X = 0$ **b.**

vertex: (0, 0)
foci: (1, 0)
directrix: $X = -1$

29. a. $X^2 + 4Y^2 = 25$ **b.**

vertices: $(\pm 5, 0)$
foci: $(\pm\frac{5\sqrt{3}}{2}, 0)$
minor axis endpoints: $(0, \pm\frac{5}{2})$

31. $336 > 0$; hyperbola; $\cos(2\beta) = \frac{7}{25}; \frac{4}{5} = \cos\beta; \frac{3}{5} = \sin\beta$

33. a. parabola **b.** $\beta = 45°; 2Y^2 = 5$ **c.** verified

35. a. circle or ellipse **b.** $\beta = 60°; \frac{9}{2}X^2 + \frac{5Y^2}{2} + 2X - 2\sqrt{3}Y = 1$ (ellipse) **c.** verified

37. f **39.** g **41.** h

43. parabola

$(1, \frac{\pi}{2})$ $(2, 0)$ $(4, \frac{7\pi}{6})$

45. ellipse

$(2, 0)$ $(1\frac{1}{3}, \frac{3\pi}{2})$

47. hyperbola

$(1, 0)$ $(-3, \pi)$ $(1.5, \frac{5\pi}{3})$

49. ellipse

$(5, \pi)$ $(1, \frac{\pi}{2})$ $(\frac{5}{9}, 0)$

51. $r = \frac{3.2}{1 - 0.8\cos\theta}$

53. $r = \frac{4}{1 - \cos\theta}$ **55.** $r = \frac{7.5}{1 + 1.5\sin\theta}$

57. a. $r = \frac{12}{2\cos\theta + 3\sin\theta}$ **b.** $-\frac{r(\pi/2)}{r(0)} = \frac{-2}{3}$ and $\frac{-A}{B} = \frac{-2}{3}$

59. Jupiter: $e \approx 0.0486$, Saturn: $e \approx 0.0567$ **61.** about 2757.1 million miles **63.** Saturn: $e \approx 0.0567$ million mi/yr

65. $r \approx \frac{482.36}{1 - 0.0486\cos\theta}$ **67.** $r \approx \frac{1780.77}{1 - 0.0457\cos\theta}$

69. In millions of miles (approx): $\overline{JS}$: 405.3, $\overline{JU}$: 1298.4, $\overline{JN}$: 2310.3, $\overline{SU}$: 893.1, $\overline{SN}$: 1905.0, $\overline{UN}$: 1011.9

71. $r = \frac{0.7638}{1 \pm 0.7862\cos\theta}$ **73.** $r = \frac{0.2865}{1 \pm 0.7862\cos\theta}$ **75.** \$582.45; \$445.94; \$881.32; \$97.92 **77.** $y = \frac{3}{1 - \cos\theta}$ **79.** verified

81. Answers will vary. **83.** Answers will vary.

85. a. $r^2 = \frac{a^2b^2}{b^2\cos^2\theta - a^2\sin^2\theta}$ **b.** $r^2 = \frac{12}{2\cos^2\theta - \sin^2\theta}$

87. $21X^2 - 4Y^2 - 50X = 25$ **89. a.** $y = 1400\sin\left[\frac{\pi}{6}(x - 4)\right] + 2600$

b. about 5.7 mo **91.** $t = \frac{\pi}{6}, \frac{5\pi}{6}, \frac{7\pi}{6}, \frac{11\pi}{6}$

93. $\cos(2\theta) = \cos^2\theta - \sin^2\theta$ $\quad \sin^2\theta + \cos^2\theta = 1$
$= 2\cos^2\theta - 1$ $\quad 1 + \cot^2\theta = \csc^2\theta$
$= 1 - 2\sin^2\theta$ $\quad \tan^2\theta + 1 = \sec^2\theta$

Summary and Concept Review, pp. 497–501

1.  **2.** **3.**

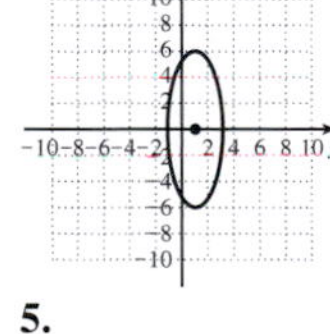

4.

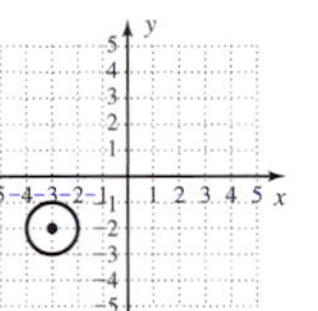

5.

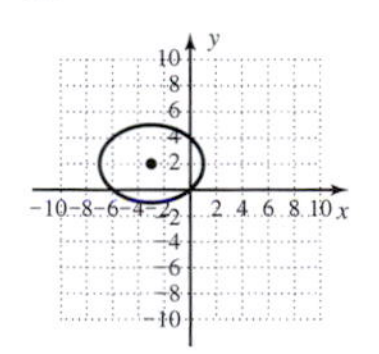

6. $(x + 1)^2 + (y - 1)^2 = 25$

7. **8.**

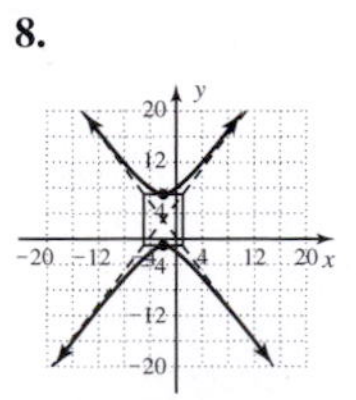

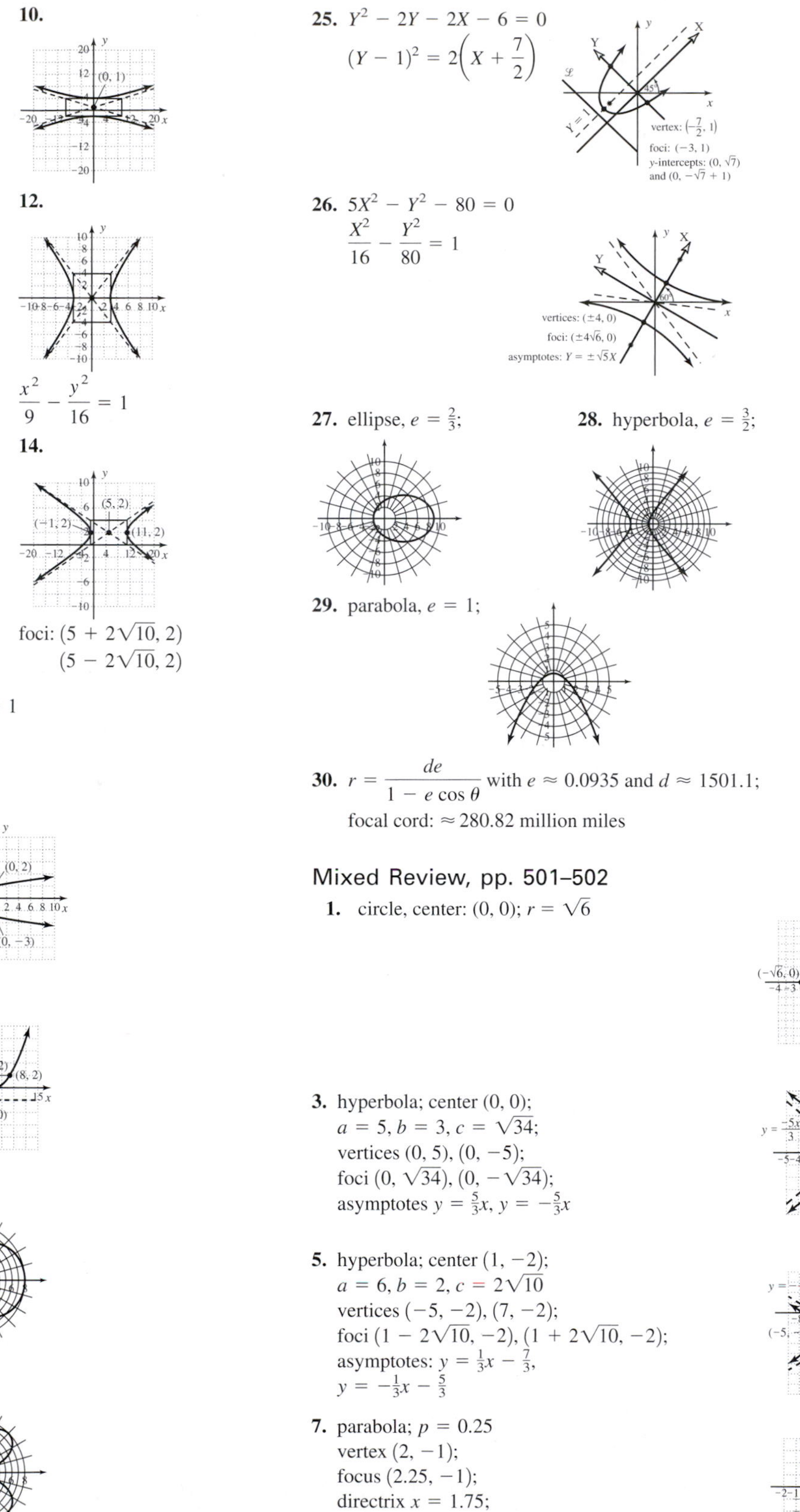

9.

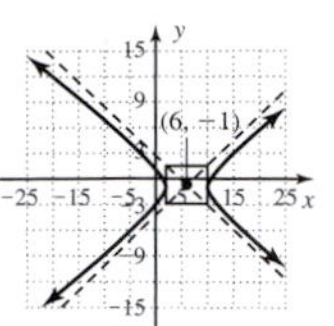

10.

11.

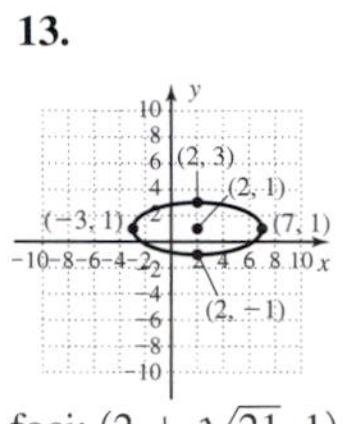

12.

$$\frac{x^2}{9} - \frac{y^2}{16} = 1$$

13.

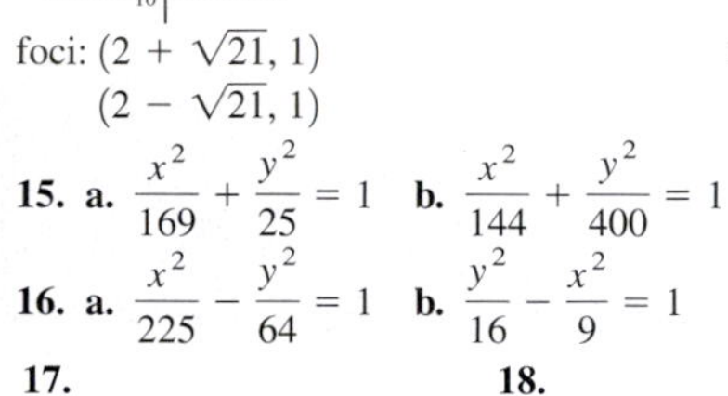

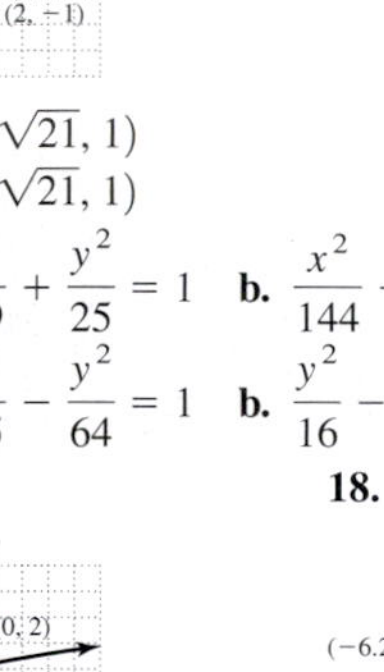

foci: $(2 + \sqrt{21}, 1)$
$(2 - \sqrt{21}, 1)$

14.

foci: $(5 + 2\sqrt{10}, 2)$
$(5 - 2\sqrt{10}, 2)$

15. a. $\frac{x^2}{169} + \frac{y^2}{25} = 1$ **b.** $\frac{x^2}{144} + \frac{y^2}{400} = 1$

16. a. $\frac{x^2}{225} - \frac{y^2}{64} = 1$ **b.** $\frac{y^2}{16} - \frac{x^2}{9} = 1$

17.

18.

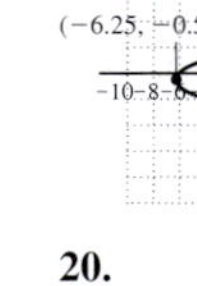

19.

20.

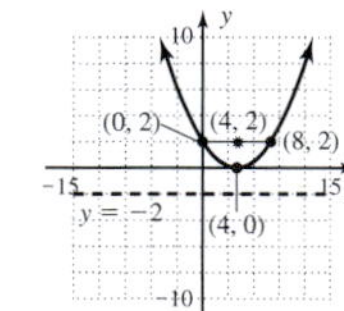

21.

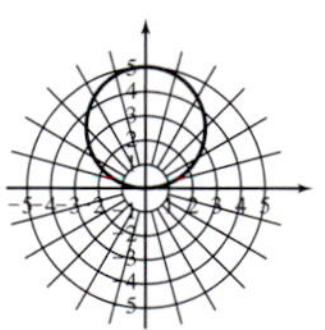

22.

23.

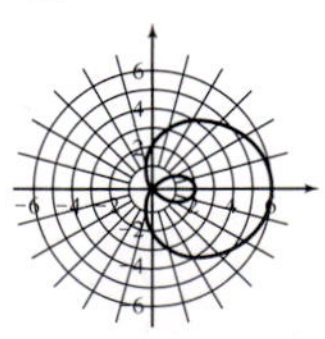

24.

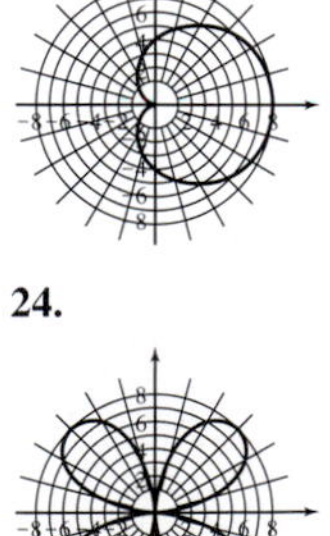

25. $Y^2 - 2Y - 2X - 6 = 0$
$(Y - 1)^2 = 2\left(X + \frac{7}{2}\right)$

26. $5X^2 - Y^2 - 80 = 0$
$\frac{X^2}{16} - \frac{Y^2}{80} = 1$

27. ellipse, $e = \frac{2}{3}$;

28. hyperbola, $e = \frac{3}{2}$;

29. parabola, $e = 1$;

30. $r = \frac{de}{1 - e\cos\theta}$ with $e \approx 0.0935$ and $d \approx 1501.1$;
focal cord: ≈ 280.82 million miles

Mixed Review, pp. 501–502

1. circle, center: (0, 0); $r = \sqrt{6}$

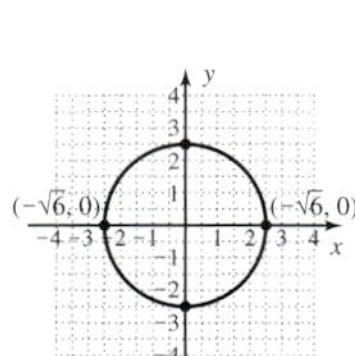

3. hyperbola; center (0, 0);
$a = 5, b = 3, c = \sqrt{34}$;
vertices (0, 5), (0, −5);
foci $(0, \sqrt{34}), (0, -\sqrt{34})$;
asymptotes $y = \frac{5}{3}x, y = -\frac{5}{3}x$

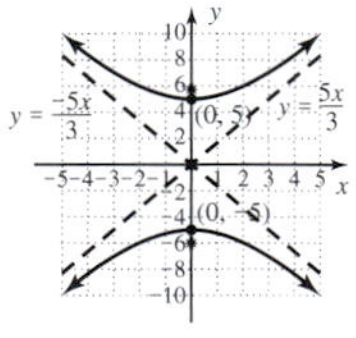

5. hyperbola; center (1, −2);
$a = 6, b = 2, c = 2\sqrt{10}$
vertices (−5, −2), (7, −2);
foci $(1 - 2\sqrt{10}, -2), (1 + 2\sqrt{10}, -2)$;
asymptotes: $y = \frac{1}{3}x - \frac{7}{3}$,
$y = -\frac{1}{3}x - \frac{5}{3}$

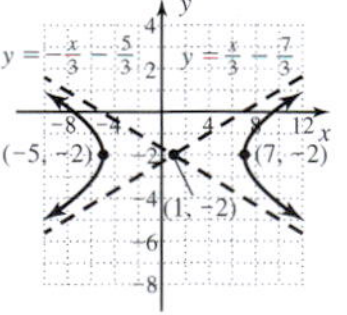

7. parabola; $p = 0.25$
vertex (2, −1);
focus (2.25, −1);
directrix $x = 1.75$;
y-intercepts: none

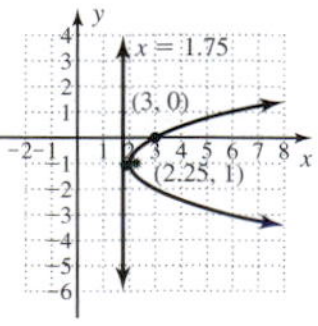

9. parabola; $p = 2$
vertex $(4, 0)$;
focus $(4, 2)$;
directrix $y = -2$;
x-intercept $(4, 0)$

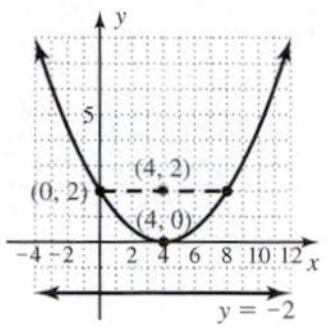

11. hyperbola; center $(3, 3)$;
$a = 5, b = 2, c = \sqrt{29}$
vertices $(8, 3), (-2, 3)$;
foci $(3 - \sqrt{29}, 3), (3 + \sqrt{29}, 3)$,
$\approx (-2.39, 3), (8.39, 3)$;
asymptotes $y = \frac{2}{5}x + \frac{9}{5}$,
$y = -\frac{2}{5}x + \frac{21}{5}$

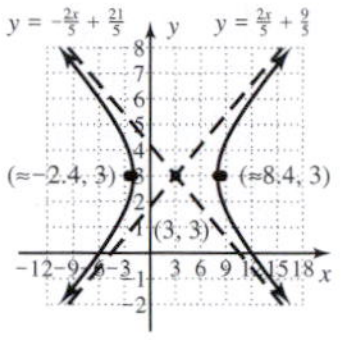

13. ellipse; center $(-2, 3)$;
$a = 1, b = 7, c = 4\sqrt{3}$
vertices $(-2, -4), (-2, 10)$;
endpoints of minor axis $(-3, 3), (-1, 3)$;
foci: $(-2, 3 - 4\sqrt{3}), (-2, 3 + 4\sqrt{3})$,
$\approx (-2, -3.93), \approx (-2, 9.93)$

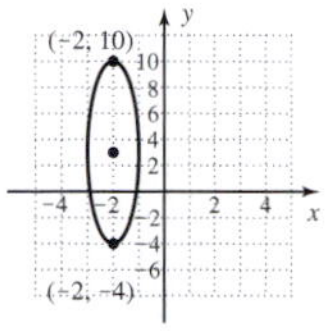

15. a. $\left(-\frac{7}{4}, \frac{7\sqrt{3}}{4}\right)$ **b.** $(2\sqrt{2}, 2\sqrt{2})$

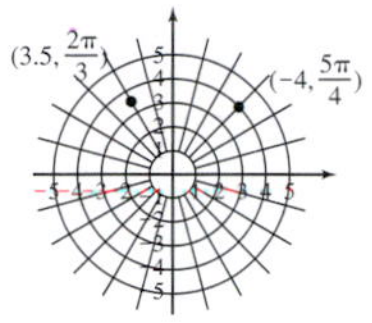

17. major axis: 116 million km; minor axis: 113.5 miilion km;
$\frac{x^2}{3364} + \frac{y^2}{3220} = 1$ (major axis horizontal)

19. $r = 5\sin(7\theta)$

Practice Test, pp. 502–503

1. c **2.** d **3.** b **4.** a

5. circle;
center $(2, -5)$;
radius 3

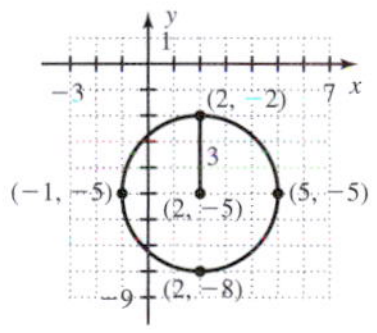

6. ellipse;
center $(-2, 1)$;
vertices $(-2, -4), (-2, 6)$;
foci $(-2, 1 - \sqrt{21}), (-2, 1 + \sqrt{21})$,
or approx. $(-2, -3, 6), (-2, 5.6)$

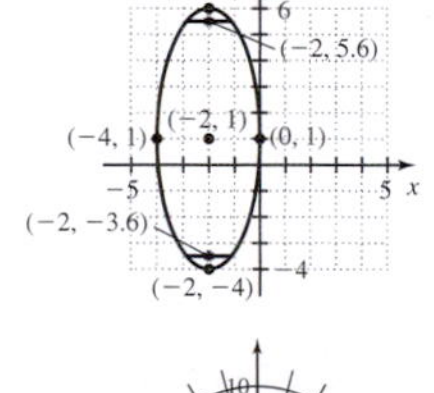

7. ellipse; center $\left(\frac{40}{9}, 0\right)$;
vertices $\left(\frac{-10}{9}, 0\right), (10, 0)$;
foci $(0, 0), \left(\frac{80}{9}, 0\right)$

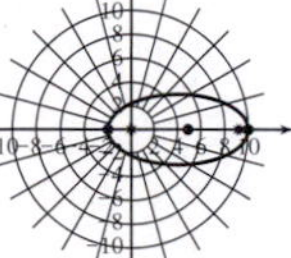

8. parabola;
vertex $(-1.2, 0)$;
focus $(0, 0)$;
directrix at $y = -2.4$

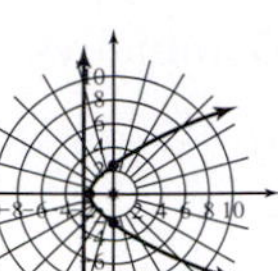

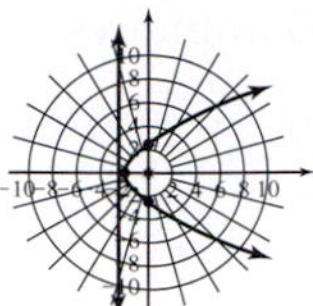

9. hyperbola;
center: $(2, -3)$;
vertices: $(2, 0), (2, -6)$;
foci: $(2, -8), (2, 2)$;
asymptotes: $y = \frac{3}{4}x - \frac{9}{2}, y = \frac{-3}{4}x - \frac{3}{2}$

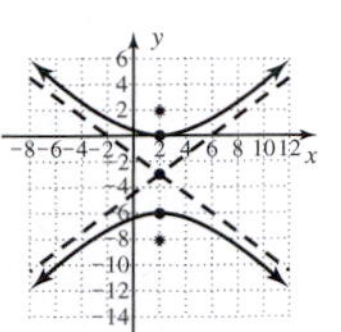

10. hyperbola;
center $(1, -2)$;
vertices $(-4, -2), (6, -2)$;
foci $(1 - \sqrt{29}, -2), (1 + \sqrt{29}, -2)$,
or approx. $(-4.39, -2), (6.39, -2)$;
asymptotes: $y = -\frac{2}{5}x - \frac{8}{5}$,
$y = \frac{2}{5}x - \frac{12}{5}$

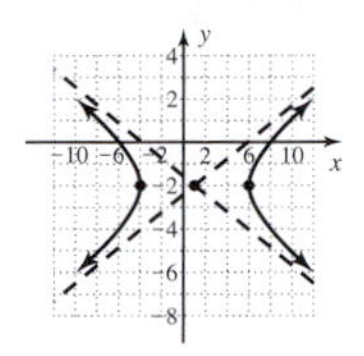

11. parabola; $\beta \approx 36.9°$; $\cos\beta \approx \frac{4}{5}$, $\sin\beta = \frac{3}{5}$

12. $Y = \frac{25}{16}X^2 - \frac{3}{4}X - \frac{11}{20}$

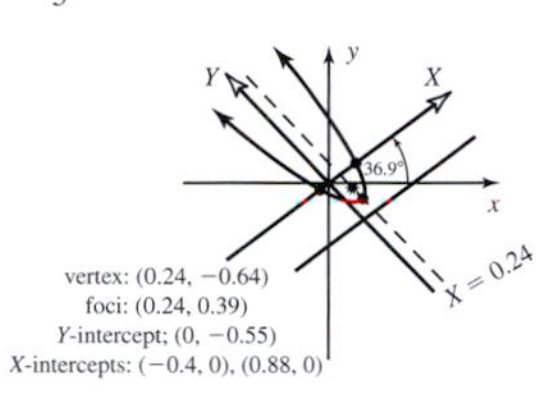

13. **14.** **15.**

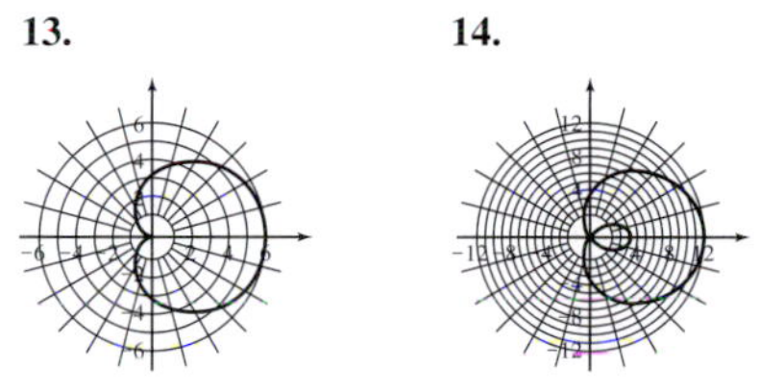

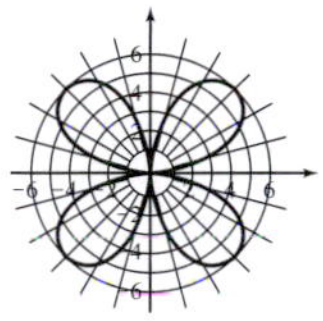

16. $r \approx \frac{1654(1 - 0.967^2)}{1 - 0.967\cos\theta}$
e is very close to 1. This makes its orbit a very elongated ellipse, where the orbit of most planets is nearly circular.

17. The ball is 0.43 ft above the ground at $x = 165$ ft, and will likely go into the goal.

18. $y = (x - 1)^2 - 4$;
D: $x \in \mathbb{R}$
R: $y \in [-4, \infty)$
focus: $(1, -3.75)$

19. $(x - 1)^2 + (y - 1)^2 = 25$;
D: $x \in [-4, 6]$
R: $y \in [-4, 6]$

20. $\frac{(x + 2)^2}{9} + \frac{(y - 1)^2}{25} = 1$;
D: $x \in [-5, 1]$
R: $y \in [-4, 6]$
foci at $(-2, 5)$ and $(-2, -3)$

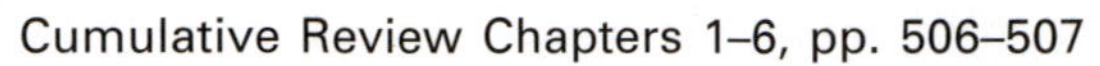

Cumulative Review Chapters 1–6, pp. 506–507

1. $\frac{5\pi}{6} + k\pi, k \in \mathbb{Z}$ **3.** $x = 1 + 6k; k \in \mathbb{Z}; x = 3 + 6k; k \in \mathbb{Z}$

5.

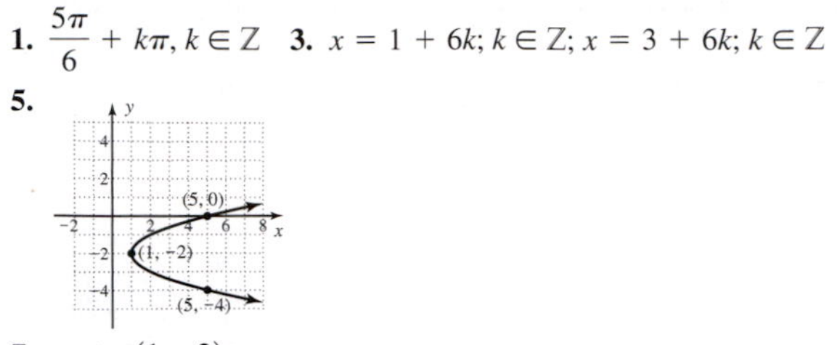

7. center $(1, -2)$;
foci $(1 + 2\sqrt{10}, -2) \approx (7.32, -2)$,
$(1 - 2\sqrt{10}, -2) \approx (-5.32, -2)$; asymptotes
$y = \frac{1}{3}x - \frac{7}{3}$,
$y = -\frac{1}{3}x - \frac{5}{3}$

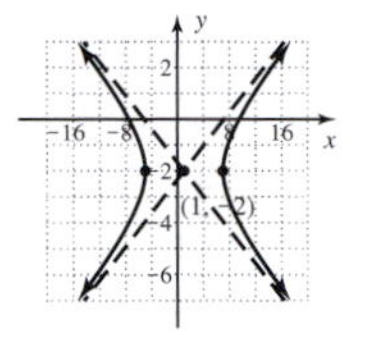

9.

11. 61.9° **13.** $A \approx 75.2\ \text{m}^2$

15. $\theta = 0, \frac{\pi}{2}, \pi, \frac{3\pi}{2}$ **17.** $\alpha = 392°, \beta \approx 7.0001$ **19.** $A \approx 39.3\ \text{ft}^2$

21. about 128 yd **23.** $2\cos^2\theta - 1, 1 - 2\sin^2\theta$ **25.** $\sin\left(\frac{\pi}{24}\right)$

27. No, it is not detected. **29.** $z^4 = -8 + 8\sqrt{3}i$

Index

D

E

F

G

H

I

Special Factorizations

$$a^2 + 2ab + b^2 = (a + b)^2$$

$$a^2 - 2ab + b^2 = (a - b)^2$$

$$x^2 + (c + d)x + cd = (x + c)(x + d)$$

$$abx^2 + (ad + bc)x + cd = (ax + c)(bx + d)$$

$$a^2 - b^2 = (a + b)(a - b)$$

$a^2 + b^2$ is prime

$$a^3 - b^3 = (a - b)(a^2 + ab + b^2)$$

$$a^3 + b^3 = (a + b)(a^2 - ab + b^2)$$

Formulas from Plane Geometry: $P \rightarrow$ perimeter, $C \rightarrow$ circumference, $A \rightarrow$ area

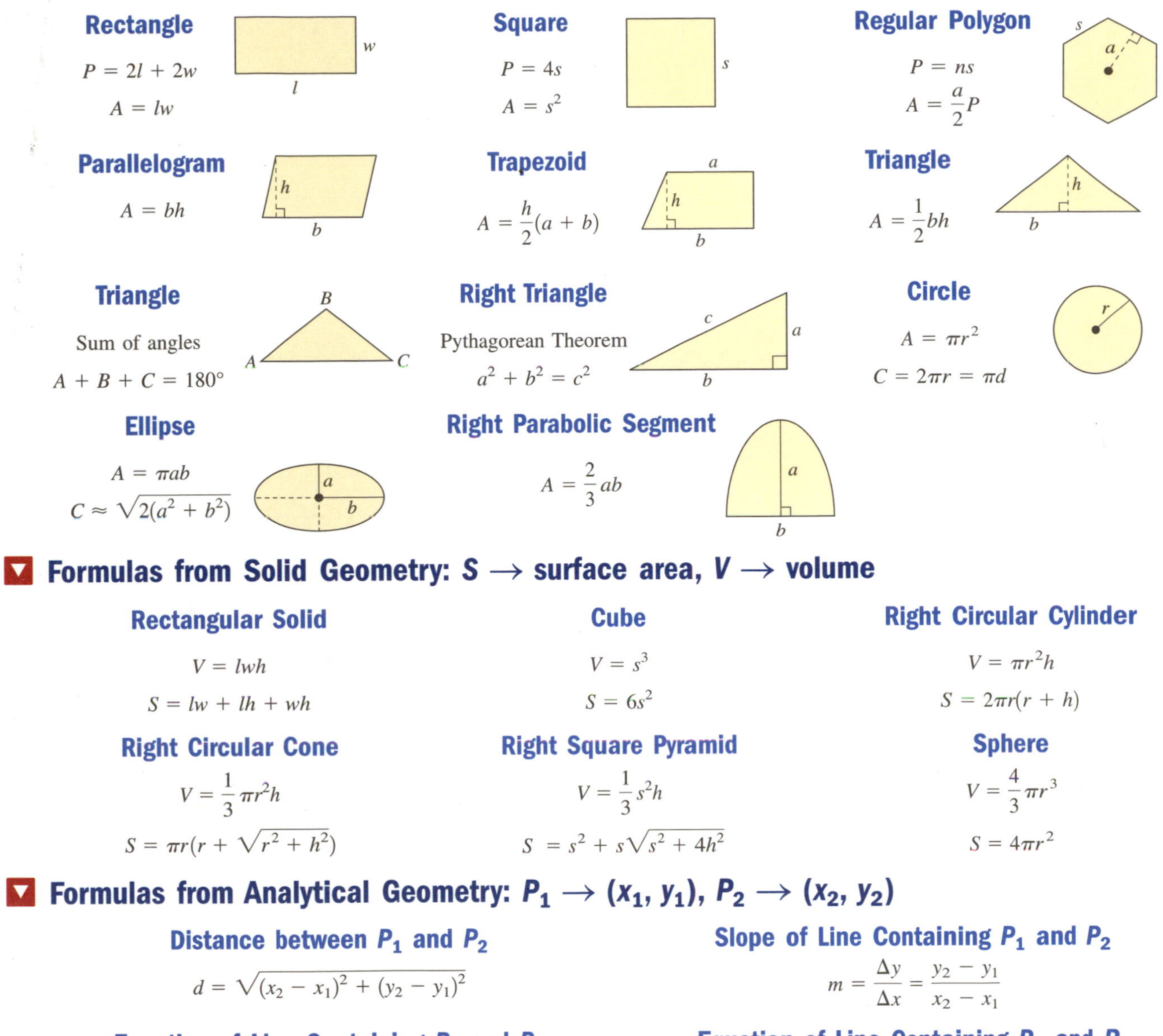

Rectangle

$P = 2l + 2w$

$A = lw$

Square

$P = 4s$

$A = s^2$

Regular Polygon

$P = ns$

$A = \frac{a}{2}P$

Parallelogram

$A = bh$

Trapezoid

$A = \frac{h}{2}(a + b)$

Triangle

$A = \frac{1}{2}bh$

Triangle

Sum of angles

$A + B + C = 180°$

Right Triangle

Pythagorean Theorem

$a^2 + b^2 = c^2$

Circle

$A = \pi r^2$

$C = 2\pi r = \pi d$

Ellipse

$A = \pi ab$

$C \approx \sqrt{2(a^2 + b^2)}$

Right Parabolic Segment

$A = \frac{2}{3}ab$

Formulas from Solid Geometry: $S \rightarrow$ surface area, $V \rightarrow$ volume

Rectangular Solid

$V = lwh$

$S = lw + lh + wh$

Cube

$V = s^3$

$S = 6s^2$

Right Circular Cylinder

$V = \pi r^2 h$

$S = 2\pi r(r + h)$

Right Circular Cone

$V = \frac{1}{3}\pi r^2 h$

$S = \pi r(r + \sqrt{r^2 + h^2})$

Right Square Pyramid

$V = \frac{1}{3}s^2 h$

$S = s^2 + s\sqrt{s^2 + 4h^2}$

Sphere

$V = \frac{4}{3}\pi r^3$

$S = 4\pi r^2$

Formulas from Analytical Geometry: $P_1 \rightarrow (x_1, y_1)$, $P_2 \rightarrow (x_2, y_2)$

Distance between P_1 and P_2

$$d = \sqrt{(x_2 - x_1)^2 + (y_2 - y_1)^2}$$

Slope of Line Containing P_1 and P_2

$$m = \frac{\Delta y}{\Delta x} = \frac{y_2 - y_1}{x_2 - x_1}$$

Equation of Line Containing P_1 and P_2

Point-Slope Form

$$y - y_1 = m(x - x_1)$$

Equation of Line Containing P_1 and P_2

Slope-Intercept Form (slope m, y-intercept b)

$y = mx + b$, where $b = y_1 - mx_1$

Parallel Lines

Slopes Are Equal: $m_1 = m_2$

Perpendicular Lines

Slopes Have a Product of -1: $m_1 m_2 = -1$

Intersecting Lines

Slopes Are Unequal: $m_1 \neq m_2$

Dependent (Coincident) Lines

Slopes and y-Intercepts Are Equal: $m_1 = m_2$, $b_1 = b_2$

Complex Numbers $z = a + bi$

Absolute Value

$$|z| = \sqrt{a^2 + b^2}$$

distance from (0, 0) to (a, b)

Trigonometric Form

$$z = r(\cos\theta + i\sin\theta)$$

where $r = |z|$

Products and Quotients

$$z_1z_2 = r_1r_2[\cos(\theta_1 + \theta_2) + i\sin(\theta_1 + \theta_2)]$$

$$\frac{z_1}{z_2} = \frac{r_1}{r_2}[\cos(\theta_1 - \theta_2) + i\sin(\theta_1 - \theta_2)]$$

Powers and DeMoivres Theorem

$$z^n = r^n(\cos n\theta + i\sin n\theta)$$

for positive integers n

Roots and the nth Roots Theorem

$$\sqrt[n]{z} = \sqrt[n]{r}\left(\cos\frac{\theta + 2\pi k}{n} + i\sin\frac{\theta + 2\pi k}{n}\right)$$

for $k = 0, 1, 2, \ldots, n - 1$

Vectors and the Dot Product

- For a position vector, $\mathbf{v} = \langle a, b\rangle$ and angle θ as shown, $a = |\mathbf{v}|\cos\theta$ and $b = |\mathbf{v}|\sin\theta$, where $\theta = \tan^{-1}\left(\frac{b}{a}\right)$ and $|\mathbf{v}| = \sqrt{a^2 + b^2}$.
- For any nonzero vector $\mathbf{v} = \langle a, b\rangle = a\mathbf{i} + b\mathbf{j}$, the vector $\mathbf{u} = \frac{\mathbf{v}}{|\mathbf{v}|}$ is a unit vector in the same direction as $\mathbf{v}$.
- Given the vectors $\mathbf{u} = \langle a, b\rangle$ and $\mathbf{v} = \langle c, d\rangle$, their dot product is denoted $\mathbf{u}\cdot\mathbf{v}$ and is defined as: $\mathbf{u}\cdot\mathbf{v} = \langle a, b\rangle\cdot\langle c, d\rangle = ac + bd$.
- Given the nonzero vectors $\mathbf{u}$ and $\mathbf{v}$ and angle θ between them, $\cos\theta = \frac{\mathbf{u}}{|\mathbf{u}|}\cdot\frac{\mathbf{v}}{|\mathbf{v}|}$.

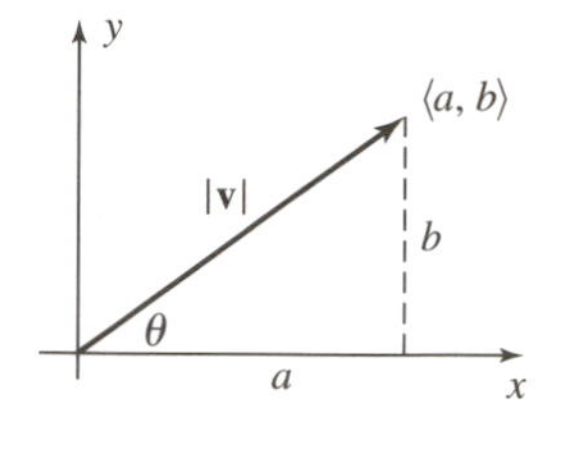

Polar Coordinates

$P(x, y)$ in rectangular coordinates can be represented as $P(r, \theta)$ in polar coordinates:

$$x = r\cos\theta \qquad y = r\sin\theta \qquad r = \sqrt{x^2 + y^2} \qquad \theta_r = \tan^{-1}\left(\frac{y}{x}\right), x \neq 0$$

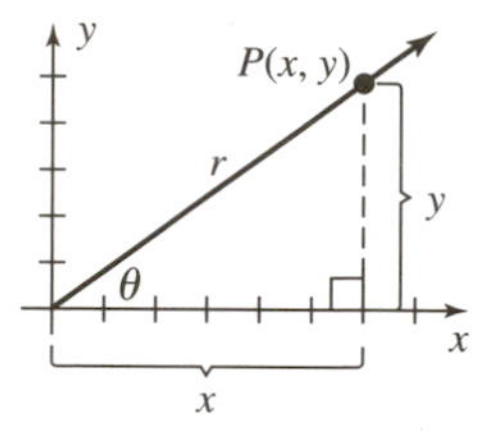

Equation of a Conic in Polar Form

Given a conic section with eccentricity e, one focus at the pole of the $r\theta$-plane, and directrix $\mathscr{L}$ located d units from this focus. Then its equation in polar form is $r = \frac{de}{1 \pm e\cos\theta}$ or $r = \frac{de}{1 \pm e\sin\theta}$ as defined by the value of e.

$e = 1$: parabola $\qquad 0 < e < 1$: ellipse $\qquad e > 1$: hyperbola

Rotation of Axes

If the x- and y-axes of the xy-plane are rotated counterclockwise by the (acute) angle β to form the X- and Y-axes of an XY-plane, the coordinates of the points (x, y) and (X, Y) are related by the formulas:

$$x = X\cos\beta - Y\sin\beta \qquad X = x\cos\beta + y\sin\beta \qquad \tan(2\beta) = \frac{B}{A - C}$$

$$y = X\sin\beta + Y\cos\beta \qquad Y = y\cos\beta - x\sin\beta$$

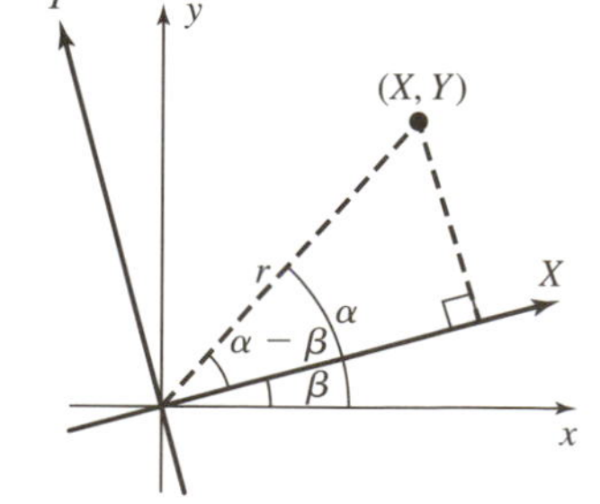

Topics from Algebra

Special Products

$$(a + b)^2 = a^2 + 2ab + b^2$$

$$(a + b)^3 = a^3 + 3a^2b + 3ab^2 + b^3$$

$$(x + c)(x + d) = x^2 + (c + d)x + cd$$

$$(a - b)^2 = a^2 - 2ab + b^2$$

$$(a - b)^3 = a^3 - 3a^2b + 3ab^2 - b^3$$

$$(ax + c)(bx + d) = abx^2 + (ad + bc)x + cd$$